MEDICINE OF AUSTRALIAN MAMMALS

MEDICINE OF AUSTRALIAN MAMMALS

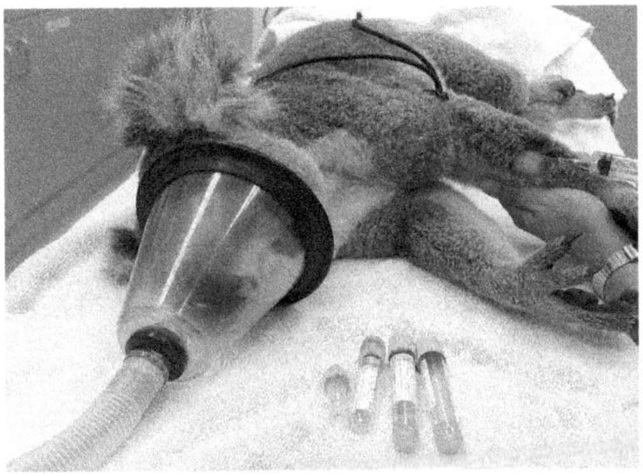

LARRY VOGELNEST • RUPERT WOODS

EDITORS

PUBLISHING

First printed 2008. This reprint 2022.

National Library of Australia Cataloguing-in-Publication entry

Medicine of Australian mammals.
Bibliography.
Includes index.
ISBN 9780643091504 (hbk.).
1. Mammals – Australia. 2. Veterinary medicine – Australia.
I. Vogelnest, Larry. II. Woods, Rupert. III. Title.

599.0994

Published by
CSIRO Publishing
36 Gardiner Road, Clayton VIC 3168
Private Bag 10, Clayton South VIC 3169
Australia

Telephone: +61 3 9545 8400
Email: publishing.sales@csiro.au
Website: www.publish.csiro.au

Front cover photographs (from top): eastern grey kangaroo, iStockphoto; x-ray of an echidna, Taronga Zoo; dingo, iStockphoto; skull of an eastern grey kangaroo, Larry Vogelnest; echidna, iStockphoto; mite, Lee Skerratt

Spine photograph: koalas, iStockphoto

Back cover photographs (from left): feeding a common wombat PY using a plastic bottle and commercial latex marsupial teat, Melbourne Zoo; taking a blood sample from a koala, Taronga Zoo; hand-rearing a pygmy possum, Robert Johnson

Set in 10/13 Minion
Edited by Adrienne de Kretser, Righting Writing
Index by Russell Brooks
Cover and text design by James Kelly
Printed by Ingram Lightning Source

Nov25_RP_ILS

Contents

Preface

Over the last 20 years there has been a dramatic increase in interest in the conservation and welfare of Australian native wildlife. Veterinarians are now frequently asked to treat these animals and there is increasing veterinary involvement in conservation programs for threatened and endangered species. Free-ranging and captive Australian mammals are now commonly used in research. Native mammals can be kept as pets in some Australian states and territories and some native Australian mammals are commonly kept as pets in other countries. Australian mammals are also popular display and educational animals in many zoos and fauna parks around the world.

Captive and free-ranging wildlife are frequently affected by both natural and anthropogenic events, with many cases requiring veterinary investigation or intervention. Australia and the rest of the world also face a number of challenges associated with emerging diseases, climate change and human population expansion, the impacts of which are often seen in wildlife.

Although the medicine of Australian mammals is covered in some textbooks, this is the first comprehensive text devoted to this subject. It is hoped that this text will be of value to veterinarians (private practitioners, zoo and wildlife veterinarians), biologists, zoologists and wildlife workers around the world. It presents the current state of knowledge on the medical management of all taxa of Australian native mammals. It is intended to be the single most important and useful, stand-alone text on the medical management of Australian native mammals. It presents the most up-to-date information available. It is specific and detailed enough to be of use to experienced zoo and wildlife veterinarians (with information on aetiology, epidemiology, pathology, clinical signs and treatment of specific diseases), but is also set out in such a way that it can be of use as a quick reference guide for veterinary practitioners that may not have much experience with Australian native mammals. The book is clinically oriented with the emphasis on providing a practical, usable, one-source reference and problem solving guide. It will assist veterinarians in making sound decisions on diagnosis, prognosis and treatment.

We have brought together Australia's leading wildlife veterinarians, who have exhaustively reviewed the literature and contributed their own knowledge in the writing of their chapters. Although there have been many peer reviewed publications and scientific studies on the diseases and medicine of Australian native mammals much of the information cited in this text is based on the author's own experiences and anecdotal reports from other people. Much of what is cited has been empirically extrapolated from the medicine of domestic animals. It is essential that clinicians and other wildlife professionals using the information from this text keep this in mind. Assessment of patients and decisions on prognosis and treatment must be made on sound clinical judgement. Recommendations from the text, particularly with respect to therapeutics must be used with caution. In most cases this will be off-label use and without specific pharmacokinetic studies carried out for the drug in the particular species being treated.

Unless stated otherwise, the values for haematology and biochemistry presented in this volume are not reference ranges. Reference ranges are not available for the majority of Australian native mammals. Where proprietary product names are used, the manufacturer or supplier's name is cited in parentheses after the first time the product is mentioned in a chapter. The full details of the manufacturer or supplier are provided in Appendix 3. The scientific name for species listed in the text is included in parentheses after the common name when first mentioned in the text for each chapter. The common names (and their construction) and scientific names for mammals used in this volume have been standardised throughout. A complete checklist of the mammals of Australia and its territories is provided in Appendix 1. Wherever possible we have used terminology that is in common usage. In some instances these terms may not be acceptable to all or scientifically accurate. One example is the term 'macropod' which is used throughout this volume to describe members of the Super-family Macropodoidea. Although not strictly correct, it is in common usage and generally recognised as the term that encompasses kangaroos, wallabies, potoroos, bettongs and the musky rat-kangaroo. Strictly speaking the derivative 'macropodoids' should be used for the Super-family and 'potoroids' for the Family Potoroidae and 'macropodids' for the Family Macropodidae.

Finally we hope that this book will ultimately contribute to improving the health, welfare and conservation of both free-ranging wild animals and those in our care. To this end, all royalties generated from the sale of this book will be donated to the Taronga Foundation's Field Grants Program and the Australian Wildlife Health Network.

Larry Vogelnest and Rupert Woods

Acknowledgments

Many people made significant contributions to this textbook. We thank the authors for their efforts in the preparation of their chapters and for their tolerance during the editing process. Your patience, dedication and attention to detail were remarkable and without you this text would not have been possible. We also thank the many veterinarians, researchers and wildlife carers who were approached for information.

We would like to thank our families for putting up with us during the writing and editing of the book, particularly the many lost weekends and nights over the three or so years it took to complete. To Linda, Dan and Rianne (Vogelnest): thank you for your support, patience and tolerance. To the long suffering Evelyn, Kaelie and Jade (Woods): thank you again and hopefully we can start to spend some more time together.

Thank you to Vanessa DiGiglio for assistance with preparation of manuscripts, to Jane Hall for assistance with providing information from the Australian Registry of Wildlife Health to authors, to Beth Croce for the brilliant illustrations and to Kaye Humphreys and the nurses and keepers at the Taronga Wildlife Hospital for assistance with the provision of information and photography.

We thank the Taronga Conservation Society Australia and the Australian Wildlife Health Network for their support, and CSIRO Publishing, John Manger and Briana Elwood for their guidance and patience.

Contributors

Paul Andrew BSc (Hons)
Curator, Taronga Zoo, Taronga Conservation Society Australia
Sydney, New South Wales, Australia

Julie Barnes BVSc (Hons) (Melb), MSc Wild Animal Health (Lond)
Clinical Veterinarian, Los Angeles Zoo
Clinical Assistant Professor, College of Veterinary Medicine, Western University of Health Sciences
California, USA

Ian Beveridge BVSc (Hons) (Melb), PhD (Melb), DVSc (Melb)
Professor of Veterinary Parasitology, University of Melbourne
Melbourne, Victoria, Australia

Wendy H. Blanshard BVSc (Qld)
Records/Special Projects Officer, Sea World
Gold Coast, Queensland, Australia

David J. Blyde BVSc (Hons) (Syd), MACVSc (Zoo Animal Medicine), MBA (UNE), Dip Wild Med (Syd), Grad Cert Captive Animal Management (CSU)
Veterinarian, Sea World
Gold Coast, Queensland, Australia

Kate B. Bodley BSc(Vet), BVSc (Syd), MVS (Melb)
Veterinarian, Melbourne Zoo, Zoos Victoria
Melbourne, Victoria, Australia

Rosemary J. Booth BVSc (Qld)
Principal Conservation Officer Threatened Species, David Fleay Wildlife Park
Conservation Services, Environmental Protection Agency
Queensland, Australia

Andrew C. Breed BSc (Murd), BVMS (Murd), MSc Wild Animal Health (Lond), MRCVS
Research Scholar, School of Veterinary Science
Australian Biosecurity CRC for Emerging Infectious Disease, University of Queensland
Queensland, Australia

Benn R. Bryant BVSc (Syd), MVS (Syd)
Senior Veterinarian, Taronga Western Plains Zoo, Taronga Conservation Society Australia
Dubbo, New South Wales, Australia

Joanne H. Connolly BSc, BVSc (Hons), PhD (Syd), PGCertTerTchg (Massey), MRCVS
Senior Lecturer in Veterinary Science
Course Co-ordinator Captive Vertebrate Management Program
School of Animal and Veterinary Sciences, Charles Sturt University
Wagga Wagga, New South Wales, Australia

Paul A. Eden BVSc (Melb), MSc Wild Animal Health (Lond)
Associate Veterinarian, Perth Zoo
Adjunct Lecturer/Tutor, Conservation Medicine Program, Murdoch University
Perth, Western Australia, Australia

Nicholas J. Gales BVMS (Murd), PhD (Murd)
Leader, Australian Centre for Applied Marine Mammal Science
Principal Research Scientist, Australian Antarctic Division
Hobart, Tasmania, Australia

Rachael B. Gray BVSc (Syd), PhD (Syd), GradCertEdStud (Syd)
Lecturer in Veterinary Anatomy, Faculty of Veterinary Science, University of Sydney
Sydney, New South Wales, Australia

Susan Hemsley BVSc (Hons), MVSc, PhD (Syd)
Senior Lecturer Veterinary Anatomy and Histology, Faculty of Veterinary Science, University of Sydney
Sydney, New South Wales, Australia

Damien P. Higgins BVSc (Qld), MVS (Syd), PhD (Syd)
Lecturer, Faculty of Veterinary Science, University of Sydney
Sydney, New South Wales, Australia

Peter H. Holz BVSc (Melb), DVSc (Guelph), MACVSc (Zoo Animal Medicine), Diplomate, American College of Zoological Medicine
Associate Veterinarian, Healesville Sanctuary, Zoos Victoria
Healesville, Victoria, Australia

Frances A. Hulst BVSc (Hons) (Syd), MVS (Syd)
Veterinarian, Taronga Zoo, Taronga Conservation Society Australia
Sydney, New South Wales, Australia

Robert Johnson BVSc (Syd), MACVSc (Feline Medicine), CertZooMed, BA
Veterinarian, Taronga Zoo and South Penrith Veterinary Clinic
Sydney, New South Wales, Australia

Philip W. Ladds MVSc (Qld), PhD (Kansas State), FACVSc, FRC Path, Diplomate, American College of Veterinary Pathologists
Specialist Veterinary Pathologist
Adjunct Professor, Graduate Research College, Southern Cross University
New South Wales, Australia

Michael J. Lynch BSc (Monash), BVSc (Hons), MVS (Melb), MACVSc (Epidemiology)
Veterinarian, Melbourne Zoo, Zoos Victoria
Melbourne, Victoria, Australia

Helen E. McCracken BSc (Vet) (Hons) (Syd), BVSc (Hons) (Syd), MVS (Melb)
Senior Veterinarian, Zoos Victoria
Senior Fellow, Faculty of Veterinary Science, University of Melbourne
Melbourne, Victoria, Australia

David R. Middleton BSc (Hons) Zool (La Trobe), BVSc (Hons) (Melb)
Former Senior Veterinarian, Healesville Sanctuary, Zoos Victoria
Healesville, Victoria, Australia

Cree L. Monaghan BSc (Murd), BVMS (Hons) (Murd), MVS (Melb)
Specialist Veterinary Manager, Perth Zoo
Perth, Western Australia, Australia

Annabelle R. Olsson BVSc (Syd), MSc (JCU Townsville), MACVSc (Avian Health)
Director, Wildlife Conservancy of Tropical Queensland
Principal Veterinarian, Boongarry Veterinary Services for Wildlife
Cairns, Queensland, Australia

Timothy Portas BVSc (Syd), MACVSc (Zoo Animal Medicine)
Senior Veterinarian, Australia Zoo
Beerwa, Queensland, Australia

Andrea Reiss BVSc (Hons) (Syd), MVS (Melb), MACVSc (Medicine of Australasian Wildlife)
Veterinarian, Perth Zoo
Perth, Western Australia, Australia

Karrie A. Rose DVM (Saskatoon), DVSc (Guelph)
Veterinary Pathologist, Australian Registry of Wildlife Health, Taronga Conservation Society Australia
Sydney, New South Wales, Australia

Lee F. Skerratt BAnSc, BVSc, PhD (Melb), MACVSc (Epidemiology)
Research Fellow, Biosecurity and Tropical Infections Diseases Research Group, School of Public Health, Tropical Medicine and Rehabilitation Sciences, James Cook University
Townsville, Queensland, Australia

Richard Speare BVSc (Qld), MB BS (Qld), PhD (JCU)
Deputy Director, School of Public Health and Tropical Medicine
Professor, Human Parasitology, James Cook University
Townsville, Queensland, Australia

David Spratt BSc (Toronto), MSc (Toronto), PhD (Qld)
Honorary Research Fellow, CSIRO Sustainable Ecosystems
Canberra, Australian Capital Territory, Australia

Simone D. Vitali BSc (Murd), BVMS (Murd), PhD (Murd)
Senior Veterinarian, Perth Zoo
Perth, Western Australia, Australia

Larry Vogelnest BVSc (Hons) (Syd), MVS (Syd), MACVSc (Zoo Animal Medicine)
Senior Veterinarian, Taronga Zoo, Taronga Conservation Society Australia
Sydney, New South Wales, Australia

Rupert Woods BVSc (Murd), BVMS (Murd), MVS (Syd), PhD (Tas), MACVSc (Zoo Animal Medicine)
Manager, Australian Wildlife Health Network
Sydney, New South Wales, Australia

EDITORS' BIOGRAPHIES

Larry Vogelnest

Dr Vogelnest graduated from the University of Sydney in 1984. He has been employed as a veterinarian by the Zoological Parks Board of NSW (now Taronga Conservation Society Australia) since January 1990. He has been Senior Veterinarian at Taronga Zoo since July 1994 and is responsible for the health and welfare of the zoo's animal collection and wildlife admitted to the Taronga Wildlife Hospital. In 1990 Dr Vogelnest completed a Masters in Veterinary Studies in Wildlife Medicine and Husbandry. In 1996 he gained membership to the Australian College of Veterinary Scientists by examination in Zoo Animal Medicine. Dr Vogelnest is on the Clinical Editorial Committee of the Australian Veterinary Journal and is the Oceania Representative of the American Association of Zoo Veterinarians.

Rupert Woods

Dr Woods manages the Australian Wildlife Health Network, a national initiative of the commonwealth government managed under the Wildlife Exotic Disease Preparedness Program (Dept of Agriculture, Fisheries and Forestry). He is Regional Co-ordinator for Australasia for the International Union for the Conservation of Nature, Species Survival Commission Veterinary Specialists Group, he sits on the National Animal Disease Surveillance Steering Committee, Chytridiomycosis Threat Abatement Team, Devil facial tumour disease steering committee and he represents the Australian wildlife community to the National Animal Health Information System (Animal Health Australia). In 1994 Dr Woods completed a Masters in Veterinary Studies in Wildlife Medicine and Husbandry and in 1995 gained membership to the Australian College of Veterinary Scientists by examination in Zoo Animal Medicine. His PhD was in elephant seal anaesthesia, completed while working for the Australian Antarctic Division between 1989 and 1992. He is interested in risk management and how systems organise.

1. Veterinary considerations for the rescue, treatment, rehabilitation and release of wildlife

Larry Vogelnest

The rescue, treatment, rehabilitation and release of wildlife serves three important purposes. Firstly, it provides an opportunity to learn about wildlife and our environment. Secondly, it may contribute to the conservation of species. Finally, and probably most importantly, it addresses the welfare of an animal that may be suffering from injury or illness or has been orphaned. It is our duty as decent human beings (and certainly as veterinarians and wildlife professionals) to relieve the suffering of animals. This may be as simple as providing first aid or facilitating euthanasia. It is essential, however, that attempts at relieving suffering do not perpetuate an animal's suffering. For a wild animal, simply being in the presence of a human being, or being placed in a confined foreign environment, may induce suffering. This is an important consideration and one of the key factors that influence differences in ones approach to management of injured wildlife compared with domestic animals.

Caring for wild animals provides an opportunity to learn more about their biology, natural history, habitats, abundance, behaviour, husbandry and care. It inspires people to develop compassion and an understanding of our wildlife and our natural environment. It educates people, and it is perhaps through this that wildlife rescue and rehabilitation contributes most to the conservation of wildlife and the environment. Despite the many benefits, it is rare that the rehabilitation and release of an individual animal contributes directly to the conservation of a species. However, this may well be the case in situations where a population is so critically endangered that each individual is crucial to its survival. Rehabilitated animals may contribute to reintroduction programs, reinforce or supplement existing populations and can be a powerful tool for conservation (IUCN 2000; Woodford 2001).

Veterinarians have several key responsibilities with respect to the rescue, treatment, rehabilitation and release of wildlife. These include animal welfare, disease risk assessment and management, human health protection and provision of euthanasia and appropriate legal advice. They must also maintain accurate records, assist with wildlife emergencies and become involved in policy development. They also have a responsibility to provide sound, rational and scientifically based advice to various groups, from members of the public to politicians. Frequently, wildlife agency personnel are trained biologists who generally consider the dynamics and welfare of populations and ecosystems as a greater priority than the individual. They therefore frequently dismiss the perspectives and expertise of wildlife carers who focus on the individual. Conversely, wildlife carers frequently do not appreciate the mandate and responsibility of wildlife agencies toward population management. This may result in strained relationships between these groups (Clark 2002). Veterinarians are well placed to act as mediators in these situations.

Decisions to treat and rehabilitate wildlife must be based on a sound practical knowledge of veterinary medicine, individual and population biology and ecology, disease risks and legal requirements. Other issues must also be considered to ensure the ongoing welfare of the individual (during rehabilitation and after release), to avoid disappointment after emotional attachment to individual animals by carers and to protect wild populations and ecosystems (Sikarskie 1992).

1 ANIMAL WELFARE

The welfare of an injured, sick or orphaned wild animal must be considered from the time of rescue until well after release. A wild animal is already likely to be suffering a degree of stress or pain when it is first discovered and determined to be in need of assistance. To be approached and handled by a human invariably compounds this. It is crucial that appropriate and humane methods are employed to accomplish the capture of the animal. In some cases veterinary assistance may be required to facilitate capture, for example, if the animal needs to be darted. Initial care and first aid may take place *in situ* as, for example, with stranded, entangled or trapped marine mammals or other species. Immediate intervention, with or without first aid, may be all that is required and the animal can be released or, if indicated, euthanased immediately. Even if it is obvious that the animal requires euthanasia it is important to ensure that it is not exposed to additional stressors such as large numbers of onlookers, people touching and talking to it, domestic animals, noise (helicopters and other vehicles) and extremes of temperature. The animal must be made as comfortable as possible. This may require moving it a short distance, providing shade or keeping it cool or moist.

If an animal is to be transported to a care facility or veterinary clinic, care must be taken to ensure that the method of transport is appropriate and as stress-free as possible. Most animals travel well in a dark, quiet, well-ventilated and cool environment. Cardboard boxes or bags (pillowcase, hessian sack) in an air-conditioned vehicle are adequate for most species. It is generally not recommended to offer the animal food or water, however this depends on the time between rescue and delivery to a care facility. The type of food and the method of administration should be appropriate to the species. The duration of the trip should be as short as possible. Animals should be checked frequently.

The ultimate aim of rescue and treatment of wildlife should be to rehabilitate and release the animal as quickly and effectively as possible. When first presented with a rescued wild animal, it is important to make an early assessment of its prognosis and suitability for rehabilitation and release. This is essential, to avoid unnecessary suffering and stress. It also avoids disappointment after substantial effort and emotional attachment to the animal. When large numbers of animals are involved in an incident such as an oil spill, bushfire or mass cetacean stranding, it is critical to triage animals. Triage ensures that maximum resources are put into animals that are most likely to survive. Human nature leads people to assist animals that are in the worst condition. The worst-affected animals—those with the least chance of survival—would thus take up time and resources that could be spent saving animals with a more realistic chance of survival. Effective triage avoids prolonged suffering in animals that clearly require immediate euthanasia.

Wild animals are very susceptible to stressors such as confinement, close proximity to other animals (particularly predators such as dogs and cats), noise and handling. If the prognosis is poor or the animal is unsuitable for release, a decision to euthanase must be made quickly. It is important that the decision to treat and rehabilitate an animal is not made to satisfy personal gratification, which may be to the detriment of the animal. If an animal is unsuitable for release but has legitimate educational or scientific value if kept in captivity, approval to do so can be sought from the relevant wildlife authority.

During initial treatment, careful consideration should be given to the judicious use of analgesics, sedatives or neuroleptic drugs to ensure that the animal is as comfortable and stress-free as possible. After initial treatment, most wild animals will require a period of rehabilitation prior to release to ensure they are restored to their former capacity. The treatment and rehabilitation period should be kept as short as possible and the environment in which the animal is held should be stress-free and provide for all its physical and psychological needs. Stress and stress-related illness are the most common causes of failure of treatment or death when treating and rehabilitating wild animals. Veterinarians are now fortunate that there is a network of experienced wildlife carers and rehabilitators who can assist with care and rehabilitation. This reservoir of knowledge and experience must not be ignored. It is important, however, that veterinarians have input into the rehabilitation process to ensure the ongoing health and welfare of the animal. This may be in the form of providing advice on nutrition and other preventative medicine or regular veterinary examinations.

Considerable time, effort and expense often goes into the treatment and rehabilitation of animals, but significantly less attention is often given to preparation for release, the release and post-release monitoring (Hall 2005). When making a pre-release assessment, veterinarians must keep in mind that the release of a native animal back into the wild must always be in the best interest of the animal. It must have recovered completely, both physically and behaviourally, to its former capacity to survive (Walraven 1994; Hall 2005). If an attempt has been made to treat animals by the application of prostheses these animals should not be released, as it is unlikely that any prosthesis will remain attached for the natural expected life of the animal. Some countries have laws and policies relating to the release of animals with disabilities, with strong recommendations not to do so (Hall 2005). Even though animals with disabilities are seen in the wild, apparently doing well, the long-term effect of the disability is difficult to assess. If an animal is released with a minor disability that appeared not to affect it during rehabilitation, it may be being committed to a life of long-term suffering and struggle for survival.

In many cases, the ultimate effect of release may be suffering and death. Many studies have demonstrated that mortality is high in animals released after rehabilitation (Brown & Tribe 2001). Despite this, it is considered by some that returning wildlife to the wild is more humane than a life in captivity. However, if not done properly animals may succumb to predation or disease or die from starvation or maladaption to an unfamiliar or inappropriate environment (IUCN 2000)—something that may not be justifiable. With our increasing knowledge and improved treatment, rehabilitation and release techniques, post-release survival has now improved and there is often minimal impact on animal welfare. In many cases this has been due to an assessment of treatment, rehabilitation and release techniques against information gained from long-term post-release monitoring studies.

Animals should always be released as close as possible to the rescue site. Many animals adapt to a particular environment, habitat or geographic area. Releasing them elsewhere may result in suffering due to maladaption. Similarly, the release of new individuals to a geographical area may pose genetic and disease risks to the resident population (Brown & Tribe 2001). The introduction of non-endemic disease or infection by the released animal may also result in considerable suffering.

2 ANIMAL HEALTH

The emergence of new infectious diseases in humans and other animals has been highlighted in recent years. Diseases such as severe acute respiratory syndrome (SARS), Australian bat lyssavirus, Hendra virus, Nipah virus, West Nile virus, monkeypox, avian influenza, rabbit haemorrhagic disease, pilchard herpesvirus, sarcoptic mange in various species, canine distemper in lions and seals, Tasmanian devil facial tumour disease and chytridiomycosis, to name a few, have had a devastating impact on animal and human populations. Wildlife populations play an important role in initiating and maintaining many of these diseases (Bunn & Woods 2005). A number of risk factors and processes have been identified, including geographic translocations of hosts and pathogens, new and or intensified contacts between hosts and pathogens, genetic change and environmental change (Daszak et al. 2005; Leighton 2005). The rescue, rehabilitation and release of wildlife, if not done properly, may contribute to these processes and lead to the disruption or disturbance of the natural balance between the host, environment and pathogens. When releasing rehabilitated wildlife, the health of the wild population and the ecosystem into which that animal is being released must take precedence over the welfare of the individual animal (IUCN 2000). Great care must be taken to maintain the balance of organisms naturally found within an animal and not introduce foreign organisms during the rehabilitation process. Each individual animal, rather than being a single organism, is a package of organisms (Woodford 2001). When releasing an animal we are releasing not only that individual but all the other organisms within it and on it.

Most wildlife rehabilitation and release and wildlife translocations are undertaken with complete lack of awareness or little regard for disease risks (Kirkwood & Sainsbury 1995; Griffith et al. 1993; Woodford 1993). Animals being rehabilitated may be exposed to new pathogens while in captivity and act as carriers when released, transmitting disease into a wild population with potentially devastating consequences for the population and the ecosystem in which it exists. Pathogens may be carried by released animals without causing disease in that species but causing severe disease in other species. Infectious disease may also be introduced to wild populations when animals with endemic disease are released into an area where those diseases do not occur. The released animal may also be exposed to diseases to which it has not developed immunity (IUCN 2000). These immunologically naive animals may become infected and die, or become carriers shedding large numbers of organisms and infecting endemic animals (Kirkwood & Sainsbury 1995; Griffith et al. 1993; Viggers et al. 1993; Woodford 1993). The release of an animal back into the wild is a stressful time for that animal and others in the area, due to competition. When animals are stressed disease can have a significant impact.

It does seem surprising that the incidence of disease introduction through the release of rehabilitated or translocated wildlife or the disease-related failure of wildlife releases or translocations is not higher, considering the large number of releases or translocations. However, it is highly likely that these scenarios go undetected or are underestimated given the general lack of post-release monitoring or surveillance of wild populations into which animals are released (Griffith et al. 1993), and the absence of veterinary involvement in some of these programs. The disease risks associated with wildlife translocation are frequently overlooked by many wildlife agencies responsible for these programs.

2.1 Disease risk assessment

Adequate disease risk assessment when dealing with wildlife presents considerable difficulty for two reasons. Firstly, there are few reliable tests for the detection of many wildlife pathogens. Secondly, many species of wild animals carry infectious agents of which we have little or no knowledge (Kirkwood & Sainsbury 1995). Although there have been surveys of the health of wild populations of some species, for the majority of species we have little knowledge of the organisms and pathogens they may naturally harbour. For these reasons appropriate measures must be taken to reduce the risk of disease during the rehabilitation and release of wildlife, including initial and ongoing clinical assessment, laboratory tests (haematology, biochemistry, serology, faecal examination, microbiology), screening tests for specific diseases (e.g. tuberculin testing), prophylactic procedures and treatments, quarantine, enclosure design and husbandry (Viggers et al. 1993; Kirkwood & Sainsbury 1995; IUCN 1998). An understanding of disease processes and the measures required in preventing disease transmission and spread is essential for all veterinarians and others involved in the rehabilitation and release of wildlife.

It is important to determine if the reason for presentation of a rescued animal is an infectious disease. Wildlife may require rescuing as a result of a natural event (disease, old age, predation, weather conditions, natural disasters such as fires or floods) or some form of human activity (hit by car, dog or cat attack, oil spills, poisoning, contamination, collision with man-made objects, trapping, shooting). The latter are far more common, but an animal that has succumbed to trauma or predation may have been weakened by a pre-existing disease process. In general terms, animals suffering from an infectious disease rather than a traumatic injury alone are poorer candidates for release and possibly pose a greater risk to other patients in the rehabilitation facility and to the population if released. It is important to screen all incoming rehabilitation patients for the presence of clinical or subclinical infectious disease (see 2.1.1).

2.1.1 Clinical assessment

Ideally, the initial clinical assessment of all rehabilitation patients should be carried out by a wildlife veterinarian. In each case, however, a disease risk (and cost–risk) assessment can be made by the wildlife rehabilitator. In some cases it may be clear that there is no underlying disease process (e.g. orphaned marsupial pouch young) and a clinical assessment by a veterinarian may not be necessary. If a clinical assessment is made it should include a physical examination, haematology and serum biochemistry analysis, serology, faecal examination for bacteria and parasites, evaluation of the blood smear for parasites and other sampling or testing as appropriate to the species (e.g. chlamydophila in koalas and birds, cryptococcus in koalas, circovirus in psittacine birds, tuberculosis in pinnipeds). In some cases it may be appropriate to collect serum or plasma to be stored at −70°C for retrospective disease investigation.

Wildlife may also be infected with exotic or notifiable diseases (e.g. avian influenza, Newcastle's disease, West Nile virus, Pacheco's disease, monkeypox). It is quite possible that an animal rescued by a wildlife carer may be the first case of one of these diseases in a country. Monitoring wildlife health may facilitate early detection of threats to biosecurity and rescued wildlife should be included as part of wildlife health surveillance systems. Monitoring the emergence of wildlife diseases is an important contribution that wildlife veterinarians and wildlife carers can make (Clark 2002; Woods & Bunn 2005). Consideration must also be given to the possibility that the animal may be harbouring a zoonotic disease such as Australian bat lyssavirus, Hendra or Menangle virus, avian influenza, chlamydophilosis, mycobacteriosis, ringworm or salmonellosis.

If pathogens are detected in the absence of clinical disease this state may be assessed as typical for a particular species, thus increasing our knowledge of wildlife health and helping us to determine what is normal for a species and what is not. Screening and clinical assessment must continue throughout the rehabilitation process as diseases that may not initially have been evident may become evident later. The stress of rehabilitation may trigger diseases, and animals may be exposed to new pathogens during the rehabilitation process (Daszak et al. 2005; IUCN 1998).

Once the disease status of an incoming rehabilitation patient has been assessed and a disease is diagnosed, a decision must be made as to whether the disease can or should be treated or whether the animal may pose a risk to other animals in the rehabilitation facility or once released. If the risk is assessed as too high (taking into account the conservation value of the animal), the animal should be euthanased. If the treatment option is chosen, we must consider if the pathogen can be completely eliminated or whether the animal may remain a carrier. It may not be necessary or beneficial in every case to eliminate all infectious agents. Wild animals harbour many organisms as part of their normal flora. All wildlife have viruses, bacteria, fungi and parasites within and on their bodies that under most circumstances do not cause disease and may in fact be beneficial to the animal. Certainly, when animals become stressed or succumb to disease, numbers of organisms may increase and cause disease or debilitate the animal even further. For example, most sick animals have increased numbers of parasites. This may be due to immunosuppression allowing parasite numbers to increase, and less grooming or preening by the animal to remove these parasites. It is beneficial in these cases to reduce the numbers of, or eliminate, the parasites in order to aid recovery. If the parasite load is low it may not be necessary to treat the animal, but simply monitor the parasite load. Many parasites are difficult to eliminate. However, testing and/or treatment for parasites should be considered in all animals presented for rehabilitation. This may need to be performed a number of times during the rehabilitation process (Zsivanovits 2004).

2.1.2 Pre-release assessment

Prior to release all animals must undergo a thorough clinical examination and health assessment as well as assessment by an experienced rehabilitator. There must be no evidence that the animal is harbouring an infectious agent that may be detrimental to it or other animals once released. All animals being released will be carrying infectious agents, but if these are endemic to the species and to the area or population into which the animal is being released there should be no concern. The presence of parasites should not preclude an animal from release providing it is known that these parasites naturally occur in that species and are endemic in that species in the area where the animal will be released. Animals should not be released into an area where diseases to which they are not immune are endemic. Testing prior to release is not totally reliable in preventing the introduction of new pathogens to the wild. This may be the case when we have little knowledge of disease or are unable to test for specific pathogens (Woodford 2001). Tests cannot be conducted on unknown pathogens and there are probably many more than the known number (Spielman 2001). In some situations and with some species it may be preferable not to release animals. This has been highlighted with examples such as the complete moratorium on the release of all rescued marine mammals in Canada, and the strong recommendation by the Scientific Committee on Antarctic Research in Australia not to release any Sub-Antarctic or Antarctic species of seals after rescue. This has been adopted as policy by the Department of Environment and Conservation in NSW. Ideally, a detailed clinical assessment similar to the initial assessment on arrival should be performed by a wildlife veterinarian. If possible, serum or plasma should be stored at −70°C. This is useful for retrospective disease investigations.

2.2 Disease risk management

While in captivity for treatment and rehabilitation, animals may be infected with pathogens from other animals, from the environment or from food or water. The risk of transmission of pathogens can be minimised by following basic principles of preventative medicine—quarantine, enclosure design, husbandry, nutrition, routine testing and treatments, pest prevention and control and post mortem examination.

The majority of wildlife rehabilitation in Australia is undertaken by individuals who care for and accommodate the animals in their own homes. Most have an area for more intensive care within the home and aviaries or similar enclosures outside for longer-term rehabilitation. No matter how large or small a facility, these basic principles of preventative medicine apply.

2.2.1 Quarantine

Quarantine is the isolation and health screening of animals with the objective of preventing or controlling introduction or spread of infectious diseases. During the quarantine period, incubating diseases may be detected. Animals may also shed infectious organisms. Quarantining animals will minimise the risk of transmission of pathogens between animals, either from a newly introduced animal or from other animals within or near the facility to animals being rehabilitated. The quarantine period is often arbitrary, given our lack of knowledge of many diseases. However, it is preferable to keep an animal in captivity for as short a period as possible.

The quarantine of wildlife undergoing rehabilitation and release serves two purposes. Firstly, it is a period of isolation during which subclinical disease may become clinical or be detected before new arrivals enter a facility and contact other rehabilitation animals. Secondly, all animals destined for release should be quarantined from animals not of the same health status or other species not destined for release. Animals that are not to be released in the same location should be kept separate from each other to avoid the transmission of pathogens that may not be endemic to a particular area. Ideally, rehabilitation animals should not come into contact with domestic animals.

All rehabilitation facilities should have an isolation area. Depending on the species and reason for presentation, barrier nursing should be practised between animals and enclosures. Barrier nursing is the use of dedicated protective clothing (gloves, gowns or overalls, gumboots), cleaning and feeding utensils and the use of footbaths and disinfection.

2.2.2 Enclosure design

Rehabilitation enclosures should be carefully designed to prevent disease transmission and minimise stress on animals. Animals should be sheltered from noise, extremes of weather and exposure to domestic animals. There should not be large numbers of enclosures in close proximity as this may facilitate transmission of pathogens. Exposure to sunlight is not only important for animal well-being (vitamin D_3 synthesis) but also helps to kill bacteria, viruses and fungi. Dark and damp areas should be minimised as these tend to harbour pathogens (e.g. aspergillus spores, mycobacteria). Enclosures and furniture must be easy to clean and disinfect. Disposable furniture (e.g. branches) is ideal. Plants can be kept in pots so they can be easily removed to aid cleaning. Plants should be rotated to limit the build-up of faecal matter and pathogens and to provide exposure to sunlight.

Certain substrates may harbour pathogens and spores of pathogens. For example, straw and some types of mulch may harbour aspergillus spores, grass and dirt may harbour coccidia and other parasites. These should be avoided in susceptible species. Good drainage and regular substrate changes are important as pathogens build up very quickly in damp or wet conditions.

Enclosures must be designed to minimise stress, as stressed animals are more susceptible to infection due to immunosuppression. Enclosures should be suitable for the species, equipped with furniture that provides shelter, privacy, visual barriers and appropriate perches (type, height and quantity), and with appropriate substrate and water bodies for aquatic or semi-aquatic species. The enclosing structures and materials must also be appropriate.

2.2.3 Husbandry

Good animal husbandry techniques are essential for maintaining healthy animals and minimising the risk of disease transmission. High standards of hygiene and sanitation are crucial for the prevention of disease transmission. Hand-washing facilities with soap (with or without antiseptic) should be available. This is not only important in preventing transmission of pathogens between animals and enclosures but in preventing zoonotic infections. Appropriate and proper use of disinfectants is important. All surfaces to be disinfected must be clean, as the presence of organic material inactivates many common disinfectants. Disinfectants must be used in accordance with the manufacturer's recommendations, i.e. dilution rates (stronger does not mean better) and how long they are effective once made up. A regular schedule of cleaning and disinfection must be maintained. Once an enclosure is vacated it should be thoroughly cleaned and disinfected before it is used by a new patient.

Sanitation is the removal of faeces and other waste material, e.g. food scraps, to prevent the build-up of pathogens in the environment. Leftover food and scraps also attract pests, some of which may carry disease. In some cases substrates may also need to be changed. Waste and water should not be flushed from one enclosure into or through another. Heavily contaminated waste material or substrates should be burned or buried to prevent exposure of free-ranging wildlife to pathogens. Though at least daily cleaning is recommended, the level and frequency of cleaning must be balanced against the stress and disturbance it may cause to the patient (Zsivanovits 2004).

If animals are housed in groups, careful consideration must be given to group make-up including age, sex and species. If animals are not compatible or there are too many, intimidation or aggression may result. Overcrowding may cause stress as well as increased contamination of the enclosure and increased risk of transmission of pathogens between animals. In general, animals should be housed individually until a veterinarian or wildlife carer is confident that they do not pose a risk to other animals.

Animals that receive appropriate treatment and optimal care in captivity will recover quickly. The rehabilitation period for any animal should be as short as possible. The longer an animal is in captivity the greater the risk of it being exposed to pathogens and the greater the chance of its natural microbial and parasitic flora changing. There are also many non-disease considerations for keeping animals in captivity for as short a period as possible (Walraven 1994).

2.2.4 Diets and nutrition

Good diet is an important component of rehabilitation. Inappropriate nutrition will lead to an increased susceptibility to infection. Food items may also introduce pathogens. Care should be taken to avoid feeding food items that may be harbouring pathogens. Rodents or birds may harbour pathogens such as salmonella, trichomonas and other parasites or viruses. Fresh fish fed to piscivorous animals may harbour various pathogens. Ideally, laboratory or farm-bred animals such as mice, rats or day-old chicks should be fed rather than wild-caught animals such as pigeons, mynahs, starlings or rabbits. Freezing prey items kills some but not all pathogens. The risk of disease transmission may be lower if birds are fed non-avian food items and mammals are fed avian food items, but this may be impractical or inappropriate depending on the species (Zsivanovits 2004).

Food should always be fresh. Spoiled food may introduce pathogens or toxins. In warm climates food spoils quickly and may need to be removed and replaced regularly. The nutrient value of food may also change if not fresh. Liquid food or gruels should be made up fresh daily or frozen immediately for later use. Frozen food should not be stored for more than three months. Prolonged freezing, inappropriate thawing and refreezing allows bacterial overgrowth, build-up of toxins and loss of nutrients.

Fresh drinking water in clean bowls should be provided at all times (even for species that may not usually drink—when they are sick or stressed their desire or need for water may change). Food and water containers must be situated to avoid contamination with faeces. They should be appropriately positioned for the species, i.e. on the ground or elevated. If animals are housed in groups there must be adequate numbers of feed stations to ensure all animals get an opportunity to feed.

2.2.5 Routine and preventative treatments

Before and during the rehabilitation process it may be necessary to provide routine treatments to eliminate or reduce the load of pathogens or to prevent animals from becoming infected with pathogens. Routine treatments are primarily for parasite control. As previously discussed, many animals naturally carry a range of internal and external parasites without harming the host, however, parasite loads may increase and cause disease when animals are injured or sick. Reducing the number of parasites may be beneficial. It is, however, not always appropriate to eliminate all parasites as it is important for the host to maintain some immunity to them. Indications for intervention for specific species and parasites are presented in the relevant chapters of this volume.

In captivity, some species become susceptible to certain infections. For example, orphaned marsupials have increased susceptibility to infections such as candidiasis, salmonellosis and coccidiosis. This may be the result of immune incompetence, immune suppression and exposure to contaminated environments or other infected animals. These diseases can be prevented by minimising stress, placing animals in clean uncontaminated environments and administering prophylactic drugs.

Although some vaccines (*Bordetella bronchiseptica*, tetanus and other clostridial vaccines, footrot vaccine [protection against lumpy jaw in kangaroos]) are used for captive Australian native mammals, it is not generally recommended to vaccinate Australian wildlife during rehabilitation prior to release.

2.2.6 Pest prevention and control

Pest species (rodents, foxes, birds, rabbits and insects) are known to transmit and harbour pathogens. Rodents can carry salmonella, leptospires, encephalomyocarditis virus and parasites. Biting insects may spread blood parasites (e.g. malaria, leucocytozoon) and viruses (e.g. pox viruses). Pests may also disturb animals, particularly at night, which may cause stress. They may also prey on them. Measures must therefore be taken to avoid attracting pests. Shelter and food are the primary attractants. Enclosures should be built to exclude pests and other wild native animals. Animal food should be stored in pest-proof containers and food scraps should be cleaned up regularly. Insect pests such as cockroaches and mosquitoes should be controlled. The judicious use of traps and baits to control pests is important, but exposure of non-target species must be avoided. Traps and baits are best set outside enclosures; if they must be placed within an enclosure, steps should be taken to ensure that non-target species cannot be trapped or poisoned. A variety of appropriate traps and bait stations are available from commercial suppliers.

3 WILDLIFE EMERGENCIES

Veterinarians are frequently called on to assist in wildlife emergencies such as oil spills, bushfires or marine mammal strandings. These events attract significant public and media attention. Dealing with wildlife in an emergency may be a significant component of an incident or a very small component of a much bigger operation. Whenever animals are involved there is usually a significant focus on that aspect of the emergency. If there is concern for the health and welfare of the animals involved, veterinary assessment and assistance will be required. Wildlife emergencies can be complex events and some understanding of your role is essential for a successful outcome. Your involvement as a veterinarian may be significant or minor. Veterinarians should concentrate on the animal issues but must consider all other aspects of the emergency. Dealing with wildlife in an emergency is often controversial and often involves philosophical issues which may overshadow practical biological issues (Sikarskie 1992). Veterinarians are frequently caught up in heated debate over these issues. Decisions to rescue and treat wildlife in an emergency must be based on sound conservation and biological principles yet take into account the welfare of the animal or animals, and the emotions of onlookers and people involved in the incident. The veterinarian should contribute in a professional manner. Authority for decision-making at emergencies is usually vested in the relevant government conservation agency representative present (usually under the Incident Control System). The veterinarian should acknowledge this authority, and appreciate that others may know a lot more than they do. If inexperienced, consult with more experienced people, use first principles, listen, learn, assess and plan. Most of all, show compassion for the animal or animals and the people involved. Where emotions are high and volunteers have invested significant effort in the rescue and care of animals, it is prudent for the veterinarian to consider all factors carefully, evaluate the situation, clinically assess animals and perhaps take diagnostic samples (even if it is clear that the outcome or prognosis is not likely to be good) before making decisions or providing advice, particularly if euthanasia is recommended.

The veterinarian's primary responsibility is animal health and welfare, but they may become involved in other aspects of the emergency. The veterinarian must keep the Incident Controller regularly informed. Statements should never be made to the media or onlookers unless the veterinarian is specifically requested to do so.

4 HUMAN HEALTH

Veterinarians have an obligation to have an awareness and knowledge of zoonotic pathogens not only to protect themselves but also to provide accurate advice and information to their staff, clients and wildlife carers. Many wildlife species may harbour zoonotic pathogens such as Australian bat lyssavirus, Hendra or Menangle virus, chlamydophila, toxoplasma, mycobacteria, dermatophytes or salmonella. The elderly, children and immunocompromised people are at greatest risk of contracting zoonotic infections. Many wildlife carers are elderly and children frequently have contact with animals in care.

Many wild animals are dangerous and can inflict serious injury through bites, scratches and blows. The greatest risk of injury is often during the initial rescue and capture. The emotional drive to rescue animals often overrides common sense. Rescuers or volunteers sometimes show little regard for their own welfare and safety, putting themselves at risk not only by exposing themselves to a potentially harmful animal but also by failing to take due care in dangerous conditions such as extremes of temperature, high seas, icy conditions, climbing cliff faces or negotiating busy roads. Such situations frequently arise during oil spills, cetacean strandings or bushfires. Veterinarians can often remove themselves from the emotion of a situation and are well placed to provide guidance and advice.

5 EUTHANASIA

Veterinarians are in the fortunate position of being able to provide and facilitate euthanasia, relieving the suffering of sick or injured animals. Where legislation allows and the operator has received appropriate training, euthanasia may also be performed by wildlife carers, wildlife authority personnel, RSPCA officers or the police.

In some cases the need to euthanase an animal will be obvious. In others a period of observation or further clinical assessment and evaluation or laboratory testing may be required. Consideration should be given to the status of the species (threatened or endangered), its biology and life history and available resources to provide proper care during treatment and rehabilitation. For example, euthanasia may be an appropriate decision in the case of a stranded healthy neonatal or infant cetacean where the prognosis for release and return to its mother is very poor and the resources required to care for it in captivity are immense. An injured sub-adult male common brushtail (*Trichosurus vulpecula*) may be unsuitable for rehabilitation because it may become ill or suffer injury during a struggle to establish territory if the only area it can be released is already saturated with possums. Releasing such an animal in such an area may just prolong its suffering, and euthanasia may be more appropriate. It should be remembered that euthanasia may be appropriate not only for sick or injured animals, for example in healthy feral animals or orphans where hand-rearing is not feasible.

5.1 Criteria for euthanasia

The following criteria can be used as a guide in the euthanasia decision-making process.

- The animal has sustained such severe injuries that it is unlikely to survive even with treatment.
- The animal has sustained injuries that may be treated but will leave it with a disability that preclude it from being released.
- The results of diagnostic tests indicate a poor prognosis for recovery with or without treatment.
- The animal is suffering from an infectious disease that may pose a risk to other animals during rehabilitation or after release.
- It may be a weak or sick animal of common species where there is no evidence of anthropogenic factors contributing to the animal's demise, e.g. flying foxes and shearwaters. Giving these animals a selective advantage through treatment and rehabilitation may be perpetuating 'weaker genes' (Hanger & Tribe 2005).
- The animal is an orphan and the prognosis for successful return to its mother or hand-rearing is poor due to its age or stage of development, the resources required and the potential for imprinting or the development of other behavioural inadequacies that would preclude survival after release.

- Lack of resources (facilities, personnel, finances, food, veterinary care, drugs) to care adequately for the animal. This may be specific to a particular species.
- Geriatric animals.
- It is illegal to release an animal, or there are recommendations not to do so based on disease risk and there is no option for permanent care in captivity.
- The animal is a feral or pest species.
- The animal is behaviourally unsuitable for release (e.g. imprinted).

Lack of interest, low priority relative to treating paying clients' animals, lack of appropriate facilities and insufficient experience or knowledge are no excuse not to attend to a wildlife patient. Many wildlife patients suffer as a result of neglect (left untreated for prolonged periods often in inappropriate housing) or an attempt at treatment and rehabilitation by inexperienced veterinarians or carers. In such cases animals should be referred immediately to an experienced wildlife veterinarian or carer, or be euthanased.

5.2 Euthanasia techniques

The objective of euthanasia is to produce immediate or rapid loss of consciousness, followed by death, with a minimum of pain, discomfort or distress. This requires rapid destruction of the brain or chemical suppression of brain activity. The technique should be able to be performed easily and safely, be reliable, preferably not affect gross and histologic necropsy findings, be readily available and cheap and be tolerable to the person performing the procedure as well as onlookers (Hanger & Tribe 2005). The IV administration of pentobarbitone is the most commonly used method and is humane and acceptable, however, there are numerous other chemical and traumatic methods for euthanasia (Cooper et al. 1989; Geraci 2000; ANZCART 2001; Hanger & Tribe 2005; Rose 2005). In some cases it may be impossible to administer euthanasia due to the size of the animal, lack of availability of a humane method or the remoteness of the animal. A good example is a very large stranded whale. In these situations it may be acceptable to allow the animal to die without intervention. After euthanasia, death must be confirmed before performing a necropsy or disposal of the carcass. If the brain has been destroyed this is usually sufficient, but in other cases cardiac arrest must be confirmed (Hanger & Tribe 2005).

Carcasses of animals that have been euthanased using pentobarbitone or prior administration of anaesthetic agents must be disposed of by deep burial or incineration. These carcasses may pose a risk to scavengers. Carcasses of animals infected with a potentially contagious disease must be disposed of in a similar fashion.

6 NECROPSIES

Every animal that dies or is euthanased during rehabilitation should undergo a necropsy. Ideally this should be carried out by an experienced wildlife veterinarian or veterinary pathologist. However, even if conducted by a wildlife carer, valuable information can be gathered (Rose 2005). A necropsy is not only of diagnostic value but provides the opportunity to learn more about the anatomy and physiology of animals. Determining the cause of death provides an opportunity to learn more about the reason the animal required rescuing and any possible underlying disease process. A necropsy may detect an infectious disease that could infect other animals in the rehabilitation facility. This allows measures to be taken to prevent further spread of the disease, thus preventing an outbreak or risk of other animals becoming infected and thus precluding them from being released. Guidelines for necropsy of Australian native mammals are presented in Chapter 4.

The majority of rescued wildlife are common species. Many of these species can be legally and routinely destroyed in pest control exercises or by hunting (Brown & Tribe 2001). The cost of treatment and the rehabilitation effort may be high. Even if the prognosis for release is assessed as good, if an infectious disease is suspected there may be considerably more value in euthanasing the animal and performing a necropsy to obtain an accurate diagnosis. The data obtained may be of significant value in wildlife disease surveillance and management.

7 RECORD-KEEPING

Accurate records should be maintained by veterinarians and carers for each animal during treatment and rehabilitation. This information provides useful data on the health of animals during the rehabilitation process and assists in establishing what is normal for a species with respect to physiological parameters and infectious agents. If post-release monitoring data is available, a useful correlation may be made between post-release survival and treatment, rehabilitation and release techniques. However, before this data can be useful on a broad scale, it must be collected in a uniform and scientifically valid way (Clark 2002). Numerous databases have been developed for wildlife rehabilitation and wildlife health. These should be standardised and adopted nationally so that data is uniform and readily retrievable (Bunn & Woods 2005). With improved documentation of the case histories of the hundreds of thousands of wild animals rescued annually, wildlife veterinarians and carers can dramatically expand the information base upon which wildlife health management is based and contribute to wildlife health surveillance (Clark 2002; Woods & Bunn 2005).

In order to maintain accurate records, each animal must be individually identified. A temporary identification can be applied during the rehabilitation process, but prior to release a permanent identification such as a passive integrated transponder, legband or eartag should be applied. Satellite or telemetry devices may be used in addition to these methods. Post-release monitoring allows retrospective evaluation of the success of rehabilitation techniques and the ability to trace potential sources of infectious disease should there be an outbreak in a wild population.

Wildlife are frequently affected by anthropogenic disasters such as oil spills or malicious activities such as poisonings and other acts of cruelty. Where animals are rescued from these situations (dead or alive) it is particularly important to maintain accurate records as legal proceedings may be taken against the offending party. Records may be required as supporting evidence.

8 LEGISLATION

Legislation pertaining to the protection, keeping, treatment, rehabilitation, release and exhibition of native and exotic non-domestic animals varies between countries, states and territories. Different wildlife and/or agricultural authorities (local, state or federal) have legislative responsibility under a variety of different acts for the protection and keeping of native fauna. Wildlife agencies also have policies and guidelines relating to the care, keeping and release of wildlife. It is important that veterinarians are familiar with this legislation so that they work within it and ensure that their clients and wildlife rehabilitators do. Veterinarians should be able to give relevant and accurate advice on the legislation, or at least know where to find it. In some countries, states or territories, registered veterinarians are required by law to relieve suffering in any animal. Such action may include euthanasia or appropriate treatment. It is unethical, immoral and unprofessional and demonstrates a lack of compassion to refuse to provide care for a sick or injured wild animal (Tribe & Hanger 2005). If treatment is the chosen course of action, in most countries the veterinarian is permitted to keep protected fauna as long as they are in need of, and receiving, veterinary care. Wildlife carers are also required to be licensed with their appropriate wildlife authority or registered with a licensed wildlife care group before being permitted to rescue and rehabilitate wildlife (Brown & Tribe 2001). This may require a demonstrated level of competency through completing an apprenticeship or attending courses before a licence will be issued. The removal and transport of certain protected species (dead or alive) such as marine mammals, migratory birds or endangered species is often highly regulated (Sikarskie 1992).

9 DEVELOPMENT OF POLICIES AND PROTOCOLS

Veterinarians should have input into the development of policies and protocols relating to the rescue, treatment, rehabilitation and release of wildlife (Sikarskie 1992). Most government wildlife agencies have developed such policies and protocols. Most relate to wildlife disasters such as oil spills, fires and marine mammal strandings, although some agencies have policies on the rescue, rehabilitation and release of wildlife in other situations. Veterinarians have been consulted in many cases and the documents and manuals that have been produced are a valuable resource for anyone involved in wildlife rehabilitation.

10 REFERENCES

Australian and New Zealand Council for the Care of Animals in Research and Teaching (ANZCART) (2001). *Euthanasia of Animals used for Scientific Purposes* (ed. JS Reilly). ANZCART: Adelaide.

Brown PR & Tribe A (2001). 'Well it sounded like a great idea at the time': wildlife rehabilitation and relocation in Australia. In *Veterinary Conservation Biology: Wildlife Health and Management in Australasia. Proceedings of a joint conference* (eds A Martin & L Vogelnest), pp. 187–197, Sydney.

Bunn C & Woods R (2005). Emerging wildlife disease—impact on trade, human health and the environment. *Microbiology Australia* **26**, 53–55.

Clark E (2002). The role of wildlife rehabilitators in emerging disease surveillance, biosecurity, and bioterrorism. *Journal of Wildlife Rehabilitation* **25**, 26–27.

Cooper JE, Ewbank R, Platt C & Warwick C (1989). *Euthanasia of amphibians and reptiles.* Report of a Joint Universities Federation for Animal Welfare and World Society for the Protection of Animals working party, UK.

Daszak P, Kilpatrick AM, Patel N, Epstein JH, Schogel L, Jones K & Cunningham AA (2005). Pathogen pollution: understanding the causes and impact of introduced wildlife diseases. In *Wildlife Health in a Shrinking World: Ecology, Management and Conservation. Proceedings of the Wildlife Diseases Association International Conference.* Wildlife Diseases Association: Cairns.

Geraci JR (2000). Euthanasia of marine mammals. In *Marine Wildlife, Proceedings 335*, pp. 85–95. Post Graduate Foundation in Veterinary Science: Sydney.

Griffith B, Scott JM, Carpenter JW & Reed C (1993). Animal translocations and potential disease transmission. *Journal of Zoo and Wildlife Medicine* **24**, 231–235.

Hall E (2005). Release considerations for rehabilitated wildlife. In *Proceedings of the National Wildlife Rehabilitation Conference*, Gold Coast, Australia.

Hanger J & Tribe A (2005). Management of critically ill wildlife. In *Proceedings of the National Wildlife Rehabilitation Conference*, Gold Coast, Australia.

IUCN (1998). *Guidelines for Re-introductions.* IUCN/SSC Re-introduction Specialist Group. International Union for Conservation of Nature and Natural Resources: Gland, Switzerland/Cambridge, UK.

IUCN (2000). *Guidelines for the Placement of Confiscated Animals.* International Union for Conservation of Nature and Natural Resources: Gland, Switzerland.

Kirkwood JK & Sainsbury AW (1995). Diseases and other considerations with wildlife translocations and releases. In *Proceedings of a Symposium on Veterinary Involvement with Wildlife Reintroduction and Rehabilitation*, pp. 12–16. World Association of Wildlife Veterinarians.

Leighton FA (2005). A global historical perspective on disease introductions. In *Wildlife Health in a Shrinking World: Ecology, Management and Conservation. Proceedings of the Wildlife Diseases Association International Conference*, pp. 42–43. Wildlife Diseases Association: Cairns.

Rose K (2005). *Wildlife Health Investigation Manual.* Zoological Parks Board of New South Wales: Sydney.

Sikarskie JG (1992). The role of veterinary medicine in wildlife rehabilitation. *Journal of Zoo and Wildlife Medicine* **23**, 397–400.

Spielman D (2001). The roles of contagious diseases in natural populations, endangered populations, captive populations, and in wildlife breeding, translocation and rehabilitation programs. In *Veterinary Conservation Biology: Wildlife Health and Management in Australasia. Proceedings of a joint conference* (eds A Martin & L Vogelnest), pp. 205–210, Sydney.

Tribe A & Hanger J (2005). The vet visit. In *Proceedings of the National Wildlife Rehabilitation Conference*, Gold Coast, Australia.

Viggers KL, Lindenmayer DB & Spratt DM (1993). The importance of disease in reintroduction programmes. *Wildlife Research* **20**, 687–698.

Vogelnest L (2005). Disease considerations for the release of rehabilitated wildlife. In *Proceedings of the National Wildlife Rehabilitation Conference*, Gold Coast, Australia.

Walraven E (1994). Rehabilitation and reintroduction of native fauna. In *Wildlife, Proceedings 233*, pp. 193–205. Post Graduate Committee in Veterinary Science: Sydney.

Woodford MH (1993). International disease implications for wildlife translocations. *Journal of Zoo and Wildlife Medicine* **24**, 265–270.

Woodford MH (2001). *Quarantine and Health Screening Protocols for Wildlife prior to Translocation and Release in to the Wild.* Office International de Epizooties, Veterinary Specialist Group/Species Survival Commission of the World Conservation Union, Care for the Wild International and the European Association of Zoo and Wildlife Veterinarians.

Woods R & Bunn C (2005). Wildlife health surveillance in Australia. *Microbiology Australia* **26**, 56–58.

Zsivanovits HP (2004). Suggestions to optimise recovery and release while minimizing the disease risks associated with raptor rehabilitation. *Journal of Wildlife Rehabilitation* **27**, 4–15.

2. Veterinary aspects of hand-rearing orphaned marsupials

Helen McCracken

1 MARSUPIAL NEONATAL BIOLOGY AND PHYSIOLOGY: CONSIDERATIONS FOR HAND-REARING

The hand-rearing of any young animal should mimic the natural situation as closely as possible to optimise its chance of normal health, growth rate and assimilation into a social group of its own species. It is therefore important to have a good knowledge of the development, growth, environment, diet and parental care of the young animal in the natural rearing situation.

1.1 Development of marsupial neonates

The characteristic of marsupials that most clearly distinguishes them from eutherian mammals is the immaturity of their young at birth (Tyndale-Biscoe & Janssens 1988). Gestation is very short in marsupials, ranging from 11–35 d (Tyndale-Biscoe & Renfree 1987) and the young are born at an early embryonic stage, weighing from 3–6 mg in the honey possum (*Tarsipes rostratus*) to approximately 900 mg in the larger kangaroos. The neonate climbs from its mother's cloaca towards her mammary gland where it attaches to a teat and completes embryonic development and growth (Russell 1982; Merchant 1989). In most marsupial species, the teats are enclosed within a bag-like pouch that supports the young as it grows, providing an environment with stable temperature and humidity. Many small dasyurid species and the numbat (*Myrmecobius fasciatus*), however, have no pouch. In some of these dasyurids, lateral folds of skin develop during gestation and progressively cover the young after attachment, partially enclosing them. As they grow, the young become exposed and hang from the mother's ventrum, supported by the *ilio-marsupialis* muscle that extends into the mammary gland and each teat. Larger dasyurids, bandicoots, wombats, marsupial moles, the koala (*Phascolarctos cinereus*), yellow-bellied glider (*Petaurus australis*), musky rat-kangaroo (*Hypsiprymnodon moschatus*) and greater bilby (*Macrotis lagotis*) all have permanent enveloping pouches that open caudally. All other macropods, possums and gliders have deep permanent forward-opening pouches. These permanent pouches may be closed and opened by contraction and relaxation of the underlying muscle, the *panniculus carnosus*. The number of teats varies between species, ranging from two to twelve. In many species the normal litter size is less than the number of teats and newly born young will attach to teats that were not used by the preceding offspring (Tyndale-Biscoe 2005). In dasyurids, particularly the smaller species, it has been observed that some species give birth to more young than can be accommodated on the teats (see Chapter 10).

The newborn marsupial is a unique composite of embryonic structures and precociously developed functional organs. The head, shoulders and fore limbs are relatively large and well-developed, enabling it to make its way to the teats unaided, whereas the hind limbs are vestigial at this stage. The neonate has closed eyes and the lips of the mouth are fused laterally, leaving only a small rostral opening. The buccal cavity, however, is large and the tongue is well-developed and muscular, allowing the young to firmly attach to a teat and commence suckling. The pinnae of the ears are fused forwards to the skin of the head. The stomach, duodenum and pancreas are in an advanced stage of development at birth, permitting digestion of milk and absorption of nutrients, but the remaining intestinal tract is poorly developed at this stage (Russell 1982; Tyndale-Biscoe 2005). The neonatal kidneys cannot concentrate urine, but water loss is compensated by the continuous suckling of the mother's dilute early-stage milk. The newborn marsupial is bright red with prominent subcutaneous blood vessels and a moist skin. While it has functional lungs, cutaneous respiration also plays a role in oxygenation during the initial climb to the pouch and in the early days of pouch life (Tyndale-Biscoe 2005). The pouch young (PY) has a form of haemoglobin adapted to the air in the pouch that is lower in oxygen and higher in carbon dioxide concentrations than ambient levels (Baudinette et al. 1985). The early PY is hairless and thyroid function is not yet developed, hence it is completely ectothermic, relying on the stable temperature of the pouch for thermoregulation (Russell 1982; Tyndale-Biscoe 2005).

As the PY grows, it undergoes a series of physical changes. The rate of growth and length of pouch life varies greatly between marsupial species, but the sequence of development of these changes is similar for all species. This development has been best studied in the tammar wallaby (*Macropus eugenii*) and the sequence in this species is indicative of the general marsupial pattern. After a gestation period of 28 d, the tammar wallaby is born at a weight of 0.3–0.4 g. On arrival in the pouch the neonate attaches to one of four teats. If the mother has been recently suckling a previous offspring, only three teats are available to the neonate, as the suckled teat is too large and elongated. For the first 100 d of pouch life, the PY remains continuously attached to that teat. There will be no milk production from the remaining teats from two days post partum, hence transfer to another teat later in lactation is not possible. During this first 100 d, the pinnae become detached from the head and change to pointing backwards. The upper and lower lips gradually separate from each other so the mouth opening becomes progressively larger. The skin gradually becomes pigmented and whiskers develop. At approximately 100 d the PY weighs 100 g, its lips are fully separated and it relinquishes the teat for the first time. Thereafter, it is intermittently attached until weaning and the process of dental eruption commences. At about 140 d, the eyes open and a fine layer of fur has formed over the body. By day 160, the kidneys are fully developed and the PY can produce concentrated urine. By day 180, the fur has thickened, thyroid function has fully developed and the PY has become endothermic (Tyndale-Biscoe & Janssens 1988; Tyndale-Biscoe 2005).

At this stage, the young marsupial is considered roughly equivalent developmentally to the newborn eutherian. In most species, this is the time of first exit from the pouch (Maynes 1976; Russell 1982). The preferred body temperature of the newly endothermic PY is several degrees lower than the temperature inside the closed pouch as this expends the least amount of energy and oxygen. It therefore needs to dissipate heat and achieves this by occasionally putting the head out of the pouch, and in some species begins to nibble grass or leaves at these times. The tammar wallaby first leaves the pouch at approximately 200 d and begins to hop around, intermittently returning to the pouch until it exits permanently at about 250 d. Herbage progressively forms a greater proportion of the young's diet between days 200 and 250 but it continues to drink milk, with peak milk intake occurring at day 240. Significant digestive physiological changes occur at this time as the diet gradually transitions from high-fat milk to grass and herbs. The gastrointestinal tracts of young herbivorous marsupials require inoculation with appropriate bacteria and protozoa to facilitate fermentation of ingesta in the forestomach in some species and caecum in others. Macropods likely acquire this microflora by grazing on herbage among the faeces of adult conspecifics and by licking saliva from their mother's lips (Tyndale-Biscoe 2005). Young common wombats (*Vombatus ursinus*) at the age of first emergence from the burrow have been observed to eat small moist faecal pellets from adults which differ from the normal dry faecal cubes (Triggs 1996; Boer 1998). The young koala feeds on a semi-liquid substance termed 'pap' produced from its mother's cloaca soon after it commences putting its head out of the pouch (see Chapter 8). Pap is unlike normal koala faeces and microbially resembles the contents of the caecum (Osawa et al. 1993).

Most young marsupials continue to suckle occasionally following permanent pouch exit. In the tammar wallaby, suckling ceases between days 300 to 350 (Tyndale-Biscoe & Janssens 1988). The period following initial pouch exit is one of accelerated growth in all species. The body weight of the young tammar wallaby increases from 500 g at 200 d to 2 kg at 300 d, then to 3 kg at 360 d (Janssens et al. 1997). Female marsupials reach sexual maturity shortly after weaning in most species. Males generally mature at a later age.

1.2 Pouch environment

The marsupial pouch provides the developing young with an environment of high stable humidity and temperature. For the hairless PY, the moist humid environment likely plays a role in cutaneous respiration immediately after birth, and subsequently prevents drying of the skin and dehydration (Fig. 2.1). Pouch temperature has been measured in several species as 34–36°C, slightly lower than marsupial adult body temperature (35.5–37°C) (Tyndale-Biscoe 2005). Young macropods, koalas, wombats and brushtail possums (*Trichosurus* spp.) will usually be solitary in the pouch, but the young of most other species generally share the pouch with one or two littermates (Russell 1982). When the neonate attaches, the teat is small (1–5 mm long) but it gradually elongates as the PY grows, reaching a length of several centimetres by late lactation, enabling access for suckling by the emerged young. Following weaning, these suckled teats gradually return to their original size, in readiness for attachment by new offspring. The pouch also expands to accommodate the growing young, then involutes following final pouch exit. Thermal intolerance is the main stimulus for the PY to commence leaving the pouch, but it appears that the mother also plays a role in determining the time of permanent emergence, in many cases denying re-entry at the time of birth of the next offspring.

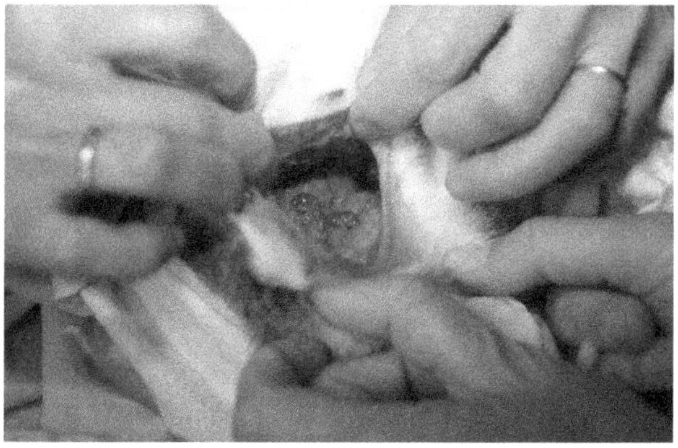

Figure 2.1 Pouch of an eastern barred bandicoot, *Perameles gunnii,* with newborn PY. Note the deep moist pouch interior with a caudally directed opening, characteristic of bandicoots.

1.3 Maternal care

Russell (1982) described three basic patterns of maternal care seen across the range of marsupial species.

Pattern A is seen in species with shallow pouches and large litters, e.g. pygmy possums (*Burramyidae*) and dasyurids. Initially the PY are well-attached to the teats, but when their lips become fully open and they can release the teat the mother leaves them in a nest when she goes out to feed. At this stage their eyes are closed, they have little fur and are probably unable to thermoregulate. While in the nest, milk is their only food. When their eyes are open and they are well-furred, they begin to leave the nest with their mother, on her back or at foot. Eventually they leave the nest alone to forage, returning occasionally to their mother to suckle, until weaning (Russell 1982).

Pattern B is seen in species that utilise nests or burrows and have deep pouches, e.g. ringtail possums (*Pseudocheirus* spp.), brushtail possums, gliders, wombats, bandicoots and the greater bilby. After they begin to release the teat, the PY remain in the pouch until their eyes are opened, they are well-furred and able to thermoregulate. They leave the pouch for the first time while their mother is in the nest and may return to it at times, although they will generally be left in the nest when their mother goes out to feed. While in the nest, milk is their only food. After a period, the young begin to leave the nest with the mother, on her back or at foot. Eventually they leave the nest alone to forage, returning occasionally to their mother to suckle, until weaning (Russell 1982).

Pattern C is seen in species that have deep pouches and don't utilise nests, e.g. macropods and the koala. When the PY is well-furred and the eyes are open, it begins to protrude its head out of the pouch occasionally. It then leaves the pouch for brief periods that gradually lengthen until it is permanently out of the pouch. It continues to suckle from the mother and follows her closely until weaning, and in some species beyond that time (Russell 1982).

In all cases, during the pouch-dwelling period, the mother keeps the PY and pouch clean by frequently licking them to remove urine and faeces. It is conjectured that this process may also stimulate the PY to urinate and defecate, hence they may be unable to do this spontaneously in the hand-rearing situation (Russell 1982). This maternal behaviour likely plays an important role in water conservation in arid-dwelling species, as the passing of dilute urine by the PY could otherwise result in severe water loss from the mother–young unit, compromising their survival. It is probable that this behaviour also plays a role in immune protection for the PY against potential pathogens in its immediate environment. Microorganisms ingested by the mother from the pouch and faeces will induce production of specific maternal antibodies that will be transferred to the PY in the immunoglobulins of the milk (Tyndale-Biscoe 2005). The mother continues to groom her young after it leaves the pouch, although it will start self-grooming at this stage (Russell 1982).

Macropods, possums, koalas and wombats, despite the considerable manual dexterity of the three former groups, do not assist young to return to the pouch. Greatest mortality occurs when young are big enough to fall out of the pouch, but not large enough to return to it unaided (Russell 1982).

1.4 Passive immunity and development of the immune system

Immunology of marsupial PY has been studied in a range of species, with most work being done on the quokka (*Setonix brachyurus*), common wallaroo (*Macropus robustus*), tammar wallaby and Virginia opossum (*Didelphis virginiana*). These studies demonstrated significant differences in the development of the immune systems of marsupials and eutherian mammals. In eutherians, thymus development largely occurs during gestation, and in most species active immune responses are almost mature at the time of birth. In the final stages of immune system maturation, the neonate is protected by a short period of passive transfer of maternal immunoglobulins, either across the placenta, as in the rabbit and human, or from the first milk (colostrum), as in horses and ruminants (Deane & Cooper 1988; Tyndale-Biscoe 2005).

In contrast with eutherians, no lymphoid tissue is present in marsupials at birth, hence the neonate is not immunocompetent (Tyndale-Biscoe 2005). Within the first few days of life lymphocytes appear in the thymus, the first lymphoid tissue to develop. All diprotodont marsupials except wombats (macropods, possums, gliders, koala) differ from most eutherians in that they possess both a cervical and thoracic thymus. Wombats have only a cervical thymus and the polyprotodont marsupials (dasyurids, bandicoots, bilby, marsupial moles) possess only a thoracic thymus. In the species with two thymi, the cervical is the larger and dominant one at all stages of development (Yadav 1973). Within the first week, the first lymph nodes appear and in the next few weeks differentiated lymphoid tissue develops in the spleen. Active immunity commences development at days 7–17 when Hassall's corpuscles are first evident in the thymus and antibody responses to antigen can be detected. Most species reach immune system maturation at the time of first release of the maternal teat, approximately halfway through pouch life (Basden et al. 1997; Old & Deane 2000).

During the period of lymphoid tissue development, PY are protected by the passive transfer of maternal immunoglobulins, essential for survival in the microbially rich environment of the pouch. Transplacental transfer from the yolk sac fluid has been demonstrated in only one species, the tammar wallaby (Deane et al. 1990). In all species studied, this protection is primarily provided through immunoglobulins secreted into the milk until the PY first relinquishes the teat, and in some cases until the time of first exit from the pouch. The immunoglobulins are absorbed unchanged across the gut epithelium of the PY throughout this period, in contrast with eutherians in which absorption of colostral antibodies is only possible in the first few days after birth. In the quokka, this passively acquired antibody has a half-life of only 8–9 d, and has virtually disappeared from the blood by 4–6 wk after cessation of suckling (Old & Deane 2000; Yadav 1971). It is also believed that secretions from the pouch epithelium and macrophages found in marsupial milk at different stages of lactation may play roles in immunological protection of the PY (Old & Deane 2000).

Hence, throughout most of pouch life the PY has two sources of immunological protection—passively acquired maternal antibodies and its own active responses. Pouch life begins with the young wholly dependent upon maternal protection and ends with it wholly dependent upon its own immune system (Deane & Cooper 1988). Hand-reared PY must be considered at risk from infectious disease because they are no longer receiving passive immunity, they will have lost most maternal immunoglobulins by about 4 wk after separation from the mother, and they are protected only by an underdeveloped active immune system.

1.5 Marsupial milk

Milk production in eutherian mammals has been studied for a wide range of species. The milk of each species is relatively uniform in composition throughout lactation once it is fully established but there are major differences between species in milk concentration and composition, reflecting the differences in lifestyles, suckling regimes and the production of altricial or precocial young. In contrast, in marsupials, milk composition is relatively uniform between species, reflecting the similarities in development and maternal care across the taxa. Notably, however, because the milk of marsupials must support the young from its embryonic state at birth until it is independent, the composition changes profoundly both quantitatively and qualitatively during the course of lactation (Merchant 1989; Tyndale-Biscoe 2005). Studies have demonstrated that the milks of numerous species show similar sequences of change through lactation (Jackson 2003; Tyndale-Biscoe 2005). These changes in the tammar wallaby and common brushtail (*Trichosurus vulpecula*) are examples of this marsupial trend (Figs 2.2, 2.3). In these species, there is a gradual increase in the carbohydrate concentration from 4–6% at birth to a peak of 11–13% shortly before first pouch emergence, after which there is a rapid decline to 1% in the tammar wallaby and 4% in the common brushtail. The peak concentration reached is very high compared with the concentration of carbohydrate in the milk of most eutherians. Protein concentration also gradually increases from 3–4% at birth to 7–8% at pouch exit, followed by a slight decrease. Changes in lipid concentration are more variable between species, especially towards the end of lactation, but the general trend is an initial gradual increase from 1–2% at birth, with a sharp rise at the time of pouch emergence up to 9% in the common brushtail and 24% in the tammar wallaby. The net effect of these changes is a gradual increase in the energy content of the milk throughout lactation. In the tammar wallaby the change is from approximately 2500 kJ/L at birth to 5000 kJ/L at day 180, then a rapid increase to 11 000–12 000 kJ/L by days 250–300 (Messer & Walker 1992). These marked changes meet the increasing energetic demands of the growing PY as it gradually becomes more active and develops endothermy.

In eutherians, lactose is the major milk sugar. It is rapidly digested to galactose and glucose by the action of the intestinal lactase, β-galactosidase, located extracellularly in the microvillus brush border of the gut epithelium (Messer et al. 1989). In contrast, marsupial milk has very low levels of lactose. The predominant

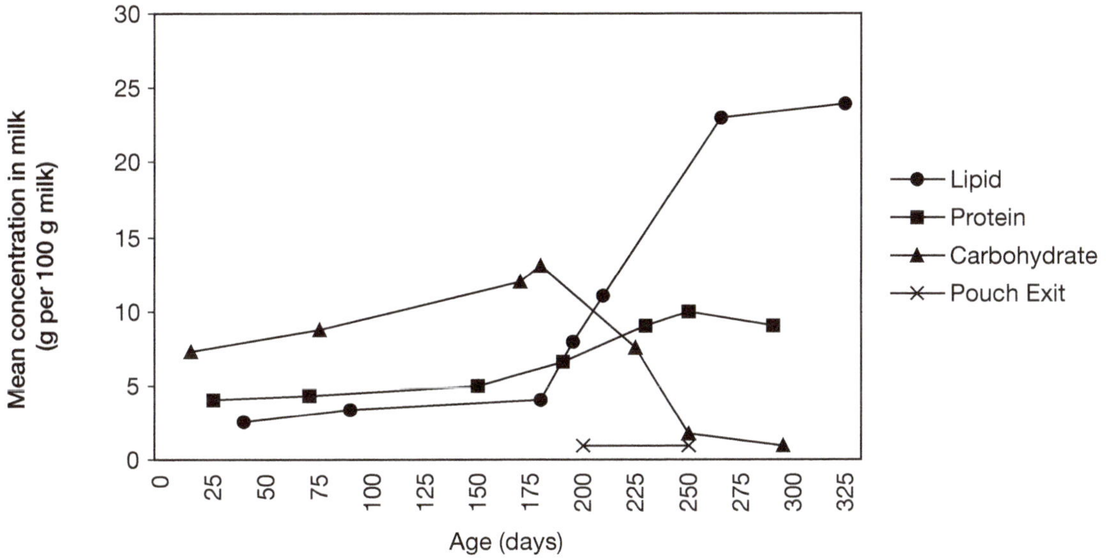

Figure 2.2 Changes in the mean concentrations of lipids, proteins and carbohydrates in the milk of the tammar wallaby throughout lactation. The pouch exit bar indicates the times of initial and permanent pouch exit. (Source: Adapted from Messer & Walker (1992).)

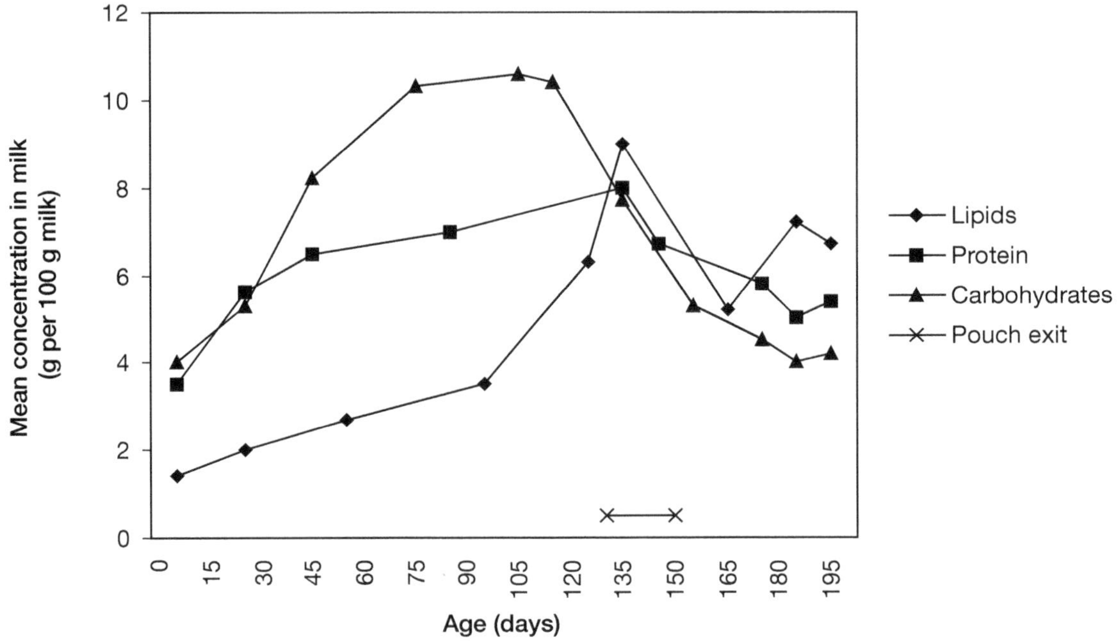

Figure 2.3 Changes in the mean concentrations of lipids, proteins and carbohydrates in the milk of the common brushtail throughout lactation. The pouch exit bar indicates the times of initial and permanent pouch exit. (Source: Adapted from Messer & Walker (1992).)

carbohydrates for most of lactation are oligosaccharides of increasing size, most composed of 1 unit of lactose and 1–8 units of galactose. Since single molecules of oligosaccharides exert the same osmotic pressure regardless of size, these increasingly larger oligosaccharides provide progressively higher levels of carbohydrate to the growing PY while maintaining the milk necessarily isotonic with plasma (Tyndale-Biscoe 2005). In contrast to the eutherian model, these oligosaccharides are transported from the gut lumen into the epithelial cells where they are digested to galactose and glucose by intracellular β-galactosidase. These monosaccharides then enter the circulation, providing the major energy source of the growing PY. Studies in the tammar wallaby suggest that the transport of sugars into the cells may occur via pinocytosis. As this is a relatively slow process, it limits the rate at which artificial formula containing lactose or oligosaccharides may be bottle-fed to hand-reared PY without causing diarrhoea. Mother-reared PY apparently suckle constantly, presumably drinking small volumes very frequently. For practical reasons, hand-reared PY are fed less frequently. Sugar accumulated in the gut lumen, due to a combination of infrequent high-volume feeds and the slow process of pinocytosis, may result in an osmotic diarrhoea (Messer et al. 1989).

The rapid decrease in carbohydrate content of marsupial milk that occurs shortly before first pouch emergence is accompanied by a change from large oligosaccharides to more simple sugars, mainly galactose in macropods, and lactose in brushtail and ringtail possums. In addition to intracellular enzyme, brushtail possums have extracellular lactase that increases in activity at the time of pouch emergence (Messer & Walker 1992). These rapid quantitative and qualitative changes in milk carbohydrates occur as the PY is starting to eat solids, and are likely related to the establishment of conditions in the gut suitable for the microbial populations required for the digestion of herbage (Merchant 1989). Hence the function of the alimentary tract is transitioning from the digestion of simple milk sugars to fermentative digestion of complex plant carbohydrates. At the time of this change, the lipid concentration of the milk dramatically increases as it becomes the major energy source, supporting the emerging PY at its time of most rapid growth and highest energetic demands as it develops locomotion and independent thermoregulation (Tyndale-Biscoe 2005).

Marsupial milk lipid is principally composed of triglycerides. The predominant fatty acid in early lactation is palmitic acid, which may be important in the synthesis of surfactants required for oxygen transport across the lungs. In the second half of lactation oleic acid is predominant, playing a key role in the synthesis of myelin sheaths around developing nerve fibres (Tyndale-Biscoe 2005).

Milk protein has been mainly studied in the tammar wallaby. Marked changes in amino acid composition occur during lactation, notably including a sudden increase in sulphur-containing amino acids, particularly cysteine, during the period of hair follicle development. This may be of significance in hand-rearing as many PY are raised from this stage of development (Messer & Walker 1992).

The concentrations of minerals in marsupial milks are similar to those in eutherians, with a few notable exceptions. Copper and iron concentrations are substantially higher in marsupial milks, particularly in early lactation when the neonatal liver has a limited storage capacity. In the early milk of tammar wallabies, copper and iron concentrations are approximately 5 mg/L and 22 mg/L respectively, compared with 0.1–0.6 mg/L and 0.3–0.5 mg/L respectively in eutherian milk. Calcium and phosphorus concentrations increase throughout lactation, doubling in levels from early- to late-stage milk, presumably reflecting the increasing requirements for these elements in the accelerated bone growth of the older PY. Sodium and potassium concentrations also change during lactation, with high concentrations of sodium in early milk decreasing to low levels at the stage of kidney maturation in the pre-emergent PY. Potassium levels follow the opposite trend (Merchant 1989; Tyndale-Biscoe 2005).

In summary, lactation in marsupials is a much more complex and changing process than that of eutherians. A major challenge in hand-rearing marsupials is offering a suitable substitute milk diet that can meet the changing needs of the developing PY.

1.6 Weaning

In species with patterns A and B modes of maternal care, weaning commences when the PY leaves the nest and accompanies its mother when she is foraging. In pattern C species, it commences when the PY first protrudes its head from the pouch, koalas eating pap and macropods nibbling a little grass, although they eat very little until they first fully emerge from the pouch. Weaning results mainly from a gradually increasing lack of co-operation by the mother when the young attempts to suckle (Russell 1982). The natural diets of adult marsupials are described in later chapters.

2 CRITERIA FOR INTERVENTION

Marsupial PY may require hand-rearing for several reasons. In Australia, the majority of hand-rearing cases are wildlife found orphaned after the mother's death by motor vehicle accident, shooting or predation. A live PY may be found in the pouch of the dead mother or emergent young may be found near the body. In some cases, emergent PY or young-at-heel may stray from the site after the mother is killed, and as they are still partially milk-dependent they will not survive alone. If pouch examination of a dead female reveals a lactating mammary gland and elongated teat, a careful search should be made of the accident scene for the dependent young. For PY at the stage of permanent attachment to the teat, removal must be performed carefully to avoid palatal and lip injuries as the tip of the suckled teat is expanded to fill the oral cavity. Alternatively, the teat may be cut from the pouch and pinned to the inside of the PY's substitute pouch. The PY will generally release it within 2–3 hr (L Dennis *pers. comm.*).

Wild PY may be found and hand-reared after they have been thrown from the mother's pouch or dislodged from her back during pursuit by a predator or in response to other stressful stimuli. It is a survival technique of some marsupial species to jettison PY in such circumstances by the simultaneous relaxation of the pouch sphincter and contraction of the longitudinal subcutaneous muscles of the pouch, effectively causing the pouch to disappear. PY that are still permanently attached to the teat are rarely thrown, as they are not a great burden to the fleeing mother. Emergent young are often safely reunited with their mother following the stressful event. The at-risk animals in these circumstances are the intermittently attached, pre-emergent PY as they are readily ejected but incapable of independently returning to the pouch. They will not survive without intervention (Booth 2002).

In captive marsupials, PY may require hand-rearing following the death of the mother. They may also be intentionally pulled for hand-rearing for a range of reasons, including failure to thrive, illness of the PY or mother, history of poor maternal care by an individual, and the conditioning of a young animal for display or educational purposes. PY may also be thrown from the pouch in captive animals, most commonly during or following capture and restraint. On these occasions, attempts may be made to return the ejected PY to the mother's pouch as an alternative to hand-rearing. This is rarely successful with PY at the stage of permanent attachment to the teat—if they have been thrown, it is generally an indication of failure of lactation or poor PY viability. Return to the pouch of pre-emergent, intermittently attached PY is best achieved following anaesthesia of the mother. The PY is warmed and rehydrated if required then returned to the pouch, and several long pieces of adhesive tape placed transversely across the pouch opening. The mother will generally groom off this tape over the next few days. It should be noted that such attempts may be unsuccessful and the PY may be re-thrown, hence it is advisable to alternatively make an immediate decision to hand-rear in the case of valuable individuals. To reduce the risk of ejection of pre-emergent PY following anaesthesia of females, care should be taken throughout the procedure to optimise the viability of the young, maintaining it at normal body temperature by preventing heat loss in the mother and closing the pouch with clothes pegs. If the female's body temperature drops significantly, a local heat source should be applied to the outside of the pouch, taking care to avoid overheating. It is advisable to tape the pouch closed prior to recovery as described above.

Fostering PY to lactating females of the same or similar species is possible only in limited circumstances and is rarely a viable alternative to hand-rearing. It is, however, utilised as a tool in endangered species breeding programs to increase the reproductive output of valuable individuals by double- and triple-clutching (Chapter 7, section 11.2.7). Several critical criteria must be met to optimise the success of this procedure: the foster mother must be of a species of similar size, pouch-life length and life habits to the PY, she must be at a very similar stage of lactation as the natural mother so that her teat size and milk composition is suitable for the PY, and her own young must be removed prior to cross-fostering (Booth 2002).

3 ASSESSMENT OF THE POUCH YOUNG

PY are very distressed if not enclosed within a pouch. On presentation, they should be immediately placed in a cloth bag, woollen hat or thick sock or wrapped loosely in a blanket. The artificial 'pouch' should then be placed in a heated box or against a person's skin beneath warm clothing until the PY can be examined.

Assessment of history for wild PY must include their exact geographic origin. If the PY is successfully hand-reared for release, most wildlife agencies require that it be released in close proximity to its site of origin. History-taking should include how the PY was orphaned, as this circumstance may influence its condition. It should also include the length of time since the PY was found, and details of its environmental conditions and nutrition since discovery.

A full clinical examination should be performed to assess the viability of the PY (sections 6.2 and 6.3 describe techniques for restraint and clinical examination). Assessment must be made of the animal's general condition, hydration status and attitude. Many PY present with some degree of hypothermia and dehydration due to a lengthy interval between orphaning and discovery, or because their rescuer did not provide adequate environmental and/or nutritional support. The healthy newly orphaned or ejected PY will wriggle vigorously, resist handling, demonstrate pouch-seeking behaviour and may vocalise loudly. Examine for problems related to the orphaning incident. Victims of vehicular accidents and arboreal species found under trees may have contusions and/or fractures. The most common fractures in macropod PY orphaned by vehicular accidents are of the tail and hind limbs (usually tibias or metatarsals), often occurring in a triad of fractures of both limbs and the tail (Speare 1988). Examine closely for this triad if one fracture is detected. Young possums, particularly ringtails, are frequently victims of cat attack and often present with small puncture wounds that may be difficult to find in their dense fur. Potentially fatal septicaemia is a common sequela to these wounds.

Species determination is crucial to determining the age, nutritional requirements and specific disease susceptibilities of the PY. Hairless and lightly furred PY can be difficult to identify, as they may not yet have the distinguishing features of their species. They are best identified from descriptions of their mothers, if available.

Age determination is essential for decision-making on the animal's prognosis and environmental and nutritional requirements. Guides to age determination for different marsupial species are provided in separate chapters in this text. More accurate age estimates may

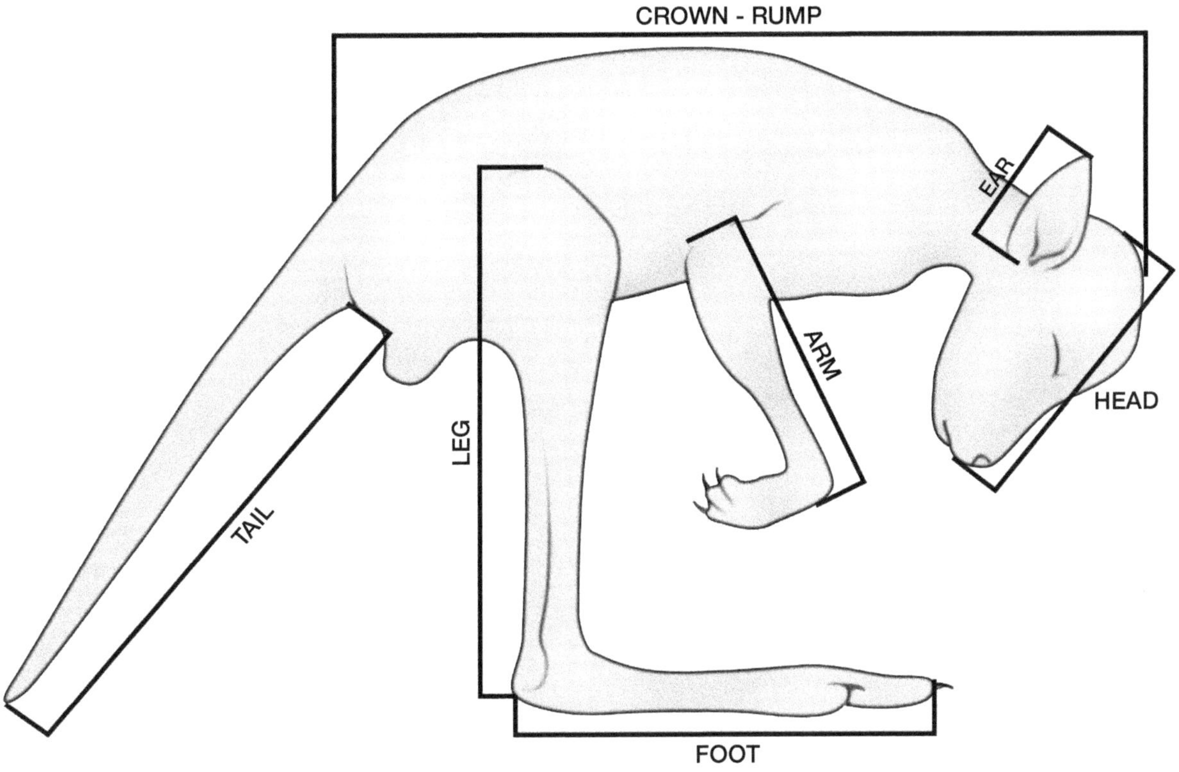

Figure 2.4 Standard measurements of marsupial PY used for age determination. (Source: Adapted from Sharman et al. (1964).)

be made from growth charts available from Wombaroo Food Products (see Appendix 3). These guides and charts require standard body measurements of the PY (Fig. 2.4). Having estimated the age of the PY, it is useful to also determine its age factor—the ratio of the present age of the PY to the age of that species at final pouch emergence (i.e. the age factor at the time of final pouch emergence is 1.0). The age factor indicates how far progressed the PY is in its development and assists in determining its prognosis for survival and its care and dietary requirements (B Rich *pers. comm.*).

A prognosis for the PY must take into consideration its health status and stage of development. It is generally considered that PY with an age factor of <0.4 (hairless with eyes closed, mouth partially closed and ears fused to the head) have a very poor prognosis. They have usually suffered significant hypothermia, are still highly dependent on passive immunity acquired constantly from their mother's milk and require an exacting rearing environment, including feeding every 2–3 hr. PY at this stage have been successfully raised by highly experienced carers but, unless they present in good condition and can be given the appropriate expert and intensive care, it is generally accepted that they should be euthanased. PY that are hairless and pigmented or with very fine fur have a better prognosis, but also require expert and intensive care. PY with a covering of fine fur have a good chance of survival with expert care, and thickly furred PY usually survive (Woods 1999). Wild PY must be assessed for their suitability for future release. If they present with injuries that would result in permanent handicap they should be euthanased or, if their problem does not affect their quality of life, considered for permanent care in a zoo or wildlife sanctuary.

4 INITIAL CARE AND STABILISATION

PY requiring hand-rearing are generally presented with hypothermia, so placement in a substitute pouch and slowly warming to normal body temperature (35.5–37.0°C) over a 2–3 hr period is the first priority. Human infant humidicribs and animal intensive care units are ideal, but a lidded box or basket with heating pad is suitable. Air temperature within these units should be 32–36°C and must be monitored, ideally with a digital thermometer, to ensure adequate warmth and to avoid overheating. The PY should not be fed until it has reached normal body temperature. Prolonged hypothermia and milk deprivation may precipitate hypoglycaemia in young animals, hence SC or IV fluid therapy with solutions containing glucose should be given in these circumstances. Ideally, blood glucose is measured before and after therapy but, given the small size of these patients, this is not always possible, hence treatment is given empirically. The recurrent tarsal veins and lateral tail veins are suitable venipuncture sites (Booth 2002). Dehydration must be corrected by IV or oral fluids, depending on its severity.

5 HUSBANDRY

The major challenges in hand-rearing marsupials are to simulate the functions of the pouch, which provides both psychological security and a stable thermal environment until the PY becomes endothermic, and to provide an appropriate milk substitute that can meet the changing needs of the developing marsupial until weaning (Booth 2002). It is also essential to be familiar with the ages of onset of key developmental stages and the type of maternal care of the species concerned, to simulate these as closely as possible. Premature or delayed pouch emergence or weaning, or failure to provide appropriate simulated maternal care (e.g. artificial nest or burrow at time of emergence or substitute mother for PY carried on the back following emergence) may be detrimental to health, growth and development of species-appropriate behaviour and activity. Care requirements for hand-reared PY at different developmental stages are presented in brief in Table 2.1.

5.1 Housing

5.1.1 Artificial pouch

The hand-reared PY requires an artificial pouch that simulates as closely as possible the security, complete enclosure and warmth of the natural state. Pouches are best made out of natural fibre, e.g. wool and cotton. For pre-emergent PY, a simple bag style with the top of the bag folded over or closed with pegs is used (Fig. 2.5). Pouches for emergent PY are designed to be suspended and have an opening at the front to give the PY independent access for exit and entry (Fig. 2.6). There should be no seam edges inside a pouch, as animals may catch toenails or teeth on them or suck and

Table 2.1 Care requirements for hand-reared marsupials at different developmental stages

Age factor	Physical features	Milk feeds per day	Temperature required	Other care requirements
<0.4	Furless pink skin, eyes closed, ears down, yellow custard faeces.	8–12	34–36°C	Prognosis poor, pouch-bound. Apply lotion to skin at least twice daily.
0.4–0.5	Furless, eyes will open, ears will become erect, thick yellow custard faeces.	6–8	32–34°C	Pouch-bound. Apply lotion to skin once to twice daily.
0.5–0.6	Colour deepening, very fine fur appearing. Starting to put head out of pouch. Mustard toothpaste consistency faeces.	5–6	30–32°C	Pouch-bound. Apply lotion to tail and feet daily. Provide with solid foods, soil and adult faeces for contact when head is out of pouch.
0.6–0.7	Short sleek fur, endothermic, starting to emerge from pouch. Olive-green toothpaste to soft pelleted faeces.	4–5	28–30°C	Provide opportunity for pouch exit and return. Give access to solid foods with constant access to pouch.
0.7–1.0	Fur becoming longer and denser. Well co-ordinated movement when out of pouch. Soft to firm dark green pellets.	3–4	20–25°C	Establish an outdoor enclosure with access to pouch, nest-box or substitute mother, depending on species. Provide natural diet.
>1.0	Fur dense. Age of permanent emergence from pouch. Firm dark green pellets.	2	Ambient	Remove pouch. Gradually reduce milk feeds to zero per day. Reduce human contact.

Source: Adapted from George (1990), Bellamy (1994) and Booth (2002).

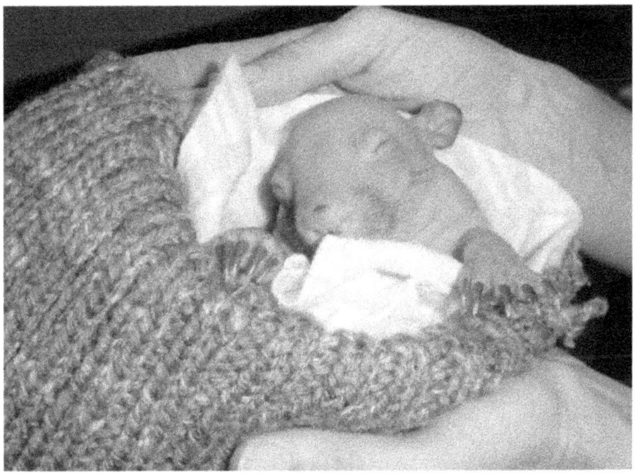

Figure 2.5 Simple bag style of artificial pouch for pre-emergent common wombat. The PY is wrapped in a soft cotton liner within the woollen bag. Photo: L Sauvarin.

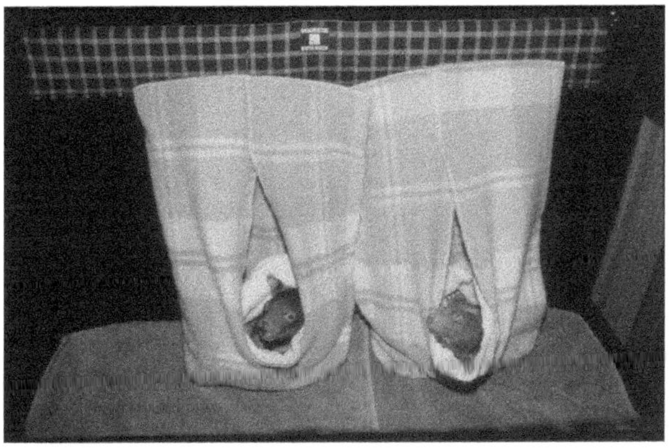

Figure 2.6 Artificial pouches for emergent PY. These are suspended by a coathanger inserted in the top, and have an opening to allow the PY to exit and enter independently. Heating is provided by an overhead heatlamp and/or a heatpad under the pouches. Where possible, PY are raised with a 'buddy' of the same species and similar age as this significantly reduces the imprinting to the human carer.

ingest loose threads. To avoid this, pouches should be either fully lined with concealed seam edges, or the seam edges should be on the outside surface. An ideal pouch for a pre-emergent PY would have a woollen or polar fleece insulating outer layer, lined with a soft cotton fabric. For hairless and finely furred PY and PY of any size with diarrhoea, a lining bag of soft cotton fabric should be used inside the pouch. This allows frequent inner linen changes without soiling the outer pouch. An ideal pouch for emergent PY has an outer layer of woollen blanket material lined with a soft cotton fabric, the edges of the opening finished with cotton bias binding and an opening in the top to allow the internal placement of a coathanger or a dowel rod for suspension. Pouches should be checked regularly for loose threads that may wind around extremities or limbs and cause avascular necrosis. PY should not share pouches with others unless this occurs naturally for their species, e.g. ringtail possums and bandicoots. For other species, however, raising a PY with a same-species 'buddy' of similar age is very beneficial, in that it promotes normal socialisation behaviour and reduces the imprinting of the PY on the human carer. Buddies are housed in separate pouches in close proximity to each other (Fig. 2.6).

The required temperature of the artificial pouch varies between stages of development (Table 2.1). The ectothermic hairless PY requires a constant 34–36°C, but as it develops hair and endothermy the temperature can be gradually decreased. Pouch temperature must be strictly controlled and monitored to prevent hypothermia and hyperthermia—common problems in hand-reared PY. It is best to use a probe thermometer, placed adjacent to the PY within the pouch. Heat may be provided by a variety of methods.

5.1.2 Hairless pouch young

These are very difficult to rear as they have exacting environmental requirements and are immune-deficient, being almost entirely dependent on passively acquired maternal antibodies. They are completely ectothermic and need to be kept at a constant temperature (34–36°C) and humidity, as in the natural pouch. This is best achieved in a human infant humidicrib or animal intensive care unit (with humidity at approximately 70%) or, less optimally, in a lidded box or basket heated by electric heatpads (Figs 2.7, 2.8). Hot water bottles are not suitable as heat sources as they cool quite rapidly and need to be monitored frequently. Electric and fan heaters are also unsatisfactory as they heat only one side of the PY. The pouch containing the PY must be wrapped in several small blankets within the unit, creating a secure enveloped environment. The pouch should be left open a little at the top to create some airflow to prevent overheating, and should be deep enough to create a narrow temperature gradient, allowing the PY to position at different levels according to its body temperature (Woods 1999). Hypothermia and hyperthermia are great risks at this stage because body temperature changes very rapidly with ambient fluxes, hence the PY should be removed from this stable environment as little as possible. Ideally

Figure 2.7 Temperature- and humidity-controlled unit for pre-emergent PY. The PY is within a bag pouch, wrapped loosely in blankets.

all feeding, cleaning and pouch changing should be done within the heated unit. The skin of these PY dries out very quickly in the artificial pouch, particularly in units without humidity control. It should be soft and smooth with a silky sheen. To maintain this, a moisturiser such as pawpaw ointment, pure sorbolene or lanolin oil should be applied frequently as a thin film over the entire body, paying particular attention to the tail and ears and flexor aspects of elbows and hocks (see Table 2.1 for recommended frequency of application). Spare pouches, cotton liners and moisturiser should be kept warm to prevent heat loss at times of linen changes and cream application (Speare 1988; Bellamy 1999; L Sauvarin *pers. comm.*)

5.1.3 Pre-emergent finely furred pouch young

These have a better prognosis than hairless PY but are also challenging to rear as they have poorly developed active immunity and are not fully endothermic. They need to be kept at constant temperature (30–32°C) and can be successfully housed in a simple heated box or basket, as described above, until close to the stage of first pouch emergence. As PY develop hair, cream can be applied less frequently. When they have a complete hair covering, only apply cream once per day to hairless areas (ears and feet).

5.1.4 Emergent pouch young

When the PY is approaching the age of first pouch emergence it should be placed in a situation that closely simulates the natural state, permitting it to commence thermoregulation by extending body parts out of the pouch then to exit and re-enter independently. From the age of first pouch exit, the PY will need to be encouraged to emerge for progressively longer periods until completely weaned from the pouch at the fully emergent body size indicated in growth charts for the species. This process must be very gradual. The initial forays from the pouch can be very stressful and the PY will require the carer's company at these times, simulating the presence of its mother. Same-age conspecific buddies and/or substitute mothers such as stuffed toys also reduce the stress associated with this stage. The emergent PY is endothermic

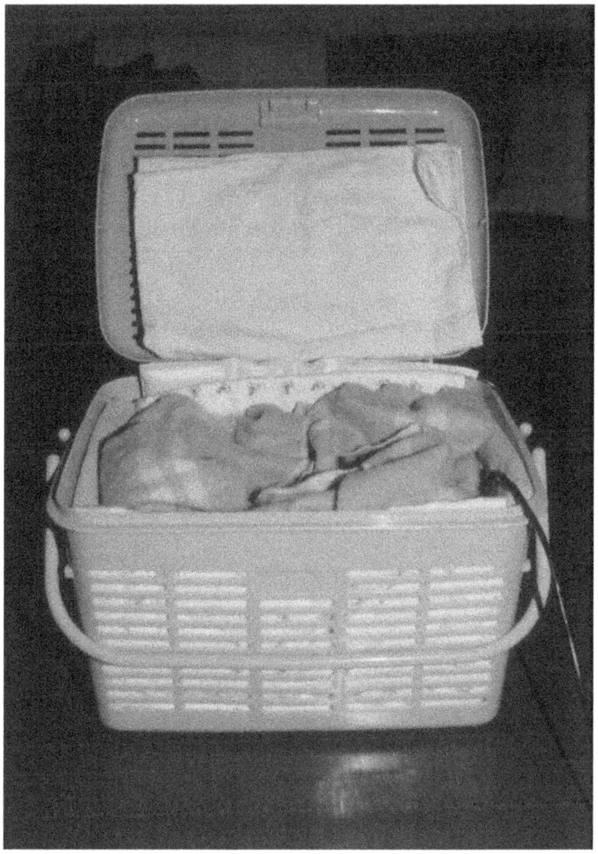

Figure 2.8 A simpler set-up for pre-emergent PY. The PY is within a bag pouch, wrapped loosely in blankets. Heating is provided by two heatpads, one positioned vertically at the rear of the basket and the other under the basket. A probe thermometer monitors temperature.

but remains in close proximity to the mother or within a nest or burrow in the natural state, hence environmental heating is still required (28–30°C).

Ground-dwelling species such as macropods and wombats are moved in their bag pouches into a front-opening pouch suspended over an electric heatpad in an indoor pen or portable infant cot (Fig. 2.9). An overhead heatlamp may also be provided. The pouch should be positioned so as to permit easy exit and re-entry. For macropods, freshly cut grass and hay are placed next to the pouch opening to encourage nibbling from the time of first emergence of the head, as in the natural situation. The pouch should be moved to an outdoor grassy yard daily (weather-dependent) for an increasing time to encourage grazing and provide necessary exposure to UV light (Fig. 2.10). It is important to provide the pouch until the age at which the species reaches permanent pouch exit.

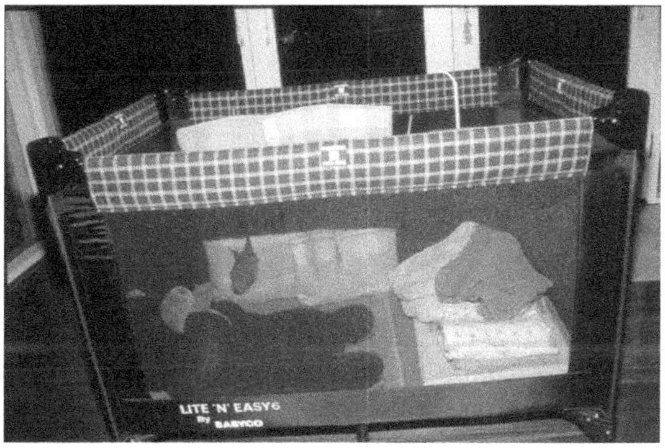

Figure 2.9 A portable infant cot provides a convenient and secure environment for the newly emergent PY. The suspended pouches each house one wombat. A toy bear acts as a substitute mother, a strategy to reduce imprinting. Replacement pouches and blankets are kept warm on a heatpad in the corner of the cot.

Figure 2.10 Emergent red kangaroo (*Macropus rufus*) PY in a suspended artificial pouch outdoors. At this age, PY are gradually introduced to the outside environment and provided with a weaning diet appropriate to their species.

Wombats first emerge from the pouch while in the maternal burrow, and for 1–2 months following permanent pouch exit they remain in the burrow while the mother leaves to graze. It is important to simulate this situation and maintain the emergent PY in a heated environment that limits its activity for this period. Seizure episodes are commonly reported in newly emergent hand-reared wombat PY (L Sauvarin & L Dennis *pers. comm.*; see also Chapter 9). These episodes generally resolve following milk feeding, and hypoglycaemia was demonstrated as the cause in several cases. Typically, affected animals have been encouraged to leave the pouch prematurely or given the opportunity to be inappropriately active, often in unheated environments (e.g. running around a room), due to a lack of appreciation of the natural environment and activity levels of the young wombat at this stage of development. It is probable that these circumstances precipitate hypoglycaemia in animals that do not yet have sufficient nutrient intake to sustain such energetic demands. This illustrates the need to be familiar with and closely simulate the maternal care and developmental milestones of each species in care.

Arboreal species such as possums and koalas are moved into a small enclosure furnished with several climbing branches and fresh browse of appropriate food species provided daily. Possums are housed in their pouches in a heated nest box that is fixed in an elevated position to encourage exit and entry. Koalas are maintained in their pouches in an open-topped box with a heatpad and a mother substitute (a furry toy or sheepskin wrapped around a thick branch). This box is placed on the floor of the enclosure, positioned to encourage climbing excursions.

When the PY is completely independent of the pouch, the environment no longer requires heating. Macropods and wombats should spend increasingly longer periods outside, ideally in a grassy yard to encourage grazing. The outdoor yard must provide adequate shade and shelter including hollow logs or other suitable substitute burrows for wombats. To prevent wombat escape by climbing or digging, the enclosure must have strong fencing or solid brick walls, and steel mesh buried approximately 10 cm below ground level. When they are competent climbers, possums and koalas are moved into larger enclosures. Possum enclosures should contain numerous sturdy branches, a nest box adjacent to a branch and an elevated feeding platform. They should be thickly foliated with branches changed weekly and a few fresh leafy branches for browse placed daily. The foliated branches are placed in steel water canisters fixed to the large permanent branches. Koala enclosures require several sturdy eucalyptus branches providing several good resting forks and thick fresh foliage placed in water canisters daily. The box containing the mother substitute may be fixed to a branch. The PY will gradually wean itself from this and begin resting in the tree forks.

5.2 Nutrition

5.2.1 Milk formulae

Numerous milk formulae have been used to hand-rear marsupials, with varying success. The most critical factor in the choice of formula for these species is that it must be low in lactose. Marsupial milk has very low concentrations of lactose, with the exception of the late lactation milk of some species, including the common brushtail (Messer & Walker 1992). Studies on hand-reared macropod PY found that they have relative lactose intolerance although they have lactase enzyme within their intestinal epithelial cells (Walker & Vickery 1989). It is hypothesised that when hand-reared PY are fed milk containing lactose, the capacity of the mechanism for transporting sugars into the cells is exceeded and unabsorbed lactose accumulates within the intestinal lumen, increasing the osmolality of the gut contents and resulting in watery diarrhoea, dehydration, malabsorption of other nutrients, malnutrition and retarded growth. Mother-reared PY suckle small amounts constantly, allowing more time for the major milk sugars, oligosaccharides, to be transported into the cells. Furthermore, these oligosaccharides are significantly larger than lactose and thus exert a lower osmotic effect on a weight basis (Messer et al. 1989).

The second factor influencing choice of formula for marsupial PY is the changing composition of natural marsupial milk from early to late lactation, meeting the vastly changing needs of the PY as it develops from its embryonic ectothermic state at birth to an animal with the energetic demands of rapid growth, locomotion and independent thermoregulation. It is generally believed that the ideal milk formula would mirror these changes.

The low-lactose formulae most commonly used for marsupial PY are Wombaroo milks (Wombaroo Food Products), Biolac milks (Biolac Milk Products) and Di-Vetelact Low Lactose Milk Replacer (Sharpe Laboratories). These formulae each use a different sugar in place of lactose. Wombaroo milks use an easily digested maltodextrin high in glucose, made from the enzymatic digestion of starch. Biolac milks use galacto-oligosaccharides, and Di-Vetelact contains lactose enzymatically converted to glucose and galactose. The compositions of these products are detailed in Table 2.2.

Wombaroo Food Products has developed a range of formulae designed to approximate the changing composition of marsupial milks. It manufactures unique products for macropods, possums, wombats and koalas, each including several different milk types formulated to simulate the natural changes in the milk of these species. There are four types of milk for macropod PY of age factors <0.4, 0.4–0.6, 0.6–0.7 and >0.7;

Table 2.2 Composition of commercial milk formulae used for marsupials

Milk type	Age factor/stage	Protein (%)	Fat (%)	Carbohydrate (%)	Energy (kJ/L)
Wombaroo	<0.4	4.6	1.9	5.6	2500
Macropod	0.4–0.6	5.8	4.0	6.3	3700
	0.6–0.7	6.5	6.2	7.2	4700
	>0.7	7.4	11.7	2.5	6200
Wombaroo	<0.4	5.0	2.2	5.0	2600
Wombat	0.4–0.6	6.2	4.5	6.0	3900
	>0.6	8.0	10.0	2.8	6200
Wombaroo	Early	4.5	8.0	7.5	5000
Koala	Mid	8.3	15.5	5.0	8200
	Late	11.0	13.5	2.0	7400
Wombaroo	<0.8	5.5	2.5	6.0	3100
Possum	>0.8	7.5	9.0	5.0	5700
Biolac	M100	5.0	5.0	5.0	3900
	M150	7.5	7.5	3.5	5210
	M200	10.0	10.0	2.0	6450
Di-Vetelact	Dilution A	3.0	4.0	5.0	2800
	Dilution B	4.8	6.0	7.6	4200

See Appendix 3 for product sources.

two types of possum milk for age factors <0.8 and >0.8; three wombat milks for age factors <0.4, 0.4–0.6 and >0.6; and three koala milks for early, mid and late lactation. The manufacturer supplies tables of data for each species that enable the carer to age the PY, determine its age factor and hence the suitable stage of milk, and determine the daily volume of milk required. The recommended daily volumes have been determined by using the energy content of the milks, and calculations of the daily calorific requirements of PY at each of the body weights listed in the tables. It is important not to overfeed. As the PY gains weight, its daily intake must be increased according to the table. The charts also instruct how to gradually transition a PY from one milk stage to the next as it grows. These milks should always be made up and fed at the concentration instructed. When using these formulae, water must be offered to pre-emergent PY between milk feeds by bottle or syringe, and provided *ad libitum* to emerged young.

Biolac has also developed milks specifically for marsupials. Three formulae are available (M100, M150 and M200), for use at the different stages of marsupial development. The manufacturer's recommendations for use of these formulae are as follows.

- Macropods: Use M100 on all hairless PY. Progress to M150 (a transition milk) when dense hair has developed. Change to M200 when the animal produces solid dark pellet droppings. When PY are fully emerged from the pouch, add canola oil to the milk, progressively increasing from 2–5 mL per 100 mL of milk.
- Possums: Use M100 only until possums are weaned.
- Koalas and wombats: For hairless PY, use M100 with 2–5 mL canola oil per 100 mL of milk. Progressively increase to M150 and M200 with age.

The manufacturer indicates that these formulae should never be diluted and should be fed at the rate of a total daily volume of 100–150 mL/kg body weight.

Di-Vetelact is a low-lactose milk formula designed for a wide range of eutherian mammals. It has been used by many Australian wildlife carers for a range of marsupial species. The manufacturer provides directions for reconstituting the milk to two dilutions, A and B. For marsupials the recommendation is to commence with dilution A, then gradually change to dilution B over a 1 wk period. The same formula is fed regardless of the age of the PY. Experienced carers recommend feeding daily volumes of 100 mL/kg body weight 'to survive' and 150–200 mL/kg to thrive and grow (L Dennis *pers. comm.*).

The compositions of these formulae is detailed in Table 2.2. The composition of the milks of the tammar wallaby and common brushtail illustrated in Figures 2.2 and 2.3 demonstrate that Wombaroo and Biolac products roughly replicate the nutrient concentrations and changes of the natural milk. The most significant difference between the formulae and natural milks is lower lipid concentration and hence energy density in the late lactation products compared with the natural milks.

Nevertheless, hand-reared PY of a wide range of species have thrived on these formulae, achieving close to normal growth rates. Di-Vetelact has significantly lower energy content than the natural milk for the later stages of pouch life and post emergence. This deficiency is largely overcome by the larger daily volumes fed of this formula than the other products, and some carers improve the protein and energy content by including egg yolk and vegetable oil (Speare 1988). Another apparent deficiency of Di-Vetelact for emergent herbivorous PY is its high sugar content compared with the very low levels in the natural milk of these species at this stage. This may be inhibitory to the establishment of the balanced gastrointestinal microflora required for digestion of herbage. Nevertheless, Di-Vetelact has been used very successfully for a very large number of marsupial PY of a wide range of species (Messer & Walker 1992). There is much passionate debate among wildlife carers over which formula is superior but, despite the apparent theoretical advantages of some products, there is no objective experimental evidence to justify the recommendation of one over another (Messer & Walker 1992; Booth 2002).

Marsupial PY are very susceptible to diarrhoea as they adjust from their mother's milk to an artificial formula, hence it is recommended to gradually introduce the formula by dilution with an oral rehydration electrolyte solution for the first few days in care. Gradually increase from half-strength on day 1 to full strength on day 4. Prolonging this process should be avoided as the sub-optimal energy intake may result in greater than acceptable weight loss in this initial period (Booth 2002).

Commercial bovine colostrum products may be advantageous for pre-emergent PY as a substitute for the maternal immunoglobulins transferred in milk throughout much of pouch life. These are added daily to formula until the age of pouch emergence (Woods 1999).

5.2.2 Milk reconstitution, storage and temperature

Water for milk reconstitution should be boiled for 10 min then cooled. The reconstituted milk can be divided into daily portions and frozen. After thawing, milk should be kept refrigerated and must be discarded after 24 hr. At each feed the target volume should be measured out and warmed to 33–36°C (close to body temperature) before feeding. Milk temperature affects acceptance, so if a PY is a difficult feeder check the milk temperature with a thermometer. It may be necessary to reheat the milk during the feed as PY may stop feeding if the milk gets too cool. After each feed, discard any remaining milk (Bellamy 1994; Booth 2002).

5.2.3 Frequency and amounts to feed

PY fed Biolac and Di-Vetelact formulas should be fed approximately 10–20% of their body weight daily. Wombaroo products should be fed at the volumes directed by the manufacturer. PY must be weighed at least daily initially, reducing to weekly at the pouch emergent stage to monitor progress and adjust milk volume accordingly. Frequencies of feeding required per day at different stages of PY development are indicated in Table 2.1. Most PY do best if a standardised feeding regime is established, with feeds evenly spaced around the clock and an equal volume offered each time. All changes to the feeding regime should be made slowly.

5.2.4 Feeding equipment and techniques

It is important to use feeding equipment that allows accurate measurement of the milk intake at each feed. For very small PY, it is best to use a syringe with a short length of infant gastric feeding tube (5–10 mm) attached as a teat. Alternatively, use a butterfly catheter with the tubing cut to 5–10 mm length. As the PY sucks, the syringe plunger is slowly depressed and the milk drizzled into the mouth. For larger PY, use a plastic feeding bottle and a commercially available artificial marsupial latex teat. A range of teats is available, suitable for macropods, wombats, koalas, possums, gliders and carnivorous marsupials (Wombaroo Food Products, Biolac Milk Products). Nipple shape and size are very important in encouraging effective feeding and avoiding deformation of the poorly calcified bone of the jaws in early PY, hence these purpose-made teats are highly recommended. A range of sizes is available, and longer teats should be used as the PY grows, simulating the lengthening of the maternal teat. The hole in the teat is made using a hot sterile hypodermic needle and should be big enough that milk drips out very slowly when the teat is inverted. The size of the hole should be such that the PY has to suck firmly, encouraging the development of normal sucking reflexes and reducing the risk of milk aspiration. After each feed, wash and rinse the equipment then sterilise in a solution or steam unit

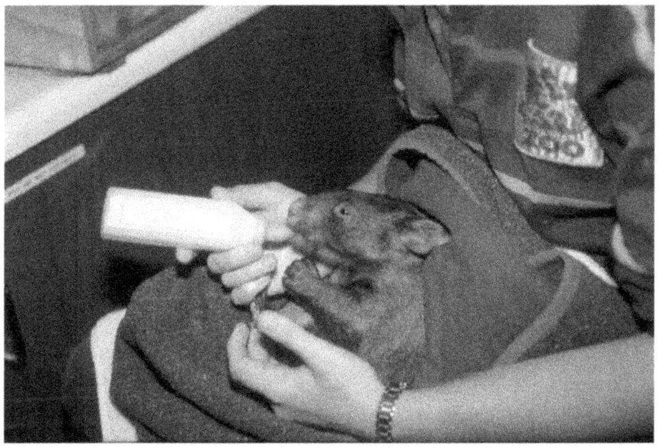

Figure 2.11 Feeding a common wombat PY using a plastic bottle and commercial latex marsupial teat. PY are fed in an upright position with the bottle only slightly tilted to reduce the risk of aspiration. They are fed within the pouch until completely endothermic.

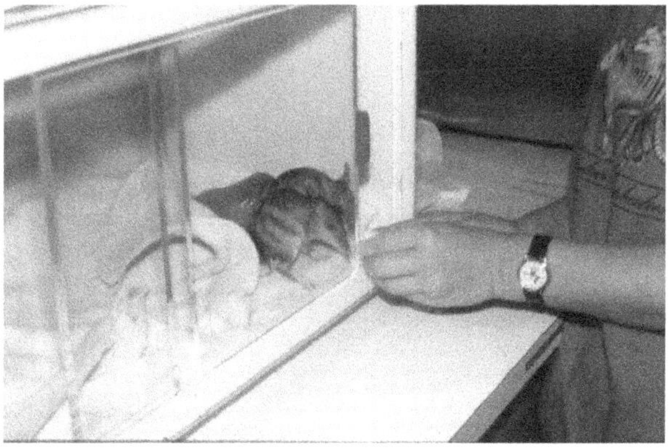

Figure 2.12 Newly emerged eastern barred bandicoot young learning to lap. As this species first emerges within the covered nest, a temperature-controlled unit provides a thermally suitable and secure environment for this stage.

designed for human infant milk bottles (Bellamy 1999; Woods 1999; Booth 2002).

Pre-emergent PY feed best in the pouch. Macropods and wombats should be held in the carer's lap in an upright half-circle position with their dorsum resting against the carer's abdomen (Fig. 2.11). One hand may be used to guide the PY to stay on the teat and to cover the eyes, preventing visual stimuli from causing distraction during feeding. Koala PY generally feed best if clinging onto their substitute mother. Possums should be kept in their pouch with only the head protruding and the pouch opening securely around the neck to prevent them escaping from, or retreating into, the pouch (Bellamy 1999; Booth 2002). Emerged PY can be fed while standing out of the pouch, clinging to a branch or on the carer's lap depending on the species, and may be taught to lap at this stage (Fig. 2.12). Changing the PY to lapping is beneficial as it will reduce contact with the carer, however, milk must not be offered in a dish *ad libitum*. A feeding regime should still be followed, intake measured and unconsumed milk removed immediately after each feed.

Newly arrived PY commonly refuse to suckle the artificial teat and formula. To train them to do so, hold the head with the hand covering the eyes to simulate the enveloping pouch environment, gently open the mouth by placing a finger between the lips, and slide the teat into the mouth. Squeeze in small amounts of milk and gently tip back the head to encourage swallowing. After a few days of frequent feeds like this, most PY start suckling well and can be established on a standard feeding regime. Aspiration is common in hand-reared marsupials and may be prevented by ensuring that the rate of milk flow into the mouth does not exceed the rate at which it is swallowed. Well-established PY may suckle very quickly, causing milk to come out through their nostrils. This is rarely associated with aspiration, but can be minimised by reducing the size of the hole in the teat and feeding more frequently to avoid frantic suckling (Bellamy 1999; Booth 2002).

Total milk intake should be assessed at the end of each day. Sub-optimal intake may be caused by a range of factors including hypothermia, incorrect milk temperature, noisy or stressful housing and/or feeding environment, unsuitable teat, teat hole too small, inappropriate feeding position or illness. The problem should be addressed and the PY offered smaller and more frequent feeds until it is consuming the daily target volume. Meanwhile, oral or SC fluids may be required to make up the deficit, to prevent dehydration. Tube feeding is rarely required, but if it is, the naso-gastric route is preferred. Measure the distances from nose to oesophagus and nose to stomach (at the level of the last rib) and mark these on the tube. Direct the lubricated tube inferiorly through the nasal cavity where there should be little resistance. After swallowing, there is minor resistance as the tube passes down the oesophagus. Aspirate to confirm that the tube is in the stomach (Speare 1988; Booth 2002).

5.2.5 Weaning

In macropods and koalas, weaning should commence when the PY first protrudes its head from the pouch. In other species, it commences at the age when the PY first emerges from the maternal burrow or nest. Herbivorous species require inoculation of the gut with appropriate bacteria at this time. This is particularly critical with koala PY. They must be assessed to determine whether they have consumed pap before coming into care; if not, they must be fed natural pap or a substitute (see Chapter 8). Macropod and wombat PY should be given access to fresh faecal pellets from an adult con-specific that have been confirmed free of parasite oocysts. Alternatively, such faeces may be mixed with water and fed as a slurry or added to the milk formula. Probiotic products have also been used. There is no strong evidence to support the comparative value of the different methods (Woods 1999; Booth 2002).

Commence the weaning process by introducing the PY to solid foods at the time of milk feeds. Thereafter, make solids continuously available and eventually the animal will eat them spontaneously. All species should be fed *ad libitum* and have fresh water available. Weaning diets should simulate natural diets for the species. As the PY begins to eat solids, the volume of milk offered at each feed and the frequency of milk feeds should be very gradually decreased. The animal's body weight should be monitored closely at this time to ensure that it is continuing to rise at the expected rate. Aim to have the PY completely weaned by the time it reaches the body size of the mother-reared young at weaning.

5.3 Toileting

PY should be stimulated to urinate and defecate before or after each milk feed. In addition to simulating the natural role of the mother, this practice allows regular observation of faecal consistency and reduces soiling of the pouch. Some carers prefer to toilet before a feed, as an empty bladder seems to improve the suckling response. Gently stroke the perineum dorsal to the cloaca with warm moist cottonwool (Fig. 2.13). Excessive stimulation may precipitate cloacal candidiasis, mucosal prolapse or urethral swelling and obstruction in males. Generally urine will be passed before faeces. PY generally urinate at each feed but will defecate only several times a day. The normal faecal consistency at different stages is shown in Table 2.1. The age of commencement of voluntary urination and defecation varies, but generally occurs following first pouch emergence (Bellamy 1994, 1999; Woods 1999).

5.4 Hygiene

Hygiene is essential for successful hand-rearing. As pre-emergent PY are not yet immunocompetent, microbial contamination of their environment should be kept to a minimum. PY self-groom by licking from an early age and will ingest bacteria from their skin and immediate environment, therefore the PY and pouch must be kept clean, simulating the natural maternal behaviour. Milk, urine and faeces should be removed by spot cleaning after each feed. If badly soiled, the PY should be bathed immediately using a gentle shampoo, then quickly blow-dried to prevent excessive cooling. After bathing, all bare skin should be moisturised with cream as described above. Pouches or pouch liners should be changed at least once a day or more frequently if soiled with urine, faeces or milk, and laundered with an anti-bacterial nappy wash solution. The humidicrib or pouch box must also be regularly cleaned and disinfected. Wash hands well using an antiseptic soap before milk preparation and feeding. Thoroughly wash and sterilise feeding equipment as described above and discard any milk remaining at the end of each feed. If caring for more than one PY, wash hands between animals and maintain separate feeding equipment for each individual. Once fur develops, the PY should be regularly brushed and encouraged to groom itself by gentle stroking under the neck and over the thighs and ribs. This action will stimulate the PY to scratch at these areas, thereby commencing self-grooming (Bellamy 1999).

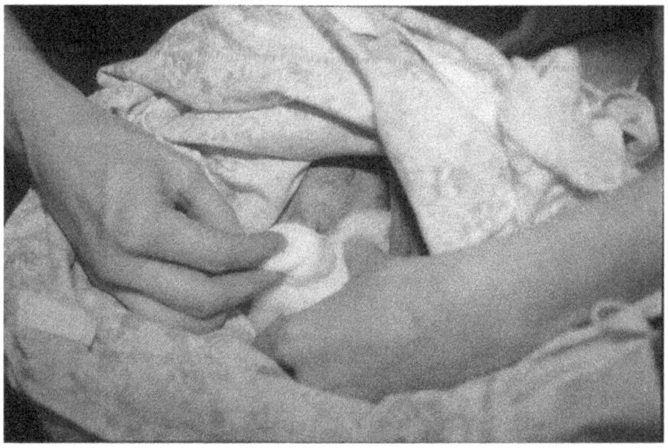

Figure 2.13 Warm moist cottonwool is used to stimulate the pre-emergent PY to urinate and/or defecate after each feed. This is done with the PY in the pouch to avoid loss of body temperature.

At weaning, food and water dishes should be washed daily, and fresh food and water provided once or twice daily. Food dishes should be put out in the shade in the late afternoon, as most species are nocturnal feeders and this will reduce spoiling. Hay should be kept in a rack off the ground. Food spillages should be removed daily so that the animals are not eating contaminated food and to minimise attracting rodents. The enclosures should be raked daily to remove faeces.

5.5 Monitoring growth

Initially PY should be weighed daily at a uniform time prior to a feed. Once a steady linear growth rate has been achieved, once or twice weekly weighing is sufficient. Measure head, tail and foot length every 2–4 weeks and plot all measurements on a graph to demonstrate trends. Compare data with species-specific growth charts to ensure that the PY is achieving the expected rate of growth. If growth rates are significantly lower than expected or if they plateau, husbandry should be re-evaluated and the animal examined for health problems (Booth 2002).

5.6 Record-keeping

Records should be kept of the animal's origin, date and cause of orphaning, age and health status on arrival, environmental conditions provided and the type of milk, solid foods and feeding equipment used. A running record should be maintained of the intake at each feed, times of urination and defecation, and faecal consistency. Such records are essential for assessment of the animal's progress, as history when health problems occur, for release site determination and for future reference to assist in the rearing of other PY.

5.7 Stress minimisation

Psychological stress is a common precipitant to disease in PY and must be avoided where possible. Stressful situations include an insecure pouch environment, excessive handling, petting and playing with children, contact with unfamiliar animals and humans, exposure to unfamiliar environments and noises, sudden changes in routine (e.g. sudden prolonged pouch deprivation during the process of pouch weaning), sudden changes in feeding schedule, and sudden withdrawal of contact with carer or buddy. It is important to remember that the mother-reared PY develops in a quiet, dark, secure and stable environment. It has a very gradual exposure to the outside world and has limited social contact with conspecifics.

5.8 Avoidance of imprinting and preparation for release

Hand-reared marsupial PY are very prone to imprinting to their carers. This is problematic even for animals that are to remain in captivity. Imprinted male macropods and wombats commonly develop sexual and non-sexual aggression towards people (Booth 2002). As the captive husbandry of these species generally involves keepers working in with the animals and macropods are commonly maintained in walk-through displays, this behaviour may present danger to both staff and visitors. Imprinted macropods commonly display tintibulation (neurotic submissive head tremoring—see Chapter 7) in response to people (Booth 2002), and wombats follow closely at a person's heels and seek human contact, inviting potentially problematic petting and feeding by zoo visitors. Furthermore, these animals may be very stressed in the absence of humans, predisposing them to injury and disease. Finally, their lack of species-appropriate social skills may result in non-participation in captive breeding programs.

To avoid imprinting during the process of hand-rearing, provide only sufficient nurturing to give a sense of security, avoid habituating the PY to stimuli not relevant to its species (e.g. dogs, cats, large groups of people), gradually wean it of human contact at the age that it would separate from its mother, and expose it to its own species from as close to the time of first pouch emergence as possible. Ideally, PY are reared together with a same age and species buddy from an early age as this significantly reduces their reliance on human carers. Considerable care should be taken to avoid imprinting in hand-reared males of the larger macropod species and to socialise them with their own species from an early age, as these animals mature to a size large enough to be dangerous to humans (Bellamy 1994; Booth 2002).

If PY are hand-reared for release into the wild, they must be well-prepared to maximise their survivability. An animal for release should be fit and healthy, meeting age-appropriate skeletal and body mass targets for its species. It should be able to recognise its own species, be familiar with species-specific social behaviour, show no evidence of being imprinted on humans and demonstrate appropriate levels of fear of humans and predators. It should be capable of selecting or catching its natural diet, exhibiting species-appropriate activity (e.g. arboreal vs terrestrial, nocturnal vs diurnal) and finding species-appropriate shelter, digging burrows or making nests. Pre-release husbandry must provide opportunities for the animal to develop these behaviours and skills (Bellamy 1994; Booth 2002; Jackson 2003).

6 GENERAL CLINICAL APPROACH TO THE SICK NEONATE

6.1 History

The most common presenting signs of illness in hand-reared marsupial PY are poor appetite and feeding response, failure to grow and gain weight at the expected rate, diarrhoea, increased respiratory rate or effort, and lethargy. These problems are most commonly caused by environmental stressors as detailed above (5.7) and sub-optimal husbandry practices including poor hygiene, unstable environmental temperature and inappropriate feeding management. Therefore, the clinical approach to the sick PY must commence with thorough history-taking regarding housing, hygiene, diets, feeding methods, volumes and frequencies and weight gain. This should include examination of feeding and care record charts (Woods 1999).

6.2 Restraint for examination

All pre-emergent PY should be examined and treated within the pouch, where possible, to maintain their body temperature and feeling of security. Emergent and fully emerged PY will usually readily enter a pouch if provided and this facilitates examination as they are generally relaxed and calm when inside with their head covered. Handlers should be seated and hold the pouch in their lap with the curvature of the animal's spine resting against their abdomen. Parts of the body are either extracted one by one for examination, or inspected within the pouch. When the head is examined, the animal will generally be calm if the eyes are covered. If the hind limbs of macropods require restraint, they should be held together, with one finger in between, at the level of the distal tibias. Never attempt to restrain these limbs by holding below the tarsus, because if the PY kicks the calcaneus may be crushed or the ligaments on the plantar aspect of the hock ruptured.

6.3 Clinical examination

6.3.1 General appearance

The animal should be alert and responsive to its environment and, unless accustomed to human handling, will seek and attempt to enter an artificial pouch. Lightly furred and older PY can stand, hop or climb (depending on the species). The PY should be accurately weighed, skeletal measurements taken and a species growth chart used to assess whether the body weight is appropriate for skeletal size. Assess hydration status by checking dryness of mouth, skin turgor, peripheral circulation and corneas and interpret the findings as for domestic animal species. To take the rectal temperature, evert the cloacal mucosa and identify the anus as the more dorsal aperture, then insert a lubricated thermometer approximately 1 cm, then grasp the cloacal skin and pull caudally. This eliminates the folds in the cloacal and rectal mucosa and facilitates deeper insertion of the thermometer. Normal temperature is 35.5–37°C. Digital thermometers are preferred because they are more reliable in this lower temperature range. Hypothermia in pre-emergent PY is indicative of inadequate temperature in their immediate environment or underlying systemic disease, and may precipitate hypoglycaemia. Hyperthermia is generally indicative of excessive temperature in immediate environment. Pyrexia due to infection rarely occurs in marsupials.

6.3.2 Integument

The skin of hairless PY should be pink, soft and smooth with a silky sheen. Furred mother-reared PY have clean sleek fur because they are regularly groomed by the dam. Hand-reared PY may have dry skin, dull tousled pelage, dried milk on face and chest, and dried faeces on hind limbs, ventral tail and around the cloaca. These signs result from failure to moisturise bare skin, clean and groom regularly, toilet at every feed and change the pouch daily, causing faecal soiling of the fur. It is important to distinguish this from faecal soiling in the well-cared-for PY, because in the latter it will have resulted from defecation in the pouch between feeds (i.e. diarrhoea).

6.3.3 Cardiopulmonary system

Normal resting heart rate ranges from 70–200 bpm. Heart sounds are as for domestic species. The cardiac impulse can be palpated and often visualised on both sides of the thorax and the femoral pulse is readily palpable. Mucous membranes (conjunctivae, gingivae, cloaca) are pink with a capillary refill time of <1 sec. Respiratory effort at rest is barely noticeable. Normal resting rate is 20–40 bpm. On auscultation, lung sounds are very difficult to hear; they are loudest just caudal to the scapulae.

6.3.4 Gastrointestinal system

Use an otoscope or laryngoscope to examine inside the mouth (many marsupial species have a narrow gape). On abdominal palpation, the stomach has a doughy feel in grazing PY, and formed faecal pellets can be palpated in the colon of weaning animals. Attempt to stimulate defecation to check faecal consistency and readiness of passage. Most PY will urinate first, then defecate after a little more stimulation. Normal faecal consistency is described in Table 2.1. Manually evert the cloaca to examine the anal mucosa that should be pink and clean. A common concern of carers of ringtail possums is apparent constipation a few weeks after commencement of weaning. The possum appears healthy and has a good appetite but no longer defecates on stimulation and no faeces are found in the pouch. If otherwise normal on examination, this is probably due to commencement of diurnal caecotrophy in this species, and the carers have failed to find the small hard pellets passed during nocturnal foraging. A normal phenomenon that concerns carers of weaning macropod PY is merycism, the equivalent of eructation and rumination in ruminants. It may be mistakenly interpreted as vomiting, which is very rare in marsupials. The PY stands upright and makes violent rhythmic jerking movements of the abdominal muscles for 10–60 sec and simultaneously progressively extends the neck, in some cases close to a vertical position. It then starts chewing rapidly, lowers its head and continues chewing, then resumes normal behaviour (Speare 1988). Teeth-grinding is not normal in PY and may be a non-specific sign of gastrointestinal disease or chronic pain.

6.3.5 Ears

The tympanic membrane is readily visualised because the external ear canal has a less acute bend than in dogs and cats. It is transparent with no vessels on its surface. PY frequently use the syndactyl claws of each hind foot to groom inside the ipsilateral ear canal. Accumulation of sebum in the external ear canal may occur when the ipsilateral syndactyl claws are bandaged to prevent sucking, or amputated. In these situations, the ears should be regularly cleaned using cottonbuds.

6.3.6 Lymphatic system

Lymph nodes are not normally palpable. The paired cervical thymi of diprotodont species are located subcutaneously on the ventral aspect of the neck. They are normally firm and bulging with their caudal borders level with the clavicles. They are prominent in PY with no or little hair, and readily palpable in those with a thick coat. Thymus size appears to be an indicator of immune status. Prolonged elevation of glucocorticoids due to chronic stress or recurrent illness may cause premature thymic involution, reducing active immunity, increasing susceptibility to infections and affecting recovery rate from illness (Speare 1988).

6.3.7 Musculoskeletal system

In PY orphaned by vehicular accident, gunshot or falls from trees, assess stance and gait and carefully palpate for fractures. Normal anatomical features of marsupials are a fibrous mandibular symphysis and hence some vertical movement and paired epipubic bones (narrow elongated bones arising from the anterior rim of the pelvis and extending cranially) that are readily palpable. Body condition is determined by an assessment of the degree of muscle-wasting and has great prognostic value. Palpate the muscles over the scapula and the dorsolateral and lateral processes of the coccygeal vertebrae just posterior to the cloaca. In the normal mother-raised PY, the scapular muscles bulge over the scapular spine, and muscles over the coccygeal vertebrae are such that the dorsolateral processes cannot be palpated and the lateral processes can only be felt on firm pressure. Few hand-reared PY are in optimal condition, but normal healthy ones are only slightly less muscled. Both processes of the coccygeal vertebrae are palpable with a little pressure and the scapular muscles are level with, or bulge slightly over, the scapular spine (Speare 1988).

6.3.8 Neurological system

Unfurred PY with eyes just opened will have difficulty standing without support. When they have a very thin coat they can stand but are unable to ambulate. As they grow, co-ordination gradually increases until at the age of first pouch emergence when they can move around well. Pupils are usually large, and less responsive to light than dogs and cats. Patellar reflex is easily elicited.

6.4 Sedation and anaesthesia

Diazepam is an excellent sedative for PY for short and minimally painful procedures such as examination of agitated individuals, administration of IV fluids, radiography and changing casts. The dose is 1.5–2.5 mg/kg IM, depending on the degree of sedation required. Sedation occurs after approximately 5 min and lasts for approximately 15 min. Full recovery occurs after about 2 hr (Speare, 1988). It can also be given slowly IV at 0.5–1.0 mg/kg to give deep sedation of rapid onset lasting approximately 20 min.

Isoflurane is the preferred anaesthetic agent for PY as it gives rapid induction and recovery. Induce at 5% using a facemask while restraining within the pouch. Maintenance (on 1.0–2.0% isoflurane) is generally via facemask because endotracheal intubation is very difficult due to the small body size and narrow gape of most marsupial PY. For long procedures, however, it is preferable to intubate. During anaesthesia of PY of all ages, ensure maintenance of body temperature by using heatpads and blankets. A handy conforming hot water bottle can be made for small PY by partly filling a surgical glove with warm water, eliminating the air, then tying a knot in the end.

7 COMMON MEDICAL PROBLEMS OF HAND-REARED MARSUPIAL POUCH YOUNG

Neonatal diseases relevant to each marsupial taxon are described in Chapters 7–13. The most common conditions occurring in all taxa are described here.

7.1 Failure-to-thrive

This is the one of the most common conditions affecting hand-reared PY. Most hand-reared infants will lose 5–10% of their body weight in the first 2 weeks of care. Thereafter their rate of weight gain and skeletal growth should be similar to that of mother-raised young. If this is not the case, there is some degree of failure-to-thrive. Speare (1988) described three categories of failure-to-thrive in marsupial PY: weight gain and skeletal growth, but at a very slow rate; steady weight with no skeletal growth; and weight loss and no skeletal growth. Animals in the last category are generally in very poor body condition and have a grave prognosis. The major causes of failure-to-thrive are malnutrition due to inappropriate milk and/or weaning diet and poor management resulting in inadequate intake. Often the carer lacks knowledge of the appropriate diet, required intake and frequency of feeds, resulting in a sub-optimal energy intake. Sub-optimal intake may also be due to a poor sucking response as a result of milk being offered at the incorrect temperature, the PY being too hot or too cold, psychological stresses or inappropriate feeding methods or equipment. Alternatively, failure-to-thrive may be caused or exacerbated by an underlying disease process so a thorough clinical examination is mandatory. Clinical signs of failure-to-thrive include generalised muscle atrophy, an unkempt coat, depression and weakness. Therapy involves evaluation and modification of the diet and aggressive treatment of any underlying disease entity, with an understanding that these animals probably have compromised immunity. Affected PY will have reduced thermoregulatory capacity, hence they should be provided with supplementary heat until body condition and growth rate are normal (Speare 1988).

7.2 Diarrhoea

Diarrhoea is a common problem of hand-reared PY, most commonly caused by inappropriate diet, poor feeding management, poor hygiene and environmental stress. Diarrhoea may also be caused by specific pathogens, but these infections generally occur as sequelae to unresolved 'management' diarrhoea as a result of altered gut flora. Chronic unresolved diarrhoea may lead to irritable bowel syndrome with resultant malabsorption and failure to thrive (Booth 2002).

The most common initial causes of diarrhoea are as follows.

- Sudden dietary changes causing changes in gut motility. These may occur in the transition from mother's milk to artificial formula, transition between milk types and at weaning.
- Poor feeding management, including excessive daily total milk volume or low-frequency feeding of large milk volumes, irregular feeding routine and inadequate milk temperature.
- Poor hygiene resulting in the ingestion of faecal organisms.
- Artificial milk formulae containing lactose.
- Environmental stressors (see 5.7).

Bacterial and yeast infections are common sequelae to diarrhoea caused by husbandry problems. A study of the faecal flora of hand-reared macropod PY found that the most common pathogens associated with diarrhoea in these animals were haemolytic *E. coli*, *Salmonella* spp. and *Klebsiella* spp. The *E. coli* were strongly associated with acute liquid diarrhoea in newly orphaned PY; *Salmonella* spp. were more commonly associated with chronic pasty diarrhoea and long-term care; *Klebsiella* spp. were also associated with long-term care, causing acute and chronic liquid diarrhoea. Non-haemolytic *E. coli* and *Streptococcus* spp. were common isolates from healthy PY and were considered to be normal faecal flora. Little antibiotic resistance was detected in most cases, however, 36% of isolates, including 100% of *Klebsiella* spp., were resistant to ampicillin/amoxycillin. Yeasts including *Candida* spp. and *Torulopsis* spp. were frequently isolated from PY with pasty diarrhoea (Cargill & Frith 1991). These infections occur most commonly following antibiotic use, and may cause oral and skin lesions in addition to diarrhoea (Booth 2002).

Coccidiosis caused by *Eimeria* spp. is a common cause of potentially fatal enteritis in some macropod species, notably the eastern grey kangaroo (*Macropus giganteus*) and less commonly in wombats. Hand-reared PY appear to be highly susceptible, hence milk-transferred maternal immunity may play a role in protection. Helminth parasites are rarely responsible for diarrhoea in hand-reared PY (Booth 2002).

The recommended approach to diagnosis and management of diarrhoea in hand-reared PY is to initially consider and correct husbandry problems. Routinely examine the faeces for parasite ova, including coccidian oocysts. In cases that are unresponsive to symptomatic treatment, faecal culture is advisable so that specific therapy may be commenced. Non-specific therapy should commence with correction of hydration deficits. Prevent further dehydration by replacing ongoing fluid losses with an oral rehydration solution offered between milk feeds. Continuation of milk feeding is recommended in order to maintain adequate energy intake. Complete replacement of milk with oral rehydration solutions often leads to weight loss, compounding the existing health problem. Good nursing, including bathing and frequent linen changes as required, is essential to prevent the occurrence of secondary infections. Apply a hydrating cream to the perineal region following each feed to prevent skin maceration caused by continual soiling and wetting. Treat specific pathogens if present. Antibiotic therapy should always be accompanied by simultaneous mycostatin therapy to prevent the development of opportunistic yeast infections (Woods 1999; Booth 2002).

7.3 Dermatologic diseases

The skin of hairless PY and the bare areas of furred PY are susceptible to drying outside the humid environment of the natural pouch. Xerosis (dry, rough, cracking scaly skin) can occur, sometimes with associated bacterial infection, if skin is not kept well-hydrated with hydrating creams (Woods 1999). Cracks may develop, particularly on the tail and across the flexor aspects of the elbow and hock joints, potentially restricting joint movement. Treatment involves application of ultra-hydration skin treatment products, resolution of infections and establishing a regular routine of moisturiser application (Table 2.1). Supplementation with oral omega-3 fatty acid products may be helpful.

Many macropod PY frequently suck a localised body part, presumably as a substitute for the teat. Frequently sucked areas include the syndactyl toes, tail tip, fore limb digits, scrotum, penis and cloaca. Some PY persistently suck the ear pinnae of their buddy. Damage results if a single structure is sucked too vigorously for too long. Clinical signs vary in severity from superficial maceration to redness, swelling and ulceration. In cases where large body parts such as the fist or tail base are sucked, deformations of the jaw and malocclusion may result (Bellamy 1994). Treatment involves preventing the sucking of damaged parts by bandaging following wound management of ulcerated lesions. If the PY has this habit, it will likely start sucking another area. Prevent further lesions by sequentially bandaging different preferred sites in order to rotate the sucked area. Alternatively, some carers provide the PY with a teat to continually suck as a dummy.

Non-specific bilaterally symmetrical alopecia involving trunk, tail and hind limbs is frequently seen in hand-reared PY. This is generally understood to be caused by environmental stress, but may also be associated with overgrooming or inappropriately high pouch temperature. It is possibly caused by a dietary deficiency of sulphur-containing amino acids, including cysteine, as these are found in high levels in tammar wallaby milk during the period of hair follicle development (Messer & Walker 1992).

7.4 Osteoporosis

This condition has been reported in numerous hand-reared PY, resulting in pathological long bone fractures. It was initially believed to be due to calcium deficiency in certain milk formulae, but as these formulae do contain adequate calcium levels several other factors are considered responsible. They include poor calcium absorption from the gut due to chronic diarrhoea, lack of exposure to UV light for vitamin D synthesis and inadequate exercise. Without vitamin D, dietary calcium is not absorbed from the gut. Some PY are housed indoors until the fully emergent stage, and the UV range of the sunlight spectrum is screened out by window glass. Mother-raised PY begin exposing body parts such as ears and feet at approximately age factor 0.6. Therefore, from this age onwards hand-reared PY should be exposed to direct sunlight for some time each day. The mother-raised PY is frequently moved around in the pouch by the mother's movements and vigorous grooming. Hand-reared PY are comparatively still in the pouch. This may be a contributory factor in the development of osteoporosis. Extension-flexion exercises and massages of limbs following each feed may be useful (George 1990).

REFERENCES

Basden K, Cooper DW & Deane EM (1997). Development of the lymphoid tissues of the tammar wallaby *Macropus eugenii*. *Reproduction, Fertility and Development* **9**, 243–254.

Baudinette RV, Gannon BJ & McAllan BM (1985). Cardiorespiratory adjustments to pouch life. In *Abstracts of the Australian Mammal Society 31st Scientific Meeting*. University of Melbourne: Melbourne.

Bellamy T (1994). Marsupial handrearing. In *Wildlife, Proceedings 233,* pp. 7–20. Post Graduate Committee in Veterinary Science: Sydney.

Bellamy T (1999). Handraising marsupials. In *Wildlife in Australia: Healthcare and Management, Proceedings 327,* pp. 273–274. Post Graduate Foundation in Veterinary Science: Sydney.

Boër M (1998). Observations on reproduction in the common wombat *Vombatus ursinus* in captivity. In *Wombats* (eds RT Wells & PA Pridmore), pp. 129–146. Surrey Beatty & Sons: Sydney.

Booth R (2002). Macropods. In *Hand-rearing Wild and Domestic Mammals* (ed. LJ Gage), pp. 64–74. Iowa State Press: Iowa.

Cargill C & Frith F (1991). Joey diarrhoea project—preliminary report. *Vetlab Newsletter*, Department of Agriculture, South Australia **36**, 10–13.

Deane EM & Cooper DW (1988). Immunological development of pouch young marsupials. In *The Developing Marsupial. Models for Biomedical Research.* Springer-Verlag: Berlin.

Deane EM, Cooper DW & Renfree MB (1990). Immunoglobulin G levels in fetal and newborn tammar wallabies (*Macropus eugenii*). *Reproduction, Fertility and Development* **2**, 369–375.

George H (1990). The care and handling of orphaned kangaroos. In *Care and Handling of Australian Native Animals, Emergency Care and Captive Management* (ed. SJ Hand), pp. 23–141. Surrey Beatty & Sons: Sydney.

Jackson S (2003). *Australian Mammals: Biology and Captive Management.* CSIRO Publishing: Melbourne.

Janssens PA, Hulbert AJ & Baudinette RV (1997). Development of the pouch young from birth to pouch vacation. In *Marsupial Biology: Recent Research and New Perspectives* (eds NR Saunders & LA Hinds), pp. 71–89. UNSW Press: Sydney.

Maynes G (1976). Growth of the parma wallaby, *Macropus parma* Waterhouse. *Australian Journal of Zoology* **24**, 217–236.

Merchant JC (1989). Lactation in macropodoid marsupials. In *Kangaroos, Wallabies and Rat-kangaroos* (eds G Grigg, P Jarman & I Hume), pp. 355–366. Surrey Beatty & Sons: Sydney.

Messer M, Crisp EA & Czolij R (1989). Lactose digestion in suckling macropodids. In *Kangaroos, Wallabies and Rat-kangaroos* (eds G Grigg, P Jarman & I Hume), pp. 217–221. Surrey Beatty & Sons: Sydney.

Messer M & Walker DM (1992). Milk substitutes for marsupials: in theory how good (or bad) are they? In *Urban Wildlife, Proceedings 204,* pp. 99–124. Post Graduate Committee in Veterinary Science: Sydney.

Old JM & Deane EM (2000). Development of the immune system and immunological protection in marsupial pouch young. *Development and Comparative Immunology* **24**, 445–454.

Osawa R, Blanshard WH & Callaghan PG (1993). Microbiological studies of the intestinal microflora of the koala, *Phascolarctos cinereus*. II. Pap, a special maternal faeces consumed by juvenile koalas. *Australian Journal of Zoology* **41**, 611–620.

Russell EM (1982). Patterns of parental care and parental investment in marsupials. *Biological Reviews* **57**, 423–486.

Sharman GB, Firth HJ & Calaby JH (1964). Growth of the pouch young, tooth eruption and age determination in the red kangaroo, *Megaleia rufa*. *CSIRO Wildlife Research* **9**, 20–49.

Speare R (1988). Clinical assessment, diseases and management of the orphaned macropod joey. In *Australian Wildlife, Proceedings 104*, pp. 211–296. Post Graduate Committee in Veterinary Science: Sydney.

Triggs (1996). *The Wombat. Common Wombats in Australia.* UNSW Press: Sydney.

Tyndale-Biscoe CH (2005). *Life of Marsupials.* CSIRO Publishing: Melbourne.

Tyndale-Biscoe CH & Renfree M (1987). *Reproductive Physiology of Marsupials.* CUP: London.

Tyndale-Biscoe CH & Janssens PA (1988). *The Developing Marsupial Models for Biomedical Research*. Springer-Verlag: Berlin.

Walker DM & Vickery K (1989). Nitrogen balance studies on pouch young *Macropus rufus* and *M. giganteus* (Marsupialia: Macropodidae): the utilisation of cow's milk and carbohydrate-free milk replacers. *Australian Mammalogy* **12**, 23–30.

Woods R (1999). Prevention of disease in hand-reared native wildlife. In *Wildlife in Australia: Healthcare and Management, Proceedings 327*, pp. 455–489. Post Graduate Foundation in Veterinary Science: Sydney.

Yadav M (1971). The transmission of antibodies across the gut of pouch young marsupials. *Immunology* **21**, 839–851.

Yadav M (1973). The presence of the cervical and thoracic thymus lobes in marsupials. *Australian Journal of Zoology* **21**, 285–301.

3. Marine mammal strandings and the role of the veterinarian

Nick Gales, Rupert Woods and Larry Vogelnest

1 INTRODUCTION

1.1 General

Marine mammal strandings, particularly those involving large numbers of whales or dolphins, attract a vast amount of public and media attention. These events are often the scene of impressive mobilisations of heavy equipment, volunteers and a variety of experts who attempt to contribute to the rescue effort. Knowledgeable veterinary expertise is a significant component of any rescue and the veterinarian at a stranding is likely to find themselves in a high-profile, challenging, operational and advisory role. Their role can be as varied as the strandings themselves and include aspects of clinical examination or necropsy and assessment, drug selection and administration, euthanasia and tag attachment. It can also be controversial. The media increasingly seek early advice on the likelihood of human causes of the stranding such as potential links to seismic or military acoustic activities that may have occurred in the vicinity. Similarly, advice to euthanase an animal may meet with polarised views. More commonly, however, veterinarians will be asked to advise or assist with health assessments and interventions for single stranded cetaceans or pinnipeds. In all eventualities, the way to maximise the positive contribution of the veterinarian is to ensure that they understand their role and the role of others, and give advice that matches their experience, knowledge and capacity in circumstances that may be novel and at times overwhelming.

Very few comprehensive texts have been produced on the wider topic of strandings; *Marine Mammals Ashore: A Field Guide for Strandings* (2nd edn) by Geraci and Lounsbury (2005) and the University of Sydney's Post Graduate Foundation in Veterinary Science *Marine Wildlife, Proceedings 335* (2000) are probably the most useful. Within the confines of this chapter it is not possible to provide a detailed clinicians' guide to diagnosis, treatment, sample collection and follow-up for all stranding scenarios. Rather, we more modestly aim to provide veterinarians, particularly those less familiar with marine mammal anatomy and physiology, with a reasonably brief and accessible summary and guide that, augmented with guides like that of Geraci and Lounsbury (2005) and the information in Chapters 17, 18 and 19 of this volume, will help them understand how and why they might best contribute to positive outcomes from marine mammal strandings. Good guides to identification and basic background biology of commonly stranded animals can be found in Baker (1999), Bryden et al. (1998) and Carwardine (1995).

We will discuss in general terms the issues of strandings of all marine mammals. Although single stranded cetaceans and pinnipeds ashore are much more common, the prognosis for single stranded cetaceans is generally poor and mass strandings often require greater veterinary involvement. For this reason, the focus of this summary will be most relevant to cetacean mass strandings.

Inexperienced veterinarians should be encouraged, once they have made an initial assessment of a stranded animal, to contact an experienced veterinarian for advice on how to proceed. This often helps enormously and can contribute to a more rapid and humane outcome. Most wildlife agencies have stranding policies and procedures and contact details for experienced marine mammal veterinarians in each state.

1.2 What is a stranding, and why do marine mammals strand?

For the purpose of this chapter we define a stranding as any of the following: any dead marine mammal on a beach or floating near-shore; any live cetacean (whale or dolphin) or dugong on a beach or in water so shallow that it is unable to free itself and resume normal activity, or any live pinniped (seal, fur-seal or sea-lion) which is unable or unwilling to leave the shore because of injury or poor health.

By far the most common form of stranding involves single animals. The causes of death or debility in these circumstances are highly varied and often include more than one factor operating simultaneously. Natural causes such as disease, parasitism, biotoxins, predator attack, animals being trapped by outgoing tides or dependent infants separated from their mothers have been widely reported, as have human factors such as vessel-strike, entanglement, shooting and pollutant contamination (Dierauf & Gulland 2001; Perrin & Geraci 2002). Typically, single strandings occur as unrelated events, but on occasions patterns of strandings may occur, indicating a larger-scale process. These patterns may be associated with the normal distribution, migration or reproduction of the species, such as patterns of strandings of juvenile fur-seals soon after weaning. Of greater concern are unusual mortality events that may manifest from human causes such as over-fishing or climate change effects, or be part of the rapid spread of an infectious agent in a marine mammal population.

The mass stranding of marine mammals is a particularly puzzling phenomenon generally restricted to cetaceans. While dugongs (*Dugong dugong*) have been recorded to mass strand in response to unusual tidal or storm events, it is the off-shore, gregarious toothed cetaceans such as pilot whales (*Globicephala* spp.), false killer whales (*Pseudorca crassidens*) and sperm whales (*Physeter macrocephalus*) that most commonly mass strand in Australia. These strandings consist primarily of apparently healthy animals, sometimes numbering in the hundreds, which strand as a group over periods of less than an hour to a few days. The media like to suggest that the cause of these events is a great scientific mystery—while it is true that the circumstances leading to a particular mass stranding are varied and extremely difficult to elucidate, the simple reason for mass strandings involves the strong social herding behaviour of the species involved. More specific contributing factors such as disease (of a subset of the animals), confusing bathymetry (ocean depth and topography), tides and rough seas are likely to be important in determining why a group of off-shore animals find themselves in close proximity to the shoreline.

While the beaching of a fully aquatic mammal like a cetacean or dugong is a clear aberration, determining when a pinniped that is hauled-out on land represents a stranding (by the above definition) can be more difficult. Extended periods of time ashore for the purpose of resting, moulting and reproduction are normal behaviours for these mammals and, while they generally frequent regular haul-out and breeding sites on relatively remote islands, they will come ashore quite regularly in areas where they may encounter people. A reluctance to leave the shore (part of the definition of a stranding for a pinniped) may also be entirely normal; e.g. bulls may be establishing a breeding territory, a southern elephant seal (*Mirounga leonina*) may be hauling out to moult, exhausted animals that have spent a long time at sea, those that may be used to humans. These animals may ignore encouragements to move or may respond aggressively. In many cases, where seals come ashore on public beaches the issues are more about management of human behaviour (and their dogs) than about management of the animal. When the reason for the seal being ashore is medically or physically based the underlying causes, as with cetaceans, are highly varied and include factors such as disease, parasitism, biotoxins, starvation (particularly following weaning) and predator attack, as well as human causes such as vessel-strike, entanglement and shooting.

1.3 What happens to marine mammals when they come ashore?

Obligate marine species such as cetaceans and sirenians (dugongs and manatees [*Trichechus* spp.]) evolved in a medium where they experience little of the gravitational consequences of mass and with a thermoregulatory system that enables them to live in waters that are usually substantially cooler than their core body temperatures. Being ashore instigates a rapid series of physical consequences that, without intervention, is certain to lead to death. The resultant physiological consequences and pathology will act additively with any potential pre-existing pathology that may have led to the stranding, and may confound and obscure a veterinarian's or pathologist's attempts to determine a primary reason for the stranding. These changes are often difficult to see and a clinical examination can be frustratingly uninformative.

For the pinnipeds, which are used to periods ashore, a disease or injury that prevents them feeding effectively at sea will ultimately lead to dehydration and starvation. A clinical examination of these animals may provide more familiar cues to the non-specialist veterinarian, but the challenge of handling these potentially dangerous creatures often limit such endeavours (Chapter 17).

1.4 Why should veterinarians attend a stranding?

There are a number of motivations why people intervene at strandings. Much of what we know about marine mammals, particularly about some of the more cryptic off-shore species like beaked whales (Ziphiidae), has been learnt by sampling and studying animals that we find on our beaches. The scientific aspects of taxonomy, anatomy, physiology, biology and pathology in particular have relied to some degree upon such opportunities. Increasingly, in recent times, the pathology and behaviour of marine mammals ashore has provided our first warning of off-shore problems of natural origin (e.g. algal blooms) or human origin (e.g. exposure to high sound sources and fisheries interactions). Thus, science can inform and serve conservation. While many veterinarians are activity involved in this contribution to science and conservation, their primary role, and hopefully motivation, is perhaps more fundamentally based on the welfare of the individual animals.

The role of the veterinarian in the clinical assessment, treatment or euthanasia of stranded animals is fundamental to optimising their welfare outcome. Something of a conundrum exists here. In most cases, the mortalities associated with strandings are a normal part of that species' life history. Wild animals die all the time from the many mechanisms that regulate their populations. Veterinarians do not spend their time at seal colonies assessing and treating the many ill and dying animals that are a normal part of any breeding season. So why attend to marine mammals that happen upon our shores in view of the public? While the scientific and conservation gains might be easy to justify—and be a critical part of the veterinarian's role—why is a welfare outcome such a product of geography and proximity to humans? The answer is perhaps two-fold. First, to some extent our intervention can be seen as in some way redressing the balance of our negative impacts on marine mammals and their environments. The second and more compelling reason for veterinarians is that we are trained and able to improve the welfare outcomes for suffering animals that are so dramatically placed before us on our coastline. To do so is part of the general public response and expectation, which is largely based on concern and empathy with the plight of these stranded mammals.

1.5 Australian stranding response networks

Tremendous advances have been made in Australia over the past 20–30 years in preparing for and responding to stranding events. Strandings occur in all states and the Northern Territory but the southern states, especially Tasmania, experience the highest frequency of strandings and mass strandings in particular. Operational responsibilities for dealing with stranding events lie with the various fauna authorities in each state or territory. Most states manage mass stranding events with a formally established Incident Control System (ICS), such as those used for bushfire response. This system implements a clear hierarchical structure with an Incident Controller (IC) in overall charge and responsible for all decisions. For the veterinarian, the IC or their designated nominee is the 'client' and represents the 'owner' of the animals, which is the state. The advice and operational role of the veterinarian is often a key component of a successful outcome, but it is essential that these roles are conducted within the ICS framework. The IC must distil advice from wildlife specialists, veterinarians and experienced volunteer groups and balance these with operational, planning and logistics priorities. Consequently there may be occasions where veterinary advice is not followed. On these occasions it is important to understand the decision that has been made, support the consensus and move on.

Co-evolving with the sophistication of the response from state agencies has been the involvement of voluntary non-government organisations. The details of these voluntary organisations differ among states, but the model of groups working within an ICS framework is similar. These volunteers generally represent an enthusiastic resource with some training and experience. They can be a great asset to the veterinarian in assisting and providing informed feedback on the animals they are responsible for. These types of voluntary groups are generally mobilised for mass stranding events and are also commonly used to monitor pinnipeds that come ashore.

Veterinarians are a key part of any stranding response. The first role of the veterinarian at a single or mass stranding event should be to identify the IC or responsible state government officer and discuss how they can most effectively contribute to the response effort.

2 HOW CAN A VET HELP? A GUIDE TO VETERINARIANS ATTENDING CETACEAN STRANDINGS

2.1 Work flow

As discussed above, the challenges facing a veterinarian and the areas in which they can contribute vary enormously with the taxa involved and whether the incident involves single or multiple individuals. This section provides a guide to help the veterinarian work through the processes of:

- initial assessment;
- stabilisation of the animals;
- clinical examinations;
- clinical pathology;
- therapeutic and other intervention options;
- considerations for immediate release, rehabilitation or euthanasia;
- marking and post-release monitoring;
- necropsy.

In conjunction with the representatives of the state authority, the veterinarian may also have a key role to play as mediator between agency staff, onlookers, volunteers and the media.

2.2 General principles

Fundamental to the correct management of a stranded marine mammal is correct identification of the species involved. This is particularly important for pinnipeds. Knowledge of the biology of the species, particularly in relation to distribution and range, seasonality of breeding, reproductive strategies, moulting, migration and social ecology will aid the decision-making process. Useful guides include Baker (1999), Bryden et al. (1998) and Carwardine (1995). If the identity of the species involved has been established, it is often useful for the veterinarian to read up on the species' biology prior to attending the animals.

Stranded marine mammals, particularly cetaceans and sirenians, are typically in a highly compromised state when they are examined. Much of the clinical picture can be as much a result of being beached as it is of any pre-existing or stranding-induced pathology. Consequently, these clinical assessments should be interpreted with caution and where possible the benefit of the doubt should be given to the animal. It is remarkable how an unresponsive beached cetacean, for which a veterinarian might determine a poor prognosis, can improve rapidly when held for a short period in shallow water.

2.3 Mass strandings: cetaceans

2.3.1 Step 1: determine your role

The essential first step for the veterinarian is to identify the IC (or equivalent) and seek advice on what role they might play. It is most important that the veterinarian's skills are best utilised, so it is important to avoid the temptation of getting involved in the general rescue effort of stabilising animals on the beach (unless there are very few people available).

Another key role is to ensure their own safety and that of others. Mass strandings in particular are extremely dangerous and it is important that this be recognised by the veterinarian. Times of highest risk for the veterinarian include clinical assessment and specimen collection (use extreme caution when working around the head or tail of the animal); moving and release of animals (rope entanglement and drowning); necropsy (slippery, often blunt knives and poor hygiene) and; at all times, exposure to the elements. People caught up in the emotions of dealing with stranded animals frequently ignore their own health and safety. Rough seas, rocks, heat, cold and the sun all pose significant risks. Although uncommon, the potential for zoonotic infections associated with marine mammals, particularly pinnipeds, must also always be considered.

2.3.2 Step 2: stabilise animals

Once human safety issues have been dealt with under the ICS, the stabilisation of animals will be the primary focus of the IC. This process need not involve the veterinarian. The first two priorities are likely to be to attempt to prevent any remaining free-swimming animals from stranding and to remove animals from the surf zone (onto the beach or out to sea).

Where possible, cetaceans should be placed into a position where they can be treated safely, both for the animal and personnel. If an animal is in shallow water and there is no significant surf it can be held gently in the water with its head facing out to sea. Such a circumstance minimises the physical trauma associated with a stranding. Where an animal is in a surf zone, on rocks or already beached, it should, if possible, be laid belly-down on reasonably level sand. Holes should be dug for its pectoral flippers.

Handling and transport of cetaceans has been reviewed (Geraci 2000). The IC and state authority at the stranding will have had experience in lifting and moving animals. Whales up to about 2 tonnes are routinely lifted with specialised slings, inflatable pontoons and heavy equipment. Live cetaceans should never be pulled backwards in the water, nor should they be rolled on land. They should never be dragged or towed out to sea by their tail stock. If animals cannot be moved, they should be made as comfortable as possible while waiting for the tide to come in. If they cannot be refloated they should be euthanased.

2.3.3 Step 3: rapidly assess and individually identify patients

The scene at mass strandings, even with well-organised responses, often appears chaotic. The number of animals may often be overwhelming and there are often many people taking what action they can with animals that are beached or beaching. For the veterinarian to make sense of much of the data they will collect during the stranding response (clinical examinations, clinical pathology, knowledge of the fate of the animals etc.) a critical first step is to ensure that all animals are temporarily marked to identify and track individuals, that their location both on the beach and relative to one another is recorded and, if known, the order in which they stranded. This is a normal part of the management of mass strandings and most state authorities have kits that include temporary tags (usually SC 'spaghetti' tags or coloured ribbon with numbers written in permanent ink, to be tied around the tail stock) and pro formas. These tasks are generally undertaken by someone other than the veterinarian, usually a marine wildlife officer or equivalent from the relevant state authority.

The rapid assessment is best conducted by a veterinarian and is used to prioritise the allocation of resources. In conducting the assessment it is important to remember that a stranded cetacean may or may not be diseased, but it is always injured or compromised to some extent by the stranding event itself. Three simple categories are useful: 1) most likely to survive; 2) less likely to survive (injured or weakly responsive); 3) dead.

Along with the individual number assigned to each animal, a colour category can be assigned for each of the three categories. The separation of the first two categories is not simple as a beached cetacean will inevitably develop respiratory fatigue and distress and will suffer superficial and musculoskeletal damage. The magnitude and rate at which this damage occurs is greater in larger animals. These changes are difficult to measure and we know little about their effects upon the animal's survival. If in doubt, assign the higher category. The combination of records of careful and systematic clinical evaluations and of the eventual fate are a logical approach to improving our ability to predict the fate of a stranded cetacean. The assessment is primarily visual and relies mainly on behavioural criteria (see 2.3.5).

Maximum resources should initially be directed and ranked towards animals in the two live categories. If, during the process of the rescue operation, a subsequent clinical investigation determines a different category from the one ascribed during the rapid assessment, the animal's colour band should be changed but its identification number must remain the same.

It is important to remember that the initial assessment is merely that. Repeated assessments of each individual are far more useful in assessing the prognosis for return to the sea. The single most important role of the veterinarian is to ensure that this longitudinal assessment occurs and the data are available for decision-making.

2.3.4 Step 4: first aid

First aid for stranded cetaceans should aim to limit further injury and maintain the animals' homeostasis (maintain it in a non-stressful comfortable position and provide supportive care) as well as possible until further action is possible (Table 3.1). It will generally be

Table 3.1 Guide to first aid for stranded cetaceans

Effect of stranding	First aid measure
Hyperthermia	Pour cool water over dorsal fin, pectoral fins, tail fluke and head region. Do not cover these areas to encourage evaporative cooling.
	Provide shade cover over the animal.
Respiratory difficulties	Clear and protect the blowhole from sand and water.
	Position in ventral recumbency.
	Move into water if possible.
Skin trauma (sun and physical)	Move from rocks or other solid sharp substrates.
	Protect from sun and wind by covering with light coloured wet cloth or similar material.
Dehydration	Provide shade and cover with wet sheet or similar material.
	Keep eyes and blowhole moist.
	Tube with oral fluids.
Pressure	Place in ventral recumbency.
	Dig depressions for pectoral flippers.
	Move to shallow water if possible.
Disorientation	Support in shallow water pointing toward the ocean.

applied by the volunteers attending a stranding, under the direction of the IC.

Once a cetacean has been stabilised (see 2.3.2), the principal physical threats are from wind and sun. The blowhole and eyes should be kept moist using sea water and protected. The blowhole region should only be flushed with water immediately after a breath has been taken (observe the breathing sequence to be sure of the timing). Care must be taken to avoid water entering the blowhole. The dorsal fin, pectoral fins, tail fluke and head region are all important areas for heat shedding for cetaceans and cool water should be applied preferentially to these parts. They should be left uncovered to encourage evaporative cooling. Ice can be used with care. The remainder of the body should be covered with light coloured, wet sheet material or something similar (note that even newspaper, seaweed or other materials can be used if nothing else is available) to protect from sun and wind. Where possible any exposed skin should be shaded. The fins can be felt regularly during first aid procedures to determine how much heat they are dumping (see 2.3.5).

2.3.5 Step 5: clinical assessment

a General

The purpose of clinical examination at a stranding is to determine where the animal is on the stranding–recovery–release pathway, and ultimately if it can be released. It is important to remember that it can be extremely difficult even for very experienced operators to determine if animals are suitable for release, based on single observations. Repeated assessments of multiple variables, preferably with the animal in the water, and how these variables change over time are crucial for informed decision-making. If in doubt, give the animal the benefit of the doubt and continue monitoring. An improving demeanour (behavioural criteria, see below) may be associated with a far better prognosis for return to sea than a worsening demeanour. Prior to decision for final release, it is crucial that the animal be assessed in the water if possible, preferably free-floating. It is quite amazing how some animals that appear about to die will improve once they are refloated.

b Behavioural criteria

Each live stranded cetacean should be classified as: 1) alert (aware, responsive to stimuli such as wave action, touch or repositioning); 2) weakly responsive (responsive only after much stimulation); 3) non-responsive (alive, but not responsive to stimuli).

c Physical and skin condition

Prior to a hands-on clinical examination of an animal, a brief general description of body condition (e.g. external injuries, abnormalities) should be made. Note that due to the general fusiform shape of cetaceans, and their firm outer blubber layer, the general nutritional status (starving vs fat) is difficult to determine visually except in extreme cases.

As cetacean skin is designed to function entirely in water it is rapidly damaged when dry, particularly when exposed to wind and direct sunlight. It is also highly vulnerable to laceration from rocks and shells etc. When exposed, the skin will wrinkle, blister and peel and can rapidly lead to critical fluid loss. Assessment of skin condition should include a description of the nature and extent (percentage of body) of the damage. Animals with substantial amounts of skin loss (>30%) are likely to be poor candidates for release and survival. It is important to differentiate between acute skin damage associated with the stranding event, and prior damage such as cookie-cutter shark (*Isistius* spp.) bites (C-shaped or oval, usually about 5–7 cm diameter) and rake (bite) marks from conspecifics, which can be quite extensive and normal in some species. There may be a normal dark reddish discharge from the rectum of pygmy and dwarf sperm whales (*Kogia* spp.), which is associated with a diet of squid and should not be mistaken for blood.

d Behaviour in water

If the animal is in the water at the time of the clinical examination, or has been moved to the water during the process of the rescue operation, a great deal of information can be derived from an assessment of its behaviour. After being stranded or rolled in the surf zone cetaceans are typically disoriented and often unable to swim and remain upright. An animal's ability and eagerness to remain in an upright position and swim are key indicators of its readiness for release. It may take some time of appropriate handling for a cetacean to regain muscle tone and orientation. This involves gentle manual support of the animal while it is fully suspended in water. A cetacean should not be released until it is able to swim strongly. Ability to swim can be difficult to assess. Various techniques include taking the animals for a 'walk' with people providing support at either side of the front. A 'break person' directly in front of the animal can also be utilised. However, caution needs to be exercised as some animals may swim off, others sink and drown, and others 'run over' the break person. Only calm animals up to about 5 m long can be 'walked', and only very experienced operators should use this technique. Conditions need to be calm, with a flat solid substrate. It has been our experience that return of tail movements associated with being moved through the water, coupled with a good righting response (ability to reposition in the water if the long axis of the body is rotated), strong and reasonably steady respiration and good reflexes generally indicate that an animal is able to swim. Animals with abnormal buoyancy, including sinking, listing to one side or circling may not be suitable for release. When animals are released as a group, it is important that all members have regained the ability to swim strongly. A single weakened animal can result in the group restranding soon after release.

e Respiration

Respiratory fatigue and distress eventually develop in all stranded cetaceans. A rapid and irregular respiratory rate, gurgling sounds while breathing, or bleeding, frothy or foul-smelling fluid from the blowhole are signs of significant respiratory distress or disease. Respiratory rates vary with species, as well as the condition and circumstances of the animal. Small cetaceans such as bottle-nosed dolphins may breathe up to 6–8 times per minute, larger whales may breath less than once per minute. The actual expiration and inspiration (when the blowhole is open) should be rapid (<1–2 sec) and forceful. Weak prolonged breaths are a poor sign. Respiratory function often improves dramatically if the animal can be stabilised in shallow water.

f Heart rate

Heart rate is of questionable use, as it is difficult to interpret in all but extreme cases. Heart rate can be palpated by holding the hand firmly under the pectoral flipper of a small to medium sized cetacean or measured directly with a stethoscope. Inspiratory tachycardia is often quite dramatic among cetaceans, thus the rate should be averaged over a few minutes to acquire a reasonable measurement.

g Body temperature

Body temperature is a measure of cardiovascular function (the animal's ability to shed heat from peripheral circulation) as well as general physiological state. An explanation and illustration of the peripheral counter current heat exchange mechanism in cetaceans is presented in Chapter 18. As water conducts heat far more efficiently than air, a stranded cetacean is enormously compromised in its ability to regulate its core body temperature, and this becomes a fundamental aspect of the first aid provided for the animal. Core body temperature can only be measured adequately with a flexible rectal probe of sufficient length. In a bottlenosed dolphin the required length is >20 cm, but it is likely to be substantially more in a sperm whale. Do not use glass thermometers. Normal temperatures are 36.5–37°C. Temperatures below 35.6°C are indicative of hypothermia or cardiovascular shock. Temperatures above 40°C are critical and life-threatening. Peripheral body temperature can be measured directly with a thermometer on the flukes and flippers. Alternatively, and more simply, a relative impression of peripheral body temperature can be acquired by holding the palm of a hand on the fins or head region of the animal, and comparing this to other skin surfaces. If the skin feels hot the animal is probably attempting to dump heat through its peripheral circulation and more cool water should be poured over these areas (and more shade provided if possible). If the skin feels neutral or cool it is not possible to draw clear conclusions of how that might relate to core body temperature.

h Reflexes

The loss of body reflexes is generally a poor sign. The blowhole should always remain tightly closed except during inspiration and expiration. The palpebral reflex can be tested by tapping gently near the eye; the reflex closure of the eye is the normal healthy response. The pupillary reflex is difficult or impossible to measure as stranded cetaceans typically have pinpoint pupils and will not respond to light sources. A fully dilated pupil is a sign of death. The corneal reflex is not useful in live animals as the lids are often held tightly shut, but can be used in assessing death. Jaw muscle and other skeletal muscle tone can be tested by attempting to open the jaw or by gently pulling the pectoral flipper forward. In both cases a healthy animal will resist the efforts. The tongue reflex, elicited by gently squeezing the edge of the tongue and observing the response, can be useful in assessing moribund animals or checking to see if animals are dead. However, it can be dangerous as some animals may resent it and should therefore only be done by experienced operators or when there is confidence that an animal may be dead.

i Maternal dependence

Small or young animals should be assessed for maternal dependence, which is an important consideration in assessing animals for release regardless of their apparent health status. It is especially important at mass strandings where a number of females may have died, and the fate of the dependent young's mother is therefore unknown. Maternal dependence is discussed in more detail under 'Criteria for euthanasia' below.

j Interpretation

A cetacean ready for immediate release should ideally be alert (responsive to external stimuli such as touch and wave action), have minimal and superficial skin damage, be able to breathe and swim, have a respiratory rate and body temperature within normal ranges, have all the measured body reflexes, and have no other evidence of major injury or disease. Issues of dependence should also have been considered.

A cetacean that is not a good candidate for immediate release has one or more of the following: lack of response to stimuli, extensive (>30%) skin damage, inability to swim without assistance, abnormal buoyancy (sinks or lists to one side), significantly abnormal respiratory function that does not improve in the water, a significantly high or low core body temperature, loss of one or more measured body reflexes, and/or evidence of major injury or disease.

Some animals fall between these categories. Where further action such as treatment or temporary holding is not possible, release and monitoring should be considered.

2.3.6 Step 6: clinical pathology

Clinical pathology can provide a major diagnostic aid to assist veterinarians with interpreting the otherwise difficult and subtle clinical signs observed in beached cetaceans. The advent of portable and rapid analysis systems mean that results from blood samples are sometimes available in meaningful timeframes, and thus inform decisions about an animal's suitability for release. As discussed earlier, the stranding event itself will ultimately lead to life-threatening pathology. The release of enzymes from muscle damage in particular can reach concentrations that may lead to fatal toxaemia. Substantial electrolyte imbalance, dehydration, changes associated with stress and any other preexisting pathology are generally dramatically evident from the blood analysis. Changes in pathology may occur as a time-series in stranded cetaceans, so serial samples, the first collected as soon as possible after the stranding event, will be most informative. Of the many variables that could be monitored, changes over time in CK, BUN, creatinine, WBC count (and differential count), PCV and TPP are probably the most useful in assisting decision-making. Techniques for venipuncture and reference values are described in Chapter 18.

Nevertheless, we know so little about the ultimate consequence of the levels of parameters in the blood that a decision not to release, or to euthanase, an animal should not be made solely on this basis. The primary assessment should always be on overt clinical appearance, with the clinical pathology informing this interpretation. There have not yet been studies relating the data from blood samples to the ultimate survival of released animals, and this step is needed to allow us to use prognostic data with more confidence.

2.3.7 Step 7: therapeutic options

The major limitations facing the veterinarian are the limited time in which to determine an appropriate regime of treatment, lack of opportunity for ongoing or follow-up treatment, the scale of treatment given the size and number of animals, and the potential cost. One-off administration of an antibiotic is probably contra-indicated as it is unlikely to reach adequate concentration for adequate duration to kill most bacteria. There is also a risk of abscess formation if injected into blubber, it may disrupt important gastrointestinal functions and could lead to potential bacterial resistance. The routine administration of corticosteroids, non-steroidal anti-inflammatory drugs, sedatives or tranquillisers is also not indicated.

The one area of therapy that may prove useful is the administration of fluids. Dehydration and electrolyte imbalance are an almost certain consequence of even relatively brief stranding events. The easiest and most rapid form of fluid administration is by mouth. This can be delivered with a horse drenching tube, or garden hose passed into the stomach. The anatomy of the nasopharynx means that there is almost no chance of introducing the tube into the larynx and any regurgitated fluid cannot be aspirated (see Chapter 18). However, if animals are in shock the fluids in the stomach may simply pool there and be poorly absorbed or regurgitated. The exact therapeutic benefit of administration of fluids PO in stranded animals is not known. Approximate volumes of fluids (oral rehydration solutions or water) that we have administered by mouth to some stranded cetacean species include false killer whale 8 L bid, long-finned pilot whale (*Globicephala melas*) 2–5 L, bottle-nosed dolphins (*Tursiops* spp.) 2 L and short-beaked common dolphin (*Delphinus delphis*) 1 L.

Intravenous administration can be achieved through a variety of veins (see section on injections for euthanasia), but generally takes more time than is available during a stranding event, particularly for multiple animals. Minimising the time before animals can be returned to open water is probably more important than attempting to redress fluid and electrolyte imbalance. Detailed information on therapeutics and veterinary procedures is given in Chapter 18.

2.3.8 Step 8: immediate or delayed release, rehabilitation or euthanasia

a Options

The IC will need to decide on a range of options for each live animal at a mass stranding. These options can be summarised as follows.

- Leave the animal with no further action when:
  - location is too isolated;
  - available logistics and resources constrain any other option;
  - in the case of large whales, natural death is the most humane option.
- Release the animal immediately (usually with a group of other animals that are good candidates for immediate release).
- Move the animal to a more logistically appropriate site for assessment and release.
- Delay (<48 hr) the release of the animal at or near the stranding site if:
  - a suitable temporary holding site is available (with volunteers holding animals in shallow water);
  - an incoming tide may assist in refloating the animal or facilitate release;
  - weather and ocean conditions are unsuitable for immediate release;
  - the animal is likely to become fit for a release if held a little longer.
- Euthanase the animal.

Sending the animal to a captive holding facility for rehabilitation and later release is not recommended. In Australia there are limited facilities available to do this. It is also extremely costly, time-consuming and generally unsuccessful. Having an animal away from its pod for any length of time is not in its best interests. The risk of disease transmission to animals in care and then to the wild population after release must also be considered. For these reasons, this practice is now discouraged by most wildlife agencies.

At many mass strandings where animals have been successfully returned to the sea, the best strategy for small to medium sized cetaceans is to move the healthy animals back to sea as a group. This can only be successfully accomplished if animals can be moved a manageable distance from their beached location to an area of relatively calm and shallow water with good access to the open sea. If sea conditions are calm, the site may be the actual stranding site. A small number of small to medium sized cetaceans can also be moved, and held, off-shore using boats.

The duration an animal is held prior to being moved off-shore can have a marked effect on its chances of survival. The period in shallow water allows the veterinarian the best chance to assess the animal's condition, as well as allowing it time to re-orientate and regain the capacity to swim effectively. Given the likelihood that the group of released cetaceans will remain together as attempts are made to herd them slowly off-shore, it is essential that seriously compromised animals not be released in the same group as apparently healthy animals. Such mixing may well lead to a further stranding event. The more compromised animals should be removed visually and acoustically from the group to be released. If attempts are made to release a second group of less-fit animals, sufficient delay should be included to minimise chances of the two groups mixing near the shore.

Holding animals at temporary beach locations for periods >48 hr is generally impractical. Under most circumstances the most compromised animals have died or been euthanased. The prognosis for any animals that remain alive depends on any pre-existing medical conditions, damage suffered during the stranding, and the quality of the care and holding location since the stranding. Each case will be different and animals that appear strong at this point may still be candidates for release. However, if the animals are still alive after more than 48 hr of holding but are still not strong enough for release, euthanasia may be the only practical option.

b Euthanasia

Euthanasia is sometimes the most humane act we can provide to a stranded cetacean. The key issues are when and how to do it. The reason a stranded cetacean might be euthanased is to end the suffering caused by the stranding event, a pre-existing pathology or injury or in circumstances where no other humane alternatives are available. The method must be humane, rapid, simple to use and safe for people and other animals in the area. Secondary concerns may be the aesthetics of the procedure (to onlookers) and the damage it may cause to valuable specimens such as the skull (this may be more important for rare specimens such as some beaked whales, or where cranial pathology might be important in determining the possibility of human-induced acoustic trauma). Large whales, particularly sperm whales, are very difficult to euthanase due to their massive size and skulls, which are difficult to penetrate using readily available firearms. In such cases a natural death, with concurrent palliative care, may be the most humane option. This approach is also preferable where there is uncertainty about the outcome of a euthanasia attempt due to inexperienced operators or inappropriate equipment. It may also be the only option in remote locations where access to the animals may be difficult and/or suitable equipment and personnel are not available.

The decision to advise the IC to euthanase a cetacean should ideally be made by a veterinarian who has clinical experience with stranded cetaceans. The veterinarian should consult extensively with appropriate biological and operational experts. In cases where several or many animals may need to be euthanased it is wise to seek a second veterinary opinion. The ultimate decision to euthanase is made by the IC in consultation with other members of the team. However, the IC will rely heavily on advice from the attending veterinarian. If it is decided to euthanase an animal, it should be carried out rapidly and with confidence to avoid unreasonable suffering and continued debate.

In cases where the decision to euthanase is unclear, it is preferable to explore other options first. The decision to euthanase an animal should be clearly explained to the public and volunteers in the area. If media are present, they will need to be informed. This is usually done through a nominee of the IC. Unless specifically requested to do so by the IC, the attending veterinarian should not discuss decisions to euthanase (or other key issues) with onlookers, volunteers or the media. People who are not immediately required for the euthanasia should be moved away from the animal. In some situations it may be necessary to screen the animal from public view. The impact of the euthanasia on other stranded whales in the vicinity should be taken into account.

Criteria for euthanasia

Clear criteria for euthanasia (modified from Geraci & Lounsbury 2005) include:

- disabling injury such as penetrating injuries of the abdomen, dislocated or broken tailstock, flukes or fins;
- substantial haemorrhage from the mouth, blowhole, genital opening or anus;
- rectal temperatures of $>42°C$;
- blistering and sloughing of a major proportion ($>30\%$) of the skin surface;
- sustained loss of major reflexes, loss of tongue withdrawal, loss of jaw tone;
- inability to move or rescue a cetacean humanely;
- certain social dependency (e.g. dependent calf stranded alone with no animals in area).

Perhaps one of the most difficult decisions regarding euthanasia is the last category, for young and potentially dependent animals. Separation from a mother is rapidly fatal for a neonate, and these animals comprise a substantial component of single stranded animals. When returned to the sea, neonates often repeatedly re-strand. Attempts to return young animals to nearby pods of conspecifics may be worthwhile, but should not be persisted with endlessly. Determining if a young animal is still dependent upon its mother becomes harder as they reach the approximate size of weaning. This is usually stated in most good cetacean field guides (e.g. Carwardine 1995), and is generally about 60% of the adult length of the mother. Animals of this size may strand because they have not managed the transition from dependence to independence—usually a period of high mortality in most wild mammal populations. If these animals still appear in reasonable body condition and are close to weaning size, attempts to return them to open water may be appropriate. If they are well below weaning size or clearly starving and in very poor health, euthanasia is indicated.

Dependent young that have stranded in circumstances where the mother is clearly identified (e.g. mother and calf stranding) can be returned to sea provided they fulfil the criteria above. However, deciding animals' fates can be extremely difficult in mass strandings with multiple dependent young and numbers of dead females. If all animals appear fit and healthy and there are minimal numbers of dead females, attempts can be made to reunite mothers and young while holding animals. The reality, however, is that this is often unsuccessful and the presence of a calf's mother cannot be guaranteed. In these circumstances euthanasia may be considered.

Methods of euthanasia

Shooting

Where practical and safe, shooting into the brain is an effective and virtually instant method of euthanasia for cetaceans less than about 8 m long. Shooting must only be carried out by experienced licensed personnel and the area around the animal must be cleared of people. Shooting should not be used in areas where ricochet is a danger.

The firearm must have a high muzzle velocity and be large-calibre (e.g. 0.308 inch), and jacketed projectiles must be used. Dorsal or lateral shots can be used, based on the position of the whale and where the projectile might exit the animal. Dorsal shots should be directed down the blowhole, aiming for an imaginary point on the midpoint of a line joining the anterior insertion of the pectoral flippers. Lateral shots should aim slightly upwards through a point between the eye and the ear opening. A well-directed shot should kill the animal instantly and cause little or no movement at the point of death. Shooting into the heart of a cetacean is not regarded as a humane method of euthanasia.

Lethal injection

Lethal injections of barbiturates are an appropriate method of euthanasia for small to medium sized cetaceans and, if there are small numbers of animals, may be an alternative to shooting. Injections can be given into a vein, the heart or the abdominal cavity. Intracardiac or intraperitoneal injections should only be administered if shooting or IV injection of a lethal solution is not practical. The amounts of solution to be administered are not well-described and the veterinarian should err towards higher doses than might be needed for terrestrial animals. A stranded cetacean is invariably in shock, which might lead to poor absorption of a drug from the abdominal cavity or from peripheral veins. It is common for a cetacean to thrash about at the point of death when euthanased with lethal drugs, particularly if the dose is low or poorly absorbed. This response can be dangerous to handlers and distressing to onlookers. The incidence of thrashing can be reduced with pre-medication using a sedative. Diazepam, xylazine, acepromazine, midazolam or pethidine may be used. Doses well above the recommended safe doses for domestic animals can be given prior to euthanasia to ensure an effect. These may be administered IM or IV.

Intravenous injections should be administered using an 18 G × 38 mm needle or 18 or 19 G butterfly catheter or scalp vein set. The vessels in the tail fluke are the most accessible (see Chapter 18). Peripheral veins can also be accessed from the pectoral and dorsal fins, but lack clear landmarks for injection sites. The heart of a cetacean can be reached by directing the needle from a point just behind the origin of the flipper to the same point on the opposite side of the body. Alternatively, it can be reached by inserting the needle either side of the sternum, at a point just posterior to a line joining the base of the pectoral flippers. A needle length of up to half the diameter of the animal's chest is required.

Because the volume of euthanasia solution required becomes prohibitive as the size of the animal increases, supersaturated potassium chloride can be administered as an adjunct. This must only be administered once the animal is unconscious. False killer whales (approximately 900–1400 kg body mass) have been euthanased by administering concentrated solutions of pentobarbitone sodium (60–100 mL of approximately 300 mg/mL solution IV) followed by a supersaturated potassium chloride solution until the heart stops (usually 300–500 mL IV). A flutter valve can be used to aid administration of large drug volumes.

Explosives

Shooting or lethal injection are not reliable methods of euthanasia in whales more than about 8 m long. Indeed, the logistics of killing an animal of this size are considerable. The careful and judicious use of explosives has been used to successfully kill large baleen whales in Western Australia, but dramatic and unfortunate mistakes with explosives have been made elsewhere in the world. The key to this technique is to force the power of the explosion down over the skull of the whale (implosion). In the absence of someone who has successfully used this technique, natural death is considered a more humane response to stranded large whales that cannot be refloated.

Unacceptable techniques

Suffocation by obstructing the blowhole is neither effective nor humane and should never be attempted. Drowning, by pouring water in the blowhole, is also unacceptable.

It is possible to kill a whale by severing the major blood vessels that supply oxygen to the brain (exsanguination). This requires cutting the vessels around the heart or the rete mirabile in the chest. Severing the carotid arteries will not lead to rapid death in a cetacean, unlike other mammals. The technique can lead to a fairly rapid death in some cases, but the risk of problems with the speed and reliability of cutting the critical vessels mean it is not recommended as an acceptable method of euthanasia.

It is important to ensure an animal is dead once a euthanasia technique has been applied. Animals should not be left or moved for burial until the veterinarian is confident that they are dead. A dead whale does not breathe, has dilated pupils and lacks corneal and palpebral reflexes and muscle tone. Mucus membrane colour and capillary refill cannot be relied upon to assess peripheral perfusion as gums are often grey or white. It is useful to either monitor individual animals until they are dead or, in the case of multiple animals, leave a capable person to watch each animal for breathing and signs of life.

Carcass disposal is at the discretion of the IC and local council. Options include burial above the high tide mark, transport and burial elsewhere, leaving animals to decompose on the beach, or towing out to sea and dumping. If animals are euthanased by lethal injection consideration must be given to scavenging of the carcasse by other animals. In these cases deep burial is recommended. Towing the carcass out to sea poses a shipping hazard and may attract sharks. The ultimate decision is made by the IC and will depend on available resources and political factors.

2.3.9 Step 9: marking and post-release monitoring

Understanding the fate of live cetaceans returned to the ocean after a stranding event is essential. It informs future decision-making by allowing assessment of the usefulness of observed clinical signs or pathology in predicting fitness for release and survivorship. If the last we see is a group of animals swimming over the horizon, we have to wonder if any or all survived once they were beyond our ability to watch them. For this reason the veterinarian should encourage the IC to ensure that as many released animals as possible are identified and monitored for as long as possible after release. There are a number of methods by which we can monitor these animals, from very simple temporary marks (that at least allow us to identify an animal if it re-strands soon after release), to sophisticated telemetry tags that can remotely track an animal at sea for weeks or months.

In several states, local fauna authorities have marking equipment such as temporary plastic ties to be put around the pectoral fins, passive integrated transponders (PIT tags) which can be detected with active scanning wands, transdermal spaghetti tags or freeze-branding equipment for brands on dorsal fins. While these marks are an important component of the stranding effort, increasingly there is interest in attaching satellite-linked radio tags to the dorsal fins of released cetaceans. The design and application of these tags requires training and substantial resources. The tags are generally attached with one or more bolts through the dorsal fin. Attachments of this type should be conducted in the presence of a veterinarian to ensure that major blood vessels are avoided, and that the tags are applied in an appropriately humane and sterile manner.

2.3.10 Step 10: necropsies

Standard necropsy techniques are appropriate for cetaceans, but differences and difficulties may be encountered. Large cetaceans, by sheer virtue of size, can be a challenge to examine. Organs the size of small cars and disgorging intestinal gases can represent very real dangers to the naïve veterinarian. The major elements of a necropsy of small to medium sized cetaceans are as follows.

- Flense the blubber back in parallel strips from a ventral or lateral mid-line incision.
- The muscle is very dark due to its high concentration of myoglobin.
- The diaphragm is often surprisingly caudal and oblique to accommodate the large thoracic cavity.
- The liver is multi-lobulated, with major hepatic sinuses that store blood during the dives.
- The kidneys are multi-lobulated, similar to many ruminants.

An excellent resource to guide veterinarians through this process is the *Guidelines on Sampling Cetaceans*, developed by and available through the Department of the Environment and Water Resources (previously known as the Department of Environment and Heritage). The guidelines are usually available from the IC at the site.

There are various priorities for sample collection at a stranding. Environmental agencies are generally interested in DNA samples, samples for residues and heavy metals, and reproductive status. Museums are often interested in carcasses, skulls or skeletons from unusual or rare species. The priority, however, for the veterinarian at a mass stranding is to perform full necropsies on key animals, specifically those that showed signs of disease prior to death or those that stranded first and died or were euthanased. Routine collection of specimens can be organised by the IC. Where possible, necropsies need to occur as quickly as possible after death. It is extremely helpful to use an experienced cetacean worker to assist with the necropsy (e.g. university student or lecturer, member of the designated authority in attendance or experienced volunteers).

2.4 Single strandings

Many of the considerations and much of what has been discussed in the veterinary approach to assessing and treating cetaceans at a mass stranding apply equally to singly stranded cetaceans. However, the level of response and the logistics of dealing with a single animal are usually far less. An important difference is that in many cases a single cetacean will strand because it is already in a highly compromised state (disease, injury or starvation). Diseased cetaceans or dependent young that have lost their mothers have a poor chance of survival and euthanasia is often the most humane action. However, it is important to remember that there will be a good deal of uncertainty about the clinical status of the animal, and indeed its exact level of maternal dependence, even after examination, and in such cases attempts to refloat and release the animal are reasonable. Single stranded cetaceans with the best prognosis for survival are those that were stranded by a rapidly falling tide or trapped in a shallow estuary or river or similar event.

2.5 Pinnipeds

This section has been adapted from Vogelnest (2000) and references therein.

Unlike the situation with a stranded cetacean, there are no direct effects on a pinniped from being on land other than interference from people and dogs, or prolonged periods ashore without food (and thus water). Clinical evaluation is usually limited to observation from a distance. If the animal can be readily approached and handled for evaluation, it usually means intervention is indicated. Blood collection as part of the evaluation is usually not an option on the beach, unless experienced handlers, equipment and veterinarians experienced in venipuncture in pinnipeds are present. First aid is also usually not an option on the beach, as any animal tolerating anything other than the most basic first aid, e.g. removal of entanglement, will require attention at a care facility.

There are five options for dealing with a live pinniped ashore, and a veterinarian should ideally be consulted in the decision-making process for each. These are: leave alone and monitor; relocate; capture for treatment on site; capture for transport to a care facility for treatment and rehabilitation; euthanasia.

2.5.1 Leave alone and monitor

If an animal appears healthy—in good condition with no or only minor wounds—and the species' biology and natural history show that it is not unusual for the species to be at that site at that time of year, it is best left alone and monitored for a day or two. Seals, particularly adults, can recover from fairly major wounds. An otherwise healthy seal with a major wound should be monitored initially to assess if the wound is having any significant effect on the animal. If there is any doubt whether a seal requires rescue, the animal should be monitored for at least 24 hr before a decision is made to intervene. The objectives of monitoring are evaluation of the animal's condition, protection from harassment and protection of the public.

There is no need to wet seals while ashore, as is the case with cetaceans. If the animal is obviously overheating and appears distressed it requires more than just cooling. A healthy pinniped that is overheating will seek shade or water. If the animal is in a public area, an exclusion zone should be established around the animal to protect it and the public. The animal's access to water should not be blocked.

2.5.2 Relocate

Healthy animals resting on a beach or moulting should not be forced back to sea or relocated unless absolutely necessary. In situations where there is danger to the public or the animal, relocation may be necessary. Even in these situations it may be preferable to leave the animal where it is and manage it by erecting barriers. The herding, capture and transport of pinnipeds requires special knowledge, skills and equipment. Herding an animal may be dangerous and is often ineffective.

2.5.3 Capture for treatment on site

It may be possible to capture and treat an animal on site and release it immediately, particularly animals that are entangled but otherwise healthy. The capture and handling of any pinniped should only be carried out by experienced people with adequate equipment to accomplish the procedure.

2.5.4 Capture for transport to a care facility for treatment and rehabilitation

Other disabilities that warrant intervention usually require treatment and rehabilitation at a care facility. Unlike cetaceans, this is a feasible option for pinnipeds. A number of zoos and oceanaria accept pinnipeds for rehabilitation. Capture, transport, treatment and rehabilitation should only be carried out by experienced people with adequate equipment, facilities and resources to care for the animal.

The decision to intervene, particularly to remove an animal to a care facility, must be based on careful veterinary assessment of the animal and discussions with the state authority representative. These decisions are complicated. The treatment and rehabilitation of pinnipeds is expensive. Most pinniped species in Australian waters are reasonably abundant and the capture, treatment and rehabilitation of any pinniped should be seen as an animal welfare rather than a conservation issue. Volunteers, onlookers and the media must be informed of reasons for a decision to capture and transport an animal to a care facility. If the on-site veterinarian is not experienced with marine mammals, they should seek advice from an experienced marine mammal veterinarian before deciding on the need for rehabilitation at a care facility. Treatment and rehabilitation at a care facility is appropriate only if:

- the animal requires it based on the assessment of a veterinarian and marine mammal experts;
- there is a good chance the animal can be restored to health;

- a suitable care facility is available, that has the necessary resources (funds, staff, food, medication) and is prepared to provide care for the necessary period;
- the care facility can adequately isolate the animal to minimise the risk of disease transmission to and from captive animals;
- the animal can be readily transported;
- due consideration is given to the likely fate of the animal after rehabilitation.

The last factor should ideally be discussed prior to capture, however, these decisions may only be possible after a more thorough assessment in a more controlled environment away from the public and media. It is important to be aware that the Scientific Committee on Antarctic Research (SCAR) discourages the release of seals after being in captivity, particularly to Sub-Antarctic islands and the Antarctic continent. Such reintroductions serve no conservation purpose and run the risk of introducing pathogens to the wild population. Only healthy animals with no disability that could interfere with their capacity to survive in the wild should be released. SCAR also recommends that rehabilitated seals be released at the site of capture to further reduce the risk of disease transmission to breeding colonies. All pinnipeds to be released must be tagged so that they can be identified if they are resighted.

2.5.5 Euthanasia

A decision to euthanase a pinniped is possibly less urgent than with a cetacean, and it may be possible to remove the animal from the public eye for euthanasia. Euthanasia can be considered after more careful evaluation by experienced people at a care facility. The principles and criteria for euthanasia are similar to those for cetaceans.

Euthanasia of seals is usually accomplished by lethal injection of barbiturates (pentobarbitone sodium) or shooting. The difficulty with injectable euthanasia of seals is that unless the animal is severely debilitated and immobile, it will need to be restrained physically or chemically for administration (see Chapter 17). Anaesthetising the animal prior to administration of euthanasia solution is usually quicker, safer to the operators, less stressful for the animal and better accepted by the public. Sites for vanous access are described in Chapter 17.

2.5.6 Criteria for intervention (treatment on site, transport to a care facility or euthanasia)

a Entanglement in netting or other debris

If the animal is healthy apart from the entanglement, it can be captured, restrained, disentangled on site and released immediately. If the entanglement has affected the animal—poor condition, strangulation, deep wounds—it will require further treatment and rehabilitation at a care facility.

b The animal is covered in oil

Oil may affect pinnipeds directly through fouling of body surfaces and orifices, or through the ingestion of the oil or inhalation of vapours. Oil is a particular threat to fur-seals as it disrupts their inner fur layer and directly compromises their ability to thermoregulate. These animals can directly ingest the oil through grooming attempts. Thorough cleaning of heavily oiled fur seals is required prior to release.

c Debilitating injuries

Some injuries, particularly non-penetrating lacerations to areas of the body covered in blubber, do not necessarily require intervention. Typically, seals can recover from quite severe injuries. Wounds can result from shark bites, territorial defence during breeding season, boat propellers, stingray barbs and gunshots. If the animal is debilitated and obviously affected by the injury—unable to move, penetrating wounds to a body cavity, fracture of a limb, stingray barb in the mouth, purulent discharge, eye trauma, fractured jaw—intervention will be necessary.

d Advanced emaciation

Emaciated animals will be dehydrated, weak and unresponsive. Experience is required to assess whether intervention is necessary. It is a normal stage of the life cycle for some animals, e.g. a young seal in the first months after weaning or an animal after a long migration, to be underweight.

e Serious illness

Free-ranging pinnipeds are susceptible to numerous infectious and non-infectious diseases. These may be primary or secondary. Diseases of pinnipeds are discussed in Chapter 17.

f Abandoned pups (otariids only)

It can be very difficult to determine if a pup is abandoned. Even if they are alone at unusual sites, they are not necessarily abandoned. Knowledge of the species biology and natural history are useful in making this decision. It can also be difficult to distinguish a 6 mo dependent pup from an independent 12–18 mo animal, as their size ranges can overlap. Emaciated animals of this age are frequently reported. These are usually animals that have been unable to catch enough prey post-weaning. They can be distinguished from animals less than 12 mo old, as they have fully erupted canine teeth. Small seals, unless there are obvious injuries, emaciation or illness or are well out of their natural ranges, should be carefully monitored before intervening. If a pup is rescued, careful consideration must be given to the hand-rearing process, prevention of imprinting, the weaning process and teaching the animal to forage. In most cases these animals are poor candidates for release; the longer the period in captive care, the less the chance of successful rehabilitation. Emaciated weanlings are better candidates for intervention, as most of these require only a short period of hand-feeding with fish in captivity until their weight and condition have improved. However, the rescue and rehabilitation of animals in both these categories must be considered carefully. Unless the animal is threatened by anthropogenic activity, these represent natural mortality events. Their rescue has little conservation value and the decision to intervene is generally an issue of animal welfare.

ACKNOWLEDGMENTS

We would like to thank Rosemary Gales and Barrie Wells, whose practical and thoughtful comments and criticisms significantly improved the manuscript, and Vanessa DiGiglio for her assistance with its preparation.

REFERENCES

Baker AN (1999). *Whales and Dolphins of Australia and New Zealand: An Identification Guide*. Allen & Unwin: Sydney.

Bryden M, Marsh H & Shaughnessy P (1998). *Dugongs, Whales, Dolphins and Seals: A Guide to the Sea Mammals of Australasia*. Allen & Unwin: Sydney.

Carwardine M (1995). *Whales, Dolphins and Porpoises: The Visual Guide to all the World's Cetaceans*. Dorling Kindersley: London.

Dierauf LA & Gulland F (2001). *CRC Handbook for Marine Mammal Medicine* (2nd edn). CRC Press: Boca Raton.

Geraci JR (2000). Handling and transport of cetaceans. In *Marine Wildlife, Proceedings 335*, pp. 97–108. Post Graduate Foundation in Veterinary Science: Sydney.

Geraci JR & Lounsbury VJ (2005). *Marine Mammals Ashore: A Field Guide for Strandings* (2nd edn). National Aquarium in Baltimore: Baltimore.

Perrin WF & Geraci JR (2002). Stranding. In *Encyclopedia of Marine Mammals* (eds WF Perrin, B Wursig & JGM Thewissen), pp. 1192–1197. Academic Press: San Diego.

Post Graduate Foundation in Veterinary Science (2000). *Marine Wildlife, Proceedings 335*. Post Graduate Foundation in Veterinary Science: Sydney.

Vogelnest L (2000). The veterinarian's approach to a pinniped ashore. In *Marine Wildlife, Proceedings 335*, pp. 181–191. Post Graduate Foundation in Veterinary Science: Sydney.

4. Wildlife health investigation and a guide to necropsy

Karrie Rose

1 APPROACH TO WILDLIFE HEALTH INVESTIGATION

1.1 Why study wildlife health?

In a global environment where approximately 75% of recently emerging and re-emerging zoonotic diseases emanate from wildlife reservoirs, there has never been more interest in wildlife health (Woolhouse 2002). Additional compelling factors driving an interest in wildlife health include public concern regarding animal welfare, zoonotic diseases and food safety issues associated with endemic and exotic disease, biodiversity protection, bioterrorism detection, climate change and its effects on animal and human health and disease, and the development of a nationally integrated biosecurity strategy (Bengis et al. 2004).

The significance of disease in wildlife populations and its impact on the conservation of biodiversity in Australia is poorly understood. Disease-causing organisms are a component of most natural ecosystems. A balance among the relationships of host, pathogen and environment most often meters the occurrence of disease. As wildlife habitats are modified through human activity and an increasing number of species are intensively managed, the possibility of disturbing these relationships increases. Wildlife health investigations are thus evolving through our increased involvement with wildlife management. The investigation of wildlife health and disease provides the opportunity to view individual species as indicators of environmental contamination, habitat degradation or climate change.

Wild animals may be subclinically infected with a variety of potential pathogens and can function as reservoirs for some disease-causing agents. Undiagnosed disease in wildlife may pose a significant risk to human health, to the agricultural industry, international trade, and to *in situ* and *ex situ* species recovery programs.

Some examples of the potential effects of wildlife disease include:

- potentially fatal respiratory illness in humans and horses in 1994, 1999 and 2004 caused by Hendra virus, a virus resident in flying fox populations (Murray et al. 1995; Breed et al. 2006);
- a 1998 outbreak of infertility and deformity in pigs caused by Menangle virus, a virus carried by bats (Philbey et al. 1998);
- the fifth Australian outbreak of highly pathogenic avian influenza in poultry occurred in Tamworth in 1997, and probably originated from free-living waterfowl (Sellek et al. 2003);
- Australian bat lyssavirus, found in a variety of bat species throughout Australia, has caused the death of two people in Queensland (Fraser et al. 1996; Allworth et al. 1996);

- a global decline in amphibian populations, including the extinction of at least one Australian species, has been associated with the presence of Chytrid fungus (Daszak et al. 1999; Lips 1999; Bonaccorso et al. 2003; Bosh et al. 2000).

Wildlife management involving capture and animal translocation in Australia includes animal research and targeted disease surveillance, the rehabilitation and release of injured or illegally held animals, the reintroduction of species that are depleted or extinct in their natural range, the creation of safe populations of wildlife where they will not be threatened by disease, harvesting or habitat degradation, the dispersal or relocation of animals deemed pests, and the movement of feral or native animals in attempts to reduce population density to control disease or protect habitat quality.

Regardless of the circumstances that initiate your involvement in a health investigation, examining a wild animal should be viewed as an opportunity to collect as much information as possible (Woodford 2001). Collecting blood and faecal samples during restraint of an animal to allow a radio-collar to be fitted provides the opportunity to screen for blood-borne and intestinal parasites, and natural food sources. Serum samples can be tested for antibodies that may reflect exposure to potential pathogens. Some nutrient concentrations can be assessed in serum, and stored serum samples can prove invaluable in defining emerging disease syndromes.

Ultimately, the collaboration of biologists, veterinarians, researchers and diagnosticians can provide cost-effective health assessment, even if this was not the primary goal of the project.

Through our involvement with wildlife health initiatives we can attempt to maximise the success of wildlife conservation activities, contribute to successful reproductive programs and participate in the banking of rare genomic materials. Each wildlife management activity can provide animal health information that is useful as a surveillance tool to help us understand the occurrence of disease and provide information to plan disease control and eradication programs (Zepeda & Salman 2003).

Information on wildlife health from all zoological parks, universities, agriculture agencies, conservation departments and wildlife rehabilitation facilities is being systematically collated by the Australian Wildlife Health Network (AWHN 2007) to contribute to animal health information systems and national disease surveillance programs.

1.2 Health screening or surveillance

A baseline health assessment can be conducted on wild animals captured for management purposes, or it can be targeted to answer specific questions regarding the health of a population.

Common goals of a baseline health assessment in live animals could include determining:

- physiological data (temperature, heart rate, blood pressure, respiratory rate, reproductive details etc.);
- morphological assessment;
- haematological and serum biochemical data and establishment of reference ranges, and identification of haemoparasites;
- identification of ectoparasites;
- the best method of external examination to detect evidence of disease;
- repeated faecal examination for evidence of internal parasites;
- bacterial fauna and potential pathogens in the upper respiratory tract, eyes, ears, lower gastrointestinal tract or lower urogenital tract;
- analysis of genetic variability using blood, skin or oral swabs;
- radiographic examination or skeletal analysis;
- stomach content or scat examination for diet analysis;
- fur or blood examination for the presence of nutrients or toxins;
- serology to determine what organisms the animal has been exposed to in the past;
- polymerase chain reaction (PCR), bacterial, fungal or viral culture to determine whether specific organisms are present within the animal (AAZV 2001).

An experienced wildlife veterinarian and a veterinary pathologist should be consulted during the planning phases of a health screening or surveillance study to ensure that the diagnostic testing is conducted in a humane fashion and that the laboratory findings will be rewarding.

1.3 Planning a health surveillance project

As with any endeavour, the key to successful wildlife health investigations is effective planning. Planning may be done very quickly by experienced personnel who are already well-equipped, but some wildlife health field projects may require more than a year of planning.

First, consideration should be given to seeking approval or guidance from the following groups:

- land owners;
- local interest groups and non-profit organisations;
- researchers involved in similar projects;
- species recovery programs (it is always best if the project is actually generated by the recovery group and designed to answer specific operational questions);
- an animal ethics committee (if live animals will be handled or observed);
- state conservation and agriculture agencies (if animal samples will be transported interstate);
- Department of Environment and Water Resources (DEWR) administers the Convention on International Trade in Endangered Species of Wild Fauna and Flora (CITES 2007) and should be contacted if wildlife samples will be transported overseas;
- the laboratories that will be receiving the samples (to determine optimum samples to collect, transport medium, transport packaging and shipping details).

The state conservation agency should be consulted before any wild animals, particularly marine mammals or migratory birds, are handled.

Projects designed to assess population health are becoming more common. Collecting samples from different populations of wildlife provides a fantastic opportunity to accumulate baseline normal data for haematology, biochemistry, parasite burdens and parameters of a general physical examination.

Baseline physiologic data determination and specific disease testing can be expensive. Collaborating with interested veterinarians and researchers can decrease the costs of investigation. However, it is important to encourage conservation agencies and recovery groups to include funding for health assessment and monitoring in their wildlife management budgets.

It can be very difficult to decide which specific disease-causing agents, toxins or nutrient concentrations to include in a health assessment. Test selection will rely on careful consideration of the species' susceptibility to a range of agents. A thorough literature and resource search of existing health information is the basis of test selection.

The sensitivity and specificity of each potential test should be evaluated. When tests have poor sensitivity, the use of several diagnostic tests to assess exposure to and infection with a single infectious agent may be required. Often, combining several tests such as culture or PCR with serology will help to differentiate:

- animals that may have been exposed to an infectious agent (and now have antibodies towards that agent) in the past, but are not necessarily still carrying the organism;
- animals that are actively infected but remain healthy;
- animals that are actively infected but have reduced function or disease;
- animals that have not been exposed to the infectious agent and could be considered susceptible due to a lack of circulating antibodies.

Always ensure that the tests you are using have been validated for the species involved in your study. If at all possible, diagnostic tests should be thoroughly evaluated in captive wildlife prior to being used in field studies. If you are not confident that the test has been validated, speak with a pathologist or microbiologists regarding means to validate new tests through your proposed study.

Carefully consider how you will interpret your findings, both positive and negative. Identifying the presence of a potential pathogen does not necessarily indicate a clinically significant infection in the animal sampled. Negative findings do not always indicate that an individual animal is free of infection. It is also wise to contact the state chief veterinary officer in advance when you plan to test for diseases listed as notifiable.

After test selection has taken place, it can take time to find laboratories that can assist with your investigation. Canvassing laboratories for the tests that they conduct, the costs involved and the optimum method of sample collection can take several weeks for a large-scale project. Some researchers can be encouraged to test samples on a collaborative basis to reduce costs. Some large commercial laboratories offer a discount for large numbers of samples shipped in bulk, if negotiated in advance.

Creating an examination checklist to be used in the field or in the clinic when each animal is examined will ensure that all required information and samples are collected and recorded. Preparing pre-labelled specimen collection kits before the field trip will reduce the time required to process samples, reduce animal handling times and decrease the likelihood of forgetting to collect a sample.

1.4 Planning a field trip

A field trip to gather data and animal samples requires a great deal of preparation. Personnel participating in a trip must be identified and trained well in advance, as this is often required in the animal ethics committee application. Generally, at least three people are required for any venture in animal handling: 1) animal handler or anaesthetic monitor; 2) examiner and sample collector; 3) recorder. The recorder can complete the examination checklist, properly label the samples, take photographs and pass any additional equipment required during the procedure (Fig. 4.1).

Planning and preparing the equipment for fieldwork can be a mind-racking experience. Obviously, it is very disappointing to be caught without a vital piece of equipment out in the field. It is important to begin the planning early so that any unusual items can be ordered and received. An equipment list is given in Table 4.1, but if this is your first trip talk to more experienced researchers and have them check your list. All equipment should be tested prior to being packed.

When samples can't be taken to a laboratory, it may be necessary to set up a remote lab. A variety of manual and electrical instruments can be taken into the field to assist with sample collection and analysis. Small generators can power most equipment. Centrifuges, microscopes and even serum biochemistry analysers can be transported quite readily. Gas anaesthetic machines can be small and portable.

Packing equipment into sealed plastic bins will protect the materials from water and pests. You may not be allowed onto some off-shore islands if your equipment is not in pest-proof cases.

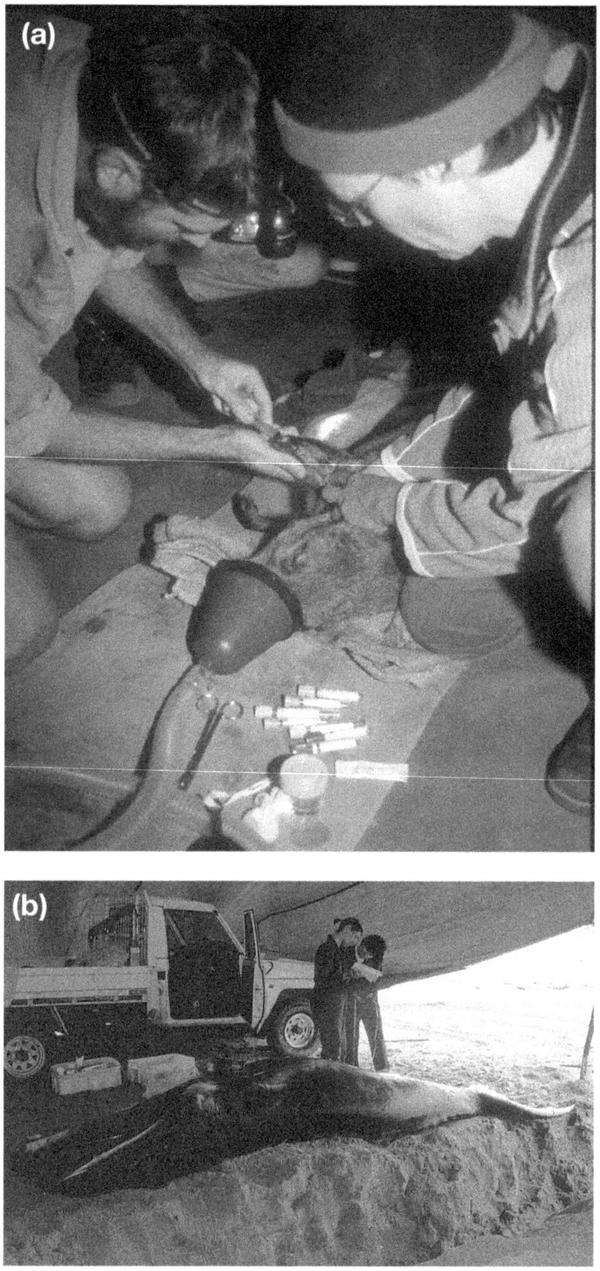

Figure 4.1 a) Sample collection from a northern hairy-nosed wombat (*Lasiorhinus krefftii*) in the field. b) Cetacean necropsy set-up on a beach.

If possible, a run-through is highly recommended to ensure that all the goals can be met with the available equipment. The run-through will also ensure that personnel are comfortable with their required tasks and can function well as a team.

When planning any project, from a single necropsy in the field to a long journey, it is important to remember

Table 4.1 Field trip equipment checklist

Personal safety equipment

- Tarps and rope to create a tent to ward off rain or sun
- Insect repellent
- Sunscreen, hat, sunglasses
- Drinking water, food
- Change of clothes
- Coveralls
- PVC apron
- Latex gloves or dishwashing gloves
- Surgical facemasks
- Rubber boots and good walking shoes
- Bucket, nail brush, antiseptic soap, paper towels
- Torch—hand-held and headlamp
- First aid kit
- Mobile phone
- GPS unit and maps
- Emergency locater beacon if on water or very remote

Carcass collection equipment

- Heavy-duty rubbish bags
- String
- Animal sample tags and pencil or indelible pen
- Wildlife incident report forms
- Mortality log forms
- Pencils and sharpener
- Clipboard with a clear piece of plastic to keep rain off

General equipment

- Good pest-proof packs for carrying equipment
- Cooler and ice packs
- Sharps disposal unit
- Camera
- Masking tape and packing tape
- Ruler
- Extra batteries

Necropsy equipment

- Scale
- Necropsy worksheet
- Knives, sharpening stone and steel
- String and manila labels
- Scalpel handle (# 4) and disposable blades (#24)
- Forceps—various
- Scissors—various
- Poultry shears or large bandage scissors
- Bone or hacksaw and spare blades
- Hammer, chisel and vice for brain removal from large skulls
- Long-handled pruning shears/ ribcutters
- Gaff hooks—useful for marine mammal necropsies

Clean-up equipment

- Tarp
- Water, scrub brush, detergent, towel
- Heavy-duty rubbish bags

Sample collection equipment

- Indelible marking pen
- Syringes—5, 10, 20 mL
- Needles—various gauges
- Butterfly catheters—various gauges
- Heparin or EDTA
- Blood collection tubes—EDTA, heparin, oxalte and serum
- Sterile plastic bottles—90 mL
- Sterile cryovials—2 mL
- Sterile plastic bags (Whirl-pak® bags)
- Large ziplock bags
- Aluminium foil
- 1 L plastic containers filled with 10% neutral buffered formalin
- 50 mL of 70–90% ethanol
- Bacterial culture swabs
- Dry sterile swabs
- Capillary tubes
- Glass microscope slides, cover slips and slide storage box
- Microscope (may require mirror as light source if no access to power)
- Centrifuge (manual serum separating devices are available)
- Saline
- 10% neutral buffered formalin
- Parasite preservative
- Any special media required—chlamydial culture medium, viral culture medium
- Methanol to fix blood films
- Faecal flotation vials and solution
- Cooler and ice bricks or liquid-nitrogen dry shipper

your own safety and comfort. Simply including a few extra tarps to erect a makeshift tent can keep you dry or shaded. Research what resources will be available on-site, and what you will have to bring.

When you finally make it to the actual field trip, take time to meet the land owners, local conservation agencies and anyone in the area who is interested in what you are doing. If these people know what you are doing and why you are doing it, the next trip will be easier to organise and you may have more help.

1.5 Sampling a wildlife population

Whenever possible, consult a veterinary epidemiologist to assist with project design prior to undertaking a wild-life health investigation. In some rare species, or those that are managed in a captive situation, it may be possible to collect samples from each animal in the population. This method of sampling is called a population census and can precisely measure the prevalence of specific diseases, characteristics or traits within the population.

Usually it is very difficult to sample the entire wildlife population, and only a small proportion of animals is examined as a study population. Prior to designing a health investigation it is important to define:

- the population at risk or of interest;
- the study population.

If you can only collect samples from a proportion or subset of the animals at risk or of interest, then you are sampling the population. Your study population will provide only an estimate of the prevalence of specific diseases or traits within the population. Sampling from a study population is not a precise measurement throughout the population at risk, and selection bias can be introduced (Wobeser 1994; Cameron et al. 2003).

When using a study population, it is very important to either limit your comments or inference to the population sampled, or attempt to collect a study population that accurately reflects the population at risk. Study populations can be affected by selection bias when sampling methods are not random. Ensuring that the age and sex distribution in the study group is similar to that in the entire population is important. Other factors such as ecosystem, geographic distribution and genetic variability should also be considered when identifying a study population.

Wildlife health sampling may be undertaken on a convenience basis rather than trying to select a study population that reflects the entire population. Animals are often sampled due to accessibility (they are already being handled or are the only animals found or trapped) or in a targeted program based on presenting signs that make them more likely to have the disease of interest. These sampling techniques, however, make it very difficult to make inferences regarding disease prevalence in the population as a whole. We can only comment with certainty on the prevalence of disease within the target group, rather than the entire population (i.e. the prevalence of lyssavirus infection in bats found with behavioural abnormalities).

Random sampling techniques are difficult if not impossible to apply to wildlife. The best option is to find a method that reduces bias and represents the population as closely as possible.

Many wildlife health surveillance projects are designed to detect specific disease. Determining the number of animals that need to be sampled to detect or rule out the presence of an infectious agent depends upon:

- an estimate of the prevalence of the disease within the population;
- the population size;
- the accuracy of the test—sensitivity and specificity;
- the level of certainty (confidence) in results, usually set by convention to 0.95 so that there is a 95% chance of detecting the disease (Wobeser 1994).

To determine the number of animals to test to detect a specific disease, use the following formula:

$$n = [1 - (1 - P)^{1/d}][N - ((d - 1)/2)]$$

where n = sample size, P = probability of detecting at least one case of the disease if it is present in the population (usually set to 95%) and d = number of detectable cases in the population. Thus, d = population size × prevalence × sensitivity of the test. N = population size.

Example

We want to determine if an army of 500 frogs is infected with chytrid fungus, which we suspect will have a 10% prevalence (approximately 10% of the frogs would be infected if the population had been exposed to the fungus). The diagnostic test is a real-time PCR with a sensitivity of 98%. We want to be 95% certain that we will detect the disease if it is present. Thus N = 500, P = 95%, d = 500 × .1 × 0.98 = 49

$$n = [1 - (1 - 0.95)^{1/49}][500 - ((49 - 1)/2)] = 28$$

At least 28 frogs would have to be tested to determine if the population was infected with the fungus, with 95% confidence, where the prevalence is 10%. The sample size required increases with level of confidence required. It also increases as the prevalence of disease decreases and estimating the sample size for very low-prevalence diseases (<0.1%) with a high level of confidence can be difficult and very expensive. For wildlife, there may also be animal and population welfare concerns, the capacity of the diagnostic laboratory may be limiting and appropriate levels of funding and resources may simply not be available. In these instances there is often a shift from sampling to detect disease at a known level of confidence, towards determining the probability of disease freedom. These studies require targeted investigation into disease events or surveillance of high-risk individuals or populations (Salmon 2003).

1.6 Prevention of disease spread among wildlife populations

The motto of fieldwork is 'Above all, do no harm'. It is important not to spread infectious agents to wildlife populations, domestic animals or humans during health investigations. Some infectious agents and parasites can be very hardy in the environment or adhere to clothes and equipment. Each site where wildlife is handled should be considered a quarantine site (CCWHC 1998; CCWHC 1999).

Protocols for the prevention of disease spread should be incorporated into written operating procedures and animal ethics and research application forms.

Outer clothing and equipment worn in the field should be removed, and placed into plastic bags in the field. Hands and boots should be washed and disinfected in the field. Animal-handling bags or boxes should be cleaned and disinfected between uses with different animals. All equipment and clothing should be thoroughly washed and sanitised prior to being used in another location. Once organic material has been removed from equipment, a solution of household bleach (1 part bleach to 9 parts water) is an effective disinfectant. This solution can also be used to scrub boats, vehicle tyres and work surfaces. The external surfaces of sample vials should be cleaned and disinfected in the field. Sample vials should be leak-proof, and should be double-bagged prior to submission to the laboratory. Carcasses should be individually bagged in leak-proof heavy-duty plastic bags to be submitted to the laboratory. The carcasses should be placed into plastic tubs, in case leakage does occur en route.

Carcass disposal should be undertaken with great care. Animal carcasses are difficult to burn, and can contaminate groundwater if buried improperly. Often the easiest way of managing excess carcasses is to double-bag them at the site and organise incineration by a commercial contaminated waste disposal company. If carcasses must be burned or buried, seek prior approval from the local council, state agriculture department and rural fire service.

2 PATHOLOGICAL APPROACH TO DISEASE INVESTIGATION

Although the procedures and tests used to investigate disease in wildlife are similar to those used in domestic animals, the approach is dramatically different.

Veterinarians are often trained that diagnosis requires them to obtain a biopsy or collect a lump of the diseased tissue into formalin and send it to a pathologist. The pathologist will examine the tissue, make observations and interpret the lesions in comparison with a large dataset of the known diseases of that species. A sample of intestine from a dog with parvovirus usually has lesions that are characteristic of that disease, so there is generally no need to pursue viral culture. The difficulty with this approach is that we know relatively little about wildlife health and there is no definitive dataset for most non-domestic species.

Specialised diagnostic testing is often required for a definitive diagnosis of illness in wildlife. It is vital to collect appropriate samples that allow conclusions in the event of a disease outbreak in wildlife. It is also important to collect samples from healthy animals to produce baseline normal health indicators that will complete the dataset. Table 4.2 gives a list of the samples that should be collected during a routine necropsy.

Table 4.2 Routine necropsy checklist—in most cases, the following samples are sufficient

Aseptically collected samples of:
liver
kidney
brain
serum
any lesion
Stored frozen pending the results of histopathology
Formalin-fixed 1 cm wide wedges of:
tonsil, axillary, inguinal, submandibular and bronchial lymph nodes
lung
heart
liver
spleen
adrenal gland
kidney
bladder
gonad
stomach, pancreas, duodenum, jejunum, ileum, large intestine and mesenteric lymph node
skeletal muscle and sciatic nerve
brain
margin of any lesion

The collection of appropriate information is just as important as the collection of diagnostic samples. Having as much information as possible about the animal and how samples were collected will greatly improve the pathologists' ability to interpret their findings.

Whenever possible, contact the pathologist or diagnostic laboratory before sample collection to ensure that you are collecting the appropriate materials and information to answer your questions.

2.1 Clinical signs and diseases of concern

Exotic diseases can affect wildlife. Australian fauna is distinct from domestic species and from wildlife in other parts of the world. Exotic diseases in wildlife could be displayed in unexpected ways.

If you encounter unusual or unexpected clinical signs or mortality in wildlife, contact the state veterinary laboratory or the Emergency Disease Watch Hotline 1800 675 888.

Categories of diseases with highest priority for investigation and surveillance in Australia include:

- mass or unexpected mortalities/morbidities of unknown causes;
- significant clusters of deaths;
- suspect livestock-associated notifiable diseases;
- undiagnosed syndromes;
- suspected human/zoonotic connection;
- diseases likely to spread and be difficult to eradicate if they become established;
- suspected exotic and OIE list diseases;
- diseases with overseas events or international drivers;
- diseases listed as key threatening processes by DEWR (AWHN 2007).

Mass mortality, behavioural abnormality, granulomas, vesicles and ulcers should trigger an investigation. If investigation of a disease syndrome in wildlife produces negative test results for relevant endemic diseases, exotic or emerging diseases should be considered.

In the face of an unusual disease event, or disease in a highly threatened species, a very detailed and thorough post mortem procedure is required. Multiple sample sets must be collected from each animal so that a wide variety of diagnostic testing can be undertaken, potentially using several different laboratories. You may be investigating a new disease. In these situations it is equally important to rule out all other possible explanations for the event as to identify the agent of concern. A complete set of samples from as many animals as possible will facilitate the investigation. Table 4.3 provides a comprehensive post mortem checklist for these situations.

2.2 Diagnostic flow

Reaching a diagnosis in wildlife involves the following:

- observers to monitor wildlife, report any abnormalities and submit carcasses to an appropriate laboratory;
- a necropsy conducted by a veterinarian, veterinary pathologist or trained layperson. Sometimes a good understanding of what killed the animal is evident at the time of examination, but this technique is not very sensitive. You might determine that an animal died due to severe trauma when it was hit by a car, but not determine that the animal was hit because it was affected by an infectious disease or exposure to toxins;
- samples collected during the necropsy are usually submitted first for histopathology. Sometimes these tissues can be examined with special stains that help to define the type of reaction in the tissues, or the type of organism. It is, however, often difficult to determine the exact organism inside a lesion using this technique;
- a pathologist most often forms a tentative diagnosis at this stage. Histopathology can take a few days to several weeks to complete;
- some diseases can be confirmed with histopathology, but additional diagnostic testing is often required for wildlife. The tentative diagnosis can be confirmed using tissues collected during the PM and kept frozen. These tissues can be submitted to specialised laboratories for bacterial, fungal or viral culture, parasite identification, electron microscopy or PCR to detect a variety of infectious agents, or toxicology testing;
- additional diagnostic tests can be as much value in ruling out some possibilities of the disease process (e.g. underlying exposure to toxins) as they are in ruling in or finding the cause of the event. These additional tests add time and expense to the diagnostic process. While most bacterial and fungal culture can be completed in 48–72 hr, it can take

Table 4.3 Comprehensive necropsy checklist

Identify species, sex and any physical forms of identification
External examination
Describe and photograph body condition/injuries
check mammary glands for evidence of lactation and collect any milk present
check the umbilicus of neonates
examine and count the teeth
examine the eyes, mouth and nose for any discharge
weigh and measure the carcass
Collect samples of any wounds or discharges for bacterial culture and histopathology
Conduct a thorough internal post mortem examination. Collect samples for genetic research, bacterial and viral culture with clean instruments as soon as the internal organs are exposed
Samples needed for a very thorough post mortem examination
Bacterial culture: Minimum of two filtering organs and any lesions. Collect samples into individual sterile containers, well labelled. Store chilled and transport immediately to a laboratory.
Viral culture: Portions of lung, kidney, liver, spleen. Place in a Whirl-pak bag or other sterile container. Store chilled or frozen.
Genetics: Small samples of skin, liver and kidney placed into sterile cryotubes and stored chilled, frozen, in RNA Later® or ethanol.
Toxicology: Fat, liver, kidney and stomach content—wrap individually in aluminium foil then place in individually labelled plastic bags and store frozen.
Histopathology: Tonsil, bronchial lymph node, lung, heart, liver, spleen, adrenal gland, kidney, bladder, gonad, stomach, pancreas, duodenum, jejunum, ileum, large intestine, skeletal muscle, brain, the margin of any lesion, in 10% buffered formalin. Maximum 1 cm thickness. Many tissues may be placed in one jar as long as the volume of 10% formalin is greater than 10 times the total volume of tissues. Store at room temperature.
Serology: Blood samples collected into red top tubes. Centrifuged and serum decanted as soon as possible. Store chilled or frozen (long-term).
Parasitology: Preserve parasites in leak-proof containers in fixative (10% formalin, glacial acetic acid, or 70% ethanol). Label with animal species, identification and site of collection. Do not freeze these samples.
An impression smear of the liver or spleen stained with Diff-Quik® and examined by oil immersion microscopy to identify blood parasites.
Skeleton: Freeze the remains, or note site of burial for a museum.

many weeks to isolate viruses or mycobacteria in culture;

- the pathologist will review the case history and the results of all examinations and tests. Their interpretation of complex findings is very important. If a sea eagle is shot, but dies of a fungal infection and other complications of being injured and in care, the cause of death is still the primary event—gunshot wound;
- after interpreting the findings, the pathologist will prepare a final report that may contain a definitive or confirmed diagnosis, but for various reasons it may not be possible to confirm the suspected diagnosis. If samples were inadequate, tests were not available, or the disease-causing agent was no longer present in the tissues etc., a definitive diagnosis may not be determined;
- when submitting samples to a veterinarian or a laboratory, ask how long they expect the tests to take and who will provide and discuss the results.

3 NECROPSY

The ideal specimen for necropsy is an ill animal that can be observed for clinical signs and then euthanased. Barbiturate administration during euthanasia should be titrated carefully to ensure that the caustic chemical is not detrimental to gross and microscopic necropsy.

A freshly dead animal is also suitable for examination. If it is not possible to bring a live or freshly dead animal to a pathologist or veterinarian for examination, a field necropsy and sample collection should be conducted. Carcasses should be chilled as soon as possible after being collected (and during any periods of storage or transportation) and frozen only as a last resort. Frozen carcasses may be examined, but are often of little use for histopathology and some other tests. In an outbreak, however, frozen carcasses may be valuable.

The lists of samples to collect during necropsy may appear lengthy and daunting, but are necessary because it is often difficult to know during the gross necropsy examination which ancillary tests will turn speculation into a definitive diagnosis. Even if you only intend to submit a portion of obviously diseased tissue for histopathology, retain a range of tissues in formalin and selected tissues frozen until you are satisfied that the investigation has concluded. Each sample collected must be properly labelled with the animal's species, identification numbers, date and organ involved. Always label the jar rather than the lid to ensure that the identification is not lost when lids are removed during sample handling. Ensure that labels are done in pencil or permanent ink which will not dissolve in the fixative (alcohol-based fixatives dissolve the ink of many permanent markers).

If you have questions or concerns regarding sample collection or transport don't hesitate to contact the clinical laboratory or pathologist prior to starting the necropsy.

It is very important to be familiar with the species that you are working with. If you are unsure of the genus and species, retain the carcass frozen until a naturalist can help with the identification.

Start to conduct necropsies on animals that die from known causes, such as trauma, to gain a good understanding of normal variations in anatomy. Allow a lot of time for the first few necropsies. With practice and a systematic approach you will get faster yet still do a thorough examination. Check the literature to help distinguish lesions from common physiological and necropsy changes (King et al. 2006). Careful observation and a systematic approach to a necropsy will prevail over years of training.

3.1 State of carcass preservation

Determining the state of preservation of a carcass is important when deciding if a full necropsy and sample collection will provide useful data. A large carcass retains heat quite well, which will accelerate the speed of tissue decomposition, which begins within hours of death. Keeping the animal cool and conducting an examination as soon as possible after death will provide the most useful information and samples. The carcass can be classified using the categories described in Table 4.4.

Table 4.4 State of carcass preservation

Preservation class	Simple description	Comments
Class 1	Live animal	Optimum specimen for examination and testing. The animal's behaviour can be monitored, blood can be collected and the animal euthanased for necropsy.
Class 2	Fresh = recently dead	Good specimen for examination and testing.
Class 3	Mild to moderate decomposition	There may be mild bloating of the carcass as the intestines fill with gas. The tissues of the mouth and eyes may be dry and wrinkled, but are not green or purple. The fur does not pull out easily. The internal organs are intact and recognisable. Examination of these animals can provide useful information, but the tissues are less suitable for additional testing than those in class 2.
Class 4	Advanced decomposition	The carcass is usually bloated and foul-smelling. The oral tissues are often green or purple. The fur is easily removed with gentle traction. The internal organs are soft or liquefied and the intestinal tract is distended with gas. A gross necropsy may provide some useful information, but tissues are generally unsuitable for further diagnostic testing other than heavy metal analysis.
Class 5	Mummified or skeletal remains	Only gross, radiographic, genetic and limited toxicological examinations are possible.

Source: CCWHC, Wildlife Health Investigation Manual

3.2 Health and safety during necropsies

Conducting necropsies on wildlife can pose the threat of direct injury or contracting a serious infection. It is important that all risks to human safety are considered prior to the initiation of an examination.

Although nearly any contact with other living organisms carries some risk of disease transmission, adopting broad principles of health and hygiene can mitigate this risk. Parasites, bacteria, fungi or viruses may cause zoonotic diseases. These organisms are routinely transmitted through direct contact during handling animals and their tissues, or through indirect contact via urine, faeces, secretions, blood, aerosols or inanimate objects contaminated with these materials.

Adequate hand-washing facilities must be available whenever there is direct contact between animals and humans. Hand-washing is the single most effective tool in preventing the transmission of zoonotic agents and thus limiting the risk of infection. Buckets, scrub brushes, fresh water, disinfectant and towels should be taken to remote locations to allow hand-washing. Alternatively, antiseptic wash or wipes may be used.

Protective clothing should be worn when conducting a necropsy. During necropsies of large animals, fluids may leak into the gloves through the wide cuffs, but this can be prevented by sealing the tops of the gloves to the wrists with tape. Hands should be washed after the procedure, even when protective clothing is worn.

Infectious diseases should be suspected whenever an animal is emaciated, has a history of respiratory signs, has a nasal or ocular discharge or when large numbers of animals are ill or dead. In these instances, a facemask and safety glasses should be worn.

Whenever necropsies are conducted there is a risk of cutting yourself with a knife or scalpel blade. Sharps disposal bins should be taken to the site of the necropsy for proper disposal of needles and scalpel blades. Designating one member of the team as the dissector can also reduce the risk of injury. When not in use, all knives, scissors and scalpels should be returned to a specified cutting-board and never left on the animal or on the ground.

Human food should not be prepared or stored near animal contact areas. Animal tissues should not be stored in containers, fridges or freezers that are used for human food.

Immunocompromised persons, such as an individual with diabetes, chronic viral disease, chronic liver or kidney disease, children, persons taking immunomodulating drugs, splenectomised people and the elderly, should be discouraged from participating in a necropsy.

Bats can carry a number of viruses fatal to humans. Anyone handling bats should be vaccinated against lyssavirus and it is important that these vaccinations are up-to-date and the operator's titre is known and appropriate (see Chapter 14). Even vaccinated persons should wear appropriate protective clothing such as long pants, long-sleeved shirts, latex gloves and puncture-resistant gloves when handling bats—a specific bat necropsy guide is included at the end of this section. Bats that exhibit neurological signs or behavioural abnormalities unrelated to trauma should be submitted to the nearest state government veterinary laboratory for lyssavirus exclusion.

Persons who have been bitten or scratched by a bat should immediately contact their local state public health service to be assessed for lyssavirus prophylaxis, regardless of their vaccination status (see Chapter 14).

Formalin and ethanol are common fixatives used in the collection of samples for further diagnostic testing. Both these agents are very caustic to exposed human tissues and ethanol can be flammable. Material safety data sheets should be obtained whenever chemicals are ordered, and should be accessible wherever the chemicals are used or transported.

A first aid kit and accident report form should be included in the equipment taken to any field site. Medical attention should be sought for any injuries sustained during the examination or illness that immediately follows the examination.

3.3 Necropsy procedure

A necropsy worksheet will help facilitate a thorough and systematic approach to a necropsy. Making the most of any animal death requires recording as much information surrounding the circumstances of death as possible. Detailed written records are one of the most valuable materials taken away from a necropsy.

Photographs or videotapes can be a valuable tool in supplementing written case reports. A set of photographs should include one image from the top, bottom, right and left sides of the animal. Additional photographs of teeth, unusual anatomical structures or possible lesions should also be taken. A general approach to necropsy is summarised in Table 4.5.

Table 4.5 General approach to a necropsy

Thorough history ⇩ External examination ⇩ Central midline skin cut (chin to anus), reflect skin ⇩ **STOP** Examine fat deposits, muscle mass, hydration ⇩ Expose internal organs ⇩ **STOP** Clean your hands and get clean instruments Look at all organs Collect clean samples for microbiology or freezing back ⇩ Collect half (or 1 cm wedge) of any lesion into formalin and place the other half into a sterile vial to be frozen for later use ⇩ Examine each organ system thoroughly, collecting 1 cm wide wedges of each organ into formalin ⇩ Complete a necropsy worksheet ⇩ Properly label, store and ship samples

3.3.1 History

Collect as much information as possible regarding the location that the animal was found, the animal's history and the person submitting the animal. A thorough history should include the following details:

- circumstances under which the animal was found;
- genus and species of the animal, natural history and state of captivity;
- diet, food and water sources;
- housing conditions—ventilation, substrate, cage type;
- exposure to toxic substances;
- recent changes in the environment, weather or husbandry;
- clinical signs, onset and progression of signs;
- treatment provided and response to treatment;
- natural death or euthanasia.

3.3.2 External physical examination

Weigh the carcass before you commence examination. Conduct a thorough external physical examination to confirm the species of animal and detect any of the following:

- ectoparasites;
- palpable fractures or dislocated joints;
- discharges from the nose, ears, mouth or urogenital tract;
- skin injuries, infections or abnormal fur or feathers;
- active mammary glands or pouched young;
- any physical identification—ear notch or tattoo;
- scan for a microchip;
- condition of teeth;
- confirm the sex.

Collect samples of any lesions for microbiological examination and histopathology.

Weigh and measure the animal. Measure the body using forms from a species or taxon advisory group.

Most Australian mammals are small to medium sized and are often easiest to examine if placed on their backs and fully opened along the ventral midline.

1 Incise the skin from the mouth to the anus and reflect the skin as far as possible off the neck, chest and abdomen. Reflect each fore limb by excising the soft tissues of the axilla, between the ribcage and scapula and, when present, disarticulate the collar bone. Disarticulate each hind limb by incising the capsule of the coxofemoral joint then cut the teres tendon as the limb is reflected.
2 Assess hydration (whether the tissues are dry and tacky or moist), fat deposits and muscle mass.
3 Make a T-shaped incision through the abdominal musculature to expose the abdominal contents. Incise the diaphragm while listening for the ingress of air and watching for the flattening of the diaphragm, to rule out pre-existing pneumothorax. Cut through the ribs or sternum along the sides of the chest, and remove any attachments of diaphragm to expose the chest cavity.
4 Without touching, observe the location of organs. Describe and collect samples of any fluids found within the chest or abdominal cavities.
5 Obtain clean instruments and collect samples for microbiological examination into sterile vials. Generally, collect samples of liver, kidney, brain and half of any lesion.
6 Blood films can be made using heart blood, or a cut section of liver or spleen, to look for blood-borne

parasites. Heart blood can also be collected for serum separation.

7 Tie the lower oesophagus with a piece of string and cut the oesophagus just cranial to the string to prevent stomach contents from contaminating the organs.
8 Remove the tongue, trachea, oesophagus, heart and lungs as one unit. Collect samples of each organ into formalin. To determine if a neonate has breathed, place a large sample of lung tissue into water or formalin. If the sample sinks, the animal did not expand its lungs with air.
9 Cut off the apex of the heart. The right side of the heart should be examined first, otherwise it will be destroyed as the left side is examined. Use scissors to cut up through the right ventricular free wall, atrioventricular valves and atrium. Examine these structures then go back to the right ventricle to locate, cut through and examine the pulmonary outflow tract, tricuspid valves and pulmonary artery. Repeat this procedure on the left side of the heart.
10 Reflect the liver and gastrointestinal tract towards the back end of the animal, but leave the rectum attached at the anus. Examining the gastrointestinal tract last minimises bacterial contamination of the tissues.
11 Examine the urogenital tract and collect samples of the kidney, bladder and gonads into formalin. The adrenal glands, located at the cranial margin of the kidney, should also be examined, measured and sampled.
12 Get clean instruments. Remove the brain by disarticulating the head and removing the skin, then using scissors, bone cutters or a vice and saw to remove a 'cap' of bone from the top of the skull. Cut the brain in half along the long axis. Place half the brain into a sterile vial and place the other half into formalin. If the animal exhibited neurological signs, collect segments of peripheral nerve (femoral and brachial), spinal cord (cervical, thoracic and lumbar) and an eye.
13 Examine the liver. Squeeze the gall bladder to ensure that it empties into the intestine and is not blocked. Remove the liver from the gastrointestinal tract and make serial sections to ensure that there are no focal parasite tracts or abscesses. Collect a sample of liver into formalin.
14 Open the gastrointestinal tract. Collect the stomach contents into a sterile container. Collect, examine and count or estimate the number of parasites. Collect 1 cm strips of stomach, duodenum, jejunum, ileum, caecum and large intestine into formalin. Describe the stomach and intestinal contents, volume and form.
15 Reflect the skin from the back of the neck, chest and abdomen to look for any signs of traumatic injury.
16 If there were any signs of anaemia or other blood disturbance, a sample of bone marrow should be collected by breaking the femur and scooping out the red marrow.
17 Open several joints to look for inflammation or degenerative changes. Collect joint fluid if it appears abnormal.
18 Write a detailed report of your findings.

In some cases, the only necropsy finding may be emaciation. If so, the teeth and claws should be examined for signs of excessive wear. The eyes should also be examined for the presence of cataracts or retinal degeneration. These findings may be indicative of old age. The carcass should also be examined for evidence of injury that might alter the ability to obtain food. If there is no obvious cause of emaciation, the possibility of heavy metal intoxication should be investigated. Generally, a diagnosis of starvation relies on ruling out other potential causes of weight loss.

3.4 Bat necropsy protocol

Staff handling bats must be educated in the zoonotic diseases carried by bats and should not handle bats unless they are effectively vaccinated against Australian bat lyssavirus (ABLV), and appropriately trained and experienced. Staff must be offered vaccination in line with the NHMRC Australian Immunisation Handbook vaccination guidelines for ABLV (see Chapter 14 and ATAGI 2007).

If ABLV is suspected or a bat has bitten or scratched people or pets, the animal should be euthanased and the entire carcass submitted to the local government veterinary diagnostic laboratory for ABLV exclusion (see Chapter 14 for signs of ABLV in bats and definition of a major contact). Specific packaging and handling requirements must be met and the laboratory consulted prior to shipping the animal.

Staff conducting or observing the examination of any bat should wear a pair of latex gloves under a pair of dishwashing gloves, an ultrafiltration facemask, eye protection, a lab coat and a PVC apron. The necropsy should be conducted within an appropriate biohazard safety cabinet.

Bats that are submitted alive for necropsy can be anaesthetised for blood and urine collection prior to euthanasia (Chapter 14). Collect 0.75–1.0 mL of blood via cardiac puncture and 1.0–2.0 mL of urine into sterile serum tubes and stored them frozen pending histopathology. Ideally, blood should be collected into a serum tube containing a gel separator, and centrifuged prior to freezing.

During the necropsy, collect the following:

1 collect into sterile containers and ship immediately or freeze: heart blood and urine (if not collected ante mortem);
2 collect into sterile containers and ship immediately or freeze: half the brain, kidney, liver, lung, spleen, uterus and foetal lung and liver if available;
3 collect into 10% buffered formalin: half the brain, salivary glands, kidney, liver, lung, spleen, uterus and foetal liver and lung if available;
4 retain the carcass in a frozen state until tests are completed (Rose 2007).

A thorough necropsy report must be created and maintained. It should include a clinical history, the origin of the bat, any details of human exposure via bite or scratch, and a list of samples taken and submitted for testing.

Note that ABLV should be included as a differential diagnosis in any native mammals showing suggestive neurological or behavioural signs of rabies-like illness (Chapter 14).

3.5 Describing lesions

Describing lesions well is essential to accurately communicate your findings with others. Describe any abnormality that you observe during the necropsy. Often the decision to identify something as a lesion is made intuitively, it 'just looks abnormal'. This decision will be easier with further experience in the appearance of normal tissues and the range of physiological and post mortem changes in animal tissues. Most internal organs are quite homogeneous, so any change in texture, colour or shape could be a lesion.

Post mortem change can alter the features of organs depending on the degree of tissue decomposition. Hollow organs often become gas-filled. Solid organs often turn green to black due to bile staining or dull red from the breakdown of red blood cells within the tissues. Describe the degree of tissue decomposition; if you are uncertain whether a tissue change is due to decomposition or a disease process, simply describe what you see. Most necropsy reports are written in full sentence format and in present tense. The first sentence of the report should be a description of the animal, including sex, age-class, weight, physical forms of identification and body condition (King et al. 2006).

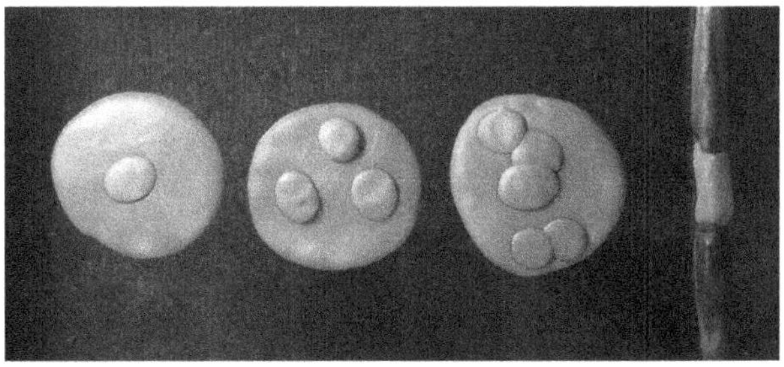

Figure 4.2 Lesion distribution: focal, multifocal, coalescing and segmental.

The body of the necropsy report includes a description of the state of decomposition, and any lesions observed. Lesions should be described in a way that allows readers to assemble a mental image of your findings. Interpretations of your findings should only be discussed in the comments section of the report.

Lesion description includes an indication of the location, distribution, size, colour, shape and texture of the change. The distribution of lesions is usually described as unilateral or bilateral, and focal, multifocal, locally extensive, coalescing, segmental or diffuse (Fig. 4.2) (King et al. 2006). Take photographs of each lesion, or draw a diagram to illustrate the size, shape and location of the abnormality.

3.6 Making a diagnosis

A morphologic diagnosis provides a description of your observations so that others can visualise the lesions and form their own conclusions. This type of diagnosis is most helpful when trying to communicate your findings to other diagnosticians.

Morphological diagnosis = chronicity + distribution + extent + organ + type of change

'Type of change' means acute, multifocal, extensive hepatic necrosis etc.

The terms peracute, acute, subacute (roughly two weeks+ duration) chronic and chronic-active are usually used to estimate the chronicity of a disease process. The chronicity of lesions can be difficult to ascribe based on gross necropsy. The severity of each lesion is usually characterised as mild, moderate or marked.

Aetiological diagnosis = describing the disease to include the name of the agent responsible for the change

Providing only an aetiological diagnosis does not describe the problem. Your report should be descriptive and include a morphologic diagnosis so that others can read it and establish their own interpretation of the findings.

4 NECROPSY SAMPLE COLLECTION

Samples are only as useful as the information collected along with them. The sample itself must be adequately labelled so that it can be correlated to the paperwork submitted. Ensure that each specimen is labelled with the date, species, organ (left or right, if appropriate) and identification number. Use only permanent markers and do a test strip to make sure that the fixative will not dissolve the ink. Ethanol will dissolve almost all inks. Pencil is a good alternative to ink for labelling cases that are not part of a medico-legal investigation.

4.1 Serum collection

Serum can be a very useful tool to complete the diagnostic picture created from the necropsy. Gross and microscopic necropsies most often provide a view of morphologic change, however, illustrating a functional change in organs may require biochemical examination of serum.

Blood may be collected immediately prior to euthanasia, or from the heart (or chest cavity) after the animal's death. If unclotted, a smear can be made or blood collected into anticoagulant for haematological analysis. Blood should also be collected into a serum tube. Serum samples can often be collected from a carcass even if a necropsy will not be conducted. Samples can be collected by removing the eye and aspirating the blood that wells up in the socket. Serum samples can be used to assess the following:

- exposure to a wide variety of infectious agents through the assessment of antibody concentrations;
- active viral infection—blood and serum samples can be submitted for viral culture;
- liver, kidney and pancreatic function;
- thyroid function, parathyroid gland function and sex hormone analysis;
- electrolyte, fluid and acid–base balance (these parameters are only reliable in blood samples collected prior to death);
- status of several micronutrients, such as vitamin D, E and A.

Serum samples are best stored frozen, preferably in a −80°C freezer but a −20°C usually suffices.

4.2 Histopathology

Formalin will only penetrate 5 mm into tissues, so tissue sections must be no thicker than 1 cm. Brain is the only exception—it should be fixed as a large organ. Always ensure that there is a 10:1 ratio of 10% neutral buffered formalin to tissue volume. Sample vials should be watertight and non-breakable, and have a wide mouth. Tissues become inflexible once fixed. Large tissues that fit through a narrow-mouthed container may be trapped once fixed. Several tissues can be placed into a single jar. Histopathology processing cassettes obtained from pathology labs can be used to hold and label individual samples (e.g. lymph nodes or lesions from different locations), rather than having many jars of formalin for one case. Formalin-fixed tissues can be stored at room temperature for long periods.

Tissues should be handled very gently. Excessive crushing, fingering or stretching of tissues can disrupt its architecture. Grasping the fascia around the organ of interest, or cutting off any tissue crushed by forceps, will avoid handling the artefact.

Tubular organs often collapse when placed in fluid, which may prevent formalin from fixing the mucosa. Tissue preservation is maximised when tubular organs are opened longitudinally, 1 cm wide sections are collected and wafted to lie flat in formalin. The intestinal tract of small mammals may be very difficult to cut open. Formalin injected through the intestine with a 25 G needle will best preserve these small samples.

Samples should represent the lesions found. Many researchers submit only the central portion of a lesion for examination, expecting that this area has the most information. The margins of the lesion, however, can be described as the advancing front and have a greater likelihood of containing an aetiological agent.

Eyes decompose very quickly. Only collect eyes into formalin if the animal has been dead for less than 12 hr. For optimal preservation for research projects, the eyes can be fixed in Bouin's solution for 24 hr then placed into 10% buffered formalin. The picric acid in

Bouin's solution can be explosive if the solution dries out. For general examination, fixation in 10% buffered formalin is usually satisfactory. Eyes from large animals should be incised along the sclera to allow the formalin to penetrate into the chambers of the eye. To orient the pathologist, place a suture in the dorsal aspect of the sclera, perhaps one suture in the left eye and two in the right.

When collecting samples from the heart, particularly if there are lesions in the myocardium, take samples of the left ventricular free wall, right ventricular free wall, interventricular septum and atrium.

The brain should be fixed as one large specimen (minus the samples collected for freezing) to allow the pathologist to orient themselves then collect the most appropriate sections. Large brains may require 7 days to fix; changing the formalin every 24 hr will accelerate the fixation process. Avoid making numerous sections through the brain, but consider making a longitudinal incision through the corpus callosum (between the cerebral hemispheres) to allow formalin to penetrate the ventricles without disrupting important anatomical landmarks.

Peripheral nerve and fresh muscle can acquire a lot of artefact if placed directly into formalin. These artefacts can be reduced by laying the tissue on a stainless steel table to cool, or on a piece of manila folder so that it does not retract as it contacts the formalin. This technique may also be useful for sections of thin skin.

In smaller animals with neurological disease, it may be wise to collect the entire head and spinal column into formalin. The dissection of such small specimens can be destructive to the tissues. You can fix the entire specimen in formalin or in demineralising solution. The demineralising solution will make the bone soft enough to section and examine under the microscope. Some demineralising solutions do not adequately fix the tissues, so it must be fixed in formalin first. Storing tissues in demineralising solutions can decrease tissue quality. Once the tissues are soft enough to cut, they should be drained and placed into 10% formalin for longer-term storage.

Formalin storage degrades the genetic material within cells. Thus, it is usually not possible to conduct a wide variety of analyses, such as PCR, on formalin-fixed tissues.

Ask how the pathologist would like the samples submitted.

4.3 Cytology

Histopathology is not always a sensitive tool for the diagnosis of haemoparasites, other protozoa and certain microbial infections. Cytological examination of tissues can often add valuable information to an investigation, and can be performed in-house to increase your own diagnostic abilities as well as save money.

Smears made from heart blood, and impressions of the liver and spleen may be stained with Diff-Quik® or Giemsa stains (Bacto Laboratories) to identify intracellular organisms. If chlamydiosis is suspected these samples should be stained with Macchiavello's technique, which stains the organism pink to purple, or a Gimeniez stain. Acid-fast stains will delineate mycobacteria as fine, beaded, pink/red intracellular bacilli.

Tissue impression smears are made by holding a wedge of tissue in a pair of forceps and pressing a cut surface of the tissue lightly against a stationary glass microscope slide. If the tissue is quite bloody it should be blotted gently with a gauze sponge or paper towel prior to making the impression smear. Several impressions are usually made on each glass slide. Always prepare 3–4 slides from each tissue so that you can apply more than one staining method to the smears. The slides are then air-dried, methanol-fixed for 5 min and stored in a slide box.

Wet mount preparations, impression smears or scrapings should be prepared from any lesion. Two smears can be made then stained using Gram's technique and Diff-Quik®, or Wright's stain, to identify microbial agents and observe cell structure. A KOH preparation may be used to digest tissue to delineate fungi or parasites.

Gastrointestinal scrapings prepared as a wet mount may be instrumental in diagnosing *Coccidia* spp. or other parasites. Modified acid-fast staining of gastrointestinal scrapings can be used to identify *Cryptosporidia* spp. Tan or white oral plaques can be scraped to identify *Candida* spp. or trichomonad parasites. Flagellate parasites are very difficult to observe in histological preparations, and are much more readily diagnosed by watching the movement of parasites within a scraping of a fresh lesion.

Fluid effusions can be collected into sterile serum and EDTA tubes to determine the following: total protein, cell concentration and differential cell count on stained smears. These tests will help differentiate transudates, exudates and haemorrhagic effusions. Urea concentration of abdominal fluids can be measured and compared with serum values when urinary tract rupture is suspected.

4.4 Bacteriology sample collection

Use sterile dissection instruments and storage containers to collect samples for microbial culture. If sterile instruments or disinfectants are not available in the field, instruments may be flame-sterilised, boiled for 15–20 min, or placed in 10% buffered formalin or 70% ethanol for 10 min. Ensure that formalin sterilisation takes place in a closed container within a well-ventilated workspace.

To determine whether bacteria are circulating in an animal's bloodstream, 2–3 filtering organs such as lung, liver, kidney or spleen can be submitted for bacterial culture. Where the carcass is mildly to moderately decomposed, bone marrow and brain are less likely to be contaminated by putrefactive bacteria than are other organs.

When microbial culture of samples from all wildlife necropsies is not economically feasible, collect and freeze the samples pending the results of histopathology.

Fluids found within the abdominal or chest cavity are quickly contaminated with bacteria leaking from the intestinal tract after death. These fluids are generally a poor choice for bacterial culture.

Often the submission of large portions of tissue in a sterile vial is much more rewarding than submitting a swab collected from a lesion. A variety of tests can be conducted on a large sample of tissue, whereas the use of swabs is limited. Bacterial swabs represent a small sample size and can dry out during transport; large tissue samples (50–100 g) are often a better option for sample collection.

If it is not possible to collect large portions of parenchymatous organs for immediate bacterial culture, the following procedure should be employed.

- Heat a scalpel blade with a gas flame.
- Sear the surface of the organ by placing the flat end of the blade against the capsule or cut surface.
- Using a fresh sterile scalpel blade, cut through the seared surface of the tissue.
- Roll a sterile swab into the cut parenchyma of the organ.

This procedure prevents the collection of putrefactive bacteria that may have colonised the tissue surface.

If lesions are observed, collect and place half the lesion into 10% buffered formalin, one quarter into transport medium or a sterile container for bacterial culture, and one quarter into a sterile container to store frozen. Small lesions may be swabbed with a microbial culture swab then placed into formalin. Bacteria are most often found at the leading edge of the lesion, on the margin of normal and abnormal tissue. Thus the lesion margin, or subcapsular region of an abscess, is the best site to collect for bacterial culture. The central region of many lesions often contains only necrotic debris.

Samples should be submitted to a laboratory as quickly as possible. Fresh tissue samples should be maintained chilled at 4°C. Swabs in transport medium should not be chilled, since this can harm anaerobic bacteria. Anaerobic bacteria survive best at room temperature. Anaerobic bacteria are best examined immediately. If this is not possible, inoculation of cooked meat agar should be considered.

Storage in liquid nitrogen (–174°C) is an ideal means to preserve microbes in tissues on a long-term basis. If liquid nitrogen is not available a –80°C freezer or –20°C freezer can be used to store samples.

Diagnosis of botulism cannot be made only by the isolation of clostridial organisms within the tissues. Samples of serum or gastrointestinal content should be analysed for the presence of botulinum toxin, usually through an ELISA test.

4.5 Toxicology sample collection

If chemical exposure is suspected or there is a significant outbreak of morbidity or mortality, a selection of specimens should be collected for toxicological examination (Table 4.6).

Generally, findings of an animal in excellent body condition, with a full gastrointestinal tract and no lesions on gross or microscopic necropsy are hallmarks of acute toxicosis. Heavy metal intoxication can also be chronic, leading to emaciation, ill thrift or neurological disease.

Heart blood, liver, kidney, brain, urine and stomach content (or vomitus) are samples most frequently collected for toxicological examinations. Lung should be collected if there is a possibility that the animal inhaled a toxin.

Table 4.6 Samples for toxicology

Toxins of concern	Tissues to collect	Comments
Heavy metals	Liver, kidney, stomach content, blood	50–100 g of tissues
		>3 mL of blood in EDTA
Organophosphates	Liver, brain, serum	50–100 g tissue
		3–5 mL serum to measure cholinesterase activity
Organochlorines	Fat, milk, brain, liver	50–100 g tissue
		50–100 mL milk
Anticoagulants	Stomach content and liver	>100 g ingesta or liver
	Blood from the live animal to assess coagulation profiles	
Snake venom	Urine or kidney	Fresh or frozen
Plant or fungal toxins	Stomach content, liver, plant of concern	
Algal or other biotoxins	Stomach content, liver, serum	Stomach content into 4–6% formalin to identify toxin-producing organism
		Stomach content, liver and serum to identify the toxins
		50–100 g tissue
		1–3 mL serum
Petrochemicals	Live animal: swab of oiled mucosa, skin, fur or environment	50–100 g tissue
	Dead animal: sections of oiled skin/fur, liver, serum	

A sample of any suspected bait or contaminated food or water should also be collected. If samples cannot be examined within 24–48 hr they should remain frozen.

Unfortunately, there is no battery of tests that will analyse for all possible toxins. It is necessary to have a high degree of suspicion of the type of toxin involved. Thus, samples are often kept at –70°C pending results of histopathology.

Large sample sizes (>100 g) are required for toxicological examination. Each specimen must be wrapped individually in aluminium foil, or placed into hexane washed glass jars that are covered with aluminium foil. Plastic containers can contain toxic chemicals that can leach into the tissues and alter toxicology test results. A large tissue sample wrapped in aluminium can be placed into a plastic bag or container pending toxicological examination.

When biotoxin exposure is suspected in aquatic animals, two sets of samples must be collected. One set of samples is used to identify the micro-organism present and the second set to identify the presence of the toxins that might have been produced. A portion of stomach content is fixed in a 4–6% solution of buffered formalin to allow specialists to identify whether any species of algae, diatoms or other micro-organisms capable of producing toxins are present. Frozen samples of stomach content, serum and liver can then be used to identify the presence of any biotoxins (freezing will rupture micro-organisms, so a sample in formalin is definitely required).

The main points to remember are that large sample sizes (>50 g) of internal organs are required for toxicological examination and that if samples cannot reach a toxicology laboratory within 24–48 hr they should be frozen.

4.6 Virology sample collection

Specimens for virus isolation need to be stored chilled or frozen. The decision to chill or freeze depends on the sample type, target agent and length of time before the sample can be shipped to a laboratory. As a general guide, if samples can be shipped to a laboratory within 3–4 d of collection (including transport time) they should be kept chilled in a refrigerator and transported on wet ice or freezer bricks. For longer periods of storage prior to shipment, all samples should be frozen at a very low (<–50°C) temperature.

Ultra-low temperatures are achieved with the use of dry ice, liquid nitrogen or special ultra-low temperature freezers. Once samples have been frozen, if they cannot be transported to a laboratory overnight (using wet ice/freezer bricks) they should be shipped in a liquid nitrogen dry shipper or on dry ice, especially during hot weather. Freezing in a standard domestic freezer (about –20°C) is often contraindicated and should only be used as a last resort.

Swabs and small tissue samples (<1 cm cube) should be placed in liquid viral transport medium at the time of collection to prevent desiccation and minimise the growth of contaminants. For storage periods longer than 6 months, larger tissue samples can also be frozen in transport medium. Unused viral transport medium is stable in a refrigerator for approximately 1 wk, but for longer periods it should be frozen (any temperature). Sample preservation is also aided by matching the size of the container to the specimen, reducing dead space.

Brain, lung, liver, spleen, intestine, tonsils or lymph nodes and sections of tissues showing lesions are the most common tissues selected for viral culture.

4.7 Parasite collection

The collection and preservation of endo- and ectoparasites is covered in Appendix 2.

4.8 Genetic material sample collection

When conducting necropsies on rare or unusual animals you may collect samples for the Australian Museum's tissue collection (Australian Museum 2007), the Animal Gene Storage & Resource Centre of Australia (AGSRC 2007) or a local museum.

The Animal Gene Storage & Resource Centre maintains a collection of frozen cell cultures, and samples of ovarian and testicular tissue. Gonads must be collected shortly after the animal has died. Gonads should be collected aseptically into phosphate-buffered saline. The tissues must then be shipped on ice packs via same-day or overnight courier. Ring the lab to ensure that it knows the sample is en route.

The Tissue Collection of the Australian Museum was established in 1988 and holds approximately 26 000 tissues from a wide variety of native and exotic fauna. Samples are available to researchers around the world.

If samples can be stored in liquid nitrogen or in a –80°C freezer, then 0.5 cm × 0.5 cm × 1.5 cm wedges of skin, liver and kidney should be collected aseptically into small cryovials. If frozen storage is not an option, similar 0.5 cm cubes of skin, liver and kidney can be placed into individual cryovials containing 70–90% ethanol. Skin samples should be composed primarily of the epidermis; the underlying fat contains scant nuclear material. Batches of samples can be collected prior to transport to the museum.

Carcasses can be very useful for educational and scientific purposes, even after a necropsy. Collaborating with a local museum can provide abundant materials for their programs and maximise the amount of valuable information collected after animal death.

4.9 Sample collection for polymerase chain reaction

When collecting samples with a view to conducting PCR to identify a new or unusual disease-causing agent, a solution called RNA*later*™ (Ambion, The RNA Company) can be very useful. PCR can be used to differentiate and characterise a wide variety of organisms. This technique, however, often examines ribosomal nucleic acid (RNA), which degrades very quickly after an animal's death. If samples can be collected immediately upon the animal's death, RNA*later*™ can be very useful, since samples can be stored at room temperature for several weeks. This makes the collection of genetic material easier in field situations where freezers may not be available.

Since PCR is such a sensitive test, aseptic sample collection is very important to ensure that the results do not reflect contamination from other organisms.

If you are unable to obtain RNA*later*™, use 70–90% ethanol as a fixative, or maintain the tissue frozen (preferably at –80°C).

4.10 Fixation for electron microscopy

Depending on the disease, electron microscopy (EM) can be undertaken on both fresh and fixed tissues. Fresh specimens should be collected when there is evidence of vesicle formation or the development of pustules or scabs. Fluid can be aspirated from vesicles with a fine needle and small syringe. All fresh fluids, scabs or other tissues should be placed in very small, sterile leak-proof vials and stored chilled after collection and until delivered to the laboratory. Do **not** submit material in syringes.

For routine EM examination, tissues can be fixed in formalin. Small cubes of tissue (5–10 mm, suitable initially for histology) should be placed as soon as possible into 10% neutral buffered formalin (at least 10 times greater than the volume of tissue) to ensure complete fixation. Histology can be completed first, to provide a provisional diagnosis and assess whether EM is required.

For optimal EM examination, Karnovsky's fixative is preferred. This fixative, or similar solution, can be obtained from an electron microscopy laboratory.

Correct tissue fixation technique should be discussed with the EM laboratory.

5 SAMPLE SHIPMENT

Always contact your chosen laboratory before shipping samples to discuss recommended transport methods. When possible, fax a copy of the courier's waybill number to the lab as soon as the parcel has left your premises so that the lab knows the samples are en route and can trace materials that fail to arrive.

Strict guidelines are in place for the transportation of samples by air and land. The information provided below is only a guide to sample packaging. Discuss with the courier and laboratory the best methods to package and transport biological specimens.

Permits are often required from state agriculture and conservation agencies prior to transporting animal samples interstate. It is a requirement that the person packing biological specimens for air transport obtain training and an accreditation certificate (Safe Transport of Infectious Substances and Diagnostic Specimens course/certificate, approved by the Civil Aviation Safety Authority of Australia).

Submission of wildlife samples to overseas laboratories without due process and management of results has the potential to result in trade, human health and biodiversity complications for Australia. Before samples are sent overseas, the state or territory Chief Veterinary Officer should be made aware of the proposal. Whenever possible, the planned examinations should be undertaken in Australia.

The following information must accompany tissues being submitted to a diagnostic laboratory:

- animal species, age, sex, location found or history in captivity;
- number of animals affected or at risk;
- clinical signs;
- any test results already known, and tentative diagnoses;
- details of any treatment and response to treatment;
- summary of necropsy findings;
- disease or organism suspected;
- tests requested;
- comment regarding zoonotic potential of the suspected pathogen, if pertinent.

Sample vials must be leak-proof. When collecting the samples, keep the outside of the vials clean. External surfaces of sample vials that may contain highly infectious agents should be disinfected prior to transport. These vials should be placed into a ziplock bag to prevent spills in the event of breakage. Heat-sealers and associated plastic bags are an excellent means of achieving a legal seal and sealing samples to prevent leakage during transport.

Dry-shippers are transport containers that have walls charged with liquid nitrogen. These containers make it possible to store and transport samples at ultra-low temperature without the risk of spilling liquid nitrogen. A handful of steel slugs placed in the toe of a nylon stocking can be chilled in the dry-shipper overnight and taken in a steel (not glass) thermos on day trips. This will allow samples to be frozen immediately after collection, even in very remote locations during field trips (but not for courier transport). This system will not meet transport safety guidelines.

Suppliers of dry ice are now difficult to find. Dry ice is classified as a dangerous good and including it in a package can significantly increase courier costs. Labelling and packaging requirements for dangerous goods are strenuous; often this is best undertaken with the assistance of a shipping company that specialises in dangerous goods handling. Carbon dioxide emitted from melting dry ice can kill several infectious and parasitic agents. Thus, the use of dry ice in sample shipment should be carefully assessed.

Blood films and cytology preparations should not be stored or transported near formalin. Formalin can cause degeneration of the cells on the smear.

If possible, retain portions of your samples. Try not to ship all of your samples to one lab in one shipment, in case the package gets lost or in case you want to confirm a diagnosis by sending samples to another lab.

5.1 Packaging fixed specimens

Formalin- or alcohol-fixed tissues can be shipped by courier or post under the following conditions. After the samples have been fixed for 24–48 hr (5–7 d for large animal brains), the fixative can be decanted. The sealed sample vial can then be placed into a ziplock biohazard bag with the case history and gross necropsy report enclosed in the bag's side panel.

Groups of these individual cases can be placed in a large ziplock or heat-sealed bag, which is then placed into an Esky containing absorbent material. The Esky is then sealed with packing tape. If the Esky is made

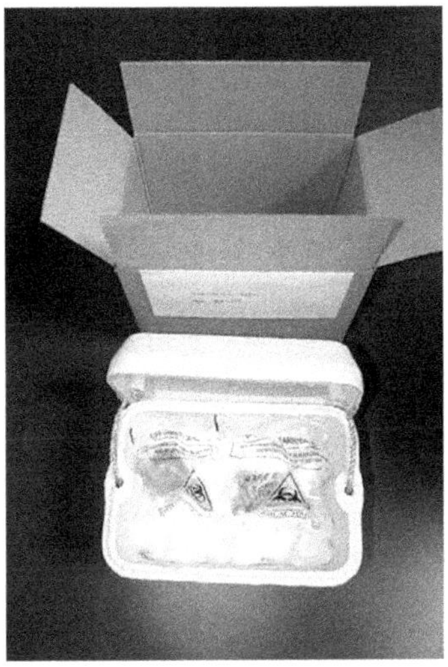

Figure 4.3 Fixed specimens for transport.

Figure 4.4 Fresh specimens for transport.

of styrofoam instead of impact-resistant plastic, it must be placed in a cardboard box for shipment (Fig. 4.3).

Generally, if more than 50 mL of formalin or 70% ethanol is being transported the shipment will be classified as a dangerous good—check with the courier. Solutions containing more than 70% ethanol are considered a dangerous good regardless of the volume shipped. Shipping dangerous goods requires additional packaging, labelling, documentation and funding.

5.2 Packaging fresh or frozen specimens

Fresh or frozen tissues could contain infectious agents and should be shipped within a three-layer packaging system produced by CSIRO's Australian Animal Health Laboratories and various other suppliers. These Biological Product and Diagnostic Transport Containers meet IATA regulations.

The Biological Product and Diagnostic Transport Containers consist of a metal tin to hold the specimen and absorbent material to pad the sample and collect any leaking fluid. The lid of the tin is sealed with two-part epoxy. The tin is placed in the polystyrene box along with two freezer bricks. The polystyrene box is placed inside a green polypropylene (hard plastic) case (Fig. 4.4). A complete set of instructions accompanies the containers. Most of the container packaging can be reused.

Fresh or frozen tissues must be shipped to the laboratory as quickly as possible. Same-day courier service is preferable, but overnight courier may be acceptable. Avoid shipping samples on Fridays. It is expensive and difficult to organise Saturday delivery. Samples that go missing in the courier system on a weekend are often of little value when found.

Whirl-pak® bags (Biolab) are very useful for sample collection in the field and for efficient use of freezer space during storage, but the bags are not leak-proof. If samples begin to melt in transit, fluids will leak out of Whirl-pak® bags. Samples contained in these bags should be transferred to a leak-proof container or placed inside a leak-proof container prior to shipping to a laboratory.

When samples are destined for overseas transport, it is important to find a reliable international courier. Try to ship the samples at the beginning of the week so that they do not get caught in customs over a weekend.

Pack all permits and sample documentation in a clearly marked envelope on the outside of the container so the paperwork can be examined without disturbing the samples. Send the receiving laboratory a copy of the documentation, courier contact details and any courier reference numbers.

Follow the shipment closely. Most courier companies allow parcels to be tracked through information posted on the internet. Check with the receiving laboratory to ensure that the materials have been delivered. It is not unusual for samples containing biological materials to be seized by customs officials at the point of export or import. If materials are seized, try to speak directly to customs officials rather than through a courier's agent. If requested, many customs officers will help by storing materials in a fridge or freezer until release of the materials can be negotiated.

ACKNOWLEDGMENTS

This chapter is reprinted with permission. *Wildlife Health Investigation Manual* © Karrie Rose, 2007. All rights reserved. Zoological Parks Board of New South Wales, PO Box 20, Mosman, NSW, 2088. www.arwh.org.

REFERENCES

AAZV (2001) *Animal Movement and Disease Risk. A Workbook*. American Association of Zoo Veterinarians: Yulee, Florida.

Allworth A, Murray K & Morgan K (1996). A human case of encephalitis due to a lyssavirus recently identified in fruit bats. *Communicable Disease Intelligence* **20**, 504.

AGSRCA (2007). Animal Gene Storage and Resource Centre of Australia, Melbourne, <www.agsrca.srivilasa.com/frozen_zoo.htm>.

ARWH (2007). Australian Registry of Wildlife Health, Sydney, <www.arwh.org>.

ATAGI (2007). Australian bat lyssavirus. In *The Australian Immunisation Handbook* (9th edn), pp. 106–117. Australian Technical Advisory Group on Immunisation, National Health and Medical Research Council: Canberra.

Australian Museum (2007). Evolutionary Biology Unit's Tissue Collection, Sydney, <www.amonline.net.au/evolutionary_biology/tissue/index.htm>.

AWHN (2007). Australian Wildlife Health Network, Sydney, <www.wildlifehealth.org.au>.

Bengis RG, Leighton FA, Fischer JR, Artois M, Morner T & Tate CM (2004). The role of wildlife in emerging and re-emerging zoonoses. *Review Scientifique et Technique Office International des Epizooties* **23**, 497–511.

Bonaccorso E, Guayasamin JM, Mendez D & Speare R (2003). Chytridiomycosis in a Venezuelan amphibian (Bufonidae: *Atelopus cruciger*). *Herpetological Review* **34**, 331–334.

Bosh J, Martinez-Solano I & Garcia-Pris M (2000). Evidence of chytrid fungus infection involved in the decline of the common midwife toad (*Alytes obstetricans*). in protected areas of centre Spain. *Biological Conservation* **97**, 331–337.

Breed AC, Field HE, Epstein JH & Dasczak P (2006). Emerging henipaviruses and flying foxes: conservation and management perspectives. *Biological Conservation* **131**, 211–220.

Cameron A, Gardner I, Doherr MG & Wagner B (2003). Sampling considerations in surveys and monitoring and surveillance systems. In *Animal Disease Surveillance and Survey Systems* (ed. MD Salman), pp. 47–66. Iowa State Press: Iowa.

CCWHC. *Wildlife Disease Investigation Manual*. Canadian Cooperative Wildlife Health Centre: Saskatoon, Saskatchewan.

CCWHC (1998). *Moving Wild Animals. Health and Environmental Issues. Assessing Risks and Benefits*. Canadian Cooperative Wildlife Health Centre: Saskatoon, Saskatchewan.

CCWHC (1999). *Management and Intervention in Diseases of Wild Animals*. Canadian Cooperative Wildlife Health Centre: Saskatoon, Saskatchewan.

CITES (2007). *Convention on International Trade in Endangered Species of Wild Fauna and Flora*, Geneva, Switzerland, <www.CITES.org>.

Daszak P, Berger L, Cunningham AA, Hyatt AD, Green DE & Speare R (1999). Emerging infectious diseases and amphibian population declines. *Emerging Infectious Disease* **5**, 735–748.

Fraser GC, Hooper PT, Lunt RA, Gould AR, Gleenson LJ, Hyatt AD, Russell GM & Katten-belt JA (1996). Encephalitis caused by a lyssavirus in fruit bats in Australia. *Emerging Infectious Disease* **2**, 327–331.

King JM, Roth-Johnson L, Dodd DC & Newson ME (2006). *The Necropsy Book* (4th edn). Charles Louis Davis DVM Foundation: Gurnee, Illinois.

Lips KR (1999). Mass mortality of the anuran fauna at an upland site in Panama. *Conservation Biology* **13**, 117–125.

Murray K, Selleck P, Hooper P, Hyatt A, Gould A, Gleeson L, Westbury H, Hiley L, Selvey L, Rodwell B & Ketterer PJ (1995). A morbillivirus that caused fatal disease in horses and humans. *Science* **268**, 94–97.

Philbey AW, Kirkland PD, Ross AD, Davis RJ, Gleeson AB, Love RJ, Daniels PW, Gould AR & Hyatt AD (1998). An apparently new virus (family Paramyxoviridae) infectious for pigs, humans, and fruit bats. *Emerging Infectious Disease* **4**, 269–271.

Rose K (2007). *Wildlife Health Investigation Manual*. Zoological Parks Board of NSW: Sydney.

Salman MD (2003). Surveillance and monitoring systems for animal health programs and disease surveys. In *Animal Disease Surveillance and Survey Systems* (ed. MD Salman), pp. 3–13. Iowa State Press: Iowa.

Sellek PW, Arzey G, Kirkland PD, Reece RL, Gould AR, Daniels PW & Westbury HA (2003). An outbreak of highly pathogenic avian influenza in Australia in 1997 caused by an H7N4 virus. *Avian Diseases* **47**(special issue), 806–811.

Wobeser GA (1994). *Investigation and Management of Disease in Wild Animals*. Plenum Press: New York.

Woodford MH (2001). *Quarantine and Health Screening Protocols for Wildlife prior to Translocation and Release into the Wild*. Office International des Epizooties, Veterinary Specialist Group/Species Survival Commission of the World Conservation Union (IUCN), Care for the Wild International and the European Association of Zoo and Wildlife Veterinarians: Paris.

Woolhouse ME (2002). Population biology of emerging and re-emerging pathogens. *Trends in Microbiology* **10** (Supplement 10), S3–7.

Zepeda C & Salman MD (2003). Planning survey, surveillance, and monitoring systems: roles and requirements. In *Animal Disease Surveillance and Survey Systems* (ed. MD Salman), pp. 35–43. Iowa State Press: Iowa.

5. Echidnas

David Middleton

1 TAXONOMY, DISTRIBUTION AND BIOLOGY

The short-beaked echidna (*Tachyglossus aculeatus*) is native to Australia and New Guinea. The larger, rare long-beaked echidna (*Zaglossus bruijni*) is confined to the New Guinea highlands. Together, these two species are the only extant representatives of their genera. Echidnas are well-represented in the Australian fossil record and, along with the platypus, are the only living remnants of the egg-laying monotremes, thought to represent the most ancient mammalian lineage surviving to the present day. Echidnas probably evolved from the platypus during the Cretaceous period (Augee & Gooden 1993) and it is possible that monotremes have been around for more than 50 million years. There are five recognised subspecies *of Tachyglossus aculeatus*: *T.a. acanthion* from the Northern Territory, northern Queensland, inland Australia and Western Australia, *T.a. aculeatus* from eastern New South Wales, Victoria and southern Queensland, *T.a. lawesii* from New Guinea, *T.a. multiaculeatus* from South Australia and *T.a. setosus* from Tasmania (Augee et al. 2006).

The name echidna derives from Ekhidna, a Greek goddess who was half woman and half reptile. Monotremes possess unique bones in their pectoral girdles: the epicoracoids and the interclavicle, both structural adaptations for digging. The monotreme pectoral girdle thus has similarities to the extinct cyanodonts and to living crocodilians. The echidna was initially identified as a relative of South America's anteater (a placental mammal) simply through lifestyle similarity, and it was not until 1802 that its close relationship to the platypus was appreciated.

The earliest record of a short-beaked echidna in captivity was from Philadelphia Zoo in 1903, an individual that was still alive in 1953, 50 yr later. This represents the longest recorded lifetime of a short-beaked echidna (Augee et al. 2006).

The short-beaked echidna has been able to exploit all terrestrial habitats including those of Tasmania and many off-shore islands and is today the most widely distributed Australian mammal. In spite of its terrestrial nature the echidna is a strong swimmer and climber. The author once encountered a healthy specimen in fresh water 1 km from the nearest land, and a captive echidna at Healesville Sanctuary (Vic.) was observed to climb a tree in its enclosure to a height of 2 m or more on a regular basis. The species is currently secure throughout most of its present range.

2 ANATOMY AND PHYSIOLOGY

2.1 External features, sensory organs, skin and associated structures

The short-beaked echidna is rotund and furred with a small head terminating in a cylindrical beak or snout. Adults can grow to a body length of 40 cm and weigh up to 7 kg. The dorsum and flanks are covered by long quills. These quills are under neuromuscular control and can be erected in defence as well as being used to grip surroundings. The quills are fixed, do not readily fall out and cannot be ejected from the skin. The body, apart from the beak, is covered in coarse hair. The density of the hair varies from heavy to sparse in different individuals. Tasmanian echidnas are smaller and have a denser hair coat. When threatened, the echidna can burrow vertically downwards quite rapidly, leaving only the spine-covered dorsum exposed. When on impenetrable substrates, the head and neck can be flexed ventro-caudally and the body curled to produce a spine-covered ball, making it very difficult for a predator to gain access to unprotected areas of skin. This is accomplished by the enormous SC muscle *panniculus carnosus*, which covers the sides and dorsum. This adaptation is most helpful for a creature that spends its foraging hours with its entire head buried in soil, searching for ants and termites.

The skin is thick, with a fat storage layer, the *panniculus adiposus*, underlying the *panniculus carnosus*. Sweat glands are restricted to the pouch area. Nipples are absent and milk is expressed directly onto the skin of the anterior pouch via specialised hair follicles concentrated into two regions (patches or areolae) of the anterior pouch skin. During feeding, the juvenile echidna sucks the milk from the skin and hair rather than licking it (Augee et al. 2006). The echidna's 'beak' is a highly refined structure of bone, cartilage and soft tissue covered in dark shiny hairless skin. Its tip is equipped with a number of sensitive electro-receptors and mechano-receptors which may be used in the location of invertebrate prey items. The beak electro-receptors can detect very small electrical impulses emitted by muscular contractions in prey (Proske 1990). The paired nostrils face upwards at the tip of the beak and the mouth opening is small. The echidna produces copious amounts of tenacious saliva from the sublingual glands. This aids in food procurement, but when combined with the narrowness of the external nares results in almost constant nasal discharge accompanied by bubbling and sniffling (Fig. 5.10). These signs are present to a degree in most normal echidnas and do not usually indicate respiratory disease.

The external ear canal is well-disguised by hair and opens as a vertical slit protected by a ridge of muscle and cartilage (Fig. 5.1). Echidnas have acute hearing aided by the arrangement of the auditory ossicles. The incus and malleus are fused, and attached to the petrosal bone. The body of the echidna conducts sound quite efficiently to the ear in particular via the snout which, when in contact with the substrate, can detect low noise and vibration (Griffiths 1989; Augee et al. 2006).

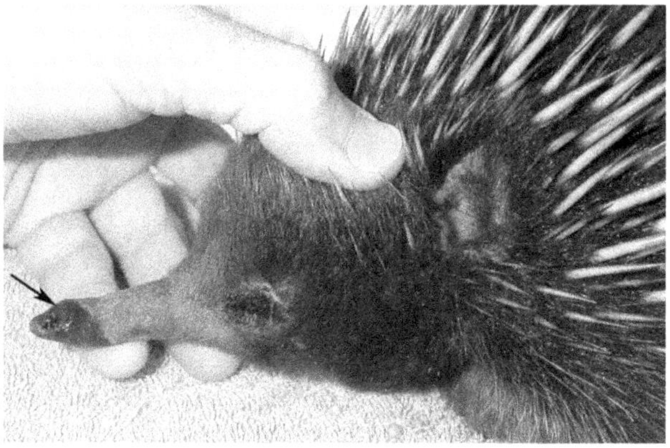

Figure 5.1 Echidnas have well-developed ears which are disguised by SC musculature as narrow vertical slits. The arrowhead indicates the site for needle insertion for blood collection from the beak sinus.

The echidna's eyes are small but well-developed. The retina is equipped with a greater proportion of cones than the human retina, and may provide good colour vision. There is almost complete decussation of the nerve fibres from each eye to the opposite side of the brain, which may limit binocular vision. In addition to electro- and mechano-receptors, echidnas probably rely heavily on olfaction for the location of prey and for seeking out oestrous females during the mating season. Blind echidnas have been reported to survive in the wild (Augee & Gooden 1993). The cerebrum of the echidna appears well-developed, deeply folded and relatively large. The frontal cortex of the echidna's brain is proportionately larger that of other mammals, though the significance of this is unclear (Griffiths 1989). This may be explained to some degree by the amount and nature of the various sensory inputs from the snout. Echidnas have been described by zoo keepers as sensitive, complex and intelligent and this may correlate with the size of the frontal cortex (Augee & Gooden 1993).

The legs and feet are stout and powerfully equipped for digging with well-developed blunt flattened claws. There are five clawed digits on the manus and four on the pes, with the most medial hind digit (digit one) greatly reduced. The second digit on the pes is elongated and, although often recognised as a useful tool for digging and excavation of foraging grounds, is also used for grooming as the shorter digits are insufficient to reach areas of the skin between the spines (Wood Jones 1923).

All male and some female echidnas possess spurs on the medial aspect of the tarsus. These are considered a remnant of an ancestral defence mechanism. A recent study concluded that the presence of spurs can be used to confirm the sex of adult echidnas, since no externally detectable spurs were present in any females in that study (Johnston et al. 2006b). However, this method may not be reliable in juvenile echidnas.

2.2 Digestive system

Echidnas have no teeth—food items are secured through the use of an elongated tongue constantly moistened with sticky saliva. The tenacious saliva, rich in glycoprotein, is produced in large amounts by the paired sublingual salivary glands which open to the floor of the mouth via multiple ducts. This 'glue' assists in the normal prehension and mastication of food items. The paired submandibular salivary glands are palpable as two distinct firm flattened mobile SC masses on the ventral neck. The tongue is long and slender, can be extended at least 18 cm beyond the tip of the snout and has a sophisticated arrangement of longitudinal and circular muscle bundles. It has a cavernous vascular network that gives it an erectile capability to forcefully probe for food in such substrates as rotten logs and termite mounds. It can protrude and retract up to 100 times per min and probe arthropod-rich crevices with great efficiency. With its strong fore limbs the echidna breaks open termite mounds, rotten logs and ant nests to probe for food. During foraging, the eyes remain closed. Prey items are crushed between backward-directed keratinous spines on the base of the tongue and above on the hard palate. This mastication can be detected audibly as a distinct grinding sound. As a result of this method of feeding, much soil and sand are ingested.

The digestive tract has a simple monogastric layout. The stomach is largely non-glandular and not involved in peptic digestion, exhibiting a relatively high pH, measured in one study at 6.2 (Griffiths 1989). It is lined by a cornified stratified epithelium which can grind invertebrate bodies with the aid of the ever-present dirt particles. The intestine of the echidna is 3.5 m long and exhibits a slow rate of throughput, with a clearance of 2 d for a termite and ant-based meal (Griffiths 1989). Faeces are solid, smooth and cylindrical and contain the undigested exoskeletons of prey items as well as large quantities of dirt. The small intestinal mucosa has high trehalase activity corresponding to the high trehalose content of termites and ants. These prey items lack lactose and sucrose, hence lactase and sucrase activity is virtually indetectable (Griffiths 1989).

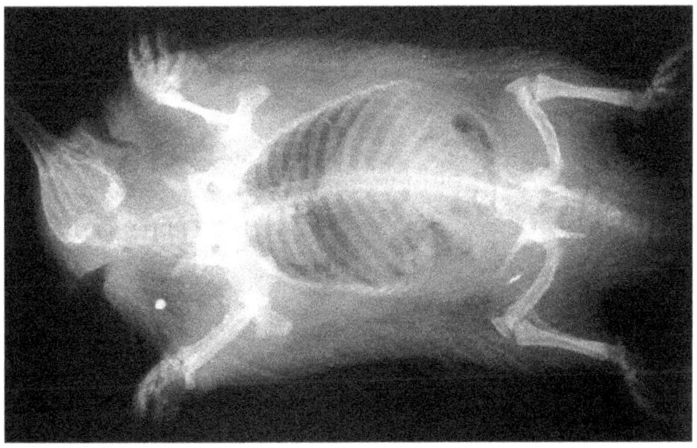

Figure 5.2 Whole-body radiograph of a sub-adult short-beaked echidna demonstrating some of its unique skeletal features.

2.3 Skeleton

The echidna's skeleton exhibits a number of interesting radiographic features (Fig. 5.2). The pectoral girdle has two additional bones, the epicoracoids and the interclavicle (structural adaptations for digging). There are cervical ribs, cranially projecting scapulae, an unfused acetabulum with a central foramen, outward-rotated hind limbs (that result in a backward-facing pes), transposition of the tibia and fibula, epipubic bones forward of the pubis and a short flattened femur. The tibia and fibula are of equivalent size and can be confused radiographically. The fore limbs are massive, with the radius and ulna easily matching the femur for size and strength. The scapulae are large. The humerus is horizontal and the glenoid broad and shallow. The paired mandibles are delicate and assist in the opening of the mouth by rotating medially. The vertebral formula of the echidna is 7C, 16T, 3L, 3S and 12Co (Griffith 1989).

2.4 Genitourinary system

The kidney of the echidna can produce concentrated urine that may be as high as 2300 mOsm/kg. The kidney's ability to conserve water in this way plays an important part in the echidna's capacity to live and thrive in arid environments.

2.4.1 Male

The reproductive tract of the male echidna is unique among mammals. The two ovoid testicles are suspended within the abdominal cavity caudal to the kidneys. There are no seminal vesicles and no discrete prostate gland. Sperm are produced by the testes and pass directly into the epididymis and on into the ductus deferens before entering the urogenital sinus and subsequently the penile urethra. The penile urethra passes only sperm. Urine is passed via the ureters to the urogenital sinus and has the opportunity to mix with sperm for a short time as the ureters and ductus deferentia join immediately prior to entering the urogenital sinus. The process of sperm being shed into the urine is known as spermatorrhoea and has been used recently as an indicator of sperm production. Urine is not conveyed by the penile urethra but enters the cloaca directly via the urinary papilla, situated on the cloacal wall at the ventral base of the penis (Johnston et al. 2007).

Figure 5.3 The echidna penis. The glans is divided into four bulbous portions by three dorso-ventral grooves. Each portion is adorned with fine epidermal projections and each receives a branch of the penile urethra, which divides laterally to produce two branches each of which divides again.

Figure 5.3 provides a dorsal view of the extruded echidna penis. The penis is securely retracted into the preputial sac inside the cloaca when not in use. The penis in adult males is around 7 cm long. Mackenzie (1919) describes the glans as being divided into four bulbous portions by three dorso-ventral grooves. Each portion is adorned with fine epidermal projections and each receives a branch of the penile urethra, which divides laterally to produce two branches each of which divides again.

2.4.2 Female

The female echidna is oviparous. The female reproductive tract consists of two functioning ovaries (unlike the platypus, in which only the left is functional), each associated with an oviduct which terminates in an infundibulum. The oviducts join directly into the urogenital sinus. The bladder also joins directly to the urogenital sinus opposite and dorsal to the entry of the ureters.

2.5 Thermoregulation and hibernation

The echidna has a relatively low body temperature among the terrestrial vertebrates (32°C) and a metabolic rate one-third that of an equivalent sized dog or cat (Griffith 1989). The body temperature is variable and echidnas are able to enter torpor and to hibernate during cold periods. Torpor is a term more applicable to short-term cold snaps after which the animal can return to activity quickly. Hibernation more correctly relates to seasonal endurance with minimal periods of awakening, a period of metabolic preparation (e.g. deposition of fat reserves) and a more protracted return to activity at the end of winter. In the colder alpine parts of their range echidnas lose 3% of body weight per month during hibernation and regularly regain activity every 12 d before re-entering their hypothermic state. Individuals engaged in hibernation can be aroused by warming gradually over a period of days (Grigg et al. 1990).

Interestingly, the echidna demonstrates a variable dependence upon hibernation in accordance with their lifecycle needs such as reproduction. Immature individuals and adults not involved in mating have been shown to hibernate for longer periods (Grigg et al. 1990). Body temperatures as low as 4°C are possible in adult echidnas and even nestlings can enter torpor for some days to endure maternal foraging absences (Griffiths 1989). All this has great relevance for the clinician as the echidna is classified as a 'heterotherm' and as such is much affected by environmental temperature. Their thermoneutral range is 20–30°C. Below 20°C muscular contraction must generate heat if the echidna is to remain active; above around 28°C heat loss is minimal and hyperthermia begins to develop (Griffith 1989). Although the echidna is unable to use sweating or panting as a means of heat loss, it is particularly adept at finding shelter in hollow logs, rock crevices or underground and restricts its activity to the cooler parts of the day and night to avoid heat stress. It will also make use of free water for cooling but this is not essential, as evidenced by the presence of echidnas in dry habitats. Daily water intake of adult echidnas held at 28°C environmental temperature has been measured at 40–80 mL/d (Griffiths 1989).

2.6 Cardiopulmonary system

The heart rate of the echidna is about 115 beats per minute, blood pressure has been measured at 123/96 mmHg and cardiac output at 59 mL/kg/min (Griffiths 1989). The heart has two unique features: coronary veins that carry blood from the heart to the right atrium (a feature not found in other mammals or in birds) and papillary muscles that insert directly onto the heart valves in the absence of chordae-tendinae. The diaphragm is broad and muscular with a tendinous centre over the apex of the heart.

Echidnas are tolerant of high CO_2 concentrations in their inspired air (to 10%), and have a metabolic requirement for oxygen that is less than half that of eutherian mammals (Schmidt-Nielsen et al. 1966). At thermoneutrality, the adult echidna consumes 0.22 mL of oxygen per gram body weight per hour. This is further evidence that the echidna is very well adapted to environmental variation and capable of exploiting low oxygen/high carbon dioxide situations.

2.7 Immune system

The general anatomical layout of the lymphatic system is the same as in other mammals (Diener et al. 1967a). Monotremes have a diffuse lymphoid system composed of nodules which are suspended by a pedicle, dispersed throughout the lymphatic vessels. The lymphoid nodules within the intestinal mesentery are grossly visible as small white raised foci near the attachment of the mesentery to the intestinal serosa (Rose 1999).

The immunological response of the echidna is also similar to the general mammalian response, however, the anamnestic response is less intense and described as similar to that demonstrated by reptiles (Diener et al. 1967b). Wild echidnas seem to enjoy relatively good health and a low incidence of infectious disease. In spite of this, echidnas in captivity are susceptible to infection, in particular bacterial septicaemia. This may be partly due to a low average body temperature (32°C), low metabolic rate and additional aspects of sluggish immunity such as a reduced level of phagocytosis as reported by Wronski et al. (2003). They demonstrated an inferior immune response in the echidna (relative to the rabbit) based upon antibody production and secondary response to antigen administration.

3 REPRODUCTION

3.1 Sex determination

Determining the sex of short-beaked echidnas can be difficult as the sex organs and associated structures are internal and the female pouch is only present when in breeding condition. Johnston et al. (2006b) describe five techniques for sex determination in adults. Only one technique—hind limb spur morphology—was suitable for use in conscious adult echidnas. Three techniques—penile palpation/extrusion, testicular palpation and ultrasound of the testes—require general anaesthesia. The latter two techniques are also subject to false negative results in young males and in males with inactive or immature testes. A fifth technique—faecal steroid analysis—requires laboratory facilities and reference ranges and is not practical for clinical use. Adult male echidnas have well-developed hind limb spurs that are evident externally while females do not (although some females have small spurs). Palpation of the penis in the cloaca with external extrusion is probably the most reliable method to use in the clinical setting, but general anaesthesia will be required to achieve an adequate level of relaxation.

3.2 Breeding and development of young

Little data is available on the breeding biology of the short-beaked echidna. This is primarily due to their cryptic nature in the wild and lack of breeding success in captivity. Echidnas show evidence of a seasonal breeding pattern. The lifecycle shows a defined period of courtship and mating during July, August and early September regardless of geographic locality (Augee et al. 2006). Although authors seem reluctant to classify the female reproductive pattern, it is likely to be seasonally polyoestrous. Females have been known to conceive twice in a single season. Sperm production is seasonal and correlates well with testicular size; production peaks in August and is lowest in late spring (Johnston et al. 2007).

The echidna is mostly solitary with adults occupying large, vaguely defined and overlapping home ranges of up to 70 ha. Mating occurs in late winter and during this time several males may be seen lined up snout-to-tail, following an oestrous female. Three weeks later, the female lays a grape sized egg directly into her pouch, which has developed into a depression in the ventral midline bounded by muscular ridges. Twins are rare. All females develop pouches during the breeding season regardless of conception. Prior to egg-laying, the female excavates a burrow which consists of a short tunnel terminating in a nesting chamber. These burrows are extremely difficult to detect as the female plugs the entrance while in residence. Hatching occurs around 10 d after laying and the young echidna, commonly referred to as a puggle, remains within the pouch sucking milk from the 'milk patch' in the anterior pouch area. At some time, prior to the development of the quills and at around 45–55 d of age (about 200 g body weight), the young exits the pouch and is deposited in a nursery burrow Fig. 5.7. It is suckled relatively infrequently, perhaps only every 5 d, when the mother returns from her foraging expeditions. The young echidna grows and develops rapidly, aided by the large quantity of milk ingested at a single suckling—up to 20% of body weight in 1–2 h. The young gain 0.4 g for each millilitre of milk ingested—an impressive conversion rate. At 100 d the puggle can weigh as much as 550 g (Griffiths 1989). Lactation lasts around 200 d and permanent nest exit is thought to occur by 240 d (Augee & Gooden 1993). In studies on Kangaroo Island, South Australia, it has been estimated that young echidnas leave the nest at between 180 and 205 d at which time they weigh 800–1300 g. Weaning is instigated by the mother failing to return to the nest. Once weaned, the mother has no further contact with her young (Augee et al. 2006).

4 HUSBANDRY

4.1 Housing

Echidnas are inquisitive and responsive to environmental stimuli and express individual personality traits through learned behaviours. In exhibit design and construction, every effort should be made to respect this most fascinating and beguiling of creatures and provide a captive environment with a rich array of possible activities based on natural behaviours.

Echidnas are enormously strong and can pull cage furniture to pieces with ease. In the hospital, echidnas are best housed in smooth plastic tubs. The vertical walls should be at least 50 cm tall to prevent escape. For short-term hospitalisation, shelter and security can be provided by large amounts of straw or shredded paper which can be easily changed. For long-term patients, consider round plastic tubs about 2 m in diameter which can be partially buried if outside. Although the beak is a delicate and sensitive organ, the echidna will use it to explore the hospital environment. Care must be taken to eliminate any sharp-edged surfaces or cavities in which the beak may become jammed and injured. Echidna enclosures should be kept dry and the base substrate should not be abrasive as it may cause ulceration of the feet. Rubber floor covering may be worked loose and the echidna may get under it. A substrate that allows burrowing, such as soil or leaf litter, should be added to provide enrichment, including the opportunity to forage for invertebrates.

Outside enclosures should have a mesh or solid impenetrable layer beneath the digging substrate to prevent escape. Alternatively, the walls can be continued vertically down to a depth of 50 cm (Jackson 2003). The perimeter walls of outdoor enclosures should be solid, smooth and vertical to at least 50 cm to prevent escape. Overhanging structures and trees should be avoided as echidnas can climb.

The thermal preference for adult echidnas is between 20–30°C. Echidnas are far more prone to heat stress than to cold stress and in most climates can be housed outside provided shelter from heat is available. This must also be considered in the hospital situation, and careful monitoring of ambient temperature is essential.

Several echidnas can be housed together in the same enclosure (including in hospital) if a minimum of 5 m^2 floor space is provided for each individual and food, shelter and foraging opportunities are sufficient to minimise direct competition. Additional food dishes and shelter locations should be provided if dominance behaviour is observed.

4.2 Feeding

In hospital and when housed without other species, food can be placed in bowls. Tubes are an alternative to bowls. PVC pipe (50 mm diameter, 150 mm long) with a cap on one end and a removable cap with a hole the diameter of an echidna's beak on the other are secured or partially buried and filled with echidna mix (L Vogelnest *pers. comm.*). When housed in mixed exhibits (e.g. with small macropods or birds), access to echidna food by other species must be prevented. PVC feeding tubes or artificial termite mounds with smooth-edged holes in the base can be used to cover and protect food bowls. Natural food items such as pieces of termite-infested wood or termite mounds will add greatly to the quality of the captive environment. Water should be available *ad lib.*

4.3 Husbandry

Hygiene is essential in maintaining the echidna's hospital environment. Daily cleaning and regular disposal of substrates is essential. The increased incidence of faecal-oral infections such as coccidiosis and salmonellosis in captive echidnas may be associated with environmental stress and hygiene (McOrist & Smales 1986).

4.4 Individual marking and identification techniques

Passive integrated transponders (PIT tags) or microchips are most commonly used to permanently mark individual echidnas. These are inserted SC on the dorsal midline between the scapulae. Coloured plastic tubing fitted over individual spines is also commonly used and has the advantage of allowing identification from a distance. Historically, tattoos on the ventral abdominal skin have been used. Paint markers can be used for temporary identification. Radio-tracking devices have also been fitted to the dorsum, anchored by various techniques including glues and ligatures.

4.5 Transport

For short-term (up to 2 hr) transport, 20 L plastic buckets or tubs with high sides (minimum 40 cm) are suitable as long as they are secured and prevented from falling over. For longer trips, strongly constructed transport boxes with smooth continuous internal surfaces free from deep corners, protrusions, cavities, holes and free edges are ideal. Some substrate material such as shredded paper should be provided to absorb fluids and enable the echidna to seek security. Provision of food and water is only necessary for extended trips (>4 hr) and then only via purpose-designed systems that prevent spillage in transit. Cardboard boxes, hessian bags and wire crates are not suitable for transporting echidnas (Jackson 2003).

5 NUTRITION

In the wild, adult echidnas are myrmecophagous—they feed almost exclusively on ants and termites. Echidnas seem to prefer termites (Augee & Gooden 1993), however, in some parts of their range (e.g. Tasmania) ants are far more common and form the greater part of the diet (Griffiths 1989). They will also eat other invertebrates such as earthworms and the larvae of beetles and moths (Booth 1999).

Invertebrate diets are characteristically high in protein and fat and most captive diets follow this basic principle. Oyarzun et al. (1996) analysed the nutritional composition of termites of *Nasutitermes* spp., which are high in protein (67%), phosphorus and iron and low in fat (2%) and calcium. Interestingly, they discovered that the dominant lipid fraction in all termite stages of development was oleic acid, which mirrors the lipid composition of echidna milk, reflecting the influence of dietary intake on milk composition (see section 8).

Termites contain considerably more water than ants (80% vs 64%). The water requirement of echidnas has been studied. It was found that at moderate humidity (9–15.6 g H_2O/m^3) in the temperature range 21–28°C echidnas feeding on termites did not require free water. However, as humidity fell the ability to maintain fluid balance without free-water declined. At 3.3 g H_2O/m^3, echidnas needed to drink 40–80 mL of water per day to remain adequately hydrated (Bentley & Schmidt-Nielsen 1967).

Due to the difficulty in maintaining a regular and adequate supply of termites and ants to captive echidnas, a wide range of alternative diets has been formulated. There has been a lot of experimentation with the formulation of captive echidna diets, with no firm indications of the ideal. Captive diets are often based on minced meat mixed with other ingredients to form a slurry. Most wild echidnas will accept a captive diet within a few days unless seriously ill. Echidnas that have suffered damage to the delicate structures of the beak may have difficulty feeding and in severe cases will be unable to survive. Formic acid (as produced by some ant species) or paw-paw (L Vogelnest *pers. comm.*) have been added to captive diets to encourage feeding, however, the real benefits of these additives are unknown.

Breeding success in captivity has usually been low and obesity is common. Although diet is certainly not the only factor in breeding success, it most probably plays an important role, suggesting that further work is required on echidna nutrition. Different zoos and fauna parks have used different diets with variable success regarding breeding and obesity (Jackson 2003). Adelaide Zoo has bred echidnas on a regular basis, an unusual achievement, and its diet is presented in Table 5.1 (D Schultz *pers. comm.*). Healesville Sanctuary has maintained echidnas in captivity for many years with good health but low breeding success. The diets offered are presented in Table 5.2.

Table 5.1 Echidna diet at Adelaide Zoo

Component	Amount
Bran	2 kg
Wheatgerm	0.1 kg
Meat and bone meal	1 kg
Blood meal	2.6 kg
Calcium carbonate	0.4 kg
Safflower oil	20 mL
Lean beef mince	2 kg

This recipe makes a bulk amount of daily feed sufficient to maintain up to 10 adults. A small animal multivitamin supplement is added to the diet as recommended by the manufacturer. The dry parts of the diet are mixed and stored separately and added to oil/meat mix daily. Water is added to make a slurry. This mix gives an energy (kj) to protein (mg) ratio of 1:34. The calcium:phosphorus ratio is 1.3:1. Faeces produced with this diet are tube-like, like those of wild echidnas.

Table 5.2 Echidna diets at Healesville Sanctuary, Zoos Victoria. Quantities are per animal per day

Component	Low-fat diet[a]	High-fat diet[b]
Minced beef	100 g	110 g
Egg	24 g	26 g
Glucose monohydrate	45 g	18 g
Wheat bran	17 g	18 g
Olive oil	9 g	14 g
Balanced Cal Powder[c]	1 g	1 g
Soluvet multivitamin supplement[d]	5 g	5 g
Fly pupae	7 g	7 g

The ingredients are blended with water (approximately 200 mL) until it is a thick consistency. *Ad lib* water is provided, as are supplementary mealworms.

a Maintenance diet fed from spring to late autumn.
b Breeding/hibernation high-fat diet is fed through winter.
c Mavlab.
d Vetafarm.

Some zoos provide different diets seasonally in an effort to support the increased nutritional requirement of breeding animals. This may take the form of providing a higher energy density by adding fat only during the winter breeding season. Manipulating the energy of the diet is also thought to lessen the incidence of obesity. Such diet variation is probably not natural in that dietary items are probably less plentiful over winter in the wild, at least in southern Australia. Johnston et al. (2007) found no seasonal variation in the body weight of wild male echidnas, which suggests adequate energy intake year-round.

Captive diets should be modelled on natural foods. If mince meat is used as the protein base it should be lean and of a high quality and digestibility. Gastric impactions have been seen when kangaroo mince that contained a high proportion of sinew was used (L Vogelnest *pers. comm.*). Lean beef mince is most commonly used. The digestion of whole invertebrate bodies is aided by the presence of soil in the gut, suggesting that soil substrates or provision of soil are an advantage in the hospital environment. If the captive diet does not contain invertebrate bodies and there is no opportunity to forage for invertebrates (not an ideal environment), soil may not be necessary. Iron is probably an important nutritional component since iron content of wild food and echidna milk is high.

Food is normally offered once daily as a semi-liquid slurry in an open bowl or tube (see 4.2). Body mass should be monitored by regular weighing (at least monthly) and food rationed accordingly (Jackson 2003). Obesity is common although no specific information regarding the health impacts of this condition is available. Water should be available *ad lib*. The amount of dietary energy required will vary according to the enclosure design and the range of physical activity. It is generally accepted that the healthiest captive environments are those that provide for natural behaviours such as digging, foraging, climbing, sheltering and the breaking apart of termite mounds and rotten logs. Dominant individuals can restrict access to food by more submissive animals; multiple feeders and covered tube feeders (see 4.2) may help avoid these problems. Echidnas have been known to feed on unusual food items such as bird seed and will go to great lengths to reach it. This may result in obesity. Access to these food sources should be prevented (L Vogelnest *pers. comm.*).

6 RESTRAINT

6.1 Capture and physical restraint

Echidnas are very strong, have sharp quills and defensively curl into a tight ball when handled. The tips of the quills can cause considerable irritation to human skin. It is not possible to grasp or palpate any part of the body that is spined, consequently, they are difficult and sometimes very awkward to capture and handle. When approached, they rapidly dig into soft substrates or can wedge themselves securely beneath an overhang or within a hollow or crevice. It is even quite difficult to extract an echidna from the corner of a simple wooden crate as it wedges itself with its quills. Excavation with a trowel or shovel may be needed to extract an echidna buried in soil. If an echidna escapes from a transport box into a car it can be virtually impossible to extract from under a seat or the boot. In some situations SC or IM administered chemical restraint (see 6.2) is the only means of retrieving an echidna.

On smooth substrates such as concrete or polythene an echidna can be picked up by one of three methods (Fig. 5.4). Using gloves or several layers of thick towel, the entire animal can be cupped with both hands and lifted up. Alternatively, an echidna can be lifted from a smooth substrate by carefully working ungloved hands, palms upwards, beneath the echidna's belly from each side. The body will curl around the hands with the head held tight against the sternum. The echidna can then be lifted and held securely with its spiny dorsum against the protected chest of the handler. Third, the hind feet may be grasped by their elongated claws and

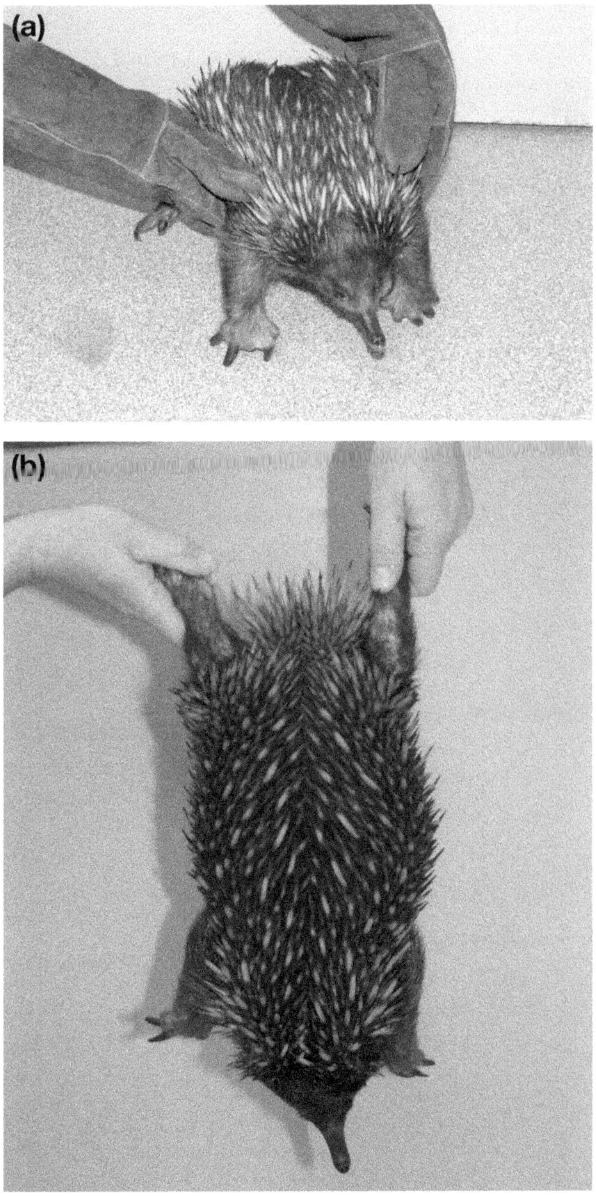

Figure 5.4 Physical restraint of an echidna. a) The use of gloves. b) Holding by the hind legs. c) Using a towel to protect the handler's forearm. d) Using bare hands placed under the spineless ventral body. Photos: Taronga Zoo.

the echidna suspended by both hind limbs. This is not a comfortable position for a conscious echidna and may result in defensive behaviour such as projectile urination and defecation. An examination of the head, neck, snout, limbs, pouch and cloaca can be performed on a conscious animal with physical restraint alone, but for a thorough examination chemical restraint will be required.

6.2 Chemical restraint

For anything other than minor procedures such as injections or transport, chemical restraint is preferable to physical restraint. Pre-anaesthetic fasting is not necessary as echidnas are not prone to regurgitation. However, as a general rule it is probably unwise to anaesthetise an echidna just after it has fed. Injectable agents have been used to good effect for examination and minor treatment but in the author's experience should not be relied upon for surgical procedures. IM injections can be given in the hind limbs or the *longissimus dorsi* and *sacrospinalis* muscles (useful in unrestrained echidnas that cannot be extracted for physical restraint).

Ketamine/xylazine mixture (5 mg/kg and 1 mg/kg), tiletamine/zolazepam (3–10 mg/kg) and ketamine/medetomidine mixture (5 mg/kg and 0.5 mg/kg) given IM provide reliable chemical restraint. Diazepam (1–5 mg/kg IM) is a useful sedative. Rissman (2000) investigated anaesthetic agents in echidnas and concluded that only two injectable regimes reliably produced surgical anaesthesia—ketamine/medetomidine mixture (5 mg/kg and 0.3 mg/kg IM) and tiletamine/zolazepam (5 mg/kg IM). Of these two mixtures, ketamine/medetomidine produced smoother inductions and shorter smoother recoveries. The medetomidine dose for echidnas is substantially higher than that used in most other mammals. Rissman (2000) considers ocular and peri-ocular reflexes and responses to be unreliable indicators of anaesthetic depth. The echidna is tolerant of increased CO_2 concentrations and responds by increasing tidal volume rather than respiratory rate. Thus respiratory rate may not be a reliable monitoring parameter during anaesthesia. An advantage of ketamine/medetomidine anaesthesia is reversibility with atipamezole at 5 times the medetomidine dose IM.

Gaseous anaesthesia is the method of choice for chemical restraint of echidnas. With appropriate physical restraint, a mask is easily placed over the beak.

Figure 5.5 Anaesthetic induction of an echidna using inhalation anaesthesia. a) In a wooden induction chamber with perspex observation panels on the lid and on one end. Photo: Healesville Sanctuary. b) Using a mask. Photo: Taronga Zoo.

Masks are usually improvised from syringe cases or barrels with rubber diaphragms cut from an examination glove (Fig. 5.5b). Care needs to be taken to avoid occlusion of the nostrils with the rubber component or the side of the mask. The inside of the anaesthetic mask must be constructed to minimise direct trauma to the

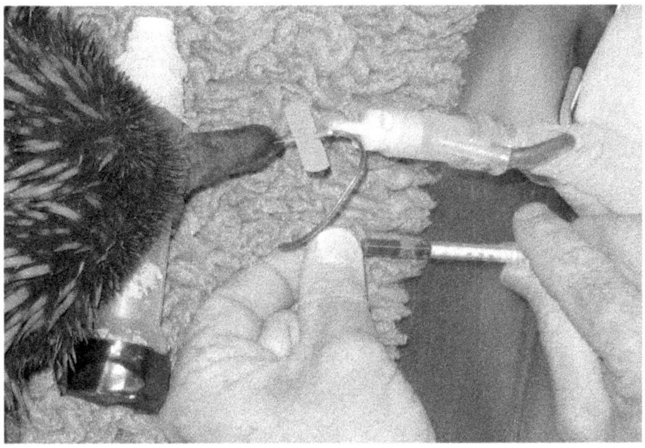

Figure 5.6 Nasal catheterisation to allow adequate ventilation and maintenance of anaesthesia in an echidna with a damaged beak. Blood is being collected from the beak sinus. Photo: Healesville Sanctuary.

beak. Mask size should also be designed to fit the beak without rubbing and abrasion. The majority of echidnas will ventro-flex the neck after mask positioning and remain reasonably immobile until induced. An induction chamber can also be used. This avoids the need to physically restrain the patient and thus may be less stressful. The chamber must be ruggedly constructed to withstand the echidna's physical strength (Fig. 5.5a).

Isoflurane in oxygen is generally used, however, other inhalation agents such as halothane or sevoflurane are also effective. 5% is used for induction and 1–2% for maintenance with an oxygen flow rate of 1 L/min. Maintenance is via a mask as echidnas are impossible to intubate (Vogelnest 1999).

There are some hazards associated with anaesthesia in traumatised individuals, especially those which have sustained injuries to the beak. The airway is narrow and subject to occlusion by blood, saliva and displaced tissue. Delaying anaesthesia until beak injuries can be more accurately assessed and stabilised may be of value, however, assessment itself usually requires chemical restraint. Gently inserting an 18 G catheter into a nostril (Fig. 5.6) may allow adequate ventilation during anaesthesia provided the patency of its lumen is maintained.

7 CLINICAL PATHOLOGY

7.1 Haematology and biochemistry

7.1.1 Sample collection

Blood can be collected from the beak sinus (Figs 5.1, 5.6), femoral vein, cephalic vein or jugular vein, but the beak sinus is by far the most accessible site (Johnston et al. 2006a). The beak sinus lies dorsally immediately caudal to the nostrils, can be seen as a bulge in the beak, and is accessed in the midline in a rosto-caudal direction. The needle should be guided sagittally and directed downwards approximately 10–15° from the long axis of the beak. If the needle is directed too deeply, penetration of the nasal cavity is possible. Sedation and/or anaesthesia greatly improve the chances of obtaining a useful blood sample from any site. When collecting blood from the beak sinus, any patient movement may lead to beak laceration. The use of a scalp vein set may minimise the risk of trauma.

7.1.2 Reference ranges and interpretation

Haematological parameters and the morphology of red and white cells of the echidna have been presented by Clark (2004). The erythrocytes of the echidna are very low in ATP but still use glucose at normal rates, presumably through alternative biochemical pathways. Platelets are variable in morphology and include both filamentous and crescent-shaped forms. The haematology and biochemistry parameters of the short-beaked echidna are more typical of other mammals compared to the platypus, which has some unique features (see Chapter 6). There is very little data on the haematological and biochemical responses of the short-beaked echidna to various physiological and pathological states, although there is adequate empirical evidence to suggest that they respond similarly to other mammals.

Table 5.3 provides haematology and biochemistry values for short-beaked echidnas.

8 HAND-REARING

The best advice for anyone contemplating hand-rearing an echidna is to speak directly with those who have done it successfully. Natural echidna milk is unique in composition (highly concentrated, high in fat and iron, low in protein and carbohydrate and low in lactose) and until recently has not been approximated by any artificially formulated products. It contains 48.9% solids and 31% lipid of which more than half (61%) is oleic acid, and the carbohydrate components are largely made up of fucosyllactose and sialyllactose (Griffiths et al. 1973; Griffiths et al. 1984).

The best milk formulas are those that most faithfully represent natural echidna milk. A variety of

Table 5.3 Haematology and biochemistry values for the short-beaked echidna

Parameter	Units	Mean	SD	n
Haematology				
Hb	g/dL	14.49	2.58	30
PCV	%	40.45	5.63	30
RBC	10^{12}/L	6.25	0.85	30
MCHC	g/dL	35.98	5.01	30
MCV	fL	64.91	5.05	30
MCH	pg	23.75	4.04	30
Platelets	10^{9}/L	414.28	125.15	18
Total WBC	10^{9}/L	11.95	5.52	30
Neutrophils	%	52.34	12.7	29
Neutrophils	10^{9}/L	6.6	3.86	29
Band neutrophils	%	0.62	1.05	29
Band neutrophils	10^{9}/L	0.08	0.15	29
Lymphocytes	%	43.69	13.14	29
Lymphocytes	10^{9}/L	5.11	2.51	29
Monocytes	%	2.66	2.14	29
Monocytes	10^{9}/L	0.30	0.27	29
Eosinophils	%	0.59	1.05	29
Eosinophils	10^{9}/L	0.08	0.17	29
Basophils	%	0.00	0.00	29
Basophils	10^{9}/L	0.00	0.00	29
Biochemistry				
Sodium	mmol/L	138.7	6.06	23
Potassium	mmol/L	3.11	0.58	21
Chloride	mmol/L	94.66	4.8	24
Bicarbonate	mmol/L	31.57	6.16	22
BUN	mmol/L	10.55	2.97	26
Creatinine	mmol/L	0.07	0.04	24
Total bilirubin	µmol/L	5.88	2.85	24
GGT	U/L	7.00	2.19	24
ALP	U/L	161.17	53.21	24
Total protein	g/L	76.19	11.51	25
Albumin	g/L	37.94	8.37	25
Globulin	g/L	38.23	5.22	24
ALT	U/L	100.00	32.93	4
AST	U/L	321.21	135.71	24
CK	U/L	79.19	37.76	21
LDH	U/L	239.75	136.28	20
Glucose	mmol/L	4.83	1.48	20
Amylase	U/L	669.74	227.84	19
Calcium	mmol/L	2.54	0.35	21
Phosphate	mmol/L	1.76	0.44	21
Cholesterol	mmol/L	4.54	Na	1
Triglycerides	mmol/L	0.39	Na	1

Sources: Booth (2003), Booth (*pers. comm.*), Healesville Sanctuary and Melbourne Zoo veterinary databases.

Table 5.4 Comparison of the constituents of monotreme milks with artificial milk replacers

Constituent	Platypus	Echidna	Wombaroo <30 d	Wombaroo >30 d	Di-Vetelact
Total solids (%w/w)	39.1	48.9	21	36	na
Lipid (%w/w)	22.2	31	8	16	4
Protein (%w/w)	8.2	12.4	7	11.5	3
Carbohydrate (%w/w)	3.7	1.6	5.5	4	5
Iron (mg/dL)	2.11	3.33	1.5	2	0.9
Calcium (mg/dL)	191	117	260	460	100
Phosphorus (mg/dL)	133	285	200	360	70
Magnesium (mg/dL)	16	16	19	34	10
Copper (mg/dL)	0.1	0.38	0.24	0.44	0.06
Zinc (mg/dL)	1.92	1.51	0.34	0.6	0.7

Platypus and echidna milk values from Griffiths et al. (1984).

formulas have been used to raise echidna puggles, with varying success. Proprietary low-lactose milks such as Di-Vetelact® (Sharpe Laboratories) made up double-strength have been used as the sole food in at least one puggle but a specifically formulated echidna milk replacer (Wombaroo Food Products) is now available. The milk is supplied in two formulations to match the age and stage of puggle development (<30 d of age and with a body weight less than 210 g and >30 d of age and with a body weight greater than 360 g). This is an attempt to mimic the changing composition of the natural milk as lactation progresses. Product supplementary information provides feeding schedules based upon developmental stage, body weight and growth rate.

Table 5.4 compares the major constituents of platypus and echidna milk with two brands of artificial milk replacer. In the wild, suckling echidnas have been shown to ingest 20% of their body weight of milk per day, presumably an adaptation to cope with the female being absent for days at a time during the nest-bound phase. In captivity, puggles are most comfortably maintained in the temperature range 20–25°C. Puggles are known to be capable of entering torpor (Griffiths 1989) and those being hand-reared may become less receptive at low temperatures, resulting in reduced intake, undernourishment and susceptibility to secondary infections (D Schultz *pers. comm.*). Due to the tendency to enter torpor, puggles kept below 20°C may not be able to digest milk. Secondary infections may be associated with high-carbohydrate diets which are, as demonstrated by analysis, unnatural and therefore not suited to puggles. Feeding is only necessary once per day and the puggle is encouraged to feed directly from the palm of the hand or out of a small flexible hand bowl or pinch pot available from kitchen supply stores. Between feeds the puggle can be housed in an unheated wooden box with a shallow layer of mulch or leaf litter on the floor.

Gavage feeding is usually unnecessary as, with patience, healthy puggles will adjust to hand-feeding within a couple of days. Gavage feeding also carries the risk of aspiration due to the mouth conformation which does not allow regurgitated milk to escape. The delicate young snout and associated structures are also at risk of injury during such procedures.

Puggles should only require feeding every 3–5 d as would happen in the wild. If the diet is roughly equivalent to wild diet in terms of composition, 10–12% of body weight is sufficient at each feed. Jackson (2003) acknowledges that puggles do not defecate as frequently

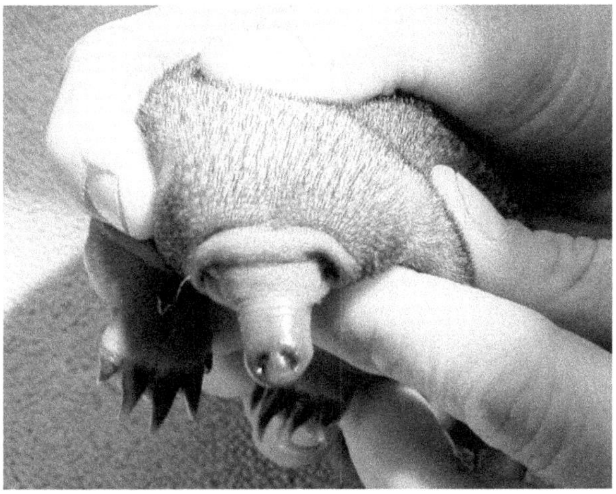

Figure 5.7 A puggle at around 40 d of age. Very small quills are beginning to appear. At this stage the young echidna would be nest-bound and fed daily at most.

as marsupial young, but detailed monitoring of dietary intake, faecal output, size and consistency of the abdomen and puggle response to palpation of the abdomen are important. An enlarging, tense and painful abdomen is an indicator of the need for further investigation and intervention. If constipation is suspected, radiographic examination of the abdomen will help to assess the severity of the condition. Treatment should include gentle massage of the faecal mass, consideration of fluid balance and the need for rehydration, faecal softeners and lubricants and, in moderate to severe cases, enemas. In rare cases surgical intervention may be indicated, however, this carries an additional risk of general anaesthesia and post-operative complications such as ileus and infection.

Jackson (2003) has summarised the process of hand-rearing and highlights several of the more common problems. These include heat stress, constipation, diarrhoea, inappetence, iatrogenic trauma through force feeding and aspiration pneumonia.

Puggles grow rapidly (see 3.2). Once weaning weight is reached (800–1300 g) they are slowly introduced to solid food such as termites, ants and fly pupae along with artificial solids such as those included in adult echidna artificial diets (see section 5). Adult food is usually presented as a semi-solid slurry and presents no difficulties in adjustment for the hand-reared puggle.

Puggles have been successfully reared from less than 40 d of age and 150 g body weight. This stage is recognised by the presence of very small (just detectable to the touch) spines (Fig. 5.7). Any abandoned echidna less than 800 g body weight should be suspected as pre-weaning and therefore may not survive without intervention in the form of artificial rearing.

9 DISEASES

9.1 Dermatologic diseases

Apart from the presence of quills, the echidna's skin is not that different from other mammals and is subject to most mammalian skin conditions. Various dermatological conditions have been reported including hyperkeratosis, proliferative dermatitis, bacterial folliculitis, mycotic epidermitis, mycotic folliculitis, acanthosis, alopecia, subcutaneous abscessation, suppurative dermatitis, pustular epidermitis, epidermal hyperplasia and ulcerative pyoderma.

9.1.1 Orthokeratotic hyperkeratosis

Whittington (1993) cites a case of hyperkeratosis in a wild juvenile echidna associated with pox virus particles. Adelaide Zoo investigated a case of spine loss and breakage in an adult. Biopsy identified a chronic proliferative dermatitis of unknown cause characterised by diffuse epidermal thickening, acanthosis, orthokeratosis, pronounced follicular keratosis, moderate dermal fibroplasia and minimal perivascular mixed round cell infiltration (D Schultz *pers. comm.*). Similar cases have been seen at Taronga Zoo and, despite extensive investigations (skin scrapings, tape preparations, biopsies, bacterial and fungal cultures, allergy skin testing and epidemiological investigation), no specific aetiology has been identified. Epidemiologically, dietary or seasonal factors were unlikely. Histological skin changes in these animals included moderate to extensive epidermal orthokeratotic hyperkeratosis, multifocal pustular epidermitis, focal suppurative dermatitis, multifocal dermatocytosis, multifocal mycotic epidermitis (*Microsporum gypseum* and *Trichoderma viride*), marked basketweave hyperkeratosis, acanthosis, bacterial folliculitis, marked epidermal hyperplasia and mycotic folliculitis (*Alternaria* sp.). Small numbers of burrowing mites were seen in some cases, but their significance was doubtful (ARWH 2007).

Numerous treatments have been used (systemic and topical antibiotics and antifungal agents, systemic and topical acaricides, antiseptic washes and rinses, dietary omega 3 and 6 fatty acid and zinc supplements). The animals recovered regardless of what was used. Some cases resolved without treatment (L Vogelnest *pers. comm.*). The skin changes are probably a final common pathway for many conditions and there are likely to be various causes associated with these cases. In spite of the numerous dermatological cases reported in captive echidnas, these conditions appear to be rare in wild echidnas.

9.1.2 Ectoparasites

Some skin diseases have been associated with ectoparasitism including mites (*Odontacarus echidnus*) and ticks (Genera *Ixodes*, *Haemaphysalis*, *Amblyomma*, *Aponomma*). The echidna tick, *Aponomma concolor*, is common and may be associated with anaemia and dermatitis (Rose 1999). Griffiths (1989) suggests tick burdens of 20 or more may result in anaemia. The external ear canal is a common site for the attachment of ticks. Ivermectin (0.2 mg/kg SC), fipronil (at domestic companion animal dose rates) and selamectin (Revolution, Pfizer Australia) are safe and effective treatments (L Vogelnest *pers. comm.*).

9.1.3 Sparganosis

Large subcutaneous swellings have been reported in echidnas from Victoria, South Australia and New South Wales (Whittington et al. 1992; ARWH 2007; AWHC 2006; Booth 1999). These masses were due to the presence of the larval stages (plerocercoids) of the pseudophyllidean cestode *Spirometra erinacei*. The plerocercoids were associated with an intense inflammatory response characterised by fibrous tissue and an inflammatory infiltrate dominated by plasma cells, lymphocytes and focal suppuration (Whittington et al. 1992). Echidnas become infected by ingesting copepods containing the intermediate first larval stage or by ingesting paratenic hosts containing a plerocercoid. Drinking water is the most likely source of such infections. Raw meat could also be a source. Unlike other animals, significant clinical lesions can develop in monotremes. Sparganosis is rarely seen in captive echidnas. Treatment can be difficult, but surgical removal of lesions can be attempted if not too far advanced (Fig. 5.8).

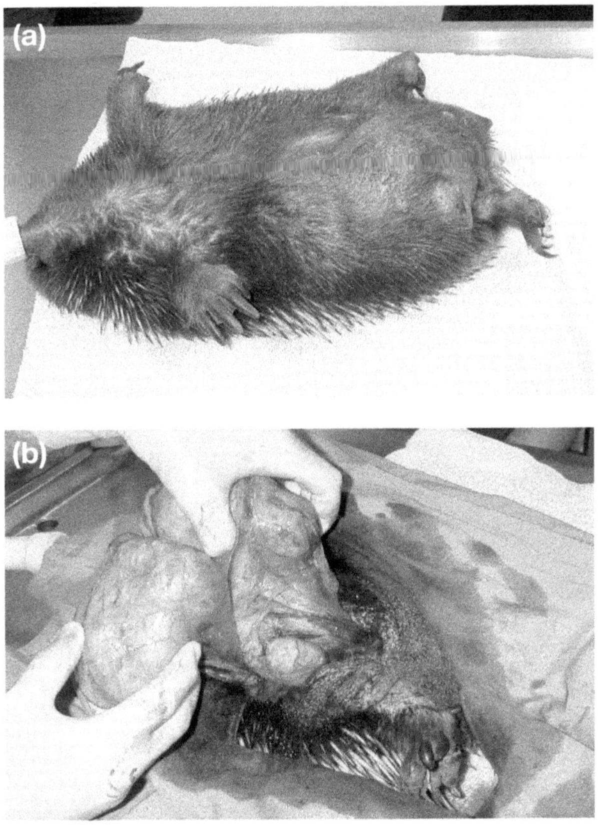

Figure 5.8 a) An anaesthetised wild echidna with a large subcutaneous sparganum. b) Surgical removal of the sparganum.

9.1.4 Trauma

This is discussed further in 9.8.1. The strength and persistence of some echidnas' attempts to escape from enclosures may lead to skin abrasions on the body, feet or beak. These frequently become secondarily infected. Treatment requires removing the animal into an environment in which it feels more comfortable and with no objects or abrasive surfaces on which it may rub. The presence of quills makes access to focal lesions difficult and in some cases quills will need to be cut at the level of the skin surface with side-cutting pliers. Quills will eventually be replaced by natural shedding and regrowth but this will take months or even years. Bandaging of feet and topical or systemic treatment may be needed.

9.1.5 Botryomycosis

Botryomycosis, a chronic granulomatous infection caused by bacteria, has been reported in an echidna at Taronga Zoo. The animal presented with soft tissue swelling and necrosis of a hind limb digit. Histologically there was ulceration and suppuration with typical Splendore-Hoeppli reaction and Gram positive cocci were present deep in the dermis. Both *Staphylococcus aureus* and a beta haemolytic *Streptococcus* were cultured, sensitive to multiple antibiotics. After a poor response to treatment with enrofloxacin at 5 mg/kg sid SC then PO for 2 wk, treatment was changed to amoxycillin/clavulanic acid at 20 mg/kg IM sid then PO bid for 4 wk. Other treatment included meloxicam at 0.2 mg/kg SC then PO and bandaging. This treatment resulted in complete resolution without recurrence (Taronga Zoo records).

9.2 Gastrointestinal diseases

A normal wild echidna's faeces is well-formed and contains visible soil and exoskeletal components of prey items. When echidnas are fed captive diets in which these environmental components are lacking, faeces are less formed, tending to be more liquid. A variety of gastrointestinal disorders have been reported including diarrhoea, constipation, motility disorders and problems associated with infectious diseases.

9.2.1 Diarrhoea and constipation

Diarrhoea has been associated with dietary imbalance, bacterial infections and the presence of endo-parasites including protozoa such as *Eimeria tachyglossi* and *Eimeria echidnae* (see 9.2.2). Acute salmonellosis is relatively common, at least in captive echidnas, with clinical signs varying in intensity from haemorrhagic diarrhoea to generalised weakness and sudden death (Whittington 1993; McOrist & Smales 1986). Cases diagnosed early may respond to oral or parenteral antibiotic therapy such as oral sulfadimidine and sulfadiazine or injectable sulfadoxine at companion animal dose rates. Additional supportive care such as parenteral and oral fluids and antidiarrhoeal preparations containing kaolin and pectin are of use.

Constipation has been seen in hand-reared puggles and is probably associated with adaptation to captive diets and their administration protocols. Whittington (1993) records that ulceration, gastritis, distension, caseation and congestion of viscera in both juvenile and adult echidnas have been associated with fungal and bacterial infections due to *Candida, Corynebacterium* and *Fusobacterium*, whose presence may be encouraged by high-carbohydrate foods and and/or overzealous antibiotic therapy.

Environmental hygiene and minimisation of captive stress such as transport and crowding may reduce the incidence of such infections.

9.2.2 Coccidiosis

Coccidiosis is one of the most significant diseases of the short-beaked echidna. It has been diagnosed in long-term captive animals, rescued wild animals under treatment and rehabilitation and wild animals.

Eimeria tachyglossi, E. echidnae and *Octosporella hystrix* are coccidian parasites of the echidna (Barker et al. 1985; Booth 1999; Whittington 1993). Whittington (1993) reported that at worst the coccidia caused a mild focal enteritis and debated the significance of these infections, however, coccidiosis is now recognised as potentially causing fatal disseminated disease in captive echidnas (Rose 1999). Enteric coccidiosis does occur in wild echidnas (AWHC 2006).

Many healthy echidnas shed coccidial oocysts in their faeces. Coccidiosis may present as a spectrum from mild focal enteritis to disseminated terminal disease. Affected animals may be depressed, lethargic and anorexic, with or without diarrhoea. Coccidiosis can therefore represent a serious and difficult clinical threat. Commonly, animals are found dead without premonitory signs.

At necropsy, in the enteric form of the disease there may be haemorrhagic and necrotising enteritis. The intestinal mucosa is often thickened due to the presence of large numbers of coccidial micro- and macrogametocytes and oocysts (Fig. 5.9), with villous atrophy

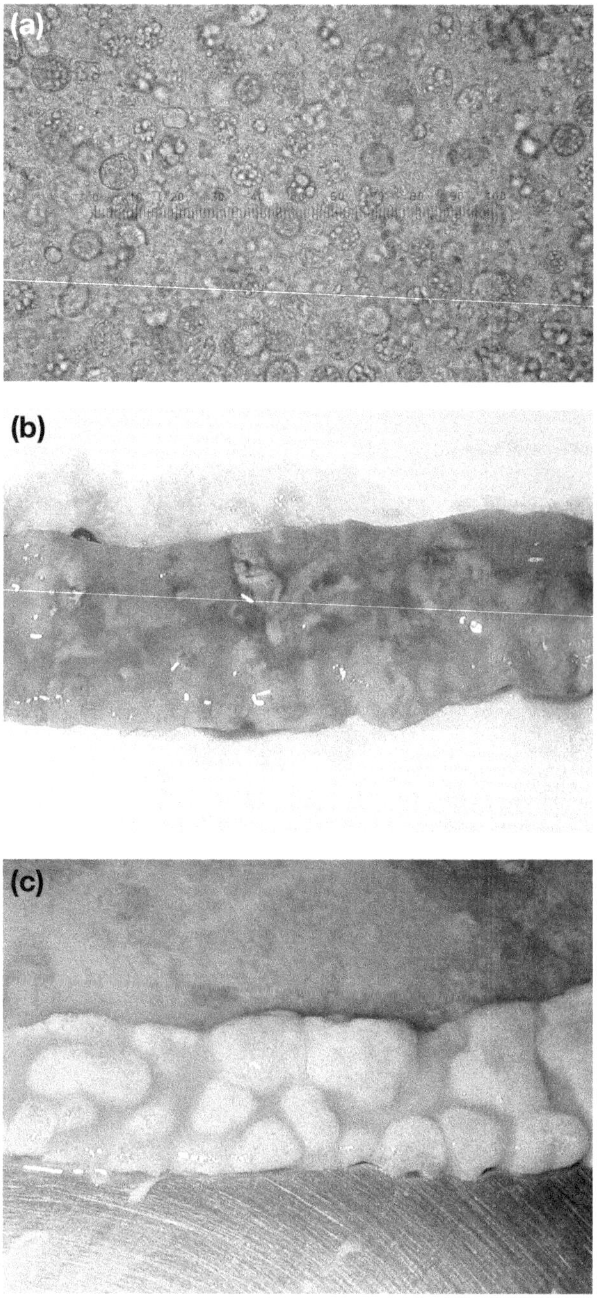

Figure 5.9 a) 400x magnification of a faecal flotation sample from a case of severe coccidial enteritis in a sub-adult hand-reared echidna. b) The lumen of the small intestine with significant epithelial damage. c) Raised areas within the small intestinal mucosa due to accumulations of coccidia.

and mononuclear infiltration of the lamina propria (Rose 1999). In the disseminated form of the disease the various coccidian life stages are found scattered throughout the endothelium and parenchyma of the lung, liver, heart, kidney, spleen and gastrointestinal tract (Dubey & Hartley 1993). There are often foci of necrosis in the spleen, liver and disseminated lymph nodules (Rose 1999). *Atoxoplasma*-like blood parasites may be seen in circulating monocytes in echidnas with disseminated coccidiosis. The association of this parasite with systemic coccidiosis is yet to be established (Rose 1999).

A number of cases of enteric coccidiosis have had concurrent bacterial septicaemia (*S. typhimurium*, *Fusobacterium* sp.), fungal pneumonia (*Cryptococcus neoformans*), fungal gastritis (*Candida* spp.) and degenerative changes in the kidney.

Diagnosis is made through the examination of faeces under light microscopy by wet preparation or faecal flotation, however, the presence of large numbers of oocytes must be interpreted in the light of clinical signs as many normal echidnas can shed large numbers of oocytes. Large numbers of coccidial oocytes in the faeces of a sick echidna must be regarded as a significant finding. Examination of blood films may be useful to detect the presence of *Atoxoplasma*-like blood parasites in monocytes (Rose 1999).

Once clinically affected, animals usually rapidly deteriorate and die. Any sick echidna shedding large numbers of coccidia in the faeces should be regarded as an emergency and treatment initiated immediately. There are no known effective treatments for coccidiosis in echidnas, but coccidial treatments that are effective in other species should be attempted. A young female echidna presenting with lethargy, depression and severe watery fetid diarrhoea was successfully treated. Faecal examination and culture revealed a heavy burden of coccidia, *Candida albicans* and a *Vibrio* spp. Treatment included toltrazuril at 20 mg/kg PO sid for 2 d, trimethoprim/sulfadiazine at 5 mg/kg of the trimethoprim component IM sid for 14 d, nystatin at 10 000 IU/kg bid for 15 d and enrofloxacin at 5 mg/kg sid PO for 15 d. A repeat faecal examination after 14 d of treatment was negative. The animal recovered and had not relapsed by 6 mo post-treatment. It was not evident which of the three organisms was the primary pathogen (Taronga Zoo records).

Although specific predisposing factors that may lead to the development of clinical coccidiosis have not been identified, it can be assumed that factors such as poor hygiene, stress and concurrent disease may be involved.

Careful attention to enclosure hygiene to reduce reinfection rates is essential. The value of pre-emptive treatment with anti-coccidial agents such as toltrazuril is yet to be established.

9.2.3 Helminthiasis

Helminth endoparasitism has been recorded in many cases. Some involved clinical disease but most are incidental findings during clinical workups or at necropsy. Segments of the cestodes *Echidnotaenia tachyglossi* and *Linstowia echidnae (tasmanica)* are frequently observed in the faeces of rescued and recently captive echidnas. McOrist (1986) detected *L. echidnae* in 9 of 36 wild echidnas examined at necropsy but failed to find cestodes of any type in 15 captive specimens. The cestode *E. tachyglossi* has also been recovered from the small intestine of subclinical cases (Whittington 1992). Fatal trichostrongyloidosis associated with *Nicollina* and *Tasmanema* spp. was diagnosed in 8 out of 36 wild echidnas. McOrist (1986) suggests that nematode burdens of 100 or less can be deemed incidental while those in the region of 500–1000 parasites are considered pathological.

Gastrointestinal helminthiasis may be clinically significant and, if diagnosed, treatment is recommended. Treatment with ivermectin at 0.2 mg/kg SC or PO, fenbendazole at 10 mg/kg and praziquantel at 10 mg/kg PO have been safe and effective.

9.2.4 Miscellaneous

Marked gastric bacterial and mycotic infection has been diagnosed in numerous echidnas suffering multisystemic disease. Emaciation and ultimately death due to massive gastric impaction with sinew from kangaroo mince has been reported in two echidnas. Despite being emaciated, the animals did not lose weight due to the mass of material in their stomachs (ARWH 2007).

9.3 Cardiopulmonary diseases

Echidnas are often erroneously diagnosed with respiratory disease because of excessive amount of froth and bubbling from the external nares. This is normal (Fig. 5.10). Tenacious saliva is produced in large amounts (see 2.2) and inevitably the nostrils, positioned as they are, attract large amounts of foreign fine particulate matter.

Figure 5.10 Normal frothy nasal discharge in a short-beaked echidna. Photo: Taronga Zoo.

Echidnas suffering beak trauma occasionally succumb to septicaemia and necrotising bronchopneumonia. This may be the result of haematogenous spread of organisms from the highly vascular tissues of the beak once secondarily infected, or the aspiration of blood, discharges or food. Pulmonary disease may also be seen in association with or secondary to other diseases.

A hand-reared sub-adult male echidna previously diagnosed with enteric coccidiosis developed a blood-stained frothing nasal discharge. The individual died soon after and histological examination of the lung revealed consolidation, proteinaceous filling of alveoli, inflammatory infiltrate consisting mainly of macrophages, large areas of necrosis associated with thrombosis of blood vessels, bronchioles packed with fibrin and large numbers of bacterial rods and fungal hyphae throughout (AWHC 2006).

Edwardsiella spp. has been isolated from pulmonary tissue of an echidna suffering from suppurative bronchopneumonia (Whittington 1993).

Larval *Ophidascaris robertsi* nematodes have been recorded from the lung of echidnas as a subclinical infection (Whittington 1992). Thoracic and pulmonary sparganosis has also been reported in echidnas (Whittington et al. 1992; ARWH 2007).

9.4 Neurological and musculoskeletal diseases

Few diagnoses of neurological conditions have been made in echidnas, though some disease entities may be associated with neurological signs. Toxoplasmosis has been diagnosed in both wild and captive echidnas (McOrist & Smales 1986) and probably causes similar disease to that seen in other susceptible species. The four individuals cited in the study by McOrist and Smales (1986) had non-suppurative encephalitis, interstitial pneumonia and focal hepatitis, splenitis and nephritis. Eliminating exposure to cat faecal contamination and dietary quality control (raw meat) is necessary to reduce the likelihood of infection.

9.5 Urinary and reproductive diseases

There is little information available on the composition of echidna urine and its relationship to clinical disease. Echidna urine is rarely available for analysis and, if voided during examination, is frequently contaminated with faeces. Kidney diseases have been reported and tubular degeneration has been seen in association with multisystemic disease and bacteremia (AWHC 2006).

9.6 Ophthalmologic diseases

The eyes of the echidna are small and well-protected with strong peri-orbital musculature. This adaptation protects the eyes during foraging but makes examination difficult. Detailed ocular examination requires sedation or anaesthesia. Conjunctivitis and keratitis are relatively common and may be related to foraging habits. Mycotic keratitis has been reported (ARWH 2007). Moderate ocular and peri-ocular trauma was seen in an echidna attempting to scavenge from an open tin can (L Vogelnest *pers. comm.*). Clinical management of ocular disease in echidnas is no different from that of other species, although application of topical preparations is difficult in all but the tamest echidna.

9.7 Infectious diseases

Table 5.5 lists the infections agents identified from short-beaked echidnas

9.7.1 Viral diseases

Herpes virus infection involving multi-systemic disease and inclusion body hepatitis has been reported by Whittington (1992) but appears to be an infrequent finding. Similarly, pox virus and adenovirus inclusion bodies have been reported but the incidence of these infections and their significance remains poorly understood (Whittington 1993).

9.7.2 Bacterial diseases

Several *Salmonella* spp. have been identified as the cause of haemorrhagic enteritis and septicaemia (Rose 1999). Bacterial septicaemia involving *Salmonella* spp. occur commonly in captive echidnas and is frequently caused by to *S. typhimurium,* although McOrist and Smales (1989) also identified *S. bovis-morbificans*, *S. dublin* and *S. saint-paul.*

Multisystemic granulomatous mycobacteriosis mostly involving the liver has been reported. *Staphylococcus* spp., *Streptococcus* spp., *Aeromonas* spp. and *Proteus* spp. have been associated with septicaemia and *Edwardsiella* spp. have been isolated from a suppurative bronchopneumonia (Whittington 1993).

9.7.3 Fungal disease

Fungal (*Candida* spp.) oesophagitis and gastritis has been diagnosed at necropsy in hand-reared juvenile echidnas or those with concurrent disease. Dermatophytes and various other fungi have been identified on the skin of echidnas either associated with pathology or as incidental findings (see 9.1.1). *Cryptococcus neoformans* has been isolated from pulmonary granulomas in an echidna (Rose 1999).

9.7.4 Parasitic diseases

Theileria tachyglossi organisms have been seen on blood smears within erythrocytes or lymphocytes and *Babesia tachyglossi* have been seen within erythrocytes. The clinical significance of these infections is unknown. *Babesia tachyglossi* presents typically with 2–8 organisms per infected erythrocyte, whereas cells infected with *Theileria tachyglossi* usually only have 1 organism per cell. These two parasites are transmitted by ticks. More recently, a protozoan parasite identified as *Hepatozoan tachyglossi* sp. nov. (Haemogregarinidae) has been reported in monocytes. Although the infected echidna was debilitated, the significance of the parasite infection was unknown (Clark et al. 2005). All other records of this genus in marsupials are from erythrocytes. *Atoxoplasma*-like parasites have been observed in an animal suffering from multisystemic coccidiosis, but the relationship between the two is not understood (Rose 1999). *Anaplasma marginale*-like organisms were reported in erythrocytes (Whittington 1993). A microfilarial nematode form, *Cercopithifilaria johnstoni,* has also been recorded from the echidna (Clark 2004).

9.8 Non-infectious diseases

9.8.1 Trauma

In the 10 years to 2006, 87 echidnas were admitted to the Australian Wildlife Health Centre (AWHC) at Healesville Sanctuary in Victoria. Seventy-three were trauma-related, and 43 of the 73 were successfully returned to the wild, which testifies to the echidna's physical strength and resilience.

Motor vehicle accidents account for most traumatic injuries. Other causes of trauma include dog attack, electrocution (electric fences close to the ground), bushfires, entrapment under objects and scavenging from open tin cans.

Fractures other than of the beak are rare, most probably as a result of the thickness of the long bones and the

Table 5.5 A catalogue of infectious agents from the short-beaked echidna

Agent	Comments
Viral	
Herpes virus	Multisystemic, hepatitis, death
Pox virus	Proliferative dermatitis
Adenovirus	Kidney, subclinical
Bacterial	
Salmonella spp.	Septicaemia, fatal, subclinical carrier
Edwardsiella sp.	Bronchopneumonia
Mycobacterium sp.	Generalised chronic fatal
Staphylococcus sp.	Generalised chronic, focal dermal, possibly fatal
Aeromonas sp.	Possible septicaemia, enteritis, peritonitis
Proteus sp.	Possible septicaemia, enteritis, peritonitis
Streptococcus spp.	Septicaemia, focal hepatitis, pulmonary oedema
Fusobacterium sp.	Oesophagitis, gastritis
Corynebacterium sp.	Oesophagitis, gastritis
Fungal	
Microsporum gypseum	Dermatopathy with quill fracture
Trichoderma sp.	Dermatopathy with quill fracture
Candida albicans	Oesophagitis, gastritis
Cryptococcus neoformans	Pulmonary granulomas
Protozoa	
Theileria tachyglossi	Incidental finding, significance unknown
Babesia tachyglossi	Incidental finding, significance unknown
Eimeria tachyglossi	Enteritis, small intestine, possibly multisystemic, fatal
Eimeria echidnae	Enteritis, small intestine, possibly multisystemic, fatal
Octosporella hystrix	Enteritis
Anaplasma marginale like	Incidental finding in normal echidnas
Hepatozoan tachyglossi	Incidental finding in debilitated individual
Atoxoplasma-like	Common incidental finding, possibly associated with multisystemic coccidiosis
Cestodes	
Spirometra erinacei	Subcutaneous and pulmonary masses of chronic nature
Echidnotaenia tachyglossi	Incidental
Linstowia echidnae	Incidental
Nematodes	
Parastrongyloides sp.	Enteritis, possibly fatal with burdens >500 worms
Nicollina spp.	Enteritis, possibly fatal with burdens >500 worms
Tachynema spp.	Enteritis, possibly fatal with burdens >500 worms
Tasmanema mundayi	Enteritis, possibly fatal with burdens >500 worms
Ophidascaris robertsi	Larvae recorded from lung, roundworm of pythons
Dipetalonema sp.	Filaroid recorded from subcutaneous tissue

(Continued)

Table 5.5 (*Continued*)

Agent	Comments
Fleas	All possibly associated with dermatological changes
Echidnophaga spp.	
Pulex irritans	
Bradiopsylla echidnae	
Stephanocircus dasyuri	
Ticks	Anaemia with burdens >20, disease vectors, dermatitis
Ixodes spp.	Possible paralysis
Haemaphysalis humerosa	Normally bandicoots, vector of Q fever
Amblyomma spp.	Mostly from Qld, 1 sp. also WA, NT
Aponomma tachyglossi	Echidna tick
Aponomma concolor	Echinda tick
Mites	Possibly associated with dermatological changes
Odontacarus echidnus	

Sources: Whittington (1992), Spratt et al. (1991), Booth (1993), McOrist (1986), ARWH (2007), Clark et al. (2006), Rose (1999).

echidna's compact muscular body shape. For an echidna to sustain a fracture (other than of the beak) the impact must have been significant and additional injuries should be expected. Observation and external examination of an echidna frequently belie the severity of the trauma it may have sustained—radiographic examination or necropsy frequently reveal several fractures and massive internal injury. Radiographic examination is indicated in all trauma cases. The spines make external examination and palpation of the limbs and abdomen difficult. The stout limbs and strong limb musculature also make palpation difficult and may increase the likelihood of distraction and displacement of bone fragments.

Due to the delicate anatomy of the echidna's beak and its specialised functions, injuries to this area are common and serious. The positioning of the nostrils at the extremity of the beak puts them at risk of damage or avulsion. Injuries to the tip of the beak, if not immediately and functionally repairable, will present feeding problems that may threaten survival. Serious injuries to the tip of the beak and the tongue are indications for euthanasia.

Beak trauma presents particular challenges. Injuries to the structures of the mouth and beak are subject to secondary infection with both bacteria and fungi, and debilitating glossitis and rhinitis have been recorded (AWHC 2006; ARWH 2007). As a general rule, traumatised beaks should be treated conservatively. The following guidelines may be useful (L Vogelnest *pers. comm.*).

- Assess prognosis for recovery. Massive soft tissue injury (including the tongue and beak sinus) and bone damage with instability of the beak, bone exposure (especially if dry or necrotic), significant deviation of the beak, tissue necrosis and infection are all indications for euthanasia. Treatment can be considered if there is minimal or minor deviation of the beak, the beak is stable, skin is intact or skin damage is minor, the tongue is intact, respiration is not compromised and there are no other significant injuries elsewhere in the body.
- Anaesthetise the animal. If beak damage is minimal, this can be done by careful mask induction. With more severe damage, an injectable induction agent should be used (see 6.2). Airway obstruction due to soft tissue swelling, blood or other fluids is a significant concern and careful continuous monitoring of respiration is essential. If airway patency is compromised, nasal catheterisation may be useful (see 6.2).
- Lateral and ventro-dorsal radiographs of the beak should be taken to assess the extent of the bone damage. Full-body radiographs are also indicated to rule out other trauma.
- Repairing the fine bones is difficult and attempts to do so are generally futile.
- Any open wounds should be flushed with copious 0.05% chlorhexidine or saline. All foreign material should be removed.

- Major skin lacerations can be sutured with fine monofilament suture material. Wound breakdown is common.
- Systemic antibiotic cover is essential. Injectable preparations are used initially and oral antibiotics can be introduced once the animal is eating.
- Soft tissue swelling, inflammation and pain must be controlled with the use of non-steroidal anti-inflammatory drugs (meloxicam, carprofen, tolfenamic acid at companion animal doses).
- In severe cases pain may need to be managed using opioids (buprenorphine or butorphanol at companion animal doses).
- Offer food (standard echidna slurry [see section 5]). The animal may not eat for a few days; this is generally not a concern in an otherwise healthy animal. If there is concern about hydration status SC fluids may be administered.
- The animal should be housed in a smooth-sided plastic tub with rounded corners (see 4.1). Echidnas still use their beaks to probe and make serious attempts to escape even if the beak is damaged, often resulting in further trauma.
- Regular monitoring and re-evaluation under anaesthesia is essential.

Fractures of the limb bones or pelvis should be carefully assessed. Significant displacement may be indication for euthanasia. Fracture repair and post-operative management can be difficult. Echidnas will not tolerate bandaging, splint application or casts, removing them quickly with their strong fore limbs. If repair is attempted, standard companion animal techniques are applicable. Open reduction of a radial fracture has been reported (Reynolds & Sommerlad 1996). Conservative treatment of uncomplicated fractures using pain relief and restricting activity is often successful (L Vogelnest *pers. comm.*)

Echidnas are frequently the victims of bushfires. Deep skin burns and beak damage are indications for euthanasia. More superficial burns are managed using standard protocols. Long-term care may be required and burnt spines take many months to be replaced (Fig. 5.11).

Figure 5.11 A short-beaked echidna with superficial burns. Photo: Taronga Zoo.

9.8.2 Neoplasia

There are few records of neoplasia in echidnas. McOrist and Smales (1986) found no cases among 73 echidnas examined at necropsy. A case of lymphoid leukaemia (Booth 1999) is the only report of haematopoietic neoplasia in a monotreme. In this echidna, the lymphocyte count rose from 2.2×10^9/L to 89.9×10^9/L during the course of the disease. There was a concurrent infection of 50% of monocytes and 10% of erythrocytes with *Theileria tachyglossi*. Whittington (1992) reported cases of leiomyosarcoma, myocardial lymphoma and fibroma of the beak. Chondroma or cartilaginous metaplasia of a fibroma, splenic lymphoma, cloacal leiomyoma and cutaneous fibroma have also been reported (ARWH 2007).

10 ZOONOSES

Zoonotic infections associated with echidnas are rare. *Salmonella* spp. are commonly identified from captive echidnas and there is potential for transmission to people. Screening for salmonella in faeces may be recommended in recently hospitalised echidnas as stress is suspected to precipitate shedding of organisms or the development of clinical salmonellosis (McOrist & Smales 1986). Echidnas are quite likely to exhibit projectile defecation if suspended by the hind limbs. Handlers risk faecal contamination if the cloaca is not directed away from people.

11 THERAPEUTICS

No specific pharmacokinetic studies have been conducted in echidnas. The empirical use of drugs based on regimes used in companion animal medicine appear effective and safe in echidnas. No specific adverse reactions have been reported. Table 5.6 lists the therapeutic agents currently in use for echidnas at the Australian Wildlife Health Centre, Healseville Sanctuary.

Table 5.6 Therapeutic agents in current use for echidnas at the Australian Wildlife Health Centre, Healesville Sanctuary

Type	Drug	Dose rate (mg/kg)	Route	Comment
Anti-bacterial	Procaine penicillin	15	IM, SC, sid	In combination
	Benzathine penicillin	10		
	Amoxycillin	12.5	IM. SC, sid	In combination
	Clavulanic acid	3		
	Sulphadiazine	50	IM, SC, sid	In combination
	Trimethoprim	10		
	Metronidazole	20	IV, SC, PO, sid	
	Oxytetracycline	20	IM, sid	
	Doxycycline	5	PO, sid	
	Amikacin			
	Ceftazidime	65	IM, sid	Whittington (1993)
Anti-inflammatory	Ketoprofen	1	IV, IM, SC, sid	
	Dexamethasone	0.2	IM, SC, sid	
	Flunixin	0.5	IM, SC, IV, sid	
	Meloxicam	0.5	IV, SC, sid	
	Meloxicam	0.2mL	PO, sid	
Anti-parasitic	Ivermectin	0.2	IM, SC q 7 d	Repeat twice
	Moxidectin	0.2	IM, SC q 7 d	Repeat twice
	Praziquantel	5	IM, PO, once	
	Fipronil	10	TOP, once	In combination
	S-Methoprene			
Anabolic	Methandriol dipropionate	2.5	IM, q 30 d	In combination
	Nandrolone decanoate	2	IM, q 30 d	
	Stanozolol	15	IM, q 30 d	
Anti-coccidial	Toltrazuril	25	PO, sid 3 d	
Anti-fungal	Itraconazole	5	PO, sid	
Combination	Nystatin	Oint	TOP	In combination
	Neomycin			
	Triamcinolone			
	Thiostrepton			
	Fusidic acid	Cream	TOP	In combination
	Betamethasone			
Analgesic	Butorphanol	0.1	IV, IM, bid	
	Buprenorphine	1	IV, IM, sid	
Dietary Supplement	Phytomenadione	0.1	PO	
	Formic acid	2%	In food at 2%	Of 1% dilution
Miscellaneous	Bromhexine HCl	1	PO tid	Mucolytic
Anaesthetics and sedatives	Isoflurane	2%	Inhalation	See text
	Tiletamine/zolazepam	5	IV, IM, SC	In combination
	Medetomidine	0.1	IM	In combination
	Ketamine	9		
	Atipamezole	0.5	IM	Reversal agent

Source: Additional information from Whittington (1993).

In conscious echidnas SC injections are most easily given on the dorsum, between the spines near the midline. IM injections can be administered in the same area by directing the needle deeper and paravertebral into the epaxial musculature or can be given in the muscles of the hind limbs. SC administration of fluids is not practical in the conscious animal, but in sedated or anaesthetised individuals it can be done via suitable sites on the ventral flanks. IV injections of small volumes can be administered into the beak sinus in anaesthetised patients but access to the cephalic, femoral or jugular veins is necessary for larger volumes. Palatable oral medications can be mixed with food, although with some echidnas appetite is sporadic and there is no guarantee of the entire dose being consumed in one sitting. Medication should not be administered directly into the mouth due to the risk of aspiration.

ACKNOWLEDGMENTS

The author thanks Larry Vogelnest for his contribution to this chapter, Frances Hulst for reviewing the manuscript, and Elizabeth Arthur for assisting with restraint photography.

REFERENCES

Archer M, Flannery TF, Ritchie A & Molnar RE (1985). First Mesozoic mammal from Australia—an early cretaceous monotreme. *Nature* **318**, 363–366.

ARWH (2007). *Australian Registry of Wildlife Health.* Zoological Parks Board of NSW: Sydney.

Augee ML (1995). Short-beaked echidna. In *The Mammals of Australia* (ed. R. Strahan), pp. 40–43. Reed Books: Melbourne.

Augee M & Gooden B (1993). *Echidnas of Australia and New Guinea.* UNSW Press: Sydney.

Augee M, Gooden B & Musser A (2006). *Echidna: Extraordinary Egg-laying Mammal.* CSIRO Publishing: Melbourne.

AWHC (2006). Australian Wildlife Health Centre, Healesville Sanctuary, Melbourne.

Barker IK, Beveridge I & Munday BL (1995). Coccidia (*Eimeria tachyglossi* nsp., *E. echidnae* nsp. and *Octosporella hystrix* nsp.) in the echidna, *Tachyglossus aculeatus. Journal of Protozoology* **32**, 523–525.

Bentley PJ & Schmidt-Nielsen K (1967). Role of the kidney in water balance of the echidna. *Comparative Biochemistry and Physiology* **20**, 285–290.

Booth R (1999). Care and medical management of monotremes. In *Wildlife in Australia: Healthcare and Management, Proceedings 327,* pp. 41–52. Post Graduate Foundation in Veterinary Science: Sydney.

Booth R (2003). Monotremata (Echidna, Platypus). In *Zoo and Wild Animal Medicine* (5th edn) (eds ME Fowler & RE Miller), pp. 278–287. WB Saunders: St Louis.

Clark P (2004). *Haematology of Australian Mammals.* CSIRO Publishing: Melbourne.

Clark P, Holz P & Spratt D (2005). *Hepatozoan tachyglossi sp.* nov. (Haemogregarinidae), a protozoan parasite from the blood of a short-beaked echidna, *Tachyglossus aculeatus. Transactions of the Royal Society of South Australia* **129**, pp. 49–52. Royal Society of South Australia: Adelaide.

Diener E, Ealey EHM & Legge JS (1967a). Autoradiographic studies on the lymphoid system of the Australian echidna *Tachyglossus aculeatus. Immunology* **13**, 339–347.

Diener E, Wistar E & Ealey EHM (1967b). Phylogenetic studies on the immune response II. The immune response of the Australian echidna *Tachyglossus aculeatus. Immunology* **13**, 329–337.

Dubey JP & Hartley WJ (1993). Disseminated coccidiosis in short-beaked echidnas (*Tachyglossus aculeatus*) from Australia. *Journal of Veterinary Diagnostic Investigation* **5**, 483–488.

Dyne GR & Walton DW (eds) (1984). *Fauna of Australia.* General articles, vol. 1A. AGPS: Canberra.

Griffiths M (1973). Observations of the comparative anatomy and ultrastructure of mammary glands and on the fatty acids of the triglycerides in platypus and echidna milk fats. *Journal of Zoology* **169**, 255–279.

Griffiths M (1978). *The Biology of Monotremes.* Academic Press: New York.

Griffiths M (1989). *Tachyglossidae.* In *Fauna of Australia,* vol. 1B (eds DW Walton & BJ Richardson), ch. 15. CSIRO Publishing: Melbourne.

Griffiths M, Green B, Leckie R, Messer M & Newgrain K (1984). Constituents of platypus and echidna milk, with particular reference to the fatty acid complement of the triglycerides. *Australian Journal of Biological Sciences* **37**, 323–329.

Grigg G, Beard L & Augee ML (1990). Thermal relations of free-living echidnas during activity and in hibernation in a cold climate. In *Platypus and Echidnas* (ed. ML Augee). Royal Zoological Society of New South Wales: Sydney.

Jackson S (2003). *Australian Mammals: Biology and Captive Management.* CSIRO Publishing: Melbourne.

Johnston SD, Madden C, Nicholson V, Cowen G, Pyne M & Booth R (2006a). Venipuncture in the short-beaked echidna. *Australian Veterinary Journal* **84**, 66–67.

Johnston SD, Madden C, Nicholson V, & Pyne M (2006b). Identifying the sex of short-beaked echidnas. *Australian Veterinary Journal* **84**, 63–65.

Johnston SD, Nicolson V, Madden C, Logie S, Pyne M, Roser A, Lisle AT & D'Occhio M (2007). Assessment of reproductive status in male echidnas. *Animal Reproduction Science* **97**, 114–127.

Mackenzie WC (1919). *The genito-urinary system in monotremes and marsupials.* Australian Institute of

Anatomical Research, Melbourne. Jenkin, Buxton & Co: Melbourne.

McOrist S & Smales L (1986). Morbidity and mortality of free-living and captive echidnas, *Tachyglossus aculeatus* (Shaw), in Australia. *Journal of Wildlife Diseases* **22(3)**, 375–380.

Menkhorst P (ed.) (1995). *Mammals of Victoria: Distribution, Ecology and Conservation.* OUP: Melbourne.

Menkhorst PW (2001). *A Field Guide to the Mammals of Australia.* OUP: Melbourne.

Oyarzun SE, Crawshaw G & Valdes E (1996). Nutrition of the tamandua: I. Nutrient composition of termites (*Nasutitermes* spp.). and stomach contents from wild tamanduas (*Tamandua tetradactyla*). *Zoo Biology* **15**, 509–524.

Proske U (1990). Electric monotreme. *Australian Natural History* **23**, 288–295.

Reynolds WT & Sommerlad SF (1996). Open reduction of a radial fracture in a short-beaked echidna. *Australia Veterinary Practitioner* **26**, 178–179.

Rissman RA (2000). Chemical immobilisation of the short-beaked echidna (*Tachyglossus aculeatus*): comparative physiological effects of seven different anaesthetic agents in echidna (alphaxalone, isoflurane, ketamine, medetomidine, medetomidine-ketamine, tiletamine-zolazepam and xylazine). MVetSci thesis (Wildlife Medicine and Husbandry). University of Sydney, Australia.

Rose K (1999). Common diseases of urban wildlife. In *Wildlife in Australia: Healthcare and Management, Proceedings 327,* pp. 365–430. Post Graduate Foundation in Veterinary Science: Sydney.

Schmidt-Nielsen K, Dawson TJ & Crawford EC (1966). Temperature regulation in the echidna (*Tachyglossus aculeatus*). *Journal of Cellular Physiology* **67**, 63–71.

Spratt D, Beveridge I & Walter E (1991). A catalogue of Australasian monotremes and marsupials and their recorded helminth parasites. *Records of the South Australian Museum Monograph Series* **1**, 1–105.

Vogelnest L (1999). Chemical restraint of Australian native fauna. In *Wildlife in Australia: Healthcare and Management, Proceedings 327,* pp. 149–187. Post Graduate Foundation in Veterinary Science: Sydney.

Whittington RJ (1988). The monotremes in health and disease. In *Australian Wildlife, Proceedings 104,* pp. 727–787. Post Graduate Foundation in Veterinary Science: Sydney.

Whittington RJ (1992a). Role of infectious diseases in the population biology of monotremes. In *Platypus and Echidnas* (ed. ML Augee). Royal Zoological Society of New South Wales: Sydney.

Whittington RJ (1992b). The impact of disease on monotreme populations. In *Proceedings of the Wildlife Disease Association, Australasian Section*, pp. 19–22. Wildlife Disease Association, Australasian section: Warrumbungle National Park.

Whittington RJ (1993). Diseases of monotremes. In *Zoo and Wild Animal Medicine. Current Therapy 3* (ed. ME Fowler), pp. 269–275. WB Saunders: Philadelphia.

Whittington R, Middleton D, Spratt DM, Muntz F, Carmel B, McCracken H, Strakosch MR, Stephanson-Shaw J, Harper PAW & Hartley WJ (1992). Sparganosis in the monotremes *Tachyglossus aculeatus* and *Ornithorynchus anatinus* in Australia. *Journal of Wildlife Diseases* **28**, 636–640.

Wombaroo (2006). Wombaroo Food Products Information Booklets. POBox 151, Glen Osmond, SA, 5064.

Wood Jones R (1923). *The Mammals of South Australia, Part 1.* RFF Rogers, Government Printer: Adelaide.

Wronski EV, Woods GM & Munday BL (2003). Antibody response to sheep red blood cells in platypus and echidna. *Comparative Biochemistry and Physiology—Part A: Molecular and Integrative Physiology* **136**, 957–963.

6. Platypuses

Rosemary Booth and Joanne Connolly

1 TAXONOMY, DISTRIBUTION AND BIOLOGY

The platypus is a highly specialised semi-aquatic, burrowing, carnivorous egg-laying mammal that is so well-adapted that it has survived relatively unchanged since the time of the dinosaurs. The platypus belongs to the Order Monotremata, the egg-laying mammals. The monotremes ('one hole') possess a common opening, the cloaca, for the reproductive, urinary and digestive systems, and the testes are abdominal (testicond). The Family Ornithorhynchidae contains one extant species, the platypus (*Ornithorhynchus anatinus*).

Platypuses are sexually dimorphic. Males are larger than females (45 – 63 cm long and weigh 1000 – 3000 g, females are 39 – 55 cm long and weigh 700 – 1750 g) and can be distinguished from adult females by the presence of a spur on the medial side of the tarsus (Grant 1989; Connolly & Obendorf 1998). Bergmann's rule applies to the platypus, with larger animals found in the cooler parts of their range (Dunn 1949).

Monotremes are long-lived for small mammals and have been maintained for up to 21 years in captivity. Grant (2004a) recorded a female platypus surviving 21 years in the wild.

Platypuses occur in freshwater streams and lakes along the eastern seaboard of Australia from Cooktown to Tasmania, mainly east of the Great Dividing Range (west only in permanent rivers) and were introduced to Kangaroo Island in 1940. They occupy a wide range of habitats, from tropical to cool temperate and pristine to degraded, and are classified as common but are potentially vulnerable to environmental perturbation (Grant & Temple-Smith 1998). Prior to their protection, thousands of platypuses were killed for the fur trade. Legal protection of platypuses was first given in Victoria in 1892, with all other states following suit by 1912 (Grant 1995).

Platypuses have a solitary nature (except when breeding or suckling) and are mainly nocturnal, but may be crepuscular during the winter. In some locations they are regularly diurnal. They depend on relatively undisturbed stream banks to support their resting and nesting burrows and show site fidelity with riverine home ranges of up to 7 km (Gardner & Serena 1995). Juvenile platypuses disperse, moving away from their natal stream.

2 ANATOMY AND PHYSIOLOGY

2.1 External features

The platypus has a streamlined body that is compressed dorsoventrally and is covered by fur except for the bill and feet. Platypuses moult seasonally with patchy loss of guard hairs. Females may also lose significant hair on the dorsal surface of the tail during the breeding season due to pugging of burrows (the process of building a mud partition within the tunnel). The platypus snout is covered by soft leathery pigmented skin (Fig. 6.1). The dorsal nostril position allows the platypus to breathe while most of its body is underwater.

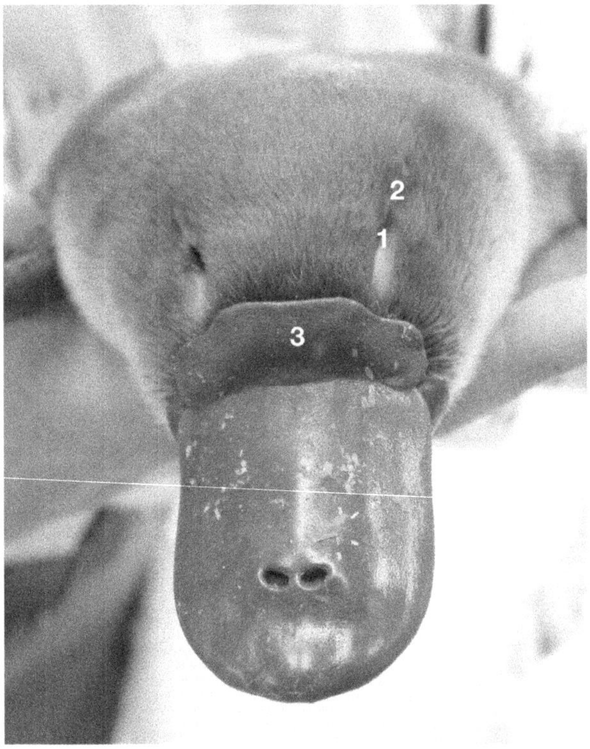

Figure 6.1 The head of the platypus. The eyes (1) and ears (2) are closed. The pores in the skin of the bill and frontal shield (3) are the sites of the electro-receptors and mechano-receptors. Photo: Taronga Zoo.

Both eyes and ears are in a groove which closes when diving. Platypuses have poor vision and their eyes are tilted up to scan for predators. A nictitating membrane is present (unlike the echidna). Their hearing is acute. Their ears have an obvious external ear canal, but little pinna development.

There are two pairs of pentadactyl limbs, with claws and webbing on the digits of the manus and pes. The pes is turned caudolaterally. Webbing on the fore limbs extends beyond the claws to form fan-shaped paddles for swimming. When walking or burrowing, the webbing extensions are folded back under with the claws out. The hind feet act as a rudder during swimming and an anchor during burrowing. The platypus has a lizard-like gait on land.

2.2 Venom gland and spurs

Male platypuses possess a crural system consisting of a venom gland on the dorsal aspect of the thigh, connected by a duct to a hollow keratinous spur on the tarsus. The spurs are used offensively by bringing them together in the ventral midline with powerful jabbing movements. Venom is pumped through a fine barely perceptible duct in the centre of the spur. The size and activity of the gland is seasonal, consistent with androgenic control. Venom appears to be produced only by sexually mature males from about 4 yrs of age. Aggressive use of the spurs in fights between males is confined to the breeding season. Reports of dogs being spurred shows that male platypuses can use their spurs to defend themselves from predators (Burrell 1927).

The toxin, of which a major component is a natriuretic peptide, can cause death in mice, dogs and captive platypuses. It causes severe and prolonged pain, oedema and localised tissue necrosis in humans, but no fatalities have been reported (de Plater et al. 1995; Fenner et al. 1992). Up to 1994 there were 15 recorded cases of human envenomation and the Commonwealth Serum Laboratory received at least one inquiry regarding treatment annually. There is currently no available antivenene (Tonkin & Negrine 1994). Supportive therapy in people has included ice, elevation, splinting, opioid IV infusion, regional nerve blocks, prednisolone and/or antibiotics. See section 8.1 for prevention by correct handling.

Juvenile females have a rudimentary spur sheath on the tarsus which is lost at around 8 mo age. There is no crural gland in females. The size and shape of the spur can be used to estimate the age of male platypuses (Fig. 6.2).

Venom can be collected from adult male platypuses (>4 yr) during the breeding season for research purposes. In some cases the venom drips from the opening in the spur. It is also possible to gently milk it down from the crural gland at the back of the hind leg. Venom can be collected into microhaematocrit tubes or via a 2 mL syringe connected to a section of drip tubing connected to the spur, to draw out up to 0.05 mL (L Vogelnest *pers. comm.*).

2.3 Skeleton

The platypus skull has a large braincase to house the well-developed forebrain (which lacks sulci and gyri, and a corpus callosum between hemispheres). A reptilian feature of the platypus skull is the ectopterygoid bones. There are no tympanic bullae. The vertebral column of the platypus comprises 49 bones (7C 17T 2L 2S 21Co). The cervical vertebrae bear cervical ribs and each thoracic rib has ossified sternal and vertebral portions, with cartilage between the two, like reptiles. The pectoral girdle of the platypus provides a sturdy base for the digging action of the fore limbs and is more like

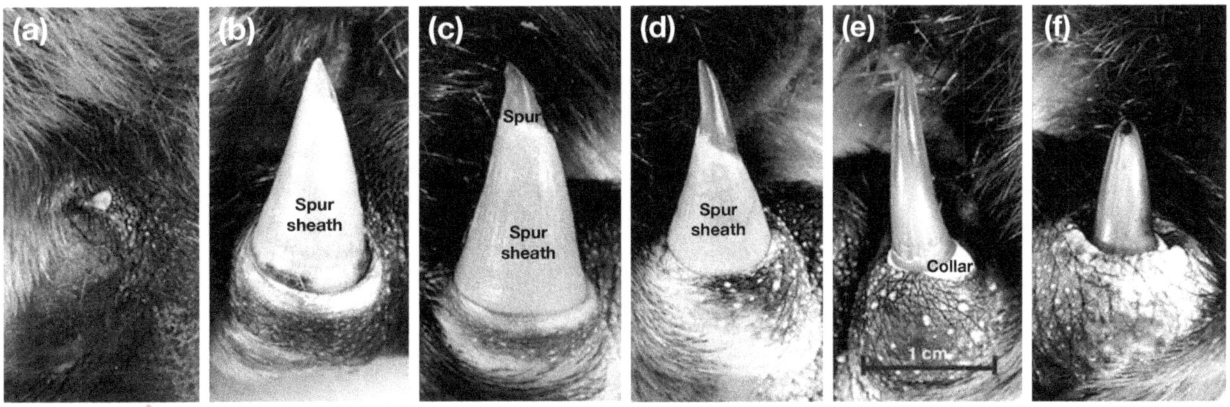

Figure 6.2 The presence of spurs in platypuses confirms age and sex. a) Subadult female, 3–8 mo. b) Juvenile male, 0–6 mo. c) Juvenile male, 6-9 mo. d) Subadult male, 9–12 mo. e) Adult male, >2 yr. f) Old adult male. Photo: P Temple-Smith.

that of lizards. It consists of paired scapulae, clavicles, coracoids, epicoracoids and a T-shaped interclavicle attached to the sternum. The pelvic girdle is typically mammalian with paired ilia, pubes and ischia, but there are also epipubic bones (as in marsupials) projecting forward from pubic bones (Griffiths 1978).

2.4 Digestive system

The most remarkable features of the platypus' digestive system are the bill, dentition, mouth and tongue, all of which allow thorough mastication of prey. The digestive tract itself is simple, reflecting the nutrient-rich food source and ample water supply consumed during feeding.

The skin of the platypus' bill contains over 100 000 electro-receptors and mechano-receptors, first reported by Scheich et al. 1986, which allows it to locate prey underwater with its eyes, ears and nostrils closed. The bill's sensory representation in the brain is nearly three-fold that of the visual and auditory representation (Pettigrew et al. 1998). The platypus bill receives its massive innervation from the well-developed trigeminal nerve (cranial nerve V). Electro-receptors are large sensory mucous glands that are distributed over the entire surface of the bill, including the frontal shields (Gregory et al. 1989) (Fig. 6.1). These receptors are extremely sensitive and allow detection of the electric field generated by the single tail-flick of a shrimp at a distance of 10 cm (Scheich et al. 1986). Mechano-receptors are pushrods which respond to tactile stimulation.

Adult platypuses do not have teeth. In early life they possess 1 premolar and 2 molars in each maxilla and 2–3 molars in each mandible. These small teeth are resorbed and replaced by keratinous grinding pads soon after the young platypuses emerge from the nesting burrow (Fig. 6.3). Despite the lack of true teeth, mastication is a significant component of monotreme digestion. Insect and crustacean prey are collected into the cheek pouches and thoroughly ground and sifted to remove much of the exoskeletons.

The tongue of the platypus has specialisations to direct food to and from the cheek pouches, from where it is repeatedly ground and filtered so that only the more digestible components of the diet are swallowed (Fig. 6.3).

The remainder of the digestive system is simple. The stomach of platypuses is lined with cornified stratified epithelium to further grind insect prey. The stomach is very small and receives ground and filtered material from the cheek pouches. This is significant during hand-rearing or tube-feeding, where volumes fed should be conservative (see 4.4). A small (2.5 cm) caecum is located at the ileocolic junction (Whittington 1988; Connollly et al. 1999a). Because of the readily assimilated ingesta, the gut transit time is short. A barium study of a mature healthy captive platypus revealed a transit time of about 5 hr (Fig. 6.4).

2.5 Cardiopulmonary system

The platypus has a four-chambered heart with a closed double circulation. The systemic aorta is derived from the left fourth arterial arch. Great cardiac veins open directly into the right atrium. A coronary vein, which is not present in other mammals, also empties directly into the right atrium. A pelvic rete mirabile is present (see 2.9).

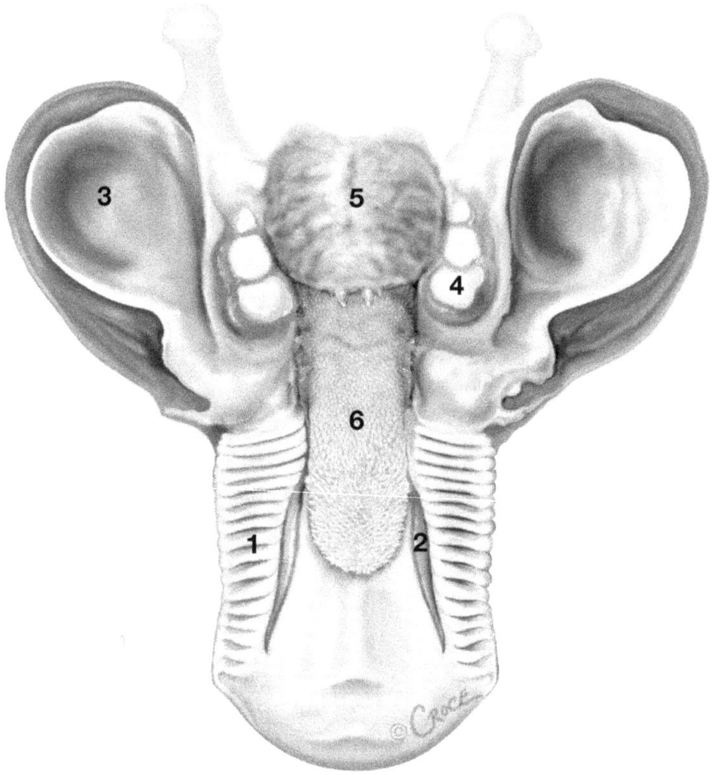

Figure 6.3 The bottom jaw and some of the oral specialisations of the platypus. 1) Filtering serrations on bottom jaw. 2) Bilateral secateuring ridges in top and bottom jaw. 3) Cheek pouches. 4) Keratinous grinding pads. 5) Torus linguae. 6) Body of tongue.

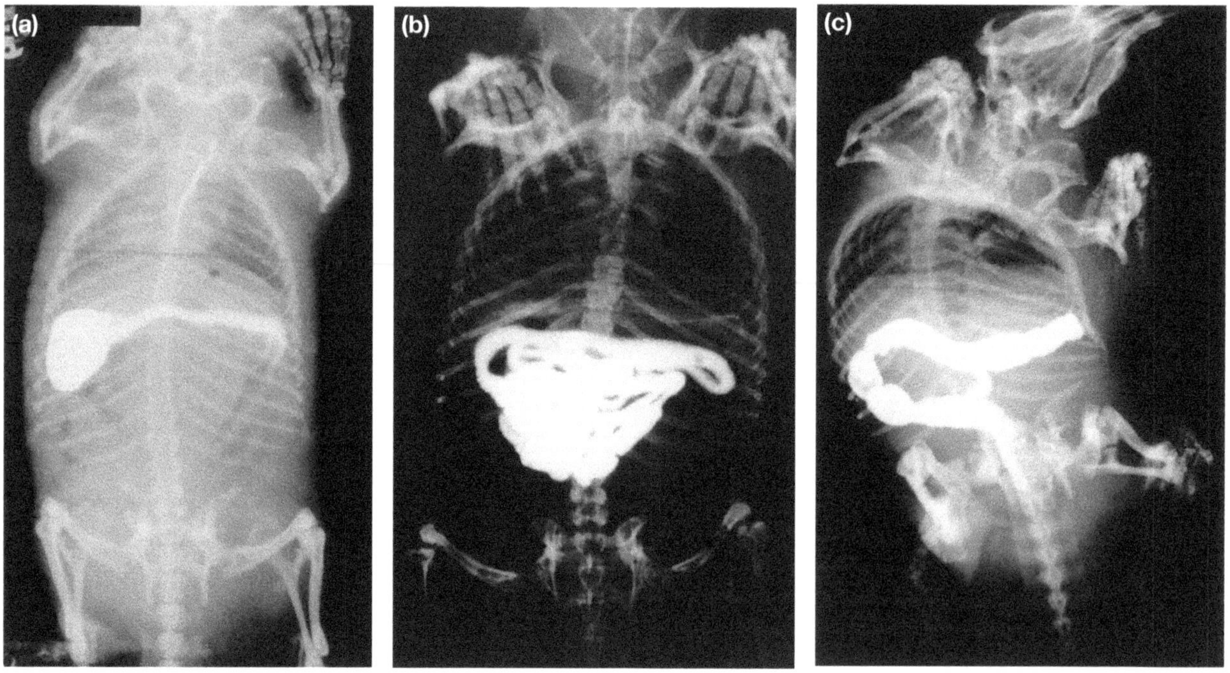

Figure 6.4 Barium study of a conscious captive platypus (DV views). a) Time 0 showing barium in stomach already entering small intestine. b) 1 hr, showing the barium in the small intestine. c) 5 hr later, showing barium in transverse and descending colon. Photos: R. Booth.

Platypus lungs are alveolar and have the typical mammalian tree-like branching system of intrapulmonary bronchi, with one lobe on the left and two lobes on the right. A diaphragm is present and is functional in respiration.

2.6 Urogenital system

Platypuses have well-developed, non-lobulated, bean-shaped ureotelic kidneys (urea, not uric acid, is the primary product of nitrogen metabolism). Ureters drain into a common urogenital sinus (opposite the neck of the bladder) which in turn drains into a cloaca, along with the rectum. A bladder is present as a diverticulum from the ventral surface of the urogenital sinus which, unlike higher mammals, does not receive the ureters.

The female platypus has paired ovaries, but only the left ovary is functional (the right is rudimentary, as in birds). The ovary bears many large follicles projecting from its surface, as in reptiles. Each ovary is enclosed by the infundibulum of the lightly convoluted oviduct. Each oviduct leads to a separate uterus, which opens independently into the urogenital sinus, thence the cloaca. The platypus is oviparous, laying 1–3 leathery eggs, 14 × 17 mm in size and weighing 1.5–2 g (Griffiths 1978; Grant 1995).

The male platypus is testicond, the abdominal testes lie caudo-dorsal to the kidneys. Seminiferous tubules in the testis drain via efferent ducts to the large epididymis and via a short vas deferens to the rostral end of the urogenital sinus. The penis lies in the caudal urogenital sinus, within a preputial sac ventral to the cloaca. The erect penis is about 7 cm long, and is extruded through the cloaca and the cloacal sphincter. The shaft of the penis bears spines and each half of the bifid glans has four evertible foliate papillae. The penile urethra, which communicates with the urogenital sinus, carries only semen (not urine). There are no ampullae, coagulating glands or seminal vesicles. Also, the prostate is disseminate rather than discrete, and bulbourethral glands are present at the base of the penile urethra (Carrick & Hughes 1978; Griffiths 1978).

2.7 Immune system

The platypus has the full repertoire of lymphoid tissues present in mammals, including a grossly visible spleen, thymus and lymphoid nodules (0.2−2 mm in cervical, thoracic, pelvic and mesentery sites), as well as histologically detectible gut-associated lymphoid tissue (tonsils, Peyer's patches and caecal lymphoid tissue) and bronchus-associated lymphoid tissue (Connolly et al. 1999a). These lymphoid tissues contain all the key elements (T and B lymphocytes and plasma cells) to mount an effective immune response against foreign antigens.

IgM, IgG (two subclasses), IgA (two subclasses) and IgE heavy chains have been isolated from the platypus (Belov & Hellman 2003). At this stage it is not known whether the platypus has IgD. Immunoglobulins have been purified from platypus serum and used to detect the immune response to *Mucor amphibiorum* (Whittington et al. 2002).

2.8 Endocrine and exocrine systems

The pituitary of the platypus is a large pear-shaped body lying adjacent to the tuber cinereum of the hypothalamus. It contains two separate endocrine glands, the pars nervosa and the pars anterior or distalis (Griffiths 1978).

The normal thyroid gland in the platypus has been described as a reddish flattened body, 2 cm × 1.75 cm, lying just proximal to the arch of the aorta in the thoracic cavity (McKenzie 1921). The parathyroids are 3 mm × 2 mm and sit either side of the proximal trachea just distal to the thyroid cartilage. The thymus lies around the base of the heart and extends onto the ventral surface of the pericardium, and is paler than the thyroid.

The paired adrenal glands sit anterior to the kidneys and measure 18 mm × 8 mm. Chromaffin tissue is located at the broader caudal end of the pear-shaped adrenal gland, with cortical tissue at the anterior end (Griffiths 1978). Platypuses have large adrenal glands in proportion to body size. The adrenal gland:body weight ratio for the platypus is 260 mg/kg compared to 40 mg/kg in echidnas (McDonald 1978). The reason is unknown, but large adrenal glands would support adaptation to change.

Platypus skin has abundant eccrine and apocrine sweat glands, but these are thought to be ineffective in evaporative cooling due to the presence of the dense fur (Griffiths 1978). Sweating from the poorly furred ventral side of the tail has been observed in the platypus in the laboratory (Dawson 1983).

There is a subcutaneous elliptical, translucent yellow, prescapular scent gland which is used for territorial marking. Captive males have been observed rubbing their neck and shoulders on tussocks in their exhibits during the breeding season. The size and activity of the gland is seasonal, consistent with androgenic control, and the gland is larger in males than females.

Platypus mammary glands lack teats but have two areas called areolae (milk patches), onto which the ducts of about 150 mammary lobules open independently and where the young suckles the milk. There are marked seasonal changes in the anatomy and histology of the mammary glands. At peak lactation, the platypus mammary glands are palpable extending from the axilla to the groin (Griffiths 1978).

Other exocrine glands noted in the platypus include paired bulbourethral and disseminate prostatic glands in the male reproductive tract; uterine glands in the female reproductive tract; Brunner's glands, intestinal glands and an exocrine pancreas in the digestive system; and crural glands in the male.

2.9 Physiological adaptations for diving and the aquatic environment

Platypuses have a number of anatomical and physiological adaptations to diving. When feeding, repeated short dives of 30 sec–3 min duration are made (Griffiths 1978; Grant 1995). Early studies of involuntary submersion in platypuses showed rapid changes and restoration of arterial oxygen and carbon dioxide. Blood pH fell along with the rise in carbon dioxide during submersion. Recovery in arterial oxygen levels occurred in 2 min, although restoration of pH to pre-submersion levels took 9–10 min.

Bradycardia develops gradually during submersion, and there is sudden release of cardiac inhibition on surfacing (Griffiths 1978). The normal heart rate when swimming on the surface is 140–230 beats/min, compared with 10–120 beats/min during forced dives (Grant 1995).

The platypus is able to thermoregulate in an aquatic environment even in alpine regions. Its thermoregulatory abilities at high temperatures are poor. It has adaptive mechanisms to reduce heat loss and thus minimise energy expenditure (Dawson 1983). The platypus has an extensive pelvic network (rete mirabile) of small parallel arterial and venous vessels that supply the muscles of the hind limbs and tail (Grant 1989). This vascular specialisation is believed to be a mechanism of counter-current heat exchange, serving to restrict heat loss and enable efficient thermoregulation. This rete is also thought to reduce the circulation and therefore oxygen supply to the hind limbs during a dive (Grant & Dawson 1978) to conserve oxygen for the brain.

The platypus has highly insulative fur and can forage for up to 12 hr in water at 0°C (Grant 1995). The guard hairs and dense underfur of the platypus trap a layer of air close to the skin, reducing heat loss.

The tail is the principal fat storage site in platypuses (about 40% of body fat). Body condition can be assessed by the tail fat index (Table 6.2).

Haemoglobin levels are also higher in platypuses (highest in nestlings) than in terrestrial mammals, believed to be an adaptation to low oxygen in the burrow and diving (see 7.1.2).

The platypus has a low average resting body temperature and a low basal metabolic rate. Physiological data for adult platypuses is presented in Table 6.1.

3 REPRODUCTION AND REPRODUCTIVE MANAGEMENT

There are still gaps in our knowledge of reproduction in platypuses. Although platypuses have been kept in captivity since the 1800s, they have been successfully bred in captivity only six times in three separate and

Table 6.1 Useful physiological values for clinically normal adult platypuses

Parameter	Value
Body temperature	32.08 ± 0.75°C.[a] Range 29.2–34.6°C (n = 2237)
Heart rate under isoflurane anaesthesia	180–220 beats/min
Heart rate during dive response under isoflurane anaesthesia	10–12 beats/min
Conscious heart rate	140–230 beats/min[b]
Conscious diving heart rate	10–120 beats/min during dives[b]
Respiratory rates under isoflurane anaesthesia	20–50 breaths/min

a Grigg et al. (1992) using implanted radiotransmitters on 5 wild platypuses in Thredbo between April and October. Cloacal temperature measured under anaesthesia is consistent with this range, although temperatures above 33°C would cause concern.
b Grant (1995).

distinctly different circumstances. The temperament of individual platypuses is the most critical factor in whether captive platypuses will breed. It would be misleading to write a 'how to breed platypuses in captivity' section based on these infrequent successes.

In captivity, success has been achieved with males and females housed with separate tunnels and nest boxes and feeding tanks, and with a communal feeding tank which allows the animals to choose to be together. They have also been bred in semi-natural situations such as a walk-through aviary and in farm dams. All successful breeding has occurred when females have had access to earth to build their own burrows. Cave-ins have occurred, so earth should have high clay content and have plant roots and logs to provide support.

3.1 Breeding season and mating

Platypuses in the wild do not live in pairs, rather, a dominant male appears to patrol a length of stream inhabited by a number of females. Mating occurs in late winter or early spring (Grant et al. 2004a). Breeding season varies with latitude, with platypuses in north Queensland commencing breeding earlier than in colder regions such as Victoria (Connolly & Obendorf 1998). Breeding usually commences in August in Queensland, September in New South Wales, and October in Tasmania (Grant et al. 2004a). In New South Wales the highest proportion of lactating females are captured in December and January. About 40% of female platypuses do not breed in consecutive seasons (Grant et al. 2004a). Female platypuses can breed in the wild as early as 2 yr of age and up to 21 yr of age (Grant et al 2004a). In captivity, the average age of females involved in any kind of reproductive behaviour was 6.2 yr (Holland & Jackson 2002).

Reproductive seasonality in the platypus has been assessed by measuring progesterone and androgen concentrations in blood, faecal and urine samples using radioimmunoassay (Jakubowski et al. 1998; New et al. 1998). Progesterone concentrations in females from plasma and excreta were higher in July–September, indicating significant ovarian activity during this period. Androgen concentrations in adult males from plasma and excreta were higher in July–August. Cortisol concentrations were higher and males lost body condition during this period, indicating stress due to rivalry during the breeding season.

The courtship ritual of the platypus involves processional swimming with the male holding the extremity of the female's tail in his bill. The two animals swim in circles and intermittently preen and nuzzle each other. From the tail-biting position, the male doubles his body under to insert the penis into the female's cloaca. Mating lasts for 10 min (Fleay 1980). The male spurs are not used in the mating ritual. The male plays no part in rearing the young and in captivity should be separated from the female after mating has been confirmed. Courtship behaviour can last several days. Mating occurred during three consecutive days in the 1998/99 Healesville Sanctuary breeding (Holland & Jackson 2002).

3.2 Gestation and incubation

The female may excavate a nursery burrow 4 mo in advance of mating. The female increases her feed intake for 9–11 d following mating then commences nest-building (i.e. lining the nursery burrow), requiring a supply of leaves, reeds and grass which must be provided in the water. Nesting material is carried in the prehensile tail. Nest-building stops abruptly after 3–5 d (Fleay 1980; Holland & Jackson 2002). In 1945, the female platypus Jill fed all day for 2 d from completion of her nest and disappeared into the nursery burrow 14 d after mating to lay and incubate her eggs. Five days later two eggs were dug out of her nest, the breeding season after the famous first breeding of Corrie (Fleay 1980). Gestation therefore is greater than 14 d and less than 19 d. Platypuses lay 1–3 eggs (Burrell 1927). The female spends 11–12 d almost exclusively on the nest, which probably represents the incubation period. In 1944, Jill did not leave the nursery burrow at all for 5 d, then left for 30 min toilet and grooming breaks every few days. Not until 12 d after presumed laying did she spend increasing time away from the nursery burrow and her hatchling (Fleay 1944). In 1998, Koorina also spent 5 d confined to the burrow before emerging for progressively longer intervals (Holland & Jackson 2002). The female instinctively blocks (pugs) the nursery burrow when she comes and goes during lactation, which emphasises the need for access to soil to facilitate captive breeding. Pugging involves using the tail to fashion multiple packed-earth blockades (pugs) in the tunnel, which causes progressive hair loss on the dorsal tail tip. On all occasions where platypuses have bred in captivity, females have had access to deep soil to construct their own nursery burrows. During the first 6 wk of lactation, the female's appetite gradually increases to almost her own body weight in daily intake.

3.3 Development of the young

From hatching, after an 11–12 d incubation period the young platypus feeds by nuzzling the mammary area of the female's abdomen. The curled posture of the mother platypus probably holds the baby firmly in place and keeps it warm. The eyes remain closed for 11 wk, and the baby platypus is fat and unco-ordinated until 15–17 wk age when it emerges for its first clumsy swim and introduction to solid food (123–136 days, n = 3) (Fleay 1980; Holland & Jackson 2002). From 17–20 wk the appetite for solids increases to maintenance levels. The young platypus continues to den with its mother until it is 5 mo old (Fleay 1944).

3.4 Lactation

Lactation occurs in October–February in New South Wales (Grant & Griffiths 1992) and November–May in Tasmania (Connolly & Obendorf 1998). Lactation can be detected by administering oxytocin at 1–2 IU/kg IM. Five minutes after injection, milk can be collected from the areolae using haematocrit tubes or pipettes. During lactation the fur covering the mammary region of a lactating female is discoloured (brownish yellow) and worn compared to fur elsewhere on the abdomen.

Platypus milk is made up of 38% total solids, mainly comprised of 22% crude lipid, 8% crude protein and 3% total hexose (n = 20) (Griffiths et al. 1984). Analysis of the fatty acid components of the triglyceride fractions of platypus milk indicated no major change in composition during lactation and the predominant fatty acids were oleic acid (25%) and palmitic acid (18%) (Grant et al. 1983). Platypus milk, like echidna milk and marsupial milk, exhibits very high concentrations of iron, up to 21 mg/kg of milk (n = 7). Given the minute size of the monotreme and marsupial liver at birth, the storage of iron would be limited and the need for iron to synthesize haemoglobin essential (Griffiths 1988).

4 PLATYPUS RESCUE

There are relatively few platypuses in captivity in Australian zoos. As platypuses are widespread and common, it is not unusual for veterinarians and wildlife carers to be presented with rescued animals. The following section aims to provide some guidance on how to assess and manage a rescued platypus.

4.1 Decision-making

1 Is the platypus an adult male and do I need to be careful handling it to avoid being spurred?

Grab the animal firmly by the tail and suspend it away from yourself and look for a spur on the heel of either hind leg (Fig. 6.2). If present, handle the animal with great care. Adult males are more likely to try to spur you during the breeding season from July–December.

If no, go to 2.

If yes, go to 5.

2 Is the platypus a lactating female?

During the breeding season, attention should be paid to whether an adult female platypus is lactating because, if so, priority should be given to returning her to the close vicinity of her young if she is healthy enough. Signs of lactation include loss of guard hairs in the mammary region of the abdomen and oozing of milk in response to gentle pressure. The lactation status of female platypuses can also be assessed using oxytocin at 1 IU IM to induce milk letdown. Lactation occurs in October–February in New South Wales (Grant & Griffiths 1992) and November–May in Tasmania (Connolly & Obendorf 1998).

If no, go to 3.

If yes, go to 5.

3 Is the platypus a milk-dependent juvenile?

Unweaned juveniles have a very short bill only as long as it is wide. If it has a square bill <30 mm × 30 mm, it is still milk-dependent (Fig. 6.5). Arrows indicate where to take measurements. A) Length of bill from tip to crease (do not include bill shield). B) Maximum bill width. Another telltale feature of a milk-dependent juvenile is that it will try to stimulate milk letdown from your hand if you put it in the position shown in Figure 6.5, and if offered milk it will slurp it up. They are generally tame and unafraid of people. Young that have not yet left the burrow will often be very fat. Body weight is not a useful indicator of age because platypuses vary enormously in weight with geographic location, age, sex and condition. A platypus weighing 600 g could

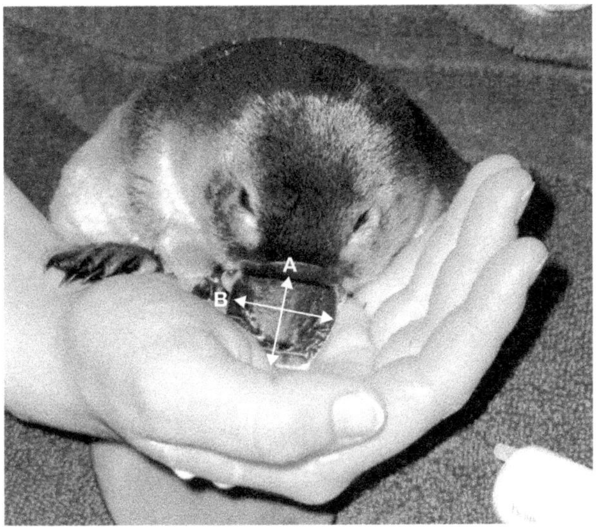

Figure 6.5 A milk-dependent juvenile Victorian platypus weighing 680 g and with a bill measuring 29 mm at A and B. This is the typical suckling posture of monotremes. Photo: R Booth.

be a subadult, a fat juvenile or an emaciated adult. A milk-dependent juvenile platypus will require intensive care for some weeks.

If no, go to 4

If yes, go to 5

4 Is the platypus an independent dispersing juvenile?

Most platypuses that come into care are dispersing juveniles (Fig. 6.6). If they are not sick or injured, rapid return (within 24 hr) to suitable habitat close to point of origin should be considered a priority.

If yes, check thoroughly for evidence of disease or injury (see 5) before release.

5 Is the platypus sick or injured?

A strap-like tail, tameness and poor waterproofing are classic signs of ill health in platypuses. The animal's condition can be assessed using the tail fat index (Table 6.2) first developed by Grant and Carrick (1978). If the animal is in poor condition it may be diseased or it may be a dispersing juvenile which has had trouble finding enough food. An experienced wildlife veterinarian should be enlisted to help determine whether a rescued platypus is sick or injured. A healthy platypus often growls on handling and wriggles constantly, and its fur dries within 10–15 min of leaving the water.

Common causes of injury include dog attack, car trauma (especially in Tasmania), motor boat injury and litter entrapment (especially look around the neck for fishing line). If the platypus exhibits any clinical signs of disease (e.g. laboured breathing, discharge, wounds), then diagnosis, treatment and hospitalisation are indicated. Treatment is based on first principles for other species (see section 9, on diseases of platypuses). Captive housing is described in section 4.2 below and feeding is described in 4.3.

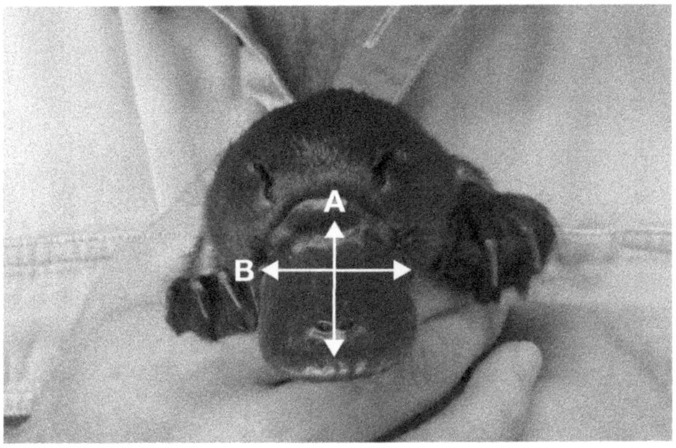

Figure 6.6 A dispersing juvenile with the bill now more rectangular than square. A) 37 mm. B) 32 mm. Photo: R Booth.

4.2 Housing

If the platypus needs to stay in care for longer than a few hours it will require housing that includes all these elements:

- a feeding tank with shallow tepid water;
- a ramp out of the water which also provides a hiding place when in the water;
- a bank, log or flat rock for grooming on;
- nest box;
- absorbent nesting material;
- short tunnel;
- minimum/maximum thermometer;
- escape-proof and predator-proof lid.

Change the water every day and don't make it too deep so that food is easy to access. Make sure the platypus cannot escape. Platypuses are extremely adept at escaping from enclosures. They are strong, can climb and squeeze through small gaps and are very determined, even when sick. Avoid ambient temperatures in excess of 27°C.

4.3 Feeding rescued platypuses

It is difficult to mimic the wild diet in captivity (see section 6).

Food to offer per day for platypuses that weigh <1 kg:

- 50 g mealworms;
- 50 g earthworms;

Table 6.2 The tail fat index of the platypus as a measure of condition

Tail fat index	Description and interpretation	Illustration
1	There is so much subcutaneous fat in the tail, it resists curling along the vertebral column. Superb condition with plenty of energy in reserve	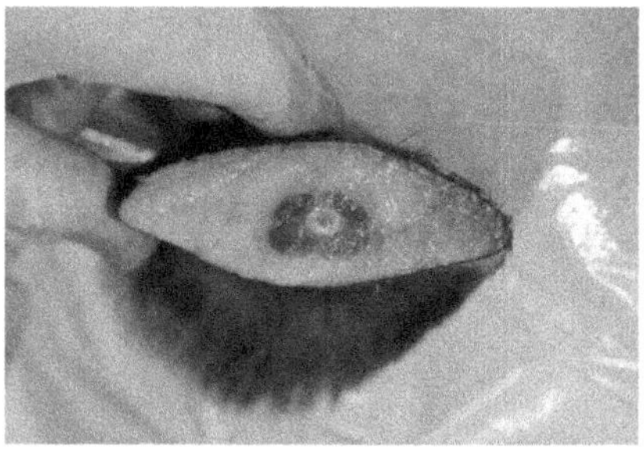
2	The tail can be curled along the midline, but the amount of subcutaneous fat restricts curling to about a third of its width. Good condition with some energy in reserve.	
3	The tail is easily curled in to about half its width. Fair condition but limited reserve. This is typically the condition of dispersing juveniles and lactating females.	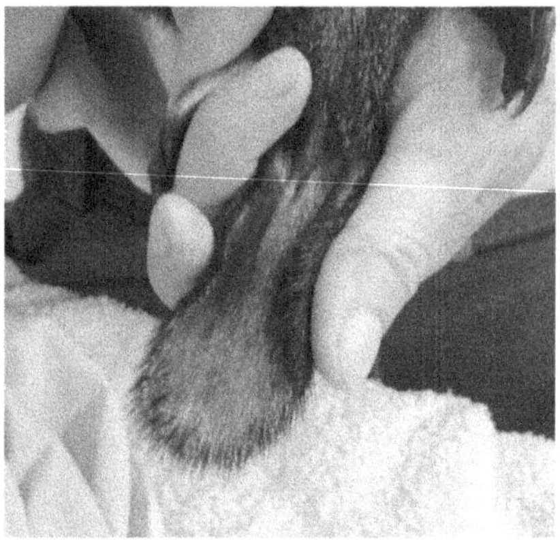
4	The tail is flat and easily curled so that the sides almost touch. Poor condition, energy intake needs to increase or the platypus may spiral downhill. There is little difference between 4 and 5. Both indicate pathology or life-threatening misadventure.	
5	The tail is strap-like and curls inwards easily and the vertebral column can be seen. Emaciated and likely to die.	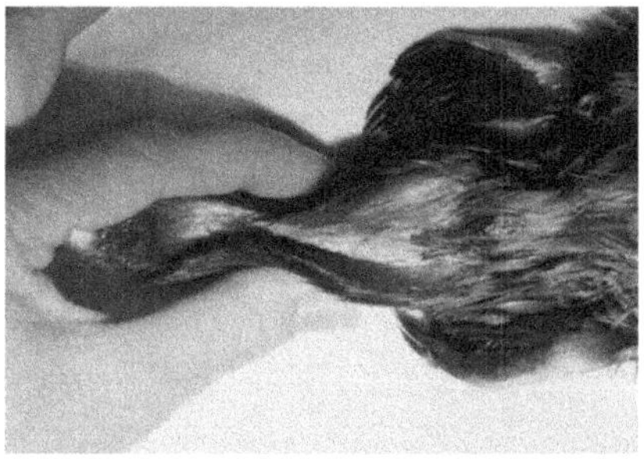

- 50 g yabby tails (sick or young platypuses cannot usually kill and break up yabbies without help);
- fly pupae, crickets and mudeyes (dragonfly larvae, which are available from good fishing bait stores).

It is important to offer more food than they require because they will never catch everything put in the water.

Weigh food (wet weight) in and out of the water each day and calculate approximate food intake, which should be 10–20% of body weight. Monitor the animal's weight daily. Animals eating less than their estimated maintenance requirement and those losing weight should receive SC fluids (2.5% glucose/saline) until a maintenance intake is achieved.

Insect or yabby milkshakes can be fed to convalescent adults and subadults to restore energy balance. An example of a suitable formula is Di-Vetelact (Sharpe Laboratories) at one scoop per 40 mL blended with 10–20 g blended mealworms and yabbies. This can be syringed very slowly into the side of the mouth or delivered by stomach tube.

In captivity, platypuses should defecate and urinate at least once daily, however, this may be difficult to detect as they most often void in the water.

4.4 Hand-rearing

Most young platypuses presented for care are dispersing juveniles that have been weaned. However, occasionally milk-dependent juveniles come into care due to misadventure such as cattle collapsing the nursery burrow or being dug up by an excavator. The most useful features to ascertain milk dependence is that the bill is square rather than rectangular (Fig. 6.5) and the unweaned young still have a pronounced suckling response if presented with milk in a cupped hand. The spur can also indicate age, but will do little to assist in the question of whether the animal is weaned. Ability to swim and forage competently in water implies that the animal is weaned.

Like captive breeding, hand-rearing has not been frequently attempted and it is not usually successful. The following advice stems from limited direct experience gleaned from hand-rearing five milk-dependent platypuses.

Because of the small size of the stomach it is important not to overfeed the animal. The small gastric volume is partially balanced by a rapid gut transit time. Do not exceed 10–15 mL per feed and feed every 6–8 hr based on the enthusiasm of feeding and the weight gain. Total daily intake should be around 20% of body weight in 3–6 doses depending on age. The general principles are feeding 10–20% body weight daily with frequency depending on age, with 3 times daily being adequate as weaning approaches. Platypuses can be raised on Di-Vetelact at one scoop per 40 mL of water. They also tolerate concentrations greater than this, and if feeding response is poor Di-Vetelact paste can be fed slowly by syringing it into the mouth.

To feed, allow the juvenile to seat itself around your hand as though it is the abdomen of the female lying on her back, and syringe small quantities of milk into a shallow well in the cupped hand. Do not submerge the nostrils. Continue to refill the well as the suckling drains it (Fig. 6.5).

If inadequate weight gain is achieved, the energy content can be increased using oil or cream. Palm oil is a rich source of palmitic acid and olive oil is a rich source of oleic acid. Best practice would be to blend the oils in the proportions found in platypus milk, but olive oil alone should provide the needed energy in a utilisable form (see 3.4).

As the platypus approaches weaning, an insectivore mix (e.g. Small Carnivore Food [Wombaroo Food Products] or Insecta-Pro [Vetafarm]) can be mixed with the milk in increasing proportions. Mealworms are the most convenient solids to introduce to a weanling. The young platypus will start grabbing at them and indicate it is ready for this next step. They can take about 10 at a time from about 14 wk. The mealworms are worked in the cheek pouches and grinding pads, sometimes for hours—manual assistance may be required to remove the flattened exoskeletons with fine forceps. At this time they should also be allowed access to water to develop swimming and foraging skills and to toilet in the water. It is not until 21 wk that you can expect the young platypus to consume enough live food to maintain itself, so the challenge is to maintain supplementary milk intake once the animal has lost its strong suckling response, which seems to happen around 14 wk. One technique is to make up a slurry of mealworms (heads removed), milk and insectivore mix and offer this in a small dish by hand.

Hand-reared platypuses should urinate and defecate daily. Cloacal stimulation with moistened cottonwool sometimes works, but a platypus will nearly always void if the cloaca and hindquarters are immersed in a basin of water (even before they would have ever entered water).

4.5 Adaptation to captivity

Not all wild animals have the temperament to adapt to a life in captivity and platypuses are at the difficult end of the scale of adaptability. It can be hard to judge whether a wild animal is adapting to captivity. Some factors which can indicate failure to adapt to captivity are:

- persistent or escalating escape behaviour;
- failure to eat normal daily food intake;
- weight loss greater than 10% of capture weight within 2 wk;
- stereotypic behaviour;
- reversal of neutrophil: lymphocyte ratio (section 7.1.2).

If animals are released back to the wild the following strategy can increase the chance of survival:

- release at point of capture;
- release within 2 wk of capture;
- release animals that are bright and alert and in good condition.

Neutrophil and lymphocyte counts can be used as an indicator of adaptation to captivity. One physiological response to an increase in cortisol in response to stress is an increase in circulating neutrophils and a decrease in circulating lymphocytes. Cortisol causes neutrophils to migrate from tissues into the circulation and causes direct lymphocytolysis (Schalm et al. 1975). This is a more useful indicator of stress than the measurement of cortisol itself, because generally restraint for blood collection produces rapid cortisol release. This paradigm has been confirmed to occur in platypuses (Booth & Whittington 1993).

Figure 6.7 shows the absolute neutrophil and lymphocyte counts for three juvenile platypuses purpose-caught for captivity. Platypus C was released due to a perception that she was failing to adapt to captivity based on changes in her white cell count combined with the development of stereotypic behaviour, increase in escape behaviour and a 5% drop in body weight. This animal was confirmed to be alive in the wild by chance recapture 106 d after release.

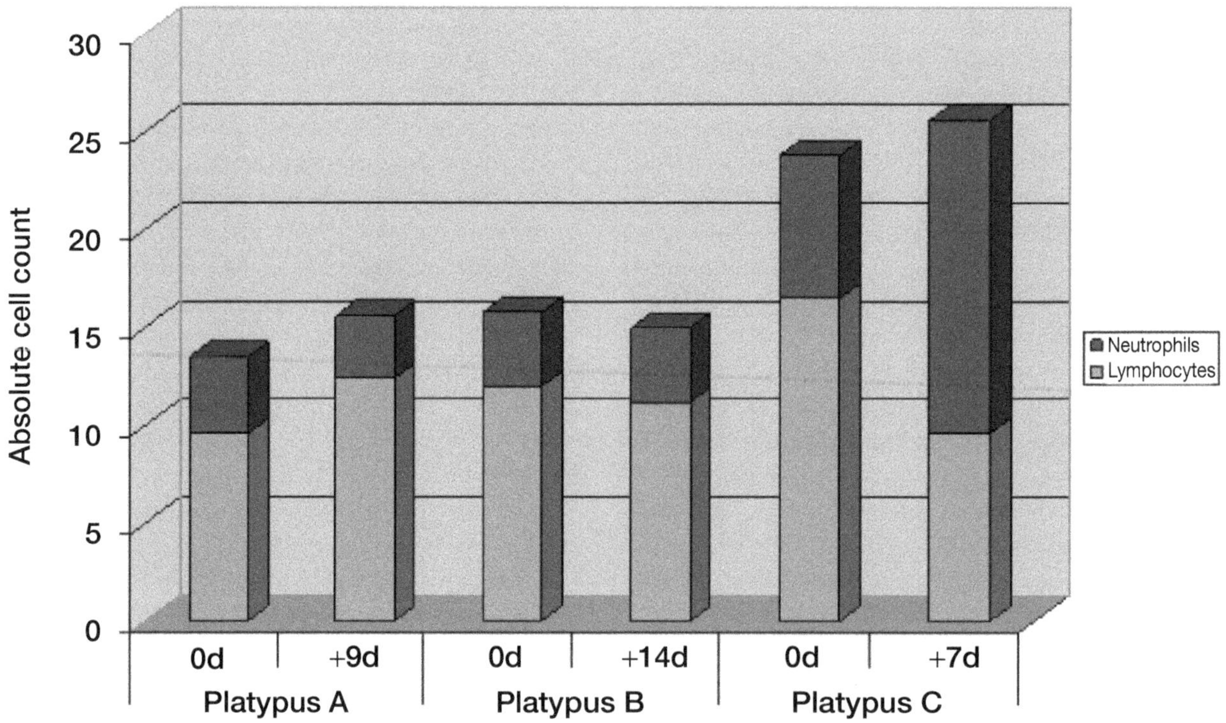

Figure 6.7 Neutrophil and lymphocyte changes observed in three juvenile captive platypuses purpose-caught for captivity in 1991. Platypus C exhibited a reversal of her neutrophil:lymphocyte ratio and a variety of non-adaptive behaviours and was released. Platypuses A and B went on to produce two litters in captivity in 1998 and 2001. The time between blood samples for each platypus varied (A = 9d, B = 14d, and C = 7d).

Another juvenile platypus released after only two nights in captivity was released solely on the basis of escalating escape behaviour and failure to eat. This platypus was confirmed to be alive 9 mo after release in his previous home range.

5 HUSBANDRY

5.1 Housing

Captive husbandry of platypuses is specialised and covered in detail by Fleay (1980), Krueger et al. (1992) and Jackson et al. (2003). Captive animals require water for feeding and exercise, and a tunnel system connected to one or more nest boxes. Tunnels have been constructed of marine ply or plastic pipe, and access hatches are required every 1 m or so to allow location and access to the animal anywhere within the system. Tunnel, nest box and tank walls should be smooth and non-abrasive. A chronic foreign body reaction in the tail of a captive platypus was attributed to incorrectly sealed fibreglass and lesions from a rough tank surface have been reported (Connolly et al. 1998). Great attention to escape-proofing is required, with clips on all latches, high smooth walls around pools and a predator-proof and escape-proof roof. Pool walls should be 1–1.2 m with at least 0.5 m above the water level at all times. Provision of a grooming platform at the tunnel entrance reduces the amount of water taken into the tunnel system. Within the water, furniture items should include partially submerged branches and overhanging ledges to shelter under or groom on. Nesting materials (grasses and leaves) should be put in the water for females during the breeding season. If soil is to be provided for tunnelling, provision must be made to prevent cave-ins by using clay soils with plentiful fibrous vegetative support. Figures 6.8 and 6.9 show an ideal hypothetical platypus facility with all the components necessary for maintenance, display, breeding and research.

Platypuses are very cold-tolerant but not heat-tolerant. Maintain the air and water temperature maximum below 27°C. If water temperature creeps above 29°C platypuses become noticeably lethargic; they have died when air temperatures exceeded 34°C. Ice cubes can be added to the water of display tanks to drop the temperature on very hot days.

Water in display and feeding tanks is typically fresh water at ambient temperature. It needs to be kept clean using filtration but there is no need to add sterilising agents such as chlorine or ozone. Weekly replacement is common, but requirements depend on whether there

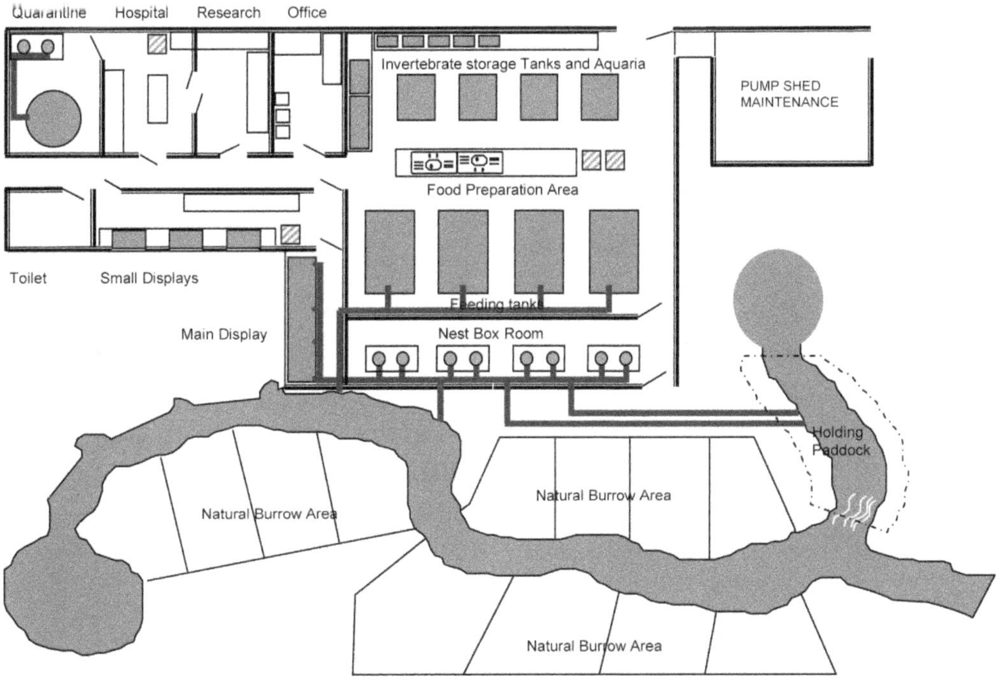

Figure 6.8 An example of the elements required for captive platypus maintenance, display, breeding and research. Source: McKee et al. (1995, unpublished).

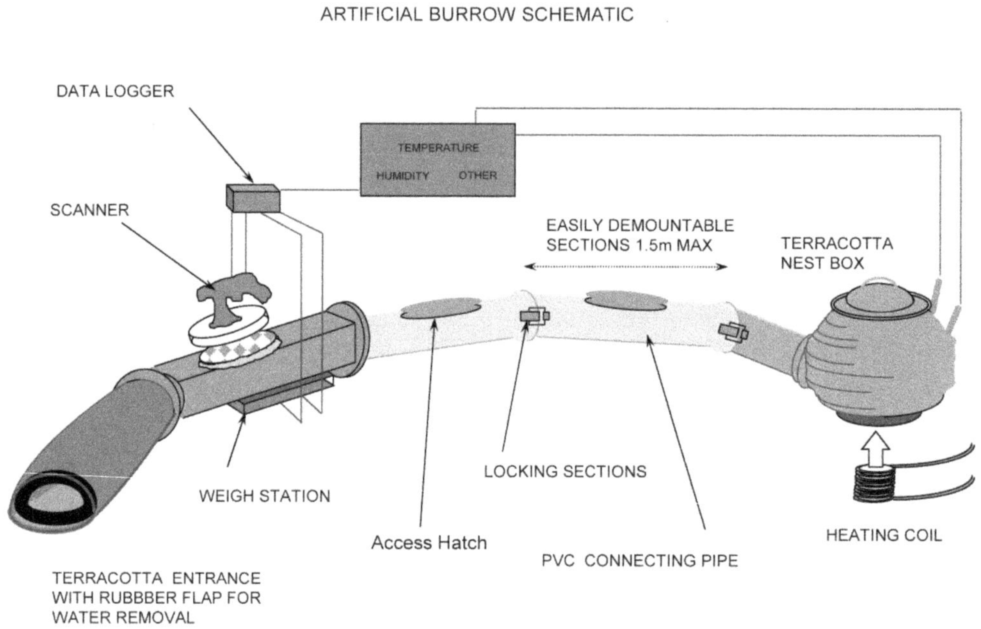

Figure 6.9 The elements of an artificial burrow system for platypuses. Source: McKee et al. (1995, unpublished).

are other occupants of the tanks (e.g. fish, yabbies, turtles).

With the sensitive bill, there is concern that pumps and filters which cause significant vibration could cause stress or interfere with feeding. It is desirable to set up noisy vibrating equipment a reasonable distance from the nesting areas and feeding tanks. There is little evidence, however, that the vibration of this equipment has a significant effect.

In the wild, platypuses are solitary when not actively courting or raising young. In captivity they have been housed in various combinations with mixed success. As a general rule, never house adult males (>4 yr) together, house male and female pairs with separate quarters so they can choose whether to be together, and female pairs are often compatible although one will always be dominant. Recent introductions must be monitored carefully for compatibility, as dominant animals may pursue subordinates until they are exhausted.

5.2 Husbandry

An important part of husbandry is the daily assessment of appetite and demeanour. Whenever platypuses are manually restrained, their condition should be routinely assessed by attitude, coat and skin condition, body weight and condition score assessed by tail fat index (Table 6.2). Frequency of weighing varies markedly between institutions holding platypuses. It is good practice to weigh captive platypuses at least monthly, however, more frequent weighing may be advantageous. Daily weighing allows for a quick close-up assessment of the animal's condition and attitude and accustoms the animal to handling, which will be an advantage when a more invasive examination is required. No adverse effects have been reported as a result of daily weighing, provided the animal can be caught quickly.

It is possible to remotely weigh platypuses without handling, but that loses the opportunity to assess demeanour. Weighbridges have been used with some success at Healesville Sanctuary. These included a data logger, a scanner, the scales and a narrowed opening before and after to slow the animal down to get an accurate reading. The scanner was positioned in the system so that the direction of travel could be determined as well. If the data logger recorded the animal's identification then weight, the animal was leaving the tunnel and entering the feeding tank. The reverse indicated that the platypus was returning from the feeding tank. Because of water retained in the coat, this measurement did not provide an accurate indication of food intake. One animal was able to escape via the weighbridge when it worked its way out through a small gap between the components.

5.3 Transport

Platypuses have been transported successfully via various means including road, sea and air. Best practice involves the acclimatisation of the platypus to a nest box which locks in and out of their regular system for at least 2 wk, which then becomes the transport box (Fig. 6.10). The transport/nest box should be a secure ventilated box packed with soft bedding (e.g. dried seagrass). Platypuses are also happy to sleep on towels or polar fleece, but artificial bedding must not have any frayed edges that could cause entanglement. The box must be smooth on the inside and free of gaps so that if the platypus tries to find a way out it will not rub its bill or feet.

During transport, careful attention should be paid to avoid temperatures exceeding 27°C. Transport should be carefully planned, and the platypus should be accompanied by a keeper to troubleshoot against delays and mishandling of the crated animal. Transport boxes can be insulated but must be adequately ventilated. Successful platypus transportation from Sydney to south-east Queensland by air involved placing the wooden nest box inside a well-ventilated Esky which was lightly packed with ice (L Vogelnest *pers. comm.*).

Fasting for several hours should not present a problem, but for transport of greater than 24 hr duration thought should be given to providing a portable feeding tank to which the animal has supervised access. It is also necessary to plan the transport of the individual's preferred live food items.

Figure 6.10 An example of a platypus nest box which converts into a transport box. Photo: de Alwys, Melbourne Zoo.

David Fleay successfully transported three platypuses from Healesville Sanctuary (Vic.) to the Bronx Zoo in New York in 1947 travelling by land, air and sea. He describes his method of creating a portable platypussary in detail in Fleay (1980). He had more difficulty in an attempt to transport a young male platypus from Victoria to London by sea in 1943 with a full portable set-up. The platypus died suddenly 4 d short of its destination, apparently caused by repeated depth charge detonations during WWII.

5.4 Individual marking and identification techniques

Individual platypuses are most commonly permanently identified using passive integrated transponders or microchips. These are inserted aseptically SC on the dorsal midline between the scapulae (Grant & Whittington 1991). Anaesthesia is generally required. The hole should be closed with tissue glue to prevent the microchip migrating out soon after insertion. Other techniques that have been used historically include tattoos on the ventral bill shield (Gardner & Serena 1995) and stainless steel bands applied to the tarsus (Grant 1995). Photographs of the individual pigmentation of the ventral bill are also useful.

6 NUTRITION

Platypuses are opportunistic benthic carnivores. In the wild they mainly eat aquatic insect larvae and small soft-shelled crustaceans, but they will also eat small fish, frogs and worms.

In captivity they can be maintained on a variety of live food including earthworms, mealworms, small yabbies, fly pupae, freshwater shrimp, prawns, goldfish, tubifex worms, trout fingerlings and aquatic insect larvae. Captive platypuses consume 15–30% body weight daily for maintenance, while lactating females consume 90–100% body weight daily (Holland & Jackson 2002). The feeding of rescued platypuses is discussed in section 4.3.

Appetite should be assessed by close monitoring of each individual's food intake on a daily basis. Best practice for platypus husbandry involves the daily weighing of both the leftover food and the animal. The wet weight of the daily ration can be measured and recorded, and feeding tanks can be fitted with coarse sieves to collect the bulk of the food waste. The waste should be partially dried on absorbent paper, then weighed, and approximate intake calculated and recorded. Platypuses can be maintained in captivity for long periods with less intensive monitoring, but detailed appetite assessment acts as an early warning system.

7 CLINICAL PATHOLOGY

7.1 Haematology and biochemistry

7.1.1 Sample collection

The most practical site for blood collection in the platypus is from a venous sinus that runs along the rostral margin of the bill. The sinus lies just beneath the skin surface. A 25 G needle or a 25 G butterfly catheter and 1–3 mL syringe are suitable. The animal must be very firmly restrained or anaesthetised. The sinus is entered parallel to the long axis of the bill at a point just lateral to the midline of the bill margin (Fig. 6.11). The needle should be rotated so that the bevel of the needle faces laterally and the needle is inserted to a depth just past the bevel. One to two millilitres of blood can usually be withdrawn without difficulty.

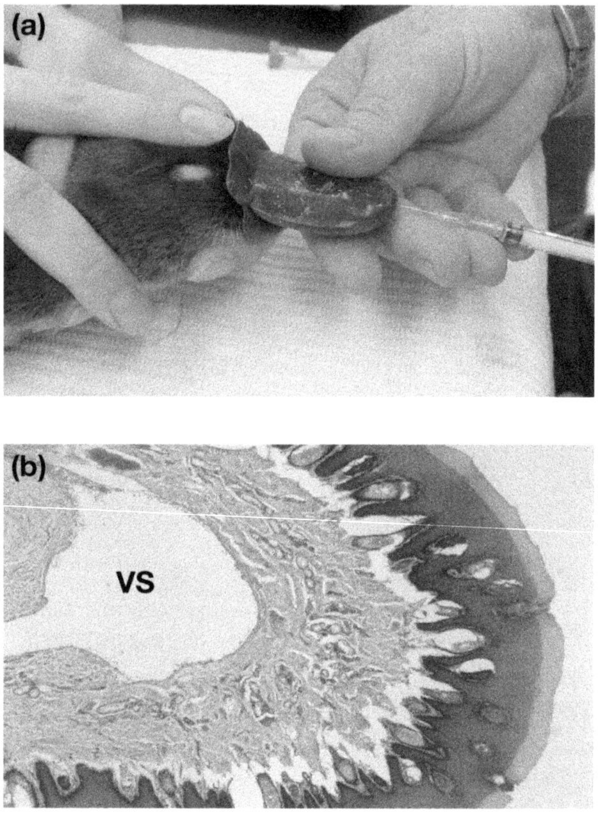

Figure 6.11 a) Blood collection from the bill sinus. b) A longitudinal cross-section of the sinus, which has been fixed and stained with H and E × 70. VS = venous sinus. Photos: J Connolly.

7.1.2 Reference ranges and interpretation

Platypuses have non-nucleate erythrocytes and platelets. Segmented and non-segmented neutrophils and eosinophils are seen in peripheral blood smears. The leucocytes of platypuses display unusual features that include the consistent presence of Döhle bodies within segmented neutrophils, and non-segmented nuclei in basophils (Canfield & Whittington 1983). Nuclei in eosinophils are ovoid (non-segmented) to tri-lobed (Clark 2004). Lymphocytes are small to medium sized with a thin rim of cytoplasm. Monocytes have a horseshoe shaped nucleus.

Haematology and biochemistry reference ranges for wild platypuses from Victoria are presented in Table 6.3. These reference ranges have been derived from blood collected from 90 healthy wild platypuses in Victoria trapped over a decade from 1989 to 1999. The program SPSS was used to identify outliers and calculate the 95% confidence intervals. Data for each parameter was examined for a normal distribution and it is on this basis that the confidence intervals were generated. The minimum and maximum values are also provided, however, 90% of healthy platypuses would be expected to have values within the confidence intervals.

The platypus has haematological adaptations for the low oxygen environments associated with diving and burrowing. These adaptations include a relatively high haematocrit, erythrocyte count and haemoglobin concentration, together with low mean corpuscular volume and high mean corpuscular haemoglobin concentration. Other unique features include a high white cell count, which mostly consists of lymphocytes and high blood urea nitrogen probably associated with the high-protein diet. There is seasonal variation in total white cell count with a pronounced lymphocytosis in spring.

Interpretation of blood results in relation to disease can be challenging, particularly with respect to leucocyte numbers which vary enormously in healthy animals. There is seasonal variation in white cell count which cannot be explained with a simple

Table 6.3 Haematology and biochemistry reference ranges for wild platypuses from Victoria (95% confidence intervals after outliers removed). The observed minimum and maximum are also provided, with outliers removed.

Parameter	Units	n	5% CL	95% CL	Min	Max
Haematology						
Hb	g/dL	87	17.47	18.32	13.4	22.3
PCV	L/L	87	0.49	0.51	0.35	0.62
RBC	$\times 10^{12}$/L	86	9.86	10.30	7.26	12.10
MCHC	g/L	90	356	363	327	399
MCV	Fl	90	48.92	50.04	42.00	56.30
MCH	Pg	84	17.60	17.84	15.00	34.40
Platelets	$\times 10^{9}$/L	57	621	817	262	2144
Total WBC	$\times 10^{9}$/L	89	22.60	26.03	9.6	40.6
Neutrophils	%	90	29.48	35.10	12.00	62.00
Neutrophils	$\times 10^{9}$/L	89	7.53	10.11	1.15	25.17
Band neutrophils	%	90	0.21	0.63	0.00	5.00
Band neutrophils	$\times 10^{9}$/L	89	0.06	0.21	0.00	1.62
Lymphocytes	%	90	60.51	66.55	32.00	86.00
Lymphocytes	$\times 10^{9}$/L	89	14.68	18.35	3.07	34.51
Monocytes	%	87	2.01	2.91	0.00	7.00
Monocytes	$\times 10^{9}$/L	87	0.58	0.91	0.00	2.74
Eosinophils	%	89	0.67	1.06	0.00	3.00
Eosinophils	$\times 10^{9}$/L	89	0.21	0.35	0.00	1.22
Basophils	%	89	0.00	0.00	0.00	0.00
Basophils	$\times 10^{9}$/L	89	0.00	0.00	0.00	0.00
Biochemistry						
Sodium	mmol/L	37	144	147	139	156
Potassium	mmol/L	37	2.21	2.61	1.40	4.00
Chloride	mmol/L	42	110	112	106	117
Bicarbonate	mmol/L	36	26.44	29.22	20.00	37.00
BUN	mmol/L	39	29.49	30.90	25.90	34.20
Creatinine	mmol/L	38	0.023	0.029	0.01	0.04
Total bilirubin	µmol/L	39	8.29	10.54	3.00	18.00
GGT	U/L	33	7.27	8.13	5.00	11.00
ALP	U/L	36	184	238	43	387
Total protein	g/L	40	64	66	57	75
Albumin	g/L	39	27	28	22	33
Globulin	g/L	40	36	39	25	46
AST	U/L	31	776	876	535	1146
CK	U/L	34	314	436	107	806
LDH	U/L	31	547	877	88	1741
Glucose	mmol/L	40	3.37	4.47	1.00	8.30
Amylase	U/L	34	100	117	54	166
Calcium	mmol/L	33	2.12	2.22	1.91	2.55
Phosphate	mmol/L	34	1.86	2.15	1.19	3.03
Cholesterol	mmol/L	41	5.79	6.60	3.10	9.41
Triglycerides	mmol/L	28	0.78	0.96	0.50	1.50

CL = confidence interval.

Source: Modified from Booth (2003).

stress lymphopenia, neutrophilia response. See section 4.5 for changes associated with adaptation to captivity and stress and section 9.1.2 for changes observed with mucormycosis.

7.2 Urinalysis

Small volumes of gelatinous urine have been opportunistically collected from platypuses under anesthesia, voided when the thermometer is inserted into the cloaca. No specific data on urinalysis in platypuses is available.

8 RESTRAINT

8.1 Physical restraint

Platypuses are surprisingly strong for their size and their loose skin can make them difficult to handle. A healthy platypus also struggles constantly. Apart from the venomous spurs in mature males (see 2.2), platypuses are otherwise poorly equipped with offensive weapons and they cannot bite or scratch with sufficient force to harm a human. However, mature males must be considered dangerous and should only be captured and handled by experienced people. If there is any doubt about the sex and maturity of a platypus it is best left alone to be dealt with by an experienced person. Mature males also use their mouth and claws to try to grab your clothing for purchase to either pull free from or spur the handler.

The most effective method of capture and restraint for superficial examination or prior to placing in a bag is to grasp the tail quickly and firmly and suspend the animal away from you (Fig. 6.12). Although the male spurs are on the heel, the danger zone is ventrally between the hind legs and along the abdomen. Never allow a mature male to get close to your body or rest on your arm. Handlers have also been spurred on the abdomen because they allowed the animal to get a grip on their clothing. If you are required to untangle a

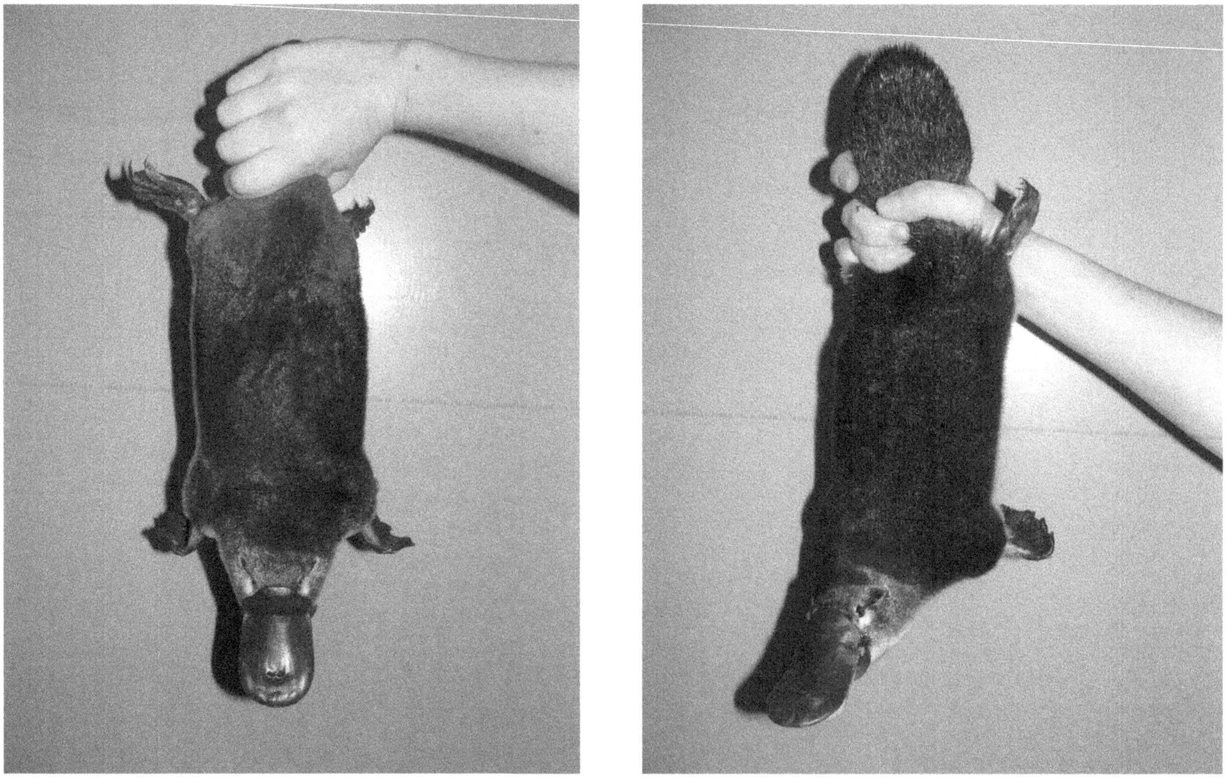

Figure 6.12 (Left) The correct way to hold a platypus, by the tail. (Right) The wrong way to hold a male platypus, where the wrist is in the danger zone for being spurred. Photos: R Booth.

mature male platypus from a net in the dark, it is wise to rest the animal on a thick towel so that it cannot gain access to any part of your body or hands while you untangle it. Once placed in a cloth bag the animals can be restrained further for very minor procedures or for anaesthetic induction using inhalation anaesthesia.

8.2 Chemical restraint

8.2.1 Sedation

Diazepam (0.5–1.0 mg/kg IM) has been used effectively for sedation for minor procedures such as radiography and transportation (Vogelnest 1999). Diazepam at low doses has also been useful in controlling anxiety and encouraging feeding in platypuses undergoing rehabilitation.

8.2.2 Anaesthesia

Regurgitation under anaesthesia has not been observed in platypuses and pre-anaesthetic fasting is not usually required. However, with a known gut transit time of around 5 hr, 1 hr fasting should be sufficient to ensure that the stomach is empty.

Isoflurane in oxygen administered by T-piece and mask is the anaesthetic of choice for platypuses. To restrain a platypus so that you can apply a face mask for inhalation anaesthesia, place the platypus into a cloth bag. Cut a hole out of one corner which is just wide enough to allow the passage of the bill. Alternatively, the bill can be directed to a corner of the bag with no hole so the animal breathes through the material (L Vogelnest *pers. comm.*). Gently restrain the animal within the bag on a foam mat while placing the mask over the bill to form a good seal and induce with 5% isoflurane and 1 L/min oxygen. Typical induction time for a healthy adult platypus is 1–2 min. Alternatively, the animal can be placed in a bag in an induction chamber. Induction takes a little longer but there is no need for physical restraint. This is particularly useful for adult males (L Vogelnest *pers. comm.*). Immediately after induction, examine the oral cavity and pharyngeal area to ensure that the upper airway is not obstructed by masticated cheek pouch contents. Long cotton swabs (15 cm) assist in clearing the airway. Transient marked bradycardia (10–12 beats/min) occurs occasionally under isoflurane anaesthesia. It usually resolves spontaneously, but may require reduction of the anaesthetic concentration. Anaesthesia is always maintained via a mask. It is very difficult to intubate platypuses because of the proximity of the larynx to the torus linguae (Fig. 6.3).

Inhalation anaesthesia and maintenance via a face mask is usually uncomplicated with good muscle relaxation, minimal salivation and regular breathing (can get down to 4–12 breaths/min). Anaesthesia can be monitored using standard small animal equipment.

In a healthy platypus, recovery from isoflurane anesthesia is rapid. Consciousness and righting reflex returns within 3–5 min of ceasing delivery of isoflurane. Normal co-ordination and behaviour return within 10–15 min. With field anaesthesia, it is advisable to retain the animal in a bag for 30–60 min after full recovery before returning it to water. Captive platypuses can be returned to their tunnel system and allowed access to water after 1 hr.

Injectable anaesthetic agents have rarely been used in platypuses and are generally not recommended or indicated.

9 DISEASES

Only a relatively small number of infectious agents and parasites have been reported from captive and free-ranging platypuses (Whittington 1992). Many are non-pathogenic and their clinical significance remains unknown. Apart from trauma and failure to thrive, non-infectious diseases are uncommon.

9.1 Infectious diseases

9.1.1 Bacterial diseases

Leptospirosis has been diagnosed in platypuses based on serology and histology. To date, *Leptospira* have not been cultured from platypuses. Approximately half of 131 platypuses captured from the upper Shoalhaven River in New South Wales were positive to *Leptospira interrogans* using microscopic agglutination (Whittington 1992). The proportion of seropositive animals increased with age. The source of exposure was thought to be from contaminated water. The *hardjo* serovar was the most common and was demonstrated in the cattle grazing at the study site. Incidental, mild chronic interstitial nephritis indicative of leptospirosis was detected in 25% of captive Victorian platypuses submitted for necropsy and examined histologically (McColl 1983; McColl & Whittington 1985). Leptospires were demonstrated in silver-stained sections of the renal cortex of a platypus that drowned in an illegal fishing net (Whittington 1992). *Leptospira* cause a subclinical infection that is not thought to be a threat to platypus populations. The condition has zoonotic potential and consideration must be given to this when dealing with platypuses.

Non-specific mixed infection with *Aeromonas hydrophila*, *Escherichia coli* and *Proteus* sp. was reported from a wild platypus that died of aspiration bronchopneumonia (Whittington & McColl 1983; Whittington 1988). The platypus presented with dyspnoea, a nasal discharge, haemoconcentration, neutropenia and lymphopenia, but was normothermic. It later died despite administration of electrolytes and amoxycillin. In a survey of pathology in captive platypuses, 45% of 20 animals showed aspiration pneumonia, with mixed growth of coliforms, *Aeromonas*, *Alkaligenes*, *Proteus*, *Staphylococcus* and *Streptococcus* (McColl 1983). This unusually high incidence of aspiration pneumonia may have been associated with other practices of husbandry at that time. Aspiration pneumonia is now rare in captive platypuses.

Salmonella enterica (including one serovar identified as *typhimurium*) has been recovered from one wild and one captive platypus that died from septicaemia (Whittington 1988). *Salmonella* was also isolated from the faeces of a captive platypus with weight loss. Excretion stopped following treatment with a potentiated sulphonamide antibiotic (trimethoprim/sulphadiazine at 5 mg/kg IM), but the animal died a year later from an unknown cause.

9.1.2 Fungal diseases

a Mucormycosis

Mucor amphibiorum is the only pathogen known to cause significant morbidity and mortality in the platypus (Whittington 1992). It causes a severe granulomatous and often ulcerative dermatitis that may progress to involve underlying muscle and disseminate to internal organs, particularly the lungs (Fig. 6.13). Affected platypuses are usually alert and display normal responses to capture and handling. The gross appearance of the skin lesions in affected platypuses varies from non-ulcerated hairless nodules and abscesses, to ulcers with underrun or

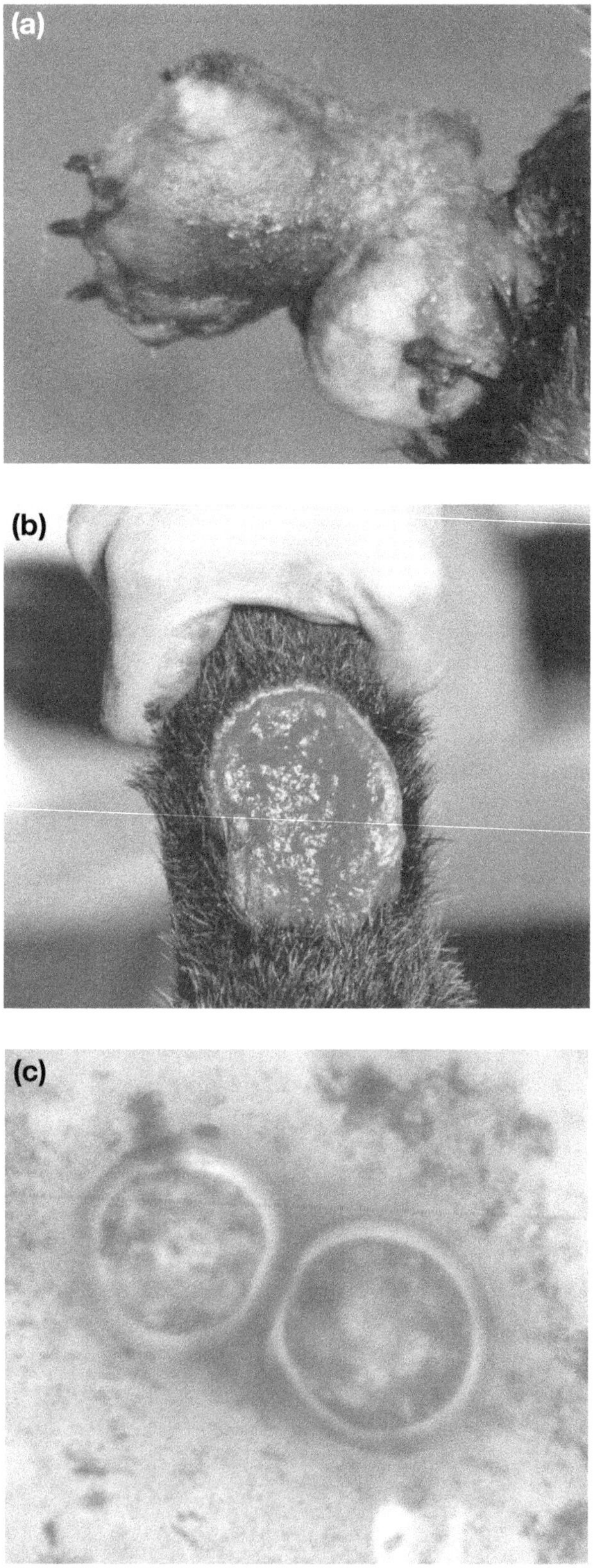

Figure 6.13 a) and b) The gross appearance of mucormycosis in the Tasmanian platypus. c) *M. amphibiorum* in DiffQuik smear preparation. Photos: J Connolly.

thickened margins, sinuses exuding pus, or exuberant granulation tissue attempting wound repair (Connolly et al. 1998). Some lesions appear as discrete entities, whereas others coalesce to form plaques. Lesions are found on haired regions including the hind limbs (38%), fore limbs (6%), tail (19%), trunk (6%) and head (6%), and unhaired regions such as the webbing of the fore limbs (13%) or bill (6%). Some affected animals have lesions at more than one site.

Histologically, skin lesions examined from seven cases were in the form of discrete poorly encapsulated granulomas or, more commonly, diffuse granulomatous or pyogranulomatous inflammation (Connolly et al. 2000). Spherules characteristic of *M. amphibiorum* infection were observed in all biopsied skin lesions. T-cells were the predominant infiltrating lymphoid cell in the diffuse lesions, indicating the importance of the cell-mediated response.

Healing mucormycosis lesions have been observed (Munday et al. 1998; Connolly unpub), however, in severely ulcerated individuals mucormycosis leads to death. It has been proposed that death could also result from secondary bacterial infections or impaired thermoregulation and mobility (Munday & Peel 1983; Obendorf et al. 1993; Connolly et al. 1998). Mortality rates in platypus populations are difficult to estimate accurately, as most animals that die are not recovered. Mucormycosis was seen in 8% (2/25) of Tasmanian platypus necropsies performed in one study (Connolly et al. 1998).

The route of entry is most likely via superficial skin wounds caused by spurring injuries, bites from water rats, eels or crustaceans, the attachment site of ticks or trombiculid mites, skin infestation by nematode larvae, or from foraging among sharp stones. Engorging ticks have been reported in association with ulcerative dermatitis in the platypus (Obendorf et al. 1993). In one case an ulcer developed at the site of a previous tick attachment and spherule-like structures consistent with *Mucor* have been observed in the haemolymph of an engorging tick found adjacent to an ulcer (Connolly et al. 1998). The respiratory route has also been proposed as a route of infection (Munday & Peel 1983; Obendorf et al. 1993; Connolly et al. 1998; Munday et al. 1998).

Platypuses with mucormycosis had significantly lower PCV, haemoglobin concentrations, lymphocyte counts, serum cholesterol and calcium concentrations, and higher serum globulin and potassium concentrations than clinically normal animals (Connolly et al. 1999b). The lymphopenia (ulcerated: mean = 11.49, SD = 3.82, n = 10; healthy: mean = 17.53, SD = 8.03, n = 26, p = 0.005) and hyperkalaemia (ulcerated: mean = 3.74, SD = 1.12, n = 10; healthy: mean = 2.69, SD = 1.08, n = 18, p = 0.027) were thought to be clinically significant. Numbers of *Trypanosoma binneyi* protozoa in blood smears were similar between the two groups. An enzyme-linked immunosorbent assay demonstrated the ability of the platypus to mount a humoral immune response and diseased platypuses had higher concentrations of serum antibody levels than clinically normal platypuses. Seroconversion was observed in a platypus coincident with the development of skin ulcers. The results also suggested that the incidence of exposure to *M. amphibiorum* was greater than the incidence of clinical disease (Whittington et al. 2002). In one study, there was no significant difference in persistent organic pollutant concentrations in *Mucor*-infected and healthy platypuses (Munday et al. 2002) and these chemicals do not appear to play a role in the epidemiology of the disease.

Limited antifungal sensitivity testing found *M. amphibiorum* to be sensitive to amphotericin B at <0.002 mg/L, but resistant to itraconazole and fluconazole (Connolly et al. 1998). If mucormycosis occurred in a captive platypus, a small lesion could be removed surgically. Possibly amphotericin B at a dose rate of 0.5 mg/kg in half-normal saline with 2.5% dextrose could be injected into the site twice weekly. Blood urea, creatinine and electrolytes should monitored for nephrotoxicity. Treatment would be discontinued if azotaemia was suspected.

The distribution of the disease appears to be expanding, with mucormycosis observed in at least ten catchments since the index cases were seen in the Elizabeth River at Campbell Town (central Tas.) in 1982. There does not appear to be any seasonality to infection and, despite the presence of *M. amphibiorum* in soil, frogs and cane toads on mainland Australia (Speare et al. 1994; Slocombe et al. 1995; Creeper et al. 1998), mucormycosis has not been reported in platypuses outside Tasmania. Differences in susceptibility to *M. amphibiorum* may result from genetic differences between Tasmanian and mainland platypuses populations. It is thought that *M. amphibiorum* has only recently been introduced into Tasmania, whereas mainland platypuses may have evolved with the organism over a longer period of time. It is also possible that the mainland and Tasmanian strains differ in pathogenicity for the platypus, as they do for the cane toad (Stewart & Munday 2004).

b Dermatophytosis

The dermatophyte *Trichophyton mentagrophytes mentagrophytes* has also been identified as a cause of tail alopecia in the platypus (Whittington 1992). Diagnosis is by the presence and morphology of characteristic arthroconidia in hairs and/or scabs from the periphery of the lesion or culture. Control measures include isolation of the affected animal, disinfection of the environment and checking in-contact animals. Treatment may include washing the area with povidone-iodine and/or applying a topical antifungal preparation. Lesions are usually self-limiting. This condition is a zoonosis. Seasonal mechanical alopecia of the tail (from tail biting or plugging) is more common than dermatophytosis.

9.1.3 Viral diseases

Cytomegalic inclusion disease in the renal collecting duct epithelium caused by an adenovirus-like infection has been reported in three of four wild and some captive platypuses from New South Wales (Whittington et al. 1990). Its significance to platypus populations is unknown. Infection was subclinical and thought to be chronic and persistent, with no inflammatory reaction. Infected cells showed dramatic enlargement and prominent intranuclear inclusion bodies, visible with low-power light microscopy. With electron microscopy, virions were non-enveloped and 80 nm in diameter, characteristic of adenoviruses. Attempted serology for adenovirus and cytomegalovirus infection was negative, but the platypus viral antigens may not have been cross-reactive with those in the test. The virus is likely to be shed in the urine with water-borne transmission.

Papilloma virus is thought to be the cause of papules seen on the webbing of front feet of platypuses from the Healesville region of Victoria (Booth 1994; Booth 1999). The histological appearance of lesions and weak positive immunohistochemical staining for human papilloma virus lead to the putative diagnosis. Most platypuses from the Healesville region have one or more small papules on the webbing of the front feet. Similar lesions have been reported in two captive platypuses in New South Wales (ARWH 2007). Wounds or abrasions to the webbing caused by foraging among litter or sharp objects, or contact with abrasive enclosure structures, could predispose to papilloma virus infection. All lesions seen in captive animals are on the dorsal surface of the front feet over the second interphalangeal joint, indicating that abrasion is a likely initiating factor (L Vogelnest *pers. comm.*).

9.1.4 Parasites

a Ectoparasites

Arthropods

Ixodes ornithorhynchi (platypus tick) is a common host-specific tick, found mainly attached to the less densely haired regions of the hind limbs and fore limbs but it can occur elsewhere (Fig. 6.14a). Large burdens (up to 200) of larvae, nymphs and engorging female ticks occur. The male tick has not been found, but is thought to inhabit platypus burrows (Roberts 1970). Dermatitis is not seen grossly (although puncture marks are visible after tick removal), but microscopic changes in the dermis are present (Whittington & Spratt 1989). Higher burdens were not associated with lower PCV. This tick is thought to be the vector of *Theileria ornithorhynchi* and *Trypanosoma binneyi*. Ivermectin at 200 µg/kg SC has been used to eliminate ticks from recently captive animals (Booth 1994). Another tick, *Amblyomma triguttatum trigutattum*, has also been recorded in platypuses (Roberts 1970).

An unnamed larval trombiculid (chigger) mite 1.1 mm ×0.5 mm in size was reported in the inner hair coat, superficial skin and external ear canal of a platypus; it caused minimal irritation without dermatitis (Whittington & Spratt 1989; Munday et al. 1998). A new species of trombiculid mite, *Guntheria (Phyllacarus) ornithorhynchi* sp. n., has recently been reported from a preserved platypus specimen collected in 1930 (Fain & Stekol'nikov 2004).

Two species of fleas, *Pygiopsylla hoplia* and *P. zethi*, have been reported in the platypus. Both are of unknown significance, though it has been suggested that they could cause skin irritation (Dunnet & Mardon 1974 in Whittington 1988).

b Endoparasites

Protozoa

Theileria ornithorhynchi (Fig. 6.14b) is a red blood cell parasite of the platypus found on average in about 1% of erythrocytes (Mackerras 1959). A small number of leukocytes containing the protozoan have also been observed (Collins et al. 1986). A high percentage of animals captured from the Shoalhaven and Kangaroo Rivers in New South Wales were infected (98% of 54 animals). The infection appeared to be rarer in platypuses from Victoria (Collins et al. 1986) and Tasmania (Connolly unpub). Diagnosis is made by examination of air-dried methanol-fixed blood smears stained with a Romanovsky stain such as DiffQuik® or Giemsa (Bacto Laboratories). Infected erythrocytes contain 1–4 0.5–2.0 μ sized, comma- or crescent-shaped parasites. Similar small piroplasms may be seen in leukocytes in infected platypuses. *Thelieria* multiplying by schizogeny form large basophilic Koch's bodies, which may also be seen in lymphocytes (Clark 2004). The parasite was previously thought to be non-pathogenic, but it was suspected as the cause of death in one emaciated juvenile platypus. The animal died from haemolytic anaemia and 12% of erythrocytes were infected with *T. ornithorhynchi* (Booth 1994). Infection may be more significant in immunosuppressed individuals, such as during weaning and dispersal.

Trypanosoma binneyi (Fig. 6.14c) occurs free in the plasma of platypuses from Tasmania and Victoria, but has not been observed in New South Wales (Connolly

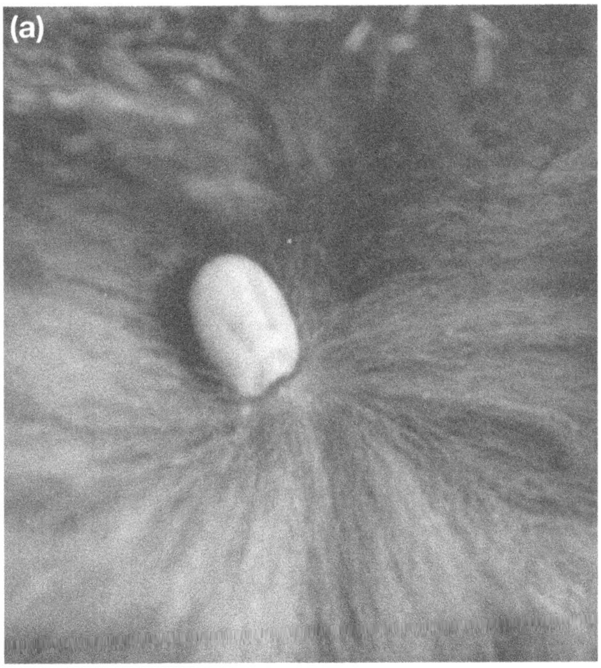

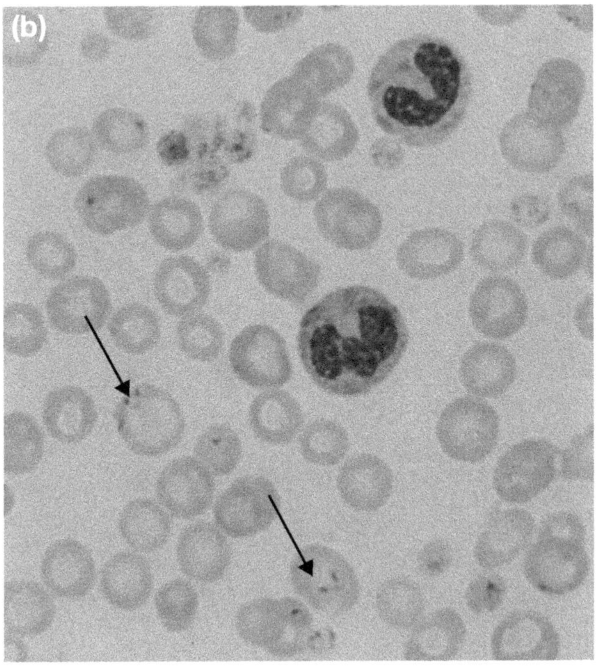

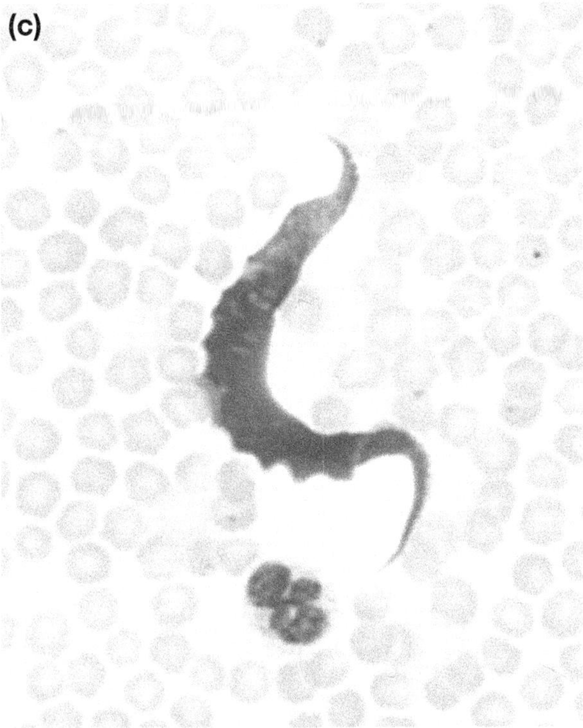

Figure 6.14 a) The platypus tick (*Ixodes ornithorhynchi*) feeding on the back of a wild platypus. This tick is thought to be the vector of the protozoan parasites b) *Theileria ornithorhynchi* and c) *Trypanosoma binneyi*. Photos: a) and c) J Connolly. b) R Booth.

et al. 1998; Munday et al. 1998). The protozoan was also reported in the blood, lung, liver and/or heart of 35% of wild platypuses from Victoria (n = 20) (McColl 1983). The pathogenicity of this parasite remains unknown. Detection is usually by observation of the parasite in DiffQuik® stained blood smears and/or histological sections of infected tissues. The organism is long and broad, with a well-developed undulating membrane and a short free flagellum (Mackerras 1959). Culture methods are difficult and often unreliable. More recently, a nested polymerase chain reaction (PCR) test has been used to detect and identify *Trypanosoma binneyi* in the platypus (Noyes et al. 1999).

Subclinical infection with *Toxoplasma gondii*-like cysts, thought to have been transmitted from faecal contamination of feed and/or bedding by cats, have been observed in the heart muscle of captive platypuses (McColl 1983). Unsporulated coccidian oocysts have also been detected in faeces and the enteric stages of *Eimeria* sp. found in the small intestine of platypuses with subclinical infections (McColl 1983; Whittington 1988).

Cestodes

Spirometra erinacei plerocercoids were found in the lung of a wild platypus and were associated with focal pneumonia and parenchymal compression (Whittington et al. 1992). There was evidence of an inflammatory and local antibody response to the plerocercoids and infection was likely due to ingestion of a host containing procercoid or plerocercoid larval stages of the parasite.

Trematodes

Several small intestinal flukes causing subclinical infection have been reported in the platypus, including *Moreauia mirabilis*, *Mehlisia ornithorhynchi* and *Maritrema ornithorhynchi* (Mackerras 1958; McColl 1983; Whittington & Spratt 1989; Whittington 1992).

Nematodes

All nematodes recorded in the platypus have been found in the skin and caused only mild dermatitis or subclinical infections. These include a larval rhabditoid nematode (Cylindrocorporidea), a larval trichostrongyloid nematode, an adult Filaroid (Lemdaninae) and *Cercopithifilaria johnstoni* (Spratt & Whittington 1989; Whittington & Spratt 1989; Spratt et al. 1991).

Faecal samples from captive platypuses can be collected opportunistically from the water in display tanks or from bags, or when examining a platypus. Coccidial oocysts, trematode eggs and pseudoparasites (ingested or contaminating large ciliates) have been found in platypus faeces (Whittington 1993). As parasitism (endo- and ectoparasites) are rare in captive platypuses, routine screening and treatment is not usually performed.

9.2 Non-infectious diseases

9.2.1 Trauma and misadventure

Many cases of trauma or misadventure in platypuses are either anthropogenic or associated with predation by native and introduced predators. Entanglement with fishing line and rubber or plastic rings has been associated with deep wounds and deaths (Booth 1994). Of 133 platypuses captured from Melbourne waterways, 11% were entangled by litter (Serena & Williams 1998). Treatment includes wound debridement and appropriate antibiotic therapy (e.g. enrofloxacin at 5 mg/kg IM sid or clavulanic acid/amoxycillin at 12.5 mg/kg IM sid). Platypuses have also been trapped in water pumps, filters, treatment plants or PVC outlet pipes (Whittington 1988; Taylor et al. 1991).

Predation by introduced predators such as the dog, fox and possibly the feral cat has been reported (Whittington 1988; Booth 1994; Grant & Temple-Smith 1998). Of 25 Tasmanian platypuses necropsied in 1994, dog attack was the cause of death in 40% of cases (Connolly et al. 1998). Native predators have included raptors such as a goshawk and eagles, and goannas, Murray cod, carpet snakes, eels and crocodiles (Richards 1986; Grant 1995; N Mooney *pers. comm.*; Connolly unpub). Platypuses have also been predated by salt water species such as single reports from a shark caught in the Kangaroo River (Whittington 1988), a fur-seal, and a leopard-seal captured near Sydney (Slip & Clippingdale).

Platypuses drown in eel, yabby and carp trap fisheries, illegally set fish nets and nets set by authorities conducting fish surveys (Whittington 1988). Design recommendations have been made to reduce platypus by-catch in the eel, yabby and carp trap fisheries in New South Wales (Grant et al. 2004b). A grid of 50–60 mm at the entrance to an eel or yabby trap was found to successfully exclude platypuses. Restricting eel traps to estuarine areas would further reduce platypus by-catch, and providing air spaces in the cod ends of traps used in dams would prevent drowning.

Platypus roadkills frequently occur in Tasmania and may occur in other states. Of 25 Tasmanian platypuses necropsied in 1994, 28% died as a result of road trauma (Connolly et al. 1998). Most platypuses were hit by cars as they crossed roads where small streams were piped under the road through a culvert (Tyson 1980; Taylor et al. 1991). Platypuses may have preferred to travel overland when the stream was flowing too rapidly through the culvert or where the step up into the culvert was too high. At Liffey in Tasmania, ramps leading up to the pipe increased usage of the culvert and reduced platypus roadkills (C Spencer *pers. comm.*). A workshop on culvert avoidance by platypuses proposed measures to ameliorate platypus roadkills where new culverts were being constructed (Tanner 2003). Measures proposed included bio-baffle installation to assist platypus passage through the culvert (successfully used with fish); provision of suitable access to culverts, including perch height; provision of an alternate dry-passage culvert adjacent to the waterway culvert; and funnel fencing installation to encourage platypuses to use the culvert rather than cross the road.

9.2.2 Toxins

Exposure of wildlife to persistent organic pollutants has been linked to a wide range of clinical signs depending on the species of animal and the type of exposure. Acute exposure may lead to death, but more commonly toxicity results from chronic exposure to residues in the environment and bioaccumulation over months. The occurrence of persistent organic pollutants in Tasmanian platypuses has recently been reported (Munday et al. 2002). Tail fat samples from 56 platypuses from six Tasmanian sites were analysed for organochlorine pesticides (DDT, its metabolites and Lindane) and polychlorinated biphenyls (PCBs). Some differences were observed in persistent organic pollutant levels in platypuses from different geographic regions. Levels of organochlorine pesticides were related to intensity of agricultural activity, while levels of PCBs were higher near industrial and/or hydroelectric developments. Paper and pigment production and transformer oil use by hydroelectric plants in the past have led to industrial pollution in some sites studied. The maximum levels of DDT and Lindane (11 857 and 96 ng/g lipid respectively) in these platypuses were the highest reported for any Australian native mammal yet tested (marsupials or monotremes) (Bolton-Grob et al. 2001 in Munday et al. 2002).

Heavy metal pollution has been detected in the South Esk River (Tas.) where tungsten and tin mines discharged mine tailings until 1959. A platypus that died of mucormycosis from the South Esk River at Perth (100 km downstream from the mines) had very high cadmium (69 ppm) and selenium (17 ppm) levels in its liver (Rounsevell 1990; Connolly et al. 1998).

A captive male platypus died as a result of being envenomated by another male platypus, which supports the idea that this is a major function for the crural system. The spurred platypus had erythematous skin on the bill and feet, SC haemorrhages at multiple spur sites, haemolysed blood and haematuria. A captive platypus died following a funnelweb spider falling into the feeding tank. No bite was witnessed, but the dead spider and the dead platypus were both found in the water (J Giles *pers. comm.*).

9.2.3 Nutritional diseases

Starvation and/or exposure are a recognised cause of winter mortalities in dispersing juvenile platypuses (Obendorf 1990). Of 25 Tasmanian platypuses necropsied in 1994, the cause of death was starvation/exposure in 16% of cases (Connolly et al. 1998). The juvenile platypuses that died of this condition typically had low body weight (<450 g in males under 6 mo), no food in the digestive tract, no fat reserves and a tail score of 4–5.

9.2.4 Thyroid hyperplasia

The normal thyroid gland in the platypus measures 2 cm × 1.75 cm and lies just proximal to the arch of the aorta in the thoracic cavity (McKenzie 1921). Diffuse thyroid hyperplasia is believed to have contributed to the death of a captive adult male platypus. The gland measured 5.5 cm × 3.5 cm, occupied almost half the pleural cavity and weighed 14 g, about 1% of total body mass. There were no observed clinical signs prior to death. Compared with values obtained from an age, sex and captive status matched platypus there was no evidence of abnormal T_3 (0.5–0.6 nmol/L, n = 2) or T_4 (19 nmol/L, n = 2) in blood samples collected coincidentally 2 mo prior to death.

The maximum ambient temperature in the platypus' tunnel system the day it died was 34°C, which exceeds the ideal for this species (see 5.1). It is surmised that the animal died of hyperthermia with the space-occupying lesion compounding the problem.

9.2.5 Hypothyroidism

One of two platypuses born in captivity in 1999 died of aplastic anaemia, hyperplastic goitre and hypothyroidism despite intensive treatment that included 500 IU SC erythropoietin and a blood transfusion of 9 mL of blood with 1:10 anticoagulant delivered into the bill sinus. The PCV was 0.21–0.22, Hb 78–80 g/L and red cells showed anisocytosis, irregular contracted cells, acanthocytes, spherocytes, fragmented red cells and mild hypochromasia. At necropsy, there was a distinct lack of erythroid component in the bone marrow, consistent with congenital hypoproliferative anaemia. The anaemia probably progressed as the foetal blood cells became senescent and removed but not replaced (P Clark *pers. comm.*).

Diagnosis of hypothyroidism was made post-mortem. The cause was unknown but possibly congenital with a defect in thyroglobulin synthesis rather than an acquired problem related to iodine intake (R Slocombe *pers. comm.*). Retrospective serum analysis revealed T_3 1.8 pmol/L and T_4 <5.0 pmol/L. The sibling of this platypus was robust and continues to thrive in captivity.

9.2.6 Developmental disorders and neoplasia

A Z-shaped bill deviation due to the absence of part of the left maxilla was reported in a juvenile platypus (Holz 1999). The platypus was underweight, unable to close its mouth and failed to thrive in captivity, having difficulty grasping its food. It was euthanased. The cause of the maxillary deformity was unknown.

9.2.7 Other

Other reported miscellaneous conditions of the platypus include focal myocardial degeneration in recently captured platypuses, hydropericardium, pulmonary oedema, renal tubular mineralisation, cerebral infarction and typhlitis (Whittington 1988). A wild adult female platypus with mucormycosis appeared to be blind and was observed to have bilateral cataracts (J Connolly unpub). The platypus was captured three times in the same area, maintaining its weight (1300 g, 1220 g and 1300 g respectively) and its moderate body condition (TVI 3–4). She appeared able to feed adequately, but was a target for predators and was found dead and scavenged 24 d after the first capture.

10 EFFECT OF HABITAT DISTURBANCE ON WILD PLATYPUS POPULATIONS

10.1 Natural occurrences

The effect of drought and flood on platypus populations has been investigated (Grant et al. 2004a). Following a 4 yr drought, both breeding success (measured by percentage of lactating females) and recruitment (measured by juveniles captured) dropped sharply. In other years, droughts had no apparent effect on breeding success or recruitment. Floods have caused significant mortality in platypus populations. No juvenile platypuses could be captured following two over-bank flood events in late December and early January. It is thought that nestlings confined to burrows may have drowned (Grant et al. 2004a). In one case, flooding dislodged a lactating platypus from its territory and it was believed that the resulting stress led to the development of a toxaemic state (neutropenia with a left shift and toxic changes) and aspiration pneumonia (Whittington & McColl 1983).

10.2 Anthropogenic impacts on populations

Agricultural practices such as allowing cattle unrestricted access to stream banks and the clearing of riverside (riparian) vegetation cause bank erosion. Sedimentation and reduced flushing flows in streams have resulted in reduced pool depths and loss of preferred foraging substrates for platypuses. Water extraction from streams also reduces the pool depth (Rounsevell 1990 unpub; Grant 1995; Grant 2004b).

The construction of a dam may affect platypus populations depending on the size of the impoundment, creating drought-like conditions downstream and pools too deep for feeding upstream (Grant 1995).

11 THERAPEUTICS

There is little published information on therapeutics in the platypus. Various antimicrobials and parasiticides have been used at small animal dose rates with apparent therapeutic effect and no adverse reactions.

Intramuscular injections need to be given into the diminutive muscles of the hind or fore legs, because the tail is mostly a fat store.

Most medications are given SC. The SC route is ideal for fluids, since the IV route is more challenging and there are no really good veins for long-term catheterisation. The skin is loose and large volumes of SC fluids can easily be administered overlying the ribcage.

There are few veins which are accessible. The bill sinus is the most accessible venipuncture site and has been used to inject volumes up to about 10 mL (see 7.1.1 and Fig. 6.11). Other potential sites for IV access include the ventral tail vein which runs along the ventral midline of the coccygeal vertebrae, and the jugular vein. The jugular vein is large, but the loose skin and short neck make it very difficult to access.

The platypus' mouth cannot be opened widely and the torus linguae of the tongue makes it impossible to see the larynx. Stomach tubing is done blind, and should be avoided if possible. It cannot be carried out on a healthy conscious platypus, and a moribund or anaesthetised platypus could easily aspirate. Sedation with diazepam at 0.5–1 mg/kg will assist the procedure if deemed necessary.

ACKNOWLEDGMENTS

The authors would like to thank Tom Grant, Peter Temple-Smith and David Middleton for unpublished information and/or permission to use their images.

REFERENCES

ARWH (2007). Australian Registry of Wildlife Health. Zoological Parks Board of NSW: Sydney.

Augee M & Gooden B (1993). *Echidnas of Australia and New Guinea*. UNSW Press: Sydney.

Belov K & Hellman L (2003). Immunoglobulin genetics of *Ornithorhynchus anatinus* (platypus) and *Tachyglossus aculeatus* (short-beaked echidna). *Comparative Biochemistry and Physiology Part A* **136**, 811–819.

Booth RJ (1994). Medicine and husbandry: monotremes, wombats and bandicoots. In *Wildlife. Proceedings 233*, pp. 395–404. Post Graduate Committee in Veterinary Science: Sydney.

Booth RJ (1999). Care and medical management of monotremes. In *Wildlife in Australia Healthcare and Management, Proceedings 327*, pp. 41–50. Post Graduate Foundation in Veterinary Science: Sydney.

Booth RJ (2003). Monotremes. In *Zoo and Wild Animal Medicine: Current Therapy 5* (ed. ME Fowler), pp. 278–287. WB Saunders: Philadelphia.

Booth RJ & Whittington RJ (1993). Indirect assessment of the health of platypus using haematology. Symposium on comparative biochemistry and physiology of marsupial and monotreme blood. Department of Biochemistry, University of Sydney: Sydney.

Burrell H (1927). *The Platypus*. Rigby: Adelaide.

Canfield PJ & Whittington RJ (1983). Morphological observations on the erythrocytes, leukocytes and platelets of free-living platypuses, *Ornithorhynchus anatinus* (Shaw). (Monotremata: *Ornithorhynchidae*). *Australian Journal of Zoology* **31**, 421–432.

Carrick FN & Hughes RL (1978). Reproduction in male monotremes. *Australian Zoologist* **20**, 211–231.

Clark P (2004). Haemoparasites of Australian mammals. In *Haematology of Australian Mammals* (ed. P Clark), pp. 147–162. CSIRO Publishing: Melbourne.

Collins GH, Whittington RJ & Canfield PJ (1986). *Theileria ornithorhynchi* Mackerras, 1959 in the platypus *Ornithorhynchus anatinus* (Shaw). *Journal of Wildlife Diseases* **22**, 19–24.

Connolly JH & Obendorf DL (1998). Distribution, captures and physical characteristics of the platypus (*Ornithorhynchus anatinus*) in Tasmania. *Australian Mammalogy* **20**, 229–235.

Connolly JH, Obendorf DL, Whittington RJ & Muir DB (1998). Causes of morbidity and mortality in platypus (*Ornithorhrynchus anatinus*) from Tasmania, with particular reference to *Mucor amphibiorum* infection. *Australian Mammalogy* **20**, 177–187.

Connolly JH, Canfield PJ, McClure SJ & Whittington RJ (1999a). Histological and immunohistological investigation of lymphoid tissue in the platypus (*Ornithorhynchus anatinus*). *Journal of Anatomy* **195**, 161–171.

Connolly JH, Obendorf DL & Whittington RJ (1999b). Haematological, serum biochemical and serological features of platypuses with and without mycotic granulomatous dermatitis. *Australian Veterinary Journal* **77**, 809–813.

Connolly JH, Canfield PJ & Obendorf DL (2000). Gross, histological and immunohistological features of mucormycosis in the platypus. *Journal of Comparative Pathology* **123**, 36–46.

Creeper JH, Main DC, Berger L, Huntress S & Boardman W (1998). An outbreak of mucormycosis in slender tree frogs (*Litoria adelensis*) and white–lipped tree frogs (*Litoria infrafrenata*). *Australian Veterinary Journal* **76**, 761–762.

Dawson TJ (1983). *Monotremes and Marsupials: The Other Mammals. Institute of Biology's Studies in Biology no. 150.* Edward Arnold: London.
de Plater G, Martin RL & Milburn PJ (1995). A pharmacological and biochemical investigation of the venom from the platypus (*Ornithorhynchus anatinus*). *Toxicon* **33**, 157–169.
Dunn LC (1949). Rules applying to warm-blooded vertebrates. In *The Systematics and the Origin of Species. No. XIII of the Columbia Biological Series 90* (ed. LC Dunn), Columbia University. OUP: London.
Dunnett GM & Mardon DK (1974). Monograph of Australian fleas (*Siphonaptera*). *Australian Journal of Zoology Supplement* **30**.
Fain A & Stekol'nikov AA (2004). A new species of the genus *Guntheria* (Acari Trombiculidae) from a duck-billed platypus collected in the New South Wales. *Belgian Journal of Entomology* **6**, 69–73.
Fenner PJ, Williamson JA & Myers D (1992). Platypus envenomation: a painful learning experience. *Medical Journal of Australia* **157**, 829–832.
Fleay D (1980). *Paradoxical Platypus Hobnobbing with Duckbills.* Jacaranda Press: Brisbane.
Fleay D (1944). *We Breed the Platypus.* Robertson & Mullens: Melbourne.
Gardner JL & Serena M (1995). Spatial organisation and movement patterns of adult male platypus, *Ornithorhrynchus anatinus* (Monotremata: *Ornithorhynchidae*). *Australian Journal of Zoology* **43**, 91–103.
Gibson RA, Neumann M, Grant TR & Griffiths M (1988). Fatty acids of the milk and food of the platypus (*Ornithorhynchus anatinus*). *Lipids* **23**, 377–379.
Grant TR (1989). *Ornithorhynchidae.* In *Fauna of Australia,* vol. 1B Mammalia, pp. 436–450. AGPS, Canberra.
Grant TR (1995). *The Platypus: A Unique Mammal.* UNSW Press: Sydney.
Grant TR (2004a). Captures, capture mortality, age and sex ratios of platypuses, *Ornithorhynchus anatinus*, during studies over 30 years in the Upper Shoalhaven River in New South Wales. *Proceedings of the Linnean Society of New South Wales* **125**, 217–226.
Grant TR (2004b). Depth and substrate selection by platypuses, *Ornithorhynchus anatinus*, in the Lower Hastings River, New South Wales. *Proceedings of the Linnean Society of New South Wales* **125**, 235–241.
Grant TR & Carrick FN (1978). Some aspects of the ecology of the platypus, *Ornithorhynchus anatinus* in the upper Shoalhaven River, New South Wales. *Australian Zoologist* **20**, 181–199.
Grant TR & Dawson TJ (1978). Temperature regulation in the platypus, *Ornithorhynchus anatinus*, production and loss of metabolic heat in air and water. *Physiological Zoology* **51**, 315–332.
Grant TR & Griffiths M (1992). Aspects of lactation and determination of sex ratios and longevity in a free-ranging population of platypuses, *Ornithorhynchus anatinus*, in the Shoalhaven River, NSW. In *Platypus and Echidnas* (ed. ML Augee), pp. 80–89. Royal Zoological Society of New South Wales: Sydney.
Grant TR & Whittington RJ (1991). The use of freeze-branding and implanted transponder tags as a permanent marking method for platypuses, *Ornithorhynchus anatinus* (Monotremata: *ornthorhynchidae*). *Australian Mammalogy* **14**, 147–150.
Grant TR & Temple-Smith PD (1998). Field biology of the platypus (*Ornithorhynchus anatinus*): historical and current perspectives. *Philosophical Transactions of the Royal Society of London B* **353**, 1081–1091.
Grant TR, Griffiths M & Leckie RMC (1983). Aspects of lactation in the platypus, *Ornithorhynchus anatinus* (monotremata), in waters of eastern New South Wales. *Australian Journal of Zoology* **31**, 881–889.
Grant TR, Griffiths M & Temple-Smith (2004a). Breeding in a free-ranging population of platypuses, *Ornithorhynchus anatinus*, in the Upper Shoalhaven River, New South Wales—a 27 year study. *Proceedings of the Linnean Society of New South Wales* **125**, 227–234.
Grant TR, Lowry MB, Pease B, Walford TR & Graham K (2004b). Reducing the by-catch of platypuses (*Ornithorhynchus anatinus*) in commercial and recreational fishing gear in New South Wales. *Proceedings of the Linnean Society of New South Wales* **125**, 259–272.
Griffiths M (1978). *The Biology of the Monotremes.* Academic Press: New York.
Griffiths M (1988). The platypus. *Scientific American* **256**, 84–91.
Griffiths M, Green B, Leckie RMC, Messer M & Newgrain KW (1984). Constituents of platypus and echidna milk with particular reference to the fatty acid complement of the triglycerides. *Australian Journal of Biological Sciences* **37**, 323–329.
Grigg G, Beard L, Grant T & Augee M (1992). Body temperatures and diurnal activity patterns in the platypus (*Ornithorhynchus anatinus*) during winter. *Australian Journal of Zoology* **40**, 135–142.
Holland N & Jackson SM (2002). Reproductive behaviour and food consumption associated with the captive breeding of platypus (*Ornithorynchus anatinus*). *Journal of Zoology* **256**, 279–288.
Holz PH (1999). Maxillary deformity in a wild platypus (*Ornithorhynchus anatinus*). *Australian Veterinary Journal* **77**, 752.
Jackson S, Serena M & Middleton D (2003). Platypus. In *Australian mammals: Biology and Captive Management* (ed. S Jackson), pp. 1–32. CSIRO Publishing: Melbourne.
Jakubowski JM, New NP, Stone GM & Jones RC (1998). Reproductive seasonality in female platypuses, *Ornithorhynchus anatinus*, in the Upper Barnard River, New South Wales. *Australian Mammalogy* **20**, 207–213.
Krueger B, Hunter S & Serena M (1992). Husbandry, diet and behaviour of platypus *Ornithorhynchus anatinus* at Healesville Sanctuary. *International Zoo Yearbook* **31**, 64–71.

Mackerras MJ (1958). Catalogue of Australian mammals and their recorded internal parasites. I–IV. *Proceedings of the Linnean Society of New South Wales* **53**, 101–160.

Mackerras MJ (1959). The haematozoa of Australian mammals. *Australian Journal of Zoology* **7**, 105–135.

McColl KA (1983). Pathology in captive platypus (*Ornithorhynchus anatinus*) in Victoria, Australia. *Journal of Wildlife Diseases* **19**, 118–122.

McColl KA & Whittington RJ (1985). Leptospiral titres in wild platypuses (*Ornithorhynchus anatinus*) in New South Wales. *Australian Veterinary Journal* **62**, 66–67.

McDonald IR (1978). Adrenocortical functions in monotremes. *Australian Zoologist* **20**, 79–85.

McKee J, Booth RJ, Spittal D & Romer E (1995). Design of a semi-free range platypus breeding and research facility. Unpublished report.

McKenzie C (1921). Further observations on some new mammalian ductless glands: the structure of the parathyroid glands. *Studies from the Australian Institute of Anatomical Research.* Jenkin, Buxton & Co: Melbourne.

Munday BL & Peel BF (1983). Severe ulcerative dermatitis in platypus (*Ornithorhynchus anatinus*). *Journal of Wildlife Diseases* **19**, 363–365.

Munday BL, Whittington RJ & Stewart NJ (1998). Disease conditions and subclinical infections of the platypus (*Ornithorhynchus anatinus*). *Philosophical Transactions of the Royal Society of London B Biological Sciences* **353**, 1093–1099.

Munday BL, Stewart NJ & Södergren A (2002). Accumulation of persistent organic pollutants in Tasmanian platypus (*Ornithorhynchus anatinus*). *Environmental Pollution* **120**, 233–237.

New NP, Jakubowski JM, Stone GM & Jones RC (1998). Seasonal pattern of androgen secretion in the male platypus, *Ornithorhynchus anatinus*, in the Upper Barnard River, New South Wales. *Australian Mammalogy* **20**, 215–220.

Noyes HA, Stevens JR, Teixeira M, Phelan J & Holz P (1999). A nested PCR for the ssrRNA gene detects *Trypanosoma binneyi* in the platypus and *Trypanosoma* sp. in wombats and kangaroos in Australia. *International Journal for Parasitology* **29**, 331–339.

Obendorf D (1990). Winter mortalities in juvenile platypus. In *Wildlife Disease Association (Australasian Section) Newsletter*, 7 September.

Obendorf DL, Peel BF & Munday BL (1993). *Mucor amphibiorum* infection in platypus (*Ornithorhynchus anatinus*) from Tasmania. *Journal of Wildlife Diseases* **29**, 485–487.

Pettigrew JD, Manger PR & Fine SLB (1998). The sensory world of the platypus. *Philosophical Transactions of the Royal Society of London B* **353**, 1199–1210.

Richards RS (1986). Predation on a platypus, *Ornithorhynchus anatinus* (Monotremata: *Ornithorhynchidae*), by a goshawk. *Australian Mammalogy* **9**, 67.

Roberts FHS (1970). *Australian Ticks.* CSIRO: Melbourne.

Rounsevell D (1990). *The Platypus in Tasmania.* Report from Tasmanian Department of Lands, Parks and Wildlife: Hobart.

Schalm OW, Jain NC & Carroll ED (1975). *Veterinary Haematology.* Lea & Febiger: Philadelphia.

Scheich H, Langner G, Tideman C, Coles RB & Guppy A (1986). Electroreception and electrolocation in platypus. *Nature* **319**, 401–402.

Schipper MAA (1978). On certain species of *Mucor* with a key to all accepted species. *Studies in Mycology* **17**, 1–52.

Serena M & Williams GA (1998). Rubber and plastic rubbish: a summary of the hazard posed to platypus *Ornithorhynchus anatinus* in suburban habitats. *Victorian Naturalist* **115**, 47–49.

Slip & Clippingdale Australian Government Antarctic Division <http://www.aad.gov.au/default.asp?casid=1769>

Slocombe R, McCracken H, Booth R, Slocombe J & Birch C (1995). Infectious diseases of captive frogs. In *Australian Society for Veterinary Pathology Annual Proceedings*, p 14. Australian Society for Veterinary Pathology: Melbourne.

Speare R, Thomas AD, O'Shea P & Shipton WA (1994). *Mucor amphibiorum* in the toad, *Bufo marinus*, in Australia. *Journal of Wildlife Diseases* **30**, 399–407.

Spratt DM & Whittington RJ (1989). Nematode parasites from the skin of the platypus *Ornithorhynchus anatinus* (Monotremata). *Systematic Parasitology* **13**, 223–229.

Spratt DM, Beveridge I & Walter EL (1991). A catalogue of Australasian monotremes and marsupials and their recorded Helminth parasites. *Records of the South Australian Museum, Monograph Series* **1**, 1–105.

Stewart NJ & Munday BL (2004). Possible differences in pathogenicity between cane toad-, frog- and platypus-derived isolates of *Mucor amphibiorum*, and a platypus-derived isolate of *Mucor circinelloides. Medical Mycology* **43**, 127–132.

Tanner Z (2003). *Culvert Avoidance by Platypuses and Subsequent Road Deaths: Workshop Aims and Outcomes.* Tasmanian Wildlife Roadkill Collective Report: Hobart.

Taylor R, Mooney N & Lange K (1991). Observations on platypus. *Tasmanian Naturalist* **105**, 1–3.

Tonkin MA & Negrine J (1994). Wild platypus attack in the antipodes: a case report. *Journal of Hand Surgery (British and European Volume, 1994)* **19B**, 162–164.

Tyson RM (1980). Road killed platypus. *Tasmanian Naturalist* **60**, 8.

Vogelnest L (1999). Chemical restraint of Australian native fauna. In *Wildlife in Australia: Healthcare and Management, Proceedings 327*, pp. 149–187. Post Graduate Foundation in Veterinary Science: Sydney.

Whittington RJ (1988). The monotremes in health and disease. *Australian Wildlife, Proceedings 104*, pp. 727–787. Post Graduate Committee in Veterinary Science: Sydney.

Whittington RJ (1992). The role of infectious disease in the population biology of montremes. In *Platypus and Echidnas* (ed. ML Augee), pp. 285–292. Royal Society of New South Wales: Sydney.

Whittington RJ (1993). Diseases of monotremes. In *Zoo and Wild Animal Medicine: Current Therapy 3* (ed. ME Fowler), pp. 269–276. WB Saunders: Philadelphia.

Whittington RJ & McColl KA (1983). Aspiration pneumonia in a wild platypus (*Ornithorhynchus anatinus*). *Australian Veterinary Journal* **60**, 277.

Whittington RJ & Spratt DM (1989). Lesions associated with metazoan parasites of wild platypuses (*Ornithorhynchus anatinus*). *Journal of Wildlife Diseases* **25**, 521–526.

Whittington RJ, Bell IG & Searson JE (1990). A viral infection causing cytomegalic inclusion disease in the renal epithelium of the platypus (*Ornithorhynchus anatinus*). *Journal of Wildlife Diseases* **26**, 55–61.

Whittington R, Middleton D, Spratt DM, Muntz F, Carmel B, McCracken HE, Strakosch MR, Stephanson-Shaw J, Harper PA & Hartley WJ (1992). Sparganosis in the monotremes *Tachyglossus aculeatus* and *Ornithorhynchus anatinus* in Australia. *Journal of Wildlife Diseases* **28**, 636–640.

Whittington RJ, Connolly JH, Obendorf DL, Emmins JJ, Grant TR & Handasyde KA (2002). Serological responses against the pathogenic dimorphic fungus *Mucor amphibiorum* in populations of platypus (*Ornithorhynchus anatinus*) with and without ulcerative dermatitis. *Veterinary Microbiology* **87**, 59–71.

7. Macropods

Larry Vogelnest and Tim Portas

1 TAXONOMY, DISTRIBUTION AND BIOLOGY

The Super-family Macropodoidea includes 19 species in New Guinea and 49 species in Australia, with three species common to both. Another six species have become extinct in Australia since European settlement, and the ranges and distribution of many others have reduced significantly. There are three extant Families: the musky rat-kangaroo (*Hypsiprymnodon moschatus*) is the only living member of the Family Hypsiprymnodontidae. All other species are placed in the Families Potoroidae and Macropodidae. In this chapter and others in this volume members of the Super-family Macropodoidea are referred to as 'macropods', a term meaning 'big foot'. Although not strictly correct it is in common usage and generally recognised as the term that encompasses kangaroos, wallabies, potoroos, bettongs and the musky rat-kangaroo. Strictly speaking the derivative 'macropodoids' should be used for the Super-family and 'potoroids' for the Family Potoroidae and 'macropodids' for the Family Macropodidae.

The Macropodoidea are a diverse group ranging in size from 0.5–90 kg body weight and are found throughout Australia. They occupy almost all terrestrial habitats including tropical rainforests, woodlands, open grasslands, deserts, mountains, islands and rocky cliffs (Strahan 1995; Menkhorst 2001). Some species are arboreal, some live in caves and two species are fossorial (Tyndale-Biscoe 2005). Most are nocturnal, while the medium and larger species tend to be crepuscular. They are essentially sedentary, occupy a persistent home range and include solitary non-social species through to gregarious species that live in well organised societies. The degree of gregariousness generally increases with body size, openness of habitat and proportion of grasses in the diet.

The teeth (diprotodont) and feet (syndactylus) of macropods place them in the Order Diprotodontia. With the exception of the tiny musky rat-kangaroo, they are distinguished from other marsupials by foregut fermentation, hopping and embryonic diapause.

The musky rat-kangaroo is considered the closest to the ancestral stock from which all kangaroos evolved. Although it superficially resembles the other rat-kangaroos, it is only distantly related. Furthermore, it bounds using all its legs rather than hopping. It is the only macropod with five toes on the hind foot, with digit one being opposable. It is omnivorous and its teeth and stomach, which is simple, resemble those of possums more than other macropods. It does not exhibit embryonic diapause (Strahan 1995; Tyndale-Biscoe 2005).

The Subfamily Potoroinae includes nine species of rat-kangaroo (including bettongs and potoroos), which range in size from 1 kg to 3 kg. Rat-kangaroos hop, but in contrast to kangaroos and wallabies can use their tails to pick up objects such as grass. An upper canine tooth is present, the premolars are very large with prominent flutings on the sides, and the four molars behind them decrease in size from front to back. The molars remain in the same position in the jaw throughout life—unlike wallabies and kangaroos, which exhibit molar progression. The rat-kangaroos have a sacculated forestomach, and all species display post partum oestrus and embryonic diapause (Tyndale-Biscoe 2005).

The banded hare-wallaby (*Lagostrophus fasciatus*) is the only species in the Subfamily Sthenurinae. It has some of the features of the potoroines and macropodines (Tyndale-Biscoe 2005). One diagnostic feature is the lower incisors that bite against the upper incisors, not behind as in other macropodids. The banded hare-wallaby is now extinct on mainland Australia and survives only on Bernier and Dorre Islands in Western Australia.

The largest Subfamily is the Macropodinae, which consists of 45 species and includes the wallabies and kangaroos. Wallabies and kangaroos are distinguished by size. The six largest species are called kangaroos and the smaller species wallabies (Dawson 1995). The kangaroos are not only united as a group by their size, they are also all grazers and, unlike many of the smaller macropod species, none are endangered (Tyndale-Biscoe 2005).

Three species, the spectacled hare-wallaby (*Lagorchestes conspicillatus*), red-legged pademelon (*Thylogale stigmatica*) and agile wallaby (*Macropus agilis*) are found in both Australia and New Guinea. Only three species are found in Tasmania: the Tasmanian pademelon (*Thylogale billardierii*), Bennett's or red-necked wallaby (*M. rufogriseus*) and the eastern grey kangaroo (*M. giganteus*) (Tyndale-Biscoe 2005).

The musky rat-kangaroo, four species of bettong, the pademelons, most of the rock-wallabies and the quokka (*Setonix brachyurus*) have 22 chromosomes. The swamp wallaby (*Wallabia bicolor*) has the smallest number of chromosomes (10). Another unique feature of the swamp wallaby is that the male has two Y chromosomes, giving a total chromosome number of 11 in the male. The male long-nosed potoroo (*Potorous tridactylus*) also has an extra Y chromosome, giving a total chromosome number of 13 (Tyndale-Biscoe 2005).

2 ANATOMY AND PHYSIOLOGY

2.1 Musculoskeletal system

Macropods have a distinctive body form which primarily relates to their locomotion. Powerful hind limbs with a long narrow hind foot and powerful fourth toe are characteristic of most species. Rock-wallabies and tree-kangaroos are an exception, having hind feet that are shorter and broader. The hind limbs and pelvic girdle are much longer, larger and more heavily muscled than the fore limbs and pectoral girdle. The animal's centre of mass is in the pelvis. The major role in propulsion and mass-suspension is performed by the hindquarters (Hume et al. 1989). The femur, tibia, fibula and pes are elongated. The tibia and fibula are closely in contact and the fibula is reduced distally to a thin splint. There is no patella. There is a strongly developed, elongate and load-bearing calcaneum. Movement of the tarsus is restricted to a back and forth motion; it cannot twist sideways. This is partly due to an expansion of the lateral process of the calcaneum. Other than in the musky rat-kangaroo, the first digit of the hind foot is absent and the second and third digits are syndactylous. They are very small and take no part in support or propulsion, but are used for grooming. The fourth digit is the largest. The fore limb is relatively small and poorly developed and has a simple manus with five digits of equal length. No digits are opposable.

The tail is long and may be slender and of uniform thickness along its length or proximally robust depending on its function. It is non-prehensile and cannot be flicked. The tail serves as a balancer when hopping or climbing trees and in the larger species it supports the hindquarters in slow 'pentapedal' locomotion. Together with the extended hind limbs, it forms a tripod to help the animal stand and move in a vertical posture. The tail of the quokka is short and does not contribute to locomotion. In nail-tailed wallabies the tail has a horny tip (Strahan 1995). Rat-kangaroos can use their tails to pick up grass and other material.

The atlanto-occipital articulation is very flexible. allowing the muzzle to remain horizontal whether the neck is horizontal or vertical. This flexibility may be associated with some weakness at this articulation, which may account for the high incidence of traumatic fractures at this site (Fig. 7.17).

Dimorphism in macropods is related to size. The smallest species are homomorphic and the largest heteromorphic. In the latter the males and sometimes the females grow throughout life, although at a decelerating rate. The males are significantly more muscular, broad-chested and wide-shouldered. In the large kangaroos the females may be a quarter to one-third the size of the male, making these one of the most heteromorphic mammals. Other than in wallaroos (*M. robustus*) antilopine (*M. antilopinus*) and red kangaroos (*M. rufus*), there are only slight differences in colouration between the sexes. Sex-specific colouration is present throughout life and is not assumed by males only upon social maturation (Hume et al. 1989).

The epipubic bones that extend from the cranial aspect of the pubis are proportionately small in macropods but are fully ossified. Since growth is persistent in macropods, the epiphyses of the limb bones remain unfused during life (Hume et al. 1989).

Some distinct features of the skull include palatal vacuities in some species and a deep mandibular masseteric fossa or canal which allows the deep masseter muscle to gain deep insertion in the mandible below the lower molar teeth (Hume et al. 1989). This is a particular feature of the potoroines and to a lesser extent the macropodines, and is not seen in other herbivorous marsupials (Sanson 1989). Posteromedially, the mandible may be concave for the internal pterygoid muscle attachment. At the inferior edge is a characteristic medially-directed inflected angle forming a shelf which is characteristic of most marsupials (Dawson et al. 1989).

In most macropods the mandibular symphysis is not fused and is flexible. This allows the lower incisors to spread apart and shear against the upper incisors. In the banded hare-wallaby and the potoroines the symphysis is more rigid and may be fused.

2.2 Dentition

Macropods have a wide range of dental adaptations reflecting their differing diets (Fig. 7.1). Most macropods have 32 or 34 teeth with the following dental formula: I 3/1 C (1)/0 P (2)1/(2)1 M 4/4 (Hume et al. 1989). In the macropodines P1 is missing. In young kangaroos P2 (a large cutting or sectorial premolar) and a large molar-like tooth, called deciduous premolar 3 (dP3), are present. These provide a shearing and grinding tooth in each jaw. Later, when M1 and M2 erupt, P2 and dP3 are shed and replaced by a single cutting tooth rostral to the molars (P3). As the animal ages the remaining molars, M3 and M4, erupt sequentially behind M2. Occasionally, old eastern grey kangaroos, red kangaroos and wallaroos may have a fifth set of molars (Jackson 2003; Tyndale-Biscoe 2005).

Dentition of macropods is typically diprotodont with a large single pair of forward-projecting lower incisors and a large diastema between these and the premolars. The second and third upper incisors, which tend to lie behind the central incisor in other diprotodonts, lie alongside these in macropodines, forming an arc with a continuous cutting edge, within which the large lower incisors fit. The lower incisors press against a tough pad on the roof of the mouth, just behind the upper incisors. This results in a shearing action when biting. The unfused mandibular symphysis also allows the lower incisors to spread apart and increases the shearing action with the upper incisors. As the mandibular symphysis in the banded hare-wallaby and potoroines is fused, the lower incisors meet the upper incisors directly. Canine teeth, while fairly well-developed in the potoroines, are absent or vestigial in the macropodines, other than tree-kangaroos which have a small canine tooth in the maxilla. As in the mandible, there is a diastema between the upper incisors and first premolar (Sanson 1989; Strahan 1995; Tyndale-Biscoe 2005) (Fig. 7.1).

The main difference between potoroine teeth and macropodine teeth is the size and structure of the molars. In all macropodines the four cusps are joined in pairs to form high transverse ridges or lophs, unlike the rounded teeth of potoroines. In the macropodines, additional ridges called links also develop at right angles to the lophs (lophodont molars). In the potoroines, lateral chewing is possible while in the macropodines lateral movement of the mandible is restricted by the position of the lower incisors inside the upper incisor arc. In macropodines, the chewing is more a forward and backward motion with a small degree of sideward motion (Tyndale-Biscoe 2005).

The dentition of macropods has been graded on the basis of degree of adaptation to processing their diets. This is based on the size of P3 and the height of the molar lophs and links, and whether the molar rows are flat or arched (Sanson 1989; Hume 1999).

2.2.1 Potoroids and basal macropodoids

These include the musky rat-kangaroo and the Potoroidea. They primarily eat invertebrates, fruits and seeds. Thus they are ingesting and masticating a variety of materials with quite different physical properties.

The premolars are well-developed longitudinal shearing blades and are used to cut or open food items to liberate the contents. The molars are therefore used primarily for crushing and grinding. The premolars are retained throughout life. The molar row is flat with all teeth in occlusion. The gape in these species is larger than other macropods, allowing the ingestion of larger food items.

2.2.2 Browsers

These include the pademelons, quokkas and swamp wallabies that browse on shrubs and soft plants, and the tree-kangaroos which feed on tree foliage. All have moderately large premolars which are used for cutting (Fig. 7.1a). They are retained throughout life. The retained P3 blocks molar progression. This may result in M1 and occasionally M2 being squeezed out from behind P3 due to molar drift. The four molars in each jaw erupt early in life and are flat with all opposing molars in occlusion. The links between lophs are small. The molars can both shear and crush. Rock-wallabies, whose diet includes more grasses, still retain P3 and some have arched molar rows. The short-eared rock-wallaby (*Petrogale brachyotis*) does not eat much or any grass and has a large P3. The nabarlek (*P. concinna*), however, sheds P3 so the molars move forward in the jaw, and because the rows are arched the molars do not all occlude at the same time. The enamel on the molars is thin and the teeth wear down quickly and are replaced by others behind them. The nabarlek is unique in that it has unlimited molar replacement. This is possibly an independently acquired adaptation to feeding on grass.

2.2.3 Intermediate browser-grazers

This includes the hare-wallabies, nail-tailed wallabies and the eight species of wallaby in the *Notamacropus* Subgenus (*M. agilis, M. dorsalis, M. eugenii, M. greyi, M. irma, M. parma, M. parryi, M. rufogriseus*). These species are predominantly grazers but will also browse. Their premolars are small and are occasionally shed, allowing some forward movement of the molars (Fig. 7.1b). The molars have higher lophs and well-developed links between them. The opposing molar rows are arched so that occlusion and wear occurs successively throughout life from M1 to M4.

2.2.4 Grazers

These are the true grazers and feed predominantly on grass which is relatively uniform in physical properties; the food items are smaller than the teeth. The gape of these animals is small. These are the six species of large kangaroos. The premolars are vestigial and are usually shed before all the molars have erupted. The molars have very pronounced lophs and well-developed links between them. Each molar row is strongly curved, with the upper and lower row curving in opposite directions so that only two pairs of molars occlude on each side at any one time (Fig. 7.1c). The molars are adapted for shearing. A feature of this group is molar progression, a phenomenon shared only with elephants, dugongs and manatees. Molar progression is the sequential eruption of the molars. Each whole row of molars moves forward in

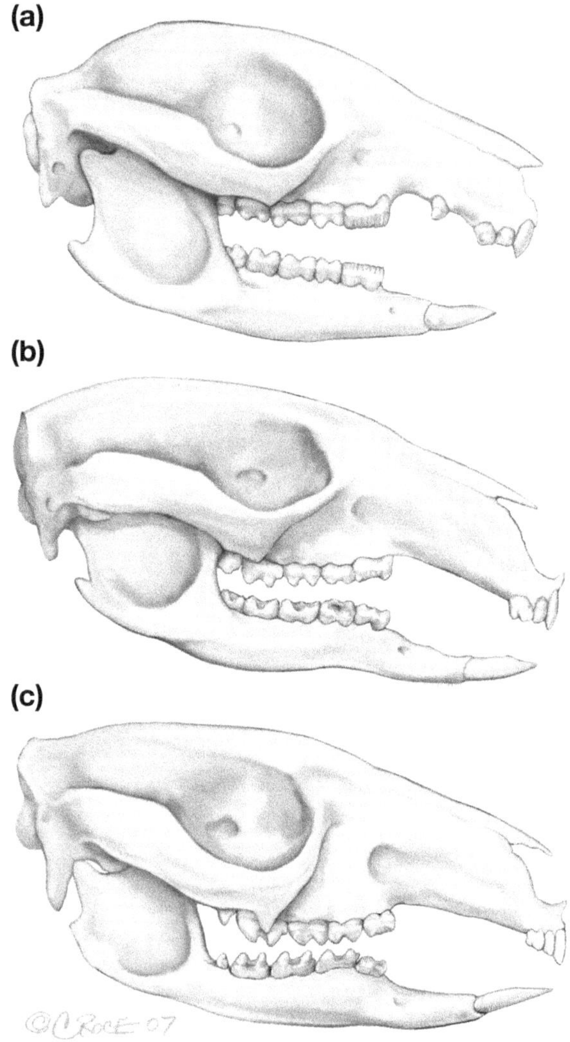

Figure 7.1 Lateral view of the skulls of macropod species showing dentition. a) A browser. b) An intermediate browser-grazer. c) A grazer.

the jaw throughout life, with the most worn-down molars being shed from the front. In old animals, only the worn-down M3 and M4 may remain.

The rate at which molar progression takes place is related to the quality of grasses eaten. Although there is variation in the rate of molar progression with the type of diet, the rate is reasonably constant within species and can be used for ageing some macropods.

Molar progression can only be used to estimate age in living macropods with the use of radiography. In the dead animal it is relatively simple. A molar index can be calculated and has been used to determine the approximate age of several macropod species (Kirkpatrick 1964; Kirkpatrick 1965; Kirkpatrick & Johnson 1969; McCauley 2003; Jackson 2003).

Molar eruption can also be used to determine the age of macropods and has been determined for several species (Jackson 2003). This is done by observing the proportion of the molar teeth that have erupted from the gum line. This system can be used in the live animal without radiography, and can potentially be used in species that do not exhibit molar progression. The limitations of this method are that it cannot be used to age animals once all molars have erupted and the gape of most macropods is small, making it difficult to examine the inside of the mouth. Most macropods would not tolerate it without chemical restraint. The mean age at which M4 is fully erupted varies between and within species and the literature is somewhat confusing in the ages quoted (Jackson 2003 and references therein). In red kangaroos, complete eruption of M4 is at the age of at least 6 yr (Sharman et al. 1964).

Both molar eruption and molar progression are highly correlated with age; however, the use of molar progression is less accurate. Sexual dimorphism occurs in both molar eruption and progression, with males acquiring their teeth earlier than females, and probably correlates with sexual heteromorphism of a particular species (Jackson 2003).

2.3 Gastrointestinal system

Wallabies of the Subgenus *Notamacropus* as well as the swamp wallaby have an unusual oesophagus lined with numerous elongate papillae (Obendorf 1984).

Macropods are foregut fermenters and the advantages of foregut fermentation are the same as those for ruminants. However, the anatomy of the macropod stomach is quite different from that of ruminants (Freudenberger et al. 1989; Hume 1999; Tyndale-Biscoe 2005). It is large and sacculated (Fig. 7.2). There are four parts that each have distinct functions. The oesophagus opens into a funnel-shaped region, which extends along the inner curvature of the stomach as a partly closed spiral groove and, like the oesophagus, is lined by squamous epithelium. It separates fine from coarse plant material and directs the fine material to the hindstomach, which is lined with an epithelium that secretes hydrochloric acid and proteolytic enzymes. Coarse plant material is directed to the forestomach, where bacterial fermentation takes place. This is the largest part of the macropod stomach and is divided into the sacciform and the tubiform forestomachs (Fig. 7.2). Both are lined with glandular epithelium that only secretes mucus. The pH in the sacciform region is maintained between 4.6 and 8.0 by the buffering action of the mucus and saliva from the parotid salivary glands, which is released during chewing. Bicarbonate, phosphate and sodium ions maintain the alkaline pH and support bacterial growth in the forestomach, which contains large populations of bacteria and ciliated protozoa. The bacteria digest the cellulose of the plant cell walls anaerobically, forming short-chain fatty acids that provide a rich source of energy (Tyndale-Biscoe 2005). Forestomach fermentation may further benefit the host by detoxifying plant chemicals and producing B vitamins. The musculature of the forestomach wall is organised into three longitudinal taeniae. Semi-lunar folds between the taeniae form haustrations that give the stomach a colon-like appearance (Freudenberger et al. 1989).

The musky rat-kangaroo has a simple stomach with low pH, no forestomach regions and no bacterial fermentation. In the Potoroidae, tree-kangaroos and pademelons, the sacciform forestomach is larger than the tubiform forestomach, which reflects their more digestible diet and lower requirement for fermentation to break down cellulose (Hume 1999) (Fig. 7.2 iii).

There is also a trend toward increasing relative size of the tubiform forestomach with increasing body size. Consequently, the browsers, which are relatively smaller in body size than the grazers, have a larger sacciform region of the forestomach (Fig. 7.2 ii). In grazers the tubiform region is larger. The forestomach contents of kangaroos can be 9–15% of total body weight (Dawson 1995).

The caecum and proximal colon of macropods is relatively small and although they contain some

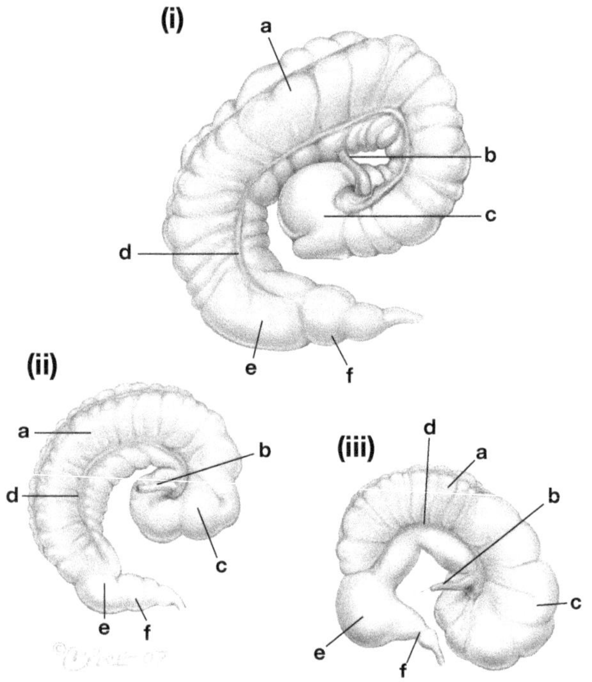

Figure 7.2 Gastric morphology of macropod species. i) Grazer. ii) Browser. iii) Potoroidae, tree-kangaroo and pademelon. a) Tubiform forestomach. b) Oesophagus. c) Sacciform forestomach. d) Taenia. e) Hindstomach. f) Pylorus. (Source: After Hume (1999) and Tyndale-Biscoe (2005).)

fermentative bacteria, they contribute little to digestion. The distal colon is an important site for the absorption of water from the faeces. The length of the colon varies with the types of habitat of the species. In the forest-welling species the distal colon is short and their faecal pellets have high water content, while arid zone species have a long colon and faecal pellets with low water content (Tyndale-Biscoe 2005).

In a process known as merycism, macropods, particularly the browsers and grazers, occasionally regurgitate food into their mouths. Merycism involves a rather violent heaving motion with vigorous movements of the fore limbs and thorax. Occasionally the bolus of ingesta may spill out of the mouth. The bolus is not generally re-chewed, unlike ruminants, and is quickly re-swallowed. The function of merycism is unknown (Dawson 1995).

In pouch young, before they start to eat herbage all regions of the stomach have a low pH and are proteolytic. It is only when the PY leaves the pouch permanently that the forestomach becomes a functional fermentation chamber with neutral pH. It also becomes colonised by bacteria and protozoa, which are presumed to come from the mother. At this time the PY is frequently seen to contact the mother's mouth, and stomach contents may be passed to the PY during merycism. This behaviour resembles and serves the same function as pap feeding of koalas and wombats and coprophagy in some possums (Dawson 1995; Jackson 2003; Tyndale-Biscoe 2005).

2.4 Skin

Macropods generally have thin skin, particularly in the inner surfaces of the limbs. This is particularly pronounced in the fore limbs, where there is significantly increased vascularity of the subcutis. This allows the animals to lose heat by evaporation of saliva, which they lick onto their forearms. The skin of the neck, shoulders, chest and abdomen tends to be thicker in males than in females. The skin on the ventral surface of the tail is also thickened.

Female macropods have four teats on the abdominal wall within the pouch. Teats are a few millimetres in length in the nulliparous animal. The teat enlarges when being suckled by a PY. At weaning it is about the length of the palate of the young-at-foot. It can be pulled out of the pouch to enable suckling by the emerging young while standing outside the pouch. After weaning the teat regresses, but it remains longer than an unsucked teat. The pouch of a multiparous female typically contains teats of different sizes. Each PY sucks on only one teat. Males lack teats.

Eccrine sweat glands are present on the hairless surfaces of the paws and feet and apocrine sweat glands are distributed over the rest of the body surface. The most developed apocrine sweat glands occur in the axillary and sternal regions. Aggregations of large apocrine glands are found around the scrotum in males and the cloaca and pouch area in females. The sternal glands are larger in males. Sebaceous glands are associated with hairs and these are particularly abundant in the sternal and axillary region in males, and around the cloaca and in the pouch of females. The oily secretions from these glands can be odorous and pigmented, particularly in males. The characteristic reddish-brown pigment of the sternal patch results from tryptophan derivatives in the apocrine secretion, which cause internal pigmentation of the hair and produce a pigment which adheres externally to the hair shafts. There are seasonal fluctuations in the amount of sternal gland secretion in some species (Hume et al. 1989; Dawson et al. 1989).

Two types of paracloacal glands are found within the cloaca: a holocrine sebaceous type and a cell-secreting type. These lie in pairs along the lateral cloacal walls and drain into the cloaca by single thin-walled ducts. These glands are small and cyst-like, and are invested in a thin striated muscle capsule. The sebaceous glands produce a pale yellow pungent secretion. The cell-secreting glands secrete an odourless cell suspension in an aqueous phase. These secretions are often released during handling and are a component of urine, are coated on faeces and used for marking (Hume et al. 1989; Dawson et al. 1989).

2.5 Locomotion

The two gaits used by macropods are pentapedal locomotion and bipedal hopping. In pentapedal locomotion the tail is used, with the fore limbs, as the third leg of a tripod to support the animal while the large hind limbs are moved forward together. This form of locomotion is used at speeds below 6 km/hr and is ungainly and energy-inefficient. Smaller macropods are able to hop even at low speeds. When hopping, the tail assists balance, but neither the tail nor the fore limbs provide support. Hopping is an economical way to travel for a large mammal, but there is less mechanical and physiological advantage for species of less than 5 kg body weight. Kangaroos can swim and when doing so the hind limbs are moved independently and alternatively (Hume et al. 1989)

At speeds of over 15 km/hr, hopping is more energetically efficient than running for similar sized species. The most economical speed for kangaroos to travel is around 20–25 km/hr. Large kangaroos can reach speeds of up to 70 km/hr (Tyndale-Biscoe 2005). Hopping rate remains constant up to about 40 km/hr with the increase in speed achieved by increasing length of stride. At speeds of 60–70 km/hr there is a noticeable increase in stride frequency and stride length, probably accompanied by a marked increase in energy consumption (Dawson et al. 1989).

2.6 Metabolism and thermoregulation

As a generalisation, macropods have a metabolic rate that is about 25% lower than that of comparably sized eutherian mammals. While this relatively lower basal metabolism impacts on macropod (and most other marsupials) physiology it does not mean that macropods are not capable of expending large amounts of energy. To compensate for the lower metabolic rate, the larger macropods have hearts one-third larger than those of comparably sized eutherian mammals. They can also move more air through their lungs with each breath (Dawson 1995).

Macropods, like most marsupials, have a resting body temperature of around 36°C. When faced with high ambient temperatures the first thing a macropod will do is try to avoid it by seeking shade (Hume et al. 1989). They may also adopt particular postures (e.g. crouching) to expose a smaller surface area for radiant heat inflow (Dawson 1995). Similarly, in cold windy conditions macropods will seek shelter. In winter some species have thicker pelts. Macropods also shiver to generate heat.

Once ambient temperatures approach or exceed body temperature, thermoregulation is by evaporative mechanisms. These mechanisms are also used to dissipate heat after exertion. This may be the result of obvious physical activity or increased muscle activity as a result of severe anxiety. Three mechanisms of evaporative heat loss are used: panting, sweating and licking. The mechanism of cooling by panting is no different from other mammalian species. Macropods also sweat. A unique feature of kangaroos is that sweating stops as soon as exercise stops, even if body temperature is still elevated and the animal is still panting rapidly. This may be related to economy in water loss (Dawson 1995; Tyndale-Biscoe 2005). The third method of evaporative heat loss is spreading saliva on the forearms. This saliva is produced by the maxillary salivary glands and has a different composition from the parotid saliva secreted during feeding. The area under the thin skin of the forearms has a dense superficial network of fine blood vessels and as the saliva evaporates it cools the blood in the veins. There is a marked increase in blood flow to the forearms in hot conditions. Antilopine kangaroos salivate on the inner thigh and this region is also highly vascularised. Males of all species also lick the scrotal region when hot. Salivating on the forearms is a common method of cooling in macropods that have undergone a period of exertion or anxiety. This can be used by the clinician as a guide to when an animal is stressed or overheating. Procedures should be aborted or cooling applied to avoid capture myopathy (see 9.10.3a).

2.7 Sensory systems

Macropod eyes are positioned high in the skull, giving them a wide field of vision, but at the same time the forward field of each eye overlaps that of the other by about 25°. This wide field of peripheral vision enables them to see movement in almost every direction, and the binocular vision enables them to have more precise close vision (Tyndale-Biscoe 2005). They have good vision both day and night. Despite this, they appear to have difficulty navigating obstacles and barriers that are not solid (e.g. wire fences), especially when alarmed and fleeing. Under these circumstances collisions are common. Colour vision has been demonstrated in the tammar wallaby (*M. eugenii*), particularly in the blue to green spectrum (Hemmi 1999).

The macropod iris is thick and uniformly brown. The pupil is circular and can be dilated easily with one drop of tropicamide. The fundus is heavily pigmented ventrally and is usually dark brown. The dorsal area of the fundus is usually lighter in colour and in some animals the choroidal blood vessels can be seen through the pigment. The optic disc sits on the junction between these two zones. The optic disc is well vascularised. Retinal vessels are not prominent and the macropod retina appears relatively avascular. All macropods have persistent hyaloid vessels. These are seen as a tuft of vessels arising from the centre of the optic disc and extending anteriorly into the vitreous toward the posterior capsule of the lens but not touching it. The prominence of the vessels varies between individuals and species. The vessels are thin-walled capillaries. In most macropods the plexus is fixed anteriorly and is not easily visualised on ophthalmoscopic examination. If not fixed, the plexus may be seen moving within the vitreous. In many kangaroos, myelination of the nerve fibre layer can be seen extending 2–3 disc diameters from around the optic disc. This myelination is generally most obvious in the lateral and medial aspects (Stanley 2002).

Most kangaroos and wallabies have large ears which can move independently through 180°. In contrast, tree-kangaroos have short rounded ears, which are not mobile.

2.8 Immune system

Passive transfer of immunity and the development of the marsupial immune system are reviewed in Chapter 2. The macropod thymus is a paired structure found SC on the ventral side of the neck. The shape and extent vary with species. In the healthy PY they are firm bulging structures with the caudal border level with or just beyond the clavicles. The medial border is close to the midline but separate from the opposite thymus. The lateral border does not extend beyond the ventral surface of the neck and the cranial limit is close to the angle of the jaw. In most species they are pear-shaped structures with the narrow part cranially. In unfurred or early furred PY the thymi can be visualised. The thymi regress fully by sexual maturity (Speare 1988; Tyndale-Biscoe 2005).

Superficial lymphnodes are not palpable in the healthy macropod. The spleen is Y-shaped and strap-like.

2.9 Reproductive system

2.9.1 Female

The anatomy of the female macropod reproductive tract is similar to that of other marsupials (Fig. 7.3). The completely separate uteri (no marsupials have a single chambered uterus) open by separate cervices into the vaginal *cul de sac* which is divided by a median septum (Dawson et al. 1989). Two separate lateral vaginae arise from the vaginal *cul de sac*. The caudal ends of the lateral vaginae open into the urogenital sinus, which also receives the urethra. Caudally, the two lateral vaginae may unite for a short distance before joining the

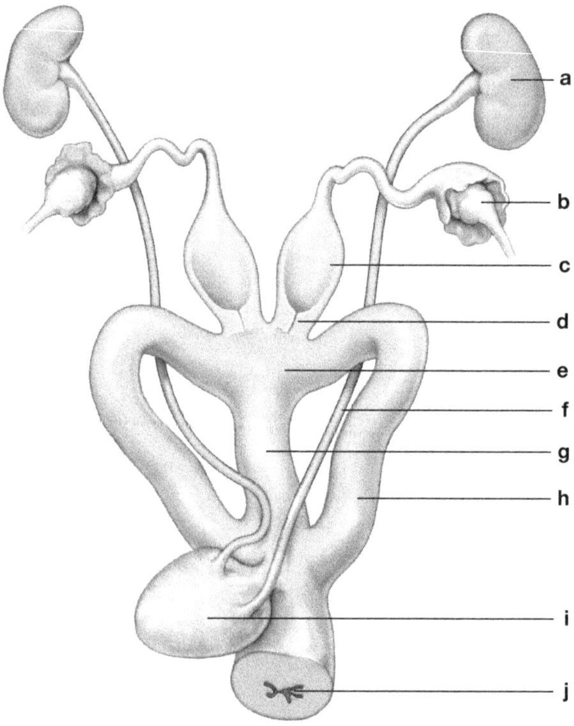

Figure 7.3 The female macropod urogenital tract. a) Kidney. b) Ovary. c) Uterus. d) Cervix. e) Vaginal *cul de sac*. f) Ureter. g) Median vagina. h) Lateral vagina. i) Bladder. j) Urogenital sinus.

urogenital sinus. The two lateral vaginae are separated by a mass of fibrous connective tissue called the urogenital strand. Prior to each parturition a pseudovaginal canal forms within the urogenital strand, connecting the vaginal *cul de sac* cranially with the urogenital sinus caudally to permit the passage of the foetus (Tyndale-Biscoe & Renfree 1987; Dawson et al.1989). The urogenital sinus, together with the rectum, opens into the common vestibule or cloaca.

In most species of Macropodidae the pseudovaginal canal becomes epithelialised after the first parturition with cuboidal epithelium of the vaginal *cul de sac* and merges with the stratified squamous epithelium of the urogenital sinus. This remains patent as a permanent median vagina. However, as is the case in most marsupials, the musky rat-kangaroo, the Potoroidea, banded hare-wallaby and the eastern grey kangaroo, the pseudovaginal canal closes rapidly after the passage of the foetus and reforms with each subsequent parturition (Tyndale-Biscoe & Renfree 1987; Dawson et al.1989). After copulation, semen is deposited in the lateral vaginae and sperm travel up the lateral vaginae, through the two cervices and uteri to the oviducts where fertilisation occurs.

2.9.2 Male

The anatomy of the male macropod reproductive tract is similar to that of other marsupials. Paired testes and epididymides are located externally in a pendulous, pedunculated prepenile scrotum. A strong active cremaster muscle encloses the spermatic cord and retracts the scrotal contents tightly against the pelvic body wall in cold weather or when stressed. When relaxed and in hot weather the scrotum hangs very low. The incidence of scrotal pigmentation is variable in adult macropods, with some heavily pigmented and others not at all. There is a 2.5–5°C temperature differential between body and testicular temperature. Kangaroos are able to maintain normal spermatogenesis at much higher ambient temperatures than eutherians of similar size. The vas deferentia join the prostatic urethra close to the neck of the bladder. Seminal vesicles and ampullae are lacking in marsupials. In macropods, three pairs of bulbourethral glands and a relatively large carrot-shaped disseminate prostate are the only accessory sex glands (Dawson et al. 1989; Hume et al. 1989). The Macropodidae are the only Australian marsupials to produce semen which coagulates to form a seminal or copulatory plug in the female's urogenital sinus and vaginae after ejaculation (Jones 1989; Dawson 1995; Paris et al. 2004). If protruding from the cloaca, the copulatory plug can be mistaken for a prolapse. Unlike a prolapse, which is rare in macropods, the plug usually disappears after a day or so.

When relaxed, the penis is sigmoid-shaped in a preputial sac in the common vestibule. The penis is rarely exposed other than during urination and erection. Unlike most other marsupials, the macropod penis is not bifid and the tip is sharply tapered.

The scrotum is discernible shortly after birth and the testes have descended completely by about 90 d in the parma wallaby (*M. parma*) and 70 d in the tammar wallaby. There is no seasonal variation in the size of the testicles or sperm production in male macropods. In prolonged severe droughts sperm production is reduced. Macropod sperm are short-lived, surviving no more than 24 hr in the female reproductive tract (Tyndale-Biscoe & Renfree 1987; Dawson et al.1989).

2.10 Urinary system

The urinary system is not significantly different from that of other marsupials. The urethra of male macropods has paired valve-like cusps approximately 2–3 cm proximal to the external orifice. Each cusp is approximately 5 mm long, with its free margin directed toward the external urethral orifice. Histologically, each cusp consists of transitional epithelium overlying a submucosa of poorly vascularised fibrous connective tissue. The epithelium and submucosa is continuous with the corresponding layer in the remainder of the urethra. A similar but larger pair of cusps lies immediately distal to the urinary bladder sphincter. The distal pair of cusps are on the lateral aspects of the urethra, are not near the ischial arch and are not associated with the bulbourethral gland. Although the functional role of these cusps is unclear, they may inhibit the retrograde movement of microorganisms that gain access to the urethra (Hatkin & Janssen 1979). These cusps can make urethral catheterisation difficult (see 9.4.1).

2.11 Tintibulation

Tintibulation is an apparent expression of anxiety or excitement in macropods. It is manifest by a wide-eyed expression and trembling of the head, neck and upper body. It is frequently seen in hand-reared animals that have a tendency for a nervous temperament. It is of no clinical significance.

3 REPRODUCTION

3.1 Oestrous cycle and pregnancy

All macropodid species that have been studied are monovular and polyoestrus with cycles ranging from 22–46 d (Table 7.1). There is a clearly defined pro-oestrus during follicular development. Ovulation is spontaneous and alternate and occurs within 2 d of oestrus. The resulting CL secretes progesterone which controls the development of the luteal phase. A rather poorly defined post-luteal phase follows as the CL regresses. In an unmated female this phase merges with the next pro-oestrus phase and ovulation. Since macropodids are monovular with alternate ovulation, in each pregnancy there is one gravid uterus ipsilateral to the CL and one non-gravid uterus ipsilateral to the developing follicle. The life-span of the CL does not extend much beyond the length of the oestrous cycle. Gestation therefore extends into the post-luteal and pro-oestrus phases without suppressing follicular growth, so that ovulation occurs within a few days before or after parturition—the post partum oestrus (Tyndale-Biscoe & Renfree 1987). If conception occurs, the resulting zygote develops to blastocyst stage where it remains in diapause as long as lactation occurs, i.e. until the first PY's final emergence from the pouch or loss of the PY. Therefore, as in all marsupials and unlike eutherian mammals, conception does not interrupt the oestrous cycle. Lactation does. The release of prolactin suppresses full development of the new CL and the next luteal phase does not occur. In tammar and red-necked (Tasmanian subspecies only) wallabies this persists for several months after weaning due to seasonality of breeding (Hume et al. 1989).

Embryonic diapause is characteristic of macropod reproduction, with the exception of the western grey kangaroo (*M. fuliginosus*), Lumholtz's tree-kangaroo (*Dendrolagus lumholtzi*) and musky rat-kangaroo, in which it does not occur. In the eastern grey kangaroo, parma wallaby and whiptail wallaby, post partum oestrus does not occur, but the females may come into oestrus and mate when the PY is older. Under favourable conditions female eastern grey kangaroos may mate when the PY is greater than 5–6 mo old, with the resulting embryo remaining in diapause due to lactational inhibition. In the swamp wallaby oestrus always occurs about 7 d prepartum.

The CL holds a central position in the control of reproduction in the Macropodidae. However, unlike other marsupials the CL does not provide the sole stimulus to the uterus to maintain gestation; the foeto-placental unit assumes this function in the second half of gestation and determines the length of gestation. The length of the luteal phase, the follicular phase and the placental influence determine whether ovulation is prepartum, post partum or absent in the different macropod species (Hume et al. 1989). The reproductive cycle of macropods differs from other marsupials in that gestation is extended from about half the oestrous cycle to occupy almost all of it, with birth occurring 1–2 d before the next oestrus (Dawson 1995). The gestation period for most macropods is 25–35 d and is shorter than the oestrous cycle in all species other than the swamp wallaby (Table 7.1).

The gestation period of macropods is considerably longer than most other marsupials. As a result, the newborn PY is relatively larger and more developed than non-macropod marsupials. This more advanced development, particularly in neurological organisation, is significant in that it reduces the risk of the single PY going astray on its journey to the pouch and attachment to the teat (Dawson 1995; Tyndale-Biscoe 2005).

In marsupials the foetus develops within an egg. The egg has a full set of shell membranes but they are much reduced compared to reptiles and birds. The yolk is also very small. In macropods the shell membranes persist for much of pregnancy, rupturing only in the last third of gestation. In macropods (and some other marsupials) there is complete loss of the allantois as the placental attachment, but it remains as a small intact vascular sac and probably functions only as an excretory store. The allantois expands rapidly near full term. The yolk sac is well developed and vascularised and is the only connection between mother and foetus. This choriovitelline placenta fulfils all the nutritive and respiratory requirements of the foetus (Dawson et al. 1989; Hume et al. 1989; Tyndale-Biscoe 2005).

Macropods are either continuous breeders or seasonal (Table 7.1). The basic reproductive pattern of macropods is controlled by lactation. However, under adverse conditions animals may become anoestrus. True seasonality, as seen in the tammar wallaby, quokka and red-necked wallaby (Tasmanian subspecies only), is controlled by day length. This occurs by either prolonging diapause until after the summer solstice, called seasonal quiescence, or by undergoing seasonal anoestrus (Tyndale-Biscoe 2005).

Table 7.1 Reproductive parameters of selected macropod species

Macropod species	Oestrous cycle (d)	Gestation (d)	Permanent pouch exit (d)	Weaning (d)	Ref.
Musky rat-kangaroo	25–26	19	147	–	5
Rufous bettong	21–36	21–30	114	155	5
Eastern bettong	17–37	20–22	109	160	5
Brush-tailed bettong	22–23	21	100	130	5
Burrowing bettong	11–35	22–23	115	165	5
Northern bettong	21–23	20–23	102–112	166–185	5
Long-footed potoroo	–	38?	140–150	106	5
Long-nosed potoroo	42	38	130	147	5
Banded hare-wallaby	–	–	180	270	5
Lumholtz's tree-kangaroo	47–64	42–48	246–275	333–515	5
Quokka	28	27	190	240	3, 5–6
Tasmanian pademelon	30	30	200	300	5
Red-necked pademelon	–	–	181	210	5–6
Red-legged pademelon	29–32	28–30	174–183	230–294	5–6
Yellow-footed rock-wallaby	32–37	31–33	190–201	210–235	5
Proserpine rock-wallaby	33–38	30–34	203–215	300–350	5
Black-footed rock-wallaby	36–38	33–35	178–197	270–350	5
Brush-tailed rock-wallaby	31	31	190–210	220–240	3, 5, 10
Unadorned rock-wallaby	30–32	30–32	189–227	290	5
Allied rock-wallaby	–	29–34	180–231	253–387	5
Purple-necked rock-wallaby	36–38	33–35	178–197	270–350	5
Nabarlek	31–36	30–32	160–180	175	5
Spectacled hare-wallaby	30	29–31	152	–	5
Rufous hare-wallaby	–	–	124	–	5
Bridled nail-tailed wallaby	30–45	21–26	119–126	190–245	5
Swamp wallaby	29–36	33–38	253–300	360–450	3–7
Western brush wallaby	–	–	180–210	–	5
Parma wallaby	41–42.5	34–35	210–212	270–320	2–3, 5–7
Black-striped wallaby	–	33–35	210	–	5
Agile wallaby	29–34	29–36	192–225	273–384	3, 5–6
Red-necked wallaby	32–33	29–30	270–300	360–517	2–3, 5–7
Pretty-faced wallaby	41–44	34–38	260–275	420–450	3, 5–6
Tammar wallaby	28–32	25–30	250–270	270–330	2–3, 5–7
Eastern grey kangaroo	36–55	34–38	310–319	540	1–7
Western grey kangaroo	30–39	28–33	300–310	540	1–3, 5–7
Common wallaroo	32–45	32	255–260	281–400	3, 5–7
Antilopine kangaroo	36–45	33–35	270	–	1, 5
Red kangaroo	34–35	33–38	235–250	330–365	1, 3, 5–7

Macropod species	Sexual maturity (months)		Postpartum oestrus	Embryonic diapause	Birth season	Ref.
	Female	Male				
Musky rat-kangaroo	>12		No	No?	Feb–Apr	5
Rufous bettong	10	12	Yes	Yes	All year	5
Eastern bettong	8–11	–	Yes	Yes	All year	5
Brush-tailed bettong	10–12	12	Yes	Yes	All year	5
Burrowing bettong	7	14	Yes	Yes	All year	5

(Continued)

Table 7.1 (*Continued*)

Macropod species	Sexual maturity (months)		Postpartum oestrus	Embryonic diapause	Birth season	Ref.
	Female	Male				
Northern bettong	6–14	8–17	Yes	Yes	All year	5
Long-footed potoroo	12?	12?	Yes?	Yes?	All year	5
Long-nosed potoroo	12	–	Yes	Yes	Jun–Aug	5
Banded hare-wallaby	12	12	Yes	Yes	–	5
Lumholtz's tree-kangaroo	24	55	No	No	All year?	5
Quokka	9–12	13	Yes	Yes	Jan–Mar	5,6
Tasmanian pademelon	14–15	14–15	Yes	Yes	Apr–June	5
Red-necked pademelon	12–26	17–20	–	Yes	All year	5,6
Red-legged pademelon	11.4	15.5	Yes	Yes	All year	5
Yellow-footed rock-wallaby	18	18	Yes	Yes	All year	5
Proserpine rock-wallaby	21	25	Yes	Yes	All year	5
Purple-necked rock-wallaby	18	22	Yes	Yes	All year	5
Black-footed rock-wallaby	18	22	Yes	Yes	–	5
Brush-tailed rock-wallaby	18	20–24	Yes	Yes	All year	3, 5, 10
Unadorned rock-wallaby	18	20	Yes	Yes	All year	5
Allied rock-wallaby	18	20–24	Yes	Yes	All year	5
Nabarlek	12–16	12+	Yes	Yes	All year	5
Spectacled hare-wallaby	11.7	15.4	Yes	Yes	All year	5
Rufous hare-wallaby	5–18	14	–	–	All year	5
Western brush wallaby	–	–	–	–	Apr–May?	5
Bridled nail-tailed wallaby	4.5–9	9–14	No	Yes	All year	5
Swamp wallaby	15–18	15–18	No#	Yes	All year	3–7
Parma wallaby	9–16	13–22	No	Yes	All year	2, 5–7
Black-striped wallaby	14	20	–	–	All year?	5
Agile wallaby	11.3–12	14–15.7	Yes	Yes	All year	5, 6
Red-necked wallaby	11–24	13–19	Yes	Yes	Jan–Aug	2, 5–7
Tammar wallaby	8–9	12–24	Yes	Yes	Jan–June	2, 5–7
Pretty-faced wallaby	24	18–24	No	Yes	All year?	5, 6
Eastern grey kangaroo	18–21	18–48	No	Yes	Oct–Mar	3, 5–7
Western grey kangaroo	14–36	29–31	No	No	Oct–Mar	2, 5–7
Common wallaroo	18–24	18–24	Yes	Yes	All year	2, 5–7
Antilopine kangaroo	–	–	No	No	All year	5
Red kangaroo	14–20	20–36	Yes	Yes	All year	5–7

#Prepartum oestrus reported.
Sources: **1** Dawson (1998). **2** Booth (2002). **3** Hume et al. (1989). **4** Holz (2003). **5** Jackson (2003). **6** Herbert & Cooper (2003). **7** McCauley (2003). **8** Grigg et al. (1989). **9** Strahan (2004). **10** Taggart et al. (2005).

Reproduction in the tammar wallaby has been studied more than any other species. They have a strictly seasonal breeding pattern. Female tammars mate approximately 1 hr after giving birth and the resulting conceptus grows to a 100-cell unilaminar blastocyst, which normally remains in diapause for 11 mo. Diapause is initially maintained by the sucking stimulus of the PY (lactational diapause) between January and late May, and by photoperiod control (seasonal diapause) between June and December. Soon after the summer solstice (22 December) the photoperiod inhibition ceases, the CL is reactivated and the blastocyst resumes development, with birth occurring in late January (Tyndale-Biscoe 2005).

3.2 Parturition, lactation and neonatal development

Maternal responsibilities in marsupials can be divided into five stages: parturition; parturition to first emergence from the pouch; first emergence to permanent emergence; permanent emergence to weaning; and the post-weaning period. Macropods typically have a reproductive sequence consisting of vacation of the pouch by one PY, birth within a couple of days, a post partum oestrus, conception and brief development of the embryo, then diapause of the embryo until the PY is within about 4 wk of permanently vacating the pouch. Embryonic development then recommences.

As the marsupial neonate is so small and fragile the mother must ensure the best chance of it reaching the pouch and attaching to a teat. One of the first signs of impending parturition is intense cleaning of the pouch. This starts 1–2 d before birth and is most intensive 1–2 hr before birth. Licking of the urogenital area also increases in intensity prior to birth. The female also adopts a birth position which is characteristic for the different species of macropod (Renfree et al. 1989). Parturition and maternal behaviour is under precise hormonal control. These include prolactin, oxytocin and prostaglandin. Progesterone declines rapidly prior to birth. The neonate climbs through the fur, taking about 3 min to reach the pouch in kangaroos. Once in the pouch the neonate attaches to a teat within a few minutes. Throughout pouch life the mother continues to clean her pouch and consumes the PY's urine and faeces. Licking of the PY's cloaca stimulates urination and defecation. Growth rate of the PY is variable between species, as is the time to permanent emergence (Table 7.1). Typical length of pouch life ranges from 150–320 d and varies between species. The PY remains permanently attached to the teat for about the first half of pouch life (Hume et al. 1989; Tyndale-Biscoe 2005).

The majority of macropods have only one young. Twins have been reported rarely in some species. The only species of macropod that routinely produces twins and even triplets is the musky rat-kangaroo (Jackson 2003). As the interval between young in the Potoroidae is short they can produce up to three young a year. All other macropods are not able to fully raise more than one young in a year. They can, however, have three young at different stages of development: one in embryonic diapause; one in the pouch attached to a teat; and one that has vacated the pouch permanently but occasionally suckles (a young-at-foot).

Lactation in marsupials is unique in that it sustains the neonate from an almost embryonic form after birth to a young-at-foot prior to weaning. This period may last over a year in some species. In all macropods there is a choice of four teats for the single young. Once attached, the PY remains on the same teat throughout pouch life. The mammary glands do not reach maximal milk production until some months after birth.

There are marked changes in milk composition throughout lactation, corresponding with changes in the sucking regime of the PY from continuous to intermittent. The most marked changes in milk composition appear to be associated with increased secretion of prolactin (at 20–24 wk in the tammar wallaby). This change in milk composition is also associated with a marked acceleration in PY growth rate. As macropods can have two young at different stages of development (a PY and young-at-foot), they can produce milk of two different kinds from adjacent mammary glands simultaneously. As there is also variable sensitivity of the mammary gland to oxytocin at different stages of lactation, milk ejection can occur semi-independently in adjacent glands in response to suckling by a tiny PY or a young-at-foot (Hume et al. 1989; Tyndale-Biscoe 2005).

Following permanent emergence from the pouch, the PEY (pouch emergent young) may spend many months getting in and out of the pouch. Once denied re-entry into the pouch (usually just before birth of the next young), the PEY still follows its mother and continues to suckle by putting its head into the pouch. This is a typical pattern for the larger species of macropod. In contrast, in the smaller species the period of getting in and out of the pouch is much shorter (a few days to weeks). Once permanently out, the PEY is often left alone under cover while the mother goes away to feed (Hume et al. 1989).

3.3 Hybridisation

Hybridisation, both within Genera and between different Genera of macropods, can occur. Most hybrids are sterile, and may have reproductive organ and pouch abnormalities and physiological and behavioural incompatibilities between mother and PY. For example, the genetically determined length of pouch life and duration of weaning and milk production and milk composition may be asynchronous between mother and PY, resulting in rejection of the PY or incompatibility with the PY's needs for growth and development. Great care needs to be taken to prevent hybridisation. It rarely occurs in the wild but may occur in captivity where macropods are held in mixed facilities with small mobs of different species. Animals preferentially choose the same species with which to breed (Jackson 2003; McCauley 2003).

4 HUSBANDRY

The husbandry of macropods requires an understanding of spatial, social and behavioural requirements and responses to stressors and disease. The veterinary management of sick or injured macropods must also take these factors into consideration as not doing so may compound the animal's problem, primarily through stress. The husbandry requirements of macropods are covered in detail by Jackson (2003) and in the Exhibited Animals Protection Act 1995; readers are referred to these sources for detailed information.

4.1 Housing and husbandry

Macropods are a diverse group of animals that have adapted to a wide range of habitats. Consequently, their housing and husbandry requirements are equally diverse.

Enclosures for macropods must take into account the unique requirements of the species and wherever possible mimic the natural environment for the species. Both the physical and psychological needs must be met. Many macropods are prone to stress and stress-related disease. Enclosures should provide a stress-free environment that includes adequate shelter, privacy, elevation, climbing structures and other furniture. All animals should be provided with a means of sheltering from wind, rain and extremes of sunlight and temperature. Although many macropods live in arid environments they do not necessarily cope well with high ambient temperatures. Most species seek refuge from heat during the day. Conversely, heating may be required for some species in cold climates. Many macropod species are nocturnal and are better suited to nocturnal displays.

Where possible, enclosures should allow for easy observation of animals from a distance (for public and staff) to minimise stress from being approached too closely. Where visitors are permitted to go into an enclosure, areas should be provided where the animals can seek refuge from unwanted visitor attention. Areas should also be provided where animals can have privacy and seek refuge and hide from other animals in the enclosure. Enclosures should be large enough to allow for the normal activity of the species and prevent overcrowding and for animals to get away from each other. Most macropods require a relatively larger area than other marsupials (Jackson 2003).

Macropod enclosures are best kept simple and free of obstacles. Fencing and other enclosing structures must be carefully designed. Macropods are prone to fence-running when stressed or being pursued and will collide with the fence or obstacles protruding from it. Fence posts should be outside the enclosure so the inside fence line is smooth and free of obstacles. Where possible, fence lines should be relatively straight and without corners. The height of the fence will depend on the species with the highest fences, about 1.8–2.4 m high, required for kangaroos at. Fences should be made of chainlink or similar mesh or solid material. Mesh size should be small enough to prevent the animals getting their heads and arms through if there are other macropods in an adjacent enclosure. Males often fight through fences, causing abrasions and even severe bite wounds on the arms (C Herbert *pers. comm.*). The fence should also prevent entry of predators such as foxes and dogs by preventing access to the enclosure through, under and over the fence. If climbing species (musky rat-kangaroo, rock-wallabies, tree-kangaroos, bettongs) are to be held in unroofed enclosures, the fences should be made of material (smooth tin, corrugated iron, timber or brick) that is unclimbable or is rimmed by a 45° outrigger 0.5 m wide facing into the enclosure. When alarmed, macropods are prone to colliding with fences if they are not clearly visible. Shadecloth or hessian attached to the fence is useful as a sight barrier and should always be in place when introducing animals to new enclosures (Williams 1990; McCauley 2003; Jackson 2003).

Enclosure furniture and planting should be appropriate for the species. Arboreal species such as tree-kangaroos should be provided with adequate climbing structures and thick shade. Rock-wallabies require a rocky mound with crevices for hiding. Macropods can be very destructive to vegetation and plants within an enclosure. Plants need to be chosen carefully for resilience and toxicity. Most plants will need protection from the animals, with barriers. Many macropods use grass tussocks, bushes and shrubs to hide or nest in (Williams 1990).

Substrates should be readily cleanable to allow regular removal of faecal material. They should be well-drained and not remain damp, to prevent the build-up of parasite oocysts and larvae, and bacteria. This is particularly important around feed and water points where animals congregate. Uneaten food must be removed daily to prevent build-up of bacteria and fungi and not attract pests. Grass, sand, soil or leaf litter are suitable substrates for most macropod species.

Feed and water containers should prevent contamination with faeces and urine, prevent access by pests and be readily cleaned and disinfected. This will aid in preventing build-up and spread of parasitic and bacterial infections and attracting pests. Feed areas should be under cover. Feed dishes, troughs, pellet hoppers and hay racks should be kept off the ground. This reduces the risk of food contamination with faeces and urine by preventing animals sitting in them. Feed areas should ideally have a concrete slab so that spilled feed and faeces can easily be picked up. Several feed areas may be required if there are a large number of animals in the enclosure. Fresh clean water should be provided in ponds, troughs or self-filling waterers. These should be cleaned regularly.

Enclosures must include suitable facilities to allow for capture, restraint and holding (off-exhibit) for sick or injured animals and for a recovery area after anaesthesia. Macropods are readily stressed and prone to trauma during capture, and to developing capture myopathy. All macropod enclosures should have a small area into which animals can be moved. This area should be away from people, enclosed with mesh fencing covered with hessian or shadecloth, have no sharp corners and have 2–3 separate yards. Animals should be allowed access to this area on a regular basis, possibly even fed in it daily so that moving animals into it is not difficult. Feed and water stations and shelter should be available.

Macropods that require hospitalisation must be housed in a quiet area away from noise, people, dogs, cats and other animals. Enclosures should be simple, solid-walled with no objects protruding from the walls and with appropriate substrate (rubber matting, wood shavings, straw). Shelter and structures in which the animal can hide and seek refuge should be provided. Rock-wallabies and tree-kangaroos should be provided with structures they can sit on top of or climb, as they generally feel more comfortable off the ground. Branches with leaves placed upside down in corners provide suitable privacy for some animals. Small species may be housed in hospital cages while larger species require a den or stall-type accommodation. Consideration should be given to the use of long-acting neuroleptic drugs in hospitalised animals (see 6.2.1).

4.2 Pest prevention and control

Pest species (rodents, foxes, dogs, cats) are known to transmit and harbour pathogens and can prey on macropods. The exclusion of cats from macropod enclosures and preventing their access to bedding and feed storage areas is essential to prevent exposure to the oocysts of *Toxoplasma gondii*. Rodents may carry *Salmonella*, *Leptospira*, encephalomyocarditis virus and parasites. Pests also disturb animals, particularly at night, causing stress or injuries associated with alarm and collision. Shelter and food are the primary attractants for pests. Food should be stored in pest-proof containers and food scraps should be cleaned up regularly. Pellet hoppers are useful in that they restrict bird and rodent access to the food. The judicious use of traps and baits to control pests is important, however, exposure of non-target species must be prevented. Enclosures should be built to exclude pests.

4.3 Individual marking and identification techniques

Accurate record-keeping is crucial to the effective management of any animals in captivity. Where possible, animals must be individually marked with a unique identifier (Jackson 2003). Techniques such as keeper familiarity are unreliable for macropods as they rarely display unique physical or behavioural characteristics that allow differentiation. Identification techniques that have been used in macropods include eartags, earmarks (notches, punches), tattoos, collars and passive integrated transponders (PIT tags) or microchips. Eartags and microchips are most commonly used. The standard site for insertion of microchips is SC on the dorsal midline between the scapulae. The point of the needle should be directed caudally. The hole should be closed with tissue glue. This, together with the caudal direction of implantation, prevents the likelihood of the microchip falling out soon after insertion.

4.4 Transport

Macropods can be transported in specially constructed boxes, pet packs or sacks (Jackson 2003; Exhibited Animals Protection Act 1995). A box or pet pack should be large enough for the animal to turn around, lie down and stand comfortably. A layer of thick cardboard or plywood should be attached to the inside of the wire door and the inside of the roof should be padded. There should be no objects or frames protruding into the box or pet pack, and feed and water bowls should be secured inside the box. Only one animal should be placed inside a box.

Figure 7.4 Suspension of a macropod in a sack for transport. Photo: Larry Vogelnest.

It is now considered preferable to transport macropods suspended in sacks. The sack should be suspended securely from a beam in a vehicle or box so that it is not touching the sides or floor. It must be large enough so that the animal can lie comfortably in it, and should be prevented from swinging by rope tied to the corners and secured to the sides of the vehicle or box (Fig. 7.4).

Females with PY still permanently attached to the teat (essentially while furless) are usually safe to transport as there is less risk of the PY being thrown from the pouch. The risk of being thrown is greater once the PY detaches from the teat. If it is necessary to transport a female macropod at this stage, the pouch may be taped closed and the tape removed prior to release. Animals should not be transported when ambient temperatures are high. All vehicles must be well-ventilated and cool. Care should be taken when releasing animals from boxes, pet packs or sacks to avoid alarming them in an unfamiliar environment (Jackson 2003). The judicious use of sedatives and tranquillisers will ensure that the animals are calm when released (see 6.2.1).

5 NUTRITION

The wild diets of macropods are diverse. Although generally considered herbivorous, many of the smaller species also eat invertebrates, roots and tubers, bulbs, fruits, seeds and fungi. Vegetation consumed by macropods is of two general types: monocotyledons, including mainly grasses and sedges; and dicotyledons such as forbs, shrubs and trees. Ferns may also be consumed by some species (Dawson 1989; Sanson 1989). Nutrition of macropods has been reviewed by Hume (1999); readers are referred to this source for detailed information.

5.1 Wild and captive diets of groups of macropods

The diets of macropods can be correlated with their dentition and gastrointestinal anatomy. The various macropod species have been categorised in groups according to dietary preferences and dentition (see 2.2) (Sanson 1989). There is also a correlation between body size and type of diet. The smaller species tend to be more specialised, feeding on more digestible energy-rich food items. The medium sized species are more generalist feeders and select a wide variety of vegetation types. These generally have relatively low fibre content. The large species, especially the kangaroos, are specialist grass-eaters. These diets are high in fibre and less digestible.

Captive diets should mimic as closely as possible the wild diet. This can be difficult, particularly when a regular supply of fresh grasses and browse is required. Most captive diets for macropods are therefore made up of various types of hay, fruits, green leafy and hard vegetables, dried corn, commercial dairy or macropod pellets and dog kibble. Lucerne and grass hays are most commonly used. Hay with a high proportion of stalk, grass awns and contamination with thistle should be avoided as these may pierce the gums, increasing susceptibility to lumpy jaw (see 9.2.2 and 9.9.26). Omnivorous species should be offered animal-based protein and insects. Most of these items are more easily digestible, and have a higher protein and energy content than the wild diet. They are also often fed in excessive quantities. Consequently, obesity is common in captive macropods. Specific captive diets for various macropod species are presented by Jackson (2003).

The diets of captive animals should not only provide for the nutrient requirements of the species but also the psychological and physical needs. Behavioural enrichment with macropods is as important as for any other species. Food and the way it is presented and the frequency of feeding can be important enrichment for captive macropods. Diet also plays an important role in dental health—food items such as bread and hard stalky items should be avoided as these promote plaque and calculus formation and gingival damage. This leads to periodontal disease, which may then progress to osteomyelitis or lumpy jaw.

5.1.1 Rat-kangaroos (*Hysiprymnodon* and the Potoroidae)

These species seek diets that are more easily digestible and energy-rich. These food items include young grasses, forbs, grass seeds, rhizomes, bulbs, tubers, swollen taproots, hypogeous fungi, plant exudates (*Acacia* gums) and invertebrates (Seebeck et al.1989). Hypogenous fungi (underground fungal fruiting bodies) are a principal food source for some of these species (Dawson 1989; Hume 1999).

5.1.2 Hare-wallabies

Although there are no detailed studies on the diets of these species in the wild, these small wallabies seem to feed on forbs, especially succulents, some green grass and grass seed heads (Dawson 1989; Hume 1999).

5.1.3 Rock-wallabies

The wild diets of these species have not been examined in detail other than the yellow-footed rock-wallaby (*Petrogale xanthopus*). Herbaceous forbs, young grasses and leaves of woody shrubs are all consumed. The diets of other rock-wallabies appear to be similar. An exception is the nabarlek. Their diet in the wet season is largely sedges and grasses and in the dry season they forage on ferns (Dawson 1989; Hume 1999).

5.1.4 Pademelons

These species eat succulent short grasses and herbage, fruits, ferns and browse (sometimes as fallen leaves). The rainforest-dwelling red-legged pademelon has a large intake of fallen fruits and leaves (Dawson 1989; Hume 1999).

5.1.5 Nail-tailed wallabies

These species primarily eat forbs, seeds, grasses and browse including fallen leaves. They particularly enjoy *Portulaca* (Hume 1999).

5.1.6 Small and medium wallabies

These species primarily graze on short grasses, sedges and browse. Quokkas are very opportunistic and eat dicotyledonous succulents, grasses, browse and sedges. The swamp wallaby is a true browser and feeds on forbs and the leaves of woody shrubs and trees. They are also opportunistic and have been known to consume grasses, fungi and ferns (Hume 1999).

5.1.7 Tree-kangaroos

Tree-kangaroos are obligate browse eaters. They eat the leaves of rainforest trees and vines and occasionally fruit (Menkhorst 2001).

5.1.8 Kangaroos

On the basis of their dentition, kangaroos are considered grazers. They will, however, opportunistically take other food items such as forbs. When available, *Portulaca oleracea* is highly selected. In times of drought kangaroos also eat the leaves of woody shrubs (Dawson 1989; Hume 1999).

5.2 Supplements

Many macropod diets are frequently supplemented with vitamin E and sometimes selenium. This is most likely based on observations of quokkas and other small macropods such as tammar wallabies and red-necked pademelon (*T. thetis*) who are prone to developing a syndrome of weakness and wasting of the hind limbs which ultimately progresses to paralysis and death (Kakulas 1961) (see 9.5.2). Histologically, lesions are typical of vitamin E deficiency. The lesions in quokkas can be completely reversed by oral dosing with 200–600 mg vitamin E daily for several days. Daily dosing with vitamin E also prevents the syndrome (Kakulas 1963). Interestingly, trace amounts of selenium in the diet (without vitamin E) had no prophylactic effect. In most but not all species, selenium has a sparing effect on vitamin E to the extent that it is an effective substitute. The quokka and tammar wallaby (and possibly other small wallabies) belong to a small group of animals in which nutritional muscular dystrophy is not prevented by trace amounts of selenium. The development of this syndrome in these small macropods is not related to diet alone. Studies have shown a correlation between size of enclosure and the incidence of muscular dystrophy in quokkas—the smaller the enclosure the higher the incidence of muscular dystrophy. Supplementation with vitamin E, however, was completely effective in preventing the development of muscular dystrophy in all animals regardless of enclosure size. It was hypothesised that the stress of overcrowding increased the vitamin E requirement (Hume 1999). This syndrome has not been seen in other macropod species, nor has a correlation between vitamin E requirements and the development of capture myopathy been demonstrated. It is certainly recommended that the diets of quokkas and possibly other small macropods be supplemented with vitamin E, however, there is no clear evidence that it is necessary for all macropod species.

6 RESTRAINT

6.1 Capture and physical restraint

The physical capture and restraint of macropods can be difficult and dangerous to both animal and people. This is particularly the case if the animals are not used to people. When approached, they often attempt to flee, and may run aimlessly into obstacles such as fences or walls resulting in wounds, fractures, ruptured organs, concussion and sometimes death. When restrained, animals may react violently, and attempts to restrain the animal further may result in contusions or fractures. Even tame animals may react in this manner when startled. In all cases, approaching, capturing and restraining a macropod can be stressful for the animal. Capture myopathy is common, particularly when capture is associated with excessive fear and/or exertion. Diazepam given immediately after capture as an anxiolytic and muscle relaxant reduces the risk of myopathy developing (see 9.5.1 and 9.10.3a).

Any capture and restraint of a macropod must be very carefully planned well in advance of the procedure if possible. Ideally, animals should be conditioned to being approached, handled and confined. Appropriate equipment and facilities such as chutes, yards, nets and bags should be available. Only personnel experienced with macropod capture should be involved. Capture and restraint of larger species is physically demanding and can be dangerous. The capture procedure must be carried out rapidly and efficiently. Prolonged pursuits will result in stress on the animal and increase potential for injury and capture myopathy. All personnel involved in the capture must understand their role in the procedure. The capture and restraint of macropods should be carried out in the cool of the morning. Procedures should not be carried out on hot days. If this is necessary, fans, air-conditioning, ice, cold water and wet towels must be available to cool the animal if necessary. A macropod's body temperature will rise rapidly even with minimal exertion or anxiety. Carrying out procedures in the morning leaves the rest of the day to monitor animals. If a capture procedure is not going to plan and animals are panting or licking their forearms, the procedure should be aborted.

During capture and restraint, female macropods with PY may eject or throw them from the pouch. It is possible to place the PY back in the pouch once the female is caught, and tape the pouch closed. Throwing may also occur as the animal is being released.

Most medium to large macropods are powerful animals. The fourth digit on the hind foot has a long sharp claw. They rarely deliberately attack (although wallaroos have been known to attack if cornered) but they often kick out with their hind legs once restrained, with the potential of causing significant injury with the claw of the fourth digit. Personnel should never position themselves in front of a restrained macropod where there is potential to be struck by the hind feet. Some animals, particularly bettongs and small wallabies, may bite while being restrained. Attempting to physically restrain the hind legs, head and neck of a macropod tends to make the animal struggle more and is dangerous to the animal and people. They are best held as depicted in Figure 7.5d or placed immediately into a bag or sack.

6.1.1 Capture equipment and techniques

A detailed description of capture techniques is provided by Jackson (2003). Large wide-mouthed hoop nets with flexible hoops are useful to capture most species. The size of the hoop, depth of the net and density of the mesh must be appropriate to the species and size of animal being captured. Captive animals can be made to run along a fence line or be cornered and rapidly scooped up in the net as they run past. Funnels or traps can also be made along a fence line to facilitate capture. If required, the animal can be transferred from the net into a thick cloth bag or sack (burlap, hessian, calico) by holding it by the base of the tail. Similarly, they can be caught by the base of the tail as they run past. Care should be taken not to hold the tail too distally as this may result in injury to the tail. The animal should be rapidly transferred into a bag or sack (Herbert 2003; Jackson 2003). If chemical restraint is to be used after capture, larger animals can be held momentarily by the tail and allowed to hop on the spot while they are hand-injected in the thigh muscles (Fig. 7.5d). Alternatively, the animal can be lifted off

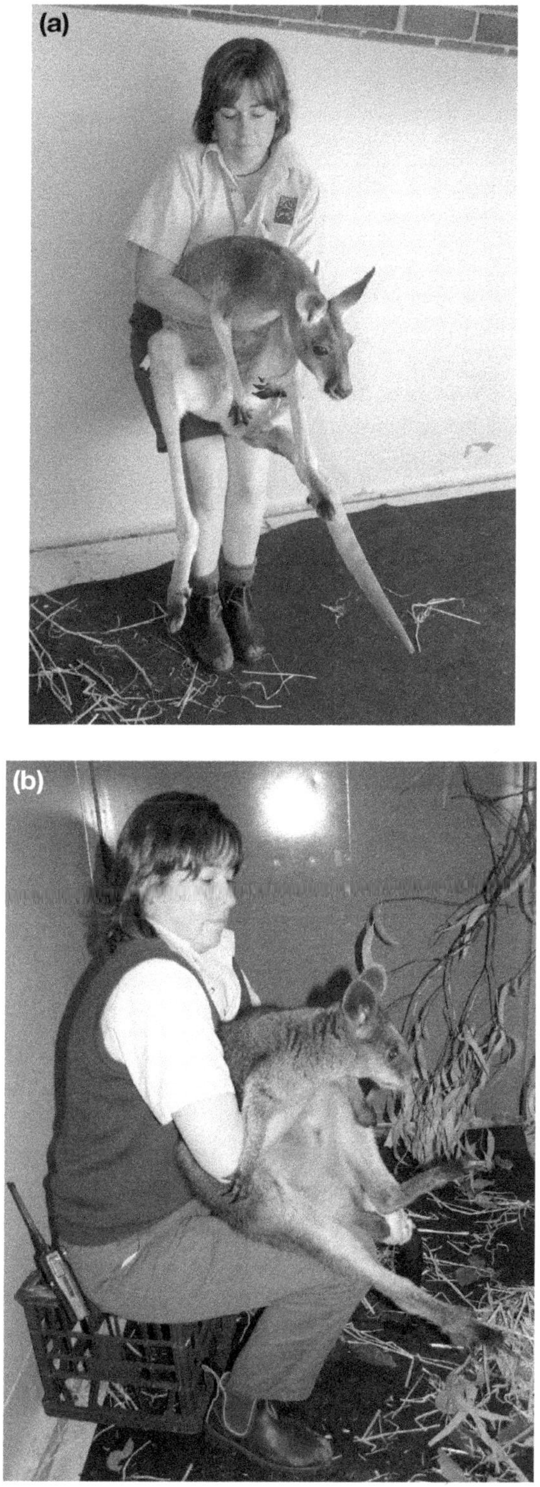

Figure 7.5 Physical restraint of a macropod. a) Standing. b) Sitting. c) Restraint in a hessian bag. d) Intramuscular injection. e) Intravenous injection. Photos: Larry Vogelnest.

the ground with one hand holding the base of the tail and the other arm around the animal's abdomen with it facing away from you, while a second person hand-injects the animal (Fig. 7.5a, b). Once injected, the animal should be released. Animals can also be readily injected IM through the material of a net or bag.

Once in the sack the animal's head should remain covered at all times to keep it calm. Minimal restraint is required while the animal is in the sack. Gentle pressure on the pelvis will keep the animal in lateral recumbency and, if the bag is open while a procedure is being carried out, the base of the tail should be held (Booth 1994). Once in a bag or sack the animal can be easily weighed using hanging scales, briefly examined (if the examination is going to be prolonged diazepam should be administered), blood sampled from the lateral caudal vein or hand-injected with chemical restraint or other drugs. If animals are to be held in bags or sacks for any length of time they should be hung from a fence, branch or beam in a building so that the bag is not touching the ground or floor. Most macropods settle and feel secure in this situation. Animals in bags must be kept cool in hot ambient temperatures and not allowed to get too cold in cool temperatures. Heavily sedated or anaesthetised animals should not be left in bags as they may suffocate if their muzzle gets wedged in a corner of the bag or their necks become kinked.

Various methods of capturing wild macropods without the use of anaesthetic drugs have been described (Hume *et al* 1989; Jackson 2003).

6.2 Chemical restraint

Injectable anaesthetic agents are either administered by hand injection after physical capture or by remote injection by darting. Injectable agents can be given IV or IM. The most accessible veins in macropods are the lateral caudal vein, cephalic and recurrent tarsal veins (see 7.1.1). IM injections are best given in the large thigh muscles. Macropods are usually darted into the large muscles of the hind limbs. Remote injection of restraint drugs is often preferable to physical capture as it reduces the risk of injury, stress and capture myopathy. Once immobilised, additional drugs can be administered IV or IM, or inhalational agents may be used.

Standard remote drug delivery systems can be used on macropods weighing more than 5 kg. Light-weight systems are available for small species. However, great care must be taken when darting these animals as there is an increased risk of fracturing bones or hitting the animal in the abdomen or thorax, which may lead to organ rupture, peritonitis and death. In these species alternative methods of capture such as traps, nets or blowpipes may be required.

It is particularly important with macropods that the animal is calm at the time of induction. Prolonged pursuit prior to darting increases the risk of injury and capture myopathy and reduces the effectiveness of anaesthesic agents. As far as possible macropods should be positioned in the shade, with the neck extended, and observed until fully recovered from the effects of chemical restraint. Animals have a tendency to stumble into obstacles or water bodies during recovery and it is important that they not be disturbed by operators or cage mates. If being recovered in a box there is a risk that they may press their head into a corner and obstruct the airway. For this reason animals are best recovered on their own in a quiet solid-walled yard. They should only be reintroduced to the larger group once fully recovered.

Even though adult macropods are foregut-fermenters they are not prone to regurgitation under anaesthesia and pre-anaesthetic fasting is not required. Hand-reared PY, however, should not be fed within 1 hr of anaesthesia because they may regurgitate and aspirate formula.

6.2.1 Sedation/tranquillisation

Diazepam is the most commonly used sedative in macropods. Doses of 0.1–2 mg/kg IM are used, but individual animal response is variable. There also appears to be some species variability. Wild brush-tailed rock-wallabies (*P. penicillata*) require as little as 0.1 mg/kg to achieve good sedation. Diazepam has a wide safety margin and it is preferable to overdose than underdose, however, a dose appropriate to the temperament and health status of the animal and the circumstances should be chosen. Debilitated animals require lower doses; wild and highly-strung animals higher. Higher doses should be used for capture and transport, to prevent capture myopathy. It should be given routinely to macropods immediately after capture if no other injectable anaesthetic agent is to be given. However, this may not be necessary for animals that are conditioned to handling or for very short procedures such as pouch checks. After IM injection it may take 15 min for sedative effects to become apparent, however, anxiolytic effects and muscle relaxation (important for the prevention of myopathy) appear to occur sooner. After injection, animals should be left quietly for 5–10 min to allow the drug to take effect. A more rapid effect can be achieved with IV administration (0.1–1.0 mg/kg). Oral administration of diazepam is very unreliable in macropods and is rarely used.

A range of short-, intermediate- and long-acting neuroleptic drugs are commonly used in macropods. The most common include azaperone, fluphenazine decanoate and zuclopenthixol acetate. They are primarily used to reduce stress and anxiety associated with transport and introduction of animals to new environments and other animals. Strategic use of one or more of these drugs can provide continuous tranquillisation for up to 3 wk, which is ideal for acclimatisation to new environments or cage mates (Table 7.2).

Despite their widespread use, there are very few studies on the safety and efficacy of these drugs in macropods. Azaperone (2 mg/kg IM) appears to have an effect within 15–20 min. The duration of effect is unknown in macropods but appears to last at least 8 hr. In other species azaperone is eliminated from body tissues within 16 hr (Swan 1993). Perphenazine enathate (5 mg/kg IM) begins to act within 12–16 hr and lasts for 7 d (Ebedes 1993). In humans, fluphenazine decanoate has an onset of action within 24–72 hr, with maximum effect reached in 48–96 h and duration of 3–4 wk. Fluphenazine has effectively been used in macropods at 2.5 mg/kg IM, however, its duration of action in macropods is unknown but appears to be at least 10 d. Haloperidol decanoate (4–6 mg/kg IM) has been administered to macropods (Vogelnest 1999). Onset of action in ungulates is 24–48 hr with a duration of up to 30 d (Ebedes 1993). Its duration of effect in macropods is unknown. Zuclopenthixol acetate (5–10 mg/kg IM) has been administered to macropods (Portas 2005). It appeared to reach peak effect after about 24 hr. The effect appears to decline 36-48 hr after injection. There appears to be species variation in sensitivity to this drug. Profound tranquillisation is seen in agile wallabies and bridled nail-tailed wallabies at these dose rates. Side effects such as peripheral oedema and urine retention have also been seen. Zuclopenthixol decanoate (10 mg/kg) combined with pipothiazine palmitate (10 mg/kg) produced

Table 7.2 Suggested chemical restraint protocols for use in macropods

Scenario	Protocol	Comments
Anaesthesia—tractable animal	Physical restraint in bag + IV tiletamine/zolazepam 1–3 mg/kg **or** IM tiletamine/zolazepam 4–5 mg/kg **or** IV alfaxalone 1.5–3 mg/kg **or** mask induction with isoflurane in oxygen	Maintain on isoflurane in oxygen via mask or endotracheal tube.
Anaesthesia—non-tractable animal	Dart with tiletamine/zolazepam 7–10 mg/kg	If required, maintain anaesthesia using isoflurane in oxygen.
Physical restraint and quick examination, e.g. pouch check or short transport	Diazepam 0.1–2 mg/kg IM	Dose used depends on species and temperament of animal. Very tractable animals may cope well without any sedation for these procedures.
Physical restraint prior to long transport and/or introduction to a new environment	Diazepam 0.1–2 mg/kg IM **or** azaperone 2 mg/kg IM + fluphenazine 2.5 mg/kg IM **or** haloperidol decanoate 4–6 mg/kg IM **or** zuclopenthixol acetate 4–6 mg/kg IM	Ideally intermediate- or long-acting neuroleptic drugs should be administered 24–48 hr prior to restraint and transport. If adequately sedated, there may be no need for additional sedatives/tranquillisers.
Anaesthesia prior to long transport and/or introduction to a new environment	As above for anaesthesia + fluphenazine 2.5 mg/kg IM **or** haloperidol decanoate 4–6 mg/kg IM **or** zuclopenthixol acetate 4–6 mg/kg IM	Lower doses of anaesthetic agents may be required due to the effect of the neuroleptic drug. Neuroleptic drugs may also deepen anaesthesia, so if being administered to anaesthetised animals it is best to do so toward the end of the procedure. They may also prolong recovery time. See also comment above.

consistent tranquillisation of recently captured red-necked wallabies for about 10 d (Holz and Barnett 1996). Onset of action was within 24 hr of injection, and peak effect was reached at 6–9 d. Compared with other species, the dose rates used in this study for both drugs was very high.

Intermediate and long-acting neuroleptic drugs are best given a few days (depending on the agent) before transport for them to be effective at the time the animal needs to be caught-up. If anaesthesia is required for capture and boxing, lower doses may be required due to the effect of the neuroleptic drug. Neuroleptic drugs may also deepen anaesthesia so, if being administered to anaesthetised animals, it is best to do so toward the end of the procedure. They may also prolong recovery time.

6.2.2 Anaesthesia

A wide range of injectable anaesthetics have been used to chemically restrain macropods. IM administration is most commonly used either by hand injection or by projectile syringe. Tiletamine/zolazepam (5–7 mg/kg IM) is the drug of choice. Lower doses (4 mg/kg) are used in tractable animals, higher doses in excited animals (10 mg/kg). It appears to be an extremely safe drug in macropods so if in doubt a higher dose rate should be used. It produces rapid and smooth induction and good muscle relaxation (Holz 1992; Shima et al. 1993; Vogelnest 1999). Duration and depth of anaesthesia after a single injection appears to be dose-dependent. Duration may be up to 2 hr and recovery may be prolonged (1–5 hr). Additional doses (1–2 mg/kg) can be administered IV, but this significantly prolongs recovery. The use of an inhalation agent after induction is thus preferable to prolong and/or deepen anaesthesia if required. Adverse reactions are rare; excessive salivation is the most common (Boever et al. 1977; Smeller et al. 1977; Shima et al. 1993). This can be controlled with atropine (0.3–0.6 mg IV or IM total dose) but is not usually necessary as laryngeal reflexes are maintained. The animal's head should be positioned to allow drainage of saliva and maintain a clear airway.

Tiletamine/zolazepam has been used in combination with α_2-agonists. Tiletamine/zolazepam (3–5 mg/kg IM) combined with xylazine (0.5–2.0 mg/kg IM), and reversal with yohimbine (0.2 mg/kg IV) or atipamezole (50–400 μg/kg IV), has been used (Blyde 1994). Tiletamine/zolazepam (2–3 mg/kg IM) combined with medetomidine (40 μg/kg IM), and reversal with atipamezole (4 times the medetomidine dose IM), was administered to red kangaroos, but did not appear to have any real advantage over tiletamine/zolazepam on its own (Vogelnest 1999).

Ketamine and xylazine (5 mg/kg each) and up to 25 mg/kg of ketamine alone have also been used to restrain macropods, but the quality of restraint is inferior to that achieved with tiletamine/zolazepam. The xylazine can be reversed with yohimbine or atipamezole. Ketamine (3–5 mg/kg) plus medetomidine (40–125 μg/kg) produces rapid smooth induction, good muscle relaxation and less salivation than tiletamine/zolazepam. Recovery to standing occurs about 30 min after reversal with atipamezole (4–5 times the medetomidine dose IM) (Jalanka & Roeken 1990; Holz 2003; Vogelnest 1999). Medetomidine (40 μg/kg) plus ketamine (4 mg/kg IM) and reversal with atipamazole (0.2 mg/kg IM) produced safe and reliable anaesthesia in eastern grey kangaroos. The atipamazole should be administered at least 30 min post-immobilisation to avoid residual ketamine effect (Pye & Booth 1998). Alfaxalone (5–8 mg/kg IM) has been used with some success in very tame macropods (S Gelis *pers. comm.*).

If IV induction is possible a number of agents can be used. Tiletamine/zolazepam (1–3 mg/kg IV) is effective and safe. Thiopentone (15–20 mg/kg), giving two-thirds of the dose as a bolus and the rest to effect, has also been used (Blyde 1994; Booth 1994). Alfaxalone (1.5–3 mg/kg IV) has also been used (S Gelis *pers. comm.*). The duration of effect is very short (5–10 min), and is useful for induction prior to inhalation anaesthesia.

Inhalation anaesthesia using isoflurane or sevoflurane in oxygen either for induction and/or maintenance is commonly used in macropods. Induction using a facemask while the animal is physically restrained is useful for smaller species and tractable animals. Premedication with diazepam may be necessary to prevent struggling and stress. The head of the animal can be kept in a bag or net and the gas fed through the material. Anaesthesia can be maintained via the mask or the animal intubated. Intubation is difficult and not usually necessary unless working on the head or in the mouth, or for prolonged procedures. The narrow gape and dental arcade make visualisation of the glottis difficult. The larynx is also a considerable distance from the oral opening. A long narrow-bladed laryngoscope, small rigid endoscope or a long curved transilluminator can be used to visualise the glottis. Long endotracheal tubes must also be used. Because of the narrow mouth, introducing the tube often obscures the view of the glottis. In this case a narrow stylet, e.g. a canine urinary catheter, can be used. Once in the trachea the endotracheal tube can be passed over the stylet and into the trachea. The stylet is then removed. Threading the tube partway onto the stylet first may make it easier and ensures that the endotracheal tube fits over it. Blind intubation can be achieved. The animal's neck must be extended and the tongue pulled out of the mouth. The larynx is held between the thumb and forefinger while the endotracheal tube is passed over the base of the tongue until the tip is resting on the glottis. As the animal takes a breath the tube is passed into the larynx and trachea. A deep plane of anaesthesia makes blind intubation easier. The cuff should be inflated (Booth 1994; Blyde 1994; Shima 1999; Vogelnest 1999).

7 CLINICAL PATHOLOGY

7.1 Haematology and biochemistry

7.1.1 Sample collection

The most commonly used site for venipuncture in macropods is the lateral caudal vein (lateral tail vein) (Fig. 7.6). It is located at the base of the tail at the midpoint of the lateral aspect. In some species it branches close to the base of the tail. It is most visible close to the tail base and tends to disappear more distally, particularly in smaller species. In species with shorter and fatter tails, such as the quokka, the vein can be more difficult to visualise. Digital pressure or a tourniquet must be used to occlude the vein. The vein is easier to visualise with the hair clipped but this is not always necessary. It is readily accessible in the conscious animal when restrained in a bag as the tail can be easily extracted without compromising restraint. It is readily accessible in the anaesthetised macropod.

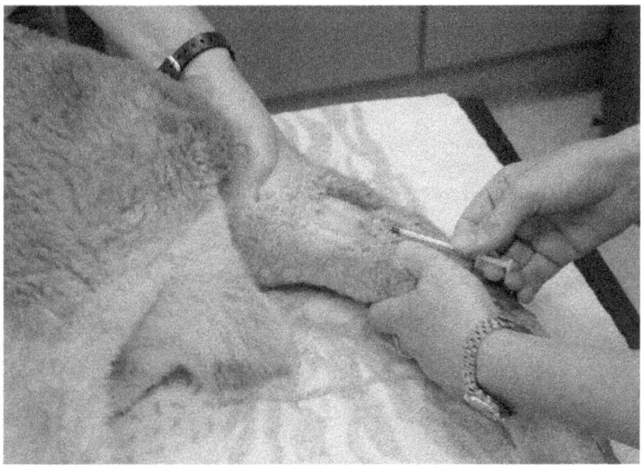

Figure 7.6 Venipuncture using the lateral caudal vein in a red kangaroo. Photo: Larry Vogelnest.

Other accessible veins include the jugular, cephalic and recurrent tarsal. These veins should only be used in a heavily sedated or anaesthetised patient as the restraint required to access them in the conscious patient may result in struggling, with the risk of injury to the patient and operator. Macropods have a short neck which makes access to the jugular vein more difficult. The cephalic vein is visible on the dorsal aspect of the antebrachium. The recurrent tarsal vein is visible on the lateral aspect of the distal half of the tibia.

7.1.2 Reference ranges and interpretation

The cytological characteristics of the blood cells of Australian macropods have been described (Clark 2004). Haematology and biochemistry values for selected macropod species are presented in Tables 7.3 and 7.4. The haematological and biochemical responses of macropods to various physiological and pathological states have been poorly studied. However, the haematological and biochemical changes in macropods frequently appear to belie the severity of the underlying pathology. It has been our experience that it is common for a macropod with severe disease (bacterial infection, nephropathy, hepatopathy) to have unremarkable haematological and serum biochemical changes.

Unlike some marsupial species, neutrophils are the predominant leukocyte in the peripheral blood of macropods. The haematological response to clinical inflammation in marsupials has been poorly studied. Macropods do not mount a marked leukocyte response to inflammation. Neutrophilia with abnormal neutrophil morphology (band neutrophils, increased cytoplasmic basophilia, Döhle bodies) was seen in red-necked wallabies with lumpy jaw (Hawkey et al. 1982). Many species of healthy marsupials have low concentrations of neutrophils. This is true for macropods. This indicates a relatively small reserve of neutrophils in the bone marrow. Therefore, in response to inflammation this reserve is rapidly depleted and neutropenia may be more likely.

When interpreting the haemogram, it is important to consider the circumstances under which the blood

Table 7.3 Haematology values for selected macropod species (range or mean ± SD)

Parameter	Long-nosed potoroo (see below)	Rufous bettong (n = 14)	Burrowing bettong (n = 3)	Lumholtz's tree-kangaroo (n = 7)	Red-bellied pademelon (n = 5)	Red-necked pademelon (n = 4)	Red-legged pademelon (n = 45)
PCV (L/L)	0.42 ± 0.05 (11)	0.50 ± 0.01	0.40 ± 0.01	0.46 ± 0.01	0.42–0.50	0.53 ± 0.04	0.51 ± 0.01
RBC (× 10^{12}/L)	8.70 ± 0.92 (22)	6.1 ± 0.2		5.7 ± 0.3	5.7–6.5		7.1 ± 0.1
Hb (g/L)	143.41 ± 11.24 (22)	165 ± 4	107 ± 6	155 ± 5	145–168	168 ± 9	176 ± 2
MCV (fL)	50.18 ± 4.03 (22)	83 ± 2 (8)		81 ± 4	74–79		72 ± 1
MCH (pg)	16.91 ± 1.34 (22)	28 ± 0 (8)		27 ± 1			25 ± 0
MCHC (g/L)	330 ± 12.5 (22)	332 ± 3 (8)		337 ± 2	330–350		344 ± 3
nRBC (/100 WBC)				2.3 ± 0.9 (4)			2.5 ± 0.4 (19)
WBC (× 10^9/L)	4.16 ± 2.27 (24)	6.49 ± 0.63 (13)		8.03 ± 0.33	4.90–8.10		4.79 ± 0.19
Neutrophils (× 10^9/L)	1.65 ± 1.48 (24)	2.07 ± 0.17 (4)		3.43 ± 0.67	0.60–1.90		1.37 ± 0.13 (39)
Lymphocytes (× 10^9/L)	2.23 ± 1.63 (24)	3.73 ± 1.18 (4)		3.63 ± 0.26	3.20–6.20		3.13 ± 0.13 (39)
Monocytes (× 10^9/L)	0.10 ± 0.11 (24)	0.18 ± 0.07 (4)		0.14 ± 0.04	0.00–0.10		0.08 ± 0.02 (39)
Eosinophils (× 10^9/L)	0.19 ± 0.16 (22)	0.13 ± 0.04 (4)		0.60 ± 0.10	0.20–0.50		0.14 ± 0.02 (39)
Basophils (× 10^9/L)	0.06 ± 0.01 (2)	0.03 ± 0.02 (4)		0.01 ± 0.01	0.00		0.01 ± 0.00 (39)
Platelets (× 10^9/L)	733 ± 164 (NS)[a]	442 ± 31 (5)		115 ± 32 (6)			314 ± 11 (40)
Reference	Perth Zoo (unpublished data)	Clark (2004) (Spencer et al. unpublished data)	Clark (2004) (Billiards et al. 1999)	Clark (2004) (Spencer et al. unpublished data)	Clark et al. (unpublished data)	Clark (2004) (Agar et al. 1986)	Clark (2004) (Spencer et al. unpublished data)

NS = not stated
a Moore & Gillespie (1968) cited in Clark (2004)
Taronga Zoo—in-house reference ranges 2006

(Continued)

Table 7.3 *(continued)*

Parameter	Yellow-footed rock-wallaby (see below)	Proserpine rock-wallaby (n = 14)	Brush-tailed rock-wallaby (n = 8)	Unadorned rock-wallaby (n = 7)	Allied rock-wallaby (n = 96)	Spectacled hare-wallaby (see below)	Rufous hare-wallaby (n = 6)	Purple-necked rock-wallaby (n = 5)
PCV (L/L)	0.33–0.55 (15)	0.46 ± 0.03	0.47 ± 0.04	0.37 ± 0.01	0.37–0.45[b]	0.56 ± 0.01 (14)	0.55 ± 0.01	0.55 ± 0.02
RBC (× 10^{12}/L)	4.6–6.4 (11)	6.8 ± 0.4	6.0 ± 0.5	5.1 ± 0.2	4.5–5.5[b]	6.5 ± 0.6 (14)	8.0 ± 0.2	5.9 ± 0.1
Hb (g/L)	111–181 (15)	149 ± 10	154 ± 17	116 ± 5	122–149[b]	192 ± 3 (12)	189 ± 3	189 ± 8
MCV (fL)	71–87 (11)	65 ± 1	78 ± 2	72 ± 1	80–85	70 ± 3 (10)	69 ± 1	94 ± 3
MCH (pg)	23–29 (11)	22 ± 0	26 ± 1	23 ± 0	26–28	30 ± 6 (12)	24 ± 0	33 ± 1
MCHC (g/L)	261–363 (15)	334 ± 2	329 ± 10	311–321	317–339	343 ± 5 (11)	345 ± 6	345 ± 4
nRBC/100 WBC		0.7 ± 0.3 (9)	1.0 ± 1.0 (5)		0.7–0.9	7.0 ± 4.0 (2)		
WBC (× 10^9/L)	3.60–12.0 (14)	7.68 ± 0.90 (13)	6.10 ± 1.80	6.23 ± 0.54	7.20–10.80	5.84 ± 0.47 (14)		7.61 ± 0.61
Neutrophils (× 10^9/L)	0.60–2.90 (10)	1.80 ± 0.27 (11)	3.28 ± 1.56		0.55–12.21	1.82 ± 0.43 (7)		2.15 ± 0.30
Lymphocytes (× 10^9/L)	1.70–10.40 (10)	5.12 ± 1.04 (11)	2.48 ± 1.45		0.32–15.14	3.56 ± 0.31 (7)		5.15 ± 0.49
Monocytes (× 10^9/L)	0.07–0.11 (4)	0.12 ± 0.06 (11)	0.13 ± 0.05 (5)		0.00–0.84	0.07 ± 0.06 (7)		0.00
Eosinophils (× 10^9/L)	0.05–0.55 (6)	0.47 ± 0.14 (11)	0.30 ± 0.22 (5)		0.00–0.68	0.52 ± 0.15 (7)		0.25 ± 0.07
Basophils (× 10^9/L)	0.04 (1)	0.05 ± 0.03 (11)			0.00–0.75	0.00 (7)		0.00
Platelets (× 10^9/L)	100–340 (8)	248 ± 25 (12)	183 ± 36 (2)		141–154	251 ± 13 (5)	154 ± 16.5	
Reference	Clark (2004) (Adelaide Zoo unpublished data)	Clark (2004) (Spencer et al. unpublished data)	ISIS (2002)	Clark (2004) (Spencer et al. unpublished data)	Clark (2004) (Spencer & Speare 1992)	Clark (2004) (Spencer et al. unpublished data)	Clark (2004) (Agar & Godwin 1991)	Clark (2004) (Spencer et al. unpublished data)

b Females were on the lower end of the range and males were at the higher ranges for PCV, RBC and Hb.

(Continued)

Table 7.3 (*Continued*)

Parameter	Northern nail-tailed wallaby (n = 7)	Bridled nail-tailed wallaby (n = 17)	Swamp wallaby (see below)	Black-striped wallaby (n = 16)	Agile wallaby (n = 12)	Red-necked wallaby (n = 47)	Tammar wallaby (n = 75)	Parma wallaby (see below)
PCV (L/L)	0.62 ± 0.01	0.51 ± 0.01	0.45 ± 0.07 (9)	0.52 ± 0.01	0.49 ± 0.02	0.40–0.56	0.47 ± 0.01	0.46 ± 0.07 (27)
RBC (× 10^{12}/L)	5.8 ± 0.1	5.7 ± 0.2	6.7 ± 0.9 (8)	6.9 ± 0.1	5.4 ± 0.2	4.4–6.6	5.9 ± 0.2	7.2 ± 0.8 (23)
Hb (g/L)	217 ± 4	171 ± 4	155 ± 30 (8)	173 ± 4	167 ± 9	140–197	159 ± 4 (74)	148 ± 22 (25)
MCV (fL)	107 ± 2	91 ± 2	70 ± 3 (8)	76 ± 2	93 ± 4	77–93	80 ± 2	64 ± 6 (21)
MCH (pg)	37 ± 1	24 ± 1	25 ± 1 (7)	25 ± 1	31 ± 1	27–33	26 ± 1	22 ± 2 (22)
MCHC (g/L)	351 ± 3	337 ± 2	346 ± 23 (8)	333 ± 3	336 ± 10	345–376	340 ± 6	330 ± 34 (24)
nRBC (/100 WBC)	0.6 ± 0.4	2.8 ± 1.3	0.0 (1)	0.2 ± 0.2 (5)	0.0 (4)	0.0–0.1		7.0 ± 10.0 (11)
WBC (× 10^9/L)	3.31 ± 0.24	9.52 = 0.76	6.67 ± 2.10 (9)	5.35 ± 0.50	6.08 ± 0.77	3.50–10.50	5.50 ± 0.30	4.73 ± 1.95 (29)
Neutrophils (× 10^9/L)	1.36 ± 0.26	1.98 = 0.18	3.55 ± 1.91 (8)	0.93 ± 0.15 (10)	2.09 ± 0.41 (7)	1.10–5.00	2.80 ± 0.40	1.77 ± 0.95 (26)
Lymphocytes (× 10^9/L)	1.47 ± 0.25	6.84 ± 0.68	2.82 ± 0.84 (8)	4.49 ± 0.54 (10)	3.66 ± 0.72 (7)	1.60–5.30	2.20 ± 0.20	2.75 ± 1.69 (27)
Monocytes (× 10^9/L)	0.04 ± 0.00	0.17 ± 0.06	0.18 ± 0.10 (8)	0.10 ± 0.04 (10)	0.08 ± 0.03 (7)	0.00–0.60	0.30 ± 0.04	0.13 ± 0.12 (25)
Eosinophils (× 10^9/L)	0.45 ± 0.05	0.46 ± 0.10	0.22 ± 0.12 (7)	0.13 ± 0.08 (10)	0.62 ± 0.19 (7)	0.00–0.50	0.20 ± 0.04	0.14 ± 0.13 (20)
Basophils (× 10^9/L)	0.01 ± 0.01	0.01 ± 0.01	0.05 ± 0.02 (2)	0.10 ± 0.04 (10)	0.00 (7)	0.00–0.20	0.00	0.14 ± 0.23 (4)
Platelets (× 10^9/L)		314 ± 35		215 ± 34 (7)	229 ± 31.1 (4)	136–485		220 ± 57 (7)
Reference	Clark (2004) (Spencer et al. unpublished data)	Clark (2004) (Spencer et al. unpublished data)	ISIS (2002)	Clark (2004) (Spencer et al. unpublished data)	Clark (2004) (Spencer et al. unpublished data)	Clark (2004) (Muir & Hawkey 1991)	Clark (2004) (Presidente 1978)	ISIS (2002)

(*Continued*)

Table 7.3 (*Continued*)

Parameter	Eastern grey kangaroo (see below)	Western grey kangaroo (see below)	Common wallaroo (see below)	Antilopine kangaroo (n = 5)	Red kangaroo (see below)	Whiptail wallaby (n = 16)	Quokka (see below)
PCV (L/L)	0.45 ± 0.06 (52)	0.47 ± 0.07 (71)	0.49 ± 0.07 (43)	0.45 ± 0.05	0.48 ± 0.07 (216)	0.52 ± 0.01	0.44 ± 0.09 (43)
RBC (× 10^{12}/L)	5.8 ± 0.8 (49)	5.2 ± 0.8 (54)	5.7 ± 0.8 (26)	7.4 ± 0.3	5.2 ± 0.8 (183)	4.9 ± 0.3	9.0 (1)
Hb (g/L)	154 ± 21 (49)	166 ± 27 (48)	169 ± 26 (26)	153 ± 18	171 ± 26 (195)	174 ± 2	141 ± 19 (42)
MCV (fL)	78 ± 7 (49)	90 ± 9 (51)	86 ± 8 (25)	61 ± 4	93 ± 7 (182)	113 ± 9	52 (1)
MCH (pg)	27 ± 3 (49)	33 ± 3 (46)	29 ± 3 (25)	20 ± 2	33 ± 3 (176)	38 ± 3	16 (1)
MCHC (g/L)	345 ± 13 (49)	361 ± 25 (47)	341 ± 15 (26)	335 ± 7	354 ± 23 (195)	333 ± 3	316 ± 39 (42)
nRBC/100 WBC	1.0 ± 1.0 (9)	2.0 ± 2.0 (20)	1.0 ± 1.0 (10)		1.0 ± 2.0 (63)	0.0	2.8 ± 1.1 (5)
Reticulocytes (%)	1.0 ± 0 (1)				0.5 ± 0.8 (15)		
WBC (× 10^9/L)	6.28 ± 3.04 (53)	6.60 ± 2.45 (73)	5.64 ± 1.80 (43)	8.50 ± 1.38	4.87 ± 1.82 (208)	4.20 ± 0.29	9.47 ± 3.95 (43)
Neutrophils (× 10^9/L)	3.57 ± 2.18 (52)	3.30 ± 1.90 (72)	2.58 ± 1.33 (43)	4.09 ± 1.31 (4)	2.71 ± 1.61 (187)	1.09 ± 0.16 (10)	4.05 ± 2.41 (43)
Lymphocytes (× 10^9/L)	2.49 ± 1.43 (53)	2.98 ± 1.32 (72)	2.51 ± 1.32 (43)	4.05 ± 0.53 (4)	1.95 ± 1.19 (195)	2.30 ± 0.24 (10)	4.48 ± 2.43 (42)
Monocytes (× 10^9/L)	0.17 ± 0.12 (50)	0.18 ± 0.15 (63)	0.29 ± 0.65 (35)	0.01 ± 0.02 (4)	0.12 ± 0.11 (154)	0.12 ± 0.06 (10)	0.31 ± 0.24 (35)
Eosinophils (× 10^9/L)	0.13 ± 0.09 (45)	0.18 ± 0.16 (54)	0.35 ± 0.64 (38)	0.28 ± 0.05 (4)	0.10 ± 0.11 (132)	0.22 ± 0.04 (10)	0.46 ± 0.28 (41)
Basophils (× 10^9/L)	0.06 ± 0.01 (6)	0.06 ± 0.04 (29)	0.03 ± 0.05 (2)	0.00 (4)	0.03 ± 0.04 (42)	0.10 ± 0.06 (10)	
Platelets (× 10^9/L)	186 ± 93 (28)	208 ± 74 (7)	212 ± 50 (12)		221 ± 108 (56)	291 ± 16 (5)	
Reference	ISIS (2002)	ISIS (2002)	ISIS (2002)	Clark (2004) (Spencer et al. unpublished data)	ISIS (2002)	Clark (2004) (Spencer et al. unpublished data)	Taronga Zoo In-house values

Table 7.4 Serum biochemistry reference ranges for selected macropod species (mean ± SD (n))

Parameter	Quokka	Red-necked pademelon	Yellow-footed rock-wallaby	Brush-tailed rock-wallaby	Bridled nail-tailed wallaby
Glucose (mmol/L)	5.6 ± 0.8 (3)	0.6 ± 0.1 (3)	6.1 ± 2.7 (19)	4.6 ± 2.0 (35)	2.9 ± 2.0 (8)
BUN (mmol/L)	6.7 ± 2.1 (3)	11.8 ± 0.4 (3)	9.9 ± 2.1 (18)	8.6 ± 1.5 (35)	15.3 ± 9.8 (8)
Creatinine (μmol/L)	80 ± 18 (3)	68 ± 8 (3)	99 ± 28 (18)	78 ± 13 (35)	137 ± 100 (8)
Calcium (mmol/L)	2.65 ± 0.08 (2)		2.53 ± 0.10 (4)	2.70 (1)	
Phosphorus (mmol/L)	1.65 ± 0.48 (2)		1.93 ± 0.62 (13)	2.10 ± 0.39 (35)	1.37 ± 0.27 (8)
Sodium (mmol/L)	140 ± 3 (2)		143 ± 4 (4)	143 (1)	
Potassium (mmol/L)	3.4 ± 0.1 (3)		3.9 ± 0.5 (9)	4.0 (1)	
Chloride (mmol/L)	97 ± 1 (2)		101 ± 3 (4)	101 (1)	
Bicarbonate (mmol/L)	21.3 ± 12.7 (2)		26.4 ± 3.3 (3)	23.7 (1)	
Cholesterol (mmol/L)	4.27 ± 0.18 (3)	3.33 (1)	2.52 ± 0.32 (4)	2.81 (1)	
Total protein (g/L C)	59 ± 8 (2)		61 ± 5 (13)	70 ± 5 (34)	59 ± 5 (8)
Total protein (g/L R)		65 ± 13 (3)	61 ± 5 (20)	67 ± 5 (34)	65 ± 4 (5)
Albumin (g/L C)	40 ± 4 (2)		48 ± 6 (13)	49 ± 6 (34)	41 ± 6 (8)
Globulin (g/L C)	19 ± 4 (2)		13 ± 6 (13)	20 ± 9 (34)	18 ± 4 (8)
T. bilirubin (μmol/L)	2.0 (2)		3.4 ± 4.2 (5)	5.2 ± 3.7 (3)	
ALP (U/L)	2827 ± 655 (2)		1141 ± 993 (14)	836 ± 595 (35)	1010 ± 290 (8)
ALT (U/L)	117 ± 17 (4)	67 ± 4 (3)	20 ± 8 (13)	21 ± 8 (20)	269 ± 207 (8)
GGT (U/L)			10.0 (1)		
AST (U/L)	37 ± 6 (4)	68 ± 9 (3)	97 ± 43 (19)	63 ± 54 (34)	100 ± 75 (8)
CK (U/L)	707 ± 550 (4)	926 ± 515 (2)	2536 ± 2902 (20)	2755 ± 2988 (34)	7704 ± 6633 (8)
Amylase (U/L)	392 ± 49 (2)		725 ± 159 (4)		
Lipase (U/L)	1.1 ± 1.4 (2)		5.5 ± 5.0 (2)		
Reference	ISIS (2002)	Taronga Zoo	Taronga Zoo	Taronga Zoo	Taronga Zoo

For bilirubin, T = total
Taronga Zoo—in-house values 2006.

(Continued)

Table 7.4 *(Continued)*

Parameter	Swamp wallaby	Parma wallaby	Agile wallaby	Red-necked wallaby	Tammar wallaby
Glucose (mmol/L)	6.8 ± 1.3 (8)	6.8 ± 2.4 (30)	4.9 ± 1.9 (9)	5.9 ± 2.2 (158)	5.5 ± 2.6 (25)
BUN (mmol/L)	8.2 ± 2.1 (8)	9.6 ± 2.1 (31)	6.2 ± 0.8 (9)	8.9 ± 2.5 (156)	7.9 ± 2.1 (25)
Creatinine (μmol/L)	133 ± 27 (8)	97 ± 27 (30)	92 ± 14 (8)	106 ± 35 (155)	106 ± 35 (24)
Uric acid (mmol/L)	0.04 ± 0.02 (1)	0.06 ± 0.02 (15)		0.04 ± 0.03 (57)	0.04 ± 0.05 (3)
Calcium (mmol/L)	2.80 ± 0.20 (8)	2.43 ± 0.18 (30)		2.58 ± 0.23 (155)	2.55 ± 0.25 (22)
Phosphorus (mmol/L)	2.58 ± 1.00 (9)	2.20 ± 0.81 (27)	2.04 ± 0.73 (8)	2.39 ± 0.78 (132)	2.13 ± 0.74 (20)
Sodium (mmol/L)	143 ± 7 (10)	145 ± 5 (23)		141 ± 5 (136)	152 ± 6 (20)
Potassium (mmol/L)	5.6 ± 2.3 (10)	4.5 ± 0.9 (23)		4.8 ± 1.2 (136)	5.0 ± 1.4 (20)
Chloride (mmol/L)	104 ± 6 (9)	102 ± 5 (22)		97 ± 5 (129)	102 ± 5 (11)
Bicarbonate (mmol/L)	26.3 ± 6.0 (2)	15.0 ± 3.6 (4)		22.1 ± 6.5 (26)	
Cholesterol (mmol/L)	2.15 ± 0.47 (7)	2.85 ± 0.54 (25)		2.20 ± 0.67 (141)	4.39 ± 2.83 (9)
Triglycerides (mmol/L)		1.00 ± 0.29 (11)		0.75 ± 0.50 (96)	1.94 ± 2.04 (4)
Total protein (g/L C)	65 ± 13 (8)	67 ± 8 (27)	63 ± 5 (8)	62 ± 12 (137)	62 ± 12 (24)
Total protein (g/L R)			65 ± 10 (9)		
Albumin (g/L C)	43 ± 10 (8)	51 ± 7 (23)	47 ± 5 (8)	41 ± 8 (114)	45 ± 8 (18)
Globulin (g/L C)	22 ± 7 (8)	16 ± 8 (24)	16 ± 3 (8)	25 ± 11 (112)	19 ± 11 (17)
T. bilirubin (μmol/L)	3.0 ± 7.0 (9)	3 ± 3 (28)		3 ± 5 (136)	3.4 ± 3.4 (20)
D. bilirubin (μmol/L)		2 ± 3 (5)		2 ± 3 (43)	1.7 ± 3.4 (5)
I. bilirubin (μmol/L)		3 ± 2 (5)		2 ± 2 (42)	3.4 ± 1.7 (5)
ALP (U/L)	3306 ± 1071 (6)	579 ± 774 (28)	1504 ± 873 (8)	1186 ± 1071 (131)	468 ± 451 (24)
ALT (U/L)	56 ± 30 (8)	77 ± 30 (29)	38 ± 5 (8)	45 ± 26 (128)	59 ± 24 (23)
LDH (U/L)	163 ± 37 (2)	624 ± 343 (15)		578 ± 531 (79)	629 ± 269 (9)
GGT (U/L)	118 ± 77 (3)	44 ± 31 (20)		46 ± 30 (73)	46 ± 34 (12)
AST (U/L)	103 ± 58 (8)	91 ± 66 (27)	29 ± 8 (8)	88 ± 71 (136)	71 ± 30 (25)
CK (U/L)	559 ± 353 (3)	824 ± 860 (20)	706 ± 327 (8)	1371 ± 1819 (48)	1060 ± 1184 (11)
Amylase (U/L)	127 ± 115 (3)	41 ± 24 (17)		68 ± 31 (75)	131 ± 63 (13)
Lipase (U/L)	51.7 ± 18.1 (3)	11.4 ± 13.1 (2)		12.5 ± 10.8 (18)	47.0 ± 60.0 (4)
Fibrinogen (g/L)	2.6 ± 0.6 (2)	3.0 (1)		1.1 ± 0.8 (10)	1.4 ± 1.3 (8)
Reference	ISIS (2002)	ISIS (2002)	Taronga Zoo	ISIS (2002)	Taronga Zoo

For bilirubin, T = total, D = direct, I = indirect
Taronga Zoo—in-house values 2006.

(Continued)

Table 7.4 *(Continued)*

Parameter	Eastern grey kangaroo	Western grey kangaroo	Common wallaroo	Red kangaroo	Long-nosed Potoroo
Glucose (mmol/L)	5.9 ± 1.7 (52)	5.1 ± 1.4 (75)	5.7 ± 1.7 (43)	7.0 ± 2.1 (181)	9.67 ± 2.70 (24)
BUN (mmol/L)	8.6 ± 2.9 (52)	9.3 ± 1.6 (74)	9.6 ± 3.9 (44)	8.6 ± 2.9 (188)	3.64 ± 1.15 (24)
Creatinine (μmol/L)	133 ± 53 (52)	110 ± 23 (73)	141 ± 44 (43)	133 ± 44 (185)	54.75 ± 7.04 (24)
Uric acid (mmol/L)	0.04 ± 0.03 (8)		0.09 ± 0.04 (10)	0.04 ± 0.04 (78)	
Calcium (mmol/L)	2.48 ± 0.23 (49)	2.60 ± 0.14 (70)	2.53 ± 0.20 (42)	2.65 ± 0.30 (184)	2.29 ± 0.17 (24)
Phosphorus (mmol/L)	2.45 ± 0.65 (44)	1.70 ± 0.50 (70)	2.29 ± 0.78 (36)	2.36 ± 0.71 (169)	1.62 ± 0.31 (24)
Sodium (mmol/L)	141 ± 4 (44)	142 ± 3 (70)	142 ± 5 (41)	144 ± 5 (170)	145.5 ± 2.12 (2)
Potassium (mmol/L)	4.2 ± 0.8 (46)	4.3 ± 0.4 (71)	4.7 ± 0.7 (40)	4.8 ± 1.0 (170)	3.20 ± 0.14 (2)
Chloride (mmol/L)	102 ± 5 (44)	98 ± 3 (70)	99 ± 4 (40)	98 ± 4 (137)	99.5 ± 0.71 (2)
Bicarbonate (mmol/L)	28.3 ± 4.5 (7)	29.5 ± 3.0 (70)	25.0 ± 10.8 (5)	25.5 ± 3.5 (22)	
Cholesterol (mmol/L)	2.25 ± 0.67 (35)	2.35 ± 0.48 (68)	2.15 ± 0.57 (37)	2.31 ± 0.67 (144)	3.99 ± 1.09 (24)
Triglycerides (mmol/L)	0.57 ± 0.53 (16)		0.97 ± 0.42 (6)	0.69 ± 0.74 (94)	
Total protein (g/L C)	60 ± 8 (45)	64 ± 7 (71)	64 ± 7 (40)	64 ± 10 (172)	55.83 ± 4.29 (24)
Total protein (g/L R)		61 ± 8 (76)			
Albumin (g/L C)	38 ± 5 (39)	35 ± 5 (71)	47 ± 10 (39)	42 ± 9 (167)	33.29 ± 3.64 (24)
Globulin (g/L C)	23 ± 6 (39)	29 ± 5 (71)	20 ± 13 (39)	23 ± 11 (162)	22.54 ± 4.04 (24)
T. bilirubin (μmol/L)	3.0 ± 2.0 (42)	0.9 ± 0.5 (70)	3 ± 3 (38)	3 ± 3 (172)	4.08 ± 2.12 (24)
D. bilirubin (μmol/L)	3 ± 2 (21)		0 ± 0 (6)	0 ± 2 (40)	
I. bilirubin (μmol/L)	2 ± 2 (21)		2 ± 2 (6)	2 ± 2 (40)	
ALP (U/L)	1037 ± 775 (46)	2311 ± 1043 (71)	532 ± 605 (39)	824 ± 648 (171)	950 ± 403 (24)
ALT (U/L)	56 ± 32 (39)	145 ± 147 (68)	50 ± 18 (41)	49 ± 28 (176)	86.42 ± 29.64 (24)
LDH (U/L)	95 ± 52 (17)		213 ± 199 (4)	155 ± 127 (88)	
GGT (U/L)	18 ± 9 (35)	13 ± 5 (69)	6 ± 3 (18)	6 ± 6 (83)	
AST (U/L)	60 ± 25 (49)	131 ± 217 (70)	75 ± 31 (41)	104 ± 65 (179)	94.08 ± 37.00 (24)
CK (U/L)	747 ± 762 (28)	1229 ± 1611 (68)	605 ± 420 (17)	638 ± 569 (89)	1483 ± 1326 (24)
Amylase (U/L)	87 ± 51 (14)	686 ± 464 (69)	65 ± 65 (16)	74 ± 34 (78)	
Lipase (U/L)	64.0 ± 130.0 (9)	8.0 ± 4.0 (11)	1.7 ± 1.9 (6)	7.8 ± 11.7 (17)	
Fibrinogen (g/L)	1.3 ± 0.5 (17)		1.5 ± 0.5 (3)	1.8 ± 1.0 (33)	
Reference	ISIS (2002)	Taronga Zoo	ISIS (2002)	ISIS (2002)	Perth Zoo (unpublished data)

For bilirubin, T = total, D = direct, I = indirect
Taronga Zoo—in-house values 2006.

samples were collected. Most macropods will require capture, physical and/or chemical restraint to obtain a sample. Animals may have been pursued or held in a trap for some time prior to sampling. Excitement, exertion, physical restraint, anxiety and pain will influence leukocyte and biochemical parameters. Leukocytosis due to neutrophilia and lymphocytosis may be seen, but this response seems less pronounced in macropods compared to eutherian mammals. Anaesthesia may negate or reverse this catecholamine-mediated response. Increased glucocorticoid concentrations (stress response) may cause mild neutrophilia, lymphopenia and, occasionally, eosinopenia.

Creatine kinase will rise to 2000–3000 U/L after physical or chemical restraint in most macropods. It is not uncommon to have levels up to 3000 U/L on routine blood samples, which are generally of little significance. Levels in the tens of thousands are significant and are usually indicative of myopathy. In these cases AST and LDH levels may also be markedly elevated. CK rises rapidly in response to cardiac and skeletal muscle damage, yet has a short half-life and only persists in high concentration in the presence of ongoing muscle damage. AST levels may rise more slowly after the initial muscle damage but has a longer half-life and high levels persist for longer than CK. Serial monitoring of AST and CK may provide insight into the duration and degree of muscle damage. Other clinical pathological changes seen with myopathy include those associated with metabolic acidosis. In advanced cases, renal function may be impaired and BUN and creatinine may be elevated (Rose 1999).

Age may also influence leukocyte concentrations. PY have different leukocyte characteristics from adults. This varies between species, with some having immature and mature myeloid cells with occasional large lymphocytes predominating; in others immature and mature neutrophils predominate. Differential leukocyte concentrations approach that of the adult by mid to late pouch life. There appears to be no differences between sexes in the species studied (Clark 2004).

There is only one report of leukaemia in a macropod. A juvenile pretty-faced wallaby (*M. parryi*) had generalised lymphosarcoma and leukaemia (Canfield et al. 1990).

Acute phase proteins such as fibrinogen can be used as a useful indicator of inflammation, particularly in macropods that do not mount a marked leukocyte response. In one study, fibrinogen concentration was increased in red-necked wallabies with lumpy jaw and was of greater diagnostic value than neutrophil concentration. In another study of red-necked wallabies and red kangaroos with unspecified bacterial infections, fibrinogen concentrations were elevated in most animals (red-necked wallaby: normal 1.3–3.6 g/L, infected 0.8–10.6 g/L; red kangaroo: normal 1.2–4.5 g/L, infected 5.9–11.7 g/L) but had similar diagnostic value to neutrophil concentration (Hawkey et al. 1982; Hawkey & Hart 1987).

Alkaline phosphatase is frequently elevated in macropods. Levels higher than 1000 U/L are common, which is significantly higher than reported reference ranges for domestic species. ALP is primarily produced in the liver, bone and intestines. In other animals, elevated levels are usually associated with hepatobiliary disease, bone disease, pregnancy, bone growth (young animals) and malignancies. The significance of these high levels in normal macropods is unknown.

Vitamin E and the selenium-containing enzyme glutathione peroxidase (GP) protect cells against oxidative damage caused by free radicals formed during cellular metabolism. Vitamin E (and not selenium) is known to be protective against development of muscular dystrophy in some small macropods, particularly the quokka (see 5.1). The diets of most quokkas in captivity are supplemented with vitamin E. If quokkas are succumbing to muscular dystrophy despite supplementation, it may be useful to measure vitamin E and GP levels (GP levels correlate directly with selenium). Although no reference ranges are available, the range in five normal quokkas from one institution (Taronga Zoo) was 2.9–3.2 mg/L vitamin E and 537–614 U/g Hb GP. Four from another institution (Melbourne Zoo) were within the range 2.0–2.6 mg/L vitamin E and 623–728 U/g Hb GP. The differences in values between the two groups are difficult to explain but may relate to the different vitamin E supplements being used. The vitamin E levels are comparable with those reported for sheep (1.8–17 mg/L), however, the GP levels in the quokkas are significantly higher than those reported for sheep (40–300 U/g Hb).

7.2 Parasitology

Endo- and ectoparasites are common in macropods. Faecal examinations should be conducted on captive macropods on a regular basis (at least twice a year) to monitor parasite loads. Direct wet preparations and faecal flotation are useful in-house techniques that can be used to diagnose endoparasite infections (Cunningham 1994). Skin scrapings with liquid paraffin or as KOH and Parker ink digests are useful for diagnosis of ectoparasite infestations.

Various haemoparasites, including haematozoa, trypanosomes, microfilariae of filarioid nematodes and larval muspiceoid nematodes may be found in the blood of macropods. These may be detected on thin blood films or by using concentration techniques (Clark 2004).

7.3 Urinalysis

Urinalysis is a useful adjunct to the diagnosis of urinary tract disease in macropods. Urine collection can, however, be difficult. With patience, free-flowing urine can be collected from tractable animals into a clean or sterile container. Urine may also be collected off the floor of enclosures but this is of less use, particularly for bacterial culture. Urine can be collected from PY while stimulating the cloaca. Male macropods are difficult to catheterise due to the presence of paired valve-like cusps approximately 2–4 cm proximal to the external urethral orifice (see 2.10 and 9.4.1) (Hatkin & Janssen 1979). The urethral opening in females is relatively inaccessible. The bladder is often difficult to palpate in macropods as it often lies within the pelvis and is between the epipubic bones. Urine can be collected by expressing the bladder if it is palpable. It is possible to collect urine by cystocentisis, however, this is best done by guiding the needle using ultrasound. Blind cystocentesis is possible by using a hypodermic needle inserted between the epipubic bones and directed caudoventrally at 45° to the body wall. The animal should be anaesthetised and placed in dorsal recumbency with the body tilted head-down. Urolithiasis has been reported in macropods (Bryant & Rose 2003) and crystals may be present in the urine (see 9.4.1). Urine sediment should be examined for crystals if urolithiasis is suspected.

Normal urine is usually pale yellow, but red urine may be seen occasionally (especially in PY) in normal animals due to the presence of a pigment. Brown urine indicates myoglobinuria, usually associated with capture myopathy. Normal urine contains no cells or bacteria. PY urine may contain traces of protein (Speare 1988). Normal pH is 5–6.

7.4 Cerebrospinal fluid (CSF) tap

CSF has successfully been collected from wallabies by inserting a 22–20 G spinal needle into the lumbar epidural space between mid- to cranial lumbar vertebrae. No specific reference data is available for CSF for macropods.

8 HAND-REARING

Macropod PY may require hand-rearing for various reasons, one of the most common being the discovery of a live PY in the pouch of a dead mother. PY may be thrown from the pouch if the mother is pursued by a predator or humans. The PY is thrown from the pouch by simultaneous relaxation of the pouch sphincter muscle and contraction of the longitudinal SC muscles of the pouch that in effect makes the pouch disappear. Very small PY that are permanently attached to the teat are rarely thrown. If a PY of this size is found out of the pouch it is usually an indication of PY failure to maintain attachment to the teat for maternal reasons (e.g. interruption of lactation) or PY weakness. Attempts at returning PY to the pouch in these situations are unlikely to succeed. If attempted, the teat must be carefully replaced in the PY's mouth. Furred PY that are already coming in and out of the pouch are more easily reunited with their mother. Unfurred, intermittently attached, pouch-dependent PY are more difficult to successfully return to the pouch (Booth 2002). Recapture of the mother and the stress of capture and restraint frequently result in the PY being thrown again. Chemical restraint may be required and the pouch may need to be closed with adhesive tape in an attempt to keep the PY inside the pouch. If attempts at replacing the PY in the pouch fail, the PY may require hand-rearing. A fully furred PY with no sign of the mother may have been thrown or fallen from the pouch or is fully emerged and has become separated from its mother. Occasionally, PY are intentionally pulled from their mother for hand-rearing. This may occur with valuable animals if a mother has a history of not rearing her own PY or if there is a need to have tame macropods for display, education or pets.

PY are permanently attached to the teat for approximately the first third of pouch life. Soon after birth the PY attaches to a teat and the commisures of the mouth close, narrowing the oral opening to a small hole. The tip of the teat expands to fill the oral cavity. Removing the teat from an attached PY can result in damage to the mouth and must be done with great care. In some cases it may be easier to cut the teat if the mother is dead, then remove the teat from the PY.

To successfully raise a macropod PY it is important to have an understanding of marsupial neonatal development, transfer of passive immunity and development of the immune system, changes in milk composition during lactation, thermoregulation, how to minimise stress, hygiene and sanitation and the diseases that affect PY. Detailed information on these is covered in Chapter 2. Features unique to macropod PY are covered here.

Determining the stage of development of macropod PY is crucial to the success of any hand-rearing attempt. The sequence of development from birth to emergence is similar for all macropods, but the length of pouch life varies significantly between species (Table 7.1). Duration of pouch life is not proportional to the size of the species. For example, pouch life in the largest macropod (the red kangaroo) is 235 d, while in the far smaller tammar and swamp wallaby it is 250 d and 265 d respectively. The timing of physiological and developmental stages in a PY are, however, related to the duration of pouch life. They vary between species and this is more important than size or age. Growth charts to determine stage of development and age of various macropod species are available (Wombaroo Food Products; Woods 1999). Growth curves based on head lengths to estimate age, rather than those based on other body parameters that may vary significantly with nutritional availability, are preferable.

The stage of physiological development of the PY is crucial to the success of hand-rearing and determines the level of commitment required for success. As a general rule, unfurred PY are less physiologically mature. They are unable to thermoregulate and are less immunocompetent. Therefore if a PY is furless, its eyes not yet open and its ears are fused to its head it has a very poor chance of survival. If the PY is unfurred but the skin is pigmented, its eyes are open or fine fur is starting to appear it may survive, but only with expert and intensive care. With fine fur the PY has a good chance of survival, but still requires significant expert care. Thickly furred PY usually survive (George 1990; Woods 1999).

Identification of the species is crucial to determining stage of development. The use of growth charts to age a PY requires its identification (George 1990). Knowledge of the species allows determinations of stage of pouch life. Some species are prone to particular diseases and knowledge of this will assist with prophylaxis (Chapter 2). Rescuers or carers may not know the species of a PY. Identification can be difficult in young PY and it may be necessary to consult an experienced macropod carer or a guide book (Strahan 1995; Menkhorst 2001). Knowledge of where the PY was found and the macropod species endemic to that area will assist identification.

A number of excellent milk formulas are available for hand-rearing macropod PY. These are tailored to different stages of development. Compared with eutherians, macropod PY have a low tolerance for lactose, which is not due to low levels of intestinal lactase activity but is the result of intracellular localisation of lactase activity. This means that lactose cannot be digested into monosaccharides until after it has been transported into enterocytes, apparently by slow pinocytosis. For this reason low-lactose milk formulas should be used for macropods (Booth 2002). Lactose (one molecule of glucose and one of galactose) is the main sugar of eutherian milk. It is a very transient component of marsupial milk within the first 4 d of lactation. Beyond this period marsupial milk contains several oligosaccharides of increasing size.

Orphaned macropods are very susceptible to dietary diarrhoea as they adjust from the mother's milk to an artificial formula. It is useful to dilute the formula with an oral rehydration solution for the first 2–3 d to allow adjustment to the formula (Booth 2002). Many wildlife carers focus on the type of milk formula as a reason for success or failure of hand-rearing. This may be true to some extent, however, other factors such as maintaining a strict feeding routine, milk temperature, toileting, hygiene, appropriate environment, lack of disturbance, minimising stress, good and consistent husbandry, patience and a natural instinct for hand-rearing are probably more important.

A normal healthy macropod PY should have prominent and readily palpable thymi (see 2.8) (Speare 1988). Prolonged elevation of glucocorticoids due to stress or chronic illness may cause premature thymic involution, thus reducing active immunity, and increase susceptibility to infections. This may be exacerbated by the lack of maternal antibodies which will disappear from the

PY's plasma within 6 wk of removal from the mother (Booth 2002).

Macropod PY should be offered solid food as soon as they are finely furred and sit with their head out of the pouch. Fresh grass with roots and dirt, chopped sweet potato, apple and carrot, lucerne pellets, hay and browse should all be offered. Exposure to parasite-free fresh adult faecal pellets may be useful in inoculating the foregut with bacteria and protozoa.

The approach to hand-rearing a PY depends on whether it will be released back into the wild or maintained in captivity. Macropods are very prone to imprinting. If the animal is to be released, every effort must be made to prevent imprinting. This can be achieved by providing only sufficient nurturing to give a sense of security, avoiding exposure to stimuli that are not relevant to the species, separating the carer and PY at a time that it would naturally separate from its mother, and exposing it to its own species as close to the time of pouch emergence as possible. A hand-reared macropod that is to be released must be both physically and behaviourally fit and completely capable of survival in the wild before being released. Sexual and non-sexual aggression in male hand-reared macropods and tintibulation (see 2.11) are problems associated with imprinting. Hand-reared imprinted male macropods, particularly larger species, can be very dangerous when they reach adult size. It has been ours and others' experience that castration and the use of GnRH agonists have no effect on this behaviour. Significant effort must be made to prevent imprinting and to socialise male macropods from an early age (Booth 2002).

9 DISEASES

9.1 Dermatological diseases

9.1.1 Papillomatosis (see 9.9.1a)

Macropod poxvirus lesions vary from single to multifocal to coalescing hyperkeratotic papillomatous proliferations often with a central umbilicated crater, typically about the head, fore limbs or tail, usually in juveniles or subadults (Fig. 7.7). Lesions may vary in size from a few millimetres up to 5 cm in diameter. Lesions are fairly characteristic, however, biopsies submitted for histopathology should demonstrate marked epidermal hyperplasia and eosinophilic intracytoplasmic inclusions displacing the nucleus of epidermal cells (McKenzie et al. 1979; Rothwell et al. 1984). Electron microscopy on sections from lesions may demonstrate intracytoplasmic inclusions in epithelial cells (Papadimitriou & Ashman 1972). Treatment is generally not indicated as lesions usually resolve over several months. Surgical excision can be undertaken if lesions are located near the eyes or mouth and are interfering with vision or the ability to eat. Strategies for preventing this disease, other than isolation of naïve PY from infected animals, have not been established as the life history of this virus has yet to be elucidated.

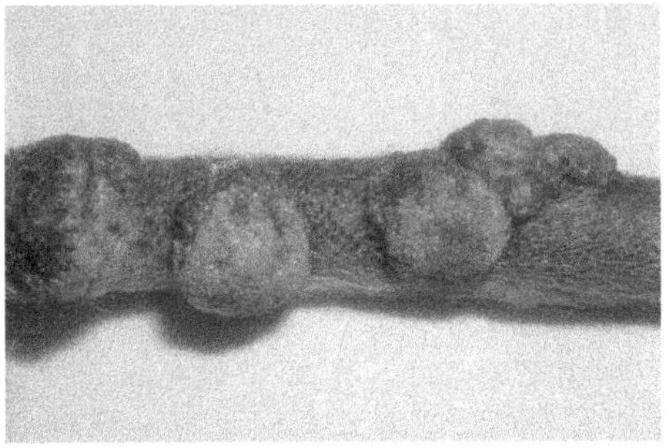

Figure 7.7 Papillomatous proliferations on the tail of a macropod.

9.1.2 Herpesvirus dermatitis (see 9.9.1f)

In addition to systemic signs of illness, localised dermatological lesions such as erythema, conjunctivitis and vesicles and ulceration of the cloaca, peri-cloacal and orofacial areas can occur in clinical outbreaks of herpesvirus-related disease (Finnie & Littlejohns 1976; Callinan & Kefford 1981).

9.1.3 Dermatophilosis

Multifocal exudative/crusting lesions and matted hair are typically present overlying erythematous and possibly bleeding dermis. Lesions can be extensive in severe cases. Squash preparations of exudate or the crusting material from lesions can be prepared with Gram's stain to demonstrate characteristic branching chains of cocci. Specialised microbial culture can be undertaken for definitive diagnosis. This disease is frequently associated with inappropriate environmental conditions and suboptimal husbandry in captive animals. Correction of the predisposing environmental factors is usually sufficient for resolution of lesions. Topical povidone-iodine solutions and/or long-acting oxytetracycline (20 mg/kg IM q 72 hr for 7–10 d) is usually effective in treating the condition.

9.1.4 Erysipelas (see 9.9.3f)

Typically infection with *Erysipelothrix rhusiopathiae* results in acute death. As well as systemic signs of illness, in more chronic cases urticarial or necrotising lesions may be present on the skin of the limbs and trunk. In some cases these may be raised and diamond-shaped. This finding is highly suggestive of *Erysipelothrix* infection. Gingivitis and oral mucosal ulceration may also be evident. Given the acute nature of this disease, diagnosis is typically at necropsy by culture of samples collected from major organs. If initiated early enough, therapy with penicillins, potentiated sulphonamides and tetracyclines may be effective. Vaccination could be considered, but as clinical cases are generally sporadic this is probably not warranted.

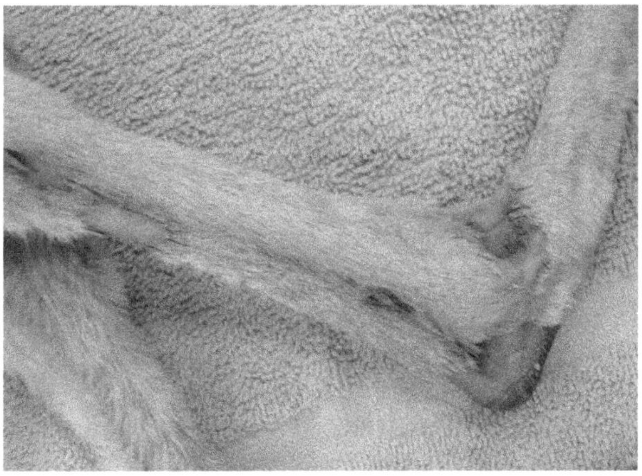

Figure 7.8 Dermatophytosis, caused by a *Trichophyton* spp., affecting the hind limbs of a juvenile red kangaroo. Photo: Tim Portas.

9.1.5 Pouch infections

The pouch of affected animals is excessively moist, greasy and foul-smelling in contrast with the relatively clean environment of the normal pouch (Johnson-Delaney 2002). The clinical appearance of this condition is highly suggestive. Swabs for bacterial culture and cytology will increase the index of suspicion. *Pseudomonas aeruginosa* has been associated with pouch infections in macropods. Treatment consists of gently cleaning the pouch with dilute chlorhexidine solutions followed by the topical application of silver sulphadiazine. In severe cases systemic antibiotics may be indicated. No specific measures have been proposed for the prevention of this condition. Macropods on long-term antibiotic therapy or subjected to chronic environmental stressors appear to be predisposed to developing pouch infections.

9.1.6 Mycobacterial dermatitis (see 9.9.2a)

Ulcerative dermatitis and/or granulomas associated with infection with *Mycobacterium avium* have been reported in free-ranging long-footed potoroos (*Potorous longipes*) (Fowler & Mitchell 2002).

9.1.7 Dermatophytosis (see 9.9.3b)

The clinical signs of dermatophytosis are highly variable and may include any or all of the following: focal, multifocal or extensive alopecia; scaling; broken hair shafts; characteristic 'ringworm' lesions. Lesions usually occur on the head, tail and extremities (Fig. 7.8). Hand-reared red kangaroo PY seem particularly susceptible to developing extensive and rapidly progressing lesions. Fungal culture of hair and scale collected from lesions should be undertaken to identify the causative organism. Histopathological examination of biopsies taken from lesions may also be useful. Wood's lamp examination for fluorescence is effective for some but not all species and strains of dermatophytes. Topical preparations, containing terbinafine or miconazole, and antifungal shampoos can be used to treat hand reared PY but are generally impractical in larger macropods or non-habituated smaller animals. Itraconazole (5 mg/kg PO sid) or griseofulvin (5–10 mg/kg PO bid) administered until 2 wk after the resolution of lesions is effective. Pulse therapy using itraconazole at 5 mg/kg sid PO, 1 wk on then 2 wk off, for 12 wk has also been used to successfully treat dermatophytosis in macropods. Adverse effects, including diarrhoea and neurological signs, have been observed in some individuals following the use of griseofulvin at 10 mg/kg PO bid for periods as short as 1 wk. In tractable animals or those that can be readily anaesthetised, weekly washes with enilconazole (Imaverol Fungicidal Wash, Austrichter) has been effective. In the captive situation, affected animals should be isolated and steps taken to prevent transmission via fomites. Steps should be taken to ensure that husbandry is optimal and environmental stressors are reduced.

9.1.8 Cutaneous candidiasis (see 9.9.3a)

Cutaneous candidiasis can occur concurrently with gastrointestinal infections or as a discrete clinical entity. Clinical signs include erythema, cracking of the skin and plaques in unfurred PY (Speare 1988). Gram stain of material obtained from skin scrapings demonstrates hyphae and budding yeasts. Culture will confirm the causative organisms. In addition to addressing predisposing factors (see 9.2.8), topical nystatin, miconazole or terbinafine preparations should be effective in resolving infections.

9.1.9 Cutaneous leishmaniasis (see 9.9.4b)

Clinically, lesions begin as papules progressing to develop areas of central ulceration. Lesions occur on the pinnae, limbs and tail of affected animals and are usually chronic in nature although they occasionally self-resolve. The lesions appeared to be most severe during the tropical wet season (Rose et al. 2004). Parasites may be evident in biopsies of lesions submitted for histopathology. Transmission electron microscopy has been utilised on formalin-fixed tissue and organisms grown in culture from lesions to demonstrate morphological features characteristic of *Leishmania* species. Infected animals exhibit variable levels of anti-*Leishmania* antibodies in serum assessed using an ELISA. PCR has also been utilised for molecular characterisation of the parasites. Treatment of this condition has not been described. Steps for preventing the disease are unknown as the source, host range and vector have not been described. In Australia, cutaneous leishmaniasis has only been diagnosed in red kangaroos held outside their normal geographic range in the Northern Territory.

9.1.10 Biting flies (*Austrosimulium pestilens* and *Simulium ornatipes*)

Infestations of *A. pestilens* are characterised by the presence of small, black, blood pool feeding flies on and about the head following periods of flooding or heavy rain. Bites are pruritic and cause localised swelling. Blood spotting occurs about the peri-ocular area and muzzle, and pronounced oedema of the peri-ocular region and muzzle can occur (McCarthy 1961a). In animals severely affected with peri-ocular oedema, vision may also be impaired. *S. ornatipes* has been observed feeding on swamp, red-necked and bridled nail-tailed wallabies, causing significant irritation and trauma to the pinnae. The presence of large numbers of these flies causes severe irritation and head-shaking. Affected animals are frequently isolated, lethargic and depressed. The clinical signs and the presence of flies are characteristic. The application of topical pyrethrin-based insect repellents is effective in providing short-term relief. The use of environmental pyrethrin-based insecticides on shelters and night houses during periods of extensive fly activity is an effective preventative strategy.

9.1.11 Pediculosis

Wild macropods are infested by a wide range of lice species but few, if any, clinical signs are apparent in mild cases of pediculosis. In more severe cases alopecia, matting of fur and pruritus with associated self-trauma are evident (Turni & Smales 2001). Lice may be removed manually or using the acetate tape impression technique for microscopic examination and identification. Eggs may be attached to the fur of the animal. Single-dose treatment with ivermectin (200–400 µg/kg) or moxidectin (500 µg/kg) topically, PO or SC, or imidacloprid (10 mg/kg) topically, is effective in eliminating infestations. Lice are most commonly seen in wild animals presenting for medical care and the routine application of insecticides listed above for newly arrived animals will eliminate infestations. Lice are rarely problematic in well-managed captive collections.

9.1.12 Ticks

Focal inflammation and craters may occur at the site of tick attachment and in some instances localised dermatitis may be present (Beveridge et al. 1985; Turni & Smales 2001). Adult ticks and nymphs can be removed manually. Single-dose treatment with ivermectin (200–400 µg/kg) or moxidectin (500 µg/kg) topically, PO or SC, or fipronil topically, is effective.

9.1.13 Mites (see 9.9.4a)

There are few reports of dermatitis in macropods associated with mites. The clinical signs of mite infestations in macropods are highly variable depending on the species of mite and the host but may include pruritus, hyperkeratosis, ulceration and localised inflammation (O'Callaghan et al. 1994; McLelland & Youl 2005). In advanced cases of sarcoptic mange in agile wallabies, thick crusting lesions with fissures are present and affected animals may be depressed, emaciated and dehydrated (McLelland & Youl 2005). Dermatitis associated with the dermanyssid mite *Thadeua serrata* has been reported in Proserpine rock-wallabies (*P. persephone*) and black-striped wallabies (*M. dorsalis*). Crateriform lesions approximately 5 mm in diameter with a central pit are present on the sparsely haired axillary and inguinal regions (Skerratt 2007). A seasonal (late summer and autumn) dermatitis caused by the same mite has been seen in brush-tailed rock-wallabies. In this species there are multiple small (<1 mm diameter) red papules, yellow crusting and alopecia extending down the medial thighs, sometimes to the foot and axilla and the medial antebrachium. Lesions are occasionally seen elsewhere on the body. Mites can be demonstrated in deep skin scrapings examined microscopically or from skin biopsies submitted for histopathology. Ivermectin (200–400 µg/kg) or moxidectin (500 µg/kg) topically, PO or SC once weekly for 4 wk is usually effective. Selamectin as per manufacturer's instructions has been used to treat sarcoptic mange in agile wallabies (McLelland & Youl 2005). There are no specific preventative measures other than isolation of infected animals during treatment.

9.1.14 Fleas

Clinical signs include pruritus, alopecia and the presence of fleas and flea faeces in the coat. Anaemia may be present in severe infestations. The presence of fleas and flea faeces in an animal's coat are diagnostic. Imidacloprid (10 mg/kg) topically (Baker & Beveridge 2001) or fipronil topically (as per manufacturer's instructions) in addition to environmental control is effective.

9.1.15 Xeroderma (excessively dry skin)

Clinical signs include dry flaking skin that is no longer smooth and supple. Cracks may be present affecting the tail, elbows and hocks. The clinical appearance in hand-reared PY is characteristic. A variety of topical moisturising lotions are routinely applied to prevent the development of xeroderma in hand-reared PY. Supplementation with an oral omega essential fatty acid formula (Megaderm, Virbac Australia) at domestic animal dose rates is also practised by some wildlife carers, though its safety and efficacy is unknown.

9.1.16 Stress or psychological alopecia

Stress or psychological alopecia can be seen in hand-reared PY raised under inappropriate conditions. There may be extensive, often bilaterally symmetrical alopecia over the trunk, tail and hind limbs. The digits of the hind limbs may be wet and macerated, secondary to excessive sucking. The clinical history is frequently suggestive, however, other potential differential diagnoses such as dermatophytosis should be eliminated. Treatment is based on correction of environmental stressors and suboptimal husbandry.

Figure 7.9 Squamous cell carcinoma on the pinna of a free-ranging parma wallaby. Photo: Tim Portas.

9.1.17 Cutaneous neoplasms

The clinical appearance of cutaneous neoplasms varies depending on tumour type and location (Fig. 7.9). Squamous cell carcinomas frequently affect the pinnae and nasal planum of albino red-necked wallabies. Definitive diagnosis is achieved by submitting biopsy tissue from the lesion for histopathology. Surgical excision can be undertaken if lesions are detected early enough.

9.2 Gastrointestinal diseases

9.2.1 Periodontal diseases

Periodontal disease occurs in both captive and free-ranging animals. Clinical signs include calculus accumulation, gingivitis, gingival recession and periodontitis (Bodley 2002). Calculus in free-ranging macropods tends to be firm, mineralised and black (Bird et al. 2002) while captive animals tend towards large amounts of soft crumbly plaque (Burton 1981). The lesions are self-evident following a detailed oral cavity examination. Removal of calculus via ultrasonic or hand-scaling and the topical application of dilute chlorhexidine solutions is indicated. Parenteral antibiotic therapy is indicated in severe cases. The provision of browse to all species, including grazing species of macropods, is believed to reduce the incidence of periodontal disease (Bodley 2002). Excessively fibrous hays and hays containing sharp awned grasses should also be avoided. Diets for captive macropods should not include bread or other rapidly fermentable carbohydrates.

9.2.2 Oral necrobacillosis (see 9.9.2b and Table 7.6)

A broad range of clinical signs may be apparent depending upon the location and extent of the lesion. The most common clinical presentation is mandibular or maxillary swelling with or without the presence of a draining sinus (Fig. 7.10). Inappetence, dysphagia, halitosis and hyperptyalism may occur, associated with the presence of an infected tooth root or discharging sinus. Other presentations include unilateral ocular discharge, injected conjunctiva and blepharospasm and unilateral nasal discharge. Signs of systemic illness may be apparent following haematogenous spread of infections and will reflect the location of secondary abscesses. The clinical signs in advanced disease are characteristic. Recurrence is common in treated animals with only 16% of treated red-necked wallabies surviving beyond 12 mo post therapy (Lewis et al. 1989). Euthanasia is frequently the most humane option in advanced cases.

A detailed oral cavity examination should be performed to identify any infected teeth or penetrating lesions. Radiography is warranted to identify retained premolars or molars, osteomyelitis and tooth root resorption (Fig. 7.11). Haematology and biochemistry may reveal a neutrophilia, elevated CK and AST. Hawkey et al. (1982) found fibrinogen levels in red-necked wallabies with oral necrobacillosis consistently elevated but a neutrophilia was variably present. A more consistent finding was abnormalities of neutrophil morphology including hypolobulation of nuclei and a basophilic cytoplasm.

Radical surgical debridement of affected soft and osteological tissues should be undertaken to achieve surgical fistulation of the lesion (Fig. 7.12). Establishment of effective drainage to prevent further progression of the osteological lesion is imperative. Copious lavage of the affected area with an antimicrobial agent such as 2% povidone-iodine to remove purulent material and achieve a localised antimicrobial effect is then performed (Fagan et al. 2005). Therapy also needs to achieve resolution of any initiating dental disease. Extraction of infected teeth and debridement of surrounding necrotic bone are required. Topical antibiotic solutions including benzyl penicillin and metronidazole are used to flush the resulting deficit, which is then packed with a dental protective paste to prevent fur-

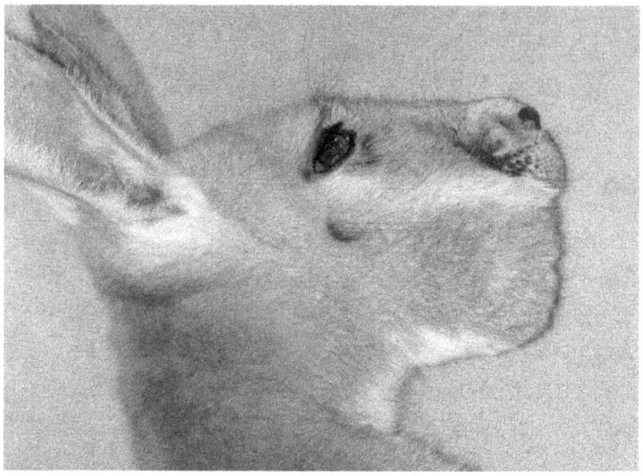

Figure 7.10 Oral necrobacillosis in a red kangaroo. Photo: Tim Portas.

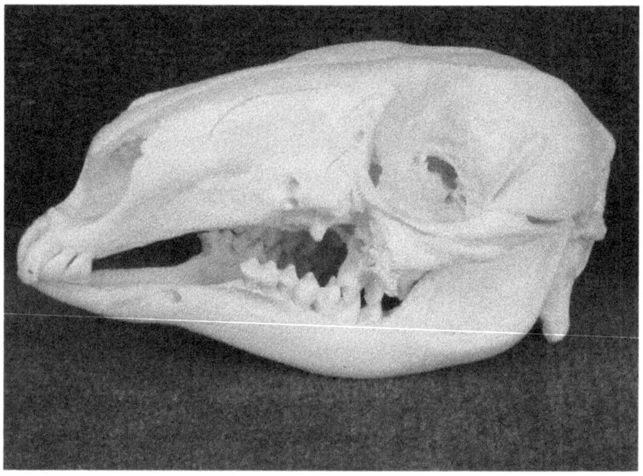

Figure 7.11 Skull of a red kangaroo showing area of osteomyelitis and bone lysis of the maxilla. Photo: Larry Vogelnest.

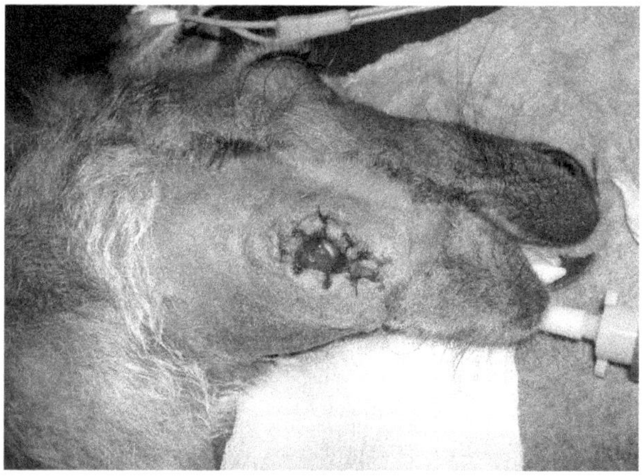

Figure 7.12 Post surgical appearance of the mandible of a western grey kangaroo with oral necrobacillosis. The surgery involved the removal of a retained premolar, the affected soft tissue was debrided, necrotic bone was curetted, antibiotic-impregnated beads were implanted and wound margins were marsupialised to the surrounding skin. Photo: Tim Portas.

ther impaction with food material (Clarke 2003). The implantation of antibiotic-impregnated polymethylmethacrylate beads following surgical debridement has been described; it has the advantage of providing a prolonged localised antibiotic effect (Hartley & Sanderson 2004). Antimicrobial therapy as dictated by the results of culture and susceptibility or clindamycin (11 mg/kg PO bid) alone or in combination with metronidazole (20 mg/kg PO bid), or oxytetracycline (40 mg/kg IM q 48 hr) should be initiated and continued until complete resolution of the lesion (Kirkwood et al. 1988). Analgesia using an NSAID such as meloxicam (0.2 mg/kg PO sid), carprofen (2–4 mg/kg SC sid) or ketopropfen (2 mg/kg SC sid) should be provided. An opioid such as buprenorphine (0.01–0.05 mg/kg IM or SC) may be used at the time of initial surgery. Skeletal scintigraphy may be useful for monitoring ongoing bone remodelling and determining the required length of treatment (Woods & Blyde 1998). Therapy for this disease is challenging as many of the oral antibiotics suitable for inclusion in therapeutic regimes are unpalatable, treatment is for a prolonged period and, unless very tractable, most macropods will not tolerate repeated manual restraint or darting for the administration of drugs. The use of long acting antibiotic preparations (long acting amoxycillin [20 mg/kg IM q 48h], long acting oxytetracycline [20 mg/kg IM q 72 hr]) and the NSAID tolfenamic acid (4 mg/kg q 48 hr) are useful for intractable animals.

Prevention of faecal contamination of feed and feeding stations is important in reducing environmental bacterial loads and minimising bacterial exposure via the oral route. Outbreaks have been associated with adverse environmental conditions, so provision of adequate shelter during periods of cold and wet weather is also important. Stocking densities should be reduced, where appropriate, to reduce overcrowding, associated environmental stress and levels of faecal contamination. Ensure appropriate diets. Avoid fibrous hays and excessive plant awns which may predispose to buccal trauma, and foods such as bread that promote plaque formation. The provision of browse encourages appropriate dental exercise and may help post-wear molars to shed more rapidly.

Vaccination with a multi serotype vaccine against *Dichelobacter nodosus* (previously *Bacteroides nodosus*) may reduce susceptibility to the disease, presumably by inducing cross-protective immunity against other *Dichelobacter* species that may act synergistically with *Fusobacterium necrophorum*. In one study, which compared the responses of tammar wallabies vaccinated once with a commercially available multi-serotype vaccine against *D. nodosus* serological responses, measured via ELISA, were not significantly different between vaccinated and unvaccinated controls. Booster vaccinations 18 mo later in a subset of vaccinated animals produced four- to five-fold increases in titres. Despite the initial lack of significant difference between vaccinated and unvaccinated animals, 4 of 42 unvaccinated controls subsequently died from oral necrobacillosis whereas no cases occurred in any of the 69 vaccinated animals during the study period (Blanden et al. 1987). A similar vaccination protocol was employed in a group of captive red-necked wallabies but was considered unsuccessful, with 2 of 65 animals developing severe oral necrobacillosis within 3 mo of vaccination (Gulland et al. 1987). Sterile abscesses, occasionally requiring surgical intervention, frequently develop at the site of vaccination. Lethargy, inappetence and pyrexia of several days duration is not uncommon in PY post vaccination. Vaccination of red-necked wallabies with killed *Fusobacterium necrophorum* in Freund's adjuvant did not protect animals against subsequent challenge (Smith et al. 1986).

9.2.3 Gastric amoebiasis (see 9.9.4b)

Clinical signs of gastric amoebiasis include progressive weight loss, diarrhoea, anorexia, bloat and, in some cases, sudden death. Chronic progressive disease has been reported in a tammar wallaby (Stedman et al. 2003); acute disease has been seen commonly in wallaroos and sporadically in red kangaroos (ARWH 2007). Diagnosis is achieved by demonstrating the presence of *Entamoeba* cysts in faeces via faecal flotation or smear. *Entamoeba* trophozoites can be identified histopathologically in the gastric mucosa from samples collected post-mortem. Indirect immunofluorescent antibody testing of formalin-fixed tissues with rabbit *E. histolytica* antiserum has been described (Stedman et al. 2003) and is required for definitive identification of the organism in histopathological sections.

Cases in wallaroos are typically acute with rapid progression to death but metronidazole (20 mg/kg PO bid) could be instituted if the disease is suspected. Apparent spontaneous recovery has been observed in mild cases of diarrhoea in wallaroos. Sulpha drugs and amprolium were ineffective in a tammar wallaby with chronic progressive disease (Stedman et al. 2003). Infected animals should be isolated during treatment.

In some cases infection appears to occur subsequent to penetration of the gastric mucosa by nematode larvae (ARWH 2007), therefore anthelmintic administration to suppress nematode parasite numbers may be a useful preventative strategy.

9.2.4 Strongyloidiasis (see 9.9.4b)

Disease associated with strongyloidiasis has only been reported in captive animals. Sudden death may occur in acute cases. In chronic cases signs include weight loss, progressive anorexia and diarrhoea (Winter 1958, Speare et al. 1982). The Baermann sedimentation technique is used to demonstrate first- or second-stage *Strongyloides* larvae in faeces. A range of anthelmintics are effective, including ivermectin (200–400 μg/kg) or moxidectin (500 μg/kg) topically, PO or SC; levamazole (10 mg/kg) PO; or thiabendazole (44 mg/kg) PO. A routine parasite surveillance program, including the Baermann sedimentation technique, should be in place for captive collections with historical or anticipated cases of strongyloidiasis.

9.2.5 Gastric trichobezoar

Gastric trichobezoars are seen principally in tree-kangaroos. Weight loss and inappetence may occur but the presence of a trichobezoar is not usually associated with clinical signs. They are more often incidental findings on abdominal palpation (Bush & Montali 1999). Contrast gastrointestinal radiography may be used to define their location and extent. Surgical removal is indicated when clinical signs are present.

9.2.6 Coccidiosis (see 9.9.4b)

The clinical signs of coccidiosis are variable and may include inappetence, abdominal discomfort, lethargy, diarrhoea (often haemorrhagic), dehydration, dependent oedema, bruxism or cloacal prolapse.

Experienced wildlife carers are often capable of detecting subtle clinical signs in the early stages of the disease and initiating treatment when it is likely to be most effective. Coccidiosis should be considered a likely differential diagnosis in any case of illness in hand-reared eastern grey kangaroos that have emerged from the pouch. Treatment should be initiated on suspicion of the disease. Diagnosis is achieved by identifying coccidial oocysts via faecal flotation in conjunction with the presence of clinical signs. Oocysts may be shed intermittently and are absent in the early stages of the disease. Their absence on faecal flotation does not preclude a diagnosis of coccidiosis. Clinical signs frequently provide a high index of suspicion.

Treatment with toltrazuril (25 mg/kg PO sid 3 d) should be initiated immediately. Antibiotic therapy with enrofloxacin (5 mg/kg IM or SC sid 5 d) is also recommended to protect against secondary bacterial infections. Fluid therapy is best administered by providing oral electrolytes or warmed SC fluids. PY with coccidiosis are typically hypoproteinaemic and administration of IV fluids is poorly tolerated. Homologous plasma transfusions (10 mL/kg IV) from healthy adults may be useful (Blyde 1999). In cases where haemorrhagic enteritis is present, phytomenadione (2.5 mg/kg IM or SC sid 5 d) may be beneficial. Pain relief is important as PY typically exhibit marked abdominal discomfort. Analgesia can be provided with flunixin meglumine (1 mg/kg IM or IV sid 3 d), meloxicam (0.2 mg/kg SC sid 3 d) or buprenorphine (0.01–0.05 mg/kg SC tid) as required.

Strict hygiene is important to prevent excessive environmental contamination of emergent PY and pre-release yards. Prevention of overcrowding and excessively moist conditions is important in captive collections. The administration of plasma harvested from exposed healthy adults to PY at the time of emerging from the pouch may confer some immunity and help prevent the disease (Blyde 1999).

9.2.7 Salmonellosis and campylobacteriosis (see 9.9.2i)

Macropods with enteric salmonellosis and campylobacteriosis present with indistinguishable clinical signs; diagnosis depends upon culture of the causative organism. Clinical signs in enteric cases include semi-formed faeces through to haemorrhagic diarrhoea, anorexia, dehydration and lethargy. Animals may also be found dead without exhibiting clinical signs. A common finding in PY is marked abdominal discomfort. Carriers are asymptomatic. Fresh faeces and/or rectal swabs should be submitted for microbial culture. Gram stains performed on faecal smears from affected animals may reveal a relatively uniform but large bacterial population and the presence of epithelial cells coated with bacteria (Speare 1988). Antimicrobial therapy is dictated by the results of microbial culture and susceptibility testing. Ceftiofur (2 mg/kg IM sid 7–10 d) can be used empirically if susceptibility results are unavailable (Holz 2003). Analgesia including buprenorphine (0.01–0.05 mg/kg SC tid) is indicated where there are signs of abdominal discomfort such as abdominal guarding and teeth grinding. Supportive care should be provided, including SC or IV fluid therapy as necessary and resolution of any predisposing stressors. Antibiotic therapy should be continued until affected animals have two consecutive negative cultures a week apart. Appropriate husbandry (including consistency in feeding times, volume, composition and temperature of milk feeds) and reduced environmental stressors help prevent the development of clinical signs in asymptomatic carriers. Affected animals should be isolated during treatment.

While the above therapeutic regimes may be effective in resolving the clinical signs associated with salmonellosis and campylobacteriosis, it must be remembered that these diseases are significant zoonoses and a carrier state can occur despite the resolution of clinical signs. Euthanasia may be the most appropriate course of action, particularly where affected animals have close and ongoing contact with people.

9.2.8 Candidiasis (see 9.9.3a)

Candidiasis may be associated with diarrhoea that is frequently yellow with a characteristic yeast-like smell. Plaques may be evident on the mucosal surfaces of the oral cavity and crusting lesions may be present around the lips. PY frequently have reduced appetite and in severe cases anorexia, presumably due to pain associated with oral cavity lesions (Obendorf 1980; Speare 1988). A Gram stain performed on faecal material or on scrapings from the oral cavity reveals the presence of budding yeasts or pseudomycelia (Fig. 7.13). Nystatin (5000–10 000 IU/kg PO tid 3–5 d) is administered separately to milk feeds to treat oral cavity, oesophageal and gastric lesions. Higher doses of nystatin may be used but may cause diarrhoea. Itraconazole (5 mg/kg PO sid 5 d) has also been used (Holz 2003). Cloacal lesions can be treated by the topical application of nystatin cream. Hygiene and husbandry practices should be reviewed, as candidiasis is often associated with suboptimal practices. A commercially available nystatin preparation containing sodium phosphate and peppermint oil has been associated with the development of intractable, and eventually fatal, diarrhoea in furless eastern and western grey kangaroo PY when used for treating candidiasis (E Latham *pers. comm.*).

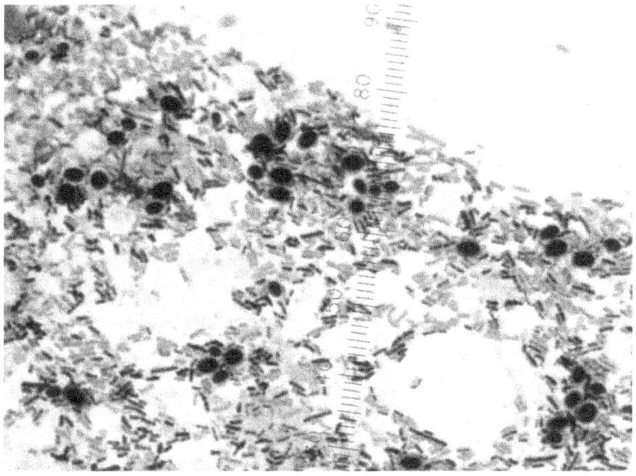

Figure 7.13 Gram stain of a faecal smear from a hand-reared red-necked wallaby PY with diarrhoea, demonstrating the presence of spherical yeasts consistent with candidiasis. Photo: Tim Portas.

Figure 7.14 Cloacal prolapse in a juvenile eastern grey kangaroo. Photo: Tim Portas.

9.2.9 Cloacal and rectal prolapse

Cloacal and rectal prolapse occurs sporadically in hand-reared PY. Prolapse of the cloacal or rectal mucosa is self-evident (Fig. 7.14). Where the mucosa has been prolapsed for an extended period it becomes congested, inflamed, traumatised and eventually dehydrated. Affected animals are anaesthetised and the prolapsed tissue reduced and replaced using gentle massage and lubrication. A purse-string suture is applied to the cloaca for 76 hr and toileting not performed during this time. Prolapses are often associated with coccidiosis, candidiasis, other causes of diarrhoea and inappropriate toileting techniques. These predisposing causes should be identified and treated where possible. Topical application of corticosteroid-based creams may be sufficient to resolve the prolapse in some cases.

9.2.10 Intestinal intussusception

Affected animals are lethargic and may develop abdominal distension and pain, defecation may cease and small amounts of blood may be passed (Blyde 1999). Diarrhoea or other clinical signs associated with enteric diseases such as coccidiosis or salmonellosis may precede the development of an intussusception in PY. Intussusceptions maybe evident on abdominal palpation but can be further confirmed by contrast intestinal radiography or exploratory laparotomy. Intussusceptions must be surgically corrected using standard surgical techniques. Supportive care including antibiotic administration and judicious fluid support is indicated. This condition occurs sporadically and there are no specific preventative measures other than those aimed at preventing potential predisposing causes.

9.3 Cardiovascular and pulmonary diseases

9.3.1 Myocardial infarction

Myocardial infarctions have been observed occasionally at necropsy in captive macropods. Dyspnoea and exercise intolerance may be evident but death without premonitory signs is the most frequent presentation. Myocardial infarction with fibrous connective tissue replacement and neocapillary formation may be seen histologically (ARWH 2007). Treatment is not usually possible given the acute nature of the disease.

9.3.2 Cardiomyopathy

Cardiomyopathy can occur secondary to *Toxoplasma gondii*-induced myocardial necrosis, exertional and nutritional myopathies and encephalomyocarditis virus infection. It also appears to occur as a primary condition in some species of macropods including antilopine, eastern grey, red and tree-kangaroos as well as wallaroos (Chineme & Njoku 1978; Bennet et al. 1993; Bush & Montali 1999; ARWH 2007). Speare (1988) cites dysrhythmias and frequent premature ventricular contractions as common clinical features in wallaroo and wallaby PY with cardiomyopathy. A systolic heart murmur, subcutaneous oedema and premature ventricular contractions evident on electrocardiographic examination were present in a female eastern grey kangaroo with a cardiomyopathy of undetermined origin (B Bryant *pers. comm.*). Acute deaths have occurred post anaesthesia in tree-kangaroos with cardiomyopathy (Bush & Montali 1999). Diagnostic approaches utilised in domestic animals are applicable in macropods including electrocardiography, echocardiography and radiography. Treatment of any underlying disease process is required. Therapy for primary cardiomyopathies is based on the principles utilised in domestic animals, although given the typically acute presentation in macropods treatment is rarely possible. There are no specific preventative measures for primary cardiomyopathies.

9.3.3 Endocarditis

A systolic murmur may be present and affected macropods may be acutely ill due to septicaemia. Cardiac enlargement as determined by thoracic radiography supports the diagnosis (Speare 1988). Echocardiography is required to establish a definitive diagnosis. Blood should be collected aseptically and submitted for culture and susceptibility testing. Antimicrobial therapy as indicated by the results of culture and susceptibility testing should be initiated and continued for 4 wk. Speare (1988) recommended the use of digoxin (10 mg/kg sid) for at least 2 mo in affected PY.

9.3.4 *Bordetella bronchiseptica* rhinitis/pneumonia (see 9.9.2c)

The clinical signs of *Bordetella bronchiseptica* rhinitis/pneumonia may include serous to mucopurulent nasal discharge, intermittent sneezing, increased respiratory effort, reduced exercise tolerance, weight loss and/or diffuse audible crackles on thoracic auscultation.

A definitive diagnosis is achieved by culture of *B. bronchiseptica* from nasal or tracheal swabs, from bronchoalveolar lavage fluid or pulmonary tissue obtained post-mortem. Thoracic radiographic findings may include an interstitial or alveolar pattern and tracheal stenosis (Fig. 7.15). Necropsy findings may include peritracheal inflammation and marked pulmonary congestion and consolidation (Ward & Blyde 2001). Antibiotic treatment with enrofloxacin (5 mg/kg sid) PO or SC or other antibiotics as indicated by culture and susceptibility testing is indicated until affected animals are culture negative. Appropriate symptomatic therapy includes nebulisation with acetyl cystine in saline or the use of bromhexine hydrochloride and the provision of thermal support. A commercially available

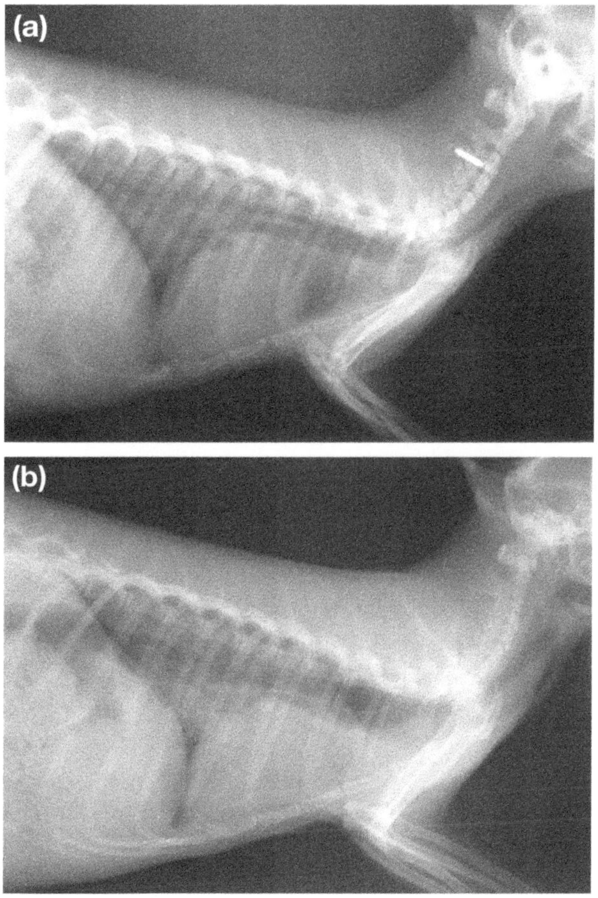

Figure 7.15 Lateral thoracic radiograph of a tammar wallaby. a) Tracheal stenosis associated with *Bordetella bronchiseptica* infection. b) A normal tammar wallaby.

canine vaccine against *B. bronchiseptica* appears to provide effective immunity in macropods, although serological responses to vaccination and challenge studies have not been undertaken (Ward & Blyde 2001; Herbert et al. 2005b). Affected animals should be isolated for treatment during outbreaks.

9.3.5 Bacterial pneumonia

Bacterial pneumonia is relatively common in macropods and a range of signs may be present including coughing, tachypnoea, dyspnoea and exercise intolerance. In advanced disease, animals are lethargic, depressed and inappetent. Clinical signs in PY may be subtle until advanced disease is present. Air movement is rarely detectable on auscultation in PY and audible respiratory sounds in sick PY should be considered significant (Speare 1988).

A thorough clinical examination including thoracic auscultation and thoracic radiography under isoflurane anaesthesia in PY or under sedation/anaesthesia in subadult and adult macropods should be performed. Ancillary diagnostic procedures include bronchoalveolar lavage for bacterial culture, including specialised mycobacterial culture, and bronchoscopy. Common significant bacterial isolates include *Escherichia coli, Klebsiella* species and *Pseudomonas* species.

Antibiotic therapy, as indicated by results of culture and susceptibility, or enrofloxacin (5 mg/kg SC sid) or ceftiofur sodium (1–2 mg/kg IM or IV sid) should be initiated and continued until clinical signs have resolved. Speare (1988) advocated a combination of gentamycin (4–7 mg/kg IM bid) and amoxycillin (10 mg/kg IM tid) for severe cases of pneumonia in PY.

Symptomatic and supportive therapy should be provided including nebulisation with acetyl cystine in saline. Hand-reared PY are particularly susceptible, so appropriate husbandry and environmental management is essential. Aspiration pneumonia is a common sequela to poor feeding technique or inappropriately shaped or sized teats. Prophylactic antibiotic therapy with enrofloxacin (5 mg/kg SC sid 3 d) may be indicated in hypothermic PY presenting for hand-rearing as pneumonia frequently develops in these animals.

9.3.6 Respiratory tract cryptococcosis (see 9.9.3c)

Clinical signs associated with cryptococcal infections of the respiratory tract vary depending upon location. They may include sneezing, coughing, dyspnoea, weight loss, inappetence and occasionally sudden death. Fine-needle aspirates from affected tissue can be stained with India ink, Gram stain, methenamine silver or methylene blue to demonstrate spherical yeasts. Histopathology of pulmonary tissue in affected animals is characteristic. Cryptococcal capsular polysaccharide antigen can be detected in serum by the latex-cryptococcal antigen agglutination test (LCAT). Treatment is with itraconazole (20–40 mg/kg PO sid) or fluconazole (10–20 mg/kg PO bid) until the LCAT is negative. Browse or exhibit furniture derived from various *Eucalyptus* species including *E. camaldulensis* may be heavily contaminated with *C. gatti* (formerly *C. bacillisporus* or *C. neoformans* var. *gatti*) and provide a source of infection for captive macropods (Vaughan et al. 2005).

9.3.7 Verminous bronchitis and pneumonia (see 9.9.4b)

Dyspnoea, coughing, exercise intolerance, depression and nasal discharge have been described in potoroid marsupials with verminous bronchitis and bronchiolitis (Love & Reddacliff 1992). *Marsupostrongylus* or *Capillaria* eggs can be found in bronchoalveolar lavage fluid and in faeces. Bronchoscopy can be used to visualise the lower airways. Capillarial nematodes and their eggs are apparent in the airways on histopathology (Love & Reddacliff 1992). *Marsupostrongylus* is evident in alveoli and bronchioles and is associated with variable degrees of bronchopneumonia (Beveridge et al. 1985). Fenbendazole (25 mg/kg PO sid 5 d) is effective for eliminating infections (Love & Reddacliff 1992). Capillarial eggs are detectable in faeces and should be considered when implementing routine parasite surveillance and control programs for potoroid marsupials. *Marsupostrongylus* has only been reported in free-ranging macropods.

9.3.8 Echinococcosis (see 9.9.4b)

Infections with *Echinococcus granulosus*, even when severe, are frequently asymptomatic. The pulmonary parenchyma is the principal site of cyst development but cysts have also been found in the liver and pleural cavity. Clinical signs observed in a small number of affected animals included dyspnoea, exercise intolerance and progressive emaciation (Johnson et al. 1999). Thoracic radiography may demonstrate cysts as radiolucent structures within the pulmonary parenchyma. In experimentally infected tammar wallabies, cysts were evident radiographically as early as 4 mo post infection (Barnes et al. 2007b). Various serological tests have been used to diagnose hydatid disease, including indirect haemagglutination test on serum to detect antibodies against larval stages, immunoelectrophoresis for antibody to antigen 5 of *E. granulosus* and Western blot test to detect antibodies to the *EpC1* protein of *Egranulosus*.

Treatment is not considered feasible, as infections are typically asymptomatic unless animals have extensive and advanced disease. Surgical resection of single lesions, if detected ante mortem, could be considered in valuable captive animals. The disease can be prevented in captive macropods by preventing the ingestion of feed contaminated with faeces from infected carnivores.

9.3.9 Encephalomyocarditis virus (see 9.9.1g)

The clinical course of disease caused by encephalomyocarditis virus (EMC) is typically acute, and sudden death is a common presentation. In less acute cases ataxia, depression, dyspnoea and other signs referable to cardiomyopathy may be evident. Myocardial necrosis associated with mononuclear cell inflammation is suggestive. Viral culture is possible on a range of cell lines. Antibody titres can be determined using a virus neutralisation test (Reddacliff et al. 1997). Treatment has not been described due to the acute nature of the disease. Wallaroos developed neutralising antibodies following vaccination with an inactivated vaccine with a booster vaccination administered 1 mo later (McLelland et al. 2005). Three of four Lumholtz's tree-kangaroos vaccinated using a similar vaccine and vaccination protocol developed antibody titres that were considered protective (Blyde 2005). Vaccination of macropods could therefore be considered in the face of an outbreak. While the role of rodents in the epidemiology of encephalomyocarditis virus has yet to be conclusively demonstrated, effective rodent control in exhibits and feed sheds is indicated (Thomson et al. 2001).

9.3.10 Haematozoan infections

Acute mortalities associated with a yet to be elucidated haemolytic anaemia syndrome occur sporadically in young eastern grey kangaroos, red-necked wallabies and swamp wallabies in coastal regions of north and mid-north New South Wales. The aetiology of this condition is uncertain, with a protozoal parasite and an omnibacteria-like organism variously postulated as the causative agent. Clinical signs are typically seen in hand-reared PY between 6 and 13 mo of age and this reflects the massive intravascular haemorrhage that occurs in affected animals. Packed cell volumes of less than 10% are common. PY are also hypoproteinaemic with total protein levels as low as 30 g/L. Mucus membrane pallor, polyuria, polydipsia and reduced appetite are variably present. Affected animals tend to bleed from tick attachment and venipuncture sites (Cook et al. 1997). Haematological examination reveals a marked, non to minimally regenerative anaemia and the presence of round to elongated micro-organisms (possibly protozoan) within small numbers ($\leq 2\%$) of erythrocytes. Microscopically, parasitised erythrocytes are evident sequestered in a range of organs. Schizont-like forms are present in blood vessels.

Treatment consists of the provision of supportive care with blood transfusions indicated in severely anaemic individuals. Little is known about the requirements for cross-matching prior to transfusion in macropods, however, reactions to single transfusions have not been reported. Imidocarb administered IV has been used with apparent success in some cases. Phytomenadione (2.5 mg/kg IM or SC sid 5 d) may be beneficial. Until this syndrome is better understood, no specific preventative measures can be recommended. Wildlife carers who have experience with this syndrome have advocated the continued exposure of hand-reared PY to low numbers of (potential) arthropod vectors throughout the hand-rearing period.

9.3.11 Pindone toxicity (see 9.10.1c)

Clinical signs of pindone intoxication in affected animals reflect internal haemorrhage and resultant organ dysfunction and may include extensive bruising, pale mucus membranes, poor capillary refill, dyspnoea and exercise intolerance. Coagulation studies reveal prolonged clotting times. Affected animals are typically anaemic, although in some cases animals that are noticeably ill have a normal haematocrit and minimal haemorrhages. Elevations in ALP and CK are common in affected animals. A presumptive diagnosis can be made on the basis of clinical signs and a history of potential exposure. Treatment involves correction of the hypovolaemia with crystalloids and/or homologous plasma or blood transfusions. Vitamin K_1 (2.5–5 mg/kg) is initially administered parentally and then orally until coagulation profiles are normal (usually 7–10 d). PY of affected females can be pulled for hand-rearing and treatment. Strategies for baiting vertebrate pests with pindone should incorporate measures that prevent exposure of non-target species such as macropods.

9.4 Genitourinary diseases

9.4.1 Urethral obstruction

Urethral obstruction in male macropods may be associated with dysuria, stranguria, haematuria, abdominal distension and discomfort, dehydration, lethargy and depression (Clark et al. 1982; Bryant & Rose 2003).

Diagnosis is by demonstration of bladder distension and/or the presence of a urolith using oblique radiographic views and/or ultrasonography. A urolith may be evident on palpation of the distal end of the penis or via transrectal urethral palpation. An anatomical barrier to urethral catheterisation is present in male macropods due to the presence of paired valve-like cusps 2–3 cm proximal to the external urethral orifice (see 2.10) (Hatkin & Janssen 1979; Bryant & Rose 2003). Attempts to treat obstructive urolithiasis in macropods must therefore rely on surgical approaches. The aetiopathogenesis of this condition in macropods is poorly understood and predisposing factors are likely to vary depending on the chemical composition of the urolith. Bryant and Rose (2003) proposed avoiding oxalate-containing feed, limiting dietary vitamin D and calcium and ensuring adequate vitamin A intake as preventative measures against the development of calcium oxalate uroliths in captive macropods.

9.4.2 Renal calculi

Renal calculi are an occasional, usually incidental necropsy finding in captive and free-ranging macropods. In some instances they may be associated with clinical signs attributable to nephritis including pyuria, haematuria, dysuria and crystalluria. Urinalysis, excretory urography and ultrasonography could all be utilised to diagnose renal calculi.

9.4.3 Renal failure

Clinical signs apparent in domestic animals with renal failure, including polyuria, polydipsia, weight loss and dehydration, are also evident in macropods. Renal failure of undetermined aetiology is seen sporadically in hand-reared eastern grey kangaroo PY. Persistent and progressive elevations in urea and creatinine are evident. No effective treatment has been determined. Cases typically progress despite supportive therapy and animals should therefore be euthanased.

9.5 Neurological and musculoskeletal diseases

9.5.1 Capture myopathy

Capture myopathy is one of the most significant diseases affecting the musculoskeletal system of macropods. This is discussed in detail under non-infectious diseases (see 9.10.3a).

9.5.2 Nutritional myopathy

Speare et al. (1982) described muscle-wasting, ataxia and infertility in captive bridled nail-tailed wallabies (*Onychogalea fraenata*) and the death of a musky rat-kangaroo with nutritional myopathy. Ataxia progressing to complete hind limb paralysis with muscle wasting ultimately resulting in death has been described in captive quokkas (Kakulas 1961). Similar lesions have been observed in captive tammar wallabies and red-necked pademelons maintained on lucerne hay (Hume 1999). Histopathological changes in affected animals included hyaline degeneration and calcification of skeletal muscle (Kakulas 1961; Munday 1972), marked fibrosis (Speare et al. 1982), swelling of myocytes (Kakulas 1961) and focal myocardial lesions. The clinical signs provide a high index of suspicion.

Treatment with vitamin E (200–600 mg PO sid) for several days reportedly resolved clinical signs in quokkas within several weeks (Kakulas 1963). Speare et al. (1982) described the use of an unspecified amount of a commercially available preparation containing vitamin E and selenium administered IV in combination with α-tocopherol for the treatment of nutritional myopathy in bridled nail-tailed wallabies. Selenium has, more recently, been demonstrated to be ineffective in preventing nutritional muscular dystrophy in tammar wallabies and quokkas (Hume 1999). Kakulas (1963) demonstrated an association between small enclosure size and the development of nutritional myopathy in captive quokkas. Oral vitamin E supplementation prevented development of disease. Daily oral dosing with vitamin E was also effective in preventing the disease in captive bridled nail-tailed wallabies, musky rat-kangaroos and Lumholtz's tree-kangaroo (Speare et al. 1982). Tree-kangaroos with muscular dystrophy were treated successfully with α-tocopherol (25 mg PO sid) (Potkay 1977).

9.5.3 Toxoplasmosis (see 9.8.3 and 9.9.4b)

9.5.4 Cryptococcal meningitis (see 9.9.3c)

Clinical signs include ataxia, head tilt, circling, blindness, weight loss, inappetence and convulsions terminally (Vaughan et al. 2005; ARWH 2007). The latex cryptococcal antigen agglutination test (LCAT) can be performed on serum or cerebrospinal fluid from suspected cases. Histopathological examination of the brains of affected animals reveals granulomatous meningitis and the presence of spherical basophilic yeasts that are PAS positive.

9.5.5 Tetanus (see 9.9.2d)

The clinical signs of tetanus in macropods include lateral recumbency, marked opisthotonos, muscular rigidity with clonic spasms (Fig. 7.16), tetanic spasms precipitated by sensory stimulation, protrusion of the nictitating membranes and hyperptyalism associated with an inability to swallow. Affected animals typically progress rapidly to death within 12–48 hr (Speare 1988; Blyde 1999). The clinical signs of this disease are highly suggestive and treatment is almost always initiated on the basis of clinical suspicion. A definitive diagnosis is achieved by culture of *Clostridium tetani*, although this rarely occurs.

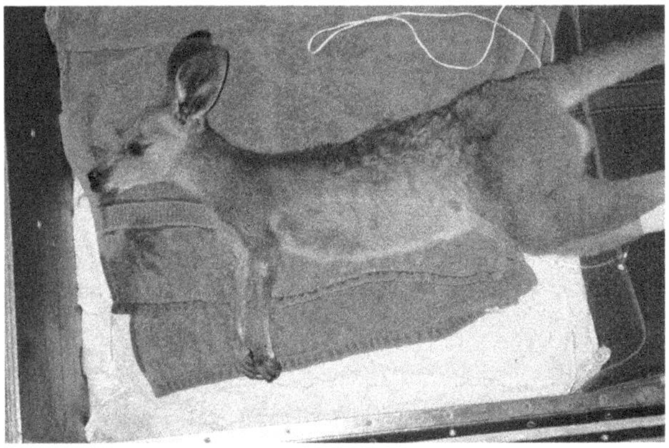

Figure 7.16 Tetanus in a juvenile eastern grey kangaroo. Photo: Tim Portas.

Treatment is unlikely to be successful. Intrathecal and IV administration of an equine tetanus antitoxin has been reported (Speare 1988). IM tetanus toxoid should also be used. Procaine penicillin (30 mg/kg) and benzathine penicillin (25 mg/kg) can be administered IM. Supportive IV fluid therapy and the use of benzodiazepines to facilitate muscle relaxation are indicated. Prevention is achieved through vaccination with either a multivalent clostridial vaccine administered SC or human or equine tetanus toxoid administered IM (Blyde 1999). An initial course of two vaccinations 1 mo apart is recommended, with subsequent annual or opportunistic vaccination. Sterile abscesses and swellings commonly occur at the injection site following administration of multivalent clostridial vaccines to macropods.

9.5.6 Neural-angiostrongyliasis (see 9.9.4b)

Migrating larvae cause multifocal neurological disease and consequently a range of clinical signs, including deteriorating motor function and proprioception. Unilateral to bilateral hind limb paresis, unilateral contraction of facial and cervical musculature, urinary stasis and excessive grooming have been reported in captive rufous bettongs (*Aepyprymnus rufescens*) infected with *A. cantonensis* (Higgins et al. 1999). Ataxia and postural abnormalities of the tail have been observed in a western grey kangaroo (ARWH 2007). Ante mortem diagnosis is difficult but demonstration of an eosinophilia in cerebrospinal fluid is suggestive. Serial brain sections can be examined histologically to demonstrate the presence of nematode parasites. Definitive diagnosis relies upon dissection of the brain to collect parasites for identification. Anthelmintic preparations are generally contraindicated as the death of the parasites incites a severe inflammatory reaction. However, fenbendazole has been used successfully to treat this condition in grey-headed flying-foxes (*Pteropus poliocephalus*) and its use could be considered in macropods (Reddacliff et al. 1999a). Treatment regimes are generally based on high-dose corticosteroid therapy. In captive macropod collections the definitive (rodent) and intermediate (gastropod) hosts need to be controlled.

9.5.7 Tick paralysis

Ascending flaccid paralysis progressing to recumbency and the eventual development of respiratory failure occurs in macropods affected by the paralysis tick (*Ixodes holocyclus*). The clinical signs and the presence of *I. holocyclus* are highly suggestive. Suspected cases of tick paralysis have also been associated with heavy burdens of other tick species (Munday 1978). Therapy follows the principles of treatment in domestic mammals and includes supportive care and the administration of hyper-immune canine serum. In high-risk areas the routine application of acaricidal preparations such as fipronil is warranted.

9.5.8 Secondary hypertrophic osteopathy

Clinical signs of secondary hypertrophic osteopathy in a wallaroo included firm, non-oedematous swelling of the fore limbs (Wayne & Nicholson 1999). Signs associated with underlying pulmonary pathology may be variably present. A presumptive diagnosis is based on radiographic demonstration of extensive smooth thickening of the radius, ulnas, metacarpals and phalanges. Given the typically advanced nature of the primary pulmonary problem when diagnosed, treatment is not usually possible. Surgical removal of a discrete neoplastic mass might be possible if detected ante mortem.

9.6 Endocrine and metabolic diseases

9.6.1 Hypoglycaemia

Hypoglycaemia is precipitated by periods of hypothermia, stress or infections in macropod PY and manifests as lethargy, muscular twitching, ataxia and bradycardia before progressing to hypoglycaemic seizures and coma (Woods 1999). Demonstration of blood glucose values less than 5 mmol/L is diagnostic. The clinical signs and history in PY are suggestive. Treatment with warmed 10% glucose (20 mL/kg IV) is initiated. Provision of a warm thermal environment and management of any predisposing problems is also indicated.

9.6.2 Hypothermia

Affected PY are cold to the touch, frequently bradycardic and lethargic. Hypothermia is extremely common in PY at initial presentation following retrieval from the pouch of dead females. It is also common where husbandry, in particular attention to maintenance of the thermal environment, is inadequate and can also occur secondary to other disease processes. The clinical signs are self-evident and rectal temperatures below 35°C are diagnostic. The PY should be placed in an environment with an ambient temperature of 37°C and allowed to warm up slowly. Warmed IV fluids will assist in raising the core body temperature. Underlying disease processes should be diagnosed and corrected. Pneumonia is a frequent sequela to hypothermia in PY and the prophylactic administration of enrofloxacin (5 mg/kg sid 5 d) is warranted. PY, particularly when furless, require constant monitoring of their environmental temperature.

9.6.3 Hyperthermia

Hyperthermia is seen in adult macropods following prolonged restraint attempts or restraint attempts during periods of high ambient temperature. The condition occurs in PY when the artificial rearing environment is excessively hot. Affected PY feel hot to touch, are tachypnoeic, lick their forearms and in severe cases may present unconscious. PY with rectal temperatures over 37°C are considered hyperthermic. Clinical signs and history provide a strong index of suspicion. Mild cases will resolve spontaneously by placing the animal in a cool environment until the rectal temperature declines to below 37°C. In more severe cases, cool (but not cold to avoid peripheral vasoconstriction) water is used to wet the skin of PY. In hyperthermic adult macropods, cooled IV fluids are administered in an attempt to reduce the animal's core body temperature in addition to the topical application of water. The artificial rearing environment for PY should be constantly monitored to ensure hyperthermia does not develop.

9.6.4 Lysosomal storage disease

Clinical signs included failure to thrive, diarrhoea, anorexia and convulsions in a juvenile eastern grey kangaroo and acute pyrexia and collapse in a juvenile red kangaroo (Rothwell et al. 1990). Histopathological examination revealed the presence of large round cells, containing abundant foamy pale eosinophilic cytoplasm throughout multiple tissues, and extensive Wallerian-type degeneration of the central nervous system. Electron microscopy was used to further characterise the lesions (Rothwell et al. 1990). Treatment has not been described.

9.7 Ophthalmologic diseases

9.7.1 Cataracts

Unilateral cataracts occur sporadically in adult macropods subsequent to a range of ocular disease processes. A syndrome of bilateral cataracts is seen sporadically in orphaned and wild macropod PY. In PY bilateral anterior and posterior subcapsular opacities that mature over several weeks to uniform opacity of the lenses are evident (Speare 1988). Sequelae observed in affected animals include lens luxation, uveitis-induced glaucoma and keratitis (Stanley 2002). Diagnosis is achieved by a thorough ophthalmic examination.

Cataract surgery in macropods is described in section 12.7.3. Glaucoma is an extremely common sequela to cataract surgery, necessitating globe enucleation or euthanasia of the PY (Stanley 2002). Speare (1988) reported the spontaneous resolution of cataracts in a juvenile agile wallaby.

Numerous theories have been proposed to account for cataracts in macropod PY. Low levels of galactokinase and galactose 1-phosphate uridyl transferase have been variably demonstrated in macropods (Stephens 1975). It was postulated that with low levels of these enzymes macropod PY reared on lactose-containing milk formulas would subsequently develop a galactosaemia due to an inability to convert galactose, a monosaccharide component of lactose. Galactosaemic cataracts were then believed to develop as a result. Given that the disease has been observed in wild macropod PY, has been seen in PY hand-reared on low-lactose milk and the fact that feeding of lactose-enriched milk formulas does not automatically result in cataract formation, this explanation seems unlikely. Speare (1988) proposed a protein or amino acid deficiency as a possible cause. Given that the aetiology of cataract formation in PY remains poorly understood, no specific preventative measures can be recommended.

9.7.2 Viral chorioretinitis (see 9.9.1c)

Viral chorioretinitis is thought to be transmitted by several species of *Culicoides* and has been reported from wallaroos and western grey, eastern grey and red kangaroos (Durham et al. 1996). The predominant clinical sign in affected animals is blindness, with all age classes of macropods except PY affected. Conjunctivitis is occasionally present. A high-stepping gait is apparent in some affected animals. A mild to moderate anterior uveitis and a poor to absent pupillary light reflex is evident on ophthalmological examination (Reddacliff et al. 1999b). The clinical signs are suggestive and disease tends to occur in outbreaks with large numbers of macropods affected (Durham et al. 1996). Virus isolation has been achieved by inoculation of a range of cell cultures including baby hamster kidney cell cultures, *Aedes albopictus* cell cultures and macropod PY kidney and lung cell cultures with homogenates of brain and retina from affected animals. A serogrouping ELISA and a microneutralisation test were used to compare Wallal virus isolates from numerous locations and confirmed a common Wallal serogroup for all isolates. Electron microscopy can be performed on retinal tissue. Indirect immunofluorescence for antigens of the Wallal serogroup can be performed on retinal tissue from affected animals. The use of a series of Orbivirus group-specific PCRs on serum from affected kangaroos demonstrated involvement of the Wallal and Warrego serogroups (Hooper et al. 1999). Histopathological examination of eyes, optic nerve and brain of affected animals reveals characteristic changes associated with this disease (Durham et al. 1996; Reddacliff et al. 1999b). Treatment has not been described. Outbreaks are associated with increased vector activity. The use of topical long-acting pyrethrins and environmentally active insecticides in enclosures and shelters may prevent or control outbreaks in captive collections.

9.7.3 Corneal ulceration

Clinical signs include blepharospasm, epiphora and corneal opacity. The lesions are typically unilateral. A detailed ocular examination should be performed under anaesthesia, if required, with fluorescein staining of the affected eye. Simple corneal ulcers are likely to resolve with topical antibiotic therapy. Approaches to the management of persistent corneal ulcers in macropods follow techniques utilised in domestic animals including corneal debridement, grid keratotomy and topical administration of 1% atropine. Corneal ulceration can occur in dehydrated PY and can be prevented by ensuring appropriate husbandry and maintaining hydration.

9.8 Generalised diseases

9.8.1 Macropodid herpesvirus (see 9.9.1f)

The clinical course of macropod herpes virus infection can be variable, with acute death in some individuals and a more protracted course of up to a week in others. Clinical signs include fever, conjunctivitis, cutaneous and mucosal erythema, incoordination, dyspnoea and oral and cloacal vesiculation and ulceration (Finnie & Littlejohns 1976; Dickson et al. 1980; Acland 1981; Callinan & Kefford 1981; Wilks et al. 1981; Britt et al. 1994). Finnie (1986) described transient infertility and genitourinary lesions in captive tammar wallabies. Multifocal necrosis of the liver and the presence of intranuclear acidophilic or basophilic inclusion bodies are suggestive. Definitive diagnosis can be achieved through the demonstration of a rising antibody titre via a serum neutralisation test. Viral culture and PCR to demonstrate herpesvirus DNA have also been used (Guliani et al. 1999). The anti-herpes viral compounds adenine 9-β-D-arabino-furanoside and (E)-5-(2'-bromovinyl)-2'-deoxyuridine have been shown to have inhibitory effects on macropod herpes virus 2 (MaHV2) *in vitro* but there are no reports of *in vivo* use in macropods (Smith 1996). Newly arrived animals should be quarantined and their herpesvirus status assessed and considered in light of the status of macropods in the collection they are to enter. Shedding of the virus following the administration of corticosteroids has been demonstrated in latently infected seropositive animals (Guliani et al. 1999) and it could be expected that disease could follow periods of stress or immunosuppression in captive or translocated animals.

9.8.2 Tammar sudden death syndrome (see 9.9.1e)

Sudden death without premonitory signs was the most common presentation during an epizootic of an *Orbivirus*-associated disease in captive tammar wallabies in eastern Australia (Rose et al. 2000). In wallabies exhibiting clinical signs lethargy, depression, ataxia, weakness, lateral recumbency with muscular fasciculations and increased respiratory rate were variably present. Viral culture of cerebrospinal fluid, cerebral tissue and parenchymatous organs with viral serogrouping and serotyping determined by virus neutralisation techniques are performed to achieve a definitive diagnosis. However, the clinical presentation coupled with gross and microscopic lesions is highly suggestive. Treatment has not been attempted, primarily due to the acute nature of the disease. At present, preventative strategies centre on control and/or limiting the vectors through the use of topical long-acting pyrethrins, environmentally active insecticides in shelters, improved drainage and rodent control in captive facilities.

9.8.3 Toxoplasmosis (see 9.9.4b)

Toxoplasmosis is a relatively common disease of macropods. Animals become infected following the ingestion of oocysts of *T. gondii* in feedstuff contaminated with felid faeces. A broad range of clinical signs has been reported from macropods with toxoplasmosis consistent with the multiple body systems affected by this disease. These signs are non-specific and while sufficient for a presumptive diagnosis cannot be relied upon for definitive diagnosis. Signs include dyspnoea, tachypnoea, coughing, depression, anorexia, lymphadenopathy and diarrhoea. Neurological signs include blindness, ataxia, circling, incoordination and dysphagia (Obendorf & Munday 1983; Patton et al. 1986; Miller et al. 2003). Ocular lesions include keratitis, uveitis, chorioretinitis, endophthalmitis and the development of unilateral or bilateral cataracts in some individuals (Stanley 2002). Death may occur without clinical signs or following a short period of non-specific illness.

A number of serological tests have been assessed for the ante mortem diagnosis of toxoplasmosis in macropods. The use of an ELISA measuring total anti-parasite immunoglobulin to *Toxoplasma gondii* has been described in Tasmanian pademelons, red-necked wallabies (Johnson et al. 1988) and tammar wallabies (Reddacliff et al. 1993a). Subsequent work by Johnson et al. (1989) suggested that the anti-kangaroo immunoglobulin used in the original ELISA was directed against IgG with little or no activity against IgM and therefore did not measure total anti-parasite immunoglobulin. The modified agglutination test (MAT) measures only IgG and false negatives can occur in cases of acute toxoplasmosis (Dubey & Odening 2001). The direct agglutination test (DAT) measures specific parasite antibody of all classes and the detection of antibodies to *T. gondii* using this test (in conjunction with a low or negative MAT) is considered to be indicative of acute toxoplasmosis (Johnson et al. 1989). Miller et al. (2003) monitored *T. gondii*-specific IgG in juvenile eastern grey kangaroos via MAT and found declining titres with increasing age. The authors concluded that this represented the possibility of transmission of maternally derived antibody via milk in juvenile kangaroos and that MAT results should be interpreted with caution in juvenile animals with suspected toxoplasmosis. Other serological tests that have been employed in macropods include the Sabin-Feldman dye test, the Latex agglutination test and the indirect haemagglutination test (Boorman et al. 1977; Dubey et al. 1988).

Impressions smears of affected tissue air-dried, fixed with methyl alcohol and stained with Giemsa may be used to identify *T. gondii* organisms although results may be variable. Histopathology alone or in combination with periodic acid-Schiff staining and/or immunohistochemical staining of parasites can also be utilised to diagnose toxoplasmosis. Histopathology is insensitive for determining if infection has occurred acutely or subsequent to the reactivation of latent infection (Reddacliff et al. 1993a). A definitive diagnosis can also be achieved by isolation of *T. gondii* following the inoculation of laboratory animals with tissues or fluids collected from animals with suspected toxoplasmosis.

Treatment is frequently unrewarding, but atovaquone (100 mg/kg PO sid) in combination with canola oil (to enhance absorption) for 30 d or clindamycin (11 mg/kg IM bid 30 d) have been recommended (Johnson-Delaney 2000). A synergistic combination of sulfadiazine and pyrimethamine has also been used to treat acute toxoplasmosis but will not eliminate infection (Dubey & Odening 2001).

Captive animals become infected following the ingestion of oocysts in feedstuff contaminated with felid faeces containing *T. gondii* oocysts. Domestic cats should therefore be prevented from accessing feed storage sheds and exhibits. Barrier-keeping may also be necessary in captive institutions housing both exotic felids and macropods to prevent the transfer of felid faecal material into macropod enclosures. Recrudescence of latent infection also occurs so environmental stressors should be minimised for captive animals. Attempts to vaccinate macropods with *Hammondia hammondi*, a related protozoal organism, were considered to provide partial protection against *T. gondii* infection. However, three of five vaccinated and subsequently challenged tammar wallabies died from toxoplasmosis (Reddacliff et al. 1993b). In another study assessing the efficacy of a commercially available attenuated strain of *T. gondii*, four of six vaccinated tammar wallabies developed fatal vaccine-associated disease (Lynch et al. 1993).

9.8.4 Mycobacteriosis (see 9.9.2a and Table 7.5)

Macropods are most commonly infected with nontuberculous mycobacteria but occasional cases of infection with *Mycobacterium bovis* have been reported in the literature (Teuscher 1957; Polema 1964). Frequently, the clinical signs of nontuberculous mycobacteriosis (Durham et al. 1996) are not evident until late in the course of the disease and may be non-specific. Clinical signs reported in tree-kangaroos include weight loss, dyspnoea, lameness, abscesses, neurological signs and blindness (Ott Joslin 1990). Mycobacterial osteomyelitis has occurred in smaller species of captive wallabies, producing lameness and neurological signs depending upon the location of the lesion (Mann et al. 1982; Richardson & Read 1986; Schoon et al. 1993). Ulcerative to granulomatous skin lesions have also been reported in several species.

Intradermal tuberculin testing produces inconsistent results and is therefore of limited diagnostic value (Buddle & Young 2003). Radiography is useful for localising the site of infection in cases where osteomyelitis or calcified lymph nodes are present. Radiographic changes include osteolysis and periosteal proliferation. Acid-fast bacteria can be demonstrated by performing Ziehl-Neelsen acid-fast stains on impression smears from affected tissue, smears made from tracheal washes, smears made utilising a tissue-concentration technique or on formalin-fixed tissue. Culture of the organism from tracheal wash samples may represent either infection or commensal colonisation. Mycobacterial culture can be performed on samples collected from lesions at necropsy and further characterised by PCR and DNA sequencing (Schoon et al. 1993; Montali et al. 1998). The histopathological lesions in infected animals are highly suggestive.

Surgical resection of localised subcutaneous abscesses and ulcerative or granulomatous skin lesions is curative (Ott Joslin 1990). Successful treatment of osteomyelitis associated with *M. avium-intracellulare* group 3 was achieved in a Matschie's tree-kangaroo using combination chemotherapy consisting of amikacin, isoniazid and ethambutol following culture and susceptibility testing of material obtained by bone biopsy (Burns et al. 1994). An epizootic of *M. bovis* in captive red-necked wallabies was apparently arrested by the administration of nicotinaldehyde thiosemicarbazon (1 mg/kg sid) in combination with isonicotinic acid hydrazide (4 mg/kg sid) for 6 wk (Teuscher 1957). Attempts to treat long-footed potoroos with a combination of rifabutin, ethambutol and clarithromycin were unsuccessful (Phelan 1996). Failure of therapy was attributed to poor patient compliance and advanced disease at the time of diagnosis. In general, infection in macropods carries a poor prognosis, recovery is rare and diagnosis is frequently at necropsy.

This is a disease of captive macropods and, while no specific preventative measures are available, attention to meticulous husbandry and limiting environmental stressors may reduce the incidence of disease. Montali et al. (1998) speculated that frequent movement of tree-kangaroos between North American institutions for breeding purposes might have been a contributory factor to the high incidence of mycobacterial disease in members of this genus through stress-induced immunocompromise. *M. avium* is ubiquitous in the environment and multiple serotypes have caused disease, complicating attempts to develop preventative strategies.

9.9 Infectious diseases

9.9.1 Viral diseases

a Macropod poxvirus (see 9.1.1)

The causative viral agent has not been characterised but dermatological, poxvirus-associated papillomatous proliferations have been demonstrated in tammar, agile and swamp wallabies, quokkas, wallaroos and red, eastern and western grey kangaroos (Papadimitriou & Ashman 1972; McKenzie et al. 1979; Arundel et al. 1979; Rothwell et al. 1984; Speare 1988). There is epidermal thickening due to acanthosis and hyperkeratosis. The surrounding epidermis is normal. Large eosinophilic inclusion bodies containing typical poxviruses are present within the cytoplasm of epidermal cells. Patchy necrosis with neutrophilic infiltration may occur (Papadimitriou & Ashman 1972; McKenzie et al. 1979; Rothwell et al. 1984). Lesions are typically seen in juvenile animals and are self-limiting (Blyde 1999). The life history has not been elucidated, but the virus is thought to occur wherever suitable hosts are present. Direct transfer via close contact or via arthropod vectors have been proposed as possible routes of transmission. There is some evidence to suggest that the virus is species-specific (Speare 1988).

b Retroviral infection in tammar and red-necked wallabies

An uncharacterised retrovirus was isolated from a captive population of red-necked and tammar wallabies in North America. It was postulated that an immunodeficiency syndrome resulting from infection with the virus led to mortalities associated with opportunistic infections, which included toxoplasmosis, histoplasmosis, aspergillosis and bacterial infections (Kapustin et al. 1999). Specific pathology associated with the postulated immunodeficiency syndrome was not described and mortalities were associated with toxoplasmosis, histoplasmosis, aspergillosis and bacterial infections. The epidemiology and significance of this virus is unknown.

c Viral chorioretinitis (see 9.7.2)

During several epidemics of blindness in large macropods in southern Queensland, New South Wales, Victoria and South Australia orbiviruses of the Wallal and Warrego serogroups were isolated from the retina and brains of affected animals (Hooper et al. 1996). Viruses from the Wallal serogroup have been most frequently isolated from affected animals, however, serological evidence suggests that infection with viruses from both serogroups occurs in affected animals. Retrospective PCR analysis of archived tissue from a blind kangaroo indicated that the disease had been present in Australia since at least 1975. The principal species affected by this disease include western and eastern grey kangaroos, red kangaroos and wallaroos.

Nucleic acid from Wallal and Warrego viruses have been identified using nested PCR in several species of *Culicoides*. The status of these species as vectors of the disease remains to be confirmed, however, epidemiological evidence is suggestive. Other insect species including mosquitoes were negative on PCR testing (Hooper et al. 1999). Outbreaks have been reported from south-eastern Australia and Western Australia but the disease could be expected to occur wherever suitable vectors and hosts are present. Epidemics of blindness occurring in the mid 1990s began in summer and subsided in winter; disease distribution closely followed water courses where macropods had congregated during drought conditions.

The principal histopathological findings in affected animals are retinitis and a severe bilateral non-suppurative panuveitis, characterised by mononuclear cell infiltration. Optic neuritis and secondary demyelination of the optic nerves also occurs. Retinal changes include extensive degeneration and necrosis with occasional retinal detachment secondary to subretinal fluid accumulation. In cases with long-standing retinal involvement, the retina may be reduced to scant atrophic remnants with small numbers of inflammatory cells. A non-suppurative meningoencephalitis may be present in some animals but is not considered to contribute to the blindness observed in affected animals. Small numbers of clinically normal macropods may also have similar histopathological lesions (Durham et al. 1996; Reddacliff et al. 1999b).

d Wallal serogroup viruses in black wallaroos (*Macropus bernardus*)

Three separate viruses of the Wallal serogroup with 83% homology with the Wallal virus causing choroid blindness have been associated with morbidity and mortality in captive black wallaroos and a black wallaroo/wallaroo hybrid housed at a wildlife park in the Northern Territory (Melville 2000). The life history of this particular group of Wallal serogroup viruses is unknown. Transmission is likely to be via arthropod vectors. Disease develops acutely and may progress rapidly to death or result in chronic illness. Clinical signs include depression, lethargy, facial oedema, conjunctivitis and focal chronic skin lesions (Melville 2000). Virus can be isolated from blood and tissue samples inoculated into *Aedes albopictus* cell culture; other cell cultures utilised for isolating Wallal viruses associated with choroid blindness should also be effective. DNA sequencing differentiates these Wallal viruses from the virus variants associated with choroid blindness. Electron microscopy can be used to characterise the viruses as Wallal viruses. Treatment consists of the provision of supportive care to chronically affected animals.

e Tammar sudden death syndrome (see 9.8.2)

During the investigation of an epizootic of sudden deaths in tammar wallabies held in zoological gardens and research facilities, an *Orbivirus* of the Eubanagee serogroup was isolated from the cerebral cortex, cerebrospinal fluid and various other organs of affected animals (Rose et al. 2000). This acute syndrome was termed 'tammar sudden death syndrome'. Koch's postulates demonstrating a causal association between the virus and this syndrome are yet to be fulfilled. Deaths associated with this syndrome occurred in New South Wales and southern Queensland in research and zoological facilities holding multiple macropod species. Outbreaks have occurred in captive facilities following periods of heavy rain and increased rodent numbers. The outbreak in the institution in which the syndrome was first noted appeared to occur over a 6 wk period with peak mortalities occurring between weeks 3 and 4. Gross necropsy lesions include pulmonary and hepatic congestion and hind limb and inguinal oedema. Consistent histopathological findings included diffuse pulmonary and hepatic congestion, and necrosis of the germinal lymphoid centres. Additional findings in a smaller proportion of animals included gastric ulceration and acute periacinar hepatocellular necrosis.

f Macropodid herpesviruses (see 9.8.1)

Macropod herpesviruses are alpha herpesviruses and are designated into two strains, MaHV-1 and MaHV-2 containing a total of nine isolated types. At least one other, as yet uncharacterised, strain has been identified (Guliani et al. 1999). Disease has only been reported in captive macropods including red, eastern and western grey kangaroos, tammar wallabies, Dorcopsis wallabies (*Dorcopsis muelleri*), parma wallabies, quokkas, brush-tailed bettongs, long-nosed potoroos and rufous bettongs. Macropod herpesviruses are assumed to occur Australia-wide but clinical disease has only been reported from captive animals. Serological studies have demonstrated higher seroprevalence in captive (41%) compared with wild populations (23%) (Webber & Whalley 1978). Cases have occurred in captive collections outside Australia (Britt et al. 1994).

Serological surveys have demonstrated the widespread prevalence of neutralising antibodies to macropod herpesviruses (Webber and Whalley 1978) and latent infection has been demonstrated in the eastern grey kangaroo (Guliani et al. 1999). It is assumed that all macropod species are potentially susceptible. The mode of transmission is unknown and dynamics of the virus are undescribed. Isolation of a macropod herpesvirus from nasal swabs and the detection of herpesviral DNA polymerase sequences from the trigeminal ganglia of eastern grey kangaroos suggest transmission via the respiratory route (Guliani et al. 1999). The reactivation of latent infection in eastern grey kangaroos following the administration of corticosteroids supports the theory that a high seroprevalence in captive populations occurs secondary to chronic stressors associated with long-term captivity (Guliani et al. 1999).

Gross findings are variable but may include pulmonary congestion and oedema, conjunctivitis, mesenteric lymph node enlargement, hepatic and splenic congestion, multiple pale hepatic foci, diphtheritic plaques on the mucosa of the oesophagus and stomach and focal ulceration and necrosis of the genitalia. Histopathological findings include low-grade interstitial pneumonia, generalised multifocal necrosis, severe hepatitis characterised by multifocal necrosis and the presence of intranuclear acidophilic or basophilic inclusion bodies within necrotic and adjacent hepatocytes.

g Encephalomyocarditis virus (see 9.3.9)

Encephalomyocarditis virus is a cardiovirus of the family Picornaviridae that has been reported to cause the deaths of captive Lumholtz's tree-kangaroos in Queensland and captive Goodfellow's tree-kangaroos at a zoo in New South Wales (Reddacliff et al. 1997; Blyde 2005). Other macropod species could be expected to be susceptible. Gross lesions in Goodfellow's tree-kangaroos included hepatic and pulmonary congestion, petechial pulmonary haemorrhage and pallor of the cardiac muscle. Histopathology revealed multifocal cardiac necrosis, pulmonary and splenic congestion and depleted follicular lymphocytes in lymph nodes (Reddacliff et al. 1997). Rodents, in which occult infections are thought to occur, are frequently implicated in outbreaks (Thomson et al. 2001). This may, however, reflect the wide array of species infected during outbreaks rather than the status of rodent species as reservoirs for the virus. The virus can be transmitted by oral, intranasal, intratracheal and aerosol routes and via contamination of wounds.

h Foot and mouth disease

Snowdon (1968) demonstrated that experimental infection of red kangaroos, Matschie's tree-kangaroo (*D. matschiei*), red-necked wallabies and long-nosed potoroos with foot and mouth disease virus was possible. Red kangaroos in this study were infectious to susceptible cattle. Vesicles on the tongue and footpads were evident in red kangaroos and Matschie's tree-kangaroo following experimental infection. A naturally occurring case of foot and mouth described from an eastern grey kangaroo housed at a zoo in India during a series of outbreaks of type O foot and mouth disease virus in the surrounding area. The animal died within 2 d of developing lameness associated with sloughing of the footpads. Focal myocardial necrosis with infiltration of inflammatory cells was evident histopathologically in the animal (Bhattacharya et al. 2003).

9.9.2 Bacterial diseases

a Mycobacteriosis (see 9.8.4)

Mycobacteriosis occurs Australia-wide in captive macropods and has been reported in macropods overseas. Infections are relatively common in marsupials and have been reported in a number of macropod species. Environmental mycobacteria are considered the source of infections in macropods (Ott Joslin 1990) and disease is most frequently associated with organisms from the *Mycobacterium avium* complex. With the exception of a number of cases of mycobacterial dermatitis in a population of free-ranging long-nosed potoroos (Fowler & Mitchell 2003), disease has only been reported from captive animals. Table 7.5 details mycobacterial species and associated pathology recorded in macropods.

As mycobacteriosis has almost exclusively been reported from captive animals, stressors of captivity or excessive environmental contamination may predispose captive animals to infection (Montali et al. 1998). Species-specific differences in susceptibility to mycobacteriosis are suspected, with Matschie's tree-kangaroo and red-necked wallabies overrepresented in the literature. The apparent high susceptibility of macropods and marsupials in general to mycobacterial infections is thought to relate to deficiencies in cellular immunity. It has been postulated that

Table 7.5 Mycobacteriosis reported in captive macropods

Macropod species	*Mycobacterial* species	Lesion	Ref.
Brush-tailed bettong	*M. avium*	Hepatic/pulmonary granulomas	ARWH 03/02/06
	M. intracellulare	Pulmonary granulomas	ARWH 03/02/06
		Lymphadenopathy	ARWH 03/02/06
		Thoracic abscessation/ osteomyelitis	Thomas (1987)
		Osteomyelitis	Richardson & Read (1986)
Rufous bettong	*M. avium*	Subcutaneous abscess	McEwan (1989)
		Pulmonary/hepatic abscesses	Lucas (1925)
	M. bovis	Pulmonary granulomas	Griffith (1928)
Long-footed potoroo	*M. intracellulare*	Pneumonia	Phelan (1996)
	M. ulcerans	Ulcerative dermatitis	Fowler & Mitchell (2002)
Long-nosed potoroo	*M. avium*	Disseminated disease	Mann et al. (1982)
Quokka	*M. avium*	Mesenteric granuloma	Munday (1978)
Grizzled tree-kangaroo	*M. avium*	Osteomyelitis	Mann et al. (1982)
Matschie's tree-kangaroo	*M. intracellulare*	Osteomyelitis	Burns et al. (1994)
	M .avium	Osteomyelitis	Mann et al. (1982)
		Spinal osteomyelitis	Kennedy et al. (1978)
		Pneumonia/osteomyelitis	Bush et al. (1995)
Goodfellow's tree-kangaroo	*M. avium*	Granulomatous pneumonia	ARWH 03/02/06
		Granulomatous pneumonia	Ott Joslin (1990)
	M. intracellulare	Granulomatous pneumonia	ARWH 03/02/06
Dendrolagus (species not specified)	*M. bovis*	Pulmonary tuberculosis	Poelma (1964)
Red-necked wallaby	*M. avium*	Hepatic/splenic abscesses	Lesslie (1958)
		Pulmonary abscess	ARWH 03/02/06
		Lymphadenopathy	Obendorf (1987)
		Spinal abscess	Tilden and Williamson (1957)
		Spinal osteomyelitis	Hime and Jones (1972)
	M. tuberculosis		Stünzi (1950)
	M. intracellulare	Disseminated disease/arthritis	Samuel *pers. comm.*
	M. bovis		Teuscher (1957)
Western grey kangaroo	*M. avium*	Disseminated disease	Lesslie (1958)
	M. avium paratuberculosis	Enteritis/lymphadenitis	Cleland et al. (2002)
	M. smegmatis	Enteric/hepatic granulomas	ARWH 03/02/06
Red kangaroo	*M. asiaticum*	Pouch mass	Pritchard (1995)
Tammar wallaby	*M. avium* paratuberculosis	Enteric/hepatic granulomas	Cleland et al. (2004)
	M. intracellulare	Osteomyelitis	Peet and Dickson (1982)

(Continued)

Table 7.5 *(Continued)*

Macropod species	*Mycobacterial* species	Lesion	Ref.
Western brush wallaby	*M. intracellulare*	Osteomyelitis	Peet and Dickson (1982)
Parma wallaby	*M. avium*	Disseminated disease	Mann et al. (1982)
		Spinal osteomyelitis	Schoon et al. (1993)
Tasmanian bettong	*M. avium*	Pulmonary granulomas	ARWH 03/02/06
	M. avium	Disseminated disease	ARWH 03/02/06
Rufous hare wallaby	*M. avium*	Cutaneous ulcer	McEwan (1989)

macropods may fail to mount an adequate granulomatous response when infected with nontuberculous mycobacteria, allowing haematogenous dissemination of the organism (Buddle & Young 2000). However, work by Young et al. (2003) found immunopathological responses consistent with those seen in placental mammals in a rufous hare-wallaby (*Lagorchestes hirsutus*) with mycobacteriosis.

Inhalation with subsequent dissemination was proposed as the primary route of infection in tree-kangaroos with mycobacteriosis on the basis of extensive pulmonary pathology (Montali et al. 1998). Other postulated routes of infection include puncture wounds and lacerations resulting in localised infection or subsequent haematogenous spread to multiple sites (Ott Joslin 1990; Richardson & Read 1986). Montali et al. (1998) cultured *M. avium* complex organisms from tracheal washes in clinically normal tree-kangaroos that remained free from disease for a further 3.5 yr, further confounding understanding the epidemiology of this disease.

M. avium subsp. *paratuberculosis* has been cultured from the gastrointestinal and lymphoid tissues of free-ranging tammar wallabies and western grey kangaroos occurring on sheep properties endemically infected with ovine Johne's disease on Kangaroo Island (Cleland 2002). Additionally, histological lesions consistent with Johne's disease were evident in eight individuals. No organisms were detected by faecal culture and the potential role of macropods as a reservoir host for this disease is unknown. A single eastern grey kangaroo returned a positive faecal culture from a farm on which ovine Johne's disease occurred in New South Wales (Abbott 2000).

Lesions are evident in a range of organs with the lungs, long bones, spinal column, liver, spleen, lymph nodes and skin most frequently involved. Pulmonary and osteological lesions in Matschie's tree-kangaroos were characterised by pyogranulomatous inflammation, necrosis and abundant acid-fast bacteria. Histiocytic hepatic granulomas were also present in some affected animals (Montali et al. 1998). Hepatic and splenic granulomas, necrosis and calcification were evident in *M. avium*-infected parma wallabies (Schroon et al. 1993). Additionally, infiltrative spinal meningeal lesions, granulomatous myelitis and multifocal tuberculous radiculitis and neuritis were uniformly present in infected animals. *M. intracellulare*-associated osteomyelitis and septic arthritis in a woylie was characterised histopathologically by the presence by extensive osteolysis and a pyogranulomatous reaction (Richardson & Read 1986).

b Oral necrobacillosis (Lumpy jaw) (See 9.2.2)

Oral necrobacillosis is one of the most common and significant diseases of captive macropods. It is characterised by chronic alveolar osteomyelitis, associated soft tissue inflammation and infection of the mandible and/or maxilla that is frequently accompanied by draining sinus tracts (Figs. 7.10, 7.11).

Multiple bacteria have been cultured from lesions in affected animals including *Nocardia* species (Tucker & Millar 1953); various *Bacteroides* species (Burton 1981; Oliphant et al. 1984), including *Bacteroides ruminicola* subspecies *brevis* (Keane et al. 1978; Taylor et al. 1978); *Actinomyces* species (Burton 1981; Samuel 1983); *Fusobacterium necrophorum* (Burton 1981; Samuel 1983; Oliphant et al. 1984); *Streptococcus* species and *Staphylococcus* species (Samuel 1983; Hartley & Sanderson 2003). It is now generally accepted that *F. necrophorum* is the principal infectious agent with several studies demonstrating it is the organism most frequently isolated from necrotic lesions (Samuel 1983; Oliphant et al. 1984). A synergistic association between *F. necrophorum* and *Bacteroides* spp., as has been described in footrot in domestic sheep, has also been postulated (Samuel 1983). Significantly, one study failed to demonstrate the presence of *F. necrophorum* on the mucosal surfaces of the oral cavity in clinically normal animals yet it was frequently isolated from the mouth of animals with oral necrobacillosis (Samuel 1983).

Oral necrobacillosis has been reported from captive macropod collections worldwide and individual cases and outbreaks have been observed in free-ranging animals during periods of environmental stress. The disease may affect any species but grazing macropods, including red kangaroos, red-necked wallabies and the eastern grey kangaroo, appear to be overrepresented, presumably due to their unique dental anatomy and the process of molar progression. A relatively high incidence of disease occurs in the bridled nail-tailed wallaby, an intermediate browser/grazer, in the captive setting.

Experimentally macropods do not exhibit increased susceptibility to infection with *F. necrophorum,* indicating a multifactorial aetiology (Smith et al. 1986). A combination of high environmental contamination, anaerobic conditions, impaired host resistance (due to chronic stress) and mucosal disruption is required for the pathogen to invade and produce clinical disease. Miller and Beighton (1979) demonstrated increased dental calculus on partially erupted and post functional molars in eastern grey kangaroos, a species exhibiting the phenomenon of molar progression. Osteolysis of the adjacent bone was evident and may have been associated with the prolonged presence of dental calculus or be a feature of the process of molar exfoliation. Retained post functional molars were frequently associated with impacted feed, suggesting that the process of molar progression may predispose macropods to the development of oral necrobacillosis. Reduced dental exercise and the ingestion of foodstuffs containing rapidly digestible carbohydrates in captive animals may also predispose to the development of the condition. Identification of individual contributory factors is therefore important in elucidating the pathogenesis of oral necrobacillosis and in developing suitable prevention and management strategies. Oral necrobacillosis can also develop following traumatic facial injury, especially where fractures of the procumbent incisors occur.

A review of necropsy records for 783 macropods at an Australian zoo revealed an overall incidence of oral necrobacillosis cases of 9.7% in captive and free-ranging macropods. The incidence in free-ranging macropods was 4.8%, compared with an incidence of 13.4% in captive macropods. Species-specific differences in the incidence of oral necrobacillosis in this review are detailed in Table 7.6. No sex predilection was evident, with approximately equal numbers of males

Table 7.6 Incidence of oral necrobacillosis in macropods presented for necropsy at an Australian zoo (1984–2005)

Species	Total number of necropsies	Oral necrobacillosis cases	Incidence (%)
Eastern grey kangaroo	256	13	5.1
Western grey kangaroo	39	1	2.6
Red kangaroo	54	16	29.6
Common wallaroo	63	11	17.5
Red-necked wallaby	97	9	9.3
Swamp wallaby	84	8	9.5
Parma wallaby	14	2	14.3
Tammar wallaby	4	0	0
Pretty-faced wallaby	59	4	6.8
Red-necked pademelon	9	0	0
Bridled nail-tailed wallaby	58	9	15.5
Yellow-footed rock-wallaby	23	3	13
Brush-tailed bettong	5	0	0
Rufous hare-wallaby	17	0	0
Quokka	1	0	0

and females affected. No seasonal trends were evident in the incidence of infection.

Outbreaks of disease have occurred, presumably the result of changing environmental and host factors leading to increased susceptibility to the disease. Eleven of 14 red-necked wallabies housed at an Irish zoo died over a 12 mo period (Keane 1977). The disease was also responsible for the deaths of 200 animals in a population of 700 red-neck wallabies over a 4 mo period during unseasonably cold weather at a British zoo (Oliphant et al. 1984). Butler (1981) reported a higher incidence in autumn and winter at Melbourne Zoo. While the incidence of the disease is highest in captive populations, sporadic cases and outbreaks have been reported in free-ranging animals including swamp wallabies, tammar wallabies (Arundel et al. 1977) and red kangaroos (Tomlinson & Gooding 1954).

Early lesions are associated with marked soft tissue inflammation and oedema, surrounding a central necrotic core. Chronic lesions are associated with extensive and expansive sub-periosteal bone proliferation. A common sequela is the presence of abscesses in visceral organs (see 9.2.2).

c *Bordetella bronchiseptica* (see 9.3.4)

Outbreaks of respiratory tract disease associated with *Bordetella bronchiseptica* have occurred in captive bridled nail-tailed and tammar wallabies (Ward & Blyde 2002; Herbert et al. 2005). A case of pneumonia associated with this organism has been identified in a single free-ranging red-necked wallaby (Western Plains Zoo archives). Reported and anecdotal cases have been limited to New South Wales but there is risk of an outbreak occurring wherever captive macropods are held at high densities.

Clinical signs reportedly persisted for up to 4 mo in tammar wallabies and up to 6 mo in a bridled nail-tailed wallaby despite ongoing antibiotic therapy. Feral European rabbits (*Oryctolagus cuniculus*) were implicated as a potential source in one outbreak. It was suggested that spill-over and establishment occurred in bridled nail-tailed wallabies held at high density during a period when rabbit numbers were unusually high (Ward et al. 2002). This organism has been cultured from the nasal cavities of clinically normal bridled nail-tailed wallabies, suggesting concurrent viral infection or other factors that may reduce immunocompetence may be involved in the development of disease and that asymptomatic carriers may act as a source of infection. Findings in tammar wallabies include suppurative inflammation, marked mucosal congestion, ulceration and necrosis of the soft cartilaginous and bony tissues of the sinus (ARWH 2007).

d *Clostridium tetani* (see 9.5.5)

Although clinical disease caused by *Clostridium tetani* is relatively uncommon, macropods appear to be highly susceptible and it is the most important clostridial disease of captive macropods. The disease appears to be more common in northern Australia and has been infrequently reported from wild macropods (Speare et al. 1989). Spores formed by the organism are a common contaminant of soil and enter the body following penetrating injuries or the contamination of wounds. *Clostridium tetani* has also been implicated as a cause of sudden death in captive kangaroos in good body condition following heavy rains (Blyde 1999). Post-mortem changes include marked opisthotonos, pulmonary congestion and rupture of the musculature of the hind limbs.

e *Disease associated with other* Clostridium *species*

Flaccid paralysis progressing to death in a single agile wallaby was thought to be associated with the neurotoxin of *Clostridium botulinum*. Clinical signs included a progressive flaccid paralysis and an absence of deep tendon reflexes (Speare 1988). Enterotoxaemia caused by *Clostridium perfringens* has also been suspected in several species of captive and semi-free-ranging macropods (English 1978; Western Plains Zoo archives).

f *Erysipelas* (see 9.1.4)

Erysipelothrix rhusiopathiae has been associated with septicaemic disease and acute death in red and western grey kangaroos. Affected animals typically die acutely or are found dead without exhibiting premonitory signs. There are limited reports of disease associated with *E. rhusiopathiae* but cases have also been seen in captive macropods held in the UK and Europe. Cases have occurred after periods of heavy rainfall and the disease typically occurs sporadically. Vaccination is generally not practised. Necropsy findings include epicardial petechial haemorrhages, subcutaneous ecchymotic haemorrhages, splenomegaly and visceral congestion. *E. rhusiopathiae* is typically cultured from all major organs.

E. insidiosa was isolated from wallabies and kangaroos dying in a number of European zoos (Poelma 1964).

g Melioidosis

The causative agent of melioidosis is *Burkholderia pseudomallei*, a common inhabitant of soil and water. This organism has been responsible for the deaths of captive and free-ranging Lumholtz's tree-kangaroos in northern Queensland (Thomas 1981; Speare et al. 1982; Martin 2005), a captive tree-kangaroo in Papua New Guinea (Egerton 1963) and several wallabies of undescribed species exported from Australia into Malaysia (Saroja 1979). *Burkholderia pseudomallei* is endemic in tropical and subtropical Australia although sporadic outbreaks have occurred in temperate regions of Australia (Low Choy et al. 2000). Infection occurs via inhalation or ingestion or though the contamination of wounds. Cases are frequently associated with periods of wet weather. Clinical signs are highly variable, ranging from an acute septicaemic form to chronic abscessation. Lungs, spleen, liver and associated lymph nodes are most frequently affected (Low Choy et al. 2000). Pulmonary emphysema, multifocal hepatic and splenic necrosis and renal, hepatic and splenic congestion were evident in infected wallabies (Saroja 1979).

h Yersiniosis

The most common causative agent of yersiniosis is *Yersinia pseudotuberculosis*, although other *Yersinia* species have been isolated from affected macropods. The disease has been identified in wallaroos, eastern and western grey kangaroos in the eastern states of Australia and Tasmania and in captive collections overseas (Munday 1978; ARWH 2007). Cases typically occur following periods of stress, including cold wet weather. Multifocal hepatic abscessation occurred in a population of captive wallaroos. Abscessation of other parenchymatous organs can also occur. Concurrent toxoplasmosis was noted in an eastern grey kangaroo dying from yersiniosis (ARWH 2007).

i Salmonellosis (see 9.2.7)

Macropods have been identified as asymptomatic carriers of a wide range of *Salmonella* serotypes throughout Australia (Speare 1988). Clinical disease, typically gastroenteritis and septicaemia, has only been described in captive animals, in particular hand-reared PY. Stressors and other factors associated with captivity predispose these animals to developing disease. Hart et al. (1985) demonstrated a marked seasonal incidence in the prevalence of *Salmonella* infections in free-ranging quokkas on Rottnest Island. High infection rates were identified in summer and low rates in winter. Multiple serotypes were frequently isolated from animals during periods of high infection. Infection was believed to be associated with seasonal changes in food supply and the consequent disruption of digestive tract physiology. Despite high levels of infection there was no evidence of *Salmonella*-associated disease in infected animals. Pretty-faced wallabies more frequently developed clinical disease compared with other macropod species held in the collection at an Australian zoo. Outbreaks in which multiple animals were affected have occurred. Necropsy findings in pretty-faced wallabies with salmonellosis include severe haemorrhagic enteritis with extensive mucosal and submucosal necrosis (ARWH 2007). Marked enteritis with fibrinohaemorrhagic exudation and enlargement and oedema of the mesenteric lymph nodes may be present in affected animals.

j Other bacterial diseases of macropods

Pasteurellosis is common in smaller species of macropods following cat attack. Treatment is frequently unrewarding (Blyde 1999). Young (1960) described an epizootic of *Pasteurella multocida* infection in a captive population of kangaroos and wallabies. Cases varied from peracute to chronic with almost 100% mortality. Extensive necrotic oral lesions were evident in many animals and histopathology revealed lymphadenomegaly, pulmonary oedema, fibrinous pleuropneumonia, hepatic congestion and ascites. Pathological findings in five red kangaroos dying during an outbreak of pasteurellosis at a Nigerian zoo included bronchopneumonia, gastrointestinal haemorrhage and hepatic and splenic congestion (Okoh 1980).

Hepatic abscessation associated with *Nocardia* species has been reported in two unidentified macropod species in New Zealand (Canfield & Hartley 1992). An extensive cavitating pulmonary abscess associated with a species of *Nocardia* was reported from a captive red kangaroo (Gray 1974). *Nocardia brasiliensis* was associated with osteomyelitis of the radius in a female red kangaroo at an Australian zoo. An undetermined species of *Nocardia* was associated with an epizootic of disease in captive wallabies and kangaroos in Queensland (Tucker & Millar 1953). Lesions included gingivitis and subcutaneous abscessation about the head.

Speare (1988) listed a range of bacterial pathogens cultured from septicaemic PY including *Klebsiella* pneumoniae, Pseudomonas, Corynebacterium, Salmonella* and *Proteus* species. Hepatic abscessation associated with *Listeria* species is an occasional finding in captive macropods (Canfield & Hartley 1992).

9.9.3 Fungal diseases

a Candidiasis (see 9.1.8 and 9.2.8)

Gastrointestinal tract infections with *Candida albicans* occur commonly in hand-reared juvenile macropods of virtually all species. *C. tropicalis* and *C. krusei* have also been associated with disease (Speare 1988). Systemic candidiasis occurs less commonly (Dixon & Baird 1984). The disease occurs Australia-wide in hand-reared macropod PY. Infections are usually associated with stress and poor hygiene and occur most frequently in PY in poor body condition. An association with the administration of antibiotics has been identified, although the association may be with the underlying bacterial infection and generalised poor cellular immunity rather than the effects of antibiotic administration (Woods 1999). The fungus affects the superficial layers of squamous epithelium and therefore affects the oral mucosa, oesophagus, squamous epithelium of the stomach and the cloaca. Gross lesions in eastern grey kangaroo PY with alimentary tract candidiasis included the presence of white plaques throughout the tract and ulceration of the gastric mucosa. Histopathologically, blastospores and mycelia were evident invading the stratified squamous epithelium of the oral mucosa, oesophagus and non-glandular stomach. A mixed leukocytic inflammation was associated with mycelial invasion of the deeper epithelial layers (Obendorf 1980). Multifocal hepatic necrosis associated with blastospores, pseudomycelia and mycelia was evident in a juvenile eastern grey kangaroo with systemic candidiasis (Dixon & Baird 1984).

b Dermatophytoses (see 9.1.7)

Trichophyton mentagrophytes, T. ajelloi, Microsporum gypseum, M. canis and *M. cookie* have been associated with dermatophytoses in a range of captive but not free-ranging macropods (Speare et al. 1988). Dermatophytosis occurs in hand-reared macropod PY Australia-wide and is a sporadic problem in captive macropod collections. In one study examining commercially harvested free-ranging animals, 6.25% of western grey kangaroos and 3.4% of red kangaroos were identified as subclinical carriers of *T. mentagrophytes* (Keep & Pericard 1983). The occurrence of clinical disease in captive but not free-ranging macropods suggests that environmental stressors and other factors encountered in captivity predispose macropods to infection and the development of clinical disease. Extensive lesions developed within several weeks in four wallaroos following translocation from a large open-range exhibit into a smaller holding facility. Lesions can be highly variable in appearance and extent. Histopathological changes include folliculitis and hyperkeratosis.

c Cryptococcosis (see 9.3.6)

Cryptococcus gatti (formerly *C. bacillisporus* or *C. neoformans* var gatti) and *C. neoformans* var *grubii* have both been associated with cryptococcosis, including cryptococcal meningitis and pulmonary and disseminated cryptococcosis in a variety of macropods. Cases of cryptococcosis have been observed in captive and free-ranging macropods. *C. gatti* has been associated with pulmonary and multisystemic cryptococcosis in red-necked wallabies, parma wallabies, quokkas (Canfield & Hartley 1992), long-nosed potoroos and a Gilbert's potoroo (*Potorous gilbertii*) (Vaughan et al. 2005). Cryptococcal meningitis has been observed in a swamp wallaby, quokkas and a spectacled hare-wallaby (*Lagorchestes conspicillatus*) (ARWH 2007). Cases of cryptococcosis could be expected wherever the causative organism is found and the necessary environmental and host factors are present to allow the establishment of infection. Cryptococcal meningoencephalitis associated with *C. neoformans* var *grubii* has been observed in long-nosed and Gilbert's potoroos. Environmental sources of *C. gatti* include a range of *Eucalyptus* species. Leaf litter and/or cage furniture derived from *Eucalyptus* species including river red gums (*E. camaldulensis*) in exhibits are considered potential sources of *C. gatti* in captive macropods. Inhalation seems the most likely route of infection with subsequent haematological spread in multisystemic cases. A consistent finding at gross necropsy in macropods with pulmonary cryptococcosis is extensive pulmonary consolidation. Small, multifocal, grey to yellow nodular lesions may also be present throughout the pulmonary parenchyma. Histopathologically, there is diffuse alveolar distension with large numbers of spherical basophilic yeasts, focal pulmonary haemorrhage, congestion and atelectasis and fibrinous exudation. Central nervous system lesions include granulomatous meningitis and encephalitis and, in some cases, optic neuritis (ARWH 2007).

9.9.4 Parasitic diseases

a Ectoparasites

Mange (see 9.1.13)

Sarcoptic mange associated with a mite consistent with *Sarcopties scabei* var *canis* has been described from free-ranging and hand-reared agile wallabies in the Northern Territory (McLelland & Youl 2005). The epidemiology in this species is poorly understood. Histopathological changes included hyperkeratosis, polymorphonuclear and mononuclear cell infiltration of the keratin layer and the presence of mites and eggs within loculi in the keratin layer. Dermatitis associated with the larvae of *Odontacarus adelaideae* has been reported from a yellow-footed rock-wallaby (*P. xanthopus*). Dermatitis associated with the mite *Thadeua serrata* has been reported in the Proserpine rock-wallaby, black-striped wallaby (Skerratt 2007) and brush-tailed rock-wallaby.

b Endoparasites

Nematodes

Labiostrongylus spp.

At least eight species of *Labiostrongylus* have been described from macropod hosts throughout Australia. Following ingestion by the host, larvae exsheath and third-stage larvae penetrate the gastric mucosa at the junction of the squamous and glandular epithelia anterior to the opening of the oesophagus in a range of host macropods (Beveridge 1978). Adults are found free-living in the stomach. Development of the parasite is seasonal in red kangaroos and tammar wallabies; eggs are laid when environmental conditions are optimal for the survival of the free-living larvae and susceptible hosts are available for reinfection (Beveridge 1978). Third-stage strongyle larvae, presumed to be *Labiostrongylus labiostrongylus*, were observed by Speare et al. (1983) in agile wallabies associated with gastric gland hypertrophy, superficial mucosal ulceration and inflammatory cell infiltration of the lamina propria of the mucosal epithelium at the junction of the squamous and glandular stomach. Nodules in the gastric mucosa of subadult eastern grey kangaroos contained *Labiostrongylus* larvae. Histopathological examination revealed epithelial ulceration and a fibrinous inflammatory reaction of the underlying mucosa (Arundel et al. 1990). Munday (1971) observed a squamous cell carcinoma associated with nodules caused third-stage *Labiostrongylus* larvae in a Tasmanian pademelon.

Globocephaloides spp.

Globocephaloides trifidospicularis has been responsible for the deaths of free-ranging juvenile eastern grey kangaroos at an enclosed reserve in Victoria (Arundel et al. 1990). Fatalities associated with *G. trifidospicularis* infection have been reported in captive western grey kangaroos housed with infected eastern grey kangaroos. Death associated with *G. macropodis* infection has been reported in a captive eastern grey kangaroo (Speare 1988) and is problematic where large groups of young macropods are maintained in captivity. *G. trifidospicularis* has been reported from eastern and western grey kangaroos, swamp wallabies, red-necked wallabies, pretty-face wallabies and tammar wallabies in South Australia, Queensland, New South Wales and Victoria (Beveridge 1978; Beveridge 1985). *Globocephaloides* species are blood feeders and produce anaemia and hypoproteinaemia, leading to death in heavy infestations. Seasonal mortality (July and August) has been observed in juvenile eastern grey kangaroos in Victoria (Arundel et al. 1990) and in Tasmania (Obendorf 1985). Juveniles emerge from the pouch in summer but do not acquire infection until autumn. Significant contamination of pasture appears to be due to infected juveniles, with adult kangaroos having low parasite burdens and minimal egg production. Moist conditions favour the survival of larvae in the environment. Mortality in juvenile kangaroos appears to be reduced following hot dry summer weather conditions that are unfavourable to the survival of larvae. Morbidity and mortality associated with this parasite appear to occur only under conditions of confinement and high population density (Arundel et al. 1990). Increasing *G. trifidospicularis* burdens in juvenile eastern grey kangaroos was associated with falls in plasma protein, haemoglobin concentration and haematocrit. Juveniles with heavy burdens of *G. trifidospicularis* have blood-stained duodenal contents and haemosiderin-laden macrophages within the connective tissue of the duodenal villi.

Hypodontus macropi

Heavy infections with *H. macropi* have been associated with enteritis and in some instances death in macropods from eastern Australian states (Arundel et al. 1977). It is a tissue feeder and attaches to the intestinal wall, causing lysis of the mucosa and haemorrhage, which is also associated with prior feeding sites. In heavy infestations, marked enteritis is present. This may include multifocal mucosal haemorrhage, focal ileal necrosis and haemorrhage and mononuclear cell infiltration of the lamina propria (Arundel et al. 1979; Beveridge 1978). Affected macropods include red and western grey kangaroos, wallaroos, swamp wallabies, red-necked pademelons and Tasmanian pademelons (Beveridge 1978). No seasonal trends in parasite numbers have been described, however, Arundel et al. (1979) demonstrated increasing numbers of *H. macropi* in red kangaroos up to 18–24 mo of age. Numbers then declined before increasing again once animals were over 6 yr of age.

Strongyloides spp. (see 9.2.4)

At least one undescribed *Strongyloides* species has been reported to cause death in captive macropods including wallaroos, red, eastern and western grey kangaroos (Arundel et al. 1977), red-necked and spectacled hare-wallabies and lesions without mortality in free-ranging animals including the agile wallaby and eastern grey kangaroo (Speare et al. 1982; Arundel et al. 1990). *Strongyloides* species have been recovered from macropods in Queensland, New South Wales and Victoria. All macropod species should be considered susceptible to infection with this parasite. Adult *Strongyloides* species live in the cardiac portion of the stomach and, in the case of the northern nail-tailed wallaby (*Onychogalea unguifera*), the caecum and upper colon (Speare et al. 1982). Adults burrow into the mucosal epithelium where they lay eggs and cause severe inflammation. The eggs rapidly develop and frequently hatch before being passed in faeces. The larvae then develop directly or indirectly to the infective stage. Arundel et al. (1990) observed an acute erosive gastritis of the epithelium of the sacculated stomach accompanied by a suppurative inflammatory reaction in subadult eastern grey kangaroos. Oedema, hyperaemia, thickening, ulceration, folding and nodulation of the gastric mucosa have also been reported (Arundel et al. 1977; Speare et al. 1982; Winter 1958). *Strongyloides* species are typically well-tolerated by the host and mortality has only been reported from captive macropod populations.

Angiostrongylus cantonensis (see 9.5.6)

Neurological disease associated with the migration of the rat lungworm *Angiostrongylus cantonensis* in abnormal hosts has been reported in captive rufous rat-kangaroos (Higgins et al. 1999), red-necked wallabies (McKenzie et al. 1978), a western grey kangaroo, a parma wallaby/tammar wallaby hybrid, purple-necked rock-wallaby (*P. purpureicollis*) and a wallaroo (ARWH 2007). The current known distribution of *A. cantonensis* in Australia is northern and south-eastern Queensland and the Sydney region. Infection with *A. cantonensis* follows the ingestion of third-stage larvae in several species of land mollusc, the intermediate hosts of the parasite. Other potential routes of infection include ingestion of paratenic hosts or possibly larvae in snail mucus trails. The precise route of infection in infected macropods remains to be demonstrated. The third-stage larvae migrate to the CNS, undergoing two moults, before reaching the subarachnoid space. In definitive hosts the fifth-stage larvae then return to the circulation and migrate to the pulmonary arteries, where they reach sexual maturity. In the abnormal host third-stage larvae migrating in the CNS cause disruption of neurological function (Prociv et al. 2000). Pathological findings may be variable and are likely to reflect the duration and severity of infection. The typical lesion is that of a meningoencephalitis, however, the intense eosinophilic inflammation seen in eutherian mammals apparently does not occur to the same degree in marsupials (Higgins et al. 1999). Nematode parasites may be evident in cross-sections of the meninges and cerebral cortex.

Capillaria spp. (see 9.3.7)

An as yet undescribed species of *Capillaria* designated as *Capillaria* sp. nov. 7 has been associated with bronchitis, bronchiolitis and pneumonia in brush-tailed and rufous bettongs and long-nosed potoroos (Love & Reddacliff 1992). Capillarial hepatitis associated with a species of *Capillaria* presumed to be *C. hepatica* has been identified in red kangaroos, brush-tailed and rufous bettongs and agile and parma wallabies (Canfield & Hartley 1992). Disease associated with *Capillaria* spp. has been described from captive collections of macropods in New South Wales. The life cycle of the *Capillaria* sp. nov. 7 is unknown. Macropods with hepatic capillariasis were presumed to have become infected following ingestion of embryonated eggs present in the soil of their enclosures (Canfield & Hartley 1992). Tenacious mucus, necrotic debris, desquamated epithelial cells and inflammatory cells occur in the bronchi and bronchioles of potoroid marsupials associated with eggs, larvae and adults of *Capillaria* sp. nov. 7. Pulmonary congestion, oedema and emphysema are also present to varying degrees. Capillarial hepatitis is associated with multifocal white nodules throughout the hepatic parenchyma. Histologically, progressive changes are evident around aggregates of eggs and occasionally adults. Early lesions are associated with minimal cellular response before becoming surrounded by connective tissue and mononuclear cells with the eventual development of calcification, giant cell infiltration and extensive fibrosis (Canfield & Hartley 1992).

Marsupostrongylus spp. (see 9.3.7)

Several species of *Marsupostrongylus* (*M. dorrigoensis, M. longilarvatus* and *M. wallabiae*) that parasitise the alveoli and bronchi of the red-necked pademelon, Tasmanian pademelon, swamp wallaby and eastern grey kangaroo have been associated with pulmonary pathology (Beveridge 1978). The parasite has been recorded from northern and eastern Australia. The life cycles of these species have not been described. An infection prevalence of 46% was reported in swamp wallabies, with a higher prevalence in animals from northern Australia (Beveridge et al. 1985). Pathology in some individuals is mild, with focal granulomas and mild bronchiolitis, and extensive in some individuals, with severe bronchopneumonia sufficient to cause respiratory compromise (Beveridge et al. 1985).

Breinlia mundayi

Mild focal peritonitis, pleuritis and pericarditis occurs in swamp wallabies associated with the presence of adult filarioid nematodes of *Breinlia mundayi.* This parasite has been identified from swamp wallabies in Queensland, New South Wales and Victoria, and from red-necked wallabies in Tasmania (Beveridge 1978; Beveridge et al. 1985). Adult nematodes are found free in the abdominal and thoracic cavities and within the pericardial sac, inducing eosinophilic infiltration and fibrosis of adjacent organs. Encapsulated calcified nematodes are found in the omentum and occasionally on the visceral surfaces. Microfilariae are found in the circulatory system and are spread from host to host via arthropod vectors. Circulating microfilariae are associated with eosinophilic infiltrations of the blood vessels of the heart, liver and renal cortex. Sequestered microfilariae cause focal granulomas in the red pulp of spleen (Beveridge et al. 1985).

Other *Breinlia* spp.

Several other species of *Breinlia* have been described in macropods, including *B. thylogali* in red-legged pademelons (*Thylogale stigmatica*) and Tasmanian pademelons associated with mild peritonitis and visceral adhesions, *B. ventricola* from the red kangaroo postulated to cause pathology similar to *Dirofilaria immitis* (Spratt & Hobbs 2003), *B. woerlei* from the nabarlek and short-eared rock-wallaby (*Petrogale brachyotis*) associated with large pulmonary cysts, and an undescribed species of *Breinlia* from the tammar wallaby associated with pleuritis and visceral and pleural adhesions (Beveridge 1978). *B. woerlei* has been observed to cause pathology similar to *D. immitis* in a captive nabarlek. Sequelae to infection in this animal included haemothorax subsequent to the rupture of a pulmonary artery caused by parasitic pulmonary thromboembolism. Valvular endocarditis and thrombus accumulation on the right atrioventricular valve were also present (ARWH 03/02/06).

Durikainema macropi

Portal vasculitis has been reported in the livers of macropods infected with *Durikainema macropi.* This nematode has been found variably in the portal veins, pulmonary arterioles and epicardial lymphatics of wallaroos, red, eastern and western grey kangaroos, agile and red-necked wallabies and the spectacled hare-wallaby (Speare & Spratt 1981). Affected macropods have been described from Queensland and one location in South Australia. The precise life history of this parasite is unknown although direct transmission form dam to young via milk has been postulated, based on the presence of larvae in the lactating mammary gland of an agile wallaby. Larvae have also been found in the skin capillaries of a male agile wallaby and transmission via hemophagous flies remains a potential mode of transmission. Gross lesions include multifocal linear 1–2 mm nodular lesions oriented perpendicular to the ventral edge of the liver. The cystic and left lateral lobes are most frequently affected. Microscopically, there is an eosinophilic portal phlebitis, often with obliteration of the terminal hepatic venules (Speare & Spratt 1981).

Pelecitus roemeri

Pelecitus roemeri is a filarioid parasite that has been found in rock wallabies, kangaroos and wallabies of the genera *Macropus* and *Wallabia*, nail-tailed wallabies and tree-kangaroos. Adult parasites are found in the SC and intermuscular connective tissue of the hind limbs and are occasionally found in the SC fascia overlying the shoulder, hips and abdomen (Spratt 1972). *P. roemeri* has been found in macropods Australia-wide. Microfilariae in the wallaroo, the normal definitive host, exhibit diurnal subperiodicity in the peripheral blood and are transmitted by several species of tabanid fly. Low-level blood microfilaraemia of relatively short duration occurs in the red kangaroo. The eastern grey kangaroo and the red-necked wallaby are considered abnormal hosts with few or no circulating microfilariae in infected animals, and delayed development of female worms and the presence of reproductive detritus in gravid females (Spratt 1974; Spratt 1975). Adult worms are normally well-tolerated in the normal definitive host. In abnormal hosts they are encapsulated within a thick fibrous tissue capsule. Portas et al. (2005) described a granulomatous splenitis in an abnormal host, an aged male western grey kangaroo, with a high-level microfilaraemia. This animal was in poor body condition although the potential role of the parasite in the condition was not clear.

Cylicospirura heydoni

Cylicospirura heydoni is a spiririud nematode inhabiting the gastrointestinal tract of a number of quoll (*Dasyurus*) species. It has been reported to cause enteritis in Tasmanian pademelons, which are considered an aberrant host (Ladds et al. 2006). Cases have only been described from Tasmania. Infection in the pademelons was thought to have occurred subsequent to the ingestion of insects carrying larvae. Multiple nodules up to 3 cm in diameter are present throughout the wall of the intestine and caecum. Abscessation of associated lymph nodes was evident in one animal. Histopathologically, the nodules consisted of fibrous tissue surrounding an area of exudation and sections of nematodes. The nodules were principally confined to the intestinal submucosa, occasionally extending into the smooth muscle layer.

Cestodes

Progamotaenia spp.

Two species of anoplocephalid cestodes, *Progamotaenia festiva* and *Progamotaenia effiga*, occur in the bile ducts of macropods invoking cholangitis but do not appear to impair the health of the host. *P. festiva* has been reported from the bile ducts of red and eastern grey kangaroos, swamp wallabies, wallaroos, northern nail-tailed wallabies, tammar wallabies, agile wallabies, antilopine kangaroos and spectacled hare-wallabies in eastern Australia (McCarthy 1961b; Presidente & Beveridge 1978; Beveridge et al. 1985). *P. effiga* has been reported from the biliary system of the western grey kangaroo (Presidente & Beveridge 1978). The life cycles of *P. festiva* and *P. effiga* are unknown but oribatid mites are considered the probable intermediate hosts. Grossly, there is dilatation and thickening of the main hepatic ducts in affected animals. In the red kangaroo there is hypertrophy of the bile duct epithelium, and fibrosis and mild eosinophilic infiltration of the duct walls. Histopathological changes in eastern and western grey kangaroos are more pronounced with more extensive hypertrophic and hyperplastic changes and inflammation (Presidente & Beveridge 1978). A severe inflammatory response with marked bile duct dilatation occurs in the northern nail-tailed wallaby (Speare et al. 1989).

Echinococcus granulosus (see 9.3.8)

Infection with the metacestode larvae (or hydatid cysts) of *Echinococcus granulosus* is widespread in macropods in eastern Australia. Cysts occur most commonly in the lungs and less frequently in the liver and pleural cavity of affected animals. Hydatid cysts have been reported from the black-striped wallaby, Mareeba rock-wallaby (*P. mareeba*), Proserpine rock-wallaby, allied rock-wallaby (*P. assimilis*), Godman's rock-wallaby (*P. godmani*), bridled nail-tailed wallaby, pretty-faced wallaby, red-necked wallaby, swamp wallaby, eastern grey kangaroo, western grey kangaroo, common wallaroo, red-legged pademelon, red-necked pademelon, tammar wallaby and Lumholtz's tree-kangaroo (Beveridge 1978; Coman 1978; Arundel 1979; Speare et al. 1989; Griffith 1998; Johnson et al. 1999; Jenkins & Morris 2003; Barnes et al. 2005, Barnes et al. 2007a, Barnes et al. 2007b). *E. granulosus* occurs in all Australian states.

Dingoes, foxes and dogs are the definitive hosts of the parasite in Australia, with the dingo the most significant (Arundel 1979). Sheep are the principal intermediate hosts with infections occurring to a lesser extent in other domestic herbivores. Macropods become infected subsequent to ingesting eggs on pasture contaminated with canid faeces. Two separate strains have been identified on the Australian mainland and a third in Tasmania. The existence of a distinct sylvatic cycle, occurring between dingoes, foxes and feral dogs and macropods, has been proposed. Recent DNA sequencing, however, suggests that extensive cross-infection occurs between the two strains and that there may be no distinct sylvatic cycle (Jones & Pybus 2001). Jenkins and Morris (2003) found a prevalence of 69% and cyst fertility of 100% for hydatids in swamp wallabies. This, coupled with an increased susceptibility of hydatid-infected swamp wallabies to predation by feral dogs, led the authors to suggest that swamp wallabies were pivotal to the maintenance of an *E. granulosus* cycle in Kosciuszko National Park. Variable prevalence of infection has been demonstrated at different sites and is likely to reflect environmental factors affecting parasite survivability (Barnes et al. 2005). Species- and sex-specific differences in prevalence of infection have been demonstrated in macropods (Barnes et al. 2005; Barnes et al. 2007a). Female eastern grey kangaroos in southern Queensland are twice as likely to be infected as males (Barnes et al. 2007a).

Despite frequent infection of macropods with *E. granulosus*, reports of morbidity and mortality are limited. Mortality has been reported in the bridled nail-tailed wallaby, the Mareeba rock-wallaby and the Proserpine rock-wallaby (Johnson et al. 1998). Macropods can remain clinically normal despite the presence of large numbers of moderately sized cysts. In animals in which death was attributed to infection with *E. granulosus*, severe respiratory compromise associated with extensive replacement and compression of the pulmonary parenchyma with hydatid cysts was evident. In experimentally infected tammar wallabies, cyst growth resulted in a 28% reduction in functional lung volume within 14 mo post infection (Barnes et al. 2007b). Sequelae to infection in this study included anaphylaxis secondary to cyst rupture and pneumothorax. Degenerative changes, including caseation and calcification of cysts, has been observed in larger macropods and is thought to be associated with an immune response to the parasite (Jenkins & Morris 2003; Barnes et al. 2005). Fewer degenerative changes have been observed in wallabies.

Trematodes

Fasciola hepatica

Cholangiohepatitis associated with the introduced common liver fluke (*Fasciola hepatica*) has been reported from numerous species of macropods in south-eastern Australia and is considered one of the most pathogenic parasites of marsupials (McManus 1979; Spratt & Presidente 1981). Cases of infection in macropods appear limited to the range of the snail, *Lymnaea tomentosa*, the intermediate host of the liver fluke. Infections have been reported in eastern grey kangaroos, red-necked wallabies, swamp wallabies and Tasmanian pademelons (McManus 1979; Spratt & Presidente 1981). Prevalence of infection is higher in macropods grazing pastures concurrently with domestic ruminants. A direct macropod snail cycle appears to occur in at least one location in the Australian Capital Territory, where the parasite is maintained in eastern grey kangaroos in the absence of domestic ruminants (Spratt & Presidente 1981). Patent infections have been observed in free-ranging eastern grey kangaroos, red-necked wallabies and Tasmanian pademelons in Tasmania and infected animals are significant sources of pasture contamination (McManus 1979). Species-specific differences in pathological responses to infection are evident. Eastern grey kangaroos are considered tolerant of moderate infections although fatalities attributed to fascioliasis have been observed (ARWH 2007). Hepatic fibrosis and severe anaemia occur in Tasmanian pademelons and red-necked wallabies, with some infections being fatal. Pathological changes include fibrosis, bile duct hyperplasia and eosinophilic cholangiohepatitis (Beveridge 1978).

Protozoa

Coccidiosis (see 9.2.6)

Enteric coccidiosis is a common disease of juvenile, captive and hand-reared macropods including red, eastern and western grey kangaroos and tammar, black-striped, pretty-faced and red-necked wallabies (Winter 1959; Presidente 1978; Speare et al. 1982; Blyde 1999). Eastern grey kangaroos appear to be especially susceptible and are most frequently affected. Various coccidial species have been found in captive and free-ranging macropods Australia-wide. Cases of coccidiosis have also been reported from captive collections of macropods held overseas (Thiruthlinathan et al. 1998; Johnson-Delaney 2000).

Coccidia of the genus *Eimeria* and less frequently *Isospora* are found in all species of macropods, although species-specific variation in prevalence of infection occurs (Mykytowycz 1964; Barker et al. 1989). Clinical disease associated with coccidial infection is usually limited to juvenile animals less than 12 mo of age and, in eastern grey kangaroos, appears to have a bimodal distribution. Anecdotal evidence suggests PY weighing 4.5–5 kg and 6.5–8 kg are the weight classes/age groups most frequently seen with clinical disease. The pathogenesis of infection is dependent on numerous factors, including the species of coccidia, and a range of host factors. Overcrowding, malnutrition and wet conditions have been associated with outbreaks in captive and free-ranging animals (Winter 1959; Barker & Harrigan 1972). Hand-reared PY are thought to be particularly susceptible, as they lack protective immunity acquired passively from their dams. Coccidial cholangitis has been described from a captive parma wallaby and tammar wallabies with concurrent enteric coccidiosis (Canfield & Hartley 1992). Mykytowycz (1964) found age-related differences in susceptibility to coccidial infection in free-ranging red kangaroos, with animals of 12 mo of age most susceptible. Sympatric eastern grey kangaroos did not demonstrate marked age-related changes in susceptibility and carried higher burdens than red kangaroos. No pathology was associated with infection in either species. An outbreak of coccidiosis in a free-ranging population of eastern grey kangaroos occurred during flooding in which large numbers of kangaroos were stranded on crowded islands with insufficient forage (Barker & Harrigan 1972). Deaths occurred in young animals and the authors postulated that environmental stressors coupled with acute exposure to large numbers of coccidial oocysts contributed to the development of disease.

Winter (1959) described an acute outbreak in a group of young captive eastern grey kangaroos and consistently found marked small intestinal changes including mild to moderate duodenal oedema progressing to marked jejunal oedema and haemorrhage. Diphtheritic changes were evident in the jejunum of the three worst-affected animals. Microscopic changes included mucosal and submucosal oedema at the level of the duodenum. At the level of jejunum, extensive haemorrhage and necrosis of the mucosa was apparent. Intestinal epithelial cells contained coccidial organisms almost exclusively in stages of schizogony. Similar pathological findings were described by Barker and Harrigan (1972) in free-ranging eastern grey kangaroos. Affected animals had haemorrhagic enteritis, with villous atrophy and oedema that was most pronounced in the jejunum. Fatal coccidiosis in captive black-striped wallabies and pretty-faced wallabies was associated with upper small intestinal haemorrhage and congestion, atrophy of the intestinal villi, epithelial sloughing and the presence of giant schizonts (Speare et al. 1982). Speare (1988) observed massive numbers of schizonts in the lamina propria associated with a lymphocytic inflammatory response in eastern grey kangaroos. Mucosal ulceration and marked villous atrophy was also present. Biliary duct hyperplasia and fibrosis were present in captive tammar wallabies with hepatic coccidiosis (Speare et al. 1989), while granulomatous lesions were evident in the hepatic portal triads of free-ranging tammar wallabies (Presidente 1978). Schizonts have occasionally been observed in the liver and mesenteric lymph nodes of black-striped wallabies.

Toxoplasmosis (see 9.8.3)

All macropod species are considered susceptible to infection with the coccidian parasite *Toxoplasma gondii* and subsequently to the development of clinical disease. Domestic cats are the definitive hosts of this parasite and macropods are infected following ingestion of sporulated oocysts in contaminated food sources. Toxoplasmosis has been reported from free-ranging macropods in Australia and virtually all locations where captive macropods are held globally. Cases in Australia are limited to locations where both the definitive hosts and macropods occur concurrently, although there is evidence to suggest disease occurs more frequently in southern Australia.

Clinical disease may be the result of primary infections with *T. gondii* following the ingestion of oocysts or recrudescence of latent disease following a period of immune compromise. A minimum oral dose of 6–50 oocysts appears to be necessary to infect macropods weighing 15–18 kg (Johnson et al. 1988). Gross lesions may not be evident at necropsy. Where lesions are present they may include pulmonary congestion, oedema and consolidation, pleural effusion, myocardial haemorrhage and pallor, lymph node and adrenal gland enlargement and inflammation, splenomegaly, haemorrhagic and ulcerative foci within the stomach and small intestine and, occasionally, the presence of fibrinous intestinal casts (Obendorf & Munday 1983; Canfield et al. 1990a; Reddacliff et al. 1993a). Histopathological findings in tammar wallabies that developed acute toxoplasmosis subsequent to experimental infection included small intestinal submucosal oedema, fibrinous exudation and focal to extensive necrosis.

Necrosis, oedema and haemorrhage of the mesenteric lymph nodes were also present. Pulmonary pathology included congestion, oedema and variable degrees of interstitial pneumonia. Focal necrosis was present in the brain, adrenal cortex, liver, spleen and cardiac and skeletal muscle (Reddacliff et al. 1993a). Free and intracellular tachyzoites were evident in H&E-stained sections of multiple tissues. Similar histopathological findings were described by Patton et al. (1986), documenting an epizootic occurring in tammar wallabies, red and eastern grey kangaroos, and long-nosed potoroos at a North American zoo. Brain lesions in affected animals were extensive with multifocal granulomatous encephalitis and meningitis. Findings in free-ranging Tasmanian pademelons and red-necked wallabies affected by an outbreak of toxoplasmosis included the presence of severe non-suppurative myocarditis and encephalitis and focal hepatic necrosis. Despite pulmonary congestion and oedema evident at gross necropsy, interstitial pneumonia was not evident on histopathological examination (Obendorf & Munday 1983). Extensive necrosis and fibrin exudation associated with *T. gondii* organisms was evident in the pulmonary parenchyma of four wallaroos dying during an outbreak at a Californian zoo (Boorman et al. 1977). A necrotising myocarditis was the predominant histopathological finding in one western grey kangaroo PY and skeletal tissue cysts in another PY dying during an epizootic of toxoplasmosis affecting a captive group of western grey kangaroos (Dubey et al. 1988). Deaths occurred only in juveniles, with subclinical infections occurring in adults. In a review of the microscopic findings in 43 cases of toxoplasmosis in macropods, Canfield et al. (1990) found an extensive range of lesions affecting the pulmonary parenchyma, cardiac and skeletal muscle, lymph nodes, spleen, gastrointestinal tract, adrenal glands, pancreas, central nervous system, liver and kidney. Gross lesions were frequently not evident in affected animals.

Cutaneous leishmaniasis in red kangaroos (see 9.1.9)

Ulcerative and papular skin lesions associated with an undescribed species of *Leishmania* have been reported from captive red kangaroos (Rose et al. 2004). The affected red kangaroos were all housed in captive facilities north of their natural geographic range in the Northern Territory. The condition has not been detected in other macropod species. The vectors, host range and source of the *Leishmania* species in the infected animals are not yet known. The host inflammatory response varies from lympho-plasmacytic through to granulomatous cellular infiltrates within the dermis. Other histopathological changes evident include extensive epidermal hyperplasia, coagulation necrosis, ulceration and suppurative inflammation. Intracellular single-celled organisms exhibiting a nucleus and frequently a kinetoplast, characteristic of *Leishmania* species, are evident.

Amoebiasis (see 9.2.3)

Chronic progressive gastritis associated with *Entamoeba histolytica* has been reported in a captive tammar wallaby (Stedman et al. 2003) and acute fatal gastritis associated with an undetermined species of *Entamoeba* has been observed in captive wallaroos and, less frequently, in captive red kangaroos (Roberts et al. 1973; Munday 1978; ARWH 2007). An undetermined species of *Entamoeba* has also been observed in the faeces of wallaroos with mild spontaneously resolving diarrhoea, although its significance in these cases is unknown. Amoebic meningoencephalitis has been reported in a captive juvenile eastern grey kangaroo and was thought to be associated with a species of *Acanthamoeba* (Norton & Hartley 1993). Cases of gastric amoebiasis have been reported from macropods in eastern Australia and captive animals in the US (Roberts et al. 1973). The epidemiology of gastric amoebiasis is poorly understood but the presence of the protozoa in the faeces of clinically normal macropods suggests clinical disease occurs subsequent to other initiating factors, including penetration of the gastric mucosa by larvae of *Labiostrongylus* species. Gross findings in wallaroos include subcutaneous oedema and focal gastric necrosis. Histopathologically, protozoa are present in the gastric submucosa, muscularis and associated blood vessels. Extensive bacterial proliferation occurs in lesions. Focal haemorrhage is present around ulcerative lesions. Trophozoites may be present in areas of mucosal necrosis accompanied by extensive invasion of the submucosa and muscularis (Roberts et al. 1973). Protozoal proliferation and invasion may occur secondary to the penetration of the gastric mucosa by larval nematodes (Munday 1978; ARWH 2007). Histopathological findings in a tammar wallaby included invasion of the gastric mucosa and displacement of gastric glands by *E. histolytica* trophozoites, protozoal invasion of the muscularis, submucosa and occasionally submucosal blood vessels and lymphatics (Stedman et al. 2003). Histological examination of the brain of an eastern grey kangaroo with amoebic meningoencephalitis revealed massive haemorrhage, suppurative acute necrotising inflammation associated with trophozoites and encysted amoeba (Norton & Hartley 1993).

Protozoal rhinitis

Rhinitis associated with a *Besnoitia*-like organism has been observed in subadult eastern and western grey kangaroos. The principal presenting clinical sign is epistaxis and infection can be fatal (ARWH 2007). A small number of cases have been observed in macropods in New South Wales. The life cycle of this organism is unknown. At necropsy macroscopic lesions are evident in the turbinates, nasal septum, larynx and trachea of affected animals. Histopathologically, a haemorrhagic rhinitis is evident associated with the presence of epithelial and subepithelial *Besnoitia*-like organisms. The organisms invade the basal cells of the pseudostratified ciliated nasal epithelium. Basal cells parasitised by the unicellular zoites become enlarged with nuclear hypertrophy. As the parasites mature the host cell becomes part of the cyst wall of the parasite.

9.10 Non-infectious diseases

9.10.1 Toxicoses

a Sodium monofluoroacetate (1080)

Sodium monofluoroacetate (1080) is a rapidly acting toxic compound that impairs cellular respiration, producing CNS anoxia and cardiovascular disturbance. It is used throughout Australia for the control of vertebrate pest species and poisoning of macropods may occur. Macropods are typically non-target casualties of baiting campaigns, although historically red-necked wallabies and Tasmanian pademelons have been specifically targeted. Species-specific differences in the LD_{50} of sodium monofluoroacetate have been demonstrated (McIlroy 1982). Marsupials have been shown to be more sensitive than eutherians to the toxic effects of sodium monofluoroacetate in experimental studies. The small body size of some macropod species such as bettongs, potoroos and pademelons renders them particularly susceptible to toxicity during baiting campaigns for European rabbits. There are few distinct pathological lesions associated with intoxication although evidence of cardiac failure may be present.

b Fluorosis

Chronic lameness and dental disease associated with high environmental fluoride levels have been observed in aged eastern grey kangaroos confined within the grounds of an aluminium smelter (Clarke et al. 2006). Lesions were confined to a sedentary population of macropods at a single aluminium smelter in south-western Victoria but could be expected in other macropods with prolonged exposure to high environmental levels of fluoride. Lesions of osteofluorosis were observed in older individuals within the affected population. Bone fluoride concentrations in affected animals were within high normal limits for bone fluoride concentrations in cattle, suggesting that eastern grey kangaroos are more susceptible to fluorosis. The lesions associated with fluorosis included osteophytosis and osteoporosis of the femur, tibia and metatarsals and osteophytosis of the proximal coccygeal vertebrae. Increased bone density, elevated fluoride levels and abnormal mineralisation and mottling of the bone matrix were also present. Dental lesions included incisor enamel hypoplasia, staining and abnormal wear.

c Pindone toxicity (see 9.3.11)

Pindone is a first-generation anticoagulant rodenticide used for controlling European rabbits. Accidental toxicity in macropods follows the ingestion of pindone-laced baits. Death secondary to coagulopathy induced by the ingestion of pindone has been observed in swamp and red-necked wallabies, eastern grey, western grey and red kangaroos. The principal finding is massive internal haemorrhage. PY appear to be particularly susceptible following ingestion of milk from affected females. During a mass toxicity episode in several species of macropods, furless PY with extensive subcutaneous and internal haemorrhages were found dead in the pouch of affected but living females. A smaller number of live PY with subcutaneous bruising and lowered haematocrits were removed from affected females, treated and successfully hand-reared.

d Mebendazole toxicity

Mebendazole at high doses has been associated with the development of a condition termed haemorrhagic septicaemic syndrome in captive macropods (see section 10).

e Toxic plants

A range of plant toxins has been associated with morbidity and mortality in macropods. Hough (1997) described gastrointestinal pain, inappetence, bradycardia and ataxia in a juvenile western grey kangaroo following the ingestion of rhododendron species. Recovery was complete following symptomatic therapy. Ellis et al. (1983) described acute death associated with oxalate nephrosis in a swamp wallaby housed in a yard containing soursob (*Oxalis pes-caprae*). Yew (*Taxus baccata*) ingestion caused fatalities in four of six red-necked wallabies (Barker et al. 1963). Hepatotoxicity and secondary photosensitisation occurred following the ingestion of *Lantana camara* by two captive red kangaroos (Johnson & Jensen 1998). One animal recovered with supportive care but the other died 1 wk later. A third animal died without exhibiting premonitory signs. An eastern grey kangaroo developed a severe diffuse chronic cholangiohepatopathy consistent with hepatotoxic mycotoxicosis (Hum 2005). The animal had been grazing perennial rye grass (*Lolium perenne*) pasture, potentially contaminated with *Pithomyces chartarum*, prior to exhibiting polydipsia and weight loss, and eventually dying.

9.10.2 Neoplasia

Neoplastic diseases reported in macropods are detailed in Table 7.7.

9.10.3 Miscellaneous non-infectious diseases

a Capture myopathy

Captive and free-ranging macropods are highly susceptible to developing capture myopathy. It is commonly seen after poorly executed capture and restraint events, during and after transport, repeated handling, in animals placed in unfamiliar environments or close confinement or exposure to or pursuit by predators such as dogs. Capture and restraint attempts during high ambient temperatures increase the risk of capture myopathy. While its development is most frequently associated with prolonged muscular exertion, it is also seen in animals experiencing fear or anxiety without physical exertion.

Development of the condition is thought to result from a combination of prolonged sympathetic nervous system stimulation causing ischaemia as a result of reduced tissue perfusion, lactic acidosis and muscular adenosine triphosphate reserve depletion. Secondary problems include cardiovascular and circulatory collapse, muscular compartment syndrome and acute renal failure subsequent to ischaemia and myoglobinuric nephrosis. Animals that survive an initial episode may die peracutely following a second episode of exertion or stress.

Affected animals may exhibit dyspnoea, tachypnoea, tachycardia, hyperthermia, muscular tremors and fasciculations, myoglobinuria, inability to hold the head in a normal elevated position, weakness and collapse. In advanced disease animals are laterally recumbent and unable to rise (Blyde 1999). In the chronic form of the disease animals have a stiff gait associated with chronic muscular lesions. Death may occur days to weeks after an event due to cardiomyopathy. The clinical signs and history are highly suggestive. Massive elevations in CK ($\geq$25 000 U/L) are evident; other biochemical abnormalities include elevated ALP, AST, urea and creatinine. Myoglobinuria may be present.

Gross pathological changes in the muscles of affected animals can be variable and may range from muscular haemorrhage, congestion and oedema through to pale and friable musculature. Chronically affected muscles may be pale and fibrosed with foci of mineralisation. Muscle groups most commonly affected include the hind limb adductors, and the psoas, cervical and dorsal musculature. Lesions may also be evident in the diaphragm and cardiac musculature. Swollen kidneys, myoglobinuria and pulmonary congestion and oedema may also be apparent. The histopathological lesions reflect the severity and duration of the insult. Microscopic changes may include myofibre degeneration, haemorrhage and oedema of the interstitium and, in chronic cases, mineralisation and interstitial fibrosis. Renal tubular necrosis, pulmonary oedema and hepatic and adrenal cortex necrosis are variably present (Rose 1999).

Affected macropods need to be effectively and quickly restrained for treatment. Benzodiazepine sedation following manual restraint or tiletamine/zolazepam anaesthesia via remote injection is indicated if animals are ambulatory at the time of presentation (see 6.2). An IV catheter is placed in the lateral caudal or recurrent tarsal vein and fluid therapy initiated to counteract shock and prevent the development of myoglobinuric nephrosis. Dantrolene sodium (1 mg/kg IV), a skeletal muscle relaxant used to treat malignant hyperthermia in humans, appears to be a useful supportive measure. Intravenous sodium bicarbonate should be administered either empirically or based on the results of blood gas analysis to correct metabolic acidosis. Prednisolone sodium succinate (5–10 mg/kg IV) may help prevent cell

Table 7.7 Neoplasms reported in some Australian macropods

Species	Neoplasm and location	Reference
Red kangaroo	Oral squamous cell carcinoma	Kosanke et al. (1979)
		Suedmeyer & Johnson (2007)
	Cutaneous melanoma	ARWH 03/02/06
	Multicentric T cell lymphoma	Suedmeyer & Johnson (2007)
	Gastric lipoma	Suedmeyer & Johnson (2007)
	Gastric squamous cell carcinoma	Lombard & Witte (1959)
	Metastatic gastric adenocarcinoma	Ratcliffe (1933)
	Metastatic gastric papilloma	Fox (1912)
	Pulmonary carcinoma	Fox (1912)
	Biliary/pancreatic squamous cell carcinoma	ARWH 03/02/06
	Mammary carcinoma with pulmonary and renal metastases	ARWH 03/02/06
	Mammary adenocarcinoma	Canfield et al. (1990c)
	Mammary adenocarcinoma with pulmonary metastases	Suedmeyer & Johnson (2007)
	Pouch basal cell epithelioma	Canfield et al. (1990c)
	Ovarian stromal tumour	Canfield et al. (1990c)
	Testicular seminoma	Wallach & Boever (1983)
Eastern grey kangaroo	Trichoepithelioma	Speare & Banks (1987)
	Subcutaneous round cell tumour	ARWH 03/02/06
	Bronchogenic adenocarcinoma	ARWH 03/02/06
	Osteosarcoma	ARWH 03/02/06
	Metastatic haemangiosarcoma	Canfield & Hartley (1992)
	Hepatic haemangiosarcoma	ARWH 03/02/06
	Cutaneous lymphosarcoma	Canfield et al. (1990c)
	Mammary adenocarcinoma hepatoma	Lombard & Witte (1959)
Western grey kangaroo	Cutaneous amelanotic melanoma	ARWH 03/02/06
	Dermal lymphosarcoma	Canfield & Hartley (1992)
Common wallaroo	Hepatic haemangioma	Canfield & Hartley (1992)
	Metastatic hepatic neoplasia	Canfield & Hartley (1992)
	Gastric carcinoma	Plimmer (1912) in Potkay
Antilopine kangaroo	Biliary adenocarcinoma	Canfield et al. (1990c)
	Thyroid adenoma	ARWH 03/02/06
Agile wallaby	Splenic lymphoma	ARWH 03/02/06
	Cutaneous squamous cell carcinoma	ARWH 03/02/06
	Renal carcinoma with hepatic metastases	ARWH 03/02/06
Pretty-faced wallaby	Lymphoid leukaemia	ARWH 03/02/06
	Blast cell lymphoma	Rothwell (1988)
Parma wallaby	Bizarre parosteal osteochondromatous proliferation	Rungsipipat et al. (1998)
	Mammary adenocarcinoma	Canfield et al. (1990c)
	Metastatic mammary carcinoma	Canfield & Hartley (1992)
	Cervical squamous cell carcinoma	Griner (1983)

(Continued)

Table 7.7 *(Continued)*

Species	Neoplasm and location	Reference
Red-necked wallaby	Aural squamous cell carcinoma	ARWH 03/02/06
	Subcutaneous adenocarcinoma	ARWH 03/02/06
	Lymphoid neoplasia	Canfield & Hartley (1992)
	Oral melanoma	Garma-Avina (1981)
	Oral adenocarcinoma	Effron et al. (1977)
Bridled nail-tailed wallaby	Thoracic haemangiosarcoma	ARWH 03/02/06
	Oral squamous cell carcinoma	ARWH 03/02/06
Tammar wallaby	Gastric fibropapilloma	Wallach & Boever (1983)
Rufous bettong	Hepatic and splenic angioma	Canfield et al. (1990a)
	Laryngeal carcinoma	Speare et al. (1982)
	Duodenal carcinoma	Canfield et al. (1990c)
Tasmanian bettong	Cutaneous melanoma	ARWH 03/02/06
	Lymphoid leukaemia	Canfield & Hartley (1992)
Brush-tailed bettong	Lymphosarcoma	ARWH 03/02/06
Tasmanian pademelon	Gastric squamous cell carcinoma	Munday (1971)
Long-nosed potoroo	Spindle cell tumour	ARWH 03/02/06
Pademelon (species unknown)	Lymphoid leukaemia	ARWH 03/02/06
Mala	Mammary adenocarcinoma	ARWH 03/02/06
Hare-wallaby (species unknown)	Nephroblastoma	ARWH 03/02/06
Quokka	Infiltrative pancreatic lipoma	ARWH 03/02/06
	Dermal lymphosarcoma	Canfield et al. (1990c)

damage through stabilising cell membranes. An alternative to corticosteroid therapy is flunixin meglumine (1.1 mg/kg IV). If the animal is hyperthermic it should be cooled using ice, wet towels, fans and cooled IV fluids. Parenteral and oral administration of vitamin E and selenium has been advocated for prevention and therapy of capture myopathy, however, reports of efficacy are anecdotal.

The capture and restraint of any macropod must be very carefully planned to prevent capture myopathy. Capture and restraint (both physical and chemical) should only be performed by experienced operators who are familiar with macropod behaviour and capture and restraint techniques. Appropriate restraint equipment must be used (both for physical and chemical capture) and operators must be familiar with their use. Performing procedures in hot weather and prolonged pursuits should be avoided. Once restrained, the animal should be kept cool. If the macropod is not successfully captured in the first one or two attempts, capture should be postponed for another day or remote chemical restraint utilised. Chemical restraint via projectile syringe is the most appropriate method for restraining free-ranging and larger captive macropods. The administration of diazepam immediately after physical capture is useful in preventing capture myopathy due to its anxiolytic and muscle relaxant effects. The use of neuroleptic agents including azaperone, zuclopenthixol acetate, haloperidol lactate, haloperidol decanoate and fluphenazine decanoate is indicated when translocating, hospitalising and intensively handling many macropod species (see 6.2).

b Arterial calcinosis

Sporadic cases of arterial calcinosis, resembling arteriosclerosis in humans, have been reported from captive macropods in Australia and overseas (Poelma 1964; Munday 1978; Griner 1983; Schoon et al. 1985). The epidemiology of this condition is poorly understood. Dietary factors have been implicated and Griner (1983) found a reduced incidence of the disease subsequent to dietary changes in a population of marsupials housed in North America. The distribution and extent of vascular lesions varies in individual animals. Histopathological findings include arterial degeneration and calcium deposition on affected vessels.

c *Trauma*

Both free-ranging and captive macropods are prone to trauma. Free-ranging animals may suffer gunshot wounds, motor vehicle accidents, collisions with fences and other objects and predation. In many of these cases, injuries are fatal. If not fatal and animals are rescued, serious consideration must be given to their welfare and the prognosis for successful rehabilitation and release. Treatment and hospitalisation of adult wild macropods is stressful for the animal. In most cases, euthanasia is indicated.

Captive macropods are also prone to trauma, most frequently the result of poorly designed facilities (see 4.1) and poorly planned and executed capture attempts (see 6.1). However, due to the often flighty nature of macropods, accidents do happen even with ideal facilities and capture procedures. Facial injuries are common, frequently resulting in fractures of the mandible and/or maxilla. This is often associated with a degree of soft tissue damage and fractured or dislodged teeth. Secondary bacterial infection of these injuries by oral flora is common and soft tissue cellulitis, necrosis and osteomyelitis are common sequelae.

In severe cases, euthanasia is often indicated. Treatment of other cases may require multiple anaesthetic procedures, fracture fixation, hospitalisation, the use of intermediate- or long-acting neuroleptic drugs, antibiotic cover (long-acting penicillins or tetracyclines—see 10.2) and pain relief. In severe cases, the use of opioids such as butorphanol tartrate (0.4 mg/kg IM or SC) or buprenorphine (0.01–0.05 mg/kg SC tid) are indicated. Non-steroidal anti-inflammatory drugs including tolfenamic acid (4 mg/kg SC q 48 hr), meloxicam (0.2 mg/kg SC) or carprofen (4 mg/kg SC) are frequently used.

A relatively common injury associated with collision with fences and other objects are fractures of the second cervical vertebra (axis), particularly of the dens (odontoid process) and cranial articular processes. This may be associated with weakness of the atlanto-occipital articulation in macropods (see 2.1). Even the most innocuous collisions can result in fracture of the axis. Animals usually collapse immediately and become unconscious. Cranial trauma is frequently suspected, however, fracture of the axis is much more likely. Lateral and ventro-dorsal radiographs of the head and neck should be taken and if a fracture is confirmed euthanasia is indicated (Fig. 7.17).

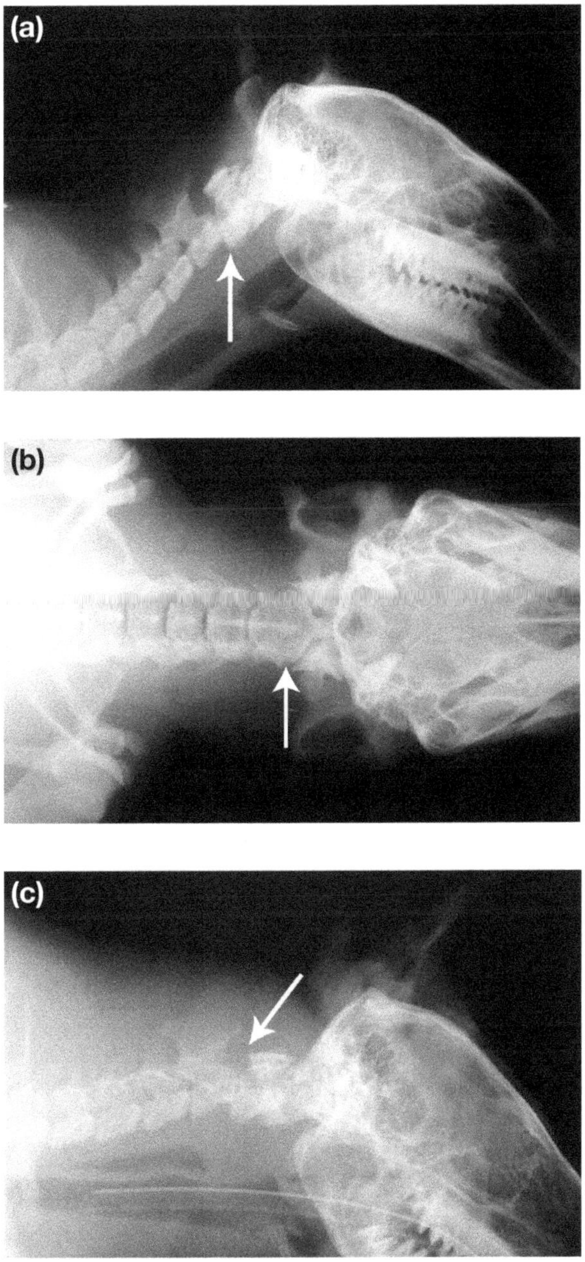

Figure 7.17 a) Lateral and (b) ventro-dorsal radiographs of an agile wallaby showing a fracture of the axis (arrow) and (c) a lateral radiograph of another agile wallaby showing a fracture of the axis (arrow). Photo: Larry Vogelnest.

10 THERAPEUTICS

10.1 Drug administration

Drugs can be administered to macropods via the same routes as in other animals. SC and IV injections can only be administered if the animal is physically or chemically restrained. IM injections can be administered by hand or by darting. In-dwelling IV catheters can be placed for continuous infusions of fluids or repeated bolus injections of large volumes. The most commonly used vein is the lateral caudal (tail) vein. In most macropods other than PY or those that are immobile it is difficult to maintain a continuous infusion. Topical preparations can be applied to animals while physically restrained or if the animal is tractable enough for application. Oral administration is also possible. With physical restraint, drugs can be administered directly to the mouth with a syringe. Drugs can also be placed in favourite food items such as bread or peanut butter and offered to the individual being treated. Many macropods will, however, refuse medicated food if the drug is not palatable. Drugs can also be administered to a mob by mixing in food such as pellets, chaff, corn or chopped vegetables. However, if the drug is not palatable compliance can be difficult. More dominant animals may also consume more of the drug than submissive animals, with the risk of over- or underdosing. Overdosing may result in toxicity and underdosing in lack of therapeutic effect and development of drug resistance. Many drugs, such as non-steroidal anti-inflammatory drugs, may also be inactivated or poorly absorbed when given orally to adult macropods (as is the case in ruminants). Orally administered drugs are commonly used in hand-reared PY. As a general rule, other than in tractable animals and with palatable drugs, oral administration of medications to macropods is considered unreliable.

When choosing a treatment regimen for a macropod, careful consideration must be given to the effect that administering the drug will have on the animal. Many macropods are prone to stress and injury when being handled or approached. The volume of a drug, frequency of dosing and palatability should all be considered. Small volumes are quicker and easier to administer orally or by injection (either by hand or remotely) while long-acting preparations can be administered every 2–3 d rather than daily. Minimising stress will also improve the chances of treatment success by reducing the secretion of endogenous glucocorticoids and thus immunosuppression, and maximising healing potential. As a general rule, where possible, physical handling of macropods for drug administration should be minimised.

10.2 Drug use

There have been few pharmacokinetic studies on macropods. In one study oxytetracycline (40 mg/kg) and penicillin G (30 mg/kg) administered IV to tammar wallabies were found to have elimination half-lives of 94 and 37 mi n respectively. Long-acting oxytetracycline (20 mg/kg) and procaine/benzathine penicillin G (30 mg/kg) administered IM to tammar wallabies were found to remain above published *F. necrophorum* MIC50 for 32 and 82 hr respectively (D McLelland *pers. comm.*).

Another study examined the pharmacokinetics of oxytetracycline in red-necked wallabies (Kirkwood & Gulland 1988). In this study, 40 mg/kg oxytetracycline was administered IV. The half-life was 11.45 hr, with plasma concentrations of 4.6 μg/ml after 24 hr and 1.1 μg/ml after 48 hr. The minimum therapeutic concentration of oxytetracycline for a wide range of bacteria including *Fusobacterium* spp. is 0.5 μg/ml. This dose rate and route of administration therefore maintains plasma concentrations at a level higher than the minimum therapeutic concentration for over 48 hr. The half-life is longer than that reported for domestic species. This may be related to the lower metabolic rate, which in many marsupials is 30% lower than the average eutherian mammal. If so, it may be likely that doses of other drugs extrapolated from domestic species may be too high when used in marsupials. However, until further pharmacokinetic studies are conducted it seems prudent to use drug doses and therapeutic regimens that have been extrapolated from domestic animals.

Being foregut fermenters, it makes sense that drug doses and therapeutic regimes used for ruminants would be most appropriate for macropods. Nevertheless, drug doses and therapeutic regimes used in dogs, cats and horses have been used in macropods with few adverse reactions. In most situations a therapeutic effect seems to be achieved. A formulary of drugs and dosing regimens provided in Table 7.8.

The use of gentamicin and clindamycin antibiotic-impregnated polymethylmethacrylate (AIPMMA) beads has been described for the treatment of lumpy jaw in a red-necked wallaby (Hartley & Sanderson 2003). Beads

Table 7.8 Drugs and doses used for treatment of macropods

Drug	Dose rate	Indications and comments
Sedatives/tranquillisers (see section 6 for more detail)		
Diazepam	0.1–2 mg/kg IM 0.1–1 mg/kg IV	Anxiolytic and muscle relaxant. Individual and species variation in response
Azaperone	2 mg/kg IM	Short duration tranquillisation. Effect in 15–20 min with approximately 8 hr duration
Perphenazine enathate	5 mg/kg IM	Intermediate duration tranquillisation. Onset of action within 12–16 hr and duration of seven days (Ebedes 1993)
Fluphenazine decanoate	2.5 mg/kg IM	Intermediate to long duration tranquillisation. Duration of action is unknown but appears to last at least 10 d
Haloperidol decanoate	4–6 mg/kg IM	Long duration tranquillisation. Duration of effect in macropods is unknown but may be up to 14 d
Zuclopenthixol acetate	5–10 mg/kg IM	Short to intermediate duration tranquillisation. Reaches peak effect after about 24 hr and declines 36–48 hr after injection. (Portas 2005)
Anaesthetics (see section 6 for more detail)		
Tiletamine/zolazepam	4–10 mg/kg IM 1–3mg/kg IV	Anaesthesia. Lower IM doses are used in tractable animals, higher doses in excited animals
Ketamine + medetomidine	3–5 mg/kg + 40–125 μg/kg	Anaesthesia. Reversal with atipamezole (4–5 times the medetomidine dose IM)
Alfaxalone	5–8 mg/kg IM 1.5–3 mg/kg IV	Anaesthesia. Useful in tame macropods. Duration of action is short
Isoflurane in oxygen	5% induction, 1–3% maintenance	Anaesthesia. Useful in tractable animals and for maintenance of anaesthesia via mask or endotracheal tube
Antibiotics		
Oxytetracycline (long acting)	20 mg/kg IM q 72 hr	Duration of treatment depends on condition being treated (7–21 d)
Clindamycin	11 mg/kg PO bid	Necrobacillosis, toxoplasmosis. Can be used in combination with metronidazole. Unpalatable. Up to 30 d treatment is required for toxoplasmosis
Metronidazole	20 mg/kg PO bid	Anaerobic infections, amoebiasis. Often unpalatable
Oxytetracycline	40 mg/kg IM q 48 hr	
Enrofloxacin	5 mg/kg SC, PO sid	
Ceftiofur sodium	1–2 mg/kg IM, IV sid 7–10 d	
Gentamycin + amoxycillin	4–7 mg/kg IM bid + 10 mg/kg IM tid	For severe cases of bacterial pneumonia in PY (Speare 1988)
Procaine penicillin/benzathine penicillin G penicillin	30 mg/kg IM q 48 hr	Bacterial infection
Tulathromycin (Draxxin, Pfizer)	2.5mg/kg SC, IM q 7 d	Registered for respiratory infections (including *B. bronchiseptica, P. multocida*) in pigs and cattle. Has been used in macropods without apparent adverse effects, efficacy still to be fully evaluated. Once a week dosing is advantageous.
Antifungals		
Itraconazole	5 mg/kg PO sid	Dermatophytosis, candidiasis. Administered until 2 weeks after the resolution of lesions. Pulse therapy (1 wk on 2 wk off for 12 wk) has been successful.
	20–40 mg/kg PO sid	Cryptococcosis; treat until LCAT is negative
Fluconazole	10–20 mg/kg PO bid	Cryptococcosis; treat until LCAT is negative
Griseofulvin	5–10 mg/kg PO bid	Dermatophytosis. Administered until 2 weeks after the resolution of lesions. Adverse effects (diarrhoea, neurological signs) have been observed at 10 mg/kg PO bid for periods as short as 1 wk

(Continued)

Table 7.8 *(Continued)*

Drug	Dose rate	Indications and comments
Antifungals		
Nystatin	5000–10000 IU/kg PO tid 3-5 d	Candidiasis. Administer between feeds to treat oral cavity, oesophageal and gastric candidiasis
Terbinafine	Topical (sid or bid)	Dermatophytosis. Useful in tractable animals
Miconazole	Topical (sid or bid)	Dermatophytosis. Useful in tractable animals
Enilconazole (Imaverol Fungicidal Wash)	Weekly washes	Dermatophytosis. Useful in tractable animals or those that can be readily anaesthetised
Nystatin cream	Topical	Candidiasis
Antiinflammatory drugs and analgesics		
Meloxicam	0.2 mg/kg PO, SC sid	
Carprofen	2–4 mg/kg SC sid	
Flunixin meglumine	1 mg/kg IM, IV sid 3 d	
Ketoprofen	2 mg/kg SC sid	
Tolfenamic acid	4mg/kg SC q 48 hr	
Buprenorphine	0.01–0.05 mg/kg IM, SC tid	
Butorphanol tartrate	0.4 mg/kg IM, SC	Only has 1–3 hr duration. Good for visceral pain
Antiparasitics		
Ivermectin	200–400 µg/kg topical, PO or SC	Pediculosis, ticks, mites, helminthiasis. Single or repeat doses may be required depending on condition being treated and response
Moxidectin	500 µg/kg topical, PO or SC	Pediculosis, ticks, mites, helminthiasis. Single or repeat doses may be required depending on condition being treated and response
Imidacloprid	10 mg/kg topical	Pediculosis, fleas. Single or repeat doses may be required depending on condition being treated and response
Fipronil	Domestic animal doses	Ticks, fleas
Selamectin	Domestic animal doses	Mites. Repeat doses may be required depending on response
Levamazole	10 mg/kg PO	Helminthiasis. Single or repeat doses may be required depending on condition being treated and response
Fenbendazole	25 mg/kg PO sid for 1–5 d	Helminthiasis. Single or repeat doses may be required depending on condition being treated and response
Thiabendazole	44 mg/kg PO	Helminthiasis. Single or repeat doses may be required depending on condition being treated and response
Toltrazuril	25 mg/kg PO sid 3 d	Coccidiosis
Atovaquone	100 mg/kg PO sid	Toxoplasmosis. Combine with canola oil to enhance absorption
Miscellaneous		
Homologous plasma transfusions	10 ml/kg IV	Coccidiosis. Harvested from healthy adults
Phytomenadione	2.5 mg/kg IM, SC sid for 5 d	In cases of coccidiosis where haemorrhagic enteritis is present and anaemia due to haematozoan infections
Acetyl cystine in saline	Nebulisation	Bordetella rhinitis/pneumonia
Bromhexine hydrochloride	2–4mg/kg PO sid	Bordetella rhinitis/pneumonia
Vitamin K_1	2.5–5 mg/kg initially parentally then orally	Pindone toxicity. Treat until coagulation profiles are normal (usually 7–10 days).
α-tocopherol	25 mg PO sid	Tree-kangaroos with muscular dystrophy (Potkay 1977)
Dantrolene sodium	1 mg/kg IV	Capture myopathy

were implanted into the affected mandible and the osteomyelitis resolved. AIPMMA provide a prolonged local concentration of antibiotic in both bone and soft tissue, while avoiding any systemic effects. In other animals, high concentrations of antibiotic penetrate only about 4 mm from the bead's surface and are higher than concentrations achieved by systemic administration. The release of antibiotics is rapid for the first 7–10 d then decreases to a low level that persists for months. The use of AIPMMA in macropods has significant advantages. It allows the delivery of high concentrations of antibiotics directly to the affected site, which parenteral administration would be unlikely to achieve. It also avoids the need for frequent administration of drugs and hence stress on the animal—a factor that may contribute to the development of lumpy jaw in the first place.

10.3 Adverse reactions

Few adverse reactions to medications in macropods have been reported. They appear to tolerate most drugs regardless of administration route. However there have been anecdotal reports of gastrointestinal upset, particularly after oral administration of antibiotics (particularly penicillins, presumably due to disruption of microbial flora within the stomach). Hand-reared PY are also prone to developing candidiasis when on antibiotics and it is wise to administer an antifungal such as nystatin to prevent this (5000–10 000 IU/kg tid 3–5 d). Probiotics may be useful once the antibiotic course is completed, to help re-establish microbial flora.

There is only one published report of an adverse drug reaction in macropods. Cases of mebendazole toxicity, termed haemorrhagic septicaemic syndrome, were described in a range of macropod species housed in a captive facility in northern Queensland (Speare et al. 2004). Mebendazole (25 mg/kg sid for 3 d) was used for the control of *Strongyloides* spp. Animals that developed haemorrhagic septicaemic syndrome were believed to have consumed higher doses in feed than conspecifics. Experimentally, mebendazole (50 mg/kg sid for 5 d) has been shown to be fatal in red-legged pademelons (Speare et al. 2004). Affected animals developed extensive haemorrhaging into organs and body cavities subsequent to septicaemia and the development of a neutropaenia secondary to mebendazole-induced bone marrow depression. High doses of mebendazole have also been reported to result in severe neutropenia and subsequent bacterial septicaemia in humans. There have been other reports of illness and death in macropods following mebendazole administration (H McCracken *pers. comm.*). Other benzimidazoles (albendazole and fenbendazole) have been linked with bone marrow toxicity in other species, however, this has not been seen in macropods with these drugs.

11 REPRODUCTIVE MANAGEMENT

11.1 Reproductive control (contraception)

The majority of macropod species breed well in captivity and in the wild and populations can quickly reach high densities. It is therefore important that captive and free-ranging macropod populations be managed carefully to prevent overcrowding, stress, disease, production of surplus animals, inbreeding, degradation of habitat, starvation and, in some cases, impacts on agriculture and other human activities (Adderton Herbert 2004). This has resulted in the development and use of a wide range of methods for controlling reproduction in captive and free-ranging macropods.

The aims of contraception are achieved by permanent or reversible prevention of fertility. The ideal contraceptive should not harm the animal, should have no effect on hormonal influence on phenotype and behaviour, have a long duration of effect, be easy to deliver and be inexpensive.

11.1.1 Permanent methods

This primarily involves the removal of the gonads—ovariectomy or ovariohysterectomy in the female and bilateral orchidectomy (castration) in the male. Tubal ligation (although rarely performed in macropods) and vasectomy can be regarded as permanent methods, as successful reversal is rare. Removal of the gonads is generally not desirable due to the loss of hormonal influence on phenotype and social behaviour. In macropods, however, these effects are generally of little consequence and castration is a particularly useful technique where mobs of mixed-sex animals are maintained for display. Castration may also be indicated in situations where it is desirable to reduce aggression, however, this is not usually successful particularly in hand-reared males. The effects on behaviour are potentially greater after tubal ligation and vasectomy than gonadectomy. This is because the former methods mean that females will be continually cycling (at least during the breeding season, or for the entire year in some species, such as red kangaroos), resulting in increased rates of aggression between males with increased potential for mating-related injuries in males and/or females. There are also potential negative side effects from increased exposure to endogenous progesterone and oestradiol. In most species, female macropods come into oestrus only once or twice a year. Therefore, the anoestrus state induced by gonadectomy is more akin to the normal status of females than are tubal ligation or vasectomy of males (C Herbert *pers. comm.*).

As surgical sterilisation is invasive, requires chemical restraint and is expensive it is a technique reserved primarily for captive or small populations of macropods and has little application to large wild populations. Targeting male fertility in polyoestrus species such as macropods is also generally an ineffective means to control fertility in free-ranging populations of animals (Adderton Herbert 2004).

Due to the anatomy of the male macropod reproductive tract, castration and vasectomy are relatively straightforward procedures (see 12.4 and 12.5). Ovariectomy or ovariohysterectomy require a laparotomy (see 12.3). These procedures are more complex and invasive and are rarely performed to control reproduction in zoo or free-ranging macropods.

The decision to permanently sterilise an individual must be made very carefully as the genetic value of an individual may change if a population experiences a catastrophic event and numerous individual are lost.

11.1.2 Reversible methods

These methods primarily involve the use of hormones and vaccines. They are generally more desirable as they have less effect on behaviour, are less invasive than surgical methods and are reversible. These methods are primarily applicable to females.

a *Oral contraceptives*

A wide range of oral contraceptives have been used in captive wild animals, but there are no reports of their use in macropods. Their efficacy and side effects have not been evaluated and compliance would be a problem.

b *Progestin injections*

Injections of synthetic progestins given at intervals consistent with the normal or expected female cyclic pattern have been used in many species. They result in deferral or suppression of oestrus. There are no published reports of their use in macropods.

c *Progestin implants*

Synthetic progestins inhibit follicular development and ovulation by reducing the concentration of FSH and LH. If ovulation does occur, changes in the cervical mucus make it impenetrable to sperm and changes in the endometrium interfere with implantation.

A number of synthetic progestins are available as implants. These include levonorgestrel (Jadelle [previously called Norplant II] Leiras Pharmaceuticals), and etonogestrel (Implanon, Organon). They are available in pre-sterilised needles and inserted using the supplied applicator. They are generally inserted SC between the shoulder blades in macropods and the hole sutured, or closed with tissue glue.

Tammar wallabies implanted with a single 70 mg levonorgestrel rod and eastern grey kangaroos implanted with two 70 mg levonorgestrel rods (Norplant II) were contracepted for 48 mo and 27 mo respectively (the duration of the studies). In humans, Norplant II provides up to 6 yr of contraception so it is highly likely that levonorgestrel implants may be effective for much longer in macropods. In tammar wallabies, follicular development continued during levonorgestrel treatment but ovulation was inhibited, suggesting that FSH concentrations were unaltered but the LH surge required for ovulation was inhibited (Hynes et al. 2007). Levonorgestrel did not affect reactivation or subsequent development of the diapausing blastocyst that had been conceived before treatment. It did not impair lactation, as young were reared to weaning in both species. The contraceptive effects of the implants were reversible, as removal of the implants from tammar wallabies was followed by a rapid return to fertility. Levonorgestrel implants do not appear to have adverse effects on behaviour or body condition, although in wild eastern grey kangaroos a small increase in body condition did occur (Nave et al. 2000; Nave et al. 2002a; Poiani et al. 2002; Nave et al. 2002b). Levonorgestrel implants (70 mg) have been used successfully in a range of other macropod species in zoos including red kangaroos, red-necked wallabies and swamp wallabies. Levonorgestrel implants therefore appear to be a safe, reliable and reversible form of long-term contraception for macropods. No specific studies have been conducted with etonogestrel implants in macropods. Insertion of the implants requires at least physical restraint and preferably chemical restraint in most macropods. Some training is required in correct insertion technique. The implants are expensive and their availability can be limited in some countries.

d Gonadotrophin releasing hormone (GnRH) agonists

GnRH is the principal hormone governing reproductive function in males and females and is secreted by the hypothalamus. It stimulates the synthesis and release of FSH and LH from the pituitary. These two hormones are essential for spermatogenesis in males, ovulation in females and production gonadal steroids in both sexes. The structure of GnRH is conserved across mammalian species. GnRH agonists are peptides that are similar to GnRH but are more potent. When administered continuously, the result is a temporary suppression of pituitary responsiveness to endogenous GnRH secretion (due to down regulation of GnRH receptors) and a subsequent decline in concentration of FSH and LH. This results in a decline in production of gonadal steroids with an inhibition of spermatogenesis in males and follicular development and ovulation in females (Adderton Herbert 2004).

A slow-release implant containing the GnRH agonist deslorelin (Suprelorin®, deslorelin 4.7 mg, Peptech Animal Health) has been used in many species of macropods and other animals (Herbert & Trigg 2005). Implants are small and usually inserted SC between the shoulder blades in macropods. The implants come in a pre-sterilised needle and inserted using the supplied applicator. The hole in the skin is best closed with tissue glue. Deslorelin has been shown to be an effective contraceptive in female macropods. In female tammar wallabies implanted with 5 mg deslorelin, contraception was achieved for approximately 1 yr. Tammar wallabies implanted with 10 mg deslorelin did not breed for 16 mo (Herbert et al. 2004a). In another study in tammar wallabies, contraception was achieved for 515 ± 87 d after implanting with 5 mg deslorelin (Herbert et al. 2005). In eastern grey kangaroos, contraception was achieved for 559 ± 111 d (minimum contraceptive duration of 7.5 mo) following implantation of one 10 mg deslorelin implant and 651 ± 21 d (minimum contraceptive duration of 19 mo) following implantation of two 10 mg implants 6 mo apart (Herbert et al. 2006). In a later study on eastern grey kangaroos implanted with 9.4 mg deslorelin (2 × 4.7 mg implants), contraceptive effect was 734 ± 65 days (n = 10) (C Herbert *pers. comm.*). The consistently longer duration of effect in the latter experiment is likely related to variations between implants used in either study (Woodward et al. 2006). The duration of effect has been shown to be dose-dependant in other species and there is considerable variation in the duration of effect between individuals, particularly when lower doses are used (Herbert et al. 2006). The duration of contraception is less predictable when lower doses are used. Contraceptive effect is estimated to occur within 1 wk of implanting and is reversible after removal of the implant or cessation of deslorelin release from the implant. In tammar wallabies, deslorelin treatment inhibits some aspect of blastocyst reactivation, gestation, parturition or initiation of lactation in the large majority of animals, as very few animals with blastocysts at the time of treatment subsequently give birth. However, deslorelin can safely be administered to a female nursing a PY with no effect on lactation or on the growth and development of the PY. It does inhibit follicular development and post partum oestrus (Herbert et al. 2004a).

No significant side effects have been reported with the use of deslorelin implants in female macropods. Changes in sexual behaviour and dominance hierarchies are likely and may have some effect in wild populations. Studies in other species have not shown any significant effect on behaviour (Adderton Herbert 2004). The behavioural effects of deslorelin implants have been studied in eastern grey kangaroos. There was no detectable effect on crepuscular activity. Body mass did not vary significantly between treated and untreated animals, suggesting that appetite and feeding strategies were not affected. Behavioural oestrus was induced approximately 3 d following combined removal of PY and deslorelin administration. Mating was observed during this oestrus, but conception did not occur and PY were not observed in any treated females (Woodward et al. 2006). Studies in other species have shown that the initial stimulatory gonadotrophin response to GnRH agonist treatment can induce oestrus. If mating and conception occurs, in most other species the pregnancy is carried to full term. The observation that treatment-induced matings in eastern grey kangaroos were unsuccessful means that deslorelin treatment can confidently be used in combination with removal of PY, without an ensuing successful pregnancy.

Deslorelin implants provide an effective, reversible and safe form of contraception for macropods. The stage of reproductive cycle, presence or absence of a blastocyst, presence or absence of a PY, removal of the PY at time of implanting, occurrence of a post partum oestrus, seasonality of the species and dose of deslorelin all influence the effectiveness and duration of treatment. Best results would be achieved if treatment was timed to coincide with the period of anoestrus (Herbert et al. 2005).

In males of a range of species (mice, sheep, pigs, humans, rhesus macaques, baboons) GnRH agonists reduce peripheral concentrations of LH and testosterone and cause the testes to regress, with the eventual induction of azoospermia. However, in other species (cattle, red deer, marmosets) males appear to be insensitive to the inhibitory effects of GnRH agonists. Testosterone and LH levels remain unchanged or increase. A study in tammar wallabies using various doses of deslorelin implants demonstrated that after acute increases in LH and testosterone concentrations, levels returned to pre-treatment levels by day 7. There was no effect on testicular size. This suggested that male tammar wallabies are resistant to the contraceptive effects of GnRH agonist treatment (Herbert et al. 2004b). A study in eastern grey kangaroos demonstrated no effect on fertility in males implanted with deslorelin, with females being impregnated at the same rate as prior to treatment (C Herbert *pers. comm.*). Deslorelin was used in male red kangaroos in a walk-through exhibit in an attempt to reduce aggressive behaviour. There was no apparent effect (K Bodley *pers. comm.*). Deslorelin is therefore likely to be an ineffective contraceptive or behavioural modifier in male macropods.

e *Immunocontraceptives*

Both passive and active immunisation against GnRH results in neutralisation of GnRH in the hypophyseal portal blood by antibodies. The antibodies bind to the GnRH, preventing it from binding with receptors in the pituitary. This results in a decline in secretion of FSH and LH and subsequent suppression of follicular development and ovulation and spermatogenesis and gonadal steroids (Short et al. 1995). Various GnRH vaccines have been developed, primarily for use in production animals. Long-term fertility control can be achieved but usually requires multiple vaccinations. Immunisation against GnRH does not have the initial stimulatory acute phase that GnRH agonists have. There is a highly variable response between individuals, with some animals not responding at all. This may result in selective pressure for non-responders or immunocompromised animals in wild populations. In a study using a GnRH vaccine in female eastern grey kangaroos given two doses 29 d apart, anti-GnRH titres peaked at 10–12 d post-vaccination and persisted for 35–40 d. All animals in the study conceived (Booth et al. 1997). Reproductive function may be maintained despite an immunological response to the vaccine (Herbert & Trigg 2005).

A study on the effect of GnRH vaccine on sexual behaviour of male western grey kangaroos concluded that immunisation against GnRH resulted in modification of male sexual behaviour and a decrease in display of aggression (Stelmasiak 1993).

The zona pellucida is the first site of attachment for sperm to the oocyte. Many regions of the zona pellucida genes are highly conserved across species. Proteins extracted from porcine zona pellucidae (PZP) have been used as antigens in vaccines to immunise various species. Many species immunised with PZP have shown reduced fertility. Tammar wallabies immunised with PZP were found to achieve significant concentrations of antibody to PZP. PZP-immunised wallabies did not produce offspring after mating or after superovulation and artificial insemination (Kitchener et al. 2002). The potential for PZP vaccines to be a viable option for contraception in captive and free-ranging macropods requires further investigation.

f *Management techniques*

This primarily involves separating sexes. This is useful in macropods, but it requires adequate facilities for animals to be housed separately.

Removal and euthanasia of early PY may also be used, particularly if there is a desire to maintain more animals of either sex. However, it is not suitable as a form of long-term population management as removal of the PY will result in recrudescence of a blastocyst in diapause, subsequent birth and post partum oestrus in most species.

11.2 Enhancing reproduction

Many macropod species are endangered and reproduction in captive and free-ranging populations may be at a very low rate. In the wild this primarily relates to the presence of various threatening processes, resulting in a loss of individuals at a rate greater than replacement. In captivity, lack of breeding may relate to environment, nutrition, health, incompatible individuals, social structure of a group, or age. In most cases, if these are identified and corrected animals will breed. However, if breeding still does not occur, methods to enhance reproduction, such as the application of assisted breeding technology, may be required to circumvent breeding problems, facilitate better genetic management of small populations or increase reproductive rate. Assisted reproduction techniques can also be used in conservation programs for demographic and genetic management of captive and wild populations. They can also be used to salvage gametes from sick, aged or dead animals.

11.2.1 Semen collection

Semen collection in macropods is invariably performed using electroejaculation with the animal under anaesthesia. Electroejaculation in macropods typically results in high volumes of seminal and prostatic fluid. Macropod ejaculates may coagulate soon after collection. Semen collection by electroejaculation in various macropod species has been described. Sperm can also be collected from the caudal epididymis by aspiration or dissection (Molinia et al. 1998; Holt et al. 1999; Paris et al. 2005).

11.2.2 Semen preservation

Semen has been collected directly into pre-warmed culture medium. Coagulation may occur to variable degrees from no coagulation to a solid plug. Samples in one study were maintained at 35°C and left for 10 min to allow sperm to migrate from the plug into the culture medium. The sperm-enriched medium was used for insemination either directly or after further processing (Paris et al. 2005).

Marsupial spermatozoa require the presence of high concentrations of cryoprotectant for successful cryopreservation and to achieve even moderate post-thaw motility. However, exposure to high cryoprotectant concentrations appears to be detrimental to sperm survival even before cooling commences. Successful cryopreservation and thawing to acceptable levels of motility of macropod semen has been difficult to achieve (Holt et al. 1999).

11.2.3 Superovulation and oestrus synchronisation

Superovulation has been achieved in tammar wallabies using various protocols with PMSG, FSH and LH with or without GnRH (Molinia et al. 1998a; Molinia et al. 1998b; Magarey et al. 2003). FSH/LH protocols result in higher ovulation success than PMSG/GnRH protocols. In one study (Magarey et al. 2003) the overall oocyte collection rate was 80%. After repeated dosing with porcine LH and FSH, detectable concentrations of antibodies were produced against the gonadotrophins. These anti-gonadotrophin antibodies did not, however, cause a significant decrease in the number of follicles or affect oocyte maturity after subsequent superovulations. All animals in this study returned to normal fertility after the study. The use of exogenous hormone stimulation in tammar wallabies resulted in high rates of fertilisation but no births. Embryos produced by AI after superovulation have a high rate of degeneration. Superovulation may affect embryo viability in monovular marsupials, possibly by altering the structural integrity of the tertiary egg coats (Molinia et al. 1998a).

Tammar wallabies predictably have a post partum oestrus within 1–6 h of birth. Ovulation usually occurs 43–60 h post partum. Removal of PY and birth can therefore be used to synchronise and time oestrus and ovulation independent of exogenous gonadotrophins (Paris et al. 2005).

11.2.4 Artificial insemination

Successful AI relies on reliable collection and preservation of semen, delivery of sperm to the appropriate site in the female and reliable detection of oestrus and hence timing of the insemination (Tribe et al. 1994). AI has only been attempted in two species of macropods: the tammar wallaby (Molinia et al. 1998; Paris et al. 2005) and the red kangaroo (Johnston et al. 1994).

As ovulation in the tammar wallaby occurs 42–60 hr post partum, birth can be used to time insemination into different sites in the female reproductive tract. Insemination is conducted with the female under anaesthesia. Four methods of insemination have been described: insemination into the urogenital sinus, insemination into the anterior vaginal *cul de sac*, insemination through the cervix (the cervices are exposed by an incision into the anterior vaginal *cul de sac* via laparotomy) and intrauterine insemination (either via laparotomy or laparoscopy) (Johnston et al. 1994; Paris et al. 2005). Successful insemination has only been achieved when semen was deposited directly into the uterus, although only two out of 22 animals inseminated by this technique gave birth. Conception occurred with semen collected by electroejaculation as well as from the epididymis (Paris et al. 2005). In the tammar wallaby, spermatozoa should be delivered by intrauterine insemination 4–6 h before ovulation to maximise the number of sperm in the oviduct (site of fertilisation) and allow sufficient time for capacitation to occur. Timing of AI in marsupials is critical as fertilisation must occur before the mucin and shell coats are laid down on the oocyte, and the passage of the oocyte through the oviduct is very rapid (Paris et al. 2005).

During natural mating, the female typically mates with several males and the total semen volume may exceed 80 mL. The coagulated ejaculate may act as a reservoir from which sperm migrate to replenish sperm in the upper tract. This is not the case with AI, and timing of AI needs to be closer to ovulation than with natural mating. Sperm deposited by AI are also lost by retrograde excretion and phagocytosis. Frequent handling, anaesthesia and surgery may all interfere with ovulation and egg and sperm transport in tammar wallabies, as seen in other species (Paris et al. 2005). The use of AI for assisted breeding in macropods needs to be developed further before it becomes a useful tool in conservation programs.

11.2.5 In vitro *fertilisation* (*IVF*)

There are very few reports of *in vitro* fertilisation in marsupials. IVF using intracytoplasmic sperm injection (ICSI) has been performed in tammar wallabies. Sperm collected by electroejaculation were injected into oocytes collected from superovulated females. Oocyte activation, sperm head decondensation and pronuclear formation occurred, but syngamy did not occur. While tammar wallaby oocytes from superovulated females can support most of the events of fertilisation, they may not be competent to complete fertilisation (Magarey & Mate 2003a; Magarey & Mate 2003b). In another study, a single epididymal sperm was injected into the cytoplasm of mature oocytes collected from Graafian follicles or from the oviduct within hours of ovulation in tammar wallabies undergoing natural cycles. Both follicular and tubal oocytes underwent cleavage. One-third of injected oocytes completed the second cleavage division, but only a single embryo reached the 8-cell stage. The presence of the mucoid coat in tubal oocytes did not prevent fertilisation by ICSI and the oocytes cleaved to a similar stage as follicular oocytes lacking a mucoid coat. The mucoid coat may be involved in maintaining the integrity of the conceptus during cleavage (Richings et al. 2004).

11.2.6 Cross-fostering

Cross-fostering is the rearing of a young by a surrogate mother of a different species. The benefit of this procedure in marsupials is that it may dramatically increase reproductive rate. Removal of a PY in most species interrupts lactational anoestrus or lactational quiescence with return to oestrus or reactivation of a blastocyst in diapause and subsequent birth and post partum oestrus. This permits the female to undergo a series of pregnancies with a minimum of intervening lactations. Despite numerous successful cross-fostering studies in macropods, only one has attempted to use the technique to increase reproductive rate in an endangered marsupial species, the brush-tailed rock-wallaby (Taggart et al. 2005). In that study, brush-tailed rock-wallabies were fostered to tammar wallabies and yellow-footed rock-wallabies.

Cross-fostering is best performed when the PY is in an immature stage of development (10–30 d for brush-tailed rock-wallabies). The procedure should be performed with both donor mother and surrogate mother under anaesthesia. The PY should be carefully removed from the teat using gentle pressure so as not to damage the teat or the PY's mouth. Reattachment of the PY to the surrogate mother's teat is best achieved using fine blunt-nosed forceps. The surrogate mother's pouch should be examined at least 20 min after the procedure to ensure that the PY is still attached to the teat (Taggart et al. 2005).

A number of criteria have been identified as important considerations for successful cross-fostering and whether fostered PY survive until weaning.

- Surrogate mothers should be quiet and tractable and of a similar size to the natural mother.
- The size (weight and head length) of the PY (regardless of age) to be fostered is similar to, or slightly smaller than that of the PY it will replace in the surrogate mother's pouch. This ensures the likelihood that sufficient milk of similar composition will be available for the PY being transferred.

- The length of pouch life in both the natural and surrogate species is similar.

Cross-fostering studies have shown that the growth rate of the transferred young reflects the growth rate of the young of the surrogate species. The time of permanent pouch exit is therefore that of the foster mother species rather than that of the natural mother. There is a close correlation between milk production, growth of the PY and maternal mass. PY transferred to surrogate mothers with a greater mass than the natural mother gain weight more quickly than those transferred to surrogate mothers of similar or smaller mass than the natural mother. Interestingly, there is little difference in head growth between young reared by their natural mothers and those reared by surrogate mothers. This probably reflects the importance of maintaining head, brain and sensory development despite nutritional fluctuations, compared with growth of other body parameters. Observations of young from the time of pouch exit suggest that behavioural traits of the fostered young are not obviously different from those of natural mother-reared animals. They readily mix with their own species once removed from the surrogate mother at weaning. Breeding behaviour and success does not appear to be affected by cross-fostering (Taggart et al. 2005).

Cross-fostering can dramatically increase the reproductive rate of threatened or endangered macropod species in captivity if adequate numbers of surrogate mothers are available. This technique has further benefits in that it should be feasible to cross-foster PY from wild animals to captive foster mothers. This would facilitate recruitment of new genetic stock from the wild for captive populations without removing adults from the wild, and would accelerate breeding in the wild population (Taggart et al. 2002).

12 SURGERY

Surgical techniques and procedures developed for use in domestic species are readily applicable to macropods. Skin, soft tissue and bone healing in macropods appear to be no different from healing in other marsupials and eutherian mammals. All the same principles of wound management are applicable. However, macropods are often highly strung and anxious animals, and careful consideration must be given to the post-operative requirements before embarking on any surgical procedure. Confinement, administration of medications and regular wound care may be difficult. Stress and activity will interfere with and impair healing. The judicious use of sedatives and neuroleptic drugs (6.2.1) should be considered as part of the post-operative management for any macropod. Although there have been no specific studies on pain management or pharmacokinetic studies conducted on analgesic drugs in macropods, it is prudent to consider the use of these drugs as part of intraoperative and post-operative management of any surgical case. The use of long-acting local anaesthesia (bupivicane) injected or splashed onto an incision site at the time of surgery (as well as the use of intraoperative systemic analgesia) may negate the need for ongoing post-operative pain management. Consideration should also be given to the use of topical antibiotic and fly-repellent spray on surgical wounds, systemic prophylactic antibiotics and tetanus prophylaxis. The use of absorbable suture material for wound closure should be considered where possible to avoid the need to recapture the animal for suture removal.

The most common indications for surgery in macropods are the repair of traumatic injuries (soft tissue and bone), repair of tissues damaged by infectious processes (particularly lumpy jaw [Remedios et al. 1996]), gastrointestinal accidents, exploratory laparotomy and reproductive procedures. Intubation and maintenance on gaseous anaesthesia is indicated for all major surgical procedures. This prevents the risk of regurgitation and aspiration, offers greater control of ventilation and facilitates treatment of emergencies during anaesthesia.

In this section various surgical procedures are described, but only specific considerations relevant to macropods are presented. The assumption is that the reader has a good knowledge of veterinary surgical techniques and procedures.

12.1 Laparotomy

Laparotomy in macropods is relatively straightforward and similar to that in most other mammalian species. Consideration must be given to the voluminous gastrointestinal tract, the presence of a pouch (with or without PY) in females and the epipubic bones. A ventral midline approach is used. The animal is placed in dorsal recumbency. If access to the caudal abdominal area is required, tilting the body with the head down to an angle of about 30° moves the gastrointestinal tract cranially in the abdomen and allows easier access to caudal abdominal organs. In the male and in the female where access to the cranial abdomen is required, the approach is straightforward. If access to the caudal abdomen is required in the female, consideration must be given to the pouch. The approach will depend on the size of the pouch and the presence/absence of a PY. If the pouch is empty or the PY can be temporarily removed, a midline incision can be made inside the pouch, midway between the teats and the cranial border of the pouch. A useful technique is to use stay sutures to hold the pouch open. A ligature using heavy-gauge suture material is placed on each side of the caudal lip of the pouch opening and retracted caudally, and secured either to the surgery table or via another bite through the skin on the proximal thigh. The pouch lip and wall should be pulled back as far as the teats (T Bishop *pers. comm.*). If a PY is present and cannot be removed, an incision can be made lateral to the pouch opening at the cranial end of the epipubic bone (Magarey et al. 2003). The skin incision is then reflected toward the midline to expose the linea alba.

12.2 Laparoscopy

Laparoscopy is generally performed to visualise the reproductive tract in females and for AI. The animal is positioned as for a laparotomy. An area on either side of the pouch is surgically prepared. A stab incision is made immediately adjacent to the lateral edges of the pouch and just caudal to the level of the pouch opening (just cranial to the cranial tip of the epipubic bones). A trochar and cannula is inserted into the abdomen on one side and the abdomen insufflated with CO_2 at a pressure of about 10 mm Hg. The trochar is then removed and a rigid laparoscope is introduced through the cannula. If a second instrument is required, a second trochar and cannula is inserted through a stab incision on the opposite side. Excellent exposure of the caudal abdominal contents is achieved. The rectum, bladder, lateral vaginae, uterine horns, oviducts and ovaries are usually clearly visible (Johnston et al. 1994).

12.3 Ovariectomy and ovariohysterectomy

The animal is prepared as for a laparotomy and tilted at an angle of about 30° with the head down. An incision is made midway between the tip of the epipubic bones and the pubis. The bladder may need to be expressed. The reproductive tract is visualised and one side exteriorised using a spay hook. For ovariectomy, one uterus is located and the corresponding ovary is exteriorised after replacing the uterus and any portion of the vaginae back into the abdomen. The fimbria and ampulla of the oviduct is teased off the ovary to enable clear visualisation of the ovarian vessels. The vessels are then clamped and ligated and the ovary removed. The procedure is repeated for the opposite side (GM Magarey *pers. comm.*). Complications such as pyometra post-ovariectomy have not been reported in macropods.

Ovariohysterectomy is technically more difficult and risky due to access and the unique anatomy of the reproductive tract (Fig. 7.3). The only indication for ovariohysterectomy in a macropod may be for therapeutic reasons. The approach is as for ovariectomy. A hole is made in the broad ligament just lateral but close to the ovaries to avoid catching the ureters (which should be visualised). The ovaries are exteriorised with gentle traction on the suspensory ligament. The vessels are clamped and ligated. Once both ovaries are ligated and detached the broad ligament is torn, staying as close as possible to the lateral margins of the blood vessels to avoid damaging the ureters. The uteri are ligated at the immediate proximal end of the cervices (Fig. 7.3). They are first ligated separately then a single ligature can be placed around both uteri. Transfixation ligatures may be required. A triple-clamp technique should be used. Once the uteri are removed, a continuos suture to close the stump may be necessary to prevent leakage into the abdomen. Large blood vessels may need to be ligated separately (T Bishop *pers. comm.*).

12.4 Castration

The animal is placed in dorsal recumbency. The scrotum and surrounding area is surgically prepared. The testicles are removed through separate incisions in each side of the scrotum at the most ventral point. The scrotal septum is very prominent and it is not possible to remove both testicles through a single incision. The testicle is held firmly in the scrotal sac and an incision made over the centre of the testicle in a cranio-caudal direction. The incision is extended through the *tunica vaginalis* into the testicle (open castration technique). The testicle and epididymis is then exteriorised. The gubernaculum is bluntly severed to allow full exteriorisation of the testicle and epididymis. The spermatic cord and cremaster muscle are then routinely clamped, ligated and cut. The scrotal incision is left open to allow drainage. The procedure is repeated on the opposite side.

A technique of scrotal ablation may also be used. For this procedure the castration is performed as described above, however, the spermatic cord is severed as high as possible. The scrotal pedicle is then cut as close as possible to the abdominal wall. The ligated spermatic cords are then pushed through the incision toward the inguinal canal. If they are too long they may need to be ligated and cut more proximally. The wound is then sutured closed.

12.5 Vasectomy

The animal is placed in dorsal recumbency. The scrotum and surrounding area is surgically prepared. A 1–1.5 cm incision is made in the skin and *tunica vaginalis* over the spermatic cord on the cranial side of the scrotal pedicle at a point about midway between the testicle and abdominal wall. A small portion of the spermatic cord is exteriorised. The vas deferens is the firmest tubular structure palpable in the spermatic cord. The vas deferens is bluntly dissected out and exposed, avoiding damaging the blood vessels and nerves. The vas deferens is a firm white tubular structure. A 1 cm piece of vas deferens is removed after ligating proximally and distally to the piece being removed. Vas deferens clips may also be used. The wound is then sutured. The procedure is repeated on the opposite side, as it is not possible to perform the procedure through a single incision due to the prominent scrotal septum. The segment of vas deferens should be examined under a microscope to confirm the presence of sperm or examined histologically to confirm that the correct tissue was removed.

12.6 Orthopaedics

Fractures are common in macropods and are usually the result of motor vehicle accidents, gunshot wounds, and accidents during capture and restraint procedures including darting accidents. All standard orthopaedic principles and techniques can be readily applied to fracture repair in macropods. However, the unique anatomy of the hind limbs and the massive mechanical stresses that are placed on the bones of the hind limbs and pelvic girdle during locomotion make adequate stabilisation of fractures of these bones very difficult. The prognosis for successful repair of fractures of the pelvic girdle and hind limbs of adult macropods is generally considered poor. It is difficult to confine and keep even the tamest macropod immobile for any reasonable length of time post-operatively. In small species and perhaps the tamest of larger species of macropods, fractures of the metatarsi may be repairable with internal or external fixation. The prognosis for fracture repair in hand-reared PY that can be confined to a pouch is much better. If fractures can be adequately stabilised by external or internal fixation and the PY can be adequately confined, bone healing is generally good. Fractures of the fore limb are less common. The prognosis for repair of these bones in adults is better, and good in PY.

12.7 Miscellaneous procedures

12.7.1 Tail amputation

Despite the macropod's obvious reliance on the tail for its pentapedal gait, tail amputation has successfully been performed in macropods. A high tail amputation was successfully performed in a male tammar wallaby (R Johnson *pers. comm.*) using standard techniques (Fig. 7.18). This captive animal adapted well to having no tail. In one case report, an adult female red kangaroo presented with necrosis, devitalisation and osteomyelitis of the distal half of the tail (Suedmeyer 1998) and 85% of the tail was amputated. A standard technique for tail amputation was used. The skin was incised in a double dorso-ventral V with the dorsal aspect extending beyond the ventral aspect. The median caudal artery and the paired lateral caudal arteries and veins were identified and ligated. Blunt dissection of the sacrocaudal musculature continued to the coccygeal vertebrae. The tail was removed after separation of vertebrae. Fascia muscle

Figure 7.18 Tammar wallaby with an amputated tail. Photo: Cathy Herbert.

sheaths were sutured together in a linear pattern with simple interrupted sutures. The skin was closed with 1/0 nylon sutures. The remaining 8–10 cm stump was bandaged and changed under chemical restraint every 6–9 d. Sutures were removed 5 wk after initial surgery. Adaptation to loss of the tail was rapid. The tail stub did not touch the ground and the animal was able to stand erect. When walking, the animal placed more weight on the tarsal joints. The muscles of the hind limb hypertrophied. Standing up from lateral recumbency was occasionally a problem.

There have been reports of wild macropods surviving with varying proportions of their tails missing (C Herbert *pers. comm.*). Although animals appear to adapt well to tail amputation, it is generally not recommended that macropods with amputated tails be released to the wild.

12.7.2 Hemimandibulectomy

An adult female Bennett's wallaby was presented with severe osteomyelitis of the left mandible that was refractory to medical treatment. A standard hemimandibulectomy was performed (Suedmeyer 1999). The entire hemimandible was removed. An *Actinomyces* spp. was cultured from the bone. The post-operative appearance was cosmetically pleasing. An moderate median shift of the right mandible occurred but did not interfere with mastication. Within 3 d of surgery the animal's appetite returned to normal using her front feet to assist in prehension of food. Twenty eight days after surgery the surgical site became infected, which progressed despite treatment. The animal was euthanased.

12.7.3 Cataract surgery

Cataract surgery in macropods is technically easy. Excellent exposure can be obtained without the need for neuromuscular relaxants. The pupils dilate well with 1% tropicamide. Lenses can be removed with short phacoemulsification times. However, in PY with galactose-induced cataracts, surgical removal of the lens often reveals extensive opacification of the vitreous, making the prognosis for return to normal vision very poor. Many cases also develop glaucoma after surgery. This is probably secondary to chronic uveitis. The prognosis even with treatment is poor and many cases require globe evisceration or enucleation or euthanasia. Because of these complications, it is difficult to justify cataract surgery in macropods (Stanley 2002).

ACKNOWLEDGMENTS

The authors thank Cathy Herbert, Karina Graham and Vanessa DiGiglio for their assistance with preparation of this manuscript.

REFERENCES

Abbott KA (2000). *Prevalence of Johne's Disease in Rabbits and Kangaroos.* Final Report Project TR.050. University of Sydney, Sydney.

Acland HM (1981). Parma wallaby herpesvirus infection. *Journal of Wildlife Diseases* **17**, 171–177.

Adderton Herbert C (2004). Long-acting contraceptives: a new toll to manage overabundant kangaroo populations in nature reserves and urban areas. *Australian Mammalogy* **26**, 67–74.

Arundel JH (1979). Hydatid disease of animals in Australia. *Australian Veterinary Journal* **55**, 126–130.

Arundel JH, Baker IK & Beveridge I (1977). Diseases of marsupials. In *The Biology of Marsupials* (eds B Stonehouse & D Gilmore), pp 141–154. Macmillan: London.

Arundel JH, Beveridge I & Presidente PJ (1979). Parasites and pathological findings in enclosed and free-ranging populations of *Macropus rufus* (Dermarest) (Marsupialia) at Menindee, New South Wales. *Australian Wildlife Research* **6**, 361–379.

Arundel JH, Dempster KJ, Harrigan KE & Black R (1990). Epidemiological observations on the helminth parasites of *Macropus giganteus* Shaw in Victoria. *Wildlife Research* **17**, 39–51.

ARWH (2007). *Australian Registry of Wildlife Health.* Zoological Parks Board of NSW: Sydney.

Baker RT & Beveridge I (2001). Imidacloprid treatment of marsupials for fleas (*Pygiopsylla hoplia*). *Journal of Zoo Wildlife Medicine* **32**, 391–392.

Barker IK & Harrigan KE (1972). Coccidiosis in wild grey kangaroos. *International Journal for Parasitology* **2**, 187–192.

Barker S, Calaby JH & Sharman GB (1963). Diseases of Australian laboratory marsupials. *Veterinary Bulletin* **33**, 539–544.

Barker IK, O'Callaghan MG & Beveridge I (1989). Host–parasite associations of *Eimeria* spp. (Apicomplexa: *Eimeriidae*) in kangaroos and wallabies of the genus *Macropus* (Marsupialia: *Macropodidae*). *International Journal for Parasitology* **19**, 241–263.

Barnes TS, McManus DP & Coleman GT (2005). Pathology and serodiagnosis of hydatid disease in macropods. In *Proceedings of the Wildlife Disease Association International Conference*, pp.107–108. Cairns, Queensland.

Barnes TS, Morton JM & Coleman GT (2007a). Clustering of hydatid infection in macropodids. *International Journal for Parasitology* **37**, 943–952.

Barnes TS, Hinds LA, Jenkins DJ & Coleman GT (2007b). Precocious development of hydatis cysts in a macropod host. *International Journal for Parasitology* **37**, 1379–1389.

Bennet RA, Browning WN & Dunker FK (1993). Dilated cardiomyopathy in an antilopine kangaroo (*Macropus antilopinus*). In *Proceedings of the American Association of Zoo Veterinarians.* St Louis, Missouri.

Beveridge I (1978). Helminth parasites of Australian marsupials. In *Fauna, Proceedings 36*, pp. 273–285. Post Graduate Committee in Veterinary Science: Sydney.

Beveridge I (1985). Marsupial parasitic diseases. In *Zoo and Wild Animal Medicine: Current Therapy 3* (ed. ME Fowler), pp. 288–293. WB Saunders: Philadelphia.

Beveridge I, Presidente PJA & Speare R (1985). Parasites and associated pathology of the swamp wallaby, *Wallabia bicolor* (Marsupialia). *Journal of Wildlife Diseases* **21**, 377–385.

Bhattacharya S, Banerjee R, Ghosh R, Biswas A & Chatterjee A (2003). Identification of foot-and-mouth disease from a captive kangaroo in a zoological garden in India. *Veterinary Record* **153**, 504–505.

Bird PS, Huynh SC, Davis D, Love DN, Blackall LL & Seymour GL (2002). Oral disease in animals: the Australian perspective. Isolation and characterisation of black-pigmented bacteria from the oral cavity of marsupials. *Anaerobe* **8**, 79–87.

Blanden DR, Lewis PR & Ferrier GR (1987). Vaccination against lumpy jaw and measurements of antibody response in wallabies (*Macropus eugenii*). *Veterinary Record* **121**, 60–62.

Blyde D (1994). Advances in anaesthesia and sedation of native fauna. In *Wildlife, Proceedings 233*, pp. 243–245. Post Graduate Committee in Veterinary Science: Sydney.

Blyde D (1999). Advances in treating diseases of macropods. In *Wildlife in Australia, Proceedings 327*, pp. 439–452. Post Graduate Foundation in Veterinary Science: Sydney.

Blyde D (2005). Encephalomyocarditis virus in captive tree kangaroos. In *Proceedings of the Australian Veterinary Association Annual Conference*, Gold Coast, Queensland.

Bodley K (2002). Dental disease in macropod species at Melbourne Zoo. In *Proceedings of the Wildlife Disease Association Australasian Section Annual Conference*, pp. 36–41, Mary River, Northern Territory.

Boever WJ, Stuppy D & Kane KK (1977). Clinical experience with Tilazol™ (C1744). for chemical restraint and anesthesia in the red kangaroo (*Macropus rufus*). *Journal of Zoo Animal Medicine* **8**, 14–17.

Boorman GA, Kollias GV & Taylor RF (1977). An outbreak of toxoplasmosis in wallaroos (*Macropus robustus*) at a Californian zoo. *Journal of Wildlife Diseases* **13**, 64–68.

Booth R (1994). Manual and chemical restraint of macropods. In *Wildlife, Proceedings 233*, pp. 443–447. Post Graduate Committee in Veterinary Science: Sydney.

Booth R (2002). Macropods. In *Hand-rearing Wild and Domestic Mammals* (ed. LJ Gage),. pp. 63–74. Iowa State Press: Iowa.

Booth R, McKee J, Pye G, Caton W, Dowsett K & Trigg T (1997). Potential immunocontraception using GnRH vaccine in eastern grey kangaroos *Macropus giganteus:* preliminary report. In *Challenging the Boundaries: Proceedings of the Australian Association of Veterinary Conservation Biologists Annual Conference*, pp 443–447. AAVCB: Brisbane.

Britt JO, Frost DF & Cockrill JM (1994). Fatal herpesviral hepatitis in a red kangaroo (*Macropus rufus*). *Journal of Zoo and Wildlife Medicine* **25**, 580–584.

Bryant BR & Rose KA (2003). Calcium oxalate urolithiasis in four captive kangaroos. In *Proceedings of the American Association of Zoo Veterinarians Annual Conference*, pp. 96–101, Minneapolis, Minnesota.

Buddle BM & Young LJ (2000). Immunobiology of mycobacterial infections in marsupials. *Developmental and Comparative Immunology* **24**, 517–529.

Burns DL, Wallace RS & Teare JA (1994). Successful treatment of mycobacterial osteomyelitis in a Matschie's tree kangaroo (*Dendrolagus matschiei*). *Journal of Zoo and Wildlife Medicine* **25**, 274–280.

Burton JD (1981). Studies into aspects of lumpy jaw in macropods. PhD thesis, University of Melbourne, Melbourne.

Bush M & Montali RJ (1999). Medical management of tree kangaroos. In *Zoo and Wild Animal Medicine: Current Therapy 4* (eds ME Fowler & RE Miller), pp 337–343. WB Saunders: Philadelphia.

Bush M, Montali RJ, Murray S, Holland S & Phillips T (1995). The diagnosis, treatment and prevention of tuberculosis in captive Matschie's tree kangaroos (*Dendrolagus matschiei*). In *Proceedings of the American Association of Zoo Veterinarians Annual Conference*, pp. 312–314. East Lansing, Michigan.

Butler R (1981). Epidemiology and management of 'lumpy jaw' in macropods. In *Proceedings of the 4th International Conference of the Wildlife Diseases Association* (ed. ME Fowler), pp. 58–60, Sydney, New South Wales.

Callinan RB & Kefford B (1981). Mortalities associated with herpesvirus infection in captive macropods. *Journal of Wildlife Diseases* **17**, 311–317.

Canfield PJ & Hartley WJ (1992). A survey and review of hepatobiliary lesions in Australian macropods. *Journal of Comparative Pathology* **107**, 147–167.

Canfield P, Bellamy T, Hartley WJ & Spielman DS (1990a). Vasoformative proliferations in a rufous bettong. *Australian Veterinary Journal* **67**, 334–335.

Canfield PJ, Hartley WJ & Reddacliff GL (1990b). Spontaneous proliferations in Australian marsupials: a survey and review. 1. Macropods, koalas, wombats, possums and gliders. *Journal of Comparative Pathology* **103**, 135–146.

Canfield PJ, Hartley WJ & Dubey JP (1990c). Lesions of toxoplasmosis in Australian marsupials. *Journal of Comparative Pathology* **103**, 159–167.

Chineme CN, Njoku CO & Evbuoma N (1978). Degenerative cardiomyopathy in a captive kangaroo (*Macropus giganteus*). *Journal of Wildlife Diseases* **14**, 52–53.

Clark P (2004). *Haematology of Australian Mammals.* CSIRO Publishing: Melbourne

Clark WT, Pass D, Biddle J & Salmon J (1982). Urinary calculi composed of magnesium hydrogen phosphate in a kangaroo. *Australian Veterinary Journal* **59**, 62–63.

Clarke DE (2003). Oral biology and disorders of chiroptera, insectivores, monotremes and marsupials. *Veterinary Clinics Exotic Animal Practice* **6**, 523–564.

Clarke E (2006). Fluorosis as a probable cause of lameness in free-ranging eastern grey kangaroos (*Macropus giganteus*). *Journal of Zoo and Wildlife Medicine* **37**, 477–486.

Cleland PC, Lehmann DR, Wright MP, Woonton TR & Phillips PH (2004). *A prevalence survey of M A paratuberculosis in Kangaroo Island macropods.* OJD Research & Development Update. Meat & Livestock Australia: Kingscote, South Australia.

Coman BJ (1978). Hydatid disease in free-living Australian mammals. In *Fauna, Proceedings 36*, pp. 145–148. Post Graduate Committee in Veterinary Science: Sydney.

Cook RW, Fraser GC & Hartley WJ (1997). Haematozoan infection in young eastern grey kangaroos. In *Proceedings of the Australian Society for Veterinary Pathology Annual Conference*, p. 42, Brisbane, Queensland.

Cunningham M (1994). Applications of clinical pathology in native wildlife. In *Wildlife, Proceedings 233,* pp. 253–295. Post Graduate Committee in Veterinary Science: Sydney.

Dawson TJ (1989). Diets of macropodoid marsupials: general patterns and environmental influences. In *Fauna of Australia,* vol. 1B Mammalia (eds DW Walton & BJ Richardson), pp. 130–142. AGPS: Canberra.

Dawson TJ (1995). *Kangaroos: The Biology of the Largest Marsupials.* UNSW Press: Sydney.

Dawson TJ, Finch E, Freedman L, Hume ID, Renfree MB & Temple-Smith PD (1989). Morphology and physiology of the metatheria. In *Fauna of Australia*, vol. 1B Mammalia (eds DW Walton & BJ Richardson), pp. 451–504. AGPS: Canberra.

Dickson J, Hopkinson WI & Coackley W (1980). Herpesvirus hepatitis in rat kangaroos. *Australian Veterinary Journal* **56**, 463.

Dixon RJ & Baird JD (1984). Systemic candidiasis in a grey kangaroo. *Australian Veterinary Journal* **61**, 308.

Dubey JP & Odening K (2001). Toxoplasmosis and related infections. In *Parasitic Diseases of Wild Mammals* (eds WM Samuel, MJ Pybus & AA Kocan), pp.478–519. Iowa State University Press: Ames, Iowa.

Dubey JP, Ott Joslin J, Torgerson RW, Topper MJ & Sundberg JP (1988). Toxoplasmosis in black-faced kangaroos (*Macropus fuliginosus melanops*). *Veterinary Parasitology* **30**, 97–105.

Durham PJK, Finnie JW, Lawrence DA & Alexander P (1996). Blindness in South Australian kangaroos. *Australian Veterinary Journal* **73**, 111–112.

Ebedes H (1993). The use of long-acting tranquillisers in captive wild animals. In *The Capture and Care Manual* (ed. AA McKenzie), pp. 71–99. Wildlife Decision Support Services/South African Veterinary Foundation.

Effron M, Griner L & Benirschke K (1977). Nature and rate of neoplasia found in captive wild mammals, birds and reptiles at necropsy. *Journal of the National Cancer Institute*, **59**, 185–194.

Egerton JR (1963). Melioidosis in a tree climbing kangaroo. *Australian Veterinary Journal* **39**, 243–244.

Ellis TM, Copland MD & Gaynor WT (1983). Oxalate toxicity in a scaly-tailed possum, a Patagonian cavy and a swamp wallaby. *Journal of Wildlife Diseases* **19**, 290–293.

English AW (1978). Enterotoxaemia in a juvenile eastern grey kangaroo (*Macropus giganteus*). *Veterinary Record* **102**, 243.

Exhibited Animals Protection Act (1995). *Standards for Exhibiting Captive Macropods (Kangaroos, Wallabies and Allies) in New South Wales.* New South Wales Department of Primary Industries. <www.agric.nsw.gov.au>.

Fagan DA, Oosterhuis JE & Benirschke K (2005). 'Lumpy jaw' in exotic hoof stock: a histopathological interpretation with a treatment proposal. *Journal of Zoo and Wildlife Medicine* **36**, 36–43.

Finnie EP (1986). Viral diseases. In *Zoo and Wild Animal Medicine* (2nd edn) (ed. ME Fowler), pp. 576–577. WB Saunders: Philadelphia.

Finnie EP & Littlejohns IR (1976). Mortalities in parma wallabies (*Macropus parma*) associated with probable herpesvirus. *Australian Veterinary Journal* **52**, 294.

Fowler A & Mitchell T (2002). Skin lesions in a free-ranging population of long-footed potoroos (*Potorous longipes*) in east Gippsland. In *Proceedings of Wildlife Disease Association (Australasian Section) Annual Conference*, p. 49, Mary River, Northern Territory.

Fox H (1912). In Potkay S (1977). *Diseases of Marsupials.* In *The Biology of Marsupials* (ed. D Hunsaker), pp. 415–506. Academic Press: New York.

Freudenberger DO, Wallis IR & Hume ID (1989). Digestive adaptations of kangaroos, wallabies and rat-kangaroos. In *Kangaroos, Wallabies and Rat-kangaroos* (eds GC Grigg, P Garman & ID Hume), pp. 179–188. Surrey Beatty & Sons: Sydney.

Garma-Avina A (1981). Oral melanoma in a captive wallaby, *Wallabia rufogrisea*. *Journal of Wildlife Diseases* **17**, 409–412.

George H (1990). The care and handling of orphaned kangaroos. In *Care and Handling of Australian Mammals* (ed. SJ Hand), pp. 123–141. Surrey Beatty & Sons: Sydney.

Gray CW (1974). Pulmonary nocardiosis in a red kangaroo (*Macropus rufus*). *Journal of Zoo Animal Medicine* **5**, 8.

Griffith AS (1928). Tuberculosis in captive wild animals. *Journal of Hygiene* **28**, 198–218.

Griffith J (1998). Parasites in pademelons. In *Proceedings of the Wildlife Disease Association (Australasian Section) Annual Conference*, p. 18, Calperum Station, South Australia.

Griner LA (1983). *Pathology of Zoo Animals.* Zoological Society of San Diego: San Diego.

Guliani S, Smith GA, Young PL, Mattick JS & Mahony TJ (1999). Reactivation of macropodid herpes virus from the eastern grey kangaroo (*Macropus giganteus*) following corticosteroid treatment. *Veterinary Microbiology* **68**, 59–69.

Gulland F, Lewis J, Kock RA & Kirkwood JK (1987). Vaccination against lumpy jaw in wallabies. *Veterinary Record* **121**, 311.

Hart RP, Bradshaw SD & Iveson JB (1985). *Salmonella* infections in a marsupial, the quokka (*Setonix brachyurus*), in relation to seasonal changes in condition and environmental stress. *Applied Environmental Microbiology* **49**, 1276–1281.

Hartley MP & Sanderson S (2003). Use of antibiotic impregnated polymethylmethacrylate for the treatment of chronic mandibular osteomyelitis in a Bennett's wallaby (*Macropus rufogriseus rufogriseus*). *Australian Veterinary Journal* **81**, 742–744.

Hatkin J & Jansen D (1979). An anatomic barrier to urethral catheterization in male *Macropodidae*. *Journal of the American Veterinary Medical Association* **175**, 1004–1005.

Hawkey CM & Hart MG (1987). Fibrinogen levels in mammals suffering from bacterial infections. *Veterinary Record* **121**, 519–521.

Hawkey CM, Ashton DG, Hart MG, Cindery RN & Jones DM (1982). Haematological findings in red-necked wallabies (*Protemnodon rufogrisea*) with necrobacillosis. *Research in Veterinary Science* **33**, 290–294.

Hemmi JM (1999). Dichromatic colour vision in an Australian marsupial, the tammar wallaby. *Journal of Comparative Physiology [A]* **185**, 509–515.

Herbert CA (2003). Handling macropodids in captivity. In *Conservation, Welfare and Handling of Australian Marsupials: A Practical Guide and Notes* (eds CA Herbert & DW Cooper), pp. 15–21. Department of Biological Sciences, Macquarie University, Sydney.

Herbert CA & Trigg TE (2005). Applications of GnRH in the control and management of fertility in female animals. *Animal Reproduction Science* **88**, 141–153.

Herbert CA, Trigg TE & Cooper DW (2004a). Effect of deslorelin implants on follicular development, parturition and post-partum oestrous in the tammar wallaby (*Macropus eugenii*). *Reproduction* **127**, 265–273.

Herbert CA, Trigg TE, Renfree MB, Shaw G, Eckery DC & Cooper DW (2004b). Effects of a gonadotrophin-releasing hormone agonist implant on reproduction in a male marsupial, *Macropus eugenii*. *Biology of Reproduction* **70**, 1836–1842.

Herbert CA, Trigg TE, Renfree MB, Shaw G, Eckery DC & Cooper DW (2005a). Long-term effects of deslorelin implants on reproduction in the female tammar wallaby (*Macropus eugenii*). *Reproduction* **129**, 361–369.

Herbert CA, Vogelnest L and Rose K (2005b). An outbreak of *Bordetella bronchiseptica* infection in a captive colony of tammar wallabies (*Macropus eugenii*). In *Proceedings of the Wildlife Disease Association International Conference*, p. 96, Cairns, Queensland.

Herbert CA, Trigg TE & Cooper DW (2006). Fertility control in female eastern grey kangaroos using the GnRH agonist deslorelin. 1. Effects on reproduction. *Wildlife Research* **33**, 41–46.

Higgins DP, Carlisle-Nowak MS & Mackie J (1997). Neural angiostrongylosis in three captive rufous bettongs (*Aepyprymnus rufescens*). *Australian Veterinary Journal* **75**, 564–566.

Hime JM & Jones DM (1972). Bone tuberculosis in wallabies (*Protemnodon rufogrisea frutica*). *Journal of Small Animal Practice* **13**, 91–96.

Holt WV, Penfold LM, Johnston SD, Temple-Smith P, McCallum C, Shaw J, Lindemans W & Blyde D (1999). Cryopreservation of macropodid spermatozoa: new insights from the cryomicroscope. *Reproduction Fertility and Development* **11**, 345–353.

Holz P (1992). Immobilization of marsupials with tiletamine and zolazepam. *Journal of Zoo and Wildlife Medicine* **23**, 426–428.

Holz P (2003). Marsupialia (Marsupials). In *Zoo and Wild Animal Medicine* (5th edn) (eds ME Fowler & RE Miller), pp. 288–303. WB Saunders: Philadelphia.

Holz P & Barnett JEF (1996). Long-acting tranquilizers: their use as a management tool in the confinement of free-ranging red-necked wallabies (*Macropus rufogriseus*). *Journal of Zoo and Wildlife Medicine* **27**, 54–60.

Hooper PT, Lunt RA, Gould AR, Hyatt AD, Russell GM, Kattenbelt JA, Blacksell SD, Reddacliff LA, Kirkland PD, Davis RJ, Durham PJK, Bishop AL & Waddington J (1999). Epidemic of blindness in kangaroos: evidence of a viral aetiology. *Australian Veterinary Journal* **77**, 529–536.

Hough I (1997). Rhododendron poisoning in a western grey kangaroo. *Australian Veterinary Journal* **75**, 174–175.

Hum S (2005). Putative sporidesmin toxicity in an eastern grey kangaroo (*Macropus giganteus*). *Australian Veterinary Journal* **83**, 678–679.

Hume ID (1999). *Marsupial Nutrition*. CUP: Cambridge.

Hume ID, Jarman PJ, Renfree MB & Temple-Smith PD (1989). Macropodidae. In *Fauna of Australia*, vol. 1B Mammalia (eds DW Walton & BJ Richardson), pp. 679–715. AGPS: Canberra.

Hynes EF, Nave CD, Shaw G & Renfree MB (2007). The effects of levonorgestrel on ovulation and oestrous behaviour in the female tammar wallaby. *Reproduction, Fertility and Development* **19**, 335–340.

ISIS (2002). *Reference Ranges for Physiological Values in Captive Wildlife*. International Species Information System: Eagan, Minnesota.

Jackson S (2003). *Australian Mammals: Biology and Captive Management*. CSIRO Publishing: Melbourne.

Jalanka HH & Roeken BO (1990). The use of medetomidine, medetomidine-ketamine combinations, and atipamazole in non-domestic mammals: a review. *Journal of Zoo and Wildlife Medicine* **21**, 259–282.

Jenkins DJ & Morris B (2003). *Echinococcus granulosus* in wildlife in and around the Kosciuszko National Park, south-eastern Australia. *Australian Veterinary Journal* **81**, 81–85.

Johnson HJ & Jensen JM (1998). Hepatotoxicity and secondary photosensitization in a red kangaroo (*Megaleia rufus*) due to ingestion of *Lantana camara*. *Journal of Zoo and Wildlife Medicine* **29**, 203–207.

Johnson AM, Roberts H & Munday BL (1988). Prevalence of *Toxoplasma gondii* antibody in wild macropods. *Australian Veterinary Journal* **65**, 199–201.

Johnson PM, Speare R & Beveridge I (1999). Mortality in wild and captive rock-wallabies and nailtail wallabies due to hydatid disease caused by *Echinococcus granulosus*. *Australian Mammalogy* **20**, 419–423.

Johnson-Delaney CA (2000). Medical problems of pet wallabies. *Exotic Pet Practice* **5**, 89–92.

Johnson-Delaney CA (2002). Reproductive medicine of companion marsupials. *Veterinary Clinics Exotic Animal Practice* **5**, 537–553.

Johnston SD, Tribe A, Perkins N, Blyde D & McGowan MR (1994). Collaborative studies in marsupial theriogenology. In *Proceedings of the Australian Association of Veterinary Conservation Biologists Annual Conference*, Canberra.

Jones RC (1989). Reproduction in male macropodidae. In *Kangaroos, Wallabies and Rat-kangaroos* (eds GC Grigg, P Garman & ID Hume), pp. 287–305. Surrey Beatty & Sons: Sydney.

Jones A & Pybus MJ (2001). Taeniasis and echinococcosis. In *Parasitic Diseases of Wild Mammals* (eds WM Samuel, MJ Pybus & AA Kocan), pp. 150–192. Iowa State University Press: Ames, Iowa.

Kakulas BA (1961). Myopathy affecting the Rottnest quokka (*Setonix brachyurus*) reversed by α-tocopherol. *Nature* **191**, 402–403.

Kakulas BA (1963). Influence of the size of enclosure on the development of myopathy in the Rottnest quokka. *Nature* **198**, 673–674.

Kapustin N, Kanitz C & Muench T (1999). Identification of a retrovirus in Bennett's wallabies (*Macropus rufogriseus frutica*) and dama (tammar) (*Macropus eugenii*) wallabies. In *Proceedings of the American Association of Zoo Veterinarians Annual Conference*, pp. 261–262, Columbus, Ohio.

Keane CT, Taylor MR, Wilson H, Smith P, Cunningham B, Devine P & English L (1977). *Bacteroides ruminicola* as a possible cause of 'lumpy jaw' in Bennett's wallabies. *Veterinary Microbiology* **2**, 179.

Keep JM & Pericard JM (1983). Dermatophytes of the larger macropods. In *Proceedings of the Wildlife Disease Association (Australasian Section) Annual Conference*, p. 11, Melbourne.

Kennedy S, Montali RJ, James AE & Bush M (1978). Bone lesions in three tree kangaroos. *Journal of American Veterinary Medical Association* **173**, 1094–1098.

Kirkpatrick TH (1964). Molar progression and macropod age. *Queensland Journal of Agricultural Science* **21**, 163–165.

Kirkpatrick TH (1965). Studies of macropodidae in Queensland: 2. Age estimation in the eastern grey kangaroo, the red kangaroo, the eastern wallaroo and the red-necked wallaby, with notes on dental abnormalities. *Queensland Journal of Agricultural and Animal Science* **22**, 301–317.

Kirkpatrick TH & Johnson PM (1969). Studies of macropodidae in Queensland: 7. Age estimation and reproduction in the agile wallaby (*Wallabia agilis* (Gould)). *Queensland Journal of Agricultural and Animal Science* **26**, 691–698.

Kirkwood JK & Gulland FMD (1988). Pharmacokinetics of oxytetracycline in clinical cases in the red-necked wallaby (*Macropus rufogriseus*). *Research in Veterinary Science* **44**, 335–337.

Kirkwood JK, Gulland FMD, Needham JR & Vogler MG (1988). Pharmacokinetics of oxytetracycline in clinical cases in the red-necked wallaby (*Macropus rufogriseus*). *Research in Veterinary Science* **44**, 335–337.

Kitchener AL, Edds LM, Molinia FC & Kay DJ (2002). Porcine zonae pellucidae immunisation of tammar wallabies (*Macropus eugenii*): fertility and immune response. *Reproduction, Fertility and Development* **14**, 215–223.

Kosanke SD, Feeback DL & Jensen JM (1979). Oral squamous cell carcinoma in a giant red kangaroo. *Journal of American Veterinary Medical Association* **175**, 974–976.

Ladds PW, Sammons J & Beveridge I (2006). Enteritis caused by *Cylicospiruria heydoni* infection in two Tasmanian pademelons (*Thylogale billardierii*). *Australian Veterinary Journal* **84**, 412–413.

Lesslie IW (1958). Tuberculosis in a wallaby caused by the avian type tubercle bacillus. *British Veterinary Journal* **114**, 358–360.

Lewis JCM, Fitzgerald AJ, Gulland FMD, Hawkey CM, Kertesz P, Kirkwood JK & Kock RA (1989). Observations on the treatment of necrobacillosis in wallabies. *British Veterinary Journal* **145**, 394–396.

Lombard LS & Witte EJ (1959). Frequency and types of tumours in mammals and birds at the Philadelphia Zoological Garden. *Cancer Research* **9**, 127–141.

Love CJ & Reddacliff GL (1992). Verminous bronchitis and bronchiolitis in potoroid marsupials associated with a new *Capillaria* sp. *Journal of Wildlife Diseases* **28**, 643–645.

Low Choy J, Mayo M, Janmaat A & Currie BJ (2000). Animal melioidosis in Australia. *Acta Tropica* **74**, 163–158.

Lucas NS (1925). Avian tuberculosis in a marsupial. *Journal of Pathology and Bacteriology* **28**, 123–124.

Lynch MJ, Obendorf DL & Redacliff GL (1993). Serological responses of tammar wallabies (*Macropus eugenii*) to inoculation with an attenuated strain of *Toxoplasma gondii*. In *Proceedings of the American Association of Zoo Veterinarians*, pp. 185–187, St. Louis, Missouri.

Magarey GM & Mate KE (2003a). Fertilisation following intracytoplasmic sperm injection in a marsupial, the tammar wallaby (*Macropus eugenii*). *Zygote* **11**, 339–346.

Magarey GM & Mate KE (2003b). Timing and ultrastructure of events following intracytoplasmic sperm injection in a marsupial, the tammar wallaby (*Macropus eugenii*). *Reproduction, Fertility and Development* **15**, 397–406.

Magarey GM, Rodger JC, Buist JM & Mate KE (2003). Effects of superovulation and surgical oocyte collection on ovarian response and natural breeding ability of the tammar wallaby (*Macropus eugenii*). *Reproduction* **125**, 701–707.

Mann PC, Montali RJ & Bush M (1982). Mycobacterial osteomyelitis in captive marsupials. *Journal of American Veterinary Medical Association* **181**, 1331–1333.

Martin RW (2005). *Tree-kangaroos of Australia and New Guinea*. CSIRO Publishing: Melbourne.

McCarthy PH (1961a). *Austrosimulium pestilens*: a source of severe irritation to marsupials. *Australian Veterinary Journal* **37**, 374.

McCarthy PH (1961b). Two parasites of marsupials in central Queensland. *Australian Veterinary Journal* **37**, 405.

McCauley D (2003). *Macropods: Their Care, Breeding and the Rearing of their Young*. Dave's Animal Farm: Texas.

McEwan D (1989). Atypical mycobacterial lesions in a mala wallaby (*Lagorchestes hirsutus*). *Veterinary Pathology Report* **25**, 8.

McIlroy JC (1982). The sensitivity of Australian animals to 1080 poison III. Marsupial and eutherian herbivores. *Australian Wildlife Research* **9**, 487–503.

McKenzie RA, Green PE & Wood AD (1978). *Angiostrongylus cantonensis* infection of the brain of a captive Bennett's wallaby (*Macropus rufogriseus*). *Australian Veterinary Journal* **54**, 86–88.

McKenzie RA, Fay FR & Prior HC (1979). Poxvirus infection of the skin of an eastern grey kangaroo. *Australian Veterinary Journal* **55**, 188–190.

McLelland DJ & Youl JM (2005). Sarcoptic mange in agile wallabies (*Macropus agilis*) in the Northern Territory. *Australian Veterinary Journal* **83**, 74–75.

McLelland DJ, Kirkland PD, Rose KA, Dixon RJ & Smith N (2005). Serologic responses of Barbary sheep (*Ammotragus lervia*), Indian antelope (*Antilope cervicapra*), wallaroos (*Macropus robustus*) and chimpanzees (*Pan troglodytes*) to an inactivated encephalomyocarditis virus vaccine. *Journal of Zoo and Wildlife Medicine* **36**, 69–73.

McManus T (1979). Native fauna and liver fluke. *Journal of Agriculture (Tasmania)* **50**, 43–45.

Melville L (2000). Wallal serogroup viruses in black wallaroos. *Veterinary Pathology Report* **55**, 12.

Menkhorst PW (2001). *A Field Guide to the Mammals of Australia*. OUP: Melbourne.

Miller WA & Beighton D (1979). Bone abnormalities in two groups of macropod skulls: a clue to the origin of lumpy jaw. *Australian Journal of Zoology* **27**, 681–689.

Miller DS, Faulkner C & Patton S (2003). Detection of *Toxoplasma gondii* IgG antibodies in juvenile great grey kangaroos *Macropus giganteus giganteus*. *Journal of Zoo and Wildlife Medicine* **34**, 189–193.

Molinia FC, Gibson RJ, Brown AM, Glazier AM & Rodger JC (1998a). Successful fertilization after superovulation and laparoscopic intrauterine insemination of the brushtail possum, *Trichosurus vulpecular*, and tammar wallaby, *Macropus eugenii*. *Journal of Reproduction and Fertility* **113**, 9–17.

Molinia FC, Gibson RJ, Smedley MA & Rodger JC (1998b). Further observations of the ovarian response of the tammar wallaby (*Macropus eugenii*) to exogenous gonadotrophins; an improved method for superovulation using FSH/LH. *Animal Reproduction Science* **53**, 253–263.

Montali RJ, Bush M, Cromie R, Holland SM, Maslow JM, Worley M, Witebsky FG & Phillips TM (1998). Primary *Mycobacterium avium* complex infections correlate with lowered cellular immune reactivity in Matschie's tree kangaroos (*Dendrolagus matschiei*). *Journal of Infectious Diseases* **178**, 1719–1725.

Munday BL (1971). Gastric squamous tumour in the stomach of a pademelon (*Thylogale billardierii*) associated with an infestation with *Labiostrongylus* sp. larvae. *Journal of Wildlife Disease* **7**, 125.

Munday BL (1972). Myonecrosis in free-living and recently captured macropods. *Journal of Wildlife Diseases* **8**, 191–192.

Munday BL (1978). Marsupial disease. In *Fauna, Proceedings 36*, pp. 335–385. Post Graduate Committee in Veterinary Science: Sydney.

Mykytowycz R (1964). Coccidia in wild populations of the red kangaroo, *Megaleia rufa* (Desmarest), and the grey kangaroo, *Macropus canguru* (Muller). *Parasitology* **54**, 105–115.

Nave CD, Shaw G, Short RV & Renfree MB (2000). Contraceptive effects of levonorgestrel implants in a marsupial. *Reproduction, Fertility and Development* **12**, 81–86.

Nave CD, Coulson G, Short RV, Poiani A, Shaw G & Renfree MB (2002a). Long-term fertility control in the kangaroo and the wallaby using levonorgestrel implants. *Reproduction Supplement* **60**, 71–80.

Nave CD, Coulson G, Poiani A, Shaw G & Renfree MB (2002b). Fertility control in the eastern grey kangaroo using levonorgestrel implants. *Journal of Wildlife Management* **66**, 470–477.

Norton JH & Hartley WJ (1993). Amoebic meningoencephalitis in a young grey kangaroo (*Macropus giganteus*). *Australian Veterinary Journal* **70**, 421.

O'Callaghan MG, Carmichael IH, Finnie JW & Conaghty S (1994). Lesions associated with infestation of a yellow footed rock wallaby (*Petrogale xanthopus xanthopus*) with larvae of *Odontacarus* (*Leogonius*) *adelaideae* (Womersley) (Acarina: Trombiculidae) in South Australia. *Journal of Wildlife Diseases* **30**, 257–259.

Obendorf DL (1980). Candidiasis in young hand-reared kangaroos. *Journal of Wildlife Diseases* **16**,135–140.

Obendorf DL (1984). The macropodid oesophagus. I. Gross anatomical, light microscopic, scanning and transmission electron microscopic observations of its mucosa. *Australian Journal of Zoology* **32**, 415–435.

Obendorf DL (1985). Intestinal parasitism in Tasmanian forester kangaroos (*Macropus giganteus tasmanicus*). *Veterinary Pathology Report* **10**, 9.

Obendorf DL (1987). Avian mycobacteriosis in captive wallabies. *Veterinary Pathology Report* **15**, 12.

Obendorf DL & Munday BL (1983). Toxoplasmosis in wild Tasmanian wallabies. *Australian Veterinary Journal* **60**, 62.

Okoh AEJ (1980). An outbreak of pasteurellosis in Kano Zoo. *Journal of Wildlife Diseases* **16**, 3–5.

Oliphant JC, Parsons R & Smith GR (1984). Aetiological agents of necrobacillosis in captive wallabies. *Research in Veterinary Science* **36**, 382–384.

Ott Joslin J (1990). Mycobacterial infections in tree kangaroos. In *Proceedings of the American Association of Zoo Veterinarians Annual Conference*, pp. 132–139, Padre Island, Texas.

Papadimitriou JM & Ashman RB (1972). A poxvirus in a marsupial papilloma. *Journal of General Virology* **16**, 87–89.

Paris DBBP, Taggart DA, Paris MC, Temple-Smith PD & Renfree MB (2004). Sperm transport, size of the seminal plug and the timing of ovulation after natural mating in the female tammar wallaby *Macropus eugenii*. *Reproduction, Fertility and Development* **16**, 811–822.

Paris DBBP, Taggart DA, Shaw G, Temple-Smith PD & Renfree MB (2005). Birth of pouch young after artificial insemination in the tammar wallaby (*Macropus eugenii*). *Biology of Reproduction* **73**, 451–459.

Patton S, Johnson SL, Loeffler DG, Wright BG & Jensen JM (1986). Epizootic of toxoplasmosis in kangaroos, wallabies and potoroos: possible transmission via domestic cats. *Journal of American Veterinary Medical Association* **189**, 1166–1169.

Peet RL & Dickson J (1982). Mycobacterium intracellulare infection in wallabies. *Australian Veterinary Journal* **58**, 215.

Phelan JR (1996). Atypical mycobacterial infections in captive long-footed potoroos (*Potorous longipes*). In *Proceedings of American Association of Zoo Veterinarians Annual Conference*, pp. 443–445, Puerto Vallarta, Mexico

Poelma FG (1964). Diseases in marsupials in captivity. *Tijdschrift Voor Diergeneeskunde* **89**, 149–154.

Poiani A, Colson G, Salamon D, Holland S & Nave CD (2002). Fertility control of the eastern grey kangaroo: do levonorgestrel implants affect behaviour? *Journal of Wildlife Management* **66**, 59–66.

Portas TJ (2005). Physical restraint, sedation and anesthesia of macropodids. *Exotic DVM* **7**, 34–37.

Portas TJ, Spratt DM & Rose KA (2005). Microfilaraemia associated with *Pelecitus roemeri* in a western grey kangaroo (*Macropus fuliginosus*). *Australian Veterinary Journal* **83**, 15–16.

Potkay S (1977). Diseases of marsupials. In *The Biology of Marsupials* (ed. D Hunsaker), pp. 415–506. Academic Press: New York.

Presidente PJ (1978). Diseases seen in free-ranging marsupials and those held in captivity. In *Fauna, Proceedings 36*, pp. 457–471. Post Graduate Committee in Veterinary Science: Sydney.

Presidente PJ & Beveridge I (1978). Cholangitis associated with species of *Progamotaenia* (Cestoda: Anoplocephalidae) in the bile ducts of marsupials. *Journal of Wildlife Diseases* **14**, 371–377.

Pritchard D (1995). *Mycobacterium asiaticum* infection in a red kangaroo. *Veterinary Pathology Report* **42**, 35.

Prociv P, Spratt DM & Carlisle MS (2000). Neuro-angiostrongyliasis: unresolved issues. *International Journal for Parasitology* **30**, 1295–1303.

Pye GW & Booth RJ (1998). Medetomidine-ketamine immobilization and atipamezole reversal of eastern grey kangaroos (*Macropus giganteus*). In *Proceedings of American Association of Zoo Veterinarians and American Association of Wildlife Veterinarians Joint Conference*, pp. 306–309.

Ratcliffe HL (1933). In Potkay S (1977). Diseases of marsupials. In *The Biology of Marsupials* (ed. D Hunsaker), pp. 415–506. Academic Press: New York.

Reddacliff GL, Hartley WJ, Dubey JP & Cooper DW (1993a). Pathology of experimentally-induced, acute toxoplasmosis in macropods. *Australian Veterinary Journal* **70**, 4–6.

Reddacliff GL, Parker SJ, Dubey JP, Nicholls PJ, Johnson AM & Cooper DW (1993b). An attempt to prevent acute toxoplasmosis in macropods by vaccination with *Hammondia hammondi*. *Australian Veterinary Journal* **70**, 33–35.

Reddacliff LA, Kirkland PD, Hartley WJ &Reece RL (1997). Encephalomyocarditis virus infection in an Australian zoo. *Journal of Zoo and Wildlife Medicine* **28**, 153–157.

Reddacliff LA, Bellamy TA & Hartley WJ (1999a). *Angiostrongylus cantonensis* infection in grey-headed fruit bats (*Pteropus poliocephalus*). *Australian Veterinary Journal* **77**, 466–468.

Reddacliff LA, Kirkland PD, Philbey A, Davis R, Vogelnest L, Hulst F, Blyde D, Deykin A, Smith J, Hooper P, Gould A & Hyatt A (1999b). Experimental reproduction of viral chorioretinitis in kangaroos. *Australian Veterinary Journal* **77**, 522–528.

Remedios AM, Sanders S, Fowler JD & Wheler C (1996). Microvascular reconstruction of a facial deficit in a red-necked wallaby (*Macropus rufogriseus*). *Journal of Zoo and Wildlife Medicine* **27**, 115–120.

Renfree MB, Fletcher TP, Blanden DR, Lewis PR, Shaw G, Gordon K, Short RV, Parer-Cook E & Parer D (1989). Physiological and behavioural events around the time of birth in macropodid marsupials. In *Kangaroos, Wallabies and Rat-kangaroos* (eds GC Grigg, P Garman & ID Hume), pp. 323–337. Surrey Beatty & Sons: Sydney.

Richardson KC & Read RA (1986). Tuberculous osteomyelitis caused by *Mycobacterium intracellulare* in the brush-tailed bettong. *Journal of Wildlife Diseases* **22**, 425–429.

Richings NM, Shaw G, Temple-Smith PD & Renfree MB (2004). Intra-cytoplasmic sperm injection in a marsupial. *Reproduction* **128**, 595–605.

Roberts ED, Williams JC & Pirie G (1973). Naturally occurring gastric amebiasis in the wallaroo. *Veterinary Pathology* **10**, 323–329.

Rose KA (1999). Common diseases of urban wildlife. In *Wildlife in Australia: Healthcare and Management, Proceedings 327,* pp. 365–427. Post Graduate Foundation in Veterinary Science: Sydney.

Rose KA, Kirkland PD, Davis RJ, Cooper DW & Blumstein DT (2000). An epizootic of sudden death in tammar wallabies (*Macropus eugenii*). In *Proceedings of American Association of Zoo Veterinarians*, pp. 166–170, New Orleans, Louisiana.

Rose K, Curtis J & Baldwin T (2004). Cutaneous leishmaniasis in red kangaroos: isolation and characterisation of the causative organisms. *International Journal of Parasitology* **34**, 655–664.

Rothwell JT (1988). Blast cell lymphoma in a pretty face wallaby (*Macropus paryii*). *Veterinary Pathology Report* **19**, 11.

Rothwell TLW, Keep JM, Xu FN & Middleton DJ (1984). Poxvirus in marsupial skin lesions. *Australian Veterinary Journal* **61**, 409–410.

Rothwell JT, Harper PA, Hartley WJ, Gumbrell RC & Meischke HRC (1990). Suspected lysosomal storage disease in kangaroos. *Journal of Wildlife Diseases* **26**, 275–278.

Rungsipipat A, Yamaguchi R, Naganobu K, Iwamoto K, Uchida K, Tateyama S, Kurogi T & Katayama N (1998). A bone tumour resembling bizarre parosteal osteochondromatous proliferation in a wallaby. *Australian Veterinary Journal* **76**, 561–564.

Samuel JL (1983). Jaw disease in macropod marsupials: bacterial flora isolated from lesions and from the mouths of affected animals. *Veterinary Microbiology* **8**, 373–387.

Sanson GD (1989). Morphological adaptations of teeth to diets and feeding in the Macropodoidea. In *Kangaroos, Wallabies and Rat-kangaroos* (eds GC Grigg, P Garman & ID Hume), pp. 151–168. Surrey Beatty & Sons: Sydney.

Saroja S (1979). Melioidosis in wallabies. *Australian Veterinary Journal* **55**, 439–440.

Schoon HA, Rosenbrunch M & Ruempler G (1985). Systemic arterial calcinosis in a grey tree kangaroo *Dendrolagus inustus*, resembling Mönckeberg type arteriosclerosis in man. *Journal of Comparative Pathology* **95**, 319–324.

Schoon HA, Schoon D, Kirpal G, Richter E, Gerdes J, Weiss R & Dressen W (1993). Enzootic *Mycobacterium avium* infection in captive parma wallabies (*Macropus parma*) with unusual spinal cord manifestations. *Journal of Comparative Pathology* **108**, 311–316.

Seebeck JH, Bennet AF & Scott DJ (1989). Ecology of the Potoroidae: a review. In *Kangaroos, Wallabies and Rat-kangaroos* (eds GC Grigg, P Garman & ID Hume), pp. 67–88. Surrey Beatty & Sons: Sydney.

Sharman GB, Frith HJ & Calaby JH (1964). Growth of the pouch young, tooth eruption and age determination in the red kangaroo, *Megaleia rufa. CSIRO Wildlife Research* **9**, 20–49.

Shima AL (1999). Sedation and anaesthesia in marsupials. In *Zoo and Wild Animal Medicine: Current Therapy 4* (eds ME Fowler & RE Miller), pp. 333–336. WB Saunders: Philadelphia.

Shima AL, McCracken H, Booth RJ & Lynch MJ (1993). Use of tiletamine-zolazepam in the immobilization of marsupials. In *Proceedings of American Association of Zoo Veterinarians,* pp 154–156.

Short RV, Flint APF & Renfree MB (1985). Influence of passive immunization against GnRH on pregnancy and parturition in the tammar wallaby, *Macropus eugenii. Journal of Reproduction and Fertility* **75**, 567–575.

Skerratt LF, Beveridge I & Johnson PM (2007). Ingurial and axillary dermatitis in wallabies in north Queensland due to the dermanyssid mite *Thadeua serrata. Australian Veterinary Journal* **85**, 510–512

Smeller JM, Bush M & Custer RW (1977). The immobilisation of marsupials. *Journal of Zoo Animal Medicine* **8**, 16–20.

Smith G (1996). *In vitro* sensitivity of macropodid herpesvirus 2 to selected anti-herpetic compounds. *Journal of Wildlife Diseases* **31**, 117–120.

Smith GR, Turner A & Cinderey R (1986). Susceptibility of wallabies to *Fusobacterium necrophorum. Veterinary Record* **118**, 691–693.

Snowdon WA (1968). The susceptibility of some Australian fauna to infection with foot-and-mouth disease virus. *Australian Journal of Experimental Biology and Medical Science* **46**, 667–687.

Speare R (1988). Clinical assessment, diseases and management of the orphaned macropod PY. In *Australian Wildlife, Proceedings 104*, pp. 211–296. Post Graduate Committee in Veterinary Science: Sydney.

Speare R & Spratt DM (1981). Pathology of a remarkable nematode parasitic in the portal veins of Macropodidae. In *Proceedings of the 4th International Conference of Wildlife Diseases Association* (ed. ME Fowler), pp. 113–117, Sydney.

Speare R & Banks D (1987). Trichoepithelioma in a wild eastern grey kangaroo (*Macropus giganteus*). *Journal of Wildlife Disease* **23**, 508–509.

Speare R, Johnson PM & Haffenden A (1982). Management of disease in captive macropods in North Queensland. In *The Management of Australian Mammals in Captivity* (ed. DD Evans), pp. 117–124. Zoological Parks Board of Victoria: Melbourne.

Speare R, Beveridge I, Johnson PM & Corner LA (1983). Parasites of the agile wallaby, *Macropus agilis* (Marsupialia). *Australian Wildlife Research* **10**, 89–96.

Speare R, Donovan JA, Thomas AD & Speare PJ (1989). Diseases of free-ranging Macropodoidea. In *Kangaroos, Wallabies and Rat-kangaroos* (eds GC Grigg, P Garman & ID Hume), pp 705–734. Surrey Beatty & Sons: Sydney.

Speare R, Skerratt LF, Berger L & Johnson PM (2004). Toxic effects of mebendazole at high dose on the haematology of red-legged pademelons (*Thylogale stigmatica*). *Australian Veterinary Journal* **82**, 300–303.

Spratt DM (1972). Aspects of the life-history of *Dirofilaria roemeri* in naturally and experimentally infected kangaroos, wallaroos and wallabies. *International Journal for Parasitology* **2**, 139–156.
Spratt DM (1974). Comparative epidemiology of *Dirofilaria roemeri* infection in two regions of Queensland. *International Journal for Parasitology* **4,** 481–488.
Spratt DM (1975). Further studies of *Dirofilaria roemeri* (Nematoda: *Filarioidea*) in naturally and experimentally infected Macropodidae. *International Journal for Parasitology* **5**, 561–564.
Spratt DM & Hobbs RP (2003). Breinlia (*Breinlia*) *ventricola* sp. nov., a nematode parasite from the heart of the red kangaroo, *Macropus rufus*, in Western Australia. *Transactions of the Royal Society of South Australia* **128**, 67–71.
Spratt DM & Presidente PJ (1981). Presence of *Fasciola hepatica* infection in native mammals in south-eastern Australia. *Australian Journal of Experimental Biology and Medical Science* **59**, 713–721.
Stanley RG (2002). Marsupial ophthalmology. *Veterinary Clinics Exotic Animal Practice* **5**, 371–390.
Stedman NL, Munday JS, Esbeck R & Visvesvara GS (2003). Gastric amebiasis due to *Entamoeba histolytica* in a dama wallaby (*Macropus eugenii*). *Veterinary Pathology* **40**, 340–342.
Stelmasiak T, Rawlin GT & Waddington JN (1993). Modification of behavior in captive kangaroos by immunization against luteinizing hormone releasing hormone (LH-RH). In *Proceedings of the Symposium on Contraception in Wildlife Management*, pp. 2.
Stephens T (1975). Nutrition of orphaned marsupials. *Australian Veterinary Journal* **51**, 453–458.
Strahan R (1995). *The Australian Museum Complete Book of Australian Mammals.* Reed New Holland: Sydney.
Stünzi H (1950). In Potkay S (1977). Diseases of marsupials. In *The Biology of Marsupials* (ed. D Hunsaker), pp. 415–506. Academic Press: New York.
Suedmeyer WK (1998). Tail amputation in a red kangaroo (*Macropus rufus*). In *Proceedings of American Association of Zoo Veterinarians*, pp. 310–312.
Suedmeyer WK & Johnson G (2007). Survey of neoplasia in red kangaroos (*Macropus rufus*), 1992–2002, in a zoological collection. *Journal of Zoo and Wildlife Medicine* **38**, 231–239.
Suedmeyer WK, Layton C & Riley SM (1999). Hemimandibulectomy to resolve *Actinomyces* sp. 'lumpy jaw' in a Bennett's wallaby (*Macropus rufogriseus fruticus*). In *Proceedings of American Association of Zoo Veterinarians*, pp. 119–122.
Swan GE (1993). Drugs used for the immobilisation, capture, and translocation of wild animals. In *The Capture and Care Manual* (ed. AA McKenzie), pp. 2–64. Wildlife Decision Support Services/South African Veterinary Foundation.
Taggart DA, Shimmin GA, Underwood G & Phillips K (2002). Survival of very small macropod pouch young following short-term isolation from the pouch at various environmental temperatures. *Animal Conservation* **5**, 275–282.
Taggart DA, Schultz D, White C, Whitehead P, Underwood G & Phillips K (2005). Cross-fostering, growth and reproductive studies in the brush-tailed rock-wallaby, *Petrogale penicillata* (Marsupialia: Macropodidae): efforts to accelerate breeding in a threatened marsupial species. *Australian Journal of Zoology* **53**, 313–323.
Taylor MRH, Wilson P, Smyth L, English L & Keane CT (1978). Laboratory investigations in the 'lumpy jaw' of macropods. *Veterinary Science Communications* **2**, 131–136.
Teuscher E (1957). In Potkay S (1977). Diseases of marsupials. In *The Biology of Marsupials* (ed. D Hunsaker), pp. 415–506. Academic Press: New York.
Thiruthalinathan R, Latha BS & Swaminathan D (1998). Coccidiosis in eastern grey kangaroos (*Macropus giganteus*). *Indian Veterinary Journal* **75**, 375–376.
Thomas AD (1981). Prevalence of melioidosis in animals in northern Queensland. *Australian Veterinary Journal* **57**, 146–148.
Thomas J (1987). Mycobacteriosis in brush-tailed bettongs (*Bettongia pencillata*). *Veterinary Pathology Report* **17**, 21.
Thomson GR, Bengis RG & Brown CC (2001). Picornavirus infections. In *Infectious Diseases of Wild Mammals* (eds ES Williams & IK Barker), pp 119–130. Iowa State University Press: Ames, Iowa.
Tilden EB & Williamson WM (1957). Unusual tuberculous lesions in a wallaby. *Journal of American Veterinary Medical Association* **131**, 526–527.
Tomlinson AR & Gooding CG (1954). A kangaroo disease: investigations into 'lumpy jaw' on the Murchison. *Journal of Agriculture Western Australia* **3**, 715–718.
Tribe A, Johnston SD & Carrick FN (1994). Artificial reproduction in marsupials. In *Wildlife, Proceedings 233*, pp. 71–105. Post Graduate Committee in Veterinary Science: Sydney.
Tucker R & Millar R (1953). Outbreak of nocardiosis in marsupials at the Brisbane botanical gardens. *Journal of Comparative Pathology* **63**, 143–146.
Turni C & Smales LR (2001). Parasites of the bridled nailtail wallaby (*Onychogalea fraenata*) (Marsupialia: Macropodidae). *Wildlife Research* **28**, 403–411.
Tyndale-Biscoe CH (2005). *Life of Marsupials.* CSIRO Publishing: Melbourne.
Tyndale-Biscoe CH & Renfree MB (1987). *Reproductive Physiology of Marsupials.* CUP: Cambridge.
Vaughan RJ, Vitali SD, Eden PA, Payne K, Warren KS, Friend T & Krockenberger M (2005). Cryptococcal infections in captive Gilbert's (*Potorous gilbertii*) and long-nosed (*Potorous tridactylus*) potoroos. In *Proceedings of the Wildlife Disease Association International Conference*, p. 106, Cairns, Queensland.
Vogelnest L (1999). Chemical restraint of Australian native fauna. In *Wildlife in Australia: Healthcare and Manage-*

ment, Proceedings 327, pp. 149–187. Post Graduate Foundation in Veterinary Science: Sydney.

Wallach J & Boever WJ (eds) (1983). *Diseases of Exotic Animals.* WB Saunders: Philadelphia.

Ward D & Blyde D (2001). An outbreak of *Bordetella bronchiseptica* infection in a captive population of bridled nailtail wallabies. In *Veterinary Conservation Biology, Wildlife Health and Management in Australia, Proceedings* (eds A Martin & L Vogelnest),. pp. 265–268. Australian Veterinary Association: Sydney.

Wayne J & Nicholson V (1999). Hypertrophic osteopathy and pneumonia in a macropod. *Australian Veterinary Journal* **77**, 98–99.

Wilks CR, Kefford B & Callinan RB (1981). Herpesvirus as a cause of fatal disease in Australian wallabies. *Journal of Comparative Pathology* **91**, 461–465.

Williams R (1990). Kangaroos in captivity. In *Care and Handling of Australian Mammals* (ed. SJ Hand), pp. 109–121. Surrey Beatty & Sons: Sydney.

Winter H (1958). Gastric strongyloidosis in kangaroos. *Australian Veterinary Journal* **34**, 118–120.

Winter H (1959). Coccidiosis in kangaroos. *Australian Veterinary Journal* **35**, 301–303.

Woods R (1999). Prevention of disease in hand reared native wildlife. In *Wildlife in Australia: Healthcare and Management, Proceedings 327,* pp. 455–489. Post Graduate Foundation in Veterinary Science: Sydney.

Woods R & Blyde D (1998). Managing 'lumpy jaw' with skeletal scintigraphy. *Veterinarian* November, 5.

Woodward R, Herberstein ME & Herbert CA (2006). Fertility control in female eastern grey kangaroos using the GnRH agonist deslorelin. 2. Effects on behaviour. *Wildlife Research* **33**, 47–55.

Young E (1960). Pasteurellosis in the kangaroo. *International Zoo Year Book* **5,** 185.

Young LJ, McFarlane R, Slender AL & Deane EM (2003). Histological and immunohistochemical investigation of the lymphoid tissue in normal and mycobacteria-affected specimens of the rufous hare-wallaby (*Lagorchestes hirsutus*). *Journal of Anatomy* **202**, 315–325.

8. Koalas

Wendy Blanshard and Kate Bodley

1 TAXONOMY, DISTRIBUTION AND BIOLOGY

The koala (*Phascolarctos cinereus*) is the only member of the Family Phascolarctidae and is included, with wombats, in the Suborder Vombatiformes. Three different 'races' or subspecies have been described, based on geographic range and morphologic characteristics: *P. cinereus cinereus*, found in New South Wales, *P. c. adustus*, found in Queensland, and *P. c. victor*, found in Victoria. Recent molecular genetic analysis has determined that morphologic variation between northern and southern populations is clinal, and might reflect adaptation to climate over an extensive range rather than a significant divergence at the DNA level (Houlden et al. 1999). The tentative conclusion was that *P. cinereus* should not be considered as having three subspecies, but should be considered as a single evolutionarily significant unit.

The koala is found in eastern Australian sclerophyll forests and woodlands, and there are populations in Queensland, New South Wales, the Australian Capital Territory, Victoria and South Australia (Melzer et al. 2000). It has a widespread distribution and is abundant in many areas. The distribution, however, is largely fragmented, with populations separated by unsuitable habitat or cleared land (Martin & Handasyde 1999).

Estimates of population size are few and inconsistent. The conservation status varies throughout its distribution, reflecting regional perceptions of current threats to koalas and to their habitat (Melzer et al. 2000). Populations in New South Wales and Queensland are subject to a number of key threatening processes, including habitat loss through clearing of native vegetation and urban development, and populations are declining in those states. Increasing urban expansion also contributes to mortality in local populations via increases in road-associated deaths and exposure to attack by domestic dogs.

In contrast, unsustainably high population levels have developed in six isolated patches across Victoria, and on Kangaroo Island (SA) (Menkhorst et al. 1998). Habitat destruction and hunting contributed to a marked decline in populations in Victoria and South Australia during the early 20th century. Koalas were reintroduced in some mainland habitats in these states, via translocation from flourishing populations that were established on off-shore islands during the late 19th and early 20th centuries. Unsustainable densities in isolated populations (including some of those established on these islands, and several mainland populations living in remnant habitat) have resulted in overbrowsing of forage trees, widespread tree death and, in some cases, starvation of koalas (Martin 1985a).

Koala populations in South Australia have largely been re-established using animals translocated from a single colony that flourished after being introduced to French Island (Vic.), during the 1880s. Three South Australian populations were found to have low heterozygosity compared with an undisturbed mainland Victorian population, and this low allelic diversity was significantly correlated with the occurrence of testicular abnormalities (Seymour et al. 2001).

Government wildlife agencies have been responding to overabundance in koala populations in south-eastern Australia for many years; management techniques have included translocating animals to less populous release sites, fertility control, and culling (Duka & Masters 2005). While culling allows immediate reduction of koala numbers to a sustainable level, and is believed by many biologists to be the most humane method for population reduction, it is not considered an acceptable management tool within the National Koala Conservation Strategy (ANZECC 1998). Translocation has become less viable as limited unoccupied habitat remains. It requires careful assessment of the health status of the animals to be moved and the habitat condition at their destination in order to avoid high mortality rates. In Victoria, development of cost-effective fertility control methods for use in large numbers of animals is considered a priority for effective management of overabundant populations (Menkhorst 2004).

Koalas in the southern part of their range (Vic. and SA) are larger and have longer darker fur than their northern (Qld) counterparts. Those in New South Wales are intermediate in size and appearance. Body weight ranges and averages given for Queensland and Victorian koalas are: Queensland females 4.1–7.3 kg (av. 5.1 kg), males 4.2–9.1 kg (av. 6.5 kg); Victorian females 7.0–11.0 kg (av. 8.5 kg), males 9.5–14.9 kg (av. 12.0 kg) (Martin & Handasyde 1995). Some large captive male koalas in south-east Queensland exceed 9.1 kg, and an exceptional individual had reached 11.1 kg when just under 7 yr old.

Free-ranging female koalas commonly live for 13–18 yr; males are capable of surviving to a similar age but tend to have a shorter life expectancy due to hazards encountered during the breeding season (Martin & Handasyde 1999). Longevity records for koalas in captivity appear to be 23 yr for a female that was supported with regular supplementary feeding (see 5.3.4d) in her later years, and 20 yr 5 mo for a female that was not supplemented (G Rawlinson *pers. comm.*).

Koalas are largely nocturnal but exhibit some activity during the day. Daily feeding typically occupies a total of about 1–4 hr, and is accomplished in several bouts of varying duration (averaging around 20 min, but ranging between 5 min and 1–2 hr). It is most commonly observed during a 4–5 hr period around dusk. About 19–20 hr of each day is spent sitting, resting or asleep, while social behaviour, grooming, and changing branches or trees occupy only a fraction of daily activity (Robbins & Russell 1978; Lee & Martin 1988; Mitchell 1990a; Martin & Handasyde 1999).

2 ANATOMY AND PHYSIOLOGY

2.1 General body form

The koala has relatively long limbs adapted for climbing. On the fore limb, digits one and two oppose digits three, four and five, effecting a forcipate grip. On the hind limb, digit one is opposed to the other digits; digits two and three are syndactylous (fused) and used for grooming. All digits other than digit one of the hind foot have sharp claws. As in most marsupials, the koala lacks a patella. Epipubic bones are present. The tail is reduced to a stump consisting of 6–7 caudal vertebrae (Cork 1987; Lee & Martin 1988).

Mature males have a sternal scent gland which produces an oily secretion with a distinctive smell, and stains the surrounding fur on the chest. Females have a pouch which opens on the ventral midline in the caudal half of the abdomen. In non-breeding females the pouch is merely a shallow triangular skin fold enveloping two tiny teats. In mated females that are nearing the time of birth, the pouch expands but the teats remain small. Pouch young (PY) do not cause a noticeable bulge in the pouch until they are about 3 mo old, but the pouch sphincter will be tightly closed. Most females resent any attempt to stretch the pouch sphincter to look inside. Scattered flecks of the young's faeces are normally evident on or around unfurred PY, and the interior of the pouch looks moist and glistening. Well-furred PY should look dry and clean. As young continue to be suckled for some months after they leave the pouch, one enlarged mammary gland with an elongated teat may be evident in an otherwise empty pouch.

The koala has relatively small, almost spherical, frontally placed eyes. The cornea occupies a relatively large segment of the globe, with the corneal diameter only slightly smaller than the equatorial diameter. A tapetum extends across the retina above the optic disc, and as in many other marsupials the retina is essentially avascular except for a small area on the optic disc itself (Schmid et al. 1992). Examination of 28 koalas determined that mean intra-ocular pressure was 24.2 ± 6 mmHg (Hirst et al. 1992a). The iris is usually brown, but very occasionally it may be blue (e.g. in heterochromia) or, in albino koalas, unpigmented. The pupil is a vertically oriented slit. Relative size and position of the nictitating membrane is similar to that in a dog or cat, and it follows the same nasal-to-temporal sweep across the eye. The lacrimal puncta are slit-like openings just inside the lid edges, 2–3 mm from the medial canthus. The 0.5–1 mm diameter opening of the nasolacrimal duct is located on the ventrolateral aspect of the nasal floor at the mucocutaneous junction (Blanshard 1994).

2.2 Digestive system

2.2.1 Dentition

The dental formula of the koala is: I 3/1, C 1/0, PM 1/1, M 4/4 (Eberhard 1972). Incisors are used to manipulate branches and leaves, but are utilised to tear off leaves only by very young animals. The first pair of incisors in either arcade is well-developed, while the other upper pairs are smaller and fairly slender. Canines are present only in the upper jaw and are small peg-like structures which probably serve no real purpose. There is a single pair of premolars in each arcade bearing a relatively sharp longitudinal ridge, and it is these teeth which slice through the stems of individual leaves (Lee & Martin 1988). Each of the four pairs of upper and lower molars has four pyramidal cusps, which interdigitate with those on the opposing arcade. This creates a series of cutting edges, between which the leaf is chewed using a cutting–shearing action (Lanyon & Sanson 1986a). There is no crushing component in the occlusal action, thus the efficiency of mastication depends upon the maintenance of cutting edges on the molars (Lanyon & Sanson 1986a).

Throughout a koala's life there is continual wear of the occlusal surfaces of the cheek teeth. The longitudinal cutting ridge of each premolar and the pyramidal cusps of each molar are gradually worn flat. Wear is more advanced on the anterior cheek teeth (Eberhard 1972; Martin 1981). The degree of tooth wear has significant implications for the koala's plane of nutrition. As the cusps of the molar teeth become worn and the total length of their available cutting edges decreases, the proportion of larger leaf fragments in the koala's stomach increases. Particle sizes in the caecum and proximal colon, however, remain similar to those in younger animals. These findings indicate that older koalas not only must excrete large particles more quickly, but also must increase their total leaf intake in order to maintain the same amount of fine digesta in the large intestine (Lanyon & Sanson 1986b). Increasing leaf intake causes an accelerated rate of tooth wear (see 7.1.12).

2.2.2 Gastrointestinal tract

Koalas are monogastric hindgut fermenters. The serosa of the stomach is off-white and the organ is usually rounded by a moderate amount of firm digesta. Part of the colon has a ligamentous attachment to the greater curvature of the stomach. The gastric mucosa normally is covered in a diphtheritic rubbery white coating. The stomach contents should be fairly dry, densely packed, finely chewed leaf material. In older animals, relatively large leaf and stem fragments, strands of fibre and cords of mucoid saliva may be evident. A large circular fleshy pink gastric gland, evident from the serosal surface, takes up much of the lesser curvature between the cardiac and pyloric sphincters, and discharges thick off-white to light brown fluid through a number of large openings on the mucosal surface.

Externally, the duodenum is greyish red and fairly thin-walled. The mucosa normally appears slightly hyperaemic, and its contents dull greyish or reddish-grey. The jejunum is a collapsed translucent thin-walled tube about 15 mm across, with a fine reticular surface pattern, and the contents are usually translucent or slightly turbid greenish-yellow. Specimens of the cestode *Bertiella obesa* may be visible through the thin wall; occasionally these migrate into the duodenum or even through the pyloric sphincter after death (Spratt 1978). Thicker-walled than the jejunum, the ileum is pinkish, opaque, and about 10 mm across. Its contents are generally pale greyish-brown.

Koalas possess the largest caecum (up to 2 m long) relative to body size for any known mammal (MacKenzie 1918). Greyish- to brownish-olive loops of the caecum and proximal colon (which are superficially similar in appearance) occupy the entire ventral portion of the abdomen. The caecum and proximal colon should be heavy with moist thick brownish-olive digesta containing few recognisable leaf fragments. Both are oval in cross-section, i.e. not distended by their contents. The mucosal surface area of the caecum and similarly capacious proximal colon is augmented by a series of longitudinal folds. Digesta evenly coats the mucosal surface and longitudinal folds, and a thin layer is retained on the mucosa after gentle rinsing under running water (McKenzie 1978). Obvious dehydration of digesta closest to the mucosal surface is abnormal. The proximal colon gradually narrows and its longitudinal mucosal folds disappear. The character of the digesta also changes, with larger leaf fragments predominating and the consistency becoming thicker and drier. Cylindrical masses of digesta are divided into pellets of increasing dryness and firmness in the distal colon, which is yellowish, tapering and relatively thick-walled.

A consistent feature in histological sections of the caecum and proximal colon is the presence of a tall fringe of mixed bacteria tenaciously adhering to the mucosa (McKenzie 1978). Culture of some of these bacteria requires very specific and selective agar media (Osawa 1992). The biochemical and metabolic characteristics of the caecal microflora and their full significance in digestion or detoxification of the koala's diet still are being investigated (Osawa et al. 1993a).

2.2.3 Liver, gall bladder and pancreas

In young koalas the liver is dull red, as in other species. With increasing age, dark pigment is accumulated and the liver becomes a uniform very deep purplish- or bluish-red. All lobe edges (except on a few of the small processes in the centre of the visceral surface) are normally very sharp. The gall bladder is a finger-like sac about 10–15 mm diameter and 5–6 cm long. Bile may be translucent and greenish, or whitish-gold with an iridescent character (Dickens 1975). The pancreas is pale or dull pink, and the distal end of the pancreatic duct is dilated to form a small flat chamber of clear fluid adjacent to where the duct enters the duodenum.

2.3 Spleen

The spleen is usually contracted and dull pink, with tapered edges. Posteriorly it is narrow and strap-like, with a slightly broader and thicker anterior end, and normally is no more than about 10 mm thick.

2.4 Urogenital system

The kidneys are simple (not lobulated). When empty, the urinary bladder is contracted, small and rather thick-walled, and lies within the pelvic cavity. The mucosa is gently undulating and usually off-white to very pale pink in colour.

The female reproductive tract is typical of other marsupials, with paired uteri and cervices. Each ovary, together with the fimbriated proximal extremity of the otherwise short convoluted oviduct, is hidden away inside a peritoneal pocket or bursa which has a small aperture (the peritoneal ostium) located in its posterior wall. The vaginal system consists of a median vaginal canal (through which the foetus passes) divided by a septum, an anterior *cul de sac* region into which each cervix protrudes, and two narrow lateral vaginal canals (through which sperm travel) connecting the *culs de sac* and the urogenital sinus. The urogenital sinus, into which the very short urethra also discharges, is approximately 40–45 mm long and opens into the cloaca, anterior to the rectum and immediately caudal to the bifid clitoris, which is a distinctly reddish triangular structure (MacKenzie 1919; Obendorf 1988).

Like other marsupials, male koalas have a prominent scrotum which is anterior to the cloacal opening. The penis has a bifid glans, each half of which is further grooved to give a four-pronged appearance. Slightly retroverted cornified spines are located on the penile shaft. The unlobed prostate encircles the urethra immediately caudal to the urinary bladder, and there are three pairs of bulbourethral glands (Young 1879; Temple-Smith & Taggart 1990).

2.5 Lymphatic system

Topography of the major superficial lymph nodes, gut-associated lymphoid tissues and efferent lymph pathways have been described (Hanger & Heath 1991, 1994).

The superficial inguinal, superficial axillary, rostral mandibular, mandibular and facial lymph nodes are palpable in healthy koalas (Hanger & Heath 1991; Blanshard 1994). The mandibular nodes are quite small and difficult to discern when not enlarged. The superficial inguinal nodes are like flat lima beans approximately 10–20 mm in length, lying in the groin near the origin of the epipubic bones. Several small axillary nodes form a tapered cord-like chain 15–20 mm long, deep in the axilla. The rostral mandibular lymph node is approximately

5–8 mm long, positioned caudal to and slightly ventral to the commissure of the lips, over the mandible at the anterior border of the masseter muscles. Facial nodes sometimes may be present over the maxilla, 10–20 mm below the eye. Popliteal and subiliac lymph nodes are absent. The ileocaecal junction is flanked by two circular caecolic lymphoid patches each about 20–30 mm diameter, and there is a cluster of small mesenteric lymph nodes, two to four of which may be 7–12 mm long (Hanger & Heath 1994).

2.6 Miscellaneous anatomical features

The linea alba is broad and almost translucent, and there is no curtain-like omentum. Koalas have an intact mediastinum, so some pathological conditions can be confined to one side of the thoracic cavity. The surface of the cerebral cortex lacks conspicuous sulci.

2.7 Fat depots

Even in healthy well-nourished captive koalas, fat depots are meagre in comparison to those of most eutherian mammals. Localised subcutaneous fat may be present in the groin or axilla, and intra-abdominal depots may be observed in animals in good body condition around the edges of the kidneys, in the splenic mesentery and, in exceptional cases, as a veneer over the lumbar muscles.

2.8 Food intake

Mean food ingested (wet matter) has been estimated in feeding trials with *Eucalyptus punctata* to be within the range 64.2–102 g/kg$^{0.75}$/d (Harrop & Degabriele 1976). Based on quoted average body weights, we could therefore expect Queensland koalas to consume approximately 220–415 g of this leaf per day, and Victorian koalas to consume approximately 320–660 g of leaf per day.

2.9 Water turnover

Free-ranging koalas rarely drink, deriving water from that preformed in eucalypt leaves and any moisture on the surface of the leaves (Lee & Martin 1988). Water balance in captive and free-ranging koalas has been studied (Degabriele et al. 1978). Five captive koalas showed mean water intakes from preformed water in eucalypt leaves of approximately 25–30 g/kg/d. Mean total water uptakes for captive animals from preformed water in leaves, metabolic water and drinking water (not always made available) were about 45–70 g/kg/d (63.3–92.3 g/kg$^{0.8}$/d). When drinking water was withdrawn, the most important contribution to maintaining water balance was a reduction in faecal water content, which decreased from 52–54% to 43%. The measured water turnover of these captive animals was about half or less than that calculated for nine free-ranging koalas, at 122 g/kg/d (179 g/kg$^{0.8}$/d). Mean total body water (5.35 L) represented 77.4% of body weight. Restrictions in activity imposed on the captive koalas by confinement in small enclosures were considered the most likely cause of the observed difference in water turnover rates.

2.10 Thermoregulation and metabolic rate

Under laboratory conditions, koalas subjected to various air temperatures from about 4.5–30°C generally maintained their body temperature between 35.5–36.5°C. When air temperature was 40°C the koalas' body temperature rose to 37.0°C. The dorsal fur of the koala is the most insulative of any marsupial examined to date. The ventral surface has half the hair density of the dorsal surface, and is more reflective of solar radiation. Postural adjustments (particularly in relation to wind direction) confer insulative flexibility (Degabriele & Dawson 1979).

In order to conserve energy, the koala leads a sedentary lifestyle and has a very low metabolic rate—only 74% of the predicted value for a marsupial (Degabriele & Dawson 1979). Panting is the major means of evaporative cooling at high ambient temperatures. Sweat glands are found on the palmar and plantar surfaces of the paws, but these areas do not have any specialised vascular adaptations for effective evaporative cooling to occur (Degabriele & Dawson 1979). Continuous panting and profuse salivation, which may wet the chest and fore limbs, have been recorded occasionally at extremes of temperature. On occasions, the palmar and plantar surfaces of the paws reportedly become encrusted with salt (Eberhard 1972). The lower critical temperature for koalas (i.e. the temperature above which their body's heat production is minimal) is about 10°C in winter and 15°C in summer, the difference due to seasonal changes in fur length and possibly tissue insulation. Basal metabolic rate was recorded at 25°C (Degabriele & Dawson 1979). Thus, if koalas are held in ambient temperatures somewhere between 10–15° and 25°C, they will be expending minimal energy on thermoregulation.

2.11 Adrenocortical function

The koala has a low adrenal weight:body weight ratio (approximately 45 mg/kg) compared to eutherian mammals. Otherwise, the adrenals have a typical mammalian structure, and all gross and (in one study) histological changes evaluated in association with disease were considered a normal mammalian response (Booth et al. 1990). Cortisol is the major glucocorticoid present in the blood of the koala. Its concentration fluctuates in irregular slow pulses throughout each 24 hr period (within the range <1–46 nmol/L), so that measurement in a single blood sample would be of little use in interpreting the effect of a presumed stressor (McDonald et al. 1990).

2.12 Immune response

The apparent susceptibility of koalas to infectious and neoplastic disease has resulted in a widespread belief that koalas are inherently immunocompromised, or 'immunologically lazy' (Wilkinson 1989). The first studies specifically designed to characterise the koala's immune response initially focused on the establishment of an *in vitro* cellular assay to examine koala lymphocyte proliferative responses to a range of mitogens. Antibody (Ab) responses and cell-mediated immunity were then investigated *in vivo* and *in vitro* following challenge with a variety of pathogenic and non-pathogenic antigens (Ags) (Wilkinson 1989, 1996). The results indicated reduced immune responses to the range of Ags classically utilised to monitor immune capability in eutherian mammals. Analysis of the kinetics and dynamics of Ab production to a wide range of protein Ags demonstrated that humoral responses were slower than those of other species (see 10.5.1e). These studies also suggested that the cell-mediated immune responses were less well-developed than in other species (Wilkinson 1989, 1996). In contrast, another major study investigated the *in vivo* immune response of koalas to innocuous Ags and determined that the koala was capable of producing a range of strong immune reactions to these Ags, including Ab and T-cell mediated responses (Emmins 1996). At least two classes of immunoglobulin (IgG and IgM) have been recognised in the koala (Wilkinson 1989, 1996; Emmins 1996).

Recent and ongoing studies aim to provide more information on the response of the koala immune system to infectious diseases (Hemsley & Canfield 1997; Higgins et al. 2005a). An understanding of the impact of retroviral infection on the immune system of koalas has only recently begun to develop (see 10.11.1). Koala retrovirus (KoRV) is believed to cause immunosuppression; increased levels of KoRV have been seen in animals with clinical chlamydiosis, and high levels of KoRV RNA in plasma have been causally associated with leukaemia/lymphoma (Tarlinton et al. 2005).

3 REPRODUCTION

3.1 Breeding season

Koalas breed seasonally, and the majority of births occur during summer. In Victoria the mating period begins in September (Lee & Martin 1988). Young are born between October and May, with most births occurring between the beginning of November and the end of January (Martin & Handasyde 1990). The breeding season is slightly earlier for northern populations than for the southern populations. At an institution in south-east Queensland holding captive koalas, the first matings for each breeding season generally occur in early August, and the last in late April (O'Callaghan & Blanshard 1992). In a captive colony in the northern hemisphere the peak breeding period is March–May (Thompson 1987).

3.2 Reproductive activity in males

Free-ranging male koalas are sexually mature at approximately 2 yr of age, however, competition from larger males restricts opportunities for mating until they are 4–5 yr old (Martin & Handasyde 1990). While male koalas are capable of mating at any time of year, reproductive activity is increased during the breeding season. Studies of males on French Island (Vic.) determined that there is a marked increase in mean plasma testosterone concentration in late July–August, with the mean concentration remaining high until January (Handasyde et al. 1990a). This increase in mean plasma testosterone concentration coincides with increased frequency of male reproductive behaviours, including 'bellowing' vocalisations, scent marking of trees with the sternal gland, and aggressive male–male interactions, prior to the commencement of reproductive activity in October (Mitchell 1990b). Adult male koalas move frequently and have larger home ranges during the breeding season (Mitchell 1990a).

3.3 Reproductive activity in females

Free-ranging females generally commence breeding late in their second or early in their third year (Martin 1981); they may produce one offspring per year and continue to breed until they are 14 yr old (Martin & Handasyde 1990). Two captive females have been recorded giving birth before 13 mo of age and at 16.5 yr old, respectively (O'Callaghan & Blanshard 1992), with successful weaning of the young in both cases.

Handasyde (1986) determined that oestradiol and progesterone are the major reproductive steroids in the peripheral plasma of the female koala. Studies of females on French Island (Vic.) demonstrated that the plasma concentration of oestradiol was highest at the beginning of the breeding period in October, and remained high until the following January. Between March and October, plasma oestradiol concentration was relatively low. The concentration of progesterone also exhibited a seasonal pattern, being low during the period of reproductive quiescence and intermittently rising markedly during November–March.

3.3.1 Oestrous cycle

If female koalas are prevented from mating, they are seasonally polyoestrous (Handasyde et al. 1990a; Johnston et al. 2000a). Peaks in plasma oestradiol occur during oestrus and immediately before birth (Johnston et al. 2000a). A number of studies have indicated that the luteal phase of the oestrous cycle is induced by the physical act of mating (Handasyde et al. 1990a; Johnston et al. 2000a, 2000b; Johnston et al. 2003). Plasma progesterone remains low during non-mated oestrous cycles, suggesting that these cycles are non-luteal and anovulatory. After mating, oestrous behaviours cease and plasma progesterone concentrations rise markedly (Johnston et al. 2000a).

Behavioural observations, morphology of the pouch and plasma concentrations of oestradiol and progesterone have been used to characterise the oestrous cycle in mated and non-mated captive koalas (Johnston et al. 2000a). The mean duration of 22 non-mated oestrous cycles (with no luteal phase) was 32.9 ± 1.1 d. The mean duration of oestrus for 24 non-mated cycles was 10.3 ± 0.9 d. The inter-oestrus interval for six mated non-parturient oestrous cycles (with a luteal phase) was 52.5 ± 0.8 d. Behavioural oestrus in captive female koalas is characterised by weight loss, vocalisation, restlessness, pseudo-male mounting of cage mates, and episodes of convulsive jerking or 'hiccoughing' behaviour, particularly in response to the call or presence of a male (Smith 1980; Thompson 1987; O'Callaghan & Blanshard 1992).

Johnston et al. (2000a) suggested that physical stimulation of the urogenital sinus is required to initiate the luteal phase, including ovulation. However, subsequent research determined that the luteinising hormone surge which occurs 24–32 hr post-insemination can be induced by deposition of semen without physical stimulation, and it was proposed that a factor (or factors) present in koala semen promotes induction of the luteal phase (Johnston et al. 2004).

3.3.2 Gestation and parturition

Smith (1979) reported a gestation period of 34–36 d. Visual confirmation of the presence of young during twice-daily pouch checking of mated captive koalas in south-east Queensland between 1989–1995, however, determined that for 103 pregnancies with known errors of 0.00–0.87 d, values for maximum possible duration of gestation were mean 34.02 ± 0.55 d, median 33.97 d, observed range 32.98–35.64 d (W Blanshard & P O'Callaghan unpub.).

Females have been observed to assume a slightly hunched position during parturition (Thompson 1987; O'Callaghan & Blanshard 1992). Parturition is rapid (approximately 30–90 sec elapses between the neonate's emergence from the cloaca and its entry into the pouch), and appears to cause some discomfort for the female (O'Callaghan & Blanshard 1992). Blood, fluid or membranes may be visible on the fur around the cloaca or pouch following parturition, but this is not always the case. Neither the degree of apparent resentment of pouch inspection nor the tightness of the pouch sphincter should be assumed to provide unequivocal evidence of parturition or the timing thereof (W Blanshard & P O'Callaghan unpub.; Johnston et al. 2000a).

Koalas usually have only a single young (Fig. 8.1). Although twins have been recorded, their survival may be compromised by the limited capacity of the pouch, which may result in premature emergence (Martin & Handasyde 1990).

Figure 8.1 Koalas usually have only a single young. Photo: W Blanshard.

4 HUSBANDRY

4.1 Housing

Structures used to house captive koalas vary considerably in design from extensive open free-range areas containing mature trees, to semi-enclosed or totally enclosed pens furnished with a network of poles and branches (Drake et al. 1990). A koala enclosure must effectively and safely contain its occupants, cater to their arboreal existence by providing freedom of movement both vertically and horizontally, offer shelter against extremes of climate, allow the animals ready access to appropriate quantities of palatable uncontaminated food, and be easy to maintain in a clean and hygienic condition.

Unless fully enclosed, walls or fences lined with material such as sheet metal are required to prevent escape from the enclosure. Walls should be a minimum of 1.2 m high, preferably with a rounded top edge, and be smooth or fully lined on the inside surface with sheet metal. Higher fences which would otherwise allow climbing (e.g. wire mesh) can be koala-proofed by affixing a strip of sheet metal at least 600 mm in height to the inside of the fence, beginning at least 1 m above ground level (P O'Callaghan *pers. comm.*). As koalas can jump horizontally from trees or poles, any climbable structure within the enclosure should be at least 1.8 m from the top of the perimeter wall or fence. The trunks of mature trees can be encircled with sheet metal guards to prevent koalas climbing to areas where they would be irretrievable (Drake et al. 1990).

4.2 Enclosure fixtures and fittings

Typically enclosures are furnished with upright natural timber poles bearing single or multiple resting forks. These may be linked by horizontal natural timber poles. It is usual to provide at least two resting forks per animal. The most common method of presenting browse to the koalas is to have tubular water containers ('water buckets' or 'leaf pots'), into which the cut ends of branches can be inserted, attached to the upright poles at an appropriate height to allow the leaves to be within reach of koalas sitting in the resting forks (Drake et al. 1990; Blanshard 1994; Jackson et al. 2003). Six millimetre thickness, 100 mm diameter PVC plumbing pipe (with an end cap) is ideal for practical and extremely durable water buckets (P O'Callaghan *pers. comm.*). Two different designs are illustrated in Blanshard (1994) and Jackson et al. (2003).

Fresh drinking water should always be available in a separate receptacle on the floor of the enclosure. Koalas have been known to eat soil occasionally. The caecal contents of six dead koalas examined specifically for such material revealed varying quantities of sand grains and small stones in each, with the total weight of gravel ranging from less than 1 g up to 7 g (Bolliger 1962). If the enclosure does not have a soil substrate, some captive facilities elect to provide a bowl of soil. Overhead sprinklers and/or shadecloth shelter sails can be used to keep animals cool during hot weather.

4.3 Enclosure maintenance

Faecal pellets and fallen leaves should be raked up and removed from under the cage furnishings every day. Urine should be hosed off concrete floors daily or every second day, and soil or sand substrate should be replaced if heavily soiled with urine. Koala urine leaves sticky residues on the enclosure furniture, and faecal pellets become trapped or trampled into crevices in the surface of poles and resting forks. Thorough physical removal of faecal material followed by scrubbing of all poles once per week is recommended. A detergent with or without a mild disinfectant (quaternary ammonium compound or benzalkonium chloride) (G Rawlinson *pers. comm.*) can be used. Poles should be rinsed before re-use. All browse should be removed and protected from contamination during cleaning procedures. Water containers should be removed, rinsed and refilled before the leaves are replaced.

Water level in the leaf containers should be checked and topped up twice daily (more frequently in very hot weather), as freshly cut eucalypt foliage takes up significant amounts of water. At least some new browse should be made available every day. Branches from the preceding day can be inspected to see if it is worthwhile leaving any of them in the enclosure, but if no young leaves ('tip' or 'tips') remain the branches should be replaced.

4.4 Hospital housing

Ideally, all but the most acutely ill or injured koalas should be housed individually in large cages or enclosures which allow them to climb freely, both vertically and horizontally. These cages should be well away from the sight/sound/smell of domestic animals. Each koala should be provided with at least two comfortable resting forks. These can be on the same large pole, or on separate poles with a horizontal pole connecting the two. A near-vertical upright pole with a side branch (≥50 mm in diameter) at an angle of 45–60°, no closer than 30–40 cm from the top of the upright, makes a suitable resting fork. Poles cut from stringybark or ironbark eucalypts are easier for debilitated or injured koalas to climb than smooth-barked species, but are more difficult to clean effectively and disinfect. A gently-inclined, rough-barked pole from the ground to a raised fork may enable a weaker koala to reach an elevated position.

Water containers for branches or leaves should be placed at the base of, or attached to, the sitting pole(s). The water container should not be so wide that the koala may inadvertently descend into it when climbing down the tree. Branches of leaves need to be securely tied or held against the upright poles, as the koala may try to sit among the branches of browse. Drinking water should be made available in a stable shallow bowl on the floor of the enclosure.

If the koala is very weak or unable to climb, a stable fork or other structure for support at ground level will be required. The use of plastic floor matting which allows urine to drain away underneath will greatly reduce soiling of the fur in koalas confined to the ground. Intensive care patients (those severely debilitated through illness or injury) can be set up in standard small companion animal hospital cages. Very thick soft padding such as a padded dog bed covered with a towel or blanket should be provided, on which the animal can lie or be propped up. It is common for wildlife carers to manage a debilitated koala indoors at home using an oval plastic laundry basket containing a roll of padding that the animal can lean against (J Starr *pers. comm.*). A second laundry basket may be inverted and tied on top of the first if extra security is required. Leaf in an appropriate water container should be secured close to the animal. Additional warmth may be provided using a heating pad, hot water bottle or heat lamp.

4.5 Moving animals and transport

Most koalas adapt quickly to routines, and can be lifted and moved short distances without much fuss if this is done on a regular basis. A sturdy bucket may facilitate moving individuals within a hospital facility. The koala is lifted by the wrists and lowered into the bucket, in which it can sit securely, holding the rim of the bucket for support. Bucket and koala can be lifted, carried or weighed without unduly disturbing the animal. Urine samples are often left behind in the bucket and may be useful if a specimen is required.

Koalas can be transported safely in hessian sacks (cushioned to prevent untoward movement if in a motor vehicle) for short periods. For longer trips a well-ventilated wooden crate, fitted with either a vertical centre timber pole to grip or a timber fork, may be used. Timber-framed, mesh-sided or slatted transport crates with the outside covered by hessian or similar material are more satisfactory, and are commonly used to transport koalas over long distances and in planes. A fork and leaf container are fitted to the centre of the floor of the crate. Australian Government specifications apply to crates used for overseas air transport (DEH 2004).

Ideally, koalas should not be transported in temperatures outside their thermoneutral range (below 10–15°C or above 25°C). Be particularly careful of higher than ambient temperatures in the interior of a motor vehicle, as both air flow and the koala's capacity to make thermoregulatory postural adjustments are restricted.

4.6 Individual marking and identification techniques

The permanent and positive identification of captive koalas is essential for effective animal management. Free-ranging koalas that are part of field studies, and contracepted or translocated animals, are also marked using various techniques. Passive integrated transponder (PIT) tags (microchips) and/or eartags are used most commonly. Tattooing the inside of the pinna has also worked well (P O'Callaghan *pers. comm.*; Jackson et al. 2003). Eartags have the advantage in field situations that animals may be identified from a distance (particularly if large coloured tags are used). A disadvantage is that tags may fall or be pulled out, so they frequently are used in conjunction with microchips.

It is essential that PIT tags be inserted correctly to prevent loss or infection. The standard site for insertion in koalas is SC on the dorsal midline between the scapulae. Directing the tip of the insertion needle caudally and closing the skin wound with tissue glue reduce the likelihood of the microchip falling out post-insertion (Vogelnest 1995).

5 NUTRITION

5.1 Browse selection and dietary preference

Koalas are strictly folivorous and exist almost exclusively on the leaves of certain *Eucalyptus* spp. The natural distribution of eucalypt species varies tremendously between and within different Australian states, and so does the koalas' opportunity to utilise them as food trees. There are several hundred different eucalypts, and koalas are known to feed on at least 40–50 species throughout eastern Australia (Southwell 1978; Warneke 1978; Drake et al. 1990; Pahl & Hume 1990). Koalas have occasionally been observed eating the foliage of non-eucalypt species, but these do not seem to be used significantly as food trees (Martin & Handasyde 1999).

While utilisation of different eucalypts may vary from one locality to another (and may change within localities from one season to another), a relatively small number of eucalypt species tend to be eaten regularly by the majority of koalas that have access to them. This tendency is recognised consistently in many localities throughout eastern Australia where the distribution of these eucalypts and koalas coincide, and they are regarded as preferred food trees or primary browse species (Fleay 1937; Robbins & Russell 1978; Warneke 1978; Hindell & Lee 1990; Martin & Handasyde 1999).

Koalas have been reported to show a degree of preference for young rather than mature foliage, but a preference for young leaves is not consistent. Animals in a socially stable population may show preferential use of some trees within their individual home ranges. While koalas may generally utilise the foliage of a particular eucalypt species, foliage from some individual trees of that species may not be eaten. Individual animals may differ in their utilisation of different eucalypt species and in their utilisation of individual trees within each species (Fleay 1937; Robbins & Russell 1978; Warneke 1978; Hindell & Lee 1990; Martin & Handasyde 1999). There have been a number of studies investigating the feeding patterns of koalas, to determine the characteristics of preferred foliage. Leaf selection has been found to correlate positively with leaf nitrogen content and water content, and correlate negatively with fibre content and total sugar content (Pahl & Hume 1990; Osawa 1993). The role of *Eucalyptus* spp. essential oils in determining koala leaf selection has also been investigated. Phenolic compounds known as formylated phloroglucinol compounds (FPC) have a strongly negative effect on feeding by marsupial arboreal folivores, regardless of other features of nutritional composition of leaves (Lawler et al. 1998). Koalas have been found to have a higher tolerance for FPC than two other arboreal folivores, the common ringtail (*Pseudocheirus peregrinus*) and the common brushtail (*Trichosurus vulpecula*), giving them a greater ability to utilise eucalypt foliage as a sole source of nutrients. While FPC may influence whether a plant is consumed, the amount eaten is probably determined by a range of factors including nutrient composition (Hume 1999; Moore & Foley 2005).

5.2 Digestion of eucalypt leaf

Eucalypt foliage is low in available carbohydrate, protein and ash but is high in lipid and phenolic compounds (including lignin). The lipid fraction includes essential oils and other plant secondary metabolites or antinutrients (Hume 1999). The fibre in eucalypt leaves makes up 30–50% of dry matter (DM) in the leaf, and tends to be highly lignified and therefore difficult to digest. Fats comprise around 17% of the DM, and much of this fraction consists of essential oils, which are toxic, and waxes, which are not digested to any great extent (Cork 1987). Proteins make up only 5–15% of the DM. Phenolic compounds may reach 25–30% of the DM and have the potential to form complexes with dietary proteins, making the latter difficult to digest (Cork 1987; Hume 1999).

Fluid and fine particles (which are potentially more digestible because of their greater surface area to volume ratio) are selectively retained in the caecum and proximal colon, while less digestible coarser particles are eliminated more rapidly. Mean retention times for digesta are 99 hr (4.1 d) for particles and 213 hr (8.9 d) for fluid (Cork & Warner 1983). Despite the size of the hindgut, studies show that the small intestine is the major site of absorption of the products of digestion, and that bacterial fermentation seems to provide only about 9% of the total energy derived from eucalypt leaves (Cork & Sanson 1990).

5.3 Feeding koalas in captivity

5.3.1 Eucalypts fed to captive koalas

Foliage from preferred food trees forms the staple diet of captive koalas. Diets are varied by providing foliage from other known food trees. The number of eucalypt plantations established by zoos and wildlife sanctuaries is testament to the fact that the same trees can be cut many times to provide browse for koalas. Regrowth, whether plantation-grown or on naturally growing *Eucalyptus* spp., is the basic food resource for most captive koala colonies (Drake et al. 1990; Blanshard 1994).

Preferred food trees utilised to feed captive koalas include:

- *E. camaldulensis*—river red gum;
- *E. globulus* spp.—Tasmanian & southern blue gum;
- *E. microcorys*—tallowwood;
- *E. ovata*—swamp gum;
- *E. punctata* spp.—grey gum;
- *E. tereticornis*—forest red gum;
- *E. viminalis*—manna gum.

Some other food trees commonly utilised to feed captive koalas include:

- *E. goniocalyx*—long-leaved box;
- *E. grandis*—flooded gum;
- *E. haemastoma*—scribbly gum;
- *E. maculata*—spotted gum;
- *E. moluccana*—gum-topped box;
- *E. nichollii*—narrow-leaved black peppermint;
- *E. obliqua*—messmate;
- *E. resinifera*—red stringybark;
- *E. robusta*—swamp mahogany;
- *E. saligna*—Sydney blue gum;
- *E. sideroxylon*—red ironbark.

Koalas also are maintained in captivity outside their natural range (in Tasmania, Western Australia, parts of South Australia, and overseas). These animals are provided with foliage grown in plantations as well as apparently palatable local eucalypt species. For comprehensive state-based lists of eucalypt species utilised to feed captive koalas, refer to Drake et al. (1990), Jackson et al. (2003) and DEH (2004).

Individual koalas vary in their preference for different eucalypts. Just because a species appears on a list of koala food trees doesn't mean that all individuals will eat it. A minimum of two or three branches of palatable browse approximately 1 m in height, bearing as much young tip as possible, are required daily to feed a healthy koala (Booth 1991). It is important to offer a choice of species each day, and to vary the species offered from one day to the next.

5.3.2 Collection and presentation of browse

Cut browse should be kept off the ground as much as practicable to prevent contamination of the leaves. Branches can be sprinkled with water to help preserve them in transit, and they should be transported in a closed vehicle or under a tarpaulin to prevent drying by the wind and exposure to exhaust fumes.

Branches should be placed upright with their cut ends in water as soon as possible after harvesting. If not destined for use on the same day they should be stored in a cool location in the shade and out of the wind, or in a coolroom.

Foliage from at least two or three known food tree species should be made available each day, although full loads of the most preferred species can be offered occasionally.

The choice of species should be changed from one day to the next. Foliage of some eucalypt species may be eaten quite enthusiastically on the first day it is offered, but may be rejected on subsequent days when its novelty value has waned.

Given the opportunity, captive koalas will selectively tend to eat tips and younger leaves before mature leaves, for the majority of eucalypt species. If large branches bearing tip are cut, some mature leaf is always available at the base of the branch anyway.

New foliage should be provided every day, and the cut ends of the branches should always be standing in water to keep the leaves as fresh as possible.

Branches on which the tip has been incompletely browsed can be rotated or repositioned within the enclosure to facilitate access to all the leaves.

Identify whether browse has been offered in excess of requirements, or whether foliage has been rejected by the animals in favour of other branches of the same or different species. This is easiest if those who put up the branches also take them down again.

5.3.3 Artificial diet

Providing fresh browse for captive koalas is labour-intensive and expensive. This prompted investigation into development of an artificial diet. Pahl and Hume (1990) reported some success feeding a combination of a thick paste (made of low-lactose milk powder and ground *Eucalyptus* spp. foliage) and a thin biscuit. The biscuit was manufactured to contain moisture, nitrogen and fibre in concentrations similar to those found in preferred leaf. While the artificial diet provided a valuable supplement to the natural diet, it did not completely replace the need for fresh foliage.

5.3.4 Feeding hospitalised koalas

a Determining food preferences

The appetite of sick koalas usually becomes more capricious than normal, and even captive animals may decline to eat species which they consume avidly when in good health.

The growth stage of leaves also becomes more critical when animals are sick. Sometimes magnificent tip will be passed over in favour of withered mature leaves at the base of a branch. Stripping and eating bark but not leaves may occur. Alternatively, the koala may bite through woody stems and eat these, allowing leaves to drop to the ground. Remember that an alternative to such bizarre eating behaviour is to eat nothing at all, so the carer should try to cater to the koala's whims however strange they seem.

b Exploiting feeding opportunities

Koalas often start eating after being disturbed from sleep or rest. Paradoxically, this can have been an unpleasant disturbance (such as receiving an injection), or something as simple as having their cage opened and fresh leaves put in or their existing leaves rustled or rearranged. Many sick koalas eat quite enthusiastically for a short time after fresh leaves are placed in their cage, then lapse into a trance-like state or fall asleep before a significant amount has been consumed. In these cases, the time spent eating and the amount of leaf eaten can sometimes be increased by offering different species of eucalypts in turn, in increasing order of preference, making a substitution each time the koala stops eating.

c Hand-feeding

Koalas generally will not eat individual leaves that have been removed from branches and offered by hand. Nevertheless, animals which are not distracted from eating by close human proximity may have their leaf intake selectively modified. Small branchlets of leaves can be held near the koala so that the leaves it might derive the most benefit from (e.g. younger, higher water content, less fibrous, easier to digest) are always the most accessible. Some koalas even stop trying to manipulate the branchlet themselves, and just expect the next leaf to be ready as soon as they are. Although very time-consuming, this process increases the number of leaves eaten in a given period and maximises consumption of younger leaves. These may be important factors in the fluid and energy balance of koalas which are spending less time eating than they should.

d Supplementary feeding

The nutrient intake of koalas can be boosted by supplementation with low-lactose high-energy milk powders mixed to a semi-liquid paste with water and administered orally via a syringe (Wood 1986, 1987; Handasyde et al. 1988; Pahl & Hume 1990; Osawa & Carrick 1990). Powders commonly used for this purpose have included (but are not limited to):

- Di-Vetelact® low-lactose animal supplement (Sharpe Laboratories);
- Infasoy® (S26 Soy) Infant Formula (Wyeth Nutritionals);
- Karicare™ Soya Formula (Nutricia Australia);
- PediaSure® Complete (Abbott Australasia);
- Portagen® (Mead Johnson Nutritionals);
- Prosobee® (Mead Johnson Nutritionals).

There is a diverse range of sources of protein, fat and carbohydrate in these products, and despite their specialised normal diet koalas seem remarkably tolerant of all these ingredients. It appears to matter little which product is used (Booth 1991; Blanshard 1994; Booth & Blanshard 1999; C Flanagan *pers. comm.*; J Hanger & G Gipp *pers. comm.*; A McKinnon *pers. comm.*; G Rawlinson *pers. comm.*).

Koalas vary in their response to attempted supplementary feeding. A proportion of animals find the taste to their liking, learn to recognise the syringe as containing something they would like to eat and actively take the tip of the syringe into their mouths. Others may be indifferent to supplementation and not voluntarily accept the syringe tip, but will swallow most of the administered paste without undue protest or evasive action. A small proportion of koalas appear to despise either the paste or the physical process involved with being supplementary fed (or both); in these few, the energy they expend trying to avoid ingesting the paste initially probably negates the beneficial effects of the minimal amount swallowed. With gentle persistence, however, very small but increasing amounts of paste may be taken during successive days until a significant quantity of extra nutrients is being accepted.

Catheter-tipped syringes are ideal for paste administration to co-operative koalas, but may deliver too large a mouthful of paste to animals still being accustomed to the supplementation process. It is common practice to give a small volume (e.g. 10 mL) of water PO by syringe following paste administration (as the paste may be sticky in the animal's mouth), and after every feed it is important to clean any residue off the koala's fur using a damp cloth or paper tissue.

Five to 8 kg koalas eating little or no leaf, when provided approximately 60 mL of paste per day (usually in 2–3 divided doses), are reported to maintain their body condition (Blanshard 1994; Booth & Blanshard 1999). Smaller quantities of paste (e.g. 20–40 mL/d) are given to animals which are eating some leaf, or those which have regained their appetite but still have to make up considerable deficits in body condition after illness. When koalas are eating nothing but paste, finely-blended eucalypt leaf can be added to the paste mixture to provide some fibre (see 10.2.6) and to ensure the specialised bacteria in the caecum and proximal colon are not totally deprived of their usual medium.

6 RESTRAINT AND ANAESTHESIA

6.1 Physical restraint techniques

6.1.1 Retrieval or capture from trees

A koala may be encouraged to back down a tree by a hand or other object placed on top of its head. Koalas which are out of reach can sometimes be 'flagged' down, using a rod or pole with a cloth on the end persistently waved over or touched against its head (Fig. 8.2b). Placid koalas can have their fore feet unhooked from the tree and slowly be pulled downwards and away from the tree to disengage their hind feet, before additional physical restraint (see 6.1.2) or placement into a sack or transport box, as appropriate. Sick or badly injured koalas are usually fairly docile and relatively easy to handle. Less tractable koalas (e.g. healthy free-ranging adults) can be very aggressive and may bite and/or scratch in attempts to escape. As these animals descend the tree they can be enveloped in or backed into a hessian sack held open against the trunk.

During field studies, koalas are sometimes captured by researchers who climb as high up the tree as feasible, then use a long pole to drape a rope noose (in which a knot has been incorporated to prevent overtightening) around the animal's neck. Traction on the rope is used to prevent the koala climbing any higher, and it is then flagged down within reach, secured in a sack and lowered to assistants waiting below. Occasionally, free-ranging koalas will leap between branches to avoid capture, or simply launch themselves into open space. It is therefore advisable to have two or more assistants, holding the corners of a tarpaulin or a blanket, under the tree to catch the koala should it fall (Fig. 8.2a).

Successful capture of free-ranging koalas from trees by remote injection of anaesthetic agents via a small dart has been reported (Lynch & Martin 2003). The method was trialled as a means of reducing the need for climbing and to shorten the duration of capture procedures, therefore decreasing stress for the animals. Darted koalas were caught using cargo nets as they fell or were dislodged from trees.

6.1.2 Physical restraint

The majority of koalas respond to gentle and adept handling, so the degree of physical restraint applied should be proportional to the task and can be scaled up according to the animal's reactions. Some useful techniques for handling and manually restraining koalas are shown in Figure 8.3.

Docile koalas can be picked up from behind, by grasping around the wrists. Lift the koala off the ground fairly slowly so that it can feel its weight is being supported. Even placid animals will only dangle patiently for a short time (Fig. 8.3a). Lateral traction on one or both arms will be necessary if it attempts to bite.

Alternatively, first the wrists then the ankles can be encircled in an initial handgrip so that the koala is restrained in a sitting position, facing away from the handler (Fig. 8.3b). After lifting, rest the animal's back against your body (Fig. 8.3c), and/or support its rump

Figure 8.2 Capture of free-ranging koalas. a) Assistants with tarpaulin positioned below the tree during a catch, in case the koala jumps or falls. b) Flagging. Photos: W Blanshard.

on some other object (e.g. table or lap). Pull the koala in closer to your body if it tries to struggle, for greater control. In co-operative animals, this is an ideal restraint position for collecting blood samples, as you can modify your grip to hold up a cephalic vein (Fig. 8.3d). Greater dexterity is required to avoid attempts to bite, so if the koala is persistent its head should be secured by another person (Fig. 8.3e).

Like all arboreal animals, koalas tend to panic when unable to grip anything with unrestrained paws. If necessary, a koala can safely be held against the ground, or picked up and carried, by grasping it with one hand around the back of the neck and the other over the pelvis, so that the claws and teeth are directed away from the handler (Fig. 8.3f). In relatively shaggy southern koalas the fur provides useful handholds. The fur of northern koalas may be too short to grasp, but their smaller size generally enables the pelvic bones and the back of the neck to be secured within each hand span. Towels, blankets and hessian sacks can facilitate the handling and restraint of aggressive animals (Fig. 8.3g).

A retrospective review of mammal bite injuries treated at emergency departments of Victorian public hospitals 1998–2004 revealed that 0.05% were inflicted by koalas (MacBean et al. 2007). Koalas should never be lifted around their chest (like a human baby). Holding or carrying an unhabituated koala clinging to the handler's body leaves the handler vulnerable to bites and scratches, especially on the face and neck. Unless habituated, only dependent juveniles (which may become more distressed if unable to cling to something for security) should be held this way.

6.2 Chemical restraint

Neither premedication with atropine nor fasting is considered essential for koalas (Blanshard 1994). Excessive salivation under anaesthesia is not usually observed, and the consistency of normal stomach contents does not lend itself to regurgitation. Some clinicians recommend pre-anaesthetic fasting for 6 hr to reduce stomach fill (Shima 1999). Anaesthesia in koalas is generally straightforward, but periods of apnoea and, in un-intubated animals, respiratory stridor due to the long soft palate are relatively common. Changing the animal's position and intubation may alleviate some of these problems.

6.2.1 Injectable agents

Diazepam at a dose of 0.5–1 mg/kg IM (Booth 1991) or 0.5 mg/kg IV (Blanshard 1994) is a useful sedative that can be given to facilitate clinical examination.

Propofol 2.5–3.0 mg/kg as a bolus IV will result in recumbency and light anaesthesia sufficient for

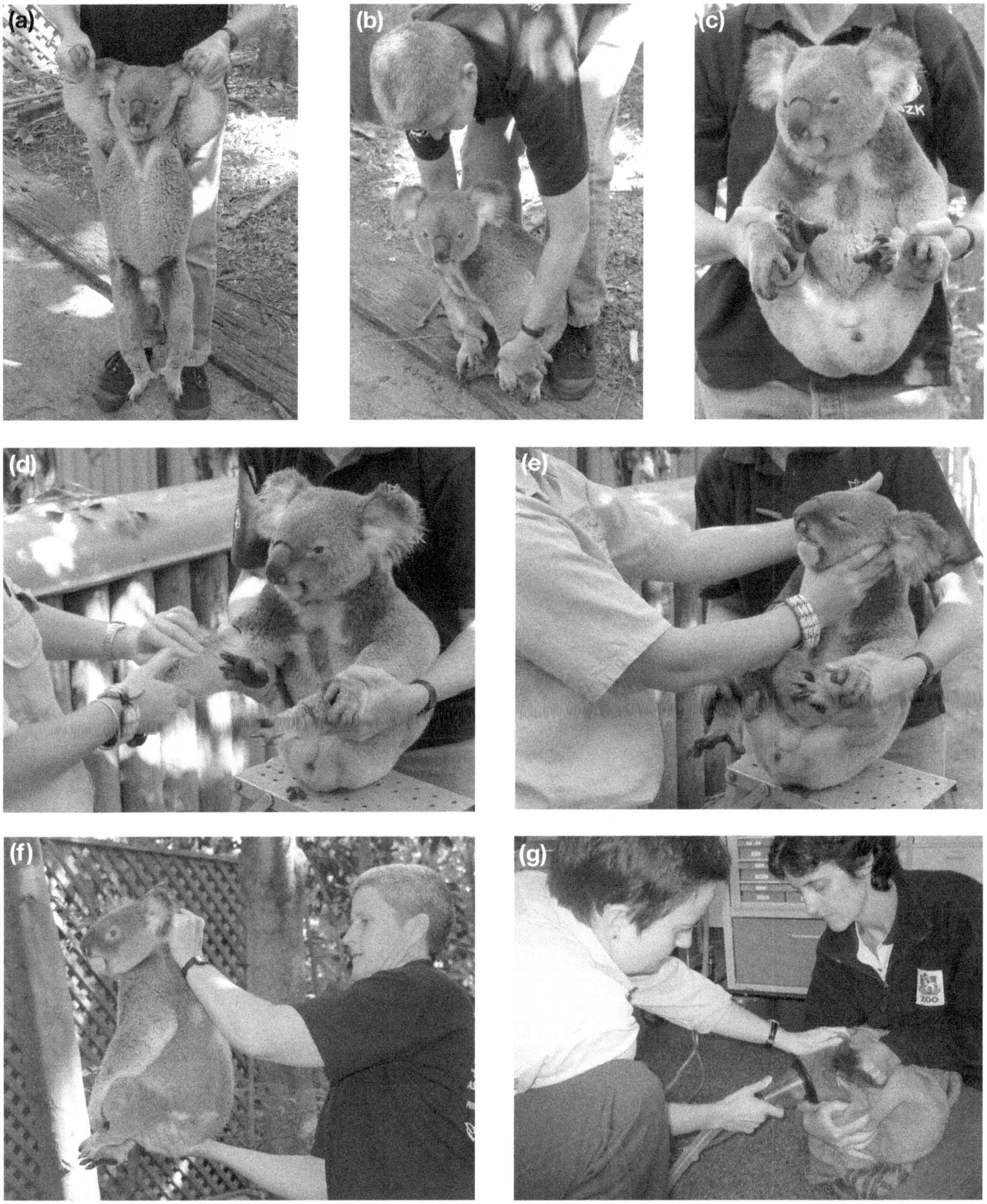

Figure 8.3 Physical restraint options for koalas vary according to task as well as demeanour, temperament and habituation to handling. a) Docile koalas may be moved short distances while dangling by their wrists. b) The hind legs can be secured in the hand-grip. c) Animals can be lifted or carried in a sitting position. d) Once body weight is supported the handler may be able to occlude the cephalic vein. e) A second person may be required to restrain the head. f) Unassisted grip for otherwise-uncontrollable struggling and biting animals. g) A towel, blanket or hessian sack can facilitate restraint, e.g. for gaseous anaesthesia. Photos: a–f) K Bodley and Lone Pine Koala Sanctuary and g) Melbourne Zoo.

intubation in placid captive animals. Larger doses (titrated to effect) may be required when animals are less accustomed to being handled (Blanshard 1993). Standard domestic animal dose rates (6–8 mg/kg IV, administered to effect, and 4–5 mg boluses given as required for maintenance of anaesthesia) have also been used (Vogelnest 1999). Propofol is ideal for very short procedures where good relaxation and fast recovery are desirable, and can be topped up if necessary. It is also useful for induction of anaesthesia prior to maintenance with a gaseous agent. Both induction and recovery are rapid and extremely smooth.

Alfaxalone at 3.0 mg/kg IM or 1.5 mg/kg IV provides sufficient anaesthesia to allow examination and/or intubation prior to supplementation with inhalant agents (J Hanger *pers. comm.*). IV administration provides approximately 10 min anaesthesia. A transient period of apnoea may occur at the time of induction when using an IV bolus.

Tiletamine/zolazepam at 5–10 mg/kg IM or 2.5 mg/kg IV is reported to give good restraint for short procedures (Blyde 1990; Shima 1999). There is some retention of muscle tone and spontaneous limb movements. At the low end of the IM dose range it produces sedation, and at the higher end of the IM dose range surgical anaesthesia. Moderate salivation has been reported with tiletamine/zolazepam anaesthesia, but caused no problems as the swallowing reflexes remained intact (Bush et al. 1990). A dose of 7 mg/kg IM has provided an appropriate level of anaesthesia for minor surgical procedures (e.g. insertion of subdermal contraceptive implants using a needle applicator) in free-ranging Victorian koalas (K Handasyde *pers. comm.*).

Dose rates in the order of 3.5 mg/kg tiletamine/zolazepam plus 55 μg/kg medetomidine given via darts fitted with 12 mm needles provided light to surgical anaesthesia in 9/19 animals, and heavy (9/19) or moderate (1/19) sedation in animals captured after being darted (Lynch & Martin 2003). Variability in response was believed to be due to variability in dart placement and drug deposition in tissues. Another three animals that were caught manually and hand-injected IM with 3.0–3.6 (mean 3.3) mg/kg tiletamine/zolazepam plus 46–59 (mean 52) μg/kg medetomidine directly into the gluteal musculature demonstrated reliable induction to light or surgical anaesthesia. Careful monitoring of anaesthesia was advocated, as medetomidine may cause alterations in heart rate, respiration and blood pressure, and may interfere with thermoregulation. Ready availability of the specific medetomidine antagonist, atipamezole, was recommended. Two of the 19 animals captured after darting with a tiletamine/zolazepam plus medetomidine combination by Lynch and Martin (2003) showed slow respiratory rates and periods of apnoea, and were given 0.5 mg atipamezole IM to restore a more regular respiratory pattern. A third animal received the same dose because its rapid induction time indicated that the dart may have delivered the drugs IV. At the end of a brief (only 5–10 min) period of handling, all animals in the study were given 2 mg atipamezole IV, and recovered to a state of light sedation within 2 min. Most of the koalas still appeared to be very slightly affected by the drugs when released 30–60 min after reversal, although all were able to climb and the releases were uneventful.

Ketamine (16 mg/kg IM) was reported to produce light restraint with muscular hypertonicity (Robinson 1981). Acetylpromazine (0.4 mg/kg IM) or xylazine (0.6 mg/kg IM) was given simultaneously with ketamine to reduce this effect (Robinson 1978, 1981). Koalas took up to 2 hr to achieve full recovery when xylazine was used. Ketamine (5 mg/kg) plus xylazine (5 mg/kg) as a single IM injection reportedly provides deep sedation. This can be reversed by giving yohimbine (0.2 mg/kg IV) (Blyde 1990). Care was advised if animals showed signs of shock or poor health. Up to 15 mg/kg ketamine plus 5 mg/kg xylazine will produce surgical anaesthesia when given IM.

6.2.2 Inhalant agents

Halothane (Dickens 1978a) or isoflurane (Wilkinson 1989; Booth 1991) in oxygen are commonly used for induction and/or maintenance of anaesthesia. Mask induction using 3.5–5.0% isoflurane in oxygen results in smooth rapid induction and rapid anaesthetic recovery. Gradually increasing the isoflurane concentration from 0.5% until anaesthesia is achieved provides a smoother and more stable anaesthetic (D Higgins *pers. comm.*).

Mask induction of hospitalised koalas can usually be achieved by an unassisted operator, without restraining the koala in a sack. Sit the koala on the table, facing away from you. Place the edge of a large towel gently around the koala's neck and secure the towel behind the neck with one hand so that it is snug but not tight. Lean forward against the koala as you apply the mask with your other hand, to prevent it backing away. The koala will probably attempt to scratch the mask off using a front leg, from which you should be protected by the folds of towel. If greater restraint is required, lean over the koala's back and use both elbows to restrict the animal's lateral movement (and to tighten the bottom edges of the towel down onto the table).

Intubation is readily achieved given a little practice and appropriate equipment. A 3.0–5.5 mm endotracheal (ET) tube is suitable for most adult Queensland koalas and Victorian animals under 10 kg body weight. The most awkward aspect of intubation, particularly in smaller individuals, is that the narrowness of the dental arcades at the oropharyngeal opening can impede introduction of a standard laryngoscope blade that is long enough to visualise the epiglottis. Use of a narrow laryngoscope blade, a perspex-tipped pen-type light source, or an angled elongated light stem from a diagnostic kit may be helpful in these cases. The long soft palate usually obscures the epiglottis from direct view until pushed dorsally using the ET tube. Intubation is possible with the animal in standard lateral recumbency, however, some clinicians prefer the koala in sternal recumbency (J Hanger *pers. comm.*), or may try dorsal recumbency so that the soft palate does not obstruct visualisation of the epiglottis (Shima 1999). It is appropriate for practitioners to begin with the patient positioning technique they prefer in more familiar species, e.g. domestic cats.

Using a stylet can assist when directing the ET tube. A standard ET tube stylet can be pre-threaded through the tube so it is protruding from the end by a few centimetres. The stylet tip is used to push the soft palate and epiglottis out of the way so the larynx can be visualised, and is then passed over the epiglottis into the larynx. The ET tube is threaded into the larynx over the stylet, and the stylet withdrawn (J Hanger *pers. comm.*). Another technique for intubating marsupials describes inserting the end of a long polypropylene catheter (e.g. a canine urinary catheter) into the larynx before passing an ET tube over the catheter and into place, then withdrawing the guide (Shima 1999). Alternatively, the ET tube can be pre-threaded over a fine rigid endoscope, which then serves both as a stylet and the means of visualising and entering the glottis (Carlisle et al. 1989; V Nicolson *pers. comm.*). Lubricating the endoscope with silicone spray before placement of the ET tube is recommended if the tube fits snugly over it.

Blind intubation can also be achieved without additional equipment (L Vogelnest *pers. comm.*). With the koala in left lateral recumbency the animal's neck is extended and the tongue pulled out of the mouth by an assistant. The larynx is held externally between the thumb and forefinger of the left hand while the ET tube is passed over the base of the tongue with the right hand, until the tip is resting on the glottis. As the animal takes a breath the tube is passed into the larynx and trachea. A deep plane of anaesthesia makes blind intubation easier.

6.2.3 Anaesthetic recovery

Koalas should be denied the opportunity to climb until fully recovered, but will be less agitated during recovery if they can grip onto something (e.g. the side of a mesh cage, a rolled-up towel or a heavy branch placed on the ground) while their co-ordination returns to normal. Either move the koala to a quiet recovery cage/area soon after extubation, or leave it undisturbed as long as possible and only transfer it to the recovery cage/area after it has spontaneously tried to sit up.

7 CLINICAL ASSESSMENT

Initial assessment of the koala should be performed with the least possible disturbance to the animal. Most koalas will adapt to the routines of captivity fairly readily, given a little time and non-threatening interaction with their carers.

7.1 Physical examination

7.1.1 Body temperature

A thermometer can be inserted several centimetres into the rectum, which opens into the cloaca dorsal to the urogenital opening. Normal body temperature is 35.5–36.5°C (Degabriele & Dawson 1979). Hypothermia is relatively common in ill or debilitated koalas, and the body temperature may fail to register on a clinical thermometer (below 32°C). Hyperthermia is relatively uncommon. Febrile states have been observed occasionally, e.g. 37.9°C in a case of septicaemia and associated inflammation, 38.7°C in a septicaemic koala which was convulsing (Blanshard 1994), and 39.4°C in a case of acute disseminated toxoplasmosis (Dubey et al. 1991).

7.1.2 Pulse rate

Normal pulse rate at rest is around 65–90 beats/min (Blanshard 1994). Sinus arrhythmia may be apparent (Rezakhani et al. 1986).

7.1.3 Respiratory rate

Normal respiratory rate at rest is around 10–15 breaths/min (Dickens 1978a). Free-ranging koalas are unlikely to be 'at rest' when being examined. Panting (rapid shallow breathing with the mouth closed) is normal at higher ambient temperatures, or in agitated animals.

7.1.4 Auscultation

Thoracic auscultation is hindered by the koala's dense fur and by the relatively small area occupied by the lungs. The chest is short and wide, and much of the lateral rib cage protects cranial abdominal organs rather than the heart and lungs. The ventral and ventro-lateral aspects of the thorax usually yield the clearest sounds. The stethoscope should be slid right up into the axillary region. Abdominal auscultation should reveal occasional peristaltic sounds. Percussion with the fingertips should generate only dull sounds, as the normal abdomen contains no accumulations of gas.

7.1.5 Lymph nodes

Numerous superficial lymph nodes are palpable in healthy koalas (see 2.5). These lymph nodes should be systematically palpated to detect enlargement and asymmetry. Unilateral or bilateral enlargement of the rostral mandibular or facial lymph nodes may cause subtle alteration in the shape of the koala's face.

7.1.6 Body weight

Regular weighing should be incorporated into the routine of any koala in care. The koala can be placed directly on the weighing platform of large scales, or transferred into a small cage, sack or strong bucket (see 4.5) in which it can be carried to and from the scales and weighed. In some zoological parks captive koalas are accustomed to being transferred directly onto a small tree fork suspended from a spring balance, in which they sit quietly while being weighed. Changes in the body condition, hydration status and gut fill (as well as the presence of PY in females) should all be considered when interpreting changes in body weight.

7.1.7 Body condition

Body condition can be assessed objectively by palpation of muscle mass over the scapulae (Dickens 1978a; Wood 1978). Although changes in muscle mass are apparent on palpation of other areas (lateral to the tibial crest, in the lower part of the arm and upper part of the forearm, and in the temporal fossa), it is more difficult to quantify progressive changes in these areas than over the scapulae. When present, subcutaneous fat depots are relatively small and confined to the axillary and inguinal areas (and around the pouch in females), so a koala's ribs will always feel prominent irrespective of whether the animal is robust or emaciated.

Place a hand across the koala's shoulders and rub your thumb over the supraspinatus and infraspinatus muscles. As body condition declines, the spine and edges of the scapula become much more prominent. Convex muscle surfaces may become flat or concave, and the infraspinatus seems to contract in overall size until muscle can only be felt clearly against the humeral end of the scapular spine. Assessment of body condition should be done on a daily basis in sick animals, as noticeable changes in body condition may precede significant changes in body weight. If the koala is dehydrated and fluid therapy is given, baseline body condition should be re-evaluated after an equilibration period, as there will be a noticeable improvement in apparent muscle mass. An example of a numerical body condition scoring system is provided in Table 8.1.

7.1.8 Hydration status

Changes in skin tone give the best guide to hydration status. Skin over the scapular area should slide smoothly backwards and forwards under your thumb. As dehydration progresses the movement of the skin becomes retarded and feels tacky. Small folds of skin pinched up over the top of the head (between the ears) usually slide briskly back into place. Koalas seem to have rather inelastic skin over much of the body, so that folds of skin twisted up over the shoulders, for example, may remain in this position for several seconds, even in healthy animals. A history which includes anorexia, rapid weight loss, polydipsia (atypically drinking from a water bowl), or the production of small dry faecal pellets should always alert the clinician that a koala may be dehydrated.

7.1.9 Gut fill

The enormous capacity and prolonged retention time of digesta in the caecum and proximal colon ensures that healthy koalas have a well-rounded ovoid body

Table 8.1 A body condition scoring system used in koalas

Condition	Score		Attributes
Excellent	5	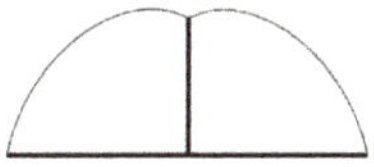	Obviously convex muscle masses on either side of the scapula. Strong muscle tone. Scapular spine apparent on careful palpation.
Good	4	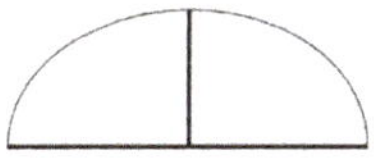	Slightly convex muscles on either side of the scapula. Good muscle tone. Scapular spine readily palpable.
Fair	3	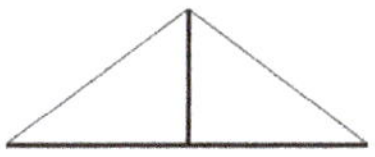	Flat to barely convex muscles on either side of the scapula. Scapular spine prominent on palpation.
Poor	2	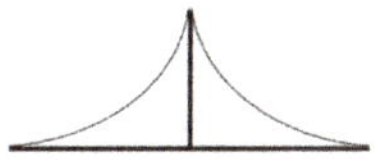	Slight dishing or concavity of muscles on either side of scapula. Scapular spine very obvious on palpation. Edges of scapula palpable.
Emaciated	1	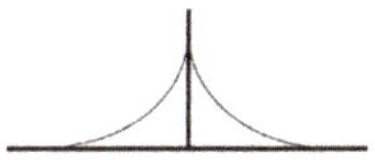	Markedly concave muscles on either side of scapula. In extreme emaciation there may be almost no muscle palpable on either side of the scapular spine; in these cases the entire scapula will be palpable.

Diagrams represent an imaginary transverse section through the scapula, with heavy lines indicating the table and vertical spine of the scapula, and fine lines outlining the supraspinatus and infraspinatus muscles.
Source: After A Reiss, in Jackson et al. (2003) and DEH (2004).

shape. In healthy juveniles and in frail but otherwise healthy aged animals, the abdomen can sometimes seem disproportionately large. After a few days of reduced food intake the abdominal girth reduces, and the animal becomes straight-sided or noticeably concave behind the costal arch.

7.1.10 Abdominal palpation

The stomach and both kidneys can be palpated in a relaxed koala. When full, the stomach occupies much of the left anterior abdomen and extends beyond the costal arch. The position and consistency of the caecum and proximal colon make identification of other organs difficult, and the epipubic bones impede palpation of the urinary bladder.

7.1.11 Gait and demeanour

Most koalas are not overly inquisitive, but when placed on the ground they usually walk or bound towards a nearby vertical structure (e.g. table leg, door jamb or even a vertical shadow or contrasting stripe of paint on a wall) in anticipation of climbing to safety. When walking slowly on the ground koalas have a rather high-stepping gait with both fore and hind legs. When agitated or alarmed they may break into a bounding gallop, or sit bolt upright with the ears erect and eyes wide open. If threatened they may raise one or both arms in the air, ready to swipe an attacker. When unwell, koalas take little or no interest in their surroundings, and adopt a ball-like resting posture with the head lowered and ears drooping, even in an unfamiliar location like the floor of a veterinary surgery.

7.1.12 Tooth wear class

Advanced tooth wear can be a limiting factor in an older koala's recovery from illness or injury, and in otherwise healthy aged koalas it may be the primary reason for debility (Eberhard 1972; Robinson 1978; Lanyon & Sanson 1986b). The cheek teeth of every adult koala coming into care should be checked as part of the initial examination. In anaesthetised animals the premolars and all four molars can be visualised. In a conscious koala you may be able to slide your index finger or a pen into the diastema between the lower incisors and premolars to pull the mouth open. It is helpful to have an assistant restrain the koala's forearms to reduce the possibility of getting scratched. Concentrate on the upper dental arcade, especially the premolars and first molars.

In very old animals, the occlusal surfaces of the upper premolars are worn down level with the hard palate and may have formed a shallow C or deep horseshoe shape. Ultimately, the two roots of each premolar may become exposed as separate islands (Martin 1981; Martin & Handasyde 1999). Perimeter notches (or even fractures) may appear in the thinning occlusal surface of the upper first molar, which by now will also be almost level with the hard palate. Gingival recedence and/or formation of periodontal pockets along the labial aspect of the upper molars may occur. If the roots of the premolar tooth on either side are exposed, it is unlikely to be in the best interests of a free-ranging koala to proceed further with treatment. Similarly, if one or both premolars are horseshoe-shaped and the surface of the first molar is worn flat so that it is level with the hard palate, euthanasia should be considered if the animal is seriously injured or already in poor body condition. The degree of wear on the second, third and fourth molars should be assessed if you are unsure about whether or not to proceed.

Figure 8.4 illustrates different stages of wear on the upper premolar and first molar teeth of 12 captive koalas in south-east Queensland between 1–15+ yr of age, as well as the full rows of upper cheek teeth of another aged, otherwise healthy individual whose degree of overall tooth wear was incompatible with effective mastication and ongoing maintenance of body condition. As there is considerable individual variation in the rate of tooth wear (R Martin *pers. comm.*), these images are presented as a reference series to assist with quantitative recording of relative tooth wear, rather than any potential usefulness in determining the absolute age of individuals. Routine husbandry practices should have biased these captive koalas' access to soft young foliage, and their rates of tooth wear may or may not reflect those typical of free-ranging animals from other localities.

7.1.13 Urine and faeces

The appearance of normal urine is described in section 8.5. Faecal pellets are khaki to olive in colour and tapered slightly at either end. Freshly passed pellets often have a glistening appearance due to a thin transparent coating of mucus. When broken open they reveal a densely-packed mass of fibre. In very aged animals with worn teeth, successive pellets are sometimes linked by strands of unchewed fibre. If food intake has been reduced for some time, shorter narrower pellets are produced. Occasionally, mauve faecal pellets are passed by healthy animals. The mauve colour seems to be confined to the mucous coating of the pellet. Segments of the cestode *Bertiella obesa* may adhere to faecal pellets (see 10.11.4a).

Many carers routinely record the number of faecal pellets produced per 24 hr period. Anecdotally, counts in the order of 90–200 pellets (or more) per 24 hr are expected in normal koalas. Pellet counts < 90 per 24 hr may reflect reduced food intake or compromised gastrointestinal function.

7.1.14 Examination of the pouch

Microflora of the pouch epithelium was investigated in female koalas in relation to current reproductive status and recent reproductive history (Osawa et al. 1992). No viable microorganisms could be isolated from swabs of the pouch epithelium in 13/17 (76%) females examined, irrespective of their reproductive status (four had no young, seven had PY and two had back young). It was proposed that the pouch epithelium usually provides a hostile environment for microbial colonisation, from which it is not possible to recover viable microorganisms under normal circumstances. Preliminary studies suggest that there are polypeptides with antimicrobial capabilities secreted into the pouch (Bobek & Deane 2002).

If doubt exists about the health status of a PY, a swab can be taken from the pouch lining for microbial culture and sensitivity testing. Isolation of any viable organisms is an indication for careful removal and physical examination of the young and the pouch. The interior of the pouch is usually pale pink. Small dark flecks of the young's faeces (from which viable microorganisms cannot normally be cultured) may adhere to the PY and to the pouch lining, but the underlying surface is relatively clean. The functioning mammary gland and teat should be pliable and pale pink.

7.1.15 Examination of PY

Healthy unfurred PY have subcutaneous fat depots visible under the skin anterior to the hind legs, in the inguinal, axillary and sometimes the cervical areas, and around the pouch edges in females. Furred PY should have rounded muscles over the scapulae and lateral to the tibial crest. Until they have begun to eat leaf, lightly furred young usually look rather concave between the ribs and pelvis. An intention tremor is normally present in PY at the time of first emergence, but disappears within the next few weeks (Blanshard 1990, 1994).

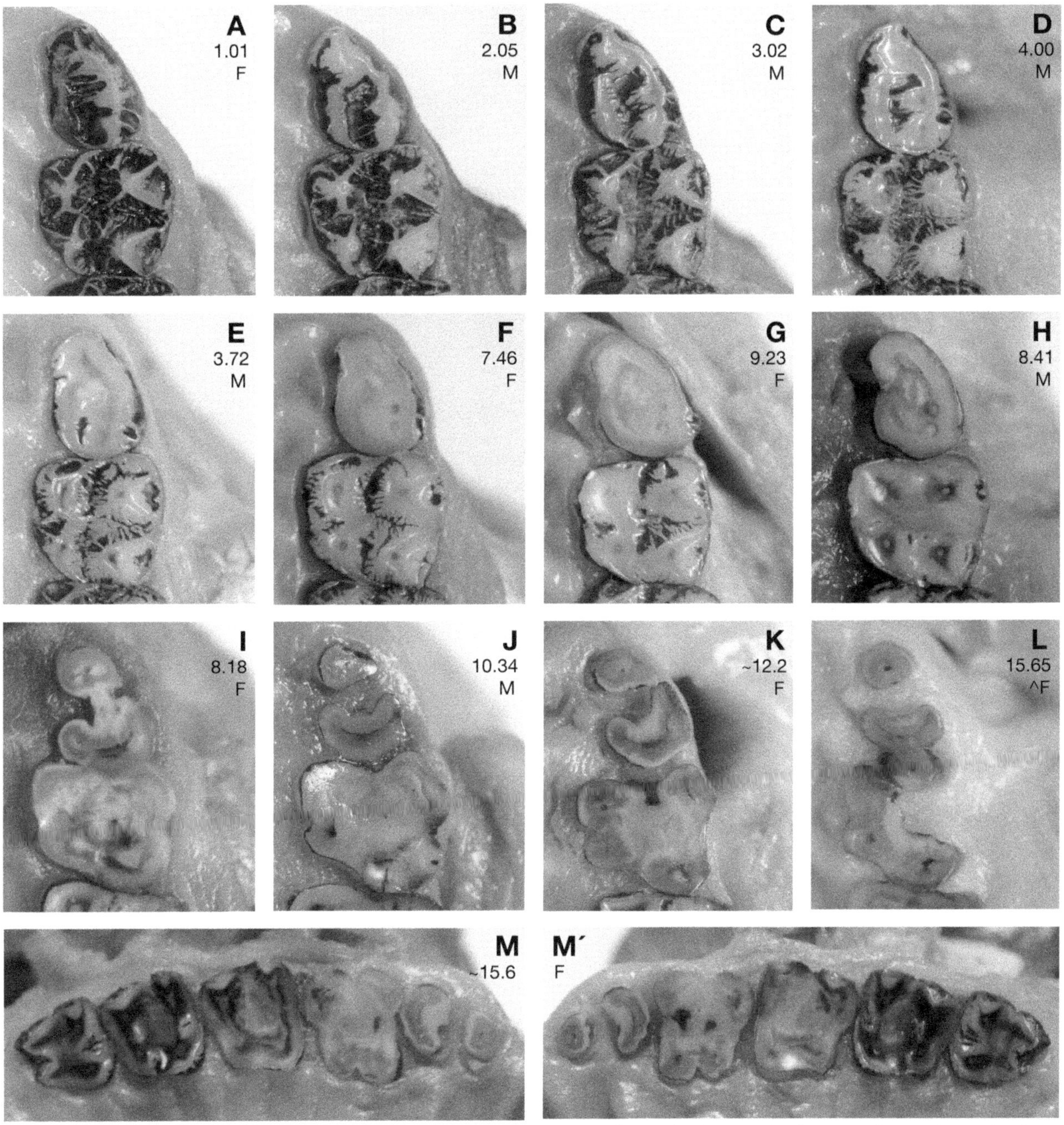

Figure 8.4 Progressive stages of wear on the upper premolar and first molar teeth of captive koalas in south-east Queensland. **A–L**. Left or right upper premolar and first molar of 12 individuals, showing variations in the degree of wear. **M, M′**. The premolar and all four molars in the upper tooth rows of one aged individual. Koala age (yr) and sex (M, F) are specified on each image. A tilde (~) indicates that age was estimated, and a caret (^) indicates an instance where the rate of tooth wear in that individual may have been reduced by virtue of regular supplementary feeding late in life. Note that neither the degree of premolar nor first molar wear could be used to rank the images in ascending order of absolute age. The full tooth rows (M, M′) are those of an aged female whose proximal colon became impacted with compact boluses of poorly masticated leaf fibre. Photos: W Blanshard and Lone Pine Koala Sanctuary.

7.2 Clinical techniques and procedures

7.2.1 Venipuncture sites

The cephalic vein is located on the cranial aspect of the forearm, as in dogs and cats. It is usually easier to palpate in females, as in adult males its position may be concealed by greater muscle development. Blood samples (1–10 mL) can be collected via 25 G, 16 mm or 22 G, 25 mm needles. In conscious animals the use of a scalp vein set may make collection easier if the animal is moving. Another accessible peripheral vein runs vertically down the caudal aspect of the tibia (Fig. 8.5b). This can be a very useful vein if the koala is restrained in a sack with only one limb exposed. Clipping the overlying fur is recommended to facilitate locating the tibial vein. Larger volumes of blood can be obtained from the femoral or jugular veins. Light anaesthesia (or serious illness) is generally required for restraint.

Cardiac puncture, when indicated for emergency administration of drugs or euthanasia solution, is accomplished using a 22 G, 25 mm needle. The koala's left arm is abducted and the intercostal spaces towards the axillary area palpated for the strongest apex beat before insertion of the needle at that point.

7.2.2 Intravenous catheterisation

A 22 G, 25 mm or a 20 G, 32 mm catheter is inserted into the cephalic vein after nicking the overlying skin with the point of a scalpel blade, and its hub secured by encircling it and the forearm with adhesive tape. If a giving set and fluid bag are then connected and secured in place with tape or a bandage, the intravenous line should exit the bandage near the underside of the elbow so that it does not get in the way while the animal is feeding. Most sick or injured koalas tolerate IV lines and associated light strapping although some bite through the IV line if it dangles in front of them. Regulation of flow rates may be difficult due to the koala's normal ball-like, flexed-elbow resting posture. The use of an infusion pump may alleviate this problem.

The koala and its IV line will become entangled if it tries to climb. Therefore IV fluid or drug administration over successive days is easiest if the hub of the catheter is plugged with an injection cap when first inserted, rather than connected directly to the giving set. The catheter and injection site are secured to the arm by a light bandage wrap and flushed with heparinised saline before and after use. Fluids can be administered intermittently by placing a needle on the end of the giving set and inserting this into the injection cap. The needle is left unsecured during fluid administration, so that if the koala climbs around its cage the needle will pull out without dislodging the catheter (Fig. 8.5a).

The vein on the caudal aspect of the tibia (Fig. 8.5b) can also be catheterised. In this case, take care to insert the catheter far enough above the tarsus such that the connection to the IV line won't be under direct pressure whenever the koala bears weight on that foot.

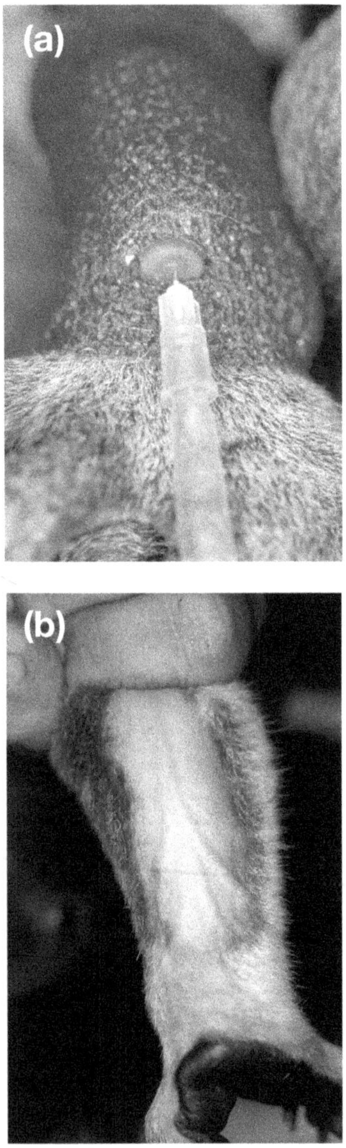

Figure 8.5 Catheterisation of peripheral veins. a) The hub of an IV catheter in the cephalic vein is plugged with an injection cap, facilitating intermittent fluid administration to a mobile patient. b) The vein on the caudal aspect of the tibia can be catheterised. Photos: W Blanshard.

7.2.3 Collecting urine samples

Koalas do not normally urinate while at rest in a fork, but may do so as they change position (when the cloaca is directed downwards). More often they pause to urinate while on the move, either in the branches or on the ground. The koala stops and raises its rump slightly so that the cloacal fur is barely off the substrate before micturating. There are a number of ways to facilitate collection of urine specimens.

Early in the morning, before otherwise disturbing the koala, lift it out of its resting fork or cage and place it on the floor. A shallow saucer is an ideal shape to swing into place under the cloaca as you see the koala lift its rump off the ground. Urine samples may be obtained fortuitously if left behind in a bucket used to hold or carry the koala for weighing. Midstream urine samples may be sterile, but all are potentially contaminated by passage through the common anogenital cloacal vestibule.

Ultrasound guided cystocentesis is possible. Note that the bladder is intra-pelvic when not distended, and the caecum and proximal colon overlie the caudal part of the ventral abdomen.

It is possible to place a urethral catheter in a male koala if required (e.g. to assess or manage bladder function after motor vehicle accident trauma); this would normally be done under anaesthesia. The penis can be extruded by applying gentle pressure on either side of the cloacal orifice. There are two regions of the male urethra which may impede catheter insertion. The first encountered is a narrowing of the lumen and some mucosal irregularity at the junction of the penile and membranous urethra, in the vicinity of the openings of the bulbourethral glands (L Vogelnest *pers. comm.*; J Hanger *pers. comm.*). The second is the prostatic sinus, a groove on either side of the urethral crest (Young 1879), just distal to the neck of the bladder. These regions are shown in Figure 8.6. Gentle traction to straighten the penis during insertion of the catheter usually facilitates passage through the first region. The urethral sinus is typically the more problematic area, and cautious

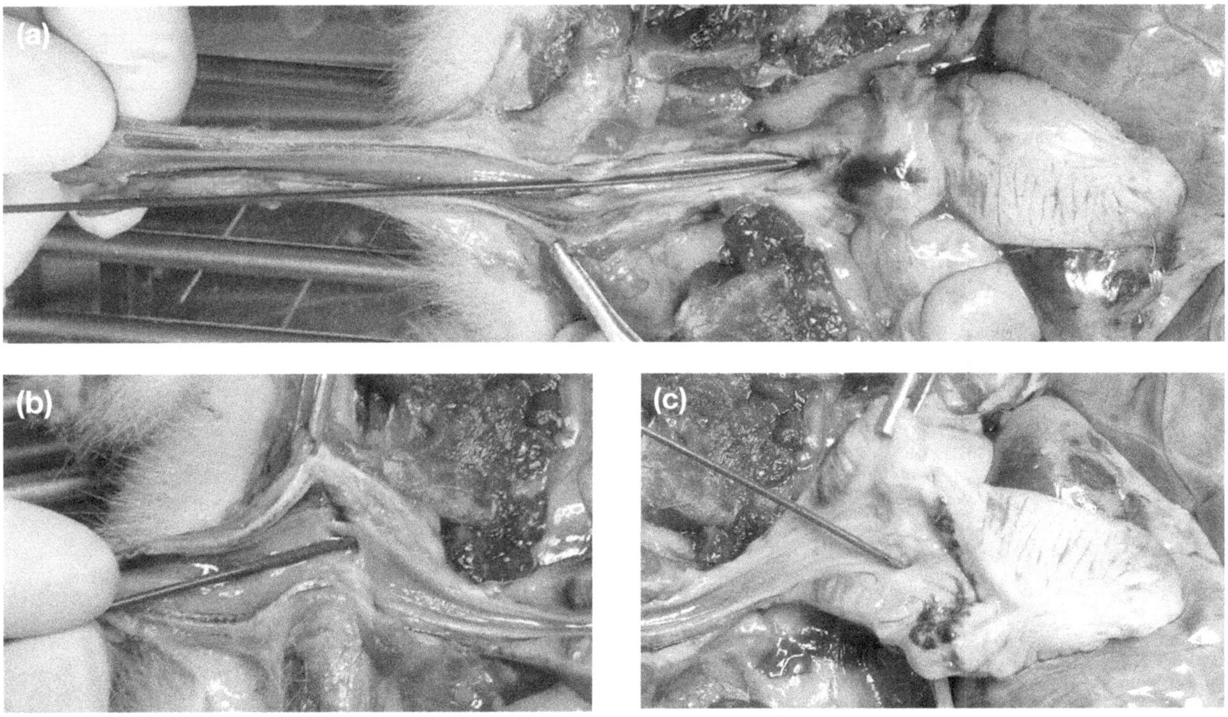

Figure 8.6 Potential impediments to successful catheterisation of the male urethra. a) The penile, membranous and prostatic portions of the urethra, exposed by removal of overlying tissue including the pubis and epipubic bones. b) Narrowing of the lumen and mucosal irregularity at the junction of the penile and membranous urethra, around the openings of the bulbourethral glands. c) The prostatic sinus, a groove on either side of the urethral crest. Note that haemorrhages in the serosa of the prostate (in a) and in the body of the prostate and mucosa of the urinary bladder (in c) are abnormalities resulting from fatal motor vehicle accident trauma. Dissection & photos: J Hanger and Australian Wildlife Hospital.

exploratory movements of the catheter tip may be required to gain entry to the bladder (J Hanger *pers. comm.*). Patience and care are required, to avoid urethral trauma. Any attempt to catheterise a female koala would be anatomically challenging (see 2.4).

7.2.4 Abdominal paracentesis

The abdominal wall is very thin. The linea alba is a semi-translucent sheet 10–30 mm in width, and the abdominal muscles are no more than 2–3 mm thick over the lateral and ventral abdominal wall. The caecum and proximal colon lie immediately below the ventral midline. With the koala restrained in a sitting position, the site used for insertion is on its right side, just ventral to the lumbar muscles, about midway between the caudal pole of the kidney and the wing of the ilium. After shaving a small area at least 20 mm square, the skin is disinfected thoroughly. A 22 G, 25 mm or a 20 G, 32 mm teflon catheter is pushed sharply, but only a short distance, through the skin and abdominal wall, and the needle immediately withdrawn slightly until the bevel is visible in the catheter at the skin surface. Once the bevel is shielded inside the catheter but is still providing support and preventing bending of the teflon, the catheter is advanced fully down the shaft of the needle until the hub of the catheter rests against the abdominal wall. The koala can then be tilted and/or the catheter rotated or withdrawn slightly until some fluid is harvested (Blanshard 1994).

7.2.5 Cerebrospinal fluid tap

CSF tap in the koala is relatively straightforward. The palpable landmarks are the occipital crest and jugular processes of the skull, the wings of the atlas (first cervical vertebra, C1) and the dorsal spinous process of the axis (C2). With the koala in lateral recumbency and the head flexed at right angles to the neck, the jugular processes of the skull are easy to palpate. Extend the neck again and identify the wings of the atlas, positioned almost within the span of the jugular processes. The point of entry is at the intersection of imaginary perpendicular lines connecting the occipital crest and the dorsal spinous process of C2, and the anterior edges of the wings of the atlas. When the head is flexed at right angles to the neck, this point is roughly midway between the occipital crest and the spinous process of C2. A 22 G spinal tap cannula is directed from the point of entry in the sagittal plane towards the angle of the jaw. Total depth of penetration is only about 20 mm (Blanshard 1994).

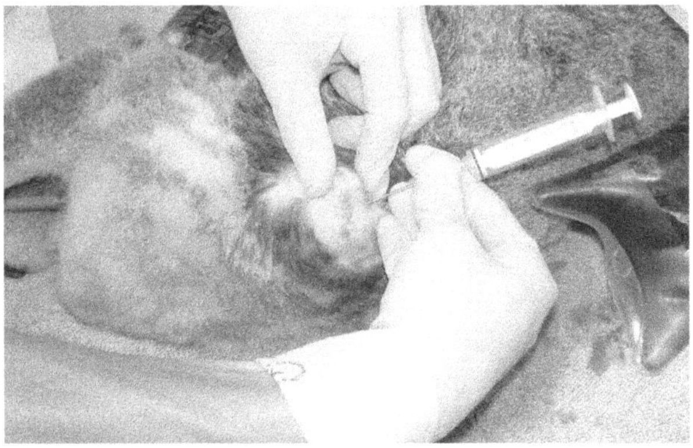

Figure 8.7 Site for bone marrow aspiration from the iliac crest. Photo: L Vogelnest and Australian Wildlife Hospital.

7.2.6 Bone marrow aspiration

Anaesthesia is required for bone marrow aspiration (see 8.2 for interpretation of aspirates). A technique for bone marrow aspiration has been described by Spencer and Canfield (1995). The wing of the ilium (the iliac crest) was found to be the most useful site for collection (Fig. 8.7), but the greater tubercle of the humerus was found to be a reasonable alternative site. A 1.25 mm Sternum Temno® bone marrow needle (N Stenning & Co.) was used to sample from both sites. In the absence of purpose-manufactured bone marrow collection needles, an 18 G needle may be used to aspirate samples for cytological assessment (J Hanger *pers. comm.*).

7.2.7 Radiography

Ideally, koalas are anaesthetised for radiography to minimise radiation exposure and improve diagnostic quality of radiographs, and should only be radiographed using physical restraint if the risk of anaesthesia is deemed too great. For thoracic views, centre the primary beam just caudal to the koala's thoracic inlet (K Reynolds *pers. comm.*). The ventrodorsal view is often much easier to interpret than the lateral view as there is less superimposition of pulmonary vessels. The scapulae will overlie part of the cranial lung lobes on both views.

In a healthy koala the entire abdomen has a ground-glass appearance and contains only scant amounts of gas, in reasonably evenly distributed small bubbles.

Individual organs cannot usually be identified. This lack of contrast is due to the absence of intra-abdominal fat, and the fact that the caecum and proximal colon occupy the ventral and ventrolateral portions of the entire abdomen (the small intestine is located dorsally). Small quantities of oral barium sulphate (e.g. 1–5 mL) given a few hours before an abdominal radiograph can be a useful marker to help assess gastric emptying, and may also enhance interpretation by providing some contrast between loops of gut. Individual faecal pellets become discernible after barium is given. Carlisle et al. (1989) recommended IV urography as a means of highlighting the anatomy of the urinary tract.

7.2.8 Ultrasonography

In contrast to radiographic examination of the koala abdomen, ultrasound provides an opportunity to identify and examine individual viscera. In preliminary studies, ultrasound using 7.5 MHz and 10 MHz transducers was used to guide cystocentesis, and to identify chronic interstitial nephritis in senile kidneys, oedema of the caecal and proximal colonic mucosa in lymphosarcoma, and increased turbidity of peritoneal fluid and urine associated with infection (L Melville *pers. comm.*). Stalder et al. (2003) proposed that ultrasonography is an effective method of detecting and grading the severity of chlamydial urogenital disease (see 10.11.2a) in both males and females, and noted the following during ultrasonographic examination of the urogenital tract.

- Epipubic bones interfered with examination of the lower genital tract.
- Kidneys and urinary bladder were sonographically comparable to those of domestic dogs and cats; cystitis was recognisable as measurable bladder wall thickening in koalas with chlamydiosis.
- Ovaries, vaginal complex and urogenital sinus were not detected in normal animals, but paraovarian cysts were detectable in females with chlamydiosis.
- Bilateral uteri were consistently visualised as circular hypoechoic structures lying dorsal to the bladder in the transverse plane.
- Testes and epididymides were sonographically comparable to those of domestic dogs and cats; lateral bulbourethral glands had a hypoechoic parenchyma of fine to medium texture.

7.2.9 Computed tomography scan and magnetic resonance imaging

Indications for CT scan and MRI are similar to those in domestic animals. CT of the head has been used postmortem to delineate the borders of a tumour of mixed cartilage and bone, and to visualise a nasopharyngeal cryptococcal granuloma (Fig. 8.18) (Woods & Blyde 1997). MRI has been used to identify cryptococcal granulomas (M Krockenberger *pers. comm.*).

7.2.10 Electrocardiogram

The normal ECG of the koala has been reported (Rezakhani et al. 1986).

8 CLINICAL PATHOLOGY

8.1 Haematology and biochemistry

Sample collection is described in section 7.2.1.

8.1.1 Reference ranges and interpretation

Haematology and biochemistry values in captive and free-ranging koalas are presented in Tables 8.2 and 8.3 (Dickens 1975, 1978a; Canfield et al. 1989a; Spencer & Canfield 1993a). The numbers and origins of koalas sampled, sampling techniques and methods of analysis of the data varied in each case, therefore the values generated by each source are not directly comparable.

A review of haematology and biochemistry in 14 diseased koalas recommended a minimum diagnostic screening panel for sick animals (Canfield et al. 1989b). This panel included PCV, total and differential WBC counts, urea, total serum protein and albumin concentrations, and AST, LDH and GGT activities, as these results would be altered in a wide variety of diseases. Blood glucose levels were not assessed in the study but should be included in an initial diagnostic screening panel.

a Red cell indices

PCV results from individual animals should be interpreted with consideration of the time of year of sample collection. Pronounced seasonal changes in haematocrit, not associated with changes in body weight, plasma protein concentration or plasma osmolarity, were observed in 15 captive koalas at a wildlife park in New South Wales where ambient temperatures ranged

Table 8.2 Haematology values for koalas

Parameter	Units	Captive and free-ranging koalas (Dickens 1978a) Adult		Juvenile		Free-ranging koalas (Canfield et al. 1989a) Adults and juveniles			
		Males	Females	Males	Females	Mean	Range	(n)	Ref values*
RBC	$\times 10^{12}$/L	3.85 ± 0.1	3.39 ± 0.1	3.46 ± 0.1	3.31 ± 0.1	3.46	2.58–4.17	(44)	2.7–4.2
Hb	g/L	130 ± 4	113 ± 2	111 ± 3	109 ± 2	114	88–139	(44)	88–*140*
PCV	%	41.5 ± 1.0	37.4 ± 0.9	37.9 ± 1.0	35.9 ± 0.8	37	28–45	(44)	29–44
MCV	fL	110.0 ± 5.2	111.7 ± 3.9	112.9 ± 6.3	108.0 ± 4.9	105	90–116	(44)	94–*117*
MCH	pg	34.4 ± 1.9	35.1 ± 1.0	33.4 ± 1.6	33.1 ± 1.3				
MCHC	g/L	313 ± 5	306 ± 6	293 ± 4	305 ± 4	314	289–332	(44)	298–330
nRBC	/100WBC	14.0 ± 12.7	17.4 ± 6.6	3.9 ± 1.2	4.4 ± 1.3		0–20	(44)	
Reticulocytes	%	1.6 ± 0.4	1.5 ± 0.2	2.6 ± 0.4	3.5 ± 1.1				
Howell–Jolly b	/100WBC	7.7 ± 2.8	7.7 ± 6.1	5.9 ± 4.8	7.1 ± 4.0				
Platelets	$\times 10^{9}$/L					307	222–558	(11)	
Total WCC	$\times 10^{9}$/L	8.5 ± 0.5	8.0 ± 0.5	7.8 ± 0.6	7.0 ± 0.5	7	3.1–10	(39)	*2.8–11.2*
Diff. (relative)									
Neutrophils	%	44.6 ± 4.2	39.6 ± 3.5	31.2 ± 3.8	39.6 ± 4.7	52	18–78		
Lymphocytes	%	49.5 ± 3.9	52.9 ± 3.3	63.9 ± 4.6	55.5 ± 4.8	40	10–57	(44)	
Monocytes	%	3.5 ± 0.6	3.8 ± 1.3	3.3 ± 0.7	3.3 ± 0.9	2.5	0–12	(44)	
Eosinophils	%	2.2 ± 0.6	3.8 ± 0.9	1.2	1.2 ± 0.3	6	0–22	(44)	
Basophils	%	0.1 ± 0.0	0.2	0.4	0.5	0	0	(44)	
Diff. (absolute)									
Neutrophils	$\times 10^{9}$/L	3.6 ± 0.3	3.5 ± 0.6	2.4 ± 0.4	2.8 ± 0.5	3.4	0.9–6.6	(39)	*0.5*–6.3
Lymphocytes	$\times 10^{9}$/L	4.3 ± 0.5	4.0 ± 0.4	5.0 ± 0.5	4.0 ± 0.4	3.1	0.7–7.3	(44)	*0.2*–5.8
Monocytes	$\times 10^{9}$/L	0.3 ± 0.07	0.3 ± 0.04	0.2 ± 0.01	0.2 ± 0.05	0.2	0.0–0.9	(44)	0.0–0.6
Eosinophils	$\times 10^{9}$/L	0.2 ± 0.02	0.3 ± 0.02	0.1 ± 0.01	0.1 ± 0.02	0.4	0.0–1.4	(44)	0.0–1.1
Basophils	$\times 10^{9}$/L	0.03	0.02	0.03	0.03	0	0	(44)	

Data from Dickens (1978a) represent >200 captive and free-ranging animals from eastern Australia, presented as the mean ± standard error by age and sex.
Data from Canfield et al. (1989a) are derived from 45 free-ranging animals from New South Wales, presented as the mean, observed range, number of samples (n), and set of reference values calculated from the mean ± 2 standard deviations.

* Reference values shown in italics fall outside the minimum or maximum value actually observed.

Table 8.3 Biochemistry values for koalas

Parameter	Units	Captive and free-ranging koalas (Dickens 1975, 1978a)	Free-ranging koalas (Canfield et al. 1989a)			
		Normal range	Mean	Range	(n)	Ref values*
Sodium	mmol/L	133–140	138	131–147	(43)	132–145
Potassium	mmol/L	5.0–5.8	5.1	3.9–7.0	(40)	*3.5*–6.8
Chloride	mmol/L	96–103	100	90–110	(43)	93–107
Bicarbonate	mmol/L		21	11–31	(41)	12–30
Anion gap	mmol/L		21.8	11.0–33.1	(38)	11.3–32.3
Glucose	mmol/L	2.2–4.4	4.5	2.7–7.2	(8)	
BUN	mmol/L	1.1–10.7	2.3	0.2–7.2	(43)	0.2–6.6
Creatinine	µmol/L	90–130	120	90–180	(43)	*80*–150
T-bilirubin	µmol/L	1.7–12	4	1–10	(43)	*0*–8
Alkaline phosphatase	U/L	30–200	98[a]	38–233[a]	(28)[a]	*25–219*[a]
			207[b]	88–352[b]	(15)[b]	88–*375*[b]
ALT (SGPT)	U/L	30	52	2–240	(39)	*0*–236
AST (SGOT)	U/L	10–50	34	1–150	(42)	*0*–134
Gamma GT	U/L		8	2–20	(40)	*0*–16
LDH	U/L	150–250	213	88–523	(40)	*79*–412
Total protein	g/L	60–75	71[a]	58–88[a]	(28)[a]	58–83[a]
			65[b]	57–77[b]	(15)[b]	*53–78*[b]
Albumin	g/L	29–38	42	33–55	(43)	34–50
Globulin	g/L		28[a]	17–38[a]	(28)[a]	18–*39*[a]
			23[b]	12–37[b]	(15)[b]	13–34[b]
Fibrinogen	g/L					0.8–3.7†
Calcium	mmol/L	2.5–2.62	2.63	2.15–3.06	(42)	2.28–2.97
Inorganic phosphate	mmol/L	1.13–1.61	1.37[a]	0.69–1.76[a]	(27)[a]	0.79–*1.96*[a]
			1.84[b]	1.20–2.34[b]	(15)[b]	1.25–*2.44*[b]
Cholesterol	mmol/L	2.07–2.84	2.12	1.07–3.5	(42)	1.15–3.09
Triglyceride	mmol/L		1.76	0.47–3.0	(42)	0.5–3.02
Osmolality (estimated)	mosm/kg		287	271–306	(40)	273–300

Data from Dickens (1975, 1978a) represent the 'range of normal biochemical values', based on an unspecified number of samples from captive and free-ranging animals.
Data from Canfield et al. (1989a) are derived from 45 free-ranging animals from New South Wales, presented as the mean, observed range, number of samples (n), and set of reference values calculated from the mean ± 2 standard deviations. They show statistically significant differences between (a) adult and (b) juvenile koalas.

* Reference values shown in italics fall outside the minimum or maximum value actually observed.

† Reference values from Spencer and Canfield (1993a).

from about 3–35°C between winter and summer (Cleva et al. 1994). In both males and females, haematocrit values were elevated in winter and low in summer, and showed statistically significant negative correlations with minimum and maximum daily temperatures (Fig. 8.8). It was proposed that increased haematocrit may be a component of the metabolic response to the greater energy demand of exposure to low environmental temperature.

Martin (1985b) described development of hypochromic microcytic anaemia affecting free-ranging female koalas during August 1979 and January 1981 at Walkerville (Vic.) and hypothesised that it resulted from dietary trace mineral deficiency (of copper and/or cobalt) rather than a seasonal effect. Recent studies suggest that improved nutritional status in free-ranging koalas (following a flush of eucalypt leaf growth after spring rainfall) resulted in significantly higher PCV (as well as increases in haemoglobin, total lymphocytes, lymphocyte:neutrophil ratio, total protein and cholesterol) compared with blood samples collected during a season when eucalypt leaf quantity and quality was lower (M Lynch *pers. comm.*).

Nucleated RBC (metarubricytes), Howell-Jolly bodies and basophilic stippling of RBC may be observed in the blood smears of koalas without any association

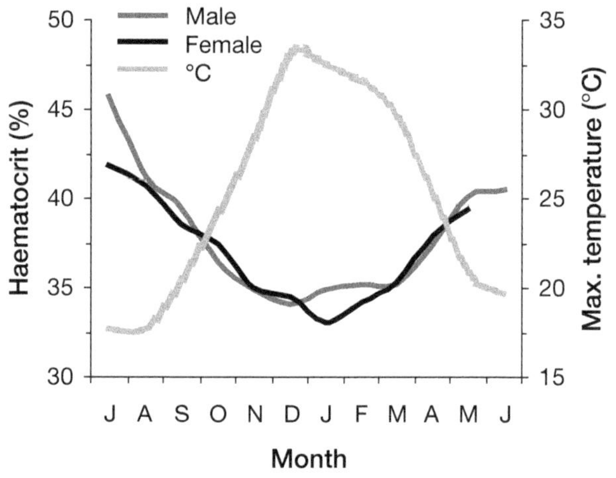

Figure 8.8 Seasonal changes in average haematocrit. Changes in average haematocrit (%) of six captive male and nine captive female koalas in New South Wales with changes in mean maximum daily temperature (°C). (Source: Modified from Cleva et al. (1994).)

with anaemia (Dickens 1978a; Clark 2004). The healthy koala also has a low level of small Heinz bodies in the blood (Spencer & Canfield 1994). Anaemia is one of the more common haematologic abnormalities detected in diseased koalas, notably those with chronic inflammatory disease or lymphoid neoplasia (Canfield et al. 1989b; Spencer & Canfield 1996). A description of erythroid regeneration following blood loss (via heavy tick infestation) has been reported by Spencer and Canfield (1993b).

b White cell indices

The morphologic characteristics of koala leukocytes have been described in detail by Clark (2004).

The lymphocyte:neutrophil (L:N) ratio is widely regarded as a useful indicator of health status in the koala (Dickens 1975, 1978a; Obendorf 1983). In differential white cell counts of healthy koalas, lymphocyte numbers exceed neutrophil numbers in most animals (L:N >1). A reversal of the L:N ratio has been reported in significantly injured or sick koalas. The L:N ratio of any individual must, however, be interpreted in light of the animal's recent history. Changes in the ratio over time may be more valuable than a single estimation. In free-ranging koalas, reversal of the L:N ratio has been shown in blood samples collected 3.5–6 hr after capture and confinement in a sack or box (Dickens 1978b). In another study, in which eight free-ranging koalas were captured and held in captivity, blood samples collected at 0 hr, 6 hr, 24 hr and 7 d after capture showed mean L:N ratios of 1.85, 0.39, 0.85 and 1.02, respectively (Hajduk et al. 1992). There were no significant differences in mean plasma cortisol levels between sampling periods.

c Serum proteins

Total serum protein values have been utilised as an adjunct to physical examination for health assessment of free-ranging adult koalas (M Lynch, N McLean & K Raymond *pers. comm.*). Total serum protein <60 g/L and/or serum albumin <32 g/L were regarded as key indicators of compromised health in free-ranging Victorian koalas. Animals demonstrating these changes, as a result of disease or reduced food availability in poor quality or overbrowsed habitat, were shown to have an increased risk of mortality when exposed to stressors such as capture, transport, anaesthesia, translocation and surgical sterilisation (M Lynch *pers. comm.*).

Dickens (1978a) and Canfield et al. (1989a) reported that young koalas may have lower total plasma protein values than adults, due to a reduced globulin fraction. Hypoalbuminaemia (with or without hypoproteinaemia) is common in koalas with lymphoid neoplasia (Spencer & Canfield 1996). Obendorf (1983) reported that six female koalas with ascending urinary tract infections (cystitis ± pyelonephritis diagnosed at necropsy) were consistently anaemic and hypoproteinaemic.

Healthy koalas are reported to have high α_1 globulin values (Dickens 1975, 1978a), with values around 10 g/L considered normal (Dickens 1978a). Quantitative estimations of normal circulating levels of immunoglobulins were performed on the sera of 40 clinically normal koalas (20 South Australian animals and 20 captive animals from south-east Queensland). The mean estimated immunoglobulin level was 5.18 g/L (range 3.5–8.6 g/L), with no significant difference in the levels of the northern or southern koala subsets sampled (Wilkinson 1996).

d Biochemistry

Marked elevation of AST, LDH and GGT levels have been observed in koalas affected by lymphosarcoma (Canfield et al. 1989b; Spencer & Canfield 1996). Dickens (1978a) reported that koalas affected by trauma, toxaemia or chronic disease frequently had elevated AST levels, while koalas affected by trauma and chlamydial cystitis had elevated LDH levels. Dickens (1978a) and Canfield et al. (1989a) recorded high ALP values in healthy juvenile koalas. As in other species, this was thought to be related to bone growth.

Koalas that are not feeding or present in a debilitated state are often hypoglycaemic. Dickens (1978a) reported blood glucose levels as low as 1.38 mmol/L in severely debilitated animals. Koalas may still be capable of clumsy ambulation when their blood glucose level is undetectable (Blanshard 1994). There have been two published reports describing diabetes mellitus in koalas (Shimizu et al. 1989; Hemsley et al. 1998).

Both pre-renal and renal azotaemia have been observed in koalas (Canfield et al. 1989b; Spencer & Canfield 1993a; C Monaghan & K Warren *pers. comm.;* R Booth *pers. comm.;* Schultz et al. 1995; Spencer & Canfield 1996; Wilkinson 1996; Haynes et al. 2004).

Unlike dogs, koalas with lymphosarcoma do not appear to commonly present with hypercalcaemia, but frequent monitoring of serum calcium levels may be required to detect transient increases in serum calcium (Spencer & Canfield 1996). Hypercalcaemia may be an early indicator of renal disease in some cases (C Monaghan & K Warren *pers. comm.*), and animals that present with unexplained hypercalcaemia should be monitored closely for development of other abnormalities that suggest renal dysfunction.

e Electrolytes

Koalas suffering from significant inanition are frequently hypokalaemic (Blanshard 1994). Periodic re-evaluation of serum potassium levels may be required in debilitated animals. Hyper- and hyponatraemia, hyper- and hypochloraemia, and hyperkalaemia have been observed in koalas secondary to acute renal failure (Spencer & Canfield 1993a; Wilkinson 1996; R Booth *pers. comm.*). Serum bicarbonate levels may become elevated in koalas that are not eating (Blanshard 1994).

8.2 Bone marrow evaluation

Bone marrow collection is described in section 7.2.6. Spencer and Canfield (1995) determined that differential counts on bone marrow aspirate smears from 10 healthy koalas were similar to those obtained for the dog and cat, however, the presence of a large number of binucleate erythroid cells in the marrow was considered normal in the koala.

Myelodysplastic disorders are common in koalas, and bone marrow aspiration can be used to confirm disease in animals that have non-specific or mild changes on blood films (J Hanger *pers. comm.*). Bone marrow cytologic and histologic examination will assist with disease staging and assessment of prognosis in cases of lymphoid neoplasia. Bone marrow may demonstrate profound hypocellularity (aplastic anaemia), hypocellularity with increased blasts and/or dysplastic haematopoietic cells, or hypercellularity with few or no clusters of haematopoietic maturation (Hanger et al. 2003). Interpretation of bone marrow abnormalities may be difficult, as changes can be subtle.

8.3 Circulating thyroid hormones

A study examining circulating thyroid hormones reported that the koala has low serum concentrations of total and free tetraiodothyronine (thyroxine; TT_4 and fT_4) and triiodothyronine (TT_3 and fT_3), but has normal mammalian levels of thyroid-stimulating hormone compared with other marsupials. These results suggest that the koala has unique features of thyroid function and tissue responsiveness that require further investigation (Lawson et al. 1996). Mean TT_4 (measured by radioimmunoassay) in 32 koalas was 3.2 ± 2.1 nmol/L (range 0.7–8.4 nmol/L), and mean fT_4 (measured by equilibrium dialysis) in 24 koalas was 3.3 ± 2.1 pmol/L (range 0.7–7.8 pmol/L). Concentrations of TT_3 and fT_3 in 22 and 14 koalas, respectively, were at the lower end of assay detection limits. It must be noted, however, that this study was not undertaken as a randomised survey of normal koalas. Blood samples were collected from 'recently injured but otherwise healthy' animals taken into care or euthanased on humane grounds (Lawson et al. 1996).

8.4 Serum insulin and fructosamine

Hemsley et al. (1998) reported serum insulin and fructosamine levels in small numbers of healthy koalas. Serum insulin range was 4.2–20.1 pmol/L (n = 4) and serum fructosamine was 420.8–566.7 μmol/L (n = 2). A koala with diabetes mellitus was found to have serum insulin of 3.1 pmol/L.

8.5 Urinalysis

Urine collection is described in section 7.2.3. Normal urine varies in appearance from transparent to turbid and from straw-coloured to golden yellow. Miscellaneous debris may be grossly visible in the sample. The odour of koala urine may be faint or very strong (Canfield et al. 1989c). Occasionally, red urine is found on the floor of an enclosure. Alternatively, a koala may be noticed with red-tinged fluid on its peri-cloacal fur. Presumably this urine contains a pigmented byproduct of eucalypt digestion (and perhaps becomes red only on exposure to the air), although not all animals offered the same leaves will produce red urine. Such urine gives a negative reaction on the blood square of reagent test strips, and should then be regarded as normal. True haematuria and imminent or recent birth are other possible causes of red-tinged fluid on the peri-cloacal fur.

8.5.1 Urine specific gravity

Urine specific gravity is usually greater than 1.060 (Canfield et al. 1989c). Although this is beyond the measuring range of most clinical refractometers, the specific gravity of concentrated urine can be determined by diluting the sample with distilled water. Turbid urine can be centrifuged before the specific gravity is determined, as the presence of crystals, cells or mucus should not be considered when estimating the concentrating ability of the kidneys.

8.5.2 Reagent strips for urinalysis

It should be noted that koala urine usually gives a slightly to very strongly false positive reaction to ketones on reagent strips (W Blanshard *pers. comm.*; Canfield 1989c). Typical readings from normal urine tested with Combur 9 Test Strips® (Roche Diagnostics Australia) or Multistix Multiple Reagent Strips for Urinalysis® (Bayer Diagnostics Business Group) are given below.

- pH—5 (occasionally 5–6)
- Protein—negative or trace up to 1.0 g/L (most often trace to 0.3 g/L)
- Glucose—negative
- Ketones—false positives up to 16 mmol/L
- Urobilinogen—normal
- Bilirubin—negative
- Blood—negative
- WBC—negative
- Nitrite—negative

Combur 9 Test Strips appear more sensitive to the presence of erythrocytes, haemoglobin and glucose in koala urine than Multistix Reagent Strips, and are less likely to give a positive reaction to ketones (Blanshard 1994). Intermittent glucosuria has been observed in a koala with a severe metabolic stressor, i.e. neoplasia causing bone destruction.

8.5.3 Sediment examination

Concentrated sediment from normal urine usually contains some unidentifiable debris, occasional squamous and transitional epithelial cells, and less than one erythrocyte and leukocyte per high-powered (× 400) field. Sperm (with characteristic hook-shaped heads) are common in the urine of males. Crystals of various shapes may be evident in many samples from normal animals, particularly after refrigeration of the specimen. Stained smears of sediment show only occasional bacteria (Canfield 1989c).

9 HAND-REARING

9.1 Normal infant koala

9.1.1 Pouch young

The term 'pouch young' (PY) describes dependent young prior to their first emergence, as well as those which emerge intermittently but still spend some time within the pouch. Young koalas first emerge from the pouch for very brief periods when approximately 6–7 mo of age. At this stage they are only lightly furred, with the fur lying fairly flat against the skin and following definite tracts. The whorl of fur over the shoulders may be slightly fluffy, but fur over the lower half of the body is often still relatively sparse, sleek and follows a distinct caudal-to-cranial direction. At the time of first emergence most PY of captive Queensland koalas weigh 300–450 g, depending on age.

9.1.2 Back young

The term 'back young' (BY) describes young which no longer use the pouch, but which are still carried by or accompany their mother and continue to suck from one of her teats. By the time they are 8–8.5 mo, normal Queensland joeys weigh at least 700 g and are physically too large to return to the confines of the pouch. At this stage they are fully furred and their coat is upright all over the body, no longer following obvious tracts. Once the young stops re-entering the pouch, the pouch reduces in size. The teat on which the PY has been feeding elongates throughout lactation and is often visible in the pouch opening, particularly after the pouch begins to involute. Weaning is a prolonged process, with leaves gradually replacing milk as the more important dietary item over 4–5 mo. Young koalas are usually physically and emotionally independent of maternal care by the time they are 1 yr old. At this stage a juvenile Queensland koala typically weighs 2.2–2.5 kg.

9.1.3 Ageing infant koalas

Initial examination of a PY or BY should include measurement of head length and body weight. Head length (in mm) is the straight-line distance between the occipital crest and the snout. It is best measured with Vernier callipers, but if these are not available it should be estimated as precisely as possible using a ruler. Head length is a better indicator of age than is body weight, which tends to be much more variable between individuals and can change rapidly with ill health. Body weight should be measured to provide a reference point for assessing progress.

Figure 8.9 illustrates the relationship between head length and body weight for 91 captive-bred, mother-reared Queensland koalas of known age, with lines of best fit for male and female data shown.

For older PY and for BY, age should be read by eye from the head length regression line for that sex. To age small PY with greater accuracy, the following formulae should be used.

- For Victorian koalas with HL <85 mm:
  $\text{Age (d)} = 0.014(\text{HL})^2 + 2.37(\text{HL}) - 17.7$
  ($r^2 = 0.992$; Martin & Handasyde 1990).
- For Queensland koalas with HL ≤60 mm:
  $\text{Age (d)} = 0.0176(\text{HL})^2 + 2.3413(\text{HL}) - 17.617$
  ($r^2 = 0.982$; 49 observations of $n = 18$ animals; W Blanshard unpub).

9.1.4 Milk composition

Marshall et al. (1990) analysed 19 milk samples from 16 koalas and determined that the pattern of alterations in carbohydrate composition were similar to those seen in other marsupials, i.e. the total carbohydrate content peaked in mid-lactation (approximately 150–290 d) and decreased as weaning approached. Higher oligosaccharides comprised most of the carbohydrate fraction, and lactose concentration was always low (<0.4%).

Krockenberger (1996) examined all milk components in 16 koalas from an estimated 140 d to 380 d of lactation, and determined that koala milk has three outstanding features compared with milk from many other marsupials.

Lipid content remains high throughout lactation (average 10.1% by weight at 140 d lactation, 16.6% at 220 d lactation then 14% at 380 d lactation). In many

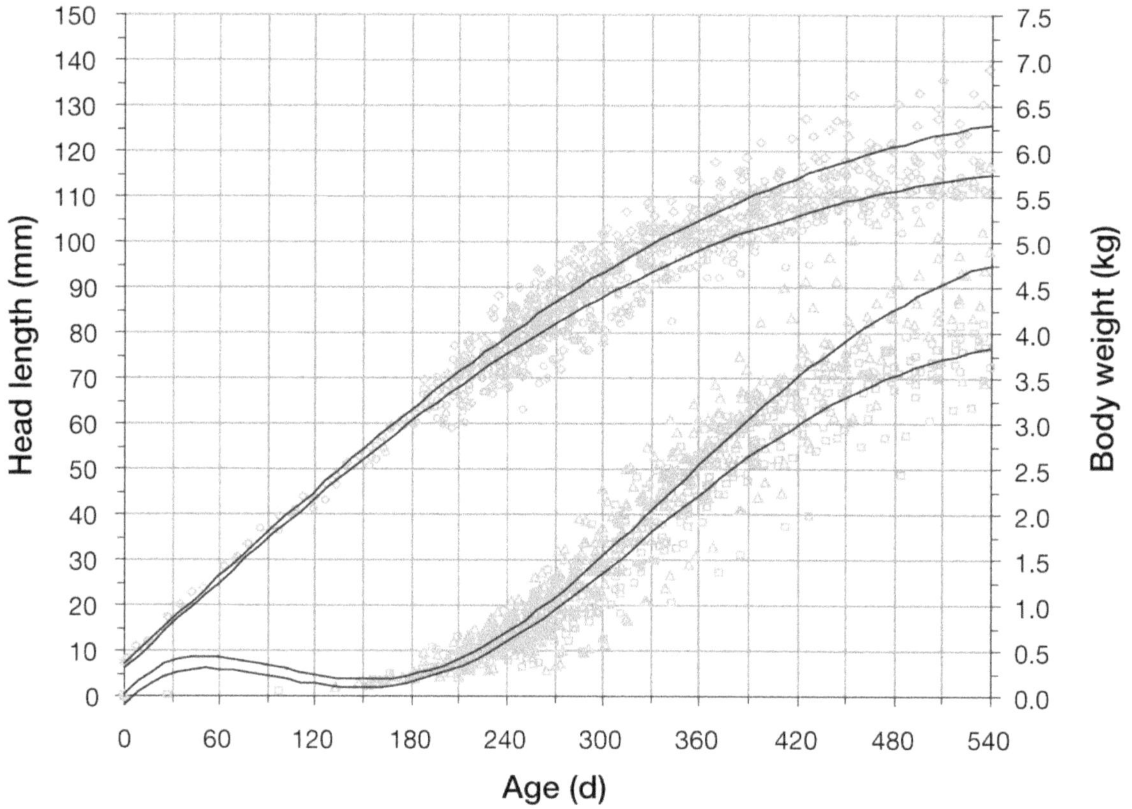

Figure 8.9 Relationship between age (d), head length (mm) and body weight (kg) for 91 captive-bred koalas in south-east Queensland during the first 1.5 yr of life. Head length = diamonds (males), circles (females). Body weight = triangles (males), squares (females). Lines of best fit (r^2 = 0.9558, 0.9595, 0.9427, 0.9569, respectively) are shown.

other marsupials, carbohydrates provide the major energy source in milk prior to pouch emergence, with lipid content rising markedly at the time of pouch emergence.

Carbohydrates in late-lactation koala milk (>270 d lactation) consist mainly of lactose and some oligosaccharides, with few or no monosaccharides, but the total carbohydrate content is low (1.1% at 380 d).

Milk solids content declines at the time of exit of young from the pouch. Milk solids content peaks at 220 d lactation (average 33.6% by weight) then declines significantly (to average 28.3% by weight) at 300 d, after which it remains constant.

9.1.5 Development of young

Young koalas progress through a series of predictable stages of physical and physiological development from pouch emergence to independence. Developmental milestones for 91 healthy, captive-bred, mother-reared koala young up to the time of assumed independence (here defined as reaching a minimum of 2.2 kg body weight and judged likely to easily adapt to separation from their mother) are summarised in Figure 8.10 (P O'Callaghan & W Blanshard unpub). These data include some atypical outlying values from a few slow developers that nevertheless grew into healthy adults.

9.1.6 Pap feeding

Around the time of first emergence from the pouch, juvenile koalas consume a special unformed maternal faeces called pap. Production of pap by female koalas seems to be initiated by the behaviour of the dependent young, which persistently nuzzles the cloaca of its mother until her normal pelleted faeces has been expelled and succeeded by thick paste or, in some cases, quite fluid material (Minchin 1937; Fleay 1937; Osawa et al. 1993b). This unformed material is ingested eagerly

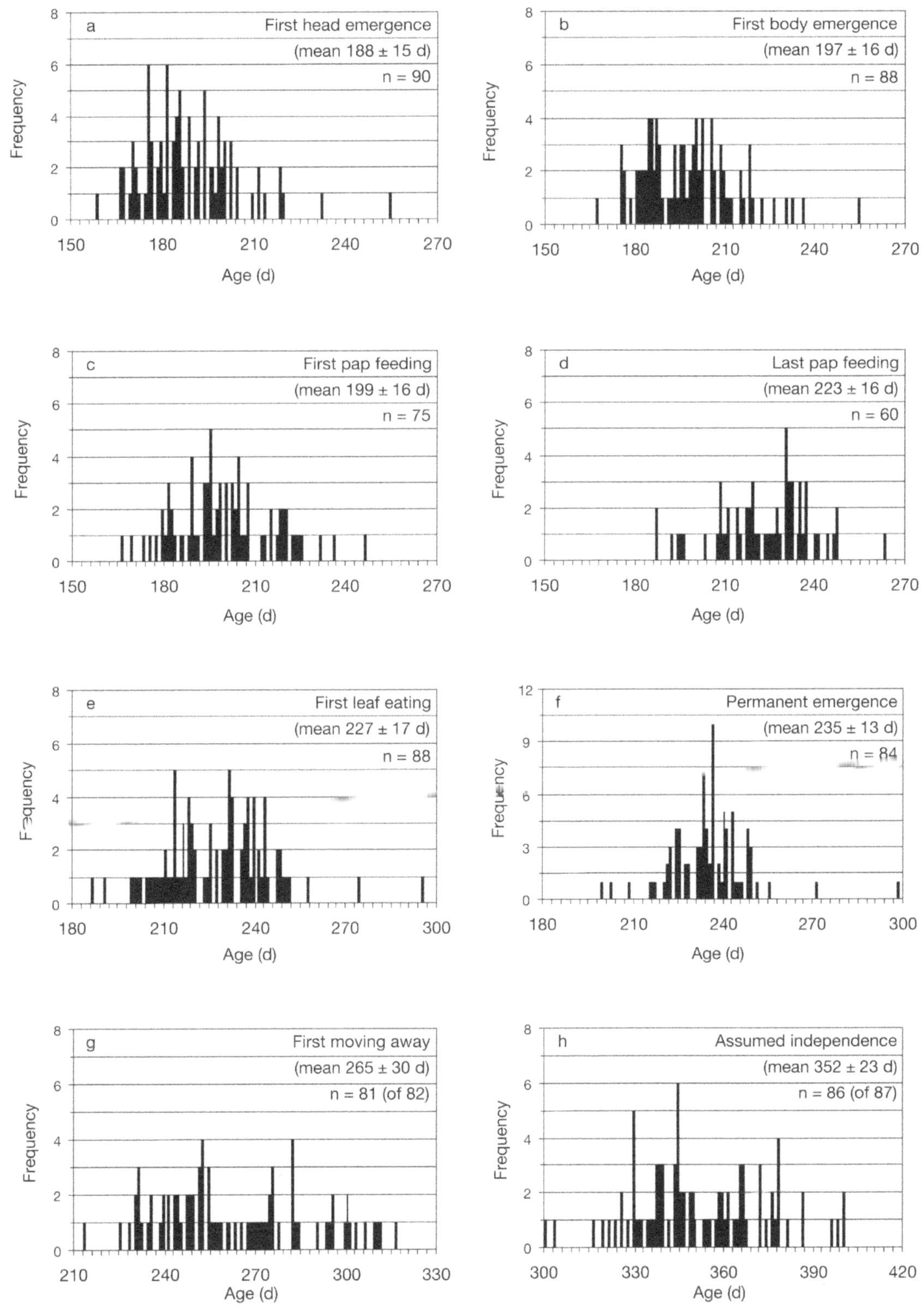

Figure 8.10 Age (d) at key developmental milestones for 91 captive-bred koalas in south-east Queensland. a) First head emergence. b) First body emergence. c) First pap feeding. d) Last pap feeding. e) First leaf eating. f) Permanent emergence. g) First moving away.* h) Assumed independence.* n = number of observations. * one outlier not shown. (Source: P O'Callaghan & W Blanshard (*unpub.*).)

Figure 8.11 Pap feeding by a 223 d old PY. Photo: W Blanshard.

by the juvenile (Fig. 8.11), which may pap feed for over an hour on each occasion. In most cases where close observations have been made in captive animals, young are known to have repeated pap feeding (2–7 times) at intervals over a 1–5 wk period.

Utilising specially formulated culture media, microbiological investigation of normal faeces and pap produced by 10 parous captive females showed that pap contained significantly higher viable counts of two bacterial groups (23–41 and 6–7 times greater, respectively) which are thought to play an important role in digestion of eucalypt leaves. It was found that viable bacteria of the former group (tannin–protein complex degrading enterobacteria) could not be recovered from the normal faeces of some healthy animals, including four of these 10 females (Osawa et al. 1993b).

These findings suggest that pap is derived from the caecal contents of the mother, and that pap feeding inoculates the juvenile koala's digestive tract with specific bacteria necessary for it to digest, metabolise and/or detoxify its natural diet. Juveniles are usually first observed eating small quantities of eucalypt leaf a few days or weeks after they have first been seen pap feeding.

9.1.7 Transfer and adoption of young

In captive colonies where female koalas with young are housed in groups, young have been observed using the pouches of, sucking from and even pap feeding from mothers other than their own. A permanently emerged BY whose mother dies or becomes ill can be adopted into a group of lactating females and their similarly aged young, and will ride on and feed from the different females. In this way, groups of three or four lactating females can raise four or five BY between them, respectively. Similarly, very old or very young mothers with BY can be housed with females whose young have recently been weaned, so the metabolic drain of lactation on the first individual is lessened.

To avoid the need for hand-rearing, an orphaned PY can sometimes be transferred to the vacant pouch of another female, if both are healthy and the female is at an appropriate stage of lactation. Unless the new teat and pouch are comparable in size to those of the natural mother's there may be too little milk or too little room available to the adoptee. The pouches of mothers with permanently emerged BY rapidly become too small for reoccupation. A 158 d old PY has been transferred successfully to a female whose 252 d old young had just stopped using the pouch and could be adopted into a group of lactating females with similarly aged young, as described above (Blanshard 1994).

On two occasions, 3 d old PY have survived and continued to develop after transfer to the unoccupied teat of a surrogate female that had just given birth herself (W Blanshard & P O'Callaghan unpub). When two young share the pouch, whether naturally or following transfer from another female, a contingency plan (either for transfer and adoption, or hand-rearing) should be devised in anticipation of further intervention being required at or approaching the time of pouch emergence, to try to ensure survival of both infants.

9.2 Hand-rearing

It is neither likely nor desirable that any carer's first attempt at hand-rearing an orphaned marsupial be undertaken with a koala. The following information assumes pre-existing experience with other species, a good working knowledge of the basic principles and a thorough understanding of the commitment and diligence required (see Jackson et al. 2003; Chapter 2 this volume).

Accurate ageing of the orphan (see 9.1.3) and assessment of its current and predicted progress through a number of key developmental milestones (see 9.1.5) are important when assessing any dependent PY or BY. There can be considerable variation between individuals in the absolute ages at which developmental milestones are reached (Fig. 8.10).

9.2.1 Environment

Young koalas may become distressed by sudden alterations in routine. It is important that a secure and constant environment is established, and that the routine remains structured so that potentially aversive stimuli are minimised. Ideally, one person should assume the role of primary carer. Physical and emotional bonds normally shared with the natural mother are quickly transferred to a human foster parent. They can also be transferred to inanimate objects, such as the substitute pouch arrangement and/or a plush toy or padded article of clothing (substitute mother) with which the juvenile can gradually be familiarised (Starr 1990a, 1990b; Koosmen 1991). Comfort provided by these familiar objects will enable the human foster parent to leave the koala with its inanimate babysitter for varying periods between feeds.

9.2.2 Substitute milk

Substitute milk provided for young koalas should be low in lactose. A number of low-lactose formulae have been used to successfully hand-rear koalas, including (but not limited to) those below.

- Biolac Milk Products—M100 early lactation milk for marsupials (with 2–5 mL canola oil added per 100 mL for furless koalas), M150 mid-lactation milk for marsupials, and M200 late lactation milk for marsupials.
- Di-Vetelact® low-lactose animal supplement (Sharpe Laboratories).
- Infasoy® (S26 Soy) Infant Formula—a soy-based infant formula (Wyeth Nutritionals).
- Wombaroo Food Products—Koala Early Milk Replacer (for koalas up to 160 d), Koala Mid-Lactation Milk Replacer (180–250 d), and Koala Late Milk Replacer (>270 d).

Koalas are very tolerant of a number of products of varying composition. At present, in spite of any theoretical advantages there is no compelling anecdotal reason to justify recommendation of one product over another (Marshall et al. 1990; Messer & Walker 1992). Most experienced wildlife carers have a favoured substitute milk formula, empirically based on what works best for both the carer and the animals in their care. When a carer first raises a young koala it is reasonable that their preferred formula for hand-rearing other species be used, rather than trialling a formula with which they are unfamiliar.

9.2.3 Milk delivery

The individual preference of the carer usually governs the method of milk delivery. Latex teats that have been successfully used while feeding young koalas include the Biolac T4 cat/kitten teat (Jackson et al. 2003) and several teats manufactured by Wombaroo Food Products. The Wombaroo FM is described as suitable for out-of-pouch koalas and MTM is described as suitable for in-pouch koalas. Catac Products manufactures a small tapered latex teat (ST1) which fits the smaller of its foster feeding bottles (FFB1), designed for hand-rearing kittens. The teats can be purchased in packets of three, separately from other components of the kit. At approximately 25 mm long, 4–5 mm in diameter near the base and tapering to a rounded tip, the ST1 is similar in size and shape to a mid- to late-lactation koala teat. Although the manufacturer acknowledges that the ST1 teat is sometimes used on a syringe, it does not recommend this practice. Nevertheless, the ST1 fits snugly onto the Luer hub of a standard syringe, and used this way allows the carer to quantify milk intake accurately at each feed. When using a new syringe, the sucking action of a vigorous feeder may draw the plunger down the syringe without assistance from the carer.

9.2.4 Substitute pap

By the time the upper and lower premolars erupt (after the first molars appear), most young have begun to consume some leaves. Young which have already started eating leaves before acquisition (and which can be assumed to have pap fed from their own mother) have brown staining of erupted cheek teeth and plant cell walls apparent in faecal smears.

PY that have not consumed pap prior to coming into care will require substitute pap. Maintaining natural pap samples anaerobically at 4°C in a refrigerator for up to 8 hr did not affect the viability and composition of its bacterial flora (Osawa et al. 1993b), but freezing the sample had an adverse effect on these characteristics (R Osawa *pers. comm.*). As it is not advisable to freeze pap, and because fresh pap is not readily available, orphaned koalas acquired as unfurred or lightly furred PY (prior to the age at which they normally consume pap from their own mothers) should be given fresh faeces collected from a number of healthy adult koalas (males and/or females). Adult faeces should be made into a slurry and fed, or added to the milk formula, on several occasions over 1–4 wk as the PY reaches the appropriate age. This should commence at the time the first cheek teeth (first molars) begin to erupt. The lower incisors and one (or two) pair(s) of upper incisors will already be present (Blanshard 1994). Intuitive carers may initiate the pap feeding process based on the behaviour of the young rather than its dental development (G Gipp *pers. comm.*). The fresh caecal contents of healthy koalas which died acutely (road kills, dog attack) may be a valuable potential source of substitute pap for hand-reared young; it has also been used for trans-faunation of adult animals that are considered at risk of caecal dysfunction, e.g. following administration of antibiotics that alter the balance of normal caecal flora (J Hanger *pers. comm.*).

Substantially reduced mortality rates were seen at one wildlife hospital following introduction of a prophylactic antimicrobial regimen (trimethoprim–sulphamethoxazole and nystatin) administered to all pre-weaned (pre-pap feeding) koalas from the time of arrival to the time of pap feeding (J Hanger & G Gipp *pers. comm.*). Weekly faecal smears were examined to monitor intestinal bacterial flora qualitatively. This regimen appears to have reduced the rate of mortality due to typhlitis/colitis (see 10.10.1d).

9.2.5 Weaning and independence

Juvenile koalas reach independence after a prolonged period of physical and physiological dependence on their mother. The mean age at which 82 captive young were first seen not to be in continuous physical contact with their mother was around 9 mo, with a range of 7.5–10.5 mo (Fig. 8.10). Most young do not move away from their mother for some time after permanent emergence from the pouch.

The gradual process of separation happens naturally and should not be unduly hurried by the carer. Therefore, following inoculation with caecal flora (from the natural mother, or substitute pap provided by the carer), small branchlets of eucalypt leaves should be placed within easy reach of the koala from its artificial pouch. As the young starts to emerge from the pouch more frequently and for longer periods, progressively larger branches of leaves can be secured close by, but the substitute pouch arrangement and/or substitute mother must still be available for physical and emotional security. The young will begin to rest among the branches, and eventually cease returning to the substitute for comfort.

10 DISEASES

10.1 Dermatologic diseases

10.1.1 Bacterial conditions

a Otitis externa

It is not possible to visualise the tympanic membrane in a conscious koala. Acute bacterial otitis externa is most often caused by *Pseudomonas aeruginosa* (Wood 1978). Other isolates in mixed infections have included *P. fluorescens*, *Proteus* spp., *Eschericia coli* and yeasts. Some infections may develop secondary to a pinna wound, and are therefore seen in fighting males.

b Erosive dermatitis

This is an uncommon secondary *Staphylococcus* spp. infection of elbow bite wounds in males. It develops as a very shallow erosion of the epithelium over the lateral aspect of the elbow. The epithelial surface is moist, yellowish and devoid of fur. A crust forms over the area, which then cracks to reveal a small amount of thin whitish-yellow exudate. In other cases, the crust covers an annulus of fibrous skin with a discharging core. The lesion continues to remain active in the absence of further trauma, and despite diligent crust removal and application of astringent or disinfectant. *S. aureus* or *S. intermedius* have been isolated. A topical antibiotic preparation is applied bid until the lesion has completely resolved. Refractory cases may require systemic antibiotic treatment (see 12.5; Table 8.6).

c Mycobacterial infection

The first naturally acquired *Mycobacterium ulcerans* infections of free-ranging animals other than humans were reported in koalas. *M. ulcerans* was recovered from ulcer material from five cases, with 4 out of 5 cases yielding mixed isolates with *M. scrofulaceum* and *M. gordonae* (Mitchell et al. 1984). In total, 11 koalas from Raymond Island (Vic.) were found to have *M. ulcerans* infections during 1980–1985 (Mitchell et al. 1987). Single or multiple ulcers were found on the volar surface of the footpads (5 of 6 animals had footpad lesions on more than one limb), the forearm or elbow (four animals), the face (four animals), rump (one animal) or scrotum (one animal).

Cutaneous ulcers were described as deep craters 5–20 mm in diameter, with a moist grey base. Histological sections showed coagulative necrosis of dermal and epidermal structures, extending into the subcutis beneath and around the margins of the ulcer (McOrist et al. 1985). Slender acid-fast bacilli were identified in necrotic areas and within macrophages in Ziehl–Neelsen-stained tissue sections (see 10.5.3).

As the majority of mycobacteria are free-living in soil or water, it was thought that they gained entry through abrasions to the epithelium. The high prevalence of infection in mature males suggested that the organism may have been transmitted during social encounters (fighting, mating) or may have contaminated wounds sustained during such encounters (Mitchell et al. 1987). All koalas were found in poor physical condition and were euthanased; treatment was not attempted.

10.1.2 Fungal conditions

a Dermatophytosis (ringworm)

The dermatophytes *Trichophyton mentagrophytes* and *Microsporum gypseum* have been isolated from both captive and free-ranging koalas. The most common sites of infection are the face (especially adjacent to or on the nose), the dorsum of the paws or the lateral aspects of the carpus and tarsus, although lesions can occur anywhere on the body. The close physical association between young koalas and their mothers helps to spread the condition. Dermatophytes have been isolated in all age groups from the very young to the very old. On furred skin the lesions are usually characterised by partial or complete alopecia, with a varying degree of hyperkeratotic scaling of the surface (into shiny silvery-white flaking layers). Occasionally the lesion has an erythematous margin. On the black, relatively hairless nasal skin the infection often seems to underrun the epithelium, with wrinkling and peeling of the black surface. Several contrasting presentations of dermatophytosis are illustrated in Figure 8.12. Blepharitis, dermatitis and paronychial infection with *T. mentagrophytes* causing deformity of the affected claws has been reported in a captive koala (Reddacliff 1986). Pruritus is not apparent in most primary dermatophyte lesions, although pruritic dermatitis was described in the koala with paronychia. Diagnosis is via skin and hair scrapings from the periphery of the lesion collected for fungal culture.

The principles of treatment for dermatophytosis that apply to companion animals apply to koalas. Daily topical application of a 0.3–0.6% solution of chloramine (Halamid®; Veterinary Companies of Australia) is usually efficaceous. Daily topical application of terbinafine cream (Lamisil® containing 10 mg/g terbinafine hydrochloride; Novartis Pharmaceuticals) has been very effective, particularly in occasional refractory cases (G Rawlinson *pers. comm.*). Successful treatment of a paronychial infection using 100 mg of griseofulvin (total dose) sid PO for 4 mo has been reported (Reddacliff 1986). Pulse therapy using itraconazole (10 mg/mL oral liquid) at 5 mg/kg sid PO as 1 wk on, 2 wk off for 12 wk has also been used to successfully treat dermatophytosis in koalas (L Vogelnest *pers. comm.*).

b Cryptococcosis

Cutaneous manifestations of infection with the fungus *Cryptococcus neoformans* have been reported (see 10.11.3).

10.1.3 Parasitic conditions

a Tick infestation

Naïve koalas with no previous exposure may be susceptible to the paralysis tick, *Ixodes holocyclus* (McKenzie 1981; Stone 1992; Stone & Carrick 1996; Martin & Handasyde 1999). For this reason, coastal areas of eastern Victoria within the range of this tick are usually avoided when koalas are translocated from other parts of that state (Menkhorst 2004).

Exsanguination during heavy tick infestation can be an important cause of morbidity or mortality (Dickens 1978a; Starr 1990a, 1990b). The tick season on the north coast of New South Wales (August–March) coincides with the time that juvenile koalas leave their mothers (Starr 1990a, 1990b). Many juveniles and occasionally adult koalas are found debilitated or dying as a result of heavy infestation with *I. holocyclus* and *Haemaphysalis* spp. ticks. Tick paralysis has not been observed in this region. Changes to the natural habitat which force koalas to travel long distances between trees through dense scrubby undergrowth are thought to predispose toward heavy infestations (Starr 1990b).

Blood samples collected from free-ranging koalas from Walkerville, on the southern coast of Victoria,

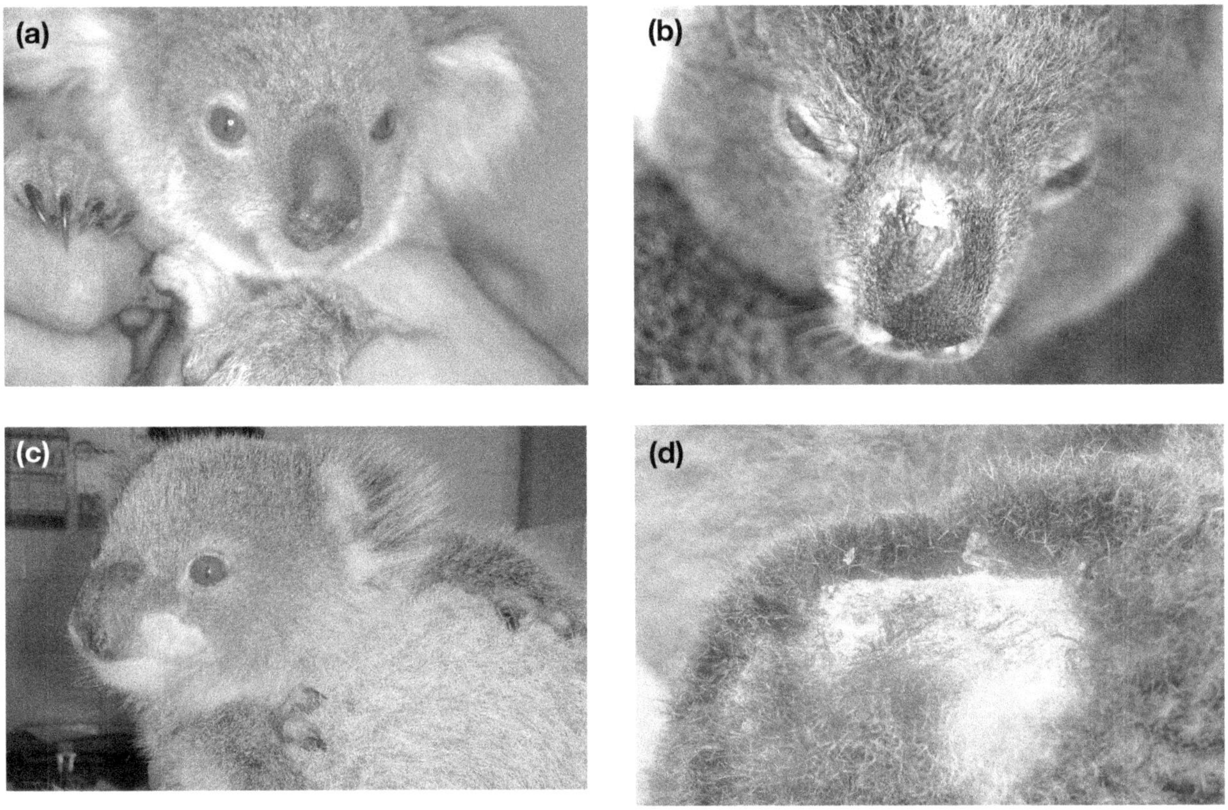

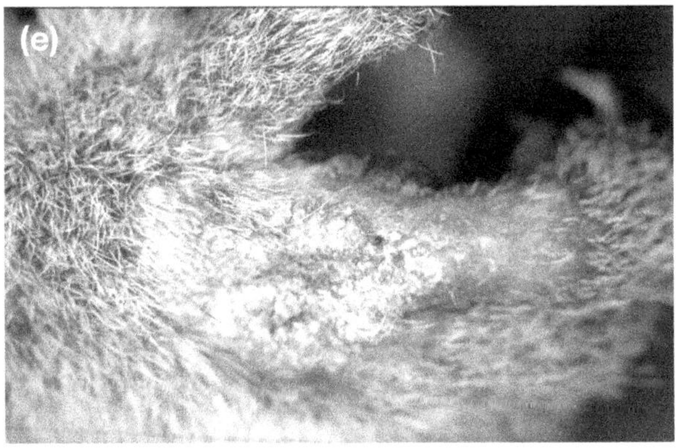

Figure 8.12 Contrasting presentations of dermatophytosis. a) An early lesion affecting the black nasal skin around the right nostril. b) More advanced lesion on the dorsum of the nose, with peeling of the epithelium. c) Localised (bilateral) alopecia on the face and first two digits of the forepaws. d) Partial alopecia and silvery flaking on the anterolateral aspect of the left stifle of a free-ranging adult, caused by *T. mentagrophytes*. e) Hyperkeratotic scaling of affected skin on the right forepaw. Photos: a), b), d), e) W Blanshard; c) Taronga Zoo.

showed mild anaemia and relative eosinophilia attributed to heavy infestations of *Ixodes tasmani* during spring and autumn (Obendorf 1983; Martin 1985b). Intense regenerative responses overcame normocytic, normochromic anaemia within 2–3 wk of tick removal from two adult and four juvenile or subadult, tick-infested free-ranging koalas brought into care from the Port Macquarie area in northern New South Wales. Recovery was characterised by high numbers of circulating reticulocytes (up to 40% uncorrected; normal 0–6%), nucleated RBC (up to 200 per 100 white cells; normal 0–10), marked polychromasia and anisocytosis, and prominent erythrocyte inclusions evident in Giemsa-stained smears. Absolute eosinophilia also was common during recovery (Spencer & Canfield 1993b).

Treatment includes physical removal of adult ticks. If many nymphs are present, bathing koalas in an acaricidal dog wash (0.025% suspension of amitraz 50 g/L) has been recommended (Starr 1990a). Topical application of a fipronyl spray is also suitable (G Gipp *pers. comm.*). An oral iron supplement may be given (Starr 1990a). Three of five treated koalas in south-east

Queensland that presented with apparent clinical signs of tick paralysis recovered after IV administration of a commercial antitoxin raised in hyperimmune dogs (Stone & Carrick 1996).

b *Sarcoptic mange*

Infestation with the mite *Sarcoptes scabiei* has been observed uncommonly in both captive and free-ranging koalas. A hand-reared juvenile koala and its human carers developed pruritic skin lesions after exposure to a wombat which had been observed to scratch constantly (Barker 1974). *S. scabiei* was recovered from skin scrapings of two juvenile free-ranging koalas from Phillip Island (Vic.) (Obendorf 1983). In another case, a large proportion of the koalas in a Queensland captive koala colony was affected during an outbreak of sarcoptic mange which resulted in several deaths over a 9 mo period (Brown et al. 1982). Clinical features in most cases included variable pruritus, small raised dry circumscribed lesions under the fur in early cases, and dry thickened encrusted hyperkeratotic lesions affecting the footpads and/or metacarpal and metatarsal regions and spreading to the skin of the thorax, abdomen and face. Skin scrapings from the lesions of all affected koalas revealed large numbers of mites.

Immersion of the individual hand-reared koala in 0.2% malathion, three times at 10 d intervals, resulted in complete clinical recovery. Prior to specific treatment of the captive koala colony, affected animals were partially immersed in a water bath and their hyperkeratotic crusts removed by vigorous scrubbing. They were then partially immersed in a 0.025% aqueous suspension of amitraz; care was taken to thoroughly wet all the fur, including sponging of the head, face and ears. Amitraz treatment was repeated after 10 d and was successful, with no adverse effects noted. Relatively high doses of oral ivermectin have been used safely to treat demodecosis in a captive koala (see below), therefore treatment of sarcoptic mange using a similar therapeutic regimen to that suggested in wombats (ivermectin 1% injection at 300 μg/kg SC, repeated at 10 d intervals for three treatments) might reasonably be used in koalas.

c *Notoedric mange*

Infestation with *Notoedres* spp. mites (described as morphologically identical to *N. cati*) has been associated with skin lesions in koalas (O'Sullivan 1949).

d *Demodectic mange*

Demodex spp. mites have been identified in skin scrapings from a captive female koala without clinical signs in south-east Queensland (Blanshard 1994), and have been confirmed as the causative agents for bilateral periocular alopecia with mild intramural lymphocytic folliculitis in a captive male koala in New South Wales (Vogelnest et al. 2000). Infection resolved following treatment with ivermectin (Ivomec® Liquid for Sheep 0.8 g/L, Merial Australia). A starting dose of 50 μg/kg sid PO was gradually increased until a clinically effective dose of 600 μg/kg ivermectin sid PO was reached. Treatment at the high dose rate was continued for 6 wk after negative skin scrape results.

e *Fur mites*

A new species of native atopomeline fur mite, *Koalachirus perkinsi* (formerly *Austrochirus perkinsi*) was described from koalas in 1958 (Domrow 1958; Halliday 2000). There are no reports of clinical signs resulting from infestation. Commonest sites of habitation are the medial surface of the thighs and the groin, with the life-cycle being completed on the host (T Warren *pers. comm.*).

f *Flea infestation*

Cat fleas, *Ctenocephalides felis felis,* have been found incidentally on the bodies of three koalas at necropsy (Griffin et al. 1983; ARWH 2007 case number 727/1). Two were free-ranging animals from northern New South Wales which became ill during the spring–summer period and were hospitalised in private dwellings (Griffin et al. 1983). The source of the flea infestations was not determined. Two juvenile koalas that were being hand-reared in a private dwelling became infested with fleas (species not identified) from a domestic cat (Blanshard 1994).

10.1.4 Miscellaneous conditions

a *Disorders of keratinisation*

Idiopathic disorders of keratinisation including diffuse seborrhoea, papillary hyperkeratosis of the volar surfaces of the paws, papular and nodular epidermal hyperplasia, and formation of cutaneous horns were reported in related ageing captive koalas (Canfield et al. 1992). The apparent familial predisposition to these dermatoses may have been coincidental, as three other cases of idiopathic disorders of keratinisation (volar hyperkeratosis, nodular dermatosis and diffuse seborrhoea) were noted in unrelated free-ranging and captive koalas. These disorders of keratinisation differed from the secondary seborrhoeic changes seen in association with sarcoptic mange and dermatophyte infection, and their aetiopathology was unknown.

b Piggy-back alopecia

After permanent emergence from the pouch a young koala rides prone on its mother's back when she moves around the branches. With the passing of time and the increasing weight of the young, the mother's fur begins to look a little the worse for wear. A saddle-shaped area over the thorax and upper back may become darker and less dense than the surrounding fur due to physical wear, loss of guard hairs and exposure of the darker finer underfur. In some cases bilaterally symmetrical circular areas of severe alopecia (which correspond to the resting position of the BY's hind feet) are apparent on the caudal flanks (Fig. 8.13). By late lactation (when the young is 10–12 mo old) the female's body condition may have declined markedly, and together with her moth-eaten appearance may be mistaken as a sign of ill health. The fur may retain this pattern for 1–2 mo after the BY ceases riding (Blanshard 1994).

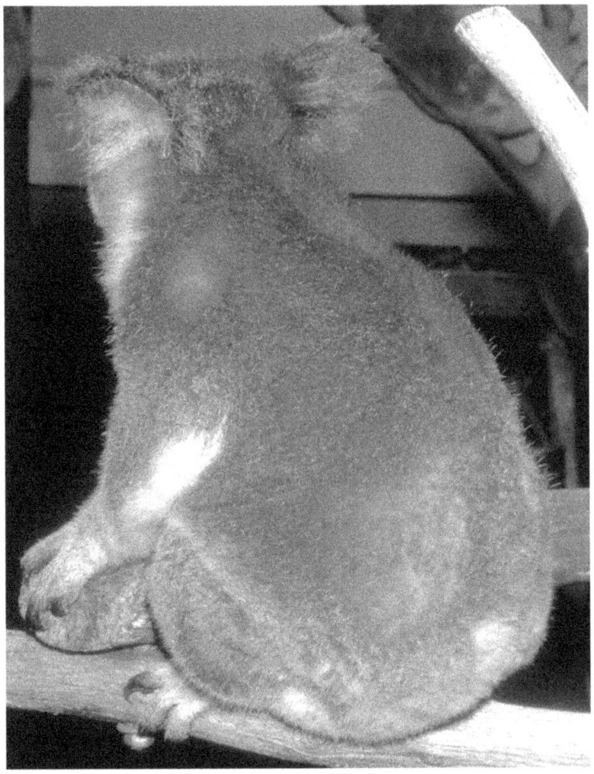

Figure 8.13 Piggy-back alopecia. This healthy female's 303 d old BY was weaned (i.e. separated) 13 d earlier. Note widespread loss of guard hairs extending onto the top of the head, and roughly circular patches of wear from the fore and (especially) hind paws of the riding young. Photo: W Blanshard.

c Claw deformities

Failure of regular weight-bearing contact between claws and branches during climbing can lead over time to elongation and deformity of the claws, e.g. after previous injury to the digit. Alternatively, a claw may regrow crookedly after traumatic detachment. After years of a relatively sedentary life-style, the claws of very old captive koalas may tend to grow in a circle back towards the pads. Their cage furniture often consists of fairly slender braches, so on many occasions the animals do not need to engage their claws at all when moving around. Treatment is to shorten and reshape the claws. Cutting of very long circular claws may require heavy plier-type nail-trimmers. The claw should be cut transversely to the same overall length as a normal claw, and a round file used to reshape the ventral aspect of the distal half to create a downward-pointing tip.

d Cutaneous myiasis (flystrike)

The sheep blowfly *Lucilia cuprina* targets moist areas in which to lay its eggs. Injured or debilitated koalas may be struck in open wounds, if exudate is present or the fur is continually moist because of soiling by urine or diarrhoea. The pouch, conjunctival sac, external ear canal and perineum are likely targets. The at-risk animal should be housed indoors or in a flyscreened enclosure during treatment if possible. If the koala must remain outdoors, fur adjacent to the affected area should be sprayed 2–3 times daily with insect repellent.

e Papillomatosis

Papillomas have been described around the face, eyes, lips and gum of young koalas (Wood 1978). Surgical removal of warts interfering with sight or eating or which become infected was reported, but most cases recovered without treatment. Papillomaviruses have been isolated from the skin of one affected and 10 clinically normal koalas (Antonsson & McMillan 2006).

10.1.5 Diseases of the pouch and mammary glands

a Pouch infection

Mortality of 3–5 mo old PY of captive koalas caused by *Pseudomonas aeruginosa*, and reported as pouch death syndrome, has been described as occurring in epidemic proportions (Wood 1978). The young were usually found dead on the floor of the enclosure. Other organisms including *Klebsiella* spp. and *Proteus* spp. were implicated in isolated cases. The epidemic was controlled when a formalised vaccine against *P. aeruginosa* was prepared and administered to mature female koalas over a 3 yr period.

In most cases, however, it is impossible to determine whether the presence of viable micro-organisms in the pouch is the cause or a consequence of morbidity or mortality of young. These examples are from Osawa et al. (1992).

- The PY of a 12 yr old captive female disappeared from her pouch within 24–40 hr of birth. A pouch swab taken 17 d later yielded a heavy pure growth of *P. aeruginosa*.
- The PY of a 5 yr old captive female disappeared from her pouch within 15 d of birth. A pouch swab taken at that time yielded a heavy growth of *P. aeruginosa*, moderate growth of *Serratia marcescens* and light growth of alpha-haemolytic *Enterococcus faecalis*.
- The pouch of a 2 yr old captive female who had given birth 61 d earlier contained an amorphous mass of thick brown paste, which may have been the highly decomposed body of her young. A pouch swab yielded heavy growth of an unidentified yeast and a light growth of *Streptococcus bovis*.

Whenever a PY is lost the pouch should be examined, and disinfected and/or treated with a topical antimicrobial preparation if necessary. The pouch should be rechecked within a few days, as inflammation of the lining may progress to true pyoderma with formation of pustules in the epithelium of the pouch, mammary gland and teat. Canfield et al. (1991) noted extensive ulcerative purulent lesions in the pouch of a captive animal. One instance of pouch pyoderma caused by *Candida albicans* was treated with ketoconazole at 10 mg/kg sid PO (seven doses); another case caused by *P. aeruginosa* relapsed after treatment with gentamicin at 2.2 mg/kg bid SC (15 doses) but responded to ceftazidime at 15 mg/kg bid SC (14 doses) (Blanshard 1994).

b Mastitis

A 173 d old PY was removed from its mother's pouch when its head was seen protruding through the pouch opening but unresponsive to touch (Fig. 8.14). The PY was very small for its age and extremely thin, and the developed mammary gland was hard and the teat swollen and shiny. Culture of milk expressed after disinfection of the teat yielded a beta-haemolytic *Streptococcus* spp. and *Serratia marcescens*. The young died 48 hr after hand-rearing was commenced, and on necropsy there was evidence of typhlitis and septicaemia. A pure growth of *S. marcescens* was obtained from the PY's heart blood, while culture of its caecal contents yielded *S. marcescens*, *Enterococcus faecalis* (β-haemolytic strain) and *Klebsiella pneumoniae*. The latter three organisms, together with *Acinetobacter lwoffi*, were also isolated from a swab of the pouch epithelium (Osawa et al. 1992).

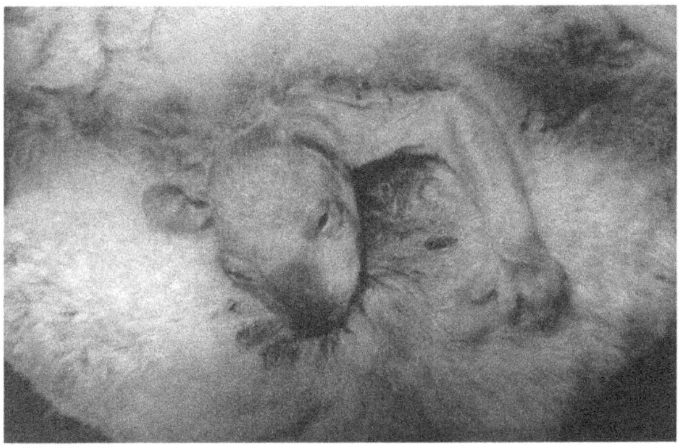

Figure 8.14 Unresponsive 173 d old PY of a female with mastitis. Note emaciation and dehydration, especially evident around the shoulder. Immediate intervention is indicated. Photo: W Blanshard.

10.2 Gastrointestinal diseases

10.2.1 Gingivitis and periodontal disease

Brown to black staining of enamel surfaces of the cheek teeth begins when leaf is first eaten. The incisor teeth usually remain an off-white colour, but as they are not mechanically cleaned by involvement in the mastication of leaf they may accumulate deposits of tartar and plaque. Periodontal pockets may form along the labial aspect of the upper molars, especially in aged animals where the height of the teeth is drastically reduced and there is mechanical abrasion of the gum by leaf fragments. A study examining 32 captive and free-ranging Queensland koalas determined that animals aged >5 yr demonstrated mild to moderate gingivitis (visible plaque and calculus, thickening of the marginal gingiva), and animals older than 10 yr had increasing gingival thickening and recedence. Despite these changes, development of severe periodontal disease was thought to be relatively uncommon (Bird et al. 2002).

Gingivitis or abscessation associated with the alveolus of the upper incisors may occur. Abscesses may drain around the roots of the teeth, or through the maxillary gum a few centimetres caudal to the incisors, or into the nose. The second and third pair of upper incisors may become loose.

Most cases of periodontal pocketing remain subclinical and are incidental findings at necropsy, but malar abscessation has been seen associated with impaction of leaf into periodontal pockets along the labial aspect of the upper molars.

Affected teeth can be cleaned under general anaesthetic and any loose incisors removed. Periodontal pockets present greater difficulty because access and exposure are difficult, and other than cleaning the exposed molar roots little can be done to correct the problem (G Seymour *pers. comm.*).

10.2.2 Impaction of cheek pouches

One or both cheek pouches may fill with chewed leaf, causing subtle swelling of the face. The leaf material can generally be easily removed by applying pressure in a caudal direction. Although impaction may occur in animals of any age for no apparent reason, particular inspection of the teeth in aged animals may reveal a predisposing cause (e.g. worn teeth).

10.2.3 Stomatitis and pharyngitis in independent koalas

Mucosal ulcerations in the soft palate and/or pharynx have been observed infrequently in recently independent juvenile or subadult captive koalas. Culture in different cases has revealed heavy growths of *Klebsiella* spp., haemolytic *Streptococcus* spp. or mixed normal flora. Failure of ulcers to resolve or their recurrence after treatment has been a feature of some cases. Severe stomatitis is included in the spectrum of diseases believed to result from the immunosuppressive effects of infection with koala retrovirus (Hanger et al. 2003; J Hanger *pers. comm.*).

Ulcerative stomatitis has been observed in free-ranging koalas. A mixed population of bacteria was seen in histological sections of extensive shallow ulcerations of the gingiva and dorsum of the tongue in a young female koala in Queensland that also had severe oedema of the submandibular tissues and face (McKenzie 1981). Three cases of stomatitis were reported in association with other disease findings in a mortality survey of 127 free-ranging koalas in New South Wales; *P. aeruginosa* was isolated from oral lesions in one case. One koala concurrently had chronic cystitis and metritis, while two had hepatic disease which appeared to be associated with toxaemia or septicaemia (Canfield 1987). Older animals with stomatitis are usually depressed and have reduced leaf intake leading to weight loss (adults) or failure to gain at an appropriate rate (subadults).

Oral candidiasis may occasionally be seen in independent koalas. White curd-like accumulations may be seen on the surface of the tongue, but in many cases only extremely subtle shallow erosions of the epithelial surface are visible in good lighting. Ovoid, occasionally budding yeast cells may be seen in stained smears from the tongue or palate. The majority of these are outside epithelial cells. It is important not to mistake the Gram-negative gliding bacteria *Simonsiella* spp. for yeast cells. This genus is an apparently non-pathogenic inhabitant of the oral cavity of warm-blooded vertebrates and has been seen in oral swabs from juvenile koalas and other marsupials (Blanshard 1994). *Simonsiella* spp. appears as multicellular filaments with rounded ends, often segmented into groups of eight cells. The long axis of each cell is perpendicular to the long axis of the filament (Larkin 1989). Most of these are seen in association with epithelial cells. An otoscope is required to examine the pharynx in young animals, and a laryngoscope or perspex-tipped pen torch is needed for adults. Juveniles with non-fungal pharyngitis show intense reddening of the caudal surface of the pharynx. Culture can be used to confirm a diagnosis of oral candidiasis, or to rule out the presence of yeasts and identify any potential bacterial pathogens.

Stomatitis and pharyngitis in independent subadult or adult animals has a guarded prognosis. Death may sometimes result from inanition and debility, or from secondary septicaemia. Microbial culture and sensitivity may reveal mixed normal flora or heavy growth of a potential pathogen.

10.2.4 Delayed gastric emptying

It is not uncommon to palpate under the left costal arch of sick koalas known to have eaten little or no leaf over previous days, only to find that the stomach is full. Delayed gastric emptying can be a significant complicating factor in the recovery of many sick koalas, irrespective of the nature of their illness. The underlying cause remains unclear in most cases, although factors known to compromise normal GIT motility in other species (e.g. dehydration and hypokalaemia) are likely to play a part.

Koalas with full stomachs may eat very enthusiastically when fresh leaves are first presented to them each day, but stop eating within a few minutes. In many cases the animal can appear deceptively bright and active, in contrast to the general picture of a debilitated koala. Passage of faecal pellets may continue despite little/no leaf intake, although pellets gradually decrease in length and diameter. In other cases a true ileus is present, with no active gut sounds apparent on auscultation and no faeces being passed. By noting how much leaf is eaten over what time, and palpating the stomach before and afterwards, an attempt can be made to gauge whether movement of the stomach contents is retarded. When assessing delayed gastric emptying (or almost any other gastrointestinal disorder) administration of a small amount of barium sulphate suspension (e.g. 5 mL/koala PO) may act as a useful marker if radiographs are subsequently taken. In some cases, all the barium remains in the stomach for days. Discernible faecal pellets on an abdominal radiograph indicate that some barium is getting through the system.

When treating cases of delayed gastric emptying, it is important to identify and correct any pre-existing fluid or electrolyte imbalances. Give some fluid orally if the koala will accept it, to soften the stomach contents. If no improvement, give metoclopramide hydrochloride at 0.5 mg/kg tid IV, IM or SC to encourage gastric emptying and accelerate intestinal transit. Metoclopramide is contraindicated in the presence of mechanical obstruction, but this can be very difficult to diagnose in the koala. If there is a likelihood of gastrointestinal pain (colic), give flunixin meglumine 1.0 mg/kg sid IV, IM or SC for a maximum of 3–4 d. Cisapride (Prepulsid® 1 mg/mL suspension; Janssen-Cilag) at 0.1–0.2 mg/kg tid PO, given 30–60 min prior to feeding, has been used to stimulate gastrointestinal motility in hospitalised animals (J Hanger *pers. comm.*).

Encourage the koala to eat leaves (even if it will only have one) at frequent intervals to stimulate gastric motility. Supplementary feed (see 5.3.4d) with paste (small amounts, often), particularly if radiographs demonstrate that barium can escape the stomach. If there is no improvement within a few days and the koala is losing weight despite supplementary feeding (or will not accept supplementary feed), a functional intestinal obstruction should be suspected and an exploratory laparotomy performed. This procedure should not be undertaken lightly, nor delayed until the koala is in such poor body condition that its chances of survival during and after surgery are negligible.

10.2.5 Gastrointestinal ulceration

Occasionally, small haemorrhages or discrete ulcerations in the gastric or intestinal mucosa are observed at necropsy. The point at which the proximal colon narrows and its longitudinal mucosal folds disappear is a common site for shallow ulceration. These haemorrhages or ulcerations may be incidental findings.

10.2.6 Diarrhoea

As in other species, a multitude of factors may cause diarrhoea of varying consistency in the koala.

Diet—observations in captive colonies suggest that there is an association between consumption of certain species of eucalypts and passing of much softer, sometimes unpelleted, faeces by some animals. In Queensland the scribbly gum (*E. signata*) has this reputation (P O'Callaghan *pers. comm.*).

Disease—enteritis may cause severe diarrhoea. A moderate to heavy growth of yeasts has been isolated from the faeces of some adult koalas with diarrhoea. In one animal, intermittent profuse diarrhoea was associated with heavy growth of a long rod-shaped yeast identified as *Candida catenulata*. Enteritis associated with *Cryptosporidium* spp. (details not specified) has been reported.

Treatment with oral nystatin has been associated with the development of watery diarrhoea in a small proportion of adult koalas.

Koalas which are being supplementary fed in the absence of significant leaf intake develop diarrhoea because the fibre content of their digesta has decreased.

Physiological diarrhoea is induced by physical stimulation of the cloacal area in maternal koalas by their PY during pap feeding.

Normal faeces is pelleted and does not soil the pericloacal fur. Abnormal faeces may vary in consistency from unpelleted semi-solid cylinders through to profuse watery puddles. In acute diarrhoea there may be precipitous weight loss associated with reduced gut contents but not necessarily any change in body condition. Faeces should be submitted for microbial culture and sensitivity testing if unexplained diarrhoea persists for more than 24–48 hr. A positive faecal occult blood test has been recorded in one adult koala with mycotic diarrhoea. Abdominal radiographs of koalas with severe enteritis may show accumulation of gas in intestinal loops. Normal barium transit would support a diagnosis of enteritis. In some cases, gastric emptying is delayed secondary to the enteritis, and radiographic distinction between enteritis and functional obstruction is not possible.

Treatment is directed at the underlying cause, if known. The severity of the diarrhoea may be temporarily improved by administering plain anti-diarrhoea mixtures containing kaolin, pectin or aluminium hydroxide (without antibiotics), or barium sulphate orally. In one koala with mycotic diarrhoea due to *C. catenulata*, heavy growth of yeast was still obtained after administration of nystatin at 10 000–20 000 U/kg tid PO for 7 d, and after administration of ketoconazole at 10 mg/kg sid PO for 7 d. Some improvement in demeanour, faecal consistency and leaf intake was evident after 3–4 d of fluconazole at 3 mg/kg sid PO (loading dose 6 mg/kg on first day). Fluconazole was given for a total of 14 d. Koalas which are being supplementary fed can have finely blended leaf added to their paste to supply fibre if necessary.

10.2.7 Colic

Clinical signs suggestive of gastrointestinal pain have been observed in captive koalas. The animal may vocalise intermittently or seem restless and uncomfortable and make serial body movements which alter the shape of the abdomen (unlike the brief contented stretching of neck and hind limbs seen in healthy koalas). Guarding of the abdomen is not apparent, probably because the abdominal muscle layers are relatively thin and incapable of palpably tensing such a large cavity. Administration of flunixin meglumine (1.0 mg/kg IV, IM or SC) seems to relieve discomfort. If clinical signs persist or recur a serious gastrointestinal disturbance should be suspected.

10.2.8 Impaction

Gross distension of the stomach and proximal small intestine with digesta, associated with apparent blockage distally by an ovoid bolus of fibrous plant material, has been described in two captive koalas (McKenzie 1981). Both were found dead the same morning in different enclosures, prompting speculation about relationship to a particular batch of feed. Rectal and colonic impaction with hard faeces and subserosal rectal haemorrhages were reported in a young captive koala (McKenzie 1981). Depression, dehydration and passage of small faecal pellets with the consistency of plasticine were noted in a 14.5 yr old captive male koala with very advanced tooth wear. At necropsy his caecum was empty and his proximal colon was completely filled by neat short compacted masses of coarse fibre. Digesta of finer consistency coated the colonic mucosal surface, and was apparently being channelled past the impacted fibre, accounting for the passage of softer and relatively non-fibrous faecal pellets.

In one captive koala, ingestion of a significant quantity of sand from the enclosure substrate resulted in decreased leaf intake, poor body condition and lethargy. No abnormality (other than reduced gut fill) was apparent on abdominal palpation. Sand had not been noted in faecal pellets but its accumulation in the caecum and proximal colon showed clearly on abdominal radiographs. The period over which the sand had been ingested was not known. Passage of the sand in faeces was facilitated by oral administration of psyllium 2 g/kg/d in divided doses in supplementary feed. Dry psyllium powder was mixed with low-lactose milk powder; the combination had to be administered promptly after the addition of water to make a paste, as it quickly became gelatinous and stiff. Overhydration was maintained with IV, SC and oral fluids to allow for the water-absorptive capacity of the psyllium. More than 100 g of sand was passed over a 2 mo period, in the latter half of which psyllium was given at a lower rate (1 g/kg/d).

10.2.9 Perforation and rupture

Duodenal rupture, and peritonitis due to ulceration and perforation of the caecum near the ileocaecal junction, have been reported in koalas. In aged animals with worn teeth, leaf stems may not be masticated efficiently and relatively long fragments may pass through the

GIT. Very occasionally such fragments may perforate the intestinal wall. The blind narrow distal end of the caecum may be the site of traumatic perforation. Asymptomatic cases are found incidentally at necropsy. The stem fragment pierces the serosa but is walled off *in situ* by fibrous tissue. In other cases there may be leakage of intestinal contents and localised peritonitis. Abdominal paracentesis should be performed if a problem is suspected, but cytology alone may not be conclusive. Microbial culture and sensitivity testing of part of the sample might demonstrate enteric bacteria. Abdominal radiography is unlikely to be diagnostic unless there has been escape of free gas into the peritoneal cavity (free gas may be present if paracentesis has preceded radiography).

10.2.10 Torsion and intussusception

Gentle torsion of the distal colon, and entrapment of caecal loops by the free end of the caecum, have been observed in a captive adult and a hand-reared juvenile koala, respectively. Both animals remained relatively bright but refused to eat. No escape of barium from the stomach of the juvenile was apparent on successive radiographs over 48–72 hr, and no faecal pellets were passed by the adult animal for several days. Both animals recovered after laparotomies to de-rotate loops of bowel that were not compromised (Blanshard 1994). Torsion or intussusception has been observed at various intestinal sites in koalas.

10.2.11 Proctitis

An aged free-ranging koala was found to have asymptomatic chronic proctitis associated with chlamydial infection (Hemsley & Canfield 1996). It was hypothesised that a primary genital tract infection was the source of an ascending infection of the rectum via the cloaca.

10.3 Hepatic diseases

Inclusion bodies in the nuclei of hepatocytes and in sloughed epithelial cells in the duodenum have been reported in histological sections from koalas, and a viral aetiology was suspected (Dickens 1975). Electron microscopy of inclusions from the liver in a similar case did not reveal any viral components. The inclusions appeared as compact amorphous masses of electron-dense material without a limiting membrane (McKenzie 1981).

Pathology suggestive of toxic liver damage has been observed in free-ranging koalas on Raymond Island (Vic.) in late spring and early summer. In one outbreak, 21 of the island's estimated population of 450 koalas (4–5%) were found dead over a 5 mo period, with liver disease identified in 17 necropsies (Mitchell 1994). The cause of the hepatopathy remains unknown.

Cholecystitis caused by *P. aeruginosa* has been described in a young koala which also had a pneumonia caused by the same organism. Death occurred 3 wk after the onset of clinical signs which included occasional vomiting, stretching of the neck and apparent dysphagia (Canfield et al. 1986a). Inflammation and calcification of the bile ducts (which in one case appeared to occlude the opening of the bile duct into the duodenum) has been reported (Dickens 1978a). A 4 mm diameter biliary calculus was recovered at necropsy from another animal (Connolly 1999).

10.4 Cardiac diseases

A koala health survey detected abnormal heart sounds under isoflurane anaesthesia in 4 of 17 captive and 9 of 50 free-ranging koalas examined (Barnes 1999). Of the 13 animals found to have audible systolic heart murmurs when anaesthetised, three of the captive animals had confirmed audible murmurs while conscious and one did not. Very few of the free-ranging koalas were successfully auscultated while conscious but 2 of 9 did not have a detectable murmur after recovery from anaesthesia. Colour-flow doppler echocardiography was carried out in one captive animal and was not conclusive of any abnormalities. In another field study, during which 25 free-ranging animals in Pine Creek State Forest (NSW) were masked down with isoflurane, arrhythmias were noted on several occasions (J Hanger *pers. comm.*). Further research is required to determine whether heart sound abnormalities might be induced by isoflurane anaesthesia in some koalas (Barnes 1999).

Constrictive pericarditis has been reported in a mortality survey of free-ranging koalas (Canfield 1987). An atrial septal defect was identified at necropsy in a captive-bred juvenile koala which had failed to grow at a normal rate and had a history of tachypnoea. The koala had a continuous cardiac murmur and profound cardiomegaly was evident on radiographic examination. At necropsy there was severe dilatation and hypertrophy of the right ventricle associated with inter-atrial shunting of blood (Atwell & Booth 1990). Pulmonic valve insufficiency resulting in congestive heart failure has been observed in a captive juvenile female koala (G Rawlinson *pers. comm.*).

10.5 Respiratory diseases

Respiratory disease may affect individual animals or groups. Captive colonies of koalas may experience outbreaks of respiratory disease (Rahman 1957; Wood 1978; McKenzie et al. 1979; Canfield et al. 1986b; Booth & Blanshard 1989; Blanshard 1992; V Nicolson *pers. comm.*; G Rawlinson *pers. comm.*; J Hanger *pers. comm.*), and episodes/cases are known to have occurred simultaneously at different sanctuaries in the same geographic region (Booth & Blanshard 1989). The aetiology of many outbreaks remains unclear.

10.5.1 Rhinitis/pneumonia complex

Rhinitis/pneumonia complex is a clinical disease syndrome with the potential for single or multiple aetiological agents, acting alone or in combination, to produce disease of varying clinical signs and severity. It is likely that there is a spectrum of clinical manifestations of upper and lower respiratory tract disease including rhinitis, pharyngitis, tracheobronchitis, pneumonia and pleuritis in which a number of micro-organisms may act as primary pathogens or secondary opportunists.

a Aetiology

In 1937 it was reported that pulmonary diseases were prevalent among Queensland koalas admitted to a New South Wales sanctuary, but practically unknown in Victorian koalas (Pratt 1937). Three types of pneumonia were mentioned– 'lobar (broncho) pneumonia, bacterial pneumonia, and staphylococcic pneumonia'—all virulent, with the last two described as 'unpreventable and incurable ... the victim does not run a temperature, so your bear is dead before you know where you are' (N Burnet cited in Pratt 1937). Retrospective necropsy findings from a number of locations were published in the proceedings of a 1976 symposium on koala biology, management and medicine. Pneumonia was reported in 12 of 227 (5.3%) koala necropsies on (mainly) Victorian koalas, and was described as a common finding in koalas dying in Victoria in late autumn and early winter (Butler 1978). An additional seven animals had cryptococcosis but distribution of their lesions was not detailed. Pneumonia was diagnosed at necropsy in 2 of 10 koalas examined in New South Wales (Finnie 1978), and was one of the pathological findings in 3 of 10 captive koalas necropsied overseas (Robinson 1978). Neither pneumonia nor other respiratory diseases were reported in any of 44 necropsies on free-ranging koalas in Victoria in a survey of morbidity and mortality conducted from 1975 to 1980 (Obendorf 1983). A 1980–86 mortality survey of free-ranging koalas from northern New South Wales found that 7 of 127 (5.5%) had pneumonia (Canfield 1987). In a 1986–87 Queensland survey of morbidity and mortality in free-ranging and captive koalas, pulmonary lesions were recorded in 11 of 75 (14.6%) animals necropsied. These were grouped in a general category of chlamydiosis, although neither chlamydial isolation nor routine microbiological culture was performed (Weigler et al. 1987).

Role of chlamydiae

An association between the order *Chlamydiales* and respiratory disease in koalas was proposed in 1984 when '*Chlamydia psittaci*' was isolated in cell culture from the nasal septum of 1 of 12 koalas with rhinitis/pneumonia complex (Brown & Grice 1984; see 10.11.2a).

Role of *Bordetella bronchiseptica*

Outbreaks of *B. bronchiseptica* pneumonia in a colony of captive koalas in Queensland were documented in detail in 1979. The organism was isolated from nasal swabs, lung tissue or bronchial exudate of seven clinically affected or dead koalas during an epizootic in March 1967 which caused 13% mortality with morbidity approaching 100%. *E. coli* and *Pseudomonas* spp. also were isolated from two animals (McKenzie et al. 1979). From 1968 there were annual outbreaks in late winter to early spring, with 10–15% morbidity and 2–3% mortality. Affected lung tissue from five koalas which died in 1976 was cultured and all yielded *B. bronchiseptica*, three in pure culture and one each with small numbers of *Proteus* spp. and micrococcus. Virus isolation was not attempted (McKenzie et al. 1979).

Microbiological culture was performed in 4 of 7 pneumonia cases described in the 1980–86 mortality survey of free-ranging koalas from northern New South Wales (Canfield 1987), and *Bordetella bronchiseptica*, *P. aeruginosa* and *Cryptococcus neoformans* were isolated (the latter from two animals). Five deaths over a 3 mo period in animals from one enclosure in a New South Wales captive colony were reported in 1986.

B. bronchiseptica was isolated from the lungs of the first two koalas. The cause of death of another two animals, whose lungs appeared normal and yielded no bacterial growth, remained undetermined although *B. bronchiseptica* was isolated from material in the nasopharynx of one. Both *B. bronchiseptica* and *C. neoformans* were isolated from the nasopharynx and lungs of the fifth animal (Canfield et al. 1986a; Canfield et al.1986b). A free-ranging juvenile male koala found dead in south-east Queensland in 1987 had severe unilateral pneumonia from which *B. bronchiseptica* was cultured (Blanshard 1994).

When a 3 yr old captive-bred male koala died within hours of showing clinical signs of lethargy, tachypnoea, open-mouthed breathing and eructation, *Bordetella bronchiseptica* isolated in heavy growth from the oedematous left lung was suspected to be the primary pathogen (Letcher et al. 1993). There was, however, a recent history of respiratory disease of unknown aetiology in two female cage-mates that exhibited sporadic sneezing and a serous nasal discharge that became mucoid.

Role of other infectious organisms

Potential primary or secondary microbial pathogens isolated at necropsy from the lungs of free-ranging koalas with pneumonia and/or other respiratory tract pathology include *Pseudomonas aeruginosa* (McKenzie 1981; Canfield 1987), *Streptobacillus moniliformis* (Russell & Straube 1979), *Staphylococcus epidermidis* (Wigney et al. 1989), *Nocardia asteroides* (Wigney et al. 1989) and *Cryptococcus neoformans* complex (Backhouse & Bolliger 1961; Canfield 1987; Krockenberger et al. 2003).

Potential primary or secondary microbial pathogens isolated or identified from captive koalas include *Corynebacterium* spp. (Rahman 1957; Blanshard 1994), *Pseudomonas aeruginosa* (Blanshard 1994), *Pseudomonas* spp. (McKenzie et al. 1979; Booth & Blanshard 1989), *Enterobacter agglomerans* (Booth & Blanshard 1989), *Proteus* spp. (McKenzie et al. 1979), α-haemolytic *Streptococcus* spp. (Booth & Blanshard 1989; Blanshard 1994), diphtheroids (Blanshard 1994), micrococcus (McKenzie et al. 1979), *Pasteurella* spp. (Blanshard 1994), *Staphylococcus aureus* (Booth & Blanshard 1989; Blanshard 1994), *E. coli* (McKenzie et al. 1979; Booth & Blanshard 1989), *Acinetobacter lwoffi* (Blanshard 1994), *Cryptococcus neoformans* complex (Canfield et al. 1986b; Krockenberger et al. 2003), *Aspergillus* spp. (Blanshard 1994), unidentified yeasts (Blanshard 1994) and *Toxoplasma gondii* (Hartley et al. 1990; Dubey et al. (1991).

b Clinical signs

Clinical signs are quite variable and may suggest involvement of the upper or lower respiratory tract, or both.

A scant serous nasal discharge may be apparent 24–48 hr prior to the onset of other signs (Wood 1978). Sneezing or coughing may be heard, and can sometimes be paroxysmal. Healthy koalas sneeze infrequently and rarely cough. There may be unilateral or bilateral, scant or profuse mucopurulent nasal discharge. Audible respiratory noises may indicate nasopharyngeal swelling, bronchitis or pneumonia. Peracute bronchopneumonia may lead to sudden death without obvious premonitory signs.

Atypical presentations, including acute shifting lameness, throat pain and dysphagia, epidermal ulceration at pressure points and inflammation of flexor tendons, have been reported as part of the spectrum of clinical signs that may occur in conjunction with upper respiratory tract disease in koalas (V Nicolson *pers. comm.*; J Hanger *pers. comm.*). This is discussed in greater detail in 10.11.2a.

c Diagnosis

In rhinitis cases, cytological examination of smears from swabs from the nasal cavity may show abundant neutrophils. Any dry yellowish powdery or flaky material adhering to the outside edges of the nostrils should be viewed as suspicious, and be cleaned away so that the nostrils can be re-examined later to check for recent discharge. Pharyngeal inflammation may be evident on examination with a laryngoscope, and the regional lymph nodes (submandibular, rostral mandibular, facial) may be enlarged. Assessment of the extent and severity of respiratory disease may necessitate radiography to confirm or rule out any lung involvement. Thoracic auscultation or percussion is made difficult by the koala's dense fur and the relatively small area occupied by the lungs.

Swabs should be taken for microbial culture and sensitivity testing to identify potential primary or secondary microbial pathogens. Viral or chlamydial isolation should be attempted, if possible, to improve understanding of the pathogenesis of this disease complex.

d Treatment

Due to the great variation in bacterial (and other) isolates from clinical cases, microbial culture and sensitivity testing is recommended to guide selection of an appropriate therapeutic agent, but the severity of the disease often dictates treatment before results are available.

A variety of oral or parenteral broad-spectrum antibiotics such as trimethoprim/sulphamethoxazole (McKenzie et al. 1979; Letcher et al. 1993), chloramphenicol (Canfield et al. 1986b), amoxycillin/clavulanic acid (Booth & Blanshard 1989) or enrofloxacin (V Nicolson *pers. comm.*) has been used in different events. Clinical responses may be slow and relapses are not uncommon (Rahman 1957; Wood 1978; Booth & Blanshard 1989; Blanshard 1992).

Local intranasal spraying of streptomycin in saline (Rahman 1957) or infusion of cloxacillin sodium (Dickens 1978a) has been utilised to treat rhinitis. Nebulisation was considered more efficacious than systemic antibiotic therapy in reducing or resolving clinical signs of rhinitis and/or bronchitis (Blanshard 1992, 1994). Pending culture and sensitivity results, gentamicin at 2 mg/kg was added to 2–3 mL saline and administered via a child-size face mask and a nebuliser pump. Affected animals were nebulised sid or bid for an average of 5 d (Blanshard 1992, 1994). Gentamicin was used in anticipation of *Bordetella* being present as a pathogen or an opportunist. Depending on culture and sensitivity results in individual cases (or from preceding cases during an outbreak), ampicillin 10 mg/kg was used occasionally instead of, or alternated with, gentamicin. As gentamicin is inactivated by penicillins *in vitro* (Laurence & Bennett 1980), the two should not be nebulised simultaneously. Koalas showing clinical signs of rhinitis or bronchitis from which no obvious bacterial pathogen was isolated were sometimes nebulised with saline alone. When pneumonia was suspected, nebulisation was used concurrently with parenteral administration of antibiotics (Blanshard 1994; Booth & Blanshard 1999).

The mucolytic bromhexine HCl at 2 mg/kg bid PO for 5 d (1 mg/kg bid thereafter), or the nasal decongestant pseudoephedrine HCl at 1 mg/kg bid or tid PO, have been given in some cases in addition to parenteral antibiotic therapy (Blanshard 1992, 1994). The process of nebulisation appears to be at least as effective as either of these drugs.

e Prophylaxis

Isolation of individual koalas at the onset of their clinical signs, and prophylactic antibiotic treatment of all animals from the same enclosure have been described (Wood 1978; Canfield et al. 1986b), but may not be feasible in outbreaks with high morbidity nor successful at preventing further cases. Establishing *in situ* quarantine conditions including mandatory disinfection of the hands and footwear of personnel, and controlling the order of entry and traffic between enclosures might contribute to confining or limiting the spread of an outbreak within a colony.

Some captive koalas have been vaccinated against *B. bronchiseptica* using a commercial vaccine, Canvac™ BB (Pfizer Australia). This is an inactivated, cell-free extract of *B. bronchiseptica* and contains an aluminium salt adjuvant. Each koala received a primary course of two SC doses given 4 wk apart, then annual boosters. Vaccination was intended to enhance the koalas' immune response to *B. bronchiseptica*, reducing the frequency with which this organism may be involved in rhinitis/pneumonia complex cases and lessening the possibility of death from peracute *Bordetella* pneumonia with septicaemia (Blanshard 1993, 1994).

Seronegative dogs are reported to develop immunity against *B. bronchiseptica* within 10–14 d of receiving their second dose of Canvac™ BB vaccine. In 1993, blood samples were collected from eight healthy 1–3.25 yr old, captive-bred male koalas in south-east Queensland prior to their first vaccination with Canvac™ BB, and at weekly intervals thereafter for 8 wk (Wk 1–Wk 8). Sera were forwarded to an immunology research laboratory in Adelaide, where the koalas' Ab responses were measured by a specifically developed indirect ELISA (Wilkinson 1996). By the time their second dose of vaccine was administered at Wk 4, three animals remained Ab negative, three had titres of 1:40, one of 1:80 and one of 1:160. Two koalas were still Ab negative at Wk 5 and Wk 6, respectively. Reciprocal Ab titres at Wk 8 were 80 (n = 2), 320 (n = 2), 1280 (n = 3) and 10 240 (n = 1). These results contrasted starkly with those of a rabbit control; 14 d after being given a single dose of the same vaccine the rabbit showed a reciprocal Ab titre >51 200 (Fig. 8.15) (Wilkinson 1996).

Based on the leisurely development of Ab titres in these eight animals, it could be assumed that in a naïve captive population a regimen of vaccination in the face

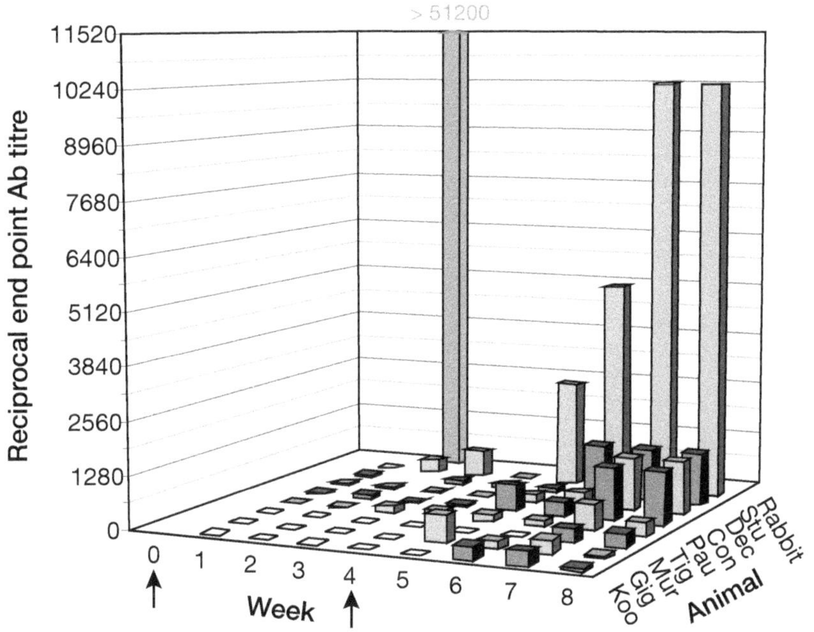

Figure 8.15 Response to vaccination. Development of indirect ELISA Ab titres to *Bordetella bronchiseptica* in eight male koalas 1–3.25 yr old that received primary and secondary vaccinations with Canvac-BB® (CSL) 4 wk apart (arrows), and a rabbit control tested 14 d after a single dose of the same vaccine. Open squares represent negative titres. (Source: Ab titres from Wilkinson (1996).)

of a respiratory disease outbreak may be too late for any effective degree of protection against *Bordetella* spp. to at-risk koalas. There is, however, strong clinical evidence for the value of ongoing vaccination in captive animals. In one large colony, in the years immediately preceding administration of the first doses of vaccine five mortalities attributed to *B. bronchiseptica* comprised 9.1%, 0%, 13.3%, 0% and 28.6% of annual deaths, respectively (Booth & Blanshard 1999). No mortalities due to infection with *Bordetella* spp. were observed in the 14 yr following institution of the vaccination program (Booth & Blanshard 1999; V Nicolson *pers. comm.*; G Rawlinson *pers. comm.*) even though the absolute incidence of clinical respiratory disease events was not obviously changed.

10.5.2 Cryptococcosis

Pulmonary and upper respiratory tract infection with the fungus *Cryptococcus neoformans* complex has been reported in free-ranging and captive koalas (see 10.11.3).

10.5.3 Mycobacteriosis

Progression of *Mycobacterium ulcerans* infection (see 10.1.1c) from deep cutaneous ulcers on the face into the upper and lower respiratory tract has been reported in two free-ranging koalas from Raymond Island (Vic.) (McOrist et al. 1985). Ulcers over the maxillae extended into the nasal turbinates, and several soft pale grey masses were apparent in pulmonary tissue in the hilar region. Acid-fast bacilli were detected in macrophages on Ziehl–Neelsen stained histological sections from both respiratory sites.

10.5.4 Pneumothorax

Unilateral pneumothorax has been observed in a captive koala. Several days after a laparotomy to correct torsion of the distal colon, the koala began to show signs of discomfort and on two occasions an arrhythmia was heard during auscultation. Radiographs showed retraction of the pleural surface of one lung from the thoracic wall. The koala made an uneventful recovery following placement of a chest drain under anaesthesia (Blanshard 1994).

10.5.5 Aspiration pneumonia

In a survey of 75 koala road kills over a 7 yr period, head injuries (59%) and haemothorax (20%) were among the most common findings. Aspiration of blood had occurred in 27% (12 of 44) of koalas with head injuries, and was regarded as a significant contributing factor in their deaths (Canfield 1991).

Aspiration pneumonia developed terminally in an adult captive koala which vomited (secondary to small intestinal torsion) and inhaled leaf material (Blanshard 1994), and was identified at necropsy as a secondary complication in a juvenile captive koala with idiopathic oesophageal neuritis (G Rawlinson *pers. comm.*).

10.5.6 Miscellaneous conditions

'Pigmentation resembling silicosis (birefringent crystals) and anthracosis', with histological evidence of a foreign body reaction, was noted in pulmonary tissue and lymph nodes of some koalas necropsied in North America (Robinson 1978). Silicosis was listed last among the diverse pathological conditions in 4 of 10 koalas examined, and is assumed to have been an incidental finding. Foreign bodies identified at necropsy in the lungs of koalas in Australia include plant material (four cases) and pneumoconiosis (one case); the clinical significance for each individual was not reported (Connolly 1999).

10.6 Urinary and reproductive diseases

Between 1980 and 1988, 235 NSW koalas were examined during a mortality survey (Canfield 1989). Various forms of urinary tract disease were identified in a total of 67 animals (61 of 187 free-ranging and 6 of 48 captive koalas). Cystitis was far more common than any other urinary tract disease and was recorded in 49 of these 67 koalas (30 females and 19 males), alone or in combination with associated renal disease (mainly hydronephrosis and pyelonephritis) or another systemic disease.

Cystitis and other diseases of the urogenital tract commonly associated with chlamydial infection are discussed elsewhere (see 10.11.2a).

10.6.1 Nephropathy

Acute and/or chronic renal failure, presumed to be non-chlamydial in origin, has been reported in free-ranging and captive koalas.

Necropsy of a koala housed at the Cleland National Park (SA) in the late 1960s indicated 'uraemia following chronic nephrosis' (Eberhard 1972). In a 1975–80 study of morbidity and mortality among free-ranging koalas in Victoria, some of the 29 koalas presented alive showed (or subsequently developed) characteristic signs of unknown aetiology including lassitude, depression, anorexia and a 'precipitous decline in metabolic function' (Obendorf 1983). Typically they were found on the ground, either wandering aimlessly, at the base of trees reluctant to climb, or prostrate and comatose. Clinical findings included unusual docility, hypothermia, bradycardia and lack of interest in eating, although some began to drink water. Progressive weight loss ensued and death invariably followed. Dehydration and loss of body condition were evident on gross necropsy. Acute tubular nephrosis was described as a common histological finding in these koalas, although its prevalence was not stated (Obendorf 1983).

Acute tubular nephrosis, chronic nephritis and membranous glomerulonephritis were recorded once each during the New South Wales disease survey (Canfield 1989). Dickens (1978a) described chronic interstitial nephritis as occurring commonly in old animals.

On the basis of clinical and pathological findings, renal failure due to acute tubular necrosis was diagnosed in a 2 yr old female Victorian koala housed at a Sydney wildlife park (Spencer & Canfield 1993c). The animal presented with depression, anorexia, polydipsia and rapid weight loss over 1 wk. Assessment of hydration status during physical examination was not mentioned. A blood sample showed azotaemia (urea 29.9 mmol/L, creatinine 581 μmol/L) and other abnormalities (see 8.1.1). A urine sample could not be obtained. Treatment consisted of SC and oral fluids, and administration of broad-spectrum antibiotics (no details provided). The following day the animal was sitting on the ground, unable to climb. Under anaesthesia a urine specimen was obtained by cystocentesis and showed low urine specific gravity (1.017), proteinuria and numerous bacteria in examined sediment. Due to the koala's deteriorating clinical state and poor prognosis, she was euthanased. Although routine bacteriological culture of the urine obtained by cystocentesis subsequently yielded *E. coli*, its isolation was considered of questionable aetiological significance in the absence of evidence of inflammation. At necropsy both kidneys were soft and mottled externally, with pale yellow cortico-medullary streaks evident on the cut surfaces. On histological examination of H&E stained sections, numerous proximal and some distal tubules were dilated and lined by flattened epithelium. Some distal tubules contained hyaline casts and some proximal tubules were filled with neutrophils.

Necrosis and early regeneration was evident in some proximal tubules (Spencer & Canfield 1993a).

a Potential aetiological and/or associated factors

Dehydration

Several morphological features of the koala kidney, including the relative medullary thickness, length of the renal papilla and difference in volume between cortical and juxtamedullary glomeruli, suggest that the koala is adapted to a habitat which is not normally limiting with respect to water. Based on the relationship between relative medullary thickness and maximum urine osmolality in eutherian mammals and other marsupials, the koala's maximum urine concentration is comparatively low (Degabriele et al. 1978). This seems contrary to accepted wisdom that the finding of highly concentrated urine (i.e. a high urine specific gravity; see 8.5.1) in healthy koalas is not unexpected considering their adaptation to a diet of eucalypt leaves (Canfield et al. 1989c).

Urine osmolality was measured as part of water turnover studies (see 2.9) conducted in captive koalas (Degabriele et al. 1978). When drinking water was available, mean urine osmolality was 1212.6 ± 79.5 mosm/kg and 1414.1 ± 89.4 mosm/kg in summer and winter, respectively. With drinking water withdrawn (in winter), urine osmolality increased to 1.2 times its former level (1691.0 ± 133.6 mosm/kg). It was suggested that although the animals were able to remain in water balance, the kidney was probably close to maximal stress under these conditions, as this value was close to the figure obtained after exogenous administration of the antidiuretic hormone vasopressin to induce production of maximally concentrated urine (1842 ± 79.3 mosm/kg). When the kidneys were concentrating maximally, there was a negative interdependence of urea and electrolyte concentrations such that as more electrolytes were excreted, less urea was excreted (Degabriele et al. 1978).

Although the koala is capable of maintaining its body temperature within reasonably narrow limits as ambient temperature rises (see 2.10), a dramatic increase in respiration rate (panting) at higher air temperatures is accompanied by a corresponding sharp increase in evaporative water loss (Degabriele et al. 1978; Degabriele & Dawson 1979). Up to an ambient temperature of 25–30°C, this evaporative loss is adequately compensated by metabolic water production. If ambient temperature increases from 30°C to 35°C, however, an 8-fold increase in respiration rate and a 2.4-fold increase in evaporative water loss can occur. Metabolic water production would then meet only about one-third the requirement to replace water lost to evaporation (Degabriele et al. 1978; Degabriele & Dawson 1979). Ambient temperature in the koala's microenvironment reportedly rarely exceeds 30°C for long periods (Degabriele 1980).

Dehydration may be a cause or an effect of renal disease (Osborne et al. 1972). It is not difficult to envisage scenarios in which an arboreal folivore without an apparent propensity to drink free-standing water and with only a modest ability to increase the osmolality of its urine might quickly become significantly dehydrated. Reduction in food intake due to ill health, injury, bushfire, habitat alienation or inappropriate captive husbandry might result in deficits in preformed and metabolic water inputs. Increased thermal load during unusually hot weather (perhaps exacerbated by low humidity) might impose excessive evaporative water losses on a species otherwise well-adapted to its specialised ecological niche.

Local or seasonal variation in foliage quality and preformed moisture content might also compromise return to normal water balance. Degabriele (1980) commented that, in times of drought, wasted koalas have become comatose and died 'with a full stomach'. Although nitrogen deficiency was suspected to be the major factor behind this phenomenon (based on the lower digestible nitrogen content of comparatively old eucalypt leaves), Degabriele (1980) noted that the preformed water content of *E. punctata* foliage varies from 65% in new growth to 40% in old fibrous leaves.

Oxalate and calcium

Oxalate nephrosis was identified in 4 of 67 koalas with urinary tract disease during a necropsy survey of 235 free-ranging and captive koalas in New South Wales, 1980–88 (Canfield 1989). The first case, reported as renal oxalosis in 1982, was in a koala approximately 7 yr old which had been held in captivity for 2 yr (Canfield & Dickens 1982). The koala became depressed and anorexic a month after treatment and apparent recovery from 'dirty tail' (see 10.11.2a) and died 2 d later. An oral trimethoprim and sulphonamide preparation had been administered for 2 mo. Gross necropsy findings were indicative of urinary tract disease and histological sections of kidney showed that many of the distal and, to a lesser extent, proximal tubular lumens were filled with radially arranged rhomboidal to triangular crystals. These crystals were birefringent under polarised light and stained positively with Pizzolato's peroxide silver method for calcium oxalate. Unstained wet preparations of bladder (and peritoneal) fluid contained characteristic octahedral calcium oxalate crystals. Morphological findings suggested that renal failure had occurred, but the source of the oxalate (exogenous or endogenous) was not apparent.

Three subsequent cases of oxalate nephrosis (at least two in free-ranging koalas) were recorded during the 1980–88 necropsy survey (Canfield 1989), and three additional cases were noted between 1995 and 1998 (Connolly 1999). A similar condition with birefringent crystals evident on histological sections of the kidneys has been observed in free-ranging and/or captive koalas in Queensland, Victoria, South Australia and Western Australia (R Booth *pers. comm.*; Booth & Blanshard 1999; C Monaghan & K Warren *pers. comm.*; Haynes et al. 2004). Renal oxalosis has also reportedly been observed in routine necropsies performed on free-ranging koalas in south-east Queensland (AD Wood cited in Lawson & Carrick 1998). Calcium oxalate precipitation in non-renal tissues of the koala has been recorded (Lawson & Carrick 1998; Krockenberger & Canfield 1999), in the follicular colloid of the thyroid gland (see 10.8) and in the substance of the prostate (see 10.6.2b). The proximate cause of these conditions has not been identified.

Aluminium

Severe renal dysfunction not associated with infection is reported as a major cause of mortality in both captive and free-ranging koalas in the Adelaide Hills (SA). As high dietary levels of aluminium had been suggested as a potential contributing factor, a regional study was undertaken to correlate the presence of aluminium in the kidneys with ultrastructural renal damage; it focused on 16 clinically affected animals (nine free-ranging and seven captive koalas). An additional free-ranging animal with no obvious signs of renal dysfunction was included as a control, and archived kidney samples from two PY from the Adelaide Hills region were also examined (Haynes et al. 2004).

Clinical signs in all 16 affected animals included lethargy, emaciation, polydipsia, polyuria and dilute urine. Clinical pathology data available for eight of the affected animals showed mean plasma urea levels of 29 ± 15 mmol/L, and mean urine SG of 1.029 ± 0.003. These contrasted with findings from the injured control animal (urea 6.2 mmol/L, urine SG 1.040). The high urea values from affected animals were similar to mean levels reported for 11 other koalas previously diagnosed with renal failure between 1995 and 2000 (30 ± 4 mmol/L). While values were not specified, elevation in creatinine and cholesterol levels were inferred to be characteristic of Adelaide Hills koalas that had succumbed to renal problems.

Clinical evidence of renal failure in the study group was confounded by an apparent absence of pathological lesions during routine histological examination of kidney sections stained with H&E, in all but one of the 16 affected koalas. In the atypical case, obvious renal pathology included focal medullary tubular damage, with elongated crystals filling the lumen of tubules in which epithelial cells were absent. Other tubules (predominantly thick segments of the loop of Henle) were blocked with acidophilic protein material and had necrotic lining cells. The crystalline material stained positively for calcium oxalate with Pizzolato's silver peroxide technique.

Kidney sections from all animals in the study were stained with solochrome azurine (for aluminium) and Perl's Prussian blue (to detect iron) and examined by light microscopy. Negative staining results in kidney sections from the adult control and the two PY indicated the absence of aluminium in these koalas. By contrast, all kidneys from the 16 affected animals showed positive staining for aluminium in some of the proximal convoluted tubules, using solochrome azurine. Positively stained areas constituted approximately 1–5% of the total renal cortical tissue. No aluminium was found in distal convoluted tubules or glomeruli. In some specimens, positive results with both stains suggested the presence of iron or iron with aluminium.

A comparison of the electron microscopic ultrastructure of renal tissue from three of the koalas which exhibited signs of renal failure showed no abnormalities in their glomeruli, basement membrane thickness or blood vessels but did reveal evidence of intracellular damage in aluminium-positive proximal convoluted tubules, which had fewer lysosomes and mitochondria than aluminium-negative tubules in the same animal. No such variability was apparent in the proximal convoluted tubules of the control animal (Haynes et al. 2004).

To date, it is uncertain whether the presence of either calcium oxalate or aluminium in the convoluted tubules is an inciting cause, or simply a consequence, of renal failure.

b Diagnosis

Clinical signs observed in koalas with nephropathy include anorexia, rapid weight loss, lethargy, obtundation, dehydration, atypical drinking, production of small dry faecal pellets and emaciation. Animals may present in a state of collapse (prostrate and comatose, hypothermic, bradycardic). An evaluation of hydration status (see 7.1.8) is important. Any retardation in the ease of skin sliding over the scapulae should be regarded as a reliable indicator of dehydration, even if PCV and/or albumin levels fall within expected reference values. Shrunken and firm, enlarged and soft, or asymmetrical sized kidneys may be palpable (A McKinnon *pers. comm.*). Potential underlying pre-renal, renal or post-renal causes (e.g. oral candidiasis, urinary tract infection, neoplasia) should be considered and addressed when investigating cases of azotaemia.

A full blood count and multiple biochemical analysis rather than a minimum screening panel are recommended. Clinical pathology results from several affected animals throughout Australia (Spencer & Canfield 1993a; Schultz et al. 1995; Wilkinson 1996; R Booth *pers. comm.*; C Monaghan & K Warren *pers. comm.*) showed a great variation in haematological and biochemical parameters, with azotaemia the only consistent finding.

Initial PCV should be evaluated carefully, taking into account the clinical assessment of hydration status (see 7.1.8), published haematological reference ranges (Tables 8.2 and 8.3), seasonal variation in haematocrit (Fig. 8.8) and any routine values previously obtained from that individual (Schultz et al. 1995).

Dickens (1978a) indicated that average values for urea in healthy koalas were around 5.4 mmol/L. Very low values were also common, while higher values (averaging 10.7 mmol/L) were seen in healthy sanctuary animals. The upper limit of Dickens' normal range for urea exceeds the values expected from healthy captive koalas by other practitioners (Canfield et al. 1989a; Booth & Blanshard 1999).

In a 1975–80 study of morbidity and mortality among free-ranging koalas in Victoria, elevated urea ($>$ 21.4 mmol/L), haemoconcentration (Hb $>$ 160 g/L, PCV 48–55%) and hypoglycaemia ($<$ 1.1 mmol/L) were nominated as poor prognostic indicators in sick or injured animals presented for veterinary appraisal and treatment (Obendorf 1983). Each abnormality was observed at least three, five and three times respectively among the 17 of 29 animals presented alive whose haematological parameters were examined, and an unspecified subset of the 17 in which blood chemistry was also measured.

In a few captive koalas from which more than one blood sample had been collected before they succumbed with evidence of calcium oxalate crystals on renal histopathology, there was a suspicion that hypercalcaemia may have preceded the development of azotaemia in some (C Monaghan & K Warren *pers. comm.*).

Neither normocalcaemia nor hypercalcaemia unequivocally predict the presence or absence of crystals consistent with calcium oxalate in histological sections of the kidneys from azotaemic koalas (C Monaghan and K Warren *pers. comm.*; R Booth *pers. comm.*).

Every effort should be made to collect a urine sample to monitor specific gravity as soon as possible after initial presentation (see 8.5.1). Interpretation of urinalysis results is discussed in section 8.5.

c Treatment

There is strong anecdotal evidence that conservative approaches to the management of azotaemic koalas result in poor patient outcomes. At least two instances of successful management are reported, with post-treatment survival of 2 yr in an adult female (Schultz et al. 1995; D Schultz *pers. comm.*) and at least 8.25 yr in a juvenile male (R Booth *pers. comm.*; M Pyne *pers. comm.*). Not surprisingly, Schultz et al. (1995) reported a more dramatic rate of decrease in serum urea (initially 22 mmol/L) in response to 3 d IV fluid therapy than the gradual decline observed simply by encouraging oral fluid intake, while R Booth (*pers. comm.*) observed urea to climb from 14 mmol/L to 18.3 mmol/L on successive days, despite administration of SC fluids. Early detection and intensive fluid therapy offer the best chance of limiting and/or reversing renal damage.

If there is clinical evidence of dehydration in the face of polydipsia and/or low urine specific gravity, re-establish positive fluid balance (see 12.1) as a matter of urgency. In small domestic animal medicine, when the concentration of urea exceeds 12.5–16 mmol/L or when the concentration of creatinine exceeds

177 μmol/L, glomerular filtration rate (GFR) is known to be diminished (Osborne et al. 1972). Therefore, a suggested indicator for therapeutic intervention in koalas is a serum urea value ≥12.0 mmol/L.

Clinicians should consider IV fluid administration. Until any electrolyte and mineral disturbances can be quantified, 5% glucose or 4% glucose/0.18% NaCl IV may be the safest short-term option. Ensure drinking water is always available, and encourage oral fluid intake by offering water or 5% glucose from a bowl brought to the animal, or from a syringe, several times per day. Select fresh browse with a high proportion of young leaves and spray them with water before feeding out. Monitor urine specific gravity as often as possible during the treatment period.

Consideration also may be given to the following principles.

- If caloric intake is inadequate, protein or amino acids may be metabolised to meet energy requirements, creating more nitrogenous waste products which may accentuate azotaemia (Osborne et al. 1972).
- Any supplementary feeding with high-carbohydrate paste should be done using small quantities, given frequently, to avoid the possibility of glucosuria and osmotic diuresis from overzealous administration.
- Anabolic steroids may be beneficial in renal failure because they promote positive nitrogen balance (only if caloric intake is adequate) and enhance skeletal deposition of calcium (Osborne et al. 1972).
- Sulphonamide antibiotics should probably be avoided in koalas with suspected renal disease because of their potential to cause crystalluria in other species (Laurence & Bennett 1980).

d Prevention

Careful attention must be paid to handling and storage of koala browse (see 5.3.2) to ensure leaf hydration. Wherever feasible, koalas in human care should be habituated to accept approach and brief physical contact by their keepers so that assessment of hydration status and body condition can become an integral part of health inspections. In intensively managed captive populations, regular (e.g. fortnightly) weighing is typically part of routine husbandry procedures for every animal that can safely be reached with a ladder. Even in extensive display enclosures, evaluation of body condition and hydration status of individuals may be undertaken opportunistically (without attempting to catch them) when animals attend feeding stations to access freshly cut browse.

In regions where koala populations have a history of nephropathy, any routine annual examinations of captive animals (including blood collection and urinalysis to screen for subclinical cases of azotaemia or unusually low urine specific gravity) might best be scheduled immediately prior to the expected onset of summer heat, to allow any abnormalities to be addressed and to provide reference data should re-evaluation of an individual be required.

10.6.2 Lithiasis

a Urinary calculi

During the 1980–88 New South Wales disease survey, two koalas with cystitis were found to have fragmented urinary calculi in the bladder. In one case the calculi were composed of ammonium magnesium phosphate (struvite), and in the other case calcium oxalate and uric acid (Canfield 1989). Bilateral renal calculi composed of 70% oxalate, 30% calcium and 0.5% ammonium were reported in 1 of 110 koalas necropsied between 1995 and 1998 (Connolly 1999).

Ultrasonographic investigation of the cause of dysuria in a 19 yr old captive-bred male koala showed a single mass free within the lumen of the urinary bladder. A cystotomy was performed and the calculus removed. Subsequent chemical analysis of the urolith revealed mainly uric acid with some calcium oxalate present. Although the koala initially recovered well after surgery, straining was observed again 16 d later. Key haematological and biochemical parameters fell within reference values. Ultrasound re-examination indicated that another two, slightly smaller, cystic calculi were now present and the koala was euthanased. Histopathology revealed mild to moderate multifocal interstitial fibrosis and lymphoplasmacytic infiltrate in the renal cortex, suggestive of scarring at sites of previous inflammation (e.g. infection), and most pronounced in a subcapsular location; there was a mild lymphocytic and neutrophilic erosive cystitis with submucosal oedema and serosal fibrosis, considered most likely secondary to the cystic calculi (G Rawlinson *pers. comm.*).

b Prostatic calculi

A free-ranging adult male koala in New South Wales presented in moderately poor body condition and was subsequently euthanased. At necropsy, more than 40 smooth black angular calculi were observed in the substance of the prostate (but not in the urethra). On chemical analysis these were found to be composed of 30% calcium and 70% oxalate. Histologically there was active chronic prostatitis with calculi present in the tubuloacinar areas and ducts, and full-thickness cystitis. Under a polarising light microscope the calculus material was birefringent; it stained positively for calcium using a von Kossa stain (Krockenberger & Canfield 1999).

10.6.3 Miscellaneous reproductive disease

A free-ranging female koala from south-east Queensland presented with a prolapse of the urogenital sinus (Brown 1979). Under general anaesthesia the exposed tissue was washed with warm saline and the prolapse reduced using slight pressure. It remained in position without requiring suturing, and the animal was released again after 4 d observation. A few similar cases in which the prolapse had been reduced without anaesthesia were also mentioned (Wood *pers comm.*, cited in Brown 1979); the pathogenesis of the condition was not known. Testicular aplasia (unilateral or bilateral failure of testicular development) has been detected in several South Australian koala populations and is believed to be the result of inbreeding depression (Seymour et al. 2001). Urethritis and urethral prolapse have been seen in an adult male koala at necropsy (ARWH 2007 case number 2210/1).

10.7 Neurologic and musculoskeletal diseases

10.7.1 Neurologic diseases

Septicaemia must be considered in any koala showing neurological signs (see 10.11.2b). Cryptococcal meningitis and encephalitis have been reported (see 10.11.3).

Hydrocephalus was diagnosed in a free-ranging emergent young estimated to be around 8 mo old (HL 75 mm), found on the ground. She exhibited a crouched posture with a tendency to circle to the left, had a domed forehead, a reduced suck reflex and vocalised abnormally (in a low whining growl). Radiographs of the skull showed enlargement of the cranial vault, abnormal angulation of the nasal and frontal bones and malocclusion of the incisor and cheek teeth. Necropsy revealed enormous dilation of the ventricular spaces, and cerebral hemispheres only 1–2 mm thick that collapsed when the build-up of CSF was allowed to escape. Increased intracranial pressure had resulted in coning of the cerebellum. A transverse section through the midbrain between the third and fourth ventricles demonstrated a transparent membrane that was occluding the cerebral aqueduct (W Blanshard unpub).

Histopathologic diagnosis of oesophageal neuritis was made in a captive 2 yr old female koala that demonstrated persistent dysphagia, coughing and weight loss. The animal had developed secondary pharyngeal ulceration and mild aspiration pneumonia (G Rawlinson *pers. comm.*).

10.7.2 Musculoskeletal diseases

Trauma is the most common cause of musculoskeletal disease (see 10.12.1).

a Degenerative joint disease

Hip joints from 81 free-ranging New South Wales koalas demonstrated a high incidence of injuries to the round ligament, which had progressed to extensive osteoarthrosis in four animals (Canfield & Spencer 1993a). The original hip joint injuries were believed to occur as a result of traumatic injury associated with an arboreal life-style.

b Scoliosis

Scoliosis has been described as an incidental finding in koalas (Hirst et al. 1992b; Canfield 1991; Markey et al. 2007) and anecdotal reports suggest that the incidence of lateral deviation of the spine may be relatively high (Morrison 2001).

c Osteomyelitis

One survey reported a high incidence of osteomyelitis lesions affecting the pre-maxilla, maxilla and mandible of free-ranging Victorian koalas (Colvin 2003). The prevalence of bony abnormalities was higher in older animals and the majority of lesions appeared be of dental origin.

d Hip dysplasia

An inherited tendency towards hip dysplasia has been identified in one captive colony. Survey radiographs demonstrate varying degrees of hip joint/femoral pathology, including shallowing of the acetabulum, flattening or loss of the femoral head, widening or loss of the femoral neck and, in some cases, curvature of the diaphysis of the femur. Radiographic changes may be evident as early as 12 mo of age. Affected koalas may or may not demonstrate gait abnormalities. Mild to severe degenerative joint disease may develop in some cases. Koalas with clinically apparent degenerative joint disease are treated using glucosamine/chondroitin sulphate and non-steroidal anti-inflammatory drugs (G Pye *pers. comm.*; A Shima *pers. comm.*).

e Myositis

Unilateral hind limb myositis and myonecrosis associated with anaerobic bacterial infection has been observed in a 3 yr old captive male koala. *Aeromonas* spp., *Klebsiella* spp. and *Clostridium* spp. were isolated from affected tissues at necropsy (Blanshard 1994).

f Neoplasia

For descriptions of neoplastic disease affecting bone, see 10.12.4.

10.8 Endocrine and metabolic diseases

Endocrinopathies have been reported infrequently in koalas. Koala thyroid and adrenal function have been investigated (see 8.3 and 2.11). There have been two cases of diabetes mellitus reported in koalas (Shimizu et al. 1989; Hemsley et al. 1998) and two other instances are known (G Rawlinson *pers. comm.*).

A captive 23 mo old female koala with a 2 mo history of failure to gain weight at a normal rate showed a sudden onset of polydipsia and polyuria. She was found to be hyperglycaemic, hypoproteinaemic and glucosuric (ketonuria was also reported, see 8.5.2) and died 3 d later. Positive staining for glycogen, noted in many pancreatic islet cells, the sub-endocardial myocardium and in vacuolated hepatic nuclei, was suggestive of persistent hyperglycaemia. Very few, scattered or single pancreatic islet and interstitial cells were positive for insulin-immunoreactivity (Shimizu et al. 1989). In the second reported case, a free-ranging 6 yr old female koala developed diabetes mellitus following hospitalisation for injuries from motor vehicle trauma (Hemsley et al. 1998). The animal presented with polydipsia while in hospital and was found to be hyperglyaemic, hyperlipidaemic and glucosuric. At necropsy, pancreatic lesions were consistent with diabetes mellitus. It was hypothesised that pathogenesis may have been related to the stress of capture, tissue damage/inflammation post trauma or administration of a parenteral corticosteroid preparation.

When histological sections were prepared from 40 pairs of thyroid glands from koalas in south-east Queensland which had died or were euthanased due to injury or disease, one instance of calcium oxalate crystals in the follicular colloid was noted (Lawson & Carrick 1998).

10.9 Ophthalmic diseases

10.9.1 Corneal ulceration

Superficial corneal ulceration is the most common eye disorder seen in chlamydial Ab-negative captive koalas. Abrasion caused by collision with a twig, the presence of a foreign body such as a fragment of bark or a scratch inflicted by another animal (the latter being the presumed cause in at least some females carrying BY) may damage the corneal surface. Diagnosis and management is as for domestic species.

10.9.2 Conjunctivitis/blepharitis

Injury to the conjunctiva or skin caused by claws or twigs may cause blepharospasm, hyperaemia and swelling of the eyelids or conjunctivae. Infection by microorganisms other than chlamydiae has been implicated in causing inflammation of the eyelids. *Enterobacter cloacae* was consistently isolated from swabs taken weekly on four occasions from both eyes of a koala with clinical signs of blepharitis (Mutimer et al. 1990). Blepharitis was reported, together with pruritic dermatitis of the foot and paronychial infection, in a koala with *Trichophyton mentagrophytes* infection (Reddacliff 1986). A case of unilateral cryptococcal conjunctivitis has been observed (R Booth *pers. comm.*).

10.9.3 Keratitis

Long-standing keratitis was noted in 22 of 70 free-ranging koalas examined on Magnetic Island (Qld) (Hirst et al. 1992b). The condition was unilateral in 12 animals and bilateral in the other 10. On slit-lamp examination affected corneas were epithelialised and not inflamed but showed a dense and usually central corneal scar. Occasionally patent blood vessels crossed the limbus and entered the scar. The anterior chamber was not inflamed and the conjunctivae were unremarkable. All koalas examined were seronegative for anti-chlamydial antibodies, and it was suggested that an external environmental agent (not identified) might be responsible for the condition. Despite significant visual impairment from corneal opacities, most affected koalas seemed to be surviving well in the wild. See also section 10.11.2a for a description of keratitis resulting from chlamydial infection.

10.9.4 Dacryocystitis

Inflammation of the lacrimal sac associated with bacterial infection has been observed in captive koalas. A pure growth of *P. aeruginosa* was isolated from the nasolacrimal apparatus in one case (Blanshard 1994).

Clinical signs include chronic intermittent scant-to-moderate ocular discharge. There may be little or no reddening of the conjunctivae and the eye may not appear to be particularly uncomfortable. Treatment is based on flushing the nasolacrimal duct. General anaesthesia and good illumination are required to locate and cannulate the nasal ostium, which is identifiable as a small dark opening on the ventrolateral aspect of the pink surface of the caudal extremity of the nostril.

10.9.5 Trichiasis

Chronic unilateral epiphora associated with upper eyelashes contacting the cornea has been observed in a captive juvenile koala. The condition was successfully overcome by performing entropion surgery to elevate the lashes away from the globe (R Smith *pers. comm.*).

10.9.6 Exophthalmos

Some koalas have unusually prominent globes and may suffer a higher frequency of traumatic corneal ulceration. When these animals look straight ahead an annulus of their sclera is visible in the palpebral opening.

10.9.7 Nuclear sclerosis and senile cataracts

Slightly bluish opacity of the lens and slowly progressive opaque white cataracts have both been seen in very old captive female koalas. Partially dilated pupils were a feature of both conditions. Intermittent luxation of a cataractous lens has been observed in an aged captive koala (G Rawlinson *pers. comm.*).

10.9.8 Optic neuritis

Optic neuritis can occur as a result of cryptococcal infection affecting the central nervous system. Animals present with acute blindness and dilated non-responsive pupils (see 10.11.3).

10.10 Neonatal diseases

10.10.1 Morbidity/mortality of emerging young

a *Misadventure*

PY have poorly developed co-ordination. If removed from the pouch prior to the time of first normal emergence, most display movements which thwart attempts to return them to the pouch—flexing their neck, grasping the mother's fur tightly with their hands and strongly extending their hind legs, effectively pushing themselves away from the pouch opening. By the time they emerge for the first time, most make a reasonably co-ordinated effort to work their way back inside. Healthy PY which fall from their mother for any reason may die of exposure, especially as their time of first emergence tends to be in winter or early spring. Until quite advanced they are incapable of climbing among branches. Even if the mother (or another female) descends to the ground in response to its distress vocalisations, the young may not be co-ordinated enough to return to the safety of the pouch (Blanshard 1994).

b *In-pouch microbial infection*

Occasionally, around the expected time of first emergence, soiling of fur around the pouch opening or visible on the young koala's head indicates that a problem exists. On extraction, the PY may be found to be partially or completely damp and soiled by its own diarrhoea. Whether the PY's diarrhoea overwhelms the pouch's sterilising system or develops secondary to a breakdown in the system is not clear, but yeasts and/or bacteria can be cultured from both the PY's faeces and from the pouch lining. There appears to be an association with the onset of pap feeding in some animals, and the causal organisms may have been ingested in maternal faeces. Two cases involving yeasts have been successfully resolved by treating the PY *in situ* with oral nystatin, with or without amoxycillin-clavulanate, as indicated by microbial culture and sensitivity results (Blanshard 1994).

c *Stomatitis, pharyngitis and gastritis*

Oral thrush caused by colonisation of the mucosa with large numbers of yeasts (usually *Candida albicans*) may occur, particularly in young animals being hand-reared. In severe cases gastritis due to yeast overgrowth also may be present. *Candida* gastritis has been seen in young koalas being reared naturally by their mothers which failed to thrive after emergence from the pouch, and it is conceivable that exposure to the yeast was at the time of pap feeding. Typically, with mild oral thrush a hand-reared animal appears hungry at feeding time but may be reluctant to take the teat into its mouth. Alternatively, it may reject the teat or spill part of its first mouthful of milk after it attempts to swallow. Regurgitation of milk through the nose may also occur. The latter is by no means conclusive of thrush, but yeast overgrowth should be ruled out in animals which start to do this on a regular basis when they were previously feeding uneventfully. Nystatin given orally at 5000–10 000 U/kg after each feed (or at least tid) for 4–7 d is effective in mild cases of oral candidiasis in hand-reared koalas. Some improvement in feeding behaviour is usually noticed within the first 24 hr of treatment.

Occasionally, generalised pharyngitis in the absence of yeasts may be noted in hand-reared young. Culture reveals only normal oral or enteric flora. Juveniles with non-fungal pharyngitis are often subdued and less interested in feeding. In these cases topical application of medical ulcer treatments (containing lignocaine and a disinfectant) 10–20 min before feeding for 24–48 hr helps to overcome reluctance to swallow milk (Blanshard 1994).

d Typhlitis

Typhlitis is a relatively common cause of mortality in hand-reared young acquired prior to or around the time of first emergence from the pouch. Paint-brush or petechial haemorrhages on the serosa of the caecum and accumulation of gas within the lumen are found on necropsy. Radiographically, the condition is difficult if not impossible to distinguish from a true obstruction causing accumulation of gas proximal to the affected area. In typhlitis it is sometimes possible to see the gas localised in intestinal loops in the ventral aspect of the abdomen. Culture of the caecal contents may reveal a heavy growth of one organism, usually a common enteric species such as *E. coli* or *Klebsiella* spp.; it is assumed that the hindgut, not yet inoculated with a balanced population of normal bacteria, may be overwhelmed by a single species (Blanshard 1994). In one wildlife hospital, administration of a prophylactic antimicrobial regimen to pre-weaned (pre-pap fed) koalas appears to have reduced the rate of mortality due to typhlitis/colitis (see 9.2.4).

10.11 Infectious diseases

10.11.1 Viral diseases

a Koala retrovirus (KoRV)

The high incidence of lymphoid neoplasia in captive and free-ranging koalas has been well-recognised for many years. Arundel et al. (1977) proposed a viral aetiology for lymphocytic leukaemia, and virus particles were first confirmed present in the bone marrow of a leukaemic koala in 1988 (Canfield et al. 1988). On the basis of their morphology, these virus particles were considered most likely to belong to the family *Retroviridae*, subfamily *Onchovirinae*. Worley et al. (1993) used PCR amplification of viral DNA and computer analysis of the sequence to confirm it as a retrovirus. Phylogenetic analysis has demonstrated strong similarities between KoRV and gibbon ape leukaemia virus (Hanger et al. 2000). A recent publication hypothesises that an intermediate vector (e.g. rodents) may have facilitated trans-species transmission of the virus (Fiebig et al. 2006).

KoRV has been isolated from both healthy and diseased koalas, and has been established as a type C endogenous gammaretrovirus. KoRV is present in the genome of all koalas tested in south-east Queensland (Hanger et al. 2000). Recent research has demonstrated an association between plasma levels of KoRV RNA and the presence of disease (Tarlinton et al. 2005). Assays for plasma viral RNA were performed in 90 northern koalas. Active viraemia was confirmed in all animals tested, and the data demonstrated a clear positive association between plasma levels of KoRV RNA and leukaemia/lymphoma in four animals. Increased levels of KoRV RNA were also observed in 20 animals with clinical chlamydiosis (Tarlinton et al. 2005). It was proposed that viral load may be a factor contributing to the severity of this disease. There appears to be a correlation between plasma levels of KoRV RNA and development of a wide range of clinical syndromes that are seen frequently in northern koalas, including neoplastic disease (aplastic anaemia, myelodysplastic disorders, leukaemia, lymphoma) and diseases suggestive of immunodeficiency such as stomatitis, glossitis and pharyngitis, fungal dermatopathies and some gastrointestinal disorders (e.g. typhlocolitis) (Hanger et al. 2003; J Hanger *pers. comm.*).

There has been limited research investigating the incidence/impacts of KoRV on southern koalas; sampling carried out in three southern populations showed a mixed prevalence of KoRV (Tarlinton et al. 2006). KoRV was not detected in any of 26 animals tested from a population present on Kangaroo Island (SA) and a review of haematology data from 225 Kangaroo Island animals found no cases of leukaemia. It was proposed that the founder animals for this population (released onto the island in the 1920s and isolated from mainland populations thereafter) were virus-negative. Other populations in south-eastern Australia demonstrated a mixed prevalence of KoRV, which was detected in 5 of 17 animals sampled from Raymond Island (Vic.) and in 3 of 5 animals sampled near Ballarat (Vic.) (Tarlinton et al. 2006).

The low prevalence of KoRV in island and other southern populations suggests that the virus was not present in the koala population in the early 20th century. Tarlinton et al. (2006) proposed only recent entry into the koala population, with ongoing infection and endogenisation spreading from northern to southern Australia.

b Other viral agents

Serological studies undertaken to investigate the involvement of wildlife in the natural cycles of the arthropod-borne Ross River and Barmah Forest viruses in the Gippsland Lakes region of coastal Victoria indicated that native mammals including the eastern grey kangaroo (*Macropus giganteus*) and koalas inhabiting areas adjacent to extensive mosquito-breeding habitat may be involved in the maintenance of these two alphaviruses in the region (Aldred et al. 1991). Of 93 samples collected from koalas on Raymond Island in 1985 and 1990, antibody to Ross River virus was detected in 22 (24%) and antibody to Barmah Forest virus in 15 (16%). These figures included 7 of 93 (8%) koalas which were seropositive for both viruses.

In another prospective study of captive marsupials maintained in urban areas of eastern New South Wales, none of 12 koalas housed in a metropolitan Sydney zoo was found to be antibody positive to Ross River virus (Old & Deane 2005).

Papillomaviruses have been isolated from the skin of koalas with and without papillomas (see 10.1.4e).

10.11.2 Bacterial diseases

a Chlamydiosis

Taxonomy of the order *Chlamydiales*

In 1999 a major revision of chlamydial taxonomy was proposed, including creation of three new families in the order *Chlamydiales* to accommodate recently discovered *Chlamydia*-like bacteria which have similar developmental cycles of replication, and the creation of a new genus within the existing family *Chlamydiaceae* (Everett et al. 1999). The genus *Chlamydia*, which previously had included all (four) known chlamydial species, was reappraised. Two genera (*Chlamydia* and *Chlamydophila*) and nine species were described. The revised taxonomy of the order (Everett et al. 1999; Everett 2000) is outlined below.

- ORDER *Chlamydiales*
  - FAMILY *Chlamydiaceae*
    - GENUS *Chlamydia*
      - SPECIES *C. trachomatis*, *C. suis*, *C. muridarum*
    - GENUS *Chlamydophila*
      - SPECIES *C. psittaci*, *C. pneumoniae*, *C. pecorum*, *C. felis*, *C. caviae*, *C. abortus*
  - FAMILY *Parachlamydiaceae*
    - GENUS *Parachlamydia*
    - GENUS *Neochlamydia*
  - FAMILY *Simkaniaceae*
    - GENUS *Simkania*
      - SPECIES *S. negevensis*
  - FAMILY *Waddliaceae*
    - GENUS *Waddlia*
      - SPECIES *W. chondrophila*

Taxonomy of the order *Chlamydiales* is in a state of flux and likely to change again in the near future, particularly with regard to generic division within the family *Chlamydiaceae* (P Timms *pers. comm.*). The trivial name 'chlamydiae' still applies to all members of the order *Chlamydiales* (Everett & Anderson 2001), but the name now encompasses a greater breadth of both pathogenic and environmental isolates and no longer precisely refers to organisms grouped within a single genus and family.

Isolation of chlamydiae from koalas: historical perspectives

When reviewing historical literature on isolation of chlamydiae from koalas, it must be remembered that any conclusions drawn depended upon the current understanding of the order *Chlamydiales*. This was based on the availability, sensitivity and specificity of different diagnostic and research methodology. If all these historical investigations could be repeated using the scope of present techniques, it is likely that the characterisation of some isolates might change substantially (P Timms *pers. comm.*).

Prior to the development of an ever-increasing array of molecular tools to study genetic differences between isolates, chlamydial diseases in koalas (and other animals) were thought to result from infection with a single species, *C. psittaci* (Cockram & Jackson 1974; McColl et al. 1984; Brown et al. 1984a). The progressive elucidation of knowledge concerning different aspects of chlamydiosis in koalas up to the mid 1990s has been documented in detail in the comprehensive review by Whittington (2001).

Current identity of *Chlamydiales* isolated from koalas

C. pneumoniae

Previously designated koala Type I ('ocular' strain) *C. psittaci* (Girjes et al. 1988, 1993) and shown by DNA homology to be closely related to *C. pneumoniae* from humans and an isolate from the respiratory tract of a horse (Girjes et al. 1994; Storey et al. 1993; Glassick et al. 1996), the koala biovar of *C. pneumoniae* is widespread in Australia's free-ranging and captive koala populations (Wardrop et al. 1999).

Analysis of partial gene sequences from more than 10 koala *C. pneumoniae* isolates obtained from widely geographically separated, free-ranging koala populations (in Qld, NSW, Vic. and SA) and one captive colony (in Qld) showed them to be 100% identical at this locus (Wardrop et al. 1999; Girjes et al. 1994; Glassick et al. 1996). Thus the koala biovar of *C. pneumoniae* appears to be monophyletic and is genetically distinct from human and equine biovars at several gene loci tested to date (Wardrop et al. 1999).

C. pecorum

Previously designated koala Type II ('urogenital' strain) *C. psittaci* (Girjes et al. 1988, 1993) and shown by DNA homology to be closely related to *C. pecorum* from a broad host range of domestic animals including cattle, sheep and pigs (Fukushi & Hirai 1993; Glassick et al. 1996), koala *C. pecorum* has been suggested as the more pathogenic of the two chlamydial species cultured from koalas (P Timms cited in Booth & Blanshard 1999; Jackson et al. 1997, 1999).

In contrast to the monophyletic nature of the koala biovar of *C. pneumoniae*, the 15 koala *C. pecorum* isolates studied by amplification and comparison of partial DNA sequences clustered into five genetically distinct subgroups, designated genotypes A–E (Jackson et al. 1997). Two of these genotypes (A and E) were koala-specific and contained multiple identical or near-identical isolates, whereas genotypes B, C and D (each represented by a single koala strain) appeared to have some relationship with non-koala *C. pecorum* isolates from domestic livestock. Koala *C. pecorum* genotype B yielded a sequence similar to a porcine isolate from Europe, koala genotype C a sequence similar to two bovine isolates from Australia, and the sequence from koala genotype D was 100% identical to two bovine isolates (one Australian, one European) and three Australian ovine isolates. These findings were regarded as strong supporting evidence for a model involving recent multiple cross-species transmission events of *C. pecorum* into koalas (Jackson et al. 1997).

Uncultured koala Chlamydiales *(UKC strains)*

A total of 94 swabs collected from 25 koalas randomly selected from a large healthy free-ranging population in New South Wales in 1999 showed prevalences of 12% (3 of 25) and 52% (13 of 25) for *C. pneumoniae* and *C. pecorum*, respectively (Devereaux et al. 2003). Although positive for *Chlamydiales* order-specific DNA, however, 10 swab samples obtained from 4 of the 25 koalas could not be speciated, being negative by both *C. pneumoniae* and *C. pecorum* species-specific molecular assays.

Direct sequencing of the available nucleic acid fragments, termed 'uncultured koala *Chlamydiales*' (UKC), confirmed the presence of non-*C. pneumoniae*, non-*C. pecorum* strains from novel *Chlamydia*-like organisms. UKC were predominantly found as co-infections with *C. pneumoniae* or *C. pecorum*. It was proposed that koalas are similar to humans in their susceptibility to the *Chlamydiales*, i.e. they may be infected with a range of *Chlamydia* and *Chlamydia*-like bacteria that vary in prevalence and ability to cause disease (Devereaux et al. 2003).

Epizootiology

It is only relatively recently that the first true population studies to use DNA-based assay procedures for the detection and speciation of koala *Chlamydiales* have been able to provide more detailed insights into the incidence and epizootiology of naturally occurring infection and disease (Jackson et al. 1999; Devereaux et al. 2003). In the first survey to use these advanced molecular techniques, ocular, urogenital and rectal swabs (a total of 170 samples) were taken from 53 koalas in two free-ranging populations in south-east Queensland and evaluated for the presence of *C. pneumoniae* and *C. pecorum* using *omp*B genus-specific PCR and species-specific DNA probe hybridisation (Jackson et al. 1999).

The level of clinical disease (17%) compared to the chlamydial infection status (85%) in one population was similar to that reported in earlier studies, confirming that infection commonly occurs without visible clinical disease. All koalas with clinical disease were found to have current *C. pecorum* infection, and no clinical cases were found to be infected with *C. pneumoniae* alone. Subjective estimates of the level of chlamydial infection (based on the DNA probe hybridisation band intensity) suggested that *C. pneumoniae* infections were low-grade while *C. pecorum* infections ranged from low- to high-grade (Jackson et al. 1999).

C. pneumoniae has been isolated from both ocular and urogenital sites in male and female koalas, with no consistent site or sex bias. Prevalence rates in different populations vary and, although *C. pneumoniae* has been detected in 5–89% of ocular and 5–62% of urogenital sites examined by PCR/hybridisation, apparent infection levels are usually of low grade (Jackson et al. 1999; Wardrop et al. 1999). *C. pneumoniae* is rarely documented in association with urogenital tract, ocular or respiratory disease in free-ranging koala populations (Jackson et al. 1997; Wardrop et al. 1999) and it remains unclear which conditions it may cause (Wardrop et al. 1999). Co-infections with *C. pecorum* can occur, making it difficult to determine specific disease associations. In one free-ranging population examined, none of the seven koalas infected with *C. pneumoniae* alone (21% of the total population of 33) had evidence of clinical disease (Jackson et al. 1999).

C. pecorum has also been isolated from both ocular and urogenital sites of male and female koalas, with overall prevalences of 10, 52 and 73% in three free-ranging populations examined (Jackson et al. 1999; Devereaux et al. 2003). In the population with the highest prevalence of infection, all 5 of the 28 koalas systematically examined that showed clinical signs of disease at the time of sampling were positive for *C. pecorum*. Much more extensive geographical surveys would be required before it could be determined whether particular genotypes of koala *C. pecorum* are disease-specific (Jackson et al. 1997).

Higgins et al. (2005b) compared samples collected at various diseased (kidney, lung, upper female reproductive tract (uFRT)) and non-diseased (uFRT) sites in free-ranging koalas from northern New South Wales and south-east Queensland using histology, immunohistochemistry for chlamydial Ag and PCR for chlamydial DNA. This study demonstrated clearly that chlamydial infection plays a predominant role in disease of the entire uFRT and a direct role in inflammation and fibrosis of the salpinges and uteri similar in appearance to that in women. A predominant (and frequently apparently primary) role for *Chlamydiaceae* in at least some cases of pyogranulomatous pyelonephritis was also established (Higgins et al. 2005b).

Sites from which different *Chlamydiales* have been isolated from the koala, and the current understanding of an association between the organisms and clinical signs of disease at these sites (P Timms *pers. comm.*), are shown in Table 8.4.

Inflammation and fibrosis within the reproductive tract may result in irreversible infertility in female koalas. However, the effect of infection on fertility rates in free-ranging populations is variable. Populations that demonstrate widespread infection do not necessarily demonstrate low fertility rates (Martin & Handasyde 1990, 1999). In the majority of infected southern populations, fertility rates (measured as the proportion of independent females that have young) range from 35–55%, but individual populations may be more severely affected (Martin & Handasyde 1999). The factors that determine the impact of infection within a population are not well understood, however, important variables include:

- the chlamydial strain involved;
- the presence of other stressors (e.g. overcrowded habitat, declining food resources);
- the presence of KoRV in the population;
- a history of previous exposure to chlamydiae.

Table 8.4 Sites at which different *Chlamydiales* have been isolated from the koala, and the current understanding of the association between these chlamydiae and clinical signs of disease

***Chlamydiales* isolates**	**Presence**			**Disease**		
	Ocul	**Urog**	**Resp**	**Ocul**	**Urog**	**Resp**
C. pecorum	+	+	?	+	+	
C. pneumoniae	+	+	+	–	–	+
Other *Chlamydiales* (UKC)	+	+		?	?	

UKC = uncultured koala *Chlamydiales*.
Ocul = ocular; Urog = urogenital tract; Resp = respiratory tract.
Source: P Timms (*pers. comm.*).

Transmission

Chlamydiales are spread by contact or aerosol and require no alternate vector (Everett et al. 1999; Everett 2000). There is some evidence that koala *Chlamydiales* might survive in the environment after being shed by infected animals. Laboratory experiments showed that koala *C. pneumoniae* can survive *in vitro* at a wide range of pH (4–8, for 4 hr) and temperature (4–25°C, for at least 28 d). Desiccation (over silica gel) was shown to inactivate koala chlamydiae within 3 d, but 1% of elementary bodies (EB) inoculated onto the leaves of nursery-size *E. tereticornis* plants kept indoors at ambient temperature (24°C) were still viable after 3 d (Rush & Timms 1996).

Faecal–oral transmission of chlamydiae during pap feeding by dependent young, and direct transfer of infected discharges from the eyes or urogenital tract of one animal to another, are other possible routes of infection. A *Chlamydiales* order-specific PCR assay used to screen semen samples collected by electro-ejaculation from eight free-ranging koalas hospitalised for treatment of cystitis detected chlamydiae in the semen of 7 of 8 of the diseased males, substantiating the presumption that infection in koalas may also be transmitted during coitus (Bodetti et al. 2002).

Over the past 70–80 years, many thousands of free-ranging koalas with overt clinical signs of ocular and urogenital chlamydial infection have been captured, examined and/or brought into care for attempted treatment. In that time there has been no obvious evidence of transmission of koala *C. pneumoniae* or koala *C. pecorum* to humans. Several veterinarians, researchers and/or carers with a history of exposure to infected and diseased koalas have been serologically assessed and found to be Ab-negative in CFT tests (Cockram 1978; AS Brown *pers. comm.*; Emmins 1996).

Clinical spectrum of infection

Ocular infection and disease (keratoconjunctivitis)

Clinical signs of ocular infection may be unilateral or bilateral. Chronic disease is usually bilateral. Although cases are seen at all times of year, a seasonal incidence is apparent. Ocular disease in affected koalas examined over a 6 yr period in New South Wales was first recorded in the summer months in 24 of 35 (69%) cases, with 22 animals having acute infections (Cockram & Jackson 1981).

Acute cases show a slight serous ocular discharge and the eye may be partly or completely closed. Slight reddening of the conjunctivae may be evident when examined. These signs increase in severity during the first few days, with the discharge rapidly becoming mucopurulent (Fig. 8.16a, b) and the conjunctivae being reddened. Swelling of the conjunctivae and subconjunctival tissues follows within 2 wk of the earliest clinical signs, with bulging (and in severe cases eversion) of the eyelids. Purulent discharge may fill the palpebral fissures and/or soil the surrounding fur. Dried discharge may gum the eyelids closed. Loss of matted fur and excoriation of the lids may follow (Cockram & Jackson 1974, 1976, 1981).

In chronic cases ('pink eye') the conjunctivae are reddened and enlarged with a granular surface (Fig. 8.16c). The palpebral conjunctivae may project beyond the lid margins and in severe cases conjunctival proliferation, especially over the nictitating membrane, may completely obscure the eye (Fig. 8.16d).

In some cases, at about the third week an associated peripheral keratitis may develop and gradually extend centrally, rendering the cornea misty bluish. In chronic severe cases pannus with increasing corneal opacity develops, so the cornea may become pinkish-white and

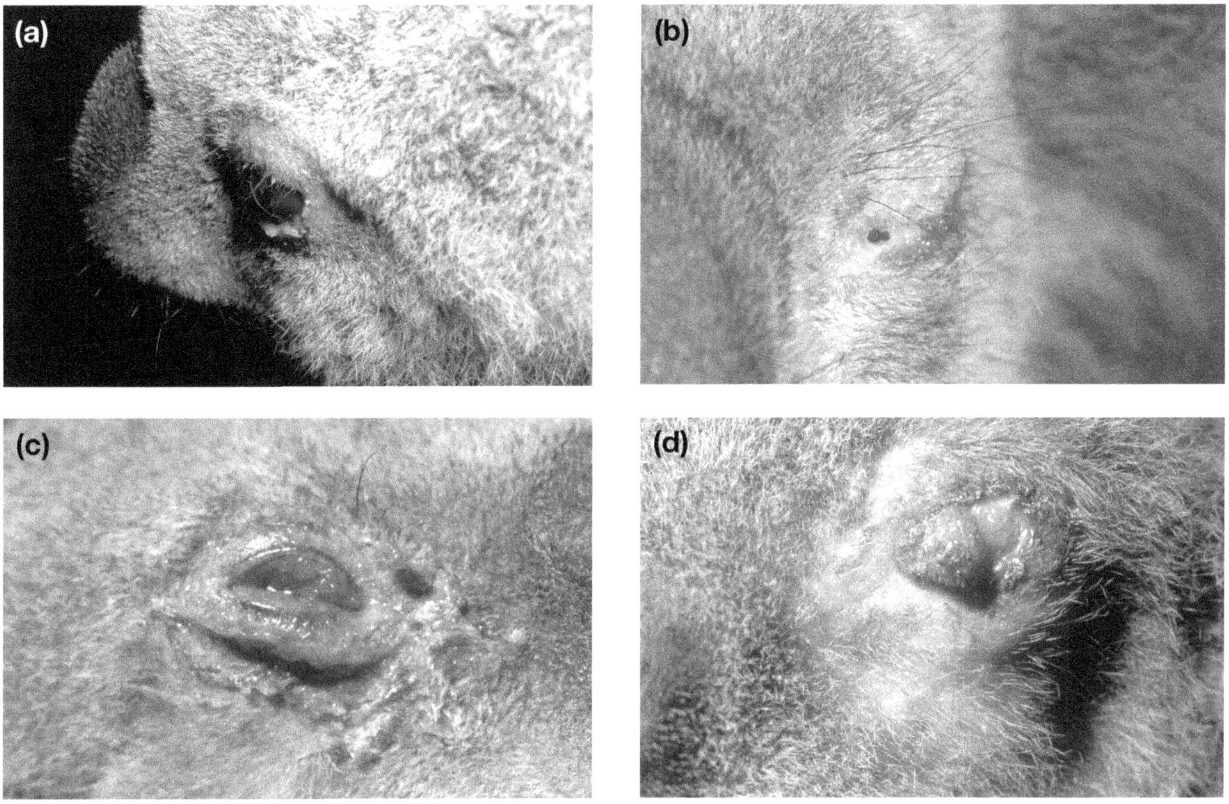

Figure 8.16 Ocular chlamydiosis. a, b) Acute conjunctivitis with mucopurulent discharge. c) More chronic conjunctivitis with chemosis and granular enlargement of the conjunctivae; mucopurulent discharge filling the palpebral fissure was rinsed away immediately prior to this photograph being taken. d) Chronic long-standing conjunctivitis with conjunctival proliferation (especially over the nictitating membrane) completely obscuring the eye. Photos: W Blanshard.

completely opaque (Cockram & Jackson 1974, 1976, 1981).

Urinary tract infection and disease

Brown et al. (1987) reported that chlamydial infection of the urinary tract may result in urethritis, cystitis, ureteritis and nephritis. Chronic cystitis (Fig. 8.17a) is commonly characterised by mucosal hyperplasia and mucosal and submucosal fibrosis around the ureteric papillae (Canfield 1989; Canfield & Spencer 1993b). Active chronic inflammation of the urogenital sinus may be seen (Canfield 1989). Upper urinary tract lesions described in koalas include hydroureter, ureteritis, hydronephrosis and pyogranulomatous and chronic interstitial nephritis (Canfield 1989; Canfield & Spencer 1993b), and it was hypothesised that renal and ureteral pathology occurred as a result of obstruction, vesicoureteral reflux and ascending infection (Canfield & Spencer 1993b). Higgins et al. (2005b) demonstrated the presence of *Chlamydiaceae* in epithelial cells and macrophages in association with pyogranulomatous pyelonephritis and focal interstitial nephritis.

Clinically, there is soiling and brownish staining of the fur around and below the cloaca and over the rump due to constant wetting with urine ('dirty tail' or 'wet bottom'). The normally pale well-groomed coat in these areas takes on the matted appearance of wet or greasy sheepskin, with clumping and pointing of small tufts of fur. The animal may not make any attempt to assume the normal posture before micturition, suggesting either urgency or true incontinence. Dysuria, signalled by tenesmus with cloacal eversion or quiet grunting vocalisation, is sometimes apparent. Gross haematuria may be evident in severe cases. Chronically affected koalas show very dark staining of fur on the rump and varying degrees of urine scalding of the skin. Ulcers and pressure sores develop over bony prominences in debilitated animals.

In cases of cystitis, urine consistently contains erythrocytes, leukocytes, bacteria and yeasts, epithelial cells,

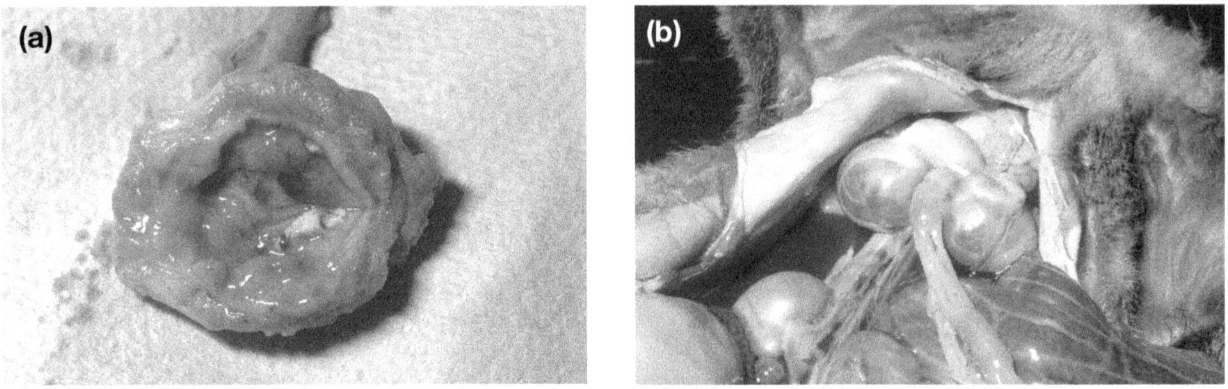

Figure 8.17 Urogenital chlamydiosis. a) Chronic cystitis. b) Bilateral cystic ovarian bursae and salpingitis. Photos: a) W Blanshard. b) W Blanshard and University of Queensland.

cellular debris and occasional renal tubular casts. There are no consistent haematological or biochemical abnormalities, although anaemia and hypoproteinaemia may occur (Canfield 1989; Obendorf 1983).

Reproductive tract infection and disease

Inflammation and fibrosis of different parts of the reproductive tract may occur, and pathology in the uFRT may be unilateral or bilateral. Cystic enlargements of the ovarian bursae (sometimes colloquially and erroneously known as 'cystic ovaries'), salpingitis (Fig. 8.17b), hydrosalpinx, metritis, pyometra, vaginitis and pyovagina have been recorded in females (Brown et al. 1987; Obendorf 1981; Canfield et al. 1983) and prostatitis in males (Canfield 1991; Canfield & Spencer 1993b). Affected females are often rendered infertile but may not show any external signs of disease. Peritonitis, with fibrinous adhesions secondary to rupture of a uterine horn, has been reported in three animals (Canfield et al. 1983; Connolly 1999).

Pathological findings and pathogenesis of urogenital tract disease in the koala have been described in detail (McColl et al. 1984; Brown et al. 1987; Obendorf 1981, 1988; Canfield 1990; Higgins et al. 2005a, 2005b). Obendorf and Handasyde (1990) described the microscopic pathology of the reproductive tracts from 17 free-ranging Victorian females grouped with reference to age, population of origin and known reproductive status. No ovarian lesions were found in 16 of the female koalas examined, however, 10 had some pathology in the reproductive tract, including five with unilateral and one with bilateral cystic ovarian bursae. All ovaries examined, whether from within cystic or normal ovarian bursae, showed some evidence of cyclic activity including one or more of the following structures: developing or maturing follicles, corpora haemorrhagica, corpora lutea or corpora albicantes (Obendorf 1981).

Clinical signs alone may provide an inadequate assessment of whether a koala has significant chlamydial inflammation of the genital tract. In the absence of cystitis, moderate to severe inflammation of the urogenital sinus, prostate or penile urethra is often clinically inapparent but asymptomatic animals may nevertheless have severe underlying disease that requires treatment (Hemsley & Canfield 1997). Unilateral or bilateral paraovarian cysts may or may not be associated with inflammation of the genital tract (Canfield 1990). Higgins et al. (2005b) reported no definitive correlation between the presence or absence of cystic ovarian bursae and histological evidence of luminal occlusion of the reproductive tract in female koalas from south-east Queensland.

Larger cystic ovarian bursae may be palpable in the region adjacent to the epipubic bones in anaesthetised females (Booth & Blanshard 1999; A McKinnon *pers. comm.*; Markey et al. 2007). A radiographic technique incorporating pneumoperitoneum has been used in a study to survey female koalas for cystic enlargements in the reproductive tract (Brown et al. 1984a). Although the epipubic bones interfere with examination of the lower genital tract in koalas, ultrasonography has been proposed as an effective method of detecting and grading the severity of chlamydial urogenital disease in both sexes (see 7.2.8). Cystitis is recognisable as measurable bladder wall thickening, and cystic ovarian bursae are detectable in females with chlamydiosis, whereas the ovaries are not detected in normal animals (Stalder et al. 2003). Ultrasound evaluation of the urogenital tract is now part of the routine clinical assessment in all free-ranging koalas admitted to some wildlife hospitals (J Hanger *pers. comm.*; C Flanagan *pers. comm.*).

Other (including respiratory) infection and disease

Atypical presentations including acute shifting lameness, throat pain and dysphagia, epidermal ulceration at pressure points and inflammation of carpal and/or patellar tendons were observed as part of the spectrum of clinical signs occurring in conjunction with upper respiratory tract disease (see 10.5.1) in two episodes in captive koalas (V Nicolson *pers. comm.*; J Hanger *pers. comm.*). *C. pneumoniae* was identified by PCR from koalas in one of these two episodes (Wardrop et al. 1999). Of the 61 koalas meeting the case definition of showing a nasal discharge on two consecutive days, 25 (41%) apparently were PCR positive (V Nicolson *pers. comm.*).

Literally thousands of compromised (traumatised or debilitated) koalas that originated from free-ranging populations in a geographical region with a high prevalence of chlamydiosis (and KoRV; Tarlinton et al. 2005) have been confined for treatment and rehabilitation without developing clinical signs of rhinitis. This suggests that, if there is an association between *Chlamydiales* and episodic respiratory disease in koalas, the presence of (an)other microbial agent(s) or epidemiological factor(s) may also be required before clinical disease results. Unequivocal evidence of a primary causal relationship between *C. pneumoniae* and clinical signs of rhinitis remains elusive.

Although chlamydiae have been cultured from the rectum of koalas, there is only one published report describing proctitis (Hemsley & Canfield 1996). It was suggested that rectal infection may represent an important reservoir of infection for other koalas, via faecal shedding of organisms.

Role of secondary (opportunistic) infections

A recent study determined that infection by non-chlamydial bacteria occasionally accompanied chlamydial infection of the female reproductive tract, but that these bacteria were probably opportunistic invaders of diseased sites (Higgins et al. 2005b). The frequency with which the *Chlamydiaceae* alone appeared to cause renal disease indicated that chlamydiae should be considered primary aetiologic agents in many cases. However, the potential for exacerbation of chlamydial disease due to opportunistic infection by normal flora should not be overlooked. Different studies have reported isolation of or secondary infection by organisms that are considered part of the normal flora in the urogenital tract (Weigler et al. 1988; Obendorf 1988; Canfield et al. 1983).

During an epidemiological survey of chlamydial infection in 65 free-ranging koalas in south-east Queensland, urogenital and conjunctival swabs for aerobic microbial culture were taken from each animal and it was reported that pathogens other than *Chlamydiales* were not isolated. *E. coli* and α-haemolytic *Streptococcus* spp. were the most common of seven bacteria or fungi isolated two or more times from urogenital swabs, comprising 40% and 32% of 64 isolates, respectively (Weigler et al. 1988).

Aerobic bacterial cultures taken at necropsy from two Victorian koalas with active pyometritis yielded pure growths of non-haemolytic *E. coli* and α-haemolytic *Streptococcus* sp., respectively. Bacteria were not isolated from cystic ovarian bursae, but non-haemolytic *E. coli* were recovered from the undistended ovarian bursae and oviducts of two other females in which the bursae contained pus and fibrous tags (Obendorf 1981). Other aerobic bacteria isolated from urine and/or swabs from the urogenital tract include coagulase-positive *Staphylococcus* spp., *Proteus* spp. and diphtheroids (Obendorf 1988).

Aerobic and/or anaerobic bacteriological procedures failed to isolate all the bacteria seen in smears of purulent material made at necropsy from four koalas with pyometra and pyovagina. Aerobic isolates included *E. coli* and β-haemolytic *Streptococcus* spp., while anaerobic isolates included at least two *Bacteroides* spp., *Actinomyces viscosus* and *Peptostreptococcus anaerobius*. The organisms that were isolated were recognised as being part of the normal vaginal flora in other species (Canfield et al. 1983).

In clinical practice, secondary infections (e.g. by yeasts) are regarded as common (J Hanger *pers. comm.*).

Confirmation of diagnosis

Studies to evaluate non-molecular methods of chlamydial diagnosis have been reviewed in detail elsewhere (Blanshard 1994; Whittington 2001).

While demonstrated evidence of (viable or non-viable) chlamydiae supports a diagnosis of chlamydiosis, apparent absence of chlamydial organisms, their nucleic acid (NA) or antigen (Ag) may reflect:

- poor sample collection or handling;
- levels of infection below the sensitivity limit of the test method employed;
- an alternative aetiology for the clinical condition observed.

Much of the literature on diagnosis of chlamydiosis in koalas is dominated by point-in-time evaluation of specimens by vastly different detection techniques, often without specific correlation to the disease status of the individual from which the sample was obtained. Epidemiological understanding may be improved or confounded when research techniques are superseded with such rapidity.

Cell culture

Usually performed only in research laboratories, cell culture is a specialised, exacting and labour-intensive test which superseded inoculation of embryonated chicken eggs for isolation of *Chlamydiales* organisms. Positive results rely on collecting viable chlamydiae from the host and providing appropriate conditions for these to infect and multiply in cultured cell lines (Grice & Brown 1985). Clinicians should contact the laboratory for details of sample collection (including swabbing technique and transport medium) before considering cell culture as a diagnostic method.

Chlamydial antigen detection

All species within the family *Chlamydiaceae* share a common lipopolysaccharide (LPS) located in the outer membrane, that can be detected by monoclonal Abs (mAbs) raised against the LPS Ag (Everett et al. 1999). Family-specific mAbs against LPS Ag are not known to recognise other members of the *Chlamdialaes*, e.g. *Parachlamydiaceae* or *Simkaniaceae* (Everett et al. 1999; Everett 2000). Currently, no information is available regarding the potential of the *Chlamydia*-like organisms from which UKC strains are derived (Devereaux et al. 2003) to be isolated in cell culture or revealed by existing Ag-detection methods.

A direct fluorescence Ab test for LPS antigen (Chlamydia Cel LPS; Cellabs) has proven to have acceptable sensitivity and specificity for testing direct smears of koala specimens, compared with detection in cell culture. A study evaluating the test in detecting *C. psittaci* in direct smears from swabs taken from koalas found that for conjunctival specimens the test was calculated to have a sensitivity of 88% and a specificity of 100%, and for urogenital specimens the sensitivity was 90% and the specificity was 84%. Overall (combined) sensitivity was calculated to be 89% and overall specificity 94% (Canfield et al. 1991).

In field trials evaluating urogenital sinus swabs from free-ranging female koalas, the Clearview® Chlamydia (Unipath Ltd) ELISA kit was estimated to have a sensitivity of 91% and a specificity of 100%. The study concluded that the Clearview® Chlamydia kit was easy to use as well as being sensitive and specific for the detection of urogenital infections in field work, and that it warranted further evaluation (Wood & Timms 1992). This test was adopted as (and the Clearview® Chlamydia MF continues to be) the most widely used in-house *Chlamydiaceae* Ag detection methodology in koala medicine.

Clearview® Chlamydia MF is intended to detect *C. trachomatis* Ag in human diagnostic specimens (female endocervical swabs or, in males, first-catch urine sediment). Its performance with specimens taken from other sites has not been established by the manufacturer. Centrifuged specimens of koala urine obtained by cystocentesis or manual expression of the urinary bladder can be tested (J Hanger *pers. comm.*).

Serological evidence of anti-chlamydial antibodies

A generic complement fixation titre (CFT) test is offered by (or can be arranged through) many larger veterinary pathology laboratories. The test detects complement-fixing Ab to *Chlamydiaceae* family-specific LPS Ag, usually in domestic livestock sera.

The CFT test has significant limitations if used as a koala screening test. Chlamydiae-naïve koalas may take at least 3–4 mo to develop detectable CF Ab levels after infection (Brown & Grice 1984, 1986). An early population study of chlamydial epidemiology in free-ranging koalas in south-east Queensland showed a high prevalence of *Chlamydiales* (in 46 of 65, 71% of animals) using cell culture isolation (Weigler et al. 1988). Only 6 of 46 (13%) infected animals had clinical signs of disease. Blood collected from each animal at the time of sampling revealed only 4 of 65 (6%) had detectable CFTs. In comparison to cell culture, the CFT test in this study was calculated to have a sensitivity of only 7% and a specificity of 95% (Weigler et al. 1988).

A koala-specific Ab ELISA developed in 1992 utilised rabbit anti-koala IgG conjugated to horseradish peroxidase in a microtitre plate assay, and was estimated to be at least 16 times more sensitive than the generic veterinary laboratory CFT test for domestic animals (Emmins & Turner 1992; Emmins 1996). The test, run on demand in Australia for koala clinicians and researchers for approximately 15 yr and incorporated in the pre-export testing protocol for koalas (DEH 2004), is no longer available (J Emmins *pers. comm.*).

Chlamydial nucleic acid (NA) detection (with or without amplification)

Theoretically, species of veterinary interest within the family *Chlamydiaceae* can be distinguished by serological and/or NA-based identification methods, however, in practice the reagents for serotyping and PCR speciation are not commercially available. Such tests can only be carried out at specialised diagnostic laboratories or in collaboration with research groups (Everett 2000).

Conventional PCR is potentially extremely sensitive as the cyclical (chain) reaction amplifies (replicates) the target NA. Viability of organisms in clinical specimens is not necessary (Gen-Probe 2000; Rybicki 2005).

Targets for the myriad different types of chlamydial PCR assays may include full or partial, conserved or variable nucleotide sequences of:

- genomic DNA—e.g. a gene coding for one of the different outer membrane proteins (omp) such as *omp*A or *omp*B, or specific variable domains (VD) within one gene such as *omp*A VD4;
- ribosomal RNA—e.g. one or more of the three molecules of rRNA (designated from their sedimentation coefficients as 23S, 16S and 5S), with or without an intermediate region such as the 16S–23S intergenic spacer.

Sample preparation, efficiency of amplification and target detection—each with different requirements for optimisation—vary between tests and between laboratories, even if the test method (PCR) sounds the same (Kaltenboeck et al. 1991; Rideout 2003; Rybicki 2005). Rigorous standardisation and validation of each clinical sampling and assay technique may be required. Potential sources of variation between tests include (but are not limited to):

- different concentrations of primers, nucleotides and other test reagents affect the efficiency of target amplification;
- different primers vary in their optimal annealing temperatures and different enzymes in their thermal stability;
- different enzymes vary in their capacity to induce random mutations during extension of hybrid NA strands;
- different numbers of amplification cycles may be performed depending on the starting concentration of the target NA.

When trying to interpret diagnostic information, especially in the case of unexpected or discordant test results, it is important for the clinician to consider *what* was being tested for, *how* and *by whom*, and thus have some understanding of the inferred limitations of the tests performed.

Suggested approach for assessing the chlamydial status of a koala

When assessing the disease status of a newly arrived animal (with no clinical signs of disease) entering a well-established captive collection of apparently healthy koalas, or an animal demonstrating clinical signs consistent with possible infection with *Chlamydiales* (including lower urinary tract disease, renal disease, reproductive tract disease or ocular disease), an appropriate diagnostic approach might include the following steps.

- Physical examination carried out under anaesthesia, with evaluation of the urogenital tract using ultrasound and blood collection for assessment of haematological and biochemical parameters.
- Clearview® Chlamydia MF kit testing of swabs collected from both eyes, nasal septum, urogenital sinus (females) or penile urethra (males) and rectum. See below for swabbing technique.
- Urinalysis including swabs of sediment (obtained by centrifugation of ≤1.5 mL of urine collected by manual expression of the bladder or by cystocentesis) for Clearview® Chlamydia MF kit testing.

If opportunities to utilise PCR are available, swabs may be submitted for this. Swabs for PCR are collected at the same sites and using the same methods as for Clearview®. The clinician should seek advice from the laboratory about sample storage and transport requirements.

As chlamydiae are spread by contact, animals should be isolated and routine quarantine measures instituted until test results are finalised. The Department of Environment and Water Resources requires that koalas intended for export from Australia must have been tested twice, with negative results, no less than 21 d apart within a 45 d pre-export isolation period (DEH 2004).

Any results must be interpreted in light of the disease status of koalas already held by an institution. The stringency of testing of new individuals should match that of existing collection animals. It is difficult to justify rigorous disease testing for an incoming individual if the disease status of collection koalas is not known.

Clinicians/zoological institutions should discuss preferences in terms of pre-transfer testing procedures prior to a planned koala transfer to reduce the likelihood of unexpected test results in newly arrived animals. If an incoming animal does present a positive test result, treatment and re-testing might reasonably be performed. An example of a treatment/testing procedure is given below.

- Following a positive test result, commence systemic antimicrobial treatment (see below), e.g. for 21 d. Maintain the animal in isolation and under quarantine conditions.
- Re-examine the animal and repeat diagnostic testing at the conclusion of treatment.
- If there is a positive test result following treatment, continue systemic therapy for a further 21 d then re-test.
- If all test results are negative following treatment, maintain the animal in isolation and perform a final examination and repeat testing 3–4 wk after the conclusion of treatment.

Procedure for swab sampling for Clearview® Chlamydia MF kit testing

Swabs should be polyester-tipped and have an aluminium or plastic shaft. If male koalas are to be tested, swab tips must be fine enough to allow entry into the penile urethra, e.g. the 170KS01 aluminium-shafted rayon-tipped dry swab (Copan Italia SPA), available from veterinary pathology laboratories in Australia.

Samples should be collected carefully and systematically, using one swab per site.

- Conjunctivae—roll the eyelid so that the swab is placed into the conjunctival fornix.
- Urogenital sinus—use an otoscope to guide the swab into the UG sinus, thus reducing contamination from elsewhere in the cloaca (see section 2.4).
- Penile urethra—extrude the penis by applying gentle pressure on either side of the cloacal orifice and insert the swab 1–2 cm into the penile urethra (see section 7.2.3).
- Rectum—evert the cloaca and establish the location of the rectal opening, then insert the swab so that it lies against the rectal mucosa.
- Nasal septum—insert the swab (directing medially) so that it lies against the mucosa of the nasal septum.

In each location, it is important that the swab is rotated/firmly smeared in a vigorous manner (5–10 times) against the mucosal surface so that epithelial cells are harvested. If not destined for immediate testing, swabs or urine specimens for Clearview® testing may be stored at 2–8°C (but not frozen) for up to 5 d after collection.

Necropsy confirmation of diagnosis

The apparent prevalence of chlamydial inclusions seen in histological sections of tissues from the same animal may vary with the staining technique employed. In one study, of 51 koala tissues histologically identified as inflamed, 16 were positive to chlamydial Ag using immunoperoxidase staining whereas only 5 and 3 had inclusions evident when stained with Gimenez and H&E, respectively. Thirteen of 21 inflamed tissue sites (including five that were negative by other Ag detection methods) were positive when examined using swab-based tests for chlamydial LPS Ag, and it was concluded that these tests appeared to have superior sensitivity than attempts to detect chlamydial Ag in formalin-fixed tissues by immunoperoxidase staining using anti-chlamydial LPS Ab (Hemsley & Canfield 1997).

Treatment

As the goal of rescue and rehabilitation programs is to return reproductively capable animals able to contribute to the sustainability of free-ranging populations, assessment of likely reproductive capability is an important part of clinical assessment.

In large wildlife hospitals which deal with hundreds of koalas each year, detection of bilateral cystic ovarian bursae is typically considered an important criterion in making a decision to euthanase (C Flanagan *pers. comm.*; J Hanger *pers. comm.*; A McKinnon *pers. comm.*). However, the lack of reliable correlation between the presence or absence of cystic bursae and evidence of reproductive tract occlusion presents a dilemma.

Anti-chlamydial antimicrobials

Systemic anti-chlamydial therapy in hospitalised, free-ranging koalas typically involves administration of enrofloxacin 10 mg/kg sid PO or SC (A McKinnon *pers. comm.*; C Flanagan *pers. comm.*) or chloramphenicol 60 mg/kg sid SC (J Hanger *pers. comm.*).

Current approaches to the treatment of ocular chlamydiosis differ considerably between three large koala hospitals located in northern New South Wales and south-east Queensland. Topical antimicrobial (and other) therapy is consistently utilised in the treatment of both acute and chronic ocular disease, and is combined with systemic administration of antimicrobials in some (but not all) protocols. There is a paucity of published data describing the long-term prognosis for treated koalas.

Facility A Initially, the topical medication used is a nonproprietary mixture of a bovine intramammary antibiotic suspension containing oxytetracycline HCL 18.5 mg/mL, oleandomycin phosphate 10 mg/mL and neomycin sulphate 10 mg/mL, with an inactive dye marker 'brilliant blue FCF' 25 mg/mL (Mastalone Blue®; Pfizer Animal Health) together with ofloxacin 3 mg/mL drops (Ocuflox®; Allergan Australia). In-house, these two products are combined in equal quantities by volume in a separate dropper bottle, and instilled as eye drops. In each case they progress to using Ocuflox® alone. Treatment is usually extended about 3 d beyond the resolution of clinical signs (A McKinnon *pers. comm.*).

Adult koalas are also routinely treated with systemic enrofloxacin 10 mg/kg sid PO, but may receive their first dose SC during initial examination under anaesthesia. In deference to the manufacturer's precautions for use in young domestic animals, subadult koalas ($\leq$2 kg) are typically not given enrofloxacin and may be given chloramphenicol as an alternative (A McKinnon *pers. comm.*).

Conjunctival ablation is regarded as the treatment of choice for badly affected eyes showing conjunctival proliferation of all three eyelids, crusting and discharge and/or corneal scarification. In some severe cases, conjunctival ablation is required before the cornea can be visualised and evaluated. Occasionally, a decision to euthanase is made only as and after that surgery has been performed; thus it can be regarded as a method of exposing the eye for examination. At one large wildlife hospital where the procedure is used frequently, approval for routine use of conjunctival ablation was based on post-operative examination of several animals by an ophthalmologist that apparently revealed no evidence of 'dry eye' or other pathology induced by an appropriate surgical technique (A McKinnon *pers. comm.*).

Under general anaesthesia, the third eyelid is grasped with forceps and scissors are used to trim away as much of the nictitating membrane as can be exposed easily. When trimming the palpebral conjunctivae, the eyelid is rolled outwards using the thumb or fingers and, once lifted away from the globe, the excess tissue is trimmed away with scissors. It is essential that care be taken to maintain the integrity of the eyelid margins, to avoid the possibility of scarring and distortion of the palpebral fissure after surgery (A McKinnon *pers. comm.*).

Post-operative management generally includes anaesthetising the animal every other day to clean clots and debris from the conjunctival surfaces and fornices. For a severely affected animal this intensive follow-up might extend about 6 d post-operatively, i.e. the process usually involves three (or fewer) general anaesthetics for cleaning. In some chronic conjunctivitis cases where koalas admitted in good body condition are given systemic and topical treatment in conjunction with surgical ablation, the turnaround time between presentation and release may be as short as 2 wk (A McKinnon *pers. comm.*).

Facility B This koala hospital imports a preparation containing oxytetracyline HCl equivalent to oxytetracycline 5 mg/g and polymyxin B sulphate 10 000 units/g (Terramycin® Ophthalmic Ointment; Pfizer Animal Health). The ointment is used topically in the eyes bid, generally for 4–5 d beyond resolution of clinical signs. Koalas with overt signs of ocular disease alone are not treated concurrently with systemic antimicrobial drugs (C Flanagan *pers. comm.*).

In chronic conjunctivitis cases, the antimicrobial ointment is used topically for 7–10 d before conjunctival ablation surgery is performed. The surgical technique tends to be relatively conservative, although the third eyelid is sometimes excised. A topical corticosteroid preparation may be commenced after, but not prior to, conjunctival surgery. Follow-up anaesthetics are not required, although saline irrigation may occasionally be indicated for post-operative cleaning. Terramycin® ointment is applied for a variable period, generally up to about 1 mo (C Flanagan *pers. comm.*).

Facility C Koalas with ocular disease are treated with two topical preparations: an ointment consisting of chloramphenicol 10 mg/g and hydrocortisone acetate 5 mg/g (Chloroptsone®; Delvet) and drops containing dexamethasone 1 mg/mL (Maxidex® 0.1% Eye Drops; Alcon Laboratories (Australia)). Typically, both preparations are applied bid, the dexamethasone drops being instilled first followed about 15 min later by the ointment (J Hanger & C Phillips *pers. comm.*). All cases of chlamydiosis, including those with clinical signs of ocular disease only, are treated systemically with chloramphenicol 60 mg/kg sid SC for 45 d, as a standard protocol (J Hanger *pers. comm.*).

Primary medical management of chronic conjunctivitis cases is preferred, with surgery reserved for individuals with an incomplete response after approximately 4 wk of medical therapy. Topical application of ocular corticosteroid preparations is considered very effective at non-surgically reducing proliferated conjunctival tissue by at least 90–95%, usually leading to marked resolution of conjunctival enlargement within 2–4 wk. Occasionally, some cases have small tags of proliferated conjunctival tissue remaining after medical treatment. When indicated, the surgical technique aims to preserve the maximum amount of normal conjunctival tissue by excising only these residual tags and leaving the third eyelid relatively intact (J Hanger *pers. comm.*).

Other Considerations A different therapeutic regimen using parenteral chloramphenicol, at 30 mg/kg q 12 hr SC, was reported to give encouraging empirical results in small numbers of captive koalas with clinical signs of chlamydiosis (Blanshard 1994). Treatment for a minimum period of 10–14 d for acute conjunctivitis and 14 d for cystitis was advised. Concurrent topical therapy was not employed in the four animals with acute conjunctivitis.

In one case in 1995, a captive 13 yr old female koala that developed clinical signs of dysuria and urine soiling of peri-cloacal fur 39 d after mating with a CFT-positive male was treated with chloramphenicol 30 mg/kg q 12 hr SC based on the presumptive diagnosis of acute chlamydial cystitis. Species-specific PCR assays on ocular and urogenital swab samples collected immediately prior to initiation of antibiotic therapy detected *C. pecorum* in the conjunctivae and co-infection with *C. pecorum* and *C. pneumoniae* in the urogenital sinus. A urogenital sinus swab for standard microbial culture and sensitivity yielded a heavy pure growth of *Klebsiella pneumoniae* sensitive to gentamicin, neomycin and chloramphenicol, and resistant to 11 other antimicrobial agents tested *in vitro*. The female was free of evidence of soiled fur and dysuria by the eighth or ninth day of treatment, which was continued for 10–14 d beyond the resolution of clinical signs, i.e. for a total of 22 d. Follow-up conjunctival and urogenital sinus swab samples collected 48 hr after the cessation of chloramphenicol therapy were PCR-negative for chlamydiae, and the female remained free of clinical signs of cystitis for the remaining 7 yr of life (M Jackson et al. unpub).

Markey et al. (2007) reported on the use of real-time PCR to quantify the apparent level of chlamydial infection in 14 free-ranging koalas with a range of clinical presentations, and to determine the effect of treatment on the shedding of chlamydiae during and after therapy in some animals considered appropriate candidates for ongoing treatment.

Where possible, sampling included swabs of sediment obtained by centrifugation of urine specimens (≤1.5 mL) collected by manual expression of the bladder or by cystocentesis. Irrespective of whether samples were taken from koalas with ocular or urogenital tract infections, normalisation of 16S rRNA copy numbers tended to result in significantly increased values from swabs of urine sediment compared with swabs from other sites, highlighting its potential importance as an adjunct to diagnosis.

Ten isolates from eight koalas were sequenced and all determined to be *C. pecorum*. Chlamydial shedding from various anatomical sites in four of the animals was followed closely over periods of 8–9 wk. In three koalas treated with parenteral chloramphenicol 60 mg/kg q 24 hr SC for 45 d, chlamydial copy number was generally no longer detectable by the end of the second week of treatment, and no evidence of relapse of infection was detected 2 wk after the cessation of antibiotic therapy. In contrast, topical (only) chloramphenicol treatment of the eyes of the fourth koala required a longer period of treatment and had little effect on chlamydial shedding from other sites (Markey et al. 2007).

Collaborative and systematic multi-disciplinary approaches to therapeutic trials of anti-chlamydial antibiotics which correlate clinical, microbiological and pharmacokinetic responses to therapy are underway (D Higgins *pers. comm.*). Their results will take understanding of the epizootiology and treatment of chlamydiosis in koalas to a new level.

Analgesic and anti-inflammatory drugs

Mild pain associated with cystitis has been managed with children's oral liquid paracetamol preparations. For slightly more painful cystitis cases, an oral preparation combining paracetamol 24 mg/mL with codeine 1 mg/mL (Painstop® for Children Day-time Pain Reliever; Paedpharm) has been administered. For both preparations, the dose rates in koalas are based on the delivery of paracetamol 15 mg/kg, up to q 6–8 hr PO (J Hanger & C Phillips *pers. comm.*). Meloxicam 0.2 mg/kg PO on the first day followed by a maintenance dose of 0.1 mg/kg q 24 hr PO thereafter is also used (A McKinnon *pers. comm.*). In cases of severe pain buprenorphine HCl (0.01 mg/kg bid SC or IM) or methadone (0.25–0.5 mg/kg q 4–6 hr IM) are used in addition to NSAIDs (J Hanger *pers. comm.*).

The acutely painful shifting lameness atypically occurring with respiratory disease was reportedly relieved (at least partially) by carprofen, whereas administration of pentosan polysulphate sodium or meloxicam had not resulted in any clinically apparent improvement (V Nicolson *pers. comm.*; J Hanger *pers. comm.*). Carprofen has been used at a dose rate of 4 mg/kg sid SC for 3 d (V Nicolson *pers. comm.*) and at an initial dose rate of 4 mg/kg sid SC on the first day then at 2 mg/kg on the second and third days, followed by at least 1 d break before reassessment and, on occasion, cautious continuation of therapy at the lower dose rate (J Hanger *pers. comm.*). Alternatively, a 3–4 d withdrawal or washout period without carprofen treatment is observed before repeating the tapered therapeutic course as above (C Phillips *pers. comm.*).

Paracetamol (with or without codeine) or an opioid may be used to provide interim analgesia in the washout period between successive courses of NSAIDs, or when switching between administration of a corticosteroid and an NSAID (J Hanger *pers. comm.*; C Phillips *pers. comm.*).

Other medications

Hexamine (methenamine) hippurate (Hiprex® Urinary Tract Antibacterial; 3M Pharmaceuticals) is administered to koalas with cystitis at one facility, at a dose rate of 250 mg per 'standard koala' sid PO, ground up into a suspension with water (A McKinnon *pers. comm.*).

Supportive care

Koalas with signs of chlamydial disease that are in poor body condition have a more guarded prognosis for clinical recovery. In these cases, supportive treatment and supplementary feeding (see 5.3.4) must be a high priority. Irrespective of the treatment regimen chosen, the principles of hygiene and nursing remain the same. Any ocular discharges should be regularly cleaned away with saline-soaked cottonwool or tissues. Soiled fur around the cloaca and rump should be washed and dried as necessary to prevent urine scalding.

Release following treatment

It is critical that the exact point of capture of each free-ranging koala be recorded when it is brought into care for any reason. Release in the same location ensures that an animal is returned to its familiar home range and to a population of comparable chlamydial status. There is little data on the post-release survival or recurrence of disease in koalas returned to the wild after treatment for chlamydiosis.

b Septicaemia

Septicaemia has been recognised in both free-ranging and captive koalas. In a mortality survey of free-ranging koalas from the north coast of New South Wales, 6% (7/127) of koalas had apparent septicaemias or toxaemias, with hepatic and/or renal microabscessation or necrosis common findings in the former group (Canfield 1987). In captive koalas septicaemia may be recognised as a primary condition, or secondary to another illness.

Primary septicaemia signalled by peracute illness or sudden death has been observed in apparently healthy animals. There is no age or sex predilection. Gram-negative bacteria can be isolated in pure culture from ante-mortem or post-mortem blood samples. Pathogens identified include *Salmonella typhimurium*, *S. sachsenwald*, *Morganella morganii*, *E. coli*, *Klebsiella pneumoniae*, *K. oxytoca* and *Pseudomonas aeruginosa* (Blanshard 1994; ARWH 2007 case number 2477/2; ARWH 2007 case number 5388/1). *P. aeruginosa* and

Aeromonas hydrophila have been isolated in mixed culture with *E. coli* (Blanshard 1994), and a mixed culture of *P. aeruginosa* with *Klebsiella oxytoca* has been identified (ARWH 2007 case number 5388/1). *Salmonella bovismorbificans* has been incriminated as a cause of sudden deaths in a captive colony (Dickens 1978a). In captive animals the most likely route of infection is ingestion of leaves inadvertently contaminated with bacteria. Cut browse should be prevented from touching the ground wherever possible. *E. coli* septicaemia in emerging PY has been observed occasionally, apparently in association with recent pap feeding (Blanshard 1994).

Septicaemia can occur as a secondary and terminal event in another illness. Primary pathogens or opportunists may invade the bloodstream in conditions such as peracute pneumonia (*Bordetella bronchiseptica*), oropharyngeal ulceration (*Klebsiella* spp., *Streptococcus* spp.) and leukaemia (Blanshard 1994).

The course of illness is very short (hours to days). Individuals have been found dead in the early hours of the morning, having appeared and behaved quite normally the previous afternoon. Dependent young may be found dead in the pouch or on the floor of an enclosure. Affected animals seen alive have consistently shown neurological signs suggestive of disturbances of balance and/or motor control (Blanshard 1994). Nystagmus is common. One koala had a 'rag doll' appearance. Almost all climb, slide or fall to the ground, presumably because they are unsteady and experiencing difficulty staying aloft. One free-ranging koala found dead showed extensive shallow grazing of the medial aspect of both hocks, consistent with having slid down a tree trunk to the ground shortly before her death.

When on the ground, affected animals may occasionally jerk their head and upper body backwards (as though they felt they were overbalancing forwards), even though sitting still. Some fall onto their side and have difficulty rising again, or adopt a low crouched posture as though the ground is moving underneath them. Localised tremor (e.g. of hind limb muscles) or generalised convulsions may occur. Vocalisations suggesting distress have been made by some animals. Septicaemia should be considered as a possible differential diagnosis in any koala showing neurological signs. Body temperature may be increased, normal or decreased. Leukocyte counts are sometimes profoundly decreased (as low as 0.1×10^9/L).

Due to its acute course, where septicaemia is suspected treatment should be instituted without waiting for the results of blood culture or other tests to confirm the diagnosis.

The suggested management protocol is as follows.

- Collect blood for microbial culture and sensitivity testing, haematology and biochemistry before or during placement of an IV catheter.
- Start an infusion of isotonic electrolyte solution.
- Give appropriate antimicrobial therapy while awaiting microbial culture and sensitivity testing results. Suggested antimicrobials include ticarcillin sodium at 45–50 mg/kg IV q 4–6 hr or gentamicin sulphate at 2–4 mg/kg IV (first dose only) plus ampicillin at 5–10 mg/kg IV then gentamicin at 2–4 mg/kg IM, SC q 8 hr and ampicillin at 5–10 mg/kg IV q 6 hr. Note that IV administration of ampicillin and gentamicin should be staggered. Alternatively, a combination of metronidazole plus a cephalosporin (e.g. ceftazidime) plus enrofloxacin may be used while awaiting microbial culture and sensitivity results (J Hanger *pers. comm.*).
- Consider NSAID therapy to manage fever and/or inflammation (e.g. flunixin meglumine at 1.0 mg/kg IV). Repeat dose after 24 hr.
- Use diazepam at 0.25–1.0 mg/kg IV or IM if necessary to overcome exaggerated physical movements and to reduce distress. Repeat as necessary.
- Consider using mannitol 1–2 g/kg as a slow infusion IV if neurological signs worsen.
- Monitor body temperature and blood glucose frequently. Modify the environment to provide warmth or to enhance air circulation around the koala's body as required.
- Try to sustain a moderate diuresis throughout the treatment period.

c *Melioidosis*

Post mortem of an adult koala found dead on Magnetic Island (Qld) revealed pronounced ecchymotic and petechial haemorrhages on most serosal surfaces, especially in the thorax. Many lymph nodes were congested and haemorrhagic. Bacteriological culture yielded *Burkholderia pseudomallei* (previously *Pseudomonas*) and other isolates, and a diagnosis of melioidosis was made (Ladds et al. 1990). It was postulated that the koala may have become infected by eating contaminated soil. Serological examination of this individual for melioidosis using a complement fixation titre (CFT) test was negative, but 8 of 50 (16%) of free-ranging koalas from the Magnetic Island colony were found to be seropositive.

d Other bacterial agents

In a prospective survey of the prevalence of anti-leptospiral agglutinins in wildlife in south-eastern Australia, sera from 25 species of native and exotic mammals were tested by a microscopic agglutination test for antibodies to 12 serovars of *Leptospira interrogans*. Eighteen of 419 (1.9%) sera from seven species were positive (titre ≥ 1:32), indicating past or present infection with leptospires. This included 1 of 3 free-ranging koalas tested, which demonstrated a 1:32 titre to serovar *hardjo* (Milner et al. 1981).

10.11.3 Fungal diseases

a Cryptococcosis

Cryptococcosis is caused by the *Cryptococcus neoformans* complex, including *C. neoformans* and *C. gattii*. These are encapsulated saprophytic yeast-like fungi which cause disease in animals and humans throughout the world. Disease is mostly sporadic, but there are occasional outbreaks (Tell et al. 1997; Fraser et al. 2003). PCR-based DNA fingerprinting is the main typing technique in the ongoing molecular epidemiologic survey of *C. neoformans*, dividing more than 400 clinical and environmental isolates into eight major molecular types (Meyer et al. 2003). Nomenclature of the *C. neoformans* complex is shown in Table 8.5.

Cryptococcosis was first definitively reported in koalas in 1960 (Backhouse & Bolliger 1960), although there are historical records of similar disease described in the early 20th century (Gordon & Hrdina 2005). The incidence of clinical cryptococcosis in koalas is higher than the incidence in domestic animals or humans in Australia. A review of all 43 documented cases of cryptococcosis in captive and free-ranging koalas was carried out by Krockenberger et al. (2003). Formalin-fixed tissues and/or mycological data were available for 29 of 43 cases, and in all 29 instances *C. gattii* was the causative agent. Of the remaining 14 cases, histological description and/or colony morphology suggested that at least nine were caused by *C. gattii*. The relatively high incidence of disease caused by *C. gattii* in koalas is believed to be the result of exposure to high environmental burdens of the organism.

C. gattii has a well-recognised association with a number of *Eucalyptus* spp. trees, particularly the river red gum (*E. camaldulensis*) (Ellis & Pfeiffer 1990). Krockenberger et al. (2002a) used clinically normal koalas as a biological sampler to investigate associations between *C. gattii* and host trees, and verified a strong and consistent relationship between *E. camaldulensis* and *C. gattii* over a wide geographic range in eastern Australia. *C. gattii* has been isolated from *E. tereticornis*, *E. gomphocephala*, *E. blakelyi*, *E. rudis*, *E. microcorys*, *E. grandis*, *Angophora costata* and *Syncarpia glomulifera* and it seems likely that other environmental sources of *C. gattii* exist in south-eastern Australia (Krockenberger et al. 2002a). The presence of dead wood in tree hollows was found to be an important substrate for the organism.

In cases of cryptococcosis, the most likely route of infection is through inhalation of airborne cryptococcal organisms from the environment. Fungal

Table 8.5 Nomenclature of *Cryptococcus neoformans* complex

Current nomenclature	Previous nomenclature	Molecular type	Environmental niche
C. neoformans var. *grubii*	*C. neoformans* var. *neoformans*		
	Serotype A	VN I, VN II	Pigeon guano
C. neoformans var. *neoformans*	*C. neoformans* var. *neoformans*		
	Serotype D	VN IV	Pigeon guano
	Serotype AD	VN III	Unknown
C. gattii	*C. neoformans* var. *gattii*	VG I	Eucalypts and other trees
		VG II	Unknown
		VG III	Unknown
		VG IV	Unknown

Sources: Fraser et al. (2003), Ellis et al. (2000), Chen et al. (1997), M Krockenberger (*pers. comm.*).

particles enter the respiratory tract, where they may be transient inhabitants or colonise the mucosa (Connolly et al. 1999). The most common presentation in koalas is a localised nasal cavity lesion, which may spread to the lungs via the airways (M Krockenberger *pers. comm.*). Dissemination from the respiratory tract to other areas may occur by haematogenous spread or by local extension through the cribriform plate to the central nervous system. Another common presentation is the development of a primary skin lesion (affecting the face and lips) that may extend locally and/or disseminate haematogenously (M Krockenberger *pers. comm.*).

C. gattii can be cultured from the nasal cavity in many koalas without evidence of invasion of the epithelium or an antigenic response to the organism. Nasal colonisation may be transient or persistent. Disease testing of healthy koalas using nasal swabs and culture has revealed that the incidence of nasal colonisation in certain koala populations is very high. The frequent presence of *C. gattii* in the nasal cavity of koalas is thought to be related to the close ecological association between koalas and host plants for the organism, and the importance of olfaction in feeding behaviour of the animal (Connolly et al. 1999).

Subclinical infection has been identified as an important intermediate in the progression from nasal colonisation to clinical disease (Krockenberger et al. 2002b, 2003). Opportunistic necropsies conducted on 10 clinically normal Ag-positive koalas resulted in detection of small focal respiratory tract nodules in two animals.

The tentative conclusion from current research (Krockenberger et al. 2002b; M Krockenberger *pers. comm.*) is that contact with *C. gattii*:

- frequently has no clinical effect on koalas (there is nasal colonisation only);
- sometimes leads to subclinical tissue invasion (probably manifested as focal respiratory tract lesions);
- relatively rarely progresses to overt disease (but the frequency of progression to overt disease may be underestimated).

The presence of high levels of fungal particles in the environment is believed to be the key factor in aetiopathogenesis of cryptococcosis in koalas. While immunosuppression (e.g. caused by malnutrition, concurrent respiratory pathogens or infection with KoRV) may contribute to the development of disease, it appears that colonisation is acquired from an environment that is heavily contaminated by the organism and perpetuated by the presence of heavily colonised koalas (Krockenberger et al. 2002a, 2003). Functional immunological studies are required, to determine the role of immunosuppression in the pathogenesis of cryptococcosis in koalas (M Krockenberger *pers. comm.*).

Clinical signs

A review of 43 cases of clinical cryptococcosis (Krockenberger et al. 2003) revealed that 31 cases showed respiratory tract involvement, including pneumonia (16 cases) and upper respiratory tract disease (13 cases). Seven cases had involvement of upper and lower respiratory tract. Sixteen cases demonstrated meningitis, 10 of which included cryptococcal encephalitis. The lymphoid system was affected in eight cases, the gastrointestinal tract in five cases, the urinary tract in five cases and the skin in three cases.

In some koalas, clinical cryptococcosis is manifested as inappetence and weight loss. When the nasal cavity is affected, nasal discharge and occasionally facial distortion are evident. Involvement of the caudal nasal cavity and nasopharynx may result in stertor and dyspnoea (Krockenberger et al. 2002a). Pulmonary disease is usually manifested as severe dyspnoea. Neurological signs have included blindness, nystagmus, a whole-body tilt and seizures (Spencer et al. 1993; Krockenberger et al. 2003). Cutaneous lesions have presented as SC cervical nodules and as dry scab-covered skin on the lateral aspects of the hind limbs.

Diagnosis

Definitive diagnosis of cryptococcosis requires isolation of the organism from a normally sterile site or histological demonstration of tissue invasion. Nasal swab samples can be collected from gently restrained conscious koalas using a sterile swab moistened with sterile saline. A nasal mucus sample is collected from the rostral 25 mm of the left and right nasal vestibule of the koala by rotating the swab against the nasal mucosa (Connolly et al. 1999). Swabs may be submitted in transport medium. In the absence of clinical signs, however, a positive culture does not necessarily imply tissue invasion.

Smears prepared from nasal discharges, cerebrospinal fluid, fine needle aspirates, impression smears, exudates from lesions or macerated tissue dissolved in potassium hydroxide may allow rapid diagnosis of cryptococcosis. Yeast-like organisms surrounded by a clear capsule, occasionally showing narrow-necked budding, are observed with suitable stains such as new methylene blue, Gram stain or India ink (Fig. 8.18a). If India ink is used, characteristic unstained capsules will be silhouetted against a black background. Particular care must be taken not to mistake fat droplets, lymphocytes or India ink particles for organisms (none of the three should be budding).

Necropsy should include examination of the brain and serial sectioning of the head to examine the nasal cavity and paranasal sinuses (Krockenberger *pers. comm.*). Grossly, lesions have been described as pale firm nodules or as pale gelatinous foci (Backhouse & Bolliger 1960; L Vogelnest *pers. comm.*). Cutaneous lesions may present as nodules containing tenacious yellow-green exudate.

Histologically, cryptococcal organisms in tissues taken by biopsy or at necropsy appear as masses of budding encapsulated yeast cells. There may be little or no surrounding inflammatory reaction, but when present it may range from an early neutrophilic inflammatory response to chronic granulomatous lesions with multinucleated giant cells and scattered lymphocytes and plasma cells (M Krockenberger *pers. comm.*). Periodic acid-Schiff and Mayer's mucicarmine stains can be used to preferentially stain the cryptococcal polysaccharide capsule for easier examination.

An immunohistochemical method for differentiating *C. neoformans* from *C. gattii* in formalin-fixed paraffin-embedded tissues has been developed, and was validated using veterinary patients including koalas (Krockenberger et al. 2001).

The latex-cryptococcal antigen test (LCAT) can be used to detect soluble cryptococcal capsular polysaccharide antigen in body fluid samples. Serial dilutions of serum, urine or cerebrospinal fluid are mixed with a suspension of latex particles coated with hyperimmune rabbit serum. Clumping of the latex particles indicates a positive test. The LCAT is quantified by repeating the test procedure using serial two-fold dilutions of the test sample until agglutination is no longer evident (Malik et al. 1996a). In species other than the koala, false positive results are considered uncommon but may occur in infections caused by *Klebsiella* spp. bacteria or *Trichosporon* spp. fungus. False negative results may also occur, especially with localised or isolated infection (particularly bone lesions).

The LCAT can be used to screen koala populations for subclinical cryptococcosis. Differentiating subclinical disease from clinical disease based on serum antigen titre is difficult, and there is no reliable cut-off LCAT titre that distinguishes between the two states. Krockenberger et al. (2002b) determined that titres $\geq$128 indicated clinical disease and titres $\leq$64 were likely subclinical cases (and therefore might resolve spontaneously). Any titre $\geq$2 is considered positive and requires clinical assessment. All koalas with positive antigen titres should be re-tested periodically, and sustained or rising titres justify further investigation or initiation of antifungal therapy.

One dyspnoeic animal has been reported with thoracic radiographic changes consistent with pneumonia (Ley 1993). Advanced diagnostic imaging techniques such as CT scans (Woods & Blyde 1997) and magnetic resonance imaging (M Krockenberger *pers. comm.*) may assist in location and characterisation of focal lesions in koalas (Fig. 8.18b).

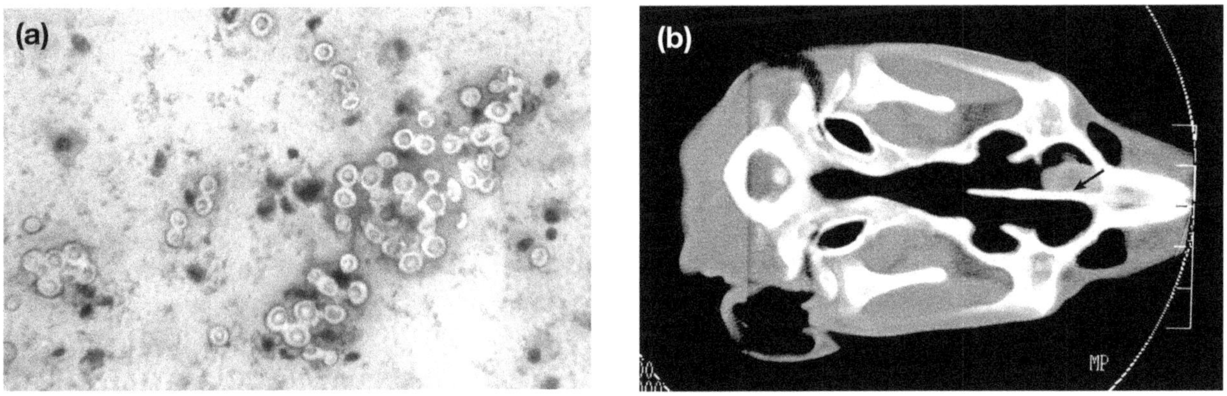

Figure 8.18 Cryptococcosis. a) Impression smear from affected lung, stained with new methylene blue. b) CT scan of the head of a koala with a nasopharyngeal cryptococcal granuloma (arrow). Photos: a) W Blanshard and b) R Woods.

Treatment

Itraconazole has successfully resolved cryptococcal lesions limited to the skin of koalas after 90 d of therapy, with no apparent side effects (Booth & Blanshard 1999; M Krockenberger *pers. comm.*). However, in human medicine it is considered less effective than fluconazole in cases of acute cryptococcosis (Andriole 1999).

Amphotericin B is considered the drug of choice for treatment of cryptococcal meningitis in humans (Andriole 1999). Nephrotoxicity is considered a serious side effect. Development of a technique for treatment of nasal and CNS cryptococcosis in domestic cats and dogs, using SC administration of amphotericin B in dextrose saline, has reduced the risk of nephrotoxicity while resulting in therapeutic efficacy (Malik et al. 1996b). Nephrotoxicity has not been observed in koalas treated using this protocol, but renal function should be monitored during therapy (M Krockenberger *pers. comm.*).

In koalas affected by cryptococcal lesions of reasonable size, combination therapy using amphotericin B plus an oral triazole is recommended (M Krockenberger *pers. comm.*). Combination therapy with amphotericin B (0.5 mg/kg in 300 mL 2.5% dextrose and 0.45% NaCl, given SC twice weekly) and either itraconazole at 100 mg/day PO or fluconazole at 50–100 mg bid PO has had some success in treatment of koalas with nasal disease, lower respiratory disease and CNS disease (Krockenberger et al. 2003). In animals with refractory cryptococcosis, monitoring of antifungal levels (and dosage adjustment if required) should be carried out.

Successful treatment depends on early diagnosis and requires long periods of therapy by a dedicated clinician and keeper. Depending on the severity of disease, the period of treatment required may range from 2–3 mo to 18 mo (M Krockenberger *pers. comm.*). CNS involvement is an extremely poor prognostic indicator. Treatment progress should be monitored by monthly LCAT and continued until the Ag titre is negative for at least 1 mo.

b *Other fungal diseases*

See 10.1.5 and 10.10.1 (pouch infection), 10.1.2 (dermatophytosis), 10.2.3 and 10.10.1 (candidiasis) and 10.2.6 (mycotic diarrhoea).

10.11.4 Parasitic diseases

a *Bertiella obesa*

The koala is the specific host of the anoplocephalid cestode *Bertiella obesa* (Spratt 1978). Adult tapeworms live in the small intestine and tapeworm segments are passed with faecal pellets. Proglottids are transversely grooved, rectangular white or yellowish structures 3–6 mm in length. Detached tapeworm segments are weakly motile and have the ability to dissociate themselves from faecal pellets. They then desiccate, shrivel and rupture, releasing hundreds of eggs. There appears to be an association between free-ranging soil mites (Orabatidae), *B. obesa* and the koala. Under experimental conditions, mites were observed actively crawling over and apparently ingesting eggs from ruptured *B. obesa* proglottids (T Warren *pers. comm.* 1993). Mites are presumed to be accidentally ingested when koalas are on the ground and actively eating soil. Other soil-dwelling arthropods or annelids could be hosts to the larval tapeworms, but this has not been investigated.

The number and volume of parasites per koala can be striking, and at necropsy the worms are sometimes found through the pyloric sphincter and into the stomach, or in the caecum (possibly following post-mortem migration) (Spratt 1978). A disease survey of free-ranging koalas in New South Wales reported that while *B. obesa* was not restricted to animals with gastrointestinal disease it tended to be common in debilitated koalas, and cestode burdens appeared to increase during the summer period (Canfield 1987). General ill-thrift in captive koalas has been attributed to large numbers of tapeworms (Dickens 1978a).

Administration of praziquantel at 5 mg/kg PO seems to be effective in eliminating an infection. At least one prophylactic treatment of debilitated free-ranging koalas hospitalised for any reason seems advisable.

b *Nematodes*

Numerous nematodes 3–4 mm in length and identified as *Marsupostrongylus* spp. were observed in histological sections and fixed lung tissue from a free-ranging male koala in Victoria (McColl & Spratt 1982). These were subsequently considered to be *M. longilarvatus* (Spratt 1984). The koala was moribund when presented, but clinical examination and necropsy revealed only a suppurative arthritis in one stifle. Routine histological examination of tissue sections showed a mild parasitic interstitial pneumonia, with nematodes in bronchioles, respiratory bronchioles, alveolar ducts and alveoli. The most prominent inflammatory response, with infiltration of neutrophils, eosinophils and mononuclear cells, occurred around the bronchioles and alveoli and in the walls of a few arterioles, with occasional necrotic foci in the pulmonary parenchyma. No respiratory distress had been noted in the koala while alive and no gross lesions were observed in the lungs at necropsy, suggesting the condition had limited clinical significance.

During 1981–96, 246 koalas were examined at the Wollongbar Veterinary Laboratory in coastal New South Wales; nine were found harbouring *Durikainema phascolarcti* in pulmonary tissue (Spratt & Gill 1998). Eight of the nine cases described occurred in koalas from the Lismore district of New South Wales. Adult worms were detected in pulmonary arteries and arterioles, and in the hepatic portal vein in several animals. Larval parasites were found in renal arteries and glomerular capillaries, in veins and sinusoids in the liver and in blood vessels and free in tissues in lung, bladder, cervix, cerebellum and medulla oblongata. In all animals larval *D. phascolarcti* were present in pulmonary capillaries, arterioles and arteries. In one cerebellar fold, which had larvae present, there was a mild vascular proliferation and loss of Purkinje cells. While no other lesions were attributable directly to the presence of the parasite, individual transverse sections of pulmonary vessels contained as many as 135 sections of nematode occupying >95% of the vessel diameter. It was surmised that such levels of infection might result in vascular and respiratory compromise (Spratt & Gill 1998).

Three thin helminths approximately 7 cm long were an incidental finding in the peritoneal cavity of a free-ranging female koala in Victoria in 1991 (R Booth *pers. comm.*). These were subsequently identified as filaroid nematodes *Breinila* cf *mundayi* (D Spratt *pers. comm.*). Another single helminth was found inside a heart chamber of the same animal and was identified as a filaroid nematode *Johnstonema* spp. (D Spratt *pers. comm.*).

c *Protozoa*

Toxoplasmosis

There are few reports of toxoplasmosis in koalas and all refer to acute or peracute illness and death in captive animals (Hartley et al. 1990; Canfield et al. 1990a; Dubey et al. 1991). The source of infection in each case was unknown, but it has been suggested that koalas might have become infected by eating soil contaminated with *Toxoplasma gondii* oocysts excreted in domestic cat faeces. The spectrum of reported clinical signs included pyrexia (39.4°C), acute onset dyspnoea, leucocytosis with absolute lymphocytosis, radiographic changes suggestive of acute pneumonia and bronchitis, and sudden death.

Cryptosporidium

Cryptosporidium spp. has been mentioned as both a transient infection in one institution (Worley et al. 1993) and as a cause of four mortalities in koalas in a second captive institution (P Whitely & J Phelan, cited in Booth & Blanshard 1999).

Coccidia

A single elliptical unsporulated oocyte was an incidental finding on faecal flotation in a captive koala (Booth & Blanshard 1999).

d *Miscellaneous*

Larvae of the kangaroo bot fly *Tracheomyia macropi* were found attached to the tracheal mucosa during necropsy of a one koala in south-east Queensland (Brown 1979). A mild tracheitis was noted (Canfield et al. 1986a).

e *Ectoparasites*

For descriptions of disease caused by ectoparasites, refer to 10.1.3.

10.12 Non-infectious diseases

10.12.1 Trauma

Trauma has been identified as the most frequent cause of death in free-ranging koalas presented for necropsy during disease surveys in Victoria (19/55 or 35%; Victorian figures include animals presented dead or alive), Queensland (35/58 or 60%) and New South Wales (48/127 or 38%) (Obendorf 1983; Weigler et al. 1987; Canfield 1987). The commonest cause of injury was being hit by a motor vehicle, followed by dog attack, falling from trees (often as trees were being felled), intraspecific fighting, gunshot wounds and drowning. The surveys in Queensland and New South Wales reported that deaths caused by motor vehicles (and in Queensland, by dogs) were more common in the last 4–6 mo of the year and that male koalas were killed more often than females, especially during this period. The seasonal variation in deaths and bias towards one sex was assumed to be related to increased activity of males associated with the breeding period (searching for females, being displaced by other males, dispersing from their mother's home range) in spring and summer.

Although any injury can occur during a motor vehicle accident, some injuries can be expected to occur more commonly than others. This seems particularly true of injuries to the head and thorax in the koala. In a survey of 75 koala road kills over a 7 yr period, the main types of injuries were head injuries (59%), appendicular skeletal injuries (21%), haemothorax (20%) and spinal injuries (9%). Koalas surviving after motor vehicle impact could be expected to show similar types of injuries. Aspiration of blood had occurred in 27% (12/44) of the koalas with head injuries, and was regarded as a significant contributing factor in their deaths (Canfield 1991).

During intraspecific fighting, the majority of bites are directed at the pinnae and the lateral aspects of the elbows. Aural haematomas, and seromas or haematomas over the distal humerus, occasionally result from fighting. Abrasions, lacerations, punctures of the skin, shredding of the tips of the claws or partial or complete detachment of the claw may be noted after trauma.

Initial treatment should follow the same basic principles as in all other species.

- Treat for shock and stabilise the koala's condition if required.
- Provide appropriate analgesia. To avoid respiratory depression titrate doses based on anticipated severity of pain, e.g. for mild to moderate pain give buprenorphine HCl 0.01 mg/kg bid SC or IM or pethidine 1.0 mg/kg q 4–8 hr IV, IM or SC. For severe pain give methadone 0.25–0.5 mg/kg q 4–6 hr IM or fentanyl bolus 5.0 µg/kg IV then 3.0 µg/kg/hr IV as a continuous rate infusion e.g. intra-operatively (J Hanger *pers. comm.*).
- Perform a thorough physical examination, remembering to assess the degree of tooth wear.
- Provide systemic antibiotic cover if indicated by the nature or severity of the injuries, especially in the case of dog attack.
- Anaesthetise once stabilised, clean and debride wounds and establish drainage if necessary.
- Consider the effect of fractures or other injuries on the koala's short-term and long-term ability to climb and eat. Repair if the prognosis is considered good (see 14.1).

10.12.2 Thermal burns

Victims of bushfires or careless prescribed/controlled fires may present with thermal burns. The nature of the fire determines the likelihood of initial survival and the severity of injuries. When danger threatens, koalas tend to climb higher up the trees, thus many koalas perish during crown fires in dense bushland. If the flames are mainly confined to the understorey and ground vegetation koalas may survive the fire, if not its aftermath. The fur, highly insulative against wind, rain and solar radiation, can protect much of the body during a fire. Koalas may present with obviously singed fur and/or mild to severe burns to the epithelium of the palms and soles of the paws, genitals, nose and ears. The claws may be charred away. Systemic signs may include shock, pain, dehydration, hyperthermia and pulmonary damage from heat and smoke inhalation.

Systemic signs must be treated first and analgesia provided if indicated. Nebulisation with saline may be helpful in cases of pulmonary damage. Antibiotic or antiseptic creams should be applied liberally and frequently to burnt areas. If the hands are burnt, particular attention should be paid to the animal's willingness and ability to manipulate browse when feeding. Animals with severe burns to the palms or soles of the paws should not be given the opportunity to climb until epithelial damage has healed. Long-term care may be required if burns are severe. Charred claws may take more than 6 mo to regrow to a functional length for a free-ranging existence.

A realistic prognosis for humane rehabilitation should be based on an assessment of the surface area, severity and location of the burns. A wildlife care group has produced guidelines for assessment of burns cases (Fowler et al. unpub 2006). It suggests that euthanasia is indicated when burns cover >50% of total body surface area and/or there are severe burns affecting the face and/or genitals.

10.12.3 Toxins

Suspected eastern tiger snake (*Notechis scutatus*) envenomation has been reported in a koala. The animal was euthanased following 72 hr hospitalisation and treatment (Booth 1992). Naïve koalas with no previous exposure are reported to be susceptible to the toxin of the paralysis tick, *Ixodes holocyclus* (see 10.1.3).

10.12.4 Neoplasia

a Lymphoid neoplasia

Lymphoid neoplasia in the koala was first documented in 1961, when two cases of leukaemia were reported independently (Backhouse & Bolliger 1961; Heuschele & Hayes 1961). In a review of marsupial diseases, it was suggested that the presence of two cases of leukaemia among the low numbers of koala necropsies described in the literature at that time (about 50) indicated that the condition may have a relatively high prevalence (Arundel et al. 1977). Lymphoid neoplasia was subsequently recorded in 3.8% (13/344) of all koala necropsies conducted by veterinary pathologists at the Universities of Sydney and Queensland over a 5 yr period (1981–85) and in 7.0% (4/57) of koala necropsies at Yanchep National Park (WA) over an 18 mo period (Canfield et al. 1987a).

A disease study conducted over a 9 yr period in northern New South Wales showed that the overall prevalence of neoplasia was comparable in free-ranging (15/199) and captive (4/54) koala necropsies, being 7.5% and 7.4% respectively. When trauma was excluded as a cause of death, the relative incidence of neoplasia in the free-ranging group was 11.9% (15/126). Most of the koalas affected by neoplasia had lymphosarcoma with associated leukaemia (Canfield 1990). It has been suggested that up to 80% of mortalities in fauna parks in south-east Queensland may be attributed to haematopoietic neoplasia (Hanger et al. 2000). Koala retrovirus is believed to be causally linked to neoplastic disease and contributes to the high incidence of lymphoid neoplasia in koalas from south-east Queensland/northern New South Wales (see 10.10.1). Immunophenotyping was carried out in 51 cases of lymphoid neoplasia and a predominance of T-cell lymphoid neoplasms was demonstrated (Connolly et al. 1998).

Clinical signs

Most cases of lymphoid neoplasia occur in middle-aged koalas (3–7 yr) without sex predilection (Spencer & Canfield 1996; Connolly et al. 1998). A review of 31 cases of lymphoid neoplasia in New South Wales and Queensland koalas (Spencer & Canfield 1996) classified cases into four anatomic groups:

- multicentric lymphosarcoma—disseminated neoplasia with bilateral and symmetrical involvement of lymph nodes, with other organs commonly affected;
- alimentary lymphosarcoma—main lesions are in the gastrointestinal tract and regional lymph nodes;
- primary lymphoid leukaemia—malignant cells are present in blood and bone marrow;
- miscellaneous forms—may involve one or more sites.

Eighteen of 31 cases were classified as multicentric lymphosarcoma. Koalas with multicentric lymphosarcoma had enlarged moist homogenous superficial lymph nodes, however, only 11 animals had clinically palpable lymphadenomegaly. Other organs were commonly diffusely affected (due to secondary leukaemia), particularly the liver and spleen. A case of thymic lymphosarcoma has been reported (Canfield & Hemsley 1996).

Variation in the body condition of koalas with lymphosarcoma suggests that acute and chronic forms may occur. Clinical signs may be extremely variable depending on the organs affected (Fig. 8.19). Depression, inappetence, lethargy, lymphadenomegaly (Fig. 8.19a) (generalised or localised, unilateral or bilateral), solitary masses, dysphagia, panting, swelling of the face and neck (Fig. 8.19c, d), nodular conjunctival swelling with ocular discharge, pale mucus membranes, abdominal pain, diarrhoea, muscle weakness and posterior paralysis have been reported (Canfield et al. 1987a). Spontaneous haemorrhage around the roots of the incisor teeth, swelling of the hind limbs with abdominal enlargement and acute shifting lameness (Fig. 8.19b) have been observed in other cases (Blanshard 1994). The most frequently reported clinical signs in the 31 cases reviewed by Spencer and Canfield (1996) were weight loss/poor body condition (48% of cases), lethargy (32%) and lymphadenomegaly (35%).

Diagnosis

Generalised lymphoid neoplasia in the koala often has a leukaemic manifestation, with leukocytosis and anaemia being common (but not invariable) findings. WBC counts as high as 820 $\times$ 10^9/L have been reported in cases of leukaemia. Alternatively, the total WBC count may be normal but either lymphoblasts or a relative lymphocytosis may be present. Absolute neutrophil numbers may be increased, normal or decreased. Anaemia is generally non-regenerative (Spencer & Canfield 1996). Thrombocytopaenia may be noted.

Hypoalbuminaemia, with or without hypoproteinaemia, is the most frequently reported abnormality. Mildly to moderately elevated urea (6.9–28.5 mmol/L), elevated LDH (597–1887 U/L), elevated AST (171–860 U/L) and elevated GGT (20–24 U/L)

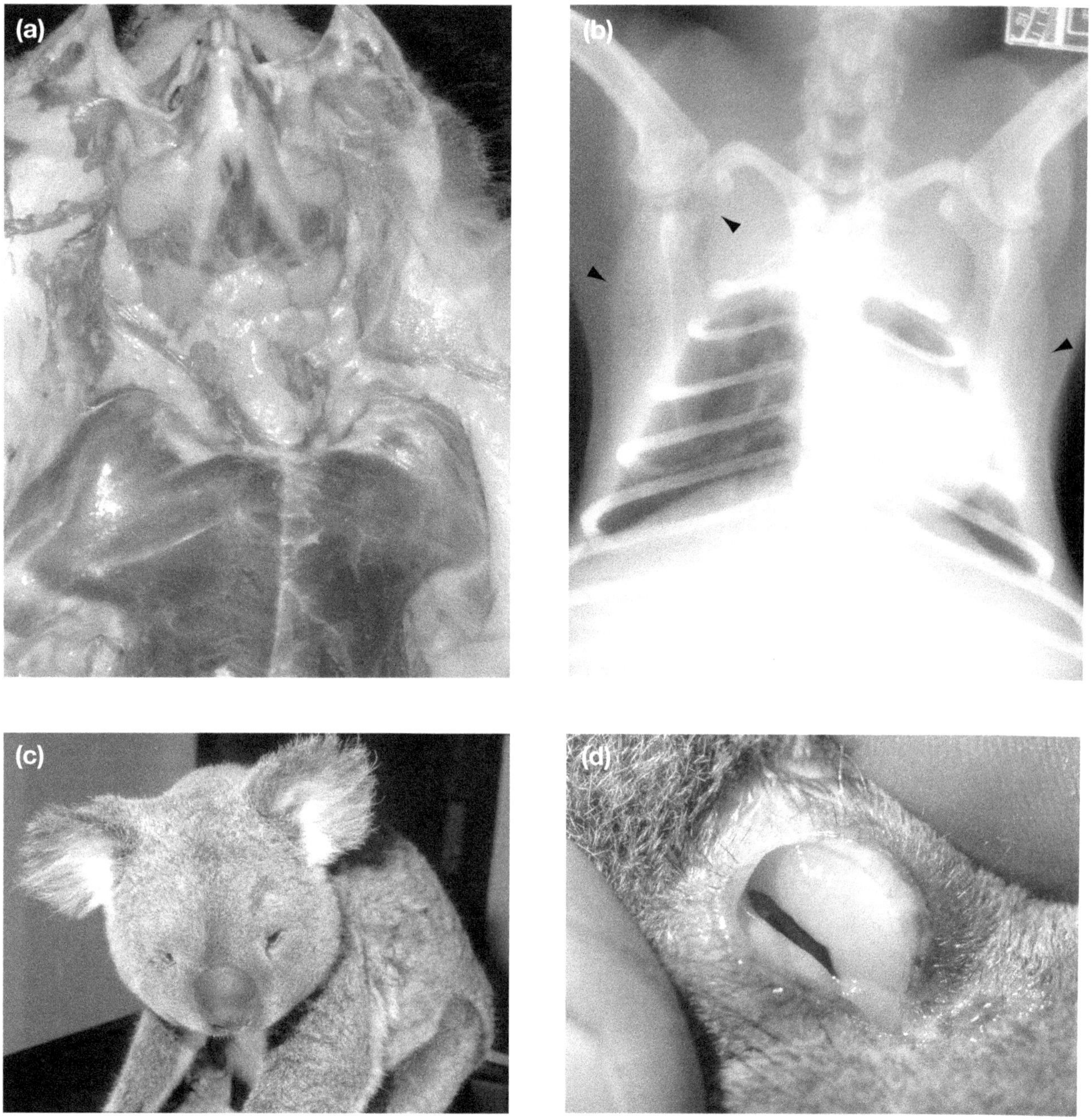

Figure 8.19 Three differing presentations of lymphoid neoplasia. a) Generalised lymphadenomegaly. b) Scapular bone necrosis (especially in the right subglenoid area), causing excruciating lameness of apparently sudden onset. c) Lymphoreticular infiltrate of facial soft tissues in a free-ranging male koala. d) Conjunctivae of the koala in c). Photos: W Blanshard.

have been observed in affected animals (Canfield et al. 1989b; Spencer & Canfield 1996). Elevations in GGT and AST are associated with liver disease, and increased blood urea and creatinine may be found in animals with renal infiltration. Prerenal azotaemia (presumably related to protein catabolism and reduced renal perfusion) is also reported (Spencer & Canfield 1996). Hypercalcaemia has not been routinely observed (see 8.1.1).

Smears made from fine needle aspirates of peripheral lymph nodes or solitary masses, or from fluid obtained by abdominal paracentesis, may reveal neoplastic cells. Radiographs of an acutely lame koala revealed multiple bilateral lucent areas in the scapulae, reminiscent of radiographic signs normally associated with multiple myeloma (Fig. 8.19b) (Blanshard 1994).

Treatment

Two instances of apparent remission following surgical removal of solitary cervical masses are known; one was an aged animal (Hanger et al. 2003) and the other a subadult female (G Rawlinson *pers. comm.*). Connolly (1999) reported that an animal with lymphoid leukaemia treated using a combination of cytarabine, prednisolone and L-asparaginase rapidly deteriorated and died.

The association between leukaemia/lymphoma and KoRV suggests that the use of immunosuppressive and cytotoxic chemotherapeutic regimens in cases of haematopoietic neoplasia in koalas is not indicated. High doses of prednisolone may result in transient amelioration of clinical signs (Hanger et al. 2003). It has been suggested that maternal lineages with a history of lymphoid neoplasia should be removed from breeding programs (Hanger et al. 2003).

Necropsy findings

Grossly at necropsy, affected lymph nodes may be enlarged, moist and homogeneous. Pale nodules or masses may be present in many different organs. Areas of necrosis may be obvious in affected bone. Histologically, affected lymph nodes show loss of normal architecture and diffuse infiltration of lymphoid cells (Canfield et al. 1987a; Canfield & Hemsley 1996). In other affected organs, lymphoid cells may be diffusely infiltrated or present in discrete scattered foci. Extensive lymphoid infiltration may be present in tissues and organs which appeared grossly normal during necropsy.

b Cranio-facial tumours of mixed cartilage and bone

In 1987 the gross and histological findings from cranio-facial tumours in two mature male koalas (one captive and one free-ranging) in New South Wales were described, together with a retrospective histological examination of two original (1981) Queensland cases. Typically, but not exclusively, the tumour involves bone of the nasal cavity and paranasal sinuses. In some cases there are palpable firm swellings or distortions of the contours of the face or hard palate. Grossly the periphery of a mass usually has a smooth glistening white nodular appearance (Fig. 8.20a). The cut surface may appear to be lobulated. Histologically the masses consist of scattered areas or irregular compartments of hypertrophying chondrocytes embedded in or surrounded by vascular and connective tissue. These areas show variable degrees of calcification and ossification (Canfield et al. 1987b).

Nasal or naso-ocular discharges and/or facial distortion are common but not consistent findings in advanced cases. The tumours seem expansive rather than invasive. Radiographic examination reveals a dense mass with a variably speckled appearance and reasonably distinct rounded outlines (Canfield et al. 1987b). In a captive 2 yr old female, the only clinical manifestation was intermittent mild epistaxis. A fine spray of fresh or dried blood was observed on the fur of the forearms and chest on three occasions at weekly intervals (Blanshard 1994). Surgical removal has been attempted, but the prognosis is generally poor (Canfield 1987b; Blanshard 1994).

c Serosal proliferations (mesothelioma)

Retrospective review of necropsy and histopathology findings in eight koalas (five free-ranging and three captive) suggested that a distinct serosal proliferative neoplasm exists in koalas. The lesions, previously described as nodular or granulomatous peritonitis, fibrosarcoma and myxofibrosarcoma, were reclassified as mesotheliomas (Canfield 1986; Canfield et al. 1990b). At necropsy all affected animals were emaciated, presumably as a result of prolonged illness. All but one koala had abdominal distension due to excessive viscous turbid red-tinged peritoneal fluid. One animal had SC oedema produced by cardiac dysfunction.

The serosal surfaces of affected body cavities bore white thickened areas, which in some cases extended up to 10 mm into underlying organs. In addition there were large numbers of smooth glistening white or red nodules, 1–10 mm in diameter, firmly attached to the serosa (Fig. 8.20b). The peritoneal cavity was affected in most cases (7/8). In two koalas, either the pelvic cavity or the pericardial sac was affected in addition to the peritoneal cavity. In the remaining case, pathology was confined to the right pleural cavity. When examined histologically, nodules consisted of collagen fibres, ground substance, plump spindle cells and sparsely distributed macrophages, lymphocytes and neutrophils covered by a single or multiple layers of uniform round mesothelial cells. Other nodules were highly vascular or contained areas of haemorrhage. The nodules were

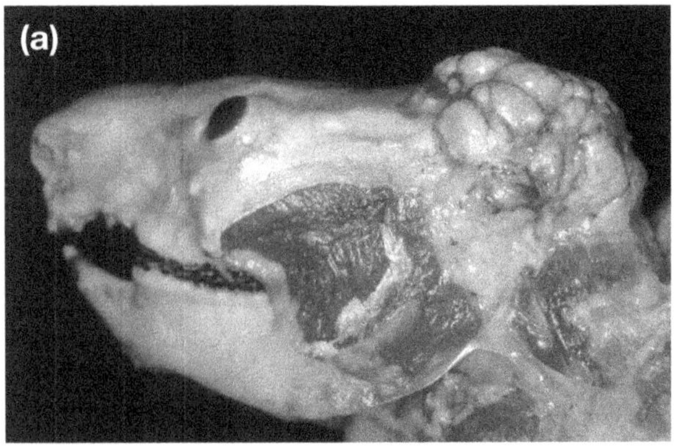

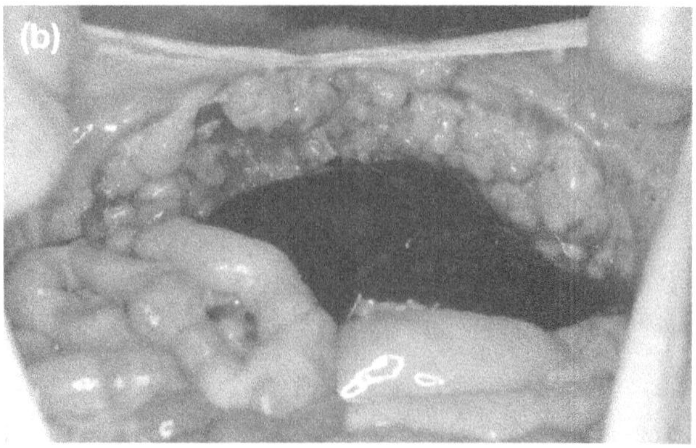

Figure 8.20 Other forms of neoplasia in the koala. a) Cranio-facial tumour of mixed cartilage and bone. b) Serosal proliferations (mesothelioma). Photos: W Blanshard.

fibrous with low cellularity. No evidence of an infective agent or irritant foreign material was identified in sections from the cases described (Canfield et al. 1990b).

d Other tumours

A variety of other tumours has been observed in koalas, including reports of:

- adenocarcinoma—an adenocarcinoma of the bile duct, with abdominal metastases (Canfield 1987b); adenocarcinoma of the genitourinary tract (C Sloan cited in Dickens 1978a); a hepatic adenocarcinoma (Canfield 1990);
- fibrosarcoma—peritoneal and pleural fibrosarcomas (AS Brown cited in Canfield 1990); a sternal fibrosarcoma (Canfield 1990); multiple oral fibrosarcomas (Barnes 1999);
- osteosarcoma—'osteo-sarcoma' was listed among diseases prevalent in koalas from Queensland (N Burnet cited in Pratt 1937); osteosarcomas of the long bones (Worley et al. 1993);
- haemangiosarcoma—splenic haemangiosarcoma (G Rawlinson *pers. comm.*);
- non-lymphoid leukaemia—a case of round cell leukaemia, with infiltration of organs including the myocardium and conjunctival involvement (Connolly 1999);
- leiomyoma—a suspect leiomyoma in the wall of the urinary bladder (Stalder et al. 2003); an intestinal leiomyoma (ARWH 2007 case number 3791/1); uterine fibroleiomyomas (see ARWH 2007 case number 3267/1, below);
- sebaceous adenoma—sebaceous adenoma of the pouch region (Reddacliff 1984); sebaceous adenoma adjacent the sternal gland (Antonsson & McMillan 2006);
- rhabdomyoma—one instance of rhabdomyoma (Connolly 1999);
- teratoma—a testicular teratoma (AS Brown cited in Canfield 1990);
- fibrovascular proliferation secondary to a transponder implant—a 5 yr old female koala was reported with exuberant fibrovascular proliferation affecting the skin and SC tissues surrounding a transponder implant (Vogelnest et al. 1997). There was evidence that the fibrovascular mass may have undergone neoplastic transformation to a myxofibroma;
- squamous cell carcinoma—a squamous cell carcinoma affecting the skin on the neck (ARWH 2007 case number 3089/2); squamous cell carcinoma on the nictitating membrane (D Blyde *pers. comm.*);
- pituitary chromophobe adenoma—an adult female koala was diagnosed with uterine fibroleiomyomas, cystic endometrial hyperplasia and ovarian granulosa cell proliferation following ovariohysterectomy. After the surgery, there was persistent behavioural oestrus with elevated blood oestrogen levels. The koala's condition deteriorated and it was euthanased. At necropsy, adrenal cortical hyperplasia and pituitary hyperplasia (considered most likely a chromophobe adenoma) were detected. It was proposed that stimulation of the adrenal gland by the atypical cells in the pituitary resulted in persistent hyperoestrogenism (ARWH 2007 case number 3267/1).

10.12.5 Nutritional issues affecting free-ranging koalas

The health of free-ranging koala populations strongly depends upon the quality of available habitat. When koalas are overabundant (as a result of habitat destruction/fragmentation or introduction into a novel site such as an off-shore island) or when drought conditions occur, defoliation of preferred food trees results in nutritional stress. At its most extreme, this may be reflected as starvation and death of animals, however, a number of physiological changes directly attributable to poor habitat quality have been reported. They include nutritionally induced anaemia, poor growth rate, reduced fertility of females, poor body condition, reduced body mass and low total serum protein and albumin levels (Martin 1985b; McLean 2003; M Lynch *pers. comm.*).

10.12.6 Congenital/developmental conditions

Various congenital abnormalities have been recorded or observed in koalas. These include atrial septal defect (see 10.4), testicular aplasia (10.6.3), hydrocephalus (10.7.1), scoliosis (10.7.2b), absence of one digit, intersex (K Nilsson and J Hanger *pers. comm.*), cystic pancreatic duct (Connolly 1999) and iris cysts (G Rawlinson *pers. comm.*).

11 ZOONOSES

There are a number of infectious agents which can affect humans, to which koalas may also be exposed. Despite this, in the 80 yr that koalas have been continuously maintained in captivity in Australia, including the past 50 yr in which systematic disease investigations in free-ranging and captive animals have increasingly occurred and dedicated wildlife hospitals for sick and injured koalas have been established, there have been no confirmed reports of transmission of disease agents between koalas and humans.

12 THERAPEUTICS

12.1 Fluid and electrolyte therapy

Many koalas do not seem to recognise drinking water, so never assume that a dehydrated koala will correct its own hydration status just because a bowl of water has been placed in its cage. Koalas that do not lap from a bowl may accept water or rehydration solutions orally if given slowly via a syringe. If the koala is eating, its fluid intake can be increased by spraying or hosing the eucalypt leaves with water before placing them in its cage. Branches should be placed upright with their cut ends in water, as soon as possible after harvesting from the tree.

Parenteral fluid therapy may be administered safely to koalas using small animal principles for rate and volume (Booth & Blanshard 1999; J Hanger *pers. comm.*). As in other species, care must be taken if serum protein levels are low, as exogenous parenterally administered fluids may not be redistributed normally within the body (see 8.1.1).

Laboratory evaluation enables specific water, electrolyte, glucose and protein deficits to be identified and therefore allows objective selection of the fluid type to be used. Low serum protein and PCV are often found in debilitated koalas and conspire to mask dehydration when interpreting laboratory results. Blood urea may be elevated in severe dehydration, and blood will be more difficult to collect from peripheral veins. Particular attention should be paid to glucose, protein and potassium levels, which may fall well below the normal values.

Koalas that have been eating little or nothing may run critically low in energy as well as water. In these cases, 4% glucose/0.18% sodium chloride or 5% glucose IV may be of more immediate benefit than balanced electrolyte solutions. In the longer term, balanced electrolyte solutions (e.g. Ringers solution or Hartmann's solution) are indicated, with particular attention to the koala's nutrient intake. Serum bicarbonate levels often seem to rise in koalas that are not eating. If frequent laboratory evaluation of electrolytes is not practicable, choose Ringers solution for fluid therapy. Amino-lite® 34X Injection (Boehringer Ingelheim) containing amino acids has been administered in IV or SC fluids, at dose rates comparable to those recommended for domestic animals.

If serum potassium levels remain below normal values despite administration of balanced electrolyte solutions containing potassium ions, potassium chloride has been given orally by dissolving effervescent tablets (Chlorvescent®; Aspen Pharmacare Australia) in water and administering fractional amounts (usually 1–2 mmol K^+) using a syringe. Frequent re-evaluation of serum potassium levels may be required.

12.2 Blood transfusion

A total of 10 mL/kg of blood drawn from the femoral vein of a female donor into acid citrate solution was administered through a filter to a male recipient. Major and minor cross-matching was performed to rule out gross incompatibility. No adverse responses were noted in either animal, and the transfusion was deemed clinically beneficial (Robinson 1978).

12.3 Drug therapy

12.3.1 Disclaimer

No pharmacokinetic studies have been published for any therapeutic drug given to koalas. With few exceptions (see below), drugs administered to koalas at dose rates extrapolated from those used in domestic animals and humans seem to be well-tolerated, and can help improve clinical condition when administered under appropriate circumstances. Despite their low metabolic rate, koalas have been shown to have a significantly faster rate of bromosulphthalein (BSP) clearance than macropods and sheep (Pass & Brown 1990). The BSP half-life in 10 koalas was 1.2 ± 0.09 min (range 0.8–1.5 min), significantly shorter than in marsupial and eutherian herbivores. Hepatic metabolism of therapeutic drugs may not be retarded, and metabolic scaling of dose rates is not practised. All drugs should be administered with consideration of their potential toxicity (anaphylactic or dose-dependent) and indirect effects in other species.

12.3.2 Adverse drug reactions

The use of two specific antibacterial agents in koalas deserves special caution. Wasting and death within 2 wk and 6 wk of administration of parenteral oxytetracycline and oral erythromycin, respectively, were reported after attempted treatment of chlamydiosis using these drugs (see 10.11.2). Koalas treated with orally administered erythromycin for 17 d became emaciated over 6 wk then died. No histologic lesions were noted at necropsy (Brown et al. 1984b). More promising results were reported when supplementary feeding with low-lactose milk paste was instituted concurrently with (modified) oxytetracycline treatment and continued for some time after cessation of treatment (Osawa & Carrick 1990). Caution in choosing to use erythromycin or oxytetracycline in koalas should probably be extended to other macrolides and tetracyclines which may share similar pharmacokinetic properties.

Nystatin may induce diarrhoea in a small proportion of adult koalas. NSAIDs are frequently used but may result in gastrointestinal ulceration in certain circumstances, e.g. dehydrated animals or those recently treated with corticosteroids (J Hanger *pers. comm.*).

12.3.3 Methods of administration

Liquid medications for oral administration may be given via syringe. Insert the tip of the syringe into the diastema between incisor and premolar teeth, and as the koala opens its mouth to eject the foreign object swing the syringe tip caudally so that the liquid is deposited at the back of the tongue. A few commercial palatable drug formulations (e.g. amoxycillin/clavulanate palatable drops) may be taken from the syringe voluntarily by some placid animals. Capsule contents or crushed tablets can be mixed thoroughly with a small amount of low-lactose milk powder, and made into a thick paste with water. Single doses may be drawn up into individual syringes and refrigerated, ready for administration on subsequent days.

Subcutaneous injections are usually given over the dorsolateral aspects of the thorax. The cranial aspect of the thigh is usually used for IM injections, as painful injections into the triceps muscles on the caudolateral aspect of the upper arm might interfere with feeding behaviour. Intravenous preparations are usually given into the cephalic vein, typically through an indwelling catheter or the administration port of a drip set (see 7.2.2).

Nebulisation can be achieved using a child-sized nebuliser mask, or by placing the animal in a nebulisation cage.

12.4 General approach to the sick koala

The nutritional status of free-ranging koalas can be adversely affected by factors such as advanced tooth wear, late lactation, drought, bushfire or clearing of large tracts of native vegetation for development. Otherwise healthy animals may therefore be presented in a state of starvation. It is well-known in other species that when injury, infection or neoplasia is combined with food deprivation, the result is an accelerated form of starvation (Lewis et al. 1987). A koala suffering from significant inanition, whatever the aetiology, may become severely dehydrated, hypoglycaemic, hypothermic and/or hypokalaemic. It then reaches a point of no return whereby any of these factors acting alone may render the animal incapable of consuming enough of its nutrient-poor eucalyptus diet to recover unaided. Koalas may still be capable of clumsy ambulation when their blood glucose level is undetectable in laboratory tests.

12.4.1 Emergency treatment

If presented with a koala which is profoundly depressed or in a state of collapse for no obvious reason:

- conduct a rapid physical examination, remembering to measure body temperature and examine the degree of tooth wear;
- clip both forearms and insert an IV catheter in one;
- collect blood for analysis but begin treatment while awaiting results;
- treat known or likely critical metabolic abnormalities (dehydration, hypoglycaemia, hypothermia) while awaiting test results. Give 5% glucose solution or 4% glucose/0.18% NaCl IV, employ a heatpad, hot water bottle or heat lamp and use warmed IV fluids if severe hypothermia is apparent.

Koalas in poor body condition should have their fluid and nutrient intake supplemented. To minimise any risk of aspiration (Booth 1991), oral administration of liquids or milk paste must be undertaken carefully in patients likely to have obtunded swallowing reflexes.

12.4.2 Suggested daily clinical assessment of a hospitalised koala

- Assess body condition (see 7.1.7) and hydration status (see 7.1.8) at least once daily;
- Assess food intake and gastrointestinal function in consultation with the primary carer:
  - amount/type of leaf consumed ± daily faecal pellet counts (see 7.1.13);
  - if the koala is eating little or no leaf, palpate under the left costal arch to assess gastric emptying (see 10.2.4);
- collect blood to assess blood glucose, total protein and electrolytes whenever indicated (see 7.2.1 and 8.1).

12.4.3 General principles of therapy

The single most important factor in successful treatment and rehabilitation of koalas is to minimise any limiting factors in their nutritional status.

No medical or surgical procedure will overcome the problems induced or compounded by inappropriate feeding and husbandry. The natural eucalypt diet adequately meets the water and energy needs of a healthy koala under normal conditions, but may provide insufficient resources to cope with increased metabolic requirements during illness or after injury. If food intake is reduced for any reason, both water and energy levels may become critically low.

There are several ways to directly or indirectly manipulate the environment of a hospitalised koala to help restore and maintain its water and energy balance (Blanshard 1994; Booth & Blanshard 1999).

Encourage positive water balance by:

- feeding younger leaves;
- ensuring all leaves are as fresh as possible (ensure branches are standing in water);
- spraying leaves with water before offering;
- encouraging oral intake (offering water from a bowl or via syringe);
- administering fluids SC or IV if indicated.

Encourage positive energy balance by:

- feeding younger leaves;
- observing individual preferences;
- positioning leaves within easy reach;
- hand feeding or supplementary feeding as required;
- administering fluids containing glucose (± amino acids) if indicated;
- establishing a routine and minimising aversive stimuli;
- maintaining the koala in a thermoneutral environment (15–25°C);
- identifying and treating specific illness or injuries;
- using anaesthesia judiciously.

12.5 Formulary

Drugs listed throughout the text and in Table 8.6 have been administered to varying numbers of captive and hospitalised free-ranging koalas, at the dose rates specified, without apparent adverse effects (sources indicated in parentheses). These dose rates are extrapolated from use in domestic animals or in humans.

Table 8.6 Drugs and doses used for treatment of koalas (see reference numbers at end of table and refer to appropriate sections of the text for additional detail)

Drug	Dose rate	Refs*	Indication	Comments
Anaesthetics				
Alfaxalone	3 mg/kg IM	11	Anaesthetic induction	Alfaxan®–CD RTU: alfaxalone 10 mg/mL (Jurox)
	1.5 mg/kg IV	11		
Atipamezole	2 mg per (Vic.) koala IV	15	Reversal agent	Specific antagonist for medetomidine
Diazepam	0.5–1.0 mg/kg IM	5	Sedative, anxiolytic	
	0.5 mg/kg IV	3		
Ketamine	10–15 mg/kg IM	26	Anaesthetic induction	Light anaesthesia with muscle hypertonicity
	5–15 mg/kg IM in combination with xylazine 5 mg/kg IM	4	Anaesthetic induction	Xylazine can be reversed using yohimbine 0.2 mg/kg
Medetomidine	55 μg/kg IM in combination with tiletamine–zolazepam 3.5 mg/kg	15	Anaesthetic induction	Medetomidine can be reversed using atipamezole
Propofol	2.5–3.0 mg/kg IV bolus, then titrate to effect	2	Anaesthetic induction	Rapid and smooth induction and recovery
	6–8 mg/kg IV, administered to effect	31		4–5 mg boluses can be given as required for maintenance of anaesthesia
Tiletamine/ zolazepam	5–10 mg/kg IM	4, 28	Anaesthetic induction	Zoletil® 100 injection: tiletamine 250 mg/mL plus zolazepam 25 mg/ mL (Virbac (Australia))
	2.5 mg/kg IV	4		
Xylazine	5 mg/kg in combination with ketamine 5–15 mg/ kg IM	4	Anaesthetic induction	Xylazine can be reversed using yohimbine
Yohimbine	0.2 mg/kg IV	4	Reversal agent	Antagonist for xylazine
Antibacterials				
Amoxycillin	10 mg/kg bid PO, SC, IM	3	Infection	
Amoxycillin/ clavulanic acid	12.5 mg/kg bid PO, SC, IM	3	Infection	Clavulox™ Palatable Drops or Injectable Suspension (Pfizer Australia). Some clinicians avoid giving PO.
Ampicillin	5–10 mg/kg tid–qid IV	3	Septicaemia	Use in combination (staggered) with gentamicin
Ceftazidime	15 mg/kg bid SC, IM	3	Infection	Particularly useful for *P. aeruginosa*
Chloramphenicol	30 mg/kg bid SC	3	Chlamydiosis	Minimum course 10–14 d for acute conjunctivitis, 14 d for acute cystitis
	60 mg/kg sid SC	11, 18	Chlamydiosis	Given for 45 d
Ciprofloxacin	10 mg/kg bid PO	6	Chlamydiosis	
Enrofloxacin	6 mg/kg sid–bid IV	11	Infection	Increased frequency of dosing used for serious infections. Can also be given in combination with metronidazole plus a cephalosporin
	5 mg/kg sid PO	8, 23	Chlamydiosis	
	10 mg/kg sid PO	9, 21, 22	Chlamydiosis	
Gentamicin sulphate	2–4 mg/kg IV, then 2–4 mg/kg bid–tid SC, IM	3	Serious infection	Monitor indicators of renal function during therapy

(Continued)

Table 8.6 (*Continued*)

Drug	Dose rate	Refs*	Indication	Comments
	2 mg/kg added to 2–3 mL saline, nebulised	3	Respiratory tract infection	Administered via a child-sized face mask using a nebuliser pump
Metronidazole	20–25 mg/kg bid given slowly IV	11	Anaerobic infection	For severe infection likely to include anaerobes
Procaine penicillin/ benzathine penicillin G	26.5 mg/kg (combined) q 72 h SC, IM	3	Infection	
Ticarcillin sodium (± clavulanic acid)	50 mg/kg q 4–6 hr IV, IM	3, 24, 27	Serious infection	Timentin® 3.1 g vial containing 3 g ticarcillin sodium plus 100 mg clavulanic acid (GlaxoSmithKline Australia)
Trimethoprim–sulphadiazine	30 mg/kg (combined) bid PO	3	Infection	
	17 mg/kg bid SC	20		
Trimethoprim–sulphamethoxazole	15 mg/kg (combined) bid PO	3	Infection	Septrin® Paediatric suspension: trimethoprim 40 mg/5 mL and sulphamethoxazole 200 mg/5 mL (Sigma Pharmaceuticals)
Chloramphenicol with hydrocortisone acetate	Topical ocular instillation	11, 18	Chlamydial keratoconjunctivitis	Chloroptsone® Eye Ointment: chloramphenicol 10 mg/g and hydrocortisone acetate 5 mg/g (Delvet)
Ofloxacin	Topical ocular instillation	21	Chlamydial keratoconjunctivitis	Ocuflox®: ofloxacin 3 mg/mL drops (Allergan Australia)
Oxytetracycline HCl, oleandomycin phosphate and neomycin sulphate	Topical ocular instillation	21, 33	Chlamydial keratoconjunctivitis	Mastalone Blue®: oxytetracycline HCl 18.5 mg/mL, oleandomycin phosphate 10 mg/mL and neomycin sulphate 10 mg/mL, with an inactive dye marker 'brilliant blue FCF' 25 mg/mL (Pfizer Animal Health)
Oxytetracyline with polymyxin B	Topical ocular instillation	9	Chlamydial keratoconjunctivitis	Terramycin® Ophthalmic Ointment: oxytetracyline 5 mg/g with polymyxin B 10 000 units/g (Pfizer Animal Health; imported from US)
Antifungals				
Amphotericin B	0.5 mg/kg in 300 mL 2.5% dextrose/0.45% NaCl given SC twice weekly	14, 17	Cryptococcosis	
Chloramine	0.3–0.6% solution applied once daily	3	Dermatophytosis	Halamid® contains chloramine 955 g/kg (Veterinary Companies of Australia)
Fluconazole	50–100 mg/koala bid PO	14, 16, 34	Nasopharyngeal cryptococcosis	Contents of Diflucan®: fluconazole 50 mg (Pfizer New Zealand) capsules can be mixed with low-lactose milk paste and given orally
	6 mg/kg (loading dose) then 3 mg/kg sid PO	3	Mycotic diarrhoea	

(*Continued*)

Table 8.6 (*Continued*)

Drug	Dose rate	Refs*	Indication	Comments
Griseofulvin	100 mg sid PO for 4 mo	25	Dermatophytosis	Prolonged treatment of paronychial infection causing deformation of the claws on one paw
Itraconazole	100 mg/koala sid PO	14, 16	Cryptococcosis	
	5 mg/kg sid PO	30	Dermatophytosis	Pulse therapy (1 wk on, 2 wk off) for 12 wk
Ketoconazole	10 mg/kg sid PO × 7 d	3	Antifungal	
Nystatin	5000–10 000 u/kg tid PO	3	Oral candidiasis	
	10 000–20 000 u/kg tid PO × 7 d	3	Mycotic diarrhoea	
Terbinafine	Topically on lesion sid	24	Dermatophytosis	Lamisil® Cream: terbinafine 10 mg/g (Novartis Pharmaceuticals Australia)
Anti-inflammatories/analgesics				
Buprenorphine hydrochloride	0.01 mg/kg bid SC, IM	3, 11	Mild to moderate pain	
Butorphanol tartrate	0.2 mg/kg IM	20		
Carprofen	4 mg/kg sid–bid PO for 24 hr then 2 mg/kg sid PO	11	NSAID	Some advise maximum 4 d duration of therapy to reduce risk of GIT ulceration; do not give NSAIDs with corticosteroids
Dexamethasone 0.1% ophthalmic suspension	1–2 drops into eye bid	11	Chlamydial keratoconjunctivitis	To reduce proliferative inflammation of conjunctivae; given concurrently with appropriate antibiotic therapy. Maxidex® 0.1%: dexamethasone ophthalmic suspension (Alcon Laboratories)
Fentanyl	5.0 µg/kg IV bolus, then 3.0 µg/kg/hr IV infusion	11	Severe pain	e.g. intra-operative and post-operative pain management
Flunixin meglumine	1 mg/kg sid SC, IM, IV	3	NSAID	Duration of therapy 1–3 d
Meloxicam	0.1–0.2 mg/kg sid PO	11, 21	NSAID	
Methadone	0.25–0.5 mg/kg q 4–6 hr SC, IM	11	Moderate to severe pain	Titrate dose according to response
Methyprednisolone sodium succinate	30 mg/kg given slowly IV	11	Spinal trauma	Give over 30 min; useful if given within 8 hr of trauma
Paracetamol (± codeine)	15 mg/kg bid–tid PO	11	Analgesic, antipyretic	Use as an alternative to carprofen for mild pain
Pethidine	1 mg/kg q 4–8 hr SC, IM, IV	3, 11	Mild to moderate pain	
Antiparasitics				
Amitraz	Thoroughly wet all fur using a 0.025% suspension	7, 29	Sarcoptic mange; nymphal tick infestation	Ectodex® EC Acaricidal Dog Wash: amitraz 50 g/L (Intervet Australia)
Fipronyl	Topical spray	10	Ticks	Frontline® Spray Treatment: fipronyl 0.29% (Merial Australia)
Ivermectin	600 µg/kg sid PO	32	Refractory demodecosis	Ivomec® Liquid for Sheep: ivermectin 0.8 g/L (Merial Australia)

(*Continued*)

Table 8.6 (*Continued*)

Drug	Dose rate	Refs*	Indication	Comments
Malathion	Bathe in 0.2% solution q 10 d × 3 times	1	Sarcoptic mange	
Praziquantel	5 mg/kg single dose PO	3	*Bertiella obesa*	
Miscellaneous				
Amino-lite® 34X	0.25 mL/kg q 8–24 hr IV, SC	3	Amino acid supplementation	Amino-lite® 34X injection: contains a number of amino acids (Boehringer Ingelheim)
Bromhexine hydrochloride	2 mg/kg bid PO × 5 d then 1 mg/kg bid PO	3	Mucolytic	
Cisapride	0.1–0.2 mg/kg tid PO	11	Prokinetic for delayed gastric emptying	Prepulsid®: 1 mg/mL suspension (Janssen–Cilag)
Frusemide	5 mg/kg IV, SC	3	Diuretic	
Hexamine (methenamine) hippurate	250 mg per (adult Qld) koala sid PO	21	Cystitis	Hiprex® Urinary Tract Antibacterial: hexamine hippurate 1 g/tablet (3M Pharmaceuticals); grind tablets and suspend in water
hCG	250 iu IM on Day 2 of oestrus	12	Induction of ovulation	
Mannitol	0.5–1 g/kg given slowly IV	3	Lowering intracranial pressure	Give over 15–20 min
Metoclopramide	0.5 mg/kg tid IV, IM, SC	3	Prokinetic for delayed gastric emptying	
Nandrolone laurate	1 mg/kg SC	3	Anabolic	
Oxytocin	1 iu/kg IM, SC	13, 19	To stimulate milk let-down	Used during research investigating milk composition
Potassium chloride	1–2 mmol K^+ sid–tid PO	3	Oral potassium supplement	Chlorvescent®: potassium chloride 595 mg plus potassium carbonate 152 mg plus potassium bicarbonate 384 mg in each tablet (Aspen Pharmacare Australia); provides potassium 14 mmol or 548 mg, and chloride 8 mmol or 283 mg
Psyllium	1–2 g/kg/d PO in divided doses	3	Sand impaction	Mix with low-lactose milk powder and administer promptly after the addition of water to make a paste

Sources: **1** Barker (1974). **2** Blanshard (1993). **3** Blanshard (1994). **4** Blyde (1990). **5** Booth (1991). **6** Booth & Blanshard (1999). **7** Brown et al. (1982). **8** Connolly (1999). **9** C Flanagan (*pers. comm.* 2007). **10** G Gipp (*pers. comm.* 2006). **11** J Hanger (*pers. comm.* 2006/07). **12** Johnston et al. (2003). **13** Krockenberger (1996). **14** Krockenberger et al. (2003). **15** Lynch & Martin (2003). **16** R Malik cited in Connolly (1999). **17** Malik et al. (1996b). **18** Markey et al. (2007). **19** Marshall et al. (1990). **20** Mathews et al. (1995). **21** A McKinnon (*pers. comm.* 2006). **22** V Nicolson (*pers. comm.* 2006). **23** E O'Connor cited in Booth & Blanshard (1999). **24** G Rawlinson (*pers. comm.* 2006). **25** Reddacliff (1986). **26** Robinson (1978). **27** Schultz et al. (1995). **28** Shima (1999). **29** Starr (1990a). **30** L Vogelnest (*pers. comm.* 2007). **31** Vogelnest (1999). **32** Vogelnest et al. (2000). **33** A Wood cited in Blanshard (1994). **34** Woods & Blyde (1997).

13 REPRODUCTIVE MANAGEMENT

13.1 Managed breeding

Three management systems may be used for breeding in captive koala colonies, and the advantages and drawbacks of each have been described (O'Callaghan & Blanshard 1992):

- random breeding—a number of males and females are housed together in a single enclosure;
- harem breeding—a single male is held in an enclosure with a group of females;
- selective breeding—the sexes are housed separately and a specific male is introduced to a female as she comes into oestrus.

13.2 Assisted reproduction

Injection of 250 IU of human chorionic gonadotrophin (hCG) on day 2 of oestrus followed by artificial insemination using fresh semen has resulted in pregnancy (Johnston et al. 2003). Koala semen has been collected by electroejaculation (Johnston et al. 1993) into an artificial vagina (AV), with use of the AV reported to produce higher-quality semen samples than those yielded by electroejaculation (Johnston et al. 1997; Johnston et al. 2003). Successful artificial insemination has been reported in six koalas, all using fresh semen collected via AV. Investigation into cryopreservation techniques for koala semen has determined that koala spermatozoa respond to short-term liquid storage in a similar way to those of eutherian mammals (Johnston et al. 2000c).

13.3 Contraception

13.3.1 Deslorelin

A small study investigating the efficacy of the gonadotrophin-releasing hormone (GnRH) agonist deslorelin on male and female koalas demonstrated that the deslorelin controlled-release implant, injected SC, resulted in approximately 12 mo contraceptive activity in four females but had no effect on the fertility of two males. Female koalas remained infertile for a period of at least 1 yr (n = 7) following administration of a single 5 mg dose (Herbert et al. 2001). In animals that could be monitored long-term, resumption of breeding occurred between 18 and 25 mo post-treatment (n = 3) (C Herbert *pers. comm.*).

13.3.2 Oestradiol and progestin implants

Short-duration exposure to high doses of oestradiol has been found to result in uterine, cervical and oviductal pathology in female koalas (Handasyde et al. 1990b), however, low-dose oestradiol implants (silastic implants releasing 2–3 pg/d of oestradiol–17β) were found to significantly reduce production of PY in 28 free-ranging koalas, without evidence of deleterious impacts on animal health (Middleton et al. 2003). An implant containing 70 mg levonorgestrel was trialled in 30 free-ranging koalas with similar results. Both implant types exhibited prolonged duration of action, with reduced production of PY for at least four breeding seasons following treatment. Reversibility of the implants has not been demonstrated.

13.3.3 Surgical sterilisation

Surgical sterilisation of male and female koalas has been performed as part of programs for management of overabundant populations (Menkhorst 2004; Duka & Masters 2005). Sterilisation of females is accomplished via laparoscopic transection and cautery of the oviducts, and of males by vasectomy (see 14.4.1).

14 SURGERY

In general, surgical management of disease/injury in koalas is performed as it would be in domestic small animal practice. However, the patient's ability to survive a major surgical intervention is dictated largely by the resources available for intra-operative and post-operative care (J Hanger *pers. comm.*).

If there is to be a successful outcome, consideration must be given to:

- careful monitoring of cardiopulmonary parameters throughout anaesthesia;
- provision of adequate analgesia and IV fluid support during and after surgery;
- appropriate husbandry and feeding (often including supplementary feeding) in the post-operative period.

14.1 Fracture repair

As in other species, internal fixation of fractures will facilitate rapid healing and an earlier return to normal mobility than external coaptation or cage rest. However, the success of surgical fracture repair in koalas depends as much on the ability of the hospital/carer to provide intensive post-operative care as on the availability of surgical expertise and equipment (J Hanger *pers. comm.*).

A small proportion of koalas coming into care already have old healed limb fractures, probably sustained as a result of falls, which may be incidental findings on radiographs of other areas. Closed fractures of long bones may heal with cage rest. Rotational mobility of the hip and tarsus or shoulder and carpus will help to compensate for imperfect alignment of fracture fragments, and rapid remodelling has been apparent radiographically in BY and juvenile animals. The koala should be provided with rough-barked poles to minimise involuntary use of the injured limb when climbing. Some animals start using their fractured limb almost immediately, and many can be released within 4–6 wk.

Koalas with fractures of the maxilla or mandible require careful assessment, as appropriate alignment of the cheek teeth is required for the molars to interdigitate and form effective cutting edges. Successful surgical repair of selected cases, with re-establishment of an adequate occlusal pattern, has been reported (J Hanger *pers. comm.*).

14.2 Laparotomy

Extreme care is required when opening the abdominal cavity through a ventral midline incision. There is no SC fat; the linea alba is a semi-translucent sheet 10–30 mm in width; the abdominal muscles are no more than 2–3 mm thick over the ventrolateral abdominal wall; the caecum and proximal colon lie immediately below the incision site.

Exploratory laparotomy has been described in detail by Mathews et al. (1995). A ventral midline incision that deviated laterally at the level of the pouch was made and the vascular mammary tissue was then undermined, allowing visualisation of the linea alba to the level of the pubis. In a pre-reproductive female, the ventral midline incision has been extended through the pouch, which was then reconstructed during closure (G Rawlinson *pers. comm.*).

14.3 Urinary bladder

Cystotomy for removal of cystic calculi or cystic masses has been performed using techniques similar to those described in domestic species (G Rawlinson *pers. comm.*).

14.4 Reproductive tract

Mathews et al. (1995) described exteriorisation of a polycystic mass arising from the cervices, hysterectomy, partial vaginectomy and bilateral ovariectomies in a 12 yr old female koala. The ovarian pedicles were short, and this factor plus the presence of the ovarian bursae meant that mobilisation of the ovaries and uteri was difficult. Ligation and transection of the oviducts greatly facilitated mobilisation of the uteri and exposure of the ovaries. The ureters were difficult to visualise within the urogenital strand, a mass of connective tissue that lies mesial to the lateral vaginae and caudal to the vaginal *cul de sac*. A transverse incision into the vaginal *cul de sac* enabled visualisation of the openings of the lateral vaginae, and transection of the vaginal tissue was carried out without damage to the ureters.

14.4.1 Surgical sterilisation

a Females

Sterilisation is performed via laparoscopic transection and cautery of the oviduct. Females are anaesthetised using isoflurane in oxygen and are positioned in dorsal recumbency on a tilted surgical table with the pelvis elevated above the head. Following routine surgical preparation (the clipped and prepared area extends from the epipubic bones to the sternum), the abdomen is insufflated with carbon dioxide such that the intra-abdominal pressure reaches 16–20 mm Hg. Three small (1–2 cm) incisions are made adjacent to the epipubic bones, allowing insertion of grasping forceps, a laparoscope and cautery scissors into the caudal abdomen. Each oviduct is visualised, and is manipulated using the forceps so that transection and cautery can be achieved without damage to blood vessels or other organs (G Johnsson *pers. comm.*).

b Males

A routine surgical vasectomy is performed via a single 10–15 mm incision on the cranial side of the pedicle of the scrotum (G Johnsson *pers. comm.*). Castration is rarely indicated and a technique similar to that described for other marsupials (see other chapters) would be appropriate.

14.5 Conjunctival ablation

Conjunctival ablation may be indicated (as an adjunct to topical and/or systemic antibiotic treatment) in selected chronic cases of ocular chlamydial disease, when proliferative conjunctivae obscure vision (see 10.11.2).

ACKNOWLEDGMENTS

A large number of our colleagues and associates have contributed directly or indirectly to making this chapter as comprehensive and (we hope) comprehensible as possible. We are sincerely grateful to them for sharing their experiences, observations, insights, opinions and understanding, and for facilitating access to reference material, unpublished data, archived clinical records and/or biological specimens. Special thanks to Rosie Booth, Jacqui Brumm, Cheyne Flanagan, Gail Gipp, Kath Handasyde, Jon Hanger, Mark Krockenberger, Michael Lynch, Roger Martin, Allan McKinnon, Vere Nicolson, Karen Nilsson, Paul O'Callaghan, Che

Phillips, Geoff Pye, Michael Pyne, Galit Rawlinson, David Schultz, Amy Shima, Peter Timms and Ray Wilkinson for their generosity, time, patience and expertise. Lone Pine Koala Sanctuary and Denzel the koala graciously facilitated our creation of Figure 8.3.

REFERENCES

Aldred J, Campbell J, Mitchell P, Davis G & Elliott J (1991). Involvement of wildlife in the natural cycles of Ross River and Barmah Forest viruses. In *Proceedings of the Wildlife Disease Association (Australasian Section) Annual Conference*, Mallacoota, pp. 25–26. WDA (Australasian Section).

Andriole VT (1999). Current and future antifungal therapy: new targets for antifungal agents. *Journal of Antimicrobial Chemotherapy* **44**, 151–162.

Antonsson A & McMillan NAJ (2006). Papillomavirus in healthy skin of Australian animals. *Journal of General Virology* **87**, 3195–3200.

ANZECC (1998). *National Koala Conservation Strategy*. Australian and New Zealand Environment and Conservation Council, Environment Australia: Canberra. <http://www.environment.gov.au/biodiversity/publications/koala-strategy.html>.

Arundel JH, Barker IK & Beveridge I (1977). Diseases of marsupials. In *The Biology of the Marsupials* (eds B Stonehouse & DP Gilmore), pp. 141–154. University Park Press: Baltimore.

Atwell RB & Booth R (1990). Cardiac failure in a koala secondary to an atrial septal defect. *Australian Veterinary Journal* **67**, 272–273.

Backhouse TC & Bolliger A (1960). Cryptococcosis in the koala (*Phascolarctos cinereus*). *Australian Journal of Science* **23**, 86–87.

Backhouse TC & Bolliger A (1961). Morbidity and mortality in the koala (*Phascolarctos cinereus*). *Australian Journal of Zoology* **9**, 24–37.

Barker IK (1974). *Sarcoptes scabiei* infestation of a koala (*Phascolarctos cinereus*), with probable human involvement. *Australian Veterinary Journal* **50**, 528.

Barnes J (1999). Koala health survey at Taronga Zoo. In *Proceedings of the Wildlife Disease Association (Australasian Section) Annual Conference*, Jervis Bay, pp. 20–22. WDA (Australasian Section).

Bird PS, Huynh SC, Davis D, Love DN, Blackall LL & Seymour GJ (2002). Oral disease in animals: the Australian perspective. Isolation and characterisation of black-pigmented bacteria from the oral cavity of marsupials. *Anaerobe* **8**, 79–87.

Blanshard WH (1990). Growth and development of the koala from birth to weaning. In *Biology of the Koala* (eds AK Lee, KA Handasyde & GD Sanson), pp. 193–202. Surrey Beatty & Sons: Sydney.

Blanshard W (1992). Aspects of the epidemiology and treatment of respiratory disease in captive koalas in Queensland. In *Proceedings of the Wildlife Disease Association (Australasian Section) Annual Conference*, Warrumbungle National Park, p. 25 (abstract). WDA (Australasian Section).

Blanshard W (1993). Emergency and critical care of koalas. In *Proceedings of the Australian Veterinary Association (Queensland Division) Annual Conference*, Brisbane. AVA (Qld Division): Indooroopilly, Qld.

Blanshard WH (1994). Medicine and husbandry of koalas. In *Wildlife, Proceedings 233*, pp. 547–623. Post Graduate Committee in Veterinary Science: Sydney.

Blyde D (1990). Immobilisation and anaesthesia in koalas. In *Control and Therapy Series, Mailing 157, No. 2945*. pp. 263–264. Post Graduate Committee in Veterinary Science: Sydney.

Bobek G & Deane EM (2002). Possible antimicrobial compounds from the pouch of the koala, *Phascolarctos cinereus*. *Letters in Peptide Science* **8**, 133–137.

Bodetti TJ, Johnston S, Pospischil A, Knox C & Timms P (2002). Screening semen from koalas (*Phascolarctos cinereus*) for *Chlamydia* species by PCR. *Veterinary Record* **151**, 147–149.

Bolliger A (1962). Gravel in the caecum of the koala (*Phascolarctos cinereus*). *Australian Journal of Science* **24**, 416–417.

Booth RJ (1991). Husbandry, handling, anaesthesia and common diseases of Australian native animals. In *Kendall Hall Seminars for Veterinarians, Recent Advance Series No. 57: Diseases of Exotic and Zoo Animals*, University of Melbourne, pp. 8–38.

Booth R (1992). Snake bite in a koala. *Wildlife Disease Association (Australasian Section), Newsletter* April 1992, 7.

Booth RJ & Blanshard WH (1989). Rhinitis in koalas. In *Proceedings of the Wildlife Disease Association (Australasian Section) Annual Conference*, Lake Eucumbene. WDA (Australasian Section).

Booth RJ & Blanshard WH (1999). Diseases of koalas. In *Zoo and Wild Animal Medicine: Current Therapy 4* (eds ME Fowler & RE Miller), pp. 321–332. WB Saunders: Philadelphia.

Booth RJ, Carrick FN & Addison PA (1990). The structure of the koala adrenal gland and the morphological changes associated with the stress of disease. In *Biology of the Koala* (eds AK Lee, KA Handasyde & GD Sanson), pp. 281–288. Surrey Beatty & Sons: Sydney.

Brown AS (1979). A disease survey of native fauna in south-east Queensland. BSc (Vet) thesis, University of Queensland, Australia.

Brown AS & Grice RG (1984). Isolation of *Chlamydia psittaci* from koalas (*Phascolarctos cinereus*). *Australian Veterinary Journal* **61**, 413.

Brown AS & Grice RG (1986). Experimental transmission of *Chlamydia psittaci* in the koala. In *Chlamydial Infections, Proceedings of the 6th International Symposium on Human Chlamydial Infections, 15–21 June 1986*. Sanderstead, Surrey (eds JD Oriel, GL Ridgway, J Schachter, D Taylor-Robinson & M Ward), pp. 349–352. Cambridge University Press: Cambridge.

Brown AS, Seawright AA & Wilkinson GT (1982). The use of amitraz in the control of an outbreak of sarcoptic mange in a colony of koalas. *Australian Veterinary Journal* **58**, 8–10.

Brown AS, Carrick FN, Gordon G & Reynolds K (1984a). The diagnosis and epidemiology of an infertility disease in the female koala *Phascolarctos cinereus* (Marsupialia). *Veterinary Radiology* **25**, 242–248.

Brown S, Wood T & Dickens R (1984b). Dirty tail and other chlamydial diseases in koalas. In *Control and Therapy Series, No. 1792.* Post-Graduate Committee in Veterinary Science: Sydney.

Brown AS, Girjes AA, Lavin MF, Timms P & Woolcock JB (1987). Chlamydial disease in koalas. *Australian Veterinary Journal* **64**, 345–350.

Bush M, Graves J, O'Brien SJ & De Wildt (1990). Dissociative anaesthesia in free-ranging male koalas and selected marsupials in captivity. *Australian Veterinary Journal* **67**, 449–451.

Butler R (1978). Patterns in koala mortality. In *The Koala. Proceedings of the Taronga Symposium on Koala Biology, Management and Medicine*, 11–12 March 1976, Sydney (ed. TJ Bergin), pp. 174–175. Zoological Parks Board of NSW: Sydney.

Canfield PJ (1986). Chronic peritonitis and pancreatitis in a koala, *Phascolarctos cinereus*. *Journal of Comparative Pathology* **96**, 469–472.

Canfield PJ (1987). A mortality survey of free range koalas from the north coast of New South Wales. *Australian Veterinary Journal* **64**, 325–328.

Canfield PJ (1989). A survey of urinary tract disease in New South Wales koalas. *Australian Veterinary Journal* **66**, 103–106.

Canfield P (1990). Disease studies on New South Wales koalas. In *Biology of the Koala* (eds AK Lee, KA Handasyde & GD Sanson), pp. 249–254. Surrey Beatty & Sons: Sydney.

Canfield PJ (1991). A survey of koala road kills in New South Wales. *Journal of Wildlife Diseases* **27**, 657–660.

Canfield PJ & Dickens RK (1982). Oxalate poisoning in a koala (*Phascolarctos cinereus*). *Australian Veterinary Journal* **59**, 121–122.

Canfield PJ & Hemsley S (1996). Thymic lymphosarcoma of T cell lineage in a koala (*Phascolarctos cinereus*). *Australian Veterinary Journal* **74**, 151–154.

Canfield PJ & Spencer AJ (1993a). Secondary degenerative arthropathy (osteoarthrosis) of the hips in ageing, free-living koalas. *Australian Veterinary Journal* **70**, 394–395.

Canfield PJ & Spencer AJ (1993b). Renal complications of cystitis in koalas. *Australian Veterinary Journal* **70**, 310–311.

Canfield PJ, Oxenford CJ, Love DN & Dickens RK (1983). Pyometra and pyovagina in koalas. *Australian Veterinary Journal* **60**, 337–338.

Canfield PJ, Brown AS & Dickens RK (1986a). Disease status of the koala (*Phascolarctos cinereus*). *Avian/Exotic Practice* **3**, 21–26.

Canfield PJ, Oxenford CJ & Lomas GR (1986b). A disease outbreak involving pneumonia in captive koalas. *Australian Veterinary Journal* **63**, 312–313.

Canfield PJ, Brown AS, Kelly WR & Sutton RH (1987a). Spontaneous lymphoid neoplasia in the koala (*Phascolarctos cinereus*). *Journal of Comparative Pathology* **97**, 171–178.

Canfield PJ, Perry R, Brown AS & McKenzie RA (1987b). Cranio-facial tumours of mixed cartilage and bone in koalas (*Phascolarctos cinereus*). *Australian Veterinary Journal* **64**, 20–22.

Canfield PJ, Sabine JM & Love DN (1988). Virus particles associated with leukaemia in a koala. *Australian Veterinary Journal* **65**, 327.

Canfield PJ, O'Neill ME & Smith EF (1989a). Haematological and biochemical reference values for the koala (*Phascolarctos cinereus*). *Australian Veterinary Journal* **66**, 324–326.

Canfield PM [sic], O'Neill ME & Smith EF (1989b). Haematological and biochemical investigations of diseased koalas (*Phascolarctos cinereus*). *Australian Veterinary Journal* **66**, 269–272.

Canfield PJ, Gee DR & Wigney DI (1989c). Urinalysis in captive koalas. *Australian Veterinary Journal* **66**, 376–377.

Canfield PJ, Hartley WJ & Dubey JP (1990a). Lesions of toxoplasmosis in Australian marsupials. *Journal of Comparative Pathology* **103**, 159–167.

Canfield PJ, Hartley WJ, Gill PA, Miller R, Obendorf DL & Brown AS (1990b). Serosal proliferations in koalas. *Australian Veterinary Journal* **67**, 342–343.

Canfield PJ, Love DN, Mearns G & Farram E (1991). Chlamydial infection in a colony of captive koalas. *Australian Veterinary Journal* **68**, 167–169.

Canfield P, Spencer A, Hartley W, Spielman D, Vogelnest L & Hulst F (1992). Disorders of keratinisation in a group of related, captive koalas (*Phascolarctos cinereus*), with a review of other skin conditions in koalas. *Journal of Zoo and Wildlife Medicine* **23**, 414–421.

Carlisle CH, Brown AS, Filippich LJ, Reynolds K & Reynolds WT (1989). Intravenous urography in the koala (*Phascolarctos cinereus*). *Veterinary Radiology* **30**, 34–40.

Clark P (2004). Haematological characteristics of Australian mammals. In *Haematology of Australian Mammals* (ed. P Clark), pp. 163–207. CSIRO Publishing: Melbourne.

Cleva GM, Stone GM & Dickens RK (1994). Seasonal changes in haematocrit in captive koalas (*Phascolarctos cinereus*). *Australian Journal of Zoology* **42**, 233–236.

Cockram FA (1978). Investigations into kerato-conjunctivitis of koalas. In *The Koala. Proceedings of the Taronga Symposium on Koala Biology, Management and Medicine*, 11–12 March 1976, Sydney (ed. TJ Bergin), pp. 177–182. Zoological Parks Board of NSW: Sydney.

Cockram FA & Jackson ARB (1974). Isolation of a chlamydia from cases of keratoconjunctivitis in koalas. *Australian Veterinary Journal* **50**, 82–83.

Cockram FA & Jackson ARB (1976). Chlamydial keratoconjunctivitis in koalas. *Australian Veterinary Practitioner* **6**, 36–38.

Cockram FA & Jackson ARB (1981). Keratoconjunctivitis of the koala, *Phascolarctos cinereus*, caused by *Chlamydia psittaci*. *Journal of Wildlife Diseases* **17**, 497–504.

Colvin J (2003). A survey of skull abnormalities in the Victorian koala, *Phascolarctos cinereus victor.* BAnSci thesis, University of Melbourne, Australia.

Connolly JH (1999). Emerging diseases of koalas and their medical management. In *Wildlife in Australia: Healthcare and Management, Proceedings 327*, pp. 13–40. Post Graduate Foundation in Veterinary Science: Sydney.

Connolly JH, Canfield PJ, Hemsley S & Spencer AJ (1998). Lymphoid neoplasia in the koala. *Australian Veterinary Journal* **76**, 819–825.

Connolly JH, Krockenberger MB, Malik R, Canfield PJ, Wigney DI & Muir DB (1999). Asymptomatic carriage of *Cryptococcus neoformans* in the nasal cavity of the koala (*Phascolarctos cinereus*). *Medical Mycology* **37**, 331–338.

Cork S (1987). Form and function in the koala. In *Koala: Australia's Endearing Marsupial* (ed. L Cronin), pp. 31–55. Reed Books: Sydney.

Cork SJ & Sanson GD (1990). Digestion and nutrition in the koala: a review. In *Biology of the Koala* (eds AK Lee, KA Handasyde & GD Sanson), pp. 129–144. Surrey Beatty & Sons: Sydney.

Cork SJ & Warner ACI (1983). The passage of digesta markers through the gut of a folivorous marsupial, the koala *Phascolarctos cinereus. Journal of Comparative Physiology B* **152**, 43–51.

Degabriele R (1980). The physiology of the koala. *Scientific American* **243**, 94–99.

Degabriele R & Dawson TJ (1979). Metabolism and heat balance in an arboreal marsupial, the koala (*Phascolarctos cinereus*). *Journal of Comparative Physiology B* **134**, 293–301.

Degabriele R, Harrop CJF & Dawson TJ (1978). Water metabolism in the koala (*Phascolarctos cinereus*). In *The Ecology of Arboreal Folivores* (ed. GG Montgomery), pp. 163–172. Smithsonian Institution Press: Washington DC.

DEH (2004). Conditions for the overseas transfer of koalas. Effective from May 2004. Department of Environment and Heritage, Commonwealth of Australia, Canberra., <http://www.environment.gov.au/biodiversity/trade-use/publications/export-conditions/koalas/index.html>.

Devereaux LN, Polkinghorne A, Meijer A & Timms P (2003). Molecular evidence for novel chlamydial infections in the koala (*Phascolarctos cinereus*). *Systematic and Applied Microbiology* **26**, 245–253.

Dickens RK (1975). The koala (*Phascolarctos cinereus*) past, present and future. *Australian Veterinary Journal* **51**, 459–463.

Dickens RK (1978a). The koala in health and disease. In *Fauna, Proceedings 36*, pp. 105–117. Post Graduate Committee in Veterinary Science: Sydney.

Dickens RK (1978b). A comparison of haematological values from koalas in different locations and the effect of stress on those values. In *The Koala. Proceedings of the Taronga Symposium on Koala Biology, Management and Medicine*, 11–12 March 1976, Sydney (ed. TJ Bergin), pp. 183–197. Zoological Parks Board of NSW: Sydney.

Domrow R (1958). A summary of the Atopomelinae (Acarina, Listrophoridae). *Proceedings of the Linnean Society of NSW* **83**, 40–54.

Drake B, Miller M & Morley NW (1990). Management of koalas in captivity. In *Biology of the Koala* (eds AK Lee, KA Handasyde & GD Sanson), pp. 323–329. Surrey Beatty & Sons: Sydney.

Dubey JP, Hedstrom O, Machado CR & Osborn KG (1991). Disseminated toxoplasmosis in a captive koala (*Phascolarctos cinereus*). *Journal of Zoo and Wildlife Medicine* **22**, 348–350.

Duka T & Masters P (2005). Confronting a tough issue: fertility control and translocation of over-abundant koalas on Kangaroo Island, South Australia. *Ecological Management and Restoration* **6**, 172–181.

Eberhard H (1972). Ecology of the koala, *Phascolarctos cinereus* (Goldfuss), on Flinders Chase, Kangaroo Island. PhD thesis, University of Adelaide, Australia.

Ellis DH & Pfeiffer TJ (1990). Ecology, life cycle, and infectious propagule of *Cryptococcus neoformans. Lancet* **336**, 923–925.

Emmins JJ (1996). The Victorian koala: genetic heterogeneity, immune responsiveness and epizootiology of chlamydiosis. PhD thesis, Monash University, Australia.

Emmins JJ & Turner SJ (1992). A new enzyme immunoassay for the diagnosis of chlamydiosis in the koala. In *Proceedings of the Wildlife Disease Association (Australasian Section) Annual Conference*, Warrumbungle National Park, pp. 54–56. WDA (Australasian Section).

Everett KDE (2000). *Chlamydia* and *Chlamydiales*: more than meets the eye. *Veterinary Microbiology* **75**, 109–126.

Everett KDE & Andersen AA (2001). Letter to the editor: radical changes to chlamydial taxonomy are not necessary just yet—reply. *International Journal of Systematic and Evolutionary Microbiology* **51**, 251–253.

Everett KDE, Bush RM & Andersen AA (1999). Emended description of the order *Chlamydiales*, proposal of *Parachlamydiaceae* fam. nov. and *Simkaniaceae* fam. nov., each containing one monotypic genus, revised taxonomy of the family *Chlamydiaceae*, including a new genus and five new species, and standards for the identification of organisms. *International Journal of Systematic Bacteriology* **49**, 415–440.

Fiebig U, Hartmann MG, Bannert N, Kurth R & Denner J (2006). Transspecies transmission of the endogenous koala retrovirus. *Journal of Virology* **80**, 5651–5654.

Fleay D (1937). Observations on the koala in captivity. Successful breeding in Melbourne Zoo. *Australian Zoologist* **9**, 68–80.

Fowler A, Wood C, Blankley M & Flanagan C (2006). Assessment and treatment of burnt wildlife. Unpublished notes for Victorian wildlife carers.

Fraser JA, Suburan RL, Nichols CB & Heitman J (2003). Recapitulation of the sexual cycle of the primary fungal pathogen *Cryptococcus neoformans* var. *gattii*: implications for an outbreak on Vancouver Island, Canada. *Eukaryotic Cell* **2**, 1036–1045.

Fukushi H & Hirai K (1993). *Chlamydia pecorum*: the fourth species of genus *Chlamydia*. *Microbiology and Immunology* **37**, 515–522.

Gen-Probe (2000). New directions in molecular diagnostic testing. Gen-Probe: San Diego, CA. <http://www.gen-probe.com/>.

Girjes AA, Hugall AF, Timms P & Lavin MF (1988). Two distinct forms of *Chlamydia psittaci* associated with disease and infertility in *Phascolarctos cinereus* (koala). *Infection and Immunity* **56**, 1897–1900.

Girjes AA, Ellis WAH, Carrick FN & Lavin MF (1993). Some aspects of the immune response of koalas (*Phascolarctos cinereus*) and in vitro neutralization of *Chlamydia psittaci* (koala strains). *FEMS Immunology and Medical Microbiology* **6**, 21–30.

Girjes AA, Carrick FN & Lavin MF (1994). Remarkable sequence relatedness in the DNA encoding the major outer membrane protein of *Chlamydia psittaci* (koala type I) and *Chlamydia pneumoniae*. *Gene* **138**, 139–142.

Glassick T, Giffard P & Timms P (1996). Outer membrane protein 2 gene sequences indicate that *Chlamydia pecorum* and *Chlamydia pneumoniae* cause infections in koalas. *Systematic and Applied Microbiology* **19**, 457–464.

Gordon G & Hrdina F (2005). Koala and possum populations in Queensland during the harvest period, 1906–1936. *Australian Zoologist* **33**, 69–99.

Grice RG & Brown AS (1985). A tissue culture procedure for the isolation of *Chlamydia psittaci* from koalas (*Phascolarctos cinereus*). *Australian Journal of Experimental Biology and Medical Science* **63**, 283–286.

Griffin DL, Canfield PJ, Collins GH & Dickens RK (1983). *Ctenocephalides felis felis* infestation of koalas. *Australian Veterinary Journal* **60**, 275.

Hajduk P, Copland MD & Schultz DA (1992). Effects of capture on haematological values and plasma cortisol levels of free-range koalas (*Phascolarctos cinereus*). *Journal of Wildlife Diseases* **28**, 502–506.

Halliday B (2000). Acarina. In *Australian Faunal Directory*. Australian Biological Resources Study, Canberra. <http://eriss.erin.gov.au/biodiversity/abrs/online-resources/fauna/afd/ACARINA/tree.html>.

Handasyde KA (1986). Factors affecting reproduction in the female koala (*Phascolarctos cinereus*). PhD thesis, Monash University, Australia.

Handasyde KA, Martin RW & Lee AK (1988). Field investigations into chlamydial disease and infertility in koalas in Victoria. In *Australian Wildlife, Proceedings 104*, pp. 505–515. Post Graduate Committee in Veterinary Science: Sydney.

Handasyde KA, McDonald IR, Than KA, Michaelides J & Martin RW (1990a). Reproductive hormones and reproduction in the koala. In *Biology of the Koala* (eds AK Lee, KA Handasyde & GD Sanson), pp. 203–210. Surrey Beatty & Sons: Sydney.

Handasyde KA, Obendorf DL, Martin RW & McDonald IR (1990b). The effect of exogenous oestrogen on fertility and the reproductive tract of the female koala. In *Biology of the Koala* (eds AK Lee, KA Handasyde & GD Sanson), pp. 243–248. Surrey Beatty & Sons: Sydney.

Hanger JJ & Heath TJ (1991). Topography of the major superficial lymph nodes and their efferent lymph pathways in the koala (*Phascolarctos cinereus*). *Journal of Anatomy* **177**, 67–73.

Hanger JJ & Heath TJ (1994). The arrangement of gut-associated lymphoid tissues and lymph pathways in the koala (*Phascolarctos cinereus*). *Journal of Anatomy* **185**, 129–134.

Hanger JJ, Bromham LD, McKee JJ, O'Brien TM & Robinson WF (2000). The nucleotide sequence of koala (*Phascolarctos cinereus*) retrovirus: a novel Type C endogenous retrovirus related to gibbon ape leukaemia virus. *Journal of Virology* **74**, 4264–4272.

Hanger J, McKee J, Tarlinton R & Yates A (2003). Cancer and haematological disease in koalas: a clinical and virological update. In *Proceedings of the 2003 Conference of the Australian Association of Veterinary Conservation Biologists*, Cairns (eds A Tribe & R Booth), pp. 19–30. AAVCB: Gatton, Qld.

Harrop CJF & Degabriele R (1976). Digestion and nitrogen metabolism in the koala, *Phascolarctos cinereus*. *Australian Journal of Zoology* **24**, 201–215.

Hartley WJ, Dubey JP & Spielman DS (1990). Fatal toxoplasmosis in koalas (*Phascolarctos cinereus*). *Journal of Parasitology* **76**, 271–272.

Haynes JI, Askew MJ & Leigh C (2004). Dietary aluminium and renal failure in the koala (*Phascolarctos cinereus*). *Histology and Histopathology* **19**, 777–784.

Hemsley S & Canfield PJ (1996). Proctitis associated with chlamydial infection in a koala. *Australian Veterinary Journal* **74**, 149–150.

Hemsley S & Canfield PJ (1997). Histopathological and immunohistochemical investigation of naturally occurring chlamydial conjunctivitis and urogenital inflammation in koalas (*Phascolarctos cinereus*). *Journal of Comparative Pathology* **116**, 273–290.

Hemsley S, Govendir M, Canfield PJ & Connolly J (1998). Diabetes mellitus in a koala (*Phascolarctos cinereus*). *Australian Veterinary Journal* **76**, 203–208.

Herbert CA, Webley LS, Trigg TE, Francis K, Lunney DH & Cooper DW (2001). Preliminary trials of the GnRH superagonist deslorelin as a safe, long-acting and reversible contraceptive for koalas. In *Conference on the Status of the Koala in 2001*, Canberra. Australian Koala Foundation: Brisbane. <https://www.savethekoala.com/conf2001pto.pdf>.

Heuschele WP & Hayes JR (1961). Acute leukaemia in a New South Wales koala (*Phascolarctos cinereus cinereus*). *Cancer Research* **21**, 1394–1395.

Higgins DP, Hemsley S & Canfield PJ (2005a). Association of uterine and salpingeal fibrosis with chlamydial hsp60 and hsp10 antigen-specific antibodies in *Chlamydia*-infected koalas. *Clinical and Diagnostic Laboratory Immunology* **12**, 632–639.

Higgins DP, Hemsley S & Canfield PJ (2005b). Immunohistochemical demonstration of the role of *Chlamydiaceae* in renal, uterine and salpingeal disease of the koala, and demonstration of *Chlamydiaceae* in novel sites. *Journal of Comparative Pathology* **133**, 164–174.

Hindell MA & Lee AK (1990). Tree preferences of the koala. In *Biology of the Koala* (eds AK Lee, KA Handasyde & GD Sanson), pp. 117–121. Surrey Beatty & Sons: Sydney.

Hirst LW, Brown AS, Kempster R & Winney N (1992a). Ophthalmologic examination of the normal eye of the koala. *Journal of Wildlife Diseases* **28**, 419–423.

Hirst LW, Brown AS, Kempster R, Hall J & Woolcock JB (1992b). Keratitis in free-ranging koalas (*Phascolarctos cinereus*) on Magnetic Island, Townsville. *Journal of Wildlife Diseases* **28**, 424–427.

Houlden BA, Costello BH, Sharkey D, Fowler EV, Melzer A, Ellis W, Carrick F, Baverstock PR & Elphinstone MS (1999). Phylogeneic differentiation in the mitochondrial control region in the koala, *Phascolarctos cinereus* (Goldfuss 1817). *Molecular Ecology* **8**, 999–1011.

Hume ID (1999). Hindgut fermenters: the arboreal folivores. In *Marsupial Nutrition* (ed. ID Hume), pp. 149–204. CUP: Cambridge.

Jackson M, Giffard P & Timms P (1997). Outer membrane protein A gene sequencing demonstrates the polyphyletic nature of koala *Chlamydia pecorum* isolates. *Systematic and Applied Microbiology* **20**, 187–200.

Jackson M, White N, Giffard P & Timms P (1999). Epizootiology of *Chlamydia* infections in two free-range koala populations. *Veterinary Microbiology* **65**, 255–264.

Jackson S, Reid K, Spittal [sic] D & Romer L (2003). Koalas. In *Australian Mammals: Biology and Captive Management* (ed. S Jackson), pp. 145–181. CSIRO Publishing: Melbourne.

Johnston SD, McGowan MR, Carrick FN & Tribe A (1993). Preliminary investigations into the feasibility of freezing koala (*Phascolarctos cinereus*) semen. *Australian Veterinary Journal* **70**, 424–425.

Johnston SD, O'Callaghan P, McGowan MR & Phillips NJ (1997). Characteristics of koala (*Phascolarctos cinereus adustus*) semen collected by artificial vagina. *Journal of Reproduction and Fertility* **109**, 319–323.

Johnston SD, McGowan MR, O'Callaghan P, Cox R & Nicolson V (2000a). Studies of the oestrous cycle, oestrus and pregnancy in the koala (*Phascolarctos cinereus*). *Journal of Reproduction and Fertility* **120**, 49–57.

Johnston SD, McGowan MR, O'Callaghan P, Cox R & Nicolson V (2000b). Natural and artificial methods for inducing the luteal phase in the koala (*Phascolarctos cinereus*). *Journal of Reproduction and Fertility* **120**, 59–64.

Johnston SD, McGowan MR, Phillips NJ & O'Callaghan P (2000c). Optimal physicochemical conditions for the manipulation and short-term preservation of koala (*Phascolarctos cinereus*) spermatozoa. *Journal of Reproduction and Fertility* **118**, 273–281.

Johnston SD, McGowan MR, O'Callaghan P, Cox R, Houlden B, Haig S & Taddeo G (2003). Birth of koalas *Phascolarctos cinereus* at Lone Pine Koala Sanctuary following artificial insemination. *International Zoo Yearbook* **38**, 160–172.

Johnston SD, O'Callaghan P, Nilsson K, Tzipori G & Curlewis JD (2004). Semen-induced luteal phase and identification of a LH surge in the koala (*Phascolarctos cinereus*). *Reproduction* **128**, 629–634.

Kaltenboeck B, Kousoulas KG & Storz J (1991). Detection and strain differentiation of *Chlamydia psittaci* mediated by a two-step polymerase chain reaction. *Journal of Clinical Microbiology* **29**, 1969–1975.

Koosmen A (1991). Rearing, management, release and relocating of orphaned, sick and injured koalas. In *Caring for Koalas: Today and Tomorrow. Proceedings of the Koala Carers Conference*, 14–15 June 1991, Armidale, NSW. Australian Koala Foundation: Brisbane.

Krockenberger AK (1996). Composition of the milk of the koala, *Phascolarctos cinereus*, an arboreal folivore. *Physiological Zoology* **69**, 701–718.

Krockenberger M & Canfield P (1999). Prostatic calculi in a free-ranging koala. In *Proceedings of the Wildlife Disease Association (Australasian Section) Annual Conference*, Jervis Bay, p. 38. WDA (Australasian Section).

Krockenberger MB, Canfield PJ, Kozel TR, Shinoda T, Ikeda R, Wigney DI, Martin P, Barnes K & Malik R (2001). An immunohistochemical method that differentiates *Cryptococcus neoformans* varieties and serotypes in formalin-fixed paraffin-embedded tissues. *Medical Mycology* **39**, 523–533.

Krockenberger MB, Canfield PJ & Malik R (2002a). *Cryptococcus neoformans* in the koala (*Phascolarctos cinereus*): colonization by *C. n.* var. *gattii* and investigation of environmental sources. *Medical Mycology* **40**, 263–272.

Krockenberger MB, Canfield PJ, Barnes J, Vogelnest L, Connolly J, Ley C & Malik R (2002b). *Cryptococcus neoformans* var. *gattii* in the koala (*Phascolarctos cinereus*): serological evidence for subclinical cryptococcosis. *Medical Mycology* **40**, 273–282.

Krockenberger MB, Canfield PJ & Malik R (2003). *Cryptococcus neoformans* var. *gattii* in the koala (*Phascolarctos cinereus*): a review of 43 cases of cryptococcosis. *Medical Mycology* **41**, 225–234.

Ladds PW, Thomas AD, Speare R & Brown AS (1990). Melioidosis in a koala. *Australian Veterinary Journal* **67**, 304–305.

Lanyon JM & Sanson GD (1986a). Koala (*Phascolarctos cinereus*) dentition and nutrition. I. Morphology and occlusion of cheek teeth. *Journal of Zoology, London (A)* **209**, 155–168.

Lanyon JM & Sanson GD (1986b). Koala (*Phascolarctos cinereus*) dentition and nutrition. II. Implications of tooth wear in nutrition. *Journal of Zoology, London (A)* **209**, 168–181.

Larkin JM (1989). Family Simonsiellaceae Steed 1962, 615[AL]. In *Bergey's Manual of Systematic Bacteriology* (ed. JT Staley), vol. 3, pp. 2107–2112. Williams & Wilkins: Baltimore.

Laurence DR & Bennett PN (1980). *Clinical Pharmacology* (5th edn). Churchill Livingstone: Edinburgh.

Lawler IR, Foley WJ, Eschler BM, Pass DM & Handasyde K (1998). Intraspecific variation in *Eucalyptus* secondary metabolites determines food intake by folivorous marsupials. *Oecologia* **116**, 160–169.

Lawson VJ and Carrick FN (1998). Morphology of the thyroid in coastal and noncoastal populations of the koala (*Phascolarctos cinereus*) in Queensland. *General and Comparative Endocrinology* **110**, 295–306.

Lawson VJ, Curtis AJ, Michelangeli VP, Stockigt JR, Carrick FN & Barlow JW (1996). Concentrations of thyroid hormones and iodothyronine binding proteins in the serum of the koala (*Phascolarctos cinereus*). *Australian Veterinary Journal* **74**, 55–58.

Lee A & Martin R (1988). *The Koala: A Natural History*. UNSW Press: Sydney.

Letcher J, Weisenberg E & Jonas A (1993). *Bordetella bronchiseptica* pneumonia in a koala. *Journal of American Veterinary Medical Association* **202**, 985–987.

Lewis LD, Morris ML & Hand MS (1987). Anorexia, inanition and critical care nutrition. In *Small Animal Clinical Nutrition* (3rd edn) (eds LD Lewis, ML Morris & MS Hand), pp. 5.1–5.43. Mark Morris Associates: Topeka, KS.

Ley C (1993). Cryptococcosis in the koala (*Phascolarctos cinereus*). Unpublished fifth year essay, Faculty of Veterinary Science, University of Sydney.

Lynch M & Martin R (2003). Capture of koalas (*Phascolarctos cinereus*) by remote injection of tiletamine-zolazepam (Zoletil®) and medetomidine. *Wildlife Research* **30**, 255–258.

MacBean CE, Taylor DM & Ashby K (2007). Animal and human bite injuries in Victoria, 1998–2004. *Medical Journal of Australia* **186**, 38–40.

MacKenzie WC (1918). *The Gastro-intestinal Tract in Monotremes and Marsupials*. Critchley Parker: Australia.

MacKenzie WC (1919). *The Comparative Anatomy of Australian Mammals. Part iv. The Genitourinary System in Monotremes and Marsupials*. Jenkin, Buxton & Co: Melbourne.

Malik R, McPetrie R, Wigney DI, Craig AJ & Love DN (1996a). A latex cryptococcal antigen test for diagnosis and monitoring of therapy for cryptococcosis. *Australian Veterinary Journal* **74**, 358–364.

Malik R, Craig AJ, Wigney DI, Martin P & Love DN (1996b). Combination chemotherapy of canine and feline cryptococcosis using subcutaneously administered amphotericin B. *Australian Veterinary Journal* **73**, 124–128.

Markey B, Wan C, Hanger J, Phillips C & Timms P (2007). Use of quantitative real-time PCR to monitor the shedding of chlamydiae in the koala (*Phascolarctos cinereus*). *Veterinary Microbiology* **120**, 334–342.

Marshall V, Carrick F, Doherty MD & Maclean DJ (1990). Aspects of the composition of koala milk. In *Biology of the Koala* (eds AK Lee, KA Handasyde & GD Sanson), pp. 229–241. Surrey Beatty & Sons: Sydney.

Martin RW (1981). Age-specific fertility in three populations of the koala *Phascolarctos cinereus* Goldfuss, in Victoria. *Australian Wildlife Research* **8**, 275–283.

Martin R (1985a). Overbrowsing and decline of a population of the koala, *Phascolarctos cinereus*, in Victoria. III. Population dynamics. *Australian Wildlife Research* **12**, 377–385.

Martin R (1985b). Overbrowsing and decline of a population of the koala, *Phascolarctos cinereus*, in Victoria. II. Population condition. *Australian Wildlife Research* **12**, 367–375.

Martin R & Handasyde K (1990). Population dynamics of the koala (*Phascolarctos cinereus*) in southeastern Australia. In *Biology of the Koala* (eds AK Lee, KA Handasyde & GD Sanson), pp. 75–84. Surrey Beatty & Sons: Sydney.

Martin RW & Handasyde KA (1995). Koala *Phascolarctos cinereus* (Goldfuss, 1817). In *The Mammals of Australia* (2nd edn) (ed. R Strachan), pp. 196–198. Australian Museum/Reed Books: Sydney.

Martin R & Handasyde K (1999). *The Koala: Natural History, Conservation and Management* (2nd edn). UNSW Press: Sydney.

Mathews KG, Wolff PL, Petrini KR, Rivers WJ, Johnston GR, O'Leary TP & Hayden DW (1995). Ultrasonographic diagnosis and surgical treatment of cystic reproductive disease in a female koala (*Phascolarctos cinereus*). *Journal of Zoo and Wildlife Medicine* **26**, 440–452.

McColl KA & Spratt DM (1982). Parasitic pneumonia in a koala (*Phascolarctos cinereus*) from Victoria, Australia. *Journal of Wildlife Diseases* **18**, 511–512.

McColl KA, Martin RW, Gleeson LJ, Handasyde KA & Lee AK (1984). Chlamydia infection and infertility in the female koala (*Phascolarctos cinereus*). *Veterinary Record* **115**, 655.

McDonald IR, Than KA, Handasyde KA & Michaelides J (1990). Factors affecting plasma adrenocortical hormone concentrations of koalas. In *Biology of the Koala* (eds AK Lee, KA Handasyde & GD Sanson), pp. 289–294. Surrey Beatty & Sons: Sydney.

McKenzie RA (1978). The caecum of the koala, *Phascolarctos cinereus*: light, scanning and transmission electron microscopic observations on its epithelium and flora. *Australian Journal of Zoology* **26**, 249–256.

McKenzie RA (1981). Observations on diseases of free-living and captive koalas (*Phascolarctos cinereus*). *Australian Veterinary Journal* **57**, 243–246.

McKenzie RA, Wood AD & Blackall PJ (1979). Pneumonia associated with *Bordetella bronchiseptica* in captive koalas. *Australian Veterinary Journal* **55**, 427–430.

McLean N (2003). Ecology and management of overabundant koala (*Phascolarctos cinereus*) populations. PhD thesis, University of Melbourne, Australia.
McOrist S, Jerrett IV, Anderson M & Hayman J (1985). Cutaneous and respiratory tract infection with *Mycobacterium ulcerans* in two koalas (*Phascolarctos cinereus*). *Journal of Wildlife Diseases* **21**, 171–173.
Melzer A, Carrick F, Menkhorst P, Lunney D & St John B (2000). Overview, critical assessment, and conservation implications of koala distribution and abundance. *Conservation Biology* **14**, 619–628.
Menkhorst P (2004). *Victoria's Koala Management Strategy.* Victorian Government Department of Sustainability and Environment: Melbourne.
Menkhorst P, Middleton D & Walters B (1998). Managing over-abundant koalas (*Phascolarctos cinereus*) in Victoria: a brief history and some potential new directions. In *Issues in Marsupial Conservation and Management: Managing Marsupial Abundance for Conservation Benefits* (ed. P Cowan), pp. 19–29. Co-operative Research Centre for Conservation and Management of Marsupials, Macquarie University: Sydney.
Messer M & Walker DM (1992). Milk substitutes for marsupials: in theory how good (or bad) are they? In *Urban Wildlife, Proceedings 204*, pp. 99–124. Post Graduate Committee in Veterinary Science: Sydney.
Meyer W, Castaneda A, Jackson S, Huynh M & Castaneda E (2003). Molecular typing of IberoAmerican *Cryptococcus neoformans* isolates. *Emerging Infectious Diseases* **9**, 189–195.
Middleton DR, Walters B, Menkhorst P & Wright P (2003). Fertility control in the koala (*Phascolarctos cinereus*): the impact of slow-release implants containing levonorgestrel or oestradiol on the production of pouch young. *Wildlife Research* **30**, 207–212.
Milner AR, Wilks CR, Spratt DM & Presidente PJA (1981). The prevalence of anti-leptospiral agglutinins in sera of wildlife in southeastern Australia. *Journal of Wildlife Diseases* **17**, 197–202.
Minchin AK (1937). Notes on the weaning of a young koala (*Phascolarctos cinereus*). *Records of the South Australian Museum* **6**, 1–3.
Mitchell P (1990a). The home ranges and social activity of koalas: a quantitative analysis. In *Biology of the Koala* (eds AK Lee, KA Handasyde & GD Sanson), pp. 171–187. Surrey Beatty & Sons: Sydney.
Mitchell P (1990b). Social behavior and communication of koalas. In *Biology of the Koala* (eds AK Lee, KA Handasyde & GD Sanson), pp. 151–170. Surrey Beatty & Sons: Sydney.
Mitchell P (1994). Hepatopathy in koalas. *Wildlife Disease Association (Australasian Section), Newsletter* April 1994.
Mitchell PJ, Jerrett IV & Slee KJ (1984). Skin ulcers caused by *Mycobacterium ulcerans* in koalas near Bairnsdale, Australia. *Pathology* **16**, 256–260.
Mitchell PJ, McOrist S & Bilney R (1987). Epidemiology of *Mycobacterium ulcerans* infection in koalas (*Phascolarctos cinereus*) on Raymond Island, southeastern Australia. *Journal of Wildlife Diseases* **23**, 386–390.
Moore BD & Foley W (2005). A review of feeding and diet selection in koalas (*Phascolarctos cinereus*). *Australian Journal of Zoology* **48**, 317–333.
Morrison R (2001). Can we straighten out Blinky Bill? *The Adelaidean, News from Adelaide University* **10**, 7.
Mutimer MD, Woolcock JB & Brown AS (1990). Aerobic bacterial flora of the conjunctivae of *Phascolarctos cinereus* (koala). *Australian Wildlife Research* **17**, 189–193.
Obendorf DL (1981). Pathology of the female reproductive tract in the koala, *Phascolarctos cinereus* (Goldfuss), from Victoria, Australia. *Journal of Wildlife Diseases* **17**, 587–592.
Obendorf DL (1983). Causes of mortality and morbidity of wild koalas, *Phascolarctos cinereus* (Goldfuss), in Victoria, Australia. *Journal of Wildlife Diseases* **19**, 123–131.
Obendorf DL (1988). The pathogenesis of urogenital tract disease in the koala. In *Australian Wildlife, Proceedings 104*, pp. 649–655. Post Graduate Committee in Veterinary Science: Sydney.
Obendorf DL & Handasyde KA (1990). Pathology of chlamydial infection in the reproductive tract of the female koala (*Phascolarctos cinereus*). In *Biology of the Koala* (eds AK Lee, KA Handasyde & GD Sanson), pp. 255–259. Surrey Beatty & Sons: Sydney.
O'Callaghan P & Blanshard W (1992). Breeding koalas in captivity. In *Proceedings of the Australasian Society of Zookeeping 1991 Annual Conference*, Dubbo, pp. 97–101. ASZK.
Old JM & Deane EM (2005). Antibodies to the Ross River virus in captive marsupials in urban areas of eastern New South Wales, Australia. *Journal of Wildlife Diseases* **41**, 611–614.
Osawa R (1992). Tannin-protein complex-degrading enterobacteria isolated from the alimentary tracts of koalas and a selective medium for their enumeration. *Applied and Environmental Microbiology* **58**, 1754–1759.
Osawa R (1993). Dietary preference of koalas *Phascolarctos cinereus* (Marsupialia: Phascolarctidae) for *Eucalyptus* spp. with a specific reference to their simple sugar content. *Australian Mammalogy* **16**, 85–87.
Osawa R & Carrick FN (1990). Use of a dietary supplement in koalas during systemic antibiotic treatment of chlamydial infection. *Australian Veterinary Journal* **67**, 305–307.
Osawa R, Blanshard WH & O'Callaghan PG (1992). Microflora of the pouch of the koala (*Phascolarctos cinereus*). *Journal of Wildlife Diseases* **28**, 276–280.
Osawa R, Bird PS, Harbrow DJ, Ogimoto K & Seymour DJ (1993a). Microbiological studies of the intestinal microflora of the koala, *Phascolarctos cinereus.* I. Colonisation of the caecal wall by tannin-protein-complex-degrading enterobacteria. *Australian Journal of Zoology* **41**, 599–609.

Osawa R, Blanshard WH & O'Callaghan PG (1993b). Microbiological studies of the intestinal microflora of the koala, *Phascolarctos cinereus*. II. Pap, a special maternal faeces consumed by juvenile koalas. *Australian Journal of Zoology* **41**, 611–620.

Osborne CA, Low DG & Finco DR (1972). *Canine and Feline Urology*. WB Saunders: Philadelphia.

O'Sullivan PJ (1949). In *Minutes of the Entomological Society of Queensland Special Meeting, 12 December 1949*, pp. 46–47.

Pahl LI & Hume ID (1990). Preferences for *Eucalyptus* species of the New England tablelands and initial development of an artificial diet for koalas. In *Biology of the Koala* (eds AK Lee, KA Handasyde & GD Sanson), pp. 123–128. Surrey Beatty & Sons: Sydney.

Pass MA & Brown AS (1990). Liver function in normal koalas and macropods. *Australian Veterinary Journal* **67**, 151–152.

Pratt A (1937). *The Call of the Koala*. Robertson & Mullens: Melbourne.

Rahman A (1957). The sensitivity of various bacteria to chemotherapeutic agents. *British Veterinary Journal* **113**, 175–188.

Reddacliff G (1984). Neoplasia in Australian marsupials. In *Clinical Oncology (Cancer: Diagnosis, Treatment and Prognosis), Proceedings 69*, pp. 137–143. Post Graduate Committee in Veterinary Science: Sydney.

Reddacliff G (1986). Griseofulvin treatment for fungal dermatitis in a koala. In *Koala Management. Proceedings of the Australian Koala Foundation Conference*, Currumbin, August. Australian Koala Foundation: Brisbane.

Rezakhani A, Atwell RB & Webster JD (1986). The electrocardiogram of the koala (*Phascolarctos cinereus*). *Australian Veterinary Journal* **63**, 264–266.

Rideout BA (2003). What does a polymerase chain reaction (PCR) result really mean? The dirty secrets of molecular diagnostics. In *Proceedings of American Association of Zoo Veterinarians*, Minneapolis, pp. 143–145. AAZV.

Robbins M & Russell E (1978). Observations on movements and feeding activity of the koala in a semi-natural situation. In *The Koala. Proceedings of the Taronga Symposium on Koala Biology, Management and Medicine*, 11–12 March 1976 (ed. TJ Bergin), pp. 29–41. Zoological Parks Board of NSW: Sydney.

Robinson PT (1978). Koala management and medicine at the San Diego Zoo. In *The Koala. Proceedings of the Taronga Symposium on Koala Biology, Management and Medicine*, 11–12 March 1976 (ed. TJ Bergin), pp.166–173. Zoological Parks Board of NSW: Sydney.

Robinson PT (1981). A review of thirty-three Queensland koala (*Phascolarctos cinereus*) immobilization procedures. *Journal of Zoo Animal Medicine* **12**, 121–123.

Rush CM & Timms P (1996). *In vitro* survival characteristics of koala chlamydiae. *Wildlife Research* **23**, 213–219.

Russell EG & Straube EF (1979). Streptobacillary pleuritis in a koala (*Phascolarctos cinereus*). *Journal of Wildlife Diseases* **15**, 391–394.

Rybicki E (2005). A manual of online molecular biology techniques. Department of Molecular and Cell Biology, University of Cape Town, Cape Town, South Africa. <http://www.mcb.uct.ac.za/manual/MolBiolManual.htm>.

Schmid KL, Schmid LM, Wildsoet CF & Pettigrew JD (1992). Retinal topography in the koala (*Phascolarctos cinereus*). *Brain, Behaviour and Evolution* **39**, 8–16.

Schultz DJ, Conaghty S & Spanner AM (1995). Azotaemia in a koala. In *Proceedings of Wildlife Disease Association (Australasian Section) Annual Conference*, Kangaroo Island, pp. 1–5. WDA (Australasian Section).

Seymour AM, Montgomery ME, Costello BH, Ihle S, Johnsson G, St John B, Taggart D & Houlden BA (2001). High effective inbreeding coefficients correlate with morphological abnormalities in populations of South Australian koalas (*Phascolarctos cinereus*). *Animal Conservation* **4**, 211–219.

Shima AL (1999). Sedation and anesthesia in marsupials. In *Zoo and Wild Animal Medicine: Current Therapy 4* (eds ME Fowler & RE Miller), pp. 333–336. WB Saunders: Philadelphia.

Shimizu T, Yasuda N, Kono I & Sakamoto H (1989). Diabetes mellitus in a koala (*Phascolarctos cinereus*). *Veterinary Pathology* **26**, 528–529.

Smith M (1979). Notes on reproduction and growth in the koala, *Phascolarctos cinereus* (Goldfuss). *Australian Wildlife Research* **6**, 5–12.

Smith M (1980). Behaviour of the koala, *Phascolarctos cinereus* (Goldfuss), in captivity. V. Sexual behaviour. *Australian Wildlife Research* **7**, 41–51.

Southwell IA (1978). Essential oil content of koala food trees. In *The Koala. Proceedings of the Taronga Symposium on Koala Biology, Management and Medicine*, 11–12 March 1976 (ed. TJ Bergin), pp. 62–74. Zoological Parks Board of NSW: Sydney.

Spencer A, Ley C, Canfield P, Martin P & Perry R (1993). Meningoencephalitis in a koala (*Phascolarctos cinereus*) due to *Cryptococcus neoformans* var. *gattii* infection. *Journal of Zoo and Wildlife Medicine* **24**, 519–522.

Spencer AJ & Canfield PJ (1993a). Renal failure in a koala. *Australian Veterinary Journal* **70**, 309–310.

Spencer AJ & Canfield PJ (1993b). Haematological characterisation of heavy tick infestations in the koala (*Phascolarctos cinereus*). *Comparative Haematology International* **3**, 225–229.

Spencer AJ & Canfield PJ (1994). Enhanced Heinz body formation, cell lysis and anaemia in a koala (*Phascolarctos cinereus*). *Comparative Haematology International* **4**, 114–117.

Spencer AJ & Canfield PJ (1995). Bone marrow examination in the koala (*Phascolarctos cinereus*). *Comparative Haematology International* **5**, 31–37.

Spencer AJ & Canfield PJ (1996). Lymphoid neoplasia in the koala (*Phascolarctos cinereus*): a review and classification of 31 cases. *Journal of Zoo and Wildlife Medicine* **27**, 303–314.

Spratt DM (1978). Note on the cestode *Bertiella obesa* in the koala. In *The Koala. Proceedings of the Taronga Symposium on Koala Biology, Management and Medicine*, 11–12 March 1976 (ed. TJ Bergin), p. 200. Zoological Parks Board of NSW: Sydney.

Spratt DM (1984). Further studies of lung parasites (Nematoda) from Australian marsupials. *Australian Journal of Zoology* **32**, 283–310.

Spratt DM & Gill PA (1998). *Durikainema phascolarcti* n. sp. (Nematoda: Muspiceoidea: Robertdollfusidae) from the pulmonary arteries of the koala *Phascolarctos cinereus* with associated pathological changes. *Systematic Parasitology* **39**, 101–106.

Stalder K, Allen G & Krockenberger M (2003). Further investigations into diseases of the koala (*Phascolarctos cinereus*). In *Proceedings of Wildlife Disease Association (Australasian Section) Annual Conference*, Healesville, p. 68. WDA (Australasian Section).

Starr J (1990a). Koalas. In *Care and Handling of Australian Native Mammals* (ed. SJ Hand), pp. 83–90. Surrey Beatty & Sons: Sydney.

Starr J (1990b). Management of koalas in an urban environment. In *Biology of the Koala* (eds AK Lee, KA Handasyde & GD Sanson), pp. 319–321. Surrey Beatty & Sons: Sydney.

Stone BF (1992). Ticks and wildlife: interactions, particularly of the Australian paralysis tick *Ixodes holocyclus* with koalas, bandicoots and fruit bats. In *Proceedings of Wildlife Disease Association (Australasian Section) Annual Conference*, Warrumbungle National Park, p. 17. WDA (Australasian Section).

Stone BF & Carrick FN (1996). Some effects of ticks on koalas, especially tick paralysis. In *Koalas: Research for Management. Proceedings of the Brisbane Koala Symposium*, 22–23 September 1990 (ed. G Gordon), pp. 65–69. World Koala Research: Corinda, Qld.

Storey C, Lusher M, Yates P & Richmond S (1993). Evidence for *Chlamydia pneumoniae* of non-human origin. *Journal of General Microbiology* **139**, 2621–2626.

Tarlinton R, Meers J, Hanger J & Young P (2005). Real-time reverse transcriptase PCR for the endogenous koala retrovirus reveals an association between plasma viral load and neoplastic disease in koalas. *Journal of General Virology* **86**, 783–787.

Tarlinton RE, Meers J & Young PR (2006). Retroviral invasion of the koala genome. *Nature* **442**, 79–81.

Tell LA, Nichols DK, Fleming WP & Bush M (1997). Cryptococcosis in tree shrews (*Tupaia tana* and *Tupaia minor*) and elephant shrews (*Macroscelides proboscides*). *Journal of Zoo and Wildlife Medicine* **28**, 175–181.

Temple-Smith PD & Taggart DA (1990). On the male generative organs of the koala (*Phascolarctos cinereus*): an update. In *Biology of the Koala* (eds AK Lee, KA Handasyde & GD Sanson), pp. 33–54. Surrey Beatty & Sons: Sydney.

Thompson VD (1987). Parturition and development in the Queensland koala *Phascolarctos cinereus adustus* at San Diego Zoo. *International Zoo Yearbook* **26**, 217–222.

Vogelnest L (1995). Use of transponder identification implants in koalas (*Phascolarctos cinereus*). *Australasian Regional Association of Zoological Parks and Aquaria, Newsletter* **22**, 4–5.

Vogelnest L (1999). Chemical restraint of Australian native fauna. In *Wildlife in Australia: Healthcare and Management, Proceedings 327*, pp. 149–187. Post Graduate Foundation in Veterinary Science: Sydney.

Vogelnest L, Canfield P, Hartley WJ & Reece RL (1997). Suspected transponder implant associated mesenchymal tumourigenesis in two feathertail gliders (*Acrobates pygmaeus*) and a koala (*Phascolarctos cinereus*). In *Proceedings of Wildlife Disease Association (Australasian Section) Annual Conference*, Flinders Island, pp. 1–5. WDA (Australasian Section).

Vogelnest LJ, Vogelnest L & Mueller RS (2000). An undescribed *Demodex* sp. and demodecosis in a captive koala (*Phascolarctos cinereus*). *Journal of Zoo and Wildlife Medicine* **31**, 100–106.

Wardrop S, Fowler A, O'Callaghan P, Giffard P & Timms P (1999). Characterization of the koala biovar of *Chlamydia pneumoniae* at four gene loci—*ompA* VD4, *ompB*, 16S rRNA, groESL spacer region. *Systematic and Applied Microbiology* **22**, 22–27.

Warneke RM (1978). The status of the koala in Victoria. In *The Koala. Proceedings of the Taronga Symposium on Koala Biology, Management and Medicine*, 11–12 March 1976 (ed. TJ Bergin), pp. 109–114. Zoological Parks Board of NSW: Sydney.

Weigler BJ, Booth RJ, Osawa R & Carrick FN (1987). Causes of morbidity and mortality in 75 free-ranging and captive koalas in south east Queensland, Australia. *Veterinary Record* **121**, 571–572.

Weigler BJ, Girjes AA, While NA, Kunst ND, Carrick FN & Lavin MF (1988). Aspects of the epidemiology of *Chlamydia psittaci* infection in a population of koalas (*Phascolarctos cinereus*) in southeastern Queensland, Australia. *Journal of Wildlife Diseases* **24**, 282–291.

Whittington R (2001). Chlamydiosis of koalas. In *Infectious Diseases of Wild Mammals* (3rd edn) (eds ES Williams & IK Barker), pp. 423–434. Iowa State University Press: Ames, Iowa.

Wigney DI, Gee DR & Canfield PJ (1989). Pyogranulomatous pneumonia due to *Nocardia asteroides* and *Staphylococcus epidermidis* in two koalas (*Phascolarctos cinereus*). *Journal of Wildlife Diseases* **25**, 592–596.

Wilkinson R (1989). Initial characterisation of the immune response of the koala (*Phascolarctos cinereus*). BSc (Hons) thesis, University of Adelaide, Australia.

Wilkinson R (1996). Characterisation of the immune responses of the koala. PhD thesis, University of Adelaide, Australia.

Wood AD (1978). The diseases of the captive koala. In *The Koala. Proceedings of the Taronga Symposium on Koala Biology, Management and Medicine*, 11–12 March 1976 (ed. TJ Bergin), pp. 158–173. Zoological Parks Board of NSW: Sydney.

Wood AD (1986). The management of captive koala colonies in Australia, USA and Japan. In *Koala Management. Proceedings of the Australian Koala Foundation Conference*, Currumbin, August, pp. 27–33. Australian Koala Foundation: Brisbane.
Wood AD (1987). The use of supplements in breeding captive koalas. In *Programme Abstracts, 3rd Koala Management Conference*, Currumbin, 9–10 May. Australian Koala Foundation/Currumbin Sanctuary & Koala Centre: Qld.
Wood MM & Timms P (1992). Comparison of nine antigen detection kits for diagnosis of urogenital infections due to *Chlamydia psittaci* in koalas. *Journal of Clinical Microbiology* **30**, 3200–3205.
Woods R & Blyde D (1997). Cryptic nasal disease revealed. *Veterinarian* November, 6.
Worley M, Rideout B, Shima A & Janssen D (1993). Opportunistic infections, cancer and hematologic disorders associated with retrovirus infection in the koala. In *Proceedings of American Association of Zoo Veterinarians*, St Louis, pp. 181–182. AAZV.
Young AH (1879). On the male generative organs of the koala (*Phascolarctos cinereus*). *Journal of Anatomy and Physiology* **13**, 305–317.

9. Wombats

Benn Bryant and Andrea Reiss

1 TAXONOMY, DISTRIBUTION AND BIOLOGY

Wombats are diprotodont marsupials in the Family Vombatidae, and are most closely related to the koala. They occur only on mainland and off-shore islands of Australia, including Tasmania. There are three species in two Genera. The common wombat (*Vombatus ursinus*) lives in temperate forests and forest edges of south-eastern Australia, where habitat is suitable for burrowing. Its density is limited by food availability. There are three subspecies, which are found on the mainland, Flinders Island and Tasmania (Strahan 1995). The southern hairy-nosed wombat (*Lasiorhinus latifrons*) inhabits semi-arid grasslands and is limited to areas of South Australia, western Victoria and far eastern Western Australia. When habitat and climate are suitable, it can be found in significant numbers. The northern hairy-nosed wombat (*Lasiorhinus krefftii*) also lives in semi-arid grasslands and is now restricted to Epping Forest National Park in central Queensland. It is critically endangered with only about 100 animals remaining (IUCN Species Survival Commission 2001). The range of all three species has decreased significantly since European settlement (Strahan 1995).

Common wombats and southern hairy-nosed wombats are often kept in zoos and wildlife parks. Wild and orphaned wombats of these two species are often presented for veterinary care. No northern hairy-nosed wombats are kept in captivity (Horsup 1998).

The three wombat species are broadly similar in appearance. The common wombat has bare skin on the nose and a coarse short coat. The ears are small and rounded. In contrast, hairy-nosed wombats have longer silky fur, soft furred muzzles and longer pointier ears. Northern hairy-nosed wombats are similar in appearance to the southern species, with a larger square muzzle (Strahan 1995). Adult wombats of all three species weigh about 20–35 kg. Males are larger than females except in the northern hairy-nosed species where females are larger in every dimension except neck girth (Johnson & Crossman 1991).

All three species of wombats were viewed as pests by European settlers and have been subject to persecution by land owners. Perceived problems include degradation of the environment through burrowing, damage to fencing and grazing competition with stock. In more recent times, burrows of southern hairy-nosed wombats have interfered with underground telecommunications cables.

Until the mid 20th century, common wombats were declared vermin with a bounty on their heads and in some cases there were attempts to eradicate populations. In most areas, wombats are now protected but land holders may obtain permits to cull if they are causing damage (Triggs 1996).

Other threats to wombats include land clearing, introduced and endemic diseases, motor vehicle trauma, inadvertent poisoning and predation by dogs and foxes. Changing pasture composition (from native perennial to annual and exotic plants) as a result of domestic stock or rabbit grazing may have a significant negative effect on wombat populations (Strahan 1995).

Wombats, in particular the hairy-nosed species, have a great ability to survive in a harsh drought-susceptible environment due to their energetically frugal lifestyle (Evans et al. 2003). They are dependent on their burrows and spend most of their time underground, emerging in darkness to graze. Burrows provide protection against predators, bushfire and, most importantly, extremes of temperature and humidity (Tyndale-Biscoe 2005). Common wombats build fairly simple burrows, whereas hairy-nosed wombats create extensive multi-entrance warren systems which may be occupied by several animals at the same time. All wombats are expert burrowers, using their fore limbs for digging and their incisors for cutting and removing obstacles. The hind limbs move soil away from the excavation site (Wells 1989).

Wombats occupy relatively small home ranges (up to 25 ha). Common wombats are solitary and defend their individual territories during contact with other wombats in areas where population densities are high. Hairy-nosed wombats may form loose social communities, sharing warren systems. Wombat territories may overlap extensively. Individuals stake their territories by scent-marking with dung and urine, scratching in dirt wallows and rubbing pheromones on posts, tree trunks and similar landmarks (Johnson 1998; Strahan 1995). Olfaction is the primary communication method; they also communicate through a variety of vocalisations including growls, grunts, huffs and squeals.

2 ANATOMY AND PHYSIOLOGY

Wombats are large, herbivorous, mainly nocturnal burrowing marsupials. They have stocky powerful bodies with short muscular limbs, well-suited for digging. They have a number of anatomic and physiological specialisations which allow them to thrive in their environmental niche.

2.1 Musculoskeletal system

Wombats have large flattened heads, thick skulls, solid jaws and a short neck. Both pectoral and pelvic girdles are strong. The short muscular legs and strong clawed feet are ideally suited for burrowing. There are five digits on both front and hind feet. The first digit of the hind foot is reduced to a small nub and the second and third toes are syndactylus. Wombats walk on the soles of their feet with a characteristic rolling gait. They walk, trot or bound and are capable of moving at high speed (40 km/hr) for short distances when necessary (Wells 1989). Wombats have a very short vestigial tail.

2.2 Skin

The skin of wombats is considerably thicker than that of other marsupials and quite inelastic. Wombats of all three species have a characteristic sacral plate on the rump which is almost as rigid as bone. The plate is used to help compact burrow walls and may also provide protection against predators.

2.3 Digestive system

Wombat dentition is specially adapted to cope with grazing high-fibre high-silica grasses. All teeth are rootless and continue to grow throughout life. Wombats have one pair of incisors, one pair of premolars and four pairs of molars in both upper and lower jaws. Wombats are diprotodont (one pair of lower incisors) but are unique in that they are the only diprotodont marsupial with one pair of upper incisors. The cheek teeth are hypsodont and the incisors have enamel over the labial surface only. The differential wear between the harder anterior enamelled surface and the softer dentine covering the posterior surface of the incisors results in the maintenance of a chisel-like biting edge (Murray 1998). All species of wombats have a split upper lip. This allows the wombat's incisor teeth to delicately pick individual blades of grass close to the ground. They use their molars and specialised jaw musculature to grind food to an extremely fine particle size, much finer than any other grazing marsupial (Tyndale-Biscoe 2005).

Wombats are hindgut fermenters with a simple gastrointestinal system, a simple stomach (with a cardiogastric gland), a relatively short small intestine, a very large sacculated colon and a very small caecum. Proteins and soluble carbohydrates are digested in the stomach and absorbed in the small intestine. The majority of energy is derived from bacterial fermentation in the colon. Wombats have a slow gut passage time (averaging 3 d but up to 8 d). This, along with the fine particulate size of ingesta, enables them to extract maximum energy and moisture from their generally poor-quality natural diet. Wombat faeces are among the driest of any animal (Hume 1999).

2.4 Reproductive system

2.4.1 Female

Female reproductive anatomy is similar to that of other marsupials. The ovaries occupy an ovarian bursa. There are paired lateral uteri each of which opens into a vaginal *cul de sac* through its own cervix. The paired lateral vaginae both open separately into the urogenital sinus at the level of the urethral opening and are separated along their length by a soft tissue median septum. During parturition the foetus tunnels through this septal tissue to form a median vagina. The urogenital sinus enters the common vestibule or cloaca at the level of the termination of the rectum (Tyndale-Biscoe & Renfree 1987).

There are two mammary glands located over the caudal ventral abdominal wall of the pouch, although wombats only raise one PY at a time (Wells 1989). The wombat pouch is rear-opening with a well-developed sphincter muscle. Pouch condition varies with reproductive status. In the juvenile animal the two teats are small and inverted and the pouch is little more than a shallow pocket held closed by the strong sphincter. Teats become everted at sexual maturity, at 2+ yr of age. As the wombat grows the pouch deepens and is often observed to be soiled, containing an exuded dry brown scale, dust and other debris (Wells 1989).

2.4.2 Male

The male has a pre-penile non pendulous scrotum, a relatively large carrot-shaped prostate and three pairs of bulbourethral glands. There are no seminal vesicles or ampullae (MacCallum 2004). The non-erect penis has a sigmoid flexure and occupies a preputial sac adjacent to the cloacal opening. The glans is bifurcate and, proximal to the bifurcation, there are a series of caudally directed spines arranged circumferentially around the penile shaft. The size of the testes, prostate and paired bulbourethral glands (located either side of the cloaca) increases in sexually mature males during the breeding season, most notably in the hairy-nosed wombats (Wells 1989).

2.5 Metabolism, water balance and ecological adaptations

All wombats, in particular the hairy-nosed species, have extremely low energy, protein and water requirements. They are able to maintain condition on a low-quality and often low-quantity herbivorous diet. Hairy-nosed wombats have a very low basal metabolic rate, only 64% of the average marsupial, or 44% of the average placental mammal. The metabolic rate of common wombats is close to that of the average marsupial (Hume 1999).

Hairy-nosed wombats receive sufficient moisture from their diet and do not need to drink. Wombats recycle urea, releasing it back into the hindgut where it aids bacterial proliferation. Because less urea is excreted in the urine, water turnover is further reduced (Tyndale-Biscoe 2005). The kidneys of the common wombat are proportionally much larger than those of the hairy-nosed wombat, however, the urine-concentrating ability of the hairy-nosed species appears to be far greater (McAllan et al. 1995). Under most conditions, wombats do not produce highly concentrated urine as they maintain water balance primarily through ecological adaptations and the ability to resorb moisture from the colon (Wells & Green 1998).

The low body temperature of wombats (34–35°C) also conserves energy and water. In the hairy-nosed wombat, body temperature falls below this level during burrow resting times (Strahan 1995). Plasma concentrations of thyroid hormones in wombats are the lowest recorded for any mammal (see Table 9.3) (Barboza et al. 1993).

All wombats are more tolerant of cold than heat. Wombats cannot sweat, but in warm environments they lie on their backs, shedding heat by exposing their thinly furred axillae and inguinal areas. Wombats may also salivate on their fore legs and chest to aid cooling (Wells 1989).

Wombats alter their behaviour patterns in response to environmental conditions. They spend much of their time resting or sleeping in their burrows, which maintain remarkably stable environmental conditions despite extreme fluctuations in temperature and humidity on the ground surface. The average daytime temperature in a southern hairy-nosed wombat burrow is 26°C in summer (compared to 38°C outside) and 14°C in winter (2°C outside). The thermoneutral range of the hairy-nosed species is 14–26°C (Tyndale-Biscoe 2005). High humidity in the burrow minimises water loss (Wells 1989).

On average wombats spend 2–6 hr a day grazing, but if conditions are unsuitable they may not emerge from the burrow at all (Tyndale-Biscoe 2005). In winter, they may emerge from burrows to graze in late afternoon, avoiding the falling temperatures of night.

In hot months they graze closer to dawn, seeking the coolest temperatures and maximising moisture intake through dewfall. Throughout the day, wombats will move to different areas of the burrow to micromanage their ambient temperature and humidity (Tyndale-Biscoe 2005).

Wombats, like most burrowing mammals, are physiologically tolerant of low oxygen and high carbon dioxide levels (Tyndale-Biscoe 2005). Because burrowing requires large amounts of energy and time, newly independent young wombats occupy burrows already constructed and vacated by adults (Johnson 1998).

Olfaction and hearing are the best developed and most important senses in wombats. Vision is better developed and more important for wombats than traditionally assumed (Strahan 1995).

3 REPRODUCTION

Study of reproductive behaviour and physiology in wild wombats has been hampered by the fact that they are nocturnal, fossorial and cryptic. Lack of breeding success in captive animals has limited the opportunity to study wombat reproduction in zoos.

Reproduction conforms to the marsupial pattern. There is a brief *in utero* phase, birth of altricial young and an extended and sophisticated lactational phase of which a significant proportion is spent in the pouch. Wombats have a relatively long intergenerational interval due to a long pouch life and period of association with the dam before independence. Under optimal conditions, female wombats raise only one offspring every 2 yr. In drought conditions, ovulation in the female southern hairy-nosed wombats and sperm production in the male may cease (Strahan 1995). Successful reproduction depends on good-quality forage near the burrow, to support both the lactating female and the weaned young (Tyndale-Biscoe 2005).

3.1 Reproductive physiology

Female wombats are monovular and polyoestrus. There are some notable differences in the reproductive strategy of common and hairy-nosed wombats. Southern hairy-nosed wombats have a defined breeding season with births occurring between November and January (Gaughwin & Wells 1978).

Free-ranging common wombats have seasonal peaks of breeding with variations, dependent on latitude and altitude, which coincide with optimum pasture availability (Strahan 1995). Births can occur at any time of year but are most frequent around spring (Moritz et al. 1998).

An early report based on urogenital cytology estimated oestrous cycle length in common wombats to average 33 d (Peters & Rose 1979). More recent studies have estimated the cycle to be significantly longer than this. One study based on behaviour and the measurement of plasma and faecal progestogens reported a mean oestrous cycle length of about 55±11 d (Paris et al. 2002a). Another study based on vaginal cytology and plasma progestogen analysis reported oestrous cycle length in the common wombat to range between 45 and 51 d (West et al. 2004). Ovulation is likely to be spontaneous in wombats.

There are no reliable estimates of the length of gestation in common wombats. Estimates of southern hairy-nosed wombat gestational length are based on observations of a mating and subsequent appearance of a newborn PY 20–22 d later in a single individual (Crowcroft & Sunderland 1997).

3.2 Breeding and life stages

One study reported that the female wombat pouch became dirty 10 d after oestrus and clean again 9–6 d prior to oestrus (Peters & Rose 1979). The periparturient pouch is also observed to be clean and moist (Triggs 1996).

Cyclic increases in activity, vocalisation and aggression have all been reported in female wombats thought to be in oestrus, as has swelling of the urogenital opening (Peters & Rose 1979). However, there have been no reports validating these indicators of oestrus by correlation with breeding behaviour, endocrine changes or visualisation of gonads. One study revealed that a male common wombat displayed increased flehmen behaviour in the period before the female's luteal phase, as defined by measured elevations in female faecal pregnanes, suggesting this male behaviour may be useful as an indirect indicator of oestrus (Paris et al. 2002).

In southern hairy-nosed wombats there may be an increase in aggressive behaviours during the breeding season. Males in particular may show evidence of bites, usually to the ears, rump and flanks. Vocalisations increase, including a cough-like vocalisation from the female when in oestrus (Wells 1978).

Courtship behaviour and mating has been observed in free-ranging common wombats. A report described chasing of the female by the male in circles and a figure-eight pattern. The female periodically slowed and allowed the male to catch up, whereupon he delivered a powerful bite to her hindquarters. She stopped in response to this and allowed the male to roll her on her side and mount. During copulation, which lasted approximately 3.5 min, the female moved from lateral to sternal recumbency. After mating the female broke from coitus and initiated another chase sequence during which several unsuccessful attempts were made by the male to bite her before she slowed again and the copulation/chase sequence was repeated. This occurred seven times over a 25 min period (Marks 1998). A report describing courtship and mating in a captive pair of common wombats divided the process into five phases:

- attraction—during which the female followed the male, made nose-to-nose contact when they passed and frequently stopped in front of and perpendicular to him presenting her cloaca by extending her hind limbs (3–4 min);
- interim—during which the animals called and responded with a series of grunts (2–3 min);
- chasing—during which the male chased the female attempting to bite her from behind (30 min);
- copulation—during which the male grasped the female around the hips, rolled onto his side and mated her (approximately 30 min);
- post-copulatory—during which the animals separated and lay in lateral recumbency (several minutes) (Boer 1998).

Courtship and mating have not been observed in free-ranging southern hairy-nosed wombats although captive mating events have been observed. In these animals copulation occurred in lateral recumbency and was preceded by the female's resistance to mating attempts and the male biting the female's hindquarters.

The life stages of the wombat are similar for all three species (Table 9.1). Early pouch life, where the young is altricial and pouch-dependent, continues up to 6 mo of age. After this time the PY's head begins to emerge occasionally from the pouch. At 7 mo the PY begins to exit the pouch, in the safety of the burrow. Around 9 mo the young emerges from the burrow to graze beside the mother. At this time, common wombat young may feed on a special soft faecal pellet produced by the dam. It is thought that this fulfils the same requirement as pap feeding in koalas, to inoculate the developing herbivorous intestine with the necessary bacterial flora (Tyndale-Biscoe 2005). Wombats are usually weaned at 12–15 mo and remain at foot until independence, around 18 mo (Tyndale-Biscoe 2005). Adult size is reached at 2–3 yr, depending on food availability and growth rates. Sexual maturity occurs at 2–4 yr of age, with males reaching maturity later than females (Wells 1989). The average life span of a wombat is 12–15 yr, although individuals have lived up to 23 yr in the wild and 30 yr in captivity (Jackson 2003).

4 HUSBANDRY

4.1 Captive housing

Wombats are strong powerful diggers and can chew through most materials. They are adept at climbing over barriers, swimming through water moats and crossing electric fences. Housing for wombats needs to be robust and escape-proof.

A solid walled den or stable can be used as a temporary holding area for a wombat. Wombats may damage themselves on mesh or chainlink fences and tend to pace and fret when there are no visual barriers. Floors should be solid (e.g. concrete) or have mesh buried under soil to prevent escape. Alternatively, fencing can be buried into the ground to a depth of 1 m. Wombats digging on concrete floors can damage their feet. Soil, sand, straw, leaf litter and wood mulch can be used as substrate. Damp floors may lead to foot problems. Wombats held in captivity for short periods do not need to have the opportunity to dig. When debilitated wombats first enter captivity, it is often advisable to limit their opportunity to dig as they can expend a large amount of energy on this behaviour. All wombats need a burrow substitute to provide both security and protection against temperature extremes. Suitable items include a sturdy wooden box, large diameter concrete pipe or large secure hollow log. Straw can be used as bedding in the shelter area.

Wombats held in long-term captivity should be provided with as much space as possible and the opportunity to dig. A minimum area of 45 m^2 has been recommended for two wombats, with an extra 9 m^2 for each additional wombat if housed in groups. Ideally, enclosures should be much larger than this to reduce

Table 9.1 Wombat PY developmental parameters

Age	**Common wombat**		**Southern hairy-nosed wombat**			**Wombaroo formula**	**Wombaroo volume per day (mL)** *	**Developmental land-marks;** ***Activity patterns***
	Weight (approx) g	**Body length (approx) mm**	**Weight (approx) g**	**Arm length mm**	**Leg length mm**			
90 d (3 mo)	150	150	130	44	40	<0.4[c]	30	Hairless, vibrissae developing, ears unfolded, lips partially open
120 d (4 mo)	390	170	290	50	50	0.4[c]	55 (45)	Beginnings of fine fur, eyes open, lips separating
150 d (5 mo)	800	188	530	55	60	0.4[d]	90 (60)	Fine fur; lower incisors erupting
180 d (6 mo)	1430	210	880	60	70	>0.6[d]	85 (65)	1 mm long fur, upper incisors erupting, *head emerging from pouch*
210 d (7 mo)	2250	230	1350	70	85	>0.6	120 (85)	Fur thicker, *emerging from pouch into burrow*
240 d (8 mo)	3350	250	2200	75	100	>0.6	160 (120)	Fully furred, *nibbling on grass, out of burrow occasionally*
270 d (9 mo)	4600	270	3250	80	110	>0.6	200 (160)[b]	*Leaves pouch permanently at 9–10 mo*
12–15 mo	12–19 kg	N/A	[e]	N/A	N/A	>0.6	[b]	*Permanently weaned at 12–15 mo*
15–18 mo	[a]	N/A	[e]	N/A	N/A	—	—	*Independent*

* First figure denotes volume for CW PY, second figure denotes volume for SHNW PY.
a Average growth (CW) is now 40 g per day.
b Milk volume depends on other foods consumed.
c Between estimated ages 100 d and 120 d milk should be gradually changed from <0.4 to 0.4 over a 7–10 d period.
d Between estimated ages 160 d and 180 d milk should be gradually changed from 0.4 to >0.6 over a 7–10 d period.
e Average growth (SHNW) is now 35 g per day

Source: Wombaroo Growth Charts, Triggs (1996), H George (*pers. comm.*), L Sauvarin (*pers. comm.*), A Reiss & B Bryant (*pers. observation*).

the likelihood of escape and stress behaviours (Jackson 2003). Care must be taken with the type of soil substrate provided and the use of heavy enclosure furnishings such as boulders and logs, as burrow collapse has resulted in the death of captive wombats. Enclosures are typically surrounded by solid smooth lined fencing (concrete, building blocks, glass panels or sheet metal) to a minimum height of 1.2 m (Jackson 2003). Water-filled moats and electric fences are not reliable barriers for wombats. Outdoor enclosures should be secure against predators, including cats, which may transmit toxoplasmosis (Booth 1999). Shade and shelter from rain and wind should be provided over part of the main enclosure. Wombats held long-term may be housed indoors; reversed daylight cycles may be used to allow visitors to observe the animals during their more active hours. Wombats are generally only active for a few hours each day and spend the remainder of the time frugally conserving energy. Behavioural enrichment for long-term captive wombats should include the provision of an adequate depth of soil for digging; browse, logs and branches (with bark attached) and native grasses (planted in enclosures or provided in clumps with roots and dirt attached) for animals to chew and dig up (Jackson 2003) along with a high point (such as a dirt mound or boulder) for surveying the territory and posts for scent-marking (Fig. 9.1).

Wombats are susceptible to heat stress and captive wombats need to be housed in an appropriate thermal environment. Common wombats may suffer heat stress at temperatures above 25°C whereas hairy-nosed wombats

Figure 9.1 A purpose-built captive wombat research and breeding facility. The area under the shelter is the burrow system.

tolerate temperatures up to 30°C (Tyndale-Biscoe 2005). Air-conditioning, misting sprinklers and climate-conscious exhibit design, including a well-insulated burrow system and a well-ventilated outdoor area, can be used to maintain appropriate environmental temperatures. Debilitated animals may need supplemental heat in the form of a heat lamp. When ambient temperatures are cooler, wombats like to bask in the sunshine.

Wombats in temporary care should be housed individually. In long-term captivity, hairy-nosed wombats may be housed in compatible groups. Common wombats are often housed individually in captivity, although hand-reared animals are frequently paired up as juveniles (same or opposite sex pairs) and may remain compatible for life. More details on short-term and long-term housing of wombats including complex captive burrow systems can be found in Booth (1999) and Jackson (2003).

4.2 Care in captivity

Although they appear to be stoic and robust animals, wild wombats often adapt poorly to the captive environment. Human interference and other disturbances (loud noises, domestic animals) should be kept at a minimum.

Wombats may not show obvious signs of stress. Indicators include loud vocalisations ('screams'); teeth gnashing; aggression (such as biting, scratching or charging without prior threat displays); pacing and escape behaviour. Escape attempts and pacing may not be readily observed and may only occur at night. Signs of pain in wombats include listlessness, teeth grinding and pawing at the abdomen. Chronic stress may manifest as bilateral alopecia or generalised immunosuppression with other disease processes developing (Spielman 1994).

Sick wombats should be disturbed as little as possible. Fresh food and water should be provided daily (although wombats are seldom seen to drink) and enclosure hygiene must be well-maintained. Health observations should be made at least daily, or more frequently if indicated. It may be necessary to encourage the wombat to move around in order to make a meaningful health assessment.

4.3 Transport

Transport boxes for wombats need to be strong and well-ventilated. Specially constructed wooden crates

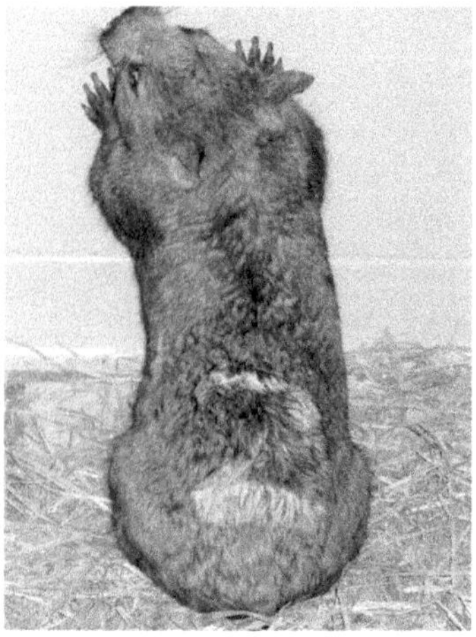

Figure 9.2 A southern hairy-nosed wombat marked with a number 2 using hair bleach. Photo: Taronga Zoo.

with mesh ventilation panels and a sliding door are recommended. A large dog transport crate (plastic and wire) may be suitable in an emergency for a debilitated wombat. Heavy-duty hessian or canvas sacks can be used to hold wombats for short periods only, as animals easily become overheated. Most wombats travel well and do not require sedation (Vogelnest 1999).

4.4 Individual marking and identification

The most common technique used for permanent individual identification of wombats is passive integrated transponders (PIT tags) or microchips. The standard site for insertion of microchips is SC on the dorsal midline between the scapulae. The needle hole should be closed with tissue glue immediately after insertion. Temporary identification methods such as stock mark paints, hair dyes and bleaches have been used. Such methods offer the ability to identify individuals from a distance and are a useful monitoring aid during introductions and field studies (Fig. 9.2).

5 NUTRITION

All wombats are strict herbivores. In the wild they principally graze on native grasses, but also eat roots, fungi, forbs and other bushes when green grass and surface water is not available (Wells 1989). They will also eat introduced grass species. Wombats actively create a 'lawn' of fresh green shoots by closely grazing the grasslands in the proximity of their burrow. This allows them to selectively consume high-nutrient high-moisture pastures, in an otherwise low-water and low-nutrient environment (Hume 1999). Generally their wild diet has a low nutrient value. The captive diet should mimic the natural diet as closely as practical.

Captive wombats should be fed high-roughage low-quality palatable hays and native grasses (preferably in clumps with soil attached). Some institutions feed specially formulated high-roughage (15%) low-protein (12.5%) pellets. A small amount of vegetables such as sweetcorn, sweet potato or carrot can be offered. In general, large amounts of vegetables and rich hays such as lucerne and muesli or rolled oats should be avoided as they may lead to bloat, other gastrointestinal diseases and obesity. Wombats should never be fed dog food. In the past, this was a common practice in some institutions. It is believed to be a contributing factor in the development of calcification disorders (Booth 1999) (see 10.6.1b). Wombats in long-term captivity need non-toxic branches or logs to chew on to encourage tooth wear. Water points and food stations are generally suspended (20 cm) off the ground. In temporary situations, heavy ceramic, concrete or metal dishes and bowls may be used. Wombats may destroy or ingest plastic bowls.

An example of a suitable daily ration for a captive wombat is:

- meadow hay and/or oaten hay (*ad lib*);
- fresh grasses, preferably native (*ad lib*);
- small amount carrot or sweet potato (200 g);
- fresh cut eucalyptus or wattle branch;
- dried maize (50 g) or fresh sweetcorn (100 g) or cut apple (100 g);
- lucerne hay (fed occasionally, small biscuit only).

Convalescing or debilitated wombats may benefit from a higher-energy diet, with greater quantities of foods such as sweet potato, sweetcorn, carrot, apple and the addition of higher-protein herbivore pellets and judicious amounts of bread. Alterations and additions to diets during convalescence are generally empirical, based on the food preferences and health status of the individual wombat. In all cases, diets should be altered gradually in order to minimise disturbances to gut flora and function. Favoured foods, including bread, may be used as a vehicle to administer oral medications.

Wild adult wombats, especially hairy-nosed wombats, often refuse to eat for days or weeks when first brought into captivity. Freshly cut native grasses may be the only food these animals will recognise or eat. Grated carrot, apple and sweet potato may also tempt reluctant feeders. Due to their low metabolic requirements, wombats (in particular hairy-nosed wombats) appear better able to tolerate extended periods of inappetence or anorexia than other marsupial species. It may be difficult to quantify the energy and fluid needs of the individual and to judge whether dietary intake is inadequate. Body weight fluctuations must be interpreted alongside gut-fill and hydration in inappetant animals, as wombats may dramatically decrease in body weight through decreased gut-fill alone.

On occasion, debilitated wombats have been force-fed (via stomach tube) under general anaesthesia or assist-fed (using a 60 mL syringe placed into the commissure of the mouth) under manual restraint. An energy-rich gruel of marsupial milk replacer powder, pureed vegetables and psyllium husk for additional fibre has been used. Adult anorexic wild-caught southern hairy-nosed wombats have been force-fed under anaesthesia, twice weekly, with 750 mL of the following pureed mix:

- 2 cups soaked kangaroo pellets;
- 200 g sweet potato;
- 2 apples;
- 2 carrots;
- 600 mL oral rehydration solution.

In the author's opinion, the usefulness of such intervention is debatable as the stress of force-feeding appears to negate the benefits of increased energy and fluid consumption. Diazepam (0.1 mg/kg SC, given daily 30 min before offering food) with or without the addition of vitamin B complex (1 mL/kg SC sid for 3 d) has also been used to stimulate appetite in debilitated and recently captive wombats (R Woods *pers. comm.*).

6 RESTRAINT

6.1 Capture and physical restraint

Wombats are extremely strong animals; their teeth and claws can cause serious injury. Some wombats (usually common wombats) may charge if cornered. All wombats, in particular hand-reared males, may become very aggressive, sometimes without apparent provocation.

Wombats may be physically restrained for very minor procedures or for induction of anaesthesia. Most clinical procedures on adult wombats (e.g. physical examination, blood collection) will require heavy sedation or general anaesthesia.

Wombats in burrows or burrow substitutes can be difficult to access and restrain. Wombats often present their rear end towards the handler as a natural defence mechanism. When restraint is attempted, they may attempt to crush the handler against the roof of the burrow with their rump.

With adequate experience and strength, all wombats, even large adults, can be picked up and held for a few minutes. This is done by approaching the wombat from behind and quickly grasping it just behind the front legs with both arms and lifting it off the ground in a bear hug. The wombat should be held firmly against the handler's chest (Fig. 9.3). Depending on the procedure, it is often more comfortable if the handler sits down once the animal has been restrained (Vogelnest 1999). Care should be taken when releasing the animal so that it does not have the opportunity to turn around and bite. Once restrained, the animal can then be placed in a sack or transport crate or injected by hand or an anaesthetic mask placed over its face (Fig. 9.5). Some animals can be run into a large open sack for restraint or transport. A blanket thrown over a sick or injured wombat may aid capture.

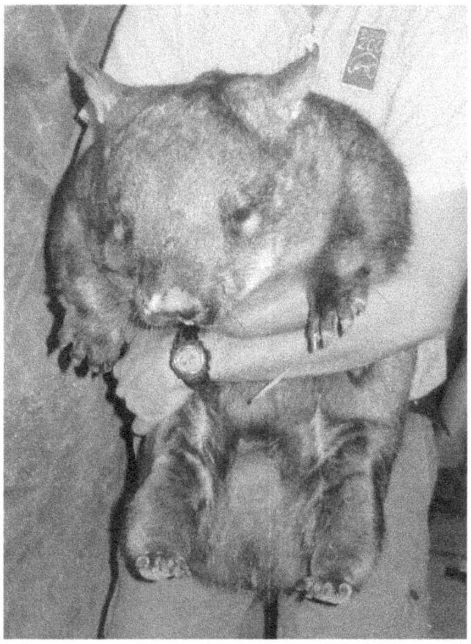

Figure 9.3 Physical restraint of an adult southern hairy-nosed wombat. Photo: Taronga Zoo.

Figure 9.4 Purpose-built slide-door trap used for northern hairy-nosed wombats. Photo: J Augustyn.

Wombats are innately trap-shy but have been successfully caught in purpose-built metal cage traps. These traps are usually of a walk-through design, positioned at burrow entrances (Fig. 9.4). Wombats attempting to exit burrows must traverse the trap and trip a wire which releases a slide, so capturing the animal (Horsup 1998).

6.2 Chemical restraint

In adult wombats sedation or anaesthesia is commonly achieved using injectable agents, given by hand, following brief physical restraint. IM injections can be given in the lateral thigh, shoulder or lumbar area. The hard dermal plate over the rump makes it impossible to give injections in this area. Wombats can be darted with injectable anaesthetic agents in the same sites.

6.2.1 Sedation/tranquillisation

Diazepam at 0.5–1.0 mg/kg IM provides adequate sedation for several hours (Booth 1999; Vogelnest 1999). Higher doses may be needed in agitated individuals. Midazolam at 0.5 mg/kg IM provides sedation of shorter duration than diazepam (Booth 1999).

The use of long-acting neuroleptic agents in wombats requires further evaluation and cannot be recommended at this time. Evidence suggests that wombats may be very sensitive to this class of drugs. Fluphenazine decanoate at 2 mg/kg IM used in a captive southern hairy-nosed wombat resulted in profound and prolonged (approximately 4 wk) tranquillisation with inappetence (Taronga Zoo records). A similar response was seen in a recently captured northern hairy-nosed wombat (Western Plains Zoo records).

6.2.2 Anaesthesia

Anaesthesia in wombats is generally uncomplicated. No pre-anaesthetic fasting is necessary in adult wombats. In milk-dependent juveniles, a minimum 1 hr fasting is recommended (Vogelnest 1999).

The injectable agent of choice in wombats is tiletamine/zolazepam. This can be given by IM injection and is an effective induction agent for all wombat species. Typically, a dose of 2 mg/kg IM provides heavy sedation, with sufficient relaxation to allow minor procedures or to facilitate induction of inhalation anaesthesia via a face mask. A dose of 3–5 mg/kg IM provides light anaesthesia in most captive wombats. Paddling and continued jaw movements may be seen when lower dose rates of tiletamine/zolazepam are used and supplementation with inhalation agents is generally required to achieve surgical anaesthesia. Very agitated wombats may require tiletamine/zolazepam doses as high as 9 mg/kg IM. Tiletamine/zolazepam doses up to 15 mg/kg IM have been used safely in wombats (Evans et al. 1998). Anaesthetic induction is rapid and smooth with good muscle relaxation at higher dose rates. Respiration and heart rate remain unchanged and palpebral, conjunctival and corneal reflexes are maintained (Evans et al. 1998).

The duration of tiletamine/zolazepam anaesthesia is dose-dependent. The more rapid recovery following low dose rates can be useful for field procedures. At higher dose rates recovery may be several hours; during this time animals are unco-ordinated and may paddle and roll. To avoid injury, recovering wombats should be placed in a secure quiet area such as a transport box or darkened den (Booth 1999).

Medetomidine at 0.125 mg/kg plus ketamine at 2 mg/kg IM has been used in common wombats (Woods 1999). Reliability and effect are less optimal than with tiletamine/zolazepam. This combination may be useful when reversibility of anaesthesia is a priority. Alfaxalone at 3-5 mg/kg IM has been used successfully in wombats. The level of anaesthesia achieved is dose-dependent (J Hanger *pers. comm.*).

Inhalation anaesthesia using isoflurane in oxygen is widely used in wombats. Isoflurane may be used as the primary induction agent (via face mask) for smaller, tractable or sedated animals and is commonly used to supplement tiletamine/zolazepam anaesthesia in all age classes (Fig. 9.5). Halothane has also been safely used in wombats.

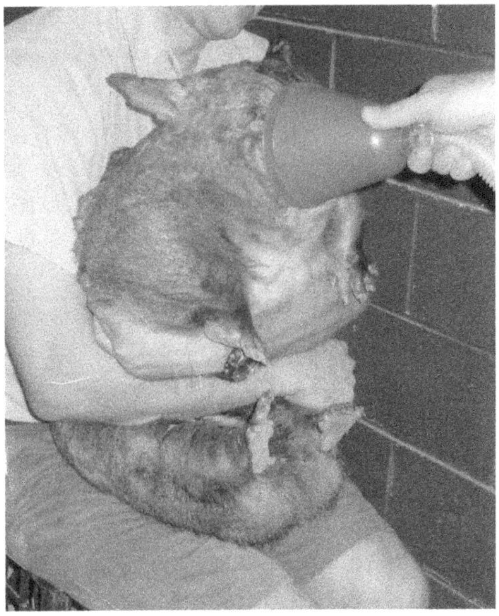

Figure 9.5 Mask induction in a southern hairy-nosed wombat. Photo: Taronga Zoo.

Intubation of wombats is difficult due to the narrow gape and dental arcade making visualisation of the glottis difficult. The larynx is also a considerable distance from the oral opening. With the animal in lateral recumbency and an assistant maintaining full extension of the head (using a tie looped around the upper incisors and a fist pressed into the dorsal neck behind the head) a long-bladed laryngoscope can be used to allow visualisation of the glottis. An alternative is to pass a stylet (a urinary catheter can be used) into the trachea via the oral cavity and guide the endotracheal tube over this (Booth 1999). Blind intubation is also possible. The larynx is palpated externally and an assistant extends the head and applies traction to the tongue. The tube is then passed over the base of the tongue until it touches the glottis. The tube is advanced as the animal takes a breath. A cuffed endotracheal tube of 5–7 mm diameter is usually suitable for an adult wombat. In many cases, wombats are maintained on face masks for the duration of anaesthesia.

6.2.3 Anaesthetic monitoring

Normal vital signs for wombats under anaesthesia are heart rate 60–120 bpm, respiratory rate 12–30 bpm and body temperature 32–36.5°C. Femoral and other peripheral pulses are easily palpated. Pulse oximetry probes may be placed on the pouch, penis, cloacal mucosa, lip or tongue although readings may be unreliable.

7 PHYSICAL EXAMINATION

Wherever possible, the physical examination of a wombat should begin with a distant observation of the undisturbed animal. Demeanour, responsiveness, behaviours, signs of disease and relevant environmental conditions should be noted. Although wombats are often active during the early morning and late afternoon hours, a free-ranging wombat seen grazing or moving around in the heat of the day is usually sick.

In most wombats a thorough physical examination will require general anaesthesia. Particular points to note when examining wombats include the following.

- Body condition is determined by assessing muscle and fat coverage of the ribs, pelvis and backbone.
- Gut fill gives an indication of recent feeding and digestion. In the healthy animal the abdomen feels full and doughy. Because of the slow passage time and large volume of the wombat gastrointestinal system, changes in body weight should be assessed in conjunction with gut fill (especially in inappetant animals) (see section 5).
- Hydration is difficult to assess clinically because wombat skin is very thick and inelastic. The oral and conjunctival mucosa appear dry even in a well-hydrated animal.
- Body temperature is taken rectally, via the cloaca. Because of wombats' very low body temperatures, regular digital thermometers may be inadequate. Mercury thermometers may measure lower temperatures.
- Integument is often unhealthy in appearance, with scarred skin and damaged coat. Ectoparasites are common. Grass seeds, other foreign matter and dried mucus are often found in nostrils and ear canals. Damaged claws and nail beds, sometimes with resultant pedal osteomyelitis, are not uncommon.
- Musculoskeletal injuries and disease can be difficult to detect due to the solid build of wombats. Radiographs are often indicated.
- Oral cavity and dental problems are often seen in wombats due to their continually growing teeth. Thorough oral examination for malocclusion, dental spurs and oral trauma will require general anaesthesia and a long-bladed laryngoscope or transilluminator.

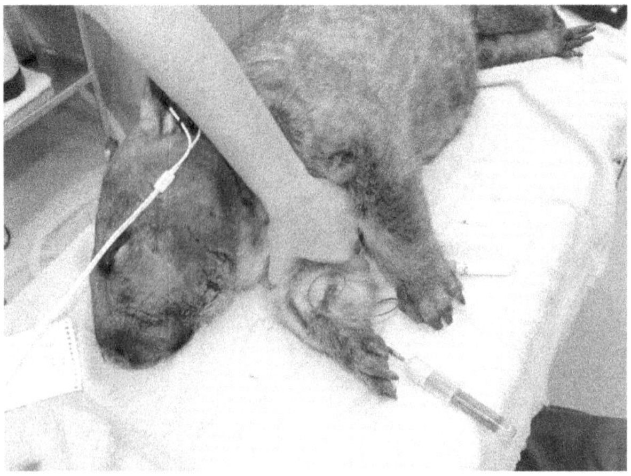

Figure 9.6 Venipuncture of the radial vein in a southern hairy-nosed wombat. Photo: Taronga Zoo.

8 CLINICAL PATHOLOGY

8.1 Haematology and biochemistry

8.1.1 Sample collection

Blood collection usually requires sedation or anaesthesia in wombats. Venipuncture sites include the radial, cephalic, brachial, medial metatarsal, caudal tibial, femoral or jugular vein (Booth 1999; Holz 2003; Clark 2004). The thickness of wombat skin can make visualisation and palpation of the veins difficult. Veins are most easily accessed after clipping the fur and use of digital pressure or a tourniquet to occlude the vein proximally. The radial vein on the medial aspect of the fore limb is most commonly used and is less mobile than the cephalic vein (Fig. 9.6). Blood flow may be slow. The femoral vein is accessed close to inguinal triangle, just caudal to the palpable femoral pulse. Large volumes may be drawn quickly from this site.

8.1.2 Reference ranges and interpretation

There are no well-developed haematology and biochemistry reference ranges for common or southern hairy-nosed wombats, despite the abundance of samples collected. The commonly published values are derived from small numbers of animals and most have not been statistically validated (Presidente 1978; Gaughwin & Judson 1980; Presidente 1982; Gaughwin et al. 1984; Booth 1999; Skerratt et al. 1999; ISIS 2002; Skerratt 2003b; Clark 2004). Reference ranges for northern hairy-nosed wombats have been statistically validated (Reiss et al. 2008).

Reference intervals for northern hairy-nosed wombats and suggested ranges for common and southern hairy-nosed wombats are provided in Tables 9.2 and 9.3.

Although more information is required, there is some evidence to suggest there are significant differences in reference ranges between captive and free-ranging wombats. Captive wombats may have higher

Table 9.2 Haematology central 90% confidence intervals or 5–95% results for northern hairy-nosed wombats and values (mean ± SD) for southern hairy-nosed wombats and common wombats

Parameter	Northern hairy-nosed wombat[a] (n = 34)	Southern hairy-nosed wombat[b] (n = 14)	Common wombat[c] (n = 31)
PCV (%)	32–42	40 ± 4	36 ± 6
Hb (g/L)	110–138	142 ± 14	119 ± 19
RBC (10^{12}/L)	3.7–4.8	5.1 ± 0.6	5.2 ± 0.7
MCV (fL)	82–92	79 ± 4	68 ± 4
MCH (pg/L)	27.3–30.5	27.8 ± 1.4	22.7 ± 1.3
MCHC (g/L)	317–364	350 ± 15	335 ± 17
WBC (10^9/L)	3.9–12.2	11.4 ± 3.3	8.4 ± 2.9
Neutrophils (10^9/L)	3.0–8.9	6.5 ± 2.7	6.1 ± 3.0
Lymphocytes (10^9/L)	0.2–3.4	4.3 ± 2.0	5.2 ± 3.0
Monocytes (10^9/L)	0.1–0.6	0.4 ± 0.4 (n = 11)	0.5 ± 0.5
Eosinophils (10^9/L)	0–0.7	0.3 ± 0.2 (n = 11)	0.15 ± 0.2
Basophils (10^9/L)	0–0.2	0.1 ± 0.02 (n = 6)	0.01 ± 0.05

a Derived from Reiss et al. (2008).
b Derived from ISIS (2002).
c Derived from Booth (1999).

Table 9.3 Biochemistry central 90% confidence intervals or 5–95% results for northern hairy-nosed wombats and values (mean ± SD) for southern hairy-nosed wombats and common wombats

Parameter	Northern hairy-nosed wombat[a] (n = 34)	Southern hairy-nosed wombat[b] (n = 14)	Common wombat[d] (n = 31)
Glucose (mmol/L)	5.7–10.7	5.5 ± 1.7	5.3 ± 1.4
BUN (mmol/L)	6.4–9.9	10.0 ± 3.9	3.7 ± 5.0
Creatinine (mmol/L)	0.23–0.36	0.23 ± 0.04 (n = 11)	0.08 ± 0.03
Total protein (g/L)	57–73	62.0 ± 5 (n = 13)	59.1 ± 10.8
Albumin (g/L)	27–36	34.0 ± 3.0 (n = 10)	29.7 ± 6.0
Globulin (g/L)	26–41	30.0 ± 6.0 (n = 9)	29.7 ± 7.5
Total bilirubin (μmol/L)	1.6–4.4	5.1 ± 3.4	5.2 ± 3.0 (n = 26)
ALP (U/L)	77–220	266 ± 125 (n = 10)	342 ± 214 (n = 28)
AST (U/L)	23–57	39 ± 23	156 ± 197 (n = 19)
ALT (U/L)	18–33	69 ± 32 (n = 13)	23.7 ± 9.2 (n = 3)
CK (U/L)	27–213	641 ± 587 (n = 9)	233 ± 454 (n = 19)
Cholesterol (mmol/L)	2.1–3.7	2.5 ± 0.6 (n = 11)	2.1 ± 0.7 (n = 12)[c]
Calcium (mmol/L)	2.2–2.6	2.3 ± 0.13 (n = 11)	2.4 ± 0.4
Phosphate (mmol/L)	0.9–2.0	1.36 ± 0.36 (n = 10)	1.64 ± 0.44
Sodium (mmol/L)	131–144	142 ± 2 (n = 11)	134 ± 3.2
Potassium (mmol/L)	3.2–7.1	4.1 ± 0.5 (n = 11)	4.7 ± 1.1
Chloride (mmol/L)	86–103	101 ± 4 (n = 11)	94 ± 5
Bicarbonate (mmol/L)	25.3–37.8	28.3 ± 2.7 (n = 2)[c]	33.6 ± 4.1
Thyroid hormone (T4) (μg/dL)	–	0.49 ± 0.15 (n = 4)[e]	0.44 ± 0.24 (n = 4)[e]

a derived from Reiss *et al.* (2008)
b derived from ISIS (2002)
c derived from Booth (1999)
d derived from ISIS in-house, Taronga Zoo (2006)
e Barboza *et al.* (1993)

WBC counts, serum protein and creatinine values and lower BUN values. These differences probably reflect better nutritional status and increased muscle mass in captive wombats (Gaughwin & Judson 1980; Skerratt et al. 1999). Immature wombats would be expected to have lower total serum protein, lower serum globulins and higher BUN, CK and ALP levels than adult wombats (Reiss et al. 2008).

The morphological appearance of southern hairy-nosed and common wombat haematocytes follows general marsupial characteristics (Clark 2004). The morphology of northern hairy-nosed wombat haematocytes has not been characterised but is similar to that of the southern hairy-nosed wombat.

There are few published reports of haematological and biochemical changes in wombats in response to disease or environmental stress. In common wombats, changes seen as a result of systemic infectious processes (e.g. generalised mange) may include increased neutrophil count, decreased lymphocyte count, presence of activated lymphocytes (detected by increased basophilic intensity in the cytoplasm), increased numbers of large lymphocytes (the majority of lymphocytes are small to medium in a healthy wombat), increased monocyte count and the presence of band neutrophils and toxic change within the neutrophil (detected by basophilic granulation of the cytoplasm) (Skerratt et al. 1999; Skerratt 2003b; Clark 2004). The extent of increased neutrophil count may give a good indication of the severity of the disease process (Skerratt 2003b). A change in the neutrophil:lymphocyte ratio, without an overall increase in leucocyte numbers, may also be an indication of inflammation.

Other changes indicating disease may include decreased MCV, MCH, total serum protein, serum albumin and creatinine levels. Increased levels of BUN, phosphorus, AST and CK may also be seen in diseased wombats (Skerratt et al. 1999). Changes seen in free-ranging wombats during drought conditions include decreased PCV, RBC count, Hb, total plasma protein and WBC counts (Gaughwin et al. 1984). These changes are typical of those seen in malnourished wombats.

8.2 Other sample collection

Urine may be collected via free catch or manual expression of the bladder in an anaesthetised wombat. Male wombats can be catheterised under anaesthesia. Cystocentesis, using a 21 or 23 G × 38 mm needle can be performed on anaesthetised wombats in dorsal recumbency. The needle is inserted perpendicular to the skin, just cranial to the pubic bones.

In most instances, wombats do not produce highly concentrated urine, conserving water instead through colonic resorption and ecological adaptations (Wells & Green 1998). Hairy-nosed species have a greater concentrating ability than common wombats (Hume 1999). Urine specific gravity in captive wombats averages 1.020–1.035 and pH ranges from 6–9. A study in free-ranging southern hairy-nosed wombats found urine specific gravity ranging from 1.024–1.077 (Wells & Green 1998). Urine varies in colour and may be turbid or deeply coloured, presumably due to food pigments. Phosphate crystals are not uncommon on sediment examination. Urine volumes are typically low (Wells & Green 1998).

Cerebrospinal fluid has been collected from wombats using the standard techniques for small domestic animals (Martin & Vogelnest 2004).

9 HAND-REARING

9.1 Rescue and considerations for rehabilitation and release

Young wombats are often rescued after the mother has been killed in a motor vehicle accident. Caring for wombat PY is a long-term investment. An orphaned wombat rescued at age 6 mo can be expected to be milk-dependent for a further 9 mo (until approximately 15 mo) and will need about 12 mo of care in total (see Table 9.1). Juvenile and adult wombats are charismatic, robust and adaptive, but can also be exuberant in their exploration and play, exhausting to care for and extremely destructive to their environment. Whenever a young wombat is rescued, consideration must be given to its long-term future. Unless circumstances preclude, orphaned PY should be reared with a view to release into the wild. All wombats, especially young males, can become extremely aggressive towards humans, particularly as they reach sexual maturity, and are often unsuitable for captivity. Levels of aggression and inappropriate behaviours vary in severity according to the individual nature of the animal and, most crucially, the experience and approach of the carer. To minimise unfavourable outcomes, wombat PY should only be raised by experienced wombat carers.

Many carers advocate constant human–wombat contact to settle a recently orphaned wombat. The carer quickly becomes the surrogate mother, which can result in inappropriate bonding or imprinting on humans. The human–wombat bond must be allowed to naturally weaken as the animal matures and becomes more independent. A suitable environment for hand-rearing (see 9.2) assists this process. Raising wombats in pairs (same or opposite sex) may reduce imprinting on humans.

Very young unfurred PY are generally poor candidates for hand-rearing. Larger unfurred and finely furred PY are often successfully raised by experienced carers and PY 5 mo and older (Table 9.1) are the best candidates for hand-rearing. Pouch-dependent PY may be well-protected by the mother's body and often suffer surprisingly little trauma during the motor vehicle accident. Injuries may range from skin grazes, lacerations and contusions through to facial trauma and skeletal fractures, especially of the long bones.

When an adult female wombat is killed on the road, pouch-emergent and older young-at-heel generally escape into the bush. If pouch examination of a wombat recently killed on the road reveals a lactating mammary gland and elongated teat, a careful search should be made of the accident scene for a juvenile. As orphans, these young will generally not survive alone. Some of these wombats are found wandering days later, near the mother's body. By the time they are rescued they may be malnourished, dehydrated and in need of critical care (Booth 1999).

9.2 Particulars of hand-rearing

All orphaned wombats should receive a thorough physical examination and close monitoring for the first week of care. Most need stabilisation with fluid therapy and warmth before they are offered milk formula and/or solid food. Artificial milk formulas should be introduced gradually. One of three milk-replacement formulas is used for wombats, depending on the experiences and preferences of the carer: Wombaroo wombat milk <0.4, 0.4 or >0.6 (Wombaroo Food Products); Biolac M100, M150 or M200 (Biolac); Di-Vetelact (Sharpe Laboratories). Regardless of milk choice, furless and newly orphaned PY should be fed approximately every 2 hr until stabilised. Furred young should be fed every 4 hr.

Wombat orphans have a reputation of being hardier than other marsupial PY. They appear more robust and less susceptible to many of the common diseases of hand-reared marsupials. Wombats tend to be stoic in nature and careful observation is needed to ensure subtle signs of stress are not overlooked. The general principles of stress minimisation, excellent hygiene and husbandry apply to all hand-reared wombats. Newly orphaned wombats generally take longer to adapt to artificial teats and formula than other marsupials and can be strong and determined in their objections. Both patience and perseverance with feeding are necessary. Some animals may feed better initially if the eyes are covered by the carer's hand. Wombats should be fed slowly, especially smaller or recently rescued animals. It may be 7–10 d (or longer) before PY settle into a good feeding pattern. Very young wombats need to be fed with a syringe and short piece of feeding tube. Older PY can be transferred onto a feeder bottle with a specialised wombat or macropod teat.

Immature wombats require supplemental heat, but have a greater tendency to overheat than other marsupials due to their lower metabolism and body temperatures. Red feet in hand-reared wombat PY may indicate heat stress (sometimes mild) or emotional distress. Wombats gradually acquire the ability to thermoregulate when their fur develops (approximately 600–800 g in common wombats, Table 9.1). Hand-reared wombats should be toileted at each feeding (as for other marsupials), although they may defecate as infrequently as once a week, especially in the first few weeks of hand-rearing. Mother-reared wombats back their rear end out of the pouch to urinate and defecate and some orphaned wombats retain this behaviour.

When the wombat begins to emerge from the pouch at around 8 mo it needs an area in which to explore and exercise. An outdoor escape-proof area with grass and dirt for digging is ideal. At pouch-emergent stage, wombats should be offered fresh grass (in clumps with dirt attached to the roots), meadow hay, herbivore pellets and small amounts of carrot, sweet potato and sweetcorn. Rich foods such as lucerne and fresh clover should be avoided as they may result in bloat. Some carers advocate the feeding of muesli and rolled oats. These should only be offered in restricted quantities if at all.

Weaning wombats need items to chew on (non-toxic branches and logs with bark attached, or pieces of wood). They enjoy chewing on pieces of charred wood and may benefit from products which help to establish the necessary gut flora for effective digestion. Mother-reared wombats are exposed to maternal faeces during weaning (although it is uncertain if the dam produces pap such as koalas do). Some carers advocate adding fresh pureed adult wombat faeces to the PY's milk formula. It may be safer to allow the pouch-emergent wombat access to fresh wombat faeces in its environment—it will sniff and probably chew some of these. The addition of a probiotic such as Protexin (International Animal Health Products) at empirical doses to the animal's diet around weaning time is practised by many carers and may be beneficial.

Valuable additional information on hand-rearing wombats can be found in the excellent notes of H George, G Parker & P Coote (1995), 'Common wombats: rescue rehabilitation release' and L Sauvarin, 'A guide to handrearing the common wombat'. Although unpublished, these notes are in common circulation and most wombat carers and wildlife care groups have ready access to copies.

10 DISEASES

10.1 Dermatological diseases

10.1.1 Sarcoptic mange

Sarcoptic mange (scabies) was first recognised in common wombats in Australia early in the 19th century. The wombat mange mite was later designated *Sarcoptes scabiei* var. *wombati* despite being morphologically indistinguishable from sarcoptic mites found on a range of other mammalian species (Skerratt et al. 1998). Sarcoptic mange is regarded as the major debilitating infectious disease of free-ranging common wombats. Southern hairy-nosed wombats are less commonly affected and sarcoptic mange has not been reported in northern hairy-nosed wombats. It is not known whether *S. scabiei* var. *wombati* is an endemic wombat parasite or was introduced, either with the arrival in Australia of the dingo (*Canis lupus dingo*) or with European colonialists and the feral pests they introduced.

S. scabiei is a cosmopolitan obligate parasite of the skin and has been reported in seven taxonomic orders of Mammalia (Skerratt et al. 1998). Sarcoptic mange occurs through most of the range of the common wombat and is enzootic in some populations. Anecdotal evidence suggests that epizootics of sarcoptic mange are associated with overpopulation, habitat degradation, drought and nutritional stress and that reduction in the local abundance of wombats can result (Martin et al. 1998). The host specificity of *S. scabiei* var. *wombati* is unknown although humans, domestic dogs and koalas have all been known to develop scabies after contact with infected wombats (Skerratt 1998).

Despite the obligate parasitic nature of *S. scabiei*, transmission of infection between wombats probably occurs most commonly after indirect contact between animals that share burrows. Wombat burrows are typically moist and cool and mites can survive off the host for variable periods of time in these conditions. *S. scabiei* var. *canis* can survive for up to 9 d off the host (at 15–25% and 25–95% relative humidity). Mites remain infestive and even display host-seeking behaviour after periods off the host (Arlian 1989).

In the various mammals in which it has been reported, sarcoptic mange can vary from mild to severely debilitating. Cross-infection of an unnatural host with a heterologous variety of *S. scabiei* tends to result in mild and self-limiting disease. The presence of mites, their eggs and excretory products within the superficial layers of the epidermis immediately causes stimulation of epidermal growth and increased production of keratinocytes resulting in hyperkeratosis. Wombats appear to manifest an induction phase typical of infestation in other mammals characterised by a lack of dermal infiltration by inflammatory cells for at least 12 d (Skerratt 2003a). Generally, the nature of the immune response by the host, particularly the cell-mediated dermal response, tends to dictate the nature and severity of disease in the individual. The host inflammatory response to infestation with *S. scabiei* typically encompasses immediate and delayed hypersensitivity inflammatory processes progressing to a phase of desensitisation (Davis & Moon 1989). This appears to also be the case for wombats, in which histological and immunohistochemical studies on sections prepared from affected skin demonstrated features consistent with both immediate and delayed hypersensitivity responses (Skerratt 2003a). Severely affected free-ranging wombats in that study also showed a reduced humoral response to the presence of mites consistent with a degree of desensitisation.

Severe crusting scabies ('Norwegian scabies') in humans is typically associated with immunosuppressive therapy or concurrent disease. Common wombats appear to be susceptible to this severe hyperkeratotic chronic form of mange in the absence of concurrent disease, possibly due to the typically relatively mild dermal cellular inflammatory response observed in affected animals which fails to limit mite abundance (Skerratt 1998). Desensitisation may also reduce the capacity of the immune response to control numbers of mites.

Most wombats experimentally infected with *S. scabiei* developed a series of clinical signs beginning with erythema then hyperkeratosis in the initial weeks. Progressive alopecia, excoriation and fissuring of affected skin became evident several weeks later (Skerratt 2003b). The progression of clinical signs was faster in animals receiving a higher infective dose. Lesions progressed from the site of exposure (the shoulder) in these animals to become bilaterally symmetrical and progressively involve the contralateral shoulder, face and flanks. Hyperkeratosis thickened to form a crust in these areas. Erythema spread to the ventral body and medial legs. Pruritis was a feature of mange in these study animals. These findings are consistent with observed signs of mange in naturally infected animals which vary from mild localised crusting lesions to a generalised severe crusting hyperkeratotic dermatosis with alopecia, excoriations, deep fissures and secondary pyoderma (Fig. 9.7). Severe skin lesions appear to impair vision and movement and affected animals are often emaciated. These animals are often observed to be active during the day and eventually succumb to starvation or misadventure. Free-ranging wombats with severe sarcoptic mange often manifest anaemia and a mature neutrophilia. Changes in serum biochemistry such as increased BUN, hypoglycaemia and hypoproteinaemia consistent with a catabolic state may be seen (Skerratt et al. 1999). At necropsy, in addition to typical skin lesions, fatty change may be seen in the liver and bacterial infection may be observed in internal organs including lymph nodes. Histological changes in the skin of wombats with severe mange have been described (Skerratt 1998).

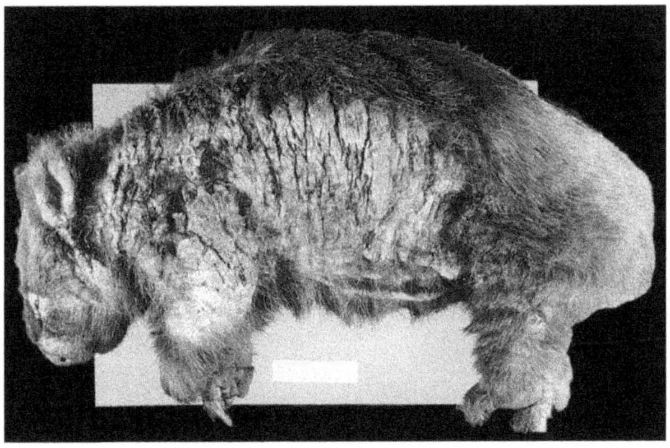

Figure 9.7 Sarcoptic mange in a common wombat. Photo: L Skerratt.

Typical severe generalised skin lesions are pathognomonic. Milder infections can be confirmed by the demonstration of mites in skin scrapings or crusts. Even in mild cases large numbers of mites are usually present. Relatively more mites are seen in samples taken from the anterolateral aspects of the body. Epizootics in free-ranging populations often become evident when usually nocturnal fossorial wombats are observed above ground during the day, with typical clinical signs. Early detection of epizootics may be achieved by observing, via spotlight, anterolateral areas of alopecia in wombats grazing at night, particularly where predisposing environmental stressors are present (Skerratt 2001).

The conservation benefits of treating severely affected free-ranging wombats are dubious. These individuals usually relapse after therapy and act as a source of mites for the contamination of burrows and infection of other animals. Euthanasia should be considered in these cases. Euthanasia should also be considered in severely debilitated and emaciated animals, particularly those with severe chronic skin damage.

Topical and systemic acaricidal therapies have been used for the treatment of sarcoptic mange in individual wombats. Whole body washes with amitraz and pour-on phosmet were used effectively in the past but are no longer recommended (Perry 1983; Booth 1999). The efficacy of topically applied therapies may be reduced by the presence of crusts, requiring that adjunctive keratolytic treatments be applied.

The use of systemic acaricidal therapies administered orally or by injection circumvents these problems. Macrocyclic lactones such as ivermectin and moxidectin are effective at 0.2–0.3 mg/kg PO or SC. Three treatments 10 d apart will eliminate clinical signs in most cases. This should be repeated with a second course after 2 wk to totally eradicate mites (Skerratt 2003b). An alternative is to treat at weekly intervals for up to 2 wk beyond the time that animals return negative skin scrapings and are pruritus-free. Topical formulations of these drugs may be useful where extensive scale does not impair their uptake. Bacterial sequelae of severe sarcoptic mange require administration of appropriate antibiotic therapy (see section 11).

10.2 Gastrointestinal diseases

10.2.1 Dental problems

Wombat teeth continue to grow throughout life. Consequently, malocclusion in wombats can result in significant recurring dental problems due to lack of normal appositional tooth wear. Inadequate provision of chewing opportunities to captive animals can result in overgrowth of incisors and malocclusion. Irregular and abnormal growth of molar teeth due to inappropriate diets may result in malocclusion (Fig. 9.8). Where a cheek tooth is fractured or lost the resulting excessive growth of the apposing tooth can also cause malocclusion. Sharp points can form on the buccal margin of the cheek teeth, resulting in lacerations. Malalignment of the jaws and subsequent malocclusion may develop in juvenile wombats as a result of trauma or bottle-feeding predominantly from one side (see 10.4.1) (Fig. 9.9).

Affected animals require regular oral examination and tooth rasping to ensure asymmetrical tooth growth does not result in orodental disease and inanition. Access to the cheek teeth for these procedures is challenging due to the narrow dental arcade and limited scope for opening the mouth. Where elected, these procedures may be facilitated by the use of gags and dental equipment developed for use in lagomorphs. Rigid endoscopy may help. Malocclusion due to jaw misalignment in juvenile wombats has a poor prognosis and, depending on the severity of the changes, euthanasia may be recommended.

Fractured incisors are common in both free-ranging and captive animals. These do not appear to predispose to alveolus infection and are usually managed by being filed even and left to grow (Booth 1994).

10.2.2 Gastrointestinal accidents

There are anecdotal reports of bowel obstruction due to intussusception in captive common and hairy-nosed wombats. Intestinal obstruction due to gastrointestinal

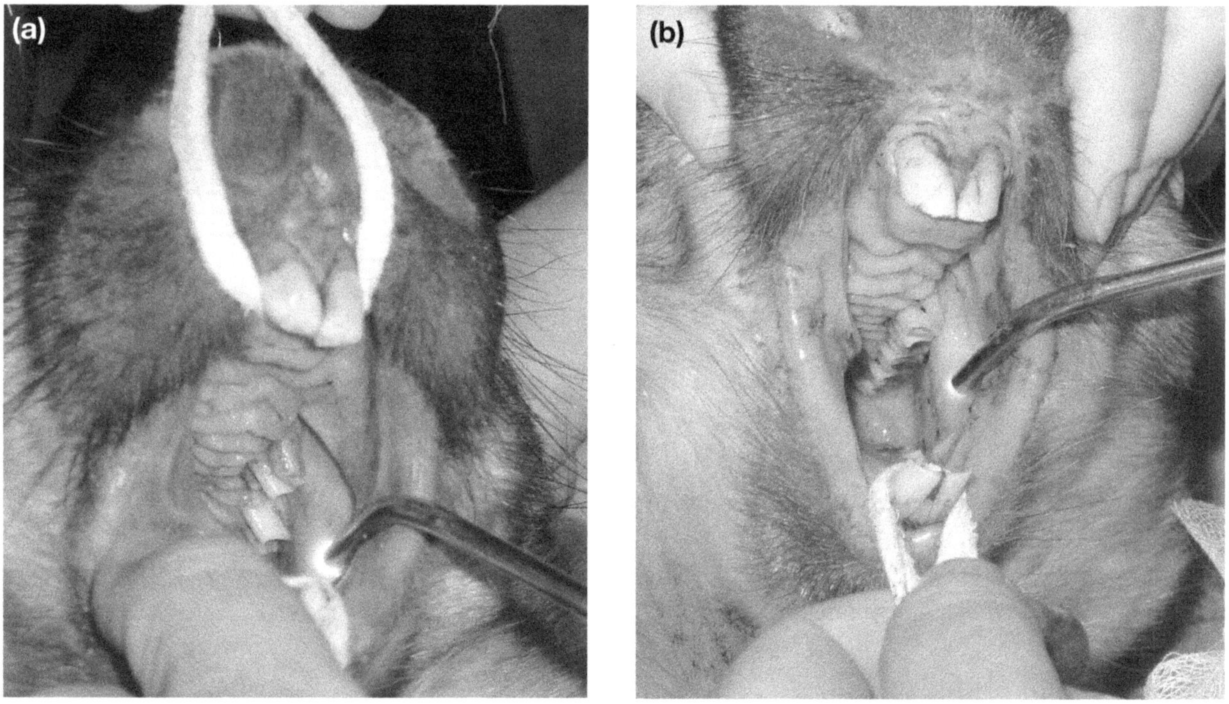

Figure 9.8 Molars in southern hairy-nosed wombats showing (a) uneven wear and (b) even wear. Photo: Taronga Zoo.

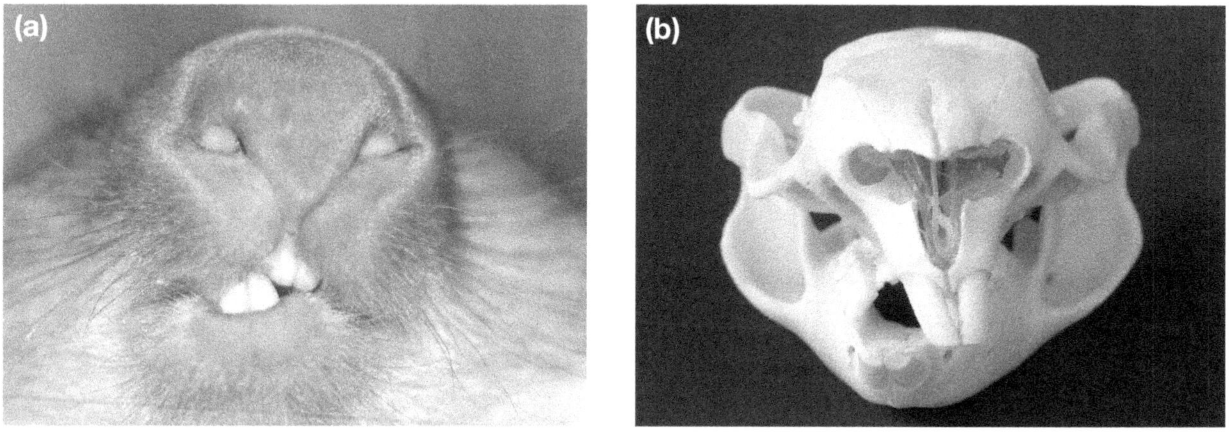

Figure 9.9 a) A hand-reared common wombat with malocclusion. b) The skull of a hand-reared common wombat showing malocclusion. Photos: Taronga Zoo.

accident should be considered in any wombat with evidence of discomfort, abdominal distension and reduced passage of faeces. One animal presented with a prolapse of the intussusceptum from the anus. Exploratory laparotomy revealed the intussusception to be at the level of the distal colon. Resection and anastomosis was performed due to significant loss of intestinal viability. The animal died shortly after surgery. Another animal was observed tooth grinding, restless and inappetant and died 24 hr later. Necropsy revealed an intussusception at the mid colon with significant loss of viability of the intussusceptum. Neither vomiting nor unequivocal behavioural evidence of abdominal pain were features of either case.

Investigation and management should broadly proceed according to the principles applied to evaluation and treatment of acute abdomen in domestic carnivores, where decision-making aims to differentiate surgical from medical causes of acute abdomen before irreversible gastrointestinal damage and shock supervenes. Blood parameters should be monitored for evidence of progressive cardiovascular compromise, although in wombats results may be misleading. Physical obstruction should be suspected where plain radiography or ultrasonography demonstrates gaseous dilation of loops of bowel. Interpretation of contrast radiography may be hampered by the normally prolonged gastrointestinal transit time of wombats. An associated loss of intestinal viability may be suspected on the basis of results of abdominocentesis. Where results of investigations are equivocal, early intervention by exploratory laparotomy should be considered.

10.3 Cardiopulmonary diseases

There is a single report of hypertrophic cardiomyopathy in a 10 yr old female captive common wombat (Machida 1997). The animal became depressed and anorectic and died 1 wk after the development of clinical signs. At necropsy the lungs were congested and the heart was enlarged with left atrial dilation. There was diffuse symmetrical hypertrophy of the left and right ventricular walls, ventricular septum and left ventricular papillary muscles which significantly encroached on the left ventricular chamber. Myocardial histopathology was consistent with classical hypertrophic cardiomyopathy in humans with bizarrely arranged myocardial cells which were hypertrophied and disordered. There was loss of the normal parallel arrangement of cells, resulting in the formation of irregular whorls of myocytes.

Pneumonia is relatively common in hand-reared juvenile wombats (see 10.4).

10.4 Neonatal diseases

10.4.1 Gastrointestinal disease

Wombats beginning to eat solids and those undergoing weaning occasionally develop bloat. This is often due to inappropriate diet. These animals are inappetant with distended abdomens and various degrees of abdominal discomfort, depression and diarrhoea. Dietary modification, supportive care and locally acting anti-bloat remedies such as Mylanta (Parke Davis), Tympanyl (Intervet [Australia]) or proprietary gripe water at empirical doses have been recommended (Rose 1999). Buscopan (Boehringer Ingelheim), at small animal dose rates, has also been used. If the animals have been fed an inappropriate diet for extended periods, it may impossible to reverse the damage to the digestive system. Common sequelae of bloat and intestinal dysfunction include torsion, intussusception, rectal prolapse and possibly clinical coccidiosis.

Young wombats will chew any number of foreign materials in the environment. Care must be taken with carpet, newspapers, furry toys and other items which can be accidentally ingested and lead to intestinal impaction, constipation or colic.

Orphaned wombats are very susceptible to dietary diarrhoea during the period of adjustment to artificial formula. Environmental stressors may also precipitate diarrhoea (see Chapter 2). If diarrhoea persists beyond 24 hr or is associated with systemic signs of ill health (inappetence, lethargy, signs of discomfort or pain), other causes must be investigated. Coccidiosis is uncommon in hand-reared wombats and is more common in subadults during rehabilitation.

Hand-raised wombats may suffer from constipation. This can be distinguished from infrequent but normal defecation by the presence of straining and abnormally firm faeces. Both diarrhoea and constipation may lead to rectal or cloacal prolapse. Constipation is generally treated empirically with increased oral fluid intake, oral administration of faecal softeners and enemas if required.

10.4.2 Jaw misalignment and dental malocclusion

Young wombats have a tendency to habitually suck on items; persistent sucking behaviours have been implicated in developmental misalignment of the jaws and subsequent dental malocclusion. Misalignment of the jaws may also be attributed to poor feeding technique, when carers fail to alternate sides of the mouth during bottle feeding. It is more likely that undiagnosed trauma at the time of orphaning (i.e. motor vehicle accidents) leads to mandibular misalignment. The resultant dental malocclusion may be difficult to treat and often carries a poor prognosis (see 10.2.1).

10.4.3 Respiratory disease

Young wombats often have dried mucus and dirt partially obstructing their nostrils. This may require gentle moistening to facilitate removal. Nasal foreign bodies such as grass seeds may be seen. In these cases there may be foul-smelling nasal discharge with associated irritation. Examination of the nasal passages (using an otoscope or endoscope) and removal of the foreign body will require anaesthesia (Woods 1999).

Pneumonia in hand-reared wombats may be a result of aspiration or primary bacterial infection and is more common in very young or debilitated individuals. Signs of respiratory disease may include nasal discharge, dyspnoea and audible respiratory sounds, and may be accompanied by inappetence and lethargy. Diagnosis of pneumonia is often problematic as auscultation is unreliable and other clinical signs may be subtle or absent. Thoracic radiographs may be required to reach a diagnosis and may also be used to evaluate response to therapy.

Treatment includes parenteral broad-spectrum antibiotics at dog dose rates for a minimum of 7 d and supportive care, including maintenance of fluid requirements, warmth and rest. Nebulisation with mucolytics, humidifiers or antibacterial agents may be of use.

Evaluation of total white cell count and differential counts on a weekly basis may assist diagnosis and efficacy of treatment of chronic cases of pneumonia. In most acute cases, the leukogram remains normal.

10.4.4 Hypoglycaemia

Hand-reared wombats occasionally suffer from single or repeated episodes of convulsions, accompanied by agitation, mental confusion or dullness. The convulsions are most frequently seen in wombats of pouch-dependent or early pouch-emergent age. These wombats may be lightly furred and not yet thermoregulatory and may have inappropriate feeding regimes (suboptimal milk formula and feeding frequency). In the natural situation these young would be constantly ingesting nutrients in a thermally supportive and stress-free environment (the mother's pouch). In a hand-reared wombat expending excessive physical activity (premature pouch-emergence) hypoglycaemia may result. Signs are most commonly seen several hours after feeding and following periods of exercise. Diagnosis is presumptive and is confirmed by blood glucose readings of <4 mmol/L. Treatment to correct the hypoglycaemic episode is empirical—bolus IV or oral glucose (10% or 50%, 500–2000 mg/kg). Management and prevention focuses on improved hand-rearing protocols and transition to Biolac or Wombaroo formula of correct strength and feeding frequency.

10.4.5 Skin

Wombats being raised on milk replacers often develop dry and scurfy skin, which is usually non-infectious and responds to topical preparations such as pawpaw ointment, lanolin or sorbolene creams. Oral fatty acid supplementation may also be beneficial.

Hand-reared wombats have presented with evidence of acute allergic skin reactions to unknown substances. Welts, weals and blisters appear on the skin and the animal may be pruritic. These cases have been treated empirically (topical soothing agents and antihistamines) with uncomplicated resolution.

10.4.6 Other

Toxoplasmosis is a common disease in wombats, including hand-reared juveniles (see 10.7.1). Salmonellosis may present with a respiratory (pneumonia) or gastrointestinal focus and rapidly spread to septicaemia and probable death if not diagnosed and treated appropriately.

Cataracts in hand-reared wombats have been linked to the feeding of lactose-based milk formulas (Speare 1988). Sporadic cases of cataracts in young wombats may be a sequel to other intra-ocular disease (e.g. toxoplasmosis).

10.5 Infectious diseases

10.5.1 Bacterial diseases

a Leptospirosis

Wild common wombats have been shown to be seropositive for a number of serovars of *Leptospira interrogans* including *pomona*, *hardjo* and *balcanica* (Munday & Corbould 1973; Durfee & Presidente 1979a). Many of these animals were clinically normal but with focal interstitial nephritis typical of subclinical leptospirosis. Experimental infection of common wombats with *L. pomona* resulted in depression, anorexia, jaundice and death (Munday & Corbould 1973). Common wombats experimentally infected with *L. balcanica* remained clinically normal but developed profuse leptospirurea, suggesting that wombats might be an effective maintenance host for this serovar (Durfee & Presidente 1979b).

In the author's experience leptospirosis appears to be an uncommon entity in captive wombats but could occur where animals are exposed to organisms in the urine of rodents or other maintenance host species. Demonstration of a rising titre to a pathogenic serovar and/or detection of leptospirurea concurrent with clinical signs consistent with leptospirosis are diagnostic. Prevention is by reducing wombat access to contaminated water and eliminating contact with rodents and other potential maintenance host species. Screening wombats for leptospirurea might be considered as part of a preventative health program, particularly where there is evidence for chronic renal failure. Implementing the above measures and avoiding contact with wombat urine by maintaining high standards of personal hygiene should safeguard staff health. Treatment with penicillins or aminoglycosides at domestic animal dose rates is indicated for treatment of acute sufferers or for the elimination of the carrier state.

b Tyzzer's disease

Tyzzer's disease is an acute fulminating fatal septicaemic disease caused by the flagellated Gram-negative rod *Clostridium piliforme* which has been described in numerous domestic, laboratory and zoo mammals. Young captive marsupials may be particularly susceptible to Tyzzer's disease, which has been reported in several species of possum, a koala (*Phascolarctus cinereus*), a wombat (species unrecorded) and an antechinus (*Antechinus* spp.) (Canfield & Hartley 1991).

10.5.2 Viral diseases

There is a single report of a mortality likely to be due to herpesvirus infection in a juvenile male orphaned common wombat (Rothwell & Canfield 1988). Intranuclear viral particles morphologically consistent with Herpesviridae were observed in hepatocytes by electron microscopy. Attempts to isolate virus in cell culture, however, were unsuccessful.

10.5.3 Fungal diseases

Spherules resembling chlamydospores of *Chyrosporium parvum* have been observed microscopically in the lungs of common and southern hairy-nosed wombats (Mason & Gaughwin 1982). The organisms were not associated with clinical illness or gross pathology.

10.5.4 Protozoal diseases

a Toxoplasmosis

Like other Australian marsupial taxa, wombats appear particularly susceptible to toxoplasmosis due to exposure to the coccidian parasite *Toxoplasma gondii*. There are reports of clinical disease in three captive common wombats (Skerratt et al. 1997; Hartley 2006) and a report describing the histopathological findings associated with toxoplasmosis in two common wombats (Canfield et al. 1990). Two of the clinical cases described were juvenile animals being hand-reared. A wild adult wombat with neurological signs was found to have toxoplasmosis (Skerratt 1998) and a serological survey revealed 23% of 27 free-ranging animals screened had been exposed to *T. gondii*. None had clinical signs (Hartley & English 2005).

Of the three clinical cases described, the juvenile animals developed acute disease characterised by dyspnoea, tachypnoea, anorexia and shock over a period of 24 hr, progressing to death in one animal. A second animal became moribund and was euthanased. Disease in the adult animal had a more protracted course characterised by lethargy, anorexia, reduced responsiveness and weight loss over a period of 36 d. Tachycardia and tachypnoea developed later in the course of disease, by which time thoracic radiographs revealed alveolar pattern pulmonary radio-densities. This animal developed hind limb ataxia late in the course of illness, became comatose and was euthanased.

Reported gross necropsy findings in wombats with toxoplasmosis include patchy congestion and consolidation of the lungs and a mottled and flaccid or streaked myocardium. Adrenal haemorrhage has been noted. One animal had no gross lesions at necropsy. Histopathological findings include pulmonary congestion and oedema with alveolar histiocytosis, fibrin accumulation and nuclear debris and myocardial necrosis with mononuclear aggregates. Focal extensive necrosis associated with neutrophilic infiltrates, gliosis and mononuclear cell perivascular cuffing have been observed in brain tissue. Focal necrosis of skeletal muscle with mononuclear and histiocytic infiltrates have been seen with low numbers of schizonts present. Lymphoid infiltration and necrosis has been described in smooth muscle, as has focal necrosis of adrenal glands with neutrophilic infiltrates.

Ante mortem diagnosis of acute infection relies on serological testing. Anti-toxoplasma IgG levels may remain elevated for years post exposure so demonstration of a rising titre in paired samples collected at least 2 wk apart is necessary for diagnosis where IgG is the predominant class of antibody measured. A four-fold increase is diagnostic.

Anti-toxoplasma IgM levels in eutherian mammals generally increase within 2–4 wk of infection and return to baseline by 12 wk post infection (Wolfe 2003). Consequently, a single elevated result for IgM suggests recent infection. A number of workers have described the use of a combination of direct agglutination testing (DAT) and modified agglutination testing (MAT) for the detection of acute toxoplasmosis in marsupials in which titres of greater than 1:64 are regarded as positive (Hartley & English 2005). This test combination potentially allows the definitive diagnosis of acute toxoplasmosis with a single sample, facilitating specific medical intervention early in the disease process and maximising the chances of a good outcome. The DAT measures anti-toxoplasma antibody of all classes. The MAT measures all anti-toxoplasma antibody classes except IgM. Presumably, an animal with a high DAT and lower or negative MAT can be considered to have had recent acute toxoplasmosis. Combined DAT and MAT testing should be interpreted with caution in wombats, however, as the single report of their application to an acutely infected common wombat describes the lack of increase in specific IgM antibodies by 22 d post appearance of clinical signs (Skerratt et al. 1997).

Serum biochemistry may be reflective of muscle and liver damage. CSF may reflect non-suppurative CNS inflammation. Tachyzoites may be seen in smears prepared from broncho-alveolar lavage, CSF or other tissues. Thoracic radiography may reveal a diffuse interstitial or alveolar pattern.

Treatment of choice in humans is the selective pharmacological inhibition of the protozoal folic acid biosynthetic pathway by a combination of pyrimethamine (0.5 mg/kg sid PO) and sulphadiazine (20 mg/kg tid PO). In these cases folinic acid supplementation (1 mg/kg sid PO) should be given and blood parameters monitored to ensure that host bone marrow suppression is not clinically significant. Clindamycin (15–25 mg/kg bid or tid PO or IM) is sometimes used but does not cross the blood brain barrier so is probably not indicated for neurological toxoplasmosis. Trimethoprim sulphonamide combinations (15 mg/kg PO bid) have also been used. Treatment should be continued for at least 4 wk.

Prevention is aimed at reducing wombat exposure to infective oocysts by imposing effective cat control and secure storage of animal foodstuffs. Vaccination of susceptible species may be a future prophylactic option, although trial use of a commercially available toxoplasma vaccine marketed for use in domestic sheep resulted in vaccine-induced disease in a group of tammar wallabies (*Macropus eugenii*) (Lynch et al. 1993).

b Coccidiosis

Southern hairy-nosed wombats are the natural host for *Eimeria ursinus* and *E. wombati* and common wombats the natural hosts for *E. arundeli* (Barker et al. 1979). Although coccidial oocysts are frequently identified in captive wombat faeces, clinical coccidiosis is uncommon. Disease can result where the infective dose is large, as might occur under insanitary or overcrowded conditions or where there is a degree of host immuno-incompetence or gastrointestinal dysfunction. Coccidiosis in juvenile captive common wombats has been described (Hum et al. 1991). Affected animals had diarrhoea or soft mucoid faeces and one animal had significant progressive weight loss. At necropsy there were extensive areas of raised indurated lesions within the small intestine. Histologically, intestinal villi were distended by hypertrophied cells of the lamina propria which contained masses of gametocytes. A neutrophilic inflammatory response occurring in affected areas of small intestinal mucosa and intestinal villous attenuation was marked in one animal. Oocysts consistent with *E. arundeli* were present in flotations prepared from the faeces of affected animals. Gross and histopathological changes ascribed to intestinal coccidiosis have been documented in southern hairy-nosed wombats but these were not associated with clinical signs (Hum et al. 1991).

Diagnosis of coccidiosis is by demonstration of a significant number of faecal oocysts by a flotation method in an animal with clinical signs in the absence of other potential gut pathogens. Toltrazuril has been used safely at 25 mg/kg PO sid for 3 d in many southern hairy-nosed wombats. In common wombats there are anecdotal reports of adverse reactions (including deaths) in a small number of animals (H McCracken *pers. comm.*).

10.5.5 Helminthiasis

The majority of nematode parasites of wombats occur in the colon and may be present in large numbers with no associated pathology. These include six strongyle species from three genera—*Oesophagostamoides* spp., *Phascolostrongylus* spp. and *Macropostrongyloides* spp. (Smales 1998). *Strongyloides spearei* have been described from the small intestine of the common wombat, where they are capable of causing mild enteritis (Presidente 1982). Larvae of the ascarid nematode *Baylisascaris tasmaniensis* have been isolated from granulomatous lesions in multiple viscera in common wombats, which are presumed to have the potential to act as intermediate hosts for this parasite. The definitive hosts are the Tasmanian devil (*Sarcophilus harrissii*) and spot-tailed quoll (*Dasyurus maculatus*). The angiostrongylid nematode *Marsupostrongylus coulstoni* has been described from the lungs of the common wombat (Spratt 1979).

Nine cestode species have been recorded in wombats (Smales 1998). Of these the cyclophyllideans *Phascolataenia* spp. and *Paramoniezia* spp. are found in the small intestine and are not pathogenic. The cyclophyllidean cestodes *Progamataenia diaphana* and *P. festiva* occur in the bile ducts of southern hairy-nosed and common wombats respectively, where they have been associated with a mild subclinical cholangitis or biliary fibrosis. Larval stages of several taeniid tapeworms have been demonstrated in common wombats. Migrating *Taenia hydatigena* larvae were associated with eosinophilic granulomatous tracts in the liver of common wombats, although peritoneal cysticercoids have not been seen (Smales 1998). A single report of *Echinococcus granulosus* has been recorded but, due to the failure of numerous other surveys to demonstrate hydatids in free-ranging wombats, it is thought that wombats are an unsuitable host (Smales 1998).

The introduced liver fluke *Fasciola hepatica* parasitises common wombats where they coincide with the occurrence of the intermediate host snail. Adult trematodes are found in the bile ducts and may cause marked biliary and hepatic fibrosis (Smales 1998).

Diagnosis of endoparasitism is by conventional means, i.e. faecal flotation techniques for nematodes and sedimentation techniques for trematodes. A Baermann technique may be necessary to demonstrate *Strongyloides* spp. larvae in faeces. These methods are relatively insensitive for the diagnosis of cestodiasis; direct visualisation of proglottids in the stool may be necessary for ante mortem diagnosis. Conventional anthelmintic drugs may be given where endoparasitism is thought to be significant. Levamisole, ivermectin, fenbendazole and praziquantel have all been used at domestic animal dose rates.

10.5.6 Ectoparasites

Aponomma aeruginans is the most common tick found on free-ranging common wombats. It appears to be a wombat-specific parasite and most commonly occurs on the underside of the limbs and in the inguinal region where it causes localised swelling (Skerratt 1998). Other ixodid ticks collected from common wombats include *Ixodes cornuatus, I. victoriensis* and *I. tasmani* (Booth 1999).

Various species of mite have been identified from common and southern hairy-nosed wombats. *S. scabiei* (see 10.1.1) is a significant pathogen of wombats. *Acaroptes vombatus* (from the skin) and *Raillietia australis* (from the external ear canal) appear to be non-pathogenic (Doube 1981).

Lycopsylla nova is a wombat-specific flea found relatively frequently on free-ranging common wombats. Various *Echidnophaga* spp. ('stick-fast' fleas) have been identified from all three species of wombat. *Boopia tarsata*, a chewing louse, is also a relatively common external parasite of the common wombat. Generally, these insect parasites are not associated with skin disease.

Conventional insecticidal and acaricidal therapies may be used to treat ectoparasitism in wombats. Moxidectin and ivermectin have been used for their acaricidal effect at domestic animal dose rates. Ivermectin should be similarly appropriate for eliminating stick-fast fleas. Topically applied acaricidal/insecticidal compounds such as amitraz and phosmet have also been used for external parasite control in wombats, as have pyrethrin-based insecticides (Booth 1999).

10.6 Non-infectious diseases

10.6.1 Nutritional diseases

a Copper toxicity

There is a single report of chronic copper toxicosis in a captive southern hairy-nosed wombat (Barboza 1990). The animal was one of a group of seven common and eight hairy-nosed wombats being held for research. The wombats were maintained on a pelleted diet incorporating a commercially available supplement formulated for growing pigs containing 30 ppm copper as $CuSO_4$.

Necropsy revealed generalised jaundice and haemoglobinuria. The kidneys were dark brown and the liver had a lobular pattern. Microscopically there was haemoglobin present in the glomerular spaces and incorporated into casts within the collecting tubules. Liver copper levels were extremely high (1166 ppm dry matter basis) relative to reference ranges for domestic livestock.

Nutritional studies were undertaken on the remaining animals in the research group, which demonstrated that similar amounts of dietary copper were retained by animals fed the original high copper ration and by animals fed a low copper diet containing 7.2 ppm copper (dry matter basis). Analyses of pastures grazed by wild common and southern hairy-nosed wombats revealed copper levels of 7.2–8.7 ppm (dry matter basis) (Barboza 1990).

Nutritional data collected from the affected animal revealed that a significantly greater amount of dietary copper was retained by that individual. None of the remaining animals developed copper toxicosis. These results suggest that copper toxicosis developed in the affected animal due to an impaired ability to limit copper retention by either limiting uptake or increasing excretion. Despite this, it remains possible that wombats are intolerant of high levels of dietary copper and it is recommended to avoid high copper rations in captive animals.

b Calcinosis cutis

A syndrome of multisystemic pathological organic deposition of calcium salts has been observed in a number of captive common wombats (Booth 1999). The syndrome has been variously termed calcinosis circumscripta and hypervitaminosis D due to the clinical and pathological similarities it shares with these conditions. Despite this, the pathogenesis of the syndrome in wombats remains unclear. It is thought to be a syndrome of metastatic calcification subsequent to episodic or persistent elevations in serum calcium levels, most likely due to hypervitaminosis D. It has been speculated that marsupials may have a lower vitamin D_3 requirement than eutherian mammals and that captive conditions may result in pathologically high levels leading to excessive gastrointestinal uptake of dietary calcium. In one study, captive animals demonstrated higher levels of 25-hydroxy cholecalciferol than free-ranging wombats and all the wombats sampled had lower levels than eutherian mammals (Fowler & Fraser 1993). Excessive uptake of vitamin D may occur where commercial dry dog foods which are supplemented with vitamin D and calcium are fed. Also, excessive exposure to ultraviolet light, as might occur in captive wombats with an enforced diurnal lifestyle, could result in excessive cutaneous 25-hydroxy cholecalciferol synthesis leading to hypervitaminosis D.

The condition is usually progressive. Animals may initially present with lameness due to inflamed nodular or plaque-like lesions in the skin of the footpads. Signs of renal failure including polydipsia, polyuria, anorexia, dehydration and weight loss usually follow. Serum biochemistry may reveal azotaemia and hypercalcaemia. Reference ranges for serum parathyroid hormone have not been established for common wombats but levels would be expected to be low in affected animals. Radiography of pad lesions demonstrates single or multiple lobulated radiodensities. Necropsy findings include variably granulomatous lesions composed of gritty mineral deposits in the kidneys, major arteries, endocardium, gastric mucosa and SC tissues of the feet. Skin lesions may ulcerate and discharge a white pasty gritty material. Uroliths may be present. Histologically, skin lesions are focal masses of granular amorphous material associated with dermal collagen and elastin fibres surrounded by zones of granulomatous inflammation. Renal histopathology may encompass interstitial nephritis, renal fibrosis and renal mineralisation (Booth 1999). Treatment is dietary modification and supportive care where renal failure is compensated. Euthanasia is indicated where renal failure is decompensated. Surgical excision of skin lesions has been attempted in a number of cases but recurrence has been common, as has progression to renal failure (Booth 1999). This condition is now rarely seen, presumably as a result of an improved understanding of the dietary and environmental needs of captive wombats.

10.6.2 Neoplastic disease

Although wombats are relatively long-lived marsupials they appear to experience a low prevalence of neoplasia in captivity. There are anecdotal reports of lymphoid neoplasia in a captive southern hairy-nosed wombat and fibropapilloma in a common wombat. There is a single report of a likely lymphosarcoma in a free-ranging common wombat (Skerratt 1998). This animal was concurrently affected by severe hyperkeratotic sarcoptic mange. Histopathology revealed large inflammatory infiltrates with many undifferentiated inflammatory cells and abnormal mitotic figures in the lungs, liver, kidney and adrenal glands. A predominantly lymphocytic mononuclear infiltrate was present in the pulmonary interstitium.

10.6.3 Trauma

a Fight wounds

Fighting between captive common wombats is not uncommon. This often appears to be associated with periods of generally increased animal activity and is frequently presumed by animal managers to be associated with oestrus. Conflicts often occur at night (the animals' most active period). Lesions vary from focal areas of hair loss to deep lacerations and are usually distributed over the dorsal areas of the body, particularly the caudal dorsal back and rump. Differentiating hair loss due to conspecific conflict from other causes of alopecia (e.g. sarcoptic mange, dermatophytosis) is straightforward where fighting is observed and lesions are accompanied by bite wounds. Lesion distribution can help—bite wounds tend to be distributed dorsally, sarcoptic mange tends to be anterolateral. Where doubt exists, alopecia should be investigated by standard dermatological investigative techniques so that parasitic and infectious causes may be ruled out.

Significant bite wounds should be assessed under general anaesthesia. Standard wound care including debridement of non-viable tissue and lavage with tissue antiseptics should be undertaken and empirical broad-spectrum parenteral antibiotic therapy should be initiated where indicated. Consideration should be given to the permanent separation of incompatible animals where fight injuries are severe and recurrent.

b Digit trauma

Escape behaviour by captive wombats often results in footpad and claw trauma which can progress to paronychia, suppurative pododermatitis, digital osteomyelitis and/or tenosynovitis. *Staphylococcus aureus*, beta-haemolytic *Streptococcus* and *Corynebacterium* spp. have all been involved. There is one report of an untreated infection progressing to involve local lymph nodes and eventually resulting in embolic pneumonia (Presidente 1982).

Traumatic foot lesions should be promptly assessed and treated and affected animals should be relocated onto clean substrate. Prophylactic antibiotic therapy should be implemented empirically. Clindamycin is a reasonable choice due to its Gram-positive spectrum of activity and ability to achieve effective concentrations within bone. Where there is clinical evidence for infection, foot radiographs should be taken to identify potential phalangeal osteomyelitis. Samples should be taken for culture and sensitivity testing to guide antibiotic choice. Intractable osteomyelitis may require digit amputation.

Prevention encompasses housing captive wombats appropriately with ample provision of refuges to reduce escape attempts. Adequate opportunity should be provided for the expression of normal digging behaviour. Husbandry practices should aim to minimise the bacterial load of enclosure substrates.

c Road trauma

Common and southern hairy-nosed wombats are frequently the victims of road trauma, often resulting in mortalities. Extensive and multiple injuries are usually incurred. PY are often orphaned by these events and may also suffer traumatic injuries.

Management and triage of wombats suffering road trauma should proceed according to the general principles of emergency management of domestic animals. Decision-making should be informed by the prognosis for rehabilitation consistent with release, where this is the goal. Shock should be detected and managed as for domestic animals. Opioid analgesics including buprenorphine and butorphanol have been used for pain management in wombats at domestic dog dose rates.

11 THERAPEUTICS

The general principles of therapy in wombats are extrapolated from those of domestic animal medicine, with modifications based on practical constraints, clinical success or failure and anecdotal evidence. No pharmacological studies have been conducted on wombats. As a starting point, therapies, dose rates and frequencies are drawn from domestic dog medicine. In most cases no consideration has been paid to the specialised metabolic and gastrointestinal physiology of wombats, which would feasibly alter pharmacokinetics.

Medications are given orally or parenterally in wombats, with the choice most often dictated by logistical constraints. If all other considerations are equal, it would appear logical to choose a parenteral route for medication, given the uncertainties surrounding gastrointestinal absorption in wombats. Wombats appear to be less susceptible to gastrointestinal disturbance following antibiotic therapy than other marsupial species.

Drug therapies that have been used widely in southern hairy-nosed and common wombats in recent times, at dog dose rates, with apparent clinical success and no recorded adverse effects include (but are not restricted to):

- antibiotics—amoxycillin (with or without clavulanic acid), enrofloxacin, long-acting penicillin, trimethoprim sulphonamide combinations, clindamycin;
- non-steroidal anti-inflammatory drugs—meloxicam, carprofen, tolfenamic acid;
- opioids and steroids—buprenorphine, butorphanol, dexamethasone;
- parasiticides—ivermectin, moxidectin, fenbendazole, levamisole, praziquantel.

Toltrazuril (25 mg/kg PO sid for 3 d) has been used safely in many southern hairy-nosed wombats. In common wombats there are anecdotal reports of adverse reactions (including deaths) in a small number of animals (H McCracken *pers. comm.*).

Long-acting neuroleptic agents have been used rarely in wombats (see 6.2.1) but have been associated with excessive and prolonged tranquillisation. If used, great caution should be exercised and considerably reduced dose rates are suggested.

Oral medications may be disguised in a favoured food item if the individual is eating well. Adult and juvenile wombats may be physically restrained for oral medication. Small volumes of liquid medication (or tablets crushed and dissolved in water) may be syringed directly into the commissure of the mouth. Debilitated or anorexic wombats are sometimes anaesthetised and treated by gavage.

Parenteral medications are generally given by IM injection. The lateral thigh, triceps muscle of the fore leg and dorsal lumbar muscles are ideal IM injection sites. The dorsal rump should be avoided due to the hard cutaneous plate. Subcutaneous injection or fluid administration is not recommended in wombats because of their thick skin and absence of a significant SC space.

Intravenous catheters may be placed in the brachial, cephalic, caudal tibial or medial metatarsal vein. Wombats requiring ongoing IV fluid administration or repeated IV medication may be managed by placing an in-dwelling catheter and delivering a bolus through an injection port several times a day (Booth 1999).

12 REPRODUCTIVE MANAGEMENT

Although there are no northern hairy-nosed wombats in captivity, common and southern hairy-nosed wombats are common in Australian zoos and fauna parks. Despite this, few institutions have attempted to implement strategic ongoing breeding programs. This is because juvenile wombats often become available to zoos after being orphaned by road trauma. The availability of these long-lived wild-born individuals precludes the need for the generation of captive-bred replacement animals.

Male and female hand-reared animals are often paired up as juveniles and permanently housed together to ensure compatibility for display purposes. Although there are anecdotal reports of animals breeding under these circumstances, reproductive failure appears more likely. Permanent pair bonding is an unnatural situation for the solitary common wombat and it is speculated that enforced habituation of individuals to one another could result in suppression of reproductive behavioural cues. If this is the case, wombats in breeding programs should be held separately and introduced for breeding at oestrus, detected by a combination of means including behavioural observation, vaginal cytology and endocrine monitoring (see 3.1, 3.2 and 12.1). In one captive breeding program, two successful breeding pairs were housed as separate individuals off-display. Introductions for breeding were timed to coincide with behavioural evidence of oestrus, particularly increased activity and evidence of seeking behaviour between animals separated by a single visual barrier (MacCallum 2004). It should be noted that courtship behaviour is aggressive and violent in wombats. Animals introduced for breeding should be monitored closely and separated if aggression is judged inappropriate and dangerous for either individual. Breeding introduction areas should have adequate scope for chase and escape.

Exhibited common and southern hairy-nosed wombats are often required to adopt a diurnal lifestyle with limited access to burrow analogues to ensure they are visible to the visiting public. This situation, contrary to the fossorial nocturnal nature of wombats, is potentially stressful and may partially explain the low rate of reproductive success in zoos. Other potential causes of reproductive failure include suboptimal nutrition and inadequate space to perform courtship and other natural behaviours. One report describes the successful breeding of a pair of captive common wombats on three occasions. The animals were housed together permanently and given access to a 600 m^2 holding area with opportunity to burrow between 6 pm and 9 am (Boer 1998).

12.1 Assisted reproduction

Semen has been successfully collected and cryopreserved from southern hairy-nosed and common wombats (Taggart et al. 1998; MacCallum 2004). Semen was collected from anaesthetised animals by electroejaculation. Semen volumes collected were up to 7.5 mL, sperm concentrations were approximately 8×10^6/mL and there was greater than 90% motility. A conventional TCF (tris, citrate, fructose) egg yolk diluent with the addition of relatively low concentrations (4–8%) of added glycerol was recommended to maximise the motility (70–90%) of sperm post-thaw in one study in which semen was frozen rapidly in liquid nitrogen vapour (Taggart 1998). A second study confirmed the suitability of the TCF diluent for the cryopreservation of wombat semen, but advocated the addition of glycerol to achieve a final concentration of 14% and a slow cooling rate of 6°C/min. Marsupial sperm are generally relatively resistant to cold shock and these studies revealed wombat spermatozoa to be particularly tolerant of the freeze/thaw process. Common wombat sperm have been retrieved from the caudal epididymis subsequent to post-mortem storage of the testis within the scrotum at 4°C for up to 3 d, cryopreserved in TCF with glycerol to a concentration of 14% and thawed with excellent recovery of motility and percentage of live sperm (MacCallum 2004).

A superovulation regime of eight injections of porcine FSH q 12 hr followed by a single injection of porcine LH 6 hr after the last LH dose resulted in the collection of 244 oocytes from 11 wombats subsequent to euthanasia (West 2003). Only 33 oocytes were defined as mature (i.e. in possession of a polar body). Incubation of the remaining viable oocytes in TCM 199 maturation media at 35°C and 5% CO_2 in air resulted in the maturation of a further 33 oocytes after 30 hr. Another study describes the maturation of 34% of oocytes collected from unstimulated common wombat ovaries after 60 hr of culture (Cleary et al. 2003).

Excreted progesterone metabolites have been measured by enzyme immunoassay for the investigation of reproductive status in female common and southern hairy-nosed wombats (Paris et al. 2002). Although the progesterone metabolites excreted by wombats have not been identified, this study revealed that, of the three anti-pregnane antibodies used in the enzyme immunoassay, the greatest cross-reactivity for the progestogens present in wombat faeces was with those raised against 5β-20α-OH-pregnane. Excreted androgens have been measured in male southern hairy-nosed wombat faeces by radioimmunoassay for the investigation of reproductive status (Hamilton et al. 2000). Results suggested that faecal testosterone provides the best indicator of reproductive status in male southern hairy-nosed wombats.

There have been no reports of successful artificial insemination in any wombat species.

13 SURGERY

The decision to apply surgical solutions to health problems in wombats is made with due consideration to the challenges inherent in the anaesthesia, hospitalisation and unique anatomical features of the taxon. Preoperative assessment should be undertaken judiciously. Inhalation anaesthesia allows more precise control over anaesthetic depth and should be elected where more complex procedures are planned, particularly where the clonic jaw and limb movements typical of injectable anaesthetics such as tiletamine/zolazepam need to be abolished (e.g. dental procedures and orthopaedic surgery). Convalescent wombats should be housed in confined areas with minimal substrate, and nest boxes to promote security and limit escape and digging behaviour.

Fractures are common sequelae to motor vehicle trauma and the approach to their management in wombats is based on the principles of fracture management in domestic animals. A significant challenge is posed by the difficulties inherent in restricting the activity of wombats in the convalescent period. Consequently, fracture stabilisation techniques that provide rigid stabilisation and a rapid return to function (e.g. orthopaedic plates and external fixateurs) are preferred. Juvenile animals that can be confined to the pouch during convalescence are better candidates for fracture repair. The efficacy of external coaptation in the management of wombat long bone fractures is undermined by the significant muscle mass between the device and the affected bone. Surgical exposure of proximal long bones is also complicated by this thick musculature (Webb 2006). Wombat skin is relatively thick and inelastic, which can interfere with retrograde placement of intramedullary pins, particularly in the femur. In these cases the proximal end of the pin may need to be left exposed and the resultant area of pressure necrosis managed (Webb 2006). Pelvic fractures in adult animals have successfully been managed conservatively with rest and confinement. Surgical access by a dorsal approach to the femoral head and neck for excision arthroplasty in one case was restricted. A ventral approach may be more appropriate for this procedure in wombats. Indications for analgesic management in wombat surgical patients are as for domestic animals.

REFERENCES

Arlian LG (1989). Biology, host relations and epidemiology of *Sarcoptes scabiei. Annual Review of Entomology* **34**, 139–161.

Barboza PS (1990). Copper toxicity in captive wombats (Marsupialia: Vombatidae). In *Proceedings of American Association of Zoo Veterinarians Annual Conference,* pp. 229–231.

Barboza PS, Hume ID & Nolan JV (1993). Nitrogen metabolism and requirements of nitrogen and energy in the wombats. *Physiological Zoology* **66**, 807–828.

Barker IK, Munday BL & Presidente PJA (1979). Coccidia of wombat: correction of host-parasite relationships. *Eimeria wombati* (Gilruth & Bull, 1912) comb. Nov. and *Eimeria ursine* Superer, 1957 from the hairy-nosed wombat and *Eimeria arundeli sp*.n. from the common wombat. *Journal of Parasitology* **65**, 451–456.

Boer M (1998). Observations on reproduction in the common wombat (*Vombatus ursinus*) in captivity. In *Wombats* (eds RT Wells & PA Pridmore), pp. 129–146. Surrey Beatty & Sons: Sydney.

Booth R (1999). Wombats: care and treatment of sick, injured and orphaned animals. In *Wildlife in Australia, Proceedings 327,* pp. 1–11. Post Graduate Foundation in Veterinary Science: Sydney.

Canfield PJ & Hartley WJ (1991). Tyzzer's disease (*Bacillus piliformis*) in Australian marsupials. *Journal of Comparative Pathology* **105**, 166–173.

Canfield PJ, Hartley WJ & Dubey JP (1990). Lesions of toxoplasmosis in Australian marsupials. *Journal of Comparative Pathology* **103**, 159–167.

Clark P (2004). *Haematology of Australian Mammals.* CSIRO Publishing: Melbourne.

Cleary M, West M, Shaw J, Jenkin G & Trounson A (2003). In vitro maturation of oocytes from non-stimulated common wombats *Reproduction, Fertility and Development* **15**, 303–310.

Crowcroft P & Sunderland R (1997). Breeding of wombats in captivity. *Zoolische Garten* **47**, 313-322.

Davis DP & Moon RD (1989). Density of itch mite, *Sarcoptes scabiei* (Acari: Sarcoptidae) and temporal development of cutaneous hypersensitivity in swine mange. *Veterinary Parasitology* **36**, 285–293.

Doube LJ (1981). Diseases of wombats. In *Wildlife Diseases of the Pacific Basin and Other Countries* (ed. ME Fowler), pp. 63–75. Wildlife Disease Association: California.

Dubey JD & Beattie CP (1988). *Toxoplasmosis of Animals and Man.* CRC Press: Boca Raton, Florida.

Dubey JP & Odening K (2001). Toxoplasmosis and related infections. In *Parasitic Diseases of Wild Animals* (eds WM Samuel, MJ Pybus & AA Kocan), pp. 478–519. Manson Publishing/Veterinary Press: London.

Durfee PT & Presidente PJA (1979a). A serological survey of Australian wildlife to leptospires of the hebdomadis subgroup. *Australian Journal of Experimental Biology and Medical Science* **57**, 177–189.

Durfee PT & Presidente PJA (1979b). Experimental infections of brush-tailed possums, common wombats and water rats with *Leptospira interrogans* serovars *balcanica* and *hardjo. Australian Journal of Experimental Biology and Medical Science* **57**, 231–240.

Evans M, Atkinson S & Horsup A (1998). Combination of zolazepam and tiletamine as a sedative and anaesthetic for wombats. *Australian Veterinary Journal* **76**, 355–356.

Evans M, Green B & Newgrain K (2003). The field energetics and water fluxes of free-living wombats (Marsupialia: Vombatidae). *Oecologia* **137**, 171–180.

Fowler AM & Fraser DR (1993). Vitamin D metabolism in nocturnal marsupials. In *Proceedings of the Wildlife Disease Association (Australasian section) Annual Conference.*

Gaughwin MD & Judson GJ (1980). Haematology and clinical chemistry of hairy-nosed wombats (*Lasiorhinus latifrons*). *Journal of Wildlife Disease* **16**, 275–279.

Gaughwin MD & Wells RT (1978). General features of reproduction of the hairy-nosed wombat in the Blanche town region of South Australia. *Australian Mammal Society Bulletin* **5**, 46–47.

Gaughwin MD, Judson GJ, McFarlane WV & Siebert DB (1984). Effect of drought on the health of wild hairy-nosed wombats, *Lasiorhinus latifrons. Australian Wildlife Research* **11**, 455–463.

Hamilton RA, Stanton PG, O'Donnell L, Steele VR, Taggart DA & Temple-Smith PD (2000). Determination of seasonality in southern hairy-nosed wombats *Lasiorhinus latifrons* by analysis of faecal androgens. *Biology of Reproduction* **63**, 526–531.

Hartley MP (2006). *Toxoplasma gondii* infection in two common wombats (*Vombatus ursinus*). *Australian Veterinary Journal* **84**, 107–109.

Hartley M & English A (2005). A seroprevalence survey of Toxoplasma gondii in common wombats (*Vombatus ursinus*). *European Journal of Wildlife Research* **51**, 65–67.

Holz P (2003). Marsupialia. In *Zoo and Wild Animal Medicine* (5th edn) (eds ME Fowler & RE Miller), pp. 288–303. WB Saunders: St Louis.

Horsup A (1998). A trapping survey of the northern hairy-nosed wombat. In *Wombats* (eds RT Wells & PA Pridmore), pp. 147–155. Surrey Beatty & Sons: Sydney.

Hum S, Barton NJ, Obendorf D & Barker IK (1991). Coccidiosis in common wombats. *Journal of Wildlife Diseases* **27**, 697–700.

Hume ID (1999). Hindgut fermenters: the wombats. In *Marsupial Nutrition* (ed. ID Hume), pp. 125–148. CUP: Cambridge.

ISIS (2002). *Reference ranges for physiological values in captive wildlife.* International Species Information System: Eagan, Minnesota.

ISIS Taronga Zoo (2006). *In-house reference ranges for physiological values in captive wildlife.* International Species Information System: Taronga Zoo, Sydney

IUCN Species Survival Commission (2001). *IUCN Red List Categories.* IUCN: Gland, Switzerland.

Jackson SM (2003). Wombats. In *Australian Mammals: Biology and Captive Management* (ed. S. Jackson), pp. 183–203. CSIRO Publishing: Melbourne.

Johnson C & Crossman D (1991). Sexual dimorphism in the northern hairy-nosed wombat, *Lasiorhinus krefftii* (Marsupialia: Vombatidae). *Australian Mammalogy* **14**, 145–146.

Johnson CN (1998). The evolutionary ecology of wombats. In *Wombats* (eds RT Wells & PA Pridmore), pp. 34–41. Surrey Beatty & Sons: Sydney.

Lynch MJ, Obendorf DL & Reddacliff GL (1993). Serological responses of tammar wallabies (*Macropus eugenii*) to inoculation with an attenuated strain of *Toxoplasma gondii*. In *Proceedings of American Association of Zoo Veterinarians*, pp. 185–187. St Louis, Missouri.

MacCallum CA (2004). The reproductive biology of the common wombat (*Vombatus ursinus)*: studies towards the development of an artificial insemination protocol. M.Phil thesis. University of Queensland, Australia.

Machida N, Hirose M, Nakamura T & Kiryu K (1997). Hypertrophic cardiomyopathy in a common wombat (*Vombatus ursinus*). *Journal of Comparative Pathology* **117**, 91–94.

Marks CA (1998). Courtship and mating in a pair of free ranging common wombats *Vombatus ursinus.* In *Wombats* (eds RT Wells & PA Pridmore), pp. 125–128. Surrey Beatty & Sons: Sydney.

Martin A & Vogelnest L (2004). Use of magnetic resonance imaging to investigate neurological dysfunction in a southern hairy-nosed wombat (*Lasiorhinus latifrons*). In *Proceedings of AAZV, AAWV and WDA Joint Conference*, pp. 152–154. American Association of Zoo Veterinarians, American Association of Wildlife Veterinarians and Wildlife Disease Association: San Diego.

Martin RW, Handasyde KA & Skerratt LF (1998). Current distribution of sarcoptic mange in wombats. *Australian Veterinary Journal* **76**, 411–414.

Mason RW & Gaughwin M (1982). Adiaspiromycosis in South Australian hairy-nosed wombats *Lasiorhinus latifrons. Journal of Wildlife Diseases* **18**, 3–8.

McAllan BM, Roberts JR & Barboza PS (1995). The kidney structure of the common wombat (*Vombatus ursinus*) and the hairy-nosed wombat (*Lasiorhinus latifrons*). *Australian Journal of Zoology* **43**, 181–191.

Moritz K, Green RH & Selwood (1998). A histological analysis of the reproductive cycle of the common wombat *Vombatus ursinus.* In *Wombats* (eds RT Wells & PA Pridmore), pp. 75–85. Surrey Beatty & Sons: Sydney.

Munday BL & Corbould A (1973). *Leptospira pomona* infections in wombats. *Journal of Wildlife Diseases* **9**, 72–73.

Murray PF (1998). Palaeontology and palaeobiology of wombats. In *Wombats* (eds RT Wells & PA Pridmore), pp. 1–33. Surrey Beatty & Sons: Sydney.

Paris MCJ, White A, Reiss A, West M & Schwarzenberger F (2002). Faecal progesterone metabolites and behavioural observations for the non-invasive assessment of oestrus cycles in the common wombat (*Vombatus ursinus*) and the southern hairy-nosed wombat (*Lasiorhinus latifrons*). *Animal Reproduction Science* **17**, 245–247.

Perry RA (1983). Successful treatment of sarcoptic mange in the common wombat (*Vombatus ursinus*). *Australian Veterinary Practitioner* **13**, 169.

Peters DC & Rose RW (1979). The oestrus cycle and basal body temperature in the common wombat (*Vombatus ursinus*). *Journal of Reproduction and Fertility* **57**, 453–460.

Presidente PJA (1978). Diseases seen in free ranging marsupials and those held in captivity. In *Fauna, Proceedings 36*, pp. 457–471. Post Graduate Committee in Veterinary Science: Sydney.

Presidente PJA (1982). Common wombat *Vombatus ursinus*: maintenance in captivity, blood values, infections and parasitic diseases. In *The Management of Australian Mammals in Captivity: Proceedings of the Scientific Meeting of the Australian Mammal Society* (ed. DD Evans), pp. 133–143. Zoological Board of Victoria: Melbourne.

Reddacliff GL, Hartley WJ, Dubey JP & Cooper DW (1993). Pathology of experimentally induced acute toxoplasmosis in macropods. *Australian Veterinary Journal* **70**, 4–6.

Reiss A, Portas T & Horsup A (2008). Hematologic and serum biochemical reference values for free-ranging northern hairy-nosed wombats (*Lasiorhinus krefftii*). *Journal of Wildlife Disease* **44**, 65–70.

Rose K (1999). Common diseases of urban wildlife. In *Wildlife in Australia, Proceedings 327,* pp. 365–427. Post Graduate Foundation in Veterinary Science: Sydney.

Rothwell JT & Canfield PJ (1988). Death due to a probable herpesvirus infection in a common wombat (*Vombatus ursinus*). *Australian Veterinary Journal* **65**, 360–361.

Skerratt LF (1998). Diseases and parasites of the common wombat (*Vombatus ursinus*) in the Healesville area of Victoria. In *Wombats* (eds RT Wells & PA Pridmore), pp. 317–328. Surrey Beatty & Sons: Sydney.

Skerratt LF (2001). Management of sarcoptic mange in wombat populations. In *Veterinary Conservation Biology, Wildlife Health and Management in Australia, Proceedings* (eds A Martin & L Vogelnest), pp. 271–276. Australian Veterinary Association: Sydney.

Skerratt LF (2003a). Cellular response in the dermis of common wombats (*Vombatus ursinus*) infected with *Sarcoptes scabiei* var. *wombati. Journal of Wildlife Diseases* **39**, 193–202.

Skerratt LF (2003b). Clinical response of captive common wombats (*Vombatus ursinus*) infected with *Sarcoptes scabiei* var *wombati. Journal of Wildlife Diseases* **39**, 179–192.

Skerratt LF, Phelan J, McFarlane R & Speare R (1997). Serodiagnosis of toxoplasmosis in a common wombat. *Journal of Wildlife Diseases* **33**, 346–351.

Skerratt LF, Martin RW & Handasyde KA (1998). Sarcoptic mange in wombats. *Australian Veterinary Journal* **76**, 408–410.

Skerratt LF, Middleton D & Beveridge I (1999). Distribution of life cycle stages of *Sarcoptes scabiei* var *wombati* and effect of severe mange on common wombats in Victoria. *Journal of Wildlife Diseases* **35**, 633–646.

Smales LR (1998). Helminth parasites of wombats. In *Wombats* (eds RT Wells & PA Pridmore), pp. 312–316. Surrey Beatty & Sons: Sydney.

Speare R (1988). Clinical assessment, diseases and management of the orphaned macropod joey. In *Australian Wildlife, Proceedings 104,* pp. 211–296. Post Graduate Committee in Veterinary Science: Sydney.

Spielman D (1994). First aid emergency care for Australian native mammals. In *Wildlife, Proceedings 233,* pp. 339–366. Post Graduate Committee in Veterinary Science: Sydney.

Spratt DM (1979). A taxonomic review of the lungworms (Nematoda: Metastrongyloidea) from Australian marsupials. *Australian Journal of Zoology* **7**, 1–45.

Storts RW (1995). Central nervous system. In *Protozoal Infections: Thomson's Special Veterinary Pathology* (eds WM Carlton & MD McGavin), pp. 380–382. Mosby: St Louis.

Strahan R (1995). *The Mammals of Australia* (2nd edn). Reed New Holland Publishing: Sydney.

Taggart DA, Steele VR, Schulz D, Dibben R, Dibben J & Temple-Smith P (1998). Semen collection and cryopreservation in the southern hairy-nosed wombat *Lasiorhinus latifrons:* implications for the conservation of the northern hairy-nosed wombat *Lasiorhinus krefftii.* In *Wombats* (eds RT Wells & PA Pridmore), pp. 180–191. Surrey Beatty & Sons: Sydney.

Triggs B (1996). *The Wombat: Common Wombats in Australia.* UNSW Press: Sydney.

Tyndale-Biscoe H (2005). Wombats: vegetarians of the underworld. In *Life of Marsupials* (H Tyndale-Biscoe), pp. 269–285. CSIRO Publishing: Melbourne.

Tyndale-Biscoe H & Renfree MR (1987). *Reproductive Physiology of Marsupials.* CUP: Cambridge.

Vogelnest L (1999). Chemical restraint of Australian native fauna. In *Wildlife in Australia, Proceedings 327,* pp. 149–188. Post Graduate Foundation in Veterinary Science: Sydney.

Webb S (2006). Fracture repair in the wombat. *Control and Therapy no. 4717.* Post Graduate Foundation in Veterinary Science: Sydney

Wells RT (1978). Field observations of the hairy-nosed wombat (*Lasiorhinus latifrons*). *Australian Wildlife Research* **5**, 299–303.

Wells RT (1989). Vombatidae. In *Fauna of Australia.* Vol. 1B: Mammalia. AGPS: Canberra.

West M, Paris MCJ, Cleary M, Shaw J, Galloway D & Trounson AO (2003). In vitro maturation (IVM) of oocytes collected following hormonal stimulation of common wombats (*Vombatus ursinus*). *Theriogenology* **59**, 405.

West M, Galloway D, Shaw J, Trounson A & Paris MCJ (2004). Oestrus cycle of the common wombat *Vombatus ursinus* in Victoria, Australia. *Reproduction, Fertility and Development* **16**, 339–346.

Wolfe BA (2003). Toxoplasmosis. In *Zoo and Wild Animal Medicine* (5th edn) (eds ME Fowler & RE Miller), pp. 745–749. WB Saunders: St Louis.

Woods R (1999). Prevention of disease in hand reared native wildlife. In *Wildlife in Australia, Proceedings 327,* pp. 455–490. Post Graduate Foundation in Veterinary Science: Sydney.

10. Dasyurids

Peter Holz

1 TAXONOMY, DISTRIBUTION AND BIOLOGY

The Order Dasyuromorphia are the carnivorous marsupials. They are characterised by having three pairs of approximately equal sized lower incisors and an absence of fusion between the second and third digits of the hind foot (Strahan 1995). The Order is divided into three Families: Thylacinidae, which contains the extinct thylacine (*Thylacinus cynocephalus*); Myrmecobiidae, which contains the numbat (*Myrmecobius fasciatus*); and Dasyuridae, which contains the remaining 66 species. This group includes the Tasmanian devil (*Sarcophilus harrisii*), quolls, antechinus, dunnarts, kowari (*Dasyuroides byrnei*) and phascogales (Krajewski & Westerman 2003). Its members are found throughout Australia and New Guinea occupying all environments ranging from high alpine regions (agile antechinus [*Antechinus agilis*], spot-tailed quoll [*Dasyurus maculatus*]) to coastal forests (quolls, Tasmanian devils, antechinus, phascogales) and deserts (mulgara [*Dasycercus cristicauda*], little red kaluta [*Dasykaluta rosamondae*], kowari).

Dasyurids range in size from the narrow-nosed planigale (*Planigale tenuirostris*) (4–9 g) to the Tasmanian devil (up to 10 kg). They were initially thought to be predominantly nocturnal, however, continued investigation has demonstrated that many species also forage during the day. Many species of dasyurids are arboreal while others are both arboreal and terrestrial and some are completely terrestrial.

The eastern quoll (*D. viverrinus*) and Tasmanian devil, once widespread throughout the Australian mainland, are now found only in Tasmania. Due to a combination of habitat destruction and introduced predators, such as the dingo and the fox, most dasyurid populations have declined since European settlement.

Dasyurids are not long-lived. Tasmanian devil life expectancy is up to 6 yr in the wild and 8 yr in captivity. Quolls can be expected to live up to 6 yr in captivity and 3 yr in the wild. The smaller dasyurids generally live for no more than 5 yr in captivity, usually up to 2 in the wild and occasionally up to 3. Males of some groups, such as the antechinus, all die in the wild after their first year (Table 10.1).

2 ANATOMY AND PHYSIOLOGY

Similar to the situation with eutherian carnivores, apart from size there is little variation in the body shape and anatomy of dasyurids. The most extreme variation in body shape of dasyurids is found in the kultarr (*Antechinomys laniger*). This species has elongated hind limbs and a long tufted tail.

2.1 Dentition

Dasyurids are polyprotodont marsupials as they have three pairs of lower incisors, unlike the diprotodonts which have only a single pair. Dasyurids also have four pairs of pointed upper incisors, well-developed upper and lower canines, two or three pairs of upper and lower blade-like premolars and four pairs of upper and

Table 10.1 Reproductive characteristics for a range of dasyurids

Species	Gestation (d)	Litter size	Weaning (d)	Sexual maturity (mo)	Mating	Birth	Life span (yr)
Agile antechinus	35	6–8	90–110	9–10	July–Aug	Aug–Sept	1–2
Brown antechinus	27	6–8	90–95	9–10	July–Sept	Aug–Oct	1–2
Dusky antechinus		8–10	90–95	9–10	May–Sept	June–Oct	1–2
Fawn antechinus	28–35	<10	120	9–10	Aug	Sept	1–2
Swamp antechinus	26–30	6–8	80–90	9–10	June–Aug	July–Sept	1–2
Tasmanian devil	31	1–4	250	24	Mar–April	April–May	7–8
Dibbler	44–53			10–11	Mar	May	1–4
Fat-tailed dunnart	13–16	7–10	70	4	July–Jan	July–Feb	1–2
Stripe-faced dunnart	12–13	<8	70	3–12	June–Jan	June–Feb	1–2
Red kaluta	38–62	8	110	10	Sept	Nov	1–2
Kowari	30–35	6	110	9	May–Nov	June–Dec	<5
Kultarr	12	<6	90–100	11.5	July–Jan	Aug–Dec	1–2
Mulgara	35–44	2–6	120		May–June	June–Sept	<6
Wongai ningaui	14–21	5–7	70		Sept–Jan	Sept–Feb	1–2
Pygmy planigale	20	5–11	70	9	All year	All year	2
Eastern quoll	21	6–8	135–150	12	May–June	May–Aug	3–4
Northern quoll		6–10	125–150	10–12	May–June	June–Jul	2
Tiger quoll	21	4–6	150	12	June–July	Jul–Aug	3–4
Western quoll	16–18	4–6	150	12	May–June	May–Sept	3

Source: McAllan (2003).

lower molars with sharp shearing cusps (Fig. 10.1). Antechinus are often mistaken for mice but are easily differentiated on the basis of dentition (Fig. 10.2).

2.2 Gastrointestinal system

Dasyurids are all carnivorous or insectivorous, although antechinus, phascogales and quolls also consume plant material and nectar, particularly in winter when invertebrates are scarce. Diets are highly digestible, high in protein, water, vitamins and minerals and low in carbohydrates. Therefore, glucose and energy requirements are met primarily from amino acid metabolism. The small intestine contains high levels of trehalase, an enzyme required to digest trehalose, a disaccharide found only in insects. Dasyurids have a short gastrointestinal tract consisting of a simple stomach, small intestine, large intestine and no caecum. Brunner's glands, which secrete an alkaline fluid that protects the proximal duodenal mucosa from the ulcerating effects of acid-pepsin secretions from the stomach, appear as a thick collar in the duodenum immediately distal to the pylorus. The glands form large lobes that drain into the duodenal lumen. Digesta transit times are brief with 100% excretion occurring 3–4 hr after ingestion in planigales and up to 13 hr in eastern quolls (Hume 1999).

2.3 Musculoskeletal system

The fore feet have five clawed toes, the first toe on the hind foot is either short or absent, and the second and third toes of the hind feet are not fused (didactyly) unlike the diprotodonts where they are fused (syndactyly). The presence of the first digit on the hind foot is usually indicative of an arboreal habit (Fig. 10.2). Two epipubic bones (rudimentary in the thylacine) extend from the anterior aspect of the pubis and support the pouch. As is the case in all marsupials other than bandicoots, dasyurids have no ossified patella. The tail of most species is shorter than the body length. Fat is stored in the tail of some species (*Pseudantechinus* spp.).

2.4 Skin

The number of teats ranges from four to 12. Many species lack a pouch. In phascogales eight teats are situated on a circular patch of skin on the abdomen. In other species low lateral folds develop at the sides of the mammary area of lactating females. These folds do not constitute a pouch in the strict sense and provide very little protection for the young which, as they develop, dangle from the mother's belly between her fore and hind legs. A permanent pouch is present in most ningauis (*Ningaui* spp.), the kultarr, spot-tailed quoll and Tasmanian devil. Dasyurids have aural, circumanal, interdigital, sternal, paracloacal and labial glands, which are used for scent marking.

2.5 Nervous and sensory systems

Dasyurids, like all marsupials and monotremes, have no corpus callosum. Arachnoid villi have been found in the lumbo-sacral spinal cord of the brown antechinus (*A. stuartii*), mulgara, kowari, fat-tailed dunnart (*Sminthopsis crassicaudata*) and white-footed dunnart (*S. leucopus*) (Attwood & Woolley 1974).

Dasyurids have a vascular retina. The arteries do not branch but form end arteries giving off discrete capillary loops. Marsupials have double cones, which are not found in eutherians. Dunnarts exhibit trichromacy, i.e. there are three separate cone visual pigments sensitive to long (LWS 535 nm), medium (MWS 509 nm) and ultraviolet (UVS 350 nm) wavelengths. Primates are the only other mammalian group to exhibit trichromacy. Trichromacy allows discrimination across a greater section of the visual spectrum to facilitate both diurnal and nocturnal activity. The LWS and MWS pigments are found in single and double cones predominantly in the central retina, while the UVS pigments are only found in single cones predominantly in the dorsotemporal retina. LWS cones are maximally sensitive to the green-yellow range of the spectrum, allowing discrimination of cryptically camouflaged prey such as insects and reptiles. Dunnarts have no tapetum and an intermediate rod:cone ratio of 40:1, which is compatible with an arrhythmic lifestyle. Each cone contains an oil droplet which filters incoming light, thereby enhancing contrast and reducing glare (Arrese et al. 1999; Arrese et al. 2002).

2.6 Metabolism and thermoregulation

Basal metabolic rate (BMR) for dasyurids can be calculated using the following equation:

$$\text{BMR} = 2.39 \times \text{body mass (BM)}^{-0.262}$$

Basal metabolic rates are highest in the small insectivores such as planigales, antechinus and phascogales, which are similar to eutherian rates. Rates for other

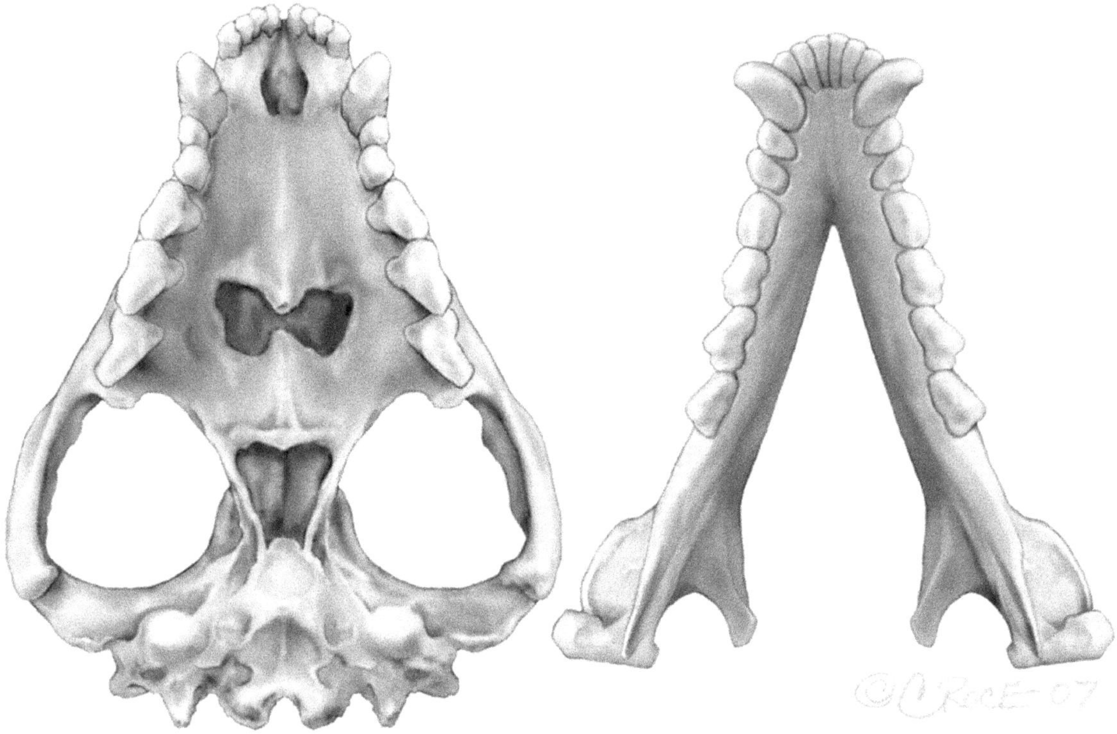

Figure 10.1 Dentition of a Tasmanian devil (I4/3, C1/1, P2/2, M4/4).

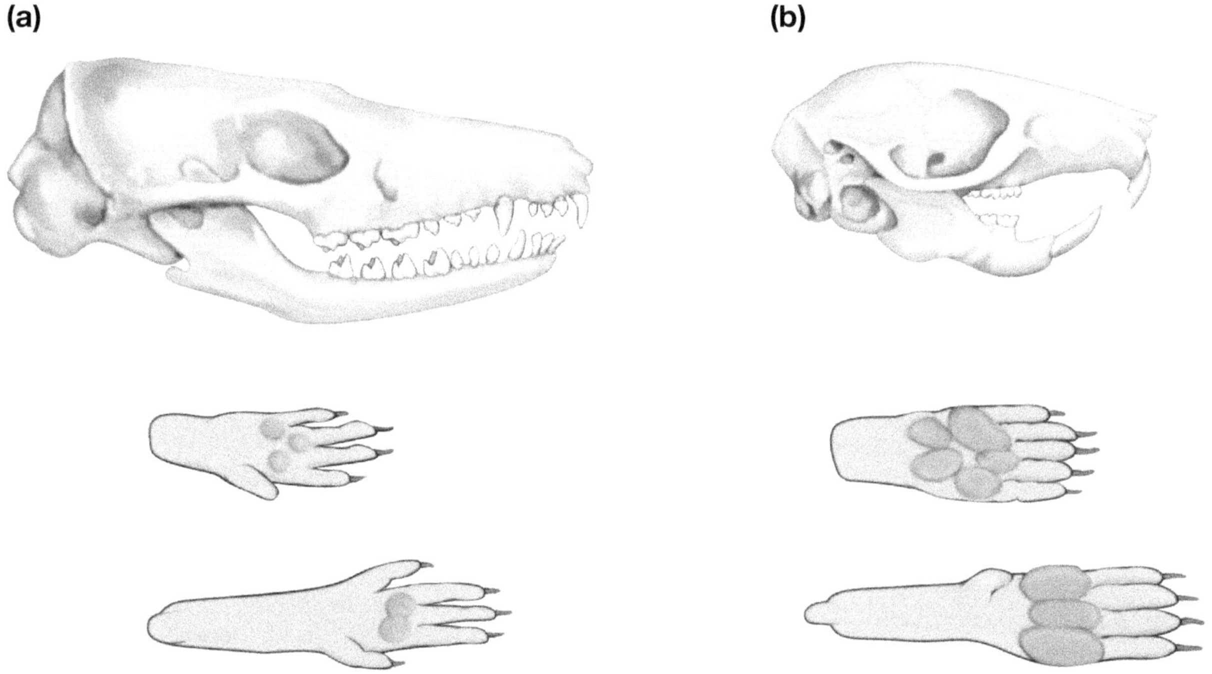

Figure 10.2 Brown antechinus (*Antechinus stuartii*) (a) are frequently mistaken as mice (b) but are easily differentiated on the basis of dentition and anatomy of the feet.

dasyurids, however, are less than half those for eutherian carnivores of similar size. Field metabolic rate (FMR), or the energy cost of free existence, is calculated by the following equation:

$$FMR = 8.36\ BM^{-0.148}$$

This ranges from six times BMR for 10 g dasyurids to twice BMR for 10 kg dasyurids (Geiser 2003).

Thermoregulatory ability is slow to develop. Kowaris cannot thermoregulate until about 100 d after birth. Body temperatures of adult dasyurids range from 32°C in planigales up to 38°C in the northern quoll (*D. hallucatus*) with most falling in the 34–36°C range. To combat cold, dasyurids produce heat by shivering. Brown adipose tissue, which is used by eutherians to generate heat, has not been found in marsupials. Dasyurids do not sweat and cope with excessive heat by licking and panting to increase evaporative cooling.

As many dasyurids are small they have a high surface area to volume ratio, which results in high heat loss. Therefore, most if not all species undergo periods of torpor to lower energy expenditure. They do not hibernate but enter shallow periods of daily torpor where the body temperature drops to varying degrees depending on species, ranging from 11°C in the kultarr to 28°C in the dusky antechinus (*A. swainsonii*). Metabolic rates drop to 10–60% of BMR during torpor bouts, which can last up to 19 hr but usually extend 2–8 hr (Geiser 2003).

The larger dasyurids are better able to conserve body heat due to their smaller surface area to body weight ratio. An eastern quoll weighing 1 kg has a daily energy requirement in summer of approximately 793 kJ. This increases to 1169 kJ/day in winter. Metabolisable energy content is approximately 3.5 kJ/kg for insects and 5 kJ/kg for wallaby flesh. In summer the quoll needs to consume 159 g of mammal flesh (two rat sized mammals) or 227 g of insects (340 insects). In winter this increases to 262 g of mammal flesh or 334 g of insects. In contrast, an 8 kg Tasmanian devil requires 518 g of prey in summer and 578 g of prey in winter, i.e. only twice as much as the quoll despite being eight times as large. The situation is even more extreme for the brown antechinus, which weighs approximately 26 g and has an FMR of 4218 kJ/kg in winter. It must consume 21 g of insects, almost its own body weight each day to survive (Tyndale-Biscoe 2005).

3 REPRODUCTION

3.1 Reproductive anatomy

Most dasyurid species exhibit sexual dimorphism with males being larger than females. Exceptions are the paucident planigale (*Planigale gilesi*) and the dunnarts, where there is no significant size difference, and the long-tailed planigale (*Planigale ingrami*) where the female is larger.

3.1.1 Female

Female marsupials have two uteri and two lateral vaginae. Between the vaginae is a mass of fibrous tissue called the urogenital strand. A birth canal forms within this strand during parturition. This canal becomes permanent after the first parturition in most macropods but in dasyurids the cleft closes rapidly after birth so that no indication of the canal remains.

3.1.2 Male

All dasyurids have a carrot-shaped prostate and bulbourethral glands, two pairs in Tasmanian devils and fat-tailed dunnarts, and three pairs in eastern quolls and brown antechinus.

3.2 Oestrous cycle, mating and gestation

Dasyurids are polyovular and most are monoestrus. Quolls, kultarrs, kowaris, dunnarts, ningauis and planigales are polyoestrus. They have an oestrous cycle which ranges from 24 d in the fat-tailed dunnart up to 60 d in the kowari. Oestrus usually lasts 2–3 d but can persist for up to 14 d in the brown antechinus. Gestation length ranges from 13 to 31 d. Post partum oestrus and ovulation are suppressed during lactation. Ovulation is spontaneous (McAllan 2003). Smaller species have more young in a litter than the larger dasyurid species. It has also been observed that some species give birth to more young than can be accommodated on the teats. However, due to the tiny size of these supernumerary neonates the extent of this phenomenon is poorly understood (Strahan1995).

Females will mate with more than one male, multiple paternity having been demonstrated in the agile antechinus and brush-tailed phascogale (*Phascogale tapoatafa*). Mating lasts from 1 hr in Tasmanian devils to 12 hr in agile antechinus. Sperm storage, which does not take place in other marsupial groups, occurs in the isthmus of the oviduct. Mating can occur up to 2 wk before ovulation in antechinus and up to 3 d in dunnarts. After mating, the sperm become immobile and reside in the oviduct until ovulation. At ovulation the sperm migrate to the site of fertilisation. At the same time the sperm head moves from its position parallel to the tail to one perpendicular to it, to maximise contact with the egg. The egg is about twice the size of a eutherian egg but the zona pellucida is much thinner. Biased sex ratios have been recorded in semelparous dasyurids, i.e. those that produce offspring in one season only, such as antechinus with first-year mothers producing more females than males. The converse occurs in iteroparous species, i.e. those that produce offspring over multiple seasons, with males more prevalent in the first year (Breed et al. 2003; Taggart et al. 2003).

In antechinus there is a 1–3 wk mating period after which all the males die before the females give birth. The mating period is governed by photoperiod and is longer in duration and later in the year the further north the populations are found. Brown antechinus mate when the rate of change of day length is 90–100 sec/d, while agile antechinus mate when the rate of change is 120–130 sec/d (Tyndale-Biscoe 2005).

Despite increasing testosterone levels, spermatogenesis ceases in mid June in the southern hemisphere, possibly due to photoperiod effects. Mortality occurs because of elevated plasma cortisol levels, which rise from 2.9 ± 0.51 μg/100 mL before mating up to 5.2 ± 0.6 μg/100 mL just prior to death, and a testosterone-dependent decrease in plasma corticosteroid binding globulin from 7.51 ± 0.53 μg/100 mL before mating to 1.96 ± 0.38 μg/100 mL during mating. While in the short term this glucocorticoid increase is beneficial as it enhances the mobilisation of energy reserves for a once-only reproductive effort, in the long term it leads to a range of ultimately fatal syndromes including hippocampal neuronal apoptosis, anaemia, anorexia, reduced splenic follicle size, heavy *Babesia* spp. parasitaemia, *Listeria monocytogenes*-induced hepatic necrosis, glomerulonephritis, renal tubular necrosis and gastrointestinal haemorrhage due to gastric and duodenal ulcers. If males are removed or castrated before the mating process begins they can survive for up to 5 yr (Barker et al. 1978; Bradley 2003).

Synchronous mating and complete post mating male mortality also occurs in phascogales and red kalutas, with some mortality reported in mulgaras, northern quolls, dibblers and ningauis. Mortality tends to occur in high-density populations, but not in low-density ones due to decreased competition between males for mates.

4 HUSBANDRY

A review of dasyurid husbandry can be found in Jackson (2003) and in the Exhibited Animals Protection Act 1995. Recommended dimensions for dasyurid enclosures can be found in Holz (2003) and in the code of practice for the welfare of wildlife during rehabilitation (AG0976 2001).

4.1 Housing and husbandry

Devils and quolls will climb, so a smooth metal skirt at least 1.2 m high around the enclosure is necessary to stop them escaping. Small dasyurids can be maintained in a glass-fronted enclosure made of solid wood at least 10-mm thick, as phascogales in particular are great chewers. Wood or metal nest boxes filled with shredded paper, sea grass or wood shavings should be provided. Arboreal dasyurids such as antechinus and phascogales require roofed enclosures with a maximum gap in the wire of 5–10 mm, depending on the species. Optimum temperature range is 15–30°C, so supplemental heating may be necessary. Enclosures should be spot cleaned daily and small enclosures should receive a full substrate change at least weekly.

Hospital requirements are similar to small eutherian carnivores. Small dasyurids can be maintained in enclosures similar to those described above or in stainless steel wire hospital cages. The wire mesh must be small enough to prevent escape of small species. A nest box should always be provided for security. Quolls and devils need to be held in sturdier accommodation with metal walls. Supplemental heat may need to be provided.

Dasyurids are just as prone to developing stereotypical behaviours as eutherian carnivores. Quolls and devils in particular are prone to stereotypic pacing. Free-ranging dasyurids spend a significant amount of time searching for food. If feeding routines are not varied, animals pace in anticipation of being fed just prior to their usual feeding time, spend a short time eating the meal then spend the remainder of their time sleeping or pacing. It is therefore important that the animal's housing and husbandry routines provide not only for their physical needs but also their behavioural needs. Enclosures should be interesting and varied with natural materials such as leaf litter, mulch, bark, branches and hollow logs. Enrichment activities should be provided and aimed at encouraging natural behaviours and stimulating senses. These may include the use of different feeding methods, varying and randomising feeding times and providing non-food related enrichment. These may include the use of bungee feeders, random scatter feeds of insects, fruits and nuts, hiding feed in the substrate and exhibit furniture, hanging papier mache balls filled with food and frozen blocks of blood. Prey species and other scents are very stimulating to dasyurids and these can be provided in the form of macropod faeces scattered in the exhibit, use of browse or nesting material from glider or possum enclosures or aromatherapy oils. Observational data has shown that macropod faeces provide the greatest stimulation for Tasmanian devils (Schaap 2002). The risk of disease (e.g. mycobacteriosis) transmission when using macropod faeces must always be carefully considered.

4.2 Individual marking and identification techniques

Dasyurids over 10 g can be permanently identified by inserting a passive integrated transporter or microchip SC between the scapulae. Tissue glue is used to seal the entry wound to prevent the microchip from tracking out along the injection site. Ear notching has been used in smaller individuals, and ear tattoos are another option.

4.3 Transport

Small dasyurids can be transported short distances in pillowcases or other cloth bags. Alternatively, they can be placed in a nest box with the entrance plugged up. Larger species can be moved in plastic crates designed for holding cats and small dogs. For longer-distance transport they must be placed in a wooden box with nesting material, a water bottle to prevent spillage and a small dish of food (Fig. 10.3).

Figure 10.3 Transport box for a Tasmanian devil.

5 NUTRITION

All dasyurids are carnivorous or insectivorous although antechinus, phascogales and quolls will eat some plant material and nectar. Small dasyurids are generalists and opportunists consuming predominantly arthropods such as beetles, spiders and ants, the size of the prey increasing with the size of the predator. Large dasyurids consume mostly vertebrates, prey size again varying with predator size. Tasmanian devils eat wallabies, wombats and a significant amount of carrion including introduced species such as rabbits and sheep. Spot-tailed quolls consume mostly rabbits and possums, while eastern quolls ingest small mammals, birds, lizards, insects and fruit such as blackberries. Western quolls (*D. geoffroii*) eat small mammals, lizards, frogs, insects and freshwater crustaceans (Hume 1999). Northern quolls consume large quantities of fruit, their favourite being the wild grape (*Ampelocissus asetosa*) and they enjoy nectar from eucalypt and grevillea flowers (Oakwood 2004).

In captivity small dasyurids such as planigales, dunnarts, antechinus, phascogales and kowaris have been successfully maintained on a mixed diet containing dog chow, hard-boiled eggs, cheese, day-old chicks and mice. This can be supplemented with assorted invertebrates such as mealworms, earthworms, fly pupae, crickets and moths, depending on what is available. A commercial insectivore mix can also be used, and pollen grains and blossoms can be fed when available. Tasmanian devils and quolls are fed dog chow and whole animals such as guinea pigs, rabbits, rats, mice, pilchards and day-old chicks (Table 10.2). Quoll diets can also be supplemented with mixed invertebrates and they may enjoy a small amount of soft fruit such as bananas, grapes and melon.

Anorexic dasyurids may be tempted to eat using favourite food items or live food. If the problem is one

Table 10.2 Sample diets for captive dasyurids

Tasmanian devil	
Ad lib	Water
Daily diet/ animal	
Monday	4 day-old chicks (35 g), 1 egg (60–70 g)
Tuesday	½ guinea pig (150–200 g)
Wednesday	1 rabbit, wallaby, sheep piece (1600 g)
Thursday	4 day-old chicks (35 g), 1 egg (60–70 g)
Friday	1 rat (250 g)
Saturday	Starve day
Sunday	1 rat (250 g) or ½ adult chicken
Kowari	
Ad lib	Water
Daily diet/ animal	½ day-old chick or mouse
	20 g moist dog chow
	1 g fly pupae
	5 mealworms
Supplement	0.2 g pollen grains once/week
	6 g egg and cheese 3–4 times/week
	2 crickets 3–4 times/week
	2 moths 3–4 times/week
Antechinus	
Ad lib	Water
Daily diet/ animal	6 g moist dog chow
	10 g egg or cheese
	2 mealworms
	1/8 day-old chick or mouse
	1 g fly pupae
Supplement	2 crickets 3–4 times/week
	2 earthworms 1–2 times/week
	2 moths 3–4 times/week
	3 pollen grains once/week
	Blossoms as available

of aggression, multiple feeding stations or the separation of warring individuals may be required. Alternatively, individual dasyurids that do not eat after being placed in a new environment may benefit from the inclusion of a companion.

Daily energy requirements for late lactation are double those for non-breeding females. For the eastern quoll to double its food intake it must consume half its body weight daily, while the antechinus needs to ingest more than its own body weight. If this cannot occur the female either draws on her own reserves and loses weight or selectively reduces her litter size by culling some of the young. To prevent this, lactating females should be fed 3–4 times maintenance.

Tasmanian devils and quolls tend to become obese and should be fed 5–10% of their body weight daily. Devils will benefit from a starve day once a week.

Metabolic bone disease has occurred in small dasyurids fed on a strict mutton diet. Affected animals had soft translucent bones. The problem was rectified by the addition of 1 tablespoon of calcium carbonate to 1 kg of meat. A laboratory colony of fat-tailed dunnarts died of iodine deficiency. Affected animals had enlarged thyroids containing follicles lined by a cuboidal to columnar epithelium that were devoid of colloid; iodine supplementation of 60 μg/d prevented mortalities (Obendorf 1993). If a dasyurid's diet is varied and balanced there should be no need to supplement with vitamins or minerals. Dental disease is uncommon in dasyurids and may be associated with the practices of feeding whole prey items and insects. Obesity is also common; energy intake should be monitored and controlled if required and activity encouraged through behavioural enrichment.

6 CAPTURE AND RESTRAINT

6.1 Capture and physical restraint

Physical restraint is only used for minor procedures or to induce general anaesthesia as it will result in considerable struggling and attempts to bite. This will produce a minor degree of pain and inconvenience when caused by an antechinus or dunnart, and the larger members of the group can inflict severe injuries.

When restraining or transporting a dasyurid it is advantageous to place it in a bag. This can range from a small calico bag or pillowcase for smaller animals up to a large hessian sack for Tasmanian devils. Devils must be closely monitored as they can chew their way out of bags. Animals in bags tend to relax as outside stimuli and visual threats are removed. The chance of injury is decreased as they cannot lash out against solid objects.

Small dasyurids can be caught in their nest box or with an Elliott trap. They are gripped by the scruff of the neck and placed in a bag. Alternatively, they can be held firmly around the body. It is important not to clasp them too firmly, as suffocation is possible.

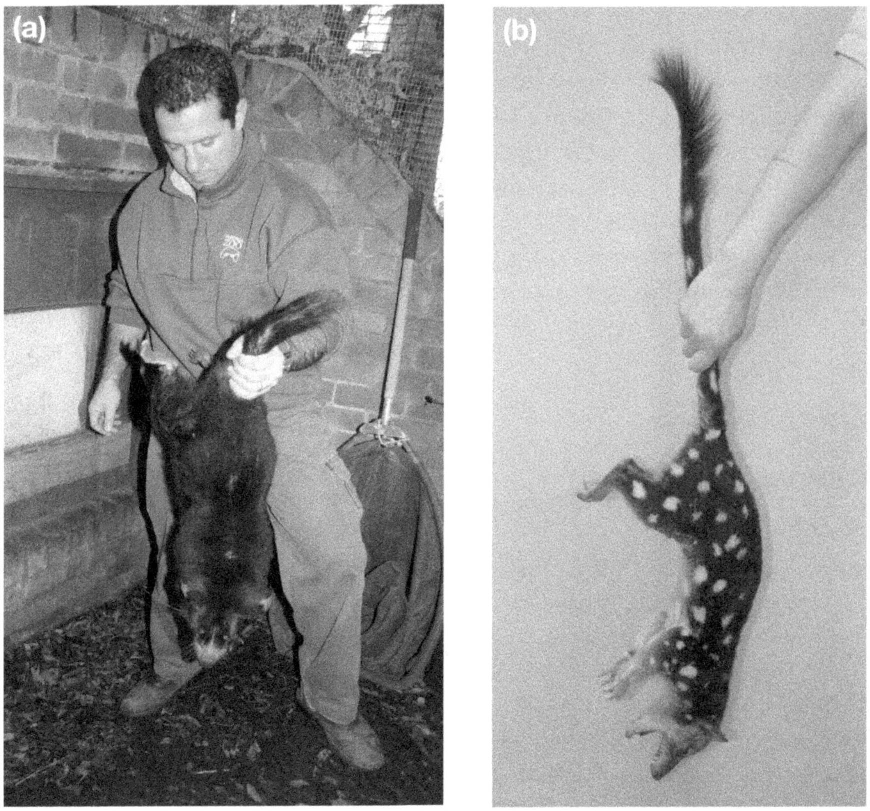

Figure 10.4 Capture of (a) a Tasmanian devil and (b) a spot-tailed quoll. Photo: Larry Vogelnest.

Larger dasyurids, such as quolls and devils, can be caught in a net or restrained by the tail and lowered into a bag. Devils and quolls can easily be held for short periods by the tail (Fig 10.4). To avoid injuring the tail it should be grasped toward its base. Some quolls are agile enough to turn and bite the handler. Dasyurids should not be carried long distances held only by the tail. Smaller dasyurids, such as antechinus, dunnarts and phascogales, should not be restrained by the tail, as severe degloving injuries and tail fractures can occur.

6.2 Chemical restraint

Most dasyurids are not prone to regurgitation under anaesthesia, but pre-anaesthetic fasting for 6–8 hr is recommended (Vogelnest 1999).

Sedation for transport is generally not required but very nervous animals may benefit from the administration of 1–2 mg/kg diazepam IM. Effects should be obvious after approximately 20 min.

Inhalation anaesthesia is the method of choice for dasyurids. If the animal can be manually restrained the safest way to induce general anaesthesia is via a non-rebreathing circuit (e.g. Ayre's T-piece) and a mask. For animals in bags anaesthesia can be induced through the bag or by extracting the head and placing it in the mask. It is important to grasp quolls and devils very firmly behind the head, as they will attempt to pull back out of the mask and potentially bite the operator.

5% isoflurane delivered in oxygen at a flow rate of 200 mL/kg/min with a minimum of 1 L/min is used for induction. Maintenance of anaesthesia generally requires 2% isoflurane, but this varies between species and individuals. Tasmanian devils and quolls have a wide gape and are easily intubated for more prolonged procedures. If isoflurane is unavailable halothane is an acceptable substitute, but is not as safe and requires close monitoring.

Injectable agents can also be used if gaseous anaesthesia is not available. To induce general anaesthesia tiletamine/zolazepam can be administered at 7–10 mg/kg IM (Vogelnest 1999). The advantage of this combination is its low volume and rapid effect. However, relaxation is variable with hypersalivation and constant limb and jaw movement sometimes occurring in Tasmanian devils. Recoveries can also be prolonged, in excess of 6 hr in one devil (Holz 1992). Xylazine 4 mg/kg combined with ketamine 20 mg/kg IM provides light anaesthesia with better muscle relaxation (Vogelnest 1999). Although likely to be useful, there are no reports of ketamine or tiletamine/zolazepam and medetomidine combinations or alfaxalone being used in dasyurids.

7 CLINICAL PATHOLOGY

7.1 Haematology and biochemistry

7.1.1 Sample collection

Anaesthesia is usually required for successful venipuncture in dasyurids. The ventral coccygeal, femoral, medial metatarsal, lateral saphenous, cephalic and jugular veins are suitable for blood collection. Access to the ventral coccygeal vein is by inserting the needle perpendicular to the tail, in the ventral midline, and advancing it until the vertebrae are reached. Withdraw the needle slightly and blood should enter the needle hub. This vein is useful for smaller dasyurids. No more than 10% of total blood volume (approximately 1% of body weight) should be collected at any one time. Access to the femoral vein or artery is by directing the needle at the pulse felt in the inguinal region (Fig. 10.5). Arterial blood is often obtained and digital pressure is required to prevent haematoma formation. The medial metatarsal vein is a small vein running along the medial aspect of the hind leg. The cephalic vein is present on the dorsal surface of either fore leg. The lateral saphenous vien is on the lateral aspect of the tibia just proximal to the tarsal joint. The latter 2 veins are useful in Tasmanian devils and quolls. The jugular vein is suitable for blood sampling in all dasyurids (Fig. 10.6). It is best accessed with the animal lying in dorsal recumbency with the neck extended. It can be found along the length of the neck in larger species with access relatively easy just at the thoracic inlet. At this point the needle can be directed almost perpendicular to the skin (L Vogelnest *pers. comm.*). Blood collection from the periorbital sinus is not acceptable.

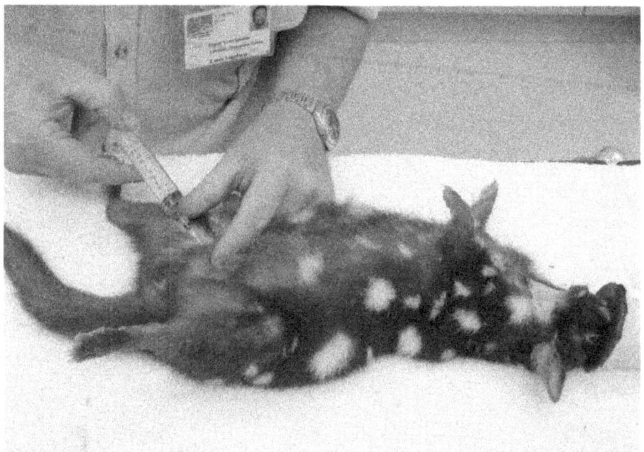

Figure 10.5 Femoral venipuncture in an eastern quoll. Photo: Taronga Zoo.

7.1.2 Reference ranges and interpretation

Dasyurids are born at an early stage of development and PY still have foetal haematological characteristics, e.g. fat-tailed dunnarts have 100% nucleated erythrocytes at 1 d of age decreasing to 15% by 4 d of age. Howell-Jolly bodies can be found in normal blood from eastern quolls, western quolls, fat-tailed dunnarts and stripe-faced dunnarts (*S. macroura*). Heinz bodies have been found in eastern quolls. Annular leucocytes occur in Tasmanian devils, eastern quolls and dunnarts. Western quolls tend to produce a lymphocytosis, rather than a neutrophilia, in response to inflammation

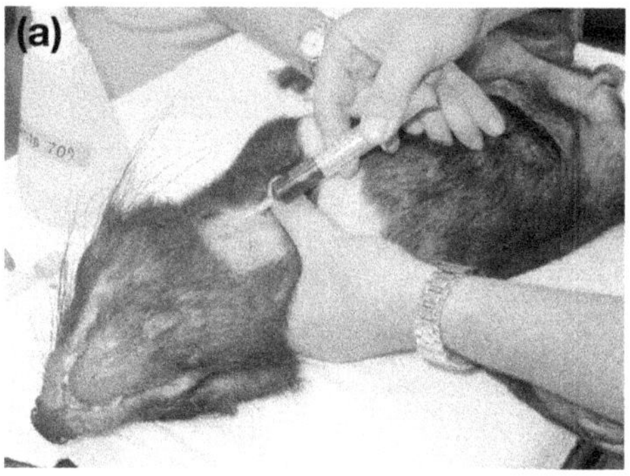

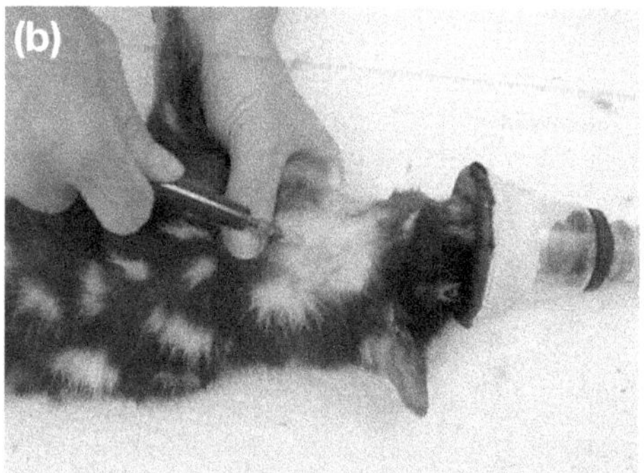

Figure 10.6 Jugular venipuncture in (a) a Tasmanian devil and (b) an eastern quoll. Photos: Taronga Zoo.

(Clark & Boardman 2005). Brown antechinus erythrocytes produce high levels of lactate from glucose. Dasyurid haemoglobin has a low oxygen affinity and a high oxygen-carrying capacity (24 mL of oxygen in 100 mL of kowari blood compared with 20 mL of oxygen in 100 mL of blood for many placental mammals) (Hallam et al. 1995). Dasyurids also have a high haematocrit, large tidal volume and high ventilation perfusion ratio. This is consistent with smaller animals having a higher metabolic rate and greater oxygen demands. Dasyurid spleens contain areas of extramedullary haematopoiesis with megakaryocytes visible histologically. A range of protozoa can be found in dasyurid blood (see 9.3.4). For a complete review of dasyurid haematology see Clark (2004).

Tasmanian devils have high plasma levels of acid phosphatase. The reason is unknown. Highest levels occur in the reticulo-endothelial system and reproductive organs. The ratio of cortisol:corticosterone in the plasma is 10:1 in quolls and Tasmanian devils. Dasyurids have high normal blood urea levels compared with domestic mammals. The reason is unknown. Table 10.3 provides haematology and biochemistry values for various dasyurid species.

8 HAND-REARING

Dasyurids have relative lactose intolerance due to low levels of intestinal lactase and should be reared on low-lactose milk. In Tasmanian devils and eastern quolls solids, lipids, protein and calcium peak in late lactation, carbohydrates peak in mid lactation and iron peaks in early lactation. Di-Vetelact (Sharpe Laboratories), Biolac and Wombaroo are all satisfactory milk substitutes for dasyurids. Upon presentation the orphaned PY should be assessed for injuries, hydration status and hypothermia. It should only be hand-raised if it presents in good condition, does not have any injuries resulting in future permanent handicaps and can be released at maturity. After assessment it should be warmed up and, if there is a delay before delivery to the intended carer, given an electrolyte solution at 100–150 mL/kg/24 hr. Orphaned PY should be held in artificial pouches made from soft material such as wool with a cotton lining. Hand-rearing furless PY has a poor success rate as they are normally attached to the mother's teat, have an extremely immature immune system, are heavily dependent on passive maternal immunity and are very susceptible to infection. Strict attention to hygiene and sterilisation of feeding utensils is important. Their skin must be treated regularly with an emollient such as sorbolene or lanolin to prevent it drying out and cracking. Feeding should be every 2 hr and the PY maintained at 34–36°C until well furred, when the temperature can be reduced to 30°C. Once fine fur has appeared feeding frequency can decrease to every 3–4 hr. Daily milk consumption varies from 10% to 20% of body weight. PY require stimulation after each feed to urinate and defecate. Rub the cloaca gently with a damp piece of cottonwool. Phascogales, antechinus and other small dasyurids are difficult to feed as they have small mouths. A very fine teat is required, or they may suckle from a fine paintbrush. Older animals may lap from the end of a syringe or eye ointment dispenser. Once the teeth have emerged and the young starts leaving the pouch, solid food such as chicken pieces and live mealworms and crickets should be offered. Live food is usually very popular. Feeding frequency can decrease to every 4–6 hr. Milk quantity should be gradually reduced and solid food increased until milk is totally eliminated from the diet. Animals for release must have minimal human contact to avoid imprinting, and access to natural prey items. Release should occur shortly before dusk and additional food may need to be provided until hunting skills are developed (Jackson 2003).

9 DISEASES

9.1 Dermatological diseases

9.1.1 Parasitic diseases

Demodex spp. have been reported from the brown antechinus, agile antechinus, spot-tailed quoll, southern dibbler (*Pseudantechinus apicalis*), red kaluta, northern quoll and hairy-footed dunnart (*S. hirtipes*) causing skin nodules, which represent a granulomatous dermatitis and folliculitis histologically (Fig. 10.7). In affected antechinus one to three solid nodules were found on the hind legs, head, tail base, cloaca or near the pouch. Nodules were filled with mites and had a honeycomb appearance on cut surface. Giant cells were present. Spontaneous regression occurred in two cases. Surgical removal in one case resulted in recurrence a month later. In quolls nodules were found on the feet and abdominal wall. Mites parasitised hair follicles causing hypertrophy

Table 10.3 Haematology and biochemistry reference ranges for selected dasyurids

Parameter	Tasmanian devil	Eastern quoll		Western quoll	Spot-tailed quoll	Kowari	Brush-tailed phascogale[a]	Fat-tailed dunnart		Brown antechinus	
		Male	Female							Male	Female
Hb (g/L)	137 (22)	176 (23)	159 (28)	156.6 (13.7)	175 (10)	187 (16)	145–178	107 (6)		161 (8)	150 (11)
PCV (%)	37.3 (6.8)	43 (13)	40 (6)	45 (4.1)	49.3 (5.4)	52.9 (4.2)	38–54	32.5 (1.9)		45 (2)	41 (2)
RBC (× 10^{12}/L)	6.25 (2.19)	10.1 (3.0)	9.8 (1.6)	8.47 (0.68)	8.05 (0.74)	10.10 (1.04)	7.4–10.8	6.8 (0.4)		10.76 (0.38)	9.54 (0.53)
MCV (fl)	64.5 (15.3)	40.6 (5.9)	40.4 (2.8)	55.2 (2.07)	61.0 (4.1)	51.7 (2.9)	47–55	48.2 (1.4)		42 (2)	43 (0.5)
MCH (pg)	23.7 (7.0)	15.6 (0.5)	15.2 (3.1)	18.4 (1.1)	21.8 (2.3)	18.4 (1.0)	15.7–18.5			15 (0.8)	15.7 (0.5)
MCHC (g/L)	366 (38)			335 (9)	358 (40)	357 (18)	313–349	330 (5)		354 (10)	365 (10)
Reticulocytes (× 10^9/L)		130 (80)	144 (96)				–	600 (100)		70 (19)	45 (6)
WBC (× 10^9/L)	6.0 (2.5)	5.7 (3.8)	5.4 (4.4)	4.9 (2.5)	4.5 (3.0)	2.9 (1.6)	1.7–4.6	1.9 (0.4)		7.6 (1.3)	5.9 (0.9)
								Male %	Female %	%	%
Neutrophils (× 10^9/L)	3.7 (2.2)	3.1 (1.3)	2.4 (1.3)	2.9 (1.9)	1.5 (0.6)	1.2 (0.7)	26–59	43.1 (2.5)	34.4 (3.4)	16 (1)	17 (3)
Bands (× 10^9/L)	0.3 (0.2)				0.0 (0.0)	0.2 (0.1)	0	5.7 (0.6)	4.5 (0.5)		
Lymphocytes (× 10^9/L)	2.2 (1.1)	3.2 (0.7)	2.2 (0.9)	1.5 (0.8)	1.6 (0.9)	1.5 (1.1)	24–65	49.6 (2.6)	60.4 (3.4)	78 (1)	78 (3)
Monocytes (× 10^9/L)	0.1 (0.1)	0.3 (0.01)	0.3 (0.05)	0.1 (0.1)	0.2 (0.2)	0.1 (0.1)	0–23	1.5 (0.3)	0.7 (0.1)	6 (1)	6 (1)
Eosinophils (× 10^9/L)	0.2 (0.3)	0.2 (0.05)	0.3 (0.04)	0.3 (0.2)	0.2 (0.2)	0.1 (0.1)	0–1	0.0	0.0		
Basophils (× 10^9/L)	0.1 (0.0)				0.0 (0.0)	0.0 (0.0)	0	0.0	0.0		
Platelets (× 10^9/L)	226 (26)				342 (69)	368 (141)	442–555				
Fibrinogen (g/L)	0.03 (0.00)			2.3 (1.5)	1.5 (1.0)	1.0 (1.1)					
BUN (mmol/L)	11.8 (3.6)	14.3–19.4		19.9 (2.4)	15.4 (1.4)	26.4 (8.2)	17.7–25.3				
Creatinine (μmol/L)	62 (18)	40–70		54 (10)	88 (18)	27 (18)	10–30				
Phosphorus (mmol/L)	1.7 (0.5)	1.6–2.0		2.3 (0.5)	2.3 (0.4)	1.8 (0.4)	0.9–2.4				
Calcium (mmol/L)	2.3 (0.2)	2.3–2.6		2.4 (0.1)	2.3 (0.3)	2.1 (0.2)	2.0–2.3				
Glucose (mmol/L)	6.3 (1.3)	5.0–7.3		6.2 (1.7)	5.9 (0.9)	8.0 (3.6)	4.0–13.7				

(Continued)

Table 10.3 *(Continued)*

Parameter	Tasmanian devil	Eastern quoll		Western quoll	Spot-tailed quoll	Kowari	Brush-tailed phascogale[a]	Fat-tailed dunnart	Brown antechinus	
		Male	Female						Male	Female
Total protein (g/L) colourimetry	60 (6)	51–58		59.5 (7.1)	64 (14)	48 (7)	50–63			
Albumin (g/L) colourimetry	33 (3)	33–36		31.5 (2.1)	40 (2)	28 (5)	34–42			
Globulin (g/L) colourimetry	27 (7)				21 (5)	18 (2)				
CK (U/L)	1152 (1210)	102–828		1423 (1323)	288 (0)	895 (1883)	112–860			
ALP (U/L)	104 (66)	198–232		469 (231)	46 (11)	53 (24)	98–423			
ALT (U/L)	26 (9)	25–51		47 (22)	24 (14)	123 (66)				
AST (U/L)	87 (50)	53–60		115 (65)	150 (85)	152 (96)	10–261			
Sodium (mmol/L)	148 (5)	134.5			14 (2)	150 (0)	143–150			
Potassium (mmol/L)	5.2 (2.1)	5.3			4.4 (0.6)	4.4 (0)	2.7–4.6			
Chloride (mmol/L)	116 (4)	113–118			115 (3)	112 (0)	110–120			
Bicarbonate (mmol/L)		15–22					15–28			
Cholesterol (mmol/L)	3.4 (0.8)			4.4 (1.1)	3.8 (0.8)	2.0 (0.2)				
Total bilirubin (μmol/L)	214 (14)	2–8		2.7 (1.5)	3 (2)	10 (3)	2–11			
Direct bilirubin (μmol/L)	5 (5)				7 (0)					
Indirect bilirubin (μmol/L)	19 (21)				0 (0)					

[a]Range

Source: Cheal (1976), Melrose et al. (1987), Haynes & Skidmore (1991), Svensson et al. (1998), ISIS (2002), Healesville Sanctuary records (2006). Sample sizes are available in the original references.

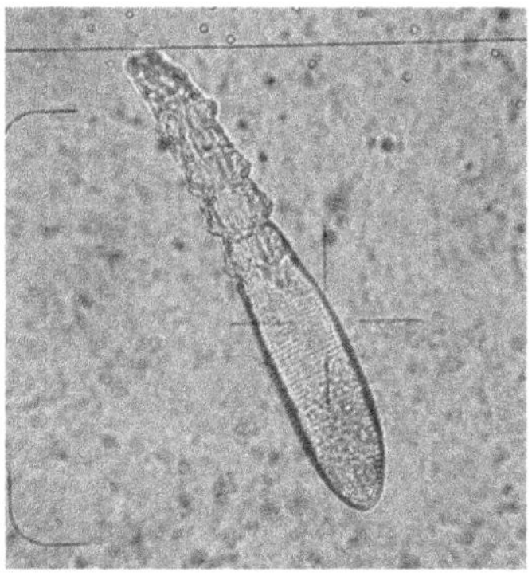

Figure 10.7 *Demodex* from an antechinus.

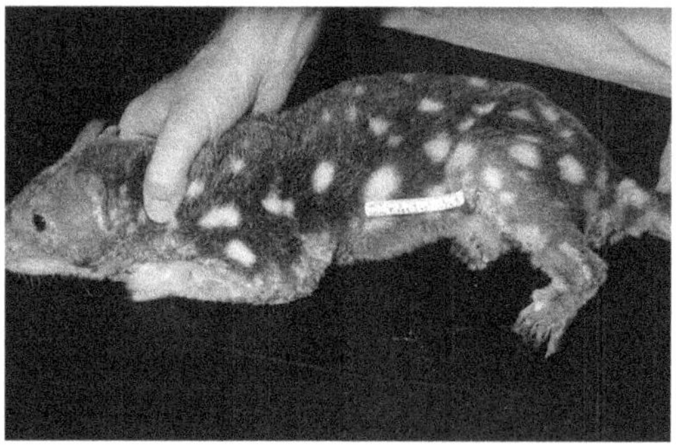

Figure 10.8 *Uropsylla tasmanica* infection in a spot-tailed quoll.

and eventual destruction of the follicle, leading to a predominantly mononuclear cell infiltrate and ulceration. Unlike the antechinus, surgical excision was curative. In the dibblers nodules appeared on the face. Seven animals were treated 2–4 times with 0.1 mg/mL ivermectin diluted in propylene glycol and dabbed onto the lesions at 7 d intervals, resulting in resolution in five individuals (Nutting & Woolley 1965; Holz 1998; Eden et al. 2004; Desch & Holz 2006).

Sarcoptiform mites belonging to the genus *Diabolicoptes* and *Satanicoptes* have been found on large dasyurids, associated with alopecia, hyperkeratosis and erythema.

There are six species of ticks, from the genera *Ixodes* and *Haemaphysalis*, recorded on dasyurids. They include the paralysis tick, *Ixodes holocyclus*, which also transmits the filarioid *Cercopithifilaria johnstoni*. This nematode occurs in the subcutaneous tissues of Tasmanian devils. No clinical signs have been associated with any of the ticks.

There have been 32 species of fleas reported from dasyurids. *Uropsylla tasmanica* occurs on Tasmanian devils and quolls. The adults are found mainly on the scrotum, lower limbs, face and ears. They attach their eggs to the basal portion of hairs. After hatching, the larvae burrow into the skin and feed on the subcutaneous tissues of the host until pupation, causing irritation leading to the formation of pustules, self-trauma and hair loss (Fig. 10.8). Treatment is with topical parasiticides such as pyrethrin spray 1 mg diluted in 10 mL water, fipronil 5–10 mg/kg or selamectin 6 mg/kg. Changing nest material and washing nest boxes frequently helps decrease ectoparasite numbers (Obendorf 1993). One study found imidacloprid to be safe and effective at a dose of 10.4 mg/kg in eastern quolls and 26.3 mg/kg in fat-tailed dunnarts (Baker & Beveridge 2001). In situations where reintroduction of fleas into an environment is limited, lufenuron has been used successfully to control fleas (L Vogelnest *pers. comm.*).

Dasyurid lice are all biting lice belonging to the genus *Boopia*.

Ophidascaris robertsi is a python ascarid whose large (up to 15 cm) larval stages have been found in the viscera and subcutaneous tissues of the brown antechinus, brush-tailed phascogale, quolls and dusky antechinus (Fig. 10.9). Plerocercoids of *Spirometra erinacei* (spargana) have been found subcutaneously in numerous dasyurid species.

9.1.2 Cutaneous mycobacteriosis

Cutaneous mycobacteriosis has been reported in spot-tailed quolls and Tasmanian devils. The causative agents are saprophytic mycobacteria such as *Mycobacterium chitae, M. fortuitum, M. smegmatis* and *M. ulcerans.* Lesions consist of single to multiple thickened areas of skin that range from focal plaques to abscesses with fistulous tracts. Histologically affected areas consist of a

Figure 10.9 *Ophidascaris robertsi* larvae in the thoracic cavity of a spot-tailed quoll. Photo ARWH.

pyogranulomatous dermatitis containing acid-fast bacilli, neutrophils, mononuclear cells, giant cells, necrosis and mineralisation. Most commonly affected sites are the dorsal neck, thorax and axilla occurring secondary to bite wounds associated with breeding. Treatment is invariably unsuccessful but can be attempted using a combination of surgical debridement and prolonged treatment with antimicrobial drugs such as amikacin 3 mg/kg bid, enrofloxacin 2.5 mg/kg bid and rifabutin 20 mg/kg sid (Raymond et al. 2000). Euthanasia is generally recommended if cutaneous mycobacteriosis is diagnosed.

9.1.3 Tasmanian devil facial tumour disease (DFTD)

DFTD was first recognised in Tasmania in 1997. None of the biologists trapping Tasmanian devils between 1964 and 1996 reported any lesions resembling DFTD (Hawkins et al. 2006). Tumours present as large solid soft tissue masses that ulcerate, first appearing on the head and/or neck regions (Fig. 10.10). The cut surface is firm, pale and slightly translucent with visible fibrous septa, often with a necrotic core. Histologically they form sub-epithelial expansile masses of round cells (8.08 ± 1.43 μm in diameter) with abundant eosinophilic cytoplasm and a high nuclear to cytoplasmic ratio (1:1.2) encased within a pseudocapsule. Mitotic figures range from zero to 12 and average four per high-power field. Inflammation is minimal, with only 7% of cases having lymphocytic infiltrates. Based on immunohistochemical stains the tumour appears to be an undifferentiated sarcoma of neuroendocrine origin that is locally aggressive and metastasises in 57% of cases to draining lymph nodes. In 41% of cases metastases are found further afield – 47% in the lung, 12% in the spleen, 6% on the heart, ovary and serosal surface of the rib and 5% in the kidney, mammary, pituitary and adrenal glands. Lymphocytes with a moderate to high mitotic index are commonly seen infiltrating other organs such as the liver. Transmission electron microscopic examination of tumour cells has not identified a viral cause (Loh et al. 2006a; Loh et al. 2006b).

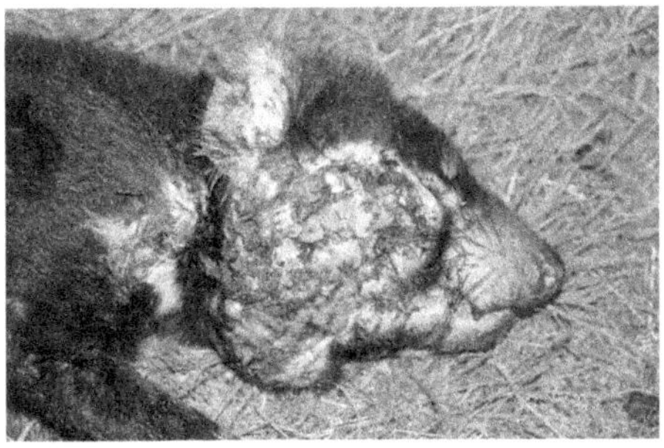

Figure 10.10 Tasmanian devil facial tumour disease.

Tumours are rarely seen in animals less than 2 yr of age (17% of cases seen occurred in 2 yr olds, 52% in 3 yr olds, 26% in 4 yr olds and 5% in devils older than 4) with males and females equally represented. Once clinical, the course of DFTD is rapid with the tumour enlarging over the course of weeks. Death usually occurs within a few months. Mortality is 100%, mostly due to starvation as the tumour destroys facial bones and dental arcades. Mature adults are scarce in regions that have been affected for a long time, with most animals being <2 yr old. Devil numbers peaked at 150 000 in the mid 1990s but mean devil sightings fell by 41% from 1992–95 to 2002–05 as a result of DFTD (Hawkins et al. 2006).

After having first been observed in the north-east of Tasmania at Mount William National Park, the distribution of DFTD has progressively increased and affected devils are now found over much of the eastern half of Tasmania. There are two apparently unaffected areas, on the north-west and east coasts, surrounded by diseased populations (Hawkins et al. 2006).

Tasmanian devils have low genetic diversity with a karyotype of 14XX or 14XY. They differ from other dasyurids by having a pericentric inversion of chromosome 5. However, 4 of 100 devils examined had both inverted and non-inverted forms of chromosome 5. Because of the characteristic chromosomal rearrangement found in the cells of every tumour (all tumours examined lack both sex chromosomes, both chromosomes 2 and one chromosome 6) direct exposure to tumour cells is regarded as necessary for the development of DFTD but that alone is not sufficient for disease to occur. Damage to the skin or mucus membranes around the head and neck, as occurs with fighting, scratching and biting, is also required before DFTD can develop. It is noteworthy that lesions are rarely observed on other parts of the body that are also subject to trauma. One diseased animal had a pericentric inversion of chromosome 5 throughout its normal tissues, but not in the cells of the facial tumour. Because of the differences in chromosome 5 between the tumour and the animal, the tumour could not have developed from the animal's own tissues but acted as an allograft (Pearse & Swift 2006). This provides solid evidence for the transmissibility of DFTD. It is possible, but not certain, that animals whose chromosome 5s are both of the non-inverted form might be resistant to development of the tumour.

Incubation period is unknown but one animal developed DFTD after 10 mo in captivity without apparent exposure to tumour cells during that time. To date, spot-tailed quolls and eastern quolls have not been diagnosed with the disease.

Currently there is no treatment. Management options are limited to attempting to separate diseased animals from disease-free susceptible animals. As the incubation period may be long and there is no preclinical diagnostic test, confidently identifying disease-free animals is difficult (AusVet 2005).

9.1.4 Miscellaneous conditions

Pedunculated papillomas of unknown aetiology have been described from the foot and head of quolls and Tasmanian devils. Lesions resembling epitrichial sweat gland cystomatosis have been described in a spot-tailed quoll (L Vogelnest *pers. comm.*; ARWH 2007 case number 5178). Many skin neoplasms have been reported in dasyurids (Table 10.4).

Dasyurids moult during or after the breeding season and animals may present with patchy fur loss. This is normal, and affected animals will regrow their fur over the following months. Aged dasyurids can also develop progressive fur loss, which will not regrow. Adult Tasmanian devils frequently have patchy alopecia, reddened skin, small skin wounds (particularly in wild animals and those kept in pairs), minor crusting skin lesions and occasionally pustules. They are frequently pruritic. Ectoparasites, trauma from bite wounds and pyoderma may be implicated in some cases, however, a cause is frequently inapparent (L Vogelnest *pers. comm.*). Multifocal to coalescing, exudative and crusting skin lesions, particularly affecting the hind quarters, are frequently reported in wild spot-tailed quolls (K Firestone *pers. comm.*). The cause of this condition is yet to be determined but is most likely the result of a combination of parasitic, environmental and social factors.

Table 10.4 Neoplasms recorded from dasyurids

Squamous cell carcinoma	Tasmanian devil × 9
	Eastern quoll × 3
	Tiger quoll × 3
	Kowari × 6
	Common planigale × 2
	Fat-tailed dunnart × 3
	Brown antechinus × 1
	Brush-tailed phascogale × 1
	Red-tailed phascogale × 1
Trichoepithelioma	Tasmanian devil × 2
	Eastern quoll × 1
	Kowari × 1
	Brown antechinus × 1
	Brush-tailed phascogale × 2
Lymphosarcoma	Tasmanian devil × 2
	Northern quoll × 2
	Fat-tailed dunnart × 1
	Fat-tailed false antechinus × 5
	Dibbler × 1
	Sandstone false antechinus × 1
Haemangioma/ haemangiosarcoma	Tasmanian devil × 1
	Tiger quoll × 1
Fibrosarcoma	Northern quoll × 1
	Kowari × 2
	Fat-tailed dunnart × 3
Hepatoma	Tasmanian devil × 2
	Eastern quoll × 1
	Tiger quoll × 1
	Northern quoll × 1
Adenocarcinoma	Tasmanian devil × 1
	Eastern quoll × 6
	Tiger quoll × 3
	Kowari × 2
	Common planigale × 1

Source: Canfield et al. (1990), ARWH (2007), Healesville Sanctuary records (2006).

9.2 Gastrointestinal diseases

9.2.1 Gastrointestinal parasites

There are 23 species of trichostrongyles infecting dasyurids, mostly occurring in the small intestine with one, *Peramelistrongylus skedastos*, found in the stomach. Of the spiruroids, which utilise invertebrate intermediate hosts, *Spirura aurangabadensis* occurs in the oesophagus of antechinus, planigales and dunnarts, while *Cyathospirura seurati*, *Cyathospirura dasyridis* and *Cyclicospirura heydoni* occur in tumour-like nodules in the stomach of quolls (Munday 1988). *Antechiniella suffodiax* and *Synhimantus australiensis* are found in antechinus stomachs. They use arthropod intermediate hosts. *A. suffodiax* burrows into the stomach wall, producing fibrous nodules. *Physaloptera sarcophili* occurs in the stomach of Tasmanian devils and quolls and uses a coleopteran intermediate host. The filarioid nematode *Sprattia capilliforme* is found in the mesenteric and portal veins of the liver in northern quolls. *Breinlia dasyuri* occurs in the spot-tailed quoll. *Capillaria rickardi* is found in tunnels in the epithelium of the stomach and duodenum of antechinus. It can cause ulceration and haemorrhage. A *Capillaria* spp. has also been recognised histologically in the tongue of northern quolls. There was no associated inflammatory reaction. *Trichinella pseudospiralis* is found in spot-tailed quolls and Tasmanian devils. Larvae occur in the muscle and adults in the intestines. *Anatrichosoma* spp. occur in the paracloacal glands of antechinus. *Baylisascaris tasmaniensis* is a large ascarid (males 24–68 mm and females 26–128 mm) found in the stomach and small intestine of Tasmanian devils and quolls. It causes visceral granulomas in wombats and pademelons, the intermediate hosts. *Ophidascaris robertsi* is a large (up to 150 mm) python ascarid that forms liver cysts in dasyurids. Larvae may also be found in the abdominal cavity of dasyurids (Beveridge & Spratt 2003) (Fig 10.9).

Anoplotaenia dasyuri are cestodes found in the Tasmanian devil small intestine. Larval stages produce 1–2 mm white foci in the heart and skeletal muscle of wallabies, pademelons, potoroos, spot-tailed quolls and eastern quolls. *Dasyurotaenia* spp. infect the small intestine of Tasmanian devils and quolls. The scolex is embedded in the muscle layer, resulting in nodular projections visible in the serosa. *Potorolepis* spp. infect insectivorous dasyurids using arthropods as intermediate hosts. Plerocercoids of *Spirometra erinacei* (spargana) have been found subcutaneously and in the body cavity of quolls, Tasmanian devils, brush-tailed phascogales, fat-tailed false antechinus (*Pseudantechinus macdonnellensis*), sandstone dibblers (*Pseudantechinus bilarni*), fawn antechinus (*Antechinus bellus*) and pygmy planigales (*Planigale maculata*). Although usually regarded as non-pathogenic, spargana have been implicated in the demise of northern quolls and antechinus.

Both acanthocephalan and pentastomid cysts have been found without reaction in antechinus livers. However, pentastomids have caused multiple mesenteric and hepatic granulomas in a northern quoll. The larval stage of the acanthocephalan *Australiformis* spp. is found occasionally in the abdominal cavity of antechinus and phascogales.

Many of the above parasites cause few if any clinical signs and are often only discovered at necropsy. A routine faecal flotation can be used to confirm the presence of helminths within the gastrointestinal tract. Treatments that have been used with no adverse effects include 0.2 mg/kg ivermectin SC, 0.2 mg/kg moxidectin SC and 10 mg/kg praziquantel PO. Canine and feline anthelmintics used at standard dose rates appear to be safe and effective in dasyurids. As many of the parasites use invertebrate intermediate hosts, the life-cycle can be disrupted by denying dasyurids access to wild invertebrates. A complete list of dasyurid parasites is found in Spratt et al (1990) and Appendix 2.

Protozoa have also been found in dasyurid gastrointestinal tracts. Unknown species of *Isospora* and *Eimeria* have been reported incidentally and a *Cryptosporidium* spp. has been found on histology in the brush border of the small intestinal crypts of brown antechinus. Associated villi were blunted.

9.2.2 Salmonellosis

Salmonella spp. are frequently isolated from dasyurid faeces. *S. mississippi* is commonly found in Tasmanian devils and eastern quolls, *S. potsdam* and *S. fremantle* are frequent isolates from dibblers, *S. charity*, *S. chester*, *S. muenchen*, *S. rubislaw*, *S. orientalis*, *S. arizona*, *S. eastbourne*, *S. wandsbek* and *S. birkenhead* have been isolated from western quolls and *S. fremantle* from slender-tailed dunnarts (*Sminthopsis murina*). These isolates were incidental findings in otherwise healthy animals (Obendorf 1993; Haigh et al. 1994).

9.2.3 *Helicobacter* and *Campylobacter*

Helicobacter spp. has been detected by PCR in the stomach, ileum, colon and rectum of Tasmanian devils and quolls. The devils also cultured positive for *Helicobacter* in the stomach and rectum but the quolls cultured positive for an isolate most closely related to *Campylobacter jejuni* in the colon (Coldham 2004). *Campylobacter coli* has been isolated from western quoll faeces. No clinical signs were associated with the organisms (Haigh et al. 1994), but they could be a potential cause of enteritis and diarrhoea. Treatment for the same organism in eastern grey kangaroos (*Macropus giganteus*) with 10 mg/kg clarithromycin PO bid was effective.

9.2.4 Tyzzer's disease

There is a report of Tyzzer's disease, caused by *Clostridium piliforme*, in an antechinus and a Tasmanian devil (ARWH 2007). Signs include depression, diarrhoea and sudden death. Grossly the liver contains diffuse white foci corresponding to areas of hepatocellular necrosis histologically. Packets of slender bacilli, whose appearance is enhanced with silver stains, are visible in the necrotic sites. The disease is usually secondary to poor husbandry (Rose 1999).

9.2.5 Pancreatitis

Interstitial pancreatitis has been diagnosed histologically in asymptomatic kultarr, fat-tailed false antechinus, brown antechinus, mulgara, kowari, fat-tailed dunnarts and white-footed dunnarts (*S. leucopus*). Lesions consist of necrosis with a lympho-plasmacytic infiltrate (ARWH 2007). Vacuolation of islet cells was found in a red kaluta that died suddenly. It was suggested that the animal had diabetes mellitus but blood glucose levels were not measured (Attwood & Woolley 1980).

9.2.6 Hepatic lipidosis and amyloidosis

Hepatic lipidosis with necrosis has been recorded in the Tasmanian devil, dunnarts, brown antechinus, dusky antechinus, sandstone dibbler, brush-tailed phascogale, eastern quoll, spot-tailed quoll and kowari. Amyloidosis has been seen histologically in spot-tailed quoll livers (ARWH 2007).

9.3 Cardiopulmonary diseases

9.3.1 Pneumonia

Bacterial pneumonia is particularly common in all dasyurid species but especially in dunnarts. Clinical signs are often vague but include anorexia, depression, weight loss, dyspnoea and sudden death. While a range of bacteria can be responsible, *Pasteurella multocida* and *Pseudomonas* spp. appear to be the most common isolates. Endogenous lipid pneumonia has been seen in dunnarts, Tasmanian devils and kowaris (ARWH 2007).

9.3.2 Mycobacteriosis

Mycobacteriosis is common in the fat-tailed dunnart but has also been recorded in other species such as the Tasmanian devil and brush-tailed phascogale. The causative agent is usually *Mycobacterium avium*, but *M. abscessus* and others have also been reported. Clinical signs include weight loss, dyspnoea, lameness, abscesses, neurological signs and blindness. Grossly multiple abscesses occur in numerous organs including lung, liver, spleen, kidney, spine and other bones, and bronchial and mesenteric lymph nodes. Histologically there is necrosis and a predominantly mononuclear cell response with minimal or absent mineralisation and encapsulation. Variable numbers of acid-fast bacilli are found within the lesions. Diagnosis in the live animal is difficult but should be high on the list of differentials in individuals with a history of chronic weight loss. Tuberculin testing is unreliable and organisms can be cultured from clinically normal animals. Treatment is also difficult, prolonged and usually unsuccessful. Options include amikacin 3 mg/kg bid, rifabutin 20 mg/kg sid, myambutol 20 mg/kg sid, ethambutol 20 mg/kg sid, enrofloxacin 2.5 mg/kg bid and azithromycin 20 mg/kg sid. Treatments have been continued for more than 3 yr, at which point mycobacteria could still be cultured from tracheal washes. Control of the disease is also problematic, as asymptomatic animals are hard to identify and the organisms are found normally in soil and water (Raymond et al. 2000; ARWH 2007).

9.3.3 Cryptococcosis

Pulmonary cryptococcosis has been diagnosed in a dusky antechinus. The antechinus was asymptomatic but was euthanased because of severe skin trauma. No lesions were seen at gross necropsy. Histologically there were foamy macrophages in alveoli. Thick-walled yeasts, compatible with *Cryptococcus* spp., were seen adjacent to the macrophage foci under the pleura (Healesville Sanctuary records 2006).

9.3.4 Parasitic diseases

Cardiopulmonary nematodes include *Nasistrongylus antechini* (formerly *Nicollina antechini*), which is a 1 cm red trichostrongylid nematode found in the nasal cavity of the agile antechinus, brown antechinus and dusky antechinus that causes erosion and metaplasia of the nasal mucosa (Munday 1988). Of the metastrongyles (lungworms) *Filaroides pilbarensis* is found in the lungs of the red kaluta, *Marsupostrongylus fragilis* occurs in the bronchi of the agile antechinus and

Marsupostrongylus lanceolatus is found in alveoli of the agile antechinus. *M. fragilis* uses the wild gastropods *Nitor pudibundus* and *Helicarion virens* as intermediate hosts. The metastrongyles can cause an eosinophilic bronchopneumonia. Treatment is with 0.2 mg/kg ivermectin SC or 0.2 mg/kg moxidectin SC.

A range of protozoa can be found in dasyurid blood. *Haemogregarina dasyuri* and *Hepatozoon dasyuri* are found in quolls and *Hepatozoon dasyuroides* is found in kowaris. None are known to be pathogenic. *Babesia* spp. causes a fatal anaemia and haemoglobinuria in antechinus (Clark 2004).

9.3.5 Arteriosclerosis

Arteriosclerosis of unknown cause consisting of medial degeneration and mineralisation of the thoracic aorta has been reported in the Tasmanian devil (ARWH 2007).

9.4 Urinary and reproductive diseases

Pyelonephritis, glomerulonephritis, membranous glomerulopathy, nephrocalcinosis, renal amyloidosis, cystic endometrial hyperplasia and endometritis have all been diagnosed at necropsy from a number of dasyurid species. Clinical signs are generally vague and non-specific. Urinary bladder stones composed of magnesium ammonium phosphate (struvite) were found at necropsy in a brush-tailed phascogale. *Klossiella* spp. has been found in the renal epithelial cells of the kowari. It is not pathogenic (Munday 1988; ARWH 2007).

Enlarged epithelial cells containing single large intranuclear inclusion bodies have been identified in the prostates of dusky antechinus, brown antechinus and brush-tailed phascogales. The hypertrophied cells decrease or occlude the lumen. Inclusions contain cytomegalovirus particles. There are no clinical signs but histologically some of the brown antechinus had a marked mixed inflammatory cell infiltrate and necrosis associated with the inclusions. Prevalence of infection increases dramatically with corticosteroid treatment (Barker et al. 1981).

Brown antechinus, fawn antechinus, sandstone false antechinus, red kaluta, kowari and stripe-faced dunnarts have also been found with cytomegalovirus intranuclear inclusion bodies in hypertrophied epithelial cells, with no associated inflammation, predominantly within the renal collecting ducts (Attwood & Woolley 1982).

Testicular torsion was described at necropsy in two fat-tailed false antechinus, one ningbing false antechinus (*Pseudantechinus ningbing*) and one red kaluta. Grossly the testes appeared reddened or blackened and histologically there was haemorrhagic necrosis within the testicular tubules. The lesions were not associated with any clinical signs (Attwood & Woolley 1982).

9.5 Neurological and musculoskeletal diseases

9.5.1 Toxoplasmosis

Toxoplasmosis is one of the most common causes of neurological disease in dasyurids (Canfield et al. 1990b). It is caused by the protozoan parasite *Toxoplasma gondii*. All dasyurid species are susceptible. Clinical signs include blindness, neurological signs, lethargy, anorexia and sudden death. Often there are no gross necropsy findings. If present, lesions include pulmonary congestion and gastric ulceration. Histologically multiple foci of necrosis with lymphoid infiltrates may be found in the lung, skeletal muscle, lymphoid tissue, central nervous system, adrenal, gut, pancreas, liver and kidney. Periodic Acid Schiff (PAS) positive tissue cysts are most commonly found in the central nervous system, muscle and adrenal. Direct agglutination test (DAT), modified agglutination test (MAT), latex agglutination test or indirect fluorescent antibody test can be used for diagnosis in the live animal. DAT measures IgG and IgM. Titres usually rise above 1:64 7–10 d after infection. MAT measures only IgG and records minimal levels early in the disease. Repeat testing of an infected animal shows both DAT and MAT titres rising. However, animals may die before the titre has a chance to rise. Cats are the definitive host of *T. gondii* with marsupials being extremely sensitive to infection. They contract the disease through exposure to cat faeces, food or water that has been contaminated by cat faeces, or by eating the flesh of another animal that contains the parasite encysted in its muscles. Oocysts can survive up to 18 mo in the environment and can be mechanically transmitted by infected earthworms, flies and cockroaches. Treatment is usually unsuccessful but can be attempted using clindamycin 10 mg/kg bid for 4 wk, trimethoprim-sulphonamide 15 mg/kg sid or atovaquone 100 mg/kg sid for at least 30 d. Control focuses on keeping cats out of feed sheds and preventing access to cat faeces. Meat should be cooked to at least

66°C or frozen for at least 2 wk prior to feeding. Equipment can be sterilised by autoclaving or heating to 70°C for at least 10 min. *T. gondii* is killed by soap and water (Canfield et al. 1990b).

9.5.2 Other parasitic diseases

Sarcocysts have been found in the skeletal muscles, tongue and diaphragm of a range of dasyurid species. They did not evoke any tissue reaction.

Angiostrongylus cantonensis has caused eosinophilic meningoencephalitis in Tasmanian devils (ARWH 2007). Clinical signs include ataxia, paresis, depression, seizures and death. Rats are the definitive host. They shed first-stage larvae in their faeces, which are consumed by snails or slugs. Devils contract the condition by ingesting the intermediate host. Parasites are visible histologically, associated with an eosinophilic meningitis and malacia, in sections of brain and spinal cord. Eosinophilia within the cerebrospinal fluid is strongly suggestive of the condition but systemic eosinophilia rarely occurs. Corticosteroids are used to decrease inflammation. Fenbendazole has cured the condition in a flying fox and could be tried in dasyurids. Prevention focuses on rodent control (Rose 1999).

9.5.3 Degenerative diseases

Leukoencephalopathy and myelopathy occurs commonly in quolls and Tasmanian devils aged 3 yr and older. Clinical signs include blindness, and gradually worsening ataxia and paresis. Histologically there is extensive Wallerian degeneration particularly in the spinal cord, optic tracts and cerebellum. There is no treatment and the syndrome is believed to represent age-related degeneration (Holz & Little 1995).

Aged quolls and devils frequently suffer from degenerative intervertebral disc disease. Prolapsed and ruptured discs can occur anywhere, but predominantly in the lumbo-sacral region, causing spinal cord compression and Wallerian degeneration. Spondylosis of vertebral bodies is often present. Diagnosis is by clinical signs that usually relate to hind end paralysis and paresis and by radiography. Treatment is generally unrewarding but can be attempted using corticosteroids. It can be difficult to differentiate this disease from myelopathy and, as both conditions are often present concurrently, euthanasia is usually indicated.

9.5.4 Otitis media and meningitis

Otitis media and meningitis have been described in dunnarts and quolls with head tilts. *Pseudomonas aeruginosa* is commonly isolated. Treatment is generally unrewarding. A case of *Pseudomonas aeruginosa* otitis externa with purulent discharge into the external auditory canal was diagnosed at necropsy in an agile antechinus.

9.6 Ophthalmologic diseases

Dasyurids do not suffer any unique ophthalmologic diseases but there is quite a high incidence of cataracts, particularly in northern quolls. The reason is unknown. In captivity, quolls with cataracts appear to cope well.

9.7 Non-infectious diseases

9.7.1 Neoplasia

Dasyurids suffer from a disproportionately large number of neoplasms (Table 10.4). One study found an incidence of 46% (Canfield et al. 1990a) compared with 2–10% for other mammal groups (Montali 1980). The reasons are not known. The most common tumours are squamous cell carcinomas, trichoepitheliomas, mammary adenocarcinomas and lymphosarcomas. Affected animals frequently present suffering from multiple neoplasms. Any ageing dasyurid that presents with vague non-specific signs of weight loss, anorexia, depression and body or skin masses should have neoplasia at the top of its differential diagnosis list (Canfield et al. 1990a) (Fig. 10.11).

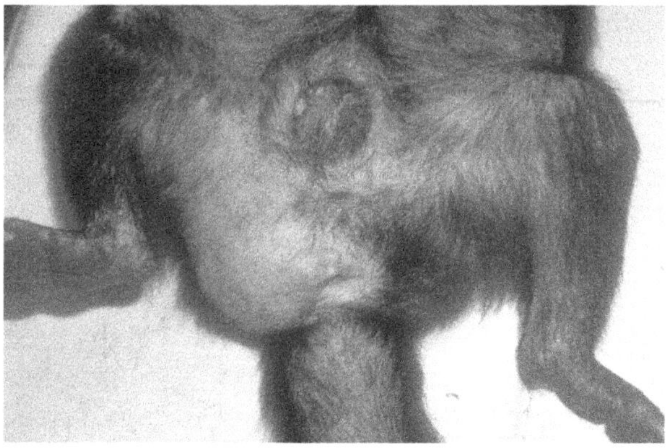

Figure 10.11 Anal gland adenocarcinoma in an aged Tasmanian devil. Photo: ARWH.

9.7.2 Toxins

a *Fluoroacetate*

In south-western Western Australia a large number of shrubs belong to the Gastrolobium and Oxylobium genera. These plants contain up to 2.65 g/kg of leaf and 6.5 g/kg of seeds of fluoroacetate (the active ingredient in 1080), which is used as a vertebrate poison. Most mammals are poisoned by less than 1 mg/kg of body mass. However, the western quoll, yellow-footed antechinus (*Antechinus flavipes*) and red-tailed phascogale (*Phascogale calura*), which are found in the area and feed on insects that consume Gastrolobium and Oxylobium leaves, are relatively resistant to the poison. Their LD_{50} ranges from 7.5–17.5 mg/kg. The eastern quoll, Tasmanian devil and yellow-footed antechinus from south-eastern Australia, where there is no natural exposure to fluoroacetate, have an LD_{50} ranging from 1.5–4.2 mg/kg (Hume 1999). This is still considerably higher than eutherians, which range from 0.06 mg/kg for dogs up to 0.4 mg/kg for cats. However, it is possible for dasyurids to be affected by consuming sufficient bait or poisoned animals such as rabbits. Signs can appear up to 23 hr after ingestion and include depression, increased sensitivity to stimuli, restlessness, rapid respiration, trembling, vomiting, straining to defecate and convulsions. Deaths occur 1.5–147 hr after ingestion with survivors showing signs of recovery after 2–60 hr. There is no antidote and treatment is supportive only using sedation, fluids and activated charcoal (McIlroy 1982).

b *Bufotoxins*

The cane toad (*Bufo marinus*) was introduced to Australia in 1935 in a failed attempt to control sugarcane pests. It was first introduced near Cairns and has since spread south to northern New South Wales, north to the tip of Cape York and west across Queensland and into the Northern Territory, reaching Kakadu National Park in April 2001. Cane toads secrete a mixture of bufotoxins and other compounds from the large parotid glands on their neck. When consumed by other animals, these toxins can be fatal. Cane toads have had a devastating effect on numerous populations of Australian native fauna. Quolls are particularly sensitive to bufotoxins and animals will die after simply mouthing a toad without actually consuming it. Death appears to be rapid and there have been no opportunities to treat any animals. Since the cane toad's arrival in the Northern Territy and Kakadu there has been a significant decline in northern quoll populations (Oakwood 2004). Cane toad toxicity has been reported in captive Tasmanian devils in Queensland. Most animals recover with supportive care, however, the deaths of two young Tasmanian devils recently imported from Tasmania were highly suspicious of cane toad toxicity (M Pine *pers. comm.*).

10 ZOONOSES

Several diseases that affect dasyurids are also potential zoonoses. Atypical mycobacteria can cause pneumonia in immunocompromised people. Exposure to organisms is by inhalation or ingestion of faeces and discharges. Many dasyurids culture positive for *Salmonella* spp. and other enteric pathogens that have caused gastrointestinal disease in people. Vigorous hand-washing with soap is enough to kill *Salmonella*.

11 THERAPEUTICS

There have been no pharmacokinetic studies done on any dasyurids. Consequently, small domestic animal therapeutics are used at small animal dose rates. There are no known contraindications for the use of any therapeutic agent in dasyurids. However, contraindications for use in domestic animals are likely to be applicable to dasyurids and this should be considered prior to the initiation of treatment.

12 REPRODUCTIVE MANAGEMENT

12.1 Reproductive control

To prevent breeding, males and females can be maintained separately or males can be vasectomised or castrated. There is no indication for the use of hormonal contraception in dasyurids.

12.2 Enhancing reproduction

Male and female dasyurids are generally maintained separately outside the breeding season. To maximise breeding success, it is important to be able to detect oestrus when it occurs. Smaller dasyurids, such as kowaris, gain weight as oestrus approaches. The pouch skin of quolls, antechinus and planigales becomes moist and reddened and the vulva oedematous. The Tasmanian devil pouch becomes deep, wet and glandular. The cells of the urogenital sinus become predominantly cornified. These can be detected by inserting a swab into the urogenital sinus. In species too small to swab, urine is usually released when the animal is picked up. These samples are examined microscopically. A predominance of cornified epithelial cells is indicative of oestrus in quolls, kowaris, phascogales and devils. Devils also experience a rise in plasma progesterone levels from 0.1 ng/mL to 2.5 ng/mL at least 7 d before mating. Levels remain elevated for about 10 d. When large numbers of clumped cornified epithelial cells are seen and the progesterone level peaks, the female should be mated with the male. Successful mating can be determined by the presence of sperm in the urine of the female (Tyndale-Biscoe & Renfree 1987; Jackson 2003).

Progesterone levels in pregnant Tasmanian devils increase to a sustained level of 20 ng/mL. Birth occurs immediately following the progesterone decline.

Various mating strategies have been used successfully. A male quoll placed adjacent to the female's enclosure will elicit a clucking response from the female if she is in oestrus. He can then be released into her enclosure. Breeding kowaris have been maintained successfully in pairs, trios (one male and two females) and larger groups. Males often need to be removed following the birth of young because of aggression from the females. A phascogale male can be placed together with a female and left until PY are observed. Alternatively, a round robin system can be used with males rotating among females on a biweekly basis. Tasmanian devils should be paired at weaning. If this is not possible, individuals should be gradually introduced with visual and olfactory contact for at least 7 d before having physical contact. Males should be removed when PY are detected. A detailed description of captive breeding management of Tasmanian devils is provided by Faulkner (2008). Dasyurids need to be closely monitored and the male removed either following mating to prevent serious and possibly fatal injury or after the arrival of the PY to prevent infanticide. Siblings should be separated prior to their first breeding season for the same reason.

Because many of the smaller species such as kowaris and phascogales breed for only one or two seasons captive spaces rapidly fill up with reproductively senescent individuals, necessitating rehousing or culling to keep the colony reproductively active (Roberts et al. 1993).

13 SURGERY

Surgery of dasyurids is similar to surgery of small animals. Most surgeries involve lump removals or traumatic wound repair. Absorbable sutures are used to avoid recapturing the animal for suture removal. Some individuals may try to remove their sutures post-operatively and it is important to be aware of this tendency. Intradermal sutures can help reduce the incidence of this.

13.1 Vasectomy and castration

Dasyurid testicles are quite pendulous. To perform a vasectomy, incise the skin over the spermatic cord proximal to the scrotum. Incise the vaginal tunic and isolate the vas deferens by blunt dissection. Ligate the vas deferens in two places and remove a small segment from between the sutures. Suture the skin and repeat the process on the contralateral spermatic cord. Castration is performed by incising the scrotum and then the vaginal tunic. Exteriorise the testicle, ligate the vascular cord and cut above the ligature, removing the testicle. Leave the scrotal skin incision open to drain and repeat the procedure on the contralateral testicle. In smaller dasyurids with particularly pendulous scrotums it may be easier to completely ablate the scrotum.

ACKNOWLEDGMENTS

The author would like to acknowledge the assistance and valuable information provided by Cathy Arrese, Ian Beveridge, Adrian Bradley, Phillip Clark, Paul Eden, Janine Ferguson, Robert Johnson, Richmond Loh, Karrie Rose, Cathy Shilton, Larry Vogelnest and Rupert Woods.

REFERENCES

AG0976 (2001). *Code of Practice for the Welfare of Wildlife during Rehabilitation*. Department of Primary Industries, Victoria. <http://www.dpi.vic.gov.au>.

Arrese CA, Dunlop SA, Harman AM, Braekevelt CR, Ross WM, Shand J & Beazley LD (1999). Retinal structure and visual acuity in a polyprotodont marsupial, the fat-tailed dunnart (*Sminthopsis crassicaudata*). *Brain, Behaviour and Evolution* **53**, 111–126.

Arrese CA, Hart NS, Thomas N, Beazley LD & Shand J (2002). Trichromacy in Australian marsupials. *Current Biology* **12**, 657–660.

ARWH (2007). *Australian Registry of Wildlife Health.* Zoological Parks Board of NSW: Sydney.
Attwood HD & Woolley PA (1974). Arachnoid villi in the spinal cord of dasyurid marsupials. *Journal of Comparative Neurology* **155**, 343–353.
Attwood HD & Woolley PA (1980). Pancreatic pathology in dasyurid marsupials. *Journal of Wildlife Diseases* **16**, 245–249.
Attwood HD & Woolley PA (1982). Histopathology of captive dasyurid marsupials. In *The Management of Australian Mammals in Captivity* (ed. DD Evans), pp. 27–30. Zoological Board of Victoria: Melbourne.
AusVet Animal Health Services (2005). Tasmanian devil facial tumour disease response: technical workshop. Department of Primary Industries, Water and Environment, Tasmania. <http://www.dpiwe.tas.gov.au>.
Baker RT & Beveridge I (2001). Imidacloprid treatment of marsupials for fleas (*Pygiopsylla hoplia*). *Journal of Zoo and Wildlife Medicine* **32**, 391–392.
Barker IK, Beveridge I, Bradley AJ & Lee AK (1978). Observations on spontaneous stress-related mortality among males of the dasyurid marsupial *Antechinus stuartii* Macleay. *Australian Journal of Zoology* **26**, 435–447.
Barker IK, Carbonell PL & Bradley AJ (1981). Cytomegalovirus infection of the prostate in the dasyurid marsupials, *Phascogale tapoatafa* and *Antechinus stuartii*. *Journal of Wildlife Diseases* **17**, 433–441.
Beveridge I & Spratt DM (2003). Parasites of carnivorous marsupials. In *Predators with Pouches: The Biology of Carnivorous Marsupials* (eds M Jones, C Dickman & M Archer), pp. 383–396. CSIRO Publishing: Melbourne.
Bradley AJ (2003). Stress, hormones and mortality in small carnivorous marsupials. In *Predators with Pouches: The Biology of Carnivorous Marsupials* (eds M Jones, C Dickman & M Archer), pp. 254–267. CSIRO Publishing: Melbourne.
Breed WG, Taggart DA & Moore HDM (2003). Sperm maturation and fertilization in Australian and American insectivorous marsupials. In *Predators with Pouches: The Biology of Carnivorous Marsupials* (eds M Jones, C Dickman & M Archer), pp. 133–146. CSIRO Publishing: Melbourne.
Canfield PJ, Hartley WJ & Reddacliff GL (1990a). Spontaneous proliferations in Australian marsupials – a survey and review. 2. Dasyurids and bandicoots. *Journal of Comparative Pathology* **103**, 147–158.
Canfield PJ, Hartley WJ & Dubey JP (1990b). Lesions of toxoplasmosis in Australian marsupials. *Journal of Comparative Pathology* **103**, 159–167.
Cheal PD, Lee AK & Barnett JL (1976). Changes in the haematology of *Antechinus stuartii* (Marsupialia), and the association with male mortality. *Australian Journal of Zoology* **24**, 299–311.
Clark P (2004). *Haematology of Australian Mammals.* CSIRO Publishing: Melbourne.
Clark P & Boardman WS (2005). Haematological response to naturally occurring disease in the western quoll (*Dasyurus geoffroyi*). *Comparative Clinical Pathology* **13**, 182–185.
Coldham T (2004). The detection and characterisation of Helicobacter species in Australian marsupials. PhD thesis. University of NSW, Australia.
Desch CE & Holz PH (2006). *Demodex tortellinioides* n. sp. (Acari: Demodecidae) from the agile antechinus, *Antechinus* agilis Dickman, Parnaby, Crowther and King, 1998 (Marsupialia: Dasyuridae) in Australia. *International Journal of Acarology* **32**, 81–84.
Eden P, Vitali S, Monaghan C & Payne K (2004). An investigation into demodecosis in southern dibblers (*Parantechinus apicalis*). In *Proceedings of Wildlife Disease Association (Australasian Section) Conference,* Kinchega.
Exhibited Animals Protection Act (1995). *Standards for Exhibiting Australian Mammals in New South Wales.* New South Wales Department of Primary Industries. <www.agric.nsw.gov.au>.
Faulkner T (2008). Sympathy for the devil. In *Proceedings of The Australasian Regional Association of Zoological Parks and Aquaria.* <http://www.arazpa.org.au/2008-ARAZPA-Conference-Proceedings/default.aspx>
Geiser F (2003). Thermal biology and energetics of carnivorous marsupials. In *Predators with Pouches: The Biology of Carnivorous Marsupials* (eds M Jones, C Dickman & M Archer), pp. 238–253. CSIRO Publishing: Melbourne.
Haigh SA, Gaynor WT & Morris KD (1994). A health monitoring program for captive, wild and translocated chuditch (*Dasyurus geoffroii*). In *Proceedings of Australian Association of Veterinary Conservation Biologists Conference*, Canberra, pp. 52–66.
Hallam JF, Holland RAB & Dawson TJ (1995). The blood of carnivorous marsupials: low hemoglobin oxygen affinity. *Physiological Zoology* **68**, 342–354.
Hawkins CE, Baars C, Hesterman H, Hocking GJ, Jones ME, Lazenby B, Mann D, Mooney N, Pemberton D, Pyecroft S, Restani M & Wiersma J (2006). Emerging disease and population decline of an island endemic, the Tasmanian devil *Sarcophilus harrisii*. *Biological Conservation* **131**, 307–324.
Haynes JI & Skidmore GW (1991). Haematology of the dasyurid marsupials *Sminthopsis crassicaudata* and *Sminthopsis macroura*. *Australian Journal of Zoology* **39**, 157–169.
Holz P (1992). Immobilization of marsupials with tiletamine and zolazepam. *Journal of Zoo and Wildlife Medicine* **23**, 426–428.
Holz PH (1998). Nodules caused by demodex-like mites in the skin of tiger quolls (*Dasyurus maculatus*). *Australian Veterinary Journal* **76**, 59–60.
Holz P (2003). Marsupialia (Marsupials). In *Zoo and Wild Animal Medicine* (5th edn) (eds ME Fowler & RE Miller), pp. 288–303. WB Saunders: St Louis.

Holz PH & Little PB (1995). Degenerative leukoencephalopathy and myelopathy in dasyurids. *Journal of Wildlife Diseases* **31**, 509–513.

Hume ID (1999). Carnivorous marsupials. In *Marsupial Nutrition* (ed. ID Hume), pp. 35–74. CUP: Cambridge.

ISIS (2002). *Physiological Data Reference Values*. International Species Information System, Minnesota. <http://www.isis.org>.

Jackson S (2003). Carnivorous marsupials. In *Australian Mammals: Biology and Captive Management* (ed. S Jackson), pp. 53–98. CSIRO Publishing: Melbourne.

Krajewski C & Westerman M (2003). Molecular systematics of Dasyuromorphia. In *Predators with Pouches: The Biology of Carnivorous Marsupials* (eds M Jones, C Dickman & M Archer), pp. 3–20. CSIRO Publishing: Melbourne.

Loh R, Bergfeld J, Hayes D, O'Hara A, Pyecroft S, Raidal S & Sharpe R (2006a). The pathology of devil facial tumor disease (DFTD) in Tasmanian devils (*Sarcophilus harrisii*). *Veterinary Pathology* **43**, 890–895.

Loh R, Hayes D, Mahjoor A, O'Hara A, Pyecroft S & Raidal S (2006b). The immunohistochemical characterization of devil facial tumor disease (DFTD) in the Tasmanian devil (*Sarcophilus harrisii*). *Veterinary Pathology* **43**, 896–903.

McAllan B (2003). Timing of reproduction in carnivorous marsupials. In *Predators with Pouches: The Biology of Carnivorous Marsupials* (eds M Jones, C Dickman & M Archer), pp. 147–168. CSIRO Publishing: Melbourne.

McIlroy JC (1982). The sensitivity of Australian carnivorous mammals to 1080 poison. In *Carnivorous Marsupials* (ed. M Archer), pp. 267–271. Royal Zoological Society of New South Wales: Sydney.

Melrose WD, Pearse AM, Jupe DMD, Baikie MJ, Twin JE & Bryant SL (1987). Haematology of the Australian eastern quoll, *Dasyurus viverrinus*. *Comparative Biochemistry and Physiology* **88A(2)**, 239–241.

Montali RJ (1980). An overview of tumors in zoo animals. In *The Comparative Pathology of Zoo Animals* (eds RJ Montali & G Migaki), pp. 531–542. Smithsonian Institution Press: Washington DC.

Munday BL (1988). Marsupial diseases. In *Australian Wildlife, Proceedings 104*, pp. 299–365. Post Graduate Committee in Veterinary Science: Sydney.

Nutting W & Woolley P (1965). Pathology in *Antechinus stuartii* (Marsupialia) due to *Demodex* sp. *Parasitology* **55**, 383–389.

Oakwood M (2004). Case of the disappearing spots. *Nature Australia* Spring, 26–34.

Obendorf DL 1993. Diseases of dasyurid marsupials. In *The Biology and Management of Australasian Carnivorous Marsupials* (eds M Roberts, J Carnio, G Crawshaw & M Hutchins), pp. 39–46. Metropolitan Toronto Zoo: Toronto.

Pearse AM & Swift K (2006). Allograft theory: transmission of devil facial-tumour disease. *Nature* **439**, 549.

Raymond JT, Tell L, Bush M, Nichols DK, Schulman FY & Montali RJ (2000). Subcutaneous atypical mycobacteriosis in captive tiger quolls (*Dasyurus maculatus*). *Veterinary Pathology* **37**, 137–142.

Roberts M, Carnio J, Crawshaw G & Hutchins M (eds) (1993). *The Biology and Management of Australasian Carnivorous Marsupials*. Metropolitan Toronto Zoo: Toronto.

Rose K (1999). Common diseases of urban wildlife. In *Wildlife in Australia: Healthcare and Management, Proceedings 327*, pp. 365–430. Post Graduate Foundation in Veterinary Science: Sydney.

Schaap D (2002). Enriching the devil: the Tasmanian devil. *Shape of Enrichment* **11**, 1–4.

Spratt DM, Beveridge I & Walter EL (1990). A catalogue of Australasian monotremes and marsupials and their recorded helminth parasites. *Records of the South Australian Museum, Monograph Series* **1**, 1–105.

Strahan R (1995). *The Australian Museum Complete Book of Australian Mammals*. Reed New Holland: Sydney.

Svensson A, Mills JN, Boardman WSJ & Huntress S (1998). Hematology and serum biochemistry reference values for anesthetized chuditch (*Dasyurus geoffroii*). *Journal of Zoo and Wildlife Medicine* **29**, 311–314.

Taggart DA, Shimmin GA, Dickman CR & Breed WG (2003). Reproductive biology of carnivorous marsupials: clues to the likelihood of sperm competition. In *Predators with Pouches: The Biology of Carnivorous Marsupials* (eds M Jones, C Dickman & M Archer), pp. 358–375. CSIRO Publishing: Melbourne.

Tyndale-Biscoe H & Renfree M (1987). Breeding biology of marsupials by family. In *Reproductive Physiology of Marsupials* (H Tyndale-Biscoe & M Renfree), pp. 14–94. CUP: Cambridge.

Tyndale-Biscoe H (2005). Predatory marsupials of Australasia: bright-eyed killers of the night. In *Life of Marsupials* (H Tyndale-Biscoe), pp. 139–163. CSIRO Publishing: Melbourne.

Vogelnest L (1999). Chemical restraint of Australian native fauna. In *Wildlife in Australia: Healthcare and Management, Proceedings 327*, pp. 149–187. Post Graduate Foundation in Veterinary Science: Sydney.

11. Numbats

Simone Vitali and Cree Monaghan

1 TAXONOMY, DISTRIBUTION AND BIOLOGY

The numbat (*Myrmecobius fasciatus*) is the only species in the Family Myrmecobiidae, in the Order Dasyuromorphia. Its closest living relatives are the carnivorous dasyurid marsupials (DeJose 1988). They are one of only two exclusively diurnal marsupial species, the other being the musky rat-kangaroo (*Hypsiprymnodon moschatus*), and are most active from mid-morning to late afternoon. Their diurnal nature reflects the activity patterns of their termite prey, which ascend to upper soil layers during the warmer part of the day (DeJose 1988; Friend 1997a). They spend nights in logs or in burrows, which they dig themselves and also use for shelter during the day. They are excellent climbers and diggers and are highly agile (Friend 1989; Power & Monaghan 2003).

Numbats were originally widespread across southern semi-arid and arid Australia (Maxwell et al. 1996). Today there remain only two remnant populations and four self-sustaining reintroduced populations, all in the south-west of Western Australia. They are threatened by predation, particularly by foxes, and by habitat destruction. Key factors necessary for survival of numbats in the wild include an abundance of their only food source, subterranean termites, and sufficient ground-level cover to afford protection from predators (Friend 1997a, b).

Numbats can live for up to 5 yr in the wild, although most do not achieve this age (Friend 1997a). In captivity they live longer. Males have been known to live up to 11 yr and females up to 7 yr (Power & Monaghan 2003).

2 ANATOMY AND PHYSIOLOGY

Adult numbats weigh about 500–700 g and have a body length of about 30 cm, with a 20 cm tail. Dorsoventral flattening of their hind quarters affords them manoeuvrability when moving in hollow logs or burrows. Both fore and hind limbs have sharp claws for digging. There are five digits on the manus, of which the fourth is the most prominent. The pes has four digits, the first digit being represented only by a skin pad (Friend 1989).

The pelage is short and stiff, which probably reflects their burrowing habits (Friend 1989). The coat is sparse compared with other marsupials, and has a low reflectivity and high absorptivity for solar radiation. This maximises solar heat gain, thereby minimising the energy expenditure required for thermoregulation. This may be an important adaptation for the species, given the poor nutritive quality of its termite diet (Cooper et al. 2003b).

Adult numbats have a sternal sebaceous gland located on the ventral midline at the cranial aspect of the sternum. The gland is hairless, pigmented and approximately 1 cm in diameter. It is more prominent in males than in females, and is used in territorial marking (Friend 1989; Power & Monaghan 2003). Numbats are unusual among termite and ant-eating mammals in that their dentition is relatively well-developed (Cooper & Withers 2004a). They have 50–52 teeth. Their long thin tongue, which can protrude several centimetres beyond the snout, reflects their dependence on termites as a sole food source (Friend 1989, 1997a). Other anatomical adaptations to their termite diet include an elongated snout, greatly enlarged submaxillary salivary glands and an unspecialised gastrointestinal tract similar to that of dasyurids (Friend 1989; Hume 1999). Their slow gut passage time of 20–30 hr optimises digestion of termites, which are of low energy density and digestibility (Cooper & Withers 2004a).

The external sex organs of the male consist of a prepenile pedunculated scrotal sac and a penis with a cleft tip and a keratinised spiny shaft. Accessory sex glands consist of a prostate and three pairs of bulbourethral glands (Fordham 1928 in Friend 1989).

Female numbats have four nipples, arranged on the caudal abdomen so that the posterior pair is closer together than the anterior pair. There is no pouch present, however, in pregnant animals skin folds may be observed on the lateral margins of the mammary area and swelling develops just cranial and caudal to the mammary area prior to birth, so that the teats are enclosed within a depression (Power & Monaghan 2003).

The ovaries are located ventral to the anterior end of the two uteri. The uteri terminate in two median vaginae close to the upper end of the urogenital sinus. Two lateral vaginal canals emanate from the median vaginae and open via separate apertures into the urogenital sinus. The cloaca is a typical distinct structure into which the rectum opens dorsally and the urogenital sinus ventrally (Friend 1989).

Like all marsupials, numbats have a relatively low basal metabolic rate and core body temperature (Friend 1989; Cooper et al. 2003a). Basking in the sun is an important feature of their thermoregulation. The relatively sparse coat has poor insulative and low reflective properties, enabling rapid and efficient solar heat gain (Cooper et al. 2003b). Numbats are able to go into torpor and reduce their core body temperature to as low as 19°C (Cooper & Withers 2004a). While torpor is an induced response to food shortage in other dasyurid species examined to date, thermal telemetry studies of numbats, both in captivity and in the wild, indicate that they are able to go into torpor spontaneously. Numbats may go into torpor on virtually a daily basis through the colder months, entering torpor around 3.30 am and achieving normal body temperature around 8 am (Cooper & Withers 2004a).

3 REPRODUCTION

Numbats are solitary and territorial, with individual animals occupying distinct home ranges. Females reach breeding age in their first year of life and males in their second. Females appear to remain fertile until they die, with the oldest recorded breeding female being 7 yr old (Power & Monaghan 2003).

Numbats have a short breeding season from December to February, with most mating occurring in January and early February (Power & Monaghan 2003). Females are seasonally polyoestrus and must be mated within 48 hr of oestrus for conception to occur (Friend 1997a). In captivity, the timing of pairing is determined by examining vaginal swabs and samples of voided urine from females to determine vaginal epithelial cell morphology. Samples from females are collected weekly from late December. Urine samples from males are collected fortnightly from December onwards to detect sperm, which usually appear in the urine from mid-December to the end of February. Males with viable sperm are introduced to females when the females are in early proestrus. The pair is kept together until PY are found (Power & Monaghan 2003).

Duration of mating may vary from less than a minute up to an hour. The only evidence of a mating may be wet fur on the scruff of the female's neck, where she has been grasped by the male during copulation (Power & Monaghan 2003). Gestation length is 14 d and four young are usually born in January or early February (Friend 1989).

Numbat young are pink and hairless and have a total body length of approximately 6–7 mm. The young attach directly to one of four nipples on the female's caudal abdomen and hold on by entwining their fore limbs in the hair around the mammary area. They remain attached to the mother for up to 6 mo. In late July they are deposited in a nest or burrow and continue to suckle at night. The young weigh approximately 70 g when deposited. Their growth rate is extremely slow and they are not weaned until about 9 mo of age when they weigh 200–300 g. This slow rate of development may reflect the poor nutritional quality of their termite diet. Dispersal occurs soon after weaning, in late November or early December (Friend 1997b, 1998).

4 HUSBANDRY AND HOUSING

Numbats have relatively specific husbandry requirements if they are to breed, display well and remain healthy. Enclosures are generally designed to provide an environment most suited to successful breeding. Numbats are usually housed individually in 5 × 3 × 2 m enclosures that are fully meshed to prevent predator attack. A mesh floor buried 1 m deep allows excavation of burrows (Power & Monaghan 2003).

The wild habitat of numbats is an arid or semi-arid environment and they do not tolerate damp or wet conditions. River sand is an ideal enclosure substrate because it drains well. Shrubs, tussocks and hollow logs are provided for natural protection. Ponds and running water are not provided as moist enclosures are believed to predispose to skin disease. Enclosures are best placed facing north to allow basking. There must be shade protection from the strong sun. Numbats can tolerate moderately cold weather but require hollow logs or nest boxes in winter to protect against severe cold. Enclosures should be cleaned every 1–2 d to remove faecal matter, and water and food dishes washed daily. Cage furniture should be replaced annually (Power & Monaghan 2003).

Numbats are very inquisitive and in captivity may be inclined to hang on wire mesh barriers to investigate adjoining enclosures. If they are startled while hanging on the wire they may become injured as they take flight, breaking nails or injuring their tail or a limb. The provision of climbing branches, anchored to the side of the enclosure, allows animals to climb to elevated positions without needing to climb mesh walls. This strategy has significantly decreased the incidence of traumatic injuries in numbats in captivity.

Numbats are solitary animals and are generally housed individually except at breeding time. Wherever possible, animals to be paired are housed in adjacent enclosures so that introductions may be effected simply by opening a door between the two enclosures. Visual barriers between enclosures (e.g. shadecloth) are important during the breeding season to prevent males becoming stressed and aggressive due to the proximity of competing males (Power & Monaghan 2003). Stress can be reduced by not housing males immediately adjacent to one another.

Pairs are introduced when the female is in proestrus, as determined by assessment of vaginal epithelial cytology (see section 3). If the pair is housed next to one another they can be introduced by opening a gateway between the enclosures. If not, the male is transferred to the female enclosure in his own nest box. Generally, little aggression is observed during introductions but occasionally a male will chase a female rather than show reproductive behaviour. If this persists for more than 24 hr the male should be removed and replaced with another male.

Breeding females require protection from heavy rainfall to prevent collapse of underground nests. For this reason, all breeding females are moved indoors prior to the onset of wet weather. Breeding and lactating females become stressed when there is too much noise or human presence, and there is significant risk of losing young if appropriate visual and noise barriers are not used. It is important to minimise movement of personnel as much as possible during the breeding months.

Provision of natural sunlight is of great importance for maintaining normal metabolism and thermoregulation in numbats, particularly breeding females. Even when ambient temperature is adequate (e.g. in numbats housed indoors) numbats actively seek out natural sunlight, and basking is their first activity when they emerge from their nest boxes. Numbats emerge from torpor around 8 am in the colder months and are usually slower and less reactive early in the morning, even during summer. Handling numbats before they leave their nest boxes in the morning does not appear detrimental to their well-being. In fact, it is often ideal to handle numbats at this time because they are less likely to become stressed or flighty and do not resist manual restraint. This facilitates captive management activities such as weighing, vaginal swabs and nail trimming.

A very sick or debilitated numbat might appear similar to a numbat in torpor. For this reason, it is preferable to wait until after numbats emerge from their nest boxes in the morning to assess their level of activity. Sick numbats can be housed in a hotbox set at 26–32°C if confinement or intensive care is required, or in a small heated room set at 16–26°C with leaf litter or woodchip substrate and a heat lamp for basking. Numbats are nervous animals and a nest box or some form of visual protection should be provided. Numbats generally head directly for the nest box if they are disturbed. This is helpful when handling is necessary, as they can be shut into the nest box then removed directly from the box rather than having to net or chase them.

Numbats can be transported for short periods using a cotton bag, pillowcase or secure nest box. For longer periods of transport, numbats should be placed in a pillowcase, which is then put into a 350 × 400 × 200 mm wooden box. The box should be lined with sea grass or similar absorbent cushioning material; only one animal should be placed in each box (Power & Monaghan 2003). Wherever possible, numbats should be transported during the cool hours of the night and early morning, for periods of less than 24 hr.

Passive integrated transponders are used for individual identification. They are inserted with the animal under general anaesthesia and are placed SC in the dorsal interscapular region (Power & Monaghan 2003). The insertion site may be clipped and surgically prepared, but chips have been inserted without clipping or surgical preparation of the site. The very thick skin of the numbat's dorsal neck region tends to resist closure using tissue glue, so to prevent transponder loss the insertion site should be closed using an absorbable suture. In the past, numbats have been identified by tattooing the inner thigh, but this technique is now obsolete due to the ease and reliability of transponder implants. The experienced observer may identify numbats by the banding pattern of their fur, which is unique to each individual (Power & Monaghan 2003).

5 NUTRITION

Numbats are myrmecophagous (eat termites). In the wild numbats feed exclusively on a range of *Isoptera* termite species. They eat approximately 15 000–20 000 termites per day and ingest large quantities of associated dirt (at least 0.33 g for every gram of organic matter ingested) (Cooper & Withers 2004a; Power & Monaghan 2003).

A reliable termite supply is essential to captive husbandry of numbats. Perth Zoo numbats are supplied with termites harvested from wild nests of *Nasutitermes exitiosus* and *Coptotermes acinaciformis* termites. The trapping, harvesting, removal and storage of termites is a complex process; see Monaghan and Power (2003) for diagrams and a detailed description. Briefly, an open-topped drum packed with slats of karri (*Eucalyptus diversicula*) wood is placed over the opening of wild termite nests. Once the termites have infested the wood, the drum is removed and the termites are separated from the wooden slats by hitting the slats with a solid metal rod (e.g. a heavy screwdriver) so that the termites fall off onto a tray. The termites may be frozen for future use or fed out live. Live termites are chilled before feeding to slow them down and are placed in stainless steel bowls so that they cannot climb out and escape. The wood remaining in the drum ('termite crumb') also contains significant numbers of termites and is valuable for activity feeds and supplementary termites during the breeding season.

To achieve breeding success in captivity, it is necessary to provide a 60–100% live termite diet to breeding females from November to March (Power & Monaghan 2003). On a 100% termite diet females are fed 40 g live termites in the morning (around 8.30 am) and another 30 g in the early afternoon (1.30 pm). They are also provided with termite crumb throughout the day.

Despite the development of reliable termite trapping techniques, the provision of large numbers of termites is difficult in captivity so nutrition is supported with an artificial diet (Table 11.1). The artificial diet is a mix of eggs, milk powder and water, cooked to a thick custard consistency before adding vitamin and mineral supplements and sterilised termite mound material (Power & Monaghan 2003). The preparation of the artificial diet is a complex but necessary aspect of captive management and numbats have not been maintained well on simpler substitutes.

6 RESTRAINT

Physical restraint of numbats is relatively simple; it involves restraining the neck and shoulder area while supporting the hind legs (Fig. 11.1). In captivity, the stress of physical restraint is minimised by capturing animals in the morning before they leave their nest boxes (Power & Monaghan 2003). They do not tend to bite, but can scratch with their sharp claws. A small soft towel or cotton bag facilitates capture and restraint.

Isoflurane is the anaesthetic of choice for numbats (Power & Monaghan 2003). Induction is achieved with 5% isoflurane in oxygen administered via face mask. Numbats often become apnoeic during induction. A unidirectional anaesthetic circuit (e.g. Bain's T-piece) is used. Although readily maintained on a mask for short

Table 11.1 Numbat artificial diet recipe

No. animals	Water (mL)	Eggs (g)	Di-Vetelact® (g)	$CaCO_3$ (g)	SF50® (g)	Termite mound powder (g)
1	57	30	12.3	0.38	0.08	5
4	228	120	49.2	1.5	0.3	20
10	570	300	123	3.75	0.75	50
15–16	910	480	198	5.63	1.5	100
20	1140	600	246	7.5	1.5	125

1 teaspoon $CaCO_3$ weighs 7.5 g
1 teaspoon SF50® weighs 3.0 g

Add eggs to water and whip with an electric beater. Add Di-Vetelact® (Sharpe Laboratories) and blend. Place mix into a stainless steel bowl and place onto a double boiler on a stovetop. Allow to cook for 20 min on low temperature, stirring regularly. The consistency should be of thick custard. Do not overcook, as the mixture will caramelise and become unpalatable. Once cooked, transfer to a stainless steel bowl and store in the refrigerator. Use within 24 hr. Immediately before feeding, add SF50® multivitamin powder (Glen Forrest Stockfeeders), $CaCO_3$ and termite mound powder (see below) and blend to a smooth consistency.

Termite mound powder preparation

Collect and store spent termite mound discarded from traps. Place mound material into a pot with a small amount of water. Boil the mound for 30 min until most of the water has evaporated. Tip the mix onto a baking tray and bake at 150°C for 1 hr. Crush the mound into a powder using a hammer and sifter. Store in an airtight container. Powder must be thoroughly dry before storage or it will rapidly become mouldy.

Source: After Power & Monaghan (2003).

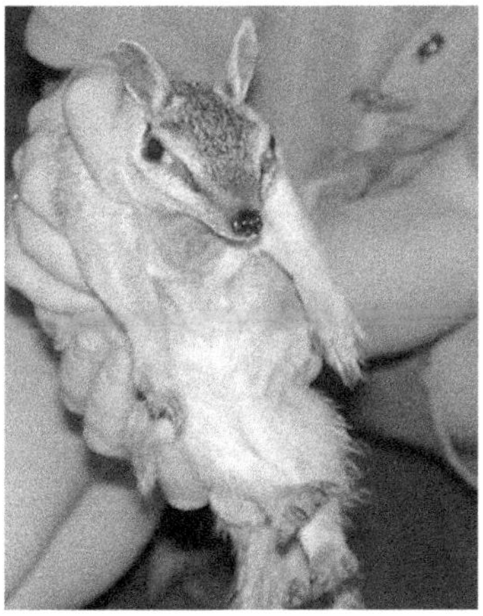

Figure 11.1 Manual restraint of a numbat.

procedures, numbats are also easily intubated using an uncuffed 2 mm endotracheal tube (Fig. 11.2). They are inclined to sudden rapid recovery during anaesthetic procedures and should be maintained at 1.5–3% isoflurane. Tiletamine/zolazepam has been used at 5–9 mg/kg IM and could be used if gaseous anaesthesia is not available (Vogelnest 1999). However, its use is difficult to justify given the ease and safety of isoflurane anaesthesia in this species. Sedation of numbats has not been undertaken although presumably the principles would be similar to those for small dasyurids. Pre-anaesthetic fasting is not essential, but an 8 hr fasting period is recommended for elective procedures.

7 CLINICAL PATHOLOGY

General anaesthesia is recommended for blood sample collection in numbats. Even in an anaesthetised animal blood collection is difficult due to the thick skin, lack of visible vessels and poor blood flow. The most reliable venipuncture sites are the femoral, jugular and brachiocephalic veins.

The landmarks for femoral and jugular venipuncture are similar to other small mammals. The site for

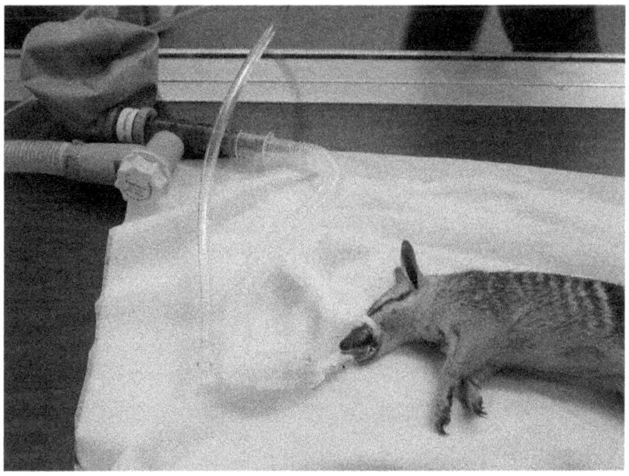

Figure 11.2 Numbat intubated and maintained under isoflurane anaesthesia.

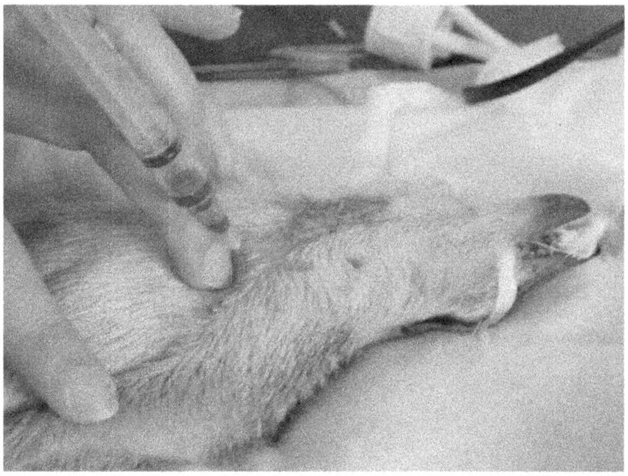

Figure 11.3 Brachiocephalic venipuncture technique in an anaesthetised numbat.

brachiocephalic venipuncture lies at the point where the vein emerges from the thoracic inlet. The animal is positioned in dorsal recumbency, with its forearm positioned alongside the body. A 23–25 G, 5/8–3/4″ needle is introduced perpendicular to the skin, either just cranial or just caudal to the clavicle, into the deep notch palpable at the medial extremity of the clavicle adjacent to the sternum. While applying suction, the needle is advanced slowly until it enters the vein and blood appears in the syringe (Fig. 11.3). If no blood appears, suction should be maintained while slowly withdrawing the needle as the vein can often be entered on withdrawal if it is missed when advancing the needle. This technique is more difficult in males than in females because the sternal gland is more developed in males, particularly during the breeding season, and this limits access to the landmarks.

Table 11.2 presents selected haematology and biochemistry values for a small number of numbats. Limited information suggests that haematological responses may not necessarily reflect the nature and severity of disease. For example, a relative neutrophilia is often seen without any evidence of inflammatory disease and polychromatic red blood cells are common in normal animals. Mild anisocytosis and nucleated erythrocytes may also be seen in clinically healthy animals (Clark 2004). White cell morphology is similar to that of other marsupial species. Serum biochemistry may be somewhat more reliable as an indicator of disease. Elevated BUN and creatinine have been observed in individuals with renal dysfunction. Low TSP has been associated with cases of chronic inappetence.

Urinalysis is most easily conducted on voided samples. Samples are collected by placing the numbat in a wire cage with a mesh floor over a tray (Power & Monaghan 2003). Although manual expression of the bladder and cystocentesis are possible under anaesthesia, the bladder is frequently difficult to palpate and many individuals urinate during induction. Dipstick urinalysis parameters are typical of most mammals. Isosthenuria and proteinuria have been noted in several individuals with renal disease.

8 HAND-REARING

Young numbats have only been successfully hand-reared when older than 6 mo, fully furred and close to being deposited (Power & Monaghan 2003). Unfurred PY which are abandoned are probably best euthanased. In the absence of specific numbat artificial milk formulas, other marsupial hand-rearing formulas have been modified to more closely resemble numbat milk composition. Wombaroo® >0.7 Kangaroo Milk Replacer (Wombaroo Food Products) has been used at 150% concentration to successfully hand-rear numbats. Other milk replacers have been used for supplementary feeding of juvenile numbats (e.g. Di-Vetelact®, Sharpe Laboratories). However, these are not as concentrated as the Wombaroo formulation and it is more difficult to achieve an adequate concentration of nutrients within an acceptable volume.

If PY appear dehydrated or debilitated on initial presentation it may be necessary to rehydrate with oral electrolytes and subcutaneous fluids (2.5% glucose) for 12–24 hr before feeding milk replacers.

If hand-reared PY are unable to suckle they may be fed using a 1 mL syringe to drop milk directly into the mouth. If they are able to suckle a very fine-tipped possum teat may be used. PY are encouraged to lap as soon as possible, as lapping minimises human imprinting and allows the gradual introduction of artificial diet into the milk. The transition from syringe feeding to termite and artificial diet feeding has been achieved in 4 wk by increasing the amount of artificial diet in the milk formula by 5 g every 3 d (Power & Monaghan 2003).

Numbat young have been fostered successfully onto other lactating numbats whose own litters were recently weaned (Power & Monaghan 2003). This technique may be attempted in deposited young as an alternative to hand-rearing.

Table 11.2 Selected haematology and biochemistry values for numbats

Parameter	Units	Mean	SD	Range	n
Haematology					
RBC	$\times 10^{12}$/L	6.19	0.54	5.43–7.35	8
Hb	g/L	143	9.77	130–157	7
PCV	L/L	0.43	0.03	0.39–0.47	7
MCV	L/L	72.13	3.23	66–76	7
MCH	pg/cell	22.0	1.6	18.6–24.1	8
MCHC	g/L	329	12	309–342	7
WBC	$\times 10^9$/L	24.7	0.54	1.7–3.2	7
Segmented neutrophils	$\times 10^9$/L	1.38	0.54	0.58–1.95	7
	%	54.57	14.01	34–74	7
Lymphocytes	$\times 10^9$/L	1.03	0.31	0.55–1.46	7
	%	43.00	14.74	21–63	7
Monocytes	$\times 10^9$/L	0.04	0.03	0–0.08	7
	%	1.86	1.21	0–3	7
Eosinophils	$\times 10^9$/L	0.02	0.02	0–0.05	7
	%	0.71	0.95	0–2	7
Basophils	$\times 10^9$/L	0.00	0.00	0	6
	%	0.00	0.00	0	6
Biochemistry					
Calcium	mmol/L	2.45	0.18	2.15–2.73	6
Phosphorus	mmol/L	1.99	0.33	1.59–2.43	6
BUN	mmol/L	15.86	2.27	12.8–17.9	6
Creatinine	µmol/L	65	8.13	53–74	6
Total bilirubin	µmol/L	3	2	2–7	6
Glucose	mmol/L	6.72	1.89	4.22–9.77	6
CK	U/L	1058	291	615–1500	6
ALP	U/L	45	10.78	30–60	6
ALT	U/L	77	63.33	30–215	6
AST	U/L	86	19	63–117	6
TSP	g/L	58	3.48	53–63	7
Albumin	g/L	41	3.64	36–47	7
Globulin	g/L	17	2.19	14–20	7

SD = standard deviation
n = number of animals
Source: ISIS (2002 and unpublished data).

9 DISEASES

9.1 Dermatologic diseases

Dermatitis has been reported in numbats. Obesity and excessive moisture in enclosure substrate are believed to be important factors in the aetiopathogenesis of dermatitis in numbats. Many cases resolve spontaneously, but unresponsive cases warrant thorough investigation including biopsy and culture.

Bacterial dermatitis has a variable presentation which may include exudative axillary lesions, alopecia, scurf and scale production. Cases usually respond to oral or topical antibiotic preparations (Table 11.3). Pododermatitis, ventral pustular dermatitis, patchy alopecia and superficial erosive lesions have been associated with *Mycobacterium intracellulare* and *M. chelonae abscessus* infection in two captive numbats (Gaynor et al. 1990). One animal responded to parenteral antibiotic treatment with amikacin at 7.5 mg/kg IM bid for 6 d with a repeat 10 d course 6 wk later. The second died during treatment and was found to have underlying pulmonary cryptococcosis. In other mammals, atypical mycobacterial infections are often opportunistic and it was suggested that both cases may have been secondary to skin trauma. Mycobacterial infection should be considered in any unusual or unresolving cases of dermatitis in numbats.

Numbat mating behaviour can be aggressive. Males may restrain the female by biting them on the scruff of the neck, a practice which may cause alopecia, crusting and ulceration in the dorsal cervical region. These cases respond spontaneously after the mating season and generally do not require veterinary intervention. However, on rare occasions severe or repeated trauma has necessitated early removal of the male.

Perivascular dermatitis, possibly related to hypersensitivity, was observed in one numbat which presented with scurfy skin and alopecia of the feet.

9.2 Gastrointestinal diseases

Sand impaction has been diagnosed in numbats and is a normal consequence of the numbats' termite diet. Termite mound is provided as part of the standard artificial diet and should not lead to impaction *per se*. It has been the authors' experience that impaction occurs when numbats maintained on sandy soils develop a concomitant predisposing disease. Lower motor neuron injury with subsequent defecation difficulties, intermittent *Salmonella* infection and chronic proctitis have all been implicated. Clinical signs include salivation, inappetence, cloacal prolapse or swelling, vomiting and tenesmus. Sand can be palpated in loops of bowel in the abdomen and can be visualised radiographically. Treatment has included dietary supplementation with psyllium, administration of palatable oral laxative gel (Cat-Lax®, Pharmaderm Animal Health), warm water enemas administered under general anaesthesia, and supportive care. Timely treatment is warranted as sand impaction has occasionally been fatal.

Diarrhoea has been occasionally observed in numbats. No specific aetiologies have been confirmed although *Salmonella* species have been cultured in animals with mucoid or malodorous faeces (Power & Monaghan 2003; see also 9.8.2). Most cases of diarrhoea are transitory and resolve without treatment. *Salmonella* culture is recommended for all recurrent or persistent cases.

9.3 Cardiopulmonary diseases

Cardiopulmonary conditions tend to be isolated incidents and have included pleurisy with pyothorax (possibly secondary to oesophageal perforation), cryptococcal pneumonia (Gaynor et al. 1990), focal granulomatous mycobacterial pneumonia, epicarditis and possible carcinoma leading to chylothorax, and atherosclerosis of the aorta and carotid vessels.

An upper respiratory tract condition, characterised by noisy respiration and seromucoid bubbling nasal discharge, has been observed occasionally. Cases have resolved without treatment. The cause is unknown and, in the absence of other clinical disease, these signs are believed to represent a stress or handling response.

9.4 Urinary and reproductive diseases

Few urogenital diseases have been observed in numbats. Isosthenuria and proteinuria were seen in a 1 yr old animal which eventually died of renal failure. At necropsy it was found to have bilateral cystic renal dysplasia, which may have been a congenital anomaly. Cases of penis prolapse have been observed after mating. These have responded to lubrication, topical application of a concentrated sugar solution to reduce oedema and manual reduction under anaesthesia. A preputial tear, which manifested as cloacal bleeding, was diagnosed in one male after mating. The condition responded to single injections of carprofen and clavulanic acid-amoxycillin, and application of topical antibiotic ointment (Tricin Eye and Ear Ointment®, Jurox). A case of unilateral testicular capsular fibroma diagnosed in a mature animal was characterised by increased size and firmness in the affected testis. Hemicastration was performed and the animal recovered uneventfully.

9.5 Neurologic and musculoskeletal diseases

Numbats are avid climbers and most sudden-onset neurologic and/or musculoskeletal conditions result from climbing injuries. Loss of tail function (suggestive of nerve damage) and degenerative arthritic changes have been observed. Anti-inflammatory drugs including dexamethasone, carprofen and ketoprofen have been administered with moderate success. Tail injuries do not usually warrant amputation, which should be avoided if possible as the tail appears to be important for balance.

9.6 Ophthalmologic diseases

Corneal ulcers due to trauma have been observed in several individuals. These have responded to topical ophthalmic medications.

9.7 Neonatal diseases

Neonatal disease is uncommon in numbats and is usually secondary to feeding difficulties or complications of hand-rearing such as aspiration pneumonia and hypothermia. Cases of poor lactation in dams have led to weight loss and starvation in PY with ensuing secondary complications such as pneumonitis. Affected PY die very quickly and intervention in these cases has generally been unsuccessful. Poor lactation appears to be associated with environmental stress, so it is critical that lactating females be maintained in a quiet warm environment and provided with excellent nutrition.

9.8 Infectious diseases

9.8.1 Parasites

Numbats are the definitive host of the acanthocephalan parasite *Multisentis myrmecobius* (Smales 1997). Acanthocephalan larvae develop in an invertebrate intermediate host, which in this case is assumed to be the numbat's staple food source, the termite. The pathogenicity of *M. myrmecobius* has not been established and many numbats appear to be asymptomatic carriers. While the parasite has been detected in association with pathology such as intestinal torsion in some wild individuals (Haigh & Friend 1999) its presence may have been coincidental. As a precaution, all numbats entering the Perth Zoo collection are currently treated with ivermectin at 200 μg/kg PO during quarantine, with a repeat dose given 3 mo after leaving quarantine. These doses should theoretically be effective in breaking the cycle of infection since the worm takes 5–12 wk to mature in the numbat and they are fed uninfected termites in captivity. However, due to the difficulty of detecting the parasite consistently in faecal samples and its dubious pathogenicity, it is not clear whether this treatment is effective or warranted.

All numbats entering and leaving the Perth Zoo collection undergo screening for faecal parasites. Unspeciated strongyle eggs have been observed in the faeces of clinically normal numbats. Two species of trichostrongyloid nematode, *Beveridgiella calabyi* and *B. inglisi*, are known to occur in numbats (Beveridge & Spratt 1996) but have not been associated with disease. Acanthocephalan eggs are occasionally detected.

Ticks (*Ixodes vesitus*, *I. holocyclus* and *Amblyomma triguttatum*), mites (*Mesolaelaps australiensis*) and stick-fast fleas (*Echidnophaga myrmecobii* and *E. perilis*) (Calaby 1960) are occasionally seen on numbats. They are rarely of clinical significance. In captivity, mild stick-fast flea infestations have been observed in the periocular and periauricular regions. These have been treated by forceps removal under general anaesthetic.

An unspeciated Prostigmata ear mite (not of the Otodectes Family) caused bilateral grey crusting of the ears in one captive numbat. The problem responded well to topical application of a pyrethrin-based ear medication.

Haemoparasites have not been identified in routine smears of numbat blood. The potential role of numbats as vectors of arbovirus has not been explored.

9.8.2 Bacteria

Salmonella spp. have been detected incidentally on routine faecal culture in the absence of clinical signs. However, *Salmonella* has been cultured in animals with mucoid or malodorous faeces and shedding may be associated with periods of stress (Power & Monaghan 2003). Treatment with oral antibacterials, including trimethoprim sulfamethoxazole and amoxycillin-clavulanic acid, has been attempted on occasions when known pathogenic strains have been isolated. However, due to intermittent shedding and lack of clinical evidence of disease the success of treatment has been difficult to ascertain. Species which have been isolated from faecal culture in numbats include *S. cholerae-suis* (var. *Australia* and var. *decatur*), *S. wandsbek* and *S. indiana*.

Campylobacter coli has been detected on faecal culture from a number of clinically normal numbats. No treatment or control was implemented.

9.8.3 Protozoa

There are no reports of protozoal infections in numbats. Toxoplasmosis should be included as a differential diagnosis in cases of central nervous system disease. However, over an 8 yr period approximately 50 numbats were tested with no positive results (Perth Zoo, unpublished).

9.8.4 Fungi

Although numbats are probably capable of contracting opportunistic fungal infections, actual incidences of disease are rare. Cryptococcal pneumonia was diagnosed at necropsy in one individual (Gaynor et al. 1990). However, the species of *Cryptococcus* was not identified. Dermatophytosis has not been diagnosed in any individuals to date.

9.9 Non-infectious diseases

9.9.1 Trauma

In captivity, numbats are prone to toenail avulsions due to their burrowing and climbing behaviour. Such injuries rarely cause significant problems and may be assessed and treated according to first principles. Numbats have presented with signs of shock and asphyxia after becoming trapped following collapse of their burrows. Treatment was symptomatic and included clearing the airway, intubation, oxygen therapy and administration of SC fluids (2.5% glucose).

A numbat presenting with loss of tail tone and hind limb ataxia was diagnosed radiographically as having ventral luxations of the first and sixth coccygeal vertebrae. This injury was thought to result from jumping behaviour, which has been observed in captivity associated with climbing fences (see section 4). The sacrococcygeal trauma may have contributed to the animal's tendency toward chronic sand impaction and other signs of poor colon motility (see 10.5).

9.9.2 Nutritional diseases

No specific nutritional diseases have been identified in numbats. It is possible that their termite diet renders them prone to vitamin K deficiency, as has been described in other myrmecophagous species (Morford & Meyers 2003). However, there have been no confirmed cases of coagulopathy which would support this diagnosis and supplementation with vitamin K is not included in the captive diet.

9.9.3 Toxins

Intoxication of unknown aetiology has been suspected in one case of sudden death in a captive numbat. Presentation included polydipsia and abdominal distension. Cause of death was acute glomerular necrosis and haemoglobinuric nephrosis consistent with ongoing intravascular haemolysis. No toxin was identified and no other numbats demonstrated similar signs, however, the pathology was considered to be consistent with exposure to a toxin.

9.9.4 Degenerative diseases

A reducible unilateral perineal hernia, probably a degenerative change associated with old age, has been observed in one wild-caught female numbat. This condition was detected incidentally and did not appear to impede defecation or cause any other ill effects. The animal died of an unrelated illness 3 mo later and at necropsy there was no evidence of complications or tissue injury associated with the hernia.

Osteoarthritis of the hips and hocks has been observed in an 11 yr old captive numbat. The animal succumbed to bile duct carcinoma shortly after diagnosis of the osteoarthritis so treatment was not attempted. Radiographic evidence of degenerative joint disease of the stifle and elbow has also been observed in captive numbats. Treatment with anti-inflammatory drugs, including dexamethasone, carprofen and meloxicam, has occasionally been attempted in animals with signs of stiffness or abnormal movement but the success of such treatments has been equivocal.

9.9.5 Neoplasia

Neoplastic conditions diagnosed in captive numbats include eyelid papilloma, squamous cell carcinoma, adenocarcinoma, bile duct carcinoma with pulmonary metastasis and testicular fibroma (Power & Monaghan 2003).

10 THERAPEUTICS

Drugs that have been administered to numbats without observable side effects are given in Table 11.3. Choice of agent and their dosage, route, frequency and duration of administration are based on those administered to domestic cats and dogs. There have been no pharmacokinetic studies in numbats. Pentosan polysulphate was used at standard canine dose rates in one numbat that subsequently died from extensive generalised haemorrhage. A bile duct carcinoma with pulmonary metastasis was diagnosed on histopathology. Haemorrhage has been reported as a possible side effect of pentosan polysulphate in animals with concomitant malignancy (Arthropharm 2001). It may be prudent to avoid this drug in numbats, particularly when neoplasia is suspected.

11 REPRODUCTIVE MANAGEMENT

In captivity, numbats are paired for mating when sperm is detected in the male's urine and cornified epithelial cells are detected on vaginal smears in the female (see section 3). Numbats are paired until PY are identified

Table 11.3 Pharmaceuticals administered to numbats without apparent side effects

Drug	Dose rate and route
Amikacin	7.5 mg/kg IM bid 6 d, then 7.5 mg/kg IM bid 10 d 6 wk later
Amoxycillin and clavulanic acid	12.5 mg/kg PO bid added to food; 8.75 mg/kg SC sid 3–5 d
Apex Ear Drops® (polymyxin B 10 000 U/mL, nitrofurazone 2.6 mg/mL, pyrethrins 0.5 mg/mL, neomycin sulphate 5 mg/mL, lignocaine hydrochloride 2.5 mg/mL; Apex Laboratories)	Topical application for ear mites, bid for 7 d
Carprofen	3 mg/kg SC single dose, or sid 3 d
Cat-Lax® (Pharmaderm Animal Health)	Add 0.5 cm to food sid to help clear sand accumulation
Dexamethasone	2 mg/kg SC sid 3 d
Ivermectin	200 µg/kg PO single dose
Ketoprofen	1 mg SC sid 5 d
Meloxicam	0.2 mg/kg SC single dose, then 0.1–0.2 mg/kg PO sid 6 d
Psyllium	Add a small pinch to food sid to clear sand impaction
Trimethoprim-sulfamethoxazole	10 mg/kg trimethoprim component PO bid 14 d
Tricin Eye and Ear Ointment® (zinc bacitracin 500 U/g, neomycin sulphate 5 mg/g, polymyxin B sulphate 10 000 U/g; Jurox)	Topical application to eyes and external wounds, sid for 3–7 d

Source: unpublished data.

(approximately 2 wk after mating). If no PY are found at this time the male may remain with the female for another cycle or may be replaced with another male (Power & Monaghan 2003).

As numbats are housed individually unless breeding is intended, there has been no use or investigation of contraceptive methods in numbats.

12 SURGERY

Standard small animal surgical principles apply in this species. Surgical characteristics of interest include a poorly defined linea alba and very tough skin which resists microchip implantation and incision. Numbats do not appear prone to post-surgical infections. Intra-abdominal telemetry devices have been implanted using a standard midline approach with routine closure (Cooper & Withers 2004a). Hemicastration has been performed in one numbat using a standard open technique.

REFERENCES

Arthropharm Pty Ltd (2001). *Safety of Cartrophen Vet in Clinical Practice.* <www.arthritis.au.com/dogsafety/saf_01.htm>.

Beveridge I & Spratt DM (1996). The helminth fauna of Australian marsupials: origins and evolutionary biology. *Advances in Parasitology* **37**, 135–254.

Calaby JH (1960). Observations on the banded ant-eater *Myrmecobius F. fasciatus* Waterhouse (Marsupialia), with particular reference to its food habits. *Proceedings of the Zoological Society of London* **135**, 183–207.

Clark P (2004). *Haematology of Australian Mammals.* CSIRO Publishing: Melbourne.

Cooper CE & Withers PC (2004a). Patterns of body temperature variation and torpor in the numbat *Myrmecobius fasciatus* (Marsupialia: Myrmecobiidae). *Journal of Thermal Biology* **29**, 277–284.

Cooper CE & Withers PC (2004b). Termite digestibility and water and energy contents determine the water economy index of numbats (*Myrmecobius fasciatus*) and other myrmecophages. *Physiological and Biochemical Zoology* **77**, 641–650.

Cooper CE, Withers PC & Bradshaw SD (2003a). Field metabolic rate and water turnover of the numbat (*Myrmecobius fasciatus*). *Journal of Comparative Physiology B* **173**, 687–693.

Cooper CE, Walsberg GE & Withers PC (2003b). Biophysical properties of the pelt of a diurnal marsupial, the numbat (*Myrmecobius fasciatus*), and its role in thermoregulation. *Journal of Experimental Biology* **206**, 2771–2777.

DeJose J (1988). Numbat conservation in WA. In *Proceedings of the 5th World Conference on Breeding Endangered Species in Captivity*, pp. 501–509. Cincinnati, Ohio.

Friend JA (1989). Myrmecobiidae. In *Fauna of Australia*, vol. 1B Mammalia (eds DW Walton & BJ Richardson), pp. 583–590. CSIRO Publishing: Melbourne.

Friend JA (1997a). Ecology and management of the numbat *Myrmecobius fasciatus*. In *1997 Fauna Conservation Course*, pp. 7–10. Department of Conservation and Land Management: Batalling Field Study Centre, Western Australia.

Friend JA (1997b). Numbats on a junk food diet. *Nature Australia* **25**, 40–49.

Friend JA (1997c). Recovery plan for the numbat (Myrmecobius fasciatus): endangered species program, project number 351: review of the recovery process. Department of Conservation and Land Management, Western Australia.

Gaynor WT, Cousins DV & Friend JA (1990). Mycobacterial infection in numbats. *Journal of Zoo and Wildlife Medicine* **21**, 476–479.

Haigh SA & Friend JA (1999). Veterinary aspects of several threatened western Australian mammal species. *Australian Association of Veterinary Conservation Biologists Newsletter* **16**, 5.

Hume ID (1999). *Marsupial Nutrition*. CUP: Cambridge.

ISIS (2002). *Reference Ranges for Physiological Values in Captive Wildlife*. International Species Information System: Eagan, Minnesota.

Maxwell S, Burbidge A & Morris K (eds) (1996). Action plan for Australian marsupials and monotremes for the IUCN/SSC Australasian marsupial and monotreme specialist group. Department of Environment and Heritage, Canberra.

Morford S & Meyers MA (2003). Giant anteater (*Myrmecophaga tridactyla*) health care survey. *Edentata* **5**, 5–19.

Power V & Monaghan C (2003). Numbats. In *Australian Mammals: Biology and Captive Management* (ed. S Jackson), pp. 99–125. CSIRO Publishing: Melbourne.

Smales LR (1997). *Multisentis myrmecobius*, gen. et. sp. nov. (Acanthocephala: Oligacanthorhynchidae), from the numbat, *Myrmecobius fasciatus*, and a key to the genera of the Oligacanthorhynchidae. *Invertebrate Taxonomy* **11**, 301–307.

Vogelnest L (1999). Chemical restraint of Australian native fauna. In *Wildlife in Australia: Healthcare and Management, Proceedings 327,* pp. 149–181. Post Graduate Foundation in Veterinary Science: Sydney.

12. Gliders and possums

Robert Johnson and Susan Hemsley

1 TAXONOMY, DISTRIBUTION AND BIOLOGY

There are 28 species of Australian possums and gliders in six Families. The smallest have an adult body size of less than 10 g while the largest may weigh up to around 4.5 kg (Smith & Winter 1984). All gliders and nearly all possums are arboreal and live in forest or heathland, but some are semi-terrestrial including the mountain pygmy possum (*Burramys parvus*), rock-haunting ringtail (*Petropseudes dahli*) and scaly-tailed possum (*Wyulda squamicaudata*) (Strahan 1995). Due to this habitat requirement, most species are found along the coast and associated ranges. Only a small number of species are found in Tasmania. Many species have a restricted range but some, such as the common brushtail (*Trichosurus vulpecula*), common ringtail (*Pseudocheirus peregrinus*) and sugar glider (*Petaurus breviceps*) have extensive ranges (Smith & Winter 1984). In addition, common brushtails and common ringtails often live in urban areas (How & Kerle 1995; McKay & Ong 1995). These two species are the most common seen in veterinary practice in Australia. Sugar gliders are commonly kept as pets in other countries and the common brushtail was introduced to New Zealand in 1858; it is now a major agricultural and ecological pest there (Tyndale-Biscoe 2005a).

Possums and gliders are nocturnal although some, such as the common spotted cuscus (*Spilocuscus maculatus*), the feather-tailed glider (*Acrobates pygmaeus*), the honey possum (*Tarsipes rostratus*) and the eastern pygmy possum (*Cercatetus nanus*), may sometimes be active in daylight hours (Renfree 1995; Turner & Ward 1995; Winter & Leung 1995; Woodside 1995). During the day they shelter in constructed nests or pre-existing structures such as tree hollows or rocky outcrops (Menkhorst 1984). Common ringtail and feather-tailed glider nests are known as dreys. These are spherical bark or leaf nests made in tree hollows or suspended in undergrowth (McKay & Ong 1995; Woodside 1995). The basic social unit of possums and gliders may consist of an individual or stable, often family-based group depending on species (Russell et al. 1989). All possums and gliders have cutaneous scent glands and para-cloacal glands that are

Table 12.1 Body weights of selected possum and glider species

Common name	Weight
Mountain pygmy possum	30–80 g
Eastern pygmy possum	15–38 g
Striped possum	245–520 g
Leadbeater's possum	100–160 g
Yellow-bellied glider	450–700 g
Sugar glider	90–150 g
Squirrel glider	190–300 g
Greater glider	900–1700 g
Herbert River ringtail possum	800–1500 g
Common ringtail	660–900 g
Honey possum	5–10 g
Feather-tailed glider	10–14 g
Mountain brushtail	2.5–4.5 kg
Common brushtail	1.5–4.0 kg
Scaly-tailed possum	1.3–2.0 kg

Source: Menkhorst (2001).

important for communication. Saliva, urine and faeces are also used for scent marking (Biggins 1984; Dawson et al. 1989; McKay 1989). Some species, such as the yellow-bellied glider (*P. australis*), common brushtail and sugar glider are very vocal (Biggins 1984; Suckling 1995).

2 ANATOMY AND PHYSIOLOGY

The limbs and tails of possums and gliders reflect their generally arboreal niche (McKay 1989; McKay & Winter 1989; Russell & Renfree 1989; Turner & McKay 1989). The pads of the digits and feet are often granulated or corrugated, which enhances friction. The fore feet are variably arranged between species. Some species such as ringtail possums have a grasping hand with the first two digits opposable to the remaining three (McKay 1989). The striped possum (*Dactylopsila trivirgata*) uses an elongated fourth fore limb digit and long tongue to extract ants and termites (Smith 1982). Digit one of the hind foot is laterally directed, clawless and opposes the remaining digits. Digits two and three of the hind foot are reduced in size and united by skin (syndactyly) (Smith & Winter 1984). The tails of possums and gliders are long, at least moderately prehensile and often have a hairless friction pad on the underside. Gliders use the tail as a rudder. The tail of the feather-tailed glider has distinctive lateral fringes of long stiff hairs, giving it a feathery appearance, and pygmy possums can store fat in their tail base (McKay 1989; McKay & Winter 1989; Russell & Renfree 1989; Turner & McKay 1989).

Gliders are distinguished from possums by the presence of a gliding membrane (patagium) extending from the hind limb to the fore limb. The gliding membrane consists of a fold of skin, connective tissue and muscle. In *Petaurus* spp. it extends from the fifth digit of the fore foot to the distal metatarsus or first digit of the hind limb. In the greater glider (*Petauroides volans*) and feather-tailed glider it extends from the elbow to the distal tibia and stifle respectively. There is elongation of the long bones of the larger gliders (McKay 1989; Suckling 1995; Woodside 1995).

Gliding is a more efficient form of movement than running, climbing or jumping. It enables animals to move quickly from tree to tree, facilitating efficient foraging. Gliding also reduces the need to come to the ground, ensuring that animals remain clean and dry. Scent trails are broken by gliding, which reduces the risk of predation. The gliding membrane provides warmth when wrapped around the body (Lindenmayer 2002).

The pouch opens cranially, has 2–6 teats and may have an incomplete median septum in *Petaurus* spp. and Leadbeater's possum (*Gymnobelideus leadbeateri*) (McKay 1989; McKay & Winter 1989; Russell & Renfree 1989; Turner & McKay 1989). Eccrine sweat glands are restricted in distribution to hairless areas. Cutaneous scent glands are found in all animals and include sternal, frontal, lip and tail glands. Para-cloacal glands also appear to be present in all species (Biggins 1984; Dawson et al. 1989; McKay 1989). The sternal gland in brushtail possums (*Trichosurus* spp.) is an obvious brown patch which is best developed in males and appears most prominent during the breeding season (McKay & Winter 1989).

The pollen and nectar feeding pygmy possums and the honey possum have a long brush-like tongue with long fine filiform lingual papillae. These probably increase the uptake of nectar. Pollen is mainly collected on the fur and ingested during grooming. In the honey possum, nectar and pollen are removed from the tongue with a series of transverse ridges on the hard palate, some of which have spines (Richardson et al. 1986).

Possums and gliders have diprotodont dentition (Smith & Winter 1984). They have lower incisors reduced to a single functional pair of large procumbent (reclining) teeth, three pairs of upper incisors, variably present canines, one or two pairs of upper and lower premolars and four upper and lower molars. The premolars and molars are well-developed with cusped grinding surfaces. In contrast, the honey possum has few teeth which, with the exception of the lower incisors, are reduced to peg-like structures (Richardson et al. 1986; Russell & Renfree 1989; Turner & McKay 1989).

Most possums and gliders digest plant material by microbial fermentation in the caecum and colon and have a simple stomach. An expanded caecum is the main site for fermentation in most species. The intestines of the striped possum and honey possum, however, are simple due to the lack of plant material in their diet. The stomach of the honey possum has a large diverticulum, which may have a role in nectar storage (Hume 1982; Smith 1982; Richardson et al. 1986). Coprophagy of caecal contents occurs in at least the common and green ringtails (*Pseudochirops archeri*) (Strahan 1995; Winter & Goudberg 1995).

Pygmy and honey possums and the feather-tailed, sugar and yellow-bellied gliders enter torpor in response to cold or decreased nutrient intake. They also huddle to conserve energy (Fleming 1980; Russell & Renfree 1989; Turner & McKay 1989; Lindenmayer 2002). Torpor usually last for less than a day. During winter the mountain pygmy possum is able to undergo extended periods resembling true hibernation, with intermittent short periods of normal body temperature and activity (Broome 1995; Geiser & Broome 1993; Geiser 1993). The feather-tailed glider can hibernate for up to 5 d during which its body temperature may decrease to 2°C (Fleming 1985; Lindenmayer 2002).

3 REPRODUCTION

3.1 Reproductive anatomy

3.1.1 Female

The anatomy of the female reproductive tract is typical of most marsupials. There are two lateral vaginae and a median vagina with a temporary central canal through which the young is born. The morphology of the reproductive tract changes with the stage of the breeding cycle. For example, in female sugar gliders the ovaries, lateral vaginae and uteri increase in size during oestrus and early pregnancy. Sugar gliders have elongated lateral vaginae compared with other species of possums and gliders (McKay 1989).

3.1.2 Male

The anatomy of the male reproductive tract of possums and gliders is similar to that of most marsupials. Testes are located in a pre-penile scrotum. Accessory sex glands consist of only a prostate and one or more pairs of bulbourethral (Cowper's) glands (Temple-Smith 1984). The glans penis is commonly bifid for a short distance terminally, e.g. in ringtail possums and pygmy possums, or may be tapered with (e.g. common brushtail) or without (e.g. honey possum) a urethral process and may have short backward projecting spines (Russell & Renfree 1989; Turner & McKay 1989). When not erect the penis forms an S-shaped curve and is withdrawn into the common vestibule. Sperm production and testes size vary seasonally in the greater glider and the common ringtail (Temple-Smith 1984). The prostate gland and bulbourethral glands of the male sugar glider increase in size during the breeding season, as does the prostate gland but not the testes of the common brushtail (Temple-Smith 1984; Lindenmayer 2002). In some species such as the common brushtail, males may be fertile all year and sperm production and testes size do not vary (McKay 1989). In brushtails a post-copulatory plug that forms from prostatic fluid after mating may prevent subsequent insemination by other males (Kerle 2001).

3.2 Reproductive physiology

Most possums are polyoestrus and polyovular, except for some of the larger species which may be monovular (Kerle 2001). Ovulation occurs on the first 1–2 d of the oestrous cycle. Ovulation is usually suppressed by lactation but females will return to oestrus 8–9 d after losing a PY. Length of gestation is usually less than 60% of the length of the oestrous cycle. However, in mountain pygmy possums, pygmy possums and the honey possum, gestation is longer (Table 12.2).

All gliders are polyoestrus, except for the greater glider which is monoestrus. It also has a short breeding season (March–June). It normally produces one young per year, but will give birth to a second young if the first is lost early in the breeding season. Males produce sperm from March to May, after which the testes reduce in size (Kerle 2001).

Embryonic diapause has traditionally been recognised as a characteristic of macropods. However, diapause has also been described in the honey possum (Oates et al. 2007), the feather-tailed glider and the eastern pygmy possum (Renfree 1981).

Monogamy is reasonably common in most of the larger species of possums and gliders. The smaller species that live in colonies, such as the honey possum, feather-tailed glider, eastern pygmy possum and mountain pygmy possum, are usually polygamous.

Body size is closely related to reproductive parameters such as age at weaning, litter size and growth rate. Larger species produce fewer eggs, have smaller litters and longer lactation periods and their young have lower growth rates. Twinning is common in sugar gliders and common ringtails and occurs occasionally in common brushtails (Hughes & Hall 1984).

Diet, particularly protein content, influences litter size and growth rates of possums and gliders (Lindenmayer 2002). Low protein in the diet of the greater glider may contribute to the production of only one offspring annually, a long lactation period and slow growth rate. The short breeding season of the greater glider corresponds with a short period during which protein levels in eucalypt leaves increase. In comparison, the yellow-bellied glider has a more varied diet and a much longer breeding period of up to 9 mo

Table 12.2 Reproductive data of selected species of possums and gliders

Species	Oestrous cycle (d)	Gestation period (d)	Annual reproductive rate (young born pa)	Season of birth	Age at weaning (mo)	Age at dispersal (mo)	Age at birth of first offspring (mo)	Refs
Common brushtail	26 (25)	16–18	1–2	Seasonal Mar–May and Sept–Nov	6–7	8–18	12–24	2, 4, 5
Short-eared possum	25–27	15–17	1	Mar–May	8–9	18–36	24–36	2, 4
Common ringtail	28	14–16	1–2.4 (1–8)[a]	May–July and Sept–Nov	6–7	8–12	10–14	1, 2, 4
Greater glider	—	—	1	Apr–June	7–9	10–11	12–24	2, 3, 4
Yellow-bellied glider	—	—	1–2	Aug–Apr	6–8	18–24	18	3, 4
Sugar glider	29	16	1–4	Aug–Dec (nth Qld all year)	4	7–10	12	1, 3, 4
Feather-tailed glider	—	—	4–8	July–Jan	3–4	—	12	3, 4

Sources: **1** Kerle (2001). **2** Jackson (2003). **3** Tyndale-Biscoe (2005b). **4** McKay (1989). **5** Lindenmayer (2002).

(Table 12.2). The Herbert River ringtail (*Pseudocheirus herbertensis*) has a bi-modal breeding season. A major breeding season occurs in early winter and a second shorter period in summer. Leadbeater's possums may give birth in any season, with peaks in autumn and spring (McKay 1989). These two species are indigenous to more tropical regions of Australia; the more tropical species tend to breed year round. In the sugar glider, spring breeding coincides with increased availability of protein-rich insect prey.

3.3 Emergence from the pouch

Postnatal growth has been described in the sugar glider, squirrel glider (*Petaurus norfolcensis*), common ringtail and Herbert River ringtail (McKay1989). The general pattern of development is similar to that of other marsupials. After leaving the pouch, the offspring of *Pseudocheirus* spp. may remain with their mother, clinging to her fur as a back young. In the Herbert River ringtail and *Petaurus* spp. they may be left in the nest. The age at dispersal varies with species (Table 12.2).

4 HUSBANDRY

The principles of exhibit design and husbandry for possums and gliders have been reviewed (Jackson 2003). Due to their nocturnal habits the use of reverse day/night cycles is common in the display of most species. Adequate space and height for climbing, exploring and escape from conspecifics is required due to the increased chance of intraspecific aggression in some species in captivity. Common brushtails, common ringtails, eastern pygmy possums, sugar gliders, squirrel gliders and feather-tailed gliders are the most commonly kept exhibit animals. Yellow-bellied gliders and Leadbeater's possums are kept to a lesser extent in zoological collections in Australia. Sugar gliders are popular pets in North America and the United Kingdom. They are social animals and need to be housed in groups. A typical captive colony comprises one dominant male, two subordinate males and four females. Young should be removed at or soon after weaning to avoid being killed. Adult animals should be checked daily for evidence of fight wounds (Booth 2000). If sugar gliders are kept as pets at least two animals are required (entire males housed together will fight). Unfortunately, in many cases they are kept in isolation as a single pet and frequently develop behavioural problems such as stereotypies. Feather-tailed gliders live in large colonies.

4.1 Hospital and rehabilitation housing

Housing requirements vary according to species, age and sex. If indoor hospitalisation is required, conventional dog or cat cages can be used as long as a nest box and other cover such as leaves are provided and the wire fronts of the cages are covered in mesh fine enough to prevent escape. Animals should always be housed in a quiet environment physically separated from dogs and cats. Animals undergoing rehabilitation can be kept outdoors in secure aviaries in more temperate climates. More solitary species such as common brushtails should be housed singly. More social species such as sugar gliders and common ringtails can be successfully group-housed outdoors in wire mesh pens with dirt, grassed or concrete floors. They should be placed in groups of two or three before being moved into outdoor aviaries. Pens should be well-constructed and vermin- and bird-proof. Outdoor pens should include adequate shelter from extremes of temperature, wind and rain (ANZCCART 2006). Sufficient numbers of nests or nest boxes should be supplied to provide a safe retreat should animals be threatened by each other. Two hanging wire baskets tied together and lined with coconut husk fibre, fine twigs, leaves or other plant material create a comfortable nesting drey for common ringtails (Fig. 12.1). Branches for climbing should also be provided. Gliders should have adequate space for gliding.

4.2 Individual marking and identification techniques

Identification techniques vary for different species, ranging from eartags and ear notching to implanting of passive integrated transponders (microchips). The use of microchips for the identification of individual animals is the most common and practical method. Implants are inserted SC between the scapulae. In smaller species this is best performed with the animal under anaesthesia. The needle hole should be closed using tissue glue to prevent the microchip falling out soon after implanting. Ear notching has been used in feather-tailed gliders and small species such as pygmy possums. Metal eartags are also commonly used.

4.3 Transport

Nest boxes removed from an enclosure with the animal secured within it can be used for transport. The nest box should be placed in a cat carry cage or pet pack

Figure 12.1 a) Artificial drey for common ringtails. b) Typical possum or glider nest boxes. Photo: Robert Johnson.

for added security. Most species are readily transported in cloth bags (calico or hessian) which are then placed in a secure and well-ventilated container for transport over longer distances. For transport over long distances animals may be placed in secure wooden boxes with adequate nesting material (e.g. shredded paper) for comfort and cushioning during the journey. Gentle encouragement from one container or enclosure to another is preferable to direct handling.

5 NUTRITION

Possums and gliders may be herbivorous, insectivorous, folivorous, nectivorous or omnivorous. Foods include leaves, fruit, flowers, seeds, sap, gum, insect exudates, insects and occasionally meat (brushtail possums, common spotted cuscus). Some species are specialist feeders and rely entirely on honey and pollen (honey possum), ants (striped possum) or leaves (greater glider) (Smith 1982; Russell et al. 1989; Russell & Renfree 1989).

Common brushtails are cosmopolitan in their dietary habits and eat native foliage, fruits and flowers, some animal protein and insects. They have adapted well to urbanisation and scavenge food scraps and eat exotic flowers and fruits.

Common ringtails eat a wide variety of native flora including the leaves of eucalypts and melaleucas. Although primarily folivorous they also eat native fruits and flowers, especially grevilleas and wattle. They also eat introduced plants such as roses and cultivated fruits. Common ringtails are coprophagic. They produce two types of faeces. Caecotrophs (large soft moist faecal pellets originating from the caecum) are produced during daylight hours when the possum is less active and are eaten immediately. Coprophagy ensures more efficient absorption of nutrients. Harder faecal pellets are passed during the more active nocturnal period and are not usually eaten.

Gliders tend to be more selective. Sugar and squirrel gliders are sap eaters but also eat seeds, including wattle and casuarina, nectar, pollen and insects. They particularly favour psyllids and lerps (or scale insects) and the waxy sweat scale-like covering they produce. They do not eat leaves, flowers or buds. Feather-tailed gliders feed on nectar, pollen, sap, small insects, lerps and flowers. The yellow-bellied glider, the largest of the *Petaurus* genus, eats the sap from a large variety of eucalypts. Like other members of the genus they also eat insects and nectar. Greater gliders are almost exclusively folivorous, preferring the leaves of a wide range of eucalypts. They also eat the fruit of turpentine (*Syncarpia* spp.) trees.

Captive animals or those undergoing rehabilitation should be offered a diet as close as possible to that of free-ranging animals. Specific diets for captive possums and gliders have been reviewed in the literature (Jackson 2003; Booth 1994). Suggested daily diets for captive common brushtails are based on a mixture of fruit and vegetables, for captive ringtail possums a mixture of fruit, vegetables, leaves and flowers and for captive sugar gliders fruit and invertebrates such as moths or crickets. This basic diet should be supplemented less frequently with other dietary components (Table 12.3).

The intake of the captive diet can be assessed by monitoring weight, skin and body condition and faecal consistency. Obesity is common in captive possums and

Table 12.3 Dietary supplements for captive possums and gliders: recommended frequency per week

Common brushtail	Common ringtail	Sugar glider
Peanuts/almonds (3×)	Peanuts/almonds (3×)	Flowers (at least 2×)
Bread (2×)	Bread (2×)	Leaves (at least 2×)
Sunflower seeds (2×)	Sunflower seeds (2×)	Soaked dry puppy food (2×)
Dry dog food (3×)	Boiled egg (2×)	Sunflower seeds (2×)
Boiled egg (2×)	Mineralised salt lick (ad lib)	Egg yolk (2×)
Invertebrates (3×)	Possum nectar mix (1×)	Bread (2×)
Leaves and flowers (3×)		Endive (4×)
Mineralised salt lick (*ad lib*)		Possum nectar mix (1×)
Possum nectar mix (3×)		*Leadbeater mix (1×)*
200 g baby rice cereal (e.g. Farex®)[1]		21 g high-protein baby cereal[1]
1 tspn glucose powder		1 hard-boiled egg white
1 tspn nutritional supplement powder (e.g. Sustagen®)[2]		150 mL honey (preferably Eucalypt based)
4 drops infant multivitamin drops (e.g. Pentavite®)[3]		1 level measure Sustagen®[2]
Combine the above ingredients and make into a slurry with a 1:4 honey:water mixture		Blend the above ingredients and add water to make a total volume of 750 mL
		Store refrigerated for up to several days

Sources: Booth (1994); M Lynch (*pers. comm.*).
Manufacturers: **1** Heinz Watties. **2** Mead Johnson Nutritionals. **3** Roche.

gliders due to a plentiful food supply and reduced exercise (Booth 2000). For example, captive sugar gliders are often fed an excess of fat and simple sugars. Mealworms are high in fat and should be avoided or fed only in low quantities. Moths, beetles and crickets provide behavioural enrichment. The inclusion of pine nuts, casuarina seeds and acacia seeds in the diet contributes to good dental health.

Disease syndromes related to poor or inadequate nutrition are common in pet sugar gliders. Owners commonly offer a wide choice of foods, which may lead to nutritional imbalances. Animals tend to select soft sweet fruits which tend to be low in calcium and high in simple sugars. Feeding fruit and a high-sugar diet to folivores such as the common ringtail can lead to bloating with gas or fluid. This condition is frequently irreversible and may be fatal.

6 RESTRAINT

6.1 Capture and physical restraint

Capture of free-ranging possums and gliders can be difficult; various techniques have been tried (Goldingay & Kavanagh 1991; Harris & Goldingay 2005; Wayne et al. 2005). Common brushtails are readily trapped in cage traps (Fig. 12.2).

Figure 12.2 Common brushtail in a cage trap. Photo: Larry Vogelnest.

Restraint techniques vary according to species, size and the degree of injury or pain suffered. Pillowslips, small linen bags or sacks for larger species are useful for restraint and transport. Nest boxes or logs with removable lids are useful as the animals are frequently in the box or log and can be readily transferred from box to bag. Most species can be readily picked up, with

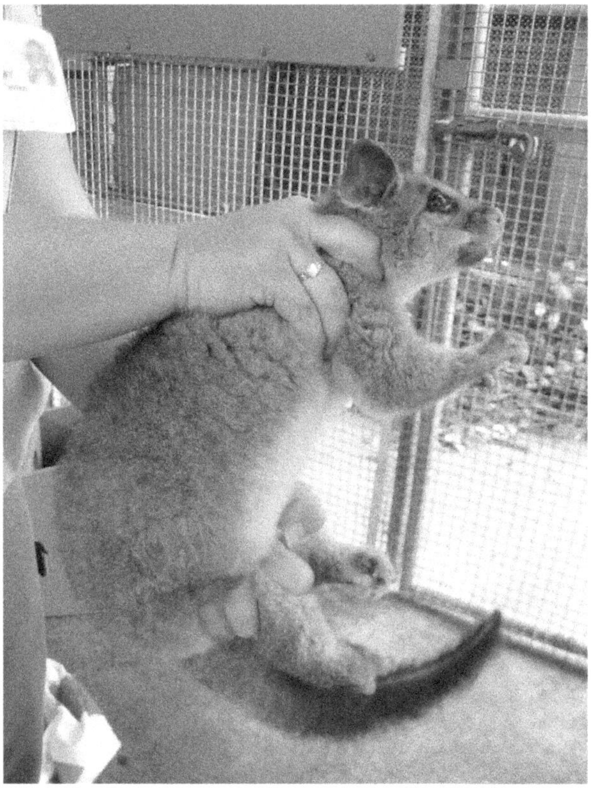

Figure 12.3 Physical restraint of a common brushtail. Photo: Robert Johnson.

a towel or cloth bag to cover the hands. Even the larger gliders can be handled this way. Gloves should not be used except perhaps with brushtails. Possums can be held by the tail, but they may turn around and bite. Brushtails and common ringtails can be grasped with one hand on the tail base or rump and the other around the neck and shoulders. This technique should only be used by experienced operators in brushtails as there is a risk of being bitten and of compromising the animal's breathing (Jackson 2003; Booth 1994) (Fig. 12.3). It is relatively straightforward in ringtails. Physical restraint can be used for minor procedures such as brief examination, medicating or handling prior to induction of anaesthesia. If in a bag, injections can be given through the cloth. Anaesthetic masks may be held over the muzzle of the animal while it is restrained in a bag or nest box. Cat squeeze cages may be used to restrain more fractious animals such as common brushtails for injection. Hollow logs or plastic plumbing pipe can also serve as restraint devices allowing access to the rear of the animal. Captive possums and gliders can become very tame, particularly if handled regularly.

6.2 Chemical restraint

6.2.1 Sedation

Sedation may be useful prior to inhalation anaesthesia or for minor procedures and transport. Diazepam (0.5–1 mg/kg IM) or butorphanol tartrate (0.4–1 mg/kg SC or IM) have been used (A Olsson & J Roffey *pers. comm.*).

6.2.2 Anaesthesia

Pre-anaesthetic fasting for 1–2 hr prior to anaesthesia is adequate as regurgitation under anaesthesia is uncommon in possums and gliders. Pre-anaesthetic handling should be minimised in order to decrease common complications of anaesthesia, such as breath-holding and apnoea. It can be difficult to assess the anaesthetic risk of a patient prior to induction; body mass and demeanour may aid in assessing risk. Approximate body weights of selected possums and gliders are presented in Table 12.1.

Inhalation anaesthesia using isoflurane in oxygen delivered via a face mask and Ayre's T-piece provides rapid induction in most cases and is the method of choice for induction and maintenance of anaesthesia (Vogelnest 1999; Holz 2002; Booth 2003). Induction with 4–5% isoflurane and oxygen flow rate of at least 1 L/min (200 mL/kg body weight) reducing to 2% isoflurane for maintenance is recommended. In smaller species such as the feather-tailed glider and the eastern pygmy possum an induction chamber may be used with a mask used for maintenance. However, most clinicians prefer masking as animals may become distressed in induction chambers (A Olsson *pers. comm.*).

Maintenance is usually via mask, however, larger species are easily intubated with the aid of a narrow-bladed laryngoscope. In some cases a stylet may be required. Intubation of sugar gliders has been described using a 1 mm endotracheal tube with stylet and a narrow-bladed laryngoscope (Booth 2000). Small-diameter tubes should be removed at regular intervals during the procedure to check for patency due to the risk of blockage with mucus.

Injectable anaesthetic agents are rarely required but may be useful if inhalation anaesthesia is not available, in field situations or with animals that are difficult to restrain. Ketamine has been used on its own or in combination with other drugs. Duration of anaesthesia and recovery time appear to be dose-related and the

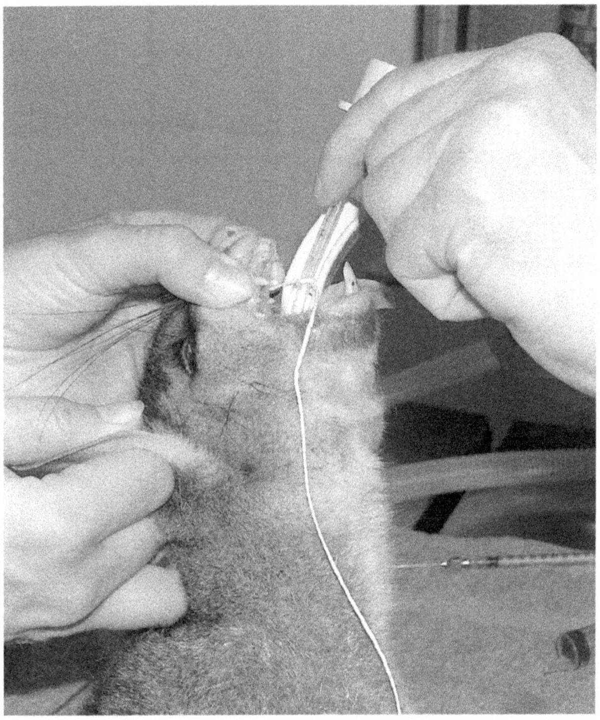

Figure 12.4 Endotracheal intubation of a common brushtail. Photo: Robert Johnson.

swallowing reflex appears to be preserved. Used alone, doses of 10–30 mg/kg are recommended for restraint of common ringtail and brushtail possums depending upon the desired level of restraint (T Bellamy *pers. comm.*). Surgical anaesthesia was induced within 2 min and lasted about 19 min in common ringtails and cuscuses administered 9–17 mg/kg IM (Salas & Stephens 2004). No adverse effects were observed.

More commonly, ketamine is combined with xylazine or medetomidine, which provides smoother induction and recovery. Doses vary. In common brushtails, ketamine (2–3 mg/kg IM) can be combined with xylazine (2–5 mg/kg IM) and antagonised with yohimbine (0.2 mg/kg IM). However, others have recommended much higher doses of ketamine (30 mg/kg IM) combined with xylazine (6 mg/kg) without apparent ill effects (Holz 2002; Holz 2003). Ketamine (1–3 mg/kg IM) and medetomidine (0.02–0.1 mg/kg IM) have also been used successfully (Vogelnest 1999). Restraint is antagonised with atipamezole (0.05–0.4 mg/kg IV).

Tiletamine/zolazepam is commonly used in larger species of both captive and free-ranging possums (4–10 mg/kg IM). Induction is usually rapid and smooth with good muscle relaxation. Reported adverse effects include tachycardia, ptyalism, respiratory depression, apnoea and muscle rigidity (Vogelnest 1999; Holz 2002; Viggers & Lindenmayer 1995). Three deaths have been reported in squirrel gliders (Holz 1992). Recovery can be prolonged. It is dose-related and recovering animals must be monitored closely to ensure patency of the airway. One report suggested that the dose required (up to 20 mg/kg) for restraint and anaesthesia of possums is higher than that for other species (Booth 1988; Viggers & Lindenmayer 1995). This is not the author's experience and lower doses are usually adequate. An advantage of using tiletamine/zolazepam is the low volume required and the rapid induction. Tiletamine/zolazepam can be combined with an α2-agonist such as medetomidine to provide smoother induction and recovery. Tiletamine/zolazepam does not have a specific antagonist but when combined with medetomidine the medetomidine can be reversed with atipamezole at 5 times the medetomidine dose (0.05–0.4 mg/kg IV or IM). Recovery is rapid (Vogelnest 1999).

Alfaxalone (5–8 mg/kg IM) has been used effectively to anaesthetise common brushtails (R Johnson *pers. obs.*; S Gelis *pers. comm.*). Using 5 mg/kg IV provides rapid short-acting anaesthesia in common brushtails and common ringtails, which is useful for brief and relatively non-invasive procedures or for surgical induction. Induction is smooth and recovery rapid and uneventful. Tiletamine/zolazepam (1–3 mg/kg IV) or propofol IV (6 mg/kg) have also been used (Vogelnest 1999).

6.2.3 Local anaesthesia

Bupivacaine 0.5% administered directly onto traumatic or surgical wounds while under general anaesthesia complements other analgesic medications and allows lower levels of chemical restraint. This technique is useful when debriding fight wounds, burns and other skin lesions.

7 CLINICAL PATHOLOGY

7.1 Haematology and biochemistry

7.1.1 Sample collection

Possums and gliders can be physically or chemically restrained for venipuncture. Chemical restraint is preferred as it has the advantages of providing immobility and reducing the effects of excitement on some haematological parameters (Jain 1993). Common sites for blood collection are the jugular vein (under general anaesthesia), lateral saphenous vein, caudal tibial artery and lateral caudal (tail) vein. Brushtail possums, ringtail possums and larger gliders can be bled from the jugular or lateral saphenous vein using a 22 or 23 G needle; for smaller gliders and possums a 25 G or 27 G needle should be used. A scalp vein or butterfly catheters may facilitate collection in smaller species. The caudal tibial artery, located superficially on the medial aspect of the stifle, has been used in species sugar glider size or smaller as an alternative to the lateral saphenous vein (Booth 1999; Clark et al. 2004). The lateral caudal vein is superficially located and suitable for blood collection even in small species.

Other sites for vascular access include the cephalic, femoral, ventral caudal and ear veins, the heart and the periorbital sinus (Presidente & Correa 1981; Hemsley 1992; Viggers & Lindenmayer 1996, 2001; Wells et al. 2000; Clark et al. 2004). The ventral caudal vein is located in the ventral midline of the tail and is found by inserting the needle perpendicular to the skin surface close to the tail base until bone is encountered. The needle is then withdrawn slowly while gentle negative pressure is applied, until blood enters the syringe and a blood sample can be withdrawn (Clark et al. 2004). This site may be useful in smaller species. In sugar gliders, venipuncture can be attempted blindly between the thoracic inlet and the angle of the mandible (Booth 2000). While these sites can be used in most species, more superficially located vessels, which can be visualised, are preferred. Very small volumes of blood can be collected into capillary tubes from incised or pricked superficial veins such as an ear vein or lateral caudal vein or the heel. Cardiac puncture allows collection of relatively large volumes of blood from any species but risks of pericardial haemorrhage, pulmonary laceration and haemorrhage or damage to the heart indicate that it should be used only in anaesthetised animals just prior to euthanasia (Clark et al. 2004; Hemsley 1992). Orbital sinus bleeding using a pipette or capillary tube might be considered in very small species such as the honey possum but it is an invasive procedure which can result in significant ocular damage (Nagy et al. 1995; Van Herck et al. 1998). Both cardiac puncture and orbital sinus bleeding should be avoided in favour of less invasive techniques.

7.1.2 Reference ranges and interpretation

Haematology and biochemistry values for possums and gliders are limited (Tables 12.4 and 12.5). Common brushtails have been the most extensively investigated. In this species age, sex and reproductive status all influence values. Haemoglobin concentration, haematocrit, erythrocyte counts and serum total protein concentrations are lower in immature animals and haemoglobin concentration, haematocrit and erythrocyte counts in adult males are higher than females (Barnett et al. 1979; Presidente & Correa 1981; Viggers & Lindenmayer 1996). Mothers with large young have lower haematocrit and haemoglobin and serum total protein concentrations than non-lactating females or those with small young (Presidente & Correa 1981). Seasonal effects have been reported in short-eared possums (*T. caninus*), with higher levels of urea and serum protein in spring and summer respectively, possibly due to variation in dietary protein intake (Barnett et al. 1979; Viggers & Lindenmayer 1996). Occasional circulating nucleated erythrocytes have been reported in apparently normal common brushtails, sugar gliders, striped possums and yellow-bellied gliders (ISIS 2002; ISIS in-house, Taronga Zoo 2005). There is little published information on the haematological and biochemical responses of possums and gliders to physiological and pathological processes. Until this becomes available, clinical judgment based on responses as seen in eutherian mammals may be appropriate.

7.2 Urinalysis

Urine can be collected by cystocentesis from anaesthetised animals or by opportunistic collection of voided samples. There are no published reference ranges for normal possum and glider urine. Samples from three common brushtails and two common ringtails without evidence of urinary dysfunction were light to moderately yellow. It was clear to slightly turbid in the common brushtails and moderately turbid in the common ringtails, possibly due to amorphous sulphate and sometimes urates, which have been reported as making the urine of the latter species turbid (Cunningham 1994; Hemsley 1992). Urine specific gravity was 1.011–1.030 in the brushtails and 1.034–1.040 in the ringtails and pH was 6.0–6.5. Trace protein and glucose levels, indicated by commercial dipstick tests, were most likely spurious results. It should be noted that normal

Table 12.4 Selected haematology values for possums and gliders

Parameter	Common brushtail[1]	Common brushtail[2]	Mountain brushtail[3]	Common ringtail[4]	Striped possum[5]	Sugar glider[5]	Squirrel glider[6]	Yellow-bellied glider[6]	Greater glider[7]
PCV (L/L)	*Male* conscious 0.47–0.59 anaesthetised 0.38–0.48 *Female* conscious 0.44–0.55 anaesthetised 0.35–0.45	*Male* 0.40–0.60 (n = 25) *Female* 0.31–0.51 (n = 23)	0.30–0.42	0.40–0.48	0.30–0.45 (n = 8)	0.34–0.50 (n = 18)	0.43–0.57 (n = 7)	0.38–0.53 (n= 10)	0.33–0.41
Hb (g/L)	*Male* 124–160 *Female* 108–150	*Male* 137–197 (n = 25) *Female* 104–162 (n = 23)	105–141	134–150	91–146 (n = 6)	122–185 (n = 18)	120–182 (n = 5)	114–193 (n = 8)	108–135
RBC ($\times 10^{12}$/L)	*Male* 5.4–7.8 *Female* 4.0–6.9	*Male* 5.8–8.0 (n = 25) *Female* 5.5–6.9 (n = 23)	4.0–5.9	4.5–6.6	7.3–8.9 (n = 3)	6.3–9.8 (n = 17)	—	—	4.6–6.2 *Male* 5.0–6.2 *Female* 4.5–5.9
MCV (fL)	53–87	58–76	68–80	68–89	48–53 (n = 3)	48–70 (n = 17)	—	—	59–73
MCH (pg/cell)	17.8–32.6	18.3–26.3	23.2–26.7	22.3–29.8	14.7–17.4 (n = 3)	17.8–22.3 (n= 17)	—	—	21–24
MCHC (g/L)	300–400	290–366	320–350	312–335	295–336 (n = 6)	315–394 (n= 18)	279–330 (n= 5)	285–364 (n = 8)	319–369
Retic (%)	—	0–3.8	—	—	—	—	—	—	—
WBC ($\times 10^9$/L)	2.2–11.9	—	2.1–6.8	4.0–9.6	3.5–24.9 (n = 8)	1.3–19.8 (n= 18)	4.7–9.1 (n= 7)	3.4–11.3 (n = 10)	1.3–6.6

(Continued)

Table 12.4 (*Continued*)

Parameter	Common brushtail[1]	Common brushtail[2]	Mountain brushtail[3]	Common ringtail[4]	Striped possum[5]	Sugar glider[5]	Squirrel glider[6]	Yellow-bellied glider[6]	Greater glider[7]
Neutrophils ($\times 10^9$/L)	0.30–2.74	—	0.5–4.8	0.7–2.7	1.39–13.2 (n = 8)	0.18–4.4 (n = 18)	0.80–1.62 (n = 7)	0.54–1.65 (n = 10)	0.4–2.0
Lymphocytes ($\times 10^9$/L)	0.90–9.56	—	0.6–3.4	1.5–6.6	1.23–12.2 (n = 8)	1.12–17.8 (n = 18)	3.08–6.46 (n= 7)	2.22–9.74 (n = 10)	0.4–5.5
Monocytes ($\times 10^9$/L)	0–0.19	—	0–0.5	0.2–1.0	0.05–3.24 (n = 8)	0.06–0.69 (n= 14)	0.22–1.22 (n = 6)	0.05–0.51 (n = 8)	0–0.3
Eosinophils ($\times 10^9$/L)	0–0.20	—	0–0.5	0.2–0.6	0.05–1.25 (n = 4)	0.3–0.99 (n = 13)	0.06–0.28 (n= 7)	0.10-0.60 (n = 9)	0–0.1
Basophils ($\times 10^9$/L)	0–0.06	—	0	0	0	0–0.04 (n = 13)	0	0–0.04 (n = 8)	0
Platelets ($\times 10^9$/L)	—	—	—	—	—	530–860 (n = 3)	—	—	—

Values have been derived from small numbers of samples and reporting style varies among authors of published papers. In addition, samples have often been collected from different populations, using different methods, stored differently and/or processed in different laboratories which may use different techniques. Age, sex and season can also affect values. For these reasons values presented should be regarded as indicative only.

Sources:
1 Fitzgerald et al. (1981). Derived from 9 male and 10 female captive adults (aged 2–9 yr). Values are observed ranges and are for combined sexes except where indicated. Possums were held in captivity in New Zealand. Animals were under barbiturate anaesthesia except where indicated.
2 Presidente & Correa (1981). Derived from adult (>2 yr) recently captured Victorian possums under ketamine-acetylpromazine or halothane anaesthesia. 79 male and 87 female possums were used to obtain values except: a) where numbers in parentheses indicate differently, and b) for serum globulin and albumin concentrations which included values from some immature animals, in which globulin levels are lower than adults. Values are calculated based on mean + 2SD and are for combined sexes except where indicated.
3 Viggers & Lindenmayer (1996). Derived from 13 male and 20 female free-living possums under tiletamine/zolazepam sedation. Values are for combined sexes and are expressed as the 5th–95th percentiles.
4 Presidente (1982b) Derived from 2 male and 4 female possums wild-caught but being held in captivity under halothane anaesthesia. Plasma electrolytes, urea and glucose values are for 2 males and 3 females. Values are observed ranges for combined sexes.
5 ISIS (2002). Observed ranges for unstated sex and age.
6 ISIS in-house, Taronga Zoo (2005). Oobserved ranges for unstated sex and age.
7 Viggers & Lindenmayer (2001). Derived from 8 male and 11 female free-living gliders under tiletamine/zolazepam sedation. Values are for combined sexes except where indicated otherwise and are observed ranges except for sex-specific ranges which are expressed as mean + 2SD.

Table 12.5 Selected serum or plasma biochemistry values for possums and gliders

Parameter	Common brushtail[1]	Mountain brushtail[2]	Common ringtail[3]	Striped possum[4]	Sugar glider[4]	Squirrel glider[5]	Yellow-bellied glider[5]	Greater glider[6]
TSP/TPP (g/L)	*Male* 59–79 (TSP) (n = 25) *Female* 51–89 (TSP) (n = 23)	55–65 (TSP)	43–67 (TPP)	—	48–71 (TPP) (n = 12)	58.0–64.0 (TPP) (n = 7)	54–77 (TPP) (n = 10)	62–72 (TPP)
Globulin (g/L)	10–40 (n = 56, mixed age) 20–30 (n = 3)[5]	19–29	14–33	—	15–34 (n = 6)	23.0 (n = 1)	29–32 (n = 3)	22–30
Albumin (g/L)	20–50 (n = 56, mixed age) 33–39 (n = 3)[5]	34–42	24–49	—	32–48 (n = 7)	37.0 (n = 1)	34–37 (n = 3)	34–50
Ca (mmol/L)	2.35–3.55 (n = 9)	2.12–2.53	1.7–3.6	1.78–2.28 (n = 6)	0.3–2.4 (n = 6)	—	—	1.78–2.5
P (mmol/L)	1.56–2.57 (n = 3)[5]	0.8–2.2	1.4–2.4	1.6–3.0 (n = 6)	1.62–3.29 (n= 6)	2.36 (n = 1)	1.04–1.67 (n = 3)	0.8–2.69
Na (mmol/L)	136–164 (n = 8)	141–148	136–146	133–152 (n = 6)	137–147 (n = 5)	—	—	—
K (mmol/L)	2.3–5.9 (n = 9)	2.7–4.9	2.8–3.6	3.1–4.3 (n = 6)	2.7–4.4 (n = 5)	—	3.94 (n = 1)	—
Cl (mmol/L)	94–108 (n = 7)	94–108	95–103	—	100–107 (n = 5)	—	—	—
Mg (mmol/L)	0.6–1.5 (n = 5)	—	2.8–3.2	—	—	—	—	—
Bicarbonate (mmol/L)	—	24–34	—	—	21.3 (n = 1)	—	—	—
CO_2 (mmol/L)	—	—	—	—	10.0–19.0 (n = 4)	—	—	—
BUN (mmol/L)	0.8–6.1 (n = 8)[5]	5.8–15.8	2.8–3.2	—	1.3–6.5 (n = 13)	2.2 (n = 1)	1.0–2.5 (n = 9)	1.3–8.4
Creatinine (umol/L)	80–142 (n = 7)[5]	50–100	57–80 (n = 8)[5]	—	27–106 (n = 7)	42.0 (n = 1)	40.0–68.0 (n = 9)	40–60
Total Br (μmol/L)	12.1–13.6 (n = 2)[5]	2–18	15.8–17.4 (n = 20)[5]	—	2–10 (n = 12)	—	—	—
Glucose (mmol/L)	2.2–9.3[7]	5.7–9.0	5.6–8.0	—	—	9.9 (n = 1)	4.9–10.5 (n = 8)	—

(Continued)

Table 12.5 (*Continued*)

Parameter	Common brushtail[1]	Mountain brushtail[2]	Common ringtail[3]	Striped possum[4]	Sugar glider[4]	Squirrel glider[5]	Yellow-bellied glider[5]	Greater glider[6]
Chol (mmol/L)	3.0 (n = 1)[4]	1.5–3.6	2.7 (n = 1)[4]	—	2.6–6.4 (n = 6)	—	—	—
Trig (mmol/L)	—	0.47–1.61	1.24 (n = 1)[5]	—	—	—	—	—
CK (U/L)	—	103–1676	1010 (n = 1)[4]	—	224–1463 (n = 5)	697 (n = 1)	126–984 (n = 7)	-
LDH (U/L)	41 (n = 1)[4]	15–627	—	—	210–274 (n = 3)	—	—	—
ALP (U/L)	65 (n = 1)[4]	838–2977	233–1590 (n = 4)[5]	—	148–432 (n = 6)	157 (n = 1)	90–654 (n = 5)	252–664
ALT (U/L)	8.5–63.5 (n = 21)	13–73	25–65	—	28–187 (n = 13)	45 (n = 1)	13–99 (n = 7)	36–140
AST (U/L)	1–151 (n = 22)	79–240	48–77	—	19–275 (n = 15)	1.0 (n = 1)	0–43 (n = 7)	-
GGT (U/L)	2 (n = 1)[4]	13–39	—	—	7–15 (n = 4)	—	—	5–11
Amylase (U/L)	—	242–672	1280 (n = 1)[5]	—	120 (n = 1)	—	—	—
Lipase (U/L)	—	—	—	—	3 (n = 1)	—	—	—

Values have been derived from small numbers of samples and reporting style varies among authors of published papers. In addition, samples have often been collected from different populations, using different methods, stored differently and/or processed in different laboratories which may use different techniques. Age, sex and season can also affect values. For these reasons values presented should be regarded as indicative only.

Sources:

1 Presidente & Correa (1981). Derived from adult (>2 yr) recently captured Victorian possums under ketamine-acetylpromazine or halothane anaesthesia. 79 male and 87 female possums were used to obtain values except where numbers in parentheses indicate differently, and except for serum globulin and albumin concentrations which included values from some immature animals, in which globulin levels are lower than adults. Values are calculated based on mean ± 2SD and are for combined sexes except where indicated.

2 Viggers & Lindenmayer (1996). Derived from 13 male and 20 female free-living possums under tiletamine hydrochloride and zolazepam sedation. Values are for combined sexes and are expressed as the 5th–95th percentiles.

3 Presidente (1982b). Derived from 2 male and 4 female possums wild-caught but being held in captivity under halothane anaesthesia. For plasma electrolytes, urea and glucose values are for 2 males and 3 females. Values are observed ranges for combined sexes.

4 ISIS (2002). Observed ranges for unstated sex and age.

5 ISIS in-house, Taronga Zoo (2005). Observed ranges for unstated sex and age.

6 Viggers & Lindenmayer (2001). Derived from 8 male and 11 female free-living gliders under tiletamine/zolazepam sedation. Values are for combined sexes except where indicated otherwise and are observed ranges except for sex-specific ranges which are expressed as mean ± 2SD

7 Barnett et al. (1979). Derived from 19 free-living male possums sampled from the ear vein while conscious. Values are expressed as mean ± 2SD.

common brushtail urine is sometimes red, presumably due to the presence of pigments.

8 HAND-REARING

Hand-rearing of orphaned possums and gliders can be difficult. Most available information relates to the common brushtail, common ringtail and sugar gliders.

8.1 Reasons for orphaning

Orphaning of free-ranging possums and gliders is commonly a result of the mother being preyed on by dogs and cats or suffering motor vehicle trauma. In captivity, mothers have been known to abandon PY due to a range of stressors.

8.2 Formulas

Low-lactose formulas available for possums and gliders in Australia include Di-Vetelact (Sharpe Laboratories), Wombaroo (Wombaroo Food Products) and Biolac (Biolac Milk Products). Biolac offers several formulas within the one brand, increasing the total solids (fat and protein) content as dietary requirements change from the furless or just furred PY to those emerging from the pouch. Two stages of Wombaroo possum milk are available, <0.8 and >0.8, the latter with a higher fat and lower protein content than the former. The formula required depends on the PY's stage of development. According to the manufacturer, the possum milk replacers are suitable for all species of possums and gliders.

8.3 Prognosis for different age classes

Generally, furless PY are much more difficult to rear than furred ones and even in the hands of experienced carers hand-rearing can be difficult. Factors to consider before commencing the task of hand-rearing a furless PY include the increased chance of problems arising later in the hand-rearing process (failure to thrive, diarrhoea) and the time, expertise and expense involved, which may be used more productively in caring for animals with a better prognosis. More importantly, undue distress to the PY and emotional attachment may be avoided if the decision not to rear is made sooner rather than later. Furless PY should be checked for palatal and lip injuries as they are often pulled off the teat, causing trauma to the oral cavity. PY should be carefully examined for the presence of any cat bite wounds.

Experienced carers recommend that common brushtails under 70 g body weight should not be hand-reared (Bishop 2005; Smith 1995). The survival rate of furless common ringtails is poor (Stanvic 1992). Possums considered for hand-rearing should, at the very least, have a fine velvety covering of hair. Common ringtail PY as small as 30 g have been raised successfully (Bishop 2005), but the survival rate improves rapidly once they reach the 60–80 g weight range (Smith 1995). Sugar gliders less than 10 g cannot be reared successfully but PY are relatively easy to rear once furred (Stanvic 1992).

8.4 Rearing techniques

8.4.1 Housing

Due to the high incidence of multiple births and their social habits, common ringtails and sugar gliders are best raised in groups. Substitute pouches for hand-rearing can be made from a variety of materials including calico or wool with a woven cotton liner. Hairless possums can be wrapped in a cotton handkerchief then the end of a stocking. Pouch young are unable to thermoregulate adequately and require supplementary heating at earlier stages of rearing. Furless PY should be kept at 34–36°C gradually reducing to 30°C as fur develops. After emerging the PY can be moved to a more natural environment, with the pouch suspended over a heating pad. As the animal matures the heating pad can be removed during the day and later at night as well (Tables 12.6–12.8). Heaters should be thermostatically controlled to avoid both hyperthermia and hypothermia. The pouch is a high-humidity environment and furless PY are subject to drying of the skin. Moisturisers such as lanolin should be applied frequently. Temperature and humidity should be regulated, e.g. with a humidicrib, but an alternative is commercially available bird hotboxes with a thermostatically controlled heating source and humidity provided via an open container of water in the box (Stanvic 1992; Bellamy 1994).

8.4.2 Feeding

Formula should always be fed at a comfortable temperature (36°C) (Bellamy 1994). Guidelines for volumes required at different growth stages are mentioned below, but manufacturers' recommendations for use should be referred to. Syringes or eyedroppers with attached custom-made teats (longer than dog and cat teats)

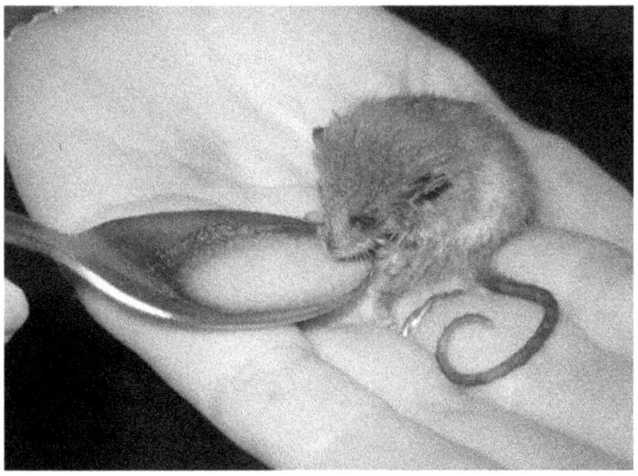

Figure 12.5 A young eastern pygmy possum (*Cercartetus nanus*) lapping formula from a spoon. Photo: Robert Johnson.

are used. For very small animals bicycle tyre valve rubber, a plastic IV catheter, infant gastric feeding tube or trimmed infusion set tubing may be used instead of a teat to achieve small and uniform droplet flow (Bellamy 1994; Smith 1995). During the feeding process it may be easier to keep the PY within the pouch, limiting movement. Older furred young can be encouraged to lap from a spoon or dish (Fig. 12.5). After each feed the perineal area should be gently wiped with a warm moistened tissue to stimulate defecation and urination. Overzealous wiping will cause inflammation and dermatitis and, occasionally, prolapse of the cloaca. This process should be continued until at least the 80 g stage in common ringtails and 300 g in common brushtails (Smith 1995). Furred PY should be brushed with a soft brush (Stanvic 1992). If reared in isolation from adults, inoculation of the gut with micro-organisms is recommended as digestion relies on hindgut fermentation. This is most easily accomplished by allowing access to faeces of adult conspecifics (Finnie 1976), but donor faeces should first be smear tested to ensure that yeasts are not present.

Frequency of feeding depends upon the PY's stage of growth (Tables 12.6–12.8). Furless PY should be fed every 2–3 hr, which reduces gradually to around 5 times per day when fur is developing and to 2–3 times per day at full emergence from the pouch. It is important to keep good records and measure and weigh the PY at least 3 times per week.

The total amount of formula fed daily depends on the composition of the formula, but should generally be 10–20% of body weight per day except in very small PY when up to 50% of body weight may be needed (Bellamy 1994). The volume of milk to be fed varies according to manufacturer recommendations.

The following suggested volumes are based on Wombaroo® product information (E Arthur *pers. comm.*).

Common ringtails should be drinking 2–3 mL/day at 6 g body weight, increasing to 4 mL at 10 g and 11 mL at 50 g. At this stage milk should be offered in a bowl to encourage lapping. When ringtails start to emerge from the pouch, at approximately 75 g, they should be drinking 15 mL daily. This volume should not be increased, to encourage feeding on solids. Full emergence occurs at 120 g body weight.

Common brushtails should be drinking up to 6 mL/day at 20 g body weight, increasing to 11 mL at 50 g. At 80 g body weight the PY should be offered milk in a bowl to encourage lapping. At this stage they should be consuming 18 mL daily. Once they reach 120 g the

Table 12.6 Guide to common ringtail development and requirements

Approximate age (d)	Body weight (g)	Number of feeds of formula per day[1]	Artificial heat (°C)[1, 2]	Stage/event
<7[3]	<2[3]			Newborn young
40	20		34–36	Independent of teat
95–105	35–50	6	30–32	Eyes may open[a4]
105–110	45–50	6	30	Fine fur. Solids can be introduced from around 60 g body weight[4]
125	70–90	4–5	28	Young may permanently leave pouch
150–220	130–400	1–2	No	Young may be weaned by mother
200	350	1	No	May be independent
300	500	Nil	No	Ready for release

a Unfurred ringtails have little chance of being reared successfully
Sources: Smith (1995). **1** Wombaroo feed charts. **2** Bellamy (1994). **3** Presidente (1982b). **4** B Clarke Orphan Native Animal Rear & Release Association Qld: www.onarr.com.au/guide_brushtail_possums.htm.

Table 12.7 Guide to common brushtail development and requirements

Approximate age (d)	Body weight (g)	Number of feeds of formula per day[1]	Artificial heat (°C)[1, 2, 3]	Stage/event
<20	<3			Newborn young[a]
40	10–15			
50	20			
65	25			
80	40	6	32–36	Independent of teat
90	70	6	32	Minimum body weight for good chance of successful rearing
100–110	100–150	5–6	30	Fine fur
110–120	150–200	4–5	30	Eyes may open
120	200	4	28	Young may leave pouch. Solids can start to be introduced at around 150 g body weight[4]
180–250	500–1200	1–2	No	Young may be weaned
190–230	800	1	No	May be independent
300	1500	Nil	No	Ready for release

a Pouch young less than 70 g bodyweight have little chance of being reared successfully.[4]
Sources: Gemmell & Hendrikz (1993), Presidente (1982a), Smith (1995). **1** Wombaroo Food Products. **2** Bellamy (1994). **3** Stanvic (1992). **4** B Clarke Orphan Native Animal Rear & Release Association Qld: www.onarr.com.au/guide_brushtail_possums.htm.

Table 12.8 Guide to sugar glider development and requirements

Approximate age (d)	Body weight (g)	Number of feeds of formula per day[1]	Artificial heat (°C)[1]	Stage/event
<26				Newborn[a]
26	2–3	8	32	Finely furred
40–60	6	8	32	Independent of teat
60	10	4	32	Fur developing
70	20	4	30	Fully furred. Eyes may open
80–100	40–50	3	28	In nest. Solids can start to be introduced at around 40 g body weight[2]
100–110	50–60	2	No	May leave nest
110–125	70–80	1	No	May be weaned
120–130	80–85	1	No	May be independent
	>85	Nil. Now totally on adult diet	No	Ready for release. Natural dispersal from the family group occurs at around 7 mo age[3]

a Unfurred PY have little chance of being reared successfully (Stanvic 1992). If attempted, feeding and heat requirements can be based on those for a 26 d PY.
Sources: Bach (1998), Smith (1979). **1** Wombaroo Food Products. **2** Clarke B & Schulz K, Orphan Native Animal Rear & Release Association Qld; www.onarr.com.au/guide_gliders.htm. **3** Tyndale-Biscoe (2005c).

volume increases to 20 mL. Pouch emergence begins at 130 g, the PY being fully emerged at 400 g when it should be consuming about 46 mL daily.

At 6 g body weight sugar gliders should be drinking 3 mL/day, increasing to 4 mL daily at 10 g. At this stage native browse should also be offered. PY heavier than 12 g usually lap from a shallow spoon. At the start of pouch emergence (20 g) PY should be consuming 5 mL daily. At 50 g they take up to 9 mL daily, and are fully emerged from the nest. Sugar gliders are ready for release once they reach 80 g body weight, when they should be consuming up to 10 mL of formula daily.

8.4.3 Solid food

During the hand-rearing process a variety of solid food as similar as possible to the natural diet should be offered (Smith 1995). For common ringtails this includes the tips and flowers of eucalypts, tea trees (*Leptospermum* spp.), bottlebrush (*Callistemon* spp.) and paperbark (*Melaleuca* spp.) as well as grevillea and wattle flowers. Hand-reared common ringtail PY should not be offered fruit as the high sugar content can predispose to bloat (see 9.8.3). For common brushtails, bark, the tips and flowers of eucalyptus, grevillea, bottlebrush and wattle can be offered. They also occasionally eat insects, such as beetles or crickets. Some fruit and vegetables such as apples, carrots, pears, rockmelon, grapes and sweet potato can be offered. Sprinkling food with a commercially available high-protein baby food supplement is recommended. For sugar gliders, natural dietary constituents such as insects, pollen, nectar, lerps, thrips, borers and the sap of eucalypts and other native trees should be offered if possible. In addition, apple, rockmelon, grapes, grevillea, banksia and wattle flowers can be offered. A high-protein supplement is recommended.

A common concern is apparent constipation in common ringtails a few weeks after weaning commences. The possum appears healthy and has a good appetite but no longer defecates on stimulation and faeces are no longer found in the artificial pouch. The animals are always normal on examination. It is thought that this must be the stage when they commence caecotrophy during the day; carers fail to find the small hard pellets passed during nocturnal foraging. After becoming aware of this situation carers generally look for and find these pellets (McCracken 2001).

8.4.4 Transition to release

At approximately 100–200 g body weight hand-reared common ringtails should be transferred to a large bird cage (approximately 900 × 900 × 900 mm). There should be a nest box hanging inside the cage and the wire mesh should be no larger than 19 × 19 mm. Multiple cages can be placed side by side to encourage socialisation. At 300–450 g body weight they should be moved to a release aviary. A suitable release weight for common ringtails is 700–800 g. Common brushtails can be moved to an aviary at 700–800 g body weight. It is important to reduce contact with the animal at this stage in order to facilitate release, which occurs at around 1400 g body weight in common brushtails. Sugar gliders need to be transferred to an aviary at 60–70 g to encourage jumping and gliding to build muscle mass and fitness. They are ready to release at 80–90 g.

9 DISEASES

9.1 Dermatological diseases

A variety of skin conditions have been diagnosed in possums and gliders. Alopecia occurs frequently in most species and may be discrete patches, diffuse and extensive and symmetrical or asymmetrical. Causes include epilation after an attack by a predator, stress, nutrition, hormones, PY being maintained at too high a temperature or infectious causes (see 9.9.3). Investigation to rule out infectious causes should occur in all cases, but in many cases no specific aetiology is determined. Despite this, identification and correction of potential underlying causes frequently results in resolution.

Bacterial and fungal skin disease is common. Mycotic dermatopathy involving yeasts (*Candida* spp.) and mixed bacterial infections, often including *Pseudomonas aeruginosa*, has been seen in feather-tailed gliders (ARWH 2007 case numbers 1468/1, 1468/3, 1468/5, 2192/1, 3650/1, 4674/1). Animals affected ranged from juveniles to adults of both sexes. Accompanying disease conditions included pneumonia, pleuritis and sinusitis. Dermatitis, epidermitis, cellulitis, pyoderma, panniculitis and severe ulceration of the skin have been recorded in common ringtails, sometimes in association with fungal infection (ARWH 2007 case numbers 730/1, 2110/1, 2221/1, 2537/3, 2793/1, 4094/1, 4133/1, 4470/1).

Ectoparasites, viruses and systemic diseases have also been associated with skin problems. A case of exudative dermatitis in a common ringtail was associated with a heavy infestation of the mite *Trichosurolaelaps striatus* (ARWH 2007 case number 730/1). Poxvirus was suspected as the cause of papillary dermatitis in a common ringtail (ARWH 2007 case number 4133/1). Thyroiditis was diagnosed at necropsy in two common ringtails with skin disease (ARWH 2007 73/1, 2793/1). One animal had bilaterally symmetrical alopecia (73/1), the other had marked ulcerative dermatitis (2793/1). No thyroid function tests were performed on these animals prior to death.

9.1.1 Exudative dermatitis in common brushtails

Exudative dermatitis, also known as lumbosacral dermatitis or rumpiness, is common in common brushtails

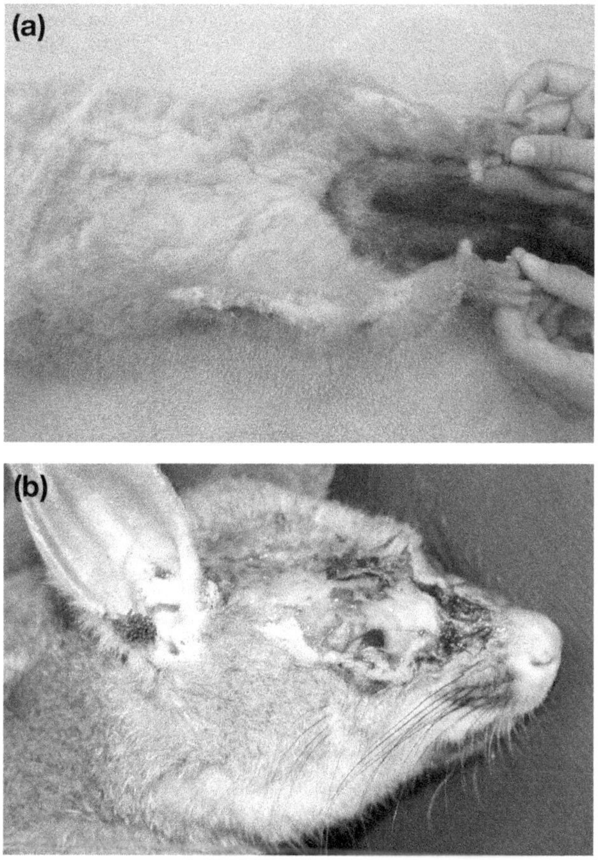

Figure 12.6 Exudative dermatitis (a) on the legs of a common brushtail and (b) on the face of a common brushtail. Note the large numbers of stick-fast fleas, *Echidnophagia myrmecobii*, at the base of the ear. Photos: a) Robert Johnson and b) Larry Vogelnest.

(Reddacliff 1981; Hemsley 1992; Hemsley & Canfield 1993b; Hemsley 1994). There is chronic ulcerative or proliferative dermatitis and a neutrophilic leukocytosis (Hemsley 1994). Lesions typically include alopecia, exudation, crusting and in severe cases ulceration in the lumbosacral or tail base region; they can also occur on the face, trunk or limbs (Fig. 12.6). An offensive odour is often present and the fur is matted and easily depilated. The actiology of the condition is unclear but is likely to be multifactorial and may involve hypersensitivity, a variety of ectoparasitic, bacterial and fungal agents, trauma and stress (Finnie 1978; Reddacliff 1981; Munday; 1988; Hemsley 1994). Mixed bacteria such as *Staphylococci, Streptococci* and *Corynebacteria* are commonly present (Reddacliff 1981; Hemsley 1994) but are likely to represent secondary infection. *Trichosurolaelaps crassipes* mites are the common ectoparasite associated with lesions. However, a recent study monitoring a wild population of mountain brushtails (*T. cunninghami*) indicated that there is no obvious correlation between pelage condition of the rump, ectoparasite loads, general health parameters or inflammation of the skin. Damage to the pelage of the rump in that population of mountain brushtails seemed to be influenced by season, gender and habitat. *Trichosurolaelaps dixoa, Atellana papilio* and *Haemolaelaps penelope* were the most commonly collected mites. Ticks identified included *Ixodes tasmani, I. trichosuri* and *I. cornuatus* and one *Acanthopsylla spp.* was found (Hufschmid 2005). The rat mange mite, *Notoedres muris*, has been recorded as the cause of crusty lesions affecting the ears, tail, snout and eyelids of brushtail possums in Victoria (Booth 1994). The species of possum was not mentioned.

Possible predisposing factors may include prolonged periods of rain, high humidity and overcrowding (Reddacliff 1981). The condition is most common in subadult and adult males and where population density is high, possibly indicating an association with territorial disputes and social stress (Munday 1988). It is frequently seen in dispersing subadult males. Some cases have been associated with underlying disease processes such as anaemia, encephalitis, hepatitis, lymphadenopathy and pneumonia (Hemsley 1994). Many animals that present with exudative dermatitis are also victims of trauma such as dog attack, possibly being more susceptible because of debilitation.

When deciding if cases of exudative dermatitis should be treated consideration must be given to the animal's fate once it has recovered. Releasing the animal into an area where it will be re-exposed to the environmental and social stressors that led to its condition may not be in its best interests. Euthanasia should be considered in severe cases, particularly young males (L Vogelnest *pers. comm.*). Treatment may be successful when there are no significant confounding factors present. A systemic acaracide such as ivermectin 200 µg/kg SC or a topical preparation such as selamectin (6 mg/kg topically once a month) should be used when mites are present (J Roffey *pers. comm.*). Topical antibacterial treatment of skin lesions is frequently required in addition to parenteral therapy although mild lesions may heal after a single injection of long-acting penicillin (Reddacliff & Spielman 1990). More severe cases may require a longer course of oral or parenteral antibiotics such as amoxycillin or amoxycillin/clavulanic acid (20 mg/kg sid SC or

IM or bid PO for up to 4 wk) (L Vogelnest *pers. comm.*). Possums frequently resent applications of medications, especially on raw or ulcerated lesions. Spray preparations of chlorhexidine 0.05% can be used, or compliant animals may be placed in a pillowslip and gently lowered into a shallow bath of chlorhexidine 0.05% for 1–2 min, ensuring that the head is above water at all times. In severe cases, anaesthetised animals may need to be clipped and bathed using antiseptic shampoo. Ointments containing silver sulphadiazine and chlorhexidine digluconate (Silvazine®, Smith & Nephew) or other commonly available topical antibiotic, antifungal and corticosteroid preparations may be applied.

9.1.2 Swollen paw syndrome

Common ringtails in the Sydney region of New South Wales can be affected by a condition known as swollen paw syndrome, which primarily involves swelling and gangrene of the distal limbs. It was first diagnosed in free-ranging animals in 1990 and has occasionally been diagnosed in captive animals. Juveniles, subadult and adult possums of both sexes can be affected. Oedema of the fore feet is the earliest clinical sign. More chronic cases may exhibit ulceration, swelling and gangrene of the feet, tenosynovitis, moist ulcerative dermatitis of the dorsal nose, moist dermatitis or alopecia, thickening, contraction and necrosis of the ear tips and crusting or ulceration of the tail tip. Some animals present with one or more feet missing. The animals are usually bright and alert and continue to eat and drink. It is speculated that the underlying process is one of avascular necrosis resulting in severe acute necrosis of the extremities, with coagulation necrosis, epidermal ulceration and dermal oedema. A range of bacteria have been isolated from affected sites, with *Staphylococcus aureus* the most common (Rose 1999). Other bacteria isolated include *Aeromonas sobria, E. coli, Klebsiella pnemoniae, Pseudomonas aeruginosa* and *Alcaligenes faecalis*. The aetiology has not been determined but photosensitisation, bacterial or viral infection, thermal injury, electrocution, mycotoxicosis or plant intoxication may be involved. Treatment is ineffective, although corticosteroids and/or antimicrobial agents have been used. Euthanasia is usually indicated.

9.1.3 Ectoparasites

Fleas, ticks and mites of numerous species infest gliders and possums (see also 9.1.1).

The stick-fast flea *Echidnophagia myrmecobii* frequently infests common brushtails and ringtails; heavy infestations may cause anaemia in juveniles. A common ringtail infested with stick-fast fleas suffering from anaemia also had pulmonary oedema and diffuse hepatopathy (ARWH 2007 case number 731/1). The paralysis tick *Ixodes holocyclus* is a frequent parasite of the common brushtail. Fleas and ticks are treated with selamectin (6 mg/kg topically once). Ticks can be removed physically.

The *Petaurus* genus is parasitised by mites of the *Guntheria* and *Petauralges* Genera (Jackson 2003). *Sarcoptes scabei* has been known to cause mange in common brushtails and common ringtails (Munday 1988; Skerrat et al. 1998). The mite *Trichosurolaelaps crassipes* causes irritation and alopecia of the lower back of common brushtails but can also be found on normal possums. Hair loss around the head, ears and extremities also occurs (J Roffey *pers. comm.*). Systemic acaracides, such as ivermectin (200 μg/kg SC) weekly for up to 3 wk, or with a topical preparation such as selamectin (6 mg/kg topically once a month) (J Roffey *pers. comm.*) are effective. Pyrethrin sprays and malathion washes have also been used (Presidente 1982a, 1982b).

Cutaneous myiasis is a common presenting condition in free-ranging common brushtails and common ringtails that have been subjected to trauma and have open wounds. Maggots should be manually removed, associated open wounds treated topically and systemic antibiotics administered. Extreme caution should be used in the application of commercial fly-strike products. Anecdotal reports suggest that death may follow application and we recommend they not be used in debilitated animals.

Subcutaneous cysts (coenuriasis) associated with infection by the larval stage of *Teania serialis* has been reported in the common ringtail (Hough 2000). If necessary, these cysts can be removed surgically.

9.1.4 Miscellaneous conditions

Pemphigus foliaceus has been diagnosed histopathologically in a free-ranging subadult common brushtail. A vesicopustular dermatitis and hyperkeratosis of the nailbed, epidermis and adjacent skin was described (Mayne Health, case no VO3/882, 28/2/2003). A similar case was described in a common brushtail that had crater-like lesions involving the tail and feet, extending to the caudal ventral abdominal skin (A Olsson *pers. comm.*). That case responded briefly to corticosteroid therapy, before euthanasia.

Burns are common in free-ranging common brushtails and ringtails as a result of bushfires or direct contact with hot surfaces such as stovetops and fireplaces. Euthanasia should be considered if the burns are extensive and may affect the animal's survival in the wild.

Pouch infections have been reported in common brushtails and common ringtails and can be associated with maternal and PY mortality if systemic infection occurs (Munday 1978b, 1988; Speare et al. 1984). Clinical signs include moist matted fur adjacent to the pouch opening and greasy malodorous liquid within the pouch, which has a friable lining. It has been recommended to administer parenteral antibiotics as well as antiseptic irrigation of the pouch, as local treatment alone has been found to have poor results (Speare et al. 1984). The cause is unknown.

9.2 Gastrointestinal diseases

9.2.1 Dental disease

Tooth root abscesses and periodontal disease occur in captive and pet gliders and possums, more commonly when young are in the transition stage from milk to solid food. They are also common in adult sugar gliders, and some clinicians believe that the disease is associated with poor husbandry, overcrowding and stress (R Woods *pers. comm*). In possums and gliders the processes do not usually extend beyond periodontal disease, unlike the more extensive condition of lumpy jaw (necrobacillosis) seen in macropods (L Vogelnest *pers. comm*). The primary pathogens involved in possums have not been identified. Periodontal disease may be associated with sinusitis in gliders, leading to a decreased appetite and weight loss. Prevention of periodontal disease is largely based on feeding a diet based on natural feedstuffs, but avoiding abrasive material. Treatment is based on cleaning of affected but essentially sound teeth, extraction of more severely affected teeth and the administration of systemic antibiotics, e.g. clindamycin 11 mg/kg bid for 2–3 wk or until signs resolve.

Severe tooth wear may lead to emaciation in elderly animals, particularly common ringtails (K Bodley *pers. comm.*).

9.2.2 Endoparasites

Endoparasites rarely cause clinical disease in possums and gliders (ARWH 2005). *Bertiella trichosuri*, an anoplocephalid cestode parasitising the intestine of common brushtails is often an incidental finding at necropsy. There appears to be increased susceptibility to infection with this parasite in suboptimal habitats (Booth 1994). Parasites from the *Bertiella* genus have also been detected in the common ringtail (Beveridge 1985). Infestation with the liver fluke *Fasciola hepatica* has been reported in common brushtails (Whittington 1982). A variety of nematode parasites including *Marsupiostrongylus minesi*, *Trichostrongylus axei* and *T. colubriformis* inhabit the gut of possums and gliders and rarely cause disease. Although routine worming of captive and rehabilitating animals is not generally required, regular faecal checks and worming as indicated is a useful control strategy. A range of anthelmintics can be used if indicated (see 11.3).

9.2.3 Salmonellosis (see 9.9.2c)

Rapidly fatal haemorrhagic enteritis and septicaemia are the manifestations of salmonellosis in young possums (ARWH 2005). Common ringtail possums appear to be most susceptible to infection (ARWH 2007 case numbers 1031/1, 1031/2). This condition needs to be differentiated from a similar syndrome caused by *E. coli* (see 9.2.5).

9.2.4 Yersiniosis (see 9.9.2d)

Acute yersiniosis is manifested as rapid death associated with enteritis, diarrhoea, dehydration and septicaemia (Rose 1999). Chronic disease is associated with multisystemic abscessation and wasting.

9.2.5 Colibacillosis

E. coli infection occurs frequently in common ringtails. Pure growths and haemolytic strains are probably significant. Signs and lesions are similar to those of salmonellosis (9.2.3). Severe granulomatous lymphadenitis was associated with *E. coli* infection in a common ringtail (ARWH 2005).

9.2.6 Tyzzer's disease (see 9.9.2g)

Tyzzer's disease is the clinical syndrome caused by infection with the bacterium *Bacillus piliformis*. The disease most commonly occurs as a fatal necrotising hepatitis and myocarditis in young possums (Canfield & Hartley 1991). Death is usually sudden or after a short illness lasting 1–2 d.

9.2.7 Miscellaneous conditions

Abdominal distension relating to gastrointestinal disease appears to be common in ringtail possums. Apart from caecal stasis in young common ringtails (see 9.8), other causes of bloat include gastric torsion and bacterial infection (Bellamy 1999). Enteritis occurs often in the common ringtail, frequently in conjunction with other diseases such as gastritis, duodenitis (*E. coli*), typhlitis, pneumonitis and hepatopathies (ARWH 2007). Colonic intussusception has been reported in two common brushtails (Hemsley 1992). Typhlitis occurs frequently in common ringtails of all ages (ARWH 2007 case numbers 1767/1, 2101/1, 2433/1, 2625/1, 3061/1, 3061/2, 3061/3). *E. coli* is a common pathogen. *Candida albicans* and *E. coli* were cultured from one case of haemorrhagic typhlitis (ARWH 2007 case number 3061/3) and *Streptococcus faecalis* from another (ARWH 2007 case number 2625/1). It is not known why gastrointestinal disease is so common in captive ringtail possums, but it is likely to be associated with inappropriate husbandry and nutrition. A juvenile male common ringtail suffered from hepatitis which yielded a pure growth of *Pseudomonas aeruginosa* (ARWH 2007 case number 2101/1).

Gastric yeast overgrowth is not uncommon in debilitated feather-tailed gliders (ARWH 2007 case number 5038/1) and hand-reared possums and gliders, particularly under suboptimal husbandry conditions or while on antibiotic therapy (see 9.8.4). *Candida* spp. predominately invade squamous epithelium, consequently, in the gastrointestinal tract the oral cavity, oesophagus, stomach and cloaca are mainly affected. Candidiasis is implicated in some cases of diarrhoea (see 9.8.4) (McCracken 2001). Prevention of candidiasis associated with antibiotic therapy is discussed in 11.1.

9.3 Cardiorespiratory diseases

9.3.1 Cardiomyopathy

Primary cardiac conditions appear uncommon in possums and gliders. A case of non-suppurative myocarditis has been diagnosed in an adult female common brushtail (ARWH 2007 case number 647/1). A second case of non-suppurative myocarditis in a common brushtail occurred in conjunction with non-suppurative interstitial nephrosis (ARWH 2007 case number 2343/1). Cardiomyopathy and signs of acute congestive heart failure were seen in an adult female common ringtail (ARWH 2007 case number 2979/1). In all cases the cause could not be determined. We are not aware of any reports of encephalomyocarditis virus in possums and gliders.

9.3.2 Pneumonia

Primary acute pneumonia has been diagnosed in juvenile and adult common ringtails. *E. coli* was isolated in one case (ARWH 2007 case number 2142/1). Acute aspiration pneumonia is a frequent finding in hand-reared animals. A case in an adult female common ringtail was associated with acute multifocal suppurative meningitis (ARWH 2007 case number 830/1). Staphylococcal pleuropneumonia occurred in an adult male common brushtail. *Staphylococcus aureus* and *Streptrococcus milleri* were isolated (ARWH 2007 case number 4602/1). *S. milleri*, F group, was isolated from the lungs of a common brushtail with bronchopneumonia. The organism was also cultured from an abscess on the fore limb and the animal was heavily infested with ticks and stick-fast fleas (ARWH case number 4420/1).

Pulmonary infections occur frequently in feather-tailed gliders (ARWH 2007 case number 1822/1, 1275/1, 4313/1, 1468/1). Two animals also suffered concomitant sinusitis (ARWH 2007 case numbers 1275/1, 1468/1). A feather-tailed glider with a transponder-induced SC sarcoma was also suffering granulomatous pneumonia from which *Mycobacterium avium* was cultured (ARWH 2007 case number 4125/1).

Verminous pneumonia occurs in common brushtails. *Marsupiostrongylus minesi* is a common parasite of possums and gliders. It rarely causes clinical disease, but pulmonary consolidation or bronchopneumonia occasionally occur (Booth 1994). Four cases in free-ranging animals from the same area of southern Sydney have been reported (Bellamy 1993). There was no apparent seasonal incidence. Two common brushtails presented with signs of severe respiratory disease including cyanotic mucus membranes and collapse. Necropsy revealed multiple pale foci within the lungs. Verminous pneumonia was confirmed by histopathology. A third common brushtail was seen in very poor condition and had cyanotic oral mucus membranes. Moist rales were detected upon auscultation of the thorax. Radiographs showed multiple focal densities. A wet preparation from a tracheal aspirate showed multiple motile larvae, which were later identified as *Marsupiostrongylus* spp. The possum was treated with avermectin (0.3 mg/kg SC weekly for 3 wk), gained weight slowly and was eventually released. A fourth young male common brushtail was presented with a fractured femur. During surgical repair the animal died; necropsy revealed multifocal densities and oedema in the lungs. A wet preparation from the cut surface of the lung showed motile larvae. Histopathology confirmed verminous pneumonia. The parasite was identified as *M. minesi*.

9.3.3 Sinusitis

Sinusitis occurs frequently in some glider species, predominantly in captive animals. It is more frequent in older animals and in conjunction with other disease processes. Dental attrition, periodontal disease and lower respiratory tract disease are also common findings. Two cases of bilateral bacterial sinusitis in yellow-bellied gliders have been diagnosed at ARWH (ARWH 2007 case numbers 2825/1, 4793/1). One animal was infected with *Pseudomonas aeruginosa* and *Bacteroides fragilis* and had a hepatic sarcoma (ARWH 2007 case number 4793/1). Cases have also been seen in feather-tailed gliders (ARWH case numbers 1275/1, 1468/1) one of which included fungal sinusitis, most likely caused by a strain of *Cryptococcus* spp. (ARWH 2007 case number 3299/1). A yeast, most likely *Candida* spp., was found within periodontal lesions.

A pentastomid parasite *Rileyella petauri* has been isolated from the lungs and nasal sinus of the sugar glider. It appears to have a direct life-cycle (Spratt 2003).

9.4 Genitourinary diseases

A variety of conditions affecting the urogenital tract have been recorded in possums and gliders (ARWH 2007). Obstructive urolithiasis, urethritis and prostatitis have been reported in an adult male common ringtail (ARWH 2007 case number 3875/1) and suppurative and non-suppurative nephritis in common brushtails, often in association with systemic disease (ARWH 2007 case numbers 2343/1, 2966/1, 4036/1, 4353/1). Cystitis, pyelonephritis and renal calculi were diagnosed in a subadult male common brushtail (ARWH 2007 case number 4036/1) and unilateral chronic pyelonephritis with thoracolumbar spondylosis in a captive adult female common brushtail (ARWH 2007 case number 4353/1). The chemical composition of the calculi was not determined. Nephritis has been diagnosed in gliders. A captive yellow-bellied glider was suffering from chronic unilateral hydronephrosis, chronic bilateral non-suppurative interstitial nephritis and a chronic membranous glomerulopathy (ARWH 2007 case number 156/1). Suppurative nephritis caused by *Serratia fonticola* and *E. coli* occurred in a squirrel glider (ARWH 2007 case number 3486/1).

Endometritis and marked oviductal cystic hyperplasia occurred in a common brushtail (ARWH 2007 case number 1971/1). The infective agents isolated were *β haemolytic Streptococcus* spp. and *E. coli*. Vaginal papillary hyperplasia and chronic interstitial nephritis were also present. Metritis and endometritis has been recorded in a captive Leadbeater's possum associated with *E. coli* septicaemia (ARWH 2007 case number 2470/1) and hydronephrosis and hydroureter in an adult male common brushtail (ARWH 2007 case number 564/1). Asymptomatic infection with *Klossiella* spp. in common brushtails has been noted as an incidental finding (Hemsley 1992).

Pet sugar gliders commonly suffer from lower urinary tract disease, particularly urolithiasis (type undetermined) and penile prolapse when fed inappropriate diets comprising mainly fruit and commercial pet food. Bladder infection, bladder rupture, peritonitis and renal disease occur subsequent to urethral blockage. Surgical treatment involving urethrostomy or penile amputation is usually unsuccessful. The condition has not been seen in pet animals on appropriate diets such as Leadbeater's mix and insectivore supplements (Johnson-Delaney 2000).

9.5 Neurological and musculoskeletal diseases

9.5.1 Choriomeningoencephalitis (CME)

Choriomeningoencephalitis primarily affects adult common brushtails. Clinical signs of disease may progress from a few weeks up to several months (Rose 1999). Depression, blindness and ataxia are the most consistent findings. Some possums have shown signs of concurrent disease including intestinal coccidiosis, vaginal oedema, dermatitis, gastrointestinal disease and pulmonary disease (ARWH 2007 case numbers 1411/1,

1834/1, 2338/1, 2545/1, 2905/1, 3046/1, 3501/1, 3619/1, 4182/1, 4231/1, 4601/1). In 30 cases of common brushtails exhibiting depression and blindness presented to the Taronga Zoo Wildlife Clinic between 1985 and 1993 none recovered sufficiently for release (Rose 1999). Tissues from those animals and 12 others from New South Wales, Victoria and Tasmania were examined histologically. Twenty-three had a chronic non-suppurative CME. Necropsy of animals with CME does not usually reveal significant gross lesions. A viral agent is considered to be the most likely aetiology of CME in common brushtails in Australia.

9.5.2 Chronic meningoencephalitis (Wobbly possum syndrome)

Chronic meningoencephalitis and optic neuritis occur in common brushtails, most commonly adults, on the eastern Australian mainland and Tasmania. It is commonly referred to as wobbly possum syndrome (WPS). Clinical signs are progressive and include depression, ataxia and blindness associated with dilated pupils. Histopathological lesions consist of non-suppurative meningoencephalitis, with the optic tract, retina and the cerebellum often involved (Reece & Hartley 1994). Viral particles have been seen in association with lesions. A similar disease syndrome has been recorded in captive and free-ranging common brushtails in New Zealand (Mackintosh et al. 1995). The disease is associated with a viral meningoencephalitis as well as mononuclear inflammation of a range of other organs including the liver, kidneys, lungs and muscle and is manifested as wasting and progressive neurological signs such as ataxia, docility and inco-ordination (Mackintosh et al. 1995; O'Keefe et al. 1997). Elevated circulating nucleated erythroid cells are common, along with eosinopaenia, decreased creatinine, increased total serum protein with decreased albumin and elevated globulin. Proteinuria, haematuria and glucosuria may be present (O'Keefe et al. 1997). Studies have attempted to elucidate the aetiology and pathogenesis of the New Zealand form of WPS and to identify the virus associated with disease in Australia (O'Keefe et al. 1997; Rose 1999). The New Zealand studies demonstrated that the disease could be experimentally transmitted between possums using intraperitoneal inoculation with tissue homogenates. Despite their similarities, it is thought unlikely that the Australian and New Zealand conditions have the same aetiology (Rose 1999).

9.5.3 Toxoplasmosis (see 9.9.4a)

Toxoplasmosis occurs frequently in common ringtails and common brushtails. Infection is often subclinical. Clinical signs depend on the organ system affected. Severity of the illness is variable, ranging from mild malaise to peracute mortality. Affected possums may be depressed, weak, anorexic, pyrexic, dyspnoeic, ataxic, hemiplegic, quadriplegic, comatose or convulsing, or may exhibit muscle stiffness, diarrhoea, emesis, uveitis, retinitis or cataract formation.

9.5.4 Angiostrongylosis

Angiostrongylus cantonensis, the rat lung worm, is a metastrongylid nematode that causes an eosinophilic meningoencephalitis in mammals and birds along the east coast of Australia. Infection follows the ingestion of third-stage larvae in several species of land mollusc, the intermediate hosts of the parasite. Other potential routes of infection include ingestion of paratenic hosts or possibly larvae in snail mucus trails. The condition has been diagnosed in common ringtails and common brushtails in Queensland and New South Wales. Central nervous system signs including ataxia, paresis, tremors, profound central depression, coma and seizures have been described (Rose 1999). Gross necropsy changes are usually visible. Occasionally nematode parasites are seen within the subarachnoid spaces. Histopathology of serial sections of the spinal cord and brain reveal nematode parasites, often without an inflammatory response. Ante mortem diagnosis of angiostrongylosis is difficult. Systemic eosinophilia rarely occurs; an eosinophilia may be present in the CSF. There is no ELISA test available for use in marsupials.

The prognosis for infected animals is poor. Corticosteroids may reduce inflammation but anthelmintics are not recommended as the dying nematodes elicit a greater inflammatory response than when they are alive. Possums are usually euthanased. Prevention involves the control and removal of the definitive host (the rat) and the intermediate hosts (slugs and snails) from the environment of captive possums.

9.5.5 Tick paralysis

The paralysis tick *Ixodes holocyclus* occurs along the east coast of Australia in tropical, subtropical and temperate regions. It is not uncommon to find multiple engorged ticks on an otherwise healthy possum and common brushtails have a demonstrated resistance to infestation by *I. holocyclus* (Doube 1979). However, tick paralysis has been diagnosed in subadult common brushtails and effectively treated by the intraperitoneal injection of specific canine paralysis tick antitoxin serum and supportive therapy (J Roffey *pers. comm.*). Doses and regime are based on those administered in domestic cats.

9.5.6 Miscellaneous conditions

Neurological conditions seem less frequent in common ringtails. A neuropathy was diagnosed in an adult female common ringtail with a multifocal intestinal haemorrhage from which *Streptococcus bovis* was cultured (ARWH 2007 case number 1637/1). Mild non-suppurative meningoencephalitis in a male PY common ringtail was characterised by docility, inco-ordination, loss of balance and wasting (ARWH 2007 case number 4710/1). One of the authors (RJ) has diagnosed a progressive ataxia in twin hand-reared common ringtail PY. No lesions were found in the CNS on gross and histopathological examination. The cause of the profound ataxia in both animals was presumed to be metabolic rather than morphological. The condition presented in a similar fashion to CME in common brushtails.

Severe head tilts in squirrel gliders, common ringtails, yellow-bellied gliders and sugar gliders that failed to respond to treatment have been reported (P Holz *pers. comm.*). Necropsy indicated a purulent otitis with meningoencephalitis. *Pasteurella multocida* was the most commonly cultured pathogen. Subacute meningitis has been reported in an adult feather-tailed glider in conjunction with extensive epidermal candidiasis, sinusitis and pneumonia (ARWH 2007 case number 1468/1). Suppurative arthritis secondary to trauma frequently occurs in common brushtails. Osteopaenia and rickets have been diagnosed in a male common brushtail PY (ARWH 2007 case number 2462/1). Multifocal discospondylitis occurred in an adult male common brushtail, however, no bacteria were cultured (ARWH 2007 case number 2786/1).

Injuries due to motor vehicles and predation by dogs, cats and conspecifics are common in free-ranging possums. These animals frequently present with neurological signs.

Hypertrophic osteopathy (Marie's disease) was diagnosed in an adult male common brushtail (Bellamy 1993). The initial sign was lameness affecting all four limbs. The animal was in poor condition and the limbs were swollen, hot and painful. Radiographs confirmed periosteal bone formation but no thoracic lesion was evident. An exploratory thoracotomy revealed a firm nodular mass encircling the distal oesophagus. The animal was euthanased. The mass was a primary osteosarcoma of the oesophageal wall and there was secondary pulmonary hypertrophic osteopathy.

Emaciation secondary to competition for food resource in an overcrowded habitat can occur in captive and free-ranging common ringtails (K Bodley *pers. comm.*). This condition has also been noticed in other species, particularly free-ranging populations of common brushtails. Cryptococcosis has been reported in possums (see 9.9.3). ABLV has not been diagnosed in Australian possums and gliders.

9.6 Endocrine and metabolic diseases

Hyperplastic or benign neoplastic masses in the adrenal glands and liver are common in possums (Canfield et al. 1990b). Thyroiditis has been diagnosed PM in two common ringtails suffering from skin conditions (ARWH 2007 case numbers 73/1, 2793/2). One animal had symmetrical alopecia and no other lesions (ARWH case number 73/1). The second animal was suffering from a multifocal eosinophilic thyroiditis, eosinophilic dermatitis and ulceration of the tail (ARWH case number 2793/2). Nutritional osteodystrophy leading to hind limb paresis is a commonly reported disease of pet sugar gliders. It has not been reported in zoo animals. Radiographic examination may reveal signs of osteoporosis of the long bones, pelvis and vertebrae. Animals in the early stage of disease may respond to cage rest and dietary correction, ensuring a calcium: phosphorus ratio of 2:1 and adequate vitamin D_3. Sugar gliders, like many nocturnal mammals, appear to rely more on gut absorption of vitamin D_3 than on the conversion of vitamin D_2 in the skin to vitamin D_3, which requires exposure to ultraviolet light (Booth 2000).

9.7 Ophthalmologic diseases

Lenticular cataracts occur in possums and gliders. Cataracts are commonly seen in captive and free-ranging common ringtails (A Fowler *pers. comm.*), brushtails and sugar gliders. In most cases the aetiology is not known, except when associated with toxoplasmosis. Previously it was thought that diet-induced galactosaemia led to osmotic changes in the lens, but this has been disproved (Speare 1988).

The most frequently seen ocular disorders in free-ranging common brushtails are injuries due to trauma or predation, which often become infected. Keratitis, suppurative keratitis, corneal ulceration and oedema and granulomatous conjunctivitis are common. Retinal atrophy, non-suppurative iridiochoroiditis, non-suppurative choroiditis, optic neuritis and bilateral atrophic lympho-plasmocytic chorioretinitis are also frequent in common brushtails with CME (see 9.5.1) (ARWH 2007 case numbers 1834/1, 2905/1, 3046/1, 3501/1, 3619/1, 4231/1). Corneal lipidosis has been observed in captive feather-tailed gliders. Serum cholesterol in these animals ranged from 19.11 to 23.72 mmol/L, which is high relative to other mammals. Serum cholesterol in normal feather-tailed gliders is not known, but in 14 healthy captive Leadbeater's possums values ranged from 1.3–3.9 mmol/L and was 2.4 mmol/L in an animal with cutaneous xanthomata (Lynch 1994).

Microphthalmia was recorded in two underweight juvenile common brushtails. One case was unilateral and the other bilateral. The cause was unknown (ARWH 2007 case numbers 1858/1, 1858/2). This condition has also been seen in common ringtails (L Vogelnest *pers. comm.*).

9.8 Neonatal diseases

9.8.1 General considerations

As for any marsupial, it is likely that many diseases seen in hand-reared possums and gliders can be ultimately attributed to poor management and husbandry practices (poor hygiene, excessive handling, noise or exposure to children or pets) (Woods 1999; see also Chapter 2 this volume). Passage of maternal immunoglobulin through the neonatal gut wall takes place over much of pouch life, although some capacity for active immunity begins to develop quite early (Yadav 1971; Belov et al. 2002). Thus hand-reared PY may lack the passive immunity that occurs in naturally reared animals, which appears to be particularly important in the first half of pouch life. Inappropriate husbandry or feeding (including inappropriate volume or composition of milk replacers) can lead to failure to thrive, hyperthermia, hypothermia, dietary diarrhoea, vitamin and mineral deficiencies or hypoglycaemia (Munday 1988; Speare 1988). Treatment is based on addressing the underlying cause and specific treatment where appropriate, e.g. rehydration and withdrawal of milk substitute for 24 hr for dietary diarrhoea, for hypoglycaemia the administration of parenteral 10% or 50% glucose (500–2000 mg glucose/kg IV) or dextrose (2–5 mL/kg 20% dextrose) along with high-energy foods and warmth (Munday 1978a; Speare 1988; Woods 1999). The skin of unfurred PY is susceptible to drying and xerosis (dry rough cracking scaly skin) can occur, sometimes with associated bacterial infection, if skin is not kept well-hydrated with hydrating creams (Woods 1999). Cataracts have been observed in artificially reared marsupial young and have been variously attributed to an inability to metabolise lactose, galactose or protein, or to amino acid deficiency. They have also been seen in free-ranging and captive possums, so should not necessarily be attributed to nutritional causes. Spontaneous resolution can sometimes occur (Stephens et al. 1975; Speare et al. 1984; Speare 1988; Rose 1999). Cloacal prolapse or inflammation may occur due to overstimulation by carers. Toileting is essential after feeding in hand-raised animals, however care must be taken not to irritate the sensitive cloacal mucus membranes.

9.8.2 Failure to thrive

Failure to thrive is the most common condition affecting orphaned PY and is discussed in Chapter 2.

9.8.3 Gastrointestinal disorders

Common ringtail PY appear particularly susceptible to gastrointestinal problems. For example, cloacal (common vestibule)/rectal prolapse possibly due to local irritation or infection or overstimulation by carers is relatively common (Hemsley 1992; Woods 1999), as is intestinal intussusception. Simple prolapses can be reduced and may be retained using a purse-string suture for 3–5 d, in addition to addressing any underlying causes (Woods 1999). Treatment of intussusception is unlikely to be successful. Cloacal stenosis has been diagnosed in a hand-raised juvenile common brushtail. The animal presented with a bloated abdomen and reduced appetite. Corrective surgery was only successful in the short term.

Bloat, manifested as abdominal distension and pain, is frequently seen in common ringtails, especially around weaning and in subadults (Fig. 12.7). The cause is unknown but it has been suggested to be dietary (high-carbohydrate, fruit-based diets) or related to antibiotic therapy (disturbances of gut flora) (Woods 1999). Caecal stasis and associated bloat is often seen

Figure 12.7 Bloat in a juvenile common ringtail. Photo: Monica Bando.

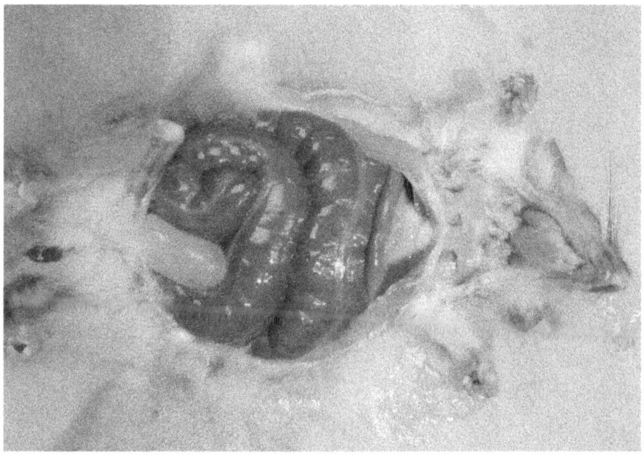

Figure 12.8 Distended caecum of the common ringtail in Fig. 12.7 above. Photo: Monica Bando.

in common ringtails weaned onto inappropriate diets, notably diets high in fruit content. It is essential to introduce a primarily folivorous diet when weaning. These foods appear to be important in establishing normal caecal function.

The hindgut of the common ringtail is very large relative to body size. In bloated animals the caecum becomes distended with fluid and loses its normal sacculated structure (Bellamy 1999) (Fig. 12.8). The distended abdomen is usually quite spongy to touch. Additional clinical signs may include a low faecal volume, small faecal pellets, increased appetite and thirst, muscle wastage, acute abdominal pain and restlessness. Treatment is symptomatic and mostly unrewarding. Despite the absence of substantial gas build-up in the gut, surfactants are commonly used by carers. Human paediatric colic treatments and papping are other popular treatments. Papping is the ingestion of the caecal contents by the PY. Bacteria are introduced into the bowel to facilitate digestion. Metoclopramide used at standard small animal dose rates (0.5 mg/kg q 8 hr) may be indicated to facilitate gastrointestinal motility. Possums are generally in very poor health and body condition by the time of presentation and usually die (McCracken 2001). As treatment is unlikely to be successful and the condition is painful, euthanasia is recommended.

Hypoproteinaemia is frequently seen in common ringtails weaned onto inadequate diets of mainly vegetables and fruit. In many cases other foods are offered but are rejected in favour of the fruits. Animals show signs of failure to thrive and evidence of hypoproteinaemia. Common ringtails of weaning age presenting with abdominal distension generally have ascites, due to hypoproteinaemia, and caecal stasis. Clinical signs are similar to those in failure to thrive, but include ascites. Diagnosis is by abdominal paracentesis revealing a transudate very low in total protein. Serum albumin is also low. Treatment is similar to that for a possum failing to thrive. The diet should be corrected and supplemented with a high-protein formula, e.g. Wombaroo Flying Fox Supplement sprinkled on the food (McCracken 2001).

Prevention of caecal stasis and hypoproteinaemia requires feeding of a natural diet including flowers and the young foliage of native trees, particularly eucalypts. Rehabilitated animals should be fed tree species which are commonly found at the intended release site. Provision of daily fresh foliage is very important as this should form the majority of the diet. Such diets better prepare the animal for release. A selection of fruits and vegetables should also be offered, including banana, apple, pear, capsicum, tomato, sweet potato, carrot and corn. The amount of fruit should be limited, otherwise it is likely to become the favoured and major dietary item, resulting in poor body condition and dental health (McCracken 2001).

9.8.4 Candidiasis

Gastrointestinal candidiasis due to *Candida albicans* is common in hand-reared marsupial PY, including possums (Munday 1988; Presidente 1982; Woods 1999). Concurrent disease, less-than-strict hygiene or antibiotic therapy may be predisposing factors. Reddening of the affected area along with the presence of grey-white plaque-like or ulcerative lesions are characteristic and diarrhoea is common (Munday 1988; Speare 1988; Woods 1999). Lesions of the mouth and common vestibule are most obvious. Affected animals frequently have a characteristic yeasty smell (L Vogelnest *pers. comm.*) Diagnosis is made by microscopic examination of scrapings and culture of lesions or faeces for the presence of yeasts (Speare 1988). Nystatin up to 50 000 IU/kg PO tid for 3–5 d has been recommended for treatment; the drug is frequently given concurrently with antibiotics prophylactically (Blyde 1999a; Woods 1999).

9.8.5 Aspiration pneumonia

Acute or chronic aspiration pneumonia can occur in hand-reared marsupial PY due to impatient feeding, too large a teat or underlying weakness (Presidente 1978; Munday 1988; Woods 1999) and has been reported in common brushtails (Presidente 1982). Supportive and antimicrobial therapy should be the basis of treatment, but generally the prognosis is poor.

9.8.6 Hypocalcaemia

Hypocalcaemic tetany has been diagnosed in four hand-reared juvenile common brushtails (Pollock 2005). Presenting signs included tetany, clenched hands and feet and inability to walk but the possums were still alert and aware of their surroundings. All four possums responded to parenteral calcium and supportive care. One possum died 4 wk after treatment. Calcium gluconate (10% 5–15 mg/kg IV slowly over a 10 min period) has been used to treat this condition. The same preparation should be diluted with 9% sodium chloride in equal parts when injected SC.

9.9 Infectious diseases

9.9.1 Viral diseases

Viral meningoencephalitis (wobbly possum syndrome) of brushtail possums is described in 9.5.2. In Australia, serological evidence of exposure to Ross River virus and Murray Valley encephalitis has been found in common brushtails without clinical signs (Anderson et al. 1952; Gard et al. 1973; O'Callaghan & Moore 1986). Antibody to Ross River virus has also been detected in asymptomatic common ringtails and sugar gliders (Gard et al. 1973). Probable herpesvirus infection, as indicated by the presence of intranuclear inclusion bodies in hepatocytes, has been recorded in the common brushtail (Munday 1978b). Poxvirus, suspected on the basis of the presence of papillary dermatitis with intracytoplasmic inclusions, has been reported in a common ringtail (ARWH 2007 case number 4133/1). In New Zealand, the presence of viruses in common brushtails has been more extensively investigated by survey because of the potential for biological control of this noxious pest. Anti-adenovirus antibodies and adenovirus particles in faeces and intestinal contents have been found, though the significance was unknown (Rice et al. 1991; Rice & Wilks 1996; Thomson et al. 2002). Papillomavirus associated with a wart-like lesion (Perrott et al. 2000) and herpesvirus, coronavirus and coronavirus-like particles have been found in intestinal contents of New Zealand common brushtails (Rice & Wilks 1996). An endogenous type D retrovirus has been detected in the liver (Baillie & Wilkins 2001). Apart from papillomavirus, clinical disease was not reported in association with the presence of virus although it could be possible. ABLV has not been reported in Australian possums or gliders.

9.9.2 Rickettsial and bacterial diseases

a Chlamydophila and rickettsial infections

The presence of Chlamydiales organisms has been confirmed, using 16S rDNA PCR, in ocular and urogenital swabs taken from greater gliders from Tumut (NSW) and mountain brushtails from Cambarville (Vic.) (Bodetti et al. 2003). In some cases in greater gliders, conjunctivitis or urogenital discharge or crusts were present but no clinical signs were apparent in the mountain brushtails. Antibodies to *Coxiella burnetii* and *Rickettsia australis*, the causative agents of Q fever and North Queensland tick typhus respectively, have been detected in common brushtails but clinical disease was not reported (Fenner 1946; O'Callaghan & Moore 1986).

b Wound infections

A range of bacteria have been isolated from infected wounds in possums and bacterial infections should be anticipated as a sequela to traumatic injury, e.g. *Staphylococcus aureus*, *β*-haemolytic *Streptococcus* and *Corynebacterium* spp. from infected wounds of common brushtails (Presidente 1978; Munday 1988). As septicaemia and death can result, local wound treatment and prophylactic antibiotics are recommended (Presidente 1978; Munday 1988; Presidente 1982a, 1982b). Common ringtails appear particularly susceptible to wound infections and septicaemia associated with cat bites (Woods 1999). *Pasteurella multocida* is often isolated in association with fight wounds, cat bite wounds and dog bite wounds (Munday 1988; Cunningham 1994). Appropriate antibiotics such as enrofloxacin, doxycycline or amoxycillin/clavulanic acid should be administered prophylactically to bitten possums as *P. multocida* infections often become systemic (Cunningham 1994).

c *Salmonellosis (see 9.2.3)*

Despite the fact that a high carrier rate for *Salmonella* spp. exists and at least eight serotypes of *S. typhimurium* have been isolated from free-ranging possums, salmonellosis does not appear to be a major cause of disease (Booth 1994; Munday 1988; Rose 1999). *Salmonella typhimurium* infection has been reported as a cause of death in captive brushtails (Hutton 1979; Keber 1979; Booth 1994) and in common ringtails (Reddacliff & Spielman 1990) and may be seen in hand-reared PY (Rose 1999). Recent capture, overcrowding and cold conditions have been suggested as causing immunosuppression and thus aiding the expression of clinical disease. Haemorrhagic enteritis and septicaemia are seen in animals succumbing to salmonellosis (Booth 1994). Supportive fluid therapy and barrier nursing are the basis for therapy (Finnie 1976), however, the zoonotic potential of salmonellosis must be considered. Because of the potential for transmission to humans, faecal culture should always be considered in juvenile ringtail possums with diarrhoea. Often it is better to euthanase affected animals.

d *Yersiniosis (see 9.2.4)*

Yersinia pseudotuberculosis commonly inhabits the intestinal tract of animals, usually without causing systemic illness. The organism is nonetheless capable of causing severe multisystemic disease in possums and gliders, including fatal enteritis and septicaemia, or subacute to chronic multisystemic abscessation (ARWH 2005). Clinical yersiniosis has been recorded in captive common brushtails and Leadbeater's possums (Hutton 1979; Munday 1988; Booth 1994; H McCracken *pers. comm.*) with immunosuppressive factors such as cold weather, capture, intra-species aggression, stress and dietary change implicated as predisposing factors. Acute disease is manifested as rapid death associated with enteritis, diarrhoea, dehydration and septicaemia (Rose 1999). Chronic disease is associated with multisystemic abscessation and wasting. Definitive diagnosis of yersiniosis involves the isolation of the organism within lesions (ARWH 2005). Lesions may include necrotic foci in the intestine, liver, spleen, kidneys, lungs and mesenteric lymph nodes.

Prognosis is usually poor but antimicrobial therapy selected after culture and susceptibility testing can be attempted. Outbreaks may be managed through substrate and husbandry changes and appropriate antibiotic therapy. The organism is ubiquitous and is found in the intestines of a wide range of species. Wild birds and rodents are probably a major source of the organism for captive animals and prevention should be based on protecting food and water from faecal contamination (Booth 1994; Rose 1999). Prevention also involves the implementation of high standards of husbandry and hygiene and the minimisation of stress. Pest prevention and control should be carried out to the highest standard. Vaccination has not been used.

e *Leptospirosis*

Leptospira interrogans (serogroup *hebdomadis*) infection appears often in common brushtails in New Zealand. Surveys indicate that around 50% of possums overall and 80% of mature animals show evidence of exposure to serovar *balcanica* or serovar *hardjo* (de Lisle et al. 1975; Hathaway et al. 1978). Australian surveys have generally indicated lower levels of infection. For example, Milner et al. (1981) found 20% of brushtails in southern New South Wales and eastern Victoria to be seropositive for the *hardjo* serovar, while Durfee and Presidente (1979) found that 14–66% of possum populations overall were sero-positive for the *hardjo* serovar, with prevalence higher in adults. A recent survey of urban common brushtails in Sydney indicated the presence of *L. interrogans* serovar *hardjo* antibodies in 11 of 136 animals (8%) and *L. borpetserseni* serovar *ballum* in 2 of 136 animals (1.5%) (Eymann et al. 2006; Eymann et al. 2007). Transmission is by direct contact between animals through urine or during mating activity. Infections are associated with interstitial nephritis but are often subclinical or mild (Durfee & Presidente 1979b; Rose 1999). Leptospirosis is a zoonotic disease, however, the risk posed to people from possums has not been quantified.

f Tuberculosis

Common brushtails are susceptible to infection with *Mycobacterium bovis* and *M. tuberculosis*. In New Zealand naturally occurring *M. bovis* is widespread and possums act as an important wildlife reservoir of the organism (Buddle & Young 2000). Lesions in infected possums occur most commonly in the lungs and lymph nodes but can be more widespread affecting a range of visceral organs (Jackson et al. 1995; Buddle & Young 2000). The lesions contain many organisms and are not walled off by fibrous tissue, facilitating generalised disease. Discharging sinuses are common (Davidson 1976; Corner & Presidente 1980; Cooke et al. 1995). It has been suggested that the failure to wall off infection could be associated with unusual features of marsupial cellular immunity (Buddle & Young 2000). Infection has not been recorded in possums in Australia (Corner & Presidente 1980; Buddle & Young 2000) and clinicians routinely exclude the disease as a cause of death as part of routine surveillance during necropsies. Any cases of granulomas in possums should be reported to the Australian Wildlife Health Network (www.wildlifehealth.org.au).

g Tyzzer's disease (see 9.2.6)

Tyzzer's disease, caused by *Bacillus piliformis*, has been recorded in free-ranging and captive common brushtails and common ringtails (Spear et al. 1984; Munday 1988; Canfield & Hartley 1991). The disease is often acute and commonly affects young animals. Clinical signs include respiratory distress, convulsions and death. Necrotising hepatic and myocardial lesions are evident at necropsy. The organism is spread oro-faecally and survives in the environment as spores. It has been suggested that clinical disease occurs in other species when latent infection reactivates in immunocompromised hosts or subsequent to stressful situations such as weaning, poor hygiene, overcrowding or concurrent disease (ARWH 2005). It is likely that this applies to possums (Ganaway et al. 1971; Rose 1999). Definitive diagnosis is by isolation and culture of the organism from lesions. High standards of husbandry and hygiene and reduction of stressors will help prevent the disease.

h Abscessation

Miscellaneous septicaemias and abscesses have been seen in free-ranging common ringtails and common brushtails (K Bodley *pers. comm.*). The causative agent is often *Yersinia pseudotuberculosis*, frequently associated with multisystemic abscessation. Abscess formation is particularly frequent in common ringtails, affecting most organ systems (T Bellamy *pers. comm.*). Pulmonary, prostatic, hepatic, renal, retroperitoneal and mesenteric lymph node abscesses have been reported (ARWH 2007 case numbers 650/1, 2230/1, 3102/1, 3602/1). Extensive abdominal abscess formation has occurred. Animals affected were subadult or adult possums. Pathogens isolated included *Proprionibacterium avidum*, *S. aureus* and *E. coli*. Abscess formation occurs frequently in common brushtails (ARWH 2007 case numbers 281/1, 602/1, 936/1, 4945/1). Abscesses have been recorded in pulmonary tissue, carpal joints, the periorbital region and nasal cavity and the intestines. One animal had a focal spinal abscess of the thoracic vertebra and the bursa of the biceps muscle (ARWH 2007 case number 602/1). Haemolytic *Streptococcus* group C was isolated from both abscesses. Vegetative valvular endocarditis and subacute non-suppurative cardiac myositis were also noted in that case. Retrobulbar abscess associated with adrenal adenoma in a squirrel glider has been recorded (ARWH 2007 case number 1831). Abscess formation appears to occur commonly as a sequela to trauma and immunosuppression.

9.9.3 Mycotic infections

Dermatophytosis caused most commonly by *Tricophyton* spp., particularly *T. mentagrophytes*, is known to occur in brushtail possums (Carman et al. 1979; Woods 1999). Treatment follows the same principles as for dogs and cats (Woods 1999). Adiospiromycosis, due to inhaled fungal spores probably of *Emmonsia crescens* has been recorded in New Zealand common brushtail with pulmonary nodules (Smith & Lancaster 1965; Hutton 1979).

Cryptococcosis has been reported in brushtail, ringtail, Leadbeater's, striped and mountain pygmy possums with meningoencephalitis and/or pulmonary lesions, pleuritis, pericarditis and peritonitis (Munday 1966; Presidente 1982b; Booth 1994; Reece & Hartley 1994). The organisms were not typed. Diagnosis is based on the presence of typical well-encapsulated yeasts (Reece & Hartley 1994). Limited surveys of free-ranging common ringtails and common brushtails in the Sydney region using LCAT have not detected serum cryptococcal antigen, nor the presence of organisms in the nasal cavity (M Krockenberger *pers. comm.*) and clinical disease appears to be uncommon.

9.9.4 Protozoal infections

a *Toxoplasmosis*

Infection in possums is most likely due to ingestion of oocysts in food contaminated with cat faeces, but transplacental transmission or ingestion of raw meat containing cysts might also be possible as these are known to occur in other species (Frenkel 1990; Rose 1999). Disease is more common in animals that are immunosuppressed or hand-reared. Deaths due to toxoplasmosis in free-ranging urban and captive common brushtails, ringtail possums and a sugar glider have been reported (Presidente 1982a; Canfield et al. 1990a; Hemsley 1992; Rose 1999). Clinical signs include sudden death, malaise, respiratory distress and neurological dysfunction (Canfield et al. 1990a). Serum AST, CK and ALT are often elevated in animals suffering from toxoplasmosis. Ante mortem diagnosis relies on serological testing to detect rising *T. gondii* IgG titre levels. Indirect fluorescent antibody tests may be used to determine serum IgM concentrations (Rose 1999).

Gross necropsy lesions, where present, include pulmonary congestion, oedema and consolidation, myocardial haemorrhage and pale streaks, splenomegaly, gastrointestinal reddening and ulceration and brain malacia. Histopathological lesions include:

- pulmonary congestion and oedema, alveolar histiocytosis, alveolar wall hypercellularity and accumulation of fibrin and nuclear debris in alveoli;
- muscle necrosis and calcification along with mononuclear and sometimes neutrophilic infiltration of cardiac, skeletal and gastrointestinal smooth muscle;
- germinal centre enlargement or necrosis, macrophage hyperplasia and follicular depletion in the spleen and lymph nodes;
- focal necrosis of the adrenal gland;
- mononuclear infiltration and fibrosis of the pancreas;
- perivascular lymphoid cuffing in the lung, brain, spinal cord and kidney;
- gliosis, neutrophilic infiltration, necrosis and calcification of the brain;
- demyelination and necrosis of the spinal cord;
- focal hepatic necrosis and portal triad lymphoid hypercellularity.

Tissue cysts unassociated with lesions are thought to be of little significance. Free tachyzoites and intracellular aggregates of tachyzoites are associated with active lesions (Canfield et al. 1990a; Rose 1999). Most cases of toxoplasmosis are diagnosed post-mortem and effective treatment regimes have not been determined. Clindamycin (10–15 mg/kg bid PO) for up to 4 wk may be effective in early stages of the disease.

Although marsupials are regarded as highly susceptible to toxoplasmosis, subclinical infections appear to be possible, as evidenced by the presence of *T. gondii* antibodies in apparently normal common brushtails (Eymann et al. 2006). Sex did not appear to have an effect on antibody prevalence, but age and location may play a role.

In general, it is believed that cysts containing bradyzoites can remain dormant in tissues of intermediate hosts for long periods without causing disease, but can reactivate when the host is immunocompromised (Frenkel 1990). This may occur in possums. Prevention of toxoplasmosis involves decreasing the opportunity for susceptible hosts to come into contact with sporulated oocysts of *T. gondii.* Effective cat control and proper storage of foodstuffs are mandatory. Vaccines based on temperature-sensitive strains of *T. gondii* or subunit vaccines may be useful in preventing the disease in captive animals in the future (Rose 1999; Jenkins 2001).

b *Other protozoa*

Eimerian coccidial oocysts are not uncommonly found in the normal faeces of brushtail and ringtail possums suggesting that infection is usually asymptomatic (Presidente 1982a; Presidente et al. 1982; Speare et al. 1984; O'Callaghan & Moore 1986; Hemsley 1992; Booth 1994), although severe diarrhoea has been attributed to coccidiosis in common brushtails (Hutton 1979). In macropod PY, in which clinical coccidiosis is more common, treatment recommendations include toltrazuril (25 mg/kg PO sid for 3 d) in conjunction with enrofloxacin (5 mg/kg IM sid for 5 d) and fluid therapy (Woods 1999). A similar regime could be attempted in clinically affected possums. The specific coccidial species involved have not been reported.

Klossiella convolutor inhabits the renal epithelial cells of ringtail possums but is considered relatively nonpathogenic (Barker et al. 1975; Presidente 1978; Speare et al. 1984). *Sarcocystis* organisms have been detected within muscle of common and mountain brushtails but their presence was not usually associated with a host reaction (Munday et al. 1978; Presidente 1978; Presidente et al. 1982). Haemogregarine protozoa have been reported in ringtail possums and sugar gliders, usually within erythrocytes but sometimes free of cells (Mackerras 1958; Speare et al. 1984). Infections were not associated with any apparent decrease in haematocrit (Mackerras 1958).

9.10 Non-infectious diseases

9.10.1 Stress

Stress factors that have been implicated in the expression of clinical disease or behavioural alterations in common brushtails include capture, introduction of new animals, cold, wet weather, poor husbandry, territorial competition and overcrowding (Hutton 1979; Keber 1979; Peters 1979). Reported manifestations of stress in captive and free-living brushtail possums include poor body condition, depression, inappetence, diarrhoea and the presence of a typical stress leucogram, indicated by leucocytosis, neutrophilia and lymphopaenia (Presidente & Correa 1981; Presidente 1982a). In free-living Herbert River ringtails wasting, lymphoid depletion and alopecia has been attributed to stress (Speare et al. 1984). Death commonly ensues in both species once clinical signs are apparent.

9.10.2 Trauma

Trauma is an extremely common reason for presentation of possums and gliders to veterinary practitioners. In urban and urban fringe areas, motor vehicles and domestic carnivores are dominant causes of traumatic injury (Hemsley & Canfield 1993a; Finnie 1988; Rose 1999). In common brushtails, motor vehicles and dogs are more common causes, although young animals may also be injured by cats, while in common ringtails and sugar gliders cat attack injuries are more common (Hemsley & Canfield 1993a; Rose 1999). Intra-specific aggression, including biting, has been recorded in captive common brushtails, common ringtails and feather-tailed gliders (Presidente 1978, 1982b; Hemsley 1992; ARWH 2007 case numbers 1970/1, 1970/2, 1970/3). Brushtail possums are more likely to suffer traumatic injury during the breeding season or if carrying dependent young, while subadult ringtail possums appear to be at increased risk (Hemsley & Canfield 1993a; Vestjens 1973). Motor vehicle trauma is usually severe and most commonly causes injuries to the head, often in combination with other body regions. Dog and cat predation often results in injuries around the neck, shoulders and thorax, and spinal injuries or abdominal injuries (Hemsley & Canfield 1993a; Rose 1999). Internal injuries can be very severe although external signs may not reflect it. Cat bite wounds are frequently contaminated with *Pasteurella multocida* and dog bite wounds are contaminated with a range of Gram negative and anaerobic bacteria (Cunningham 1994; Rose 1999). Treatment of bite wounds includes debridement, systemic antibiotics and supportive care but prognosis is guarded (Rose 1999). Important considerations prior to antibiotic use in ringtail possums are presented in 11.1. Significant tail injuries in ringtail possums are associated with a poor prognosis for release as the tail is necessary for climbing and balance. Unilateral coxo-femoral luxation is common in ringtail possums and its treatment is usually unsuccessful (Rose 1999). Common ringtails tend to tolerate traumatic injury and treatment very poorly compared with common brushtails (Finnie 1988; Hemsley1992).

Smaller gliders may fall victim to cat predation. Injuries are frequently fatal. Occasionally animals will survive and may be presented for treatment. The larger glider species frequently become entangled in barbed wire fences, overhead wires or netting that protects fruit trees. This may result in severe lacerations to the gliding membranes (see 13.2), fractured or dislocated limbs or tail (L Vogelnest *pers. comm.*).

Bushfires are the most obvious cause of burns in free-ranging possums and gliders, but domestic fires or appliances are also possible causes for common brushtails which often live in close association with human habitation. Treatment is centred on fluid and serum electrolyte support and topical and systemic antibiotics (Rose 1999). There is a reasonable chance of recovery in animals with less than 15% of skin surface area burnt and when the head and/or joints are not affected (Spielman 1994). Power-line electrocution is likely in urban possums and is usually expected to result in death, but if an animal survives it may manifest typical thermal injuries which are progressive (Munday 1988; Rose 1999). These injuries typically include extensive deep skin ulceration exposing muscle and bone.

9.10.3 Myopathy

Focal necrosis of unknown aetiology of skeletal and/or myocardial muscle has been recorded in free-ranging

Table 12.9 Neoplastic conditions reported in possums and gliders

Species	Neoplastic conditions reported	References
Common brushtail	Mammary gland hyperplasia, lymphosarcoma, squamous cell carcinoma, aural haemangiopericytoma, pulmonary large cell sarcoma, adrenal cortical adenoma	Canfield et al. (1990b), ARWH (2006 case numbers 1625/1, 4040/1, 4178/1, 4241/1)
Common ringtail	Lymphosarcoma, metastatic adenocarcinoma of the liver, lungs and thoracic wall, mammary adenocarcinoma, anal sac related sarcoma	Canfield et al. (1990b), ARWH (2006 case numbers 4070/1, 4070/2)
Sugar glider	Lymphosarcoma, splenic erythroid hyperplasia, duodenal adenocarcinoma, subcutaneous vascular hamartoma, skin-associated leiomyoma	Canfield et al. (1990b), Hough (1992)
Squirrel glider	Lymphosarcoma, schwannoma, adrenal cortical adenoma, mammary adenocarcinoma, basal cell epithelioma of the pouch	Canfield et al. (1990b), ARWH (2006 case number 4198/1)
Leadbeater's possum	Cutaneous xanthomata possibly due to excessive cholesterol in an artificial nectar diet	H McCracken & M Lynch (*pers. comm.*)
Eastern pygmy possum	Splenic erythroid hyperplasia, sertoli cell tumour, hepatoma, lymphosarcoma	Canfield et al. (1990b)
Western pygmy possum	Hepatoma	Canfield et al. (1990b)
Feather-tailed glider	Rhabdomyosarcoma, hepatocellular adenoma, epaxial and coelomic fibrosarcoma, subcutaneous sarcoma, sarcoma on the hind foot, osteosarcoma	ARWH (2006 case numbers 2342/1, 3077/1, 4125/1, 4208/1, 4232/1, 4363/1, 4773/1, 4804/1)

Herbert River ringtails and mountain brushtails and a captive common ringtail (Griner 1983; Presidente et al. 1982; Speare et al. 1984). Focal eosinophilic myositis has also been recorded in north Queensland rainforest possums (Speare et al. 1984). Posterior paresis or paralysis has been seen in captive common ringtails and captive and free-ranging sugar gliders (Arundel et al. 1977; Presidente 1982b). In the case of the ringtail possums, the condition was responsive to vitamin E and selenium administration, though no details were given on doses and regime. Exertional (capture) myopathy appears uncommon in possums and gliders but was seen in a brushtail possum subsequent to capture and restraint (Griner 1983).

9.10.4 Neoplasia

A range of neoplastic and hyperplastic conditions have been reported in possums and gliders (Table 12.9). Surgical and palliative principles in relation to neoplasia apply as for other species, but chemotherapy has not been attempted.

Tumour formation (sarcomata) with local metastases has been reported subsequent to the implantation of passive integrated transponders in feather-tailed gliders (Vogelnest et al. 1997; D Whiteside *pers. comm.*; ARWH 2007 case numbers 4125/1, 4363/1, 4407/1, 4804/1) (Fig. 12.9). Metastasis to regional lymph nodes and viscera is rare. Infiltration into the subcutaneous and adjacent tissues may occur. Implant-associated neoplasia has been reported in a mahogany glider (*P. gracilis*) (R Booth *pers. comm.*).

9.10.5 Toxins

A number of intoxications have been reported in possums and gliders, mostly related to substances contacted in gardens, human dwellings or animal enclosures. Ingestion of camellia blossoms by a common brushtail has been associated with diarrhoea, focal hepatic

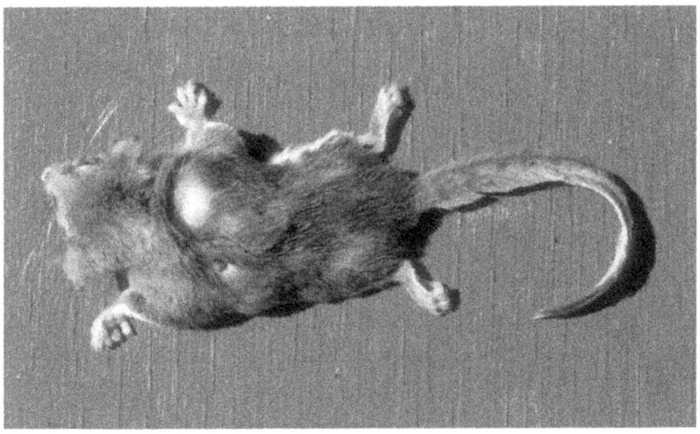

Figure 12.9 A feather-tailed glider with a transponder implant-associated tumour. Photo: Larry Vogelnest.

necrosis and death (Presidente 1978). Other common garden plants are poisonous (azaleas, rhubarb, daffodil bulbs, oleander, rhododendrons) and although specific poisonings have not been recorded in possums they could occur in urban animals (Woods 1999). Deaths in common brushtails associated with haemorrhage consistent with toxicity due to anticoagulant rodenticides (Woods 1999; ARWH 2007 case numbers 1074/1, 1188/1) have been reported. Zinc toxicosis was suspected in a colony of feather-tailed gliders (ARWH 2007 case numbers 4291/1, 4291/2, 4291/3). The affected gliders had been introduced into a holding facility fitted with new wire. Several days later and up to a week after introduction, five animals were found dead or moribund. Thoracic haemorrhage and red urine were common findings at necropsy. Pancreatic and hepatic degeneration and necrosis suggestive of zinc toxicosis were seen histologically. Toxicological screening for zinc in combined samples revealed levels of 60 mg/kg. Iron levels in the same tissue samples were 2500 mg/kg. Normal zinc and iron levels for this species are not recorded. Elevated iron levels have been noted in association with zinc toxicosis in some animals (Breitschwerdt et al. 1986; Doneley 1992). Oxalate nephrosis has been recorded in a common ringtail and a scaly-tailed possum (Ellis et al.1983; Hemsley 1992), presumably due to consumption of oxalate-containing plants (Seawright 1982).

Poisoning caused by the ingestion of anticoagulant rodenticides in common brushtails may be misdiagnosed as primary respiratory disease. Signs observed include dyspnoea, pale or cyanotic oral mucus membranes and spontaneous haemorrhage into the gastrointestinal tract and lungs. Treatment is usually unrewarding as animals are often in the terminal stage of poisoning. Mild cases can be treated with vitamin K_1, phytomenadione at an initial dose of 2.5 mg/kg SC followed by 5 mg/kg PO for at least 14 d, and supportive therapy.

9.10.6 Miscellaneous conditions

Hepatic vacuolation consistent with lipidosis has been recorded in captive and free-living ringtail possums, sugar gliders and feather-tailed gliders (Hamerton 1936; Speare et al. 1984; ARWH 2007 case numbers 4402/1, 583/1). Focal hepatic necrosis has been recorded in a captive common ringtail and a captive sugar glider (Griner 1983). Histopathological examination of PY often reveals vacuolation of hepatocytes, but this is thought to be physiological. Colonic intussusception is a relatively common finding in adult ringtail possums and in hand-reared PY (Speare et al. 1984; Presidente 1982b; Hemsley 1992). Chocolate intoxication was included as a differential diagnosis in a group of 30 brushtail possums trapped in a chocolate-waste bin, but a definitive diagnosis was not made (R Woods *pers. comm.*). Animals recovered uneventfully after sedation, washing, fluids and heat.

10 ZOONOSES

Traumatic bite wounds and scratches and associated sequelae from handling aggressive brushtail possums are common. However, with the exception of *M. bovis*, reports of infections associated with handling possums and gliders are rare. *M. bovis* infection is commonly reported in veterinarians and other workers performing necropsies on common brushtails in New Zealand. These primarily involve infections that result from contamination of cuts and abrasions on the hands (Cooke et al. 2002). Other diseases of possums and gliders such as leptospirosis, yersiniosis, dermatophytosis and salmonellosis are all potential zoonoses (Stevenson & Hughes 1980; Blyde 1999b). The role of possums and gliders in the epidemiology of other diseases that have been known to affect humans such as Ross River virus, Murray Valley encephalitis and rickettsial diseases is unknown.

11 THERAPEUTICS

There have been no pharmacokinetic studies on possums and gliders. Most drugs and dosage regimes have been extrapolated from those used in domestic species. However, administration of drugs should be approached with caution as side effects and complications have been experienced in some species.

11.1 Antibiotic therapy

Subadult common ringtails appear particularly sensitive to antibiotics (Bellamy 1999). A dysbiotic syndrome related to disturbance of normal gut flora leading to wasting and death, similar to that seen in guinea pigs and other hindgut fermenters, has been reported in common ringtails. However, unpublished observations indicate that the judicious and careful use of some antibiotics in common ringtails is without overt side effects or complications (T Bellamy, P Holz & A Olsson *pers. comm.*; Hough 2000.). Amoxycillin/clavulanic acid (12.5 mg/kg bid SC sid for 3–5 d), enrofloxacin (5 mg/kg SC sid for 5 d) and doxycycline (5 mg/kg initial dose then 2.5 mg/kg PO sid for 3–5 d) have all been used with success in some cases (Woods 1999). Nonetheless, antibiotics should be used with caution in common ringtails (McCracken 2001). In general, parenteral use is preferred to oral dosing and duration of therapy should be kept to a minimum. Animals in the transitional stage from hand-rearing and pouch to juvenile appear more sensitive to deleterious gastrointestinal side effects (T Bellamy *pers. comm.*). The use of probiotics once the course of antibiotics is complete may be of use in preventing dysbiosis.

Common brushtails appear more tolerant of antibiotics. Broad-spectrum antibiotics such as amoxycillin/clavulanic acid combinations can be given orally (12.5–20 mg/kg bid for 5 d) in this species with no apparent adverse effect (McCracken 2001). Ideally, antibiotics with a narrower spectrum of activity such as amoxycillin and penicillin should be used if possible. Long-acting penicillin (benzathine penicillin 112.5 mg/mL, procaine penicillin 150 mg/mL) can also be used in common brushtails at 1 mL/10 kg q 48 hr repeated 2–3 times. Clindamycin has been used (10 mg/kg PO bid for 7–10 d) in yellow-bellied gliders for treatment of sinusitis and periodontal disease without any observable adverse effects. Droppers and small syringes can be used to dose animals while they are manually restrained.

Nystatin (5000–10 000 U/kg tid for 5 d) may be prescribed to prevent secondary fungal or yeast overgrowth during antibiotic therapy. It is also used at the higher dose (10 000 U/kg tid for 7 d) in the treatment of oral candidiasis. As nystatin works topically it should be administered without food.

11.2 Analgesia

As with other animals, a multimodal approach to acute pain management is desirable when treating possums and gliders. A combination of analgesics (e.g. an opioid, non-steroidal anti-inflammatory and local anaesthetic) appears to provide good pain relief. Opioids such as butorphanol tartrate (0.4 mg/kg IM or SC) or buprenorphine (0.005–0.01 mg/kg SC or IV bid) have been used. Non-steroidal anti-inflammatory drugs including tolfenamic acid (4 mg/kg SC sid for 5 d), meloxicam (0.2 mg/kg SC or PO followed by 0.1 mg/kg PO sid for 5 d) and carprofen (4 mg/kg SC once only) are commonly administered to injured possums and gliders.

11.3 Anthelmintics

Although routine prophylaxis against nematodes is not required, if clinically apparent infestations are present anthelmintics can be used. However, some studies have shown that the doses of anthelmintics required to control nematodes effectively in possums appear higher than those recommended for animals for which they are currently registered (Ralston et al. 2001). Possums appear to tolerate these higher doses without adverse effects. Eprinomectin (7.5 mg/kg PO) and doramectin (0.8 mg/kg PO) have been used without side effects. Treating possums with a levamisole (37.5 mg/kg)–albendazole (23.75 mg/kg) combination (Arrest, Ancare New Zealand) orally twice, 7 d apart, reduced the faecal egg count by 99% (Ralston et al 2001). Other anthelmintics used without obvious side effect in possums and gliders include fenbendazole (20–50 mg/kg PO sid 3 d), oxfendazole (5 mg/kg PO once only) and ivermectin (0.2 mg/kg PO or SC once only) (Booth 2000). Ivermectin and praziquantel have been shown to have no significant effect on periparturient female mountain brushtail or on the subsequent survival of PY (Viggers et al. 1998).

Selamectin (6 mg/kg topically once a month) has been used for treatment of *Trichosurolaelaps crassipes* infestations in common brushtails (J Roffey *pers. comm.*) and for successful treatment of an exudative dermatitis caused by a sarcoptid mite (species unidentified) in a common brushtail (A Olsson *pers. comm.*). It also reduces parasite burdens in free-ranging mountain brushtails as soon as 48 hr after treatment (Viggers et al. 1998).

In another study, a Leadbeater's possum, a yellow-bellied glider and a common ringtail were treated with imidacloprid (10 mg/kg topically) after observation of fleas (*Pygiopsylla hoplia*) on their bodies and nest boxes. No live fleas or adverse reactions were observed for 27 d following treatment. Imidacloprid appears to be a useful drug for treating flea infestations in possums and gliders due to its ease of application, absence of toxicity and long residual effect (Baker & Beveridge 2001).

11.4 Fluid therapy routes

The route of fluid administration depends on the size, age and species of possum or glider. SC administration is most commonly used, especially in smaller species. Intraperitoneal fluids can be given to PY. IV fluids can be given to larger species but sedation will be required unless the animal is moribund. Common brushtails tolerate IV catheterisation of the lateral tail vein provided that movement is restricted. It is not usually necessary to splint the tail.

11.5 Miscellaneous treatments

Metaclopramide (0.5 mg/kg SC bid prn) has been used to enhance gastrointestinal motility. Vitamin K_1 (phytomenadione) appears to be effective in cases of rodenticide poisoning at an initial dose of 2.5 mg/kg SC, followed by 5 mg/kg PO for at least 14 d.

12 REPRODUCTIVE MANAGEMENT

12.1 Reproductive control

12.1.1 Non-surgical control

The use of gonadotrophin-releasing hormone (GnRH) analogs has been trialled in the common brushtail for population control (Jolly et al. 1996). More recent studies have involved the use of two different doses of deslorelin in common brushtails (J Eymann *pers. comm.*). Both 5 mg and 10 mg implants had a contraceptive effect in females, but no contraceptive effect on males. A dose-dependent difference between the 5 mg and 10 mg dose may exist but has not been confirmed. Results from field trials indicated that the 5 mg dose achieves infertility for 15–16 mo. The implants were injected before the main breeding season in February, and females had offspring 1 yr later. Data from the field trials involving females injected with a 10 mg implant is not available.

12.1.2 Surgical control

Either vasectomy or castration is the preferred surgical method for reproductive control (see 13.1). Ovariectomy or ovariohysterectomy are not routinely performed, however, tubal ligation has been used as a method of population control in captive common ringtails and common brushtails (H McCracken *pers. comm.*).

12.2 Assisted reproduction

Semen collection and cryopreservation, induced superovulation, artificial insemination and oocyte and embryo harvesting have been carried out successfully in common brushtails by Molinia and Rodger (1996), Molinia et al. (1998), Jungnickel et al. (1999) and Glazier and Molinia (2002). Readers are referred to these sources for more information.

13 SURGERY

Generally there are no major differences between approaches to surgical problems in possums and gliders compared with other small mammals such as dogs and cats. The use of subcuticular or absorbable sutures will reduce the need for repeat handling and the incidence of wound breakdown. Wound healing times appear no different from those in other mammals. Self-mutilation of surgical wounds appears to be inversely related to good surgical technique and effective intra-operative and post-operative pain control. Meloxicam may be given (0.2–0.5 mg/kg PO sid) for 2–3 d to provide pain relief (B Carmel *pers. comm.*). Elizabethan collars are not tolerated and contraindicated in coprophagic species such as ringtail possums.

Examples of surgeries more specific to possums and gliders are mentioned below.

13.1 Castration and vasectomy

Castration requires two separate scrotal skin incisions. To limit the risk of ligature slippage, the vas deferens is identified and ligated separately from the blood vessels using fine non-absorbable suture material. The skin wound is preferably glued closed rather than sutured in order to reduce self-trauma and subsequent wound breakdown. Splash local anaesthesia using 0.5% bupivacaine is used as part of a multimodal approach to analgesia. Alternatively, local anaesthetic may be injected into the scrotal stalk pre-operatively after induction. We have not seen toxicity associated with its use.

A scrotal ablation technique in the sugar glider has been described (Newbury et al. 2005). This method leaves no pendulous sac and requires only one small skin incision (Lightfoot 1999).

Vasectomy can be used in dominant males as a means of reproductive management. The vas deferens is easily identified even in the smaller species due to the pendulous nature of the scrotum. The technique is similar to that performed on macropods. Two small incisions are made on the cranial side of the scrotal stalk over the spermatic cord and proximal to the testes. The vasa deferentia are identified and a small portion excised, depending upon the size of the patient. Both ends of the remaining tissue are ligated with absorbable suture material. Tissue glue is used to close the skin wound, keeping the amount of suturing to a minimum. If tissue glue is unavailable, fine absorbable suture material may be used.

13.2 Glider membrane tears

Gliding membrane damage occurs more commonly in the larger gliders – the yellow-bellied glider and greater glider. Holes or complete tears may occur as a result of trauma, often due to contact with overhead wires and barbed wire. Surgical treatment involves the use of subcuticular suturing and tissue glue. The incidence of wound breakdown is quite high. Often gliders will not tolerate the sutures and quickly pull apart any suture line. Some reports indicate that good surgical repair of tears is possible if each side of the membrane skin is sutured separately using a soft fine non-absorbable suture such as 4/0 polyglactin (A Olsson *pers. comm.*). Conservative treatment—allowing the wounds to heal by secondary intention—appears to be more successful than surgery. Chlorhexidine flushing once daily followed by the application of silver sulphadiazine and chlorhexidine ointment is usually effective. Occasionally wounds will require surgical debridement.

Healed membrane wounds frequently lead to significant contraction and distortion of the membrane, rendering the animal unsuitable for release.

13.3 Orthopaedics

In general, cat and dog orthopaedic principles apply to possums and gliders. Splints may be glued to the skin of small gliders to immobilise fractures (A Olsson *pers. comm.*). Splinting of fractures in gliders can be difficult due to the constraints imposed by the flight membrane. Internal fixation is usually the treatment of choice in long bone fractures. Excision arthroplasty of the proximal head of the femur has been performed in a possum (Prescott 1972).

ACKNOWLEDGMENTS

The authors would like to thank Jane Hall for providing and collating data from the Australian Registry of Wildlife Health.

REFERENCES

Anderson SG, Donnelley M, Stevenson WJ, Caldwell NJ & Eagle M (1952). Murray Valley encephalitis: surveys of human and animal sera. *Medical Journal of Australia* **1**, 110–114.

ANZCCART (2006). *Australian Marsupials: ANZCCART Facts Sheet*. ANZCCART: Glen Osmond. <www.adelaide.edu.au/ANZCCART/publications/facts.html>.

Arundel JH, Barker IK & Beveridge I (1977). Diseases of marsupials. In *The Biology of Marsupials* (eds B Stonehouse & D Gilmore), pp. 141–154. Macmillan: London.

ARWH (2005). *Common Diseases of Urban Wildlife—Mammals*. Australian Registry of Wildlife Health: Sydney. <www.arwh.org>.

ARWH (2007). Australian Registry of Wildlife Health. Zoological Parks Board of NSW: Sydney.

Bach C (1998). *Birth Date Determination in Australasian Marsupials* (1st edn). Australasian Regional Association of Zoological Parks and Aquaria. Taronga Zoo: Sydney.

Baillie GJ & Wilkins RJ (2001). Endogenous type D retrovirus in a marsupial, the common brushtail possum (*Trichosurus vulpecula*). *Journal of Virology* **75**, 2499–2507.

Baker RT & Beveridge I (2001). Imidacloprid treatment of marsupials for fleas (*Pygiopsylla hoplia*). *Journal of Zoo and Wildlife Medicine* **32**, 391–392.

Barker IK, Munday BL & Harrigan KE (1975). *Klossiella* spp. in the kidneys of peramelid, petaurid and macropodid marsupials. *Zeitscrift fuer Parasitenkunde* **46**, 35–41.

Barnes M (2002). Sugar gliders. In *Hand-rearing Wild and Domestic Mammals* (ed. LJ Gage), p. 60. Iowa State Press: Ames, Iowa.

Barnett JL, How RA & Humphreys WF (1979). Blood parameters in natural populations of *Trichosurus* species (Marsupialia: Phalangeridae). I. Age, sex and seasonal variation in *T. caninus* and *T. vulpecula*). *Australian Journal of Zoology* **27**, 913–926.

Bellamy T (1993). Clinical case report: brushtailed possums. *Proceedings of Wildlife Disease Association (Australasian Section) Annual Conference*, p. 37. Wildlife Disease Association (Australasian Section).

Bellamy T (1994). Handrearing native animals. In *Wildlife, Proceedings 233*, pp. 7–20. Post Graduate Committee in Veterinary Science: Sydney.

Bellamy T (1999). First aid for native fauna. In *Wildlife in Australia, Proceedings 327*, p. 269. Post Graduate Foundation in Veterinary Science: Sydney.

Belov K, Nguyen M-AT, Zenger KR & Cooper DW (2002). Ontogeny of immunoglobulin expression in the brushtail possum (*Trichosurus vulpecula*). *Developmental & Comparative Immunology* **26**, 599–602.
Beveridge I (1985). The genus Bertiella (Cestoda: *Anoplocephalidae*) from Australasian mammals—new species, new records and redescriptions. *Systematic Parasitology* **7**, 241–289.
Biggins JG (1984). Communications in possums: a review. In *Possums and Gliders* (eds AP Smith & ID Hume), pp. 35–57. Surrey Beatty & Sons/Australian Mammal Society: Sydney.
Bishop P (2005). *Raising Joey Possums*. Blue Mountains WIRES training course notes. WIRES: Blue Mountains, NSW.
Blyde D (1999a). Advances in treating diseases of macropods. In *Wildlife in Australia, Proceedings 327*, pp. 439–452. Post Graduate Foundation in Veterinary Science: Sydney.
Blyde D (1999b). Update on zoonotic disease. In *Wildlife in Australia, Proceedings 327*, pp. 491–499. Post Graduate Foundation in Veterinary Science: Sydney.
Bodetti TJ, Viggers K, Warren K, Swan R, Conaghty S, Sims C & Timms P (2003). Wide range of *Chlamydiales* types detected in native Australian mammals. *Veterinary Microbiology* **96**, 177–187.
Booth NH (1988). Intravenous and other parenteral anesthetics. In *Veterinary Pharmacology and Therapeutics* (eds NH Booth & LE McDonald), pp. 212–274. Iowa State University Press: Ames.
Booth R (1994). Medicine and husbandry: dasyurids, possums and bats. In *Wildlife. Proceedings 233*, pp. 423–442. Post Graduate Committee in Veterinary Science: Sydney.
Booth RJ (2000). General husbandry and medical care of sugar gliders. In *Kirk's Current Veterinary Therapy XIII Small Animal Practice* (ed. E Bonagura), pp. 1157–1163. WB Saunders: Philadelphia.
Breitschwerdt EB, Armstrong PJ, Robinette CL, Dillman RC, Karl ML & Lowry EC (1986). Three cases of acute zinc toxicosis in dogs. *Veterinary and Human Toxicology* **28**, 109–117.
Broome LS (1995). Mountain pygmy possum *Burramys parvus*. In *The Mammals of Australia* (2nd edn) (ed. R Strahan), pp. 208–210. Reed New Holland: Sydney.
Bruner DM & Gillespie JH (1973). *Hagan's Diseases of Domestic Animals* (6th edn). Cornell University Press: Ithaca.
Buddle BM & Young LJ (2000). Immunobiology of mycobacterial infections in marsupials. *Developmental and Comparative Immunology* **24**, 517–529.
Canfield PJ & Hartley WJ (1991). Tyzzer's disease (*Bacillus piliformis*) in Australian marsupials. *Journal of Comparative Pathology* **105**, 167–173.
Canfield PJ, Hartley WJ & Dubey JP (1990a). Lesions of toxoplasmosis in Australian marsupials. *Journal of Comparative Pathology* **103**, 159–167.
Canfield PJ, Hartley WJ & Reddacliff LJ (1990b). Spontaneous proliferations in Australian marsupials—a survey and review. 1. Macropods, koalas, wombats, possums and gliders. *Journal of Comparative Pathology* **103**, 135–146.
Carman MG, Rush-Munro FM & Carter ME (1979). Dermatophytes isolated from domestic and feral animals. *New Zealand Veterinary Journal* **27**, 143–144.
Clark P, Holz P & Duignan P (2004). Collection and handling of blood samples. In *Haematology of Australian Mammals* (ed. P Clark), pp.1–20. CSIRO Publishing: Melbourne.
Cooke MM, Jackson R, Coleman JD & Alley MR (1995). Naturally occurring tuberculosis caused by *Mycobacterium bovis* in brushtail possums (*Trichosurus vulpecula*): II. Pathology. *New Zealand Veterinary Journal* **43**, 315–321.
Cooke MM, Gear AJ, Naidoo A & Collins DM (2002). Accidental *Mycobacterium bovis* infection in a veterinarian. *New Zealand Veterinary Journal* **50**, 36–38.
Corner LA & Presidente PJA (1980). *Mycobacterium bovis* infection in the brush-tailed possum (*Trichosurus vulpecula*): I. Preliminary observations on experimental infection. *Veterinary Microbiology* **5**, 309–321.
Cunningham M (1994). Applications of clinical pathology in native wildlife. In *Wildlife, Proceedings 233*, pp. 253–298. Post Graduate Committee in Veterinary Science: Sydney.
Davidson RM (1976). The role of the opossum in spreading tuberculosis. *New Zealand Journal of Agriculture* **133**, 21–25.
Dawson TJ, Finch E, Freedman L, Hume ID, Renfree MB & Temple-Smith PD (1989). Morphology and physiology of the Metatheria. In *Fauna of Australia*. Vol. 1B Mammalia (eds DW Walton & BJ Richardson), pp. 451–504. AGPS: Canberra.
de Lisle GW, Almand KB, Julian AF & Wallace J (1975). Leptospirosis in the opossum (*Trichosurus vulpecula*). *New Zealand Veterinary Journal* **23**, 215–216.
Doneley R (1992). Zinc toxicity in caged and aviary birds—'new wire disease'. *Australian Veterinary Practitioner* **22**, 6–11.
Doube BM (1979). Seasonal patterns of abundance and host relationships of the Australian paralysis tick *Ixodes holocyclus* Neumann (Acarina: Ixodidae), in southeastern Queensland. *Australian Journal of Ecology* **4**, 345–360.
Durfee PT & Presidente PJA (1979a). A sero-epidemiological study of *Leptospira interrogans* serovar *balcanica* in four brush-tailed possum populations in Victoria, Australia. *Australian Journal of Experimental Biology and Medical Science* **57**, 191–201.
Ellis TM, Copland MD & Gaynor WT (1983). Oxalate toxicity in a scaly-tailed possum, a Patagonian cavy and a swamp wallaby. *Journal of Wildlife Diseases* **19**, 290–293.
Eymann J, Herbert CA, Cooper DW & Dubey JP (2006). Serologic survey for *Toxoplasma gondii* and *Neospora caninum* in the common brushtail possum (*Trichosurus vulpecula*) from urban Sydney, Australia. *Journal of Parasitology* **92**, 267–272.

Eymann J, Smythe LD, Symonds ML, Dohnt MF, Barnett LJ, Cooper DW & Herbert CA (2007). Leptospirosis serology in the common brushtail possum (*Trichosurus vulpecula*) from urban Sydney, Australia. *Journal of Wildlife Diseases* **43(3)**, 492–497.

Fenner F (1946). The epidemiology of North Queensland tick typhus: natural mammalian hosts. *Medical Journal of Australia* **2**, 666–668.

Finnie EP (1976). Husbandry and diseases of orphaned pouch-young marsupials. *Australian Veterinary Practitioner* **6**, 19–28.

Finnie EP (1978). Marsupial therapy. In *Fauna, Proceedings 36*, pp. 619–622. Post Graduate Committee in Veterinary Science: Sydney.

Finnie EP (1988). Diseases and injuries of other Australian mammals. In *Australian Wildlife, Proceedings 104*, pp. 91–93. Post Graduate Committee in Veterinary Science: Sydney.

Fitzgerald AE, Clarke RTJ, Reid CSW, Charleston WAG, Tarttelin MF & Wyburn RS (1981). Physical and nutritional characteristics of the possum (*Trichosurus vulpecula*) in captivity. *New Zealand Journal of Zoology* **8**, 551–562.

Fleming MR (1980). Thermoregulation and torpor in the sugar glider, *Petaurus breviceps* (Marsupialia: Petauridae). *Australian Journal of Zoology* **28**, 521–534.

Fleming MR (1985). The thermal physiology of the feather-tailed glider, *Acrobates pygmaeus* (Marsupialia: Burramyidae). *Australian Journal of Zoology* **33**, 667–681.

Foley WJ (1992). Nitrogen and energy retention and acid–base status in the common ringtail possum (*Pseudocheirus peregrinus*): evidence of the effects of absorbed allelochemicals. *Physiological Zoology* **65**, 403–421.

Frenkel JK (1990). Transmission of toxoplasmosis and the role of immunity in limiting transmission and illness. *Journal of American Veterinary Medical Association* **196**, 233–240.

Ganaway JR, Allen AM & Moore TD (1971). Tyzzer's disease. *American Journal of Pathology* **64**, 717–731.

Gard G, Marshall ID & Woodroofe GM (1973). Annually recurrent epidemic polyarthritis and Ross River virus activity in a coastal area of New South Wales. II. Mosquitoes, viruses and wildlife. *American Journal of Tropical Medicine and Hygiene* **22**, 551–560.

Geiser F (1993). Hibernation in the eastern pygmy possum, *Cercartetus nanus* (Marsupialia: Burramyidae). *Australian Journal of Zoology* **41**, 67–75.

Geiser F & Broome LS (1993). The effect of temperature on the pattern of torpor in a marsupial hibernator. *Journal of Comparative Physiology B* **163**, 133–137.

Gemmell RT & Hendrikz JK (1993). Growth rates of the bandicoot *Isoodon macrourus* and the brushtail possum *Trichosurus vulpecula*. *Australian Journal of Zoology* **41**, 141–149.

Glazier AM & Molinia FC (2002). Development of a porcine follicle-stimulating hormone and porcine luteinizing hormone induced ovulation protocol in the seasonally anoestrus brushtail possum (*Trichosurus vulpecula*). *Reproduction, Fertility and Development* **14**, 453–460.

Goldingay R & Kavanagh R (1991). The yellow-bellied glider: a review of its ecology, and management considerations. In *Conservation of Australia's Forest Fauna* (ed. D Lunney), pp. 365–375. Royal Zoological Society of New South Wales: Sydney.

Griner LA (1983). *Pathology of Zoo Animals. A Review of Necropsies Conducted over a Fourteen-year Period at the San Diego Zoo and San Diego Wild Animal Park*. Zoological Society of San Diego: San Diego.

Hamerton AE (1936). Report on the deaths occurring in the Society's gardens during the year 1935. *Proceedings of the Zoological Society of London* 1936, pp. 659–686.

Harris JM & Goldingay RL (2005). Distribution, habitat and conservation status of the eastern pygmy-possum *Cercartetus nanus* in Victoria. *Australian Mammalogy* **27**, 185–210.

Hathaway SC, Blackmore DK & Marshall RB (1978). The serologic and cultural prevalence of *Leptospira interrogans* serovar *balcanica* in possums (*Trichosurus vulpecula*) in New Zealand. *Journal of Wildlife Diseases* **14**, 345–350.

Hemsley S (1992). Clinicopathological investigation of the sick and injured possum. MVSc thesis. University of Sydney, Australia.

Hemsley S (1994). Dermatitis in free-living common brushtail possums (*Trichosurus vulpecula*). *Australian Veterinary Practitioner* **24**, 147–155.

Hemsley S & Canfield P (1993a). Traumatic injuries occurring in possums and gliders in the Blue Mountains, New South Wales. *Journal of Wildlife Diseases* **29**, 612–615.

Hemsley S & Canfield PJ (1993b). Dermatitis in common brushtail possums (*Trichosurus vulpecula*). *Proceedings of Wildlife Disease Association (Australasian Section) Annual Conference*, pp. 23–24. Wildlife Disease Association (Australasian Section).

Holz P (1992). Immobilisation of marsupials with tiletamine and zolazepam. *Journal of Zoo and Wildlife Medicine* **23**, 426–428.

Holz P (2002). Restraint and anaesthesia of possums. In *Zoological Restraint and Anaesthesia* (ed. D Heard). International Veterinary Information Service: New York. <www.ivis.org>.

Holz P (2003). Marsupialia (Marsupials). In *Zoo and Wild Animal Medicine* (eds ME Fowler & RE Miller), pp. 288–303. Elsevier Science: Philadelphia.

Hough I (2000). Subcutaneous larval *Taenia serialis* in a ring-tailed possum (*Pseudocheirus peregrinus*). *Australian Veterinary Journal* **78**, 468.

Hough I, Reuter RE, Rahaley RS, Belford C, Miller R & Mitchell G (1992). Cutaneous lymphosarcoma in a sugar glider. *Australian Veterinary Journal* **69**, 93–94.

How RA & Kerle JA (1995). Common brushtail possum *Trichosurus vulpecula*. In *The Mammals of Australia* (2nd edn) (ed. R Strahan), pp. 273–275. Reed New Holland: Sydney.

Hufschmid J 2005. Ectoparasite burdens and pelage condition in mountain brushtail possums (*Trichosurus cunninghami*). *Proceedings of Wildlife Disease Association (Australasian Section) Annual Conference*, p. 104. Wildlife Disease Association (Australasian Section).

Hughes RL& Hall RS (1984). Embryonic development in the common brushtail possum, *Trichosurus vulpecula*. In *Possums and Gliders* (eds AP Smith & ID Hume), pp. 197–212. Surrey Beatty & Sons/Australian Mammal Society: Sydney.

Hume ID (1999). *Marsupial Nutrition*. CUP: Cambridge.

Hutton JB (1979). Some diseases of possums. In *Opossum Field Day* (ed. M Godschmidt), pp. 22–25. Ministry of Agriculture and Fisheries: Rangiora.

ISIS (2002). *Reference Ranges for Physiological Values in Captive Wildlife*. International Species Information System: Eagan, Minnesota.

ISIS in-house, Taronga Zoo (2005). *Reference Ranges for Physiological Values in Captive Wildlife*. International Species Information System: Taronga Zoo, Sydney.

Jackson S (2003). *Australian Mammals: Biology and Captive Management*. CSIRO Publishing: Melbourne.

Jackson R, Cooke MM, Coleman JD & Morris RS (1995). Naturally occurring tuberculosis caused by *Mycobacterium bovis* in brushtail possums (*Trichosurus vulpecula*): I. An epidemiological analysis of lesion distribution. *New Zealand Veterinary Journal* **43**, 306–314.

Jain NC (1993). *Essentials of Veterinary Haematology*. Lea & Febiger: Philadelphia.

Jenkins MC (2001). Advances and prospects for subunit vaccines against protozoa of veterinary importance. *Veterinary Parasitology* **101**, 291–310.

Johnson-Delaney C (2000). Medical update for sugar gliders. *International Conference in Exotics, Exotic DVM*, **2.3**, 91–93.

Jolly SE, Scobie S & Cowan PE (1996). Effects of vaccination against gonadotrophin-releasing hormone (GnRH). on the social status of brushtail possums in captivity. *New Zealand Journal of Zoology* **23**, 325–330.

Jungnickel MK, Harman AJ & Rodger JC (1999). Ultrastructural observations on *in vivo* fertilisation in the brushtail possum, *Trichosurus vulpecula*, following PMSG/LH superovulation and artificial insemination. *Zygote* **7**, 307–320.

Keber AW (1979). The North Island experience. In *Opossum Field Day* (ed. M Godschmidt), pp. 26–29. Ministry of Agriculture and Fisheries: Rangiora.

Kerle A (2001). *Possums: The Brushtails, Ringtails and Greater Glider*. UNSW Press: Sydney.

Lightfoot T (1999). Sugar glider orchiectomy. *Exotic DVM* **1.4**, 11–13.

Lindenmayer D (2002). *Gliders of Australia: A Natural History*. UNSW Press: Sydney.

Lindenmayer DB, Dubach J & Viggers KL (2002). Geographic dimorphism in the mountain brushtail possum (*Trichosurus* caninus)—the case for a new species. *Australian Journal of Zoology* **50**, 369–393.

Lynch M (1994). Veterinary care and management of non-domestic animals. MVetStud thesis. University of Melbourne, Australia.

Mackerras MJ (1958). Catalogue of Australian mammals and their recorded internal parasites: Part I. Monotremes and marsupials. *Proceedings of the Linnean Society of New South Wales* **83**, 101–125.

Mackintosh CG, Gill JM & O'Keefe JS (1995). A newly discovered disease of the brushtail possum: wobbly possum syndrome. *New Zealand Veterinary Journal* **43**, 126.

McCracken H (2001). Husbandry and veterinary care of orphaned marsupial pouch young: hand-rearing native fauna. In *Workshop Proceedings, July 2001, Veterinary Conservation Biology: Wildlife Health and Management in Australia Proceedings*, Taronga Zoo: Sydney.

McKay GM (1989). Family Petauridae. In *Fauna of Australia*. Vol. 1B Mammalia (eds DW Walton & BJ Richardson), pp. 665–678. AGPS: Canberra.

McKay GM & Ong P (1995). Common ringtail possum *Pseudocheirus peregrinus*. In *The Mammals of Australia* (2nd edn) (ed. R Strahan), pp. 254–256. Reed New Holland: Sydney.

McKay GM & Winter JW (1989). Phalangeridae. In *Fauna of Australia*. Vol. 1B Mammalia (eds DW Walton & BJ Richardson), pp. 636–651. AGPS: Canberra.

Menkhorst PW (1984). The application of nest boxes in research and management of possums and gliders. In *Possums and Gliders* (eds AP Smith & ID Hume), pp. 517–525. Surrey Beatty & Sons/Australian Mammal Society: Sydney.

Menkhorst PW (2001). *A Field Guide to the Mammals of Australia*. OUP: Melbourne.

Milner AR, Wilks CR, Spratt DM & Presidente PJA (1981). The prevalence of antileptospiral agglutinins in sera of wildlife in southeastern Australia. *Journal of Wildlife Diseases* **17**, 197–202.

Molinia FC, Gibson RJ, Brown AM, Glazier AM & Rodger JC (1998). Successful fertilization after superovulation and laparoscopic intrauterine insemination of the brushtail possum, *Trichosurus vulpecula*, and tammar wallaby, *Macropus eugenii*. *Journal of Reproduction and Fertility* **113**, 9–17.

Molinia FC & Rodger JC (1996). Pellet-freezing spermatozoa of two marsupials: the tammar wallaby, *Macropus eugenii*, and the brushtail possum, *Trichosurus vulpecula*. *Reproduction, Fertility and Development* **8**, 681–684.

Munday BL (1966). Diseases of Tasmania's free-living animals. *Research Bulletin No. 5*, pp. 1–29. Tasmanian Department of Agriculture.

Munday BL (1978a). Diseases of native land mammals of Australia. In *Fauna, Proceedings 36*, pp. 11–18. Post Graduate Committee in Veterinary Science: Sydney.

Munday BL (1978b). Marsupial disease. In *Fauna, Proceedings 36*, pp. 335–385. Post Graduate Committee in Veterinary Science: Sydney.

Munday BL (1988). Marsupial diseases. In *Australian Wildlife, Proceedings 104*, pp. 299–365. Post Graduate Committee in Veterinary Science: Sydney.

Munday BL, Mason RW, Hartley WJ, Presidente PJA & Obendorf D (1978). *Sarcocystis* and related organisms in Australian wildlife. I. Survey findings in mammals. *Journal of Wildlife Diseases* **14**, 417–433.

Nagy KA, Meienberger C, Bradshaw SD & Woller RD (1995). Field metabolic rate of a small marsupial mammal, the honey possum (*Tarsipes rostratus*). *Journal of Mammalogy* **76**, 862–866.

Newbury S, Hanley CS & Paul-Murphy J (2005). Sugar glider castration and scrotal ablation. *Exotic DVM* **7.1**, 27–30.

Oates JE, Bradshaw FJ, Bradshaw SD, Stead-Richardson EJ & Philippe DL (2007). Reproduction and embryonic diapause in a marsupial: Insights from captive female honey possums, *Tarsipes rostratus* (Tarsipedidae). *General and Comparative Endocrinology* **150**, 445–461.

O'Callaghan MG & Moore E (1986). Parasites and serological survey of the common brushtail possum (*Trichosurus vulpecula*) from Kangaroo Island, South Australia. *Journal of Wildlife Diseases* **22**, 589–591.

O'Keefe JS, Stanislawek WL & Heath DD (1997). Pathological studies of wobbly possum disease in New Zealand brushtail possums (*Trichosurus vulpecula*). *Veterinary Record* **141**, 226–229.

Perrott MR, Meers J, Greening GE, Farmer SE, Lugton IW & Wilks CR (2000). A new papillomavirus of possums (*Trichosurus vulpecula*) associated with typical wart-like papillomas. *Archives of Virology* **145**, 1247–1255.

Peters AJ (1979). Possums in captivity. In *Opossum Field Day* (ed. M Godschmidt), pp. 15–21. Ministry of Agriculture and Fisheries: Rangiora.

Pollock J (2005). Tetany in hand-reared common brushtail possums (*Trichosurus vulpecula*). *Proceedings of 2005 National Wildlife Rehabilitation Conference*, Gold Coast, Australia.

Prescott CW (1972). Excision arthroplasty in a possum. *Australian Veterinary Journal* **48**, 624–625.

Presidente PJA (1978). Diseases seen in free-ranging marsupials and those held in captivity. In *Fauna, Proceedings 36,* pp. 457–471. Post Graduate Committee in Veterinary Science: Sydney.

Presidente PJA (1982a). Common brushtail possum *Trichosurus vulpecula*: maintenance in captivity, blood values, diseases and parasites. In *The Management of Australian Mammals in Captivity. Proceedings of Australian Mammal Society Scientific Meeting*, 1979 (ed. DD Evans), pp. 55–66. Zoological Board of Victoria: Melbourne.

Presidente PJA (1982b). Common ringtail possum *Pseudocheirus peregrinus*: maintenance in captivity, blood values and diseases. In *The Management of Australian Mammals in Captivity. Proceedings of Australian Mammal Society Scientific Meeting,* 1979 (ed. DD Evans), pp. 75–81. Zoological Board of Victoria: Melbourne.

Presidente PJA & Correa J (1981). Haematology, plasma electrolytes and serum biochemical values of *Trichosurus vulpecula* (Kerr) (Marsupialia: Phalangeridae). *Australian Journal of Zoology* **29**, 507–517.

Presidente PJA, Barnett JL, How RA & Humphreys WF (1982). Effects of habitat, host sex and age on the parasites of *Trichosurus caninus* (Marsupialia: Phalangeridae) in northeastern New South Wales. *Australian Journal of Zoology* **30**, 33–47.

Ralston MJ, Stankiewicz M & Heath DD (2001). Anthelmintics for the control of nematode infections in the brushtail possum (*Trichosurus vulpecula*). *New Zealand Veterinary Journal* **49**, 73–77.

Reddacliff G (1981). Dermatoses: zoo animals. In *Dermatology, Proceedings 57,* p. 407. Post Graduate Committee in Veterinary Science: Sydney.

Reddacliff G & Spielman D (1990). Diseases and parasites of Australian fauna—a brief introduction. In *Care and Handling of Australian Native Animals. Emergency Care and Captive Management* (ed. SJ Hand), pp. 191–197. Royal Zoological Society of New South Wales/Surrey Beatty & Sons: Sydney.

Reece R & Hartley W (1994). The pathology registry and some interesting cases. In *Wildlife, Proceedings 233,* pp. 217–233. Post Graduate Committee in Veterinary Science: Sydney.

Renfree MB (1981). Embryonic diapause in marsupials. *Journal of Reproduction and Fertility Supplement* **29**, 67–78.

Renfree MB (1995). Honey possum *Tarsipes rostratus.* In *The Mammals of Australia* (2nd edn) (ed. R Strahan), pp. 258–260. Reed New Holland: Sydney.

Rice M & Wilks CR (1996). Virus and virus-like particles observed in the intestinal contents of the possum, *Trichosurus vulpecula. Archives of Virology* **141**, 945–950.

Rice M, Wilks CR, Pfeiffer D & Jackson R (1991). Adenovirus precipitating antibodies in the sera of brush-tailed possums in New Zealand. *New Zealand Veterinary Journal* **39**, 58–60.

Richardson KC, Wooller RD & Collins BG (1986). Adaptations to a diet of nectar and pollen in the marsupial *Tarsipes rostratus* (Marsupialia: Tarsipedidae). *Journal of Zoology (London)* **208**, 285–297.

Rose K (1999). Common diseases of possums. In *Wildlife in Australia, Proceedings 327,* pp. 385–392. Post Graduate Foundation in Veterinary Science: Sydney.

Russell EM & Renfree MB (1989). Tarsipedidae. In *Fauna of Australia.* Vol. 1B Mammalia (eds DW Walton & BJ Richardson), pp. 769–782. AGPS: Canberra.

Russell EM, Lee AK & Renfree MB (1989). Natural history of the Metatheria. In *Fauna of Australia.* Vol. 1B Mammalia (eds DW Walton & BJ Richardson), pp. 505–526. AGPS: Canberra.

Salas LA & Stephens SA (2004). Capture and immobilisation of cuscuses and ringtail possums in Papua New Guinea. *Wildlife Research* **31**, 101–107.

Seawright AA (1982). Oxalates. In *Animal Health in Australia.* Vol. 2 Chemical and Plant Poisons, pp. 74–76. AGPS: Canberra.

Skerrat LF, Martin RW & Handasyde KA (1998). Sarcoptic mange in wombats. *Australian Veterinary Journal* **76**, 408–410.

Smith MJ (1979). Observations on growth of *Petaurus breviceps* and *P. norfolcensis* (Petauridae: Marsupialia) in captivity. *Australian Wildlife Research* **6**, 141–150.

Smith AP (1982). Is the striped possum (*Dactylopsila trivirgata*; Marsupialia, Petauridae) an arboreal anteater? *Australian Mammalogy* **5**, 229–234.

Smith B (1995). *Caring for Possums.* Kangaroo Press: Sydney.

Smith JMB & Lancaster MC (1965). Adiaspiromycosis in the brush opossum, *Trichosurus vulpecula*, in New Zealand. *Sabouraudia* **4**, 146–147.

Smith A & Winter J (1984). A key and field guide to the Australian possums, gliders and the koala. In *Possums and Gliders* (eds AP Smith & ID Hume), pp. 579–594. Surrey Beatty & Sons/Australian Mammal Society: Sydney.

Speare R (1988). Clinical assessment, diseases and management of the orphaned macropod joey. In *Australian Wildlife, Proceedings 104*, pp. 211–290. Post Graduate Committee in Veterinary Science: Sydney.

Speare R, Haffenden AT, Daniels PW, Thomas AD & Seawright CD (1984). Diseases of the Herbert River ringtail, *Pseudocheirus herbertensis*, and other North Queensland rainforest possums. In *Possums and Gliders* (eds AP Smith & ID Hume), pp. 283–302. Surrey Beatty & Sons/Australian Mammal Society: Sydney.

Spielman D (1994). First aid emergency care for Australian native mammals. In *Wildlife, Proceedings 233*, pp. 339–366. Post Graduate Committee in Veterinary Science: Sydney.

Spratt DM (2003). *Rileyella petauri* gen. nov., sp. nov. (Pentastomida: Cephalobaenida) from the lungs and nasal sinus of *Petaurus breviceps* (Marsupialia: Petauridae). *Australia Parasite* **10**, 235–241.

Stanvic S (1992). *Possums: Rescue—Rearing—Rehabilitation—Release.* Wildlife Information and Rescue Service training notes. WIRES: Blue Mountains, NSW.

Stephens T, Crollini C, Mutton P, Gupta JD & Harley JD (1975). Galactose metabolism in relation to cataract formation in marsupials. *Australian Journal of Experimental Biology and Medical Science* **53**, 233–239.

Stevenson WJ & Hughes KL (1980). *Synopsis of zoonoses in Australia*. AGPS: Canberra

Strahan R (ed.) (1995). *The Mammals of Australia* (2nd edn). Reed New Holland: Sydney.

Suckling GC (1995). Sugar glider *Petaurus breviceps*. In *The Mammals of Australia* (2nd edn) (ed. R Strahan), pp. 229–231. Reed New Holland: Sydney.

Temple-Smith PD (1984). Reproductive structures and strategies in male possums and gliders. In *Possums and Gliders* (eds AP Smith & ID Hume), pp. 89–106. Surrey Beatty & Sons/Australian Mammal Society: Sydney.

Thomson D, Meers J & Harrach B (2002). Molecular confirmation of an adenovirus in brushtail possums (*Trichosurus vulpecula*). *Virus Research* **83**, 189–195.

Turner V & McKay GM (1989). Burramyidae. In *Fauna of Australia.* Vol. 1B Mammalia (eds DW Walton & BJ Richardson), pp. 652–664. AGPS: Canberra.

Turner V & Ward SJ (1995). Eastern pygmy possum *Cercatetus nanus.* In *The Mammals of Australia* (2nd edn) (ed. R Strahan), pp. 217–218. Reed New Holland: Sydney.

Tyndale-Biscoe H (2005a). Life in the trees: koala, greater glider and possum. In *Life of Marsupials* (2nd edn), pp. 219–265. CSIRO Publishing: Melbourne.

Tyndale-Biscoe H (2005b). Reproduction and development. In *Life of Marsupials* (2nd edn), pp. 37–102. CSIRO Publishing: Melbourne.

Tyndale-Biscoe H (2005c). Pygmy possums and sugar gliders: pollen eaters and sap suckers. In *Life of marsupials* (2nd edn), pp. 183–218. CSIRO Publishing: Melbourne.

Van Herck H, Baumans V, Brandt CJ, Hesp AP, Sturkenboom JH, van Lith HA, van Tintelen & Beynen AC (1998). Orbital sinus blood sampling in rats as performed by different animal technicians: the influence of technique and expertise. *Laboratory Animals* **32**, 377–386.

Vestjens WJM (1973). Wildlife mortality on a road in New South Wales. *Emu* **73**, 107–112.

Viggers KL & Lindenmayer DB (1995). The use of tiletamine hydrochloride and zolazepam hydrochloride for sedation of the mountain brushtail possum, *Trichosurus caninus. Australian Veterinary Journal* **72**, 215–216.

Viggers KL & Lindenmayer DB (1996). Variation in hematological and serum biochemical values of the mountain brushtail possum, *Trichosurus caninus* Ogilby (Marsupialia: Phalangeridae). *Journal of Wildlife Diseases* **32**, 142–146.

Viggers KL & Lindenmayer DB (2001). Hematological and plasma biochemical values of the greater glider in Australia. *Journal of Wildlife Diseases* **37**, 370–374.

Viggers KL, Lindenmayer DB, Cunningham RB & Donnelly CF (1998). The effects of parasites on a wild population of the mountain brushtail possum (*Trichosurus caninus*) in south-eastern Australia. *International Journal of Parasitology* **28**, 747–755.

Vogelnest L (1999). Chemical restraint of Australian native fauna. In *Wildlife in Australia, Proceedings 327*, pp. 149–188. Post Graduate Foundation in Veterinary Science: Sydney.

Vogelnest L, Canfield P, Hartley WJ & Reece RL (1997). Suspected transponder implant associated mesenchymal tumourigenesis in two feather-tailed gliders (*Acrobates pygmaeus*) and a koala (*Phascolarctos cinereus*). In *Proceedings of Wildlife Disease Association (Australasian*

Section) Annual Conference, pp. 50–54. Wildlife Disease Association (Australasian Section).

Wayne AF, Cowling A, Ward CG, Rooney JF, Vellios CV, Lindenmayer DB & Donnelly CF (2005). A comparison of survey methods for arboreal possums in jarrah forest, Western Australia. *Wildlife Research* **32**, 701–714.

Wells RMG, Jones A, Clout MN, Sarre SD & Anderson RK (2000). Seasonal effects on the haematology and blood chemistry of wild brushtail possums, *Trichosurus vulpecula* (Marsupialia: Phalangeridae) in New Zealand. *Comparative Haematology International* **10**, 68–73.

Whittington RJ (1982). Chronic fascioliasis in a brushtail possum (*Trichosurus vulpecula*). *Australian Veterinary Journal* **59**, 192.

Winter JW & Goudberg NJ (1995). Green ringtail possum *Pseudochirops archeri*. In *The Mammals of Australia* (2nd edn) (ed. R Strahan), pp. 244–246. Reed New Holland: Sydney.

Winter JW & Leung LK-P (1995). Common spotted cuscus *Spilocuscus maculatus*. In *The Mammals of Australia* (2nd edn) (ed. R Strahan), pp. 266–268. Reed New Holland: Sydney.

Woods R (1999). Prevention of disease in hand reared native wildlife. In *Wildlife in Australia, Proceedings 327*, pp. 455–489. Post Graduate Foundation in Veterinary Science: Sydney.

Woodside DP (1995). Feather-tailed glider *Acrobates pygmaeus*. In *The Mammals of Australia* (2nd edn) (ed. R Strahan), pp. 262–264. Reed New Holland: Sydney.

Yadav M (1971). The transmission of antibodies across the gut of pouch-young marsupials. *Immunology* **21**, 839–851.

13. Bandicoots and bilbies

Michael Lynch

1 TAXONOMY, DISTRIBUTION AND BIOLOGY

Bandicoots and bilbies are small to medium sized (150–2500 g) omnivorous marsupials found only in Australia, New Guinea and surrounding islands. They are usually nocturnal and their appearance is characterised by a pointed nose that is used in foraging for ground-living invertebrates and other food items. They are opportunistic species, exploit a wide range of food items, have a rapid reproductive rate, grow quickly and are not long-lived. There are 21 recognised species of which 11 are found in Australia. However, since European settlement the Australian species have suffered severe contractions of their natural ranges and three of the 11 species are now extinct. Overgrazing by sheep, cattle and rabbits has reduced the ground cover needed by bandicoots for nesting and increased their susceptibility to predation by foxes and cats.

The Australian bandicoots and bilbies can be divided into three Families (Peramelidae, Chaeropodidae and Thylacomyidae) within the Order Peramelemorphia. Although all species of the Order Peramelemorphia can be termed bandicoots, members of the Family Thylacomyidae are known as bilbies.

The Family Peramelidae contains eight species in three Genera, *Perameles* (long-nosed bandicoots), *Isoodon* (short-nosed bandicoots) and *Echymipera*. The extinct pig-footed bandicoot (*Chaeropus ecaudatus*) is now classified into a separate family, the Chaeropodidae (Tyndale-Biscoe 2005). The Family Thylacomyidae comprises a single genus, *Macrotis*, containing two species—the lesser bilby (*Macrotis leucura*), now extinct, and the greater bilby (*Macrotis lagotis*), commonly referred to as 'the bilby'. Table 13.1 lists the species of Australian bandicoots.

The long-nosed bandicoot (*Perameles nasuta*) is still relatively common along the eastern coast of Australia, but the ranges of the other two extant long-nosed bandicoots (eastern barred bandicoot, *Perameles gunnii*, and western barred bandicoot, *Perameles bougainville*) have dramatically reduced. The desert bandicoot, *Perameles eremiana*, once occupied central and western arid regions but is now extinct. The eastern

Table 13.1 Species of bandicoots and their weights

Common name	Weight (g)
Family Peramelidae	
Golden bandicoot	260–670
Northern brown bandicoot	500–3100
Southern brown bandicoot	400–1600
Western barred bandicoot	190–250
Desert bandicoot[a]	Weight unknown
Eastern barred bandicoot	450–900
Long-nosed bandicoot	500–1900
Rufous spiny bandicoot	500–2000
Family Chaeropodidae	
Pig-footed bandicoot[a]	Weight unknown
Family Thylacomyidae	
Greater bilby	800–2500
Lesser bilby[a]	360–435

a Extinct species.
Source: Adapted from Seebeck et al. (1990).

barred bandicoot is still common in Tasmania but the mainland population consists of a remnant wild population in eastern Victoria and three small populations in the Victorian grasslands that are the result of captive breeding and release. The western barred bandicoot was once widespread over arid central and southern Australia but is now free-ranging only on two islands off the coast of Western Australia (Dorre and Bernier Islands). Conservation efforts are currently directed at maintaining the free-ranging populations and captive breeding and release of animals into predator-proofed areas in Western Australia and South Australia.

Two of the three short-nosed bandicoots: the southern brown bandicoot (*Isoodon obesulus*) and the northern brown bandicoot (*Isoodon macrouras*) still occupy a substantial part of their former distribution. However, the golden bandicoot (*Isoodon auratus*), once distributed over a vast area in arid and tropical Australia, has undergone a huge contraction in range. It is now found only on a few islands off the northern coast of Western Australia. The cause of the rapid and dramatic decline of bandicoot and bilbies inhabiting the central part of Australia is unclear. Grazing by cattle and the introduction of foxes and rabbits may have contributed to the decline of these species. However, it appears that the major impact was the change in land burning patterns caused by movement of Aboriginal people away from their traditional nomadic life and onto cattle stations and missions. The pig-footed bandicoot formerly occurred over a wide area in central and southern Australia and is believed to have become extinct sometime in the 1950s. The rufous spiny bandicoot (*Echymipera rufescens*) is still relatively common in the rainforest areas of Cape York.

Bilbies were once distributed over most of Australia's arid zone. However, the greater bilby is now found only in an isolated area in south-west Queensland and in the northern deserts of the Northern Territory and Western Australia. Bilbies have been reintroduced to the Peron peninsula (WA) and to smaller fenced reserves in the south-west of Western Australia and the Roxby Downs area of South Australia.

2 ANATOMY AND PHYSIOLOGY

The evolutionary ancestry of bandicoots and bilbies is not clearly resolved because they share anatomical characteristics with both the other major groups of Australian marsupials. Like the dasyurids they have polyprotodont dentition possessing three or more pairs of incisors in the top and bottom jaw. However, they also display syndactyly of the second and third digits of their hind feet, a characteristic shared with kangaroos and possums. Based on analysis of mitochondrial DNA, there is general agreement that bandicoots and bilbies are a very ancient and separate group from the dasyurid or diprotodont marsupials of Australasia. The divergence of the bandicoots and bilbies from other Australian marsupials is estimated to have occurred about 60 million years ago (Tyndale-Biscoe 2005).

With the exception of the pig-footed bandicoot, bandicoots and bilbies all possess subauricular or interauricular cephalic skin gland complexes (Stoddart 1980). The copious watery secretion they produce conveys a pungent odour highly characteristic of some bandicoot and bilby species. The glandular complex increases in activity markedly during breeding periods in both sexes, but particularly in males. In bilbies the complex is located between the ears and measures 25 × 18 mm in the adult animal. Its presence can be readily palpated. Bilbies have unusual structures of a glandular appearance located at the mouth commissures (Fig. 13.1). The histological structure and function of these masses is undescribed.

Tedman (1990) described anatomical features of the northern brown bandicoot and noted that the oral cavity receives saliva from relatively small parotid but large mandibular (submaxillary) glands. Saliva produced from the mandibular glands is of a mucoid type. Bandicoots and bilbies have a simple oval stomach and the small intestine is not clearly demarcated into different regions. The size of the caecum varies between species. In the northern brown bandicoot and long-nosed bandicoot it is of moderate size but in *Permales* spp. it is only about 2 cm in length. The caecal

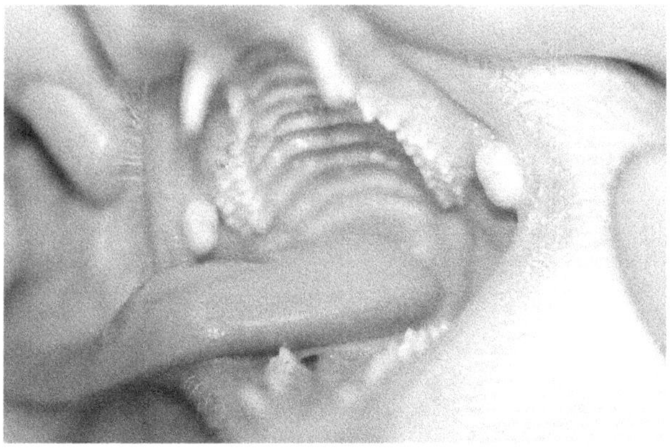

Figure 13.1 Gland-like structures in the commisures of the mouth of the greater bilby. Photo: Larry Vogelnest.

length of the extinct pig-footed bandicoot was reported to be 10 cm (Parsons 1903), suggesting that the species consumed a significant amount of plant material. In most bandicoot species the colon is short relative to the small intestine (26% of tract length in the northern brown bandicoot) but in the bilby the distal colon is much longer, presumably to maximise the resorption of water for this arid-zone species (Hume 1999).

Studies of the passage of food particles through the gut of the northern brown bandicoot and long-nosed bandicoot suggest that bandicoots maximise their use of plant material by selectively retaining small particles in the caecum and proximal colon (Hume 1999). Possession of a digestive tract that is more differentiated than that of carnivores, together with selective retention of digesta in the hindgut, allows bandicoots and bilbies to switch between plant, animal and fungal material according to the relative abundance of each food type.

A number of anatomical features of the bandicoot and bilby urogenital tract are distinctive among marsupials. The female tract of the eastern barred bandicoot is represented in Figure 13.2. The bladder is quite cranially placed, therefore the urethra is quite long. The remarkable features of the reproductive tract are the expansive vaginal caeca, very long lateral vaginae and the short median vaginal *cul de sac*, which terminates a considerable distance from the cranial end of the short urogenital sinus. Dissection of the reproductive tract of a multiparous eastern barred bandicoot revealed that the lateral vaginae open into the vaginal caecae and that each caeca connects with the median vagina. The uterine horns are separate and each has a single caudal opening into the median vagina. The median vagina is divided by a central septum. McCracken (1990) found the same anatomical relationships and external appearance of the urogenital tract in the greater bilby. The anatomy of the urogenital tract has been described in three other bandicoot species (Hill 1899; Tedman 1990) and they differ from the bilby and eastern barred bandicoot in two respects. The vaginal caecae of the northern brown bandicoot, southern brown bandicoot and long-nosed bandicoot share a median partition wall and so have an externally undivided bilobed appearance. Additionally, in the northern brown bandicoot the septum that divides the median vagina is reported present only in nulliparous individuals (Tedman 1990).

Male bandicoots have a large prostate and bulbourethral glands in relation to their body size (Tedman 1990) and these organs are usually easily palpated in the abdomen of breeding males.

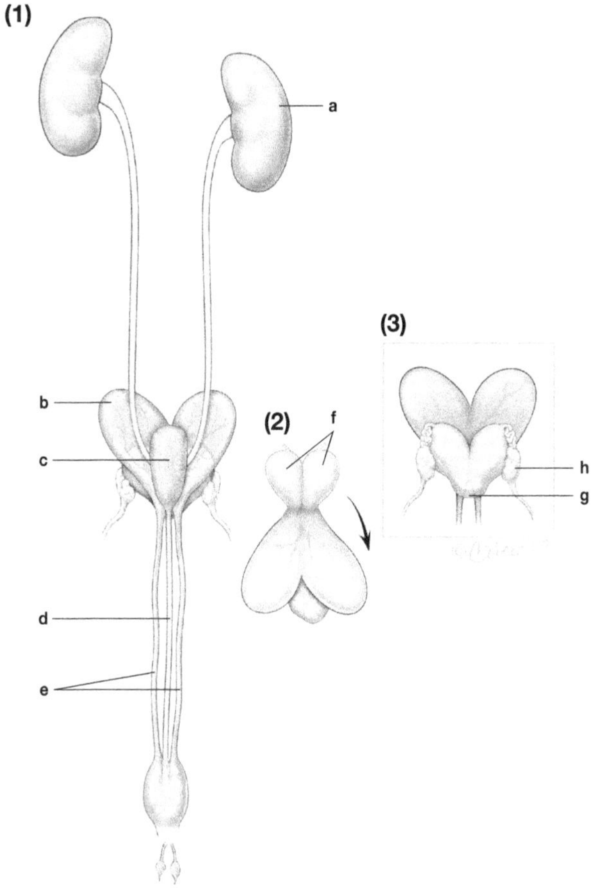

Figure 13.2 Urogenital anatomy of the female eastern barred bandicoot. 1) Ventral aspect of urogenital tract 2) Ventral aspect with bladder and vaginal caecae reflected forwards 3) dorsal aspect of urogenital tract.
a) Kidney b) Vaginal caeca c) Urinary bladder d) Urethra e) Lateral vaginae f) Uterine horns g) Median vagina h) Ovary

A number of bandicoot and bilby species occupy arid habitats and have behavioural and physiological adaptations that enable them to survive in such environments. Studies of the golden bandicoot showed that it can maintain rates of water and electrolyte turnover under conditions of extreme aridity (Bradshaw et al. 1994). Its field metabolic rate was extremely low under drought conditions and increased markedly after rain, and the volume of water required to maintain hygric balance was lower than that required by temperate-zone bandicoots (Withers 1992; Bradshaw et al. 1994). Bilbies also require less water than the predicted amount for a marsupial of their size (Gibson & Hume 2000). Arid-zone insectivorous animals often store fat in their tails because of inevitable short-term food shortages (Hume 1999). However, it appears that only one extant bandicoot species, the western barred bandicoot, has developed this capacity (Gordon & Hall 1995).

3 REPRODUCTION

Bandicoot and bilby reproductive characteristics are such that, with favourable environmental conditions, they can achieve rates of reproduction greater than any other marsupial group. All species are polyoestrus and capable of breeding throughout the year, but within their natural ranges breeding activity is influenced by food availability and climatic factors (Heinsohn 1966; Gemmell 1990; Reimer & Hindell 1996).

The oestrous cycle of bandicoots and bilbies is relatively short, from 12 d to 37 d. The length of gestation is also extremely short, ranging from 12 d to 16 d (Jackson 2003). Despite the short gestation period, bandicoot and bilby young are born in a relatively advanced stage compared to other marsupials because of the nutrient exchange efficiency of their placental structure (Tyndale-Biscoe 2005). Bandicoots and bilbies are unique among marsupials in that they form a chorio-allantoic placenta, where there is very intimate contact between maternal and foetal tissue (Renfree 1982). In all other marsupial groups the allantois remains enfolded by the wall of the yolk sac and never reaches the chorion. Eutherian (placental) mammals all have a chorio-allantoic placenta yet bandicoots and bilbies are true marsupials, i.e. their embryonic young are born and nurtured in a well-developed pouch.

The lactation period of bandicoots and bilbies is extremely short. Weaning occurs at approximately 55–75 d in bandicoots and around 3 mo in the greater bilby (Heinsohn 1966; Hall 1990; McCracken 1990). This rapid growth is possible because their milk is very energy-dense, more so than any other mammal group except the seals (Tyndale-Biscoe 2005). If conditions are favourable, females return to oestrus during late lactation so that they give birth to a new litter as the older litter is weaned. Sexual maturity in female bandicoots usually occurs around 3 mo of age, although this may be delayed depending on seasonal food availability (Lyne 1964; Gemmell 1986; Jackson 2003). Female bilbies reach sexual maturity around 5–7 mo (McCracken 1990; Southgate et al. 2000). In captivity, bandicoots and bilbies breed well if held in favourable nutritional and social conditions, and in enclosures designed to minimise stress.

The growth and development of the northern brown bandicoot has been described by Hall (1990) and is presented in Table 13.2. This species has a gestation period of 12.5 d and a pouch life of 50 ± 7 d. McCracken (1990) studied the development rate of the greater bilby from pouch life to sexual maturity (see Table 13.3).

The life-span of bandicoot species is relatively short compared with many other Australian mammals. In a study of wild southern brown bandicoots in Victoria no

Table 13.2 Development of the northern brown bandicoot from birth to weaning

Age (d)	Development features	Weight (g)
2	Umbilicus disappears	0.30
4	Deciduous claws on digits 2, 3 and 4 of fore limbs lost	0.45
5	Prominent ear bulge	0.71
7	Pouch present in females	1.25
9	Scrotum present in males	1.30
12	Teats present in female pouch	1.80
15	Claws appearing on hind limb digits	2.25
18	Facial vibrissae appear	3.5
23	Eyelids separating	7.1
25	Righting reflex present	9.5
30	All vibrissae present	16
35	Ears unplugged	23
37	Lips opening	27
39	Pelage pigment shows on dorsum	33
42	Short fur all over body	39
45	Eyes commence to open	53
48	Detaching from teat	62
50	Eyes fully open	77
52	Young leave pouch but are not fully weaned	88
60	Mostly independent	138

Source: Adapted from Hall (1990) and Bach (1998).

Table 13.3 Development of the greater bilby from birth to sexual maturity

Age (d)	Development features
17–21	Ears unfused and turned backwards
31–41	Facial vibrissae appear
37–45	Pelage pigment appears
50–58	Pelage pigment complete
59–65	Mouth fully open
63–66	Eyes fully open
66–69	Short fur all over body
78–82	Young leave pouch but are not fully weaned
94–150	Weaned
160+	Sexual maturity

Source: Adapted from McCracken (1990).

animal was found to live beyond 3.5 yr (Lobert & Lee 1990). A similar longevity was found in wild eastern barred bandicoots (M Watson *pers. comm.*). In captivity, the usual longevity of these species is 4–5 yr. Captive bilbies usually live to 7–10 yr; the average life expectancy in one wild population was 4 yr (Southgate et al. 2000).

4 HUSBANDRY

The captive management of all wild animals is greatly assisted by knowledge of their natural habitat preferences, diet and behaviour. The wide distribution of bandicoots and bilbies in Australia prior to European settlement meant that they occurred in a wide variety of habitat types ranging from tussock grasslands and heaths to open forest. They adjust their behaviour to exploit altered habitats such as regenerating grasslands after fire, and cleared pasture, as both provide an abundance of insect larvae. All Australian bandicoots and bilbies are nocturnal in behaviour. Most of the bandicoot species rest in shallow grass nests during the day, although the northern brown bandicoot uses hollow logs for daytime shelter. Bilbies construct burrows up to a few metres long for shelter during daylight hours.

4.1 Housing and husbandry

In captivity, bandicoot species are best housed in enclosures with a 10–20 cm sand or soil floor, with areas of leaf litter to allow the animals to forage and dig grass nests. Galvanised wire mesh of 10 × 10 mm must be laid under the substrate to prevent animals digging out of the enclosure. Enclosures should be planted with tussock grasses and/or small shrubs and contain branches and hollow logs. Plywood panels leaning at an angle against enclosure walls and filled with straw bedding provide additional shelter. The walls of the enclosure should be smooth-sided up to 1.5 m to prevent animals from catching their nails in mesh when jumping or climbing. At least one-third of the enclosure should be covered to protect animals from excessive rain and sun.

Bandicoots often display aggression when more than two animals are kept in areas that are too small. This is especially true if males are housed together; therefore bandicoots are often kept as individuals or male/female pairs. The breeding male bandicoot is not removed from the female while she has young in the pouch or at foot but it is important to remove independent young swiftly. Adult bandicoots often injure young animals left with them past the time of independence. A suitably sized enclosure for a pair of bandicoots is about 6 × 3.5 m, although animals are often successfully kept in smaller areas. More animals can be kept in larger enclosures, but aggression between males may be a problem. It was observed that free-ranging eastern barred bandicoots actively defended foraging areas, and that the mean home range of males was 12.94 ha and of females was 2.38 ha (Dufty 1991).

In contrast, bilbies are less aggressive towards other individuals and can more easily be kept in pairs or small groups. A study of captive bilbies suggested that there is a rigid dominance hierarchy among males, but they can be maintained without destructive fighting (Johnson & Johnson 1983). Bilby pairs can be successfully housed and bred in enclosures as small as 2.5 × 2.5 m but ideally should be housed in larger areas (J Butcher *pers. comm.*). An enclosure size of 8 × 5 m can be used to house a pair of animals or a single male and two females. Animals housed in small enclosures should be provided with a nest box that is accessed via an artificial tunnel of plastic pipe 15 cm in diameter and a few metres in length. Bilbies are usually housed on sand at least 20 cm deep, which must be kept clean and dry to prevent the development of foot problems.

Like the bandicoots, bilbies are nocturnal and shelter in cool burrows during the day. Enclosures for captive bilbies require a means of cooling when the ambient temperature approaches or exceeds 33°C. This can be achieved via overhead fans and mist sprinklers turned on for short periods so the sand substrate does not become excessively moist. Outdoor enclosures for bilbies should have solid sides up to 1 m and be meshed above that height. Bilbies displayed in nocturnal exhibits can be adequately housed in areas separated by solid smooth partitions at least 75 cm high as bilbies tend to climb rather than jump.

Sick or injured bandicoots and bilbies can be temporarily housed as solitary animals in smaller enclosures. If their medical condition contraindicates the use of enclosures with sand or dirt substrates they can be housed on a non-abrasive surface such as epoxy flooring, butcher's paper, newspaper or clean towels if a softer substrate is required. These should be changed daily. Animals should be provided with a nest box so they can take refuge during the daylight hours. Animals can be kept in such enclosures for lengthy periods while under treatment but may become stressed by frequent handling for medical treatments. Twice daily handling for short periods is usually well-tolerated.

4.2 Individual marking and identification techniques

Passive integrated transponders are commonly used to identify bandicoots and bilbies. The devices are easily inserted under the skin of the interscapular area, usually without the need for anaesthesia in suitably restrained animals. However, the skin on long-nosed bandicoots is very fragile and tears easily as the implanting needle is inserted. In this species it is preferable to have the animal anaesthetised and a stab incision made with a scalpel blade, through which the implanting needle is inserted (L Vogelnest *pers. comm.*). Skin holes made by implanting needles or stab incisions can be closed with tissue glue. Ear tattoos are a useful method to apply a permanent external identifying mark in bandicoots and bilbies. Tattoos can be applied by the use of pliers containing needle blocks or a tattoo gun. Because of the delicate nature of the pinnae in most bandicoot species, care must be taken when using tattoo guns not to over-apply them in one area. Doing so can quickly result in a full-thickness laceration of the pinna.

4.3 Transport

Bandicoots and bilbies are best transported in calico bags placed within an adequately ventilated transport box. The bag should be large enough to give the animal some freedom of movement and cushioned with grass bedding or a thick towel. The transport of bilbies in boxes without additional containment within sacks has resulted in nail damage because animals try to dig out of the box (J Butcher *pers. comm.*). It has been observed that bilbies become more agitated when transported during night-time hours when they would normally be active. This has resulted in some animals suffering nasal abrasions even when contained within a sack (J Butcher *pers. comm.*). Bandicoot species can be transported in darkened boxes without sacks as long as adequate bedding material such as grass hay is provided. This method of transport is useful when the animal is destined for field release. In that situation transport boxes can simply be opened at the field site, allowing the bandicoot to vacate when it chooses.

5 NUTRITION

Bandicoots and bilbies are omnivorous and consume a diet of insects and other invertebrates, hypogeal fungal fruiting bodies, seeds, grasses and plant bulbs. They occasionally consume small reptiles (Norman 1991). They can be considered to be opportunistic feeders in that they adjust their dietary intake depending on seasonal availability of food items. In a study of bilbies in south-western Queensland, it was found that the plant component of their diet consisted mostly of seeds and did not vary greatly across seasons. In contrast, the consumption of invertebrates varied seasonally when availability increased over summer; the favoured food items were termites and ants (Gibson 2001). Ants and termites were also found to make up a substantial portion of the diet of golden bandicoots studied on Marchinbar Island (NT) (Southgate et al. 1996). It is not surprising that bandicoots and bilbies in Australia's arid and semi-arid zones exploit these invertebrate food items, given the abundance of ants and termites in those areas. Additionally, the colonial habits of ants and termites offer a concentrated food source, minimising required foraging times and maximising rates of prey intake.

Studies of the natural diet of temperate-zone bandicoots indicate that they are opportunistic omnivores (Hume 1999). In a mixture of open pasture and eucalypt forest habitat in southern Tasmania, eastern barred bandicoots were found to have a marked variation in the seasonal consumption of invertebrates, correlating with a spring breeding peak (Reimer & Hindell 1996). The study found that the proportion of plant and fungal material in the diet was highest (58%) in autumn, equal to that of insects in winter and lowest (28%) in spring. Claridge (1991) compared the diets of southern brown bandicoots and long-nosed bandicoots in eucalypt forest in south-east New South Wales. Both species fed mainly on ants, cockroaches, beetle larvae and plant material, but with varying proportions consumed seasonally (presumably dependent on availability). It was noted that the southern brown bandicoot consumed a significant amount of hypogeal fungi in all seasons.

It is likely that a significant amount of the water needs of bandicoots are met by water contained in food items or by intake of dew on consumed plant material. Free-ranging golden bandicoots were able to maintain water and electrolyte balance despite loss of body condition during conditions of severe drought (Bradshaw et al. 1994). Nagy et al. (1991) studied the physiology of free-ranging southern brown bandicoots and concluded that the animals did not drink from water bodies in their environment. Regardless of these findings, bandicoots and bilbies in captivity should be provided with fresh water.

5.1 Diet of captive animals

Diets fed to captive animals usually consist of food items similar to those consumed by their wild counterparts, or at least approximating the nutritional composition (percentage of protein, fat etc.) of the natural diet. The approach to nutrition of captive bandicoots and bilbies in Australia has been the former—there have been no studies aimed at establishing the nutritional composition of natural diets and the actual nutritional requirements of bandicoots and bilbies are unknown. Diets fed to captive animals tend not to vary seasonally, whereas it is well-known that the diet of free-ranging animals often vary widely over the course of a year.

Diets fed to captive bandicoots and bilbies consist of a mixture of commercial dried dog kibble, seeds, insects, invertebrates, vegetables and small amounts of fruit. Some examples of bandicoot and bilby diets used in Australian facilities are reported in Jackson (2003) and the diet fed to eastern barred bandicoots at Melbourne Zoo is presented in Table 13.4. Recently captured wild bandicoots can be encouraged to eat by providing extra invertebrate food items and adding aromatic food items such as canned cat food or highly palatable supplements (e.g. Nutrigel, Troy Laboratories) to the standard diet.

Table 13.4 Diet of captive eastern barred bandicoots held at Melbourne Zoo

Average amounts offered to each animal per day
40 g dried dog kibble
20 g mixed vegetables (chosen from carrot, pumpkin, sweet potato, endive, mushroom, sweetcorn, broccoli and beetroot) and apple
15 g insects (e.g. crickets, mealworms)

These dietary modifications are beneficial in animals debilitated by illness.

The diet shown in Table 13.4 provides (on a dry matter basis) 30% crude protein, 16% fat, 0.89% calcium and 0.82% phosphorus. The fat content of the diet is high because insects are a very rich source of fat. The dried dog kibble is a high-protein puppy formulation (Advance Pedigree Puppy Rehydratable, Masterfoods). These food items are used because a substantial proportion of the natural diet of this species is insects and other invertebrates. Protein concentrations in invertebrates are relatively high, ranging from 40–70% on a dry matter basis (Bernard & Allen 1997). Crude fat content is quite variable in insects with species, season and life stage; it ranges between 10% and 58% (Dierenfield & Barker 1995). While fat is an important source of energy for insectivorous animals, the appropriateness of feeding a diet containing 16% fat year-round warrants further investigation.

6 CAPTURE AND RESTRAINT

6.1 Capture and physical restraint

Bandicoots and bilbies are fairly innocuous and tend not to bite or scratch. They are delicate and timid and great care must be taken at all times to avoid injury to the animal. Spring or gravity traps are used to capture free-ranging bandicoots with the animal attracted into the trap using bait such as a mixture of oats, peanut butter and honey. Capture of bandicoots and bilbies from pens is best attempted when they are in a nesting area so that the risk of injury is minimised. If located in a nest box, a calico bag can be easily slipped over and around the animal to scoop it up.

To capture smaller bandicoots from grass nests, a cloth catch net is often used. Mesh netting is not suitable as it will result in a higher rate of minor abrasions. The aim is to position the net at the head end and flush the animal out into the bag. The net is held with its edge on the ground at one end of the nest and a hand gently slipped under the other end of the nest to feel for the animal's rump. Prolonged attempts to capture flighty animals when they are moving around an enclosure should be avoided as the animal may injure itself while trying to evade capture. Once the animal is in the net it can be transferred to a soft cloth handling bag. This is facilitated by using a cloth net with a drawstring

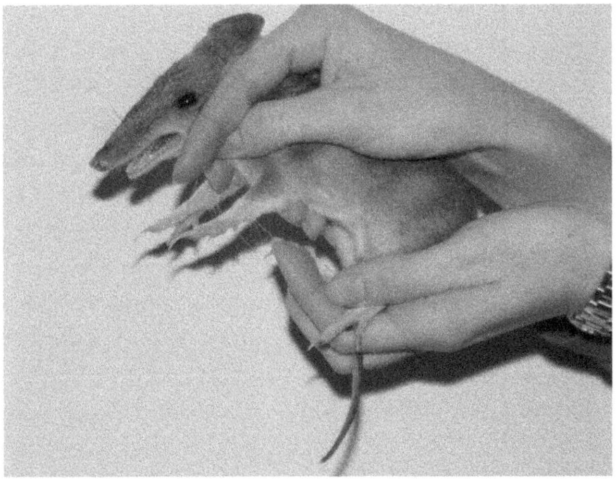

Figure 13.3 Physical restraint of a juvenile long-nosed bandicoot.

opening which can be placed inside the top of the handling bag.

Smaller species can be captured from a nest box or grass nest by hand restraint. When capturing animals by hand from a grass nest, it is wise to have an assistant holding a net against the ground in front of the nest in case the animal evades the restraint. To capture the animal, one hand is used to apply pressure to the top of the nest to discourage the animal exiting while the other hand is introduced to locate the rump of the animal. Once this is achieved, this hand is moved forward in order to restrain the neck. The animal is lifted out with the rump supported by one hand and the other hand on the back of the animal with the neck restrained between the fore and middle fingers (Fig. 13.3). Bilbies are more difficult to restrain in this manner due to their size. If capturing this species by hand it is most appropriate to firmly grasp the tail base and quickly lift them into a handling bag. Bandicoots should not be restrained by grasping the skin alone as in some species hair is easily lost and the skin can be torn. Bandicoot species should never be restrained by grasping the tail due to its fragility.

Simple examinations can be performed on animals when they are in a bag. The best method of doing this is to sit with the bag on the operator's lap and expose the areas of interest in a sequential manner while keeping most of the animal inside the bag. Small species can be turned over and gently restrained in the bag between the handler's legs to allow examination of the pouch.

When releasing animals after an examination it is best to shut them in a straw-filled nest box to allow them to settle. When the handler is leaving the enclosure the box is quietly opened to let the animal exit in its own time.

6.2 Chemical restraint

Inhalation anaesthesia using isoflurane in oxygen delivered via a T-piece is the method of choice for the chemical restraint of bandicoots and bilbies. Most species are easily restrained in a soft cloth bag with the nose exposed to allow placement of a face-mask. Alternatively, the mask can be placed over the nose while the animal is wedged in a corner of the bag. Bilbies may struggle excessively during mask inductions and for this reason midazolam at 0.1–0.2 mg/kg IM as a premedication is sometimes useful. Diazepam has also been used as a pre-anaesthetic sedative at a dose of 0.5–1.0 mg/kg IM (Vogelnest 1999). Pre-anaesthetic fasting is unnecessary in bandicoots and bilbies as they are not prone to regurgitation. Intubation is generally easily achieved with the aid of a small-bladed laryngoscope. Suitable endotracheal tube sizes are 2 mm (uncuffed) for smaller species and 3.0 mm or 3.5 mm for bilbies.

Recovery from inhalation anaesthesia is usually rapid and smooth in bandicoots, but bilbies occasionally have slightly prolonged recoveries. Animals should be allowed to recover fully in a nest box before release. If midazolam or diazepam has been used as a premedication, recovery can be aided by the administration of the benzodiazepine antagonist flumazenil at a dose rate of 0.05–0.1 mg per mg of benzodiazepine. This can be delivered by IM or IV injection. Flumazenil can be used during an anaesthetic procedure if it is judged that the midazolam sedation is significantly depressing respiration.

Injectable anaesthetic agents have rarely been used in bandicoots and bilbies as inhalation anaesthesia is simple and safe in all situations.

7 CLINICAL PATHOLOGY

7.1 Haematology and biochemistry

7.1.1 Venipuncture

Venipuncture sites most commonly used in bandicoots and bilbies are the femoral vein, the cephalic vein and, in species with a substantial tail, the lateral tail vein. In bilbies, collection of blood is most easily achieved from the jugular vein or lateral tail vein. The jugular vein is also used in the smaller bandicoot species, but more commonly blood is collected from the femoral vein. In many species, the thin skin in the inguinal area means the femoral vein can be visualised. This is particularly true for female animals of most species, where it is visible running inside the pouch. When the femoral vein cannot be directly visualised, its position is located by palpating the femoral artery pulse. The femoral vein runs parallel and just caudal to this artery and a 25 G needle is introduced to the area until blood flashes in the needle hub.

A number of other sites have been used to collect blood from bandicoot species. These include the lateral saphenous vein and medial tibial artery in eastern barred bandicoots (Booth 1994) and the orbital sinus of the northern brown bandicoot (Kemper et al. 1990) and golden bandicoot (Bradshaw et al. 1994). Collection of blood from the orbital sinus of bandicoots is not recommended by this author as other alternatives pose less risk to the animal's welfare.

7.1.2 Reference ranges and interpretation

Haematology and biochemistry reference ranges have not been established for all species of bandicoot and bilbies. Clarke (2004) presents values derived from small numbers of individuals from a number of bandicoot and bilby species. Wicks (2005) reports on the haematological characteristics of the southern brown bandicoot. Table 13.5 presents blood data collected from captive eastern barred bandicoots and bilbies and

Table 13.5 Haematology and biochemistry values (mean ± SD) of the eastern barred bandicoot,[1] southern brown bandicoot,[2] long-nosed bandicoot[3] and greater bilby[4]

Parameter	Eastern barred bandicoot (n = 50)	Southern brown bandicoot (n = 54–65)	Long-nosed bandicoot (n = 4–10)	Bilby (n = 31–46)
PCV (%)	46.0 ± 7.0	43 ± 4.0	41.5 ± 6.8	49.4 ± 6.3
RBC (× 10^{12}/L)	8.79 ± 1.40	6.96 ± 0.56	—	7.31 ± 0.95
Hb (g/L)	155 ± 25	145 ± 9.8	139 ± 21	160 ± 21
WBC (× 10^9/L)	8.14 ± 3.88	3.50 ± 1.40	8.86 ± 8.18	14.56 ± 4.66
Neutrophils (× 10^9/L)	2.72 ± 2.26	1.34 ± 0.78	3.05 ± 1.89	4.89 ± 2.03
Lymphocytes (× 10^9/L)	4.88 ± 2.56	1.74 ± 1.11	5.13 ± 8.10	8.67 ± 4.08
Monocytes (× 10^9/L)	0.21 ± −0.20	0.04 ± 0.05	0.21 ± 0.23	0.61 ± 0.44
Eosinophils (× 10^9/L)	0.25 ± 0.16	0.31 ± 0.21	0.49 ± 0.49	0.33 ± 0.31
Sodium (mmol/L)	151.2 ± 5.3	—	—	143.3 ± 2.9
Potassium (mmol/L)	5.72 ± 6.02	—	—	4.68 ± 0.68
Chloride (mmol/L)	110.2 ± 3.8	—	—	104.5 ± 3.4
Bicarbonate (mmol/L)	22.9 ± 4.1	—	—	23.7 ± 4.4
BUN (mmol/L)	14.1 ± 2.8	—	12.2 ± 1.97	12.2 ± 2.0
Creatinine (μmol/L)	70.0 ± 20.0	—	90.3 ± 17.4	64.1 ± 17.1
Calcium (mmol/L)	1.32 ± 0.78	—	—	2.51 ± 0.17
Phosphate (mmol/L)	2.31 ± 0.68	—	1.84 ± 0.48	2.16 ± 0.52
Protein (g/L)	48.1 ± 5.5	—	56.6 ± 4.83	58.1 ± 5.1
Albumin (g/L)	31.2 ± 3.8	—	29.8 ± 5.31	36.9 ± 2.9
Globulin (g/L)	16.7 ± 6.8	—	26.8 ± 3.27	20.9 ± 4.2
Total bilirubin (μmol/L)	5.94 ± 4.53	—	—	7.92 ± 6.02
ALP (U/L)	282 ± 292	—	132 ± 64	223 ± 155
ALT (U/L)	74 ± 29	—	82 ± 35	70 ± 19
AST (U/L)	44 ± 38	—	50 ± 36	67 ± 26
CK (U/L)	446 ± 650	—	327 ± 339	643 ± 668
Cholesterol (mmol/L)	3.16 ± 1.24	—	—	3.59 ± 1.17
Glucose (mmol/L)	—	—	7.37 ± 1.96	7.24 ± 2.01

Sources: **1** R Booth (unpub). **2** Wicks & Clark (2005). **3** Taronga Zoo. **4** I Smith (unpub).

from wild southern brown bandicoots. All animals were apparently healthy at the time of blood collection.

There is no published data on the haematological and biochemical responses of bandicoots and bilbies to various physiological and pathological states. There are also no unique haematological or biochemical features of bandicoots and bilbies. In most cases standard principles of interpretation of results can be applied, however, as is the case in many marsupial species, haematological and biochemical results in bandicoots and bilbies frequently belie the severity of pathological changes.

8 HAND-REARING

Bandicoots and bilbies are distinctive among the marsupials in that their young grow and develop very rapidly. Successful hand-rearing of bandicoots and bilbies requires consideration of the milk composition needed to achieve rapid rates of growth. Only three species in this group have had their milk composition studied in detail—the northern brown bandicoot, the eastern barred bandicoot and the southern brown bandicoot (Merchant 1990; Ikonomopoulou et al. 2005; R Rose unpub). These studies showed that the composition of bandicoot milk changes through the lactation period. In the first few days of lactation the milk is very dilute, with total solids measured as 8% (w/w) in the northern brown bandicoot. During lactation the milk of all three species becomes steadily more concentrated. At 3–4 wk of lactation solids have risen to around 28%, and by the end of lactation the milk of the northern brown bandicoot contains 45% solids. Eastern barred bandicoot milk contains 54% solids. Table 13.6 shows the variation in amount of carbohydrate, protein and fat in these species between mid and late lactation.

The very high fat content of bandicoot milk provides a concentrated energy source for rapid growth. Rapid growth is also achieved by a relatively high consumption of milk in the second half of lactation compared to other marsupial species (Merchant 1990). Hand-reared bandicoots should be offered 15–20% of their body weight as milk daily, compared to the 10–15% usually recommended for macropod species.

Very little information is available on the milk composition of bilbies. McCracken (1983) opportunistically collected milk samples from two bilbies when PY were removed. These samples were taken on day 37 and day 102 of pouch life, representing early and late lactation. The early lactation sample contained 12.7% solids (2.2 g carbohydrate, 4.3 g protein and 6.1 g lipid); the late lactation sample contained 49.5% solids (1.0 g carbohydrate, 15.4 g protein and 39.9 g lipid). These results indicate that bilbies, like the bandicoots, have very concentrated milk in late lactation with a high protein content and a very high lipid content.

A milk formula that has been used to hand-rear eastern barred bandicoots from around 45 d of age is Wombaroo 0.7 Kangaroo Milk Replacer (Wombaroo Food Products). If prepared according to the manufacturer's instructions for use in kangaroos, this milk would be deficient in protein and lipid throughout much of the lactation period. Four eastern barred bandicoot PY of around 50–55 d of age died from presumed suboptimal energy intake using this product (Booth & McCracken 1994). The problem was rectified by adding supplementary lipid to the Wombaroo milk product, in the form of cream or canola oil. Supplementation should commence around 30 d of age and gradually increase until day 50, when it should be double the manufacturer's instructions for use in kangaroos. Although this approach has been successful, it would be theoretically more correct to also supplement the milk with additional protein.

Another artificial milk, Di-Vetelact (Sharpe Laboratories), has been used to successfully rear many long-nosed bandicoots (A Sehlmeier *pers. comm.*). The amount of milk consumed by most individuals daily is

Table 13.6 Carbohydrate, protein and fat content of bandicoot milk (g/100 mL) in mid and late lactation

	Carbohydrate		**Protein**		**Lipid**	
	Mid	*Late*	*Mid*	*Late*	*Mid*	*Late*
Northern brown bandicoot	8.0	2.0	10.0	14.0	12.0	26.0
Eastern barred bandicoot	5.5	2.5	10.2	12.0	8.4	17.4
Southern brown bandicoot	1.7	1.1	13.0	9.2	16.1	27.8

Sources: Merchant (1990); Ikonomopoulou et al. (2005); R Rose (unpub).

approximately 15–20% of body weight, as has been found in other bandicoot species. When the PY become well furred, egg yolk is introduced into the milk and animals tend to thrive more successfully with this supplement. An initial amount of 0.5 mL/60 mL of milk is used, gradually increasing to approximately 2 mL/60 mL milk. This approximates changes in the natural composition of bandicoot milk in the later stages of lactation.

Another product, Biolac (Biolac Milk Products), can be used for hand-rearing bandicoots (Jackson 2003). Biolac produces three products—M100 for furless PY, M150, a transitional milk for when the animal first becomes furred, and M200 for the later part of PY development. The M200 product contains elevated lipid in the form of canola oil. When the joey is nearing weaning, 2–5 mL of canola oil is added per 100 mL of formula (Jackson 2003).

Toileting of hand-reared PY, by gentle stimulation of the cloaca with a moist cottonball, is required after and sometimes before feeding. Long-nosed bandicoots frequently suffer anuria despite stimulation, resulting in massive bladder distension. This is sometimes relieved by applying gentle pressure to the bladder during stimulation. If this is unsuccessful cystocentesis may be required on one or more occasions. Most cases resolve uneventfully (L Vogelnest *pers. comm.*).

Opinions vary on the minimum age and stage of development before hand-rearing is attempted. A reasonable indication for hand-rearing is if the animal has at least short fur over its body. Table 13.2 lists the developmental features of the northern brown bandicoot in relation to days of age. Most Australian bandicoot species have developed a first blush of fur by about 30 d of age and the fur is well-developed by approximately 40 d, when the animals will be 45–50 g. Orphaned bandicoots and bilbies should be kept at 28–30°C until the time they would normally emerge from the pouch. Very young PY can be fed by using an IV catheter attached to a syringe to introduce drops of milk to the mouth. PY in the later stages of pouch life quickly learn to lap milk from a small dish.

In bandicoots, weaning occurs around 1 wk after leaving the pouch. At this stage animals will be about 60 d of age and weigh 100–150 g. Bilbies will be milk-dependent for a longer period after pouch emergence. It is useful to introduce a tray of dirt containing insects (e.g. cricket abdomens and mealworms) into the environment as animals approach weaning to encourage foraging behaviour (A Sehlmeier *pers. comm.*). Weaned animals are then moved to solid-sided larger enclosures and disturbed as little as possible, to avoid imprinting. At weaning a greater variety of solid foods are introduced. The food items in the diets of adult animals can be used in weanling bandicoots and bilbies but should be presented in a form that is easily eaten by the young animal. Canned cat food mixed through the food may be useful when introducing solids as some animals find it highly palatable.

9 DISEASES

9.1 Dermatological diseases

Captive bilbies are frequently observed to have skin and hair conditions including poor coat quality, bilaterally symmetrical alopecia, hyperkeratosis and tail tip necrosis. However, the most common lesions are mild alopecia and hyperkeratosis. In males, these lesions usually occur on top of the head (not to be confused with the normal interauricular cephalic skin gland complexes) and in females they occur around the tail base. In bilbies kept in outdoor enclosures, lesions are often more severe and extensive in hotter weather (J Butcher *pers. comm.*). Nutritional supplementation with an omega-3 fatty acid source such as Megaderm (Virbac) or cold-pressed sunflower oil at 2 mL/kg sid (introduced slowly over a few days) often improves the condition but does not resolve it. Dorsal alopecia and hyperkeratosis affecting the head and tail of one animal resolved after a humidifier was introduced to its indoor enclosure. The husbandry change was prompted by a consideration that the micro-environment of a bilby burrow is cool and humid and those conditions were not being approximated in captivity.

Investigations into possible infectious and parasitic causes of dermatological conditions in bilbies have largely been unrewarding. However, *Trichophyton* spp. was isolated from bilbies in one captive colony (Wilson 1984). The mite *Ornithonyssus bacoti* was occasionally noted in another captive colony and is believed to be a contributory factor for alopecia (Buchecker & Kleinig 2002). It is commonly believed that stress is a significant contributory factor to the skin and hair problems often seen in bilbies (Buchecker & Kleinig 2002; J Butcher *pers. comm.*). Possible stressors include enclosure change, social incompatibility, frequent handling and concurrent systemic disease.

Intraspecific aggression, particularly between males, is sometimes identified as a cause of partial alopecia of the rump area, and is often accompanied by small bite wounds and scratch marks.

Ulceration and abscessation of the hind feet, primarily involving the plantar pad of the tarsus, have been observed in some captive bilby colonies (Fig. 13.4) (Blyde 1993; Buchecker & Kleinig 2002). Blyde (1993) described four bilbies from one captive colony presenting with abscessation of the feet and tail—a pure growth of *Staphylococcus aureus* was obtained from two of these cases. One animal died from suspected *S. aureus* septicaemia. Blyde suggested that maceration of plantar surfaces due to excessive moisture in the enclosure substrate may have been a predisposing factor for the condition. In a second colony, foot ulcers were occasionally observed and it was felt that frequent changing of the sand substrate (every 3 mo) assisted in preventing the condition (Buchecker & Kleinig 2002). In a third bilby colony, where the condition has not been seen, the enclosure's sand substrate is sieved daily to remove faecal matter. The managers believe that this practice and the maintenance of a very dry substrate are important in preventing foot problems (J Butcher *pers. comm.*). In another collection the condition is most prevalent in aged bilbies. Debridement, bandaging, antibiotics and anti-inflammatory treatment may result in resolution, but the condition frequently recurs (L Vogelnest *pers. comm.*).

A papilloma-like syndrome has been seen in captive and free-ranging western barred bandicoots and is considered to be a significant factor in hampering the recovery program for that species (Friend 2002). The disease manifests as multicentric proliferative lesions involving the skin and mucosal surfaces. Skin lesions first manifest as hair loss, progressing to skin erythema, thickening and the formation of nodular growths. The nodules can be situated anywhere on the body but are commonly seen around the face, feet and pouch. They are extremely debilitating around the eyes and feet, and euthanasia of captive animals is often required on welfare grounds. Grossly and histologically the smaller skin lesions resemble papillomas; the larger lesions demonstrate malignant transformation into carcinomas. Large intranuclear inclusion bodies have been observed in some cases and conjunctival lesions examined using transmission electron microscopy suggest that the inclusion bodies are papilloma virus or polyoma virus. Indirect immunohistochemistry using anti-bovine papillomavirus antibody was positive in some animals (O'Hara et al. 2004).

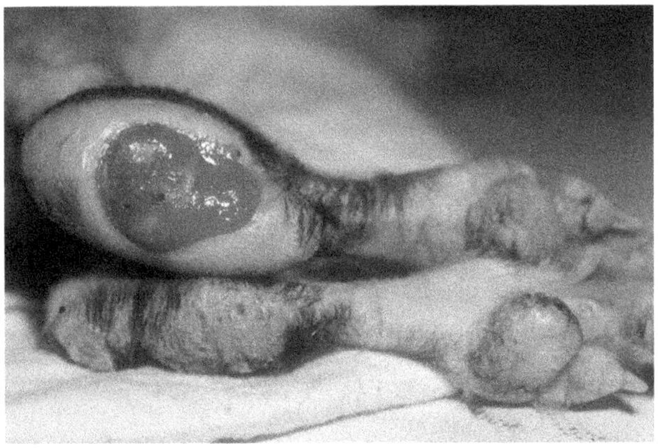

Figure 13.4 Ulceration and swelling of the hock in an aged bilby. Photo: Taronga Zoo.

9.2 Gastrointestinal diseases

Periodontal disease is often seen in captive bilbies, particularly in older animals (Buchecker & Kleinig 2002; B Bryant *pers. comm.*; J Butcher *pers. comm.*) and is presumably related to a diet that does not mimic the abrasive qualities of the diet of free-ranging animals. Periodontal disease is less common in bandicoots and is usually associated with an unsuitable diet. The author has seen severe periodontitis in a captive eastern barred bandicoot that had been predominantly fed a banana-based gruel. Wild eastern barred bandicoots fed by tourists at picnic spots were also noted to have the condition (Munday 1978).

Clinical signs of periodontal disease include calculus formation on the lateral aspects of the teeth, gingivitis, gingival recession and tooth loss (Fig. 13.5). Lenghaus et al. (1990) noted severe bronchopneumonia and pleurisy in a free-ranging eastern barred bandicoot, and considered this secondary to periodontal disease. *Actinobacillus* spp. bacteria were isolated from the lung lesions. It is not known why bilbies appear to suffer periodontal disease more frequently than bandicoots given their diet in captivity and in the wild is similar to many of the bandicoot species. Control and prevention of this condition in bilbies has been achieved by reducing soft food items and introducing a commercial dog kibble formulated to reduce periodontal disease (Hills Canine t/d diet, Hills Pet Nutrition). Another strategy used successfully to control periodontal disease in bilbies was the inclusion of browse in animal diets and

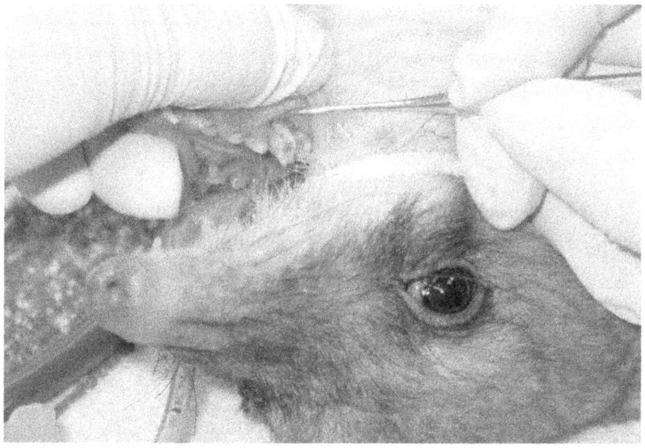

Figure 13.5 Periodontal disease in an aged bilby. Photo: Taronga Zoo.

enrichment items that require the animal to chew (e.g. *Acacia* spp. gum inside bamboo sticks) (P Eden *pers. comm.*).

A review of case material held in pathology databases held at Melbourne Zoo (MZ 2006), Healesville Sanctuary (HS 2006), Perth Zoo (PZ 2006) and Taronga Zoo (ARWH 2007) revealed a number of cases of gastric disease in bandicoots. Gastric ulceration not associated with the presence of parasites has been noted in several eastern barred bandicoots (MZ 2006) and two southern brown bandicoots (ARWH 2007 case number 94/47; HS 2006 case number HS95/140). In three cases the animals died from bacterial septicaemia, including one animal where the ulcer had obviously perforated (MZ 2006 case number MZ950617). In most cases the gastric ulceration was presumed to be associated with stress. Some of the animals were wild-caught and believed to be adapting poorly to captivity, while others were wild animals with concurrent disease conditions.

9.3 Liver and pancreatic diseases

Herpesvirus infection has been suggested as the cause of hepatic pathology observed in a captive bilby (ARWH 2007 case number 88/2384) that died after a short period of illness. The liver was grossly abnormal and on histopathological examination had universal massive acute hepatocellular necrosis associated with eosinophilic intranuclear inclusion bodies. Similar pathology and inclusion bodies were seen in a southern brown bandicoot (MZ 2006 case number MZ818).

An eastern barred bandicoot (MZ 2006 case number MON109011) was found to have a grossly distended gall bladder containing granular brown fluid at necropsy. Histologically, the animal had cholecystitis and acute cholangiohepatitis. Although no microbiological investigations were conducted it is presumed that the condition resulted from bacteria ascending the bile duct. A captive eastern barred bandicoot died following the rupture of a liver abscess (HS 2006 case number HS920371). A growth of mixed coliforms and anaerobes was cultured from the peritoneal effusion.

Significant cholelithiasis has been observed in three eastern barred bandicoots held in two captive colonies (MZ 2006). In each case the gall bladder was grossly distended with choleliths 2–4 mm in size. Choleliths were also observed in the bile ducts of these animals, however, complete bile duct obstruction was not observed and none of the animals had significant liver pathology. However, in one case (MZ 2006 case number MZ911079) the animal had been found to have a transiently elevated bilirubin level (30 µmol/L) over a year before it died of unrelated causes. The bile duct contained many choleliths and was grossly distended and it might be presumed that the transient elevation of bilirubin indicated a duct blockage that spontaneously resolved.

Choleliths were analysed in only one of the cases (MZ 2006 case number MON94367) and were found to consist predominantly of cholesterol. No investigations of the normal composition of bandicoot and bilby bile have been conducted. Cholesterol is the major constituent of choleliths in humans, and risk factors include obesity, increasing age and female sex hormones (Greenberger & Isselbacher 1998). Further investigations into the diets of free-ranging and captive bandicoots and the normal composition of their bile and choleliths are required, to develop understanding of this condition.

In two eastern barred bandicoots death was attributed to chronic fibrosing pancreatitis (MZ 2006). This condition has also been seen in a southern brown bandicoot (ARWH 2007 case number 1448/84)

9.4 Cardiopulmonary diseases

Mild pneumoconiosis resulting from the inhalation of air pollutants is a common incidental finding at necropsy of captive bandicoots held in urban areas (MZ 2006).

Multifocal pulmonary abscesses containing large numbers of *Cryptococcus* spp. yeasts have been observed in a captive bilby (ARWH 2007 case number 2408/95) and captive eastern barred bandicoots (MZ 2006 case number MZA30544; ARWH 2007 case number 5515/1). Animals presented with gradual weight loss and one animal had a facial swelling, enlarged submandibular lymph nodes and epistaxis. In the latter case (ARWH 2007 case number 5515/1), disseminated cryptococcosis involving the lung, heart, kidney, spleen, eye, brain, left maxilla and submandibular lymph nodes was diagnosed at necropsy and histologically. *Cryptococcus gatti* (formerly *C. neoformans* var. *gatti*) was cultured from multiple organs.

Acute diffuse bacterial pneumonia has been seen on a number of occasions in captive bilbies (ARWH 2007; PZ 2006 case number M230412) but causative agents and/or predisposing factors were not identified. *Pseudomonas aeruginosa* was cultured from a captive eastern barred bandicoot that died from pleuropneumonia (ARWH 2007 case number 1172/93) and an *Aeromonas* spp. was cultured from cases of fibrinous pleuritis in two other captive eastern barred bandicoots (MZ 2006 case numbers SERW351/76, SERW235/76). The occurrence of pneumonia caused by *P. aeruginosa* is not surprising as this organism is very widespread, particularly in environments that are moist and unhygienic. It can exist as a commensal in many species and high environmental load or animal stress could be predisposing factors in the development of infection. The reports of pneumonia caused by an *Aeromonas* spp. are unusual in that the organism is often pathogenic in reptiles but only rarely in mammals.

Cardiac insufficiency is occasionally diagnosed at necropsy in bandicoots and bilbies of varied age (ARWH 2007; MZ 2006) and pulmonary oedema, liver congestion and ascites may be seen. The causes of cardiac insufficiency in this group of marsupials have not been elucidated, but in one eastern barred bandicoot (where cardiac insufficiency was not diagnosed) descending aortic mural changes resembling early atherosclerosis were detected (MZ 2006 case number MZ921422). It is not known whether this observation might be related to dietary intake of cholesterol.

Myocardial necrosis was the cause of death in at least three captive eastern barred bandicoots (ARWH 2007) but the aetiology of this pathology is unknown. Considerations should include the possibility of acute viral infection (e.g. encephalomyocarditis virus) or toxic insult to the myocardium (K Rose *pers. comm.*).

9.5 Genitourinary diseases

Nephritis and pyelonephritis have been seen in bilbies, southern brown bandicoots, northern brown bandicoots, long-nosed bandicoots and eastern barred bandicoots (ARWH 2007; MZ 2006). In most cases the diagnosis of the condition was made at necropsy, and it is likely that infection in other parts of the urinary tract may not have been detected due to incomplete sampling of tissues. In one case of pyelonephritis and prostatitis in an eastern barred bandicoot (MZ 2006 case number HS980419) a *Streptococcus* spp. was isolated from the kidney. It was presumed that infection had spread to the prostate from the urinary tract. In another eastern barred bandicoot (MZ 2006 case number MZ961022) the renal medulla contained a well-circumscribed focus of mild inflammation with large numbers of *Cryptococcus* spp. yeasts.

Prostatitis has been observed in a golden bandicoot (ARWH 2007 case number 1306/1) where the condition appeared to be subacute and perhaps initiated by trauma.

Of particular interest are the deaths of at least 11 eastern barred bandicoots from five different captive colonies from severe acute bilateral nephrosis (ARWH 2007; MZ 2006). In all cases pathologists examining tissues concluded that the most likely cause of the lesion was a toxic insult but no causative toxin has been identified. In most cases animals were found dead, and at necropsy were in good condition with no pathology other than nephrosis. Six of the 11 deaths involved male and female cage-mates that died within 24 hr of each other. The deaths were investigated in detail at one facility (Booth & McCracken 1994). Two animals had tissues and gut contents analysed for phenolics (including tannins), zinc and copper and the results were compared to levels in wild eastern barred bandicoots that had died from trauma. Very low levels of phenolics were found in all animals and there were undetectable levels of zinc and copper in all tissues. Other toxins considered were manufactured chemical agents, mycotoxins, spider bites and toxic plants or invertebrates that animals may have consumed, but no source of such toxins could be identified (H McCracken *pers. comm.*).

9.6 Ophthalmologic diseases

Traumatic ocular disease has been observed in free-ranging eastern barred bandicoots captured during field studies (Booth & McCracken 1994), due to trauma from bait hooks in the trap or from the sides of the trap. Injuries were more common when animals were left in uncovered traps for extended periods. It would be expected that the nervous demeanour of most smaller bandicoot species contributes to the incidence of corneal trauma under these circumstances because animals easily panic in response to unfamiliar stimuli.

Ophthalmic examinations were conducted on 55 captive eastern barred bandicoots following observations of a high incidence of corneal and lenticular lesions in animals held in large semi-natural enclosures and under more intensive conditions (H McCracken *pers. comm.*). Twenty-two animals (40%) had at least one ocular abnormality that was considered severe enough to affect vision. Lesions included corneal scarring, cataracts and optic nerve atrophy (Miller et al. unpub). No unifying causal factor to explain the high incidence of ocular pathology in this group of animals could be established. Suggested aetiological factors included toxoplasmosis and trauma from thorny vegetation. *T. gondii* infections in other marsupial species sometimes results in choroidoretinitis (Attwood et al. 1975; Ashton 1979) but this has not been described in bandicoots or bilbies. Unfortunately, there has been no systematic histological examination of the eyes of eastern barred bandicoots, even those identified with lesions in the study, so the incidence of retinal lesions due to *T. gondii* remains uninvestigated.

Ocular disease has been reported in free-ranging western barred bandicoots (Bodetti et al. 2003; Warren et al. 2005). Eleven of 19 animals trapped for translocation on Bernier Island (WA) had at least one eye condition (Warren et al. 2005). These included corneal opacity, conjunctivitis and blepharitis. Four of five animals with conjunctivitis had chlamydial DNA detected by polymerase chain reaction (PCR), and four new types of organism were characterised using gene sequencing. Supporting evidence for the pathogenicity of these uncultured Chlamydiales genotypes was the lack of clinical response to non-chlamydiacidal ocular antibiotic therapy and rapid response to topical and systemic tetracyclines. A Chlamydiales genotype has been isolated from cloacal and conjunctival swabs in 11 of 21 bilbies tested from a wild population from south-west Western Australia, but no clinical disease was noted (Bodetti et al. 2003). It is not known whether this novel uncultured genotype is capable of causing disease.

9.7 Infectious diseases

9.7.1 Bacterial diseases

Bacterial infections isolated from specific organ systems have been considered in previous sections. Mortality resulting from generalised bacterial infections is not uncommon in bandicoots and bilbies.

Multisystemic granulomatous disease caused by saprophytic atypical *Mycobacteria* spp. is a sporadic cause of death in captive marsupials. Mycobacteriosis affecting the lungs and gastrointestinal tract was the cause of death in a captive northern brown bandicoot (ARWH 2007 case number 1546/1). A captive long-nosed bandicoot presenting in poor condition was found to have granulomatous disease of the lungs and liver caused by atypical mycobacterial infection (ARWH 2007 case number 3235/1).

Bacterial septicaemia is not an uncommon cause of death in bandicoots and bilbies. A captive eastern barred bandicoot with chronic active inflammation in the liver, kidney and prostate died from bacterial septicaemia caused by *Proteus mirabilis* (ARWH 2007 case number 184/1). This organism is a commensal of the intestinal tract and septicaemia commonly originates from endogenous infection of the urinary tract. In this animal the origin of infection was unclear, as the duration of the processes seemed to be similar in all three organs.

Bacterial septicaemia secondary to environmental contamination of wounds is not an uncommon finding in all species. Septicaemia caused by a mixed *E. coli* and a *Proteus* spp. infection were attributed to complications of traumatic toe wounds in two eastern barred bandicoots (MZ 2006). Similarly, an eastern barred bandicoot died from *Vibrio metschinkovii* septicaemia thought to be secondary to a toe wound (MZ 2006 case number HAMK/42). This is a saprophytic organism usually associated with brackish water and can cause sporadic disease (Jean-Jacques et al. 1981). Wild bandicoots surviving attacks by cats are at risk of developing *Pasteurella multocida* septicaemia from contaminated bite wounds; prophylactic treatment with amoxycillin and clavulanic acid or enrofloxacin at standard dose rates is recommended.

Septicaemia caused by *Erysipelothrix rhusiopathiae* has been observed in a captive bilby (ARWH 2007 case number 87/1910), a northern brown bandicoot (Eamens et al. 1988) and a southern brown bandicoot

(MZ 2006 case number MZ539). The serotypes of *Erysipelothrix* from the bilby and northern brown bandicoot were isolated and found to be different. They were also distinct from serotypes that are reported to cause disease in pigs, sheep, domestic fowl and wild birds in Australia (Eamens et al. 1988). Animals normally acquire *Erysipelothrix* infections from environments contaminated by carrier animals, contact with fish, or water bodies that have been contaminated.

9.7.2 Parasitic diseases

This section will highlight clinically significant parasites rather than attempt to list all species that have been identified from bandicoots and bilbies.

a Ectoparasites

Heavy ectoparasite burdens are occasionally seen in captive and free-ranging bandicoots. It is likely that the incidence of significant ectoparasite burdens is influenced by environmental conditions favourable to the parasite, animal densities and individual animal health. Lenghaus (1990) attributed the death of 4 of 134 eastern barred bandicoots presented for necropsy to heavy tick (*Ixodes tasmani* and *I. faecialis*) and flea (*Pygiopsylla zethi*) burdens. A captive-bred eastern barred bandicoot, released into a large fenced wildlife reserve in early winter, was found to have a heavy and clinically significant flea burden when recaptured a few weeks later. The flea, identified as *Pygiopsylla hoplia*, has a wide host range and the high densities of southern brown bandicoots, rufous bettongs (*Aepyprymnus rufescens*) and long-nosed potoroos (*Potorous tridactylus*) within the same enclosure may have resulted in high flea densities. Heavy flea burdens are occasionally seen in free-ranging long-nosed bandicoots and southern brown bandicoots (HS 2006).

The northern brown bandicoot is the host of at least three species of ticks, *Haemophysalis humerosa*, *I. tasmani* and the paralysis tick *I. holocyclus*. Bandicoots are the principal hosts of *I. holocyclus* and are generally thought to be immune to its toxin. However, it is suspected that heavy tick infestations on juvenile northern brown bandicoots may have adverse effects on animal health (Gemmell 1991) and occasionally result in mortalities (Stone 1992). Roberts (1970) is a useful reference for other tick species found in bandicoots and bilbies.

Other ectoparasites whose hosts include bandicoot species include the mites *Haemolaelaps marsupialis* and *Odontacarus* spp. (Jackson 2003). The mite *Ornithonyssus bacoti* has been seen on bilbies at one captive breeding institution (Buchecker & Kleinig 2002). A captive eastern barred bandicoot exhibiting alopecia of the rump area and hyperkeratotic skin over the dorsum was infested with a mite from the genus *Petauralges* (MZ 2006 case number MZ910320). The condition resolved with systemic and local anti-acaracidal treatments. A wild southern brown bandicoot with generalised alopecia and thickened and cracked exudative skin lesions was found to have a very heavy burden of a *Sarcoptes*-like mite (HS 2006 case number 92/0593). Mites of a *Demodex* spp. have been detected in clinically normal eastern barred bandicoots (Norman 1991).

Ectoparasite burdens in bandicoots have been successfully treated using products developed for use in companion animals. Fipronil 2.5 g/L sprayed lightly over entire coat and imidacloprid 100 mg/mL deposited as a 10 mg/kg dose percutaneously have both been used to treat flea burdens in bandicoot species. Ivermectin (10 mg/mL) at 200 ug/kg SC can be used for heavy tick or mite burdens. When animals are carrying light tick burdens, the parasites can simply be removed by hand.

b Endoparasites

Helminths

Spratt et al. (1991) is a useful reference detailing the helminth parasites that have been found in bandicoots and bilbies. The lungworms *Marsupostrongylus bronchialis* and *Filostrongylus peramelis* are not uncommonly recovered from bandicoots including the northern brown, long-nosed and eastern barred bandicoots. These parasites are not normally associated with significant pathology (Beveridge 1978) but unidentified lungworms were the cause of multiple focal abscessation in northern brown bandicoots (Norman 1991).

Physaloptera peramelis, a spiruroid parasite found in the stomach of most bandicoot species, causes ulcerative lesions when approximately 15+ worms are present (Lenghaus et al. 1990; Norman 1991). The parasites attach to the gastric wall by their mouthparts, causing severe granulomatous inflammation of the submucosa. The life-cycle of this parasite is probably indirect, with arthropod or insect intermediate hosts. Treatment using fenbendazole at 10–30 mg/kg sid PO for 3 d has been successful in eliminating the parasite.

A *Gnathostoma* spp. burden in a wild northern brown bandicoot was found to be clinically significant as the parasites were found deeply embedded in the stomach wall (Norman 1991). Their presence had resulted in severe granulomatous disease of the submucosa.

Several *Capillaria* spp. have been found in the epithelia of the lips, tongue and oesophagus, mucosa of the stomach and small intestine of eastern barred bandicoots (Lenghaus et al. 1990) and long-nosed bandicoots (ARWH 2007 case number 1852/1). Although often found in large numbers, and eggs are readily detected on faecal examination, these parasites are usually considered non-pathogenic. Attempts at treating *Capillaria*-positive animals with ivermectin or fenbendazole have not been successful. In one captive colony of eastern barred bandicoots, *Capillaria* spp. and *Parastrongyloides australis* nematodes were considered the cause of diarrhoea and enteritis in several animals (Lenghaus et al. 1990). In histological section, large numbers of nematodes were seen burrowing through the intestinal mucosa and were associated with an inflammatory reaction.

Migrating nematodes suspected to be *Strongyloides* spp. or *Parastrogyloides* spp. were the cause of significant enteritis in two juvenile eastern barred bandicoots. One was a free-ranging animal (Lenghaus et al. 1990) and the other was a captive individual (MZ 2006 case number MZG/75). In both animals nematodes were embedded deeply in the mucosa; in the captive animal they had caused granulomas in the submucosa with some parasites present in larger blood vessels. The animal also had multifocal hepatic abscesses that cultured a profuse growth of *Plesiomonas shigelloides*.

Other nematode parasites of the bandicoot gastrointestinal tract that occasionally cause pathology include *Peramelostrongylus* spp. (Obendorf & Munday 1990). A heavy intestinal burden of this parasite was found to be the cause of death of a free-ranging juvenile eastern barred bandicoot (MZ 2006 case number GHPG/47). *Peramelostrongylus skedastos* is found in the stomach of the eastern barred bandicoot and in one captive animal it was felt that parasitic granulomas in the lung parenchyma were due to migrating larvae of this nematode (MZ 2006 case number SERW185/78). Pulmonary parasitic granulomas have been associated with *Mackerrastrongylus* spp. infection of the small intestine (MZ 2006 case number SERW978/77). Visceral larval migrans associated with unidentified nematodes have been noted as incidental findings in a number of wild long-nosed bandicoots (HS 2006). Heavy burdens of *Echinonema* spp. have been found in the small intestine of wild northern brown bandicoots and southern brown bandicoots (Norman 1991).

Moniliformes semoni is a large acanthocephalan (up to 10 cm) that anchors to the mucosa of the stomach and small intestine with a spiny proboscis (Beveridge 1978). Heavy burdens of this parasite are likely to be pathogenic.

The nematode parasite *Angiostrongylus cantonensis* has been observed in a captive bilby (ARWH 2007, case number AWHR 4686/1). The animal presented with neurological dysfunction and the parasite was found in the brain, associated with severe haemorrhage and inflammation.

Larval stages of the reptile nematode *Ophidascaris robertsi* present in earthworms and other invertebrates can develop in the subcutaneous tissues, body cavities and liver of mammalian hosts that include the northern brown and long-nosed bandicoot. Because of their large size (up to 10 cm) they may be pathogenic (Beveridge 1993).

The filarioid worm *Cercopithifilaria pearsoni* is found in the SC connective tissues of the northern brown bandicoot and *Cercopithifilaria johnstoni* has been found in eastern barred, long-nosed, southern brown and northern brown bandicoots (Spratt 1990). Transmission is mediated by ixodid ticks and although the microfiliaria typically accumulate in the small lymphatic vessels of the skin, small numbers can be observed in peripheral blood films (Clark et al. 2004).

Bandicoots and many other Australian mammals can serve as intermediate hosts for the cestode *Spirometra erinacei* although the masses of plerocercoids found in body cavities do not seem to significantly affect animal health. The cestode *Potorolepsis peramelidarum* was an incidental finding in a wild long-nosed bandicoot killed by predators (ARWH 2007 case number 1537/1).

Protozoa

General

Parasitism of the bandicoot intestine and caecum by coccidial protozoa of the genus *Eimeria* is very common, but is rarely associated with significant pathology and therefore positive animals are not usually treated. However, in the occasional bandicoot, infection with coccidia is associated with mucosal inflammation and diarrhoea. This seems to be most common in young animals or those under presumed physiological stress (J Samuel *pers. comm.*). Treatment in eastern barred bandicoots has been attempted with toltrazuril at 10–20 mg/kg PO but has been unsuccessful. Treatment with a trimethoprim/sulpadiazine combination (at a dose rate of 5 mg/kg of the trimethoprim component PO or SC q 24 hr for 5–7 d) has also been attempted, largely unsuccessfully (L Vogelnest *pers. comm.*).

Disseminated protozoal infections with a coccidian-like organism have been described as the cause of death in a captive adult southern brown bandicoot (ARWH 2007 case number 4118/1) and two eastern barred bandicoots that had been recently weaned (MZ 2006 case numbers MZP/40, MZA30171). Vasculitis confined to the intestinal wall was a pathological feature in the southern brown bandicoot and in one of the eastern barred bandicoots. The latter animal was a recently weaned captive individual who died acutely. On gross necropsy, severe haemorrhagic enteritis was observed in the anterior small intestine. The second eastern barred bandicoot (MZP/40) was a recently weaned free-ranging animal. Histological examination showed proliferative enteritis associated with coccidian organisms. In addition, the liver and spleen contained focal areas of necrosis associated with coccidian-like intracellular protozoal schizonts containing zoites. Protozoal schizonts containing zoites were also observed in the liver and lymph nodes of the southern brown bandicoot.

The protozoan *Cryptosporidium muris* has been detected in the faeces of captive bilbies but was not associated with clinical disease (Warren et al. 2003). It is believed the animals acquired the infection from food or water contaminated by mice, and that stress associated with overcrowding may have predisposed some animals to infection.

Infection of renal tissue with *Klossiella* spp. is a common finding in bandicoots and, to a lesser extent, bilbies (Lenghaus et al. 1990; ARWH 2007; M Bennett *pers. comm.*). It is usually detected at necropsy as an incidental finding. Protozoal parasites from the genus *Sarcocystis* spp. have been seen in eastern barred bandicoots, western barred bandicoots, long-nosed bandicoots and bilbies (Obendorf 1990; ARWH 2007). They are often associated with focal inflammatory lesions in skeletal and cardiac myocytes; in one bilby protozoal zoites were also associated with necrotic foci in the adrenal gland, liver and pancreas (ARWH 2007 case number 3317/1).

Intracellular haematozoa that have been found in bandicoot species include *Babesia* spp., *Theleria* spp. and *Hepatozoon* spp. (Clark et al. 2004). Natural infections with these protozoal parasites have not yet been associated with pathologic effects, although it is expected that they might become significant in an immunosuppressed host. Trypanosome flagellates have been reported from a wide range of Australian mammals including bandicoots. They can be detected by routine examination of blood smears and occupy an extracellular position. Forms resembling the trypanosome *Trypanosoma thylacis* described in the northern brown bandicoot (Mackerras 1959) were also observed in Tasmanian eastern barred bandicoots (Bettiol et al. 1998). These forms appeared distinct from those seen in Tasmanian southern brown bandicoots, which were smaller in size, and may represent a distinct species (Bettiol et al. 1998). No disease related to infection with trypanosomes has been reported in bandicoots.

Bandicoots are susceptible to experimental infection with *Giardia duodenalis* and infection has been found in the scats of free-ranging eastern barred bandicoots and southern brown bandicoots in Tasmania (Bettiol et al. 1997). The significance of infection in wild populations and the incidence of disease from infection are unknown. A heavy burden of *Giardia* spp. was thought to be significant in the death of three 70 d old captive northern brown bandicoots (ARWH 2007 case number 87/426). It was speculated that rodents introduced the parasite to the enclosure and that nutritional deficiencies in the recently weaned animals predisposed them to infection.

Toxoplasmosis

Toxoplasma gondii is an obligate intracellular coccidian parasite that can infect a wide range of mammals and birds. Domestic cats and certain other feline species are the only known definitive hosts for *T. gondii* and shed oocysts in their faeces. While the life-cycle of the parasite can be completed if a cat consumes sporulated oocysts, they are more commonly infected by ingesting intermediate hosts that harbour tachyzoites (rapidly multiplying forms of *T. gondii*) or bradyzoites (tissue cyst forms) (Frenkel 1990).

Bandicoots and bilbies become infected with *T. gondii* by ingesting oocysts in soil or paratenic hosts such as earthworms (Bettiol et al. 2000). After infection the oocyst ruptures in the intestine and releases sporozoites that multiply locally to form tachyzoites. Tachyzoites are dispersed to the rest of the body via blood and lymph eventually encysting in brain, skeletal and cardiac muscle and liver. These cysts, containing the slowly multiplying bradyzoites, are microscopic and survive in tissues throughout the life of the host. Infection is usually subclinical in intermediate hosts that evolved in the presence of felids, but in immunologically impaired individuals or susceptible species the tachyzoite form can cause significant disease and death. Like many Australian marsupials, bandicoots and bilbies often develop clinical disease when infected with *T. gondii*. Their susceptibility is commonly attributed to their lack of past exposure to the parasite, as cats were absent from Australia until European settlement (Dubey & Beattie 1988).

Clinical signs of toxoplasmosis observed in bandicoots prior to their death include uncharacteristic docility, apparent blindness, inco-ordination and loss of ability to seek refuge (Obendorf & Munday 1990). Acute toxoplasmosis in bandicoots and bilbies is characterised by widespread focal necrosis in tissues including the lung, gut mucosa, brain, myocardium, skeletal muscle, adrenal gland, pancreas and kidney (Canfield et al. 1990; Lenghaus et al. 1990). No evidence of *T. gondii* has been found in bandicoot and bilby PY, suggesting that *in utero* infections are uncommon or do not occur (Lenghaus et al. 1990; Miller et al. 2000).

The recrudescence of latent infection is possible in all species and sometimes accompanies known or presumed immunosuppression (Frenkel 1990). In the author's experience with captive eastern barred bandicoots, encephalitis producing unusual behaviours (e.g. activity in daylight hours and docility) is the predominant manifestation of presumed recrudescence of latent infection. At necropsy, marked inflammatory lesions in the brain and the presence of many tissue cysts suggest a diagnosis of recrudescent infection.

Diagnosis of toxoplasmosis in bandicoots and bilbies is aided by measuring antibodies to *T. gondii* using direct (DAT) and modified agglutination tests (MAT). The DAT detects both M and G class immunoglobulins (IgG and IgM antibodies); the MAT detects only IgG antibodies. Experimental studies on macropods have suggested that the marsupial response to infection with *T. gondii* is the formation of IgM antibodies 7–10 d after infection, characterised by a rising DAT titre (Johnson et al. 1989; Lynch et al. 1993). Thirty days after infection, surviving macropods had increasing MAT titres indicative of an IgG-related seroconversion (Johnson et al. 1989). Macropods previously exposed to *T. gondii* had DAT and MAT titres greater than 1:64 dilutions (Lynch et al. 1993).

The impact of toxoplasmosis on bandicoot and bilby populations is measured by the mortality rate from acute infection and the incidence of recrudescence of latent infection. Populations in close association with high densities of domestic cats (e.g. urban fringes, intensive agricultural areas) are likely to be at greater risk of *T. gondii* infection than populations where cat density is relatively lower. Obendorf et al. (1996) collected sera from 150 eastern barred bandicoots in southern Tasmania and tested for agglutinating antibodies to *T. gondii* using the DAT and MAT. Seventeen (11.3%) were classified as either positive or suspicious reactors. The frequency with which these animals were retrapped (29%) over the 32 mo study was much lower than seronegative animals (68%), suggesting that although some bandicoots survive initial infection they do not act as a long-standing carrier for *T. gondii*. Many of the 134 Victorian eastern barred bandicoots examined at necropsy by Lenghaus et al. (1990) had tissue cysts containing bradyzoites, suggesting a significant number of animals survive initial infection. However, potentially fatal generalised infections were present in 10% of animals, indicating that toxoplasmosis has the potential to be a significant disease for this species and that recrudescence of latent infection may be a major manifestation of disease. The role of nutritional and weather stressors in contributing to recrudescence is unknown.

Treatment of animals showing clinical signs of toxoplasmosis is usually unrewarding and the prognosis for survival is poor. The most common treatment regimen is a combination of a sulphonamide and trimethoprim given at 25 mg/kg PO sid. This is given in conjunction with pyrimethamine at 0.5 mg/kg PO sid and folic acid 1 mg/kg PO sid.

9.8 Non-infectious diseases

9.8.1 Neoplastic diseases

As in many other marsupial species, benign and malignant neoplasms are not uncommon in bandicoots and bilbies, particularly in aged animals. Records held at the Australian Registry of Wildlife Health at Taronga Zoo, Perth Zoo, Melbourne Zoo and Healesville Sanctuary reveal 38 cases of neoplastic disease in bandicoots and bilbies (see Table 13.7).

Table 13.7 Neoplastic disease observed in bandicoots and bilbies

Species	Case number	Diagnosis
Bilby	ARWH 642/1	Mesenchymal neoplasia associated with scapula. Cellular origin unknown
	ARWH 2582/1	Pulmonary fibrosarcoma
	ARWH 3112/1	Basal cell carcinoma
	ARWH 4770/1	Haemangiosarcoma
	ARWH 740/94	Pancreatic haemangioma
	ARWH 1580/94	Osteosarcoma
Long-nosed bandicoot	ARWH 2138/2	Gastric carcinoma
	ARWH 2792/1	Squamous cell carcinoma of larynx
	ARWH 90/961	Mammary cystadenoma and hepatoma
Southern brown bandicoot	ARWH 79/682	Lymphosarcoma
	ARWH 89/4922	Renal fibrosarcoma
	ARWH 85/1487	Subcutaneous fibrosarcoma
	ARWH 83/226	Pulmonary reticulum cell sarcoma
	PZ A01234	Invasive oral tumour involving mandible. Tumour type not identified
Northern brown bandicoot	ARWH 95/1158	Mammary adenocarcinoma
	HS HS890423	Disseminated metastatic adenocarcinoma of unknown origin
Golden bandicoot	ARWH 1921/1	Neoplasia associated with cloaca. Tumour type not identified
	PZ 910234	Splenic tumour. Tumour type not identified
	PZ 950238	Hepatic adenocarcinoma
Western barred bandicoot	ARWH 2430/1[a]	Focal carcinoma of pouch
Eastern barred bandicoot	ARWH 2965/2	Metatarsal sarcoma
	ARWH 2401/96	Mammary adenocarcinoma
	ARWH 94/39974	Subcutaneous fibroma
	MZ HSG/13	Metastatic leiomyosarcoma of skin
	MZ HS7/8	Fibrohistiosarcoma on nasal planum
	MZ MZ910320	Lipoleioma of subcutaneous tissue around hock
	MZ MZ5/96	Mammary cystadenocarcinoma
	MZ MZ920647	Leiomyoma of colonic wall
	MZ HS950618	Disseminated haemangiosarcoma
	MZ HS931010	Laryngeal basal cell carcinoma
	MZ MZ980643	Bile duct carcinoma
	MZ MZ970784	Hepatic carcinoma
	MZ HS7/53	Adrenocortical adenocarcinoma
	MZ HS930498	Pedal fibrosarcoma
	MZ MZ940714	Thyroid adenocarcinoma
	MZ MZA10951	Disseminated apocrine adenocarcinoma
	MZ HS911193	Subcut malignant spindle cell tumour
	MZ HS921164	Facial histiocytoma

a This individual also had multiple papillomatous proliferations predominantly located at the muco-cutaneous junctions. This syndrome is discussed in section 9.1

9.8.2 Trauma

Road trauma is a commonly reported cause of death in bandicoots, especially in populations on the fringes of urban areas. In a study of the causes of death of 134 wild eastern barred bandicoots over a 7 yr period, 55 animals died from road trauma (Lenghaus et al. 1990). It is difficult to assess the relative importance of road trauma as a cause of mortality because road kills are often conspicuous and therefore likely to be overrepresented in sampling.

Fox, dog and cat predation is well-recognised as a contributing factor in the historical and ongoing decline of many bandicoot and bilby populations. Bandicoots are also preyed upon by owls and occasionally by diurnal raptors. Often, animals killed by predators are not consumed or are consumed only partially; necropsy to establish predator type can be useful for population management programs.

Booth and McCracken (1994) reported the necropsy findings that distinguish fox, dog and cat attacks on eastern barred bandicoot. The victims of fox predation are often found with intact skin, but on its removal a typical feature is massive thoracic trauma. Usually there is extensive SC and intra-thoracic haemorrhage and occasionally punctures in the soft tissue are paired, representing the canine teeth. A domestic or feral dog of similar size to a fox may cause the same pattern of damage. Foxes and dogs often preferentially eat the hindquarters and abdominal contents of the bandicoot.

Eastern barred bandicoots observed to be killed by cats usually lack the degree of thoracic trauma caused by canids but have sharp puncture marks through the skin as well as the soft tissue. However, in long-nosed bandicoots predated by cats the skin is usually extensively torn (L Vogelnest *pers. comm.*) and this may reflect a difference in fragility of skin. Bandicoots suspected as being killed by raptors usually do not have significant SC trauma and the pattern of claws on the dorsum may be quite clear (Booth 1994). Many raptors are scavengers and feed on bandicoots that died from other causes.

Traumatic injuries in captive bandicoots can result from incompatible pairings or delayed removal of independent young. The most common signs of an incompatible pairing are weight loss and patchy alopecia over the rump area. Young should be removed from their parents immediately after weaning because aggression from females towards offspring after this time is common, and injuries can be quite severe. Female bandicoots may occasionally cannibalise their PY in stressful environments (e.g. overcrowding, frequent disturbance).

Toe injuries and avulsion of toenails that result in bone exposure are seen in bandicoots held in enclosures where walls are not smooth to a height of 1.5 m. They can easily catch toes and nails up to that height by jumping against the wall. Another consequence of this behaviour is splitting of the tissues at the ventral tail base and, on occasion, severe nasal trauma (R Woods *pers. comm.*). If untreated, these injuries often become infected and can result in loss of the tail. Toe and nasal planum trauma is occasionally seen in bandicoots that have been trapped and become distressed. Animals may attempt to dig out of traps resulting in avulsion of multiple nails, or strike themselves against the cage wire resulting in depressed fractures of the nasal planum. These injuries can be largely prevented by provision of nesting material in the trap, wrapping a waterproof cover around its outside and ensuring animals are not held in the trap for prolonged periods.

9.8.3 Degenerative diseases

Like most captive mammals, the longevity of bandicoots and bilbies in captivity frequently exceeds that of their free-ranging counterparts. A very frequent finding in aged animals is arthritis of joints of the hind limb, particularly the stifle and tarsal joints. Intervertebral disc protrusion at L1–L2 necessitated the euthanasia of a golden bandicoot (P Eden *pers. comm.*). Bilateral cataracts are a common finding in aged captive bandicoots and bilbies.

9.8.4 Toxic diseases

A potential nephrotoxin in eastern barred bandicoots has been described in section 9.5. Free-ranging bilbies and bandicoots can be exposed to toxins introduced by human activities, including agricultural chemicals and poison baits intended for vertebrate pests.

Persistent organochlorines have been banned from use in Australia for almost 20 yr (longer in some states) and, although these compounds can persist in soil for variable periods, their current impact is negligible. However, it is of interest to consider the potential

Table 13.8 Susceptibility of bandicoots and bilbies to sodium fluoroacetate (1080) poisoning

Species	LD50 (mg/kg)	Adult weight (kg)	Adult LD50 (mg)
Southern brown bandicoot			
Western Australia	20.0	0.4–1.6	8.0–32.0
South Australia	8.0	0.4–0.8	3.2–6.4
Northern brown bandicoot	3.5	0.5–3.1	1.8–10.9
Eastern barred bandicoot	5.4	0.45–0.9	2.7–5.9
Golden bandicoot	8.94	0.26–0.67	2.3–6.0
Bilby	15	0.8–2.5	12.0–37.5
Fox	0.13	4.7–8.3	0.6–1.1
European rabbit	0.37	0.96–2.42	0.4–0.9

Source: Adapted from McIlroy (1983); Twigg et al. (1990); Booth & McCracken (1994).

historical impact of these compounds on bandicoot and bilby health. Five of six free-ranging eastern barred bandicoots whose tissues were analysed for dieldrin tested positive, with values of 1.29–6.52 ppm (Lenghaus et al. 1990). The animals were collected from the Hamilton area in western Victoria in the 1980s, where bandicoot habitat overlaps with farmland. The significance of this toxin in the animals was unknown.

Sodium fluoroacetate (1080) is a poison used widely in Australia to control populations of rabbits, foxes and dogs. Concerns about the impact of this poison on non-target species led to investigations into the sensitivity of native wildlife to 1080 (McIlroy 1983; Twigg et al. 1990). Table 13.8 shows the marked variation of bandicoot and bilby susceptibility to poisoning with 1080, depending on species and geographic location.

The variation in susceptibility to 1080 poisoning between Western and South Australian populations of southern brown bandicoots was attributed to previous exposure to toxic plant products (McIlroy 1983). Fluoroacetate occurs naturally in the genera *Oxylobium*, *Gastrolobium* and *Acacia* and most of the species known to produce this compound are confined to the lower south-west of Western Australia. Bandicoots and bilbies also appear to have a high innate tolerance to fluoroacetate compared to target species, but the amount of 1080 in poison baits places the smaller bandicoot species at risk. Baited carrots or oats for rabbit control usually contain 0.175–0.2 mg 1080/g bait. In theory, 15 g of bait can poison small bandicoots.

Intensive monitoring of an eastern barred bandicoot population during 1080 baiting for rabbits showed that two animals died from suspected 1080 poisoning. One animal presented with neurological signs consistent with fluoroacetate intoxication and subsequently died. The second animal was found dead and at necropsy was found to have stomach contents containing carrot, the vehicle of bait delivery. The diagnoses were not confirmed by toxicological analysis for 1080.

FOXOFF baits (Animal Control Technologies) contain 3 mg of 1080, sufficient to kill a number of the bandicoot species. Clinical trials using non-baited FOXOFF showed that it is within the behavioural repertoire of captive eastern barred bandicoots to dig up and consume baits (Booth & McCracken 1994). While FOXOFF baits may cause bandicoot deaths, there have been no studies into the frequency of their consumption by free-ranging animals. FOXOFF bait sizes range between 30 g and 60 g, reducing the likelihood that bandicoots of a susceptible size would consume the whole bait. In addition, the 1080 component is located in the centre of the bait, further reducing the chance of non-target species poisoning.

10 THERAPEUTICS AND SURGERY

No pharmacokinetic studies have been carried out on the use of therapeutic agents in bandicoots and bilbies, therefore treatment regimens are based on those used in other mammals and anecdotal evidence of efficacy. Table 13.9 provides a basic formulary for drugs that have been used in bandicoots and bilbies.

Most bandicoot species and bilbies can be handled twice daily for a period of 5–7 d without significant ill effect, provided they are held in enclosures from which they can easily be captured. Food intake and weight should be monitored daily while under intensive treatment regimens, and treatment suspended if the

Table 13.9 Drug formulary for use in bandicoot and bilby species

Agent	Dosage	Comments
Antibiotics		
Amoxycillin	20 mg/kg IM/SC q 24 hr or PO q 12 hr	Amoxycillin/clavulanic acid usually chosen in preference
Amoxycillin/clavulanic acid	12.5 mg/kg SC q 24 hr or PO q 12 hr	Commonly used as first choice antibiotic for minor wounds
Ceftazidime	15 mg/kg IM q 12 hr	Has been used for complicated toe infections in bandicoots
Cephalexin	20–30 mg/kg PO q 12 hr	Likely to be useful
Clindamycin	11 mg/kg PO q 12 hr	Has been used for periodontal disease and toe infections
Doxycycline	5 mg/kg PO loading dose 2.5 mg/kg PO q 12 hr for 2 doses then q 24 hr	Likely to be useful
Enrofloxacin	5 mg/kg PO or IM q 12 hr	Likely to be useful
Metronidazole	50 mg/kg PO q 24 hr	Likely to be useful
Trimethoprim/ sulphamethoxazole	15 mg/kg PO q 12 hr	Likely to be useful
Antiparasitics		
Fenbendazole	10–30 mg/kg PO q 24 hr for 3 d. Repeat in 14 d	Has been used to successfully treat *Physaloptera peramelis*
Fipronil	2.5 g/L preparation. Spray to coat animal	Has been used for treatment of flea and louse burdens in bandicoots
Imidacloprid	100 mg/mL preparation. 10 mg/kg percutaneous	Has been used for treatment of flea burdens in bandicoots
Ivermectin	200 ug/kg PO or SC. Repeat in 14 d	Used for treatment of heavy tick and mite burdens. Slight burdens of ticks can be individually removed
Moxidectin	200–400 ug/kg PO or SC. Repeat in 14 d	May be useful for treating *Physaloptera peramelis* and *Capillaria* spp. infections
Pyrimethamine	2 mg/kg PO q 24 h + sulfadiazine 20 mg/kg PO q 8 hr	Used for treatment of toxoplasmosis. Need to supplement with folinic acid 1 mg/kg PO q 24hr during treatment
Toltrazuril	2.5 mg/kg PO as single dose	Used for treatment of coccidial infections. Limited success
Trimethoprim/ sulphonamide	5 mg/kg of trimethoprim component SC or PO q 24 hr	Used for treatment of coccidial infections. Local ulceration at injection site seen in an eastern barred bandicoot
Analgesics		
Buprenorphine	0.01 mg/kg SC q 12 hr	Has been used intra- and post-operatively
Carprofen	2 mg/kg SC/PO q 12 hr	Injectable form has been used post-operatively. Oral form has been used as a 5 d course in arthritic animals
Meloxicam	0.3 mg/kg PO day 1 then 0.1 mg/kg PO q 24 hr	Routine treatment duration is 5 d

stress of treatment is perceived to be detrimental to animal health. Oral medications are most easily administered by restraining the animal in a calico bag and introducing a syringe into the side of the mouth. Palatable medications can be hidden in favourite food items. Standard small mammal sites for SC and IM injections are applicable to bandicoots and bilbies, however, access for IV injections or catheterisation can be difficult.

Surgical approaches applicable to other small mammal species are suitable for use in bandicoots and bilbies. As with other female marsupials, the pouch must be reflected for a midline approach when conducting abdominal surgery.

ACKNOWLEDGMENTS

The author very much appreciated the contributions to this chapter made by Mark Bennett, Rosie Booth, June Butcher, Benn Bryant, Paul Eden, Mandy O'Hara, Helen McCracken, Randy Rose, Larry Vogelnest, Rupert Woods and Ian Smith. The author would also like to sincerely thank Jane Hall of Taronga Zoo for her time spent retrieving data from the Australian Registry of Wildlife Health. Thanks also to Kate Bodley and Kris Warren for their reviews of the manuscript.

REFERENCES

ARWH (2007). *Australian Registry of Wildlife Health.* Zoological Parks Board of NSW: Sydney.

Ashton N (1979). Ocular toxoplasmosis in wallabies (*Macropus rufogriseus*). *American Journal of Ophthalmology* **88**, 322–332.

Attwood HD, Woolley PA & Rickard MD (1975). Toxoplasmosis in dasyurid marsupials. *Journal of Wildlife Diseases* **11**, 543–551.

Bach C (ed.) (1998). *Birth Date Determination in Australasian Marsupials* (1st edn amended). Australasian Regional Association of Zoological Parks and Aquaria: Sydney.

Bernard JB & Allen ME (1997). Feeding captive insectivorous animals: nutritional aspects of insects as food. Fact Sheet 003. American Zoological Association Nutrition Advisory Group: Silver Spring, Maryland.

Bettiol SS, Kettlewell JS, Davies NJ & Goldsmid JM (1997). Giardiasis in native marsupials of Tasmania. *Journal of Wildlife Diseases* **33**, 352–354.

Bettiol SS, Jakes K, Le DD, Goldsmid JM & Hocking G (1998). First record of trypanosomes in Tasmanian bandicoots. *Journal of Parasitology* **84**, 538–541.

Bettiol SS, Obendorf DL, Nowarkowski M, Milstein T & Goldsmid JM (2000). Earthworms as paratenic hosts of toxoplasmosis in eastern barred bandicoots in Tasmania. *Journal of Wildlife Diseases* **36**, 145–148.

Beveridge I (1978). Helminth parasites of Australian marsupials. In *Fauna, Proceedings 36*, pp. 273–285. Post Graduate Committee in Veterinary Science: Sydney.

Beveridge I (1993). Marsupial parasitic diseases. In *Zoo and Wild Animal Medicine: Current Therapy 3* (ed. ME Fowler), pp. 288–293. WB Saunders: Colorado.

Blyde D (1993). Staphylococcal infections in the greater bilby (*Macrotis lagotis*). In *Proceedings of the Wildlife Disease Association (Australasian Section)*, pp. 25–27. North Stradbroke Island, Qld.

Bodetti TJ, Viggers K, Warren K, Swan R, Conaghty S, Sims C & Timms P (2003). Wide range of *Chlamydiales* types detected in native Australian mammals. *Veterinary Microbiology* **96**, 177–187.

Booth RJ (1994). Medicine and husbandry: monotremes, wombats and bandicoots. In *Wildlife, Proceedings 233*, pp. 395–420. Post Graduate Committee in Veterinary Science: Sydney

Booth R & McCracken H (1994). Morbidity and mortality report for eastern barred bandicoots, *Perameles gunii*. Veterinary Department, Melbourne Zoo: Melbourne.

Bradshaw SD, Morris KD, Dickman CR, Withers PC & Murphy D (1994). Field metabolism and turnover in the golden bandicoot (*Isoodon auratus*) and other small mammals from Barrow Island, Western Australia. *Australian Journal of Zoology* **42**, 29–41.

Buchecker J & Kleinig S (2002). Husbandry manual for the greater bilby *Macrotis lagotis sagitta* (draft). Monarto Zoological Park: South Australia.

Canfield PJ, Hartley WJ & Dubey JP (1990). Lesions of toxoplasmosis in Australian marsupials. *Journal of Comparative Pathology* **103**, 159–166.

Claridge AW, McNee A, Tanton MT & Davey SM (1991). Ecology of bandicoots in undisturbed forest adjacent to recently felled logging coupes: a case study from the Eden woodchip agreement area. In *Conservation of Australia's Forest Fauna* (ed. D Lunney), pp. 331–345. Royal Zoological Society of NSW: Sydney.

Clark P (2004). *Haematology of Australian Mammals*. CSIRO Publishing: Melbourne.

Clark P, Adlard RD & Spratt DM (2004). Haemoparasites of Australian mammals. In *Haematology of Australian Mammals* (ed. P Clark), pp. 147–162. CSIRO Publishing: Melbourne.

Dierenfield ES & Barker D (1995). Nutrient composition of whole prey commonly fed to reptiles and amphibians. In *Proceedings of the Association of Reptilian and Amphibian Veterinarians*, pp. 3–15. Sacramento, California.

Dubey JP & Beattie CP (1988). *Toxoplasmosis in Animals and Man*. CRC Press: Boca Raton, Florida.

Dufty AC (1991). Some population characteristics of *Perameles gunii* in Victoria. *Wildlife Research* **18**, 355–366.

Eamens GJ, Turner MJ & Catt RE (1988). Serotypes of *Erysipelothrix rhusiopathiae* in Australian pigs, small ruminants, poultry, and captive wild birds and animals. *Australian Veterinary Journal* **65**, 249–252.

Frenkel JK (1990). Transmission of toxoplasmosis and the role of immunity in limiting transmission and illness. *Journal of American Veterinary Medical Association* **196**, 233–247.

Friend T (2002). Workshop on disease in western barred bandicoot populations, 16 July 2002. Department of Conservation and Land Management: Perth, WA.

Gemmell RT (1986). Sexual maturity in the female bandicoot *Isoodon macrourus* (Gould) in captivity. *Australian Journal of Zoology* **34**, 199–204.

Gemmell RT (1990). The initiation of the breeding season of the northern brown bandicoot *Isoodon macrourus* in captivity. In *Bandicoots and Bilbies* (eds JH Seebeck, PR Brown, RL Wallis & CM Kemper), pp. 205–212. Surrey Beatty & Sons: Sydney.

Gemmell RT, Cepon G, Green PE & Stewart NP (1991). Some effects of tick infestations on juvenile northern brown bandicoots (*Isoodon macrourus*). *Journal of Wildlife Diseases* **27**, 269–275.

Gibson LA (2001). Seasonal changes in the diet, food availability and food preference of the greater bilby (*Macrotis lagotis*) in south-western Queensland. *Wildlife Research* **28**, 121–134.

Gibson LA & Hume ID (2000). Seasonal field energetics and water influx rates of the greater bilby (*Macrotis lagotis*). *Australian Journal of Zoology* **48**, 225–239.

Gorden G & Hall LS (1995). Tail fat storage in arid zone bandicoots. *Australian Mammology* **18**, 87–90.

Greenberger NJ & Isselbacher KJ (1998). Diseases of the gallbladder and bile ducts. In *Harrison's Principles of Internal Medicine* (14th edn) (ed. AS Fauci), pp. 1725–1736. McGraw-Hill: New York.

Hall LS (1990). Growth and a description of the development of external features of pouch young of captive *Isoodon macrourus*. In *Bandicoots and Bilbies* (eds JH Seebeck, PR Brown, RL Wallis & CM Kemper), pp. 123–133. Surrey Beatty & Sons: Sydney.

Heinsohn GE (1966). Ecology and reproduction of the Tasmanian bandicoots (*Perameles gunii* and *Isoodon obesulus*). In *University of California Publications in Zoology* Vol. 80 (eds GA Bartholomew, JH Connell, J Davis, CR Goldman, C Hand, PR Marler & KS Norris), pp. 1–106. University of California Press: Berkeley/Los Angeles, California.

Hill JP (1899). Contributions to the morphology and development of the female urogenital organs in the Marsupialia 1. On the female urogenital organs in the *Perameles*, with an account of the phenomena of parturition. *Proceedings of Linnean Society of New South Wales* **24**, 42–82.

Hume ID (1999). *Marsupial Nutrition*. CUP: Cambridge.

Ikonomopoulou MP, Smolenski AP & Rose RW (2005). Changes in the milk composition during lactation in the eastern barred bandicoot (*Perameles gunii*) (Marsupialia: Peramelidae). *Australian Journal of Zoology* **53**, 59–65.

Jackson SM (2003). Bandicoots. In *Australian Mammals: Biology and Captive Management* (ed. SM Jackson), pp. 127–144. CSIRO Publishing: Melbourne.

Jean-Jacques W, Rajashekaraiah KR, Farmer JJ, Hickman FW, Morris JG & Kallick CA (1981). *Vibrio metschnikovii* bacteremia in a patient with cholecystitis. *Journal of Clinical Microbiology* **14**, 711–712.

Johnson AW, Roberts H, Statham P & Munday BL (1989). Serodiagnosis of acute toxoplasmosis in macropods. *Veterinary Parasitology* **34**, 25–33.

Johnson CN & Johnson KA (1983). Behaviour of the bilby, *Macrotis lagotis* (Reid), (Marsupialia: Thylacomyidae) in captivity. *Australian Wildlife Research* **10**, 77–87.

Kemper C, Kitchener DJ, Humphreys WF, How RA, Schmitt LH & Bradley AJ (1990). The biology of the northern brown bandicoot, *Isoodon macrourus* (Marsupialia: Peramelidae) at Mitchell Plateau, Western Australia. *Australian Journal of Zoology* **37**, 627–644.

Lenghaus C, Obendorf DL & Wright FH (1990). Veterinary aspects of *Perameles gunii* biology with special reference to species conservation. In *Management and Conservation of Small Populations* (eds TW Clark & JH Seebeck), pp. 89–108. Chicago Zoological Society: Chicago.

Lobert B & Lee AK (1990). Reproduction and life history of *Isoodon obesulus* in Victorian heathland. In *Bandicoots and Bilbies* (eds JH Seebeck, PR Brown, RL Wallis & CM Kemper), pp. 311–318. Surrey Beatty & Sons: Sydney.

Lynch M, Obendorf DL, Statham P & Reddacliff G (1993). An evaluation of a live *Toxoplasma gondii* vaccine in the tammar wallaby (*Macropus eugenii*). *Australian Veterinary Journal* **70**, 352–353.

Lyne AG (1964). Observations on the breeding and growth of the marsupial *Perameles nasuta* Geoffroy with notes on other bandicoots. *Australian Journal of Zoology* **22**, 303–309.

Mackerras MJ (1959). The haematozoa of Australian mammals. *Journal of Zoology* **7**, 105–135.

McCracken HE (1983). Reproduction in the female bilby, Macrotis lagotis. BSc(Vet) thesis. University of Sydney, Australia.

McCracken HE (1990). Reproduction in the greater bilby, *Macrotis lagotis* (Reid)—a comparison with other perameloids. In *Bandicoots and Bilbies* (eds JH Seebeck, PR Brown, RL Wallis & CM Kemper), pp. 199–204. Surrey Beatty & Sons: Sydney.

McIlroy JC (1983). The sensitivity of Australian animals to 1080 poison VI. Bandicoots. *Australian Wildlife Research* **10**, 507–512.

Merchant JC (1990). Aspects of lactation in the northern brown bandicoot *Isoodon macrourus*. In *Bandicoots and Bilbies* (eds JH Seebeck, PR Brown, RL Wallis & CM Kemper), pp. 219–228. Surrey Beatty & Sons: Sydney.

Miller DS, Mitchell GF, Biggs B, McCracken HE, Myroniuk P & M Hewish (2000). Agglutinating antibodies to *Toxoplasma gondii* in sera from captive eastern barred bandicoots in Australia. *Journal of Wildlife Diseases* **36**, 213–218.

Munday BL (1978). Marsupial disease. In *Fauna*, *Proceedings 36*, pp. 335–385. Post Graduate Committee in Veterinary Science: Sydney.

Nagy KA, Bradshaw SD & Clay BT (1991). Field metabolic-rate, water flux, and food requirements of short-nosed bandicoots, *Isoodon obesulus* (Marsupialia: Peramelidae). *Australian Journal of Zoology* **39**, 299–305.

Norman RJ (1991). *Physaloptera peramelis* infection in the eastern barred bandicoot (*Perameles gunii*): taxonomy and pathogenicity. Diploma of Wildlife Medicine and Husbandry thesis. University of Sydney, Australia.

Obendorf DL & Munday BL (1990). Toxoplasmosis in wild eastern barred bandicoots, *Perameles gunii*. In *Bandicoots and Bilbies* (eds JH Seebeck, PR Brown, RL Wallis & CM Kemper), pp. 193–197. Surrey Beatty & Sons: Sydney.

Obendorf DL, Statham P & Driessen M (1996). Detection of agglutinating antibodies to *Toxoplasma gondii* in sera from free-ranging eastern barred bandicoots (*Perameles gunii*). *Journal of Wildlife Diseases* **32**, 623–626.

O'Hara AJ, Warren KS, Swan RA, Sims C & Friend JA (2004). Cutaneous papillomatosis and carcinomatosis in the highly endangered western barred bandicoot (*Perameles bougainville*). *Veterinary Dermatology* **15** (Suppl. 1), 25.

Parsons FG (1903). On the anatomy of the pig-footed bandicoot (*Choeropus castanotis*). *Journal of Linnean Society of London, Zoology* **29**, 64–80.

Reimer AB & Hindell MA (1996). Variation in body condition and diet of the eastern barred bandicoot (*Perameles gunii*) during the breeding season. *Australian Mammalogy* **19**, 47–52.

Renfree MB (1982). Implantation and placentation. In *Reproduction in Mammals* (2nd edn). *Book 2: Embryonic and Fetal Development* (eds CR Austin & RV Short), pp. 26–69. CUP: Cambridge.

Roberts FHS (1970). *Australian Ticks*. CSIRO Publishing: Melbourne.

Seebeck JH, Brown PR, Wallis RL & Kemper CM (eds) (1990). *Bandicoots and Bilbies*. Surrey Beatty & Sons: Sydney

Southgate R, Palmer C, Adams M, Masters P, Triggs B & Woinarski J (1996). Population and habitat characteristics of the golden bandicoot (*Isoodon auratus*) on Marchinbar Island, Northern Territory. *Wildlife Research* **23**, 647–664.

Southgate RI, Christie P & Bellchambers K (2000). Breeding biology of captive, reintroduced and wild greater bilbies, *Macrotis lagotis* (Marsupialia: Peramelidae). *Wildlife Research* **27**, 621–628.

Spratt DM (1990). Monotremes, marsupials, murids and monkey (?). filaria. In *Proceedings of Wildlife Disease Association (Australasian Section)*, pp. 1–2. Lake Eucumbene, NSW.

Spratt DM, Beveridge I & Walter EL (1991). *A Catalogue of Australasian Monotremes and Marsupials and their Recorded Helminth Parasites*. Monograph Series Number 1, Records of the South Australian Museum. South Australian Museum: Adelaide.

Stoddart MD (1980). Observations on the structure and function of cephalic skin glands in bandicoots (Marsupialia: Peramelidae). *Australian Journal of Zoology* **28**, 33–41.

Stone BF (1992). Ticks and wildlife: interactions, particularly of the Australian paralysis tick *Ixodes holocyclus* with koalas, bandicoots and fruit bats. In *Proceedings of Wildlife Disease Association (Australasian Section)*, p. 17. Warrumbungle National Park, NSW.

Tedman RA (1990). Some observations on the visceral anatomy of the bandicoot *Isoodon macrourus* (Marsupialia: Peramelidae). In *Bandicoots and Bilbies* (eds JH Seebeck, PR Brown, RL Wallis & CM Kemper), pp. 107–116. Surrey Beatty & Sons: Sydney.

Twigg LE, King DR & Mead RJ (1990). Tolerance to fluoroacetate of populations of *Isoodon* spp. and *Macrotis lagotis* and its implications for fauna management. In *Bandicoots and Bilbies* (eds JH Seebeck, PR Brown, RL Wallis & CM Kemper), pp. 185–192. Surrey Beatty & Sons: Sydney.

Tyndale-Biscoe H (2005). *Life of Marsupials*. CSIRO Publishing: Melbourne.

Vogelnest L (1999). Chemical restraint of Australian native fauna. In *Wildlife in Australia, Proceedings 337*, pp. 149–187. Post Graduate Foundation in Veterinary Science: Sydney.

Warren KS, Swan RA, Morgan-Ryan UM, Friend JA & Elliot A (2003). *Cryptosporidium muris* infection in bilbies (*Macrotis lagotis*). *Australian Veterinary Journal* **81**, 739–741.

Warren K, Swan R, Bodetti T, Friend T, Hill S & Timms P (2005). Ocular *Chlamydiales* infections of western barred bandicoots (*Perameles bougainville*) in Western Australia. *Journal of Zoo and Wildlife Medicine* **36**, 100–102.

Wicks RM & Clark P (2005). Clinical haematology of the southern brown bandicoot (*Isoodon obesulus*). *Comparative Clinical Pathology* **14**, 56–60.

Wilson GS (1984). The diseases and parasite status of the bilby (*Macrotis lagotis*) and rufous (western) hare wallaby (*Lagorchestes hirsutus*) in captivity. Diploma of Wildlife Health and Medicine thesis. University of Sydney, Australia.

Withers PC (1992). Metabolism, water-balance and temperature regulation in the golden bandicoot (*Isoodon auratus*). *Australian Journal of Zoology* **40**, 523–531.

14. Bats

Annabelle Olsson and Rupert Woods

1 TAXONOMY, DISTRIBUTION AND BIOLOGY

The Order Chiroptera includes 202 genera and over 1116 species (Wilson & Reeder 2005). Seventy-nine species are recognised in Australia. The Family Pteropodidae (fruit-bats or pteropodids) includes 13 of these species, which also occur in New Guinea or surrounding islands. This Family is distinguished from others by external appearance and dietary preferences and includes the blossom bats, tube-nosed bats, fruit-bats and flying-foxes. In this chapter, unless otherwise specified, we use the term 'pteropodids' to refer to these species. The insectivorous and carnivorous bats (Families Megadermatidae—ghost bats, Rhinolophidae—horseshoe-bats, Hipposideridae—leaf-nosed bats, Emballonuridae—sheath-tailed bats, Molossidae—free-tailed bats and Vespertilionidae—ordinary bats) include 66 species of which 27 also occur in New Guinea or surrounding islands. Identification is made through their significant morphological (Table 14.1) and dietary diversity (Jackson 2003; Strahan 1995; Wilson & Reeder 2005).

The majority of Australian bat species probably entered from New Guinea. The high species diversity on Cape York is a reflection of this migration, its close geographical proximity and diversity of habitats. Bat diversity diminishes with increasing latitude with only six species found in Tasmania (Strahan 1995; Hall & Richards 2000).

Both the spectacled (*Pteropus conspicillatus*) and grey-headed (*P. poliocephalus*) flying-foxes are listed as vulnerable, as are several species of insectivorous and carnivorous bats. Six other insectivorous and carnivorous bat species are thought to be extinct, critically endangered or endangered (Environment Australia 1999).

Over much of their range, pteropodids are now affected by increasing urbanisation and exotic fruit cultivation which can lead to fragmentation of traditional roosting sites. Pteropodids normally move from one part of their roost site to another over the seasons, preventing significant damage to the roost trees, but the restricted size of the urban camps means there is now a greater impact on the trees and many are significantly damaged or killed before the colony moves on. However, in spite of these limitations, pteropodids often choose urban campsites where they are sheltered by surrounding buildings, protected from shooting and have ready access to flowering garden trees and cultivated fruit trees. Together with the increased media focus on the potential zoonotic disease risks of pteropodids, increased pressure is being placed on government agencies to relocate or destroy populations in urban areas.

2 ANATOMY AND PHYSIOLOGY

The anatomy and physiology of bats is similar to most other eutherian mammals. However, there are a few notable exceptions, primarily in relation to their capacity for flight.

Table 14.1 Key to the Families of Australian bats using external features

Key		Description	Identification
1		Tail absent	2
		Tail present	3
2	i	Small to large bats with dog-like facial appearance, first and second finger with terminal claw (exception is *Dobsonia magna*), ear forms complete ring, no tragus, can have tubular nostril. *Nyctimene* spp. have a tail.	Pteropodidae
	ii	Pale coloured bat with large ears joined over head, claw on thumb only, prominent tragus	Megadermatidae
3	i	Tail protrudes from membrane	4
	ii	Tail fully enclosed within membrane	5
4	i	Tail protrudes beyond end of membrane, large wrinkled lips, hairy toes, musky smell, complex thick ears	Molossidae
	ii	Tail protrudes dorsally half-way along tail membrane, prominent eyes, sometimes throat pouch	Emballonuridae
5	i	Complex noseleaf on face with a horseshoe cutaneous plate surrounding the nostrils, broad ears, no tragus, short tail	Rhinolophidae
	ii	Elaborate noseleaf, two phalanges in each toe instead of two in first toe and three in others	Hipposideridae
	iii	Simple face, sometimes rudimentary noseleaf, ears with prominent tragus, long tail	Vespertilionidae

External anatomical features used as part of this key are presented in Figures 14.1 and 14.2. The tragus is a prominent flap of skin at the entrance to the ear.
Source: After Hall (1988).

2.1 Wing

Chiropterans are distinguished from other mammalian Orders by the characteristic wing membrane (patagium) of the fore limbs and their ability to achieve sustained flight. Most insectivorous and carnivorous bats, but no pteropodids, have a uropatagium or membrane that extends between the hind legs and partially or totally covers the tail (Jackson 2003).

The wing is essentially the mammalian forearm highly modified for flight. Major skeletal adaptations for flight include a keeled sternum and lightweight but solid bones. The radius is the main forearm bone to which the ulna is fused over most of its length. A thumb (digit one) is present in all species but is larger in pteropodids. This digit controls the shape of the propatagium (the part of the wing membrane attached anterio-laterally to the arm bones between the shoulder and the carpus) during flight. For pteropodids, digit one is also an essential tool for manipulating branches and food items, hanging for defecation and urination, and for fighting. The remaining digits (two to five) are elongated rigid spokes supporting the wing membranes. Each digit is made up of a metacarpal and phalanges one and two (Hill & Smith 1986; Hall 7 Richards 2000; Heard 2003). With the exception of the bare-backed fruit-bat (*Dobsonia magna*), a distinguishing feature of pteropodids is the presence of a nail at the end of digit two (Fig. 14.1).

All bats have three flight membranes that form the main body of the wing. The membranes are essentially modified skin stretched around the framework of bones. They are the propatagium, the plagiopatagium and the dactylopatagium (Fig. 14.1). In most insectivorous and carnivorous bats the tail is connected to the legs by a fourth broad membrane (uropatagium), which is used during abrupt changes in flight direction (Hall & Richards 2000). The uropatagium is represented in pteropodids by a small flap of skin down the inside of the legs. The membranes are very elastic and have numerous parallel bands of elastin running through the plagiopatagium perpendicular to the humerus and radius (Fig. 14.1), making the wing pliable and resilient.

Although the primary function of wing membranes is flight, they also serve a number of other purposes. The membranes are well-supplied with blood vessels that dilate to promote heat loss in hot and humid environments. In pteropodids, heat loss is further assisted by wing fanning or allowing the wings to hang in a breeze. Additionally, they urinate on the wing membranes to promote evaporative cooling. In cold, windy or wet weather pteropodids wrap their wings tightly around their bodies to reduce heat loss and keep warm and dry.

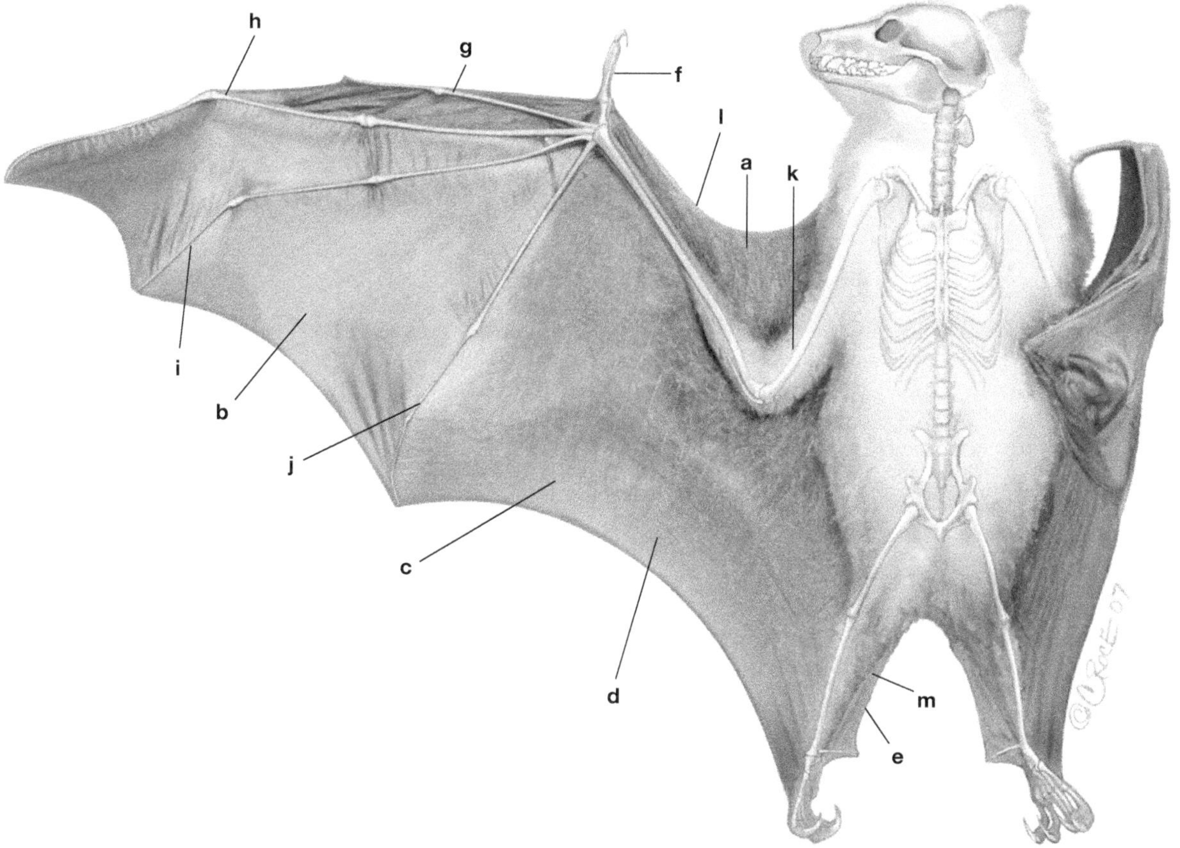

Figure 14.1 Ventral view of a grey-headed flying-fox showing key anatomical features. a) Propatagium. b) Dactylopatagium. c) Bands of elastin. d) Plagiopatagium. e) Uropatagium. f) Digit one. g) Digit two. h) Digit three. i) Digit four. j) Digit five. k) Median vein. l) Wing or cephalic vein. m) Uropatagial vein.

Tube-nose bats (*Nyctimene* spp.) have tan wings dappled with unpigmented patches that provide camouflage when wrapped around their bodies at rest.

Bats use their wing membranes during birth to assist the pup from the vulva to the nipple. Wing membranes also serve as a cradle when the pup is suckling or sleeping and insectivorous bats use their wings to help capture prey before transferring it to their mouth (Hill & Smith 1985).

Transillumination of the wing is used to age insectivorous and carnivorous bats by visual identification of the translucent epiphyseal cartilages of juveniles (Parnaby 1992). Fusion takes place over several months. Pre-flight bats have three bands of cartilage visible; bats flying for 1 mo have two visible bands; those flying 2–3 mo have one band; those flying for more than 3 mo have no visible bands and epiphyses are fused (approximately 5 mo of age). In insectivorous and carnivorous bats changes in appearance and pelage colour or texture are unreliable guides to distinguishing juveniles from adults.

2.2 Pelvic girdle and hind limbs

In insectivorous and carnivorous bats the sacrum and ilium are fused to the level of the acetabulum, whereas in pteropodids they are entirely fused (Heard 2003). In all species the hind limbs are rotated at the hip so the legs can bend anteriorly towards the ventrum. The patella is absent in all species.

2.3 Locking mechanism

Since bats perch inverted with their entire body weight suspended from the toes, their legs are adapted for pulling rather than pushing. Most bats have a locking mechanism, consisting of a small tendon in each toe that keeps the claws in a flexed position without muscular contraction. This allows them to sleep while hanging. To take off they must flap to attain a horizontal position to reduce pressure on the hind limbs and allow the hind limb claws to unlock.

The claws are compressed into blade-like hooks to assist with hanging as well as grooming (Hall & Richards 2000).

2.4 Gastrointestinal tract

Pteropodid dentition reflects a diet of fruit, nectar and leaves, with large canine teeth and flattened molars that have a smooth hollow trough and high cusps. The cheek pouches can expand to hold pieces of fruit temporarily during feeding. Food is repeatedly chewed then pressed between the tongue and well-defined palatine ridges to extract the juice. Most fibre is eventually spat out after the juice is extracted (leading to the common misconceptions that bats vomit fibre or defecate from their mouths). The gastrointestinal tract has evolved into a shortened simplified structure, designed predominantly for digesting fruit juice, nectar and pollen (Nelson 1989). Blossom bats have a brush-tipped tongue for collecting nectar and pollen from flowers. Insectivorous bats possess molars with sharply pointed cusps and clearly defined ridges. They also have a muscular callus-like papilla on the gingiva at the level of the lower premolars for breaking off and ejecting chitin pieces. In most species the upper third of the oesophagus is lined with keratinised epithelium for protection from the sharp chitin pieces (Heard 2003).

The gastrointestinal transit time is very short relative to other mammals. Insectivorous bats generally have longer transit times (35 min in fully active animals to 170 min in animals at rest) than *Pteropus* spp. (>12–34 min for cultivated fruits) (Jackson 2003). These anatomical and physiological modifications reduce the weight of gut contents while flying.

Neonates have fine recurved milk teeth (teeth that point backwards) that assist in holding onto the nipple, especially during flight. These are replaced by permanent dentition about the time pups begin to fly.

2.5 Sensory system

Pteropodids have large forward-facing eyes well-adapted for nocturnal vision, while still retaining a high degree of visual acuity in broad daylight. The retina is avascular; the central artery is absent and nourishment presumably occurs by diffusion from choroidal capillaries. As would be expected of nocturnal mammals the retina consists almost entirely of rod cells. In many pteropodids colour vision and a *tapetum lucidum* that causes the eyes to shine bright red in a spotlight are present (Hall & Richards 2000). In contrast, insectivorous and carnivorous bats have small eyes and may have poor vision, but compensate for this by their ability to image their environment acoustically via echolocation (Hall & Richards 2000). The imaging system is made up of a transmitter (larynx) and a receiver (ears and associated neural systems) (Heard 2003).

Figure 14.2 External features of the head of an insectivorous bat (eastern pipistrelle) showing the prominent tragus at the entrance of the ear.

Insectivorous and carnivorous bats have very acute hearing and most species have a well-developed tragus (a prominent flap of skin at the entrance to the ear) to help catch sound outside the normal mammalian acoustic range (Fig. 14.2). The nose may be relatively unmodified to complex with dermal appendages (e.g. nose-leaf). These structures are thought to be involved in catching and redirecting sound for navigation and hunting by echolocation.

All pteropodids have simple and relatively small external ears with no tragus. The inner ear of pteropodids shows no special modifications. Internally, the cochlea is relatively small (Hall & Richards 2000).

2.6 Integument

The majority of neonatal insectivorous and carnivorous bats are born without visible fur on their bodies. This gradually appears everywhere except the wings and the ventrum, which remains bare until 2–3 wk of age (Bogan 1972).

Most adult bats have furred bodies and unfurred wings. A healthy coat in an adult bat is thick and glossy. Most male pteropodids have modified sebaceous glands around the neck ruff. These exude sebum with a distinctive straw colour (pink in the spectacled flying-fox) and fruity odour. This is very pronounced during the breeding season and in the presence of other adult males. In females a white discharge may be noted from perivulvar glands during the periparturient period (D Heard *pers. comm.*).

Wing membranes are normally dry, supple and smooth, with numerous parallel bands of elastin visible in the patagium. The wing pits (axillae) of pteropodids may feel greasy and smell strongly of bat musk.

2.7 Cardiovascular system

The cardiovascular system is typically mammalian but because bats fly the heart is larger, primarily in the right atrial and ventricular muscle mass (Heard 2003).

2.8 Physiology

Body weight varies from 3–10 g for small insectivorous bats (e.g. free-tailed bats, *Mormopterus* spp.) to >1 kg (e.g. grey-headed, spectacled and black flying-foxes (*P. alecto*)). Monitoring body weight in hospital patients is critical in assessing adequate food intake and hydration status.

Some bats use daily torpor and hibernation to minimise their metabolic rate. Bats enter torpor by allowing their body temperature to drop almost as low as the air temperature. This lowers their respiratory and heart rate and conserves energy. Hibernation is similar to torpor but is generally much more prolonged (up to several months); torpor is usually limited to several hours or days. Hibernation allows many species of bats to survive winters on small fat stores. Disturbing bats during hibernation wastes these limited resources and may mean that they do not survive (Churchill 1998).

Pteropodid and some insectivorous and carnivorous bats' core body temperatures are 37–39°C. However, smaller pteropodids and insectivorous and carnivorous bats that undergo torpor may exhibit extreme body temperature variability. Generally, rectal temperature is unreliable as a measurement of core body temperature, especially in smaller species.

Heart rates vary between species, time of day and level of activity. Larger species have rates of 100–400 bpm while smaller species can have rates of 250–450 bpm, and up to 1000 bpm.

3 REPRODUCTION

Although there are many features common to all mammalian species the chiropterans have the greatest diversity of reproductive strategies of any mammalian Order (Crichton & Krutzsch 2000; Hall & Richards 2000).

Although most bats are polygynous, some will pair during the mating season (Nelson 1965). Breeding cycles usually commence in response to changes in natural photoperiod. There is considerable variation in the number and timing of oestrous cycles. All temperate species are monoestrus and all tropical species are polyoestrus. Some species exhibit sperm storage in females over winter. Others use delayed blastocyst implantation or embryonic diapause as a reproductive strategy for controlling the timing of births (Heard 2003). Gestation periods are long relative to those of other mammals (Table 14.2): 40–60 d in small vespertilionid or 'evening' bats (Family Vespertilionidae) up to 90–180 d in pteropodids (Heard 2003; Jackson 2003).

Table 14.2 Gestation periods and annual breeding cycles of some Australian bat species

Species	Sexual maturity (mo)	Mating months	Birthing months	Gestation (approx. d)
Vespertilionid bats (Nth America)				40–60
Free-tailed bats (*Mormopterus* spp.)	>12	Aug–Nov	Oct–Jan	40–60
Black flying-fox	18–24	Feb–Apr	Sept–Nov	
Grey-headed flying-fox	18–2	Feb–Apr	Sept–Nov	150
Little red flying-fox	18–24	Nov–Dec	May–June	90–150
Spectacled flying-fox	18–24	Feb–Apr	Sept–Dec	150
Bare-backed fruit-bat		May–June	Sept–Nov	
Eastern blossom bat			North: year-round South: Oct/Nov & Feb/Apr	
Northern blossom bat			Oct/Nov, just prior to wet season	

Source: Heard (2003).

Australian pteropodids are born in October–December, except for little red flying-foxes (*P. scapulatus*) which are born around April. Pteropodid offspring are generally weaned by about 5–6 mo (at approximately 60–70% of adult weight), whereas insectivorous bat species are generally weaned within 2 mo (usually around 80% of adult body weight).

4 HUSBANDRY

4.1 Housing

Housing requirements for chiropterans are well-described (George 1990; Barnard 1995; Fascione 1995; Lollar & Schmidt-French 1998; Jackson 2003).

Pteropodids are preferably housed outdoors in snake-proof enclosures with access to direct sunlight, natural air flow and rain. Galvanised wire may cause rubbing injuries to the flexor surfaces of joints and can be a source of zinc if animals lick either the wire or rainwater off it, which may result in toxicity. Vinyl-coated wire is preferable; polyethylene trawler netting hung from the ceiling and sides of the enclosure is a suitable alternative. Bats occasionally become tangled in mesh regardless of mesh size but heavier-gauge mesh carries a lower risk of cutting injuries if entanglement occurs. Ropes, vines and netting are appropriate perches. At least some netting or rope should extend to the ground to give grounded pteropodids a means to climb back up into the canopy. It is recommended that enclosures be double-meshed to reduce contact with free-ranging bats to minimise the risk of transmission of diseases such as Australian bat lyssavirus (ABLV).

Flight enclosures for pteropodids should be at least eight times the animal's wing span long (Fascione 1995) and 1.8 m high. Landing and hanging material such as branches, ropes or towels should be placed at either end away from the walls to reduce injury from attempts to land on wire. Towels hung vertically from the roof also provide privacy and hanging and landing opportunities. Provided the towels are not flat against a solid wall, impact injuries rarely occur.

Insectivorous and carnivorous bats are housed indoors or outdoors but, like pteropodids, have better breeding success with natural light and temperature cycles. However, microclimate control is easier with indoor enclosures. This is particularly important for tropical species housed in temperate climates. All enclosures must be snake-proof. Housing requirements differ between species but all need a hide or shelter and a white or blue light source for attracting live insects. The required area of a flight aviary depends on the species to be housed. It should be greater than 3 × 4 m for highly manoeuvrable bats such as horseshoe, (*Rhinolophus* spp.), leaf-nosed (*Hipposideros* spp.) and forest bats (*Vespadelus* spp.) and greater than 7.5 × 7.5 m for fast flying, less manoeuverable species such as free-tailed and sheath-tailed bats (*Saccolaimus* spp. and *Taphozous* spp.) (Lollar & Schmidt-French 1998; Jackson 2003). The social behaviour of any species should be taken into account before setting up an exhibit or rehabilitation enclosure to avoid overcrowding and interspecies competition (Barnard 1995; Jackson 2003; SM Barnard unpub).

4.2 Hospitalisation

A standard dog or cat hospital cage is suitable for initial assessment and treatment of pteropodids and large carnivorous bats (e.g. ghost bats, *Macroderma gigas*). The cage must have a secure perch placed near the top and a cover to limit visual stimuli. A wire-top cat cage is adequate for temporary holding of smaller pteropodid species (e.g. little red and spectacled flying-foxes) and tube-nosed bats. A towel hung vertically in the cage provides shelter and, in smaller species, perching. The cage must be tall enough (greater than 2 × body length) for an adult bat to hang without its ears touching the floor and wide enough for the bat to extend and flap both wings. Avoid cages with vertical bars that may entangle a wing.

Insectivorous and carnivorous bats and small pteropodids (e.g. eastern blossom bats, *Syconycteris australis*) can be hung on a towel in a securely closed cardboard or wooden box or a plastic tank with a mesh lid.

The ideal substrate for hospital cages is newspaper, changed regularly. Boxes in which animals can hide (hide boxes) can be made from a variety of cardboard boxes. Hotboxes or incubators maintained at 22–25°C are ideal for most hospitalised species.

4.3 Feeding and water

Diets and detailed information on feeding hospitalised and debilitated bats is given in section 5. Food is presented in shallow trays to self-feeding insectivorous and carnivorous bats. Pteropodids eat from small hanging buckets or spill-proof dishes on the cage floor (if the cage is small). Nectivorous bats are fed from a shallow dish or (preferably) a feeder bottle. To avoid bacterial and fungal colonisation, feeder bottles should be thoroughly cleaned and sterilised daily.

Water should be provided for all species. Pteropodids drink from a hanging dish or bowl. Small insectivorous bats are offered water from a shallow dish or bowl with marbles or pebbles in it to avoid bats drowning. Spectacled and black flying-foxes are known to drink sea water (A Olsson unpub; D Heard *pers. comm.*) and should be provided with a salt lick *ad lib*.

4.4 Transport

Pteropodids and ghost bats can be transported in a small pet carrier made from plastic and/or wire. It is advisable to line the carrier with fine mesh or shadecloth to prevent accidental injury to animal and people. These species require a roosting perch or trawler netting secured near the ceiling of the box from which to hang. Insectivorous and carnivorous bats <100 g body weight are safely transported on a cloth in a small plastic container, or secured inside a cloth bag.

Bats which may be injured or ill are best wrapped securely in a towel or cloth which is then placed in the transport cage. This minimises the risk of further injury to the animal, and protects the rescuer from bites and scratches. Care must be taken in hot or humid conditions that animals do not overheat.

4.5 Individual marking and identification

When a number of bats are kept together it is important to have a reliable means of identifying individuals. Hand-reared bats can be identified temporarily by painting different combinations of claws with brightly coloured nail varnish. The bat should be restrained while the varnish dries. Varnish is rarely groomed off and lasts several weeks to months. More permanent methods include passive integrated transponder (PIT) tags (or microchips), tattoos of the patagium or ears, ear notching, forearm and thumb bands, and necklaces (Jackson 2003). All banding in Australia is regulated by the Australian Bird and Bat Banding Scheme (ABBBS) and all banders are licensed under the ABBBS. PIT tags are inserted SC on the dorsal midline between the scapulae. The needle should be directed toward the head so that gravity pulls the chip into the channel created by the needle. The needle hole should be sealed with tissue glue to prevent the PIT tag falling out soon after implantation (L Vogelnest *pers. comm.*).

5 NUTRITION

Chiroptera have the greatest diversity of diet and feeding habits of any mammalian order (Heard 2003). Bats consume large quantities of food relative to their body size but process it rapidly through their modified gastrointestinal tract. They are usually fed at dusk as they are most active at night.

5.1 Wild diets

Pteropodids consume a very wide variety of native plant species including fruits, blossoms and leaves. Preference is generally given to highly aromatic and light-coloured fruits and blossoms highly visible in moonlight (e.g. *Melaleuca* blossoms) and fruits or blossoms on the periphery of the forest canopy (Hall & Richards 2000). The grey-headed flying-fox is considered a generalist feeder, while the spectacled flying-fox is a specialist frugivore. The tube-nosed bats are fig specialists. The volume of fruit ingested in all species is determined by its protein content (Hall & Richards 2000).

Nectar is the principal food source for the blossom bats (*Synconycteris* and *Macroglossus* spp.). While collecting nectar they ingest significant amounts of pollen, which is also an important source of protein.

Folivory is widespread in pteropodids (Hall & Richards 2000; A Olsson *pers. obs.*; P Tully *pers. comm.*; J Maclean *pers. comm.*). It is thought this provides protein and trace minerals. Fig and lilly pilly leaves (any species) are readily accepted and usually readily available in urban areas.

The insectivorous and carnivorous bats are primarily insectivorous. Their diet is high in chitinous roughage from wing cases of mature insects. The exceptions are the ghost bat, which feeds predominantly on rodents, birds, frogs and other bats, and the southern myotis (*Myotis macropus*) which feeds on fish (Jackson 2003).

5.2 Captive diets

Table 14.3 gives suitable captive diets for bats (Jackson 2003).

5.2.1 Frugivorous and nectivorous bats

Flying-foxes (*Pteropus* spp.) and the bare-backed fruit-bat usually readily accept commercial fruits. While larger species can hold and chew large pieces, dicing the fruit reduces wastage and is easier for smaller bats and recovering patients to manage. Most fruits are low in calcium, except figs, which are calcium-rich. Canned fruits are suitable and are convenient to store for long periods. Fruit purees can be offered as emergency nutrition but the high fibre load impairs absorption of nutrients if fed regularly (George 1990) and gastrointestinal tract problems such as rectal prolapse can result (see 9.2.3c). Fruit and nectar eaters are fed commercial lorikeet nectar mixes or Wombaroo high-protein supplement (Wombaroo Food Products) blended with fruit juice or blossom bat mix (Table 14.3). This diet can be fed alone or with Wombaroo Lorikeet and Honeyeater mix, or Wombaroo high-protein mix.

5.2.2 Insectivorous bats

Insectivores are initially fed whole mealworms or their viscera (squeezed directly into the mouth of a bat restrained between thumb and forefinger). These are usually readily accepted and high in calories. They eat 25–50% of their body weight each night (Hopkins 1990). This is equal to 6–10 mealworms for a small bat and up to 30 for a larger bat. However, for long-term maintenance mealworms alone are not adequate as they are high in fat and low in calcium, phosphorus and vitamins A, E, D_3 and B complex. Mealworms should be fed Wombaroo Insectivore, Wombaroo Small Carnivore powder or similar for 48 hr prior to feeding out so their gut is filled with nutrients (Hopkins 1990; Jackson 2003). Adult insects (moths, cockroaches, grasshoppers, crickets) should be offered *ad lib* (Table 14.3).

5.2.3 Carnivorous bats

Carnivorous bats are fed a diet that approximates the wild diet (e.g. mice, baby rats, day-old chickens, larger insects, fish). The items can be fed whole or chopped to allow easy consumption and digestion. Carnivorous bats eat 10–20% of their body weight in food per night

Table 14.3 Captive diets for Australian bat species

Species	Diet	Quantity per bat per day
Flying-foxes	Chopped fruit, e.g. banana, apple, pear, pawpaw, mango, figs, any soft fruit in season, tinned fruit	350 g
	Wombaroo[1] high-protein supplement	7–10 g
	or Complan[2]	5 g
	Fresh native flowers, e.g. bottlebrush, eucalypts	Supplement
	Fresh mulberry leaves	Supplement
Blossom bats	Blossom bat mix:	10 mL mixed with 10 mL of water
	2 bananas	
	500 mL apple juice	
	150 g raw sugar	
	150 g Glucodin (generic)	
	120 g S26 Soy[3]	
	Blend bananas and apple juice to total 500 mL.	
	Blend in sugar and Glucodin.	
	Add S26 Soy and blend for 1.5 min.	
	Pour into smaller containers and freeze (icecube trays for small numbers of bats).	
	Thaw for use as required.	
	Fresh native blossoms, e.g. banksias, bottlebrush, *Melaleuca* spp.	Supplement
Ghost bat	Day-old chick	1
	Adult mouse	1
	Mealworms	*Ad lib* twice daily
	Crickets, grasshoppers, cockroaches, moths, other large insects	Supplement
Insectivores	Mealworms	*Ad lib*
	Flies, crickets, cockroaches, other flying insects	*Ad lib*
	Wombaroo[1] Insectivore or Small Carnivore mixed with very lean mince	3:1 (finely minced heart)

Manufacturers: **1** Wombaroo Food Products. **2** Complan Foods. **3** Wyeth Nutritionals.
Source: Adapted from Jackson (2003).

(Hopkins 1990) which equates to one day-old chick, one mouse and 5–10 mealworms per animal per day.

5.3 Feeding sick animals

Sick or injured animals that are not eating require assisted feeding or gavage. Assisted feeding is maintained until voluntary food intake is adequate for maintenance and healing. Patients should be weighed daily to monitor progress.

In an inappetant bat, injectable fluids containing 5% dextrose and vitamin supplements are sufficient to maintain energy levels for short periods. However, prolonged use of simple carbohydrate therapy will induce osmotic diarrhoea and does not correct protein or fat depletion. If treatment is to be prolonged, early enteral nutritional support decreases recovery time, weight loss and morbidity.

High-energy supplements such as Poly-Aid Plus (Vetafarm) (1 g/10 g body weight q 12 hr) or any pureed or liquid diet appropriate to the species, made up to a runny paste, may be used to provide readily absorbable energy, protein and fats to the debilitated patient and are easily administered via gavage if required. Probiotics at domestic animal dose rates appear to be a useful adjunct in debilitated animals, particularly following antimicrobial therapy, to assist restoration of normal gut flora.

Debilitated or inappetant bats may be assist fed using a syringe held to the mouth of the hanging or recumbent bat, allowing the animal to lick the liquid food off the tip. If the animal does not take the food it may need to be fed by gavage. A 14–18 G stainless steel bird crop feeding needle, fine nasogastric human infant feeding tube (size 3–4) or IV cannula (14–18 G) can be used. If mouth or head injuries are present, a pharyngostomy or nasogastric tube may need to be placed. Gavage is performed with the bat in the head up position (gravity assists food retention in the stomach) in the same manner as for any small mammal. A very rough guide for gavage volumes in debilitated patients is 2–3 mL/100 g body weight, given every 3–4 hr.

The following liquid diets are ideal for initially feeding debilitated patients.

- Pteropodids—apple or pear juice blended with a high-protein supplement or Poly-Aid Plus.
- Insectivorous and carnivorous bats—a blend of mealworm viscera and Poly-Aid Plus initially, gradually replaced by Wombaroo Insectivore or Small Carnivore mix.

The liquid diet is gradually replaced by a maintenance diet appropriate to the species as the patient recovers. Maintenance diets may be pureed (with added liquid if required) for ease of swallowing if patients are reluctant or unable to chew solid food.

6 RESTRAINT

All bats are capable of inflicting bites and scratches and may harbour zoonotic pathogens. While physical restraint may be adequate for simple procedures, chemical restraint allows a safer and more thorough assessment with less stress to the animal.

Veterinarians should ensure that only people vaccinated against rabies handle bats, that the vaccinations are up-to-date with appropriate levels of seroconversion and that carers have been properly trained in handling bats (see 9.8.1a). A major contact with an ABLV bat includes not only a bite but also saliva contamination of broken skin and mucus membranes. The use of personal protective clothing is recommended during handling. This may include long pants, a long-sleeved shirt (preferably of medium-weight cloth), covered footwear, gloves, a surgical or other industrial safety mask and safety glasses (Fig. 14.3). Safe physical restraint also depends on common sense and experience—larger bats are capable of inflicting a bite even through protective clothing if given the opportunity.

6.1 Physical restraint

Most insectivorous and carnivorous bats are small and adequately restrained by grasping over the dorsoventral thorax between finger and thumb. Ghost bats and diadem leaf-nosed bats (*Hipposideros diadema*) are an exception due to their larger size and long sharp canines. They are handled as for a pteropodid.

Figure 14.3 Appropriate personal protective clothing for handling bats (little red flying-fox). Photo: Gary Tabor.

Pteropodids are restrained by wrapping them with a thick towel while they hang. Once the towel is wrapped around the body to contain the wings and envelop the head, one hand is used to grasp the feet. The other hand grips the head through the towel under the mandible or with the middle finger and thumb on either temporomandibular joint. The index finger is placed over the top of the cranium. Pteropodids accustomed to handling can be minimally restrained by the hind feet, and loosely wrapped in a towel (Fig. 14.4). To release, place the wrapped bat in a cage then gently withdraw the towel as the animal orients itself.

Some species (e.g. tube-nosed bats) are solitary and easily stressed (A Olsson *pers. obs.*; J Maclean *pers. comm.*; P Tully *pers. comm.*). Handling time should be minimised to prevent stress. Stressed bats can be placed in a dark quiet box to recover. Administration of glucose syrup (5 mL to 200 mL water) PO given at 1–2 mL/100 g body weight or 5% dextrose saline given at 5–10 mL/100 g SC at the time of handling appears to aid recovery.

Figure 14.4 Physical restraint of an adult spectacled flying-fox accustomed to handling. Photo: Andrew Breed.

6.2 Chemical restraint

Fasting is unnecessary prior to chemical restraint except immediately after feeding or drinking. This is because rapid gastric emptying and short gastrointestinal transit times, as well as a tight oesophageal sphincter (Heard 2003), make it very unlikely a bat will regurgitate while under anaesthesia. Fasting is also contraindicated in debilitated bats because of their high metabolic rates and greater likelihood of developing hypoglycaemia.

Once chemical restraint is achieved bats should be placed in ventral or lateral recumbency with the head slightly lower than the feet to avoid postural hypotension and accumulation of pulmonary secretions.

The most common IM sites for drug administration are the dorsal neck muscles (Booth 1994) and the body of the scapular and triceps muscles. Avoid the pectoral muscles in insectivorous and carnivorous bats as they largely control wing movement and flight. Poor injection technique causes muscle trauma and scar tissue formation which reduces manoeuvrability in flight. Additionally, bats do not possess a broad sternum and the needle may penetrate between the ribs and enter the thorax if the animal struggles during injection (D Heard *pers. comm.*).

Heart and respiratory rates vary with changes in anaesthetic depth. Additional responses to monitor include the palpebral reflex (disappears at light to medium depth), pupillary light response, foot grip (the locking mechanism that allows the bat to hang by a foot is lost under deep anaesthesia) and corneal reflex. Loss of the last two responses is indicative of deep anaesthesia.

Due to a high surface area to body weight ratio, hypothermia is common in anaesthetised bats and can occur rapidly. All anaesthetised bats should therefore be provided with supplementary heat. However, the patagium is very sensitive to thermal injury. Electric heating pads, heat lamps and heated water bottles and rice bags must never be used. Appropriate heat sources include circulating water blankets and forced air warmers as used in human operating and recovery rooms. The water pad should be set at 32–38°C. Recovering bats should be maintained in a temperature-controlled cage at 25–28°C. During recovery they should be wrapped to avoid wing flapping, placed in lateral recumbency with their head lower than their feet, and have the feet hooked onto the cage wire or a towel.

6.2.1 Injectable agents

Diazepam (0.5–2.0 mg/kg IM or IV) produces sedation for between 30 min and 4 hr. Highly agitated or aggressive bats require a higher dose than calm bats and duration of effect is generally shorter. Sedation is useful in larger bats to facilitate examination and prior to general anaesthesia.

In pteropodids, medetomidine (0.3–0.8 mg/kg IM) is a useful sedative. However, as with other alpha-2-adrenergic agonists it induces profound bradycardia, poor peripheral circulation and altered thermoregulation. It is antagonised with atipamezole at 5 times the medetomidine dose IM.

Propofol is used for induction or maintenance of anaesthesia in pteropodids. It can be diluted 1:4 with sterile water for accurate dosing in small animals. At 8–10 mg/kg IV it produces rapid induction and 5–15 min of anesthesia, which can be maintained by continuous infusion or repeated boluses. It appears very safe, providing rapid smooth inductions and recoveries (J Barrett & A Olsson unpub). Its disadvantage is that IV administration is difficult if the bat is excited or aggressive.

Ketamine alone is not recommended because of poor muscle relaxation, and struggling and flapping during recovery (Heard et al. 1996a). The addition of an alpha-2-adrenergic agonist (xylazine, medetomidine) improves muscle relaxation, reduces the ketamine dosage and improves analgesia. Several combinations of ketamine and xylazine have been used in pteropodids: 10–20 mg/kg ketamine with 2–4 mg/kg xylazine (Heard et al. 1996a; A Olsson *pers. obs.*) or 20–40 mg/kg ketamine with 2–4 mg/kg xylazine (J Barrett & A Olsson unpub). The preferred dose rate is 10–20 mg/kg ketamine with 2–4 mg/kg xylazine IM. Induction occurs within 2–5 min, immobilisation lasts 10–40 min and recovery is usually quiet without wing flapping. The combination can be partially antagonised with atipamezole (0.02–0.06 mg/kg IM) but this is not recommended because it may unmask the adverse effects of ketamine (Heard et al. 1996a).

Ketamine (6 mg/kg) and medetomidine (60 µg/kg) combination administered IM has been used to anaesthetise pteropodids. The eyes frequently remain open with this combination and the corneas should be protected with a lubricating ophthalmic preparation. The medetomidine is antagonised with atipamezole at 5 times the medetomidine dose (J Barrett & A Olsson unpub). Note that when dealing with very small volumes of drugs, making a 1:10 or 1:100 dilution with sterile water facilitates accurate dosing of small patients.

Tiletamine/zolazepam (10–40 mg/kg IM) (J Barrett & A Olsson unpub) produces rapid induction and has a wide safety margin. The higher dose rate is used to sedate aggressive animals whose injury or illness necessitates euthanasia. In pteropodids, tiletamine/zolazepam (20 mg/kg) sprayed into the mouth produces rapid and effective immobilisation of unrestrained aggressive animals. This is particularly useful if ABLV infection is suspected (L Vogelnest *pers. comm.*) and when rescuing pteropodids trapped in netting or barbed-wire fencing. Recovery from tiletamine/zolazepam may be prolonged and violent with wing thrashing and extreme agitation (D Heard *pers. comm.*; A Olsson *pers. obs.*).

Following IM anaesthesia, patients should be allowed to recover in dim lighting until the pupillary light reflex returns. Bats placed in a semi-vertical recovery position usually climb to a hanging position within 1–2 hr.

6.2.2 Inhalation anaesthesia

Inhalation anaesthesia is the method of choice for bats. Induction is via induction chamber for very small animals or, for larger animals, a mask using a non-rebreathing system (e.g. Ayre's T-piece) and precision vaporiser. Anaesthesia is maintained via the mask or an uncuffed (to avoid pressure necrosis of the delicate tracheal mucosa) 2–3 mm endotracheal tube (Heard 1994).

If a pteropodid is aggressive and difficult to handle, an injectable agent can be used for sedation or induction prior to inhalation anaesthesia. If a bat is in shock (e.g. hypovolaemia, endotoxaemia), inhalation anaesthesia is safer than an injectable agent. Ideally, these animals should be stabilised prior to induction.

Isoflurane (induction 4–5%, maintenance 2–2.5%) in oxygen (1–2 L/min) is presently the inhalant of choice because of the wide safety margin and rapid smooth induction and recovery. The bat respiratory system is very efficient and induction (2–3 min) and recovery times are rapid. Although it renders the animal unconscious, isoflurane is a poor analgesic. Opioids or non-steroidal anti-inflammatory drugs are used as part of a balanced anaesthetic regimen for surgery and other painful procedures (Table 14.7).

Atropine (0.02 mg/kg SC) administered at or following induction reduces the profuse pharyngeal secretions occasionally encountered under inhalation anaesthesia in bats, but this is usually unnecessary.

7 CLINICAL PATHOLOGY

7.1 Haematology and biochemistry

7.1.1 Sample collection

In insectivorous and carnivorous bats, total circulating blood volumes range from 4.8–13.0 mL/100 g (Kallen 1977). The volume of blood that can be safely collected is 0.5–1% of body weight, which is roughly equivalent to 5–10% of the total blood volume.

Physical restraint for venipuncture is adequate for some species. However, chemical restraint is recommended for safety and greater control during venipuncture. Keegan (1979) devised a plastic cylinder with a slit through which a wing could be extended, and blood drawn from the cephalic (wing) vein.

Venipuncture in bats <100 g can be performed with either a 0.5–1 mL heparinised syringe and a 25 G needle or a heparinised microhaematocrit tube to draw blood from the needle hub by capillary action or with a calibrated modified micropipette under gentle controlled suction (Gustafson & Damassa 1985). Alternatively, a 25 G needle with the hub snapped off is inserted into a vein and the blood allowed to drip into an open-topped tube containing EDTA or heparin (R Speare *pers. comm.*).

Cardiocentesis is only performed in anaesthetised bats prior to euthanasia. Blood collection from the infra-orbital sinus is traumatic and unnecessary and therefore not recommended (Baer 1966). The external jugular veins are large in insectivorous and carnivorous bats and are used for collection in some larger species (Heard 2003). In pteropodids they are relatively small vessels and are difficult to access for blood collection (Heard 2003; A Olsson *pers. obs.*).

Peripheral limb vessels are preferred for venipuncture. These include the median, cephalic (wing) vein and uropatagial veins (Fig 14.1). The median vein is located on the medial aspect of the humerus and yields relatively large volumes of blood rapidly in pteropodids (Heard 1998). It is difficult to access in the conscious animal. The wing or cephalic vein (located along the leading edge of the wing membrane) is readily accessible in all bat species but collapses when even light negative pressures are applied (Figs 14.1, 14.5).

In insectivorous and carnivorous bats the uropatagial (interfemoral) vein is accessed in the tail membrane, close to and parallel with the long bones of the hind limb. In pteropodids, the uropatagial vein is visible across the posterior aspect of the hind foot. To access the vessel for blood collection the bat is placed on a towel so that it grips with its feet. The uropatagial vein is useful in conscious pteropodids (Fig 14.1).

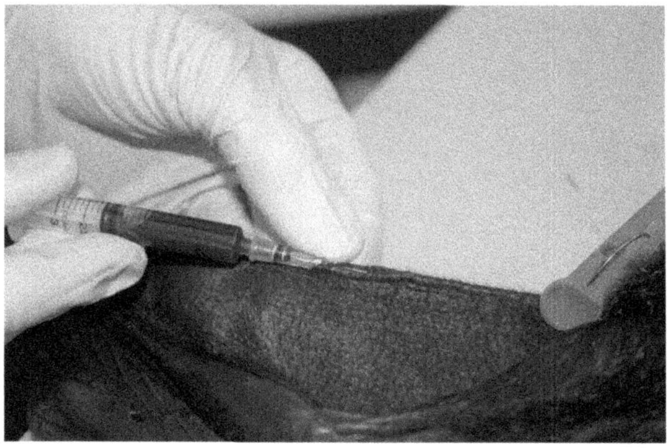

Figure 14.5 Blood collection from the wing or cephalic vein in an anaesthetised adult pteropodid. Photo: Andrew Breed.

7.1.2 Reference ranges and interpretation

Chiropteran haematology and biochemistry values are similar to domestic carnivore species. Hyperkalaemia (in excess of 8 mEq/L) in otherwise healthy animals is due to prolonged collection time and/or poor technique, or due to collection into EDTA (Heard et al. 2006; Olsson 2000). Bats, particularly pteropodids, have low blood cholesterol, urea and creatinine values (Heard 2003). Table 14.4 provides haematology and biochemistry values for some Australian bats.

7.2 Urinalysis

Bats usually urinate when restrained. A midstream voided sample can be collected by placing the bat head up and collecting urine in a container (with a funnel) placed under the animal. Gentle stimulation of the urogenital region with a stroking motion using a finger or soft cloth stimulates micturition. Alternatively, cystocentesis can be performed as for small domestic species, with the bat restrained in dorsal recumbency. Urinalysis and urine specific gravity can be interpreted as for domestic species.

The urine of fruit-eating bats may be clear, or range in colour from yellow to pink. Urine composition and odour are influenced by diet. The occasional fruity odour is a normal physiological finding associated with ketone production in an otherwise healthy animal.

Table 14.4 Haematology and biochemistry values for some Australian bat species (range)

Parameter	Little bent-winged bat (n = 6)	Gould's wattled bat (n = 7)	Lesser long-eared bat (n = 8)	Grey-headed flying-fox (n > 25)	Little red flying-fox (n = 4)	Spectacled flying-fox (n = 63)
PCV (L/L)	0.47–0.49	0.38–0.66	0.53–0.56	0.44–0.56	0.46–0.56	0.35–0.55
RBC ($\times 10^{12}$/L)	10.7–11.1			8.2–9.4	9.0–11.5	7.5–10.7
Hb (g/L)	192–196			157–199	151–207	133–180
MCV (fl)	44.0–44.8			54–59	44–56	48–50
MCH (pg)	17.6–18.0			18–22	15–20	15–19
MCHC (g/L)	398–414			344–376	303–407	302–396
WBC ($\times 10^{9}$/L)				11–22	10–16	9–35
Neutrophils (%)				33–45	43–63	23–65
Lymphocytes (%)				49–58	19–43	15–61
Monocytes (%)				3–10	2–13	1–12
Eosinophils (%)				3–6	3–12	0–24
Basophils (%)				0–3	0–1.0	0–1
Total protein (g/L)				73–77	74–87	54–86
Glucose (mmol/L)						0–6.8
BUN (mmol/L)						0.2–3.4
Creatinine (mmol/L)						0–0.08
Albumin (g/L)						25.6–48.4
Globulin (g/L)						32.9–33.1
A/G ratio						0.7–1.5
Total bilirubin (μmol/L)						0–24
ALP (U/L)						<2333
AST (U/L)						7.9–167.1
ALT (U/L)						5.8–31.8
CK (U/L)						<1884
Cholesterol (mmol/L)						0–1.6
Calcium (mmol/L)						2.2–2.6
Phosphorus (mmol/L)						1.3–4.3
Ca:P ratio						0.5–2.1
Sodium (mmol/L)						132–147
Potassium (mmol/L)						2.7–14.3
Chloride (mmol/L)						94–113
Bicarbonate (mmol/L)						4.2–23.8

Source: Modified from Clark (2004) and Olsson (2000).

7.3 Faecal analysis

Faecal smears, wet preparations and flotation are prepared and interpreted as for domestic species. Stained faecal smears of pteropodids contain relatively low numbers of Gram-positive rods and cocci. Budding yeasts are abnormal. Pollens are frequently seen. It has been our experience that insectivorous and carnivorous bats have predominantly Gram-negative bacteria.

8 HAND-REARING

Hand-rearing of bats is well-described in numerous publications (Barnard 1995; Barnard & Sachs 1992; SM Barnard unpub; George 1990; Lollar & Schmidt-French 1998). Indications for hand-rearing include abandonment of the pup by its mother due to disturbance of roost sites or creches and illness or injury in mother or pup. Prognosis for successful rearing of both pteropodid and insectivorous and carnivorous bat pups is good even from birth if proven protocols are followed. Pups with significant injuries (including but not limited to open humerus fractures, open phalangeal fractures, compromised blood supply to more than one-third of the patagium, extensive wing tears, loss of thumbs), congenital defects or serious illness should be euthanased. A limiting factor in the decision to hand-rear bats in some areas is availability of rabies-vaccinated carers.

8.1 Pteropodids

8.1.1 Neonates

In the first 4 wk of life, pteropodids have a bare ventrum and difficulty thermoregulating. Naked premature pups have been successfully reared when strict hygiene principles are applied. Neonatal pups are kept wrapped snugly in a triangular cloth (Fig. 14.6). They usually suck the edge of the cloth or a dummy teat. A folded tissue used as a nappy keeps the pup clean. Care is taken to ensure urine and faeces are removed from the skin and wing membranes at every feed to prevent fungal and bacterial infections (see 9.1). Alternatively, the 'Mumma technique' is used—wrapping pups to give them more freedom and keep them cleaner (Parry-Jones 2000). With this technique the pup is given a roll of towelling to cling to. The whole roll is then wrapped snugly in a cloth, resembling the mother's wings. The pup can crawl up and down the roll beneath the covering cloth, and can move away from its urine and faeces.

8.1.2 Housing

Initially pups are kept wrapped up in a small pet carrier or basket at 28°C. It is important to expose pups to 10–15 min of sunlight every day during feeding or cleaning as exposure to ultraviolet light is important for drying the skin and may be important in vitamin D synthesis and calcium metabolism (see 9.9.3a). After 4 wk they are encouraged to hang from towelling-covered walls or towels hung on a plastic-coated clothes-horse. Towels are hung as hammocks and curtains to provide a secure environment. Most pups flap vigorously by 4–6 wk and are flying at 8–12 wk. They are housed at this time in a small snake-proof aviary. Towels, foliage and perches are placed at either end with ropes or netting extending to the floor so bats have a way of returning to the canopy if they fall. Bats are encouraged to fly by calling from one end of the aviary (George 1990).

Pteropodids are released in groups once all are flying. Ideally the release cages are winched high into the canopy adjacent to an existing colony (J Maclean *pers. comm.*). The juvenile bats are allowed to come and go for support feeding for several weeks before the cage is removed.

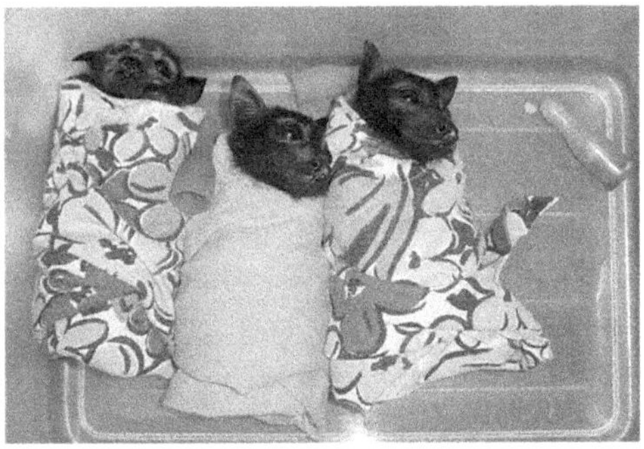

Figure 14.6 Neonatal pteropodid pups wrapped in a triangular cloth. Photo: Andrew Breed.

8.1.3 Feeding

A number of milk formulae have been used to successfully raise pteropodids (George 1990; Wilson 1988). These include Wombaroo Flying-fox Milk, Complan and Nan Number One (Nestle Australia) human infant formulae and Di-Vetelact (Sharpe Laboratories).

Although flying-fox teats (Wombaroo) are best, little red flying-fox and blossom bat pups require a smaller teat (e.g. a possum teat). For pups less than 4 wk of age, a 10 mL syringe fitted with a teat allows better control of milk flow. This reduces milk aspiration and ensures adequate intake, particularly if the pup has a weak sucking reflex. By 2–4 wk of age, when the sucking reflex is strong, the syringe is replaced with a small glass bottle and by 6 wk pups can lick from a guinea pig self-feeder bottle. Milk or feeding formula is changed every 4 hr and the self-feeder thoroughly cleaned and disinfected.

Newborn pups are fed every 2–3 hr. This is reduced to every 4 hr by 1 wk of age, to every 6 hr by 6 wk of age.

Once pteropodids are hanging and flapping, fruit juice and small pieces of soft fruit are offered and the amount of fruit increased weekly until release. Pteropodids in the wild suckle for at least 6 mo, so milk formula or a high-protein source (e.g. Wombaroo high-protein powder supplement) is added to the fruit, and bottle feeds are withdrawn over this time.

8.1.4 Care of pteropodid pups

The infant must be warm and active (George 1990) before feeding, and the formula should be warmed to body temperature. Care is taken to ensure the milk is evenly warmed to prevent oral and oesophageal burns. It is much easier to feed a bat that is securely wrapped to contain the wings and feet. The head is held slightly lower than the feet and the animal turned to one side. These precautions reduce the risk of milk aspiration.

Before each feed the bat is inverted (head up) and the anogenital area stimulated using a moist finger or soft cloth to induce urination and defecation. This area, as well as the wing pits and membranes, is then thoroughly cleaned with a soft cloth and warm water or baby change lotion. The latter sanitises but does not oil the skin. A full shower under a running tap is usually unnecessary unless major soiling has occurred. After any bathing or sponge-down, the bat is thoroughly patted dry with an absorbent cloth or paper towel. It is then hung in the sunlight for at least a few minutes and encouraged to flap and groom by gently extending each wing in turn. Tickling under the wing will usually produce a wing stretch.

8.2 Insectivorous and carnivorous bats

Insectivorous and carnivorous bat hand-rearing techniques have not been as widely published as those used for pteropodids (Hopkins 1990; Barnard & Sachs 1992; Barnard 1995; SM Barnard unpub).

8.2.1 Neonates

Most species breed in colonies containing large numbers of bats which keep neonatal pups warm. Consequently, an artificial heat source is required to keep the temperature at 27–38°C (Barnard 1995; Lollar & Schmidt-French 1998). Humidity should be kept at 55–80% (Barnard & Sachs 1992).

8.2.2 Housing

Newborn pups are housed in a small bag or box (Hopkins 1990), styrofoam container (Barnard 1995; Barnard & Sachs 1992), or bird brooder (Lollar & Schmidt-French 1998). Whatever the housing, a maximum/minimum thermometer should be used and humidity monitored to prevent hyperthermia and dehydration. As pups become well-furred and more active they are moved to larger cages.

8.2.3 Feeding

Two formulae are regularly used: Wombaroo Insectivorous Bat Milk and Biolac Puppy Milk (Biolac Milk Products). Insectivorous and carnivorous bats grow and develop very rapidly, and all require high-energy milk with a high fat and protein content in early life.

8.2.4 Care of young insectivorous and carnivorous bats

Bats are initially hung upside down to feed to prevent aspiration of milk. Small eyedroppers, IV catheters and other fine tubing apparatus are used to feed pups (Hopkins 1990; Barnard & Sachs 1992; Lollar 7 Schmidt-French 1998). Feeds are small (0.1–0.5 mL depending on species and body weight) and frequent (1–4 hr).

As for pteropodids, the bat should be stimulated to urinate and defecate before every feed. After the first few days on a new formula, faeces are black and firm (Barnard 1995).

In most species pureed mealworms are added to milk formula by 6 wk of age (Barnard 1995; Lollar & Schmidt-French 1998). Alternatively, some carers remove the mealworm head and squeeze the gut contents into the pup's mouth (Barnard 1995). Feeding over a dish of mealworms will encourage the bat to self-feed. It is offered mealworms on forceps at this time.

Milk is continued for several weeks after the bat is fully furred and has begun eating insects. It is then creched with others of its species and encouraged to start flying (Hopkins 1990) and catching insects. Some carers gently throw young bats into the air to make them fly, but the natural instinct to fly can be encouraged safely by placing the pup on a shelf or hanging towel at least 2–3 m high within the enclosure and releasing insects at dusk, the time wild bats are becoming active. Most pups will be stimulated to take off and start short unassisted flights at this time of evening in response to the available insects. If prepared food is withheld at this time the bats are more likely to attempt to catch insects. Food dishes can be placed in the enclosure later in the night to ensure nutritional requirements are met. A nightlight in the enclosure of weanling bats attracts insects and provides the opportunity for bats to practise feeding on the wing during the night.

When bats reach adult weight and are self-feeding on the wing they are released at dusk without a recent feed (otherwise they become sleepy and will be reluctant to leave). They should be released from an elevated area with no nearby obstacles.

8.3 Inappetant pups

Inappetence in hand-reared pups may be due to:

- formula fed too frequently, too much or of the wrong type;
- missed feeds and subsequent dehydration;
- colic;
- inappropriate ambient temperature and/or relative humidity;
- incompatible or too many nursery mates;
- systemic disease.

9 DISEASES

9.1 Dermatologic diseases

9.1.1 Physiological alopecia

Alopecia in wild bats is rare. When present it is usually related to ectoparasitism (see 9.8.4d) or is a normal physiological change associated with reproductive behaviour or moulting (Hill & Smith 1985).

9.1.2 Nutritional alopecia

Alopecia in captive bats is often the result of malnutrition. This may result from inadequate dietary supplementation or leaving infant bats on milk formula beyond normal weaning (A Olsson *pers. obs.*; SM Barnard *pers. comm.*). Hand-reared insectivorous pups are susceptible to alopecia from protein (essential amino acid) (Olsson & Barnard unpub) and fat-soluble vitamin deficiencies. The skin is unaffected, but patchy or symmetrical hair loss occurs. Once the diet is corrected or the bat is weaned, the hair regrows.

Bats fed nectar diets without protein supplement experience weight and fur loss and fail to thrive; many die (Howell 1974). Wombaroo high-protein supplement, Poly-Aid Plus, powdered milk, bee pollen (from health food stores) and high-protein baby cereal are good protein sources.

9.1.3 Allergies

Possible allergic reactions in captive bats include acute bilateral conjunctival and facial swelling with associated respiratory distress, cutaneous ulcerations and pruritus (D Heard *pers. comm.*). Suspected allergens include flowering plants and insect bites (SM Barnard unpub). Treatments include antihistamines or corticosteroids at small animal dose rates.

9.1.4 Trauma

Skin trauma occurs when overcrowding leads to fighting among cage mates. Overcrowding can lead to urine or faecal scalding, initially apparent as fur bleaching. Thermal burns from power lines and heatpads may permanently destroy hair follicles and melanocytes, resulting in alopecia, depigmentation and scarring. Wing membrane wounds are common (see 13.3) and frequently difficult to treat successfully.

9.1.5 Fungal infections

Cutaneous candidiasis caused by *Candida albicans* is common in poorly raised or chronically ill juvenile pteropodids. It may be observed first on the wing membranes as areas of erythema, which develop into a greyish pseudomembranous slime. The infection may spread to other body areas, including the mouth. Furred areas are wet and sticky, and fur falls out in sticky clumps. Recurrent infections suggest underlying immunosuppression. Budding yeasts are seen on stained impression smears.

Treatment involves holding the bat securely in a head up position (preferably wrapped in a towel with only the affected part free) and washing the affected areas thoroughly under warm running water. The cleansed areas are patted dry several times with a soft cloth or paper towel and a 1–3% chlorhexidine solution (non-soap based) applied and left on. Access to direct sunlight for at least 10 min, twice daily, will aid recovery. Topical application of antifungal creams (e.g. nystatin or miconazole) may be required, but is usually unnecessary with good cleaning protocols and is usually licked straight off unless an Elizabethan collar is fitted to the bat. Denuded areas of fur regrow, but will take up to 4 wk after the commencement of treatment. Good hygiene is essential for prevention.

Isolated lesions suspected to be dermatophytosis are common on hand-raised pups. Diagnosis is usually made following a skin scraping stained with gentian violet demonstrating fungal hyphae. Culture of the lesions should be performed to confirm the diagnosis. Treatment includes application of a topical antifungal cream such as miconazole. An Elizabethan collar is usually unnecessary if the cream is rubbed into the lesion well, although distracting the bat with a bottle feed helps prevent medication being licked off. Generalised dermatomycosis is rare and usually secondary to other problems such as poor husbandry. It should be cultured and is treated with systemic antifungals (see Table 14.7).

9.1.6 Otitis externa

Ear infections or head injuries are associated with head tilts, ear scratching, asymmetry and altered balance. Otitis and other ear problems are managed as in domestic mammals. Bat pinnae are very delicate and their vascular supply is easily disrupted by trauma (such as pups sucking on one another), resulting in necrosis and sloughing.

9.2 Gastrointestinal diseases

9.2.1 Oral cavity

The cheek pouches of larger species may become impacted with food. Oral ulcers suggest ingestion of irritant substances. If infectious agents are suspected, cultures are performed on swabs taken from the lesions. Bats rescued from barbed-wire fences may have holes or lacerations in the mouth due to chewing the wire barbs in an attempt to free entangled skin and wing membranes. Oral burns occur in suckling pteropodids due to maternal electrocution or microwave-overheated liquids. Burns to more than 30% of the dorsal surface of the tongue is an indication for euthanasia.

Oral candidiasis is common in hand-reared bats. Although reluctant to suckle or feed, they appear hungry. There may be a rust-coloured salivary stain at the commissures of the mouth. Classic white plaques visualised in the oral cavity are often associated with marked halitosis. When the plaques are removed the underlying tissue is ulcerated and bleeds easily. Diagnosis is based on history, clinical signs and microscopic examination of a smear for the characteristic yeasts and pseudohyphae. Treatment includes nystatin, fluconazole or itraconazole (Table 14.7).

Dental plaque, calculus and associated gingivitis are very common in captive insectivorous and carnivorous bats. Dental staining and enamel etching are common in pteropodids. Teeth are examined for premature loss, degree of wear, trauma and movement. It has been our experience that calcium-deficient diets may be associated with brittle or soft teeth. Gingivitis is usually secondary to dental or systemic disease.

Tooth wear occurs at different rates depending on diet, age and other factors such as hardness of food. Captive frugivorous and nectivorous bats feed well and maintain body condition with extremely worn teeth, whereas carnivores and insectivores do not. Severe bacterial gingivitis with underlying osteomyelitis or tooth root abscesses will result in anorexia and possibly bacteraemia/septicaemia. Surgical intervention and long-term antibiotic therapy may be of benefit but euthanasia is recommended in complicated or refractory cases.

9.2.2 Pharynx and oesophagus

Many oral diseases (e.g. burns and candidiasis) extend to the pharynx and oesophagus. Examination of the pteropodid oropharynx usually requires sedation or anaesthesia and a light source. Most insectivorous and carnivorous bats, however, gape in a threat posture if tapped lightly around the head with a tongue depressor or similar, allowing examination. The most obvious sign of disease is dysphagia.

9.2.3 Gastrointestinal tract

Abdominal palpation of conscious bats is difficult because of their well-developed rectus muscles. The gastrointestinal tract is thin-walled and usually contains little to no material. The anus should be examined (especially in infants) for signs of diarrhoea, tenesmus or prolapse. Soft stools are often passed by bats during physical restraint. When available, faeces are examined for intestinal parasites and their ova. In pteropodids, ascarid worms and occasionally eggs (*Toxocara pteropodis*) may be observed in faeces (and occasionally vomited up) around the time of weaning (see 9.8.4a). Melaena is always abnormal in pteropodids, and in insectivorous and carnivorous bats should be viewed with suspicion if other signs of illness are also present.

a Stomach

Signs of disease include inappetence, anorexia, bloat, diarrhoea or constipation and upper abdominal pain. Vomiting is uncommon but will occur after ingestion of noxious substances and starfruit by pteropodids. Vomiting is often a terminal sign in a sick animal.

In debilitated and dehydrated insectivorous and carnivorous bats, gastric impaction can occur if they are fed dry artificial diets or insects. Gastric impaction is characterised by inappetence, vomiting and occasionally constipation. In large bats, cranial abdominal palpation will identify a hard mass and elicit a pain response. Treatment includes oral and/or parenteral fluids (see section 11). A small amount of vegetable oil may be administered orally or by feeding tube. The animal is monitored closely and fluid therapy continued until the ingesta passes. If surgery is indicated, oil is discontinued more than 24 hr beforehand.

Gastric dilatation is common in hand-reared bats on inappropriate milk formulas or commercial milk replacers in which the ingredients have been altered. It is also associated with aerophagia during feeding if the teat hole is too large or too small, or if the bat is a greedy feeder. Aerophagia can be minimised by having the infant firmly restrained in the correct position and by controlling the flow of milk, either by pinching the teat or fitting the teat over a 10 mL syringe and depressing the plunger at the appropriate rate. It is advisable to remove the teat from the infant's mouth once the bottle is empty to prevent aerophagia through the empty teat. Chronic feeding of highly nutrient-dense diets may cause digestive disturbances. These diets lead to bacterial overgrowth and associated gastroenteritis, which may also lead to bloat. Both *Clostridium perfringens* and *Candida albicans* have been implicated as a cause of gastritis in pteropodids fed inappropriate foods (SM Barnard *pers. comm*; Fox 2004; Wood 2004; A Olsson *pers. obs.*). The bloating in these cases may be due to inflammation causing gastric ileus or peristaltic dysrhythmia. Bloat caused by obstruction is rare. Heard et al. (1995) describe gastric dilatation and volvulus caused by toxocariasis in a variable flying-fox (*P. hypomelanus*).

The onset is usually rapid and, depending on species, occurs 3–10 d from beginning hand-rearing (SM Barnard *pers. comm.*). If undiagnosed, it can become life-threatening. Gas must be eliminated from the gastrointestinal tract as quickly as possible. In pteropodids, it is sometimes necessary to pass a stomach tube and aspirate the contents, including fluid. In small bats, decompression is relatively successful with drug therapy. Simethicone infant gas relief (0.01 mL/g) alternated with metoclopramide syrup (0.01 mL/g) is often effective if the bat is still passing faeces. However, if no faeces have been passed for 12 hr discontinue treatment as use in an obstructed animal may lead to rupture of the gastrointestinal tract. Both drugs are diluted (1 drop/4 drops water); 2 drops of this mixture are added to each mL of milk formula. This therapy is continued for a week after resolution of abdominal distension (Barnard 1995; Lollar & Schmidt-French 1998; SM Barnard *pers. comm*; Olsson & Barnard unpub). A probiotic is administered concurrently at domestic animal doses rates.

When bacterial overgrowth occurs, fluids and electrolytes are administered. Where possible, culture and sensitivity testing should be performed. If the bat is depressed and pyrexic a broad-spectrum antibiotic is indicated. Otherwise, supportive therapy alone may be adequate. Gastric candidiasis occurs particularly in debilitated and/or immunosuppressed pups, and often follows prolonged antibiotic therapy. Diagnosis is based on identification of the organism in stomach washes. Faecal examination may identify increased numbers of yeasts and pseudohyphae. Gastrointestinal yeast infection in pteropodids is treated with nystatin 50 000 U/100 g, PO tid for 5–7 d, bid for 2 d and sid for 3 more days (Wood 2004). Fluconazole (10 mg/kg PO sid for 10–14 d) is also effective (D Heard *pers. comm.*). After antimicrobial therapy, probiotics are administered for over a week, preferably until weaning, to restore and maintain intestinal flora.

b Intestines

Inappropriate diet may lead to gastrointestinal microbial overgrowth, which leads to diarrhoea.

Diarrhoea may be acute or chronic. Treatment of acute diarrhoea is based on resting the gastrointestinal tract for up to 24 hr to facilitate healing. Parenteral fluids are essential to restore and maintain hydration. Oral fluids are withheld for 24 hr then introduced in small volumes every 3–6 hr, supported by parenteral fluids until diarrhoea is resolving. Very young or weak bats may require parenteral or enteral emergency nutrition such as Poly-Aid Plus or similar (see section 5). Antimicrobial therapy is usually unnecessary for simple acute enteritis.

Chronic diarrhoea results when clinical signs have been overlooked or inappropriately treated. This is relatively common in hand-reared animals. Malabsorption is a common cause of chronic diarrhoea, and is suspected when there is a combination of diarrhoea, weight loss, alopecia and anaemia (PCV < 20–25%). Management initially involves correcting husbandry, providing appropriate therapeutics and ensuring a readily digestible diet appropriate to the species and age group. Diets for patients with maldigestion/malabsorption are highly digestible and low in fibre and contain readily available fats. Care must be taken to avoid excessive amounts of fats that are poorly absorbed and will result in osmotic diarrhoea in adult bats. Increased protein is indicated for protein-losing enteropathies. Refer to section 5 for specific dietary management.

Colitis is occasionally seen and is usually associated with frequent defecation and blood or mucus in the stool. In hand-reared animals it may be associated with generalised gastrointestinal tract microbial overgrowth due to inappropriate diet. Medical and dietary management depends on cause but there do not appear to be any contraindications to using small animal treatment regimens.

c *Rectum and anus*

Total or partial rectal prolapses usually occur secondary to tenesmus caused by constipation. This is often seen in bats fed blended diets and unable to separate fibre from juice (Snell 1994). Rectal prolapses can also be associated with lower bowel irritation, diarrhoea and endoparasitism. Severe prolapses are more likely to present in bats hand-reared by unobservant or inexperienced carers unable to recognise early signs of gastrointestinal tract disease such as change in volume, appearance, colour, odour or frequency of faeces.

A partial prolapse often disappears between straining episodes. It frequently self-corrects once straining is identified and treated. If intervention is warranted, the problem is corrected by applying firm pressure on either side of the anus, using a finger and thumb. A lubricant is applied in combination with a local anaesthetic spray to prevent further straining.

Total prolapses always require intervention. If the prolapsed portion appears devitalised or torn, surgical resection and anastamosis via a laparotomy or euthanasia are indicated. Uncomplicated prolapses can be reduced after application of sugar or an electrolyte powder to reduce swelling; small sections at a time are inserted back into the body. A well-lubricated cottonbud is carefully inserted into the centre of the prolapse and gently rolled backwards and forwards to assist the return of the prolapsed tissue. Once replaced, a purse-string suture is placed around the anus (using non-absorbable 2/0–4/0 suture material) to reduce its diameter while still allowing the passage of faeces. The suture is removed after 2–4 d. An Elizabethan collar is used to prevent self-trauma. Antibiotics and non-steroidal anti-inflammatory drugs may be indicated (see 11.1).

Conditions involving the anus include imperforate anus (see 9.9.5) and trauma. Traumatic lesions are common with cat attacks and in hand-reared pups overstimulated to induce defecation. Left untreated, ulcers may develop. A painful anus will make the animal reluctant to defecate, leading to constipation. In severe cases, the anus will heal closed requiring surgery or chemical cautery to regain and maintain patency of the opening. Lotagen ointment (Schering-Plough Animal Health), which cauterises granulation tissue and facilitates healing, can be used. Care must be taken to use Lotagen only for 2–3 d or until ulcers are cauterised, to avoid strictures forming.

9.2.4 Liver

An outbreak of hepatitis was reported in a captive spectacled flying-fox colony in Cairns (Qld). It was associated with haematuria, myoglobinuria, abdominal pain and death within 1–3 d. No histopathology or microbiology was performed and the cause was unknown.

9.2.5 Pancreas

Pancreatic insufficiency results in pale greasy faeces (bats fed diets containing fat), chronic diarrhoea or unformed faeces, and weight loss and altered weight. The disease is uncommon and has been reported anecdotally in obese captive little free-tailed bats (*Mormopterus loriae*) fed almost exclusively on mealworms. A pancreatic enzyme supplement was reported to improve the character of the faeces.

9.3 Cardiovascular diseases

Nutritional cardiomyopathies are relatively uncommon in bats (Duncan et al. 1996; Heard et al. 1996b). Dilated cardiomyopathy induced by hypovitaminosis E has been described in a captive collection of Australasian and south-east Asian flying-foxes (*P. poliocephalus, P. hypomelanus* and *P. giganteus*) (Heard et al. 1996b).

Mucus membrane and tongue colour are altered by decreased haemoglobin oxygen saturation and reduced peripheral perfusion associated with circulatory failure. Dependent oedema associated with cardiac disease is observed in the head and neck because of bats' perching position (Heard et al. 1996b). This is also seen with hypoproteinaemia due to glomerulonephropathy (Deem 1999) and protein-losing enteropathy. The most common cause of head and neck swelling, however, is abscessation of bite wounds.

At low body temperatures supraventricular arrhythmias and second-degree heart blocks are common in bats, but these generally resolve with warming (Heard 2003).

9.4 Lymphatic diseases

Peripheral lymph nodes are found in the same locations as in other small mammals. Peripheral lymphadenopathy has been associated with septicaemic conditions and, in one case, a suspected allergic reaction in a hand-reared spectacled flying-fox that had been fed chocolate.

9.5 Respiratory diseases

The minimum metabolic requirement in level flight is much greater than the maximum aerobic capacity of a non-flying mammal of the same size (Maina et al. 1991). Consequently, bats at rest are not likely to manifest overt clinical signs of respiratory disease until dysfunction is well-advanced (Heard 1998). Bats with respiratory disease are reluctant to fly, have decreased exercise tolerance, may lose weight and fail to thrive. Bats rarely breathe through their mouth so any impediment to nasal breathing will cause the animal to puff through its cheeks. This must be differentiated from teeth grinding (bruxism), which is indicative of pain. Sneezing associated with a clear nasal discharge is relatively common in pteropodids and appears to be of no clinical significance.

Radiography is an important tool in diagnosis of respiratory disease. Interpretation is as for other mammals. Some bats are too small for respiratory diagnostic tests. However, if respiratory disease is suspected nasal and tracheal washes are performed, preferably under inhalational anaesthesia, by instilling 0.3–1 mL sterile saline/100 g body weight into the nasal passages or trachea and aspirating it immediately to collect samples for culture and cytological examination. Free fluid identified in the chest with radiographs or ultrasonography can be aspirated for cytology and culture.

Aspiration of milk is the most common cause of pneumonia in hand-reared bats. All bats are at risk of aspiration when they are fed with their head higher than their feet. Treatment depends upon the cause, but is aimed at alleviating pulmonary inflammation and controlling oedema and associated infection. Non-steroidal anti-inflammatory drugs are indicated (Table 14.7). Pulmonary oedema is controlled with furosemide (2–8 mg/kg IM, PO). Secondary bacterial infection is treated with broad-spectrum antibiotics (e.g. amoxycillin/clavulanic acid combination, 12.5 mg/kg of each component PO bid; enrofloxacin 5 mg/kg PO or SC sid; or trimethoprim/sulfadiazine 30 mg/kg PO bid). Postural drainage occurs with the bat in a hanging position. Parenteral fluids and warmth are critical in debilitated and hypothermic patients.

9.6 Urinary and reproductive diseases

Occasionally white sandy material, composed of calcium deposits, protein and inflammatory cells, is passed in the urine of young hand-raised bats. It may be an indication of ascending urinary tract infection caused by bacteria transferred from the anus during stimulation to toilet. Other causes include mild dehydration during times of unseasonably high environmental temperatures, and high levels of minerals or other dietary supplements in captive diets. If it persists over several feeds it should be investigated further by urinalysis, cytology and culture and sensitivity.

Vaginal bleeding occurs around the time of parturition, and during oestrus in healthy captive bats. Penile bleeding is uncommon. Unexplained bleeding in either sex requires further investigation.

Occasional 'abortion storms' are reported in the media. These are usually in colonies close to human habitation and therefore attract public attention, and where factors such as heat stroke, toxin exposure (e.g. paralysis tick toxin) and natural disasters (e.g. cyclones) may be significant. If mass abortion is discovered in a colony, appropriate action should be taken to submit samples to relevant authorities and to report it to the Australian Wildlife Health Network (www.wildlifehealth.org.au).

Differential diagnoses for haematuria include urinary tract obstruction, trauma, neoplasia and infection. Treatment is as for other mammals.

9.7 Neurological and musculoskeletal diseases

Central nervous system disease in pteropodids has been investigated by Barrett (2004). One hundred consecutive submissions of wild pteropodids with a history or clinical signs suggesting central nervous system disease were investigated over a 4 yr period. The diagnoses and number of cases were ABLV (32), spinal and head injuries (29), neural angiostrongylosis (18), congenital hydrocephalus (2), bacterial meningoencephalitis (2), vascular hepatopathy (1), chronic injury with mydriasis (1) and no diagnosis (16). There were no cases of lead poisoning. Note that the total number of diagnoses is 101, as one flying-fox had both neuro-angiostrongylosis and a spinal injury.

Because of its relatively high prevalence in bats displaying signs of neurological disease ABLV should be suspected and ruled out in any bat presenting with such signs. Veterinarians presented with bats must use appropriate precautions in handling and management (see 9.8.1a).

Examination for neurological disease includes an assessment of behaviour and observation of neurological signs. General demeanour (allowing for species differences) is important. Some bats become aggressive when handled or restrained, while others such as tube-nosed bats often become limp.

Osteosarcomas or fibrosarcomas have been suspected in aged pteropodids. Generalised muscle soreness can be caused by overexertion in high heat and humidity. Bats shift position from limb to limb while hanging, hang from all four limbs and appear stiff when climbing. They are reluctant to fly. Transient fever and neuralgia accompany active trypanosomiasis and leptospirosis in pteropodids (R Speare *pers. comm.*). Muscle fasciculations can be associated with toxicoses such as lead poisoning, central nervous system disease, septicaemia and nutritional deficiencies and require further investigation.

Joint swelling occurs with inappropriate nutrition, infection, trauma, drug toxicoses and immune-mediated and degenerative joint diseases. One or more joints may be involved, depending on the cause. The main causes of multiple joint swelling are malnutrition and infection. A single swollen joint is usually due to trauma or foreign body penetration. In young bats and those weighing less than 100 g the affected joint can be transilluminated to reveal the injury without anaesthesia and radiographs. Treatment is directed at the cause (e.g. antibiotics for bacterial infection). Non-steroidal anti-inflammatory drugs and opiods (e.g. buprenorphine 0.03 mg/kg SC or IM bid or tid) for pain, joint immobilisation and occasionally surgery to clean and stabilise an infected joint may be indicated. Joint infections must be treated promptly and aggressively since compromised joint function may result in the bat being unreleasable.

9.8 Infectious diseases

There are an increasing number of important infectious diseases being identified in free-ranging and captive bats. Some of these are pathogens of major significance to humans. These include diseases such as Nipah virus, Hendra virus, lyssaviruses, Ebola viruses and the SARS-like coronaviruses (Breed et al. 2006; Calisher et al. 2006; Messenger et al. 2003).

9.8.1 Viral diseases

a Australian bat lyssavirus (ABLV)

ABLV is closely related to the rabies virus. The first recognised case of ABLV, in a black flying-fox, was described to the author (Olsson) in 1996. It has been identified by virus isolation or PCR in four flying-fox species (black, little red, grey-headed and spectacled flying-fox) and one insectivorous bat, the yellow-bellied sheathtail bat (*Saccolaimus flaviventris*) (Allworth et al. 1996; Tidemann et al. 1997; Field 2000; Thompson 1999). However, there is serological evidence of exposure to lyssaviruses from most Australian microchiropteran Familes (Field 2005). ABLV causes fatal encephalitis in both bats and humans (Allworth et al. 1996; Fraser et al. 1996; Field & Ross 1999; Rupprecht 1999) and has caused the deaths of two people in Australia (for a review of potential human exposure see McCall et al. 2000). The Australian Technical Advisory Group on Immunisation recommends that rabies vaccine should be given prophylactically to people who are at risk of occupational exposure to the virus, and therapeutically to people who have been bitten or scratched by Australian bats of any species (ATAGI 2007; www.health.gov.au).

There is an Australian Veterinary Emergency Plan (AUSVETPLAN) disease strategy for ABLV in domestic animals and captive bat colonies, which gives detailed information on ABLV and provides guidance for its management (Animal Health Australia 2006). Much of the information presented here is from that source.

Clinical signs of infection in bats are varied but commonly include paresis or paralysis of the hind limbs, inability to fly or general paresis, dysphagia or abnormal function of the mouth and/or larynx, vocalisation and overt aggression (Field & Ross 1999). Other signs include fitting, twitching and tremors, and weakness with respiratory difficulties (which can be mistaken for pneumonia) (Barrett 2004). ABLV should be suspected in any bat found dead or moribund, those with

Table 14.5 Clinical recognition of bats likely to be ABLV-positive

Clinical signs and other characteristics	Ease of recognition of ABLV infection	Comments
Dead or moribund	Difficult	Indistinguishable from other moribund bats.
Poor or nonspecific clinical description	Difficult	The number of bats in this category can be minimised if veterinarians make detailed and specific inquires about the bats' physical and neurological abilities.
Overt aggression	High index of suspicion	Wild bats are usually not aggressive. Most bat bites and scratches are clearly defensive. Any suggestion by experienced wildlife carers that a bat is uncharacteristically aggressive or angry should be considered very seriously. These bats are particularly likely to attempt to bite or scratch people and pose a high risk of human infection. Both human lyssavirus deaths involved aggressive bats.
Paresis, paralysis of hind limbs	High index of suspicion	While there are other causes of paresis few have a good prognosis, and all bats that have hind limb paresis and/or are unable to hang should be submitted for ABLV testing.
Generalised paresis	High index of suspicion	Bats found with unexplained generalised weakness or with progressive weakness should be submitted for ABLV testing.
Abnormal function of mouth or larynx	High index of suspicion	These clinical signs are uncommon in ABLV-negative bats. All bats with a history that includes difficulty swallowing, abnormal salivation or tongue movements, odd yawning etc. should be submitted for testing.
Fitting, twitching and tremors	High index of suspicion	All bats showing evidence of CNS and/or cranial nerve involvement should be submitted for ABLV testing.
History or clinical signs suggesting a specific non-ABLV diagnosis	Low index of suspicion	Perhaps the most difficult ABLV-positive bat to recognise. Many of the >90% of rescued bats that are ABLV-negative have histories and clinical signs, e.g. being caught in barbed-wire fences, electrical burns, found as orphans etc. that suggest a specific non-ABLV reason for rescue. ABLV-positive bats masquerading as one of the common ABLV-negative bats are unlikely to be recognised.
Weakness, with respiratory difficulties	Difficult	The presence or development of respiratory abnormalities in a bat with lyssavirus-like weakness should not reduce the otherwise high index of suspicion for ABLV.
Yellow-bellied sheath-tail bats	High index of suspicion	Due to the extraordinarily high prevalence of ABLV in this species (62.5%) and the difficulty associated with safely examining this small animal, all sick, injured or orphaned yellow-bellied sheathtail bats should be submitted for ABLV testing.

Source: Modified from Barrett (2004).

non-specific clinical signs, or yellow-bellied sheath-tailed bats. Importantly, it should also be considered even when the history or clinical signs suggest a non-ABLV diagnosis (Barrett 2004). Bats as young as 8 wk may be infected and exhibit clinical signs (Field & Ross 1999). Death in bats occurs a few days to a week after first clinical signs. Bats exhibiting the aggressive form are usually dead within 24–48 hr. Table 14.5 presents clinical recognition of bats likely to be ABLV-positive.

For veterinarians in practice it is crucial to recognise the importance of ABLV as a potentially fatal disease of humans and its varied clinical presentation in bats, and take appropriate steps to minimise risk to themselves and their staff.

When the clinical signs and histopathology suggest lyssavirus infection, specific diagnostic tests are necessary to confirm infection (Barrett 2004).

Lyssavirus infection should be immediately excluded whenever a bat has bitten, scratched or otherwise potentially exposed a person to ABLV. The ABLV status of any bat having direct contact with companion animals should be determined if the bat is available.

Suspect lyssavirus infection should be confirmed by laboratory examination when any animal has acute progressive neurological signs, particularly if the signs include behavioural changes, paresis and paralysis. Specimens should initially be sent to the state or territory veterinary reference laboratory (or other appropriate

Table 14.6 Tests currently available for ABLV

Test	Specimen	Test detects	Time taken to obtain result
FAT	Fresh brain	Viral antigen	0.5 d (same day)
TaqMan assay (separate assays for pteropid and insectivorous ABLV)	Fresh brain	Viral genome	1 d (same day)
PCR for generic lyssavirus (nested RT-PCR and sequencing)	Fresh brain	Viral genome	2–3 d
Virus isolation using neuroblastoma cell cultures	Fresh brain	Live virus	5 d
Immunohistochemistry	Formalin-fixed brain	Viral antigen	2 d
Serum neutralisation test (rabies RFFIT)	Serum	Rabies neutralising antibody	3 d
Serum neutralisation test (ABLV FAVN)	Serum	ABLV neutralising antibody	5 d

FAT = fluorescent antibody test; FAVN = fluorescent antibody virus neutralisation test; RFFIT = rapid fluorescent focus inhibition test; RT-PCR = real-time polymerase chain reaction.
Note: Several of the tests are offered by state animal and public health laboratories in Australia (check with local laboratories).
Source: Information provided by CSIRO Australian Animal Health Laboratories (2006). Refer to AAHL for the most up-to-date information.

laboratory), from where they may be forwarded to the CSIRO Australian Animal Health Laboratory for testing or confirmation of positive or suspicious test results. If a human exposure has occurred, the state or territory laboratory will immediately notify the relevant public health department.

A positive result must be notified immediately to the Chief Veterinary Officer of the state or territory concerned, who will also immediately notify their health department equivalent.

Before shipping specimens, contact the receiving laboratory to discuss arrangements for sampling, transport and sample reception.

Whole unpreserved brains, severed heads or whole animals should be chilled and forwarded on ice to the testing laboratory. Freezing should be avoided unless chilling is not possible. If possible, animals should not be shot through the head due to loss of diagnostic material and the potential for dissemination and exposure to potentially infected material. If no brain material is available, the trigeminal ganglion or spinal cord should be submitted. Unless the operator is vaccinated and experienced, the head or brain should not be removed before submission because of the risk of self-inoculation. In the much smaller insectivorous and carnivorous bats, considerable architectural damage may occur to brains during removal; these species should therefore be submitted whole to the laboratory.

Care must be taken to identify the bat species as this information is extremely valuable. This may be difficult and requires experience, particularly for some insectivorous and carnivorous bats. If accurate identification is not possible, the whole carcass should be submitted to the laboratory so that it can be retained and properly identified. Digital photographs can aid identification.

Both unpreserved and formalin-fixed samples of other tissues could be collected if a differential diagnosis is required. Serum may be required for other purposes, such as the serum neutralisation test. The tests currently available for ABLV diagnosis are listed in Table 14.6.

Fluorescent antibody test (FAT) is the initial test of choice for the diagnosis of lyssavirus infections in domestic or wild animals because it is the most rapid and reliable of the tests used. It involves applying, to a smear of brain tissue, of specific fluorescein-labelled antibodies directed against the viral nucleocapsid proteins. Further tests depend on whether the case was associated with human exposures. To distinguish ABLV from rabies and other lyssaviruses, specific polymerase chain reaction (PCR) tests are used in conjunction with nucleotide sequencing (Animal Health Australia 2006). Methods for diagnosing ABLV infection are similar for samples from humans, bats or other animals and similar to those used for classical rabies. Appropriate laboratories for assessing samples include the CSIRO Australian Animal Health Laboratories (CSIRO Division of Animal Health, 5 Portarlington Rd, Geelong Vic. 3212 [contact the duty veterinarian]) and Queensland Health Scientific Services (39 Kessels Road,

Cooper Plains Qld 4108). These laboratories provide advice on sample submission and transport procedures. Further information on testing is available on the website of the Australian Government Department of Health and Ageing (www.health.gov.au/).

The available data indicate that the incubation period for ABLV is similar to that for rabies. The incubation period for rabies is typically 10 d to several months, and periods of several years have been reported. Once clinical signs develop death occurs within <1–2 wk (Animal Health Australia 2006).

Infected bats have been reported from the northern and eastern coastal areas of Australia. Inland, ABLV has been found in bats at Narromine (NSW) and near Mt Isa (Qld) (Field & Ross 1999; Garner & Bunn 1997). Serological evidence suggests a wide geographical and taxonomic distribution in bats in Australia (Field 2005). It should be assumed that all Australian bat species have the potential to be infected with, and transmit, lyssaviruses.

The prevalence of infection, and therefore risk to people, differs between clinically normal bats and sick, injured and rescued bats. The infection prevalence (by PCR or virus isolation) of ABLV infection in clinically normal wild-caught Australian bats is very low (<1%) (Field 2005). However, in sick, injured and rescued bats the prevalence of infection of ABLV (by PCR or virus isolation) in pteropodids varies from approximately 5–20% depending on the species (7.8% in black flying-fox, 16.9% in little red flying-fox, 4.6% in grey-headed flying-fox, 1.0% in spectacled flying-fox). Adult bats of some species with CNS signs appear to have a higher prevalence than comparable young bats. Prevalence also appears higher in adult animals of some species with CNS signs. In one study, sick and injured adult little red flying-foxes presenting with CNS signs had a prevalence of more than 60% (Barrett 2004). Prevalence in sick, injured and rescued insectivorous and carnivorous bats is lower (up to 5%) but also appears to vary with species. One study found that the yellow-bellied sheath-tail bat had significantly higher antibody prevalence (up to 62.5%) than other species, suggesting that it plays an important role in the ecology of ABLV (Field 2005). Sick and injured animals pose the greatest public health risk, at least to those who come in contact with them, e.g. veterinarians, veterinary staff and wildlife carers (Field & Ross 1999; Animal Health Australia 2006).

Because of the variability of nervous signs and the relatively high prevalence of infected bats coming into contact with humans through illness or accident, all humans who are likely to be in contact with bats should receive pre-exposure rabies vaccination. The current indications for pre- and post-exposure rabies vaccination, recommended dosage and regimen are described in the *Australian Immunisation Handbook* (ATAGI 2007). The handbook includes information about vaccination for people who work with live bat lyssavirus in research laboratories, other laboratory workers (e.g. those performing ABLV diagnostic tests) and veterinarians and their assistants who come into contact with bats.

Contact with bats should be strictly avoided by the general public. Further information on preventative measures can be found in the *Information Leaflet on Australian Bat Lyssavirus* (www.health.gov.au/), produced by the Australian Department of Health and Ageing, the Australian Department of Agriculture, Fisheries and Forestry and the Communicable Diseases Network Australia. Bats should only be approached by rabies-vaccinated, experienced bat handlers and animal health professionals. Only they should attempt to capture and care for sick or injured bats.

Only a pathologist who has been vaccinated against rabies should perform necropsies on bats. If the necropsy cannot be carried out in a biological safety cabinet, the pathologist should wear goggles, facemask, P2 respirator, thick rubber gloves and overalls or a waterproof apron. Carcasses should be incinerated, and used instruments should be soaked in disinfectant then boiled or steam sterilised.

If accidental exposure occurs—e.g. a person is bitten by a bat, or bat saliva contacts the hands or face—first-aid measures should be applied immediately to remove the pathogen. These include:

- immediately scrubbing the wound thoroughly with soap and water;
- rapidly applying an antiseptic—either a phenolic, alcoholic, halide or quaternary ammonium compound.

Proper cleaning of the wound is regarded as the single most effective way to reduce transmission of lyssaviruses. The bat should be retained and the local public health department should be immediately contacted for advice (Animal Health Australia 2006).

Double-wire enclosures are used to reduce the risk of transfer of this and other viral disease from free-living pteropodids to captive animals. It is worth remembering that ABLV should also be included as a differential diagnosis in any domestic animals showing suggestive neurological or behavioural signs of rabies-like illness in Australia.

Bats are frequently held in zoos and wildlife sanctuaries in Australia for display and education. The housing varies from double-mesh 'off exhibit' to large walk-through aviaries with public access. The following conditions apply in most of the major zoos: bats are not vaccinated against rabies; keepers who are responsible for the daily care of the animals are vaccinated against rabies and their level of protection is monitored. These arrangements are currently under review.

b Hendra virus (HeV)

In 1994 an outbreak of an acute respiratory illness occurred in Hendra, a suburb of Brisbane (Qld). Twenty-one horses and two people (the trainer and a stable hand) were infected; one of the people died. Six additional outbreaks (1994, 1999, two in 2004 and two in 2006) infected seven horses and two people, killing one person (Field et al. 2007). A virus (Family Paramyxoviridae, Genus Henipavirus) was shown to be the aetiological agent of the disease (Calisher et al. 2006; Field & Ross 1999; Murray et al. 1995; Philbey et al. 1998; Young et al. 1996).

Based on antibody surveys, all Australian flying-fox species appear to be reservoirs throughout their natural range (Halpin et al. 1996; Field & Ross 1999). Pteropodids appear to be well-adapted to the virus and no clinical disease syndrome has been identified. However, humans, dogs, cats, guinea pigs and horses are known to develop severe and frequently fatal illness following infection. Horses, cats and guinea pigs can excrete virus in their urine (Calisher et al. 2006). Field (2005) proposed three alternative models for the maintenance of infection:

- infection is enzootic in all *Pteropus* species throughout their distribution;
- infection is enzootic in a particular *Pteropus* species with a periodic epizootic pattern in the other species;
- infection is periodically epizootic in all these species, persisting in a spatial or temporal mosaic across their distribution.

He contends that the pattern of spillovers from pteropodids to horses fits better with the second and third hypotheses. That is, a periodic outbreak in a local population of pteropodids results in an increased probability of spillover to horses in a specific locality during a limited period (Calisher et al. 2006). Further research is necessary to understand the biology of the virus and pathogenesis of the disease (Field & Ross 1999).

HeV does not appear to be highly contagious in experimentally infected domestic species (Williamson et al. 1998). A sero-survey (Selvey et al. 1996) of Australian bat handlers failed to detect evidence of exposure to HeV, suggesting humans in regular close contact with pteropodids are at minimal risk of HeV infection. Although it is not certain how HeV could be transmitted from pteropodids to horses, the isolation of the virus from uterine fluid and aborted foetal tissue of pteropodids suggests that horses could ingest the virus on feed or pasture recently contaminated by birth products. Ingestion of pasture contaminated with urine or the ingestion of saliva-laden spats (expelled masticated food pulp) are other plausible modes of transmission. The available evidence indicates that transmission of the virus from horses to people, albeit rare, occurs through physical contact with nasal and oral secretions emanating from very ill, dying or dead horses (Hanna et al. 2006). HeV is a notifiable disease in Australia.

HeV is diagnosed post-mortem by a range of laboratory methods including serology, PCR and virus isolation on a range of tissues including lung, liver, kidney and spleen (Murray et al. 1995; Halpin et al. 1996; Hooper et al. 1996). Captive populations can be regularly tested for antibodies to HeV, but further research is needed to determine the relationship between exposure, shedding and titres.

c Menangle virus

Menangle virus is another paramyxovirus (genus *Rubulavirus*) isolated from stillborn piglets during an outbreak of reproductive disease at a piggery at Menangle (NSW) in 1997 (Ross et al. 2001). Two of 250 people in contact with the infected pigs had high titres of antibodies to the virus and both reported a febrile illness with a measles-like rash, but neither had direct exposure to pteropodids. Although attempts to isolate virus from pteropodids were unsuccessful, paramyxovirus-like virions labelled with antibody to Menangle virus from a convalescent sow were seen by electron microscopy in flying-fox faeces collected beneath the roost near the piggery (Calisher et al. 2006). Bats from a nearby colony were also antibody-positive for Menangle virus (Hall & Richards 2000; Ross et al. 2001). A sero-survey of pteropodids in far north Queensland also found evidence of exposure to this virus (Hall & Richards 2000). The clinical presentation in bats is unknown. Ante-mortem diagnosis relies on detecting circulating antibodies.

9.8.2 Bacterial diseases

The normal bacterial flora of the gastrointestinal tract of pteropodids and other frugivorous species is predominantly Gram-positive and aerobic (Heard et al. 1997). In contrast, insectivorous and carnivorous bats have been found to be predominantly Gram-negative and aerobic (*Enterobacteriacae*).

Primary bacterial disease is uncommon (Heard 2003). Most infections are sequelae to interspecific (cats and dogs) and intraspecific bites, often caused by *Pasteurella multocida*. Most bats heal well and severe infections associated with bites are uncommon. However, young or stressed animals are more likely to develop significant cellulitis, osteomyelitis and abscesses, with possible septicaemia and bacteraemia. Treatment, as in other mammals, is based on wound cleaning, debridement if indicated and antimicrobial therapy preferably based on culture and sensitivity. Blood culture may be useful in identifying the source of a systemic bacterial infection.

In pteropodids, puncture wounds from barbed-wire fencing or shot pellets may cause septic arthritis. Onset of clinical signs (hot painful swollen joints) is usually rapid. Medications that penetrate the synovial capsule include the cephalosporins, extended-spectrum penicillins, fluoroquinolones, azithromycin, aminoglycosides and lincosamides (Table 14.7). Septic arthritis unresponsive to antimicrobials is surgically explored, cultured and flushed. This may need to be repeated several times, and arthrodesis may occur during healing. This may be an indication for euthanasia.

Bacterial placentitis and metritis can be a cause of or sequel to dystocia (D Heard *pers. comm.*). Bacterial tooth root abscesses occur in pteropodids and are usually associated with pulp cavity exposure. Bacterial gingivitis is common in captive insectivorous and carnivorous bats due to malnutrition and soft diets, and may progress to debilitating tooth loss and osteomyelitis (see 9.2.1).

Antibodies to *Leptospira* spp. serovars have been identified in the serum of many bat species (Constantine 1970). Serological studies have identified pteropodids as carriers of leptospirosis (Smythe et al. 2002). Leptospiral antibodies were detected in 28% of sera of 271 pteropodids sampled across northern and eastern Australia. *Leptospira kirschneri* serovar *australis* was most frequently identified. It was assumed, based on distribution and potential contact with other flying-fox species, that Asian bats are also reservoirs for this bacterium. Cox et al. (2005) found 11% of 173 kidney samples from four species of Australian pteropodids were positive for leptospiral DNA. Positive individuals were identified in each species. Significant differences were not detected between species, between species within the same geographic area or between geographically separated samples from the same species. Of 46 urine samples tested, 18 (39%) tested positive by PCR, confirming that pteropodids are carriers of pathogenic *Leptospira* spp. and shed them into the environment.

9.8.3 Fungal diseases

Fungal infections are not uncommon in bats and have been discussed elsewhere in this chapter (see 9.1.5). Histoplasmosis is caused by the dimorphic fungal soil saprophyte *Histoplasma capsulatum*. Infection is usually asymptomatic or a benign respiratory disease. Bats have been implicated as a source of the organism but the risk of infection is considered low (Heard 2003).

9.8.4 Parasitic diseases

Chiropterans are hosts to numerous parasites, primarily ectoparasites, most of which are non-pathogenic or insignificant to bat population health. Trematodes are the main macro-endoparasites of insectivorous bats (Heard 2003).

a Toxocara pteropodis

Toxocara pteropodis is a large ascarid found in all Australian and south-east Asian pteropodids (Prociv 1990). The adults and ova are found only in suckling pups. When the pups wean the adult nematodes die and may be seen in the faeces. The ova are ingested by adult bats in the colony during mutual grooming and eating of contaminated fruit and leaves. The larvae hatch in the gastrointestinal tract then migrate to the liver, where they remain dormant. In lactating females the larvae migrate to the mammary glands and pass out in the milk to the pups. The larvae in the male livers eventually die and are encysted. This parasite rarely causes morbidity and mortality (Heard et al. 1995). Death associated with *T. pteropodis* obstruction of the upper airway with worms in a grey-headed flying-fox pup has been reported (Prociv 1990). In hand-reared grey-headed (L Vogelnest *pers. comm.*) and spectacled flying-fox pups ill thrift is commonly associated with a significant ascarid infestation in the stomach and intestines (see 9.2.3).

Pyrantel (11 mg/kg PO) is the drug of choice. Although anthelmintics are administered at the same dosages as in other small mammals, care should be taken when calculating doses and administering drugs with low therapeutic index to small species (DeMarco et al. 2002).

b *Angiostrongylus*

Neuro-angiostrongylosis in pteropodids has been reviewed (Barrett 2004). All reported cases of angiostrongylosis in pteropodids in Australia have been attributed to *Angiostrongylus cantonensis.*

Histological evidence of infection has been found in grey-headed, little red, black and spectacled flying-foxes (ARWH 2007; Barrett 2004; Reddacliff et al. 1999). The majority of cases have been from Queensland and New South Wales. It appears to be a relatively common cause of neurological disease. In the most comprehensive study to date (Barrett 2004), 18 of 100 pteropodids (18%) with clinical CNS signs had gross or histological evidence of neuro-angiostrongylosis involving the brain. The clinical picture was dominated by paresis, particularly of the hind limbs, and depression. None were aggressive and 12 of the 18 were well-nourished, with ample fat reserves.

Diagnosis is confirmed histologically. There is eosinophilic and granulomatous meningoencephalitis. Sections of nematodes may be found in brain sections. The meningitis is typically most severe around the brain stem and cerebellum, with macrophages predominating in the infiltrate, often with large numbers of lymphocytes and eosinophils, some plasma cells and occasionally multinucleated giant cells. Nematode sections are most frequently seen in the subarachnoid space of the cerebral sulci and/or cerebellar folds (Barrett 2004).

Presumably, pteropodids acquire infection by ingesting intermediate or paratenic hosts such as slugs, snails and planarians, or their slime (although slime has not been confirmed to be a vehicle of infection). These hosts could readily be concealed among fruit. It might seem unlikely that pteropodids would eat these invertebrates, but faecal analysis has shown that pteropodids regularly ingest insects, spiders and mites. Slugs and snails do climb high into trees, especially in wet weather, and might prove attractive to a hungry flying-fox or be ingested unintentionally (Barrett 2004).

The circumstances in which pteropodids with neuro-angiostrongylosis are found, and the initial clinical signs, closely resemble those of pteropodids with the paretic form of ABLV infection (see 9.8.1a). However, whereas most bats with ABLV rapidly deteriorate within 24–48 hr of being rescued and die or are euthanased *in extremis,* most pteropodids with neuro-angiostrongylosis have a comparatively prolonged (>3 d) and stable clinical course, and are more often euthanased as a result of failure to improve and an assessment that the bat is unsuitable for release, rather than *in extremis.* Some pteropodids with neuro-angiostrongylosis have periods of temporary improvement in CNS functions (e.g. appearing stronger or regaining the ability to hang), which never occurs in bats with ABLV (Barrett 2004).

However, it needs to be remembered that ABLV should be suspected and ruled out in all pteropodids with evidence of neurological disease. Because of this, and the fact that treatment is unrewarding, pteropodids with suspected angiostrongylosis should be euthanased. Precautions outlined for personal protection against ABLV should be followed when handling these animals.

c Haemoparasites

Hepatocystis spp. are common in free-ranging tropical and subtropical bats and may be confused, on stained blood smears, with more pathogenic *Plasmodium* spp. *Hepatocystis levenei,* identified in grey-headed flying-foxes from Queensland, is transmitted by *Culicoides nubeculosus* (Boulard et al. 1985; Landau et al. 1985). *Hepatocystis* parasitaemias are self-limiting in captivity (D Heard *pers. comm.*).

d Ectoparasites

Nycteribiid and streblid flies (spider-like wingless flies) are ubiquitous non-pathogenic ectoparasites of bats worldwide. The former parasitises primarily bats of the Old World tropics. Streblid flies are common on both New and Old World tropical bats. Other external parasites include fleas, mites and ticks. These may cause alopecia if pruritus develops and bats scratch excessively.

Since the mid 1990s spectacled flying-fox maternity colonies in northern Australia have been repetitively affected by the paralysis tick *Ixodes holocyclus*, with high morbidity and mortality (A Olsson *pers. obs.*; P Tully *pers. comm.*; J Maclean *pers. comm.*). It is the commonest cause of illness in wild bats from that region. Clinical signs include congested or cyanotic mucus membranes, laboured breathing with grunting, and exaggerated abdominal heaving as the diaphragm is progressively paralysed.

In addition to respiratory effects, bats with tick toxicity have prolonged corrected QT intervals, bradycardia and rhythm disturbances including sinus bradydysrhythmia, atrial standstill, ventricular premature complexes and idioventricular bradydysrhythmia (Campbell 2003). The QT prolongation is thought to reflect delayed ventricular repolarisation and predisposes to polymorphic ventricular tachycardia and cardiac arrest in response to sympathetic stimulation.

Animals are most commonly presented between October and December each year, which coincides with the months in which bats give birth. Coincidentally, during mortalities of birthing mothers in 1998 and 2001 large numbers of orphans were found with palatine defects (see 9.9.5).

Diagnosis of tick paralysis is based on recognition of clinical signs associated with progressive respiratory and peripheral nervous system paresis and paralysis, and finding the tick *in situ*. Mid-thoracic spinal injury may cause similar signs. Treatment includes canine tick antiserum from hyperimmune dogs (given at 1–2 mL/kg) and symptomatic management with sedatives, atropine, bronchodilators and diuretics as needed (Atwell 2001; Campbell 2003). As for other mammals, tick paralysis is progressive and frequently fatal if not treated. Euthanasia is considered if the paralysis is advanced and respiratory failure is inevitable.

9.9 Non-infectious diseases

9.9.1 Neoplasia

Neoplasia has been reported in non-Australian bats overseas (Barnard 1995; Andreasen & Duimstra 1996). Craniofacial tumours, including soft tissue sarcomas, carcinomas (e.g. haemangiosarcoma and squamous cell carcinoma) and osteosarcomas have been identified in aged pteropodids (A Olsson unpub; D Heard *pers. comm.*) In Australia, neoplasia in bats appears rare. A malignant mesothelioma of the peritoneal cavity has been reported in a grey-headed flying-fox (R Johnson *pers. comm.*).

9.9.2 Toxins

a Lead poisoning

Lead poisoning in Australia's urban-dwelling pteropodids is likely due to inhalation and ingestion of lead particles from industrial and vehicle emissions or lead arsenate in insecticides applied to fruit (Barrett 2004; Hariono et al. 1993; Skerratt et al. 1998; Sutton & Wilson 1983). Pteropodids that feed in fruit orchards or those that roost in areas where they are not wanted are often shot (Hall et al. 2001; Parry-Jones 2001; Pierson 1984). However, lead in tissue rarely causes poisoning.

Clinical signs of lead poisoning have been described as an inability to fly, loss of appetite, weakness, emaciation, unco-ordinated movements, muscle tremors and anaemia, very high lead concentrations in the kidney (370 ppm) and moderately high in the liver (16 ppm) (Skerratt et al. 1998). The animal described by these authors later tested positive to ABLV, which was then believed to have been the primary cause of death (Barrett 2004). The threshold above which blood or tissue levels of lead in pteropodids are associated with clinical toxicity are unknown (Barrett 2004). Lead poisoning in pteropodids could be suspected when blood lead concentrations exceed 5 ppm (μmol/L) and blood protoporphyrin concentrations are elevated. Liver lead levels from 50 pteropodids in apparent good health captured from camps in urban Brisbane (Qld) were <5 ppm wet weight and consistent with levels in healthy individuals of other species (Barrett 2004). Hariono et al. (1993) suggested that levels of >10 ppm wet liver and >25 ppm wet kidney indicated toxicity in urban bats.

Lead poisoning in pteropodids can be treated with calcium disodium edetate (Calsenate, Parnell Laboratories), 200 g/L diluted to 10 mg/mL, 100–200 mg/kg SC sid 7 d (Booth 1994). Concurrent supportive therapy includes parenteral fluids, nutrition and benzodiazepines to control convulsions. Blood lead concentrations should be monitored for 2–3 wk after therapy (Rose 1999).

The incidence of lead poisoning may be decreasing in pteropodids in Australia. All new cars in Australia are required to use lead-free petrol, which may explain the lower levels of background lead in recent studies on pteropodids (Barrett 2004). Barrett (2004) did not diagnose lead poisoning in 100 sequential cases of neurological disease in pteropodids.

b Rodenticide poisoning

Bats do not eat rodenticides. However, those that roost in buildings are poisoned when rodenticides are fumigated into wall spaces and attics (Greenhall 1982; Tuttle 1988). Non-targeted wildlife are poisoned when crops are sprayed (Solymar 2001). Clinical signs include lameness (reluctance to hang by feet alone, shifting from leg to leg while hanging), swollen joints, ventral haematoma, dyspnoea, respiratory depression, nasal and rectal bleeding and vomiting of blood. Necropsy reveals internal bleeding. Treatment for anti-coagulant rodenticide poisoning is as for other species, using vitamin K.

c *Fluorosis*

Fluorosis-induced multicentric hyperostosis has been described in three captive pteropodid species—Indian fruit-bat (*P. giganteus*), grey-headed flying-fox and Egyptian fruit-bat (*Rousettus aegyptiacus*)—from the same facility (Duncan et al. 1996). All affected animals were more than 2 yr old. The fore limb bones were most frequently involved. Lesions began as discrete eccentric asymmetric bone proliferations that frequently coalesced into large raised firm areas. Characteristically, articular bone surfaces were not involved and tendons and ligaments diverted around the nodules. Skin over the nodules was swollen, taut, warm, smooth, hairless and predisposed to abrasions. Necropsy bone specimens had markedly raised fluoride levels (2200–4700 ppm) compared with similar samples from another collection (290–320 ppm). Diet analysis revealed shrimp meal (1400 ± 200 ppm) and dicalcium phosphate (1900 ± 300 ppm) to be the main fluoride sources.

9.9.3 Nutritional disorders

a *Metabolic bone disease*

Dietary calcium may be the major limiting factor in reproductive success, growth and adult size in pteropodids (Barclay 1994). This is because of their large size at birth and their rapid growth to free-flying individuals.

Although uncommon in free-ranging bats, metabolic bone disease is readily induced in malnourished captive animals, particularly young rapidly growing animals with a high calcium and vitamin D requirement. The disease is due to inadequate dietary calcium, a calcium and phosphorus imbalance, or a lack of exposure to dietary vitamin D (see 8.1). In the wild, ingestion of native figs and folivory are a significant source of calcium, essential micronutrients (e.g. vitamin E) and protein (Kunz & Ingalls 1994). It is also possible that vitamin D activation associated with exposure to ultraviolet light is important in normal calcium dynamics in pteropodids (Heard 2003).

Clinical signs include bone folding and pathological fractures. These usually involve the long bones but in severe cases the spine and pelvis may be affected. Collapse of the pelvis will result in dystocia in females that survive to maturity. Swollen joints occur in juveniles hand-reared on inappropriately formulated milk replacers, as in other mammals (Speare 1988). Finger joints, particularly the terminal ones, are most commonly affected. Swellings are non-painful and hard, with shiny reddened smooth skin.

Diet and husbandry correction is effective unless deformities are severe. Oversupplementation may occur, especially in small species. Thickened bone cortices and epiphyses are seen.

b *Vitamin B_{12} deficiency*

Vitamin B_{12} (cobalamin) deficiency is induced in frugivorous and folivorous bats fed an unsupplemented diet of washed and peeled fruits (bananas, papayas, pears and oranges) (Green et al. 1975). Clinical signs of abnormal flying and ataxia are primarily related to the development of a demyelinating myelopathy. Free-ranging pteropodids obtain Vitamin B_{12} from the bacteria in their gastrointestinal tract and by consuming insect-infested fruit. Chronic antibiotic therapy may destroy the gastrointestinal tract flora and increase the likelihood of deficiency. In captivity, vitamin B_{12} is provided by adding multivitamins to the fruit diet, supplementing with low-fat powdered milk or Wombaroo high-protein supplement, or feeding a complete balanced diet including at least some native fruits.

c *Iron storage disease*

Haemochromatosis (iron storage disease) has been described in captive Egyptian fruit-bats but has not been reported in Australian species (Farina et al. 2005).

d *Protein deficiency*

Protein deficiency is usually caused by malabsorption exacerbated by stress. Other causes include malnutrition, chronic diarrhoea, protein-losing nephropathy (Deem et al. 1999), infection, haemorrhage, trauma, burns and critical illness. In addition to anaemia, protein malnutrition is associated with marked hypoalbuminemia, oedema, muscle atrophy, delayed wound healing and an impaired immune system. Although there are often no signs of emaciation, growth is depressed in young animals. Hepatic lipidosis may also occur. The normally low total plasma protein (54–87 g/L) makes it difficult to interpret deficiency.

Treatment includes providing an adequate well-balanced diet and correcting and treating any underlying disease.

e Hypervitaminosis A

Hypervitaminosis A in carnivorous bats fed excessive amounts of liver will cause painful and swollen joints. This is corrected by feeding killed whole mammal or bird prey.

9.9.4 Electrocution

Electrocution from contact with power lines is a significant cause of injury and death in pteropodids in suburban areas. In some fruit-growing areas deterrent electric grids have been used to electrocute frugivorous bats. Insectivorous and carnivorous bats may receive similar damage to the extremities when they come in contact with exposed electrical wires inside buildings.

Burns to the pinnae, wing membranes, uropatagium, toes and/or thumbs are common with electrocution. These may require surgical debridement or amputation if ischaemia occurs. If sparks are produced on contact, the membranes of both wings are frequently pitted with small holes. Occasionally, electrocuted bats have pronounced muscle fasciculations, seizures and cardiac arrhythmias.

Due to the insulating property of fur, infant pteropodids frequently survive the electrocution that kills the mother. However, the mouths of infant bats should be examined for burns sustained from scalding milk when the electric current courses through the mother's body (Hall & Richards 2000).

Electrocuted bats are treated symptomatically. Seizures and fasciculations are controlled with diazepam (2 mg/kg IV or IM). Treatment is repeated (1–2 mg/kg qid to tid) as indicated. Burns are cleaned >2 × daily with sterile saline solution, and Silvazine cream (silver sulphadiazine 10 mg/g and chlorhexidine digluconate 2 mg/g, Smith & Nephew) applied. Parenteral antimicrobial drugs (e.g. enrofloxacin 5 mg/kg SC, PO sid) may be required. Parenteral administration of amikacin (5–10 mg/kg SC, IM, IV tid) is used in severe burns associated with marked cellulitis or bacteraemia and septicaemia. The animal must be hydrated using physiological saline or Hartmann's solution (see 11.2, Tables 14.8 and 14.9) to reduce renal tubular injury. Surgery is delayed until the full extent of the damage is evident (usually over 1 wk). Even apparently unharmed bats are kept for observation for at least 1 wk post-electrocution since necrosis and internal injuries may not be evident immediately. Ischaemic necrosis of the patagium may only become apparent a week or more following electrocution.

9.9.5 Congenital defects

Portosystemic shunts in neonatal pteropodids have been reported. Signs include ill thrift, ascites, jaundice and depression. Euthanasia is recommended in these animals.

In 1998 and 2001 several north Queensland spectacled flying-fox colonies produced 35–45 neonates per season with congenital palatine defects, ranging from pinpoint holes to complete rostrocaudal separation (J Maclean *pers. comm.*; A Olsson *pers. obs.*). Craniofacial abnormalities were noted in some, although facial symmetry remained. Some had poor fur growth or alopecia on the face and head. Most had some form of orthopaedic abnormality, particularly length and shape of toes and claws, with associated hanging difficulties. Postulated causes include exposure to toxic compounds (plant alkaloids) (Hall & Richards 2000) or some infectious agent (e.g. virus) during the first trimester of pregnancy. Surgical correction was successfully attempted on some bats with minimal palatine defects, but most were euthanased.

Imperforate anus is occasionally seen in neonatal bats. Absence of faeces and a bulging perineum accompanied by abdominal distension and discomfort are obvious indicators. Surgical correction under local anaesthesia to resect the persistent membrane is usually successful. In small bat species, this procedure is difficult to perform and euthanasia is recommended. Regardless of species, however, as these animals clearly would not have survived in the wild without intervention it can be argued that they should be either euthanased or kept in captivity, and not returned to the wild.

9.9.6 Degenerative diseases

Degenerative joint disease is occasionally seen in ageing bats. Non-steroidal anti-inflammatory drugs (Table 14.7) are used for pain relief. In small insectivorous and carnivorous bats copper indomethacin (Avigesic,

Vetafarm) at 0.1 mL/200 g, diluted to 1:10 for bats <100 g, can be used. Alternatives include pentosan polysulphate (3 mg/kg SC weekly for 4 wk) and tramadol (0.5–2 mg/kg PO sid).

10 ZOONOSES

The main zoonoses associated with Australian bats are ABLV and HeV (see 9.8.1a and b). Avoiding the exposure of horses to pteropodids, by not placing food troughs under trees in which pteropodids feed, may be useful in preventing HeV. Veterinarians should be vigilant regarding sudden acute respiratory disease with pulmonary oedema in horses and include HeV in the list of differential diagnoses. Guidelines for veterinarians examining horses have been presented (Queensland Department of Primary Industries and Fisheries 2005). Other potential zoonoses include leptospirosis, histoplasmosis, bite wound infections and occasionally ringworm. Melaka virus and Nipah virus have not been reported in Australia.

11 THERAPEUTICS

11.1 Drug administration

Intravenous and IM drug administration in bats has been described (section 7). Many drugs can be given SC at the scruff of the neck using a similar technique as for small domestic mammals. Intraperitoneal injections, especially for fluid administration in bats weighing <100 g, are given using a similar technique for small domestic species, using a 25 G needle inserted paramedially from the ventral surface of the bat craniodorsally into the peritoneal cavity, taking care to just penetrate the abdominal muscle wall. Tenting the abdominal skin and muscles prior to injection assists correct placement of the needle.

When administering drugs orally to bats, care is required to minimise the risk of being bitten. Many bats will lick pleasant-tasting oral medications from a syringe while hanging in a cage, or medication can be hidden in a small amount of food or in whole prey for carnivorous bats. If physical restraint is required for administration, wrap the bat in a towel or cloth. Always place bats in a head up position to medicate orally to prevent medication running out of the mouth.

11.2 Drug use

Pharmacokinetic studies on some anaesthetics such as xylazine and ketamine (Heard et al. 1996a) have been conducted on bats, but generally empirical use of drugs in bats is based on dosage regimens used in small companion animals, particularly dogs. Table 14.7 lists some of the therapeutic agents successfully used in bats. Sulphonamides routinely used in captive bats (Lollar & Schmidt-French 1998; Barnard 1995) rarely cause bone marrow hypoplasia and swollen joints. Ivermectin is avoided in bats due to potential toxicity (DeMarco et al. 2002).

11.3 Fluid therapy

The goal of fluid therapy is to replace deficits within the first 24 hr, as well as provide for maintenance fluids (Tables 14.8, 14.9). Unless contraindicated, oral nutrition is also provided. In large bats (e.g. pteropodids) fluids are administered IV through the cephalic vein or uropatagial veins. A bolus equivalent to 10% circulating blood volume may be given initially. However, it is often impractical to maintain an intact IV line unless the bat is moribund or sedated. Fluids may be administered via an intraosseous catheter placed in the distal radius or proximal tibia. In very small bats a small-gauge hypodermic needle is used, while in a large bat a spinal needle is used to prevent a bone plug. Intraperitoneal and SC fluids are administered to bats of all sizes, but are particularly useful in fluid delivery to small bats <100 g.

12 REPRODUCTIVE MANAGEMENT

Separating the sexes is the most common option for reproductive control in all bat species. Where indicated in larger bat species, surgical sterilisation is a straightforward procedure in both sexes (see section 13). Surgical sterilisation is appropriate for captive bats where breeding is not desirable, especially for common species or commonly represented genetic lines. Other indications include wild individuals in permanent care, and females with pelvic abnormalities likely to result in dystocia. Melengestrol acetate contraceptive implants have been used successfully in female Rodrigues fruitbats (*Pteropus rodricensis*) with minimal effect on social behaviour (Hayes et al. 1996). However, they were associated with weight gain and retarded hair growth above

Table 14.7 Therapeutic agents used in bats

Drug	Dose, route and interval	Comment
Antibiotics		
Amoxycillin	10 mg/kg IM, SC, PO bid	
Amoxycillin/clavulanic acid	12.5 mg/kg SC, IM sid	
	12.5 mg/kg PO bid	
Enrofloxacin	5 mg/kg SC or PO sid	
Metronidazole	25 mg/kg PO sid	
Trimethoprim/sulfadiazine	30 mg/kg IM, SC, PO bid	
	30 mg/kg PO sid or bid	
Antifungals		
Fluconazole	10 mg/kg PO	
Miconazole	Topical 10–14 d	
Nystatin	0.5 mL/100 g PO qid	Can use topical ointment
Anthelmintics		
Fenbendazole	25 mg/kg PO once and repeat in 14 d	Not in bats <100 g
Praziquantel	5–20 mg/kg PO as required	Cestodes, trematodes
Pyrantel	11 mg/kg PO once	
Anaesthetics and sedatives		
Diazepam	0.5–2 mg/kg IM, IV	Sedative, anticonvulsant
Isoflurane	5% induction, 2.5% maintenance	
Medetomidine	0.3–0.8 mg/kg IM	Antagonise with atipamezole 5 times the medetomidine dose (mg/kg) IM
Medetomidine 1 mg/mL/ketamine 100 mg/mL	0.0125 mL/kg combination IM	
Propofol	8–10 mg/kg IV	
Tiletamine/zolazepam	10–40 mg/kg IM	Absorbed across mucus membranes
Xylazine 20 mg/mL/ketamine 100 mg/mL	0.2–0.4 mL/kg combination IM	
Atipamezole	5 times medetomidine dose (mg/kg) used (IM)	
Analgesics and antiinflammatory drugs		
Buprenorphine	0.03 mg/kg SC, IM TID or bid	
Carprofen	2–4 mg/kg PO sid or bid	
Meloxicam	0.08 mL/kg SC sid	
Tramadol	0.5–2 mg/kg PO sid	
Miscellaneous		
Calcium disodium edentate	100–200 mg/kg SC sid	
Canine tick paralysis antiserum	1 mL/kg SC, IP, slow IV if premed with dexamethasone	With symptomatic management
Dexamethasone	0.1–2 mg/kg IV, SC	Dose depends on condition being treated
Furosemide	2–8 mg/kg IM, PO	Pulmonary oedema
	2–8 mg/kg PO sid for 3–5 d	
Metoclopramide HCl syrup	2 drops of diluted mix PO qid	Dilution 1 drop/4 drops water
Pentosan polysulphate	3 mg/kg SC weekly for 4 wk	
Lignocaine 2% jelly	Topically	

Table 14.8 Fluid therapy for bats <100 g body weight

Weight of bat (g)	Dosage[1] (mL/g)	No. injections in 24 hr or as needed[2]
1.0–1.1	0.03–0.04	3–5
1.2–1.5	0.04–0.05	3–5
1.6–1.9	0.05–0.06	3–5
2.0–2.4	0.06–0.07	3–4
2.5–2.9	0.07–0.08	3–4
3.0–3.4	0.08–0.09	3–4
3.5–3.9	0.09–0.10	3–4
4.0–4.4	0.10–0.15	3–4
4.5–4.9	0.15–0.20	2–3
5.0–5.4	0.20–0.25	2–3
5.5–5.9	0.25–0.30	1–2
6.0–6.5	0.25–0.30	1–2
6.5–6.9	0.30–0.35	1–2
7.0–7.4	0.35–0.40	1–2
7.5–7.9	0.40–0.45	1–2
8.0–8.4	0.45–0.50	1–2
8.5+	0.50+	Volume and no. injections calculated at maintenance of 80 mL/kg/d (0.8 mL/100 g/d)

1 The higher dosage is the maximum safe amount of fluid that can be delivered in one injection.
2 It is not necessary to inject bats throughout the night. Injections can begin at approximately 6 am and end at midnight.
Source: Modified from Barnard (1995).

Table 14.9 Fluid therapy for bats >100 g body weight

Body weight (g)	mL/24 hr	mL/hr	mL/6 hr injections
100	8	0.33	2
150	12	0.5	3
200	16	0.67	4
250	20	0.83	5
300	24	1.0	6
350	28	1.17	7
400	32	1.33	8
450	36	1.5	9
500	40	1.67	10
550	44	1.83	11
600	48	2.0	12
650	52	2.2	13
700	56	2.3	14
750	60	2.5	15
800	64	2.67	16
850	68	2.83	17
900	72	3.0	18
950	76	3.17	19
1000	80	3.33	20

Fluids can be administered IV or SC.
Source: Adapted from Ilkew (1984).

the inplant site. There was also a 22% rate of rejection or loss of implants, suggesting improper surgical placement or removal via grooming.

13 SURGERY

13.1 Surgical considerations

During surgical preparation care should be taken to avoid chemical damage to the skin and patagium. Chlorhexidine solution is non-irritating in comparison to iodine-based compounds and alcohol (Heard 2003).

Pteropodids are notorious for chewing sutures, licking wounds and exposed bone ends incessantly and removing bandages. Perioperative analgesics (Table 14.7) are essential to reduce self-trauma. Intradermal suturing can be used but, even so, self-trauma to the suture line can be expected if the bat can physically reach the site. Bats, especially pteropodids, can be fitted with Elizabethan collars or home-made neck braces. However, pteropodids will work at removing collars and braces. Vigilance is required to ensure wounds remains relatively intact while healing. In some cases it is preferable to leave wounds open to heal by secondary intention as suturing or application of medications seems to increase a bat's attention to the wound.

13.2 Orthopaedics

Fractures to larger limb bones occur as a result of shooting injuries, collisions with vehicles, power lines and trees and occasionally predator attacks. Frequently, fractures are so severely comminuted that euthanasia or amputation are the only management options. Internal fixation using pins, hypodermic needle shafts or spinal needles, or Type 2 external fixation devices (Heard 2003) are used if the fracture is simple. Bandaging and splinting fractures is not recommended because they are difficult to apply, can damage patagial membranes and are very poorly tolerated by bats, which will constantly chew at the dressings.

Metacarpal and phalangeal fractures are common. Small distal fragments can be amputated without reducing flight ability. However, wounds generally heal poorly over digits, and constant licking by the patient often results in exposure of the bone and ischaemic necrosis of the dactylopatagium. Therefore, digit and thumb injuries require careful consideration. Loss of any digit is usually accompanied by loss of a significant portion of patagium and resultant loss of flying ability, which in a wild bat is an indication for euthanasia. It has been our experience that many pteropidids cope with part or all of one thumb missing and can be released. Similarly, there are reports of wild and apparently healthy pteropodids with no thumbs. However, it is likely that these types of injuries compromise the animals and their release should be approached with caution. Many veterinarians consider loss of both thumbs an indication for euthanasia.

13.3 Wing membrane surgery

Most large membrane injuries carry a guarded prognosis, especially when the edge of the membrane is lacerated. Forces keep the wound distracted and dehiscence is common regardless of repair technique. Most bats with severe wing lacerations are not releasable. Blood is supplied to the membrane via a network of very fine vessels which are readily damaged during injury and surgery, leading to ischaemic necrosis of often significant areas of tissue around the injury. For this reason it is preferable to leave lacerations unsutured for a few days until all the damage is evident. Once devitalised areas are debrided an attempt can be made to suture. A simple interrupted pattern using 3/0–9/0 non-absorbable suture tacks the wound together, and tissue glue can be used to hold wound edges in apposition. Sometimes the spot-tacking holds wound edges together long enough for the resultant gaps to heal closed by secondary intention.

Heard (2003) describes a technique to stabilise large wing lacerations, where the edges are debrided then pinned to a corkboard for alignment. Tissue glue is then applied to adhere the edges together.

Scar tissue usually follows large defects in the wing membrane, and this can cause flexion of the digits. Smaller holes and tears, however, often heal without treatment.

13.4 Soft tissue surgery

Intra-abdominal surgery is usually performed through a midline longitudinal incision. The *rectus abdominus* muscles are large and should be avoided when making a paramedian incision (Heard 2003). Otherwise, surgery in this region is as for other small mammals.

13.4.1 Vasectomy

Vasectomy is performed either by incising the ventral scrotal sac and identifying the vas deferens as it exits the epididymis (LaFortune et al. 2004) or by incising in the prescrotal area. Good-quality surgical loupes with halogen headlamp and microsurgical instruments facilitate the surgery. A small segment of the vas deferens is removed and the ends ligated. Histological confirmation that the correct tissue was excised is useful.

13.4.2 Castration

A closed castration technique should be used because of the open inguinal ring. A prescrotal incision as in dogs can be used.

13.4.3 Caesarean section

Caesarean section is performed through a midline abdominal incision and a longitudinal incision in the uterine body wall. Once the foetus and placenta is removed, closure is as for other mammals. Tubal ligation or ovariohysterectomy may be performed at the time if indicated. Abdominal sutures are buried and an Elizabethan collar is fitted to prevent the surgical site from being self-traumatised. Close observation of the bat for wound dehiscence or self-trauma following abdominal surgery is essential.

ACKNOWLEDGMENTS

The authors thank Sue Barnard, Andrew Breed, Hume Field, Darryl Heard and Rick Speare for commenting on the manuscript and contributing personal observations; and bat rehabilitators Jenny Maclean (Tolga Bat Hospital, Tolga, Qld), Pam Tully (Batreach, Kuranda, Qld) and Robin Wingrave (Far North Queensland Wildlife Rescue Association Inc.) for sharing their personal experiences.

REFERENCES

Allworth A, Murray K & Morgan J (1996). A human case of encephalitis due to a lyssavirus recently identified in flying-foxes. *Communicable Diseases Intelligence* **20**, 504.

Animal Health Australia (2006). *Disease Strategy: Australian Bat Lyssavirus* (Version 3.0). Australian Veterinary Emergency Plan (AUSVETPLAN) (3rd edn). Animal Health Australia: Canberra.

Andreasen CB & Duimstra JR (1996). Multicentric malignant lymphoma in a pallid bat. *Journal of Wildlife Diseases* **32**, 545–547.

Arguin P, Murray-Lillibridge K, Miranda M, Smith J, Calaor A & Rupprecht C (2002). Serologic evidence of lyssavirus infections among bats, the Philippines. *Emerging Infectious Diseases* **8**, 258–262.

ARWH (2007). *Australian Registry of Wildlife Health*. Zoological Parks Board of NSW: Sydney.

ATAGI (Australian Technical Advisory Group on Immunisation) (2007). Australian bat lyssavirus. In *The Australian Immunisation Handbook* (9th edn). National Health & Medical Research Council: Canberra.

Atwell RB & Campbell FE (2001). Reactions to tick antitoxin serum and the role of atropine in treatment of dogs and cats with tick paralysis caused by *Ixodes holocyclus*: a pilot survey. *Australian Veterinary Journal* **79**, 394–397.

Baer GM (1966). A method for bleeding small bats. *Journal of Mammalogy* **47**, 340.

Barclay RMR (1994). Constraints on reproduction by flying vertebrates: energy and calcium. *American Naturalist* **144**, 1021–1031.

Barnard SM (1995). *Bats in Captivity*. Wild Ones Animal Books: Springville, CA.

Barnard SM & Sachs N (1992). The successful hand rearing of two orphaned vampire bats *(Desmodus rotundus)*. *Animal Keepers Forum* **19**, 280–288.

Barrett JL (2004). Australian bat lyssavirus. PhD thesis. School of Veterinary Science, University of Queensland. <http://eprint.uq.edu.au/archive/00002417/>. Accessed 26 September 2006.

Barrett JL & Olsson AR (in prep). Anaesthesia of bats. In *Maintenance of Bats in Captivity* (ed. SM Barnard). Krieger Publishing: Malabar, Florida.

Bogan M A (1972). Observations on parturition and development in the hoary bat, *Lasiurus cinereus*. *Journal of Mammalogy* **53**, 611–614.

Booth R (1994). Medicine and husbandry: dasyurids, possums and bats. In *Wildlife, Proceedings 233*, pp. 423–441. Post Graduate Committee in Veterinary Science: Sydney.

Boulard Y, Landau I, Miltgen F, Baccam D & Humphrey-Smith I (1985). Ultrastructural study of the sporogony of *Hepatocystis levinei* (Haemoproteidae), a parasite of Australian bats, in *Culicoides nubeculosus*. *Prostistologica* **21**, 93–105.

Breed AC, Field HE, Epstein JH & Daszak P (2006). Emerging henipaviruses and flying foxes—conservation and management perspectives. *Biological Conservation* **131**, 211–220.

Calisher CH, Childs JE, Field HE, Holmes KH & Schountz T (2006). Bats: important reservoir hosts of emerging viruses. *Clinical Microbiology Reviews* **19**, 531–545.

Campbell FE, Atwell RB & Smart L (2003). Effects of the paralysis tick, *Ixodes holocyclus*, on the electrocardiogram of the spectacled flying-fox, *Pteropus conspicillatus*. *Australian Veterinary Journal* **81**, 328–331.

Churchill S (1998). *Australian Bats*. New Holland: Sydney.

Clark P (2004). Haematological characteristics of Australian mammals. In *Haematology of Australian Mammals* (ed. P Clark), pp. 163–207. CSIRO Publishing: Melbourne.

Constantine DG (1970). Bats in relation to the health, welfare, and economy of man. In *Biology of Bats* (ed. WA Wimsatt), pp. 319–449. Academic Press: New York.

Cox TE, Smythe LD & Leung LK-P (2005). Flying foxes as carriers of pathogenic *Leptospira* species. *Journal of Wildlife Diseases* **41**, 753–757.

Crerar S, Longbottom H, Rooney J & Thornber P (1996). Human health aspects of a possible lyssavirus in a black flying-fox. *Communicable Disease Intelligence* **20**, 325.

Crichton EG & Krutzsch PH (eds) (2000). *Reproductive Biology of Bats*. Academic Press: Sydney.

Deem SL, Heard DJ, Clippinger TL & Buergelt CD (1999). Cranial edema associated with a protein-losing nephropathy in a golden-mantled flying-fox (*Pteropus pumilus*). *Journal of Zoo and Wildlife Medicine* **30**, 126–131.

DeMarco JH, Heard DJ, Fleming GJ, Lock BA & Scase TJ (2002). Toxicosis associated with topical administration of ivermectin in dog-faced fruit bats (*Cynopteris brachyotis*). *Journal of Zoo and Wildlife Medicine* **33**, 147–150.

Duncan M, Crawshaw GJ, Mehren KG, Pritzker KPH, Mendes M & Smith DA (1996). Multicentric hyperostosis consistent with fluorosis in captive fruit bats (*Pteropus giganteus, P. poliocephalus, and Rousettus aegyptiacus*). *Journal of Zoo and Wildlife Medicine* **27**, 325–338.

Environment Australia (1999). *Action Plan for Australian Bats*. Australian Department of Environment and Water Resources: Canberra. <http://www.environment.gov.au>.

Farina LL, Heard DJ, LeBlanc DM, Hall JO, Stevens G, Wellehan JFX & Detrisac CJ (2005). Iron storage disease in captive Egyptian fruit bat (*Rousettus aegyptiacus*). *Journal of Zoo and Wildlife Medicine* **36**, 212–221.

Fascione N (1995). *Flying-foxes Husbandry Manual*. American Zoo and Aquarium Association, Chiroptera Advisory Group.

Field HE (2000). A fatal case of Hendra virus infection in a horse in north Queensland: clinical and epidemiological features. *Australian Veterinary Journal* **78**, 279–280.

Field HE (2005). The ecology of Hendra virus and Australian bat lyssavirus. PhD thesis. School of Veterinary Science, University of Queensland.

Field HE & Ross T (1999). Emerging viral diseases of bats. In *Wildlife in Australia: Healthcare and Management*,

Proceedings 327, pp. 501–512. Post Graduate Foundation in Veterinary Science: Sydney.

Field HE, Breed AC, Shield J, Hedlefs RM, Pittard K, Pott B & Summers PM (2007). Epidemiological perspectives on Hendra virus infection in horses and flying foxes. *Australian Veterinary Journal* **85**, 268–270.

Fox S (2004). Bloat in flying-foxes. Flying-fox Information and Conservation Network (FFICN). listserve.

Fraser GC, Hooper PT, Lunt RA, Gould AR, Gleeson LJ, Hyatt AD, Russell GM & Kattenbelt JA (1996). Encephalitis caused by a lyssavirus in flying-foxes in Australia. *Emerging Infectious Diseases* **2**, 327–331.

Garner G & Bunn C (1997). Update on surveillance for Australian bat lyssavirus. *Australasian Epidemiologist* **4**, 27–30.

George H (1990). Grey-headed flying-foxes. In *Care and Handling of Australian Native Animals* (ed. SJ Hand), pp. 159–169. Surrey Beatty & Sons/Royal Zoological Society of NSW: Sydney.

Green R, van Tonder GJ, Oettle G, Cole SV & Metz J (1975). Neurological changes in fruit bats deficient in vitamin B_{12}. *Nature* **254**, 148–150.

Greenhall AM (1982). *House Bat Management.* US Department of the Interior, Fish and Wildlife Service Resource Publication 143. US Government Printing Office: Washington DC.

Gustafson AW & Damassa DA (1985). Repetitive blood sampling from small peripheral veins in bats. *Journal of Mammalogy* **66**, 173–177.

Hall L & Richards G (2000). *Flying-foxes: Fruit and Blossom Bats of Australia.* UNSW Press: Sydney.

Hall L, Richards G & Markus N (2001). Why culling will fail to solve the flying-fox problem at Melbourne Botanic Gardens. <http://www.austrop.org.au/ghff/home.htm>.

Halpin K, Young P & Field HE (1996). Identification of likely natural hosts for equine morbillivirus. *Communicable Disease Intelligence* **20**, 476.

Hanna JN, McBride WJ, Brookes DL, Shield J, Taylor CT, Smith IL, Craig SB & Smith GA (2006). Hendra virus infection in a veterinarian. *Medical Journal of Australia* **185**, 562–564.

Hariono B, Ng J & Sutton RH (1993). Lead concentrations in tissues of fruit bats (*Pteropus* sp.) in urban and non-urban locations. *Australian Wildlife Research* **20**, 315–320.

Hayes KT, Feistner ATC & Halliwell EC (1996). The effect of contraceptive implants on the behaviour of female Rodrigues flying-foxes, *Pteropus rodricensis. Zoo Biology* **15**, 21–36.

Heard DJ (1994). Endotracheal anesthesia in five species of flying-fox (*Pteropus pumilus, P. rodricensis, P. hypomelanus, P. poliocephalus,* and *P. vampyrus*). *Bat Research News* **34**, 112.

Heard DJ (1998). Medical management of Megachiropterans. In *Zoo and Wild Animal Medicine: Current Therapy 4* (eds ME Fowler & RE Miller), pp. 344–353. WB Saunders: Philadelphia.

Heard DJ (2003). Chiroptera. In *Zoo and Wild Animal Medicine* (eds ME Fowler & RE Miller), pp. 315–333. WB Saunders: St Louis, Missouri.

Heard DJ, Garner M & Greiner E (1995). Toxocariasis and intestinal volvulus in an island flying-fox (*Pteropus hypomelanus*). *Journal of Zoo and Wildlife Medicine* **26**, 550–552.

Heard DJ, Beale C & Owens J (1996a). Ketamine and ketamine:xylazine ED_{50} for short-term immobilization of the island flying-fox (*Pteropus hypomelanus*). *Journal of Zoo and Wildlife Medicine* **27**, 44–48.

Heard DJ, Buergelt CD, Snyder PS, Voges AK & Dierenfeld ES (1996b). Dilated cardiomyopathy associated with hypovitaminosis E in a captive collection of flying-foxes (*Pteropus* spp.). *Journal of Zoo and Wildlife Medicine* **27**, 149–157.

Heard DJ, DeYoung JL, Purich B & Ellis GA (1997). Comparative rectal bacterial flora of four flying-fox (*Pteropus sp.*) species. *Journal of Zoo and Wildlife Medicine* **28**, 471–475.

Heard DJ, Ruiz MM & Harr KE (2006). Comparison of serum and plasma biochemical values in Malaysian flying foxes (*Pteropus vampyrus*). *Journal of Zoo and Wildlife Medicine* **37**, 245–248.

Hill JE & Smith JD (1985). *Bats: A Natural History.* University of Texas Press: Austin.

Hooper PT, Gould AR, Russell GM, Kattenbel GA & Mitchell G (1996). The retrospective diagnosis of a second outbreak of equine morbillivirus infection. *Australian Veterinary Journal* **74**, 244–245.

Hopkins CS (1990). Carnivorous and insectivorous bats. In *Care and Handling of Australian Native Animals* (ed. SJ Hand), pp. 171–184. Surrey Beatty & Sons: Sydney.

Howell DJ (1974). Bats and pollen: physiological aspects of the syndrome of Chiropterophily. *Comparative Biochemistry and Physiology* **48A**, 263–276.

Jackson S (2003). Bats. In *Australian Mammals: Biology and Captive Management* (ed. S Jackson), pp. 297–350. CSIRO Publishing: Melbourne.

Kallen FC (1977). The cardiovascular system of bats: structure and function. In *Biology of Bats.* Vol. III (ed. WA Wimsatt), pp. 289–483. Academic Press: New York.

Keegan DJ (1979). Restraining cage and method for bleeding fruit bats (*Rousettus aegyptiacus*). *Laboratory Animal Scientific* **29**, 402–403.

Kunz TH & Ingalls KA (1994). Folivory in bats: an adaptation derived from frugivory. *Functional Ecology* **8**, 665–668.

Lafortune M, Canapp SO, Heard DJ & Farina LL (2004). A vasectomy technique for Egyptian fruit bats (*Rousettus aegyptiacus*). *Journal of Zoo and Wildlife Medicine* **35**, 104–106.

Landau I, Humphery-Smith I, Chabaud AG, Miltgen F, Copeman B & Boulard Y (1985). Description and experimental transmission of the haemoproteid

Hepatocystis levinei n.sp., a parasite of Australian Chiroptera. *Annales de Parasitologie Humaine et Comparee* **60**, 373–382.

Legendre AM (1976). Estrogen-induced bone marrow hypoplasia in a dog. *Journal of American Animal Hospital Association* **12**, 525–527.

Lollar A & Schmidt-French B (1998). *Captive Care and Medical Reference for the Rehabilitation of Insectivorous Bats*. Bat World: Austin, Texas.

Maina JN, Thomas SP & Hyde DM (1991). A morphometric study of the lungs of different sized bats: correlations between structure and function of chiropteran lung. *Philosophical Transactions of the Royal Society of London B* **333**, 31–50.

McCall BJ, Epstein JH, Neill AS, Heel K, Field HE, Barrett J, Smith GA, Selvey LA, Rodwell B & Lunt R (2000). Potential human exposure to Australian bat lyssavirus: Brisbane South and South Coast, Queensland, 1996–1999. *Emerging Infectious Diseases* **6**, 259–264.

McCoy RH (1974). Bacterial diseases of bats: a review. *Laboratory Animal Science* **24**, 530–534.

Messenger SL, Rupprecht CE & Smith JS (2003). Bats, emerging virus infections, and the rabies paradigm. In *Bat Ecology* (eds TH Kunz & MB Fenton), pp. 622–679. University of Chicago Press: Chicago.

Murray K, Rogers R, Selvey L et al. (1995). A novel morbillivirus pneumonia of horses and its transmission to humans. *Emerging Infectious Diseases* **1**, 31–33.

Nelson JE (1965). Behaviour of Australian Pteropodidae (Megachiroptera). *Animal Behaviour* **13**, 544–547.

Nelson JE (1989). Pteropodidae. In *Fauna of Australia*. Vol. 1B Mammalia (eds DW Walton & BJ Richardson), pp. 836–844. AGPS: Canberra.

Olsson AR (2000). Normal haematological and serum biochemical values and the effects of anaesthesia and parasites in the spectacled flying-fox, *Pteropus conspicillatus*. MSc thesis. James Cook University of North Queensland, Australia.

Parnaby H (1992). An interim guide to identification of insectivorous bats of southeastern Australia. *Technical Reports of the Australian Museum* **8**, 1–33.

Parry-Jones G (2000). *The Grey-Headed Flying-fox Care & Rescue*. Wambina Flying-fox Research Centre: Gosford, NSW.

Parry-Jones G (2001). The killing of flying-foxes. <http://www.austrop.org.au/ghff/home.htm>.

Philbey AW, Kirkland PD, Ross AD, Davis RJ, Gleeson AB, Love RJ, Daniels PW, Gould AR & Hyatt AD (1998). An apparently new virus (Family *Paramyxoviridae*) infectious for pigs, humans and fruit bats. *Emerging Infectious Diseases* **4**, 269–271.

Pierson ED (1984). Can Australia's flying-foxes survive? *Bats* **1**, 1–3.

Prociv P (1990). Aberrant migration by *Toxocara pteropodis* in flying-foxes: two case reports. *Journal of Wildlife Diseases* **26**, 532–534.

Queensland Department of Primary Industries and Fisheries (2005). Handling suspect Hendra virus cases in all equines: guidelines for veterinarians. <http://www2.dpi.qld.gov.au/health/16503.html>. Accessed June 2006.

Reddacliff LA, Bellamy TA & Hartley WJ (1999). *Angiostrongylus cantonensis* infection in grey-headed fruit bats (*Pteropus poliocephalus*). *Australian Veterinary Journal* **77(7)**, 466–468.

Rose K (1999). Common diseases of urban wildlife. In *Wildlife in Australia: Healthcare and Management, Proceedings 327*, pp. 365–427. Post Graduate Foundation in Veterinary Science: Sydney.

Ross T, Philbey A, Kirkland P & Field H (2001). Megachiroptera and Menangle virus in Australia. In *Veterinary Conservation Biology Wildlife Health and Management in Australasia. Proceedings of International Joint Conference* (eds A Martin & L Vogelnest), pp. 241–244. Taronga Zoo: Sydney.

Rupprecht CE (1999). Rabies: global problem, zoonotic threat, and preventative management. In *Zoo and Wild Animal Medicine: Current Therapy* (eds ME Fowler & RE Miller), pp. 136–146. WB Saunders: Philadelphia.

Selvey L, Taylor R, Arklay A & Gerrard J (1996). Screening of bat carers for antibodies to equine morbillivirus. *Communicable Disease Intelleligence* **20**, 477–478.

Skerratt L, Speare R, Berger L & Winsor H (1998). Lyssaviral infection and lead poisoning in black flying-foxes from Queensland. *Journal of Wildlife Diseases* **34**, 355–361.

Smythe LD, Field HE, Barnett LJ et al. (2002). Leptospiral antibodies in flying-foxes in Australia. *Journal of Wildlife Diseases* **38**, 182–186.

Snell J (1994). *Standards Recommended for the Care and Exhibition of Flying-foxes*. Ku-Ring-Gai Bat Colony Committee: Sydney.

Solymar B (2001). Rodent management on farms and in new tree planting to prevent accidental poisonings of raptors and other non-target wildlife. *Bird Studies of Canada*. Ontario. <http://www.bsc-eoc.org/download/Rodent%20Management%20on%20Farms.pdf>.

Speare R (1988). Clinical assessment, diseases and management of the orphaned macropod joey. In *Australian Wildlife, Proceedings 104*, pp. 211–296. Post Graduate Committee in Veterinary Science: Sydney.

Steece R & Altenbach J (1989). Prevalence of rabies specific antibodies in the Mexican free-tailed bat (*Tadarida brasiliensis mexicana*) at Lava Cave, New Mexico. *Journal of Wildlife Diseases* **25**, 490–496.

Strahan R (1995). Bats. In *The Mammals of Australia*. Reed Books, Sydney.

Sutton RH & Wilson PD (1983). Lead poisoning in grey-headed fruit bats (*Pteropus poliocephalus*). *Journal of Wildlife Diseases* **19**, 294–296.

Thompson GK (1999). Veterinary surgeons' guide to Australian bat lyssavirus. *Australian Veterinary Journal* **77**, 710–712.

Tidemann CR, Vardon MJ, Nelson JE, Speare R & Gleeson LJ (1997). Health and conservation implications of Australian bat lyssavirus. *Australian Zoologist* **30**, 369–376.

Tuttle M D (1988). *America's Neighborhood Bats*. University of Texas Press: Austin.

Williamson MM, Hooper PT, Selleck PW, Gleeson LJ, Daniels PW, Westbury HA & Murray PK (1998). Transmission studies of Hendra virus (equine morbillivirus) in fruit bats, horses and cats. *Australian Veterinary Journal* **76**, 813–818.

Wilson P (1988). Bats (order Chiroptera). In *Australian Wildlife, Proceedings 104*, pp. 105–109. Post Graduate Committee in Veterinary Science: Sydney.

Wilson DE & Reeder DM (eds) (2005). *Mammal Species of the World: A Taxonomic and Geographical Reference*. Johns Hopkins University Press: Baltimore.

Wood J (2004). *Bloat in Flying-foxes*. Flying-fox Information and Conservation Network (FFICN). listserve.

Young PL, Halpin K, Selleck PW et al. (1996). Serological evidence for the presence in *Pteropus* bats of a paramyxovirus related to equine morbillivirus. *Emerging Infectious Diseases* **2**, 239–240.

15. Rodents

Andrew Breed and Paul Eden

1 TAXONOMY, DISTRIBUTION AND BIOLOGY

All of the approximately 60 extant species of Australian rodents belong to the family Muridae, the largest and most widespread of all mammalian families (Musser & Carleton 2005). Various species groupings have been proposed for Australian rodents at different times and further changes to the taxonomy are likely to take place in the future (Watts 1973; Strahan 2002; Musser & Carleton 2005). The most recent comprehensive taxonomic review is by Musser and Carleton (2005), in which these species are placed in six Divisions (an informal classificatory rank between Subfamily and genus).

- *Pseudomys* Division. Comprises about 50 species in at least 10 genera including:
  - five species of hopping mouse (genus *Notomys*);
  - five species of rock-rat (genus *Zyzomys*);
  - two species of tree-rat (genus *Mesembriomys*);
  - one species of rabbit-rat (genus *Conilurus*);
  - greater stick-nest rat (*Leporillus conditor*).

In older literature, this group was often treated as a distinct Subfamily (Pseudomyinae or Conilurinae) or Tribe (Conilurini) within Muridae. All but two members of this group are endemic to Australia, with the brush-tailed rabbit-rat (*Conilurus penicillatus*) and the delicate mouse (*Pseudomys delicatulus*) also found in New Guinea (Flannery 1995).

- *Uromys* Division—mosaic-tailed rats. Comprises six species—four *Melomys* spp and two white-tailed rats (genus *Uromys*). Related species and three related Genera are found in New Guinea (Flannery 1995). This group was formerly treated as a distinct Tribe (Uromyini) within Muridae.
- *Hydromys* Division—water rat (*Hydromys chrysogaster*). Other members of the genus Hydromys and related genera are found in New Guinea (Flannery 1995). This group, together with the next, was often treated as a distinct Subfamily (Hydromyinae) or Tribe (Hydromyini) within Muridae.
- *Xeromys* Division—water mouse (*Xeromys myoides*). Other related genera are found in New Guinea (Flannery 1995).
- *Pogonomys* Division—prehensile-tailed rat (*Pogonomys mollipilosus*) (scientific name in doubt, see Appendix 1). Six other *Pogonomys* species and related genera are found in New Guinea (Flannery 1995). This genus was formerly included in a Tribe Anisomyini, which otherwise included many New Guinean genera.
- *Rattus* Division—true rats. Eight species are found in Australia.

In recent literature the first five groups are often referred to as the Old Endemics and sometimes treated as a distinct Subfamily, the Hydromyinae. The term 'Old Endemics' is a useful one for the various groups that entered Australia at the very end of the Miocene period, approximately 5 million yr ago. The true rats are often referred to as the New Endemics due to their more recent arrival in Australia, probably around 1 million yr ago (Strahan 2002). There are also three rodent species, the house mouse (*Mus musculus*), brown rat (*Rattus norvegicus*) and black rat (*Rattus rattus*) that were introduced to Australia during the early years of European settlement.

Despite the extinction of at least seven Australian rodent species since the arrival of Europeans, several species, previously thought to be extinct, were rediscovered in the latter part of the 20th century. These include the central rock-rat (*Zyzomys pedunculatus*), New Holland mouse (*Pseudomys novaehollandiae*), Hastings River mouse (*Pseudomys oralis*) and heath mouse (*Pseudomys shortridgei*). Sixteen of the extant Australian rodent species are currently threatened with extinction (IUCN 2006), all of which are Old Endemics.

The Old Endemics have adapted to live in a wide range of habitats from rainforest to desert. The New Endemics have had less time to adapt to Australian conditions and most are found in forested areas. In general, they maintain a typical Asian rat appearance and lack the morphological adaptations of the Old Endemics.

2 ANATOMY AND PHYSIOLOGY

Australian rodents have compact bodies with various adaptations to allow proficient climbing, jumping, digging, burrowing or swimming. Various tail adaptations have a range of functions, including fat storage in rock-rats (*Zyzomys* spp.), a long tail that provides balance when hopping in hopping mice (*Notomys* spp.), the ability to shed the tail skin when seized by a predator (or inexperienced handler) in hopping mice and prehensility to aid climbing in the prehensile-tailed rat. They also vary greatly in size from less than 10 g (e.g. delicate mouse) to over 1 kg (e.g. giant white-tailed rat (*Uromys caudimaculatus*)).

A feature that readily distinguishes Australian rodents from all marsupials of similar size is their teeth. Rodents have one pair of large upper incisors and one pair of large lower incisors and broad molars with a gap (diastema) between the incisors and molars. Small dasyurid marsupials have continuous rows of sharp pointed teeth (Fig. 10.2). The diprotodont marsupials have a single pair of lower incisors which are much more forward-pointing than those of rodents, and most diprotodont marsupials have more than one pair of upper incisors. Rodent incisor teeth have an open apical pulp that results in continuous growth, essential to cope with the gnawing behaviour of most species. The labial aspect of the incisors has an enamel coating, with softer dentine on the lingual surfaces. This difference in the structural surfaces of the incisors results in the teeth wearing to a chisel shape that is ideal for cutting. The molars do not grow continuously and are termed brachyodont (short-crowned). Differential use of incisors and molars is facilitated by a large amount of rostro-caudal movement of the temporomandibular joint. When the mandible is in the forward position the incisors occlude while the molars do not; when the mandible is moved back the molars are in occlusion and the incisors are not. Rodents do not possess canine teeth and the Australian species also lack premolars. The dental formula for Australian rodents is I1/1 C0/0 P0/0 M2–3/2–3.

The digestive tract is generally simple, with modifications to the stomach and large intestine in some species. Carnivorous species such as the water rat have a relatively small stomach, with most (74%) of the stomach surface area composed of glandular tissue. This is in contrast to the herbivorous species (e.g. giant white-tailed rat, greater stick-nest rat and fawn-footed melomys (*Melomys cervinipes*)), which have a large stomach composed primarily of a non-glandular forestomach (Breed & Ford 2007). Some herbivorous species possess an enlarged caecum, where bacterial and protozoal fermentation aids in the digestion of cellulose.

Differentiation of native Australian rodents from introduced species is not always easy but can usually be achieved by examining several characteristics (Table 15.1). The species most likely to be confused are some of the small native mice (particularly some members of the genus *Pseudomys*) and house mice. Female house mice have five or more pairs of teats while female native mice have only two pairs of teats, located in the inguinal region. House mice also have a characteristic, often pungent, musty smell. The upper incisor of most house mice bears a small notch or ledge on the inner surface of the upper incisor teeth (Fig. 15.1), a wear facet that is rarely observed in native mice. The physiological parameters for native rodents are likely to be similar to those for domestic rodents (Table 15.2).

Table 15.1 Characteristics for differentiating native and non-native Australian rodents

Characteristic	House mouse (*Mus musculus*)	Native mice (*Pseudomys* spp.)	Brown rat (*Rattus norvegicus*)	Black rat (*Rattus rattus*)	Native true rats (*Rattus* spp.)
Coat	Extremely variable	Variable[a]	Grey-brown above, white-grey below. Coarse fur	Grey-brown above, white-cream below. Sleek fur	Variable[a]
Body shape and size	Variable. Head and body length 60–95 mm. Body tapers to tail. Pointed nose. Smaller eyes relative to body size.	Variable.[a] Larger eyes relative to body size.	Robust. Larger than native *Rattus*. Head and body length 180–255 mm. Body tapers to tail.	Slender. Head and body length 165–205 mm. Body tapers to tail.	Variable[a]
Ear	Relatively small	Relatively large in comparison to house mouse.	Does not reach past middle of eye when bent forwards.	Long and thin. Reaches past middle of eye when bent forwards.	Does not reach past middle of eye when bent forwards.
Tail	≈Head and body length. Generally uniform in colour. Tail length 75–95 mm.	Usually bicoloured. Paler ventral surface. Generally more hairy than house mouse tail.	≤Head and body length. Stout and scaly tail. Tail length 150–215 mm.	>Head and body length. Tail length 185–245 mm	≤Head and body length.
Number of nipples (adult female only)	≥5 pairs	2 pairs	6 pairs	≥5 pairs	3–6 pairs
Disposition	Often aggressive	Usually not aggressive	Often very aggressive	Usually not aggressive	Usually not aggressive
Habitat	Almost all including farmland, deserts and areas of human habitation.	Usually less disturbed habitat.	Generally close to urban centres. Seldom far inland. Watercourses.	Usually disturbed or degraded habitat. Often urban buildings. Watercourses	Usually less disturbed habitat.
Other useful characters	Notch on inner surface of upper incisor (Fig. 15.1). Distinctive musty odour.	Absence of notch on inner surface of upper incisor (Fig. 15.1)	Typical rat odour.		

a See references for individual species characteristics.
Source: Menkhorst (2004); Strahan (2002); Watts & Aslin (1981); W Gleen (*pers. comm.*).

Table 15.2 Physiological parameters for domesticated rodents

Parameter	Brown rat	House mouse
Respiratory rate (breaths/min)	70–150	100–125
Tidal volume (mL)	1.5–1.8	0.3–0.5
Heart rate (bpm)	260–450	500–600
Blood volume (mL/kg)	70	80
Body temperature (°C)	38	37.5

Source: Orr (2002).

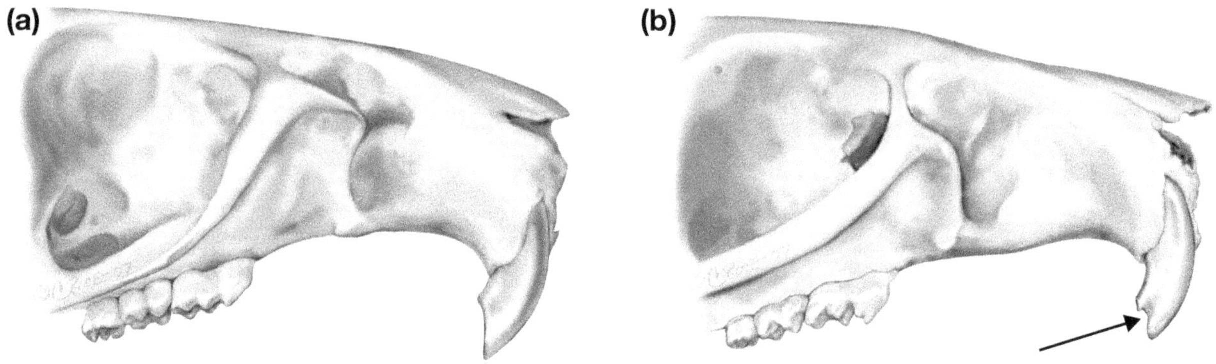

Figure 15.1 Skulls of (a) *Pseudomys* sp. and (b) a house mouse (*Mus musculus*) showing the small notch or ledge on the inner surface of the upper incisor in the house mouse (arrow).

Table 15.3 Reproductive parameters for native and domestic rodents

Parameter	Spinifex hopping mouse	Plains rat	Water rat	Bush rat	Brown rat	House mouse
Lifespan in captivity (yr)	3–5.5	2–3	2–6	2–3	2.5–3.5	1.5–2.5
Sexual maturity (d)	60–84	84–180	130	68–120	42–56	42–49
Oestrous cycle (d)	4–10	5–21	11	4–6	4–5	4–5
Duration of oestrus (hr)	4–8	5–21	N/A	4–6	14	14
Gestation (d)	32	30	35	23	20–22	19–21
Litter size (avg)	1–6 (3.5)	1–7 (4)	1–7 (4)	3–8 (5)	6–16 (10)	6–12 (8)
Number of nipples	4	4	4	10	12	≥10
Birth weight (g)	3	3–4	—	—	4–7	1–1.5
Adult body weight (g)	27–45	50–80	340–1275	40–225	220–500	20–60
Eyes open (d)	—	—	—	—	12–14	12–14
Weaning age (d)	30	28	28–30	31	21	21

Source: Orr (2002); Watts (1978); Watts & Aslin (1981).

3 REPRODUCTION

In comparison to other mammals, most rodents have a short generation time with a relatively short gestation, large litter size and rapid rate of maturation. This is exemplified in the three rodent species introduced to Australia and may have contributed to their successful invasion of many parts of Australia.

Female Old Endemic rodents have two pairs of teats and in most species (except hopping mice) the young are continuously attached to the teats for about the first 2 wk of life. Female New Endemics have three to six pairs of teats and hence larger litter sizes than the Old Endemics (Table 15.3). They also rear their young in a nest, leaving them there to feed and returning to suckle them. New Endemics have shorter gestation times than Old Endemics and their higher reproductive rate facilitates the periodic population explosions that occur in the long-haired rat (*Rattus villosissimus*) and dusky rat (*Rattus colletti*) (Yom-Tov 1985). Oestrous cycle lengths have been recorded for several Australian rodent species (Table 15.3) but, as for many other mammals, an oestrous cycle is an artefact of captivity and does not occur regularly in wild populations. There is also variation in male reproductive anatomy and physiology, with New Endemics having relatively large testes and producing many more sperm than some Old Endemic species (e.g. hopping mice) (Breed 1997). The accessory sex gland arrangement of Australian rodents is highly variable (Breed 1986). Males of most species have a pair of large seminal vesicles and coagulating glands as well as typical ventral and dorsolateral prostates and ampullary glands. Hopping mice have very large ventral prostates while their seminal vesicles, coagulating glands, dorsolateral prostates and ampullary glands are variably reduced. An os penis is present in all male Australian rodents (Breed 1986).

Determining sex in Australian rodents is not always straightforward. Both sexes have a genital papilla that covers the penis in males and the clitoris in females. The testes of juvenile male rodents are located intra-abdominally and they enlarge and descend into the scrotum as the animal matures. In female rodents the vagina is located immediately caudal to the genital papilla. The entrance to the vagina is often difficult to identify in juvenile animals but can be obvious in animals that have recently mated or given birth. Similarly, the presence of nipples can be difficult to ascertain in juvenile female animals. Sex determination of adult rodents can be achieved by locating the presence of teats in the inguinal region of females, or the scrotal and epididymal sacs of males. In juveniles, the anogenital distance of males is much greater than that of females and there is usually a distinct unfurred line between the vagina and anus in juvenile females. A direct comparison with known individuals of both sexes may be required for certainty when sexing is performed by inexperienced handlers (Fig. 15.2).

4 HUSBANDRY

Australian rodents may be arboreal, terrestrial, aquatic and/or burrowing and hence have a diverse range of housing requirements. Most are largely nocturnal or crepuscular so reversed light cycles are often used in enclosures for display purposes, although some rodent species show a partial shift from nocturnal to diurnal activity in captivity. Although a wide range of Australian rodent species have been kept in captivity, the spinifex hopping mouse (*Notomys alexis*) and plains rat (*Pseudomys australis*) are by far the most commonly kept native rodent species in Australia.

Rodents generally thrive in captivity. The design and materials used to construct their enclosure depend to some extent on the purpose of the facility, however, it is recommended to replicate their natural environment where possible. This varies considerably with the species; guidelines are outlined below. Most rodents are fast and agile, jump high, burrow and gnaw through many materials. Enclosure design and husbandry practices (particularly during cleaning) must take these into consideration to prevent escape or injury of animals. Jackson (2003) provides more detail on husbandry and housing of Australian rodents.

In designing and constructing an enclosure for Australian rodents, the following should be considered.

- Enclosure design: materials such as glass, perspex, metal and plastics are recommended for enclosure construction. Most rodents will gnaw through wood, some plastics and fine metal mesh. Enclosure lids need to be well-fitted and can be made from flywire providing they are out of animal reach; hopping mice can jump up to 70 cm high. Heavy rocks or other large objects should not be placed on top of sand in the enclosure as animals may dig under them and become trapped. Figure 15.3 shows an enclosure set up for central rock-rats.
- Enclosure size: this should reflect the natural behaviours of the species (e.g. terrestrial or arboreal). For smaller terrestrial species a floor space of 50 × 40 cm is adequate, whereas for larger terrestrial species (e.g. greater stick-nest rat) the minimum recommendation is 200 × 200 cm. Vertical space of up to 2 m is recommended for some arboreal species, and water rats require space of at least 300 × 300 cm and a pond 100 × 100 × 50 cm deep (Jackson 2003).
- Sufficient shelter and nest boxes: this may include purpose-built nest boxes but can include items such as hollow logs, pieces of bark, rock shelters, tussock grasses and plastic or cardboard tubes or boxes.
- Absorbent substrate: this may include sawdust, sand, newspaper, leaf litter and soil. Sawdust or sand makes a good substrate for energetic burrowers. Sand should be fine-grained as coarse sand can be abrasive.

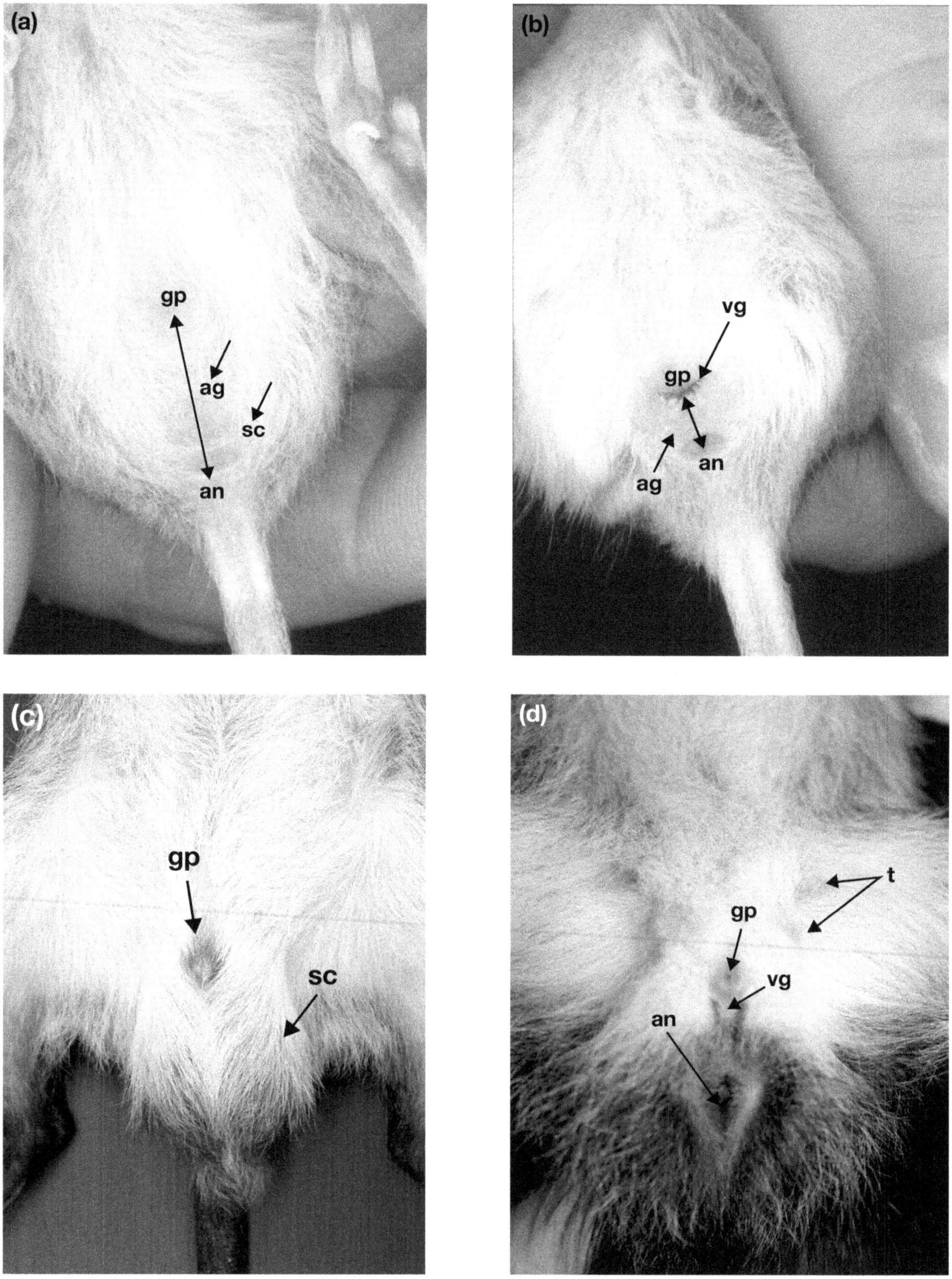

Figure 15.2 (a) Adult male hopping mouse. (b) Adult female hopping mouse. (c) Adult male black-footed tree-rat. (d) Adult female black-footed tree-rat. t = teats; sc = scrotum; gp = genital papilla; an = anus; vg = vagina; ag = anogenital distance.

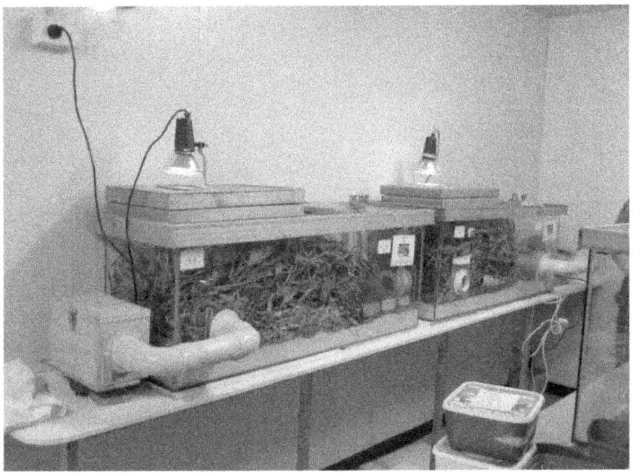

Figure 15.3 Off-display enclosures for central rock-rats.

- Nesting material: this may include shredded paper, dried grasses and straw. Most species are very competent nest-makers.
- Environmental enrichment: this may include pieces of native vegetation (e.g. saltbush, *Atriplex* spp.; grasses, *Triodia* spp.; daisy bush, *Olearia* spp.) for rodents to gnaw on, a network of branches for arboreal species, rodent activity wheels and cardboard or paper to tear up for nesting.
- Enclosure location: consider environmental temperature and ventilation, e.g. ensure enclosures are not placed in direct sunlight on the inside of a window, as overheating may occur.

Enclosures should be well-ventilated and easy to clean to avoid build-up of ammonia formed by the action of bacterial urease on urea in urine. Ammonia is a potent respiratory irritant which may predispose to respiratory infections. Frequency of cage cleaning should vary according to the density of animals in the enclosure, substrate type and level of stress experienced by animals during the cage cleaning procedure. Spot cleaning can be done where areas of substrate soiled by urine and faeces are cleaned on a daily basis, with cleaning of the entire cage and complete substrate changes undertaken less often.

Fresh food and water should be provided daily and careful attention paid to hygiene of food and water containers. Desert-adapted species may require little water but water should always be provided, particularly if breeding is intended.

Temperature in the enclosure should remain moderate (15–25°C) for most species (Jackson 2003); even desert-adapted species are mostly burrow-dwellers and not adapted to extreme heat or cold. Prevention of drafts and provision of plenty of bedding helps to avoid exposure to cold temperatures.

The social organisation of many Australian rodent species is relatively poorly understood but there are some marked differences in behaviour between species when held in captivity (Watts & Aslin 1981; Jackson 2003). Species from arid habitats tend to nest communally, unlike those from southern Australia and the tropical forests where most species live in smaller groups or are solitary (Happold 1976; Breed & Ford 2007).

4.1 Hospital care

Debilitated rodents are likely to have reduced homeostatic abilities, hence careful attention to temperature and hydration status is required. If possible a hospital cage should:

- be located away from natural predators to reduce stress from olfactory and auditory stimuli;
- be thermostatically controlled, usually offering a warm environmental temperature (25–28°C);
- be free from draughts, i.e. solid walls and floor rather than mesh;
- have provision for supply of supplemental oxygen for cases of shock and respiratory disease. Care should be taken to avoid inadvertently chilling the animal while providing supplemental oxygen.

Rodents may lose their appetite with illness, and it can be a challenge to encourage feeding in these animals. High-calorie pastes such as Nutrigel (Troy Laboratories) may assist in stimulating appetite and improving energy intake. Other techniques include feeding high-fat foods (e.g. sunflower seeds, mealworms) and warming food prior to feeding.

4.2 Transport

As many rodent species are nocturnal, capture for transport or restraint is often easier to achieve during the day when the animal may be resting in its nest box. It is useful to design a nest box such that entrances may be sealed with the animal inside (e.g. using a slide or locking door, or stuffing with a towel or calico bag) and the nest box used as the transport container. Alternatively, the animal may be transferred to a calico bag. It is necessary to ensure that the bag is adequately tied shut to prevent escape. As some rodents may chew through bags it is wise to place the bag in a more solid container during transport. Larger species may be trained to enter a pet carrier.

4.3 Individual marking and identification

The use of passive integrated transponders (microchips) for individual identification is now common. Insertion is best performed under general anaesthesia to provide sufficient restraint and minimise stress. The recommended location for placement of microchips in rodents is interscapular (Miller 1992). The needle should be directed caudally and inserted to the depth of the notch on the needle. This ensures that the microchip is deposited some distance from the hole in the skin. The hole in the skin should be closed with tissue glue or sutured to prevent the microchip falling out. Other reported methods of identification include ear tags, ear tattoos and ear notching (Jackson 2003). However, these methods are becoming less favoured for ethical reasons.

5 NUTRITION

Australian rodent species show variation in the types of food they eat as well as in their degree of dietary specialisation. Most are omnivorous, eating a range of seeds, fungi, invertebrates and other plant material, but some have quite specialised requirements (Strahan 2002; Jackson 2003).

Some examples of natural diets are:

- water rats—carnivorous (fish, yabbies, prawns, molluscs);
- greater stick-nest rat—herbivorous (especially succulent plants);
- giant white-tailed rat—omnivorous (fruits, nuts, insects, small vertebrates);
- *Melomys* spp.—herbivorous (leaves, grass, fruits);
- central rock-rat—granivorous (Nano et al. 2003);
- bush rat (*Rattus fuscipes*)—great dietary variation (fungi, leaves, seeds, insects);
- hopping mice—omnivorous (seeds, insects, fungi, roots, leaves).

Specific captive diets for Australian rodents are described in Jackson (2003).

Detailed nutritional requirements of laboratory rodents have been published by the National Research Council (1995) and can be used as a guide for the requirements of many Australian rodents. To meet nutritional and behavioural requirements for a particular species in captivity, it is best to provide a varied diet that reflects the natural diet. Diets generally consist of a mix of fresh vegetables and fruit, seed mixes, other plant material, invertebrates and commercial rodent pellets. Excessive levels of high-energy foods (e.g. sunflower seeds) can lead to obesity and other nutritional imbalances, while excessive levels of some fruit and vegetables (e.g. citrus fruit) can lead to gastrointestinal disturbances in some species. Vitamin and mineral supplements are not required if a varied diet is supplied. Rodents should be provided with suitable objects on which to gnaw to allow normal behaviour and reduce the likelihood of dental disease. The water rat and water mouse can be fed fish, yabbies and prawns, supplemented with commercial dog or cat food.

6 RESTRAINT

6.1 Capture and physical restraint

Rodents that are handled frequently often become tame, making handling an easy and relatively stress-free procedure. Rodents in unfamiliar surroundings or those not used to handling will try to avoid capture and often bite. Improper handling can result in injury to the animal and/or handler. Unlike domestic mice and rats, grasping the tail of many Australian rodents may result in the tail skin slipping off or degloving of the tail (Fig. 15.4). This method of restraint is not recommended for Australian rodent species. Some species (e.g. central rock-rat) may also become stressed and collapse during prolonged handling, so minimising handling time is recommended.

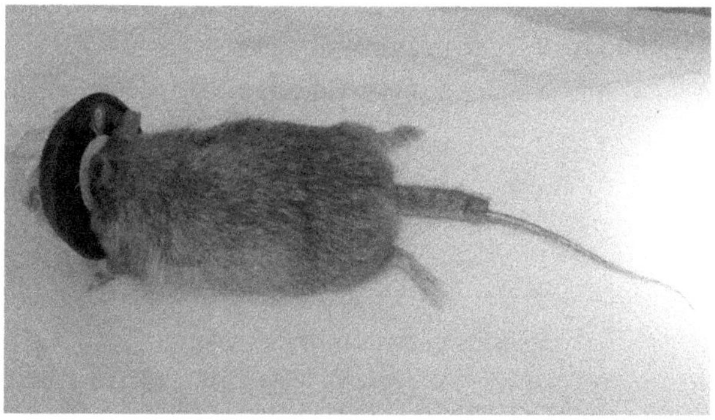

Figure 15.4 Grasping the tail of some Australian rodent species may result in slipping or degloving of the tail, as seen in this central rock-rat.

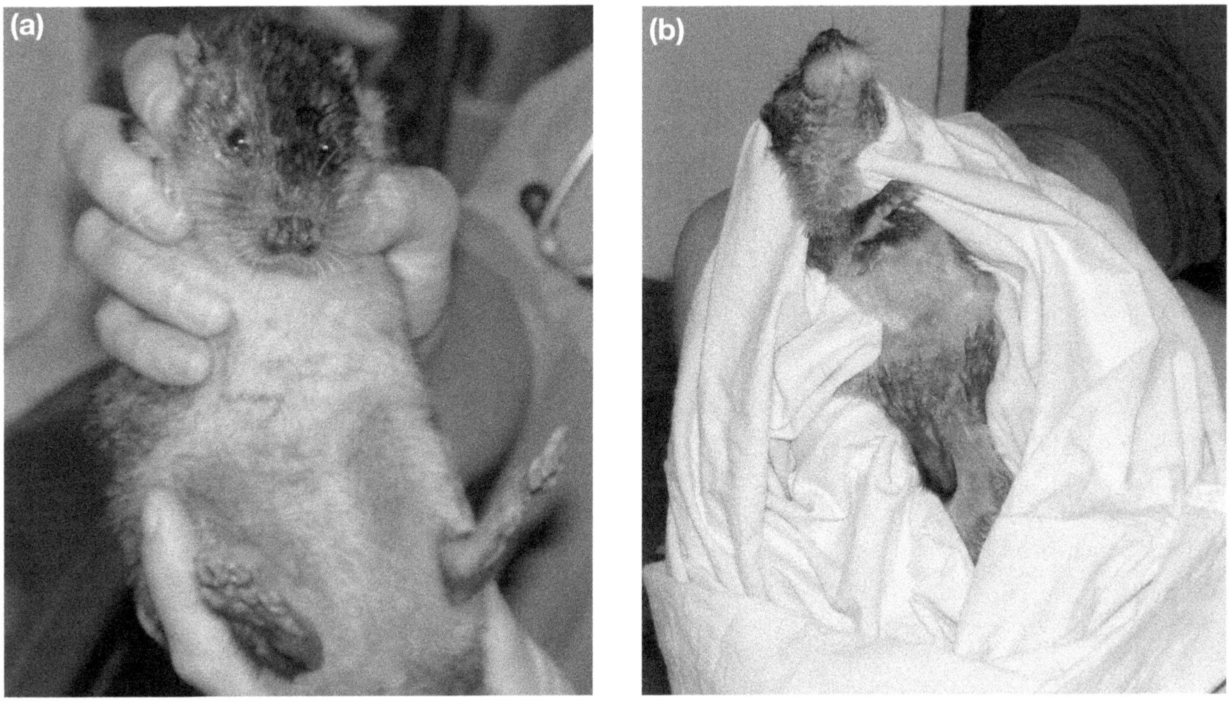

Figure 15.5 Physical restraint of a water rat using (a) bare hands or (b) protected with a cloth.

Grasping the skin over the shoulder blades (the scruff) between the forefinger and thumb and supporting the body with the other fingers of the same hand is generally successful in restraining Australian mice. For rats, a two-handed approach is used. One hand either holds the scruff as in a mouse or grasps with the forefinger and thumb around the shoulders with the thumb under the mandible to prevent biting, while the other hand supports the hind quarters (Fig. 15.5a). The use of thick gloves may be preferable in some situations but they decrease tactile sensation, reducing the handler's ability to finely adjust pressure applied to the animal. Medium to large rats can easily bite through most leather gloves. Rats and mice can also be manually restrained through a cloth bag, then examined by turning the bag inside out while still maintaining a hold on the animal through the bag (Fig. 15.5b). This minimises the risk of escape by avoiding the need to adjust grip on an animal that is presented in a bag. Clear plastic bags can also be successfully used to restrain, measure and sex rodents with minimal need for direct handling.

6.2 Chemical restraint

Chemical restraint is preferable for wild-caught or rarely handled individuals if a thorough examination or other prolonged procedure is necessary. When performed with appropriate equipment and by experienced personnel, this method of restraint reduces stress and the risk of injury.

Rodents cannot vomit and so do not require pre-anaesthetic fasting although checking and clearing the oral cavity immediately after induction is recommended to avoid accidental aspiration of food material. Inhalation anaesthesia is the preferred method for chemical restraint. It has the advantage of providing oxygen, resulting in a relatively rapid and smooth induction and recovery, and allows safe and rapid titration of the anaesthetic dose to the required effect. Isoflurane or sevoflurane are the preferred inhalational agents, with a recommended concentration of 3–5% for induction and 1.5–3% for maintenance of anaesthesia. Induction can be achieved using a face mask with the animal physically restrained, in a cloth bag or in an induction chamber (Fig. 15.6). Anaesthesia is usually maintained using a face mask as endotracheal intubation is difficult. Blind intubation is possible but difficult (L Vogelnest *pers. comm.*).

A range of injectable tranquillisers, sedatives and anaesthetic agents have been reported for use in non-native pet rodent species (Table 15.4). The use of these

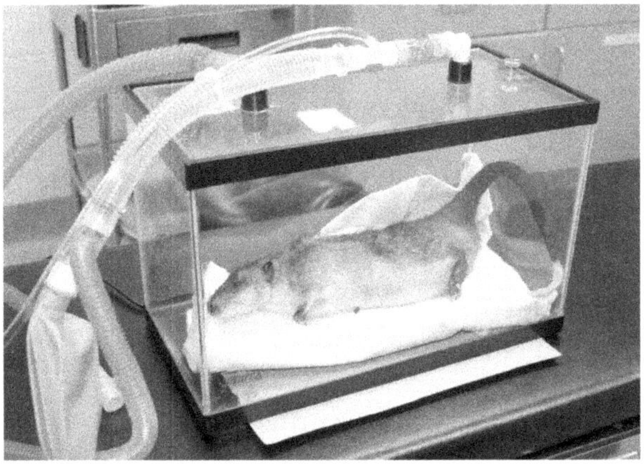

Figure 15.6 Anaesthetic induction of a water rat in an induction chamber.

agents in Australian rodents has not been reported. Butorphanol has been used as a pre-anaesthetic agent in rodents at 0.1–0.2 mg/kg SC (R Johnson *pers. comm.*). When using injectable agents, animals should be weighed to ensure accurate dosing.

7 CLINICAL PATHOLOGY

Rats and mice have a blood volume of approximately 75 mL/kg body weight and a maximum of 10% of that volume (0.75 mL from a healthy 100 g animal) can be safely removed at any one time (Orr 2002). Obtaining a blood sample from rodents <100 g is often challenging but the following sites and procedures are recommended: lateral tail vein, saphenous vein, cephalic vein, jugular vein, femoral vein and cardiac puncture (recommended for terminal blood collection only and performed under anaesthesia). Use of the sublingual vein has also been reported for larger laboratory rodents (Zeller et al. 1998).

Venipuncture is easier if the animal is anaesthetised and when its body is warm, or at least if the tail is warm when the lateral tail vein is to be used. This can be achieved by placing the animal in an incubator or hot box for 10–15 min at 35°C, placing it on a warm hot-water bottle or immersing the tail in a bowl of warm water at approximately 35–40°C for 5–15 min. For larger animals a tourniquet can be applied to a fore leg to aid blood collection from the cephalic vein. A 25–27 G needle attached to a 0.5–1.0 mL syringe is generally appropriate. The use of a pre-heparinised syringe or microhaematocrit tube can facilitate blood collection from smaller animals. Due to the small size of many rodent species and therefore the small volume of blood that can be safely collected, recommended minimum tests that should be requested may include packed cell volume, total plasma solids and blood smear analysis for estimated white blood cell count and differential white cell count.

Haematology and biochemistry of blood samples from rodents is similar to domestic small animals, with some differences in cell morphology (Clark 2004). Generally, rodents tend to show a relatively high level of reticulocytes and often the lymphocyte count outnumbers the neutrophil count (Sainsbury 2003). Alanine aminotransferase (ALT) may not be present in all rodent species so may be of limited diagnostic use (Sainsbury 2003). Gamma glutamyl transferase (GGT) is present in the liver of rodents (Sainsbury 2003). Rodents with chronic renal failure may show evidence of increased serum BUN levels (Quesenberry & Carpenter 2004). Tables 15.5–15.7 provide haematology and biochemistry values for selected Australian and domestic rodent species, and Table 15.8 provides basic urinalysis data for domestic rodents.

Table 15.4 Injectable chemical restraint agents used in domestic rodents

Agent	Dose rate	Comments
Ketamine	22–44 mg/kg IM	
Ketamine/medetomidine	50–75 mg/kg ket + 1 mg/kg med IP	House mouse
Tiletamine/zolazepam	20–40 mg/kg IM	Brown rat
Butorphenol	1–2 mg/kg SC q 4 hr	Brown rat; analgesia
	1–5 mg/kg SC q 4 hr	House mouse; analgesia
	0.1–0.2 mg/kg SC[a]	Native rodents; pre-anaesthetic sedation

a R Johnson (*pers. comm.*).
Source: Morrisey & Carpenter (2004).

Table 15.5 Haematology values for Australian rodents

Parameter (sample size)	Common rock-rat September (34–47) Mean ± SD	Common rock-rat January (37–48) Mean ± SD	Swamp rat male (3) Mean ± SD	Spinifex hopping mouse Male (27) Mean ± SD	Spinifex hopping mouse (3) Mean ± SD	Spinifex hopping mouse (3) Range	Plains rat (5) Mean ± SD	Plains rat (5) Range	Central rock-rat (1–4) Mean ± SD	Central rock-rat Range
PCV (L/L)	0.46 ± 0.04	0.46 ± 0.05	0.48 ± 0.1	0.54 ± 0.01	0.38 ± 0.1	0.37–0.38	0.42 ± 0.03	0.39–0.45	0.51 ± 0.04	0.37–0.42
RBC (× 10^{12}/L)			7.1 ± 1.0		7.8 ± 0.05	7.4–8.3	7.3 ± 0.4	6.8–7.9	6.2 ± 0.6	5.65–6.8
Hb (g/L)	147 ± 16	142 ± 20	189 ± 11		122 ± 1.5	126–129	133 ± 7	126–144	122 ± 2	120–121
MCV (fl)			70 ± 10		49 ± 3	46–52	58 ± 2	56–60	64 ± 4	60–65
MCH (pg)			27.2 ± 3		16.7 ± 1.2	16–18	18.2 ± 1.3	16–19	19.3 ± 1.5	18–21
MCHC (g/L)			393 ± 15		339 ± 4	336–343	317 ± 20	281–329	306 ± 21	287–328
WBC (× 10^9/L)	4.3 ± 2.0	9.2 ± 5.6	9.7 ± 2.5		3.2 ± 1.3	2.1—4.6	6.1 ± 1.7	4.6–8.1	8.6 ± 3.8	5.6–13.8
Neutrophils (× 10^9/L)			0.8 ± 0.3		0.83 ± 0.42	0.5–1.3	0.10 ± 0.07	0–0.2	3.11 ± 0.94	1.8–4.0
Neutrophils (%)	57.2 ± 13.2	49.0 ± 9.8								
Lymphocytes (× 10^9/L)			8.9 ± 2.6		1.83 ± 0.55	1.3–2.4	5.78 ± 1.51	4.5–7.7	4.87 ± 3.39	1.74–8.97
Lymphocytes (%)	39.9 ± 11.6	46.4 ± 8.6								
Monocytes (× 10^9/L)					0.47 ± 0.81	0–1.4	0.02 ± 0.04	0–0.1	0.45 ± 0.15	0.3–0.6
Eosinophils (× 10^9/L)					0	0	0	0	0.12 ± 0.1	0.06–0.28
Basophils (× 10^9/L)					0	0	0.18 ± 0.45	0–0.9	0	0
Platelets (× 10^9/L)					534 ± 74	482–586	281 ± 27	252–317	117	117
Reference	Clark (2004)				Old et al. (2005)					

Source: Clark (2004); Old et al. (2005).

Table 15.6 Biochemistry values for Australian rodents

Parameter (Sample size)	Plains rat (5)		Spinifex hopping mouse (3)		Central rock-rat (1–4)	
	Mean ± SD	Range	Mean ± SD	Range	Mean ± SD	Range
Total protein (g/L)	60.6 ± 3.65	55–64	47 ± 5.66	43–51	54.75 ± 4.03	49–58
Albumin (g/L)	48 ± 3	44–51	29 ± 2.83	27–31	33.75 ± 3.3	30–38
Globulin (g/L)	12.6 ± 1.52	11–15	18 ± 2.83	16–20	21.33 ± 2.52	19–24
Calcium (mmol/L)	2.32 ± 0.24	2.05–2.58	1.82 ± 0.2	1.68–1.96	2.55 ± 0.07	2.46–2.62
Phosphorus (mmol/L	2.09 ± 0.53	1.44–2.55	2.65 ± 0.44	2.34–2.96	3.37 ± 1.27	1.53–4.43
Magnesium (mmol/L)	1.12 ± 0.14	0.97–1.3	1.28 ± 0.06	1.24–1.32	1.21 ± 0.32	0.73–1.39
Sodium (mmol/L)	145.8 ± 3.83	141–150	137.5 ± 7.78	132–143	152 ± 2.58	149–155
Potassium (mmol/L)					9.4 ± 2.63	5.5–11.3
Chloride (mmol/L)					115.75 ± 1.5	114–117
BUN (mmol/L)	9 ± 2.44	7.2–13.3	6.7 ± 0.14	6.6–6.8	9.08 ± 2.01	6.9–11.7
Creatinine (mmol/L)	0.02 ± 0.01	0.02–0.03			0.06 ± 0.02	0.02–0.07
Glucose (mmol/L)					3.85 ± 1.34	2.9–4.8
Cholesterol (mmol/L)	1.92 ± 0.4	1.36–2.3	2.96 ± 0.81	2.38–2.53	3.13	3.13
Triglycerides (mmol/L)	1.79 ± 0.41	1.28–2.24	1.51 ± 0.88	0.88–2.13	1.88	1.88
Total bilirubin (µmol/L)	3.2 ± 1.3	2–5	1.5 ± 0.71	1–2	<2	
ALP (U/L)	50.4 ± 42.17	18–112	770.5 ± 514.7	407–1134	107	107
ALT (U/L)	50.8 ± 23.47	27–84	12.5 ± 7.78	7–18	86	86
GGT (U/L)	3 ± 1.58	1–5				<1–2
GLDH (U/L)					5.37 ± 3.66	1.3–8.4
ßOHB (mmol/L)					0.34 ± 0.28	0.16–0.2
AST (U/L)	165.6 ± 24.74	136–195	178 ± 147.08	74–282	262 ± 65.3	168–319
CK (U/L)					390.97 ± 350.16	168–319

Source: Old et al. (2005).

Table 15.7 Haematology and biochemistry values for domestic rodents (range)

Parameter	House Mouse	Brown Rat
PCV (L/L)	0.37–0.46	0.37–0.49
RBC (× 10^{12}/L)	7.9–10.1	5.4–8.5
Hb (g/L)	11.0–14.5	11.5–16.0
WBC (× 10^9/L)	5.0–13.7	4.0–10.2
Neutrophils (%)	10–40	6–17
Lymphocytes (%)	55–95	9–34
Monocytes (%)	0.1–3.5	0–5
Eosinophils (%)	0–4	0–6
Basophils (%)	0–0.3	0–1.5
Platelets (×10^9/L)	600–1200	450–885
Total protein (g/L	59–103	59–78
Albumin (g/L	25–48	33–46
Globulin (g/L)	6	22–35
Calcium (mmol/L)	1.2–2.4	1.9–3.2

Parameter	House Mouse	Brown Rat
Phosphorus (mmol/L)	1.8–3.0	1.7–2.7
Magnesium (mmol/L)	0.6–1.3	1.1–1.3
Sodium (mmol/L)	143–164	142–150
Potassium (mmol/L)	6.3–8.0	4.3–6.3
Chloride (mmol/L)	105–118	100–109
BUN (mmol/L)	6.4–11.1	4.3–7.9
Creatinine (mmol/L)	0.04–0.1	0.03–0.07
Glucose (mmol/L)	4.1–10.2	4.1–9.0
Cholesterol (mmol/L)	1.5–2.7	1.1–3.6
ALP (U/L)	43–71	40–191
ALT (U/L)	44–87	4–144
AST (U/L)	101–214	54–192
CK (U/L)	155	111–334

Source: Quesenberry & Carpenter (2004).

Table 15.8 Urinalysis values for domestic rodents

Parameter	House mouse	Brown rat
USG	1.060	1.022–1.050
pH	5.01	5–7
Protein (mg/L)	Males proteinuric	<300

Source: Bihun & Bauck (2004).

8 HAND-REARING

Hand-rearing of Australian rodents has been reported for a stick-nest rat, black-footed tree-rat (*Mesembriomys gouldii*) and water rat (Jackson 2003). The process followed similar procedures as for other native mammals, although it can be extremely challenging and in many cases is unsuccessful when orphaned young are only a few days old (Saunders 2003). Neonatal rodents have poor thermoregulation and careful attention needs to be paid to the ambient temperature at which they are held. Providing an appropriate environment which is hygienic, minimises stress and prevents risk of escape is also important. Feeding apparatus is frequently determined by trial and error, but may include items such as an eye dropper, 1 mL syringe or a catheter attached to a syringe. For larger sized rodents a teat designed for smaller possums may be suitable (Jackson 2003). Rodent milk has relatively high protein (35%) and dry matter, moderate fat (47%) and low sugar (12%) content (Sainsbury 2003), although there is variation with species and period of lactation (figures in parentheses are for the house mouse) (Jackson 2003). The quantity of milk required can be calculated according to the formula $E = 225\ W^{0.83}$ where E is the energy intake in kilocalories per day and W is the body weight in kilograms. A minimum of 10–20% of body weight should be consumed on a daily basis (Jackson 2003). Various milk replacers (e.g. Wombaroo Dog Milk Replacer (Wombaroo Food Products) and Di-Vetelact (Sharpe Laboratories)) have been used for hand-rearing rodents; those produced for carnivores are thought to be better suited for rodents than those produced for herbivores. Orphaned rodents are likely to be dehydrated and malnourished at initial presentation, and correction of this prior to starting supplementary feeding is important. Age of weaning varies with species but weaning tends to occur rapidly (Jackson 2003). More detailed guidelines are given below and in Jackson (2003).

- Supplementary heating is important until the animal is fully furred and active. Ambient temperatures of 32–40°C for unfurred animals and 28–32°C for more developed young are recommended, with a temperature gradient to allow thermoregulation.
- Correct dehydration—5 g glucose or 1 g electrolyte replacement in 100 mL water. Warm pup prior to feeding to reduce risk of aspiration. Alternatively, SC fluids may be administered.
- Weigh daily—adjust feed volume accordingly.
- Toilet animal at feeding as for marsupial young.
- Volumes of milk provided at each feed vary with stage of development. Higher volumes (50–150% body weight) may be required for very small animals; volumes ranging from 30–50% body weight may be required for larger animals.
- Frequency of feeding can be determined by assessing daily volume required and maximum volume of each feed. Typical ranges are 4–6 times daily.
- Record data—gender, date, time, body weight (to nearest 0.1 g for smaller species and 1 g for larger species), general activity and demeanour, faecal and urine output, food offered, food consumed.
- Approaching weaning, provide variety of solid foods, including browse to maintain dental health.

9 DISEASES

Healthy rodents spend a large amount of time grooming and the first signs of ill health are often reflected in the external appearance. The coat often appears dull and unkempt, activity decreases, animals eat and drink less and lose weight. They may eventually display a hunched posture, remain isolated from enclosure mates and show no interest in surroundings.

9.1 Dermatological diseases

9.1.1 Bite wounds and other traumatic injuries

Rodents can be particularly aggressive towards each other. When kept in colonies, dominant animals frequently pick on submissive animals. Barbering as seen in domestic mice is not common in Australian rodents. Wild rodents may be caught and injured by predators such as cats. Degloving/skin slip injuries of the tail may occur during handling. Clinical signs include skin wounds and fur loss, subcutaneous abscesses and loss of skin from the tail. Submissive animals are frequently in poor condition if there is a history of aggression between individuals.

Wounds are usually treated under general anaesthesia (debridement, irrigation, topical and/or systemic antibiotics, wound dressings). It is important to rectify the underlying cause of aggression, e.g. unsuitable social grouping for particular species, overcrowding, insufficient nesting areas. Introducing submissive animals back into a colony after treatment can be difficult and euthanasia should be considered if alternative social groupings cannot be provided. Bites from predators such as domestic cats and conspecifics usually require systemic antibiotic therapy even for minor injuries as bacteria are often seeded deep into muscles or body cavities. *Pasteurella multocida* has been isolated from the mouth of a water rat (Obendorf & Smales 1985).

Prevention of injury includes good husbandry with particular attention to stocking density (Jackson 2003) and appropriate handling and restraint techniques. Aggression and bite wounds can occur as a normal part of mating behaviour in some species (e.g. central rock-rat), therefore separation may not be warranted.

9.1.2 Mites

Mites known to infect Australian rodents include members of the genera *Laelaps, Mesolaelaps, Murichirus, Radfordia, Myocoptes leptotrombidium* and *Cheyletiella* (Old et al. in press; Perth Zoo records). Stress due to overcrowding or suboptimal husbandry such as poor hygiene may predispose Australian rodents to increased mite populations. The clinical significance of mite infestations is not clear and mites may be seen without evidence of clinical disease. Signs of mite infestation may include pruritus, alopecia and evidence of self-trauma, e.g. excoriations, scabbing, ulceration. Diagnosis is based on clinical signs with observation of significant numbers of mites on skin scraping and microscopy.

Treatments that have been used in clinical cases of mite infestations include topical selamectin at 6 mg/kg as a single dose and ivermectin at 0.2 mg/kg q 10–14 d PO, SC or topically (Carpenter et al. 2001). Ivermectin may also be used topically if diluted 1 part 1% ivermectin to 100 parts of an equal mix of water and propylene glycol (Table 15.9). An aerosol insecticide for aircraft cabin spraying (Top of Descent [active constituent d-Phenothrin] Callington Haven) has been used successfully to eliminate and control a mite infestation in a colony of spinifex hopping mice and laboratory mice (L Vogelnest *pers. comm.*). Prevention includes quarantine of new introductions to colony, appropriate husbandry and stocking densities. Routine treatments with acaricides are not usually required.

9.2 Gastrointestinal diseases

9.2.1 Dental malocclusion

Any disease or disorder that disrupts the normal attritional movements of rodent incisors will lead to malocclusion. In most cases malocclusion is caused by trauma to the mouth, but may also be related to poor nutrition or an inappropriate diet that does not promote normal teeth wear. Osteomyelitis, oral fractures, oral infections and loss of teeth have been associated with malocclusion (Sainsbury 2003). Nutritional imbalances, such as excessive selenium and insufficient vitamin C, have also been linked to malocclusion as these imbalances affect the synthesis of collagen, which may cause weakening of periodontal ligaments and allow teeth to move from their normal position.

Clinical signs include visible overgrowth or distortion of teeth, food dropping from the mouth, ptyalism, oral trauma, weight loss and deterioration in body condition. Diagnosis and treatment often require general anaesthesia. Identification and treatment of the underlying cause is desirable. The crowns of affected teeth can be trimmed with a dental bur or diamond disc. The incisors can be clipped. Exposed pulp can be capped with calcium hydroxide paste to avoid infection. Extraction of affected teeth is an alternative to trimming. Prevention involves provision of an appropriate and balanced diet for the species and provision of suitable plant material for gnawing to allow normal tooth wear.

9.2.2 Gastroenteritis

Viruses such as parvoviruses and others have caused diarrhoea in laboratory rodents (Pritchett & Corning 2004). Bacteria (Salmonellae, *Clostridum piliforme*), fungi (*Candida* spp.) and parasites (Oxyurids, Mastophora, cestodes, protozoa) may or may not be pathogenic. Clinical signs include ptyalism, diarrhoea, wetness or soiling of the perineal region, dehydration, lethargy, emaciation and sudden death. Diagnosis is based on observation of clinical signs, faecal flotation and examination for parasite eggs and cysts as well as faecal culture and susceptibility testing.

Treatment includes hospitalisation and general supportive care—fluids (oral and/or parenteral), warmth, nutritional support, antiparasitic medications (e.g. ivermectin 0.2–0.4 mg/kg SC q 10–14 d; piperazine citrate 4–7 mg/mL drinking water for 3–10 d) and antibiotics as indicated.

Prevention involves species-appropriate husbandry and stress minimisation. Oxyurid nematodes (e.g. *Aspicularis tetraptera*) can be a common finding on faecal examination of healthy rodents and their presence in a healthy animal is not necessarily an indication to administer antiparasitic medications (Hill et al. 2007). Routine worming is not generally required unless indicated by the presence of clinical disease or on routine faecal flotation examinations.

Oesophagitis and gastritis caused by *Candida albicans* was diagnosed in three greater stick-nest rats. Clinical signs included weight loss and ptyalism. Diagnosis was confirmed on swabs taken from the oesophagus at necropsy of two animals. Treatment with nystatin was unsuccessful; one animal recovered after treatment with itraconazole at 5 mg/kg sid PO for 3 wk (Taronga Zoo records, ARWH 2007). There were no apparent predisposing factors that may have contributed to the animals developing the disease.

9.2.3 Tyzzer's disease

Tyzzer's disease is caused by *Clostridium piliforme* (previously *Bacillus piliformis*) and has been observed in a number of Australian rodent species including Mitchell's hopping mouse (*Notomys mitchellii*), spinifex hopping mouse and common rock-rats (*Zyzomys argurus*) (Perth Zoo records). Clinical signs include mucoid to haemorrhagic diarrhoea, dehydration, lethargy, weight loss and sudden death (Pritchett & Corning 2004). Diagnosis is usually made at necropsy, with typical findings including multiple white foci (<1–3 mm diameter) through the hepatic parenchyma, and evidence of ulcerative to haemorrhagic enteritis. There may also be evidence of myocarditis. Histopathologic findings reveal intestinal epithelial (particularly ileal and caecal) and hepatic necrosis. Definitive diagnosis is achieved through identification of the organism in associated lesions using silver staining (Wobeser 2001).

Treatment includes hospitalisation and supportive care as for enteritis but mortality of affected individuals is often high. Systemic antibiotics such as tetracycline, doxycycline or enrofloxacin (Table 15.9) may reduce morbidity and mortality. Medication via drinking water may be more practical in an outbreak.

Tyzzer's disease is often the result of overcrowding and/or poor hygiene. Hygienic conditions and appropriate husbandry are therefore important in preventing the disease. The source of infection is often food or bedding contaminated with bacterial spores. Spores are resistant to many disinfectants and can survive for over a year in bedding or soil. Reduction of environmental load of spores is also important in an outbreak, and may include regular changing of substrate and disinfection of the enclosure using 0.5% sodium hypochlorite.

9.3 Cardiopulmonary diseases

9.3.1 Respiratory infection

Various infectious agents including viruses, bacteria (e.g. *Pasteurella* spp., *Mycobacteria* spp. and *Mycoplasma* spp.) and fungi (e.g. *Cryptococcus neoformans*) may cause respiratory disease in Australian rodents (ARWH 2007 case numbers 2896/1, 3748/1, 4166/3, 4281/1, 4899/1; Mackie et al. 2001). Poor husbandry such as inadequate ventilation or inappropriate humidity may predispose animals to infection. Clinical signs include dyspnoea, sneezing, nasal discharge, inappetence, weight loss, lethargy and dehydration. Diagnosis is based on observation of clinical signs, radiography, necropsy findings and histopathology.

Treatment includes hospitalisation and supportive care, systemic antibiotics as indicated by culture and susceptibility testing from necropsy samples if possible. Fluoroquinolones, tetracyclines, tylosin, cephalosporins and potentiated sulphonamides have been reported to be effective in non-native rodent species (Table 15.9).

Prevention involves quarantine of new introductions to a colony, appropriate husbandry ensuring adequate ventilation, humidity and temperature control and minimising dust from the enclosure substrate.

9.4 Urinary and reproductive diseases

9.4.1 Chronic progressive nephropathy

Infectious agents such as bacteria and protozoa (Klossiella-like organism) and diet (oxalates) may cause renal disease. Some species, particularly the greater stick-nest rat, are thought to have a genetic predisposition to developing chronic tubular nephropathy, possibly secondary to disorders in oxalate metabolism (S Bigwood *pers. comm.*; C Monaghan *pers. comm.*). Clinical signs include polydipsia/polyuria, wetness around the perineal region and weight loss.

Diagnosis is based on serum biochemistry and urinalysis (proteinuria, isosthenuria, azotaemia). Renal interstitial and/or cortical fibrosis, tubular dilation and pseudodilation, tubular protein casts, glomerulonephritis and oxalate nephrosis are seen histologically (Perth Zoo records; ARWH 2007 case numbers 1457/1, 2041/1, 2144/1, 2299/1, 2991/1, 4023/1, 4067/1). Fluid diuresis may slow the progression to renal failure (see section 11 for more details). Dietary caloric and protein restriction and administration of anabolic steroids has been described for non-native rodent species with the condition (Orr 2002). Food items high in oxalates (e.g. saltbush, *Atriplex* spp; beetroot, *Beta vulgaris*) should be restricted in the diet of susceptible species such as the greater stick-nest rat.

9.4.2 Metritis/pyometra

Uterine infections have been reported in the spinifex hopping mouse (ARWH 2007 case number 1819/1, unpublished data), a golden-backed tree-rat (*Mesembriomys macrurus*) (ARWH 2007 case number 3309/1) and domestic rodent species. Clinical signs include vaginal discharge, abdominal distension and inappetence. Diagnosis is based on clinical signs, ultrasonography and radiography. Treatment requires ovariohysterectomy and supportive care including fluid therapy, heat, supplemental nutrition, analgesia and systemic antibiotics.

9.5 Ophthalmological diseases

Cataracts have been observed in a number of Australian rodent species, particularly the greater stick-nest rat and black-footed tree-rat (*Mesembriomys gouldii*), and has frequently been associated with glucosuria (Perth Zoo records; ARWH 2007 case numbers 2041/1, 2064/1, 4159/1). The possibility of an underlying condition such as diabetes mellitus has been investigated, without conclusion (Perth Zoo records). Lens opacity, corneal oedema, scleritis, serous ocular discharge, exophthalmos, visual deficits, reduced appetite and weight loss may be seen in animals with cataracts (Perth Zoo records). When investigating cataracts in Australian rodents, diagnostic investigations should include urinalysis to check for glucosuria and ketonuria, and blood glucose measurement. There are no reports of cataract removal in Australian rodents.

9.6 Neonatal diseases

Cannibalism of neonates is common in large colonies and may be provoked by poor husbandry, stress, excessive handling and overcrowding.

9.7 Infectious diseases

9.7.1 Viruses

Little is known about the range of viruses infecting Australian rodents, and even less about the effects of these viruses on individual animals and populations. Antibodies to a number of viruses have been found in the house mouse in Australia (Smith et al. 1993; Moro & Lawson 2003). These include murine cytomegalovirus (MCMV), murine rotavirus (epizootic diarrhoea of infant mice), mouse adenovirus, mouse parvovirus, mouse hepatitis virus, mouse reovirus, lymphocytic choriomeningitis virus (LCMV), Sendai virus and Theiler's murine encephalomyelitis virus.

Limited studies have not found these viruses to be present in any Australian rodent species. MCMV is of interest as a potential biological control agent for house mice in Australia. Since house mouse plagues cause millions of dollars of damage in the agricultural industry there is keen interest in controlling their numbers by humane and ethical methods. In house mice, MCMV appears to cause no overt disease and is carried asymptomatically in many populations. This makes it a potentially suitable candidate to be genetically modified for immunocontraception and fertility reduction, and thus control of population size.

Mossman virus (Campbell et al. 1977) has been found in the Cape York rat (*Rattus leucopus*) and bush rat in northern Queensland. The virus has been found in *Laelaps echidninus* mites, suggesting they may play a role in transmission. No pathology has been reported in Australian species but pneumonia has been recorded in experimentally infected house mice. One study has shown neutralising antibodies to J virus in 28% of wild house mice, 14% of brown rats, 12% of pigs and 2% of humans in northern Queensland (Jun et al. 1977). Severe haemorrhagic pneumonia has been reported in house mice associated with this virus.

Antibodies to encephalomyocarditis virus (EMCV) have been found in several rodent species in Australia including the water rat, canefield rat (*Rattus sordidus*), grassland melomys (*Melomys burtoni*), brown rat and black rat (Pope & Scott 1960). Infection with the virus is usually asymptomatic in rodents but it is a significant pathogen of captive wildlife, having caused mortalities in primates, elephants, lions, ungulates and marsupials (McLelland et al. 2005; Thomson et al. 2001). It is a potential zoonosis and evidence of human infection has been detected in rural New South Wales, but it has not been found to cause clinical disease in people in Australia (Kirkland et al. 1989).

9.7.2 Bacteria

Streptobacillus moniliformis has been reported to infect captive spinifex hopping mice and wild house mice from south-eastern Queensland (Hopkinson & Lloyd 1981; Taylor et al. 1994). This is the cause of rat-bite fever in humans and transmission is primarily through bite wounds (Quesenberry & Carpenter 2004). It is a commensal in the nasopharynx of the brown rat and may be endemic in wild house mice. It may result in disease at high population densities and during times of stress, e.g. mouse plagues. Pathology includes single or multiple swollen joints, subcutaneous abscesses and hepatomegaly.

Various Salmonella species have been reported in Australian rodents (Glazebrook et al. 1978; How et al. 1983). These include the brush-tailed rabbit-rat, Lakeland Downs mouse (*Leggadina lakedownensis*), grassland melomys, western chestnut mouse (*Pseudomys nanus*), common rock-rat, large rock-rat (*Zyzomys woodwardi*), Tunney's rat (*Rattus tunneyi*), giant white-tailed rat, canefield rat and bush rat. Studies have confirmed Salmonellae in Australian rodents from northern Western Australia and north-eastern Queensland but they are likely to be present in many other areas.

A study conducted in northern Western Australia reported a marked seasonal variation in prevalence of Salmonella, with maximum prevalence in the wet season (January) (How et al. 1983). Numerous serotypes of Salmonella exist and all are considered potentially pathogenic. Determination of serotype is recommended if Salmonella is isolated from a clinical case. Transmission is primarily faeco-oral or via fomites. Chronic carrier states can develop, acting as a potential reservoir for infection of other animals. Pathology in Australian rodents was not reported in these studies. Clinical signs of infection in domestic rats may include lethargy, weight loss, conjunctivitis, poor coat condition and diarrhoea (Pritchett & Corning 2004).

Leptospira species have been reported in the bush rat, giant white-tailed rat, water rat and brown rat. Many rodents have asymptomatic infections. In one study 15% of bush rats surveyed in northern Queensland were found to be excreting *Leptospira celldoni* and none showed a detectable antibody response (Glazebrook 1978). No pathology associated with Leptospira has been reported in Australian rodents. Domestic rodents may carry and shed this organism without overt signs of disease.

Cilia-associated respiratory (CAR) bacillus and Mycoplasma species have been reported in captive spinifex hopping mice and laboratory rats and mice (Mackie et al. 2001). The organism is an inhabitant of the respiratory tract and mild to severe suppurative bronchopneumonia has been seen in spinifex hopping mice.

Burkholderia pseudomallei (melioidosis) has been reported to cause suppurative pneumonia in the bush rat and giant white-tailed rat in northern Queensland. One survey found 14.5% of rodents (mainly black rats) had antibodies to *B. pseudomallei* (Glazebrook et al. 1978).

9.7.3 Rickettsial diseases

Orientia tsutsugamushi (scrub typhus) has been reported in the canefield rat, fawn-footed melomys, grassland melomys, bush rat, black rat, bandicoots and humans in north-eastern Australia including Cape York Peninsula. It has also been reported from the Northern Territory and northern Western Australia (Glazebrook et al. 1978; Domrow 1966; Odorico et al. 1998). The organism is passed from adult mites (e.g. *Leptotrombidium deliense*) to their offspring by trans-ovarial transmission (infected female mites lay eggs infected with the bacterium). No pathology associated with the organism has been reported in Australian rodents.

Rickettsia australis (Queensland tick typhus) and *Rickettsia honei* (Flinders Island spotted fever) have been reported in the giant white-tailed rat, bush rat, swamp rat (*Rattus lutreolus*), black rat, brown rat, house mouse, marsupials (macropods, antechinuses, bandicoots, possums), domestic dogs, possibly reptiles and humans (Graves et al. 1993). *R. australis* is found in eastern Australia (Cape York to Gippsland) and *R. honei* on Flinders Island and possibly Tasmania (Odorico et al. 1998). These organisms are thought to be transmitted by Ixodid ticks (*Ixodes holocyclus, I. cornuatus, I. tasmani*). One study showed 89% of bush rats from Gippsland (Vic.) to have antibodies to spotted fever group rickettsia. No pathology associated with these organisms has been reported in Australian rodents.

9.7.4 Ectoparasites

External parasites such as ticks, mites and fleas have been found on many species of Australian rodents but have not been shown to cause disease under natural conditions. However, the role of ectoparasites as agents for transmitting pathogens between rodents is well-established. Australian rodents may host a diverse range of ectoparasites, with 22 species of fleas recorded from the bush rat and 14 from the swamp rat (Dunnet & Mardon 1974). A large number of mite species have been found living on Australian rodents. Most inhabit the skin but mites have been found inhabiting the nasal passages and oral cavity (Spratt 2002).

9.7.5 Endoparasites

While there have been no endoparasites recorded from most Australian rodent species, helminths of the bush rat and water rat have received considerable attention (Smales 1997; Warner 1998). Studies have also been published on the endoparasites of the canefield rat (Smales 1992), fawn-footed melomys (Smales et al. 2004) and Cape York rat (Smales & Spratt 2004). The bush rat shows a helminth community structure dominated by nematodes, as in most other terrestrial rodents. The water rat, in contrast, hosts predominantly trematodes which require an aquatic mollusc for a first intermediate host and hence are usually found in animals that live in or near water. These parasites may then move to a second intermediate host which can be an insect, crustacean, fish or amphibian. The water rat's aquatic life-style, together with its varied diet, make it particularly susceptible to trematode infection with over 30 species recorded in water rats (Smales 1997).

Spratt (2002) reviews parasites of the respiratory tract of Australian mammals including the lungworm, *Angiostrongylus cantonensis*, found in black rats and brown rats. Various mollusc species act as an intermediate host. This parasite is of particular interest due to its ability to cause fatal neurological disease in abnormal hosts, including humans and numerous other species. A related lungworm, *Angiostrongylus mackerrasae*, has been found in Australian rodents but is not known to infect other species.

The lungworm, *Gallegostrongylus australis*, is found in the lungs of the bush rat where it generally causes only mild pathological changes. In the swamp rat it can be responsible for debilitating lung damage (Spratt et al. 2001). It may be that the lungworm originally evolved in the bush rat and has only recently infected swamp rats, hence causes more significant disease in the latter species which has had less time to evolve a balanced host–parasite relationship.

Metacestode pleroceroids of *Spirometra* sp. (spargana) have been found in the connective tissue of water rats, causing inflammation, fibrosis and occasionally severe inflammation and muscle atrophy resulting in leg weakness (Obendorf & Smales 1985).

Helminths occasionally cause clinical disease in captive rodents and regular faecal parasitology and necropsy of all dead animals is recommended. The cestode *Rodentolepis fraterna* has been associated with enteritis and mortality in a captive spinifex hopping mouse where approximately 200 adult worms were found in the intestinal lumen (Hill et al. 2007).

If high levels of endoparasites are discovered, treatment of all animals in the enclosure may be undertaken using praziquantel (10 mg/kg orally) for cestodes, or ivermectin (0.2–0.4 mg/kg PO or SC) for nematodes.

Haemoparasites that have been found in rodents include *Trypanosoma lewisi* (Mackerras 1959 cited in Clark 2004), *Hepatozoon* spp. (O'Donoghue & Adlard 2000 cited in Clark 2004), *Cercopithifilaria johnstoni* (Mackerras 1962 cited in Clark 2004) and other microfilaria. The presence of these parasites rarely appears to cause clinical disease, but in severe infections *Hepatozoon muris* may cause anaemia and splenomegaly (Clark 2004).

9.7.6 Protozoa

Rodents are an intermediate host of *Toxoplasma gondii* and it has been reported in the muscle, kidney, heart and brain tissue of many rodent species without inflammatory response. However, granulomatous inflammation has been reported in the heart and brain of the water rat (Smales & Obendorf 1996).

Sarcocystis murotechis has been reported in the bush rat, swamp rat, canefield rat, giant white-tailed rat, fawn-footed melomys, long-tailed mouse (*Pseudomys higginsi*), water rat, broad-toothed rat (*Mastacomys fuscus*), brown rat and black rat in eastern Australia (Munday et al. 1979; Munday et al. 1978; Munday & Mason 1980). It has a two-host life-cycle. The definitive host appears to be snakes, where sexual reproduction of the parasite occurs. Sporocysts are shed in the snake's faeces. The intermediate host is usually a herbivorous animal, in which asexual reproduction of the parasite takes place. The life-cycle can be completed when the intermediate host is eaten by a snake. Tiger snakes (*Notechis* spp.) appear to be the definitive host for *S. murinotechis* in south-eastern Australia and Tasmania. In tropical and subtropical Australia the parasite has a life-cycle that involves bush rats and carpet pythons (*Morelia spilota*). Infection rates in wild-caught bush rats are often high, with up to 95% infected in some areas. Cysts are found in skeletal muscle but have not been reported to be associated with pathology.

Klossiella hydromyos has been reported in the water rat (Winter & Watt 1971). It probably occurs throughout the water rat's range in mainland Australia but has not been reported from Tasmania. Direct transmission is likely, with oral ingestion followed by urinary excretion. Klossiella is thought to have contributed to protozoal nephritis in a greater stick-nest rat and degenerative changes to kidney tubules have been reported to occur in other species infected with *Klossiella* spp. (Smales & Obendorf 1996; Winter & Watt 1971).

9.8 Non-infectious diseases

Most reports of non-infectious diseases in Australian rodents are from the few species that have been maintained in captivity. Dental malocclusion and fight wounds are the most frequently encountered while chronic progressive nephropathy and ocular cataracts are also significant (see 9.4.1, 9.5).

A number of neoplasms have been reported in Australian rodents including lymphoma (ARWH 2007 case numbers 1338/1, 2404/1, 3129/1), hepatic sarcoma (ARWH 2007 case number 1457/1), hepatic adenocarcinoma (Perth Zoo records), islet cell adenoma (Perth Zoo records), trichoblastoma (basal cell tumour) (ARWH case number 5617/1) and pheochromocytoma (ARWH 2007 case number 2035/1).

Intussusception (ARWH 2007 case number 2991/1), uterine torsion (ARWH 2007 case number 2164/1) and hepatic lipidosis have been reported from captive Australian rodents.

Bite wounds and other traumatic injuries are relatively common in both wild and captive Australian rodents (see 9.1.1). Bone fractures are not uncommon. Clinical signs include lameness and loss of appetite. Some fractures can be treated with external stabilisation using splints, slings and analgesia. Antibiotic therapy is indicated for treatment of open fractures (Table 15.9). Small rodents generally adapt very well to limb or tail amputation if necessary.

Metabolic bone disease and pathological fractures have been reported in exotic rodent species fed diets primarily comprising sunflower seeds (Keeble 2002). Appropriate nutrition is essential to prevent this developing in captive Australian rodents.

10 ZOONOSES

Rodents worldwide have long been associated with the transmission of zoonotic diseases and Australian rodents are no exception. There are many pathogens that infect rodents in Australia that can also infect people, but the importance of rodents in transmission of disease to humans is often unclear. Most of the pathogens can infect other animals, which may be the source of human infections, while many human cases may result in only mild symptoms and hence go unreported.

Singleton et al. (2003) reviewed rodent-borne zoonotic diseases in Australia and reported the detection of leptospirosis, various rickettsia, *Streptobacillus moniliformis*, *Angiostrongylus cantonensis*, lymphocytic choriomeningitis virus and cryptosporidia.

Leptospirosis is a significant disease in Australia with over 200 human cases recorded each year, over half in Queensland (Glazebrook et al. 1978). In 1986 a case of leptospirosis was reported in a zoologist who almost certainly contracted the disease while handling wild bush rats (Looke 1986). The leptospire bacteria were probably present in the rat's urine and entered the handler's bloodstream when rat urine came in contact with bite wounds on his hands acquired while handling the rats.

Rickettsial organisms carried by rodents (see 9.7.3) are recognised as an important cause of fever in tropical

Queensland. Spotted fever has an annual incidence of approximately 1% in humans on Flinders Island (Graves et al. 1993; Odorico et al. 1998).

Rat-bite fever, caused by *Streptobacillus moniliformis*, is a potential zoonosis associated with Australian rodents. The causative organism has been isolated from captive spinifex hopping mice and wild house mice (Hopkinson & Lloyd 1981; Taylor et al. 1994). The disease is transmitted primarily through bite wounds, with an incubation period of 3–10 d. Clinical signs of zoonotic infection include relapsing fever and chills, vomiting, myalgia, regional lymphadenopathy and a maculopapular rash affecting the extremities (Mitchell & Tully 2004). Bite wounds from rodents should be cleaned and antiseptic applied immediately to reduce the risk of this zoonosis.

Several of the potential zoonoses carried by Australian rodents may be of little significance due to low prevalence or other features of their epidemiology. For example, there have been no reported human cases of lymphocytic choriomeningitis virus in Australia despite its presence in the feral house mouse population (Singleton et al. 2003); the presence of cats is a much more significant factor in the risk of toxoplasmosis to humans than Australian rodents; *Angiostrongylus cantonensis* has not been reported from Australian rodents and is only known to infect introduced rodent species.

11 THERAPEUTICS

Due to the small size and high metabolic rate of rodents, dose rates of medications are often higher than those for domestic carnivores. Recommended dose rates for common pet rodent species are frequently used when medicating Australian rodents (Table 15.9). Alternatively, allometric scaling may be utilised to assist in determining the dose rate for a particular medication. Allometric scaling relates a drug dose to the metabolic rate of an animal rather than its body weight, and accounts for the physiological phenomenon of metabolic rate being inversely related to body size. This technique utilises dose rates for a species in which the pharmacokinetics of a drug are known (e.g. domestic dog) as a reference as follows:

1. Determination of the specific minimum energy cost for the known (control) and unknown (patient) animals.
   Specific minimum energy cost, SMEC = $k(W_{kg})^{-0.25}$ where k = 70 (for placental mammals), W = body weight in kg.
2. Determination of the ratio SMEC (patient):SMEC (control animal).
3. Dose rate (patient) = ratio × dose rate (control).
4. Daily frequency (patient) = ratio × daily frequency (control) (this step determines the number of treatments per day).
5. Hours between treatments = 24/step 4.

Systemic antibiotic treatment of rodents can result in clostridial enterotoxaemia, particularly antibiotics with a Gram-positive spectrum and when given orally (Carpenter et al. 2001). No other specific contraindications (e.g. species susceptibility to adverse reactions) for the commonly used medications have been reported in rodents.

Administration of medications to rodents can be via parenteral, topical or oral means. Oral medications may be administered directly or may be included in food or water. Direct administration is preferred to administration in food or water as it guarantees an accurate dose, but it is not always practical (e.g. medicating a large group of rodents). For IM injections a maximum volume of 0.05 mL should be given at any one site to a mouse-sized rodent (Saunders 2003), and rotation of sites of administration is recommended. Larger volumes may be administered via the SC or IP route. For IP injection the animal should be well-restrained or anaesthetised and in dorsal recumbency. The site for injection should be aseptically prepared. The needle is passed through the skin and body wall at an angle of approximately 20° to the body, parallel to the midline in a cranial direction. The right caudal quadrant of the abdomen is generally used to avoid perforation of the caecum or bladder (Hrapkiewicz et al. 1998). The syringe should be aspirated before injection to check for visceral perforation. Table 15.9 provides recommended dose rates for commonly used medications in rodents.

Supportive therapy is often an important component of treating sick rodents and may include supplementary heat, fluid therapy and provision of oxygen. Fluid therapy may be administered to correct dehydration or as a supportive measure where fluid intake may be reduced. Rates of administration are frequently extrapolated from those used for domestic small mammals. Fluids may be administered IV where feasible, although SC administration is utilised more

Table 15.9 Frequently used medications in rodents

Drug	Dose rate	Comments
Antibiotics		
Cephaloridine	10–25 mg/kg SC/IM sid[1]	
Doxycycline	2.5–5 mg/kg PO bid[1]	Do not use in young or pregnant animals
Enrofloxacin	5–10 mg/kg PO/IM bid[1]	Dilution of injectable enrofloxacin is recommended to minimise the risk of muscle necrosis
	0.05–0.20 mg/mL[1]	Administered in drinking water; change water every 24 hr
Tetracycline	10–20 mg/kg PO q 8–12 hr[2]	
	5 mg/mL drinking water[3]	Change daily
Trimethoprim-sulpha	15–30 mg/kg PO, SC q 12 hr[2]	
Tylosin	10 mg/kg PO, SC q 12 hr[2]	
Analgesics and antiinflammatory drugs		
Butorphanol	1–5 mg/kg SC q 2–4 hr[1]	For analgesia
	0.1–0.2 mg/kg SC[4]	For pre-anaesthetic sedation
Carprofen	5–10 mg/kg PO[1]	
Meloxicam	1–2 mg/kg SC, PO sid/bid[2]	
Antiparasitics		
Fipronil	7.5 mg/kg topically q 30–60 d[2]	
Ivermectin	1–3 drops topically[2]	Use 1% ivermectin solution diluted 1:100 with a 1:1 solution of propylene glycol and water Used for mite control
	0.2 mg/kg SC, PO q 10 d[2]	Used for mite control
	2 mg/kg PO q 10 d[2]	For enteric nematodes
Piperazine citrate	4–7 mg/mL in drinking water for 3–10 d[2]	Change medicated water daily
Praziquantel	6–10 mg/kg PO[1]	For cestodes
Other medications		
Dexamethasone	0.5–2.0 mg/kg SC, IM, IV[2]	
Diazepam	3–5 mg/kg IM[2]	
Frusemide	2–10 mg/kg SC, PO q 12h[2]	
Oxytocin	0.2–3.0 IU/kg SC, IM, IV[2]	
Prednisone	0.5–2.2 mg/kg PO, SC, IM[2]	

Source: **1** Carpenter et al. (2001). **2** Quesenberry & Carpenter (2004). **3** Flecknell (1991). **4** R Johnson (*pers. comm.*).

frequently. Other alternative routes of administration include oral and intraperitoneal. Isotonic crystalloid solutions are usually recommended. These may include Hartmann's solution, 5% glucose or a mix of 1:1 Hartmann's and 5% glucose.

12 REPRODUCTIVE MANAGEMENT

Watts (1980) provided the first review of reproductive success of over 30 Australian rodent species in captivity. Many were found to breed well initially but nearly all had significant reduction in production of young over several generations. The spinifex hopping mouse, plains rat and western chestnut mouse are notable exceptions and have been successfully bred in captivity for many generations (Breed 1989; Watts 1980). Various factors may affect the reproduction of Australian rodents in the wild including food availability, temperature and photoperiod.

Prevention of breeding of captive Australian rodents usually involves management of social groupings (separation of males and females). Castration has also been used. Facilitation of breeding is generally attempted by trying to imitate the environmental conditions that favour breeding in the wild as closely as possible, e.g. plentiful food and water, appropriate photoperiod and low stocking density.

Captive breeding for conservation purposes and/or reintroduction to the wild has been undertaken for Shark Bay mice (*Pseudomys fieldi*), greater stick-nest rats and central rock-rats.

13 DIAGNOSTIC IMAGING

Chemical restraint is often advisable prior to diagnostic imaging to reduce stress of the animal and facilitate the procedure. The abdomen of Australian rodents appears disproportionately large in relation to the thorax, compared to that in domestic carnivores. The caecum is usually visible in the left caudal abdomen and frequently contains gas.

Due to the small size of most Australian rodents mammography films or dental films are often required for sufficient detail. A radiographic unit that allows variation of the focal film distance allows magnification of areas to be examined, which is particularly useful for small species.

Ultrasound probes of a higher frequency (e.g. 7.5 MHz) are appropriate for most rodents as they provide better detail, and the decrease in depth penetration is not important in very small animals. The use of a stand-off pad is often beneficial for mice due to their small size. Ultrasonography of rodents has proven useful for the diagnosis of pyometra and abdominal masses.

14 SURGERY

Rodents have a relatively high metabolic rate and high surface area:volume ratio and hence lose body heat and fluids rapidly. Maintaining body temperature during anaesthesia and surgery is essential. The use of a heatpad and/or bubble wrap, and clipping the minimum amount of hair, are strongly recommended, to avoid hypothermia. Warm parenteral fluids and analgesia should be given pre-operatively. Monitoring rectal temperature for long procedures and for animals that do not recover rapidly from anaesthesia is advisable. Pre-oxygenation is recommended but care should be taken not to excessively cool the patient with anaesthetic gases. Blood loss should be kept to a minimum as a seemingly small volume of blood can comprise a significant percentage of the circulating blood volume.

Common surgical procedures performed on rodents include castration, ovariohysterectomy, tail amputation, fracture repair and treatment of other traumatic injuries. These procedures are described in detail for domestic rodents in Quesenberry and Carpenter (2004).

Castration of rodents can be performed using an open or closed technique; closed castration is preferred due to the potential risk of inguinal herniation with open castration. Male rodents can retract their testes into the abdomen. A gentle rolling movement over the abdomen from cranial to caudal will help move the testes back into the scrotum. The surgical incision may be midline or directly over each testis. It is recommended that the incision be made as dorsal as possible to prevent contamination of the surgical wound with enclosure substrate. The skin incision may be left open or may be closed with a tissue adhesive (Quesenberry & Carpenter 2004).

Ovariohysterectomy is performed using a ventral midline approach, following a similar technique to that used for domestic carnivores. Due to the small size of these animals microsurgery instruments may be valuable, as may the use of haemaclips. Ligation of the uterus cranial to the cervix is recommended, to prevent urine leakage into the abdominal cavity. Gentle tissue handling is also important due to the fragility of tissues and potential for adhesions. A three-layer closure of the incision site is recommended. The risks associated with ovariohysterectomy are greater compared to castration, thus castration is the preferred technique to prevent breeding (Quesenberry & Carpenter 2004).

REFERENCES

ARWH (2007). *Australian Registry of Wildlife Health.* Zoological Parks Board of NSW: Sydney.

Breed WG (1986). Comparative morphology and evolution of the male reproductive tract in the Australian hydromyine rodents (Muridae). *Journal of Zoology* **209**, 607–629.

Breed WG (1989). Comparative studies on the reproductive biology of three species of laboratory bred Australian conilurine rodents (Muridae: Hydromyinae). *Journal of Zoology* **217**, 683–699.

Breed WG (1997). Interspecific variation of testis size and epididymal sperm numbers in Australasian rodents with special reference to the genus *Notomys*. *Australian Journal of Zoology* **45**, 651–669.

Breed B & Ford F (2007). *Native Mice and Rats.* CSIRO Publishing: Melbourne.

Campbell RW, Carley JG, Doherty RL, Domrow R, Flippich C, Gorman BM & Karabatsos N (1977). Mossman virus, a paramyxovirus of rodents isolated in Queensland. *Search* **8**, 435–436.

Carpenter JW, Mashima TY & Rupiper DJ (2001). *Exotic Animal Formulary* (2nd edn). WB Saunders: Philadelphia.

Domrow R (1966). Mite parasites of small mammals from scrub typhus foci in Australia. *Australian Journal of Zoology* **15**, 759–798.

Dunnet GM & Mardon DK (1974). A monograph of Australian fleas (Siphonaptera). *Australian Journal of Zoology Supplement Series* **30**, 1–273.

Flannery TF (1995). *Mammals of New Guinea.* Reed Books: Sydney.

Glazebrook JS (1978). Rodent diseases in tropical Queensland. *Fauna, Proceedings 36*, pp. 719–733. Post Graduate Committee in Veterinary Science: Sydney.

Glazebrook JS, Campbell RSF, Hutchinson GW & Stallman ND (1978). Rodent zoonoses in North Queensland: the occurrence and distribution of zoonotic infections in north Queensland rodents. *Australian Journal of Experimental Biology and Medical Science* **56**, 147–156.

Graves SR, Stewart L, Stenos J, Stewart RS, Schmidt E, Hudson S, Banks J, Huang Z & Dwyer B (1993). Spotted fever group rickettsial infection in south-eastern Australia: isolation of rickettsiae. *Comparative Immunology, Microbiology and Infectious Diseases* **16**, 223–233.

Happold M (1976). Social behaviour of the conilurine rodents (Muridae) of Australia. *Zeitschrift für Tierpsychologie* **40**, 113–183.

Hill NJ, Rose K, Deane EM & Old JM (2007). *Rodentolepis fraterna*: the cause of mortality in a new host, the spinifex hopping mouse (*Notomys alexis*). *Australian Veterinary Journal* **85**, 62–64.

Hopkinson WI & Lloyd JM (1981). *Streptobacillus moniliformis* septicaemia in spinifex hopping-mice (*Notomys alexis*). *Australian Veterinary Journal* **57**, 533–534.

How RA, Bradley AJ, Iveson JB, Kemper CM, Kitchener DJ & Humphreys WF (1983). The natural history of Salmonellae in mammals of the tropical Kimberley region, Western Australia. *Ecology of Disease* **2**, 9–32.

Hrapkiewicz K, Holmes D & Medina L (1998). *Clinical Medicine of Small Mammals and Primates* (2nd edn). Manson Publishing: London.

IUCN (2006). *2006 IUCN Red List of Threatened Species.* International Union for the Conservation of Nature and Natural Resources. <www.iucnredlist.org>.

Jackson S (2003). Rodents. In *Australian Mammals: Biology and Captive Management* (ed. S Jackson), pp. 351–380. CSIRO Publishing: Melbourne.

Jun MH, Karabatsos N & Johnson RH (1977). A new mouse paramyxovirus (J virus). *Australian Journal of Experimental Biology and Medical Science* **55**, 645–647.

Keeble E (2002). Gerbils. In *BSAVA Manual of Exotic Pets* (4th edn) (eds E Meredith & S Redrobe), pp. 34–46. British Small Animal Veterinary Association: Quedgeley.

Kirkland PD, Gleeson AB, Hawkes RA, Naim HM & Boughton CR (1989). Human infection with encephalomyocarditis virus in New South Wales. *Medical Journal of Australia* **151**, 176–178.

Looke DFM (1986). Weil's syndrome in a zoologist. *Medical Journal of Australia* **144**, 597–601.

Mackie JT, Booth R, Caton W & Stevenson R (2001). Concurrent infection with cilia-associated respiratory bacillus and mycoplasmas in spinifex hopping-mice (*Notomys alexis*) with pneumonia. *Australian Veterinary Journal* **79**, 502–504.

Menkhorst PW (2004). *A Field Guide to the Mammals of Australia* (2nd edn). OUP: Melbourne.

McLelland DJ, Kirkland PD, Rose KA, Dixon RJ & Smith M (2005). Serologic responses of barbary sheep (*Ammotragus lervia*), Indian antelope (*Antilope cervicapra*), wallaroos (*Macropus robustus*), and chimpanzees (*Pan troglodytes*) to an inactivated encephalomyocarditis virus vaccine. *Journal of Zoo and Wildlife Medicine* **36**, 69–73.

Mitchell MA & Tully TN (2004). Zoonotic diseases. In *Ferrets, Rabbits and Rodents: Clinical Medicine and Surgery* (2nd edn) (eds KE Quesenberry & JW Carpenter), pp. 429–434. WB Saunders: St Louis.

Moro D & Lawson M (2003). Pathogen study of house mice on islands reveals insights into viral persistence in isolated populations. In *Rats, Mice and People: Rodent Biology and Management* (eds GR Singleton, LA Hinds, CJ Krebs & DM Spratt), pp. 51–54. Australian Centre for International Agricultural Research: Canberra.

Munday BL & Mason RW (1980). Sarcocystis and related organisms in Australian wildlife: III. Sarcocystis murotechis sp. n. life cycle in rats (Rattus, Pseudomys and Mastocomys spp.) and tiger snakes (Notochis ater). *Journal of Wildlife Diseases* **16**, 83–87.

Munday BL, Mason RW, Hartley WJ, Presidente PJA & Obendorf D (1978). Sarcocystis and related organisms in Australian wildlife: I. Survey findings in mammals. *Journal of Wildlife Diseases* **14**, 417–434.

Munday BL, Hartley WJ, Harrigan KE, Presidente PJ & Obendorf DL (1979). Sarcocystis and related organisms in Australian wildlife: II. Survey findings in birds, reptiles, amphibians and fish. *Journal of Wildlife Diseases* **15**, 57–71.

Musser GG & Carleton MD (2005). Rodents. In *Mammal Species of the World: A Taxonomic and Geographic Reference* (3rd edn) (eds DE Wilson & DM Reeder), pp. 21–42. Johns Hopkins University Press: Baltimore.

Nano TJ, Smith CM & Jeffreys E (2003). Investigation into the diet of the central rock rat, *Zyzomys pedunculatus*. *Wildlife Research* **30**, 513–518.

Obendorf DL & Smales LR (1985). The internal parasites and pathological findings in *Hydromys chrysogaster*

(Muridae: Hydromyine) from Tasmania. *Australian Journal of Zoology* **33**, 33–38.
Odorico DM, Graves SR, Currie B, Catmull J, Nack Z, Ellis S, Wang L & Miller DJ (1998). New Orienta tsutsugamushi strain from scrub typhus in Australia. *Emerging Infectious Diseases* **4**, 641–644.
Old JM, Hill NJ & Deane EM (2007). Isolation of the mite *Myocoptes musculinus* Koch from the spinifex hopping-mouse (*Notomys alexis*). *Laboratory Animals* **41**, 292–295.
Orr HE (2002). Rats and mice. In *BSAVA Manual of Exotic Pets* (4th edn) (eds E Meredith & S Redrobe), pp. 13–25. British Small Animal Veterinary Association: Quedgeley.
Pope JH & Scott W (1960). A survey for antibodies to encephalomyocarditis virus in man and animals. *Australian Journal of Experimental Biology and Medical Science* **38**, 447–450.
Pritchett KR & Corning BF (2004). *Biology and Medicine of Rodents*. International Veterinary Information Service: Ithaca. <www.ivis.org/advances/Reuter/corning/chapter_frm.asp>.
Quesenberry KE & Carpenter JW (2004). *Ferrets, Rabbits and Rodents: Clinical Medicine and Surgery* (2nd edn.). WB Saunders: St Louis.
Sainsbury AW (2003). Rodentia (Rodents). In *Zoo and Wild Animal Medicine* (5th edn) (eds ME Fowler & RE Miller), pp. 420–442. Elsevier: St Louis.
Saunders R (2003). Other insectivores and rodents. In *BSAVA Manual of Wildlife Casualties* (1st edn) (eds E Mullineaux, D Best & JE Cooper), pp. 75–94. British Small Animal Veterinary Association: Quedgeley.
Singleton GR, Smythe L, Smith G, Spratt DM, Aplin K & Smith AL (2003). Rodent diseases in Southeast Asia and Australia: inventory of recent surveys. In *Rats, Mice and People: Rodent Biology and Management* (eds GR Singleton, LA Hinds, CJ Krebs & DM Spratt), pp. 25–30. Australian Centre for International Agricultural Research: Canberra.
Smales LR (1992). A survey of the helminths of *Rattus sordidus*, the canefield rat, together with a description of *Ancistronema coronatum* n. g., n. sp. (Nematoda: Chabertiidae). *Systematic Parasitology* **22**, 73–80.
Smales LR (1997). A review of the helminth parasites of Australian rodents. *Australian Journal of Zoology* **45**, 505–521.
Smales LR & Obendorf DL (1996). Protozoan parasites and pathological findings in *Hydromys chrysogaster* (Muridae: Hydromyinae) from Queensland. *Journal of Wildlife Diseases* **32**, 344–347.
Smales LR & Spratt DM (2004). Helminth community structure in *Rattus leucopus* (Gray) (Muridae) from Australia, Papua New Guinea and Papua. *Australian Journal of Zoology* **52**, 283–291.
Smales LR, Heinrich B & McKillup SC (2004). The helminth parasites of *Melomys cervinipes* (Rodentia: Muridae: Hydromyinae). *Australian Journal of Zoology* **52**, 65–80.
Smith AL, Singleton GR, Hanseen GM & Shellam G (1993). A serologic survey for viruses and *Mycoplasma pulmonis* among wild house mice (*Mus domesticus*) in south-eastern Australia. *Journal of Wildlife Diseases* **29**, 219–229.
Spratt DM (2002). Parasites and pathology of the respiratory tracts of native and feral mammals in Australia: a review. *Australian Mammalogy* **24**, 177–192.
Spratt DM, Haycock P & Walter EL (2001). Life history and pathogenesis of *Gallegostrongylus australis* (Nematoda: Angiostrongylidae) in muridae. *Parasite* **8**, 115–125.
Strahan R (2002). *The Mammals of Australia* (2nd edn). Reed New Holland: Sydney.
Taylor JD, Stephens CP, Duncan RG & Singleton GR (1994). Polyarthritis in wild mice (*Mus musculus*) caused by *Streptobacillus moniliformis*. *Australian Veterinary Journal* **71**, 143–145.
Thomson GR, Bengis RG & Brown CC (2001). Picornavirus infections. In *Infectious Diseases of Wild Mammals* (3rd edn) (eds ES Williams & IK Barker), pp. 119–130. Iowa State Press: Ames.
Vogelnest L (1999). Chemical restraint of Australian native fauna. In *Wildlife in Australia: Healthcare and Management, Proceedings* 327, pp. 149–188. Post Graduate Foundation in Veterinary Science: Sydney.
Warner LR (1998). Australian helminths in Australian rodents: an issue of biodiversity. *International Journal for Parasitology* **28**, 839–846.
Watts CHS (1973). *Handbook of South Australian Rodents and Small Marsupials*. Field Naturalists Society of SA: Adelaide.
Watts CHS (1978). Eutherian mammals: rodents and bats. In *Fauna, Proceedings 36*, pp. 197–219. Post Graduate Committee in Veterinary Science: Sydney.
Watts CHS (1980). Australian hyrdromyine rodents: maintenance of captive colonies. In *International Zoo Yearbook* (ed. PJS Olney), pp. 180–184. Zoological Society of London: London.
Watts CHS & Aslin HJ (1981). *The Rodents of Australia*. Angus & Robertson: Sydney.
Winter H & Watt D (1971). *Klossiella hydromyos* n. sp. from the kidneys of an Australian water rat (*Hydromys chrysogaster*). *Veterinaria Pathologia* **8**, 221–231.
Yom-Tov Y (1985). The reproductive rates of Australian rodents. *Oecologica* **66**, 250–255.
Zeller W, Weber H, Panoussis B, Bürge T & Bergmann R (1998). Refinement of blood sampling from the sublingual vein of rats, *Laboratory Animals* **32**, 369–376.

16. Dingoes

Frances Hulst

1 TAXONOMY, DISTRIBUTION AND BIOLOGY

The dingo, *Canis lupus dingo*, is closely related to the domestic dog, *Canis lupus familiaris*, both being subspecies of the grey wolf, *Canis lupus*. In general terms, the veterinary management of the dingo is based on that of the domestic dog. This chapter will focus on aspects specific to the dingo.

Dingoes are found wild in Australia and parts of south-east Asia (Corbett 2001). They are thought to have evolved from the Indian wolf, *Canis lupus pallipes*, 6–10 000 yr ago and probably arrived in Australia about 4000–5000 yr ago with south-east Asian seafarers (Savolainen et al. 2004). The dingo is Australia's largest terrestrial predator, is distributed over most of the mainland and can be found in diverse environments from tropical to alpine, and coastal to arid inland. However, it has never reached Tasmania. European arrival allowed great increase in dingo numbers due to pastoralism and the use of artesian bores. As a result of dingo predation on livestock a dingo fence was built, that has partly excluded the dingo from the main sheep and cereal-growing areas of south-eastern Australia. The dingo is afforded protection in some parts of its range (e.g. national parks), while in other parts it is considered a pest and, along with other wild dogs, is subject to control measures. Poisoning with 1080 (sodium fluoroacetate) in shallow-buried or helicopter-dropped meat baits is the main method of control. Targeted padded-jaw traps, exclusion fencing, opportunistic shooting and livestock guard dogs are also used (Fleming et al. 2001). Increasing hybridisation with domestic dogs, however, poses the biggest threat to conservation of the dingo (Corbett 2001; Fleming et al. 2001; Woodall et al. 1996) and it has recently been listed as vulnerable on the IUCN Red List of Threatened Species (Corbett 2004).

The dingo is held in zoos and animal parks in Australia and overseas, particularly in Europe and North America. In New South Wales it can be kept as a pet without a licence (Corbett 2004; Davis 2001; Fleming et al. 2001; Oakman 2001). However, dingoes are by nature highly sensitive, wary and elusive. Even when well-socialised with humans they remain independent and unpredictable, and liable to revert to aggressive behaviour when frightened or threatened. For this reason, they do not generally make good pets. Dingoes can live for 12–14 yr in captivity, but rarely live longer than 10 yr in the wild (Corbett 2001).

2 ANATOMY AND PHYSIOLOGY

The dingo is a medium sized short-coated dog with a bushy tail. Average shoulder height is about 60 cm. Body weight ranges from 10–20 kg, with males being heavier than females. Dingoes from arid central Australia tend to be smaller than those from cooler or tropical areas. Skull measurements, body size and coat colour show a cline between tropical, alpine and desert dingoes. Ginger with white points is the predominant coat colour, but black, black and tan, and white dingoes also occur. While these coat colours can occur in domestic dogs and hybrids, patches, sable and brindle patterns are not seen in pure dingoes (Corbett 2001).

Many wild dogs and dingo hybrids are visually indistinguishable from pure dingoes, making differentiating pure dingoes difficult. Dingoes can, however, be differentiated from wild dogs and hybrids by a composite score based on a series of skull measurements. Unfortunately, these are best performed at necropsy (Corbett 2001). In the live animal, microsatellite studies have attempted to distinguish pure dingoes from those that have undergone hybridisation with dogs. Past hybridisation events, particularly recent ones, can be detected by this method but a reliable source of pure dingo DNA, such as pelts collected by early explorers, is needed to make these studies definitive (Wilton 2001).

Dingoes can be classed as pups, juveniles, yearlings or adults based on characteristics of canine teeth (Thomson & Rose 1992). An accurate method of estimating the age of adult dingoes and wild dogs uses pulp cavity as a proportion of tooth width, measured 15 mm from the root tip and averaged over both upper canines. These measurements are made from lateral radiographs and applied to a formula to give an age estimate (Kershaw et al. 2005).

The anatomy of the dingo is essentially identical to that of the domestic dog. There is no evidence of special physiological adaptations to the Australian environment (Newsome & Coman 1989). The dingo does, however, differ from the domestic dog in social behaviour and reproductive pattern.

3 BEHAVIOUR

Wild dingoes are essentially crepuscular, particularly in hotter parts of their range, while captive dingoes are more active during the day. They are generally not nomadic and confine their activity to a home range, which varies in size from less than 20 km^2 to more than 100 km^2 depending on region and prey abundance (Corbett 2001; Green & Catling 1977; Harden 1978). Dingoes are often seen alone but are usually part of a small flexible socially integrated pack that comes together for breeding or co-operative hunting. Solitary animals are more common where small to medium sized prey such as rodents or rabbits abound, since these are easily caught by one animal. Increasing social cohesion and larger packs form with greater reliance on larger prey such as wallabies and kangaroos (Fleming et al. 2001). Dingoes communicate by scent marking and a variety of howls (Corbett 2001). Barking is rare (Newsome & Coman 1989). Within packs, dingoes display a strong dominance hierarchy between and within sexes. The frequency of aggressive interactions within a pack varies with the pack stability but is generally higher between males; the intensity of these interactions may increase during the breeding season (Corbett 1988). The dominant male and female dingo are generally the only pair to raise offspring. In captivity, it is difficult to establish a stable social group with unrelated animals from different sources. Dingoes are therefore best kept in pairs with their offspring (Jackson 2003). Desexed animals may be kept in groups, but this is not always successful.

4 REPRODUCTION

In common with most wild canids, dingoes are seasonal breeders and produce one litter each year; domestic dogs and dingo/dog hybrids may have two oestrous periods per year. Most matings take place in April and May, with most births in June or July (Catling et al. 1992; Thomson 1992a). A copulatory tie occurs as in other canids. Behavioural data suggests a pro-oestrus period of 1–2 mo (Thomson 1992a), however, vaginal cytology is more indicative of a 10–12 d pro-oestrus and oestrus (Corbett 2001). Gestation averages 63 d with one to eight pups per litter (average three to five) (Harden 1978; Newsome & Coman 1989). Only a small percentage of females breed in their first year (Corbett 2001). Male dingoes in central Australia show a marked seasonal variation in reproductive parameters. There is an increase in testis weight, prostate weight and semen volume, which peak in April and May. Males from cooler areas of south-east Australia do not show such seasonal changes and are capable of breeding year-round. In response to drought, there is a reduction and delay in reproductive response (Catling et al. 1992). All oestrous females mate (usually with the dominant male) and some become pregnant (Corbett 1988). In a pack, offspring of subordinate females are usually killed by the dominant female—the subordinate females then help to rear her pups. Thus only one litter per pack is raised (Corbett 1988). Whelping takes place in a den using available rocky outcrops, hollow logs or burrow entrances. Preference is shown for elevated sites near water. In addition to the mother, other pack members may regurgitate food for the pups. Pups leave the den at about 8 wk of age and increasingly accompany adults on feeding forays (Thomson 1992a), becoming independent from as early as 4 mo (Corbett 2001). Mortality is highest in recently independent pups that do not remain associated with their parents as subadults (Harden 1978). Captive dingoes may show an increase in intensity of aggression during the breeding season, especially between members of the same sex (Green & Catling 1977), which presents challenges for management.

5 HUSBANDRY

5.1 Identification

Individual dingoes can be identified by a combination of visual characteristics, ear tags, ear tattoos or passive integrated transponders (PIT tags). PIT tags are usually implanted under the skin between the scapulae and provide the most reliable permanent form of identification. Ear tags are easily lost during fighting or other trauma and may result in permanent tears and notches in the ear pinna.

5.2 Exhibit design

Exhibit design and holding facilities have been well-described (Jackson 2003). The suggested minimum area for a compatible pair of dingoes is 220 m^2 with 6.6 m^2 per additional animal (EAPA 2003). Fencing should be constructed to prevent dingoes from climbing, jumping or digging their way out. As long as adequate shelter is available, additional heating or cooling is generally not required. A holding area with a den should be available, allowing the dingoes to be held off-exhibit singly or in pairs and to be fed individually. Visual barriers or solid fencing may be needed between enclosures if there is a high level of aggression between animals.

5.3 Hospital housing

Where possible, animals undergoing veterinary treatment should be maintained in their familiar areas. If a dingo's condition warrants it being taken off-exhibit or isolated from other dingoes, the animal may be restricted to its holding area. This allows continued visual, auditory, olfactory and limited tactile contact with conspecifics. If a serious contagious disease is present or the dingo requires intensive treatment, hospitalisation may be necessary. Removing a dingo from its companions for a prolonged time (as little as 12–24 hr for some individuals) may involve a lengthy reintroduction period to avoid extensive fighting (see 5.4).

Suitable hospital facilities depend on the nature of the dingo and how ill it is. Tame or severely ill animals may be kept in a hospital cage suitable for domestic dogs, though not in a room with other animals. Wild dingoes will need a larger holding pen with a den open on one side. Two adjacent holding pens with a slide between is ideal—the dingo can be moved from one to the other to facilitate cleaning without the handler having to go in with the animal. The enclosure should allow the dingo privacy while still giving access to capture it by hand, net or dart.

5.4 Reintroduction to a group

If prolonged separation cannot be avoided, the animal to be reintroduced should be placed in visual, auditory and olfactory contact with the group while remaining protected from attack. If the dingo is not subject to aggressive advances from group members, it can be allowed access to the group. Reintroduction should take place at the beginning of the day so there is opportunity to watch the group for a long period. Plans should be made to separate animals quickly if fighting occurs. It may be wise to feed the group prior to reintroduction, to avoid food-motivated fighting. In some situations, introduction may occur with a feed so that animals are distracted.

The neuroleptic drug zuclopenthixol dihydrochloride at 3.5 mg/kg PO twice daily was given to a group of dingoes to facilitate reintroduction of an animal. The group and the animal to be introduced were all medicated for 1 wk before and 1 wk after the introduction, which went without incident (Vogelnest 1999).

6 NUTRITION

6.1 Natural diet

Dingoes consume prey items including insects, lizards, rodents, rabbits, birds, macropods, sheep, cattle and horses. Mammals predominate, and small to medium sized macropods and rabbits are particularly important prey species. Domestic stock may be killed by dingoes (Fleming et al. 2001) but do not make a significant contribution to the diet (Corbett 2001). Free-ranging dingoes consume about 7% of their body weight per day in food, which equates to approximately 1 kg per day of prey for an average sized dingo. Dingoes in desert regions require 100 mL/kg/d of water in summer, reducing to 70 mL/kg/d in winter. Lactating bitches require 100 mL/kg/d, though extra fluid is gained by ingesting urine and faeces of pups elicited during licking. It has been suggested that ingested mammalian prey may meet the daily water needs at times of lower requirement (Green & Catling 1977).

6.2 Captive diet

Captive diets should match wild diets in nutritional value, texture and variety. Obesity may become a problem in captive dingoes and food intake should be adjusted to maintain a lean body condition. Food intake of 4–5% of body weight per day is suggested as a guide. Captive diets should be based on whole prey items to maintain dental health and contribute to behavioural enrichment. Whole rabbits or chickens, and kangaroo, beef or horsemeat with hide and bone included are good options. Rabbit carcasses should be frozen for ≥1 wk prior to thawing and feeding out to kill tissue stages of the rabbit tapeworms *Taenia pisiformis* and *T. serialis*, since the dingo can act as definitive host. Meaty bones such as shank bones should be offered regularly. Good-quality dry dog food may replace some of the prey items. Dingoes should be fed individually once or twice a day; complete starve days are unnecessary. If a varied and balanced diet is offered, supplements are unnecessary. Where diets are based on meat alone, a calcium supplement will be necessary to correct the calcium:phosphorus ratio, as for domestic dogs. Energy requirements increase by up to 60% during late pregnancy and lactation. Drinking water should be available at all times. Medications may be hidden in items such as unsalted butter or liverwurst, which are generally well-accepted.

7 RESTRAINT

7.1 Capture and physical restraint

Due to their unpredictable nature, dingoes that will allow handling should always be muzzled and firmly restrained for examination regardless of temperament. Manual restraint methods do not differ from those for domestic dogs. More tractable dingoes, once muzzled, can be picked up off the ground and examined, treated or injected while being held by the handler. This may also be done with the animal standing on the ground or on a table. Dingoes that will not tolerate handling may be confined to a small pen and caught in a net as they run past. Animal handlers must be experienced, so that the dingo is not chased for a prolonged period prior to capture, as hyperthermia may result. IV or IM anaesthetic agents may be administered through the net or, in the case of IV agents, by withdrawing one hind leg to access the lateral saphenous vein. Care must be taken to restrain the dingo's head to avoid being bitten during injection and to prevent the animal from traumatising its gums and fracturing teeth on the net. The dingo's eyes should be covered with a cloth or towel during restraint to minimise visual stimuli. Darting is preferred if facilities and experienced personnel for netting are not available (see 7.2).

Cage traps may be used for semi-habituated animals, though wild dingoes are too wary to enter (Corbett 2001). As with excessive physical exertion, prolonged time in a trap may lead to hyperthermia. Wild dingoes may be caught with padded foothold traps. One study reported a 10% mortality rate (Thomson 1992b), however, over 100 wild dogs or dingoes have been successfully caught without incident by the following method. The animal is caught in a foothold trap and, as soon as possible, its head is restrained by a person using a dog-catching noose (pole with rope noose at the end). A second handler takes hold of the hind legs and the animal is carried, with trap still attached, to a restraint board. The animal is strapped to the board with three straps (one just anterior to the pelvis, one slightly posterior to the front legs and one over the neck), its head is covered with a cloth and the trap is removed. Dingoes will lie quietly without struggling for procedures such as blood collection, skin biopsy or radio-collaring without requiring sedation (D Jenkins *pers. comm.*). If chemical restraint is needed for a dingo in a leghold trap, drugs can be administered by hand or pole syringe after noosing or netting.

7.2 Chemical restraint

Chemical restraint agents may be administered via hand-held syringe, pole syringe or dart. Dingoes may be darted using a blowpipe or CO_2 powered pistol or rifle. More powerful projectors, such as those that use powder charges, are less suitable. Darts should be air-inject and lightweight with fine short barbless needles, e.g. 20 G × 1″ (0.9 × 24 mm) for blowpipes,

18 G × ¾″ (1.5 × 20 mm) for pistol or rifle, to avoid dart impact injury. The large muscle mass of the thigh is the preferred site for dart injection. The weight of the dart and its velocity at impact both play a part in determining any impact injury. Even lightweight darts fired at high speed may result in broken bones or severe muscle trauma.

As with other carnivores, food and water should be withheld for 12 hr prior to general anaesthesia.

Sedatives and tranquillisers used in domestic dogs would be suitable for use in dingoes. The use of neuroleptic agents other than zuclopenthixol (see 5.4) has not been reported in dingoes. In a hospital situation, acepromazine or diazepam may sedate a dingo sufficiently for intensive treatment.

Dingoes can be anaesthetised using any drug or drug combination at dose rates for domestic dogs, providing they can be safely and correctly administered. IV access may be difficult in less tractable animals, so IM drugs are often more useful. Where drug dose rates are quoted as a range, it is preferable to choose the higher end of the range rather than the lower end to ensure adequate immobilisation, particularly in the field where body weights must be estimated and animals are frequently active or apprehensive. If supplemental doses are required to achieve adequate restraint, an additional half-dose of the primary immobilising agent may be given. In the field, or in captive situations where prompt return to the pack is desirable, use of drugs or drug combinations that can be antagonised is an advantage. This also allows the option of daily immobilisations for treatment with immediate return to the pack, and minimises potential problems associated with prolonged recovery.

IV agents such as propofol, alphaxalone and thiopentone are all suitable for use in dingoes at standard canine dose rates. Tiletamine/zolazepam at 1.5–2 mg/kg IV or 8–10 mg/kg IM is usually sufficient for short procedures or to facilitate intubation. This combination has a wide safety margin and smooth induction and recovery, though recoveries can be prolonged (up to 8 hr) with higher doses. A single dose of 120 mg ketamine and 60 mg xylazine was suggested for immobilisation of wild dingoes (13–18 kg) by helicopter-darting (Thomson 1992a). A standard dose of 0.5 mg etorphine plus 2.5 mg acepromazine was also considered a useful reversible combination for immobilising wild dingoes (Green 1976). Though their use has not been reported in dingoes, ketamine 2.0–5.0 mg/kg with medetomidine 0.04–0.1 mg/kg IM or tiletamine/zolazepam at 1–3 mg/kg combined with 0.03–0.08 mg/kg medetomidine IM provides light to surgical level of anaesthesia in wild canids, depending on the dose (Kreeger 1999; Kennedy-Stoskopf 2003), and could be considered in dingoes. The medetomidine may be reversed using atipamazole at 4 times the medetomidine dose IM. Butorphanol/xylazine can be used on relatively calm captive canids (Kreeger 1992). Dose rates given for grey wolves are 2 mg/kg xylazine and 0.4 mg/kg butorphanol IM, antagonised by naloxone 0.05 mg/kg and yohimbine 0.125 mg/kg IV (Kreeger et al. 1989). These combinations could be considered for use in dingoes.

Whenever a dingo is removed from its pack for more than a few hours, there is the potential for a change in group dynamics and resultant fighting. Even in a relatively stable pack, the animal being reintroduced may be severely injured or killed. Elective procedures should be carefully timed to avoid peak aggression times such as breeding season. Animals should be fully recovered from anaesthetics before being reintroduced. There should be adequate daytime left to observe behaviour of the group, and sufficient staff and facilities to quickly separate animals if serious fighting occurs.

8 CLINICAL PATHOLOGY

Collection and analysis of samples for diagnostic purposes are similar to those used for domestic dogs. Clinical pathology values, including haematology and serum biochemistry, fall within reference ranges for domestic dogs. The morphology of the blood cells in dingoes is similar to those of domestic dogs (Clark 2004) and they share many of the same blood groups (Symons & Bell 1992).

9 REPRODUCTIVE MANAGEMENT

Dingoes breed readily in captivity so assisted reproduction techniques are usually not required. Permanent contraception or injectable or implantable forms of temporary contraception are generally preferred as compliance with oral dosing is often difficult.

9.1 Permanent contraception

In a non-breeding group, it is preferable that all females be ovariohysterectomised and all males castrated to avoid more intense fighting and loss of condition associated with the breeding season. Vasectomy or castration of the male will not prevent potential adverse effects in females from prolonged cyclic exposure to endogenous steroids associated with the obligate hormonal pseudopregnancy that follows ovulation. Elective procedures such as these are best undertaken outside the breeding season. Surgical sterilisation may be performed prior to sexual maturity; this may assist in pack stabilisation although desexed dingoes still fight.

9.2 Temporary contraception

9.2.1 GnRH agonist implants

Deslorelin (Suprelorin®, Peptech Animal Health), a GnRH agonist implant, is considered the safest reversible contraceptive for female canids (AZA Wildlife Contraception Centre 2006). A dose of 4.7 mg is recommended for female dogs up to 30 kg and the loose skin between the scapulae is the recommended site. As a GnRH agonist, deslorelin stimulates the reproductive system, which may result in oestrus and ovulation within the first 2 wk even if out of breeding season. Positive feedback then causes suppression of the reproductive endocrine system, providing contraception until the hormone content of the implant is depleted. Duration of action is 6–12 mo, with individual variation. The initial oestrus and ovulation seen with deslorelin can be suppressed with oral progestin treatment for 7 d prior and 8 d post implant if separation of the sexes is not possible. Deslorelin is also an effective contraceptive in males but its use for this purpose is not recommended, since it is more difficult to ensure continuous absence of sperm over a variable period. Deslorelin is useful, however, in suppressing aggression in males, although an initial period of heightened aggression corresponding with raised testosterone levels may occur before down-regulation. Deslorelin should not be used in pregnant animals and may prevent epiphyseal closure of the long bones in juveniles, resulting in taller animals. There is a lack of data on prepubertal animals and long-term effects on fertility (AZA Wildlife Contraception Centre 2006).

9.2.2 Progestins

Medroxyprogesterone acetate, megestrol acetate (PO or injection) and melengestrol acetate implants are recommended only for short-term use of not more than 2 yr without an intervening pregnancy (AZA Wildlife Contraception Centre 2006). Prolonged use of progestin contraceptives is associated with the development of endometrial hyperplasia, mineralisation and inflammation, endometrial and mammary neoplasia, weight gain and diabetes mellitus. Progestins should not be used in pregnant animals. Proligestone (Covinan, Intervet Australia) is available as an injection marketed for use in domestic dogs for permanent or temporary suppression or postponement of oestrus and claims to have little risk of undesirable effects on the endometrium. This drug may be useful for suppression of oestrus until next breeding season. In situations where a bitch can be reliably orally medicated, oral megestrol acetate given in early pre-oestrus can suppress oestrus. Mibolerone, a synthetic androgen, is sometimes used in female domestic dogs. It is not recommended in other canids as it can increase aggression (AZA Wildlife Contraception Centre 2006).

9.3 Breeding

Captive dingoes usually breed readily. If a large group is kept in a pack situation, natural breeding behaviour can be expected with the dominant male and female generally the only animals to successfully raise pups. If dingoes are kept in stable pairs, they can be allowed to breed naturally. It is safer to separate the bitch for whelping, keeping the male in visual and olfactory contact and introducing him when the pups are 3–4 wk old. Alternatively, an oestrous female (detected by behaviour, vulval swelling and vaginal discharge as in domestic dogs) can be introduced to a male for mating and separated when copulation has occurred. This introduction is best done in neutral territory and plans made for quick separation if necessary. A quiet whelping area with a den is needed for the bitch to give birth and rear her pups.

10 HAND-REARING

Dingoes can be hand-reared using any commercially available puppy-rearing formula. Puppy-feeding teats and bottles are available through pet shops although a syringe or dropper can be used for very young pups. Standard hygiene practices for hand-rearing orphans should be followed. It is wise to offer a balanced rehydration solution at the first feed. A staged increase to full-strength formula can then be instituted over the next 24–36 hr.

Pups should be bottle-fed in ventral recumbency with the head slightly elevated. Feeding frequencies and volumes are the same as for domestic puppies, starting at two-hourly and extending to every 4–6 hr depending on the stage of growth. Pups should be encouraged to lap at about 3–4 wk of age, when soft or finely chopped solid foods can be introduced. Captive dingo pups of this age that are parent-reared generally weigh 1.25–1.75 kg (Dingo Sanctuary, Australian Native Dog Conservation Society *pers. comm.*). Weaning is completed by 8 wk of age.

Before attempting to feed formula, hypothermia, hypoglycaemia and dehydration should be corrected if present. This can be achieved using SC fluids or by passing a stomach tube. Feed volumes should be reduced by 30% if given by stomach tube to prevent regurgitation. Body temperature in pups less than 4 wk of age should be 35.5–36.1°C. Newborn pups require an ambient temperature of 30–32.2°C for the first week of life, gradually reducing to 23.9°C by 4 wk of age (Barrette 2002). Heat can be provided using warm hot-water bottles and blankets. Burns or overheating can result from improperly monitored heatpads.

The critical period for socialisation of dingo pups is 3–12 wk. A good relationship between the pup, other dingoes and humans should be fostered during this time. Where possible, the pup should spend time with other pups of similar age and have regular positive handling experiences with human carers.

11 DISEASES

Dingoes can develop any disease seen in domestic dogs. As a rule, clinical signs, diagnosis and disease prevention do not differ from domestic dogs. Treatment is also based on that used in domestic dogs, however, route of drug administration, dosage frequency or even choice of treatment may have to be modified depending on the ability to handle the animal or its social situation. See current canine veterinary texts for specific diagnostic tests, treatment regimes and drug dose rates. A suitable preventative medicine calendar can be based on current recommendations for dogs held in the same geographic area. Generally, products are chosen that allow minimal animal-handling at greater intervals.

11.1 Non-infectious diseases

11.1.1 Primary lymphangiectasia

Few published reports describe non-infectious diseases in the dingo. Primary lymphangiectasia has been reported in a 3 yr old male dingo, which was successfully maintained with nutritional management for more than 2 yr. The dingo presented with chronic diarrhoea. After other causes were ruled out, a diagnosis of lymphangiectasia was made based on low serum total protein and albumin, combined with increased faecal levels of alpha-1 protease inhibitor. The diagnosis of primary lymphangiectasia was confirmed by intestinal biopsy revealing dilated lacteals and submucosal lymphatics without associated inflammation. The dingo was treated by feeding a fat-restricted diet with medium-chain triglyceride supplementation (Suedmeyer et al. 2004).

11.1.2 Persistent corneal erosion

Persistent corneal erosion has been reported in a 6 yr old dingo that presented with blepharospasm, epiphora and corneal ulceration with no predisposing cause. The ulcer failed to heal after third eyelid flap and topical antibiotics. The corneal erosion was then debrided, a multiple striate keratotomy was performed, a scarified third eyelid flap placed and antibiotic drops and atropine drops instilled in the eye for 4 d. Healing took place over 2–3 wk. The appearance of the corneal erosion was consistent with primary corneal disease due to epithelial basement membrane disease (Turner & Phelan 1996).

11.1.3 Trauma

Bite wounds are common when dingoes are housed together and there is disruption to the social group. More dominant group members may kill subordinate animals. Aggressive interactions between dingoes are common year-round with changes in group dynamics, though fight injuries are more common during the breeding season. Males show more aggressive encounters than females (Green & Catling 1977; Corbett 1988). In the wild, dingoes may receive injuries from larger prey such as buffalo or cattle (Fleming et al. 2001). Trap injuries may also be significant, particularly if the dingo has been in the trap for some time.

11.1.4 Toxins

a Cane toad poisoning

Predators including dingoes are highly susceptible to poisoning by cane toad (*Bufo marinus*) secretions; in dogs, mortality rates of 20–100% may be seen in untreated cases (Merck Veterinary Manual 2003). Domestic dogs can die within 15 min of mouthing a cane toad due to the effects of the bufotenine toxin. Diagnosis is based on history and clinical signs, with acute onset of ptyalism, hyperaemic mucus membranes and neurologic abnormalities progressing to collapse (Reeves 2004). Treatment is symptomatic and supportive including washing the oral cavity with water. A study on the effects of cane toads on native fauna showed that after invasion of an area in the Northern Territory the abundance of cane toads had a negative effect on dingo numbers, at least in the short term (Catling et al. 1999).

b Sodium fluoroacetate (1080)

1080 (sodium fluoroacetate) is the preferred poison for wild dog control in Australia. Canids are particularly sensitive to 1080 compared to other mammals, while birds and reptiles are less sensitive. Average baits contain 6 mg 1080 (Fleming et al. 2001), with a lethal dose for dogs of 0.06–0.20 mg/kg (Goh et al. 2005). LD50 for the dingo is 0.1 mg/kg (McIlroy 1981). 1080 acts to disrupt cellular energy metabolism by blocking the Krebs cycle. Clinical signs of 1080 poisoning in canids appear 30 min to 2 hr after ingestion depending on the amount ingested and are characterised by central nervous system excitation and gastrointestinal hypermotility. In the initial stages the dog becomes excited or frenzied, has running fits, howls and is hyperaesthetic. This is followed by salivation, vomiting, urination/defecation and tenesmus, progressing rapidly to convulsions, coma and death within 2–12 hr. Diagnosis is based on clinical signs and history of exposure. Treatment aims to minimise absorption and provide supportive care. Recommended treatment includes inducing emesis or, if not fully conscious, gastric lavage and oral activated charcoal/colestipol. Control seizures with pentobarbital sodium or gaseous anaesthetics. Commence IV fluid therapy with 0.9% NaCl. Give 300 mg/kg sodium bicarbonate (8.4% w/v) over 15–30 min. Monitor serum calcium and potassium and supplement as needed. Treatment may need to be continued for 12–18 hr. Acetamide can be used instead of sodium bicarbonate: dissolve 15 g in 1 L 5% glucose. Give 10 mL/kg IV over 5 min then 8 mL/kg/hr for first litre. Continue at 5 mL/kg/hr until clinical signs resolve (Goh et al. 2005).

c Strychnine

Strychnine can still be used in baits in South Australia, Queensland and Western Australia and can be used to poison the jaws of traps in South Australia, Queensland, Western Australia and New South Wales (Fleming et al. 2001). LD50 for dogs is 0.05–0.1 mg/kg. Strychnine causes central nervous system excitation by blocking the inhibitory effects of glycine at post-synaptic glycine receptors. Clinical signs commence 10–20 min after ingestion and include seizures, opisthotonos, hyperreflexia and respiratory paralysis. Diagnosis is by history of exposure and clinical signs. Signs may be confused with 1080 poisoning, although dogs poisoned with strychnine usually remain conscious until near death. Treatment includes gastric lavage and activated charcoal to try to minimise absorption, together with supportive treatment including the use of diazepam or phenobarbital to control seizures (Merck Veterinary Manual 2003).

11.1.5 Neoplasia

Lymphosarcoma, lipoma, pancreatic adenocarcinoma, fibromatous epulis, squamous cell carcinoma, perianal adenoma, cutaneous sebaceous gland adenoma and canine eosinophilic granuloma have all been reported in dingoes (ARWH 2007; Western Plains Zoo records 2006; Taronga Zoo records 2007).

11.1.6 Miscellaneous conditions

Common dog diseases such as flea and mosquito allergy dermatitis and degenerative joint disease are seen in captive dingoes. There are no published reports of hip dysplasia in wild dingoes, but bilateral hip dysplasia requiring femoral head osteotomy has been seen in captive dingoes (Taronga Zoo records 2007).

11.2 Infectious diseases

11.2.1 Viral diseases

Modified live vaccines against canine distemper, infectious canine hepatitis, canine parvovirus and canine parainfluenza marketed for domestic dogs have been used routinely in captive dingoes without untoward effects (Taronga Zoo and Western Plains Zoo records 2006). Vaccination programs follow those used in domestic dogs.

Parvovirus has been reported in captive dingoes (Hooper et al. 1983).

Infectious canine hepatitis has been reported from wild dingoes in central Australia (Corbett 2001). Canine distemper virus epidemics have been recorded in dingoes in the Barkly Tableland in central Australia in the 1960s and 1970s (Corbett 2001).

Rabies is exotic to Australia, although the dingo, wild dog or fox could serve as a reservoir host. Rabies, or other lyssaviruses such as the closely related Australian bat lyssavirus, should be included in the differential diagnosis of any dingo that presents with neurological disease, especially in northern Australia. Highly immunogenic inactivated cell culture vaccines for dogs are widely available for parenteral use in countries with endemic rabies. Oral vaccination projects using modified live or recombinant vaccines have proved successful in reducing the incidence of rabies in foxes and coyotes and other species in North America and Europe (World Health Organization 2005). Vaccination of dingoes in Australia is not presently recommended.

11.2.2 Bacterial diseases

Australian wild dogs and dingoes are reported to be susceptible to anthrax (Cormack Gates et al. 2001) although no reports could be found of disease or deaths in dingoes. Carnivores are generally highly resistant to anthrax and often recover without treatment. Dingoes and wild dogs may acquire infection from scavenging contaminated livestock carcasses.

11.2.3 Fungal diseases

Ringworm due to *Microsporum gypseum* (Taronga Zoo records 2006) and *M. canis* (Corbett 2001) has been seen in captive dingoes.

11.2.4 Parasitic diseases

The impact of parasites in captive dingoes is analogous to that in kennelled domestic dogs. Regimes developed for the control or elimination of parasites in domestic dogs are equally applicable to dingoes. Parasitic disease epidemiology may be different in free-ranging dingo populations. In captive dingoes, faecal examination for gastrointestinal parasites should be performed at least once or twice a year, more often if there is a history of parasite infestation. In endemic areas, heartworm prevention should be prescribed. Flea prevention may be necessary in some captive situations.

a Endoparasites

Nematodes

Infection with the canine lungworm *Oslerus (Filaroides) osleri* is relatively common in wild dingoes in south-east Australia although infection is unusual in domestic dogs in Australia (Dunsmore & Spratt 1979). 83% of dingoes sampled in the south-east highlands were infected, compared with 6% in central Australia (Corbett 2001). Ingested hatched or un-hatched first-stage larvae undergo further larval development stages as they invade the wall of the duodenum, then migrate to the lung parenchyma and alveoli before forming nodules in the trachea and major bronchi which contain adult worms measuring 5–12 mm. The submucosal nodules measure up to 10 mm diameter and are mostly found around the tracheal bifurcation. The pre-patent period is 18 wk and animals older than about 7 mo appear resistant to infection. Clinical signs in dogs are proportional to the severity of infection and the number and size of nodules. The most severe clinical cases are seen in pups <1 yr old, where heavy infestations may cause lung damage due to migrating larvae, together with partial obstruction of the airway by nodules surrounding adult worms. Coughing, respiratory distress and emaciation with up to 75% mortality may result. In older animals, an occasional non-productive cough may be the only clinical sign. Direct transmission can occur from an infected bitch to her pups during weaning while the bitch is regurgitating food for the pups. This method of transmission is thought to be the reason why *O. osleri* infection is rare in domestic dogs, where other food sources are generally made available for weanling pups (Dunsmore & Spratt 1979). Diagnosis relies on identification of first-stage larvae in tracheal washes or in faeces using the Baermann technique. The strongyloid larvae measure 325–378 μm. Treatment is with fenbendazole or ivermectin using regimes presented for domestic dogs.

Dirofilaria immitis has been reported in dingoes in Australia. 56% of dingoes in a tropical region of the Northern Territory were infected with *D. immitis* (Starr & Mulley 1988). Infected animals showed a high rate of circulating microfilaria. 73% of dingoes in Kakadu National Park were infected (Corbett 2001). Heartworm disease caused high mortality in dingo populations at Kapalga (NT) in the 1980s. The epizootics were associated with late onset of the preceding wet season, which could have encouraged high populations of the mosquito species that transmit heartworm most readily. Deaths occurred 6 mo after the start of the wet season (Corbett 2001). Heartworm infection in dingoes can be expected to follow the same geographic distribution as that seen in domestic dogs. Heartworm prophylaxis delivered by annual injection (e.g. Proheart SR-12 Injection, Fort Dodge Australia) or by monthly oral dosing may be easier to administer in captive dingoes that are difficult to handle.

The whipworm *Trichuris vulpis* infected 1% of dingoes sampled in study areas in south-east Australian highlands (Corbett 2001). *Ancylostoma caninum* and *Uncinaria stenocephalica* (hookworm) infestation has caused significant disease in wild dingoes due to enteric haemorrhage and resultant anaemia (Brown & Copeman 2003; Corbett 2001).

Cestodes

The hydatid tapeworm *Echinococcus granulosus* is frequently reported in dingoes, with the incidence of infection varying from 25–100% depending on locality (Brown & Copeman 2003; Corbett 2001; Jenkins et al. 2001; Jenkins & Morris 2003; Jenkins et al 2006). *E. granulosus* was likely to have been introduced to the Australian mainland with dogs and sheep during European settlement (Jenkins & Morris 2003). It is widely distributed in Australia, especially in sheep-rearing areas of the Great Dividing Range in eastern Australia, in south-west Western Australia and some cattle populations in northern Western Australia, and in northern Queensland (Lymbery et al. 1995; Small & Pinch 2003; Banks et al. 2006a). Wildlife plays a major role as a reservoir of infection for *E. granulosus*, especially in south-east Australia. Wild dogs including the dingo and, to a lesser extent, foxes act as definitive hosts, with macropods and sometimes feral pigs and wombats acting as intermediate hosts (Jenkins & Morris 2003). Macropods, particularly wallabies, are highly susceptible to hydatidosis. Cysts most commonly develop in the lungs and the effects may render the wallaby more vulnerable to predation, thus completing the cycle (Jenkins 2005). It was thought that *E. granulosus* from domestic sheep and dogs was a different strain from that found in wild dogs and macropods (Baldock et al. 1985), but no detectable genetic differences have been found. The parasite on mainland Australia is the common sheep strain G1 (Banks et al. 2006b) and can be transferred between wildlife, domestic animals and humans (Jenkins & Morris 2003). Hydatidosis in humans is a potentially fatal zoonotic disease (see section 12). Eradication of the disease on the Australian mainland is considered unlikely due to the sylvatic cycle in dingoes and wallabies, which often live in inaccessible areas (Jenkins & Morris 2003). Wallabies are a favoured prey and remain a source of infection for dingoes; the dingoes then defecate on pastures grazed by domestic stock, contaminating the pastures with *E. granulosus* eggs (Grainger & Jenkins 1996).

Dingoes become infected at an early age (Banks et al. 2006b) by eating protoscolices from cysts in prey animals. Worm burdens are often very heavy, commonly in the thousands, with up to 300 000 worms reported from two dingoes in the tablelands of south-east New South Wales (Jenkins & Morris 2003).

E. granulosus does not cause clinical signs in dingoes. Infections can be eliminated using praziquantel at a dose of 5 mg/kg PO. In a captive situation, ensuring raw offal from infected intermediate hosts is not fed to dingoes will prevent reinfection. The use of praziquantel baits around bush recreational areas, distributed at 4–6 wk intervals, has been suggested as a means to control infection in free-ranging dingoes in those areas, thus limiting opportunity for human infection (Jenkins 2005).

The rabbit tapeworms *Taenia pisiformis* and *T. serialis* are common in wild dogs and dingoes in central Australia, which mainly feed on rabbits (Brown & Copeman 2003; Jenkins unpublished data). *Dipylidium caninum*, *Spirometra erinacei* and *Taenia hydatigena* have also been reported (Brown & Copeman 2003; Corbett 2001). Tapeworm burdens have no discernible effect on dingo health.

Protozoa

Dingoes may be exposed to *Toxoplasma gondii* cysts in the flesh of prey items, such as native animals and sheep or pigs. 10% of dingoes sampled in south-east New South Wales had antibodies to *T. gondii* using a direct (DAT) and indirect (MAT) agglutination test (Johnson et al. 1990). The young age of dingoes positive on both MAT and DAT suggested that they were more likely to have been infected in the first year or two of their life

(Johnson et al. 1990). Some animals may succumb to fatal toxoplasmosis early in life. Pulmonary infection is most common in dogs, and affected animals may show anorexia, fever, lethargy and dyspnoea. Lungs from affected animals show focal areas of coagulation necrosis adjacent to and involving small vessels and bronchioles. It is likely that the pathology would be similar in dingoes.

Neospora caninum is a protozoan in the apicomplexa group and is a major cause of congenital infections and abortion in cattle in many parts of the world. The domestic dog and the coyote (*Canis latrans*) are the only confirmed definitive hosts of *N. caninum* (Dubey 2003; Gondim et al. 2004) although it is likely that other canid species may also act as definitive hosts. Given its close relationship to the domestic dog, it is highly likely that the dingo can act as a definitive host for *N. caninum* and thus may play a role, together with the domestic dog, in the transmission of *N. caninum* to cattle in Australia. Low numbers of unsporulated oocysts are shed in the faeces of the definitive host 5–13 d after infection; shedding persists for up to 27 d (Reichel et al. 2007). Ingested sporulated oocysts develop into tachyzoites and tissue-encysted forms in intermediate hosts (e.g. cattle, other herbivore species). As in the dog, the dingo can presumably become infected by eating infected tissues. *N. caninum* antibodies were detected in the serum of 1 of 117 dingoes from New South Wales and 14 of 52 dingoes from Queensland, giving an overall prevalence of 9% (Barber et al. 1997). However, the incidence of clinical *N. caninum* infection in dingoes is unknown. Reports of clinical cases of *N. caninum* infection in dogs are infrequent, although in one study the overall seroprevalence in domestic dogs from Melbourne, Sydney and Perth was 9% (Barber et al. 1997). Subclinically affected bitches can transmit the parasite to developing foetuses, with part or all of a litter becoming infected (Reichel et al. 2007). Neosporosis is most severe in young congenitally infected pups. Signs depend on the site parasitised, but hind limb paresis developing into paralysis, often with rigid hind limb extension, is typical. Dermatitis may also develop. *N. caninum* oocysts in dog faeces are infrequent and may be confused with those of *Hammondia heydorni,* which can cause diarrhoea, and the feline coccidians *Hammondia hammondi* and *Toxoplasma gondii* which may pass through the dog digestive tract (Reichel et al. 2007). Diagnosis is usually based on serology and clinical signs. PCR, histopathology and immunohistochemistry may also be useful. Treatment may be successful if attempted early in the disease course using current regimes for the treatment of toxoplasmosis in dogs. Clindamycin has been used successfully to treat cutaneous neosporosis in dogs (Reichel et al. 2007).

Anaplasma (Ehrlichia) platys is an obligate intracellular rickettsial organism that infects the platelets of dogs and causes infectious canine thrombocytopaenia. *Babesia canis vogeli* is an intra-erythrocytic protozoan parasite that may cause haemolytic anaemia and thrombocytopaenia in domestic dogs. *B. canis vogeli* and *A. platys* were detected in the blood of free-ranging dogs in remote Aboriginal communities in the Northern Territory and north-west New South Wales using PCR-based assays (Brown et al. 2006). Approximately 50% of dogs were infected with *A. platys, B. canis vogeli* or both, with the highest prevalence in dogs from the top end of the Northern Territory. Both organisms are transmitted by the brown dog tick *Riphicephalus sanguineus* which is endemic in the tropical areas of northern Australia (Brown et al. 2006). *B. canis vogeli* has been infrequently reported in dingoes and the consequence of infection is not known (Clark 2004). No reports were found for *A. platys* infection in dingoes, but it is likely that infection would have a similar prevalence to that of free-ranging dogs in brown dog tick endemic areas. Both *A. platys* and *B. canis vogeli* are difficult to detect on Romanowsky-stained blood films because they are usually present in low numbers. Co-infection with *B. canis vogeli* and *A. platys* resulted in lower total platelet counts than single infections with either organism (Brown et al. 2006).

Sarcocystis spp. tissue cysts are commonly seen as an incidental finding in heart and striated muscle of dingoes. Dingoes, along with other carnivores, act as vectors for sarcocystis, excreting sporocysts in faeces that are consumed by herbivores, which develop tissue cysts (Corbett 2001).

Pentastomid arthropods

Linguatula serrata is a pentastomid arthropod that inhabits the nasal sinus of dingoes and other canids (Newsome & Coman 1989; Riley et al. 1985). Incidence of infection appears low, with less than 1% of wild dingoes examined from south-east Australia infected (Corbett 2001). The dingo acts as a primary host after eating infected tissues of the secondary host, usually an insect or herbivorous mammal. The adult pentastomes can grow up to 12 cm long and live in the nasal passages, ingesting mucus. Infections in dogs are usually asymptomatic. Rarely, there may be sneezing or epistaxis. Diagnosis may require imaging technology such as rhinoscopy. Treatment may be attempted with ivermectin at a dose of 200–400 µg/kg if removal of the pentastomes is not possible.

b Ectoparasites

Lice

Trichodectes canis has been reported from wild dingoes (Corbett 2001).

Ticks

Dingoes are susceptible to paralysis by the Australian paralysis tick *Ixodes holocyclus* (Barnes et al. 2001), which occurs along the east coast of Australia from Queensland to south-east Victoria. Clinical signs, diagnosis and treatment are the same as for domestic dogs. Preventative acaricidal preparations effective against *I. holocyclus* marketed for domestic dogs and cats (e.g. fipronil and cythioate) may be used to prevent tick paralysis in dingoes. If dingoes can be handled, they should be searched daily for ticks during the tick season.

Riphicephalus sanguineus, *Haemaphysalis* spp. and *Amblyomma* spp. have been reported in dingoes. The last two are primarily macropod parasites. Heavy infestation in pups may cause anaemia and have led to deaths (Brown & Copeman 2003; Fleming et al. 2001).

Fleas

Flea infestation and flea allergy dermatitis may occur in captive dingoes. Clinical signs are similar to those seen in domestic dogs. All flea control products used on domestic dogs can be used in dingoes. Oral preparations containing lufenuron or cythioate that may be mixed in the food are easier to administer than topical or injectable preparations if animals are not readily handled. Rabbit stick-fast fleas (*Echidnophaga myrmecobii*) are common in wild dogs and dingoes in central Australia, which mainly feed on rabbits (Brown & Copeman 2003).

Mites

Sarcoptic mange due to *Sarcoptes scabiei* has been reported in wild dingo populations throughout Australia (Corbett 2001; Thomson 1992b). It causes varying degrees of alopecia, scaling and crusting, thickening, lichenification, fissuring and haemorrhage in the skin and is seldom seen as a debilitating disease on its own. The highest incidence (20%) was reported from the Fortescue River area of north-west Australia, with males twice as likely to be affected as females (Corbett 2001). In the wet–dry tropics, sarcoptic mange in dingoes is usually associated with plagues of dusky rats (*Rattus colletti*) (Corbett 2001).

Demodex canis causes localised facial alopecia in captive dingoes (Newsome & Coman 1989). Generalised demodectic mange has been reported (ARWH 2007).

12 ZOONOSES

The most significant zoonotic disease associated with dingoes is hydatidosis, which is potentially fatal in humans. Infection with the hydatid tapeworm *Echinococcus granulosus* is frequently reported in dingoes, and can be transmitted to humans during handling. Standard precautions (e.g. use of gloves, hand-washing) should be taken to minimise the risk of transmission. Dingoes may also act as a source of infection for humans by distributing eggs around bush campsites in national parks (Jenkins 2005) (see 11.2.4).

ACKNOWLEDGMENTS

The author gratefully acknowledges the expertise and assistance of Peter Fleming, David Jenkins, John Ellis and the Dingo Sanctuary, Australian Native Dog Conservation Society, in the preparation of this chapter.

REFERENCES

ARWH (2007). *Australian Registry of Wildlife Health.* Zoological Parks Board of NSW: Sydney.

AZA Wildlife Contraception Centre (2006). Contraception recommendations, carnivores. *American Zoo and Aquarium Association Wildlife Contraception Centre.* <www.stlzoo.org/contraception>.

Baldock FC, Thompson RCA, Kumaratilake LM & Shield J (1985). *Echinococcus granulosus* in farm dogs and dingoes in southeastern Queensland. *Australian Veterinary Journal* **62**, 335–337.

Banks DJD, Copeman DB, Skerratt LF & Molina EC (2006a). *Echinococcus granulosus* in Northern Queensland. 1. Prevalence in cattle. *Australian Veterinary Journal* **84**, 303–307.

Banks DJD, Copeman DB & Skerratt LF (2006b). *Echinococcus granulosus* in Northern Queensland.

2. Ecological determinants of infection in beef cattle. *Australian Veterinary Journal* **84**, 308–311.

Barber JS, Gasser RB, Ellis J, Reichel MP, McMillan D & Trees AJ (1997). Prevalence of antibodies to *Neospora caninum* in different canid populations. *Journal of Parasitology* **83**, 1056–1058.

Barnes J, Vogelnest L, Hulst F, Rose K & Bryant B (2001). Tick toxicity in zoo animals in eastern Australia. In *Proceedings of the American Association of Zoo Veterinarians, American Association of Wildlife Veterinarians, Association of Reptile and Amphibian Veterinarians and National Association of Zoo and Wildlife Veterinarians Joint Conference*. Orlando, Florida.

Barrette VT (2002). Puppies. In *Hand-rearing Wild and Domestic Mammals* (ed. LJ Gage), pp. 13–18. Iowa State University Press: Iowa.

Brown B & Copeman DB (2003). Zoonotic importance of parasites in wild dogs caught in the vicinity of Townsville. *Australian Veterinary Journal* **81**, 700–702.

Brown GK, Canfield PJ, Dunstan RH, Roberts TK, Martin AR, Brown CS & Irving R (2006). Detection of *Anapasma platys* and *Babesia canis vogeli* and their impact on platelet numbers in free-roaming dogs associated with remote Aboriginal communities in Australia. *Australian Veterinary Journal* **84**, 321–325.

Catling PC, Corbett LK & Newsome AE (1992). Reproduction in captive and wild dingoes (*Canis familiaris dingo*) in temperate and arid environments of Australia. *Wildlife Research* **19**, 195–209.

Catling PC, Hertog A, Burt RJ, Wombey JC & Forrester RI (1999). The short term effect of cane toads (*Bufo marinus*) on native fauna in the gulf country of the Northern Territory. *Wildlife Research* **26**, 161–185.

Clark P (2004). *Haematology of Australian Mammals*. CSIRO Publishing: Melbourne.

Corbett LK (1988). Social dynamics of a captive dingo pack: population regulation by dominant female infanticide. *Ethology* **78**, 177–198.

Corbett LK (2001). *The Dingo in Australia and Asia*. JB Books: Marleston, South Australia.

Corbett LK (2004). *Canis lupus ssp. dingo*. In *2004 IUCN Red List of Threatened Species*. <www.iucnredlist.org>.

Cormack Gates C, Elkin B & Dragon D (2001). Anthrax. In *Infectious Diseases of Wild Mammals* (eds ES Williams & I Barker), pp. 396–412. Iowa State University Press: Iowa.

Davis E (2001). Legislative issues relating to control of dingoes and other wild dogs in NSW. 1. Approaches to future management. In *A Symposium on the Dingo* (eds CR Dickman & D Lunney). Royal Zoological Society of NSW: Sydney.

Dubey JB (2003). Review of *Neospora caninum* and neosporosis in animals. *Korean Journal of Parasitology* **41**, 1–16.

Dunsmore JD & Spratt DM (1979). The life history of *Filaroides osleri* in wild and domestic canids in Australia. *Veterinary Parasitology* **5**, 275–286.

EAPA (2003). Draft standards for exhibiting Australian mammals in New South Wales (2003). *Exhibited Animals Protection Act 1986*, NSW Department of Primary Industries: Canberra.

Fleming P, Corbett L, Harden B & Thomson P (2001). *Managing the Impacts of Dingoes and Other Wild Dogs*. Bureau of Rural Sciences: Canberra.

Goh CSS, Hodgson DR, Fearnside SM, Heller J & Malikides N (2005). Sodium monofluoroacetate (Compound 1080) poisoning in dogs. *Australian Veterinary Journal* **83**, 474–479.

Gondim LFP, McAllister MM, Pitt WC & Zemlicka DE (2004). Coyotes (*Canis latrans*) are definitive hosts of *Neospora caninum*. *International Journal for Parasitology* **34**, 159–161.

Grainger J & Jenkins DJ (1996). Transmission of hydatid disease to sheep from wild dogs in Victoria, Australia. *International Journal for Parasitology* **26**, 1263–1270.

Green B (1976). The use of etorphine hydrochloride (M99) in the capture and immobilisation of wild dingoes, *Canis familiaris dingo*. *Australian Wildlife Research* **3**, 123–128.

Green B & Catling P (1977). The biology of the dingo. In *Australian Animals and their Environment* (eds H Messel & ST Butler), pp. 51–60. Shakespeare Head Press: Sydney.

Harden RH (1978). The biology of the dingo. In *Fauna, Proceedings 36*, pp. 451–455. Post Graduate Committee in Veterinary Science: Sydney.

Hooper PT, Reece RL & Graydon RJ (1983). Pathology of selected cases from two Melbourne zoos. In *Wildlife Diseases Association (Australasian Section) Annual Proceedings*. Wildlife Diseases Association (Australasian Section).

Jackson S (ed.) (2003). *Australian Mammals: Biology and Captive Management*. CSIRO Publishing: Melbourne.

Jenkins DJ (2005). Hydatid control in Australia: where it began, what we have achieved and where to from here. *International Journal for Parasitology* **35**, 733–740.

Jenkins DJ (2006). *Echinococcus granulosus* in Australia, widespread and doing well! *Parasitology International* (supplement) **55**, 203–206.

Jenkins DJ and Morris B (2003). *Echinococcus granulosus* in wildlife in and around the Kosciuszko National Park, south-eastern Australia. *Australian Veterinary Journal* **81**, 81–85.

Jenkins DJ, Fleming P & Cathles II (2001). The role of wildlife in the transmission of *Echinococcus granulosus* with particular reference to south-eastern Australia. In *Wildlife Health and Management in Australasia. Proceedings of Australian Association of Veterinary Conservation Biology*. Sydney.

Jenkins DJ, Romig T & Thompson RCA (2006). Emergence/re-emergence of Echinococcus spp.: a global update. *International Journal for Parasitology* **35**, 1205–1219.

Johnson AM, Phillips P & Jenkins D (1990). Prevalence of *Toxoplasma gondii* antibodies in dingoes. *Journal of Wildlife Diseases* **26**, 383–386.

Kennedy-Stoskopf S (2003). Canidae. In *Zoo and Wild Animal Medicine* (eds ME Fowler & RE Miller). WB Saunders/Elsevier Science: St Louis.

Kershaw K, Allen L, Lisle A & Withers K (2005). Determining the age of adult wild dogs (*Canis lupus dingo, C. lupus domesticus* and their hybrids). I. Pulp cavity: tooth width ratios. *Wildlife Research* **32**, 581–585.

Kreeger T (1992). A review of chemical immobilisation of wild canids. In *Proceedings of the American Association of Zoo Veterinarians and American Association of Wildlife Veterinarians Joint Meeting*, pp. 271–283. Oakland, California.

Kreeger TJ (1999). Chemical restraint and immobilization of wild canids. In *Zoo and Wild Animal Medicine: Current Therapy 4* (eds ME Fowler & RE Miller). WB Saunders: Philadelphia.

Kreeger TJ, Mandsager RE, Seal US, Callahan M & Beckel M (1989). Physiological response of grey wolves to butorphanol-xylazine immobilization and antagonism by naloxone and yohimbine. *Journal of Wildlife Diseases* **25**, 89–94.

Lymbery AJ, Thompson RCA, Constantine CC & Kruger JG (1995). The geographic distribution of hydatid infection in cattle in Western Australia. *Australian Veterinary Journal* **72**, 430–432.

McIlroy JC (1981). The sensitivity of Australian animals to 1080 poison. II: Marsupial and eutherian carnivores. *Australian Wildlife Research* **8**, 385–399.

Merck Veterinary Manual (2003). Merial Ltd and Merck & Co.: Whitehouse Station, New Jersey. <www.merckvetmanual.com/mvm>.

Newsome AE & Coman BJ (1989). Canidae. In *Fauna of Australia* Vol. 1B Mammalia (eds DW Walton & BJ Richardson), pp. 993–1005. AGPS: Canberra.

Oakman B (2001). The problems with keeping dingoes as pets and dingo conservation. In *A Symposium on the Dingo* (eds CR Dickman & D Lunney). Royal Zoological Society of NSW: Sydney.

Reeves MP (2004). A retrospective report of 90 dogs with suspected cane toad (*Bufo marinus*) toxicity. *Australian Veterinary Journal* **82**, 608–611.

Reichel MP, Ellis JT & Dubey JP (2007). Neosporosis and hammondiosis in dogs. *Journal of Small Animal Practice* **48**, 308–312.

Riley J, Spratt DM & Presidente PJA (1985). Pentastomids (Arthropoda) parasitic in Australian reptiles and mammals. *Australian Journal of Zoology* **33**, 39–53.

Savolainen P, Leitner T, Wilton AN, Matisoo-Smith E & Lundeberg J (2004). A detailed picture of the origin of the Australian dingo, obtained from the study of mitochondrial DNA. *Proceedings of the National Academy of Sciences* **101**, 12387–12390.

Small LM & Pinch DS (2003). Survey for hydatidosis in cattle bred in the northern region of the Northern Territory of Australia. *Australian Veterinary Journal* **81**, 355–358.

Starr TW & Mulley RC (1988). *Dirofilaria immitis* in the dingo (*Canis familiaris dingo*) in a tropical region of the Northern Territory, Australia. *Journal of Wildlife Disease* **24**, 164–165.

Suedmeyer WK, Ludlow C, Layton C, Dennis J & Miller M (2004). Primary lymphangiectasia in a dingo (*Canis familiaris dingo*). *Journal of Zoo and Wildlife Medicine* **35**, 572–575.

Symons M & Bell K (1992). Canine blood groups: description of 20 specificities. *Animal Genetics* **23**, 509–515.

Thomson PC (1992a). The behavioural ecology of dingoes in northwestern Australia. II. Activity patterns, breeding season and pup rearing. *Wildlife Research* **19**, 519–530.

Thomson PC (1992b). The behavioural ecology of dingoes in northwestern Australia. I. Study areas and animals. *Wildlife Research* **19**, 509–518.

Thomson PC (1992c). Capture of dingoes from a helicopter with tranquilliser darts loaded with ketamine hydrochloride and xylazine hydrochloride. *Wildlife Research* **19**, 601–603.

Thomson PC & Rose K (1992). Age determination of dingoes from characteristics of canine teeth. *Wildlife Research* **19**, 597–599.

Turner AG & Phelan JR (1996). Persistent corneal erosion in an Australian dingo. *Australian Veterinary Journal* **74**, 435–436.

Wilton AN (2001). DNA methods of assessing dingo purity. In *A Symposium on the Dingo* (eds CR Dickman & D Lunney). Royal Zoological Society of NSW: Sydney.

Woodall PF, Pavlov P & Twyford KL (1996). Dingoes in Queensland, Australia: skull dimensions and the identity of wild canids. *Wildlife Research* **23**, 581–587.

World Health Organization (2005). Rabies. *Human and Animal Rabies.* World Health Organization: Geneva. <www.who.int/rabies/en/>.

Vogelnest L (1999). Chemical restraint of Australian native fauna. In *Wildlife in Australia: Healthcare and Management, Proceedings 327*, pp. 149–188. Post Graduate Foundation in Veterinary Science: Sydney.

17. Pinnipeds

Julie Barnes, Damien Higgins and Rachael Gray

1 TAXONOMY, DISTRIBUTION AND BIOLOGY

Two Families in the Order Pinnipedia are found in Australian waters: the Otariidae (eared seals; fur-seals and sea-lions) and the Phocidae (true or earless seals). Otariid species found in Australian waters are the Australian fur-seal (*Arctocephalus pusillus*), the New Zealand fur-seal (*Arctocephalus forsteri*), the Australian sea-lion (*Neophoca cinerea*), the subantarctic fur-seal (*Arctocephalus tropicalis*) and the Antarctic fur-seal (*Arctocephalus gazella*). The southern phocids include the leopard seal (*Hydrurga leptonyx*), southern elephant seal (*Mirounga leonina*), Weddell seal (*Leptonychotes weddellii*), crab-eater seal (*Lobodon carcinophaga*) and Ross seal (*Ommatophoca rossii*). Southern elephant seals and leopard seals are seen frequently in Tasmanian waters and occur less frequently in other states. Weddell, crab-eater and Ross seals, Antarctic and sub-antarctic fur-seals and Hooker's sea-lions (*Phocarctos hookeri*) are only occasional visitors to the Australian coast.

Typical lengths and weights of Australian pinnipeds are presented in Table 17.1. Figure 17.1 displays the approximate range of Australian and New Zealand fur-seals in Australian waters, the Australian sea-lion and Antarctic and subantarctic fur-seals, with details on their breeding range.

The southern elephant seal has a circumpolar distribution with a non-breeding range extending from the Antarctic continent to the tropics (Carrick & Ingham 1962). Most breeding occurs on islands on either side of the Antarctic convergence (Bonner 1994). It is not uncommon for healthy individuals to haul out in Australia, particularly on the Tasmanian coast, and New Zealand.

The leopard seal is the most widely distributed of the Antarctic pinnipeds, ranging between 50–80°S (Hofman et al. 1977). Although primarily a resident of the pack ice where breeding occurs (Siniff & Stone 1985), there is some movement of individuals to the subantarctic islands and the southern continents including south-eastern Australia and frequently Tasmania (Gwynn 1953; Rounsevell 1988; Bonner 1994). Individuals hauling out in more northerly locations are commonly debilitated. Weddell, crab-eater and Ross seals also have a circumpolar distribution and are generally found south of the Antarctic convergence.

2 ANATOMY AND PHYSIOLOGY

2.1 Anatomy

Otariids and phocids differ anatomically. For example, otariids have prominent external pinnae, large fore flippers and smaller hind flippers that can be rotated forwards under the body to enable them to walk quadrupedally, assisting in terrestrial locomotion. Propulsion through water is primarily from the long fore flippers, while the hind flippers control direction. In contrast, phocids are unable to rotate the hind flippers under the body and therefore have a less agile, slug-like terrestrial locomotion. They employ lateral undulating movements of the hind flippers and lower body for aquatic

Table 17.1 Lengths and weights of Australian pinniped species

Species	Source	Male length (m)	Female length (m)	Male weight (kg)	Female weight (kg)
Australian fur-seal	Warneke & Shaughnessy (1985)	2.16(2.01–2.27, n = 25)	1.57(1.36–1.71, n = 122)	279 (218–360, n = 13, territorial bulls)	76 (41–113, n = 122)
Australian sea-lion	Walker & Ling (1980)		1.48 ± 0.13(1.32–1.81, n = 28)		73.9 ± 7.09 (63.0–81.6, n = 14, lactating)
	Gales & Costa (1997)			Up to 300	Up to 120
	Bonner (1994)	2.0–2.5	1.7–1.8	400	80–105
New Zealand fur-seal	Crawley & Warneke (1979)	1.45–2.50	1.25–1.50	120–185	40–70
	Croxall & Gentry (1987)			120–155	40–50
	Troy et al. (1997)	1.50–1.79 (n = 22)		78–160 (n = 22)	
Subantarctic fur-seal	Condy (1978)			Up to 140	Up to 50
	Shaughnessy & Ross (1980)	1.29–1.64 (n = 7)			
	Bester (1987)			131.3 ± 20.0 (97.0–158, n = 8)	35.6 ± 4.5 (28.0–46.0, n = 29)
	Croxall & Gentry (1987)			70–160	24–45
	Kerley (1987)			88.3 ± 15.7 (n = 18, max. 118)	34.1 ± 8.2 (n = 4)
	Bonner (1994)	1.80	1.45		
Antarctic fur-seal	Payne (1979)			130 (>9 yr)	31.2 (5 yr)–38.2 (>11 yr)
	Bonner (1982)	1.84 ± 0.08 (1.72–1.97, n = 15)	1.28 ± 0.08 (1.13–1.39, n = 21)	140 ± 14.2 (126–160, n = 4)	38.2 ± 5.9 (30–51, n = 17)
	Croxall & Gentry (1987)			110–200	22–50
	Bonner (1994)	1.65–2.0	1.15–1.40		
Leopard seal	Hamilton (1939)	2.21–3.05 (n = 18); 2.89 (>4 yr)	2.21–3.58 (n = 20); 3.29 (>4 yr)		
	Laws (1993)			Up to 600 kg (males + females)	
	Gray (2005)	2.94 ± 0.19 (2.76–3.50, n = 14)	3.07 ± 0.20 (2.80–3.42, n = 22)		
Southern elephant seal	Jones (1981)	4.88 ± 1.84 (breeding, n = 11)		2176 (breeding, n = 11)	
		4.73 ± 2.89 (bachelor, n = 19)		1973 (bachelor, n = 19)	
	Laws (1993)	4.2–4.5	2.6–2.8	3000–4000	400–900
	Woods et al. (1995a)		2.50 (2.10–2.90, n = 96)		369 (215–610, n = 96)

Australian fur-seals breed on the islands of Bass Strait (Brothers & Pemberton 1990; Goldsworthy *et al.* 1997). New Zealand fur-seals breed on the South Island of NZ, the NZ subantarctic islands, the offshore islands of South and Western Australia (Crawley & Warneke 1979; Shaughnessy *et al.* 2005), Bass Strait (Arnould *et al.* 2000), Maatsuyker Island (Brothers & Pemberton 1990), the North Island of NZ (Dix 1993) and Macquarie Island (Csordas & Ingram 1965). There are occasional reports of non-breeding NZ fur-seals in Western Australia, NSW, Victoria, and southern Qld (Shaughnessy 1999, references therein).

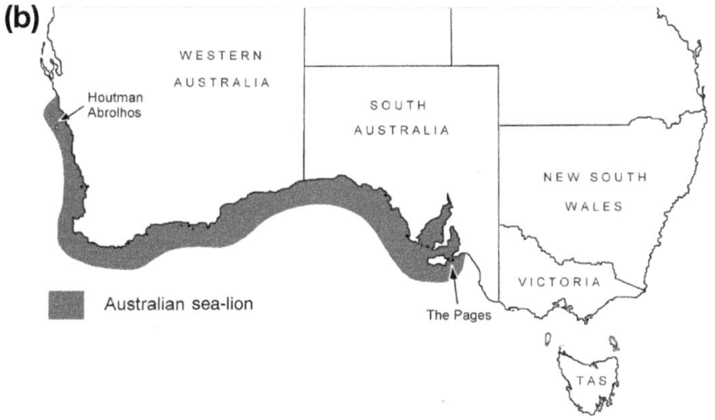

The Australian sea-lion ranges from Houtman Abrolhos in Western Australia to The Pages in South Australia (Marlow 1975; Gales *et al.* 1994) with breeding on at least 50 offshore islands within this range (Gales *et al.* 1994).

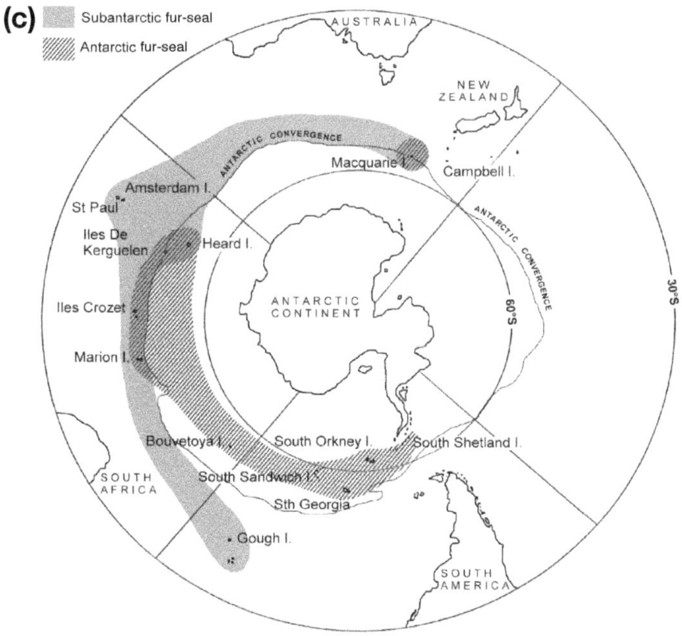

The subantarctic and Antarctic fur-seal breed on several subantarctic islands north and south of the Antarctic convergence respectively (Hofmeyr *et al.* 1997). They breed sympatrically at several islands close to the convergence and hybridization has been reported (Condy 1978; Goldsworthy & Shaughnessy 1989; Hofmeyr *et al.* 1997). Some subantarctic fur-seals range to South Africa, South America, New Zealand (Shaughnessy & Ross 1980; Taylor 1990; Nowak 2003) and Australia (Shaughnessy 1999 and references therein).

Figure 17.1 Approximate ranges a) Australian and New Zealand fur-seals in Australian waters; b) Australian sea-lions; c) Antarctic and subantarctic fur-seals and southern phocids. (Source: Csordas & Ingham (1965); Warneke (1982); Ling (1987); Shaughnessy & Warneke (1987); Brothers & Pemberton (1990); Goldsworthy et al. (1997); Irvine et al. (1997))

locomotion, steering with the fore flippers. Phocids lack external pinnae and the opening of the ear canal is barely visible just behind the eyes.

The flexible cartilaginous ribcage of pinnipeds extends more caudally than that of terrestrial mammals, so surgical access to some abdominal organs can be difficult. The skeletal muscle is dark red, almost black, because of the relatively high myoglobin concentration. Pinnipeds have a simple stomach and considerably longer intestinal tract than terrestrial mammals. The large intestine is relatively short, the small intestine can be up to 20 times longer than the large intestine (Richardson & Gales 1987; Rommell & Lowenstein 2001) and the caecum is vestigial (Rommel & Lowenstein 2001). The liver is large and multi-lobed and lies mostly to the left of the midline (Rommel & Lowenstein 2001). The kidneys are multi-lobed and venous drainage forms a superficial stellate plexus, visible on the kidney surface (Harrison & Tomlinson 1956).

The anatomy of the reproductive tracts is generally similar to that of terrestrial carnivores (Boyd et al. 1999). The male urogenital opening is found on the ventral abdomen caudal to the umbilicus and the female urogenital opening is found underneath the tail just ventral to the anus. Male pinnipeds have a baculum or os penis. The testes are scrotal in otariids, but are para-abdominal (deep to the blubber in the inguinal area) in phocids (Bryden 1967; Harrison 1969). The uterus is bicornuate.

The anatomical adaptations of the pinniped eye maximise visual sensitivity and visual acuity in air and in water at low light intensity. The eye is characterised by a large globe, high density of photoreceptors, well-developed tapetum, large rounded lens and a slit-like pupil, which is generally vertical in orientation (Wartzok & Ketten 1999). There is a central circular flattening of the cornea, which should not be mistaken for a lesion. Tear production is constant and copious, typically producing a wet area below the eye when the animal is dry.

2.2 Integument and thermoregulation

Fur-seals have highly developed compound hair follicles consisting of a single guard hair follicle and several intermediate and underfur follicles creating an undercoat of dense fur which, by trapping air, has excellent insulating and water-repelling qualities (Scheffer 1964; Ling 1970). The blubber layer is comparatively thin in fur-seals with the pelage providing most of the insulative capacity of the integument (Rommel & Lowenstein 2001). In contrast, phocids have few or no secondary or underfur fibres (Scheffer 1964) and rely on a thick blubber layer for insulation (Ling 1965a). Sea-lions generally have a small number (1–4) of secondary or underfur fibres (Scheffer 1964) and, like phocids, rely on blubber for their primary insulation (Pabst et al. 1999). Moulting of the pelage occurs annually, generally during summer and autumn, except in the Australian sea-lion which moults on a 17–18 mo cycle; females moult approximately 4 mo after parturition (Higgins 1993). For most species, the moult involves the regeneration of the pelage only; in otariids, the moult can take several weeks to months (Rommel & Lowenstein 2001). Some phocids, including the southern elephant seal, undergo an extensive moult (Ling 1965b; Laws & Sinha 1993) in which large areas of the stratum corneum are shed with the hair. Southern elephant seals spend 4–7 wk ashore and fast during the annual moult (Ling 1965b), using blubber reserves for energy. The timing of the moult must be considered when remote monitoring devices are deployed on free-ranging pinnipeds, as the instruments are most commonly glued to the pelage and will be shed with the pelage during the moult.

2.3 Diving adaptations

Several anatomical and physiological adaptations contribute to the diving ability of pinnipeds. Foraging behaviour, and therefore diving strategy, differs between species. Fur-seals generally undertake short shallow dives (Gentry et al. 1986), whereas Australian sea-lions are benthic foragers and dive to greater depths to forage on or near the sea floor (Gales & Mattlin 1997 and references therein; Costa et al. 2001). Phocids, especially Weddell and elephant seals, generally dive deeper and for longer than otariids, which is reflected in their degree of adaptation. A post-moult female southern elephant seal from Macquarie Island recorded a maximum dive depth of 1256 m and dive duration of 120 min (Hindell et al. 1991).

Diving adaptations increase transport, storage and conservation of oxygen, and aid thermoregulation in the aquatic environment. Physiological adaptations associated with the dive response (see 6.2.3) are reviewed by Lynch et al. (1999). Anatomical adaptations include a distensible dilatation of the ascending aorta, which facilitates the maintenance of mean arterial blood pressure and perfusion of the brain and heart during diving bradycardia (Elsner 1969; Drabek 1975). A sphincter surrounding the caudal vena cava cranial to the diaphragm, an enlarged caudal vena cava and the large venous reservoir of the hepatic sinus caudal to the diaphragm allow massive pooling of blood in the venous system and control venous return to the heart, preventing volume overload of the heart during diving (Harrison & Tomlinson 1956, 1963). Venous plexuses in the cervical, lumbar and renal regions may also have a blood storage function (Harrison & Tomlinson 1956). A pericardial retia and a duplicated caudal vena cava have been reported in some species (Harrison & Tomlinson 1956, 1963). Arteriovenous anastomoses, a network of intertwining arteries and veins, create vascular counter-current heat exchange systems that dissipate and conserve heat, particularly from the flippers, to control core body temperature (Riedman 1990). The intravertebral extradural vein provides a large alternative venous pathway from the brain and spinal cord (dural sinuses) that eventually empties into the caudal vena cava (Harrison & Tomlinson 1956, 1963).

Haematological adaptations include larger erythrocytes, lower erythrocyte counts, greater haematocrit and haemoglobin concentrations (Lenfant 1969; Ridgway 1972) and greater blood volume (Bryden & Lim 1969; Ridgway 1972) than in terrestrial mammals. Among the pinnipeds, greater haematocrit and haemoglobin values are seen in phocid seals (Lenfant et al. 1970; Horning & Trillmich 1997).

There are a few adaptive features of note in the respiratory system. Pinnipeds have a flexible ribcage which enables the thoracic cavity to compress during diving (Ridgway 1972; Pabst et al. 1999 and references therein; Rommell & Lowenstein 2001). Depending on species, the cartilaginous support of the trachea varies from complete cartilaginous rings to dorsally incomplete cartilage rings to only small ventral bars (King 1983), permitting its collapse during diving and, sometimes, anaesthesia. Last, the terminal airways are reinforced with muscle in phocids and cartilage in otariids (Denison & Kooyman 1973). Under compression during diving, the alveoli collapse and air can be forced into the reinforced upper airways of the bronchial tree; reducing exposure to nitrogen during diving (Pabst et al. 1999 and references therein).

2.4 Physiological data

The normal adult heart and respiratory rates are 60–120 beats and 3–6 breaths per minute respectively, but both vary considerably during stress or physical restraint. Body temperature is also variable (35.0–38.0°C) and is most reliably measured with a deep rectal thermometer.

3 REPRODUCTION

Pinniped reproduction is characterised by synchronised annual parturition and postpartum mating within a few days or weeks of parturition (Stirling 1983). The Australian sea-lion is an exception as it has a 17–18 mo non-seasonal breeding cycle (Ling & Walker 1978; Gales et al. 1992) and the breeding season is asynchronous between colony sites (Gales et al. 1992). Southern phocids give birth on ice or land and have a 3–6 wk lactation period, during which females fast (southern elephant seal and leopard seal) or partially fast (Weddell seal) (Laws & Sinha 1993). Females remain near the pupping site and weaning is generally abrupt (Oftedal et al. 1987), resulting in a post-weaning fast of the young until they learn to forage on their own (Atkinson 1997). In contrast, otariids give birth on land and have a prolonged lactation, characterised by an initial intense nursing period followed by alternating periods of foraging at sea for 2–10 d (depending on species and individual) with terrestrial nursing. Weaning is more gradual in otariids (Oftedal et al. 1987) as pups begin to accompany their mothers on foraging trips to sea during the latter part of lactation.

Embryonic diapause is thought to be an obligate part of the reproductive cycle of all pinnipeds (Daniel 1981). The synchronicity of the seasonal pattern relies on this feature (Boyd 1991a) as mating occurs close to parturition and in most pinnipeds the period of active gestation is approximately 8 mo (Daniel 1981; Boyd 1991a). The Australian sea-lion, however, is thought to have an active gestation period of up to 14 mo (Gales et al. 1997). Embryonic diapause occurs when the embryo reaches the blastocyst stage; cell division slows and there is a delay of 2–4 mo, depending on gestation length, before the blastocyst reactivates and implants (Daniel 1981).

Sexual maturity in pinnipeds occurs at 2–7 yr of age and is dependent on species, sex and environmental factors affecting growth rate (Atkinson 1997; Boyd et al.

Table 17.2 Reproductive parameters for Australian pinniped species

Species	Age at sexual maturity, age at which reproduction possible i.e. first ovulation (females)	Age at sexual maturity (males)	Arrival prior to parturition (pre-natal attendance)	Birth to sea (post-natal attendance)	Pupping duration	Median pupping date	Birth–oestrus/mating	Length of lactation/time of weaning
Australian sea-lion	3 yr[1]		1.8 d (range 0–5)[2], 2-3 d[3]	9.8 d[2], 14 d[3]	5 mo[4], 4–5 mo[5]	Non-annual and asynchonised between sites[5]	7 d[2,3]	15–18 mo (up to 40 mo)[2], 18 mo[5]
Australian fur-seal	4 yr or later, occasionally 3 yr[8], 3–6 yr[13,14]	4–5 yr, breeding status at 9–12+[8,13], 4–5 yr, social maturity 11 yr (range 8–13 yr)[14]		8.4 d[15]	Nov–Dec[13], late Oct–late Dec[14,16]	1 Dec[14]	5–6 d[8,13,14]	11–12 mo[8,13,14], 313 ± 5.1 d[15], some to a 2nd, rarely 3rd yr[8,13]
New Zealand fur-seal	4–6 yr[18,19]	Age of first harem ~10 yr[18], 4–6 yr, socially mature 10–12 yr[19]	2.1 d[20]	10 d[11,21], 9 d[20]	Late Nov–mid Jan[21], mid Dec–early Jan[22], mid Nov–late Dec[20]	21 Dec (SA)[21], 9/10 Dec (NZ)[20,23]	7.4 d[21], 8 d[11,20]	9+ mo[22], 10–11 mo[19]
Subantarctic fur-seal	4–6 yr[24], first pupping 4–5 yr[18]	3–4 yr, socially mature 7+ yr, territorial males 8+ yr[24] 4 yr+, sexual and social maturity 8 yr, full adulthood 10–11 yr[25], harem tenure 8+ yr[18]	1.4 d[26], 6 d[27]	8.7 d[26]	Late Nov/early Dec–early/mid Jan[24,28,29], Nov–Feb[27]	11 Dec (Amsterdam Is.)[28], 17 Dec (Marion Is.)[29]	8 – 12 d[27]	10–11 mo[24,28,29], 6–8 mo (Prince Edward Is.)[27]
Antarctic fur-seal	Pup at 3, 4 or 5 yr[30] Pup at 3 yr[18], first pupping 3 yr, all pupped by 6 yr[31]	Harem tenure 7–10 yr[18]	1 d[32], 2 d[30,31,33,34]	7 d[31], 7.8 d[32]	Late Nov–mid Dec[29]	6 Dec[29]	6 d[33], 7 d[30,32]	4 mo[30], 112 d[29,31], 117 d[33]
Leopard seal	3–4 yr[10], 2–7 yr[12], 3–6 yr (mostly 4 yr)[17]	4–6 yr[10], 3–6 yr[12], 4–5 yr[17]	Unknown	Length of lactation	Late Oct–mid Nov[35], mid Oct–mid Nov[17], Nov–Dec[34]	Unknown	1 mo (post-weaning)[17], 48 d[36]	30 d[17,34]
Southern elephant seal	3, 4 or 5 yr at Macquarie Is., 2 yr at South Georgia[8], 3–6 yr[6]	Attend harems 10 yr, secure harem status 12+ yr[9], present at breeding sites at 4 yr & maintain harems at 7 yr (South Georgia)[8], 6–9 yr (Macquarie Is.)[6]	5 d[8,6], 6 d[7]	Length of lactation	Early Sept–late Oct (Macquarie Is.)[6]		3rd week of lactation [6,7,8], 19 d[36]	23 d[6,7,8]

Sources: **1** Stirling (1972). **2** Higgins & Gass (1993). **3** Marlow (1975). **4** Higgins (1993). **5** Gales et al. (1992). **6** Carrick et al. (1962). **7** Condy (1979). **8** Warneke (1982). **9** Jones (1981). **10** Laws (1957). **11** Crawley (1990). **12** Øritsland (1970). **13** Warneke (1979). **14** Warneke & Shaughnessy (1985). **15** Arnould & Hindell (2001). **16** Pemberton & Kirkwood (1994). **17** Tikhomirov (1975). **18** Croxall & Gentry (1987). **19** Crawley & Warneke (1979). **20** Miller (1975). **21** Goldsworthy & Shaughnessy (1994). **22** Brothers & Pemberton (1990). **23** Mattlin (1981). **24** Bester (1987). **25** Bester (1990). **26** Georges & Guinet (2000). **27** Condy (1978). **28** Roux & Hes (1984). **29** Kerley (1983). **30** Payne (1977). **31** McCann & Doidge (1987). **32** Goldsworthy (1995). **33** Doidge et al. (1986). **34** Laws (1984). **35** Siniff & Stone (1985). **36** Laws & Sinha (1993).

1999). The age of sexual maturity in males is more variable and, although often capable of spermatogenesis at 3–6 yr of age, many are not behaviourally capable of breeding until 8–10 yr of age (Atkinson 1997).

Otariids and southern elephant seals are highly gregarious, polygynous and sexually dimorphic; bulls defend individual terrestrial territories. Slight reverse sexual dimorphism is seen in the Antarctic ice breeding seals, which generally maintain aquatic territories (Stirling 1983).

Table 17.2 outlines reproductive parameters for each species.

4 HUSBANDRY

The housing and husbandry requirements of pinnipeds are specialised.

4.1 Housing

Captive pinnipeds should have access to dry resting areas and at least one pool of sufficient depth and size to allow the animals to perform their natural behaviours. Otariids are good climbers and high fencing is required for these species. Pools with gentle sloping sides are best for all phocids and otariid pups, allowing them easy access into and out of pools and a place to rest in shallow water. The substrate should not contain objects that can be worked loose, such as rocks, as pinnipeds often ingest foreign objects. Air and water temperature should be maintained within a range that is consistent with the natural geographical range of the species. Sweeney and Samansky (1995) recommend the following guidelines for water temperature for pinnipeds: 0–24°C for polar species, 7–27°C for temperate species and 13–29°C for tropical species. Pinnipeds are much more tolerant of cold than of high temperatures, but extremes of both should be avoided. Ambient temperatures of 26–28°C approach the upper thermal limit for most pinnipeds and they should have access to water (pools, hosing, sprinklers) under such conditions to avoid heat stress. Outdoor facilities should have a shaded area to protect from direct sunlight and heat. Lighting should mimic the natural photoperiod for the species as closely as possible and natural lighting is preferable. Ideally, off-exhibit holding areas should have a large flat dry area that can be isolated from pools. This can be used for veterinary procedures, isolation/quarantine and to isolate mothers and pups. A sloped shallow depression at one end of the dry area that can be filled with water is useful for young pups or debilitated animals. Pools should be cleaned regularly – the cleaning interval is usually determined by the number of animals, type of pool and type of life-support system used to maintain water quality.

Social species should not be housed alone unless there are specific reasons for doing so, e.g. isolation due to illness, medical treatment or quarantine. Interspecific and intraspecific aggression, sex ratios and individual compatibility also need to be considered when housing pinnipeds together. Non-social phocid species, such as the leopard seal, should not be housed with other species due to territorial aggression. They often require individual housing, however, compatible females can sometimes be housed in pairs if provided sufficient space.

Although a relatively small number of wild southern phocid species, such as the leopard seal and southern elephant seal, have been held in captivity, they traditionally have had poor long-term survivorship, most likely resulting from inadequate husbandry and poor acclimatisation. Consequently the suitability of these species, particularly the southern elephant seal, for captivity is questionable. Leopard seals have been maintained successfully for several years but few have survived long-term in captivity. Survivorship of southern elephant seals in captivity has been very poor. A thorough understanding of biology and husbandry requirements, and the ability to provide adequately sized enclosures with appropriate air and water temperatures, is critical for their health and survival in captivity.

4.2 Hospital and quarantine housing

Pinnipeds undergoing veterinary care, quarantine or rehabilitation can be held in pens for several weeks without access to water for swimming. However, they are more content if they have free access to water and ideally should be provided with a pool, at least large enough for the animal to immerse itself. The water must be changed daily or turned over by placing a slow-running hose in the pool, such that it constantly overflows. Pools must be cleaned regularly. Water does not need to be salinated for short-term holding. Where animals have no access to water, drinking water must be provided. These animals also enjoy being hosed. Pens should have non-abrasive washable floors, be well-ventilated and large enough for the animal to move around freely. A slightly raised, preferably slatted, platform should be provided as a dry resting area. Pups sometimes like to hide or sleep in a roomy pet-pack or crate within the holding area.

Pinnipeds undergoing quarantine should be housed with a separate air and water space from any other pinnipeds in the facility, to reduce the risk of cross-infection. Wastewater or filtered water from quarantine pools should not be recirculated through other marine mammal pools and may require disinfection prior to discharge, depending on local regulations.

4.3 Water quality

Salt water is preferred, although pinnipeds can be housed long-term in fresh water. Oral sodium chloride supplementation is required for all pinnipeds held in fresh water to ensure general well-being (see 5.4). When the use of natural seawater is not possible, some institutions use artificial seawater salts. Good water quality is essential for the health of captive pinnipeds. Some pools are maintained as an open seawater system, with minimal to no filtration. Semi-closed and closed systems require filtration, or pools need to be emptied and filled regularly. Chlorine, ozone and ultraviolet irradiation are commonly used for disinfection of water in marine mammal pools. Chlorine, ozone and, less frequently, copper are used to remove algae. Monitoring levels of oxidative compounds such as chlorine and ozone is critical as excessive amounts may result in corneal damage (Stoskopf et al. 1983). Bromide, a component of fresh water, seawater and organic matter, is oxidised to bromine when exposed to chlorine or ozone, and bromine in excessive amounts is irritating to tissues. Water should be tested regularly for temperature, pH, salinity, chlorine, bromine (if oxidants such as chlorine or ozone are used), any added chemicals, and bacteria. Daily water testing is preferred, although weekly testing may be adequate for established life-support systems. Pinnipeds can be maintained safely in water of pH 7.2–8.4, salinity 20–36 ppt and total chlorine below 1.0–1.5 ppm. Free chlorine should be maintained at approximately 50% of total chlorine, and combined bromine below 1.0 ppm. Total coliform bacterial counts are recommended to be less than 500 MPN (most probable number) per 100 mL of water (NSW DPI unpub), although less than 1000 MPN per 100 mL water is a commonly accepted international standard (APHIS 2002). Arkush (2001) provides more detail on water quality for marine mammal pools.

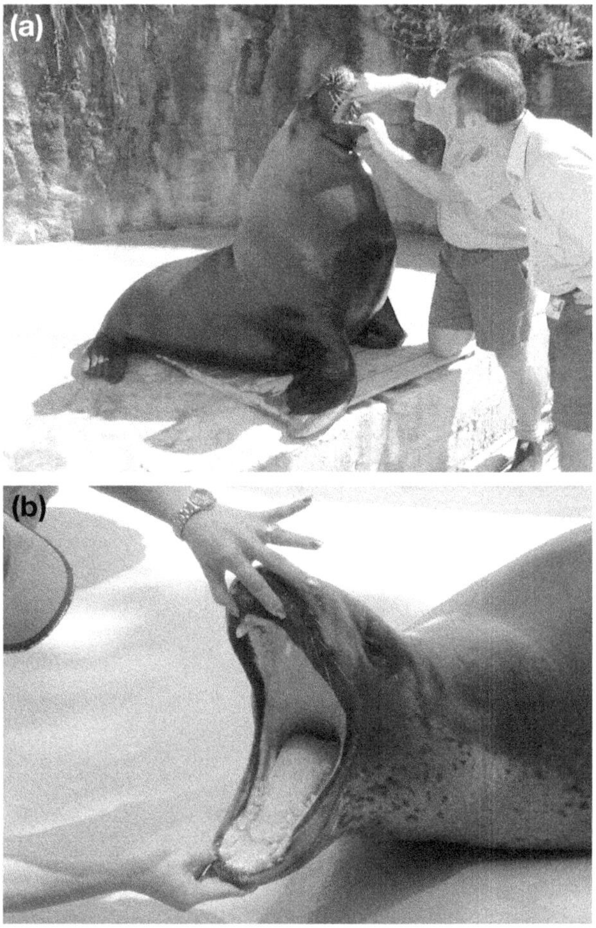

Figure 17.2 (a) An Australian sea-lion and (b) a leopard seal trained for dental examination.

4.4 Husbandry

Animals should be weighed monthly to monitor body condition and health. If there are health concerns animals should be weighed more frequently. Captive pinnipeds can and, where possible, should be trained to facilitate veterinary examination and procedures, including blood collection (Fig. 17.2). Behavioural enrichment is also important for pinnipeds and can be achieved through socialisation, training and enrichment activities such as toys and feeding activities.

4.5 Transport

Cages or crates are ideal for transporting pinnipeds. They must be well-ventilated and large enough to allow an animal to stretch full length, raise its head and turn around. Openings must be small enough to prevent the animal biting anyone through them, but large enough to allow the animal to be cooled by wetting. Internal plywood panelling can prevent the animal from damaging its teeth on cage mesh, but ventilation must not be compromised. An adequate number of handholds should be placed around the outside of the cage or crate so that it can be kept reasonably level when lifted and moved, and to prevent handlers from being bitten.

A high-sided open or closed and ventilated vehicle or trailer can be used for the transport of adult pinnipeds over long distances when purpose-built cages or crates are not available. Pinnipeds transported in open trailers must be protected from direct sunlight. Large pet-packs are ideal for the transport of small pups.

Ambient temperatures of 10–20°C are suitable for transport. Good ventilation is necessary to prevent overheating during transport and a temperature-controlled vehicle may be required in some cases. Animals must be monitored closely for signs of distress, hyperthermia or hypothermia. Cooling can be achieved by wetting the animal with a water spray, or placing ice on the top of a wire mesh cage so that water drips onto the animal as the ice melts. Although most pinnipeds are unlikely to drink during transport, they should be given the opportunity to drink every 4 hr by offering water from a hose or spray, or ice cubes can be dropped into the crate.

4.6 Individual marking and identification techniques

Individuals can be identified temporarily and prior to moulting with oil-based paint markings, by shaving or by bleaching the fur with a mixture of potassium persulphate and hydrogen peroxide. The most common methods used for permanent identification are placement of a passive integrated transponder (microchip or PIT tag) or a plastic livestock tag. Livestock tags are approximately 15 × 40 mm and are placed in the interdigital webbing approximately 25 mm from the edge of the webbing between the upper two digits of the hind flipper of phocids, or into the loose skin on the trailing edge of the upper aspect of the fore flipper of otariids. It is not uncommon for plastic flipper tags to be lost due to damage to the interdigital webbing or by being torn out, or by damage or failure of the tag itself. Freeze branding, which is commonly used for permanent identification in cetaceans, is infrequently employed as a means of identification in pinnipeds and brands may disappear over time with moulting. The use of hot branding is controversial due to the concern about the degree of pain caused by the technique.

5 NUTRITION

5.1 Diet

Although free-ranging pinnipeds eat a wide variety of fish and invertebrate species, pinnipeds in captivity are generally maintained on a limited variety of commercially available fish species and squid. Mullet, mackerel, yellow-tail, whiting, herring, pilchards and squid (5–10% of diet) are commonly fed to captive pinnipeds in Australia. Variety is important as nutritional value, particularly fat content, varies greatly between fish species and with season, age, sex and source of fish. Fat content is the major determinant of the energy value of the fish, as carbohydrate content in fish is very low. Mackerel and herring are generally considered to be high-fat fish, while mullet, yellow-tail, whiting and pilchards are mid–low fat fish. Squid has the lowest fat content. Ideally, pinnipeds should be fed a selection of high- and low-fat fish. Underweight animals are usually fed an increased proportion of high-fat fish to encourage weight gain.

Marine mammals derive most of their water from ingested fish and some from the metabolism of fat, protein or carbohydrate. The amount of water in fish is inversely proportional to the concentration of fat. Some captive pinnipeds drink fresh water from a hose or trough. Although consumption of fresh water is not frequently observed in captive pinnipeds, animals that have no access to water and those with significantly reduced food intake should be given access to fresh water in the enclosure.

5.2 Storage and handling of fish

Fish fed to pinnipeds should be of a quality suitable for human consumption. To retain its nutritive value and quality, fish should be quick-frozen, packed in containers that are impervious to air and moisture, labelled with the date of packaging and stored at or below –23°C (Chrissey 1998; Geraci 2000a). Storage time should not exceed 6 mo as, even under optimum freezing conditions, fats become rancid due to lipid peroxidation. High-fat fish, such as mackerel and herring, deteriorate more rapidly and should be stored for a maximum of 4 mo.

To minimise nutrient loss, dehydration, lipid peroxidation and rancidity, microbial contamination and loss of palatability, the preferred method of thawing is in a refrigerator at 4–6°C (Chrissey 1998; Geraci 2000a).

Fish can be thawed in cool potable water (21°C or below), preferably under running water, but this method results in increased leaching of nutrients (Chrissey 1998). Fish should not be thawed at room temperature or left soaking in water as this will reduce quality and lead to increased microbial contamination. Once thawed, fish should be refrigerated or kept on ice. Unused fish should be discarded after 24 hr. Fish should not be thawed and refrozen.

Fish are the intermediate hosts for many parasites of marine mammals, particularly helminths. Freezing is one of the most effective means of killing these parasites and the effectiveness of this method depends on several factors: temperature of the freezer, length of time the fish is held frozen, type of parasite and mass of fish in the container/box. Cestodes are more susceptible to freezing than nematodes and trematodes appear to be more resistant than nematodes. Thus the freezing time required to kill different parasitic species varies and all parts of the fish must be frozen at a low enough temperature for long enough that all parasitic larvae are killed. The lower the freezing temperature, the less time is needed to kill the parasites. The US Food and Drug Administration recommends that freezing and storing of fish at –20°C or below for 7 d, or freezing and storing at –35°C or below (commercial blast-freezing) for 15 hr, is sufficient to kill parasites (FDA 2001). All parts of the fish must reach these freezer temperatures for the recommended duration. Some bacterial pathogens found in fish can survive freezing temperatures and resume growth when thawed. This highlights the importance of using thawing techniques that do not create conditions favourable for pathogen growth. In most cases pathogen growth is very slow below 10°C, and 4.4°C is below the minimum growth temperature of most pathogens. Pathogen growth is relatively fast at temperatures above 21.1°C (FDA 2001).

5.3 Feeding

When formulating diets for pinnipeds, age, species, climatic conditions and body condition of the animal need to be considered. The optimal daily intake for young growing animals is 8–15% of body weight and for adults 4–8% of body weight, depending on the fat content of the fish. Whole fish should be fed whenever possible, as cut or eviscerated fish have lower nutritional value. The total daily ration is better tolerated if spread over the course of a day, preferably divided into three feeds. Occasional reduced intake or refusal of one or two meals is not a concern. In fact, well-blubbered animals can safely go without food for several days, or reduced food intake for several weeks. It is normal for food intake to decrease significantly during moult or the breeding season; some species, e.g. the southern elephant seal, fast during moult. Fasting also occurs during lactation in some species, e.g. leopard seals and southern elephant seals. It is very important that food type, amount eaten and feeding behaviour are recorded daily for each individual. The amount of fish eaten per feeding session is calculated by subtracting the weight of leftover fish from the total amount in the bucket prior to feeding.

Most pinnipeds eat more readily in water than on land. Captive pinnipeds are accustomed to frozen–thawed fish fed by hand, but free-living pinnipeds brought into captivity will rarely feed voluntarily. Whole fish can be thrown or offered by hand to the animal. If this is unsuccessful, fish can be left on the enclosure floor or in the pool but must be removed within 12 hr. It is common for one to several days to pass before newly captive pinnipeds voluntarily take fish. Animals that do not begin to take fish voluntarily after several days, or are severely debilitated, will require force-feeding (see 5.5). Live fish can be used to initiate feeding if an animal refuses to accept frozen–thawed fish and is considered too dangerous to force-feed. However, this method is not commonly employed due to a lack of consistent supply of appropriately sized fish.

5.4 Nutritional supplementation

Some species of fish commonly fed to captive pinnipeds contain endogenous thiaminases that destroy thiamine (vitamin B1). Vitamin E is depleted in fish by oxidation of polyunsaturated fatty acids during frozen storage. Therefore, pinnipeds fed frozen fish diets should be supplemented with 25–30 mg thiamine per kg of fish fed and 100 IU vitamin E per kg of fish fed (Geraci 2000a). Thawing of fish, particularly in water, results in leaching of vitamins, minerals and salts. Commercially available (Aquatic Mammal Chewables, Vetafarm; Vita-Zu mammal tablets, Mazuri) or compounded multivitamin-mineral supplements 2–3 times per week will typically compensate for vitamin and mineral losses in a frozen–thawed fish diet (Geraci 2000a). In addition, pinnipeds housed in fresh water must be supplemented daily with 2–3 g sodium chloride/kg of fish fed to prevent hyponatraemia (Geraci 2000a). All of these supplements are available in tablet or capsule form and can be inserted into the fish via the gill flaps. Put the daily supplement requirement, based on expected total food intake, into one or two fish and offer the medicated fish prior to the main feed.

5.5 Assist-feeding

For assist-feeding, the animal is restrained (see 6.1) and the mouth is opened using soft rope, strong towel strips or climbing tape on the upper and lower jaws. Whole fish are placed head first into the back of the mouth using tongs or forceps, not hands. The animal will usually swallow voluntarily if the fish is placed correctly. Massaging the throat can encourage and assist swallowing. Once swallowed, introduce the next fish. Start with only a few fish at each feed and gradually increase the number of fish. With successive feeds the seal will take fish more readily and the level of restraint can be reduced. Generally, pinnipeds will feed voluntarily within 1–2 wk. Assist-feeding is usually a multi-person task and force-feeding of adult pinnipeds can be difficult and dangerous.

Fluids, milk formulas and fish gruel can be administered by stomach tube. The animal is restrained and the mouth is held open as described for force-feeding. A gag with a central hole is placed in the mouth and a stomach tube is passed through the central hole into the pharynx. Allowing the animal to assist in swallowing the tube reduces the risk of passing it into the trachea. If the tube is in the oesophagus, it can usually be palpated in the neck of young animals, but is frequently not palpable in adults. Volumes of fluid that can typically be safely given by stomach tube are 100–200 mL for young pups (<10 kg), 200–400 mL for older pups (10–20 kg), 1–2 L for subadults/adults less than 100 kg and 2–4 L for adults over 100 kg.

Pinnipeds undergoing rehabilitation that are in poor body condition as a result of not having eaten for an extended period are likely to have compromised gastrointestinal function and should be slowly reintroduced to food. For the first 24–48 hr they should be stomach tubed with fluids only (water or electrolyte solution), then semi-solid food such as fish gruel or Hills a/d diet (Hill's Pet Nutrition) can be introduced, starting with a small quantity mixed with the fluids. The volume of semi-solid food per feed can be gradually increased over 48–72 hr. If no problems have arisen, solid food can be gradually introduced.

6 RESTRAINT

6.1 Capture and physical restraint

The choice of restraint technique will depend on species, size and demeanour of the seal, availability and skill of handlers, ease of access to the animal, proximity to water and other hazards, ambient temperature and intended type of procedure. Physical restraint can often be used for minor procedures, injections or assist-feeding, or for restraint for administration of chemical immobilising agents. Capture of large pinnipeds should be well-planned and performed by experienced handlers because of the difficulty and danger involved. Pinnipeds, particularly otariids, are faster and more flexible than they appear. It is essential that all handlers have a clear escape route in mind at all times and remember that their escape might be slowed by slippery, uneven or sandy substrate. Some pinniped species, particularly fur-seals, flee to available water when they are approached. Pinnipeds are susceptible to hyperthermia during prolonged manual restraint in warm conditions and to hypothermia if unable to express thermoregulatory behaviours in cold, wet or windy conditions. Excessive pressure over the thorax during physical restraint can compromise breathing.

There are several methods commonly employed for physical restraint. Herding boards are very useful to facilitate the movement of pinnipeds from one area to another or into a crate or squeeze cage while providing some protection for handlers. Injections or quick procedures can be performed while using herding boards to pin a seal against a wall, behind a gate or in a corner. Barriers should be high and smooth as pinnipeds often climb the barrier or may damage their teeth by biting it.

Physical restraint can be hazardous to handlers as pinnipeds are quick and agile and inflict severe bites. Therefore, the capture and restraint of any pinniped should be undertaken with extreme care and only by experienced handlers. Additional restraint can be achieved by placing a head bag, which is a hessian or ventilated canvas bag on a net hoop, over the head to reduce vision and, if deep enough, to pin the pectoral flippers against the body (Fig. 17.3). The head and neck can be briefly restrained further by pinning with the net hoop or a concave handlebar attachment on a pole, though care must be taken not to impede respiration.

Figure 17.3 Capture of a southern elephant seal using a head bag. Photo: Corey Bradshaw.

Seals under 50 kg can generally be restrained by throwing a towel or head bag over the head, then quickly grabbing the neck just behind the base of the skull, with both hands from behind. The head and neck are then pushed to the ground while the handler straddles the body, using the knees to hold the pectoral flippers tightly against the thorax. It is important to hold the pectoral flippers off the ground, as this markedly reduces the animal's ability to roll or lift itself. If needed, a second person can hold the pectoral flippers against the animal's side. This method of capture should only be attempted by experienced handlers due the risk of being bitten. A net can be used when it is too dangerous to capture a seal by this method or if handlers are inexperienced.

Seals up to 100 kg can be caught with a net. The ideal net design is conical, with a hole in the apex through which the muzzle can protrude, and with proportions similar to those of the seal so that the seal runs into the net and becomes firmly encased. The apex of the net, which covers the head, should ideally be made of canvas to impede the animal's vision but be open at the tip to allow clear breathing. Approaching quietly and slowly from behind, the large net or hoop net is thrown quickly over the seal and pulled in around the seal as much as possible. An additional head bag can be placed over the head, if needed. The animal must then be quickly restrained: one person grabs the neck and pins the head to the ground, another person grabs the pectoral flippers and holds them firmly against the animal's side and, if needed, a third person restrains the hind flippers. Pinnipeds of this size should not be straddled by handlers while being captured and restrained, as it can be dangerous to both the animal and handlers. Pinnipeds often struggle vigorously when initially netted, which can result in handlers being thrown and injured. Although this method can be used for pinnipeds over 100 kg it becomes more hazardous. Chemical restraint should be considered for larger pinnipeds. Administration of a sedative, such as a benzodiazepine, prior to capture can assist physical restraint in situations where full chemical restraint is undesirable.

Purpose-designed squeeze cages can be useful for larger pinnipeds. Restraint boards can also be used: the animal is placed on a board and restrained by several strong straps across the body. Restraint boards can be incorporated into the bottom of squeeze cages; once the seal is strapped to the board, the squeeze cage is opened and the seal is removed on the board. Restraint boards can also be used in field situations.

6.2 Chemical restraint

6.2.1 Practical considerations

Chemical restraint demands as much planning as physical restraint. Pinnipeds can be expected to flee or attack following drug administration, risking entry to the water or misadventure on cliffs, ledges, pool edges or abrasive substrates. For more aggressive species or individuals, pursuit of handlers can sometimes be used to lure animals away from such threats. Pinnipeds can be blocked from hazards by herding boards, or they may be physically restrained for drug administration. Pools should be emptied and covered or fenced off, or the procedure performed in the bottom of the pool. When physical restraint is not used, the risk of flight can be reduced by not crowding close to the animal, keeping a low profile, avoiding eye contact and using low-impact light-weight gas injection darting systems (Lynch et al. 1999; Higgins et al. 2002). In general, fur-seals are more easily disturbed than sea-lions and phocids. Interaction by conspecifics, especially during the breeding season, can pose a threat to the animal and to handlers, and can impede anaesthetic induction by stimulating the patient.

6.2.2 Drug administration

Inhalation agents offer the best control of anaesthetic depth but require physical restraint if used as induction agents. IV induction of southern elephant seals is routinely performed with highly reproducible results;

during manual restraint the agent is administered into the intravertebral extradural vein (see 7.1.2). IM injections are generally administered into the lumbar or gluteal muscles. Long needles are required to penetrate the blubber; minimum needle lengths of 3.8 cm are needed for smaller pinnipeds and up to 10 cm for large southern phocids. Hand injection can be performed using a needle attached to a pre-loaded 1.2 m extension tube. The needle is placed, the operator then retreats and injects the drug swiftly and flushes the tubing with saline to ensure the entire dose has been administered. This useful technique can often be performed with only herding boards for restraint. Administration by dart is the least reliable method, but has the advantage of not requiring physical restraint. It is therefore useful with intractable animals, when insufficient skilled handlers are available or if physical restraint poses a risk to the seal or handlers due to heat stress or other physical hazards. Darts should generally be fired into the lumbar or gluteal musculature, as the skin and blubber are thinner in this area. They should hit perpendicular to the body surface to avoid ricochet or deposition of drug into the blubber, and gun pressure should be moderated to prevent the dart from rebounding. Light-weight air-pressurised plastic darts elicit least response from the animal (Higgins et al. 2002).

6.2.3 Anatomical and physiological complications

Pinnipeds should not be considered as a single group for the purposes of chemical restraint. Anatomy and physiology differ between otariids and phocids and, for some drugs, response to dose and complications differ even among phocid species. With a few exceptions, chemical restraint methods for phocids should still be considered developmental, and there is little knowledge of the pharmacokinetics and physiological effects of most drugs in these species.

Cardiorespiratory emergencies can occur in any species, although predispositions vary depending on the species and chemical restraint method used. Although adaptations allow pinnipeds to survive periods of apnoea during diving (see 2.3), they should not be relied upon to function normally under the influence of anaesthetic agents. The pharmacokinetics of anaesthetic agents may also be affected by apnoea (Woods 1994). The degree of anatomical and physiological adaptation varies considerably between species—anaesthetised southern elephant seals commonly tolerate 8–20 min periods of apnoea during anaesthesia (Slip & Woods 1996; Woods et al. 1996), while shorter-diving leopard seals, crabeater seals and otariids appear no more tolerant than terrestrial mammals. However, any apnoea should be cause for concern and increased diligence, as in terrestrial mammals. As movement of the nares and thoracic wall sometimes occur without sufficient inspiratory volume, diligent monitoring of blood oxygen saturation (pulse oximetry), end tidal CO_2 (capnography) and expired air volume is essential (Lynch et al. 1999; D Higgins *pers. obs.*).

The ability of the phocine trachea to collapse dorsoventrally often predisposes these species to inspiratory obstruction, especially when the cervical musculature is relaxed. In addition, large or male phocids and otariids are prone to respiratory obstruction by pharyngeal and palatine soft tissue. The flexible thoracic structure and elastic pulmonary interstitium of pinnipeds makes larger animals particularly susceptible to progressive pulmonary collapse and ventilation–perfusion mismatch and, although the literature is equivocal, Robin et al. (1963) suggest that phocid respiration is poorly stimulated by hypercapnoea. This, and observations that end tidal CO_2 readings of otariids and phocids tend to rise throughout extended procedures despite regular respiratory rates (J Barnes, M Lynch & D Higgins, unpub), suggest that positive pressure ventilation should be seriously considered where procedures exceed 20–30 min (i.e. where the improvement of ventilation outweighs the disadvantages of prolonging the procedure by intubation, the greater depth of anaesthesia required for intubation and the risk of stimulating vomition in non-fasted free-ranging pinnipeds). In addition, end tidal CO_2 readings (without associated tachypnoea) of Australian sea-lions immediately after gaseous induction appeared to be higher in animals that had been physically restrained for longer before induction (D Higgins *pers. obs.*). Capnography and positive pressure ventilation should be considered for anaesthetised pinnipeds where manual restraint has been vigorous or prolonged.

The possible influence of physiological adaptations for diving in the management of pinniped anaesthesia remains poorly defined. These adaptations, known collectively as the dive response, are most developed in the deep-diving phocids. Literature on the phenomenon is reviewed by Lynch et al. (1999). The response comprises apnoea, bradycardia, peripheral vasoconstriction and redistribution of blood flow so that circulating blood is mostly limited to vital organs such as the heart and brain and pooling of blood in the caudal vena cava and hepatic sinus. Although bradycardia is vagally mediated and several authors have suggested the use of the parasympatheticolytic drug atropine as a preventative, its efficacy has not been formally assessed and the authors found no apparent disadvantage to omitting it. The dive response is often reported to be a cause of death during anaesthesia, however, without detailed monitoring equipment the presentation of apnoea and bradycardia is often difficult to distinguish from that of cardiorespiratory depression from other causes, e.g. profound anoxia or pharmacological depression.

All pinnipeds have low surface area to mass ratios and either subcutaneous blubber or dense fur coats. They rely on active peripheral vasodilation for cooling and anaesthetic agents can interfere with these mechanisms. Therefore, pinnipeds are susceptible to hyperthermia during anaesthesia and body temperature should be monitored by feeling the skin, especially the flippers, for hot areas. Ideally a rectal thermometer should be used to measure body temperature; in larger animals a long probe may be required for accurate measurement of core body temperature. Animals can be cooled by application of cold packs to the extremities and axillae, wetting the animal and using a fan. Extremes of cold can induce hypothermia (Gales et al. 2005), so thermal blankets and windbreaks should be kept on hand in cold environments.

6.2.4 Cardiopulmonary support and resuscitation

As apnoea and tracheal collapse are common causes of respiratory failure, particularly in phocids, in the authors' experience endotracheal intubation alone is often inadequate to restore ventilation. Interestingly, endotracheal intubation has induced apnoea in southern elephant seals (Woods et al. 1996). Essential equipment for chemical restraint should include a method for positive pressure ventilation, e.g. an endotracheal tube and breathing bag or oxygen demand valve. Ideally, oxygen should be available for emergencies.

Visual endotracheal intubation is often impeded by the location of the larynx, which is caudal relative to many carnivores and is obscured by the soft palate and pharyngeal folds. Manual placement of the tube by feeling the laryngeal opening with the hand is the most efficient means in all but small pinnipeds, although for safety the animal must be completely anaesthetised and a gag firmly placed. In otariids, the tracheal bifurcation into the primary bronchi is more cranial compared to other mammals (King 1983), occurring at the thoracic inlet and not at the pulmonary hilus, as occurs in phocids (Rommel & Lowenstein 2001 and references therein). Therefore, in otariids the endotracheal tube should be placed just beyond the larynx, the cuff inflated and bilateral ventilation confirmed by thoracic auscultation. In the authors' experience, ideal tube placement varies among phocids. In leopard seals, the collapsible nature of the trachea makes caudal placement desirable to bypass tracheal collapse during spontaneous inhalation. However, where an airtight seal is needed for positive pressure ventilation, the large tracheal diameter, relative to that of the larynx, makes placement of the inflatable cuff immediately behind the larynx more effective unless the cuff is very large.

Reports describing the use of cardiac and respiratory stimulants are uncommon and equivocal. The authors and Gales et al. (2005) have successfully stimulated respiration by administration of doxapram (1–2 mg/kg) IV in pinnipeds with adequate cardiac function. Peripheral venous access is often difficult to obtain during an emergency, particularly in otariids, and a jugular cut-down may be necessary to obtain reliable IV access. Peripheral administration of cardiac and respiratory stimulants to pinnipeds during cardiorespiratory collapse is often ineffective, although it is difficult to determine whether poor drug distribution is due to the dive response or simply to anoxic or pharmacological reduction of cardiac function. In such animals, Woods et al. (1996) and Higgins et al. (2002) report improved efficacy with intratracheal administration of doxapram (1–4 mg/kg) and adrenaline (1:1000, 0.02–0.06 mL/kg). Management of cardiopulmonary emergencies in southern elephant seals under field conditions has been described by Woods (1994).

It is not uncommon for phocids and otariids to become apnoeic during recovery. These animals generally respond well to stimulation, rocking, or intubation and ventilation provided the problem is detected before cardiac function is compromised. Close monitoring should be continued through the entire recovery phase, and extubation should be performed as late as safely practical.

6.2.5 Inhalation anaesthesia

Mask induction of anaesthesia with isoflurane in oxygen is becoming the method of choice for chemical restraint of pinnipeds that can be safely physically restrained, or sedated then physically restrained. Females of all ages and juvenile to subadult male otariids are commonly anaesthetised by this method and recent studies report induction of crab-eater seals (Gales et al. 2005), subadult leopard seals (J Barnes unpub) and Weddell seals (Bodley et al. 2005) with isoflurane 5% in oxygen 10–15 L/min. Typically, isoflurane 1.5–3% via mask or endotracheal tube is suitable for maintenance. Prior to induction, both phocids and otariids can be sedated with midazolam (0.25–0.5 mg/kg IM) to facilitate physical restraint. Significantly fewer complications are seen than in most studies using injectable agents alone. They include transient apnoea, which can be remedied by physical stimulation, and hypothermia in Antarctic conditions.

Field procedures necessitate equipment that is robust and portable and, in Antarctic conditions, heated (Gales et al. 2005). However, in the controlled environment of a zoo or rehabilitation facility, it is feasible to use a standard vaporiser and closed circuit of appropriate capacity. Although pinnipeds use a greater percentage of their tidal volume at rest than do other mammals, their tidal volume is similar to that of other mammals (10–20 mL/kg).

6.2.6 Injectable agents: phocids

Tiletamine/zolazepam has been used for anaesthesia of thousands of southern elephant seals with few complications. IV administration into the intravertebral extradural vein (0.54 mg/kg) to over 1000 seals immobilised them within 1 min and lasted 20 min without fatality and without apnoea exceeding 5 min (Field et al. 2002). Thinner seals had longer recovery times. IM administration (1 mg/kg) to 15 seals produced heavy sedation in approximately 10 min, lasting approximately 10–50 min without fatality (Woods et al. 1994b).

Adult female Weddell seals have been effectively immobilised with IV tiletamine/zolazepam (0.50–0.65 mg/kg) without fatality or pronounced apnoea, whereas there was a 25% mortality rate with IM administration. A starting dose of 0.50 mg/kg IV is recommended for other age classes of Weddell seal and other phocid species (Wheatley et al. 2006).

Tiletamine/zolazepam IM (1.2–1.4 mg/kg) has been used successfully in approximately 50 leopard seals. Higgins et al. (2002) describe much more reliable induction and better airway maintenance than during chemical restraint with midazolam/pethidine. Induction is smooth, taking 19 min, anaesthesia lasts approximately 20–30 min and recovery is smooth, with mobility returning after cognitive function—an advantage in avoiding misadventure. The authors have used tiletamine/zolazepam (1–1.25 mg/kg IM) six times for induction of captive leopard seals prior to maintenance by inhalation anaesthesia (isoflurane) for up to 3 hr with one mortality occurring during an apparently normal recovery. Mortalities in free-ranging leopard seals associated with the use of tiletamine/zolazepam alone were two during apparently normal recoveries and one following accidental IV administration. All three seals that died during recovery were partially mobile and apparently breathing well, before apnoea occurred and was followed swiftly by cardiac arrest. The three seals had moderate to marked pulmonary congestion with interlobular oedema at necropsy. The findings concur with reports of resistance to positive pressure ventilation in these animals. We speculate that signs might be due to pulmonary hypertension but the cause of this is unknown and further investigation is needed to understand the pathophysiology during anaesthesia. Accidental IV administration induces rapid anaesthesia and profound apnoea; intubation and ventilation is sufficient to maintain the animal, but must be instituted rapidly. Differentiating induction and inappropriate apnoea from sleep and normal apnoea respectively can be difficult, and it is sometimes necessary to lightly rouse the animal to determine the difference.

Midazolam/pethidine combination has been used to chemically restrain southern elephant (0.04 and 4 mg/kg IM respectively), leopard (0.22 and 1.15 mg/kg IM) and crab-eater seals (0.25–0.4 and 2.4 mg/kg IM) (Woods et al. 1994a; Higgins et al. 2002; Tahmindjis et al. 2003). Response is variable, induction is inhibited by stimulation and the margin of safety appears to be small. In leopard seals, and to a lesser extent crab-eater seals, profound relaxation of the cervical musculature often causes partial to complete airway obstruction, even in animals that are insufficiently anaesthetised to be intubated. Compulsive locomotion is not uncommon and mobility can occur beyond loss of cognition, increasing potential for misadventure during induction and recovery.

A common problem encountered when using IM agents, especially when administered by dart, is incomplete induction. Presumably this results from deposition of some of the drug between muscle planes or, in the case of darts, deposition of drug into blubber as the dart transits this layer or fails to reach muscle. Fatalities have occurred when animals are given supplemental doses IM, presumably because agent deposited in tissues is absorbed later, leading to subsequent overdose. To deepen or maintain anaesthesia a different and more controllable agent, e.g. isoflurane in oxygen, is ideal. IV ketamine (1–3 mg/kg to effect) has been successfully used in a range of phocids, albeit in small numbers (Baker et al. 1988; Woods et al. 1994a; Tahmindjis et al. 2003; J Barnes unpub). Some mortalities have occurred during maintenance of anaesthesia with incremental IV tiletamine/zolazepam, but this has not been studied thoroughly.

Midazolam alone (0.25–0.5 mg/kg IM) is used commonly to facilitate physical restraint for minor procedures, or prior to induction with inhalation agents. Major determinants of the extent of sedation are the completeness and accurate placement of the injection and the seal's degree of excitement prior to and during induction. Once restrained, if the level of sedation is insufficient, supplemental IV doses (0.1–0.2 mg/kg) have been used effectively to increase the depth of sedation (J Barnes unpub). Diazepam 0.1–0.5 mg/kg IM or IV can be used for mild sedation. Oral diazepam (0.2 mg/kg) has been used with unpredictable efficacy for light sedation in captive phocids (Gales 1989).

Benzodiazepines (midazolam and diazepam) can be antagonised with flumazenil 0.002–0.01 mg/kg IV or IM. More precise doses are yet to be investigated and determined. Opiates can be antagonised with naloxone 0.012 mg/kg or naltrexone 0.05–0.12 mg/kg IM. These drugs hasten recovery but reversal can be incomplete, restoring physical mobility without full return of cognitive function and thus increasing the potential for misadventure (Higgins et al. 2002; Tahmindjis et al. 2003). The duration of action of these compounds in pinnipeds is unknown. These antagonists are therefore best suited to hastening recovery in animals that are stable and well-oxygenated and can be observed beyond the expected recovery time for non-reversed animals, or for reversal of specific drug effects where anaesthetic control can be maintained by other means (e.g. inhalation agents or IV ketamine).

6.2.7 Injectable agents: otariids

Tiletamine/zolazepam (1.2–2.0 mg/kg IM) has been the most commonly used injectable induction agent in Australian otariids, producing reliable chemical restraint for approximately 20 min. All animals, but especially large animals, should be observed closely for pharyngeal obstruction during induction and are best intubated as soon as possible.

In recent years, medetomidine/ketamine combinations have been commonly used for induction of otariids. The authors have successfully used medetomidine (70–100 μg/kg) and ketamine (2 mg/kg) IM in Australian and New Zealand fur-seals, with successful reversal following administration of atipamezole (350 μg/kg IM). Medetomidine (140 μg/kg) and ketamine (2.5 mg/kg) IM (Haulena et al. 2000) and medetomidine (70 μg/kg) and tiletamine/zolazepam (1 mg/kg) IM (Haulena & Gulland 2001) have been used successfully in northern otariid species, using atipamezole (200 μg/kg IM) for reversal. A combination of medetomidine (13 μg/kg), midazolam (0.25 mg/kg) and butorphanol (0.4 mg/kg) administered IM produced light anaesthesia in aged California sea-lions and was effectively reversed with atipamezole (0.06 mg/kg), flumazenil (0.002 mg/kg) and naltrexone (0.1 mg/kg) IM (Spelman 2004). Although there are no reports of use of this combination in Australian otariids, it may merit consideration.

Midazolam (0.25–0.5 mg/kg IM) can be used to facilitate restraint for minor procedures or prior to induction of inhalation anaesthesia. It produces more reliable sedation following IM administration than diazepam and has a shorter duration of action.

7 CLINICAL PATHOLOGY

7.1 Blood collection

7.1.1 Otariids

Blood collection from otariids can be difficult due to the limited accessibility of peripheral veins. The most commonly used venipuncture sites are the caudal gluteal vein, the interdigital veins of the hind flippers, a brachial vein and the jugular vein.

The caudal gluteal vein is deep within the gluteal muscles and can be difficult to locate. It runs caudally, lateral to the sacrum. To sample this vein the animal is restrained in sternal recumbency with the hind flippers

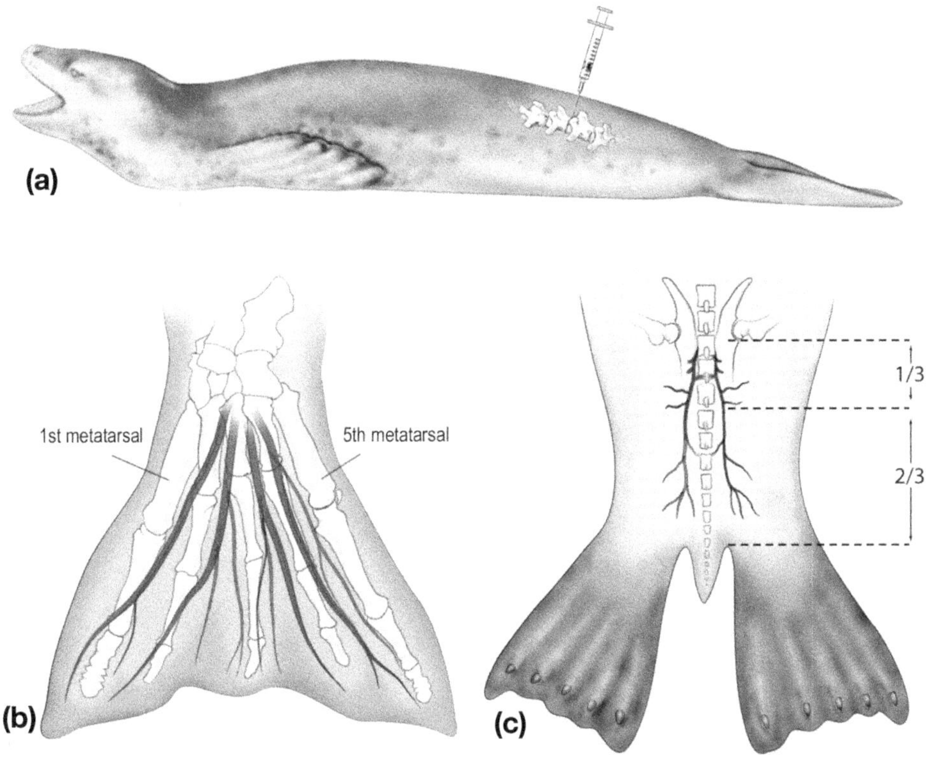

Figure 17.4 Sites for blood collection in otariid and phocid seals. a) Intravertebral extradural vein in a phocid seal. b) Dorsal view of the left hind flipper showing interdigital veins which can be used in a phocid or otariid seal. c) Caudal gluteal vein in an otariid seal. (Source: Illustrations adapted from King (1983), Geraci & Lounsbury (1993), Sweeney (1993) and Bossart et al. (2001) and based on drawings by Bozena Jantulik.)

spread. A needle is inserted perpendicularly, approximately one-third of the distance from the femoral trochanter to the base of the tail, just lateral to the sacral vertebrae (Geraci & Sweeney 1978) (Fig. 17.4). If the vein is not immediately found, the needle can be slightly withdrawn and 'walked' to either side (Clark et al. 2004). In smaller animals (<25 kg) a 20–22 G, 2.5 cm needle is used, in animals 25–100 kg an 18–20 G, 3.8 cm needle is used and in larger animals longer needles will be required.

The interdigital veins can be visualised on the dorsal surface of the webbing of the hind flipper. Application of a tourniquet around the base of the hind flipper and warming the flipper can improve visualisation of these veins. The veins are small and the blood flow slow, so a heparinised butterfly catheter can aid blood collection (Gulland et al. 2001). An alternative site in the hind flipper is just proximal to the origin of the interdigital webbing on the dorsal surface. The needle is inserted at a shallow angle centrally in the inverted V that forms the proximal point of the interdigital webbing (junction of furred and non-furred area) in any of the interdigital spaces (K Bodley *pers. comm.*). Sometimes the vessel can be palpated between the index finger and thumb, which aids location of the venipuncture site, but more often the vessel is not palpable and must be attempted blindly. A better blood flow rate is obtained from this vessel than from the small vessels within the interdigital webbing. The use of a tourniquet and butterfly catheter and warming the flipper also improve the success of venipuncture at this site.

Another useful vein is the brachial vein, which is found on the medial side of the fore flipper in the loose tissue on the trailing edge of the flipper between the elbow and carpus (Fig. 17.5). It is usually an easily visualised vessel that is useful for venipuncture, catheter placement and IV drug administration in otariids. Rolling the animal onto its side or back and placing the flipper perpendicular to the body allows access to this vein. The vessel can be easily accessed in pups and physically restrained or sedated small pinnipeds, but large otariids require anaesthesia unless animals are trained for the behaviour.

Figure 17.5 Brachial vein—site for blood collection and IV catheter placement in otariid seals.

The jugular veins, which run from the angle of the jaw to the thoracic inlet, can be used for venipuncture but are smaller in pinnipeds than in other species (Harrison & Tomlinson 1963). The jugular vein is generally not palpable or visible except in pups and sometimes in adult animals in poor body condition. The vein can usually be located in the jugular groove in the mid-cervical region on a line running between the commissures of the lips and the point of the shoulder (Gage 2003). For emergency venous access in otariids, the subclavian vein can be located by inserting a needle perpendicularly into the angle between the sternum and the first rib (Gulland et al. 2001), or the cranial vena cava can be located by inserting a needle between the first two ribs close to the midline (Hubbard 1968). Both methods require dorsal recumbency and are not recommended for routine venipuncture. The intravertebral extradural vein (see 7.1.2) is generally not a recommended venipuncture site for otariids.

7.1.2 Phocids

The preferred site for venipuncture in phocids is the lumbar portion of the intravertebral extradural vein, which is a large valveless vein located dorsal to the spinal cord within the vertebral canal (Harrison & Tomlinson 1956, 1963). The seal is placed in sternal recumbency and an intervertebral space is located by palpating along the dorsal midline of the caudal lumbar vertebrae (L3–L5). A needle is inserted into the intervertebral space perpendicular to the skin until blood is observed in the hub of the needle (Fig. 17.4). For pups and smaller phocids (<50 kg) a 20 G, 2.5–3.8 cm needle is used and for larger phocids an 18 G, 8.9 cm spinal needle is used. Large Antarctic phocids may require needles that are 12.7 cm or longer. Large volumes of blood can be readily obtained.

The plantar interdigital veins of the hind flipper form a rich vascular network in the metatarsal region just proximal to the origin of the interdigital webbing (Geraci & Lounsbury 1993) and can be used for blood collection. These veins are located by inserting a 20–21 G, 2.5–3.8 cm needle at a 10–20° angle to the skin over the second digit or medial to the fourth digit, at the origin of the interdigital webbing (Geraci & Sweeney 1978) (Fig. 17.4). Interdigital veins are not visible on the surface of the webbing in phocids as the flippers are covered with fur.

7.2 Haematology and biochemistry

Reference ranges for haematology and biochemistry in pinniped species, particularly those in the southern hemisphere, are limited by sample size, especially within age or sex cohorts. Captive animals are often sampled during periods of clinical disease or abnormality, so results cannot be used to develop species-specific reference ranges and it is best that individual baseline values be obtained for captive animals (Bossart et al. 2001). Published values for haematology and biochemistry in free-ranging pinnipeds and non-published values for clinically normal captive animals are shown in Tables 17.3, 17.4 and 17.5. For unpublished values from captive pinnipeds, results are included only for individuals considered to be of normal health status. Repeat samples from the same individuals are included if sample collection times were sufficiently spaced (>6 mo). For captive animals, the range of observed values has been provided, while the mean and standard deviation are provided when the sample size exceeds six.

Most marine mammals have larger erythrocytes, lower erythrocyte counts and higher haematocrit and haemoglobin concentrations than terrestrial mammals (see 2.3). Larger erythocytes, and their variability in size, introduces potential error when using electronic cell counters for red blood cell counts, MCV and haematocrit measurements of marine mammals unless the analyser is calibrated. Haematocrit should therefore be determined by the microhaematocrit method in addition to automated methods (Bossart et al. 2001).

Table 17.3 Published values for haematology of free-ranging and non-published values for clinically normal captive Australian pinnipeds

Parameter	Australian fur-seal (captive) Adults (n = 12) Subadults (n = 2)	Australian sea-lion[1] Lactating females (n=23)	Australian sea-lion[1] Yearlings (n = 8)	Australian sea-lion[1] Subadult males (n = 7)	Australian sea-lion (captive) Adults (n = 3) Subadults (n = 2)	NZ fur-seal (captive) Adults (n = 3) Subadults (n = 4)	Antarctic fur-seal[2] (n = 2)	Leopard seal (captive)	Leopard seals in Antarctica[5] Adults	Leopard seals in NSW Subadults at time of haul-out[5]
PCV (L/L)	0.50 ± 0.04 0.44–0.59	0.56 (0.48–0.64)	0.58 (0.54–0.62)	0.57 (0.54–0.59)	0.44–0.58 0.31–0.43	0.48–0.55 0.47–0.52	0.46–0.52		0.47 ± 0.03 (n = 28)	0.45 ± 0.07 (n = 21)
Hb (g/L)	171 ± 28.1 158–213	187 (162–210)	196 (175–209)	193 (184–203)	146–183 100–111	156–190 132–193		196[3] 178–186[4]	171 ± 13.0* (n = 14)	186 ± 30.3 (n = 12)
RBC ($\times 10^{12}$/L)		5.46 (4.77–6.08)	5.61 (5.05–5.99)	5.67 (5.71–5.88)			5.9–6.5	4.40[3] 4.37–5.21[4]		5.37 ± 1.40 (n = 6)
MCV (fl)		103 (96–112)	104 (100–109)	101 (97–103)			78–79	98.1–113[4]		83.3–120 (n = 3)
MCH (pg)								35.9–41.2[4]		32.0–42.3 (n = 3)
MCHC (g/L)	296–325 (n = 3) 357–362	33.5 (31.1–34.4)	33.5 (32.4–35.6)	34.3 (33.6–35.0)	298–382 (n = 2) 323 (n = 1)	269–302 (n = 2)		364–367[4]	358 ± 5.87* (n = 14)	387 ± 57 (n = 8)
WBC ($\times 10^9$/L)	8.33 ± 3.31 5.8 (n = 2)	11.3 (6.3–14.6)	11.4 (7.7–12.8)	11.2 (8.7–12.7)	3.96–6.82 6.71–9.90	5.83–11.2 5.28–7.59		4.41[3]	8.37(7.34, 9.55) (n = 27)	9.90 ± 3.99 (n = 21)
Neutrophils (%/$\times 10^9$/L)	65.8 ± 11.1/5.64 ± 2.75 63–70/3.7–4.1	56.9 (31–76)	51.5 (33–83)	60.1 (39–79)	40.9–66.1/ 1.62–4.5 61–75/ 4.09–7.43	56–87/ 3.79–9.76 56–79/ 3.01–5.04			59.2(55.6, 63.0)/ 4.95 (4.29, 5.72) (n = 27)	69 ± 14/ 7.01 ± 3.89 (n = 20)
Band Neutrophils (%/$\times 10^9$/L)									0.33 ± 0.62/ 0.03 ± 0.05 (n = 26)	4.30 ± 7.12/ 0.45 ± 0.90 (n = 20)
Lym-phocytes (%/$\times 10^9$/L)	25.7 ± 9.21/2.02 ± 0.73 15–32/0.9–1.9	29.6 (12–59)	39.6 (14–56)	25.7 (15–42)	28.0–51.0/ 1.3–2.02 20–32/ 1.98–2.15	10–43/ 1.12–4.58 16.9–40/ 1.08–2.88			16.6 ± 5.30/ 1.42 ± 0.58 (n = 27)	20.7 ± 11.2/ 1.93 ± 1.10 (n = 20)

(*Continued*)

Table 17.3 (*Continued*)

Parameter	Australian fur-seal (captive) Adults (n = 12) Subadults (n = 2)	Australian sea-lion[1] Lactating females (n=23)	Australian sea-lion[1] Yearlings (n = 8)	Australian sea-lion[1] Subadult males (n = 7)	Australian sea-lion (captive) Adults (n = 3) Subadults (n = 2)	NZ fur-seal (captive) Adults (n = 3) Subadults (n = 4)	Antarctic fur-seal[2] (n = 2)	Leopard seal (captive)	Leopard seals in Antarctica[5] Adults	Leopard seals in NSW Subadults at time of haul-out[5]
Monocytes (%/× 10^9/L)	4.99 ± 2.60/ 0.44 ± 0.30 1–2/0.1	2.0(0−7)	1.4(0−5)	0.9(0−2)	1.97−7.07/ 0.08−0.41 3−4/ 0.2−0.4	1−3/ 0.11−0.34 0−6/0−0.38			2.87 (2.18, 3.76)/ 0.24 (0.18, 0.31) (n = 27)	4.70 ± 5.00/ 0.47 ± 0.52 (n = 20)
Eosinophils (%/× 10^9/L)	3.42 ± 4.56/ 0.23–0.29 2–13/0.1–0.8	11.4(1−37)	7.5(1−13)	13.3(2−38)	0−1.01/ 0−0.04 1−4/ 0.09−0.27	0−2/0−0.12 (n = 2) 0−3/0−0.18			16.5 (13.9, 19.6)/ 1.35 (1.08, 1.69) (n = 26)	1.20 ± 3.40/ 0.11 ± 0.33 (n = 20)
Basophils (%/× 10^9/L)	0.09 ± 0.30/ 0.01−0.02 0–2/0−0.1				0 0	0 0−2.04/ 0−0.13			2.12 (1.42, 3.16)/ 0.19 (0.13, 0.28) (n = 27)	0.05 ± 0.22/ 0.01 ± 0.03 (n = 20)
Platelets (× 10^9/L)									343 ± 123[6] (n = 14)	

Mean ± standard deviation, range of values or mean (95% confidence interval) shown.
Sources: **1** Needham et al. (1980). **2** Fayolle et al. (2000). **3** Brown (1957) (n = 1). **4** Williams & Bryden (1993) (n = 2 captive). **5** Gray (2005) —(mean ± SD or 95% confidence interval).
* IDEXX.

Table 17.4 Published values for haematology of southern elephant seals

Parameter	Pups[1]	Pups (males; females)[2]	Young[3] (n = 5)	Female 2–3 yr[4] (n = 3)	Adults[1]	Adults (females)[2]	Adults[5] (n = 20)	Postpartum females[6]
PCV (L/L)	0.55 ± 0.04(n = 8)			0.39–0.73	0.59 ± 0.06(n = 44)	0.61–0.71(n = 5)	0.54 ± 0.04	0.56 ± 0.04(n = 12)
Hb (g/L)	207 ± 18.4(n = 8)	158 ± 22.0(n = 14) 162 ± 16.0(n = 12)	245 ± 19		233 ± 25.2(n = 43)		224 ± 14	233 ± 11(n = 6)
RBC (× 10^{12}/L)	2.75 ± 0.20(n = 3)	3.13 ± 0.68(n = 19) 3.06 ± 0.83(n = 15)		3.5–5.7	2.54 ± 0.31(n = 28)	2.70–3.72(n = 2)	12.5 ± 0.2	3.1 ± 0.6(n = 10)
MCV (fl)	197 ± 14.3(n = 3)			118–195	239 ± 27.1(n = 28)		213 ± 5.0	
MCH (pg)	72.4 ± 7.50(n = 3)				95.5 ± 10.0(n = 27)		88.1 ± 2.3	
MCHC (g/L)	37.6 ± 4.34(n = 8)				39.1 ± 1.45(n = 43)		41.3 ± 0.5	
WBC (× 10^9/L)		15.2 ± 4.38(n = 20) 14.2 ± 4.96(n = 16)				12.5–20.8(n = 4)	11.2 ± 2.9	
Neutrophils (%)							63.0 ± 9.0	
Lymphocytes (%)							17.0 ± 7.0	
Monocytes (%)							5.0 ± 2.0	
Eosinophils (%)							15.0 ± 9.0	
Basophils (%)								
Platelets (× 10^9/L)				125 (71.6–199)			241 ± 56.6	

Mean ± standard deviation or range of values shown.
Sources: **1** Lane et al. (1972). **2** Bryden & Lim (1969). **3** Seal et al. (1971). **4** Fayolle et al. (1972). **5** Melrose et al. (1995). **6** Lewis et al. (2001).

Table 17.5 Published values for serum biochemistry of free-ranging and non-published values for clinically normal captive Australian pinnipeds

Parameter	Captive Australian fur-seal Adults (n = 12) Subadults (n = 1)	Australian sea-lion[1] Female adults (n = 19)	Australian sea-lion[1] Yearlings (n = 11)	Australian sea-lion[1] Subadult males (n = 6)	Australian sea-lion (captive) Adults (n = 2) Subadults (n = 2)	NZ fur-seal (captive) Adults (n = 3) Subadults (n = 3)	Southern elephant seal[2] Young (n = 5)	Southern elephant seal[3] Postpartum (early lactation) females (n = 13)	Southern elephant seal[3] 2 d old pups (n = 13)	Leopard seal[4] (captive) Adult & subadult (n = 2)	Leopard seals in Antarctica[5] Adults (n = 29)	Leopard seals in NSW[5] Subadults at time of haul-out (n = 8)
Glucose (mmol/L)	5.02 ± 1.32	4.1 ± 1.07	4.6 ± 1.48	4.5 ± 0.53	3.7–5.0 6.83–7.12	3.11–14.1 (n = 2) 5.34–6.38		7.2 ± 1.1	7.9 ± 1.3		6.74 ± 0.57[a]	6.06 ± 1.91
BUN (mmol/L)	11.9 ± 2.18 10.4	11.3 ± 4.8	14.3 ± 7.72	7.7 ± 2.03	7.28–7.71 9.26–10.8	15.3–45.3 (n = 2) 11.7–16.6	17 ± 3	11.0 ± 1.2	11.1 ± 2.0		6.65 ± 3.07	8.25 ± 5.68
Creatinine (μmol/L)	106 ± 16.4 90	100 ± 30	80 ± 20	100 ± 20	80 65–75	90–160 50–60		138 ± 21	95 ± 9		93.4 ± 19.1	77.5 ± 38.5
Total protein (g/L)	71.2 ± 5.47 (n = 11) 63	69 ± 12.8	74 ± 3.9	73 ± 9.7	68–70 59–79	61–77 (n = 2) 61–67	76 ± 3	77 ± 6	62 ± 3		73.0 ± 8.05	73.6 ± 8.14
Fibrinogen (g/L)											1.52 ± 0.61[c]	
Albumin (g/L)	33.5 ± 1.75 (n = 11) 34				31–34 29–35	29–47 (n = 2) 34–38	38 ± 4.5	34 ± 2	39 ± 3		30.5 ± 3.04	31.6 ± 8.35
Globulin (g/L)	37.7 ± 4.80 (n = 11) 29				36–37 30–43	30–32 (n = 2) 24–30		43 ± 5	23 ± 1		42.5± 6.01	42.0 ± 13.7
Bilirubin (μmol/L)	2.11 ± 1.03 (n = 7) <5	3 ± 0.9	5 ± 4.0	3 ± 1.2	0.21–3.51	3.4 (n = 1) 0.80–0.9 (n = 2)		16 ± 9	8 ± 4		3.56 (2.98, 4.25)	6.04 ± 3.94
ALP (U/L)	59.8 ± 31.5 35	62 ± 23	157 ± 32	85 ± 26	41–77 92–105	29–215 136–412		76 ± 27	466 ± 159		75.0 (65.5, 85.8)	51.3 ± 16.7
AST (U/L)	66.3 ± 30.1 29	77 ± 41 (n = 14)	48 ± 17 (n = 9)	66 ± 9 (n = 4)	30–42 23–44	25–34 11–31		73 ± 13	63 ± 12		71.1 (58.8, 86.0)	74.4 ± 42.2
ALT (U/L)	58.2 ± 22.7 49				26–41 43–59	39.4–88 24–50		14 ± 11	31 ± 16		83.6 (60.0, 117)	50.5 ± 21.2

(Continued)

Table 17.5 *(Continued)*

Parameter	Captive Australian fur-seal Adults (n = 12) Subadults (n = 1)	Australian sea-lion[1] Female adults (n = 19)	Australian sea-lion[1] Yearlings (n = 11)	Australian sea-lion[1] Subadult males (n = 6)	Australian sea-lion (captive) Adults (n = 2) Subadults (n = 2)	NZ fur-seal (captive) Adults (n = 3) Subadults (n = 3)	Southern elephant seal[2] Young (n = 5)	Southern elephant seal[3] Postpartum (early lactation) females (n = 13)	Southern elephant seal[3] 2 d old pups (n = 13)	Leopard seal[4] (captive) Adult & subadult (n = 2)	Leopard seals in Antarctica[5] Adults (n = 29)	Leopard seals in NSW[5] Subadults at time of haul-out (n = 8)
CK (U/L)	302 ± 672 83	271 ± 263 (n = 13)	276 ± 204 (n = 8)	389 ± 139 (n = 4)	76–237 182–258	58.5–185 97–188		1096 ± 593	441 ± 383		242 (192, 306)[a]	
Cholesterol (mmol/L)	6.25 ± 0.80 (n = 8) 6.5	5.8 ± 1.4	6.7 ± 0.7	5.7 ± 0.6	7.41	3.8 (n = 1) 5.5–7.59 (n = 2)		8.8 ± 1.7	3.3 ± 0.8		6.67 ± 1.62	7.50 ± 2.17
Calcium (mmol/L)	2.41 ± 0.12 (n = 7) 2.55	2.23 ± 0.39	2.64 ± 0.11	2.39 ± 0.28	2.4	2.5 (n = 1) 2.3–2.5 (n = 2)		2.35 ± 0.12	2.71 ± 0.09		2.44 (2.36, 2.52)	2.49 ± 0.16
Phosphate (mmol/L)	1.88 ± 0.44 (n = 11) 1.48	2.14 ± 0.43	2.08 ± 0.43	2.54 ± 0.36	1.5–2.2 1.45–1.88	2.7 (n = 2) 1.6–2.05		1.61 ± 0.19	2.09 ± 0.27		2.49 ± 0.53	2.06 ± 0.34
Na^+ (mmol/L)	149 ± 3.02 (n = 7) 147	145 ± 17.5	154 ± 1.8	147 ± 9.3	148–153	149 (n = 1) 149–150 (n = 2)		146 ± 1	147 ± 2		152 ± 5.23[b]	153 ± 3.59
K^+ (mmol/L)	4.18 ± 0.54 (n = 8) 4.1	4.5 ± 0.75	4.3 ± 0.31	4.3 ± 0.32	3.5–4.4	4.2 (n = 1) 3.5–4.0 (n = 2)		4.0 ± 0.2	4.3 ± 0.5		4.21 (4.09, 4.33)	4.34 ± 0.78
Cl^- (mmol/L)	108 ± 2.19 (n = 7) 110	103 ± 10.1	109 ± 2.7	105 ± 4.1	108–114	108 (n = 1) 106–112 (n = 2)		102 ± 1	99 ± 2		112 (110, 114)[a]	106 ± 3.51
HCO_3^- (mmol/L)	26.7 ± 2.00 (n = 7) 23	15 ± 4.7	18 ± 4.4	15 ± 4.5	18.1–26.5	33.3 (n = 1) 21.7–27.7 (n = 2)				26.7 –28.3	25.7 ± 4.51	25.3 ± 5.98
Anion gap	18.3 ± 3.62 (n = 7)				17–25.3	11.9 (n = 1) 20.3 (n = 1)					19.6 (18.3, 20.9)	22.8 –33.9 (n = 4)
Amylase (U/L)	12 (n = 1)				0 (n = 1)						1.38 (0.92, 2.05)[a]	0–4.00 (n = 4)
Lipase (U/L)	51 (n = 1)				199 (n = 1)	15 (n = 1) 83–113 (n = 2)					353 (266, 467)	5.00 –58.0 (n = 4)

Mean ± standard deviation, range of values or mean (95% confidence interval) shown.
a n = 28. **b** n = 27. **c** n = 26.
Sources: **1** Cargill et al. (1979). **2** Seal et al. (1971). **3** Engelhard et al. (2002). **4** Williams & Bryden (1993). **5** Gray (2005) —(mean ± SD or 95% confidence interval).

In most marine mammal species, disease-induced peripheral blood leukocyte changes generally follow the typical mammalian pattern. Basophils are not seen commonly in peripheral blood (Bossart et al. 2001) and have not been seen in blood smears of Australian fur-seals and Australian and New Zealand sea-lions (Clark et al. 2002). Descriptions of erythrocyte and leukocyte cell morphology for a number of pinnipeds found in Australian waters can be found in Clark et al. (2002), Clark (2004) and Gray (2005).

The response of biochemical analytes to disease in pinnipeds is not well-defined. Documented differences to terrestrial mammals include higher fasting blood glucose concentrations (although there is considerable interspecies and individual variation), postulated to be due to restraint, handling and/or venipuncture and elevated levels as a diving adaptation (Ridgway 1972; Medway & Geraci 1978; Bossart et al. 2001); and higher basal BUN concentrations (Ridgway 1972; Medway & Geraci 1978; Bossart et al. 2001; Gage 2003), thought to reflect a higher metabolic rate and a high-protein diet (Ridgway 1972).

In marine mammals, quantification of globulin fractions by serum protein electrophoresis (SPE) provides clinically useful information and SPE is the preferred method for determining albumin concentration (Bossart et al. 2001). Although it is a non-specific test it can aid identification of general types of disease processes, e.g. acute and chronic inflammatory processes including changes in acute phase proteins and hepatitis, as well as neoplasia (Bossart et al. 2001).

When interpreting blood results, the effects of stress, exercise, physical restraint and anaesthesia need to be considered. For example, a decrease in haematocrit during anaesthesia has been reported in several pinniped species due to splenic expansion and pooling of erythrocytes in the spleen (Castellini et al. 1996; Costa et al. 1998). Conversely, an elevation in haematocrit due to release of erythrocytes from the spleen has been recorded during periods of disturbance, restraint, excitement or strenuous exercise (Geraci & Smith 1975; Medway & Geraci 1978; Castellini et al. 1996). Leukocytosis with neutrophilia, lymphopaenia and eosinopaenia can also be seen with stress in pinnipeds (Medway & Geraci 1978), similar to the stress-mediated adrenocortical response reported in dogs. With regard to biochemical analytes, most increases in CK are due to skeletal muscle injury including that associated with strenuous activity and handling stress or stranding (Bossart et al. 2001). AST can also increase in response to handling stress, and increases in potassium can be seen with severe physical exertion in pinnipeds (Medway & Geraci 1978).

7.3 Urine, cerebrospinal fluid and bone marrow collection

As in domestic dogs, urine can be collected in free flow or (under anaesthesia) by catheterisation, expressing the bladder or cystocentesis. Cerebrospinal fluid is collected in lateral recumbency from the epidural space at the level of the atlanto-occipital joint and bone marrow can be collected from the pelvis or femoral trochanter by techniques similar to those used in domestic dogs. No specific reference data is available for pinnipeds, so interpretation of findings is generally based upon extrapolation from domestic animal medicine.

8 HAND-REARING

Most Australian otariid species are born in breeding colonies where pups generally remain until weaned, but young pups (<1 yr old) may be found alone away from these colonies while the mother is foraging at sea. If a pup is found emaciated and/or injured it is probably abandoned and will require intervention, whereas pups in good body condition should be left alone and monitored on the beach over the next few days to see if the mother returns. The Australian sea-lion generally has shorter foraging trips (1–2 d) than the fur-seals (2–5 d) (Shaughnessy 1999). Pups removed from the beach in good body condition should be immediately returned to the same location and monitored to determine if the mother is returning and still feeding the pup. Pups that are hand-reared are in most cases unsuitable for release because they have not received normal social contact and foraging training from their mother, and are therefore unlikely to survive in the wild. Otariid pups also readily become imprinted on their carers during hand-rearing, further reducing the chance of successful return to the wild. Emaciated and/or injured weanling otariids (12–18 mo of age) found away from breeding colonies have presumably had insufficient foraging success. These animals are often mistaken for pups as they are usually of a similar size, but can be distinguished from pups by their fully erupted canine teeth. The rehabilitation and release of these animals may be more successful, however, consideration must be given to the fact that after a period in captivity the animals may pose a disease risk to the wild population.

The need for hand-rearing a southern phocid pup in the Australian region is unlikely to occur as there are no reports to date of any southern phocids being born in captivity in Australia. Hand-rearing techniques and formulas for northern hemisphere phocid pups have been described (Geraci 2000a; Townsend & Gage 2001; Duerr 2002) and should be consulted as a starting point if the need arises to hand-rear a pup, since there is no published information relating to hand-rearing of southern phocid pups. Phocid pups that are hand-reared are usually weaned within a month and require an energy-rich formula with a high fat content. An anecdotal report of an attempt to hand-rear a southern elephant seal pup on Macquarie Island was unsuccessful due to lack of experience and inability to formulate a milk substitute that provided the pup with enough calories for an adequate growth rate (R Woods *pers. comm.*).

For captive otariids, assessment of adequate suckling in mother-fed pups may take several days to a week. If the pup is strong and shows normal suckling behaviour or there is a milk beard (milk around the mouth) it should be left with the mother and its weight monitored. Pups typically lose a small amount of weight before they start to gain weight, so they should not be removed at the first sign of weight loss. If indicated, the pup can be stomach tubed with electrolyte solution and returned to the mother. If the pup appears weak and is making poor or no attempts to suckle or the mother will not allow the pup to suckle, it should be removed for hand-rearing.

Hand-rearing of pinniped pups should be performed by experienced carers in appropriate facilities, with shared advice from other institutions experienced in hand-rearing the species. Pups that are dehydrated, emaciated or ill need stabilisation with oral or SC fluids and appropriate medication prior to commencing feeding. Young otariid pups (<3 mo) require feeding with a milk formula, whereas older otariid pups (>6 mo) can be stomach tubed with fish gruel or force-fed fish to wean them early.

8.1 Formula feeding

Some otariid pups may accept a bottle within 1–2 d, while others may take several days. Human and lamb teats are typically used and preference for teat type varies between individuals. Stomach tubing will be required until the pup accepts the bottle readily. Volumes given by stomach tube are approximately 15–20% less than the amount comfortably consumed during a bottle-fed session. Oral electrolyte solution is given for the first two to four feeds, then 25% milk formula is added and the proportion gradually increased to full strength over 2–4 d.

Newborn pups should be fed formula at 30–32°C, commencing with a total daily intake of 3–5% of body weight and increasing to 10–12% over 3–4 d. The amount of formula fed varies with the type used and with each individual pup. As a guideline, pups can usually be bottle-fed 15–25 mL/kg per feed initially for five or six feeds per day. Once the pup is suckling well, the amount of formula at each feeding is gradually increased. Aim to have pups suckling up to 200–250 mL per feed five times a day by 4 wk of age and 250–300 mL per feed four times a day by 6 wk of age (Townsend & Gage 2001). Feed newborns every 4 hr, decreasing the number of feeds as the pup gets older, to four feeds per day when the pup reaches 10 kg. Pups on formula normally produce yellow to tan soft-formed faeces. If diarrhoea, constipation or vomiting occurs the quantity and composition of the formula must be altered.

Milk composition varies between species and within species depending on stage of lactation, length of foraging trips and body condition of the female. Pinniped milk has a high fat content (25–60% depending on the species) and lipids contribute to most of the gross energy of the milk. Carbohydrate content is very low (less than 1% lactose). The average lipid content for the Australian fur-seal is 42% (Arnould & Hindell 1999), similar to that recorded for other otariid species except the Australian sea-lion which has a relatively low average lipid content of 25–30% (Kretzmann et al. 1991; Gales et al. 1996). For most pinniped species the fat content is considerably lower in the very early post-partum period and tends to increase throughout lactation. Protein content remains stable at 10–12%. Water content is usually 30–55% and is inversely proportional to fat. Phocids have a much shorter lactation period than otariids and consequently produce milk with a higher fat content. Phocid pups have a significantly higher average daily weight gain than otariid pups. For example, the average lipid content for milk of the southern elephant seal and crab-eater seal are 40–52% (Hindell et al. 1994; Carlini et al. 1994) and 51% (Green et al. 1993) respectively.

Different institutions prefer different home-made or commercial formulas which they have successfully used to hand-rear pups. Commonly used powdered formulations include Di-Vetelact (Sharpe Laboratories) and Wombaroo sea-lion milk (Wombaroo Food Products). Di-Vetelact is a low-lactose (95% lactose-free) animal milk substitute and is often used as the basis of a formula. Additional fat is provided in the form of peanut, fish or vegetable oil and lecithin. Thickened cream is often added to formulas to increase the fat content; it must be treated with a lactase product, such as Lacteeze (Gelda Pharmaceutical) one to several hours before use to reduce the lactose content. Pinnipeds tolerate little to no lactose and high dietary lactose levels will cause severe diarrhoea. Additional protein can be added in the form of a whey protein concentrate. Wombaroo sea-lion milk closely resembles the composition of Australian sea-lion milk and has been used successfully to rear sea-lions and fur-seals. Formulas should be mixed slowly to avoid aeration and can be refrigerated for up to 24 hr. Formula not consumed during a feed should be discarded. Pups with access to only fresh water should be supplemented with sodium chloride; approximately 2–3 g/d (Townsend & Gage 2001).

8.2 Weaning

Fish can be introduced into the diet as early as 3 mo of age for otariid pups, either as blended fish added to the milk formula or by force-feeding with small pieces of fish. The formula feeds are gradually reduced and the amount of fish gradually increased until the pup is voluntarily eating approximately 20% of its body weight per day. Fish diets should be supplemented with vitamins and minerals (see 5.4). Add calcium supplements if the diet largely consists of blended formulas, using fish with heads and tails removed or small pieces of fish, until whole fish are being consumed. Weaning generally takes 2–6 wk depending on the method used; force-feeding is the faster method. Allowing pups to play in the pool with ice-cubes that contain chunks of fish is another way to introduce pups to fish. The animal's weight should be monitored closely during this time. Pups typically lose up to 10% of body weight during the weaning process.

8.3 Husbandry

Pups are typically weighed at the same time each day prior to feeding: daily for the first month, then twice weekly until 3 mo, then weekly. If ambient temperatures are low, young pups may need a heat source. Young pups should have access to shallow water and a pool with a gradually sloping entry, as they lack the strength to exit deep pools. Pups can swim from birth but are poorly co-ordinated (Stirling 1971) so swimming should be supervised until the pup's swimming skills have developed.

8.4 Immune status

Pinnipeds show evidence of acquisition of passive immunity via the endotheliochorial placenta and via colostrum (Cavagnolo & Vedros 1979; Ross et al. 1994; Marquez et al. 1995). The relative importance of these two routes of transfer has yet to be determined in pinnipeds and the significance of failure to receive colostrum is unclear, as some studies suggest active production of immunoglobulins *in utero* and post-partum (Cavagnolo & Vedros 1979; King et al. 1998; Marquez et al. 1998). Options for supplementation of passive transfer include oral and/or parenteral serum from the mother or another pinniped, or a powdered bovine colostrum supplement (Impact, Wombaroo Food Products), though the efficacy of the latter is unknown.

8.5 Husbandry and medical problems

Common problems during hand-rearing include gastrointestinal disorders, pneumonia, hypoglycaemia, dehydration, hypothermia and hyperthermia, drowning and (especially in otariids) imprinting. Early diagnosis and intervention is important. Presentation, diagnosis and management of most of these problems are the same as for domestic small animals. Pups that did not obtain colostrum might be predisposed to infections. Diarrhoea, constipation, inappetence and/or vomiting are generally associated with inappropriate or changing feeding regimes. Anaemia can occur in young otariids, especially free-ranging animals, due to hookworm (*Uncinaria* spp.) infestation or malnutrition. Pups housed in fresh water without sodium chloride supplementation can develop hyponatraemia, which produces vague neurological signs such as inco-ordination, muscle fasciculation or seizures. Treatment requires administration of IV and oral sodium chloride.

9 DISEASES

Most of the literature on pinniped diseases is from the northern hemisphere. Many of these diseases have not been reported in Australian pinnipeds. An overview of infectious diseases described in pinnipeds in the northern hemisphere but not yet reported in Australia is given in Table 17.6. This section focuses on diseases reported in Australian pinnipeds. In addition to referenced literature, information was obtained from the examination of 232 clinical observations of 37 pinnipeds at Taronga and Melbourne Zoos and Sydney Aquarium, all located in south-east Australia, between 1987 and 2005 and records in the Australasian Registry of Wildlife Health from 94 necropsies performed on free-ranging and captive pinnipeds between 1978 and 2005 (ARWH 2007).

Table 17.6 Infectious pathogens of pinnipeds not yet reported to cause disease in Australian pinnipeds

Aetiology	Systems affected	Significance	References
Phocine morbilliviruses (phocine, monk seal WA, monk seal G)	Respiratory, nervous, lymphoid	Northern hemisphere: mass mortalities. Southern hemisphere: seroconversion in phocids and otariids.	Kennedy-Stoskopf (2001); Duignan et al. (1997)
Canine distemper virus	Respiratory, nervous, lymphoid	Northern hemisphere: mass mortalities.	Kennedy-Stoskopf (2001); Kennedy et al. (2000)
Phocine herpesvirus 1 and 2	Respiratory and digestive, or adrenal adenitis and hepatitis. Possibly abortion.	PHV1fatal in pups. In adults probably secondary infection. PHV2 apathogenic.	Kennedy-Stoskopf (2001); Gulland et al. (2001); Duigan (2000)
Adenovirus (not characterised)	Hepatitis	Uncommon, reported in California sea-lions.	Kennedy-Stoskopf (2001)
Seal poxviruses	Cutaneous (typically head, neck, flippers), mucosal	Self-limiting. Zoonotic.	Tryland et al. (2005a); Muller et al. (2003)
Influenza A and B	Respiratory, systemic	Seroconversion. One associated mass mortality (phocids).	Ohishi et al. (2002); Callan et al. (1995); Geraci et al. (1982); Osterhaus et al. (2000)
Calicivirus (multiple serotypes)	Cutaneous, mucosal	Self-limiting vesicular disease (oral, extremities).	Smith et al. (1981)
Coronavirus	Digestive	Necrotic enteritis, harbour seals.	Bossart et al. (1990)
Brucella pinnipediae	Subclinical	Northern and southern hemispheres: potential pathogen for humans, cattle.	Tryland et al. (2005b)
Leptospira interrogans	Renal, reproductive, systemic	North America: may be pathogenic and enzootic, apathogenic strains.	Acevedo-Whitehouse et al. (2003); Colagross-Schouten et al. (2002)
Bordetella bronchiseptica	Respiratory	Probable secondary invader	Munro et al. (1992)
Burkholderia pseudomallei	Multisystemic (melioidosis)	Captive seals, Hong Kong	Hicks et al. (2000)
Coxiella burnetti	Possibly abortion	One report. Potentially zoonotic	Lapointe et al. (1999)
Toxoplasma	Nervous	Emerging disease	Gulland et al. (2001)
Cryptosporidia spp.	Subclinical	Zoonotic	Measures et al. (1999)
Sarcocystis neurona-like	Nervous	Emerging disease	Gulland et al. (2001)
Coccidia	Digestive	Pathogenic in captive seal pups	Munro et al. (1991)

9.1 Dermatologic diseases

Dermatologic conditions accounted for 20 clinical observations from 12 animals. Of these, eight animals developed blistering, erosion or ulceration of the palmar surface of the pectoral flippers. One incident was attributed to contact with hot metal; the rest were idiopathic. A male Australian sea-lion suffered recurring episodes of fore limb lameness of 1wk duration followed by multifocal sloughing of the superficial, and occasionally deep, palmar epithelium. Biopsies from this animal indicated moderate perivascular non-suppurative dermatitis and mild acanthosis, but no aetiological agent could be identified. All conditions resolved spontaneously. Observations on other animals included ventral wheals, patchy alopecia and miliary dermatitis. Of these, two were described as fungal folliculitis (one case associated with failure to moult, one attributed to *Microsporum gypseum*) and three others tested positive to a range of intradermally administered allergens, including pollens of *Pinus* spp. and *Casuarina* spp. and various weeds and grasses. In at least one case, signs resolved after removal of the animal from its pollen-contaminated environment. Response to administration of antihistamines has been variable (K Bodley *pers. comm.*).

Disturbed moults are common in captive pinnipeds but the cause in most cases is unknown.

9.2 Gastrointestinal diseases

9.2.1 Oral cavity

Heavy dark brown to black staining of the teeth is typically seen in otariids from a young age but is not typical of phocids. Chipped, worn or fractured teeth, dental calculus, gingivitis, periodontitis, bacterial or fungal tooth root abscessation and gingival trauma due to fish spines have all been observed in captive pinnipeds. Severe wearing of multiple teeth with exposure of pulp cavities and tooth root abscessation has been seen in a subadult leopard seal hauled out in New South Wales, presumably associated with defective tooth development of unknown aetiology. Self-limiting multifocal gingival, lingual and palatine papilloma was observed in one subadult New Zealand fur-seal and Gales (1984) reports excision of a fibromatous epulis from a leopard seal. An unclassified focal oral neoplasm in an Australian fur-seal caused marked multifocal gingival inflammation and ulceration, and resorption of alveolar bone.

Oral candidiasis was a common disease of southern elephant seals held at Taronga Zoo. Vogelnest et al. (1996) described gingival inflammation, ulceration and plaque formation around the teeth, particularly the incisors and canines, and *Candida albicans* was isolated in 14 of 20 episodes. Lesions described by others, in other phocid species, include inflammation of the mucocutaneous junctions of the mouth, eye, anus and vulva, and conjunctivitis and dermatitis. In these lesions, and those reported by Vogelnest et al. (1996), *Candida albicans* was isolated in conjunction with other opportunistic invaders such as beta-haemolytic streptococci, *Vibrio* spp. and other mixed flora.

9.2.2 Gastrointestinal

Idiopathic malaise and inappetence occur commonly, accounting for 30 observations from 15 otariids. Some developed diarrhoea and, less commonly, vomiting or abdominal discomfort. Signs were transient, lasting days or occasionally 1–2 wk. There was no apparent seasonality, sex or species predilection. Captive pinnipeds are known to sometimes lose their appetite during hot weather or as their body condition increases and it is normal for some species, especially southern elephant seals, to eat little or nothing during moult or the breeding season. Constipation sometimes occurs in captive pinnipeds, presumably for reasons similar to those in other carnivores, and was severe enough in one case to cause intestinal necrosis and death.

Captive pinnipeds commonly ingest foreign bodies dropped into pools or loosened from the substrate. Ingestion of rocks has been reported as a normal behaviour of free-ranging Australian sea-lions (Needham 1997). Three Australian sea-lions in the current review showed diverse signs in association with large quantities of stones in their stomach. These included inappetence, vomiting, weight loss, abdominal guarding, adoption of unusual postures and vague behavioural changes resembling neurological signs. Resolution of signs in some cases was associated with regurgitation of stones. In two events, removal of large quantities of stones by gastrotomy was necessary due to progressive weight loss. Cordes and O'Hara (1979) reported the death of a captive leopard seal from gastric impaction by 23 kg of stones and shells. Exploratory laparotomy and gastrotomy in another captive leopard seal with chronic anorexia and weight loss revealed five school hats and numerous other objects in the stomach.

Radiography and sometimes abdominal palpation are effective in detecting large radio-opaque gastric foreign bodies and should be considered for pinnipeds, especially Australian sea-lions, showing vague and otherwise inexplicable signs of clinical illness.

Fatal peritonitis was diagnosed in five otariids, attributed to hepatic disease, intestinal perforation by foreign bodies such as fish spines or fishhooks and line, or small intestinal volvulus.

Fatal acute fibrinous or necrotising gastroenteritis or colitis was diagnosed in eight free-ranging and captive pinnipeds, often with associated septicaemia. Organisms cultured from tissues in association with histological findings consistent with pre-mortem invasion included *Actinomyces naeslundi* (possibly originated from a periodontal lesion), *Clostridium difficile*, *Clostridium sordelli*, *Proteus mirabilis* and *E. coli*.

Helicobacter sp. has been found in association with ulcerative gastritis of harbour seals (*Phoca vitulina*) in the northern hemisphere (Harper et al. 2003). Some captive Australian otariids have recently been found to carry *Helicobacter* sp. asymptomatically or, as in one case, in association with chronic gastritis (Oxley & McKay 2004; Oxley et al. 2004).

9.3 Cardiopulmonary diseases

Chronic or acute bronchopneumonia are common necropsy findings in Australian pinnipeds and, even excluding mycobacterial infection (see 9.8.1b), are the most common cause of non-traumatic death in Western Australian pinnipeds (Mawson & Coughran 1999). Reported aetiological agents include *Nocardia asteroides* (Orchard 1979), haemolytic streptococci and, secondary to tuberculosis, *Proteus vulgarus*. One pup died during hand-rearing as a result of systemic aspergillosis, thought to have progressed from chronic lung infection (M Lynch *pers. comm.*).

Endocardiosis has been described in an ageing Australian fur-seal and idiopathic myocarditis and suppurative valvular endocarditis have also been observed.

9.4 Genitourinary diseases

Genitourinary diseases of Australian pinnipeds are uncommon. Dystocia has occurred infrequently in Australian zoos (see 13.2). Vaginal prolapse has been reported in an Australian sea-lion (Read et al. 1982). Multiple uterine leiomyomas have been found at necropsy in one aged Australian fur-seal and an ovarian granulosa cell tumour in another.

In the northern hemisphere, *Leptospira interrogans* has been isolated in association with renal disease. *Coxiella burnetti* (Lapointe et al. 1999), leptospires and phocine herpesviruses have been isolated from aborted placentae but their significance is unknown.

9.5 Neurologic and musculoskeletal diseases

Idiopathic epilepsy in captive Australian fur-seals has been controlled long-term with oral phenobarbitone. Meningoencephalitis, suppurative meningitis and chronic encephalitis have all been diagnosed in Australian pinnipeds. However, excluding cases resulting from bacteraemia or systemic aspergillosis, specific aetiological agents have not been identified.

Encephalitides of northern hemisphere pinnipeds have been attributed to viral, bacterial and protozoal (*Toxoplasma gondii* and *Sarcocystis neurona*) infection. Biotoxins such as domoic acid cause severe neurological disease in pinnipeds in the northern hemisphere.

Trauma, which is especially common in free-ranging pinnipeds, can result in fractures or osteomyelitis of the extremities. Arthritis has been observed in aged captive pinnipeds and has responded to treatment with injectable pentosan polysulfate sodium and oral glucosamine/chondroitin, using canine doses. Open bite wounds often heal extremely well but bite wounds, trauma or iatrogenic infection from injections (especially if administered into blubber) have resulted in deep abscesses.

9.6 Endocrine and metabolic diseases

No endocrine or metabolic diseases have been confirmed in pinnipeds in Australia. However, the prevalent patchy or delayed moults in captive pinnipeds are suggestive of endocrinological, nutritional or metabolic disturbance. The moult period is associated with reduced food intake and (anecdotally) disease and fatalities are more common in this period.

9.7 Ophthalmologic diseases

Ocular disease is common in captive and free-ranging pinnipeds (Stoskopf et al. 1983; Greenwood 1985), particularly corneal lesions. Our review of records for 37 captive pinnipeds found that ocular anterior segment disease accounted for 52 of 232 clinical observations

(22%) in Australian pinniped species, and southern elephant seals appear particularly susceptible (Greenwood 1985; Vogelnest et al. 1996). While corneal opacities such as generalised oedema, corneal ulceration or scarring are observed most frequently, chronic keratitis, descemetoceles and corneal perforation are also reported. Blepharospasm is a common presenting sign and may persist for days to months. The aetiology of corneal lesions in captive pinnipeds is not fully understood and is postulated to be multifactorial, involving water quality, chlorine, ozone and bromine concentrations, exposure to ultraviolet light and opportunistic pathogens. There have been no systematic studies to examine any of these hypotheses. Anecdotally, incidence of ocular disease differs between collections. Shading of pools and the use of non-reflective pool linings have reduced prevalence of the condition, suggesting that glare may be a significant factor in ocular disease. Our review suggested a predisposition to corneal lesions during summer months and two presentations were associated with known over-chlorination of water. Excessive amounts of oxidative disinfectants in water, such as chlorine, ozone and bromine, can result in unilateral or bilateral corneal oedema generally involving the whole cornea with roughened but intact epithelium (Stoskopf et al. 1983). An epidemiologic survey indicated an increased frequency of corneal lesions in California sea-lions (*Zalophus californianus*) maintained in fresh water (Dunn et al. 1996), however, corneal lesions are also common in most pinniped species housed in salt water.

Patchy iris depigmentation of unknown cause has been reported in several pinniped species (Greenwood 1985) and has been observed by the authors in New Zealand fur-seals. Cataracts, synechiae formation, lens rupture, lens luxation and uveitis also occur. Cataracts occur frequently in both captive and wild pinnipeds and could be associated with trauma, metabolic or infectious disease, nutritional or congenital factors, advanced age and possibly ultraviolet light (Stoskopf et al. 1983), although the cause of bilateral cataract development in pinnipeds is often undetermined. Cataracts are very common in aged pinnipeds. Unilateral strabismus has been reported in pinnipeds (Greenwood 1985) and has been observed by the authors in a captive Australian fur-seal. Unilateral fungal panophthalmitis in a captive Australian fur-seal necessitated evisceration of the eye and replacement with a prosthesis. Fungal elements were seen histologically in the eviscerated material (L Vogelnest *pers. comm.*).

9.7.1 Ocular examination

Anaesthesia is usually required for a thorough ocular examination due to the strong palpebral muscles, a prominent nictitating membrane, the ability to retract the globe and a narrow slit-like pupil which closes to a pinhole in high light intensity. Topical mydriatics such as atropine sulfate and tropicamide elicit poor pupillary dilation (Stoskopf et al. 1983; Dutton 1991; Barnes & Smith 2004). Marginally better results are sometimes obtained if topical mydriatics are applied repeatedly at intervals commencing 24 hr prior to examination. Subconjunctival atropine is only partially effective and the response tends to be delayed. Adrenaline 1:1000 injected intracamerally (into the anterior chamber) produced rapid and effective pupillary dilation in a fur-seal for cataract removal (Barnes & Smith 2004). Diagnostic procedures used for assessment of the pinniped eye are similar to those used for domestic species. Normal values for intraocular pressure, measured by tonometry, are not published for most pinniped species but Stoskopf et al. (1983) report that pressure readings for harbour seals and California sea-lions are very similar to normal human values. Braun et al. (1996) report intraocular pressures of approximately 28 mmHg in physically restrained Hawaiian monk seal (*Monachus schauinslandi*) pups, 12–15 mmHg in isoflurane-anaesthetised pups and up to 56 mmHg in physically restrained pups with active bullous keratitis. Ultrasound is a useful diagnostic tool for examining the posterior segment of the eye.

Seals with significantly impaired vision compensate quickly using tactile and acoustic cues, and cope well in a familiar environment. A reduction in visual capacity and even blindness can be difficult to diagnose due to this adaptability and the sensitivity of the vibrissae to movement when attempting to assess the menace response.

9.7.2 Treatment

Treatment of ocular disease in pinnipeds is similar to that of domestic species. Due to the strength of the eyelids, topical preparations can be very difficult to administer effectively and animals must remain out of water for a minimum of 10–15 min to allow the drug to take effect. Saline (5%) drops or head-dips into salt water appear to help decrease corneal oedema in pinnipeds housed in fresh water (Gulland et al. 2001). The authors have found that oral and/or topical non-steroidal anti-inflammatory drugs can be effective for pain relief, but corticosteroids are indicated for uveitis. Subpalpebral lavage systems, such as those used in horses, have been placed successfully in pinnipeds for extended administration of topical ophthalmic medications (Borkowski et al. 1999). The use of serum topically (bid–qid) is recommended for chronic superficial and deep non-healing ulcers. Use of autologous serum is preferred, but if that is unavailable then serum from another individual of the same species or, as the last option, serum from an individual of a different species may be used. As serum is a blood product, the risk of disease transmission must be considered when using serum derived from another animal. In the authors' experience, corneal debridement and/or superficial keratectomy have been beneficial in several cases of non-healing ulcers. Corneal flap techniques have been used with some success. Conjunctival pedicle grafts have been used to repair corneal ruptures. Cataract extraction is a viable option in suitable candidates and has been described in a northern elephant seal (*Mirounga angustirostris*) (Filer et al. 2003) and in New Zealand fur-seals (Dutton 1991; Barnes & Smith 2004). Surgical enucleation or evisceration and optional replacement with a prosthesis of a non-salvageable eye is an option only in captive pinnipeds.

9.8 Infectious diseases

9.8.1 Bacteria

a Salmonella

A range of Salmonella serotypes have been isolated from faeces of captive and free-ranging pinnipeds, including typhimurium PT101, PT1, Cerro, Derby, Enteriditis PT4, PT8 and PT untyped and typhimurium untyped (Duignan 2003). A captive New Zealand fur-seal died of salmonellosis due to *Salmonella newport*, a serotype rarely recorded in New Zealand (Cordes & O'Hara 1979) and untyped *Salmonellae* have been reported as cause of death in two free-living Australian fur-seal pups in Victoria (Norman 1995). At least some of these can be expected to be zoonotic.

b Tuberculosis

Mycobacterium pinnipedii, a member of the *M. tuberculosis* complex, causes granulomatous lesions in the lung, lymph nodes and viscera of free-ranging and captive Australian pinnipeds (Forshaw & Phelps 1991; Woods et al. 1995b; Hunter et al. 1998; Cousins et al. 2003). The incidence of the disease in pinnipeds is thought to be low, although true incidence is unknown. Infected animals may show no clinical signs or in some cases may exhibit vague signs such as anorexia, dysphagia or weight loss. Coughing is not typically observed (Forshaw & Phelps 1991). As this organism is zoonotic (Cousins et al. 2003) any live pinniped with respiratory disease or wasting, and all pinniped necropsies, should be approached with caution. In otariids, multifocal granulomas with central necrosis occur but, unlike cattle, pigs and humans, there is less caseation, and mineralisation and giant cell formation were not prominent features of lesions. As seen in horses and dogs, fibrocytic proliferation was a prominent feature in two pinnipeds with marked pleuritis and lesions resembled fibrosarcomas (Forshaw & Phelps 1991). To the authors' knowledge, there are no anecdotal or published reports of treatment of this disease in pinnipeds.

Ante mortem diagnosis of tuberculosis in pinnipeds is both challenging and limited. A serological test (ELISA) that was previously available in Australia for use in pinnipeds is no longer available. Assays for cell-mediated immunity such as lymphocyte transformation and gamma interferon assays have not been established for pinnipeds. The intradermal tuberculin test, which measures the cellular response of the immune system to mycobacterial antigens, is the only tuberculosis test currently available for screening pinnipeds in Australia (D Cousins *pers. comm.*). However, there are limitations in both sensitivity and specificity of this test and false positive reactions do occur in pinnipeds (Forshaw & Phelps 1991; G Reddacliff & L Lim unpub; D Cousins *pers. comm.*). An unpublished study by Reddacliff and Lim for the Australian National Parks and Wildlife Service (Infectious disease monitoring in captive and wild pinnipeds, Project 8889/23, July 1990) determined a preliminary estimate of high sensitivity (90–100%) and lower specificity (46–63%) for the tuberculin skin test in pinnipeds, although the sample size was small and biased as captive animals experiencing a tuberculosis outbreak at the time were included in the study. Comparative intradermal skin testing was found to have slightly higher sensitivity and specificity than single intradermal skin testing. As pinnipeds are commonly exposed to environmental mycobacteria, the comparative tuberculin test is recommended. During an investigation of a tuberculosis outbreak in a captive collection of pinnipeds in Western Australia in 1986, the results of tuberculin skin testing showed an apparent high degree of sensitivity but the results were not statistically significant due to the small number of animals tested (Forshaw & Phelps 1991). The incidence of false negative reactions has not been estimated but appears to be extremely low. Until large numbers of tested animals have been thoroughly examined at necropsy for tuberculosis, sensitivity and specificity of tuberculin skin testing in pinnipeds cannot be accurately determined.

A comparative intradermal tuberculin test should be performed on free-ranging pinnipeds coming into rehabilitation facilities or zoological collections to assess the risk to handlers and other pinnipeds. To perform this test, shave a 3–4 cm^2 area of skin on both sides of the neck. Take a pinch of skin from the centre of the shaved area and make three separate measurements, using callipers, of the double skin thickness and record the mean skin thickness. Inject 0.1 mL of avian tuberculin PPD (25 000 IU/mL) intradermally in the centre of the shaved area on one side of the neck. Repeat the same procedure on the other side of the neck with 0.1 mL bovine tuberculin PPD (1 mg/mL) and mark the injection sites with a waterproof marker. An increase in double skin thickness of $\geq$5 mm at 72 hr at either site is considered a positive reaction. If the increase in thickness at the avian site is equal to or greater than the increase at the bovine site it indicates prior exposure to atypical mycobacteria, which is not a concern. A significant increase at the bovine site, if greater than at the avian site, indicates prior exposure to mycobacteria of the *Mycobacterium tuberculosis* complex, which includes the causative agent of tuberculosis in pinnipeds, *Mycobacterium pinnipedii* sp. nov. (Cousins et al. 2003). Results from a single test should be interpreted cautiously due to the presumed low specificity of the test; a positive reaction at the bovine site may be a false positive. If an animal is a positive reactor on initial testing, without evidence of clinical disease and with low risk of exposure, quarantine and retesting no less than 60 d later is recommended. An animal with sustained positive or increased reactions on subsequent testing should be considered highly likely to be infected and is a potential risk to other pinnipeds and a significant zoonotic risk. Euthanasia and necropsy is recommended. Tracheal washes or bronchoalveolar lavage may be attempted on positive reactors to isolate the organism, but failure to culture the organism does not rule out the possibility of tuberculosis. In the case of a positive reactor with signs of clinical disease such as weight loss, anorexia, dysphagia and/or respiratory signs, euthanasia and necropsy is recommended.

An experimental ELISA test for tuberculosis in pinnipeds was developed and performed at the Australian Reference Laboratory for Bovine Tuberculosis, Department of Agriculture WA, but is no longer available (D Cousins *pers. comm.*). This test was used in the two studies mentioned above and showed promise, but again the number of animals tested was small so sensitivity and specificity of the test has not been accurately determined (Forshaw & Phelps 1991; Reddacliff & Lim unpub). More free-ranging animals brought into rehabilitation centres and zoological collections as well as captive animals were serologically tested during the 1990s, but data has not been retrospectively collated with skin test results and necropsy findings. Ten animals were tested from the captive population in Western Australia during the tuberculosis outbreak in 1986 and the serology results correlated 100% with skin test results (seven positive and three negative). However, serology samples were collected 7–22 d after skin testing and Reddacliff and Lim question whether skin testing stimulated a serological response. In Reddacliff and Lim's unpublished study nine animals were serologically tested and found negative at the time of skin testing; three of them were tuberculin skin test positive. Thus, serology samples should be collected at the time of skin testing to eliminate the possibility of skin testing influencing a serological response.

Radiographic examination is not useful, even in advanced cases of tuberculosis, as mineralisation is not a feature of the disease in pinnipeds (Forshaw & Phelps 1991). Culture of the organism should be possible from nasal or pharyngeal swabs, tracheal washes or bronchoalveolar lavage and possibly faeces, but this requires that the animal be shedding the organism at the time of sampling, which may not occur in early cases. Failure to isolate the organism from these samples would not rule out the possibility of tuberculosis. Biopsies of lesions are more likely to result in isolation of the organism by culture, but tuberculosis lesions in pinnipeds are typically not superficial. Samples should be submitted to mycobacterial reference laboratories for culture as the organism is slow-growing, difficult to culture and requires special media.

c *Other bacteria*

Mass mortalities such as those caused in 1998 by *Campylobacter* spp.-associated necrotising vasculitis and haemorrhagic pneumonia in New Zealand sea-lions in the Auckland Islands and a New Zealand fur-seal from Otago (Duignan 2003), and in 2002–2003 by *Klebsiella pneumoniae*-associated disease of pups and, less commonly, adult New Zealand sea-lions in the Auckland Islands (Duignan et al. 2004) have not yet occurred in Australia. However, the presence of potential wide-ranging carriers such as pinnipeds, fish and seabirds, the increasing potential for introduction of new pathogens to the marine environment and the potential for changing marine resources to affect the susceptibility of pinnipeds to disease make it possible that these infections may occur in Australian populations.

There is evidence that *Brucella* spp. is widespread in cetaceans and pinnipeds in the northern hemisphere and there is serological evidence of exposure of pinnipeds of the Antarctic Peninsula to *Brucella* spp. (Retamal et al. 2000). There is no evidence that Australasian marine mammals have been exposed, but evidence that marine isolates can infect humans and cattle indicate that investigation of Australasian pinnipeds for exposure is warranted.

Thornton et al. (1998) described a wide range of bacteria cultured from pinnipeds although the significance of many was not determined. *Pseudomonas mallei, Clostridium chauvoei, C. novyi, Neisseria mucosa* var *heidelbergensis, Klebsiella pneumoniae, Salmonella* spp. and *Pasteurella multocida* have been isolated from pinnipeds overseas in association with lesions and are known bacterial pathogens in terrestrial mammals (Smith et al. 1978). Common water-borne organisms are frequently cultured from lesions or faeces of pinnipeds. Their significance is unknown.

9.8.2 Viruses

Viral respiratory diseases that have been responsible for epizootics in the northern hemisphere (see Table 17.6) have not been reported in Australasia and there is insufficient data to predict the likelihood of mortalities occurring in this region. Although an idiopathic mass mortality that occurred in Antarctica in the mid 1950s (Laws & Taylor 1957) is sometimes suggested to have been viral, associated pathology more closely resembled a *Campylobacter*-associated epidemic of haemorrhagic septicaemia in New Zealand sea-lions in the Auckland Islands in 1998 (Duignan 2003). Antibodies to canine distemper virus have been detected in the sera of two species of Antarctic seal (Bengston et al. 1991), and New Zealand sea-lions and New Zealand fur-seals have been reported seropositive to a morbillivirus antigenically similar to phocine distemper virus (Duignan 2000). In addition, Weddell and crab-eater seals have been reported seropositive for phocine herpesvirus (Harder et al. 1991; Stenvers et al. 1992), in one case in association with outbreaks of respiratory disease in 1984 and 1990 although causation was not established. Avian influenza has caused deaths in northern hemisphere populations and, although it has not been reported in Australasia, there is potential for transmission from birds (Duignan 2000).

9.8.3 Parasites

Lungworms (*Otostrongylus circumlitus, Parafilaroides* spp.) and respiratory mites (*Orthohalarachne* spp.) (Duignan 2000) can cause disease when in large numbers. Heartworm (*Dirofilaria immitis*) causes disease in pinnipeds similar to that in dogs, though this parasite is becoming less common due to improved prophylaxis in dogs. Prophylaxis is recommended for captive pinnipeds in endemic regions. Oral ivermectin (10–24 μg/kg) or milbemycin oxime (0.5 mg/kg) monthly are commonly used for prevention.

Roundworms (*Contracaecum osculatum, Anisakis* spp.), tapeworms (*Diphyllobothrium* spp.) and acanthocephalans (*Corynosoma australe*) have been reported in Australian pinnipeds (Arundel 1978; Cordes & O'Hara 1979). *Anisakis simplex* and *Contracaecum* spp. have been reported from southern elephant seals (Morgan et al. 1978), sometimes in association with ulcerative gastritis. Similar parasites are common in captive and other free-ranging Australian pinnipeds.

Massive intestinal burdens of *Diphyllobothrium* spp. occur asymptomatically in free-ranging leopard seals (Higgins & Gray unpub) and probably other species, although they are not commonly detected in routine faecal examination of captive animals (K Humphreys *pers. comm.*). *Diphyllobothrium delphini* form cysts within the blubber of southern elephant seals (Morgan et al. 1978) and similar subcutaneous plerocercoids have been reported in fur-seals in New Zealand (Cordes

& O'Hara 1979) and Australia. Duignan (2000) reports that granulomatous mastitis is sometimes associated with heavy burdens of plerocercoid cysts in New Zealand sea-lions.

Hookworm (*Uncinaria* spp.) causes pup mortalities through anaemia and dehydration in some New Zealand pinniped colonies (Morgan et al. 2000) and occurs in Australian otariids (Norman 1995). Infection is acquired through suckling and possibly percutaneously. Occult infection can occur, as shedding of eggs ceases after the pup has reached a certain age.

The principles of treatment and drugs for parasitic infections in pinnipeds are the same as for domestic carnivores.

9.9 Non-infectious diseases

9.9.1 Neoplasia

Neoplasms reported in Australian fur-seals include multiple uterine leiomyomas, ovarian granulosa cell tumour, an unspecified adenocarcinoma, an axillary soft tissue sarcoma and a possible fibrosarcoma. Those in New Zealand fur-seals include a carpal osteosarcoma and a renal adenocarcinoma with cardiac, lymphoid and small-intestinal metastases (ARWH 2007). All were in captive animals, most of which were aged.

9.9.2 Toxins

a Biotoxins

Of the algal toxins, domoic acid causes sporadic mass mortalities in pinnipeds in North America (Scholin et al. 2000) and saxitoxin has been implicated in mass mortalities in West Africa (Reyero et al. 1999). Mortalities caused by biotoxins have not been reported in Australian waters although algal blooms do occur. New Zealand fur-seal mortalities coincided with toxic blooms on New Zealand's east coast in 1997–98 (Duignan 2000).

b Xenobiotics

The impact or extent of chronic exposure to marine pollutants, such as heavy metals and organohalogens, in pinnipeds has not been studied in the Australasian region.

9.9.3 Nutritional diseases

Clinical nutritional disease is uncommon in captive pinnipeds fed a complete diet with appropriate supplementation (see 5.4). Free-ranging pinnipeds commonly present in an emaciated condition, presumably due to displacement from their normal habitat or range, or impaired foraging ability secondary to other disease, depletion of fish stocks or lack of access to their usual prey items. These animals are usually malnourished.

Haemochromatosis has been documented as the cause of death in captive adult northern fur-seals (*Callorhinus ursinus*) and California sea-lions (Garcia et al. 2000). High levels of iron were measured in the liver of these captive animals with gross and histopathological hepatic lesions, but neither elevated hepatic iron levels nor lesions were found in free-ranging adult otariids or captive pups that were comparatively assessed. There is speculation that increased bioavailability of iron may result from feeding a greater proportion of fish compared to cephalopods in captivity than are consumed by free-ranging pinnipeds. Though these findings suggest that iron accumulation can be a pathological process in captive northern hemisphere otariids, ferrous deposits in hepatocytes and hepatic, splenic and adrenal macrophages (haemosiderosis) of three adult captive Australian fur-seals, two adult captive subantarctic fur-seals and a captive southern elephant seal were not associated with histopathological changes (ARWH 2007).

9.9.4 Trauma

Bite wounds between conspecifics are common in both captive and free-ranging pinnipeds particularly during the breeding season. Males can be very aggressive toward each other while defending territories and harems. Although wounds can be severe, they generally heal remarkably well in the salt water environment. In captivity, fight wounds are rarely significant and only occasionally require treatment. Wounds should always be left open and may require antibiotic therapy to prevent secondary infection.

Hauled-out free-ranging animals often present with injuries such as fishing net or boat propeller lacerations, gunshot wounds, bite wounds from cookie-cutter sharks (*Isistius brasiliensis*) or other sharks, or strangulation wounds from debris such as plastic packing straps. The same principles of treatment apply as in domestic species and pinnipeds generally heal well, including extensive deep wounds by secondary intention. Embedded stingray barbs may cause swellings or discharging sinuses, but in many cases remain undetected externally. Barbs are sometimes observed radiographically and if detected should be removed surgically if accessible.

10 ZOONOSES

Zoonotic pathogens reported from pinnipeds are listed in Table 17.7. The possibility of injury and disease transmission is always present when working with pinnipeds and a greater risk is posed when handling or examining free-ranging animals of unknown or poor health status. Trauma from bites and their sequelae are by far the most common zoonotic conditions associated with pinnipeds. New infectious organisms are continually being identified and reported for marine mammals and the zoonotic potential of many of these disease agents is unknown. *Mycobacterium pinnipedii* is zoonotic and is harboured by some Australian pinnipeds, thus handlers should be aware of this zoonotic risk when dealing with any pinniped, particularly free-ranging pinnipeds or those with respiratory disease or wasting. Gloves should be worn during examination of the oral cavity to prevent transcutaneous infection with *Mycoplasma phocacerebrale*, the cause of 'seal finger' and other potentially zoonotic pathogens. Handlers should also be

Table 17.7 Zoonoses associated with pinnipeds worldwide

Organism	Significance	Prevalence	Reference
Seal pox (parapox virus)	Papule, vesicle, pustule, scab	North, uncommon South	Hicks & Worthy (1987); Shaughnessy (1999)
Influenza A and B	Conjunctivitis, potential respiratory	North	Webster et al. (1981); Osterhaus et al. (2000)
Mycobacterium marinarum	Cutaneous	Ubiquitous aquatic	Woods & Gutierrez (1993)
M. pinnipedii sp. nov.	Respiratory	Common South	Cousins et al. (1993); Thompson et al. (1993); Cawthorn (1994); Hunter et al. (1998); Cousins et al. (2003)
Brucella spp.	Potential zoonosis	North	Brew et al. (1999)
Leptospira spp.	Fluids, tissues	North	Smith et al. (1978)
Vibrio spp.	Gastroenteritis, wounds, septicaemia	Ubiquitous aquatic	Woods & Gutierrez (1993)
Edwardsiella tarda	Gastroenteritis, wounds/cellulitis, septicaemia, meningitis	Common aquatic	Cowan et al. (2001)
Clostridium spp.	Potential via wounds	Common intestinal	K. Rose (*pers. comm.*)
Erysipelothrix rhusiopathiae	Transcutaneous, local, can become generalised	Uncommon, common in fish	Woods & Gutierrez (1993)
Coxiella spp.	Potential illness, abortion	North, isolated from one placenta	Lapointe et al. (1999)
Mycoplasma phocacerebrale	'Seal finger'; swelling locally and local joint	Common	Baker et al. (1998)
Salmonella spp.	Gastroenteric	Potentially common	Duignan (2003)
Cryptosporidium spp., *Giardia* spp.	Gastroenteric protozoa	North	Measures et al. (1999); Deng et al. (2000)
Trichinella spp.	Intestinal nematode	Where seal flesh eaten (infective larvae in muscle)	Forbes et al. (2003)
Toxoplasmosis	Doubtful significance	North; Australian coastal cetaceans	Migaki et al. (1977); Bowater et al. (2003)

North = northern hemisphere pinniped species. South = southern hemisphere pinniped species.

aware of the zoonotic risks associated with handling fish during food preparation and feeding sessions.

11 THERAPEUTICS

The principles of therapeutics in pinnipeds are similar to those used for terrestrial carnivores. Drug doses are generally extrapolated from those used in canine medicine as few pharmacokinetic studies have been performed in pinnipeds. Stoskopf et al. (2001) provide a formulary for pinnipeds based on published and anecdotal information, and advise clinicians that most drugs have only been used in a limited number of individuals and caution should be exercised with regard to potential side effects and drug interactions.

11.1 Routes for drug administration

Practical considerations frequently determine routes of drug administration. Oral (PO) administration is the easiest method in pinnipeds that are eating; medications can be placed in fish via the gills, mouth or coelomic cavity. Alternatively, medications and larger volumes of fluid can be delivered via stomach tube. The SC route is used most commonly for administration of fluids; it is important that the fluids are deposited between the blubber and muscle layer. IM administration to the lumbar or gluteal muscles, or the large muscles overlying the scapulae in otariids, is frequently used for delivery of injectable drugs. Long needles are required so that drugs are not deposited in the blubber layer, as systemic distribution from blubber is much slower and less reliable than from muscle. Injections into the blubber may also result in abscessation. Injections can be administered by hand if pinnipeds are restrained or trained to accept injections, or by dart, pole syringe or needle and extension tube if they are not (see 6.2.2). IV access in pinnipeds is limited. Veins used for otariids include the jugular, the interdigital veins on the dorsal surface of the hind flippers and brachial vein. The large intravertebral extradural vein is typically used in phocids, or the interdigital veins on the plantar surface of the hind flippers can be used (see 7.1). The extradural vein in otariids should not be used for administering drugs IV except in an emergency, as it may result in damage to the spinal cord. For extended access, IV catheters can be placed into the intravertebral extradural vein of phocids and the interdigital veins or the brachial vein of otariids (Figs 17.4, 17.5). Jugular catheters can be placed in otariid pups but surgical cut-downs are usually required to access the jugular vein in older animals. Catheters are difficult to maintain in most otariids for extended periods. Intraperitoneal (IP) administration can be used but care must be taken not to damage organs or introduce infection. Intraosseous catheters have been placed into the medullary cavity of the proximal tibia in small otariids for fluid therapy and blood transfusions (Gulland et al. 2001). Intratracheal (IT) administration can be used, most commonly for emergency drugs during resuscitation. Nebulisation for severe respiratory disease can be achieved via mask or by placing an animal into a sealed cage. Ointment and sprays can be applied topically, though pinnipeds need to be kept out of water long enough to achieve sufficient contact time. Effective application of topical ophthalmic agents may be limited by the strong palpebral muscles.

11.2 Pharmaceuticals

Most of the antibiotics, antifungals, anthelmintics, anti-inflammatory drugs and analgesics used in domestic animal medicine have been used in pinnipeds. The authors have routinely used oral and injectable carprofen at the canine dose for treatment of musculoskeletal and ophthalmic conditions, and for pain relief. In a few cases, we have used carprofen in otariids continuously for several months with no observed clinical side effects, no significant alterations in haematological and biochemical parameters and no gastrointestinal ulceration on necropsy. Lucas et al. (1999) also reported the use of carprofen in pinnipeds with no observed side effects. The authors and Lynch (1999) have routinely used meloxicam in pinnipeds at the canine dose with no observed clinical side effects. Gastroprotectants such as omeprazole and famoditine can be administered concurrently with non-steroidal anti-inflammatory drugs to reduce the risk of gastrointestinal side effects.

11.3 Euthanasia

Euthanasia of pinnipeds is most commonly accomplished by lethal injection of barbiturates (pentobarbitone sodium) or shooting. IV administration of pentobarbitone is the preferred method and physical or chemical restraint is required for administration. If IV access cannot be achieved, then less preferred but acceptable routes are intraperitoneal or intracardiac (AVMA 2001). The former produces less consistent results than IV administration. Very long needles, about half the width of the animal's chest, are required for intracardiac injections and should be inserted perpendicular to the skin just caudal to the axilla. The intracardiac route is recommended for use only in heavily sedated, anaesthetised or unconscious animals (AVMA 2001). It is important to prevent scavengers gaining access to a carcass that contains high concentrations of euthanasia solution; carcasses should be disposed of by deep burial or incineration.

Shooting may be necessary for large and dangerous pinnipeds or when the option of lethal injection is unavailable. The procedure should be performed only by an experienced operator, shooting a high-velocity bullet into the brain (Geraci 2000b). Pinnipeds do not have large frontal sinuses. Fur-seals and sea-lions can be shot 2.5 cm above the ear, perpendicular to the body. Elephant seals and leopard seals can be shot at the same point or down through the cranium at the widest point of the head (Vogelnest 2000).

12 SURGERY

The principles of surgery in pinnipeds are similar to those in other mammals, but consideration needs to be given to anatomic differences (see 2.1) and aquatic needs. While major surgical procedures are performed infrequently in pinnipeds, some conditions requiring surgery more commonly reported or seen by the authors include repair of traumatic wounds, exploratory laparotomy, removal of gastrointestinal foreign bodies or obstruction, mass or tumour removal, caesarean section, prolapsed uterus, fracture repair or removal of infected bone, tooth extraction, root canal therapy, ophthalmic surgery and castration.

12.1 Wounds

Most traumatic skin wounds heal without surgical intervention, even those with significant tissue deficits. Debridement and flushing with saline and antiseptics may be required. Wound closure is rarely required.

12.2 Laparotomy

Laparotomy is one of the more common surgical procedures performed on captive pinnipeds. Keeping some of the unique anatomical features in mind (see 2.1), the procedure and principles are the same as for domestic animals. All pinnipeds, with the exception of fur-seals, have a thick blubber layer which is well-vascularised (Pabst et al. 1999) but tends not to hold sutures well under tension (Sweeney 1990). Sweeney (1990) reported that pinnipeds have relatively thin abdominal musculature and a thin peritoneum, so abdominal wound closure tends to be weak, yet Richardson and Gales (1987) reported that the abdominal wall of Australian sea-lions is extremely muscular. The authors have not experienced problems with wound dehiscence following abdominal surgery.

12.3 Castration

Otariids have scrotal testes and castration is performed as for the dog. Phocids have para-abdominal testes that lie below the blubber layer adjacent to the abdominal musculature in the inguinal area, so castration of these animals requires two incisions just caudal and slightly lateral to the opening of the prepuce (Gage 2003). Blunt dissection is required to identify and exteriorise the testes, which can then be removed using a technique similar to a dog castration.

12.4 Post-operative management

The surgical wound and the animal's enclosure should be kept clean and dry for several days post surgery to optimise wound healing. Pinnipeds out of water for prolonged periods benefit from frequent hosing for cooling and behavioural enrichment. Activity should be minimised for several days following abdominal surgery to reduce the risk of wound dehiscence. Wound dressings and bandaging are usually not feasible, except on the flippers.

13 REPRODUCTIVE MANAGEMENT

13.1 Pregnancy diagnosis

Pregnancy diagnosis and assessment of foetal viability can be achieved during mid to late gestation using ultrasonongraphy. Radiography is also useful for pregnancy diagnosis, but not for assessing foetal viability. Abdominal enlargement and increased food consumption are usually observed in late gestation. Foetal movement can sometimes be seen in very late gestation.

Elevated progesterone concentrations can be a useful indicator of pregnancy (Atkinson 1997; Robeck et al. 2001), but current knowledge of sex steroid concentration patterns during embryonic diapause and pregnancy is too limited for most pinniped species to be used as a reliable indicator of pregnancy. In addition, serial sampling is required as individual values can be variable and often misleading when interpreting results.

Pseudopregnancy, defined as the prolonged period of high concentrations of progesterone in animals that are not pregnant, occurs in some otariid and phocid species (Atkinson 1997). During pseudopregnancy the progesterone concentration is elevated and indistinguishable from pregnant animals for at least the duration of embryonic diapause, and possibly longer in some species as the functional life of the corpus luteum in non-pregnant animals varies between species. In pregnant animals, progesterone increases at the end of embryonic diapause and remains elevated during active gestation, although the pattern of change in progesterone concentration varies between otariids and phocids (Boyd et al. 1999). Gales et al. (1997) investigated the breeding cycle of free-ranging female Australian sea-lions. The mean plasma progesterone values showed an increase from about 12 ng/mL at 3 mo postpartum to a peak of about 24 ng/mL at 5–6 mo postpartum, subsequently falling to about 12 ng/mL by 8 mo postpartum and remaining at that concentration for the next 8 mo. The progesterone profile of pregnant and non-pregnant sea-lions during the first 3 mo postpartum (embryonic diapause and pseudopregnancy respectively) were similar, with a mean concentration of 12 ng/mL. The significant rise in progesterone and oestradiol concentrations between 3.5–5 mo postpartum is believed to be when the blastocyst reactivates and implants. A similar pattern has been documented in Antarctic fur-seals (Boyd 1991b); mean plasma progesterone concentration was about 5 ng/mL throughout embryonic diapause and pseudopregnancy (unmated females), increasing to 35 ng/mL at the end of embryonic diapause followed by a slow decline through the remainder of gestation, falling to about 15 ng/mL by 8 mo postpartum. The concentration of serum progesterone was found to be on average 30 ± 13 ng/mL in free-ranging breeding female New Zealand sea-lions sampled approximately 1 mo after the peak pupping period (Dutton et al. 2000) and it was unknown which females were pregnant. The harbour seal exhibits a different progesterone profile. Progesterone rises after ovulation and remains elevated throughout embryonic diapause (mean concentration approximately 30 nmol/L) with a significant increase during the last 3–4 mo of gestation in pregnant animals (mean concentration up to 60–80 nmol/L) and a sharp decline at parturition (Reijnders 1990). In non-pregnant harbour seals, progesterone concentration during pseudopregnancy remained elevated as in pregnant animals for about 4 mo, equivalent to the duration of embryonic diapause.

Measurement of circulating steroid concentrations is invasive and serial sampling can be difficult in pinnipeds unless they are trained for venipuncture. Faecal steroid analysis is a useful non-invasive technique for monitoring the reproductive cycle. Serum or plasma steroid concentrations have been shown to correlate well with concentrations in faeces with a time lag of 1–2 d in many species, although there is no published data of faecal steroid analyses to monitor the reproductive cycle in pinnipeds. Dutton et al. (2000) measured the concentration of progesterone in faeces at a single time-point in free-ranging breeding female New Zealand sea-lions (mean of 447 ± 177 ng/g dry faeces). However, the faecal and serum samples were collected at the same time so, as expected, no correlation was found between serum and faecal concentrations. Individual faecal steroid concentrations can vary markedly, so serial analyses are required to monitor changes in the reproductive cycle. Use of this method in pinnipeds is significantly hampered by the fact that pinnipeds typically defecate in water. A non-invasive method using saliva has been adopted in the endangered Hawaiian monk seal for monitoring the oestrous cycle, with plasma and salivary sex steroid concentrations being well-correlated. This method has also been used successfully to detect pregnancy in a captive California sea-lion (Atkinson 1997).

13.2 Parturition

Approaching parturition, the female should be provided with ample haul-out space with an area of seclusion. Inappetence usually occurs within 1–2 d of parturition. There may be little obvious behavioural display other than restlessness just prior to, or during, parturition. Time between appearance of the amniotic sac and birth can range from a few minutes to close to 3 hr (Stirling 1971; Marlow 1975; Higgins & Gass 1993). Cephalic presentation and (less commonly) posterior presentation are both normal and the umbilical cord usually breaks during birth. The zonary, endotheliochorial placenta is usually expelled within an hour of birth. Twins are rare. Dystocia occurs infrequently and is managed as in other carnivores. Ultrasound is useful to assess foetal viability prior to deciding the best treatment option. Care must be taken to avoid excessive force when manually extracting a pup as this may result in tearing of the reproductive tract. A significant uterine tear with extensive bleeding caused during manual delivery of a large dead pup was the cause of death in a captive Australian sea-lion. Prostaglandin $F_{2\alpha}$ (Cloprostenol 500 mg IM) has induced abortion of dead full-term foetuses in California sea-lions; foetuses were delivered 36–40 hr later (Gulland 2000). Oxytocin (20 IU IM) has been used in pinnipeds to stimulate milk letdown postpartum but its effectiveness is variable.

13.3 Contraception and control of aggression in males

Male pinnipeds typically become more aggressive during the breeding season and inter-male fighting is common. Breeding and territorial or aggressive behaviour of males towards females and other males during the breeding season have been most commonly controlled by physical separation, castration of males and contraception of females.

Castration controls fertility and aggressive behaviour but its permanency may be undesirable. Gonadotropin-releasing hormone (GnRH) agonists such as leuprolide, buserelin and deslorelin show promise as a reversible alternative in male pinnipeds. GnRH agonists cause an initial significant increase in circulating testosterone concentration and enhancement of sperm production, followed by a decline to baseline or less by approximately 2 wk after injection (AZA 2005). The inhibitory effect lasts approximately 2 mo in Hawaiian monk seals (Atkinson et al. 1993; Atkinson et al. 1998), harbour seals and northern elephant seals (Robeck et al. 2001) when injected with a depot suspension of leuprolide acetate (7.5 mg IM in Hawaiian monk and 40 mg IM in northern elephant seals). The same drug given (0.09–0.12 mg/kg IM) at monthly intervals over the breeding season successfully controlled undesirable male-associated behaviours in California sea-lions (Calle et al. 1997). The disadvantages of injectable leuprolide are cost, the need for repeated injections and adverse injection site reactions. Buserelin (9.9 mg depot implant SC) administered twice, 2 wk apart, completely suppressed testosterone for almost 4 mo during the breeding season in harbour seals (Siebert et al. 2002). Deslorelin implants produce reversible dose-related long-term suppression of reproductive function in male domestic dogs (Trigg et al. 2001) and have been successful in suppressing testosterone and controlling aggression in male sea otters (Calle et al. 1999). Deslorelin (2×4.7 mg implants SC) is reported to have successfully decreased seasonal aggression in male Australian sea-lions, New Zealand fur-seals and California sea-lions during the breeding season (D Blyde *pers. comm.*).

13.4 Contraception in females

Since most pinniped species have a short and defined breeding season, separation of males and females is one of the most commonly used methods of contraception. When space is a limiting factor an alternative form of contraception may be required. The limited research on female pinniped contraception has focused on the use of porcine zona pellucida (PZP) vaccine in northern phocids (Brown et al. 1997a, b) and California sea-lions (Renner & Dougherty 1999). Antibodies are raised against PZP antigens, preventing fertilisation of the ovum by blocking sperm binding. A single-dose administration has resulted in effective contraception in free-ranging grey seals (*Halichoerus grypus*) and captive grey and harp seals (*Phoca groenlandica*). Captive California sea-lions given a series of three injections of PZP vaccine did not become pregnant during the subsequent breeding season, but the study is preliminary and the safety and efficacy of the vaccine has not been fully determined. PZP vaccine is not available in Australia. Proligestone, a synthetic progesterone hormone, has been effective in spotted seals (*Phoca largha*) when given as a single injection (5 mg/kg) 2 mo prior to estimated oestrus (Katsumata et al. 2003). Efficacy of deslorelin is not yet established in female pinnipeds but it produces reversible long-term suppression of oestrus in female domestic dogs (Trigg et al. 2001) and is effective in several exotic wild felid and canid species (Bertschinger et al. 2001). GnRH agonists, such as injectable leuprolide and deslorelin implants, are initially stimulatory, causing oestrus and ovulation before down-regulation of the reproductive system (AZA 2005).

ACKNOWLEDGMENTS

The authors thank Kaye Humphreys for assistance with clinical pathology data; Mikaylie Wilson for gathering product information; marine mammal keepers at Taronga Zoo for husbandry advice; David Blyde, Kate Bodley, Debby Cousins, Kaye Humphreys, Michael Lynch, Larry Vogelnest and Rupert Woods for personal communications; Taronga Zoo and Melbourne Zoo for clinical and necropsy information; Jane Hall for retrieval of information from the Australian Registry of Wildlife Health; Michael Terkildsen for assistance with seal distribution figures; Bozena Jantulik for assistance with, and initial drawings of, blood collection sites; and Michael Lynch for reviewing the manuscript.

REFERENCES

Acevedo-Whitehouse K, de la Cueva H, Gulland FMD, Aurioles-Gamboa D, Arellano-Carbajal F & Suarez-Guemes F (2003). Evidence of *Leptospira interrogans* infection in California sea lion pups from the Gulf of California. *Journal of Wildlife Diseases* **39**, 145–151.

APHIS (2002). Animal welfare act and animal welfare regulations: Part 3 – Standards; subpart E – Specifications for the humane handling, care, treatment, and transportation of marine mammals, pp. 104–127. Animal and Plant Health Inspection Service, US Department of Agriculture.

Arkush KD (2001). Water quality. In *CRC Handbook of Marine Mammal Medicine* (2nd edn) (eds LA Dieruf & FMD Gulland), pp. 779–790. CRC Press: Boca Raton, Florida.

Arnould JPY & Hindell MA (1999). The composition of Australian fur seal (*Arctocephalus pusillus doriferus*) milk throughout lactation. *Physiological and Biochemical Zoology* **72**, 605–612.

Arnould JPY & Hindell MA (2001). Dive behaviour, foraging locations, and maternal attendance patterns of Australian fur seals (*Arctocephalus pusillus doriferus*). *Canadian Journal of Zoology* **79**, 35–48.

Arnould JPY, Littnan CL & Lento GM (2000). First contemporary record of New Zealand fur seals *Arctocephalus forsteri* breeding in Bass Strait. *Australian Mammalogy* **22**, 57–62.

Arundel JH (1978). Parasites and parasitic diseases of Australian marine mammals. In *Fauna: Part B, Proceedings 36*, pp. 323–333. Post Graduate Committee in Veterinary Science: Sydney.

ARWH (2007). *Australian Registry of Wildlife Health*, Zoological Parks Board of NSW: Sydney.

Atkinson S (1997). Reproductive biology of seals. *Reviews of Reproduction* **2**, 175–194.

Atkinson S, Gilmartin WG & Lasley BL (1993). Testosterone response to a gonadotropin-releasing hormone agonist in Hawaiian monk seals (*Monachus schauinslandi*). *Journal of Reproduction and Fertility* **97**, 35–38.

Atkinson S, Ragen TJ, Gilmartin WG, Becker BL & Johanos TC (1998). Use of a GnRH agonist to suppress testosterone in wild male Hawaiian monk seals (*Monachus schauinslandi*). *General and Comparative Endocrinology* **112**, 178–182.

AVMA (2001). Report of the American Veterinary Medical Association Panel on Euthanasia 2000. *Journal of American Veterinary Medical Association* **218**, 669–696.

AZA (2005). Contraception Advisory Group 2005 recommendations. American Zoo and Aquarium Association Wildlife Contraception Centre, St Louis Zoo. <www.stlzoo.org/contraception>.

Baker JR, Anderson SS & Fedak MA (1988). The use of a ketamine-diazepam mixture to immobilise wild grey seals (*Halichoerus grypus*) and southern elephant seals (*Mirounga leonina*). *Veterinary Record* **123**, 287–289.

Baker AS, Ruoff KL & Madoff S (1998). Isolation of *Mycoplasma* species from a patient with seal finger. *Clinical Infectious Diseases* **27**, 1168–1170.

Barnes JA & Smith JS (2004). Bilateral phacofragmentation in a New Zealand fur seal (*Arctocephalus forsteri*). *Journal of Zoo and Wildlife Medicine* **35**, 110–112.

Bengston JL, Boveng P, Franzen U, Have P, Heide-Jorgensen MP & Harkonen TJ (1991). Antibodies to canine distemper virus in Antarctic seals. *Marine Mammal Science* **7**, 85–87.

Bertschinger HJ, Asa CS, Calle PP, Long JA, Bauman K, DeMatteo K, Jöchle W, Trigg TE & Human A (2001). Control of reproduction and sex related behaviour in exotic wild carnivores with the GnRH analogue deslorelin preliminary observations. *Journal of Reproduction and Fertility Supplement* **57**, 275–283.

Bester MN (1987). Subantarctic fur seal, *Arctocephalus tropicalis*, on Gough Island (Tristan Da Cunha group). NOAA Technical Report NMFS 51 Status, Biology and Ecology of Fur Seals. In *Proceedings of an International Symposium and Workshop*, April 1984 (eds JP Croxall & RL Gentry), pp. 57–60. Cambridge, England.

Bester MN (1990). Reproduction in the male sub-Antarctic fur seal *Arctocephalus tropicalis*. *Journal of Zoology, London* **222**, 177–185.

Bodley K, van Polanen Petel T & Gales N (2005). Immobilisation of free-living Weddell seals (*Leptonychotes weddellii*) using midazolam and isoflurane. *Polar Biology* **28**, 631–636.

Bonner WN (1982). The status of the Antarctic fur seal, *Arctocephalus gazella*. In *Mammals in the Seas*. Food and Agriculture Organisation, Fisheries Series 5, Vol. 4, pp. 423–430.

Bonner WN (1994). *Seals and Sea Lions of the World*. Blandford: London.

Borkowski R, Moore PA, Mumford S, Patchett K, Rice J & Rubinstein B (1999). Extended use of subpalpebral lavage systems for treatment of keratitis in a harbor seal (*Phoca vitulina*). In *Proceedings of the 30th Annual Conference of the International Association for Aquatic Animal Medicine*, pp. 39–40. Boston, Massachussetts.

Bossart GD & Schwartz JC (1990). Acute necrotizing enteritis associated with suspected coronavirus infection in three harbor seals (*Phoca vitulina*). *Journal of Zoo and Wildlife Medicine* **21**, 84–87.

Bossart GD, Reidarson TH, Dierauf LA & Duffield DA (2001). Clinical pathology. In *CRC Handbook of Marine Mammal Medicine* (2nd edn) (eds LA Dieruf & FMD Gulland), pp. 383–436. CRC Press: Boca Raton, Florida.

Bowater RO, Norton J, Johnson S, Hill B, O'Donoghue P & Prior H (2003). Toxoplasmosis in Indo-Pacific humpbacked dolphins (*Sousa chinensis*), from Queensland. *Australian Veterinary Journal* **81**, 627–632.

Boyd IL (1991a). Environmental and physiological factors controlling the reproductive cycles of pinnipeds. *Canadian Journal of Zoology* **69**, 1135–1148.

Boyd IL (1991b). Changes in plasma progesterone and prolactin concentrations during the annual cycle and the role of prolactin in the maintenance of lactation and luteal development in the Antarctic fur seal (*Arctocephalus gazella*). *Journal of Reproduction and Fertility* **91**, 637–647.

Boyd IL, Lockyer C & Marsh HD (1999). Reproduction in marine mammals. In *Biology of Marine Mammals* (eds JE Reynolds & SA Rommel), pp. 218–286. Smithsonian Institution Press: Washington DC.

Braun R, Paulson M, Omphroy C, Heath B, Gilmartin W, Ragen T, Breese M & Driscoll C (1996). Corneal opacities in Hawaiian monk seals. In *Proceedings of the 27th Annual Conference of the International Association for Aquatic Animal Medicine*, pp. 97–99. Chattanooga, Tennessee.

Brew SD, Perrett LL, Stack JA, MacMillan AP & Staunton NJ (1999). Human exposure to *Brucella* recovered from a sea mammal. *Veterinary Record* **144**, 483.

Brothers N & Pemberton D (1990). Status of Australian and New Zealand fur seals at Maatsuyker Island, southwestern Tasmania. *Australian Wildlife Research* **17**, 563–569.

Brown KG (1957). The leopard seal at Heard Island, 1951–1954. *ANARE Interim Reports* **16**, 1–34. Antarctic Division: Melbourne.

Brown RG, Bowen WD, Eddington JD, Kimmins WC, Mezei M, Parsons JL & Pohajdak B (1997a). Evidence for a long-lasting single administration contraceptive vaccine in wild grey seals. *Journal of Reproductive Immunology* **35**, 43–51.

Brown RG, Bowen WD, Eddington JD, Kimmins WC, Mezei M, Parsons JL & Pohajdak B (1997b). Temporal trends in antibody production in captive grey, harp and hooded seals to a single administration immunocontraceptive vaccine. *Journal of Reproductive Immunology* **35**, 53–64.

Bryden MM (1967). Testicular temperature in the southern elephant seal, *Mirounga leonina* (Linn.). *Journal of Reproduction and Fertility* **13**, 583–584.

Bryden MM & Lim GHK (1969). Blood parameters of the southern elephant seal (*Mirounga leonina*, Linn.) in relation to diving. *Comparative Biochemistry and Physiology* **28**, 139–148.

Callan RJ, Early G, Kida H & Hinshaw VS (1995). The appearance of H3 influenza viruses in seals. *Journal of General Virology* **76**, 199–203.

Calle PP, Stetter MD, Raphael BL, Cook RA, McClave C, Basinger J, Walters H & Walsh K (1997). Use of depot leuprolide acetate to control undesirable male associated behaviours in the California sea lion (*Zalophus californianus*) and California sea otter (*Enhydra lutris*). In *Proceedings of the 28th Annual Conference of the International Association for Aquatic Animal Medicine*. Harderwijk, Holland.

Calle PP, Raphael BL, Cook RA, McClave C, Basinger JA & Walters H (1999). Use of depot leuprolide, cyproterone and deslorelin to control aggression in an all male California sea otter (*Enhydra lutris nereis*) colony. In *Proceedings of the 30th Annual Conference of the International Association for Aquatic Animal Medicine*, pp. 42–45. Boston, Massachusetts.

Cargill CF, Needham DJ & Judson GJ (1979). Plasma biochemical values of clinically-normal Australian sea lions (*Neophoca cinerea*). *Journal of Wildlife Diseases* **15**, 105–110.

Carlini AR, Marquez MEI, Soave G, Vergani DF & Ronayne de Ferrer PA (1994). Southern elephant seal, *Mirounga leonina*: composition of milk during lactation. *Polar Biology* **14**, 37–42.

Carrick R & Ingham SE (1962). Studies on the southern elephant seal, *Mirounga leonina* (L.). I. Introduction to the series. *CSIRO Wildlife Research* **7**, 89–101.

Carrick R, Csordas SE & Ingham SE (1962). Studies on the southern elephant seal, *Mirounga leonina* (L.). IV. Breeding and development. *CSIRO Wildlife Research* **7**, 161–197.

Castellini JM, Meiselman HJ & Castellini MA (1996). Understanding and interpreting hematocrit measurements in pinnipeds. *Marine Mammal Science* **12**, 251–264.

Cavagnolo RZ & Vedros NA (1979). Serum and colostrum immunoglobulin levels in the northern fur seal, *Callorhinus ursinus*. *Developmental and Comparative Immunology* **3**, 139–146.

Cawthorn MW (1994). Seal finger and mycobacterial infections of man from marine mammals; occurrence, infection and treatment. *Conservation Advisory Science Notes No. 102*. Department of Conservation: Wellington, New Zealand.

Chrissey SD (1998). Handling fish fed to fish-eating animals: a manual of standard operating procedures. US Department of Agriculture, Agricultural Research Service, National Agricultural Library. <www.aphis.usda.gov/ac/fishfeed.html>.

Clark P (2004). Cytological characteristics of haematological cells from Australian mammals. In *Haematology of Australian Mammals* (ed. P Clark), pp. 95–146. CSIRO Publishing: Melbourne.

Clark P, Boardman WS & Duignan PJ (2002). Cytology of haematological cells of otariid seals indigenous to Australasian waters. *Australian Veterinary Journal* **80**, 161–164.

Clark P, Holz P & Duignan PJ (2004). Collection and handling of blood samples. In *Haematology of Australian Mammals* (ed. P Clark), pp. 1–20. CSIRO Publishing: Melbourne.

Colagross-Schouten AM, Mazet JAK, Gulland FMD, Miller MA & Hietala S (2002). Diagnosis and seroprevalence of leptospirosis in California sea lions from coastal California. *Journal of Wildlife Diseases* **38**, 7–17.

Condy PR (1978). Distribution, abundance, and annual cycle of fur seals (*Arctocephalus* spp.) on the Prince Edward Islands. *South African Journal of Wildlife Research* **8**, 159–168.

Condy PR (1979). Annual cycle of the southern elephant seal, *Mirounga leonina* (Linn.) at Marion Island. *South African Journal of Zoology* **14**, 95–102.

Cordes DO & O'Hara PJ (1979). Diseases of captive marine mammals. *New Zealand Veterinary Journal* **27**, 147–150.

Costa DP, Gales NJ & Crocker DE (1998). Blood volume and diving ability of the New Zealand sea lion, *Phocarctos hookeri*. *Physiological Zoology* **71**, 208–213.

Costa DP, Gales NJ & Goebel ME (2001). Aerobic dive limit: how often does it occur in nature? *Comparative Biochemistry and Physiology Part A* **129**, 771–783.

Cousins DV, Williams SN, Reuter R, Forshaw D, Chadwick B, Coughran D, Collins P & Gales N (1993). Tuberculosis in wild seals and characterisation of the seal bacillus. *Australian Veterinary Journal* **70,** 92–97.

Cousins DV, Bastida R, Cataldi A, Quse V, Redrobe S, Dow S, Duignan P, Murray A, Dupont C, Ahmed N, Collins DM, Butler WR, Dawson D, Rodriguez D, Loureiro J, Romano MI, Alito A, Zumarraga M & Bernardelli A (2003). Tuberculosis in seals caused by a novel member of the *Mycobacterium tuberculosis* complex: *Mycobacterium pinnipedii* sp. nov. *International Journal of Systematic and Evolutionary Microbiology* **53**, 1305–1314.

Cowan DF, House C & House JA (2001). Public health. In *CRC Handbook of Marine Mammal Medicine* (2nd edn) (eds LA Dierauf & FMD Gulland), pp. 767–778. CRC Press: Boca Raton, Florida.

Crawley MC (1990). Suborder Pinnipedia. In *The Handbook of New Zealand Mammals* (ed. CM King), pp. 243–280. OUP: Oxford.

Crawley MC & Warneke R (1979). New Zealand fur seal. In *Mammals in the Seas*. Food and Agriculture Organisation Fisheries Series 5, Vol. 2, pp. 45–48. FAO: Rome.

Croxall JP & Gentry RL (1987). The 1984 fur-seal symposium: an introduction. NOAA Technical Report NMFS 51: status, biology and ecology of fur-seals. In *Proceedings of International Symposium and Workshop*, April 1984 (eds JP Croxall & RL Gentry), pp. 1–4. Cambridge, England.

Csordas SE & Ingham SE (1965). The New Zealand fur seal, *Arctocephalus forsteri* (Lesson), at Macquarie Island, 1949–64. *CSIRO Wildlife Research* **10**, 83–99.

Daniel JC (1981). Delayed implantation in the northern fur seal (*Callorhinus ursinus*) and other pinnipeds *Journal of Reproduction and Fertility* **29**, 35–50.

Deng MQ, Peterson RP & Cliver DO (2000). First findings of *Cryptosporidium* and *Giardia* in California sea lions (*Zalophus californianus*). *Journal of Parasitology* **86**, 490–494.

Denison DM & Kooyman GL (1973). The structure and function of the small airways in pinniped and sea otter lungs. *Respiration Physiology* **17**, 1–10.

Dix B (1993). A new record this century of a breeding colony in the North Island for the New Zealand fur seal *Arctocephalus forsteri*. *Journal of the Royal Society of New Zealand* **23**, 1–4.

Doidge DW, McCann TS & Croxall JP (1986). Attendance behaviour of Antarctic fur seals. In *Fur Seals: Maternal Strategies on Land and at Sea* (eds RL Gentry & GL Kooyman), pp. 102–114. Princeton University Press: Princeton.

Drabek CM (1975). Some anatomical aspects of the cardiovascular system of Antarctic seals and their possible functional significance in diving. *Journal of Morphology* **145**, 85–106.

Duerr R (2002). Harbor seals and northern elephant seals. In *Hand-rearing Wild and Domestic Mammals* (ed. LJ Gage), pp. 132–142. Iowa State Press: Ames, Iowa.

Duignan PJ (2000). Diseases of cetaceans and pinnipeds In *Marine Wildlife, Proceedings 335,* pp. 431–462. Post Graduate Foundation in Veterinary Science: Sydney.

Duignan PJ (2003). Disease investigations in stranded marine mammals, 1999–2002. DOC Science Internal Series 104. Department of Conservation: Wellington, New Zealand.

Duignan PJ, Nielsen O, House C, Kovacs KM, Duffy N, Early G, Sadove S, St. Aubin DJ, Rima BK & Geraci JR (1997). Epizootiology of morbillivirus infection in harp, hooded, and ringed seals from the Canadian arctic and western Atlantic. *Journal of Wildlife Diseases* **33**, 7–19.

Duignan PJ, Castinel A, Grinberg A & Wilkinson IS (2004). *Klebsiella pneumoniae* epidemics in New Zealand sea lions: a natural phenomenon or an adverse human impact? In *Abstracts of 22nd Wakefield Fisheries Symposium, Sea Lions of the World: Conservation and Research in the 21st Century*. Anchorage, Alaska.

Dunn JL, Abt DA, Overstrom NA & St Aubin DJ (1996). An epidemiologic survey to determine risk factors associated with corneal and lenticular lesions in captive harbor seals and California sea lions. In *Proceedings of the 27th Annual Conference of International Association for Aquatic Animal Medicine*, pp. 100–102. Chattanooga, Tennessee.

Dutton AG (1991). Cataract extraction in a fur seal. *Journal of American Veterinary Medical Association* **198,** 309–311.

Dutton G, Duignan P & Wilkinson I (2000). Concentrations of progesterone in the serum, milk and faeces of the female New Zealand sea lion (*Phocarctos hookeri*) from the Auckland Islands. In *Proceedings of the Joint*

Conference of Wildlife Disease Association (Australasian Section) and Wildlife Society of New Zealand Veterinary Association, pp. 20–21. Leigh, New Zealand.

Elsner R (1969). Cardiovascular adjustments to diving. In *The Biology of Marine Mammals* (ed. HT Andersen), pp. 117–145. Academic Press: London.

Engelhard GH, Hall AJ, Brasseur SMJM & Reijnders PJH (2002). Blood chemistry in southern elephant seal mothers and pups during lactation reveals no effect of handling. *Comparative Biochemistry and Physiology Part A* **133**, 367–378.

Fayolle C, Leray C, Ohlmann P, Gutbier G, Cazenave J, Gachet C & Groscolas R (2000). Lipid composition of blood platelets and erythrocytes of southern elephant seals (*Mirounga leonina*) and Antarctic fur seal (*Arctocephalus gazella*). *Comparative Biochemistry and Physiology Part B* **126**, 39–47.

FDA (2001). Fish and fisheries products hazards and control guidance (3rd edn). Center for Food Safety and Applied Nutrition, US Food and Drug Administration. <www.fda.gov>.

Field IC, Bradshaw CJA, McMahon CR, Harrington J & Burton HR (2002). Effects of age, size and condition of elephant seals (*Mirounga leonina*) on their intravenous anaesthesia with tiletamine and zolazepam. *Veterinary Record* **151**, 235–240.

Filer R, Haulena M, Fauquier D & Gulland FMD (2003). Clear corneal cataract extraction in a northern elephant seal (*Mirounga angustirostris*). In *Proceedings of 34th Annual Conference of International Association for Aquatic Animal Medicine*, pp. 155–156. Waikoloa, Hawaii.

Forbes LB, Measures L, Gajadhar A & Kapel C (2003). Infectivity of *Trichinella nativa* in traditional northern (country) foods prepared with meat from experimentally infected seals. *Journal of Food Protection* **66**, 1857–1863.

Forshaw D & Phelps GR (1991). Tuberculosis in captive colony of pinnipeds *Journal of Wildlife Diseases* **27**, 288–295.

Gage LJ (2003). Pinnipedia (seals, sea lions, walruses). In *Zoo and Wild Animal Medicine* (5th edn) (eds ME Fowler & RE Miller), pp. 459–475. Saunders/Elsevier Science: Missouri.

Gales NJ (1984). Ketamine hydrochloride and diazepam anaesthesia of a leopard seal (*Hydrurga leptonyx*) for the biopsy of multiple fibromatous epulis. *Australian Veterinary Journal* **61**, 295–296.

Gales NJ (1989). Chemical restraint and anesthesia of pinnipeds: a review. *Marine Mammal Science* **5**, 228–256.

Gales NJ & Costa DP (1997). The Australian sea lion: a review of an unusual life history. In *Marine Mammal Research in the Southern Hemisphere*. Vol. 1 Status, ecology and medicine (eds M Hindell & C Kemper), pp. 78–87. Surrey Beatty & Sons: Sydney.

Gales NJ & Mattlin RH (1997). Summer diving behaviour of lactating New Zealand sea lions, *Phocarctos hookeri*. *Canadian Journal of Zoology* **75**, 1695–1706.

Gales NJ, Cheal AJ, Pobar GJ & Williamson P (1992). Breeding biology and movements of Australian sea lions, *Neophoca cinerea*, off the west coast of Western Australia. *Wildlife Research* **19**, 405–416.

Gales NJ, Shaughnessy PD & Dennis TE (1994). Distribution, abundance and breeding cycle of the Australian sea lion, *Neophoca cinerea* (Mammalia: Pinnipedia). *Journal of Zoology, London* **234**, 353–370.

Gales NJ, Costa DP & Kretzmann M (1996). Proximate milk composition of Australian sea lion milk throughout the entire supra-annual lactation period. *Australian Journal of Zoology* **44**, 651–657.

Gales NJ, Williamson P, Higgins LV, Blackberry MA & James I (1997). Evidence for a prolonged postimplantation period in the Australian sea lion (*Neophoca cinerea*). *Journal of Reproduction and Fertility* **111**, 159–163.

Gales N, Barnes J, Chittick B, Gray M, Robinson S, Burns J & Costa D (2005). Effective, field-based inhalation anaesthesia for ice seals. *Marine Mammal Science* **21**, 717–727.

Garcia AR, Montali RJ, Dunn, JL, Torres NL, Centeno JA & Goodman Z (2000). Hemochromatosis in captive otarids. In *Proceedings of American Association of Zoo Veterinarians and International Association for Aquatic Animal Medicine Joint Conference*, p. 197. New Orleans, Louisiana.

Gentry RL, Costa DP, Croxall JP, David JHM, Davis RW, Kooyman GL, Majluf P, McCann TS & Trillmich F (1986). Synthesis and conclusions. In *Fur Seals: Maternal Strategies on Land and at Sea* (eds RL Gentry & GL Kooyman), pp. 220–264. Princeton University Press: Princeton.

Georges J-Y & Guinet C (2000). Early mortality and perinatal growth in the subantarctic fur seal (*Arctocephalus tropicalis*) on Amsterdam Island. *Journal of Zoology, London* **251**, 277–287.

Geraci JR (2000a). Nutrition of piscivorous animals. In *Marine Wildlife, Proceedings 335*, pp. 121–132. Post Graduate Foundation in Veterinary Science: Sydney.

Geraci JR (2000b). Euthanasia of marine mammals. In *Marine Wildlife, Proceedings 335*, pp. 85–91. Post Graduate Foundation in Veterinary Science: Sydney.

Geraci JR & Lounsbury VJ (1993). Specimen and data collection. In *Marine Mammals Ashore: A Field Guide for Strandings*, pp. 175–228. Texas A&M Sea Grant Publication: Galveston.

Geraci JR & Smith TG (1975). Functional hematology of ringed seals (*Phoca hispida*) in the Canadian arctic. *Journal of Fisheries Research Board of Canada* **32**, 2559–2564.

Geraci JR & Sweeney J (1978). Marine mammals: clinical techniques. In *Zoo and Wild Animal Medicine* (ed. ME Fowler), pp. 580–587. WB Saunders: Philadelphia.

Geraci JR, St Aubin DJ, Barker IK, Webster RG, Hinshaw VS, Bean WJ, Ruhnke HL, Prescott JH, Early G, Baker AS, Madoff S & Schooley RT (1982). Mass mortality of harbor seals: pneumonia associated with influenza A virus. *Science* **215**, 1129–1131.

Goldsworthy SD (1995). Differential expenditure of maternal resources in Antarctic fur seals, *Arctocephalus gazella*, at Heard Island, southern Indian Ocean. *Behavioural Ecology* **6**, 218–228.
Goldsworthy SD & Shaughnessy PD (1989). Subantarctic fur seals *Arctocephalus tropicalis* at Heard Island. *Polar Biology* **9**, 337–339.
Goldsworthy SD & Shaughnessy PD (1994). Breeding biology and haul-out pattern of the New Zealand fur seal, *Arctocephalus forsteri*, at Cape Gantheaume, South Australia. *Wildlife Research* **21**, 365–376.
Goldsworthy SD, Pemberton D & Warneke RM (1997). Field identification of Australian and New Zealand fur seals, *Arctocephalus* spp., based on external characters. In *Marine Mammal Research in the Southern Hemisphere.* Vol. 1 Status, ecology and medicine (eds M Hindell & C Kemper), pp. 63–71. Surrey Beatty & Sons: Sydney.
Gray R (2005). A comparative assessment of selected health indices in the leopard seal, *Hydrurga leptonyx*, and Weddell seal, *Leptonychotes weddellii*, in Prydz Bay, Eastern Antarctica. PhD thesis. University of Sydney, Australia.
Green B, Fogerty A, Libke J, Newgrain K & Shaughnessy P (1993). Aspects of lactation in the crabeater seal (*Lobodon carcinophagus*). *Australian Journal of Zoology* **41**, 203–213.
Greenwood AG (1985). Prevalence of ocular anterior segment disease in captive pinniped. *Aquatic Mammals* **1**, 13–15.
Gulland F (2000). Domoic acid toxicity in California sea lions (*Zalophus californianus*) stranded along the central California Coast, May–October 1998. NOAA Technical Memorandum, NMFS-OPR-8. National Marine Fisheries Service, US Department of Commerce: Silver Spring, Maryland.
Gulland FMD, Haulena M & Dierauf LA (2001). Seals and sea lions. In *CRC Handbook of Marine Mammal Medicine* (2nd edn) (eds LA Dierauf & FMD Gulland), pp. 907–926. CRC Press: Boca Raton, Florida.
Gwynn AM (1953). The status of the leopard seal at Heard Island and Macquarie Island, 1948–1950. ANARE Interim Reports **3**, 1–33. Antarctic Division: Melbourne.
Hamilton JE (1939). The leopard seal *Hydrurga leptonyx* (De Blainville). *Discovery Reports* **18**, 239–264.
Harder TC, Plotz J & Liess B (1991). Antibodies against European phocid herpesvirus isolates detected in sera of Antarctic seals. *Polar Biology* **11**, 509–512.
Harper CG, Xu SL, Rogers AB, Feng Y, Shen Z, Taylor NS, Dewhirst FE, Paster BJ, Miller M, Hurley J & Fox JG (2003). Isolation and characterization of novel *Helicobacter* spp. from the gastric mucosa of harp seals *Phoca groenlandica. Diseases of Aquatic Organisms* **57**, 1–9.
Harrison RJ (1969). Reproduction and reproductive organs. In *The Biology of Marine Mammals* (ed. HT Andersen), pp. 253–348. Academic Press: New York.
Harrison RJ & Tomlinson JDW (1956). Observations on the venous system in certain pinnipedia and cetacea. *Proceedings of Zoological Society of London* **126**, 205–233.
Harrison RJ & Tomlinson JDW (1963). Anatomical and physiological adaptations in diving mammals. *Viewpoints in Biology* **2**, 115–162.
Haulena M & Gulland FMD (2001). Use of medetomidine-zolazepam-tiletamine with and without atipamizole reversal to immobilize captive California sea lions. *Journal of Wildlife Diseases* **37**, 566–573.
Haulena M, Gulland FMD, Calkins DG & Spraker TR (2000). Immobilization of California sea lions using medetomidine plus ketamine with and without isoflurane and reversal with atipamezole. *Journal of Wildlife Diseases* **36**, 124–130.
Hicks BD & Worthy GA (1987). Sealpox in captive gray seals (Halichoerus grypus) and their handlers. *Journal of Wildlife Diseases* **23**, 1–6.
Hicks CL, Kinoshita R & Ladds PW (2000). Pathology of melioidosis in captive marine mammals. *Australian Veterinary Journal* **78**, 193–195.
Higgins LV (1993). The nonannual, nonseasonal breeding cycle of the Australian sea lion, *Neophoca cinerea. Journal of Mammalogy* **74**, 270–274.
Higgins LV & Gass L (1993). Birth to weaning: parturition, duration of lactation, and attendance cycles of Australian sea lions (*Neophoca cinerea*). *Canadian Journal of Zoology* **71**, 2047–2055.
Higgins DP, Rogers TL, Irvine AD & Hall-Aspland SA (2002). Use of midazolam/pethidine and tiletamine/zolazepam combinations for the chemical restraint of leopard seals (*Hydrurga leptonyx*). *Marine Mammal Science* **18**, 483–499.
Hindell MA, Slip DJ & Burton HR (1991). The diving behaviour of adult male and female southern elephant seals, *Mirounga leonina* (Pinnipedia: Phocidae). *Australian Journal of Zoology* **39**, 595–619.
Hindell MA, Bryden MM & Burton HR (1994). Early growth and milk composition in southern elephant seals (*Mirounga leonina*). *Australian Journal of Zoology* **42**, 723–732.
Hofman RJ, Reichle RA, Siniff DB & Müller-Schwarze D (1977). The leopard seal (*Hydrurga leptonyx*) at Palmer Station, Antarctica. Adaptations within Antarctic ecosystems. In *Proceedings of 3rd SCAR Symposium on Antarctic Biology* (ed. GA Llano), pp. 769–782. Smithsonian Institute: Washington DC.
Hofmeyr GJG, Bester MN & Jonker FC (1997). Changes in population sizes and distribution of fur seals at Marion Island. *Polar Biology* **17**, 150–158.
Horning M & Trillmich F (1997). Development of hemoglobin, hematocrit, and erythrocyte values in the Galápagos fur seal. *Marine Mammal Science* **13**, 100–113.
Hubbard RC (1968). Husbandry and laboratory care of pinnipeds In *The Behavior and Physiology of Pinnipeds* (eds RJ Harrison, RC Hubbard, RS Peterson, CE Rice &

RJ Schusterman), pp. 299–358. Appleton-Century-Crofts: New York.

Hunter JE, Duignan PJ, Dupont C, Fray L, Fenwick SG & Murray A (1998). First report of potentially zoonotic tuberculosis in fur seals in New Zealand. *New Zealand Medical Journal* **111**, 130–131.

Irvine A, Bryden MM, Corkeron PJ & Warneke RM (1997). A census of fur seals at Montagu Island, New South Wales. In *Marine Mammal Research in the Southern Hemisphere.* Vol. 1 Status, ecology and medicine (eds M Hindell & C Kemper), pp. 56–62. Surrey Beatty & Sons: Sydney.

Jones E (1981). Age in relation to breeding status of the male southern elephant seal, *Mirounga leonina* (L.), at Macquarie Island. *Australian Wildlife Research* **8**, 327–334.

Katsumata E, Hori T & Tsutsui T (2003). Contraceptive effect of proligestone on spotted seals and crossbreeds of spotted and harbor seals. *Journal of Veterinary Medical Science* **65**, 619–623.

Kennedy-Stoskopf S (2001). Viral diseases. In *CRC Handbook of Marine Mammal Medicine* (2nd edn) (eds LA Dierauf & FMD Gulland), pp. 285–308. CRC Press: Boca Raton, Florida.

Kennedy S, Kuiken T, Jepson PD, Deaville R, Forsyth M, Barrett T, van de Bildt MW, Osterhaus AD, Eybatov T, Duck C, Kydyrmanov A, Mitrofanov I & Wilson S (2000). Mass die-off of Caspian seals caused by canine distemper virus. *Emerging Infectious Disease* **6**, 637–639.

Kerley GIH (1983). Comparison of seasonal haul-out patterns of fur seals *Arctocephalus tropicalis* and *A. gazella* on subantarctic Marion Island. *South African Journal of Wildlife Research* **13**, 71–77.

Kerley GIH (1987). *Arctocephalus tropicalis* on the Prince Edward Islands. NOAA Technical Report NMFS 51: Status, biology and ecology of fur seals. In *Proceedings of International Symposium and Workshop,* April 1984 (eds JP Croxall & RL Gentry), pp. 61–64. Cambridge, England.

King JE (1983). *Seals of the World* (2nd edn). UQP: St Lucia.

King DP, Sanders JL, Nomura CT, Stoddard RA, Ortiz CL & Evans SW (1998). Ontogeny of humoral immunity in northern elephant seal (*Mirounga angustirostris*) neonates. *Comparative Biochemistry and Physiology Part B: Biochemistry and Molecular Biology* **121**, 363–368.

Kretzmann MB, Costa DP, Higgins LV & Needham DJ (1991). Milk composition of Australian sea lions, *Neophoca cinerea*: variability in lipid content. *Canadian Journal of Zoology* **69**, 2556–2561.

Lane RAB, Morris RJH & Sheedy JW (1972). A haematological study of the southern elephant seal, *Mirounga leonina* (Linn.). *Comparative Biochemistry and Physiology* **42A**, 841–850.

Lapointe J-M, Gulland FM, Haines DM, Barr BC & Duignan PJ (1999). Placentitis due to *Coxiella burnetti* in a Pacific harbor seal (*Phoca vitulina richardsi*). *Journal of Veterinary Diagnostic Investigation* **11**, 541–543.

Laws RM (1957). On the growth rates of the leopard seal, *Hydrurga leptonyx* (De Blainville, 1820). *Säugetierkunde Mitteilungen* **5**, 49–55.

Laws RM (1984). Seals. In *Antarctic Ecology* Vol. 2 (ed. RM Laws), pp. 621–715. Academic Press: London.

Laws RM (1993). Identification of species. In *Antarctic Seals Research Methods and Techniques* (ed. RM Laws), pp. 1–28. CUP: Cambridge.

Laws RM & Sinha AA (1993). Reproduction. In *Antarctic Seals Research Methods and Techniques* (ed. RM Laws), pp. 228–267. CUP: Cambridge.

Laws RM & Taylor RJF (1957). A mass dying of crab-eater seals (*Lobodon carcinophagus*). *Proceedings of Zoological Society of London* **129**, 315–325.

Lenfant C (1969). Physiological properties of blood of marine mammals. In *The Biology of Marine Mammals* (ed. HT Andersen), pp. 95–116. Academic Press: New York.

Lenfant C, Johansen K & Torrance JD (1970). Gas transport and oxygen storage capacity in some pinnipeds and the sea otter. *Respiration Physiology* **9**, 277–286.

Lewis M, Campagna C, Uhart M & Ortiz CL (2001). Ontogenic and seasonal variation in blood parameters in southern elephant seals. *Marine Mammal Science* **17**, 862–872.

Ling JK (1965a). Functional significance of sweat glands and sebaceous glands in seals. *Nature* **208**, 560–562.

Ling JK (1965b). Hair growth and moulting in the southern elephant seal, *Mirounga leonina* (Linn.). In *Biology of the Skin and Hair Growth* (eds AG Lyne & BF Short), pp. 525–544. Angus & Robertson: Sydney.

Ling JK (1970). Pelage and molting in wild mammals with special reference to aquatic forms. *Quarterly Review of Biology* **45**, 16–54.

Ling JK (1987). New Zealand fur seal, *Arctocephalus forsteri*, (Lesson), in South Australia. NOAA Technical Report NMFS 51: Status, biology and ecology of fur seals. In *Proceedings of International Symposium and Workshop,* April 1984 (eds JP Croxall & RL Gentry), pp. 53–55. Cambridge, England.

Ling JK & Walker GE (1978). An 18-month breeding cycle in the Australian sea lion? *Search* **9**, 464–465.

Lucas RJ, Barnett J & Riley P (1999). Treatment of lesions of osteomyelitis in the hind flippers of six grey seals (*Halichoerus grypus*). *Veterinary Record* **145**, 547–550.

Lynch M (1999). Pinnipeds: anaesthesia, medicine and surgery. In *Wildlife in Australia: Healthcare and Management, Proceedings 327,* pp. 81–93. Post Graduate Foundation in Veterinary Science: Sydney.

Lynch MJ, Tahmindjis MA & Gardner H (1999). Immobilisation of pinniped species. *Australian Veterinary Journal* **77**, 181–185.

Marlow BJ (1975). The comparative behaviour of the Australasian sea lions *Neophoca cinerea* and *Phocarctos hookeri* (Pinnipedia: Otariidae). *Mammalia* **39**, 159–230.

Marquez MEI, Slobodianik NH, Ronayne de Ferrer PA, Carlini AR, Vergani DF & Daneri GA (1995).

Immunoglobulin A levels in southern elephant seal (*Mirounga leonina*) milk during the suckling period. *Comparative Biochemistry and Physiology Part B: Biochemistry and Molecular Biology* **112**, 569–572.

Marquez MEI, Carlini AR, Slobodianik NH, Ronayne de Ferrer PA & Godoy MF (1998). Immunoglobulin M serum levels in females and pups of southern elephant seal (Mirounga leonina) during the suckling period. *Comparative Biochemistry and Physiology Part A: Molecular and Integrative Physiology* **119**, 795–799.

Mattlin RH (1981). Pup growth of the New Zealand fur seal *Arctocephalus forsteri* on the Open Bay Islands, New Zealand. *Journal of Zoology, London* **193**, 305–314.

Mawson PR & Coughran DK (1999). Records of sick, injured and dead pinnipeds in Western Australia 1980–1996. *Journal of Royal Society of Western Australia* **82**, 121–128.

McCann TS & Doidge DW (1987). Antarctic fur-seal, *Arctocephalus gazella*. NOAA Technical Report NMFS 51: Status, biology and ecology of fur-seals. In *Proceedings of International Symposium and Workshop*, April 1984 (eds JP Croxall & RL Gentry), pp. 5–8. Cambridge, England.

Measures LN & Olson M (1999). Giardiasis in pinnipeds from eastern Canada. *Journal of Wildlife Diseases* **35**, 779–782.

Medway W & Geraci JR (1978). Clinical pathology of marine mammals. In *Zoo and Wild Animal Medicine* (ed. ME Fowler), pp. 604–610. WB Saunders: Philadelphia.

Melrose WD, Feast MM, Woods R, McMinn A, Jupe DML & Bell PA (1995). Haematology, red cell enzymes, and red cell metabolic intermediates of 20 wild southern elephant seals (*Mirounga leonina*) from Macquarie Island. *Comparative Haematology International* **5**, 1–6.

Migaki G, Allen JE & Casey HW (1977). Toxoplasmosis in a California sea lion (*Zalophus californianus*). *American Journal of Veterinary Research* **39**, 135–136.

Miller EH (1975). Annual cycle of fur seals, *Arctocephalus forsteri* (Lesson), on the Open Bay Islands, New Zealand. *Pacific Science* **29**, 139–152.

Morgan IR, Caple IW, Westbury HA & Campbell J (1978). Disease investigations of penguins and elephant seals on Macquarie Island. Government of Victoria Research Project Series Number 47, April 1978, pp. 1–51. Department of Agriculture: Victoria.

Morgan K, Norman RJdeB, Duignan PJ, Gibbs N & Best H (2000). Studies on hookworm in New Zealand otariid seals. In *Proceedings of Wildlife Disease Association (Australasian Section) and NZVA Wildlife Society Joint Conference*. Warkworth, New Zealand.

Muller G, Groters S, Siebert U, Rosenberger T, Driver J, Konig M, Becher P, Hetzel U & Baumgartner W (2003). Parapoxvirus infection in harbor seals (*Phoca vitulina*) from the German North Sea. *Veterinary Pathology* **40**, 445–454.

Munro R & Synge B (1991). Coccidiosis in seals. *Veterinary Record* **129**, 179–180.

Munro R, Ross H, Cornwell C & Gilmour J (1992). Disease conditions affecting common seals (*Phoca vitulina*) around the Scottish mainland, September–November 1988. *Science of the Total Environment* **115**, 67–82.

Needham D J (1997). The role of stones in the seal lion stomach: investigations using contrast radiography and fluoroscopy. In *Marine Mammal Research in the Southern Hemisphere.* Vol. 1: Status, ecology and medicine (eds M Hindell & C Kemper), pp. 164–169. Surrey Beatty & Sons: Sydney.

Needham DJ, Cargill CF & Sheriff D (1980). Haematology of the Australian sea lion, *Neophoca cinerea*. *Journal of Wildlife Diseases* **16**, 103–107.

Norman RJdeB (1995). Disease in free-living Australian seals. In *Proceedings of Australian Association of Veterinary Conservation Biologists*, pp. 41–50. Melbourne.

Nowak RM (2003). Pinnipeds. In *Walker's Marine Mammals of the World*, pp. 65–123. Johns Hopkins University Press: Baltimore.

Oftedal OT, Boness DJ & Tedman RA (1987). The behaviour, physiology, and anatomy of lactation in the pinnipedia. *Current Mammalogy* **1**, 175–245.

Ohishi K, Ninomiya A, Kida H, Park CH, Maruyama T, Arai T, Katsumata E, Tobayama T, Boltunov AN, Khuraskin LS & Miyazaki N (2002). Serological evidence of transmission of human influenza A and B viruses to Caspian seals (*Phoca caspica*). *Microbiology and Immunology* **46**, 639–644.

Orchard VA (1979). Nocardial infections of animals in New Zealand, 1976–78. *New Zealand Veterinary Journal* **27**, 159–165.

Øritsland T (1970). Sealing and seal research in the South-west Atlantic pack ice, Sept.–Oct. 1964. In *Antarctic Ecology* Vol. 1 (ed. MW Holdgate), pp. 367–376. Academic Press: London.

Osterhaus ADME, Rimmelzwaan GF, Martina BEE, Bestebroer TM & Fouchier RAM (2000). Influenza B virus in seals. *Science* **288**, 1051–1053.

Oxley APA & McKay DB (2004). Fecal shedding of *Helicobacter* spp. by co-housed Australian sea lions (*Neophoca cinerea*) and Australian fur seals (*Arctocephalus pusillus doriferus*). *Veterinary Microbiology* **102**, 235–243.

Oxley APA, Powell M & McKay DB (2004). Species of the family *Helicobacteraceae* detected in an Australian sea lion (*Neophoca cinerea*) with chronic gastritis. *Journal of Clinical Microbiology* **42**, 3505–3511.

Pabst DA, Rommel SA & McLellan WA (1999). The functional morphology of marine mammals. In *Biology of Marine Mammals* (eds JE Reynolds III & SA Rommel), pp. 15–72. Smithsonian Institution Press: Washington DC.

Payne MR (1977). Growth of a fur seal population. *Philosophical Transactions of Royal Society of London* **B279**, 67–79.

Payne MR (1979). Growth in the Antarctic fur seal *Arctocephalus gazella*. *Journal of Zoology* (London) **187**, 1–20.

Pemberton D & Kirkwood RJ (1994). Pup production and distribution of the Australian fur seal, *Arctocephalus pusillus doriferus*, in Tasmania. *Wildlife Research* **21**, 341–352.

Read RA, Reynolds WT, Griffiths DJ & Reilly JS (1982). Vaginal prolapse in a South Australian sea lion (*Neophoca nove hollandia*). *Australian Veterinary Journal* **58**, 269–271.

Reijnders PJH (1990). Progesterone and oestradiol–17β concentration profiles throughout the reproductive cycle in harbour seals (*Phoca vitulina*). *Journal of Reproduction and Fertility* **90**, 403–409.

Renner MS & Dougherty M (1999). Preliminary investigation of the safety and efficacy of porcine zona pellucida (PZP) immunocontraception in captive California sea lions (*Zalophus californianus*). In *Proceedings of 30th Annual Conference of International Association for Aquatic Animal Medicine,* p. 91. Boston, Massachusetts.

Retamal P, Blank O, Abalos P & Torres D (2000). Detection of anti-Brucella antibodies in pinnipeds from the Antarctic territory. *Veterinary Record* **146**, 166–167.

Reyero M, Cacho E, Martinez A, Vazquez J, Marina A, Fraga S & Franco JM (1999). Evidence of saxitoxin derivatives as causative agents in the 1997 mass mortality of monk seals in the Cape Blanc Peninsula. *Natural Toxins* **7**, 311–315.

Richardson KC & Gales NJ (1987). Functional morphology of the alimentary tract of the Australian sea lion, *Neophoca cinerea*. *Australian Journal of Zoology* **35**, 219–226.

Ridgway SH (1972). Homeostasis in the aquatic environment. In *Mammals of the Sea: Biology and Medicine* (ed. SH Ridgway), pp. 590–747. Charles C Thomas: Springfield.

Riedman M (1990). Adaptations for a marine existence. In *The Pinnipeds: Seals, Sea Lions and Walruses* (M Riedman), pp. 1–49. University of California Press: Berkeley.

Robeck TR, Atkinson SKC & Brook F (2001). Reproduction. In *CRC Handbook of Marine Mammal Medicine* (2nd edn) (eds LA Dierauf & FMD Gulland), pp. 193–236. CRC Press: Boca Raton, Florida.

Robin ED, Murdaugh HV Jr, Pyron W, Weiss E & Soteres JP (1963). Adaptations to diving in the harbor seal: gas exchange and ventilatory response to CO_2. *American Journal of Physiology* **205**, 1175–1177.

Rommell SA & Lowenstein LJ (2001). Gross and microscopic anatomy. In *CRC Handbook of Marine Mammal Medicine* (2nd edn) (eds LA Dierauf & FMD Gulland), pp. 129–164. CRC Press: Boca Raton, Florida.

Ross PS, de Swart RL, Visser IK, Vedder LJ, Murk W, Bowen WD & Osterhaus AD (1994). Relative immunocompetence of the newborn harbor seal, *Phoca vitulina*. *Veterinary Immunology and Immunopathology* **42**, 331–348.

Rounsevell D (1988). Periodic irruptions of itinerant leopard seals within the Australasian sector of the Southern Ocean, 1976–1986. *Papers and Proceedings of Royal Society of Tasmania* **122**, 189–191.

Roux J-P & Hes AD (1984). The seasonal haul-out cycle of the fur seal *Arctocephalus tropicalis* (Gray, 1872) on Amsterdam Island. *Mammalia* **48**, 377–389.

Scheffer VB (1964). Hair patterns in seals (Pinnipedia). *Journal of Morphology* **115**, 291–304.

Scholin CA, Gulland F, Doucette GJ, Benson S, Busman M, Chavez FP, Cordaro J, DeLong R, De Vogelaere A, Harvey J, Haulena M, Lefebvre K, Lipscomb T, Loscutoff S, Lowenstein LJ, Marin III R, Miller PE, McLellan WA, Moeller PDR, Powell CL, Rowles T, Silvagni P, Silver M, Spraker T, Trainer V & Van Dolah FM (2000). Mortality of sea-lions along the central California coast linked to a toxic diatom bloom. *Nature* **403**, 80–84.

Seal US, Erickson AW, Siniff DB & Cline DR (1971). Blood chemistry and protein polymorphisms in three species of Antarctic seals (*Lobodon carcinophagus, Leptonychotes weddelli* and *Mirounga leonina*). *Antarctic Pinnipedia* Vol. 18 (ed. WH Burt), pp. 181–192. American Geophysical Union: Washington.

Shaughnessy PD (1999). *Action Plan for Australian Seals*, pp. 1–62. Environment Australia: Canberra.

Shaughnessy PD & Ross GJB (1980). Records of the subantarctic fur seal (*Arctocephalus tropicalis*) from South Africa with notes on its biology and some observations of captive animals. *Annals of South African Museum* **82**, 71–89.

Shaughnessy PD & Warneke RM (1987). Australian fur seal, *Arctocephalus pusillus doriferus*. NOAA Technical Report NMFS 51: Status, biology and ecology of fur-seals. In *Proceedings of International Symposium and Workshop,* April 1984 (eds JP Croxall & RL Gentry), pp. 73–77. Cambridge, England.

Shaughnessy PD, Dennis TE & Seager PG (2005). Status of Australian sea lions, *Neophoca cinerea*, and New Zealand fur seals, *Arctocephalus forsteri*, on Eyre peninsula and the far west coast of South Australia. *Wildlife Research* **32**, 85–101.

Siebert U, Driver J, Rosenberger T & Sandow J (2002). Reproduction control in a group of common seals (*Phoca vitulina*) through the use of a gonadotropin-releasing hormone (GnRH, Suprefact®-3–months-depot). In *Proceedings of 33rd Annual Conference of International Association for Aquatic Animal Medicine*, p. 111. Albufeira, Portugal.

Siniff DB & Stone S (1985). The role of the leopard seal in the tropho-dynamics of the Antarctic marine ecosystem. In *Antarctic Nutrient Cycles and Food Webs* (eds WR Siegfried, PR Condy & RM Laws), pp. 555–560. Springer-Verlag: Berlin.

Slip DJ & Woods R (1996). Intramuscular and intravenous immobilization of juvenile southern elephant seals

(*Mirounga leonina*). *Journal of Wildlife Management* **60**, 802–807.
Smith AW, Vedros NA, Akers TG & Gilmartin WG (1978). Hazards of disease transfer from marine mammals to land mammals: review and recent findings. *Journal of American Veterinary Medical Association* **173**, 1131–1133.
Smith AW, Skilling DE & Latham AB (1981). Isolation and identification of five new serotypes of calicivirus from marine mammals. *American Journal of Veterinary Research* **42**, 693–694.
Spelman LH (2004). Reversible anesthesia of captive California sea lions (*Zalophus californianus*) with medetomidine, midazolam, butorphanol, and isoflurane. *Journal of Zoo and Wildlife Medicine* **35**, 65–69.
Stenvers O, Plotz J & Ludwig H (1992). Antarctic seals carry antibodies against seal herpesvirus. *Archives of Virology* **123**, 421–424.
Stirling I (1971). Studies on the behaviour of the South Australian fur seal, *Arctocephalus forsteri* (Lesson). II. Adult females and pups. *Australian Journal of Zoology* **19**, 267–273.
Stirling I (1972). Observations on the Australian sea lion, *Neophoca cinerea* (Peron). *Australian Journal of Zoology* **20**, 271–279.
Stirling I (1983). The evolution of mating systems in pinnipeds. In *Advances in the Study of Mammalian Behaviour* (eds JF Eisenberg & DG Kleiman), pp. 489– 527. American Society of Mammalogists, Special Publication No. 7: Shippensburg.
Stoskopf MK, Hirst LW & Graham D (1983). Ocular anterior segment disease in captive pinnipeds. *Aquatic Mammals* **10**, 34–44.
Stoskopf MK, Willens S & McBain JF (2001). Pharmaceuticals and formularies. In *CRC Handbook of Marine Mammal Medicine* (2nd edn) (eds LA Dierauf & FMD Gulland), pp. 703–727. CRC Press: Boca Raton, Florida.
Sweeney J (1990). Surgery. In *CRC Handbook of Marine Mammal Medicine: Health, Disease and Rehabilitation* (ed. LA Dierauf), pp. 215–233. CRC Press: Boca Raton, Florida.
Sweeney JC (1993). Blood sampling and other collection techniques in marine mammals. In *Zoo and Wild Animal Medicine* (3rd edn) (ed. ME Fowler), pp. 425–428. WB Saunders: Philadephia.
Sweeney J & Samansky T (1995). Elements of successful facility design: marine mammals. In *Conservation of Endangered Species in Captivity: An Interdisciplinary Approach* (eds EF Gibbons Jr, BS Durrant & J Demarest), pp. 465–477. State University of New York Press: Albany.
Tahmindjis MA, Higgins DP, Lynch MJ, Barnes JA & Southwell CJ (2003). Use of a pethidine and midazolam combination for the reversible sedation of crabeater seals (*Lobodon carcinophagus*). *Marine Mammal Science* **19**, 581–589.
Taylor RH (1990). Records of subantarctic fur seals in New Zealand (Note). *New Zealand Journal of Marine and Freshwater Research* **24**, 499–502.
Thompson P, Cousins DV, Gow BL, Collins DM, Williamson BH & Dagnia HT (1993). Seals, seal trainers and mycobacterial infection. *American Review of Respiratory Disease* **147**, 164–167.
Thornton SM, Nolan S & Gulland FMD (1998). Bacterial isolates from Californian sea lions, harbor seals and northern elephant seals admitted to a rehabilitation center along the central California coast. *Journal of Zoo and Wildlife Medicine* **29**, 171–176.
Tikhomirov EA (1975). Biology of the ice forms of seals in the Pacific section of the Antarctic. *Rapports et Procès-verbaux des Réunions/ Counseil Permanent International pour l'Exploration de la Mer* **169**, 409–412.
Townsend FI & Gage LJ (2001). Hand-rearing and artificial milk formulas. In *CRC Handbook of Marine Mammal Medicine* (2nd edn) (eds LA Dierauf & FMD Gulland), pp. 829–849. CRC Press: Boca Raton, Florida.
Trigg TE, Wright PJ, Armour AF, Williamson PE, Junaidi A, Martin GB, Doyle AG & Walsh J (2001). Use of a GnRH analogue implant to produce reversible long-term suppression of reproductive function in male and female domestic dogs. *Journal of Reproduction and Fertility Supplement* **57**, 255–261.
Troy S, Middleton D & Phelan J (1997). On capture, anaesthesia and branding of adult male New Zealand fur seals *Arctocephalus forsteri*. In *Marine Mammal Research in the Southern Hemisphere.* Vol. 1 Status, ecology and medicine (eds M Hindell & C Kemper), pp. 179–183. Surrey Beatty & Sons: Sydney.
Tryland M, Klein J, Nordoy ES & Blix AS (2005a). Isolation and partial characterization of a parapoxvirus isolated from a skin lesion of a Weddell seal. *Virus Research* **108**, 83–87.
Tryland M, Sorensen KK & Godfroid J (2005b). Prevalence of *Brucella pinnipediae* in healthy hooded seals (*Cystophora cristata*) from the North Atlantic Ocean and ringed seals (*Phoca hispida*) from Svalbard. *Veterinary Microbiology* **105**, 103–111.
Vogelnest L (2000). The veterinarian's approach to a pinniped ashore. In *Marine Wildlife, Proceedings 335,* pp. 181–191. Post Graduate Foundation in Veterinary Science: Sydney.
Vogelnest L, Hulst F & Woods R (1996). The veterinary management of southern elephant seals (*Mirounga leonina*) at Taronga Zoo. In *Proceedings of Annual Conference of American Association of Zoo Veterinarians,* pp. 318–325. Puerto Vallarta, Mexico.
Walker GE & Ling JK (1980). Body weights and dimensions of adult female Australian sea lions, *Neophoca cinerea.* *Journal of Mammalogy* **61**, 164–165.
Warneke RM (1979). Australian fur seal. In *Mammals in the Seas.* Food and Agriculture Organisation Fisheries Series 5, Vol. 2, pp. 41–44. FAO: Rome.
Warneke RM (1982). The distribution and abundance of seals in the Australasian region, with summaries of biology and current research. In *Mammals in the Seas.*

Food and Agriculture Organisation Fisheries Series 5, Vol. 4, pp. 431–475. FAO: Rome.

Warneke RM & Shaughnessy PD (1985). *Arctocephalus pusillus*, the South African and Australian fur seal: taxonomy, evolution, biogeography, and life history. In *Studies of Sea Mammals in South Latitudes* (eds JK Ling & MM Bryden), pp. 53–77. South Australian Museum: Adelaide.

Wartzok D & Ketten DR (1999). Marine mammal sensory systems. In *Biology of Marine Mammals* (eds JE Reynolds III & SA Rommell), pp. 117–175. Smithsonian Institution Press: Washington DC.

Webster RG, Geraci J, Petursson G & Skirnisson K (1981). Conjunctivitis in human beings caused by influenza A virus of seals. *New England Journal of Medicine* **304**, 911.

Wheatley KE, Bradshaw CJA, Harcourt RG, Davis LS & Hindell MA (2006). Chemical immobilization of adult female Weddell seals with tiletamine and zolazepam; effects of age, condition and stage of lactation (Abstract). *BMC Veterinary Research* **2**, 8.

Williams R & Bryden MM (1993). Observations of blood values, heart rate and respiratory rate of leopard seals (*Hydrurga leptonynx*) (Carnivora: Phocidae). *Australian Journal of Zoology* **41**, 433–439.

Woods GL & Gutierrez Y (eds) (1993). *Diagnostic Pathology of Infectious Diseases.* Lea & Febiger: Philadelphia.

Woods R (1994). Chemical restraint of southern elephant seals (*Mirounga leonina*). PhD thesis. School of Pharmacy, University of Tasmania, Australia.

Woods R, McLean S, Nicol S & Burton HR (1994a). Use of midazolam, pethidine, ketamine and thiopentone for the restraint of southern elephant seals (*Mirounga leonina*). *Veterinary Record* **135**, 572–577.

Woods R, McLean S, Nicol S & Burton H (1994b). A comparison of some cyclohexamine-based drug combinations for chemical restraint southern elephant seals (*Mirounga leonina*). *Marine Mammal Science* **10**, 412–419.

Woods R, McLean S, Nicol S & Burton H (1995a). Antagonism of some cyclohexamine-based drug combinations used for chemical restraint of southern elephant seals (*Mirounga leonina*). *Australian Veterinary Journal* **72**, 165–171.

Woods R, Cousins DV, Kirkwood R & Obendorf DL (1995b). Tuberculosis in a wild Australian fur seal (*Arctocephalus pusillus doriferus*) from Tasmania. *Journal of Wildlife Diseases* **31**, 83–86.

Woods R, Mc Lean S, Nicol S, Slip DJ & Burton HR (1996). Use of the respiratory stimulant doxapram in southern elephant seals (*Mirounga leonina*). *Veterinary Record* **138**, 514–517.

18. Cetaceans

David Blyde and Larry Vogelnest

1 BIOLOGY

Forty-three of the 80 known species of cetaceans live around the Australian continent. These include 35 species in the Suborder Odontoceti (toothed whales) and eight species in the Suborder Mysticeti (baleen whales) (Bryden et al. 1998; Rogers 2000; Menkhorst 2001). The baleen whales include the rorquals in the Family Balaenopteridae which have longitudinal grooves in the throat and a small pointed dorsal fin.

Although most cetacean species strand, a few species regularly strand in large numbers. These include sperm (Family Physeteridae), pygmy killer (*Feresa attenuata*), melon-headed (*Peponocephala electra*), false killer (*Pseudorca crassidens*), short- and long-finned pilot (*Globicephala* spp.) and common killer whales (*Orcinus orca*) (Rogers 2000). The role of the veterinarian at a marine mammal stranding is presented in Chapter 3.

Three species of cetaceans are held in captivity in oceanariums in Australia: the Indo-Pacific bottle-nosed dolphin (*Tursiops aduncus*), the common bottle-nosed dolphin (*Tursiops truncatus*) and the Indo-Pacific hump-backed dolphin or sousa (*Sousa chinensis*). This chapter primarily deals with these species. They will be referred to here on as 'dolphins'.

Dolphins are social animals that can live up to 50 yr. The common bottle-nosed dolphin is large, robust and uniformly mid-grey to brown without obvious stripes or colour patches. It is found throughout tropical and temperate oceans, in-shore and off-shore, in a wide variety of habitats including bays, estuaries and harbours. In Australia it is most common around the southern coast from southern Queensland to Albany in Western Australia. It is gregarious in groups of up to 20 individuals. The Indo-Pacific bottle-nosed dolphin is smaller and paler and has dark flecks on the ventral surface. It is found mainly in warm shallow in-shore waters. It is most common around the northern coast of Australia. The genetics of Australian dolphins are being investigated to determine the relationships between the many different morphometric types of bottle-nosed dolphins in Australian waters (K Charlton *pers. comm.*) The Indo-Pacific hump-backed dolphin is robust, uniformly pale, light grey above and whitish pink below. It is primarily found in coastal tropical and warm temperate waters, mostly less than 20 m deep in estuaries, tidal rivers and channels through mangroves. On open coasts it is often seen in the surf zone. It is gregarious in small groups of up to 25 individuals (Menkhorst 2001).

2 ANATOMY AND PHYSIOLOGY

Cetaceans share the general mammalian anatomical and physiological characteristics, with significant differences associated with adapting to a marine environment.

2.1 Digestive system

Odontocete teeth are closely spaced and uniform in shape and size and have growth rings in cross-section that can be used for ageing (Perrin & Myrick 1980).

Mysticetes have a series of baleen plates suspended from each side of the upper jaw. The abdominal cavity is small and the gastrointestinal tract differs from those of terrestrial mammals. The oesophagus is penetrated dorso-ventrally by the laryngeal tube. In most cetaceans food passes either side of this structure to reach the stomach. In the pygmy sperm whale (*Kogia breviceps*) the left side is a blind pouch and food must pass to the right of the laryngeal tube. Dolphins and other Odontoceti have three stomach compartments and a simple but long intestinal tract. The three stomach compartments functionally correspond to regions of the stomach of monogastric mammals. The first stomach (forestomach) is an enlargement of the distal oesophagus. Digestion starts here, aided by enzymes and hydrochloric acid that refluxes from the second chamber which is the glandular or fundic stomach and is pink to red. The third is the pyloric stomach, which secretes mucus and prepares food for digestion.

The pylorus opens into a dilated pear-shaped ampulla which begins the intestinal tract. The intestinal tract of odontocetes is not visibly organised into small and large intestines, and in small species can be up to 30 m long. Mysticetes have a distinguishable colon (Geraci 2000a). Gastrointestinal transit time is very rapid, which has implications for oral drug administration. Parasites are commonly found in the stomach chambers of wild odontocetes.

2.2 Respiratory system

The upper respiratory tract has been modified to allow cetaceans to breathe at the water surface and swallow simultaneously. Complex groups of upper airway muscles in cetaceans ensure an airtight seal of the closed glottis, which is particularly important when diving. This has implications when performing bronchoscopy as it can be difficult to pass the scope. The blowhole is on top of the head and to the left of the midline in many odontocetes. In dolphins and the common killer whale it is in the midline. The blowhole is paired in mysticetes and single in odontocetes. Immediately below the blowhole in odontocetes are four pairs of nasal sacs which function in sound production. Ventral to the nasal sacs, a bony septum divides the nasal passage into left and right internal nares. The nasal passages then open into the nasopharynx, a tubular hollow space about 4–5 cm in diameter in dolphins. Projecting into the centre of the nasopharynx is the beak-like opening of the arytenoepiglottideal tube (also referred to as the goosebeak). The space between the arytenoepiglottideal tube and the pharyngeal wall is the nasopharyngeal fornix. The glottis is rosette-shaped and not covered by an epiglottis. The larynx sits in a muscular valve and together they form the arytenoepiglottideal tube. It remains tightly shut other than during respiratory cycles. The arytenoepiglottideal tube is elbow-shaped (45–80°) and joins the trachea caudally. The trachea has complete tracheal rings (Harper et al. 2001; Tsang et al. 2002) (Fig. 18.1). This anatomical arrangement has important implications for intubation for inhalation anaesthesia. The lungs are large, adapted for deep diving and turgidly elastic and the pleura is thick and well-vascularised. Cartilaginous rings continue down to the origin of the alveoli, making the lungs rigid and resilient (Geraci 2000a).

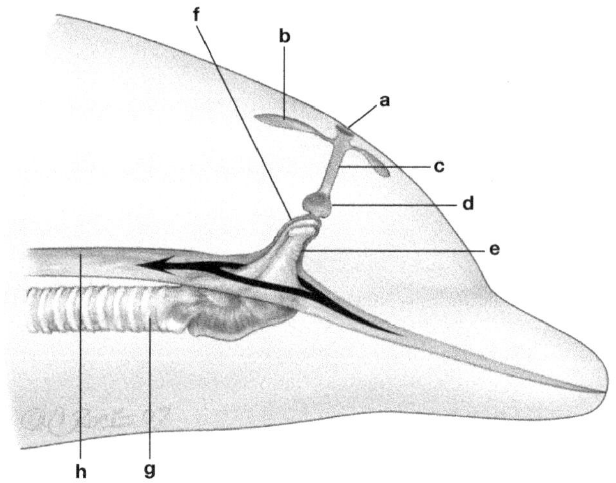

Figure 18.1 Upper respiratory and oesophageal anatomy of a dolphin. a) Blowhole. b) Nasal sacs. c) Nasal passages (paired). d) Nasopharynx. e) Arytenoepiglottideal tube. f) Nasopharyngeal fornix. g) Trachea. h) Oesophagus.

2.3 Cardiovascular system

The dolphin heart is large relative to body mass. The ventricles are thick-walled and highly trabeculated. There are numerous retia mirabile in the thorax and head. Peripheral venous retia help regulate body temperature. Surface arteries (particularly in the flukes, flippers and dorsal fin) are surrounded by a network of veins encased in a rigid channel of connective tissue underlying the dermis (Fig. 18.2). To retain body heat, arterial blood flows to the surface under low pressure and returns along the surrounding venous rete, absorbing heat from the central artery. To cool, blood

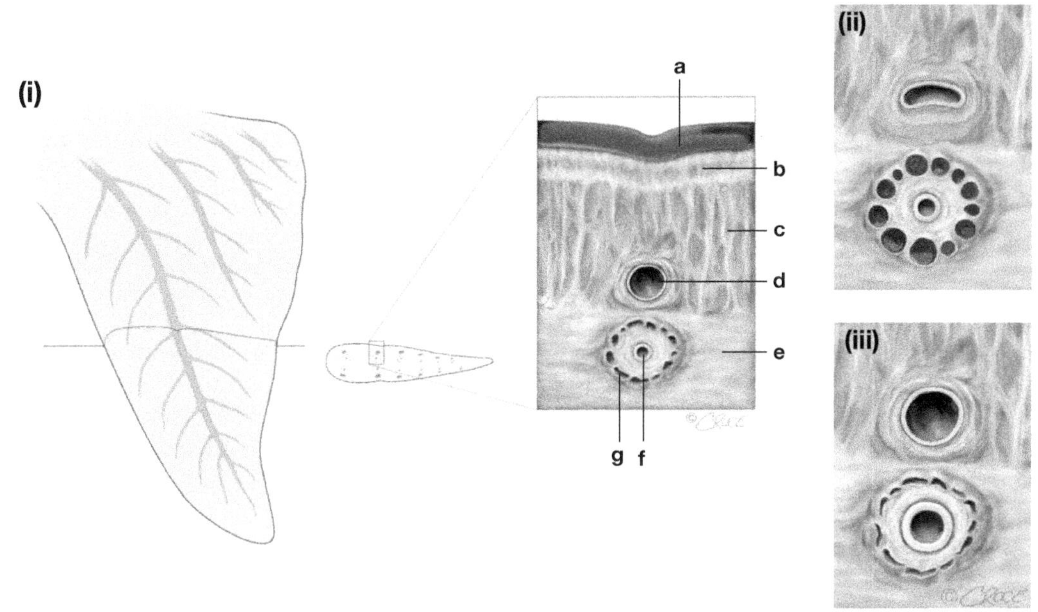

Figure 18.2 i) Blood vessels and the peripheral venous retia in the tail fluke, epidermis (a), dermis (b), connective tissue (c), vein (d), dense connective tissue core (e), artery (f), periarterial venous retia (g). ii) Conserving heat. iii) Cooling. (Source: After Geraci (1986a), Geraci & Lounsbury (1993).)

flows under high pressure, thereby collapsing the surrounding veins against the rigid tunnel walls, and returns by superficial veins that lie closer to the surface of the skin. These vessels in the flukes are useful for venipuncture (Geraci 2000a).

2.4 Viscera

The liver has a prominent dividing cleft and large intrahepatic sinuses. There is no gall bladder. The spleen is ovoid, small (4–8 cm) and firm and there may be one or more smaller accessory spleens. The kidneys are elongated and hyperlobulated (multirenculate).

2.5 External features

Cetaceans do not have external ears but have small openings in the sides of the head directly behind each eye. The large eyes have flattened corneas for underwater vision. There are no hind limbs and the fore limbs are modified into a flipper. The skin is smooth and rubbery and generally lacks hairs and glands. The row of bristle-like hairs along the snout of newborn cetaceans is lost within a few weeks of birth. Rorquals have a sparse beard along the chin (Geraci 2000a).

2.6 Physiology

Dolphins' normal respiratory rate is 1–3 breaths per minute. Normal heart rate is 50–90 beats per minute. Normal rectal temperatures range from 35°C–37°C. The temperature of cetaceans can be assessed by inserting a sensor probe 10–15 cm into the rectum. The body temperature may fluctuate and this should be considered when using body temperature to clinically evaluate a cetacean.

2.6.1 Water balance

The osmotic concentration of sea water is about four times that of mammalian body fluids, which favours loss of body fluids. Cetaceans have evolved a number of strategies to compensate for this: the skin is impermeable to sea water, little water is lost through the respiratory tract or by evaporative cooling, and water is conserved by concentrating milk and urine. Cetaceans obtain water from their diet (free-formed water and metabolic water from oxidation of fat) and do not require an additional source. However, they can and do drink water they swim in.

Marine mammals secrete large amounts of aldosterone in response to stress. Aldosterone promotes the reabsorption of sodium from the renal tubules, drawing water back into the body. This mechanism conserves water when food is not available or the animal is not eating (Geraci 2000a).

2.6.2 Thermoregulation

Most cetaceans live in an environment that is significantly cooler than their body temperature. Cetaceans have developed several mechanisms to deal with the fact that heat is conducted about 20 times faster in water than in air. The large body size and small appendages results in low surface area to volume ratios. Therefore they have a large body core to produce heat but only a small surface area to lose it. Blubber, a meshwork of connective tissue that supports fat cells, acts as a very effective insulating layer just beneath the skin.

The peripheral venous retia function as a counter-current heat exchange mechanism in both warm and cold environments (Fig. 18.2). Core body temperature is conserved by selective vasoconstriction of superficial vessels that would otherwise dissipate heat. In cold environments core body temperature of cetaceans can drop by about 2°C, thus lowering the temperature gradient between external and internal environments (Geraci 2000a).

In extremely cold water the surface tissues are kept active partly by cold-adapted isoenzymes in cells and the large unsaturated fatty acid component of blubber (Geraci 2000a).

2.6.3 Adaptations for diving

Cetaceans exhale when they dive to great depths. At about 100 m the lungs are almost totally collapsed. The residual air is forced into dead air spaces from which nitrogen cannot diffuse into the circulation. The rib-cage is flexible and can collapse without injury. The cartilaginous airways prevent crushing of the lungs and allow them to open quickly when returning to the surface. The numerous retia mirabile engorge when diving to counter the pressure on the thorax, ears and nasal sinuses. At depths there is no gas exchange in the lungs as they are collapsed. Instead the lungs are adapted for rapid and efficient extraction of oxygen when inflated. The oxygen is circulated in the large volume of haemoglobin-rich blood.

During longer dives heart rate slows significantly. There is peripheral vasoconstriction with shunting of blood away from the skin, muscle and abdominal viscera. Body temperature decreases. Stored oxygen is directed to the brain and heart. Oxygen is stored in myoglobin and can be released during long dives. Accumulation of carbon dioxide does not trigger respiration as it does in terrestrial mammals. After an anaerobic dive the animal must remain at the surface for a long time to recover.

3 REPRODUCTION

There are no external genitalia in males. The genders can be distinguished by features of the genital slits. The male has a separate anal slit behind the genital slit. The female has a longer and more prominent slit which encompasses the more closely spaced genital and anal openings. Internally, the genital slit in the female is directed cranially while in the male it is directed caudally (Geraci 2000a). The testes are internal and are very large compared to those of most terrestrial mammals of similar body size. Veins carrying cool blood from the dorsal fin and flukes are juxtaposed to arteries supplying the testes. This allows the production and storage of sperm at a cooler temperature.

Male dolphins generally reach sexual maturity at about 8–10 and females at 10–12 yr of age. Captive animals may reach sexual maturity earlier due to increased food availability. Mature sperm was found in semen obtained from a 2 yr old male bottle-nosed dolphin (H Muraco *pers. comm.*) but this is probably an exception.

Dolphins in Australian waters appear to breed year-round, with a peak of births in summer.

Dolphins in captivity in Australia are polyoestrus with peaks in sexual activity and births in summer. Ovulation is spontaneous, with an oestrous period of about 21 d. Gestation lasts 12 mo, after which a single offspring is born. Calves are normally born tail first, weigh about 10 kg and are 98–126 cm long. Lactation lasts 12–18 mo and the young become independent around 2–3 yr of age. Lactation may suppress oestrus. Female offspring stay with the mother and the maternal pod for a long time. Male offspring generally leave the maternal pod and form pairs or live singly, interacting with female pods only for breeding.

4 HUSBANDRY

4.1 Housing

Many dolphins seem to adapt well to captivity. They are active, intelligent and inquisitive and their housing and husbandry must meet their behavioural and physical requirements. Housing standards have been developed for captive cetaceans in Queensland (Queensland Wildlife Parks Association 2003) and New South Wales (Department of Primary Industries 2005). These standards cover minimum space, containment structures, water and climatic conditions.

Table 18.1 Ideal water parameters for cetacean pools

Parameter	Acceptable range
Temperature	15–30°C
pH	7.5–8.4
Salinity	1.030–1.035 ppt
Alkalinity	80–100 ppm
Coliform count	<500 cfu
Chlorine levels	<1.0 ppm

Ideally dolphins should be kept in natural sea water. Artificial salt water can be used but is expensive and there may be problems with the availability of the constituents. Dolphins fare better in natural outdoor enclosures. Indoor enclosures are associated with a higher incidence of disease, particularly respiratory disease. Pool construction can be natural or artificial. If artificial (concrete, fibreglass) the enclosure should be durable, watertight, non-porous, non-abrasive, non-toxic and easily cleaned and disinfected. Sharp projections and loose fittings should be avoided. Water quality is easier to maintain in artificial pools and there are fewer opportunities to encounter foreign objects, such as stones that may be ingested, than in natural pools. Natural pools (i.e. dug into natural substrate or a netted natural marine water body) can be constructed much larger for less cost than artificial pools. The pools should be large enough to accommodate the natural behaviours of the animals and deep enough for them to dive and avoid the effects of ultraviolet radiation. The water should be shaded and the temperature maintained between 15°C–30°C. If chemical treatment is used to maintain water quality and cleanliness, it should not harm the animals. Many different disinfection systems have been used. Chlorine and ozone are the most common. Both are suitable as long as the water quality is monitored carefully and regularly. As dolphins live entirely in water, water quality is critical to maintain healthy animals (Arkush 2001; see Chapter 17, 4.3). Table 18.1 shows important water parameters and their acceptable levels.

4.2 Training

Cetaceans are very amenable to training and a number of behaviours can be trained to assist in their management. Behaviours that can be trained include eye examinations and instillation of medications, rectal temperature recording, presentation for ultrasonographic examination, and collection of blood, urine, stomach fluid, semen or faeces. Animals can be conditioned to haul out onto scales to measure body weight regularly. This is extremely useful and animals should be weighed at least monthly (preferably weekly) to objectively assess body condition.

4.3 Estimating body weight

If weighing scales are not available, an equation can be used to estimate normal body weight (ENBW) by measuring the animal's length (Bedford 2000). For common bottle-nosed dolphins the equation is:

$$\text{ENBW} = (1.616 \times \text{body length in cm}) - 236$$

For Indo-Pacific bottle-nosed dolphins it is:

$$\text{Log ENBW} = (0.1389 \times \text{log body length}^{3.2019})$$

Normal adult body weights for the Indo-Pacific bottle-nosed dolphin are 120–170 kg for males and 110–150 kg for females; for the common bottle-nosed dolphin 180–230 kg for males and 170–220 kg for females; and for the Indo-Pacific hump-backed dolphin 220–250 kg for males and 160–200 kg for females. Body weights fluctuate seasonally with animals tending to be heaviest in winter and lightest in summer, generally corresponding with food intake. Body condition can be assessed visually by experienced observers. Severe and pathological weight loss is evident when the ribs are easily seen and the muscles along the dorsal thorax are wasted, giving a concave shape lateral to the midline. Any sudden or unexpected weight loss should be investigated.

4.4 Individual marking and identification techniques

Generally, animals are identified by their markings or other obvious physical differences, including the shape of the rostrum, the shape and outline of the dorsal fin, and visible scars. Microchips or passive integrated transponder (PIT) tags can be inserted IM to identify individuals. A suitable site for implanting is the muscle just caudal and lateral to the dorsal fin. Implanting on the left side is the standard.

4.5 Transport

Purpose-made padded stretchers with openings through which the flippers can protrude and openings in the genital region to prevent urine scalding are ideal. Stretchers without these openings should be used only for short periods (20 min) to avoid crushing the flippers and overheating. The fabric of the stretcher should be smooth and easily disinfected. Cetaceans should always be transported in ventral recumbency, never in lateral recumbency. The stretcher can be suspended in a water-filled container for long trips with the dorsal surface exposed to the air so the animal can breathe. Foam padding will help protect and support the animal. Transport time should be minimised. Keep the animal wet by lying moist towels or sheets over it or by continuously spraying with water. Never cover or fill the blowhole with water. Temperatures during transport should not exceed 30°C or be less than 15°C. Overheating is more of a problem than cold. It may be useful to have a supply of ice on hand. Always protect the animal from direct sunlight, wind and exhaust fumes. Never roll cetaceans to move them as this distresses them greatly and may damage the flippers and dorsal fin. A small crane fixed to the transport vehicle or separate is useful for lifting heavy animals.

Cetaceans should never be transported free in a water-filled container as there is a risk of drowning or hitting the side of the container and causing serious injury.

The use of sedatives when transporting cetaceans is debatable. Unless the animal is highly agitated it may be better to forgo the use of these agents, as they can cause respiratory depression and affect thermoregulation. Midazolam (0.04–0.06 mg/kg) has been used routinely at one institution when transporting potentially stressed animals, without side effects (D Spielman *pers. comm.*).

5 NUTRITION

Wild bottle-nosed dolphins eat mostly fish and squid. Indo-Pacific bottle-nosed dolphins eat fish from in-shore reefs and sandy benthic habitats while Indo-Pacific hump-backed dolphins eat mostly fish, some squid and crustaceans (Menkhorst 2001). Wild and captive Indo-Pacific hump-backed dolphins have been seen catching and eating birds.

Captive dolphins are primarily fed fish at least three times a day. A variety of species should be fed daily as different species have different nutritional content varying with sex, season, handling and storage. Whole fish rather than fillets or cleaned gutted fish should be fed. The calcium in the bones, vitamin A in the liver and fats in the gonads are all essential in a balanced diet. As a rough guide, captive animals should be fed 5–7% of their body weight daily. This should be increased in pregnant, lactating and growing animals. Captive dolphins, like other species in captivity, have a tendency to become obese and therefore body weight should be monitored and controlled.

Most fish fed to captive dolphins has been frozen, which is important in killing parasites in fish. The appropriate storage and thawing of frozen fish is important to maintain nutrient content and quality (Geraci & St Aubin 1980; Geraci 1986b). The ideal thawing conditions to maintain hydration and nutrients are in an air-tight container at 4–8°C. Thawing fish in water leaches essential vitamins and minerals from the fish. Fish contain thiaminases that destroy thiamine. Fat-soluble vitamins are also readily lost. All fish should be supplemented daily with vitamin E (50 IU/kg of fish fed) and thiamine (20 mg/kg of fish fed). These supplements can be given as tablets or capsules in the fish, and various commercial products are available (see Chapter 17). Because of the short transit time of food in the gastrointestinal tract, supplements should be easily digestible.

An artificial diet was formulated in the US using traditional fish feed industry ingredients and proved palatable in trials. However, the animals needed 25% more of the artificial diet on a daily weight basis to maintain weight (Wright et al. 2005). It could be useful if storage space is a problem or supplies of fish are disrupted.

When free-ranging cetaceans are brought into captivity they may not eat dead fish for several days and may require force-feeding if they are in poor condition. Force-feeding is easy. Restrain the animal, open the jaws using soft rope or rolled-up towels around each jaw and introduce the fish deep into the oesophagus. Because of the nature of cetacean anatomy there is no risk of food entering the respiratory tract.

See Chapter 17, section 5, for additional information on nutrition.

6 RESTRAINT

Handling and restraining captive cetaceans is frequently required to facilitate moving animals between facilities or to conduct routine husbandry or veterinary procedures. There is always potential for injury to both animal and people so it is important that the procedures be carried out only by appropriately trained handlers. Many procedures can be performed using operant conditioning, making them safer and less stressful for animal and handlers. A training program to facilitate transportation and handling should be incorporated into the husbandry program of all captive dolphins.

6.1 Physical restraint

Cetaceans are often physically restrained. They can be herded into small shallow areas where they are easier to catch. Small cetaceans can be restrained by grabbing them from above and pushing both pectoral fins against the body. Once restrained, they can be placed on a smooth flat surface. If restraint will exceed 10–15 min they should be positioned on foam pads (7.5–10 cm thick) or mattresses to alleviate pressure on the stomach and lungs. Even wild animals tolerate physical restraint and can be restrained for up to 3 hr without undue stress as long as environmental conditions are suitable and appropriate supportive care is provided. Restraint is rarely required for longer than 90 min. Always take care with the tail as it can cause significant injury to handlers if it hits them. Dolphins can bite and their teeth are sharp. They can also damage themselves if not appropriately restrained. Keep them cool and wet while restrained, but if they are being sprayed or hosed with water ensure it does not enter the blowhole. Do not subject them to long periods of direct sunlight—they can suffer sunburn in 30 min under Australian conditions. If the procedure is likely to take longer than this, erect a shade shelter.

6.2 Chemical restraint

Cetaceans are rarely chemically restrained. Many procedures such as dentistry, endoscopy and minor surgery can be accomplished by combining sedation and local anaesthesia. Sedatives that are useful in captive cetaceans include midazolam (0.095 mg/kg IM) and diazepam (0.15–0.2 mg/kg IM). Responses to both drugs vary between individuals. Sedated animals should be monitored closely and kept separate until the effects of the sedative have worn off. Both drugs can be reversed by flumazenil (0.002–0.004 mg/kg IM or IV). Butorphanol has also been used at 0.05–0.2 mg/kg IM with reasonable results (Chittick et al. 2006). The effects of butorphanol can be reversed by naltrexone (0.3 mg/kg IM).

General anaesthesia is rarely performed in cetaceans. It can be done successfully but is logistically difficult. Induction can be accomplished IM or IV (see 7.1 and 13). IV agents include thiopentone (15–20 mg/kg), propofol (3.5–4.8 mg/kg) or tiletamine/zolazepam (1–2 mg/kg). Tiletamine/zolazepam may be the agent of choice as it is not irritant perivascularly. IM agents include tiletamine/zolazepam (2–2.5 mg/kg) and medetomidine (0.04 mg/kg) plus ketamine (2 mg/kg). Medetomidine can be reversed with atipamazole (0.2 mg/kg IM). Intubation is achieved by dislocating the modified larynx (arytenoepiglottideal tube) from the nasopharyngeal sphincter of the blowhole by reaching into the mouth and intubating blindly via the oral cavity (Fig. 18.1). Laryngospasm does not occur. The diameter of the trachea is large and adult dolphins require a 24–32 mm endotracheal tube. The bifurcation of the trachea is more cranial in cetaceans and care must be taken to avoid introducing the endotracheal tube into either bronchus. Maintain gaseous anaesthesia with intermittent positive pressure ventilation, ensuring the chest makes normal excursions. Useful inhalation agents include isoflurane, sevoflurane or desflurane in oxygen. Anaesthesia should be monitored for oxygen saturation, pulse rate, end tidal carbon dioxide concentration and minimum anaesthetic concentration. End tidal carbon dioxide levels should be kept below 40 mmHg by the rate and depth of ventilation.

7 CLINICAL PATHOLOGY

7.1 Haematology and biochemistry

7.1.1 Sample collection

Numerous sites can be used for venipuncture in cetaceans (Fig. 18.3), but the preferred site in most cetaceans are vessels that run along grooves parallel to the leading edge of the tail fluke near the midline on the flukes' dorsal and ventral surfaces. The groove can be palpated or seen and is often more prominent on the ventral surface. These vessels comprise a superficial vein and a deeper artery surrounded by a periarterial venous rete (Fig. 18.2). Blood may be collected from either and it may be difficult to determine which vessel has been entered. A 23 G scalp vein set can be inserted perpendicularly or at a slight angle to the skin into the vessels and blood can be aspirated into a syringe. Ideally, captive cetaceans should be trained to present their tail flukes for unrestrained blood collection (Fig. 18.4).

Cetacean blood clots slowly so a significant sample can be collected from even a poor flow. Partial prothrombin time is long in dolphins because of a lack of

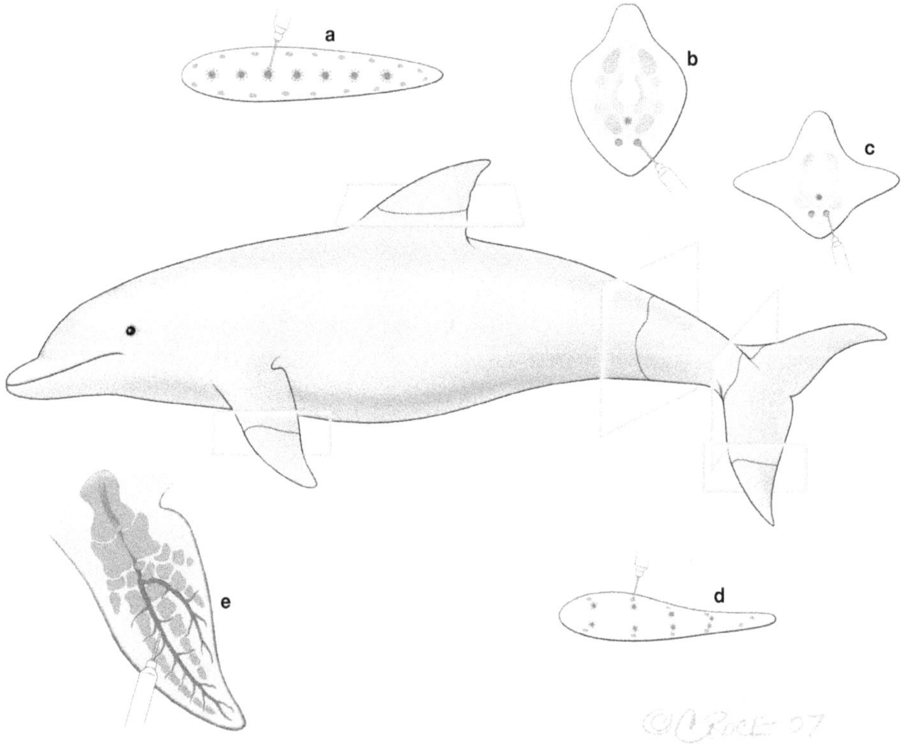

Figure 18.3 Blood sampling sites in a cetacean. a = dorsal fin, b & c = caudal peduncle d = tail fluke, e = pectoral flipper. (Source: After Geraci & Sweeney (1986), Geraci & Lounsbury (2005).)

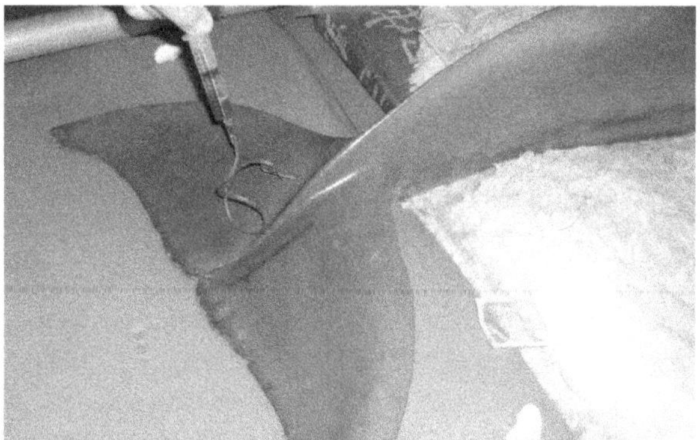

Figure 18.4 Blood collection from the tail fluke in a dolphin.

Factor XII and plasma prekallikrein, which disables the intrinsic pathway of the coagulation cascade (Bossart et al. 2001). Captive dolphins should be bled regularly (at least six-monthly, preferably monthly) for routine haematological and biochemical analyses. This provides invaluable reference data for individual animals and allows the detection of subclinical disease.

7.1.2 Reference ranges and interpretation

Haematology and serum chemistry are extremely useful in monitoring the health of captive cetaceans. The peripheral blood of cetaceans readily reflects systemic pathological processes, particularly inflammation. Changes are similar to those seen in domestic companion animals, which may be used to guide interpretation. However, a number of factors have been shown to influence haematology and biochemistry values in dolphins. These include gender, age, nutrition, stress, habitat, diving capacity and seasonality. These need to be considered when interpreting results (Fair et al. 2006). Pelagic species have been shown to have higher RBC counts, PCV and haemoglobin concentration than those residing in coastal environments. This is thought to be due to the greater physiological oxygen demand on deeper-diving off-shore dolphins (Duffield et al. 1983). In a study by Fair et al. (2006), non-pregnant female dolphins were shown to have higher mean triglyceride concentrations than males had. In another study, overall serum creatinine levels were found to be 15–38% higher in summer than in winter. Cholesterol levels were also found to be seasonally variable, with values ranging from 96–348 mg/dL. The highest levels recorded were for adult females in spring (244.9 ± 27.9 mg/dL) and males and juveniles in winter (177.5 ± 40.1 mg/dL, 219.1 ± 39.3 mg/dL). Overall free fatty acid levels peaked in summer (0.610 ± 0.191 mEq/L for adult males, 0.789 ± 0.309 mEq/L in adult females and 0.547 ± 0.151 mEq/L in juveniles) and dropped off in winter (Terasawa et al. 2002). Significantly higher fibrinogen levels have been observed in pregnant dolphins than in non-pregnant dolphins. Plasma creatinine levels can be higher in males than in females (Fair et al. 2006). In general, RBC counts are higher in neonates than in adults; they decrease during growth and as the individual starts to dive (Reidarson 2003). Younger dolphins also tend to have higher levels of CK than adults do. Adult dolphins have been shown to have higher concentrations of AST, alpha-1, alpha-2 and gamma globulins, and total protein than juveniles (Fair et al. 2006). Stress and handling time may affect blood parameters such as glucose, TCO_2 and HCO_3, with these increasing with longer duration of handling (Varela et al. 2006).

As in all mammals, an increased PCV and haemoglobin concentration generally reflects dehydration and low PCV indicates either blood loss or reduced red cell production. In cetaceans in general, mild normocytic anaemia can be associated with infection, and microcytic anaemia can be indicative of haemorrhagic gastric and duodenal ulcers. Leukocytosis is often associated with tissue damage in dolphins, and large increases above normal of up to 40% are a common finding in individuals post transport. Neutrophilia is a common finding with infection. A high neutrophil count (>10 000 × 10^9/L) may indicate fungal or bacterial infection. Lymphopaenia (<1000 × 10^9/L) may infer viral infection. Chronic infections are normally correlated with increases in monocytes and lymphocytes (Medway & Geraci 1986). Numbers of circulating eosinophils are significantly lower in captive than in free-ranging animals, most likely due to the reduced numbers of parasites in captive animals (Reidarson 2003). In general, it has been reported that red blood cell count is lower in aquatic mammals than in terrestrial mammals (Shirai & Sakai 1997).

ALP levels have been shown to be affected by age, with higher levels in rapidly growing younger animals whose bones are undergoing ossification. Consequently ALP levels can be used as an indicator of bone metabolism (Fair et al. 2006). ALP is also a useful indicator of nutritional status in dolphins and other cetaceans due to its positive correlation with increasing body weight. However, increased ALP levels are rarely associated with liver pathology in cetaceans, unlike in terrestrial mammals (Reidarson 2003). Potassium levels in dolphins can increase following rigorous exertion. Glucose levels in cetaceans are generally higher than in other mammals. This may be due to the need for adequate glucose levels in the brain during the deep and extended periods of diving exhibited by many cetaceans (Medway & Geraci 1986). Electrolyte values combined with total protein and albumin values can be used to evaluate hydration status. Dehydration should be suspected when the albumin level exceeds 40 g/L and the PCV exceeds 0.48. These should be interpreted in conjunction with other results.

Table 18.2 provides haematology and biochemistry values for dolphins.

7.2 Urinalysis

Captive cetaceans can be trained to urinate on demand. Male and female dolphins can be catheterised to collect urine. The technique is simple in males. The penis is extruded and a well-lubricated and disinfected small-gauge catheter (5 French) is introduced into the urethra and threaded proximally until urine is obtained. The technique is slightly more complicated in females. With the animal restrained in lateral recumbency, an assistant opens the genital slit and an otoscope is introduced into it. The urethral os is visible just dorsal to the clitoris, which sits cranially. A well-lubricated and disinfected medium catheter (8–14 French) is introduced and threaded proximally until urine is obtained. If the catheter is directed too far dorsally, it may enter the vagina. If the bladder contains urine the animal may urinate in response to the urogenital stimulation so the handler should be ready to collect it. Alternatively, urine can be collected via ultrasound-guided cystocentesis. Normal urine is clear and odourless. The pH is generally around 6. Specific gravity ranges between 1.040–1.060. A trace of protein is normal but there should be no blood if the urine is collected atraumatically.

7.3 Gastrointestinal sampling

Faecal analysis can be useful for culture and to detect occult blood. However, faeces can vary from a cloud in the water to a paste-like consistency, so collection can be difficult. Animals can be conditioned for rectal

Table 18.2 Haematology and biochemistry values for the common bottle-nosed dolphin

Parameter	Mean	SD	Range	n
Haematology				
RBC (10^6/µL)	3.61	0.28	2.8–4.8	59
Hb (g/dL)	14.47	1.11	11.3–18.2	59
HCT(%): automated	40.37	3.03	32–50	59
MCV (fl)	112.2	6.46	96–126	59
MCH (pg)	40.41	2.55	33–45	59
MCHC (g/dL)	35.93	1.01	32–38	59
WBC (10^3/µL)	10.31	2.58	5.8–19.5	59
Neutrophills: band (10^3/µL)	0	0.03	0–0.2	59
Neutrophils: mature (10^3/µL)	4.6	1.65	1.8–12.7	59
Lymphocytes (10^3/µL)	1.96	0.93	0.2–6.2	59
Monocytes (10^3/µL)	0.35	0.31	0–1.6	59
Eosinophils (10^3/µL)	3.35	1.27	1.3–7.1	59
Basophils (10^3/µL)	0.04	0.08	0–0.3	59
Platelets (10^3/µL)	167.12	41.68	73–281	59
Fibrinogen mg/dL	138.14	89.22	50–400	59
Biochemistry				
Glucose (mmol/L)	5.257	0.68	3.70–6.83	62
Sodium (mmol/L)	155	2.44	151–164	62
Potassium (mmol/L)	3.97	0.32	3.4–4.8	62
Chloride (mmol/L)	113.47	2.91	106–119	62
Bicarbonate (mEq/L)	20.69	4.01	9–28	62
Anion gap (mEq/L)	24.87	5.33	17–40	62
BUN (mmol/L)	23.67	3.38	16.42–31.06	62
Creatinine (µmol/L)	96.36	23.87	53.04–159.12	62
Total protein (g/L)	74.5	5.8	64–100	62
BUN:creatinine ratio (mg/dL)	65	19.21	25.56–110	62
Albumin (g/L)	44.90	3.0	32–52	62
Globulin (g/dL)	2.96	0.68	2–6.8	62
Albumin:globulin ratio (g/dL)	1.59	0.34	0.47–2.48	62
Total bilirubin (µmol/L)	1.54	0.51	0–3.42	62
Calcium (mmol/L)	2.33	0.12	2.08–2.63	62
Phosphorus (nmol/L)	1.68	0.23	1.16–2.30	62
Magnesium (mg/dL)	1.44	0.14	1.2–1.9	62
Uric acid (mg/dL)	1	0.52	0.2–2.3	26
ALP (U/L)	272.39	149.58	43–899	62
ALT (U/L)	44.42	23.39	19–122	26
AST (U/L)	255.89	91.96	142–733	62
SDH (U/L)	10.53	10.83	1–51	59
LDH (U/L)	493.38	83.94	378–706	26
CPK (U/L)	152	41	82–291	62
GGT (U/L)	27.45	3.9	17–39	62
Amylase (U/L)	1.12	0.43	1–3	26
Lipase (U/L)	10.35	4.6	4–27	26
Cholesterol (mmol/L)	3.65	0.66	2.29–4.76	26
Triglyceride (mg/dL)	83.03	32.42	41–182	62
Iron (µg/dL)	98.85	35.07	32–206	62

Source: Adapted from Goldstein et al. (2006).

swabs to be taken. Faecal occult blood can be difficult to interpret as fish blood may give false positive reactions. To avoid this, cetaceans can be fed washed fish fillets for 3 d prior to testing.

Gastric samples can be easily obtained under physical restraint or animals can be conditioned for unrestrained gastric lavage. An equine stomach tube is passed into the stomach. Material may reflux up the tube or it can be aspirated. Gastric samples should be evaluated for colour, consistency, odour, pH and cytology. Gastric samples should be collected if upper gastrointestinal pathology is suspected, such as gastric ulceration.

7.4 Respiratory tract sampling

Cytological and microbiological analyses of exhaled air are useful for examining the health of the respiratory tract. Captive animals can be trained to exhale onto agar plates or into sterile containers. A trans-tracheal wash can be performed by slowly advancing a dog urinary catheter into the blowhole to a depth of 10–15 cm. When the animal exhales, introduce 20 mL of sterile saline then apply negative pressure to aspirate material. Cytological examination of the sample should include white cells, bacteria and parasites. Both wet preparations and stained smears can be used. The presence of small numbers of epithelial cells and leukocytes is normal. Abnormal findings include red blood cells, increased numbers of leukocytes, parasite eggs, protozoa and fungi. Abnormal findings without clinical signs of illness need to be evaluated carefully and the need for treatment should be considered. A detailed guide to the cytology of cetaceans can be found in Sweeney et al. (2003). The sample may also be cultured for bacteria or viruses. Bronchoscopy has been used to examine and sample the respiratory tracts of dolphins (Harper et al. 2001; Tsang et al. 2002).

8 NEONATAL CARE AND HAND-REARING

8.1 Neonatal care

Survival of cetacean calves in captivity has increased dramatically over the last few years. This has been due to improved nutrition of the dam, size of the birthing pool and increased medical knowledge. Calves should be monitored closely for the first 4 wk after birth. Sucking from the dam should be observed within the first 24 hr or intervention may be necessary.

The date at which the calf should undergo its first physical examination is arbitrary. Some institutions catch and examine calves at 4 wk of age, or earlier if behavioural or physical abnormalities are observed. Other institutions wait until the animal is 5–6 mo old if no problems are detected. Capturing and restraining juvenile animals should be planned and carried out carefully. Respiratory and cardiac arrest has occurred in juveniles that were taken out of the water. If animals stop breathing they should be returned to the water immediately. Procedures should be quick and the calf returned to the water and its mother as quickly as possible. Always take care when restraining newborn calves as the mothers are usually very protective and may attack. Keeping the calf in sight and hearing of the mother can help keep both calm.

Calves must suckle within the first 24 hr to obtain the antibody-rich colostrum. Cetaceans have a diffuse epitheliochorial placenta and therefore neonates most likely receive passive immunity from colostrum rather than transplacentally. If a calf does not suckle in the first 24 hr it may not absorb sufficient antibodies for adequate immunity. There are, however, reports of surviving calves that did not suckle for the first 24 hr. If there is concern about adequate passive transfer from direct observations that the calf has not suckled, the dam can be manually restrained and milk collected from the mammary glands. This can be accomplished by applying negative pressure to the gland by using a neonatal face-mask and negative pump (e.g. a 60 mL syringe). The milk is given to the calf by stomach tube. Alternatively, plasma can be collected from a healthy adult animal and given to the calf IV, IP or PO at a rate of 10 mL/kg. Lactation lasts 7–18 mo and captive calves start eating whole fish at 3–6 mo of age

8.2 Hand-rearing

Orphaned wild or mismothered captive cetaceans may need to be hand-reared. Various milk formulas have been used (Table 18.3). Both formulas were developed in the US and some ingredients are not available in Australia. Di-Vetelact (Sharpe Laboratories) is readily available in Australia and has proven safe and useful in hand-rearing dolphins. It can be started at the normal dilution of 1 scoop per 60 mL and gradually increased in concentration to 1 scoop per 30 mL. The milk can be given by stomach tube initially, then the calf gradually introduced to a bottle and teat. The best initial teat

Table 18.3 Milk formulas used for dolphins

Sea World (USA) formula	April formula[a]
Herring fillets with viscera — 750 mL	Herring fillets with viscera — 750 mL
Zoological 30/55[b] — 225 g	Multimilk powder
Zoological 33/40[b] — 450 g	Safflower oil — 50 mL
Salmon oil — 50 mL	Lecithin — 15 mL
Taurine — 500 mg	Taurine — 250 mg
Lactobacillus — 1.5 tablets	Lactobacillus — 3 tablets
Osteoform[c] — 1 tablespoon	Osteoform — 1 tablespoon
Sea World marine vitamin tablets	Bottled water — 1100 mL
Whipped cream — 100 mL	Multivitamins with zinc
50% dextrose — 60 mL	
Bottled water — 1100 mL	

a April formula was named after an orphaned common bottle-nosed dolphin calf rescued at an estimated 3–4 wk of age and successfully raised at the Gulfarium, Fort Walton Beach, Florida.
b Zoological (milk substitute, Pet-Ag) 30/55 contains 30% protein and 55% fat; 33/40 contains 33% protein and 40% fat.
c Osteoform (vitamin/mineral mix, Vetamix).
Source: Reproduced from Townsend (1999).

for dolphins is a lamb teat, but experimentation may be required to find the teat that a neonate prefers. The volume given per day should be 20% of body weight, divided into five or six feeds. Calves should be weighed at least weekly to ensure they are gaining weight. Antibiotics can be given PO or IM if indicated for the first 5–7 d. Ceftiofur (2 mg/kg sid IM for 5–7 d) or amoxycillin/clavulanic acid (12.5 mg/kg bid IM for 5–7 d) are the drugs of choice.

Several wild orphaned dolphins have been successfully introduced to experienced mothers. The female does not need to be lactating at the time. Generally, experienced females begin producing milk within 24 hr of introduction to a calf and will allow the orphan to suckle. Close observation and regular weighing of the orphan are required. It is recommended to vaccinate orphaned calves against *Clostridium perfringens* toxin if the pathogen is a problem in a collection. The first dose can be given on day 2 or 3 after the introduction and a booster given 4 wk later.

9 CLINICAL APPROACH TO THE SICK DOLPHIN

Captive dolphins show few clinical signs when they are unwell. They may have decreased or no appetite, they may be listless and lethargic or they may rest in shallow water with their ventral surface on the bottom. It should be noted that clinically normal dolphins can show similar signs if they have been overfed, are unhappy or the water temperature is greater than 30°C. The individual history is a useful diagnostic indicator, e.g. whether its daily feed intake has decreased or it is losing weight. Behavioural observations are also important. Abnormal interactions with conspecifics and trainers may indicate illness.

If a dolphin is suspected to be unwell, observation from a distance is the first step in a clinical assessment. Note the position in the water. The animal may be more or less buoyant than normal or may list to one side. Listing to one side may be due to pneumonia. Monitor the respiratory rate and effort and note normal, increased, shallow, laboured or decreased respiration. Err on the side of caution—intervention is indicated at the earliest suspicion of illness. If a bacterial infection is suspected, antibiotic therapy can be commenced prior to diagnostic procedures or while awaiting results, either by adding antibiotics to the food or by IM injection. Doxycycline (1.5 mg/kg bid) is the drug of choice for oral administration in adult animals and ceftiofur (2 mg/kg sid) for IM injection. These drugs are chosen because they are safe, broad-spectrum and effective against commonly encountered bacteria. Treatment should be continued for at least 7 d or until the white blood cell count has returned to normal and the animal appears clinically well. Antibiotic treatments for 4–6 wk are sometimes required. All animals on prolonged broad-spectrum antibiotics should receive nystatin, to avoid fungal overgrowths. If leukocyte counts remain elevated despite long-term antibiotic treatment, consider changing the antibiotic. Fungal infections may also be a cause.

Haematological and serum chemical analyses are extremely useful in assessing the health of cetaceans and blood should be collected early in the investigation. Ancillary diagnostic tests can be performed, depending on the results. Urine can be collected and analysed (see 7.2). Take blowhole samples and trans-tracheal washes if respiratory disease is suspected (see 7.4). Radiography is useful in diagnosing pneumonia, gastrointestinal foreign bodies and skeletal disease. Ultrasonography can assist in diagnosing respiratory, urogenital and gastrointestinal disease. Bronchoscopy can be used to examine and sample the respiratory tract (Harper et al. 2001; Tsang et al. 2002). Gastroscopy can be used to assess the health of the upper gastrointestinal tract (Dover & Van Bonn 2001).

10 DISEASES

10.1 Dermatological diseases

Skin lesions are common in dolphins. Lobomycosis, a yeast infection in the skin caused by *Loboa loboi*, has been diagnosed in a wild dolphin stranded in south-east Queensland. Craters and mild ulcerations of the epidermis were seen. *Erysipelothrix rhusiopathiae* often causes skin lesions in dolphins (see 10.8.2b). Rake marks caused by bites from conspecifics and other traumatic injuries can be portals of entry for environmental contaminants such as atypical mycobacteria, *Nocardia* spp. and fungi. Papillomavirus lesions can occur on the genitalia and on the oral mucosa (see 10.8.1b). Poxvirus causes characteristic tattoo lesions or pinhole lesions. Squamous cell carcinomas have been seen in oral and genital mucosa (McKinnie & Dover 2003).

Nematodes of the genus *Crassicauda* are commonly observed in mammary gland tissue. They can cause mammary gland swelling and may decrease milk production in lactating females. They can be visualised on ultrasonography in the milk ducts. Eggs may be seen in the milk. There is no known effective treatment.

10.2 Gastrointestinal diseases

10.2.1 Clostridial enteritis

Clostridium perfringens can cause gastrointestinal disease such as severe enteritis, colic and sudden death in captive animals (Blyde 2003). Diagnosing clostridial disease is difficult as the clinical signs and necropsy findings are non-specific. Ante mortem diagnosis relies on recognising clinical signs and detecting the organism in the faeces of affected animals either by Gram stain or culture. Clinical signs include depression, abdominal pain, anorexia, vomiting and diarrhoea. Intestinal intussusceptions and volvulus can be complications of the disease. Animals become extremely weak and die from electrolyte and acid–base disturbances and endotoxic shock. Sudden death may be the only sign. Necropsy diagnosis includes pathologic lesions such as hyperaemic areas of the intestine, ulceration of the intestinal mucosa, a fluid-filled pericardial sac and bilateral pulmonary oedema. Detection of the organism or the toxin in tissues at post-mortem can also aid diagnosis, however, *C. perfringens* may be a normal inhabitant of the gastrointestinal tract, it is widespread in the environment and is a rapid necropsy invader. The detection of *C. perfringens* in organs is therefore not a reliable basis for a diagnosis of clostridial disease. Laboratory detection of toxin is problematic as false positive and false negative results both occur. Toxin can diffuse from the gastrointestinal tract after death. Confirming a diagnosis can be difficult and is mostly based on gross necropsy findings and excluding other causes.

Disease associated with *C. perfringens* can be prevented by vaccination (Ultravac 5 in 1, Pfizer Animal Health). Initially two doses of 1.5 mL for adults and 1 mL for calves and juveniles weighing less than 100 kg are given 4–6 wk apart, then every 6 mo. Calves should be first vaccinated at about 6 mo of age. A brief post-vaccination anorexia may be seen. Pregnant animals have been vaccinated safely, but it should be avoided.

10.2.2 Foreign bodies

Captive cetaceans often ingest foreign bodies such as plastic bags, balls, stones, straws, jewellery, coins and hardware such as nails and screws. Most foreign bodies cause no harm and are regurgitated. Some dolphins keep favourite objects as toys, repeatedly regurgitating, playing with and reswallowing them for weeks or months. However, foreign bodies may pass to the intestine and cause a partial or complete obstruction, although this is rare in cetaceans due to their unusual gastrointestinal anatomy. Galvanised foreign bodies contain zinc which can cause localised gastritis and can be absorbed, causing toxicity. Lead sinkers are potential foreign bodies and may cause toxicity (see 10.9.3b).

Potentially harmful foreign bodies can be removed from the proximal compartment of the stomach using an endoscope. Physically restrain the dolphin and introduce an endoscope through a cannula to avoid damaging the scope. Dolphins may be sedated prior to the procedure to facilitate restraint and endoscopic examination. Alternatively, with the animal's jaws held open with towel strips or rope, a long arm can be introduced down the oesophagus and into the stomach to retrieve the foreign body.

10.2.3 Gastric ulceration

Gastric ulcers have been diagnosed in captive cetaceans (Abollo et al. 1998; Harper et al. 2002). Clinical signs include anorexia, lethargy, vomiting and diarrhoea. Diagnosis can be confirmed by endoscopic examination. Bleeding gastric ulcers can cause significant anaemia. The anaemia can be regenerative or non-regenerative depending on the length of time the ulcer has been bleeding. Initially the anaemia is regenerative but as iron is gradually lost through the gastrointestinal tract it becomes non-regenerative. Treat gastric ulcers as in horses, including antibiotics such as amoxycillin/clavulanic acid (12.5 mg/kg bid PO) or clarithromycin (5 mg/kg bid PO), histamine H2-receptor antagonists such as ranitidine (0.6 g/100 kg tid PO) and gastric acid suppressants such as omeprazole (0.25 mg/kg bid PO). Dolphins that are prone to developing gastric ulcers can be treated prophylactically using gastric acid suppressants and/or mucosal protectants such as cimetidine (5 mg/kg sid PO).

10.3 Respiratory diseases

Pulmonary disease is common in free-ranging and captive cetaceans. Bacterial pneumonias are common in captive cetaceans kept in indoor facilities. It is less common in animals kept in natural outdoor conditions. Pneumonia may be acute or chronic. Clinical signs are non-specific and include decreased appetite, lethargy and unresponsiveness. Respiratory rate may be increased and coughing (forceful expiration of air through the blowhole) is sometimes seen. Rales may be auscultated and the animal's breath may smell foul. There may be a discharge from the blowhole. Body temperature may exceed 37°C. Abnormal buoyancy such as floating on one side may be seen. Haematology usually reveals a leukocytosis, with the distribution of cells varying with the chronicity and severity of the disease. Expired air from the blowhole can be directed onto agar plates for culture. Swabs and trans-tracheal washes can also be obtained for cytology and bacteriology (see 7.4). Radiography, ultrasonography and bronchoscopy (Harper et al. 2001; Tsang et al. 2002) are useful in the diagnosis of respiratory disease.

Antimicrobial therapy should commence immediately and should be administered systemically (IM or IV) in severe cases (see section 9). Ongoing antimicrobial therapy should be directed by results of culture and sensitivity testing.

10.4 Urinary diseases

Urinary tract infections are seen occasionally. Cystitis can be diagnosed by urinalysis. Renal calculi have been diagnosed in captive cetaceans using urinalysis and ultrasonography or as incidental findings at necropsy. The calculi are generally made of uric acid. Clinical signs can include haematuria and anaemia. Treatment includes frequent low-volume feeds, fluid administration and urinary alkalinizing agents such as Ural (Sigma Pharmaceuticals) at a dose rate of 4 g bid PO for 6–8 wk (G Lacave *pers. comm.*).

10.5 Neurological and musculoskeletal diseases

Meningitis is a common finding in stranded cetaceans. It may be caused by bacteria, viruses, fungi (e.g. *Aspergillus fumigatus*) or protozoa (e.g. *Toxoplasma gondii*) (Bowater et al. 2003). Ante mortem diagnosis is difficult. Treatment is directed toward the specific aetiologic agent if it can be identified.

Osteomyelitis of dental alveoli may follow tooth infections. Worn or broken teeth are prone to infection. Removing infected teeth is simple in dolphins (see section 15).

Infections caused by *Clostridium* spp. may cause acute muscular swelling, muscle necrosis and emphysema within affected muscle tissue.

10.6 Endocrine and metabolic diseases

Endocrine diseases are not commonly reported in captive cetaceans. There have been no reports of diabetes, hyperthyroidism, hyperadrenocorticism or hypoadrenocorticism in captive cetaceans. Thyroid function can be assessed by monitoring thyroid hormone levels in serum and ultrasonography of the thyroid gland. Abnormal thyroid glands diagnosed on histopathology at necropsy have been associated with stillborn bottle-nosed dolphin calves (Garner et al. 2002).

10.7 Ophthalmologic diseases

Eye examinations are best carried out on conscious unrestrained animals that have been conditioned to undergo these procedures. Physically restrained untrained animals tend to close their eyes when attempts are made to examine them. Cetaceans have very strong ocular muscles and can retract the globe into the socket, making it almost impossible to view the eye. There are no reliable methods to block the facial nerve and general anaesthesia is difficult and risky.

Ocular disease is common in captive cetaceans. The most common problem is corneal opacity, which occurs when the salinity levels falls below about 1.030 for a substantial time (usually longer than 24 hr). This resolves once the salinity level is corrected.

Speed lines on the cornea are common in captive dolphins. These appear as horizontal linear opacities across the middle of the cornea. They are thought to be the result of particulate matter rubbing against the cornea where the eyelids meet when they are closed. They are generally self-limiting and resolve spontaneously.

Foreign bodies can lodge on the cornea and cause localised keratitis. Small corneal ulcers are usually the result of trauma. Ulcers can be diagnosed using fluorescein stain applied to the corneas of trained animals. Small uncomplicated corneal ulcers can be treated by applying antibacterial drops, e.g. ofloxacin (Ocuflox, Allergan Australia), or ointment. Drops are preferable as ointment tends to run off the surface of the eye into the water. Drops should be applied at least three times a day for up to several weeks. Subconjunctival injections of antibiotics (25 mg enrofloxacin every 3–4 d) may also be useful. If topical treatment is not possible because the animal will not open its eye for drops, parenteral treatment may be necessary. Doxycycline at 1.5 mg/kg bid PO is useful as doxycycline readily penetrates the anterior and posterior chambers of the eye.

Fungal keratitis or keratomycosis has been reported (Huff 2003) and is probably more common in captive animals housed in natural pool systems. Initially, it presents as a small white corneal opacity that is a fungal colony. It should be removed physically by rubbing a swab or cottonbud over the cornea until the lesion is removed. The animal may have to be physically restrained. If the fungal colony is not removed, the eye closes in 24–48 hr followed by a severe fungal uveitis. This can be treated with fluconazole (2 mg/kg PO bid for 6–8 wk).

Cataracts occur in older animals. However, their impact in captive animals that use sonar to navigate is probably not important. Cataract-induced uveitis is a common sequel. It can be treated with topical and parenteral corticosteroids (prednisolone 0.25–1.0 mg/kg PO sid or q 48 hr). Subconjunctival injections of long-acting corticosteroids, e.g. methylprednisolone acetate (20 mg once or every 4–6 wk), may also be used to treat this condition.

10.8 Infectious diseases

10.8.1 Viral diseases

a Morbillivirus

Morbilliviruses have caused mass mortalities in a number of free-ranging species of cetaceans in North America and Europe (Kennedy 1998). These morbilliviruses are now grouped together as cetacean morbillivirus (CeMV) and include porpoise morbillivirus (PMV) and dolphin morbillivirus (DMV). Little is known about the prevalence of morbillivirus antibodies in cetacean populations, however, antibodies have been demonstrated in up to 14 odontocete species in the northern hemisphere. Pilot whales (*Globicephala* spp.) were found to have very high seroprevalence of morbillivirus antibodies and may be the reservoir in cetaceans. DMV antibodies have been detected in a few cetacean species in New Zealand and Tasmania (Duignan 2000). A serosurvey by serum neutralisation of free-ranging and captive bottle-nosed dolphins and melon-headed whales (*Peponocephala electra*) on the eastern coast of Australia revealed antibodies to both DMV and PMV (W Blanshard *pers. comm.*). A case of mortality suspected to be associated with morbillivirus in a subadult male melon-headed whale on the Queensland coast was recently seen. No gross abnormalities were detected at necropsy. Histopathology of the brain revealed demyelinating encephalitis with low numbers of intranuclear eosinophilic inclusion bodies detected in nerve cells, astrocytes and oligodendroglia consistent with morbillivirus infection. Immunohistochemistry to confirm the diagnosis is pending at this time.

CeMV is transmitted via the respiratory route. Clinical signs in free-ranging cetaceans are not usually observed. Animals usually strand dead in poor condition and there are often excessive ectoparasites, indicating that the animal was compromised for some time. The disease is generally diagnosed on histopathology and by demonstrating viral antigen with immunohistochemistry, immunofluorescence, antigen-capture ELISA or PCR. If animals are found alive they generally show abnormal behaviour and respiratory compromise. Antibody titres can confirm a clinical infection (Duignan 2000).

Treatment is supportive but always unrewarding. Ensure that diseased animals are kept in quarantine and isolated from captive animals to prevent transmission of the disease.

b Papilloma virus

Papilloma virus infects free-ranging and captive dolphins and wart-like lesions have been seen in free-ranging and captive cetaceans in Australia. Virus has not been identified in these cases to confirm a papilloma virus. Two different viruses have been isolated from cetaceans—Type 1 and Type 2 (Rector et al. 2005). The virus causes wart-like lesions around the external genitalia and the oral cavity. The genital location of the warts and high prevalence in some species suggests venereal transmission. Diagnosis is made by histopathology on biopsy samples. Virus particles can be detected by electron microscopy. Treatment is not rewarding and the lesions are generally self-limiting.

c Herpesvirus

A herpesvirus was isolated from the brain of a stranded harbour porpoise with encephalitis (*Phocoena phocoena*), from the North Sea (Kennedy et al. 1992). It has also been isolated from oesophageal ulcers and skin lesions in stranded cetaceans. There have been no reports of the virus causing disease in dolphins in Australia.

10.8.2 Bacterial diseases

a Brucellosis

Brucellosis has not been diagnosed in cetaceans in Australasia. *Brucella* spp. have been associated with abortion in captive bottle-nosed dolphins. In cetaceans, brucellosis has been shown to cause reproductive, neurologic, lymphatic, musculoskeletal and other systemic disease. Brucellosis may contribute to decreased conception rates, increased spontaneous abortion and stillbirth rates and early neonatal mortalities in captive cetacean populations. It may be transmitted venereally and is potentially zoonotic. The *Brucella* isolate from bottle-nosed dolphins is now called *Brucella delphini* (Miller et al. 1999). Serological testing for brucellosis using commercially available tests may not be accurate (Meegan et al. 2006). Diagnosis can be made on culturing the organism from the placenta and foetus at necropsy (Miller et al. 1999).

b Erysipelas

Cetaceans are extremely susceptible to infection with *Erysipelothrix rhusiopathiae* (Sweeney & Ridgway 1975). Free-ranging and captive animals have been diagnosed with the disease. Disease syndromes range from mild disease with classic skin lesions to sudden death. Animals become infected when they ingest the bacterium. It is found in many soil types and resides in the mucus layer of many marine fish. It can survive freezing and thawing and it is therefore difficult to prevent exposure in captive cetaceans.

The skin disease is characterised by grey rhomboid dermal plaques over the entire body (Sweeney 1978). The skin may ulcerate. Animals may become septicaemic and die if not treated quickly and aggressively using IV or IM antibiotics. Clinical signs include anorexia and profound lethargy.

Haematology often reveals a leukocytosis which can progress to a severe leukopaenia just prior to death. The organism can be isolated from lesions and blood culture in live animals and many organs at necropsy. Erysipelas titres can be performed using a commercially available test but the results are difficult to interpret. False negatives are common.

Mild cases can be treated successfully with appropriate antibiotics and general supportive care. Amoxycillin and clavulanic acid are the antibiotics of choice at 12.5 mg/kg bid PO, but ciprofloxacin at 20 mg/kg bid PO or doxycycline at 1.5 mg/kg bid PO may also be used. Treatment should continue for at least 21 d.

Captive animals can be vaccinated using commercially available vaccines (Eryvac, Pfizer Animal Health). Two doses (1 mL IM) are given 4–6 wk apart, then annually. Some swelling may occur at the injection site for 5–7 d following vaccination. Some animals may also develop anorexia for 5–7 d. Rare anaphylactic reactions following the administration of vaccines have been reported. Animals remain still, show muscle fasciculations and sometimes convulse. If anaphylaxis occurs, give adrenaline (1 mg IM) and remove the animal from the water. Monitor respiration and pulse rates. Supply oxygen via a mask placed over the blowhole. Ensure appropriate emergency drugs are on hand when using this vaccine in dolphins.

c *Nocardia*

Infections with *Nocardia asteroides, N. brasiliensis* and *N. transvalensis* produce variable clinical signs depending on the organs infected. Nocardia produces granulomatous pathology which may abscess. Diagnosis relies on isolating the organism from infected tissues. Cytological and bacteriological examinations of trans-tracheal washes, exhaled air and blowhole swabs are useful for diagnosis. This may be possible ante mortem but it is more likely to be diagnosed at necropsy.

Infection is by inhalation or contamination of external wounds. The bacterium is ubiquitous and found in soil in many geographic locations. The disease is therefore seen most commonly in rescued stranded juvenile animals, which are often compromised and more likely to be exposed to Nocardia. Certain weather patterns such as high wind and heavy rainfall can increase animal exposure to the organism.

Treatment involves administering appropriate antibiotics. The antibiotic of choice is amikacin at 15 mg/kg sid IM. Treatment should be continued for up to 6 mo.

d *Melioidosis*

This disease is an often fatal septicaemic infection caused by *Burkholderia pseudomallei*. It was first reported in marine mammals in Hong Kong in 1976 (Kinoshita et al. 2003). It is a common inhabitant of soil in tropical and subtropical Australia. Animals are infected when they inhale the organism in water droplets or ingest bacteria in run-off, or when external wounds become contaminated. Melioidosis is more common in warm wet seasons and outbreaks are often associated with monsoons or cyclones. There are no dolphinariums in Australia in the melioidosis geographic range but the disease should be considered in strandings in the area where *B. pseudomallei* is endemic (northern parts of Australia).

The onset of disease is acute and the diagnosis is frequently made at necropsy after sudden death. If clinical signs occur they initially include inappetence, changes in behaviour including lethargy, then anorexia and pyrexia. Treatment can be successful if appropriate antibiotics are given early and aggressively. The treatment of choice is ceftazidime at 20 mg/kg sid IM in conjunction with meropenem at 30 mg/kg tid IM or IV. Treatment should be continued for up to 12 wk (Kinoshita et al. 2003). An alternative treatment regimen requires a 6 wk course of ceftazidime (30 mg/kg tid IV) and trimethroprim sulphadiazine (25 mg/kg bid PO/IM) plus folic acid (25–50 mg/dolphin bid) for 2–4 wk followed by at least 20 wk of amoxycillin/clavulanic acid (10 mg/kg tid PO) (D Spielman *pers. comm.*).

Prevention involves isolating animals from the bacteria. No vaccines are available.

10.8.3 Fungal diseases

a *Aspergillosis*

Aspergillosis caused by *Aspergillus fumigatus* occurs in captive cetaceans. Infection is generally opportunistic in immunocompromised or stressed animals. It usually enters via the respiratory tract but can also enter via a skin wound or the opening of the ear. Clinical signs depend on the organs affected and can include dyspnoea, head tilt, depression and convulsions as it spreads to the nervous system. Haematology usually shows a leukocytosis with a neutrophilia. Aspergillus titres are not reliable as false negatives are common. Aspergillosis should be suspected with persistent leukocytosis and neutrophilia that does not respond to antibiotic therapy. Treatment should begin with fluconazole at 2 mg/kg bid PO for at least 28 d.

10.8.4 Parasitic diseases

Cetaceans harbour a diverse assemblage of parasites (Duignan 2000). Generally these are of little clinical significance and are usually incidental findings at necropsy. Parasites may occasionally be implicated in the cause of illness or death, but they are rarely the primary pathogen. Parasites are rare in captive cetaceans. The only parasite of clinical significance in dolphins in Australia is *Toxoplasma gondii*. Toxoplasmosis has been diagnosed as the cause of death in free-ranging and captive cetaceans. Indo-Pacific hump-backed dolphins seem particularly susceptible to this disease, possibly due to their ecology. They are known to prey on birds in the wild and inhabit estuaries close to human populations. Indo-Pacific hump-backed dolphins that died from toxoplasmosis stranded after periods of heavy rainfall (Bowater et al. 2003). Clinical signs are non-specific and can include lethargy, anorexia and neurological signs. Toxoplasmosis has caused stillbirths in captive and free-ranging dolphins including

Indo-Pacific hump-backed dolphins. Ante mortem diagnosis is difficult. Antibody titres can be difficult to interpret and false negatives occur. Treatment would include injectable clindamycin (Dalacin C Phosphate sterile solution, Pfizer Australia) at 20 mg/kg IM or IV tid for 10–14 d and general supportive care.

10.9 Non-infectious diseases

10.9.1 Neoplasia

Neoplasia in captive and free-ranging cetaceans is uncommon. Squamous cell carcinomas in the mouth and around the genitalia have been diagnosed in captive and free-ranging dolphins. They have been associated with papilloma virus, suggesting malignant transformation of a benign papillomatous lesion (Bossart et al. 2005).

Lymphoproliferative disease has been diagnosed in stranded dolphins. One was diagnosed with granulocytic leukaemia and another with lymphocytic lymphoma (Cowan 2003). Lymphosarcoma and haemangiosarcoma were diagnosed in an aged captive common bottle-nosed dolphin.

10.9.2 Trauma

Minor wounds are common in captive dolphins. Wounds are more frequent in groups of males. Wounds from predators may be seen in stranded cetaceans. Boat propeller injuries are also common. On the eastern coast of Australia, many of the popular beaches are netted or have baited drumlines to protect swimmers from shark attacks. Dolphins commonly get caught in these nets and suffer moderate to severe trauma when trying to escape. They also get caught on the baited drumlines while searching for food and suffer hook wounds to their mouths and other parts of their bodies. Animals also become entangled in crabpot lines.

Wounds are generally treated conservatively by local debridement of necrotic tissue and antibiotic cover. Applying topical medications is generally not warranted as, in water, they do not stay on the tissue long enough to be effective. Some wounds become infected with opportunistic environmental organisms such as atypical mycobacteria, Nocardia or fungi.

Fractures of the mandible and maxilla have been sustained when free-ranging and captive dolphins swim into objects at high speed. These can be repaired using conventional techniques under general anaesthesia.

10.9.3 Toxins

a Algal biotoxins

Marine algal biotoxins produced by dinoflagellates have been implicated in the deaths of cetaceans. Toxin-producing algae occur in Australasian waters (Arthur et al. 2006). Marine biotoxins have killed free-ranging cetaceans in North America (Duignan 2000). There have been no reported cetacean mortalities associated with these toxins in the southern hemisphere. Captive cetaceans are unlikely to be exposed to biotoxins. Animals affected by algal biotoxins are usually found dead. Diagnosis is by detecting the toxic acid in the urine, faeces and brain of affected animals. The syndrome may be suspected when animals are found dead with no premonitory signs in conjunction with algal blooms.

b Zinc and lead

Zinc and lead gastrointestinal foreign bodies can cause toxicity. Signs include lethargy, vomiting, diarrhoea and haemolytic anaemia. Radiographs help diagnose metallic foreign bodies. Blood zinc levels and blood lead levels can be determined. Blood must be collected into EDTA tubes. Blood lead levels >0.7 μmol/L indicate significant exposure and treatment may need to be implemented. Blood zinc levels >70 μmol/L may indicate abnormal exposure and may require further investigation. Treatment for heavy metal poisoning in cetaceans is primarily supportive. The foreign body should be removed if possible, by manual or endoscopic retrieval. Often the animal will regurgitate the object voluntarily. Fluids should be given PO and antibiotics and corticosteroids (prednisolone 0.25–1.0 mg/kg PO sid for 5–7 d) for potential gastritis should be considered. Calcium EDTA (Calsenate injection, Parnell Laboratories) can be given slowly IV at a dose rate of 80 mg/kg or 0.4 mL/kg. Take care as it is a perivascular irritant and can cause phlebitis and skin sloughing at the site if injected perivascularly. If this occurs the area should be treated topically with DMSO and flucortisone sid or bid for 10–14 d.

11 ZOONOSES

Zoonoses associated with cetaceans are uncommon. Mycobacterial infections, brucellosis, erysipelas and lobomycosis (a skin disease caused by the yeast *Loboa loboi*) have been diagnosed in cetaceans and are potentially zoonotic. Always take care when treating stranded animals or performing necropsies on stranded or captive animals. Appropriate personal protective equipment should always be used to avoid contracting zoonoses.

12 EUTHANASIA

Euthanasia is often required for stranded cetaceans and occasionally for captive animals. In small cetaceans such as dolphins pentobarbitone sodium can be injected IV into vessels in the tail fluke flippers, dorsal fin or caudal peduncle (Fig. 18.3). Intracardiac or intraperitoneal injection is only acceptable after heavy sedation with xylazine or midazolam. Xylazine is recommended only as a sedative prior to euthanasia. Sedation prior to euthanasia is also recommended to prevent excitement and violent movements during and after injection of the euthanasia solution. The euthanasia of larger cetaceans is difficult and should be carried out only by experienced operators. These methods are discussed in Chapter 3.

13 THERAPEUTICS

Drugs can be administered to cetaceans via a number of routes. Oral administration can be accomplished by inserting the drug into fish which are fed to the animal. If the animal is not eating, drugs can be mixed with water and given by gavage. A stomach tube is easy to pass in cetaceans and due to the arytenoepiglottideal tube there is no risk of aspirating regurgitated liquid. Gastrointestinal transit time in cetaceans is rapid and orally administered drugs with gastric protective coatings or capsules may not break down in time to be absorbed adequately. This must be considered when choosing oral medications.

Intramuscular injections are straightforward but care must be taken to use a needle long enough (21 G × 3.8 cm) to deposit the drug into muscle. Most cetaceans have a thick blubber layer which is poorly vascularised in some areas. Injection into the blubber may lead to aseptic necrosis then large abscesses which can be very difficult to treat. IM injections can be given anywhere from the dorsum to the lateral line of the animal, between the neck and the tail fluke (Fig. 18.5).

Intravenous injections can be given into the vessels of the tail flukes, flippers, dorsal fin and caudal peduncle (Fig. 18.3). Care must be taken with irritant agents

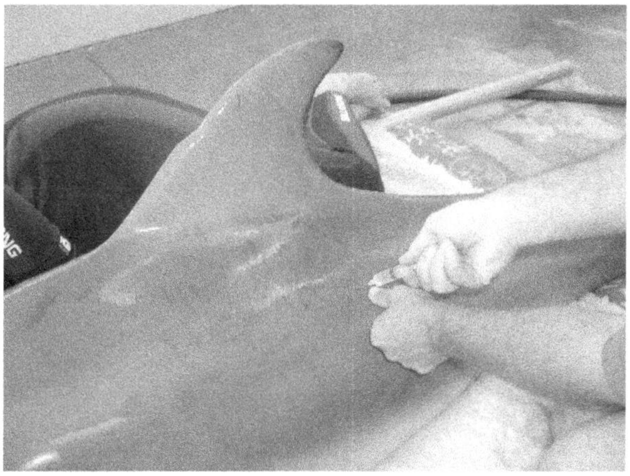

Figure 18.5 IM injection in a dolphin.

which can easily leak from the vessels into surrounding tissue and cause phlebitis and tissue necrosis.

In cetaceans, the dermis directly contacts the muscles below it and there is no SC space—this route cannot be used for injections. This has implications for vaccines which are designed to be given SC in terrestrial mammals. These vaccinations must be given IM in cetaceans, increasing the risk of anaphylactic and tissue reactions.

Non-irritant drugs can be injected intraperitoneally. Ensure the needle is long enough to penetrate the abdominal musculature and that abdominal organs are not accidentally lacerated.

Subconjunctival injections can be used to deliver corticosteroids and antibiotics to ocular structures such as the cornea and anterior chamber.

Captive cetaceans have been successfully treated with a variety of therapeutic agents. Most antibiotics are appropriate for use in cetaceans. The exception is trimethoprim/sulphonamide combinations, that have been reported to produce bone marrow suppression in cetaceans (McBain 1984). Whilst the addition of folic acid at a dose rate of 25–50 mg PO bid may reduce the likelihood of adverse reactions, it is preferable to avoid use of these drugs in cetaceans. Non-steroidal anti-inflammatory drugs have produced significant side effects in dolphins (e.g. gastric ulceration and renal impairment) and should be avoided. They should only be used in extreme circumstances such as endotoxic shock. Corticosteroids are a preferred choice to treat non-infectious inflammatory disease.

There is rarely a need to use anthelmintics in captive cetaceans fed frozen–thawed fish. Ivermectin,

Table 18.4 Therapeutic agents and dose rates used in dolphins

Drug	Dose rate (mg/kg)
Antibiotics	
Amikacin	5–10 bid IM
Amoxycillin/clavulanic acid	12.5 bid PO, IM
Amoxycillin	20 bid PO, IM
Azithromycin: loading	9.6 once PO, IV
Azithromycin: maintenance	5.2 sid PO
Cephalexin	22 tid PO
Ceftazidime	20 sid IM
Ceftiofur	2.0 sid IM
Ciprofloxcin	15–29 bid PO
Clarithromycin	5 bid PO
Clindamycin	5.5 bid PO, 10 bid IM, IV
Doxycycline	1.5 bid PO
Enrofloxacin	5 bid PO, IM
Meropenem	13 tid IM
Metronidazole	7 bid PO
Ofloxacin	5 bid PO
Rifampicin	2.5 bid PO
Tetracycline	55–65 bid PO, IM
Vancomycin	1–1.5 tid PO
Antifungals	
Fluconazole	2 bid PO
Ketaconazole	10 bid PO
Terbinafine	2 sid PO
Sedatives	
Butorphanol	0.05–0.2 IM
Diazepam	0.15–0.2 PO, IM
Midazolam	0.095 IM
Pethidine	2.0 IM
Anaesthetics	
Ketamine	2.0 IM
Medetomidine	0.04 IM
Propofol	3.5–4.8 IV
Thiopentone	15–20 IV
Tiletamine/zolazepam	2–2.5 IM, 1–1.5 IV
Reversal agents/antidotes	
Atipamezole	0.2 IM
Calcium disodium edetate	100–200 IV
Flumazenil	0.002–0.004 IM, IV
Naltrexone	0.005 IM
Corticosteroids	
Methylprednisolone acetate	0.1–1.0 IM
Prednisolone	0.25–2.0 sid, bid, PO, IM

(Continued)

Table 18.4 *(Continued)*

Drug	Dose rate (mg/kg)
Non-steroidal anti-inflammatory drugs	
Flunixin meglumine	0.5 SID PO, IM
Meloxicam	0.1–0.25 sid PO, IM
Anabolics	
Ethyloestrenol	0.05 sid PO
Norandrostenolone	1.0 IM
Antiparasiticides	
Ivermectin	0.2 PO
Gastrointestinal	
Cimetidine	5 BID PO
Omeprazole	0.25–0.3 bid PO
Ranitidine	6.6 tid PO
Reproductive	
Altrenogest	0.044 sid PO

praziquantel, febantel and pyrantel administered PO at canid dose rates have all been used safely in cetaceans. Short periods of anorexia and lethargy have been observed following administration of some anthelmintics. Levamisole should be avoided, as deaths have occurred following its use in cetaceans.

14 REPRODUCTIVE MANAGEMENT

14.1 Monitoring and enhancing reproduction

The oestrous cycle is about 30 d but long periods of anoestrus have been observed in captive cetaceans. Females may be more likely to undergo an oestrous cycle after changing the social structure or introducing a new breeding male.

Oestrus can be detected by behavioural observations, ultrasonography, hormonal levels in serum and/or urine or vaginal cytology. Oestrus can be monitored in captive female cetaceans by ultrasonography of the ovaries. A follicle can be followed by ultrasonography (Brook 2001) daily until it is seen to ovulate (from 14–20 mm in diameter depending on the individual). Mating generally occurs at ovulation. Monitoring the oestrous cycle of captive female cetaceans allows more precise timing of male–female introductions as sexes are usually separated other than for breeding, and for artificial insemination.

Pregnancy in captive females can be diagnosed in two ways. Three successive serum progesterone levels, taken 2 wk apart, exceeding 9.5 nmol/L (3 ng/mL) are diagnostic for pregnancy. Alternatively, pregnancy may be detected using ultrasonography from 14 wk of gestation. There are many advantages to ultrasonographic diagnosis and monitoring of pregnancy. Measuring the diameter of the cranium and thoracic cavity allows a reasonable prediction of the time of parturition (Lacave et al. 2004). An Excel computer program is available to predict parturition dates based on ultrasonographic foetal measurements (available from Dr Geraldine Lacave at geraldine.lacave@skynet.be). Routine ultrasonography of pregnancy also assists in monitoring foetal health.

Animals can be conditioned to stay in right or left lateral recumbency or dorsal recumbency in the water for this procedure. If animals must be physically restrained it should be in right or left lateral recumbency. The exact position of the probe depends on the size of the foetus, but the operator should begin at the genital slit and move cranially and dorsally.

Assisted reproduction techniques are used in captive cetaceans. Males can be conditioned to provide semen samples which can be collected and cryopresrved. Artificial insemination of females has produced many live calves. Female dolphins can be synchronised by treating them with altrenogest (Regumate for horses, Intervet Australia) at 1 mL/50 kg or 0.044 mg/kg sid PO. When the altrenogest is stopped, female dolphins should come into oestrus 15–27 d later (Robeck et al. 2005). Sex selection using a flow cytometer (MoFlo SX, Dako, Fort Collins, CO, US) and artificial insemination has been successful in dolphins (O'Brien & Robeck 2006). This allows the production of young of a desired sex, which will have significant implications for managing captive cetaceans.

14.2 Reproductive control (contraception)

There have been few indications for contraception in cetaceans. If facilities are available the sexes can be separated. Mature females can be contracepted using oral altrenogest. Altrenogest has been used for up to 2 yr without adverse effects. Trainers must take care to avoid the medication touching their skin when they are administering it. Deslorelin implants (Suprelorin, Peptech Animal Health) can decrease testosterone levels in mature males and reduce reproductive drive and sperm concentration. Although deslorelin may be used as a male contraceptive it does not guarantee complete sterility. Theoretically, deslorelin should be a useful contraceptive in female dolphins but its use has yet to be evaluated. In males, two 4.7 mg implants or a single 9.4 mg implant are usually inserted IM, just lateral to the ventral midline.

14.3 Dystocia

Dystocias are rare in cetaceans but have been reported in captive animals. Generally the calves are already dead by the time the dam is restrained. The calf can usually be extracted per vagina. Foetotomy may be necessary to assist extraction.

15 SURGERY

Major surgical procedures are rarely performed on captive cetaceans. Abdominal surgery is rarely performed due to the limited exposure of the organs and the difficulty of closing the inelastic skin. Surgery in captive cetaceans is limited to dentistry, draining abscesses, biopsies and wound management.

Teeth are often worn down or broken in captivity due to animals playing with objects such as rocks and hard balls. Worn or broken teeth may abscess and require extraction. Regional anaesthesia for the anterior mandible can be used by blocking the infra-alveolar nerve. This can be accomplished by palpating the mandibular foramen, lateral to the base of the tongue, through the mouth and injecting 10 mL local anaesthetic around the foramen, using a 22 G, 9 cm spinal needle. The patient's mouth should be manually opened as wide as possible by placing towels around the rostral areas of both the maxilla and mandible and pulling the jaws apart. The operator can then insert their hand into the mouth and palpate the mandibular foramen lateral to the base of the tongue on the ventral surface of each mandible. They guide the needle into this area and infiltrate local anaesthesia. All dolphin teeth are single-rooted and extraction is usually a simple procedure that requires only an extractor and an elevator (Ridgway et al. 1975).

ACKNOWLEDGMENTS

Dr Blyde would like to thank Jim McBain for his assistance and mentoring. The authors thank Vicki Knight for her contribution to the clinical pathology section, Justine O'Brien for reviewing the reproductive management section and Derek Spielman for his revision of the manuscript.

REFERENCES

Abollo E, López A, Gestal C, Benavente P & Pascual S (1998). Long-term recording of gastric ulcers in cetaceans stranded on the Galician (NW Spain) coast, *Diseases of Aquatic Organisms* **32**, 71–73.

Arkush KD (2001). Water quality. In *CRC Handbook of Marine Mammal Medicine* (2nd edn) (eds LA Dierauf & FMD Gulland), pp. 779–790. CRC Press: Boca Raton, Florida.

Arthur KE, Limpus CJ, Roelfsema CM, Udy JW & Shaw GR (2006). A bloom of *Lyngbya majuscula* in Shoalwater Bay, Queensland, Australia: an important feeding ground for the green turtle (Chelonia mydas). *Harmful Algae* **5**, 251–265.

Bedford G (2000). The use of body length, girth and weight relationships as a husbandry tool for captive bottlenose dolphins. In *Proceedings of 10th ARAZPA/ASZK Conference*, pp. 264–277. Gold Coast, Qld.

Blyde D (2003). Gastrointestinal disease in captive marine mammals associated with *Clostridium perfringens*. In *Proceedings of Annual Conference of Australian Association of Veterinary Conservation Biologists* (eds A Tribe & R Booth), pp. 87–89. University of Queensland, Cairns, Qld.

Bossart GD, Reidarson TH, Dierauf LA & Duffield DA (2001). Clinical pathology. In *CRC Handbook of Marine Mammal Medicine* (2nd edn) (eds LA Dierauf & FMD Gulland), p. 421. CRC Press: Boca Raton, Florida.

Bossart GD, Ghim S, Rehtanz M, Goldstein JD, Varela RA, Ewing RY, Fair P, Lenzi R, Joseph B, Hicks CL,

Schneider LS, McKinnie CJ, Reif JS, Sanchez R, Lopez A, Novoa S, Bernal J, Goetti M, Defran RH & Jenson BA (2005). Orogenital neoplasia in Atlantic bottlenose dolphins (Tursiops truncatus). In *Proceedings of Dolphin Neonatal and Reproduction Symposium* (ed. D Olson), p. 7. Indianapolis Zoo, Indianapolis, Indiana.

Bowater RO, Norton J, Johnson S, Hill B, O'Donoghue P & Prior H (2003). Toxoplasmosis in Indo-Pacific humpbacked dolphins (*Sousa chinensis*), from Queensland. *Australian Veterinary Journal* **81**, 627–632.

Brook FM (2001). Sonographic imaging of the reproductive tract of the female bottlenose dolphin *Tursiops truncatus aduncas*. *Reproduction* **121**, 419–428.

Bryden MM, Marsh H & Shaughnessy P (1998). *Dugongs, Whales, Dolphins and Seals: A Guide to the Sea Mammals of Australasia.* Allen & Unwin: Sydney.

Chittick EJ, Gearhart S, Dold C, Walsh MT & Dalton L (2006). Preliminary findings with butorphanol sedation in cetaceans. In *Proceedings of 37th Annual Conference of International Association for Aquatic Animal Medicine* (ed. W Van Bonn), pp. 144–145. Bahamas.

Cowan D (2003). Leukemia in the bottlenose dolphin (*Tursiops truncatus*). In *Proceedings of 34th Annual Conference of International Association for Aquatic Animal Medicine* (ed. N Young), p. 204. Hawaii.

Department of Primary Industries NSW (2005). *Standards for Exhibiting Bottle-nosed Dolphins (Tursiops truncatus) in New South Wales.* NSW Department of Primary Industries: Sydney. <http://www.agric.nsw.gov.au/reader/1218.htm>.

Duignan PJ (2000). Diseases of cetaceans and pinnipeds. In *Marine Wildlife, Proceedings 335*, pp. 431–462. Post Graduate Foundation in Veterinary Science: Sydney.

Dover SR & Van Bonn W (2001). Flexible and rigid endoscopy in marine mammals. In *CRC Handbook of Marine Mammal Medicine* (2nd edn) (eds LA Dierauf & FMD Gulland), pp. 621–642. CRC Press: Boca Raton, Florida.

Duffield DA, Ridgway SH & Cornell LH (1983). Hematology distinguishes coastal and offshore forms of dolphins (*Tursiops*). *Canadian Journal of Zoology* **61**, 930–933.

Fair PA, Hulsey TC, Varela R, Goldstein JD, Adams J, Zolman ES & Bossart GG (2006). Hematology, serum chemistry, and cytology findings from apparently healthy Atlantic bottlenose dolphins (*Tursiops truncatus*) inhabiting the estuarine waters of Charleston, South Carolina. *Aquatic Mammals* **32**, 182–195.

Garner MM, Scwetz C, Ramer JC, Rasmussen JM, Petrini K, Cowan DF, Raymond JT, Bossart GD & Levine GA (2002). Congenital diffuse hyperplastic goitre associated with perinatal mortality in eleven captive-born bottlenose dolphins (*Tursiops truncatus*). *Journal of Zoo and Wildlife Medicine* **33**, 350–355.

Geraci JR (1986a). Marine mammals: introduction and identification. In *Zoo and Wild Animal Medicine* (ed. ME Fowler), pp. 750–757. WB Saunders: Philadelphia.

Geraci JR (1986b). Marine mammals: nutrition and nutritional disorders. In *Zoo and Wild Animal Medicine* (ed. ME Fowler), pp. 760–764. WB Saunders: Philadelphia.

Geraci JR (2000a). Cetacean anatomy and physiology. In *Marine Wildlife, Proceedings 335*, pp. 377–384. Post Graduate Foundation in Veterinary Science: Sydney.

Geraci JR & Lounsbury VJ (2005). *Marine Mammals Ashore: A Field Guide for Strandings* (2nd edn). National Aquarium in Baltimore: Baltimore.

Geraci JR & Medway W (1974). Simulated field blood studies in the bottle-nosed dolphin *Tursiops truncatus*: changes in hematology and chemistry during blood and plasma storage. *Journal of Wildlife Disease* **10**, 410.

Geraci JR & St Aubin DL (1980). Nutritional disorders of captive fish eating animals. In *The Comparative Pathology of Zoo Animals* (eds RJ Montali & G Migaki), pp. 41–49. Smithsonian Institution Press: Washington DC.

Geraci JR & Sweeney J (1986). Marine mammals: clinical techniques. In *Zoo and Wild Animal Medicine* (ed. ME Fowler), pp. 771–777. WB Saunders: Philadelphia.

Goldstein JD, Reese E, Reif JS, Varela RA, McCulloch SD, Defran RH, Fair PA & Bossart GD (2006). Hematologic, biochemical, and cytologic findings from apparently healthy Atlantic bottlenose dolphins (*Tursiops truncatus*) inhabiting the Indian River Lagoon, Florida, USA. *Journal of Wildlife Diseases* **42**, 447–454.

Harper CG, Borkowski R, Hoffman AM & Warner A (2001). Development of a standardized nomenclature for bronchoscopy of the respiratory system of harbour porpoises (*Phocoena phocoena*). *Journal of Zoo and Wildlife Medicine* **32**, 190–195.

Harper CG, Feng Y, Xu S, Taylor NS, Kinsel M, Dewhirst FE, Paster BJ, Greenwell M, Levine G, Rogers A & Fox JG (2002). *Helicobacter cetorum* sp. nov., a urease-positive Helicobacter species isolated from dolphins and whales. *Journal of Clinical Microbiology* **40**, 4536–4543.

Huff D (2003). Keratomycosis in a Pacific white-sided dolphin (*Lagenorhynchus obliquidens*). In *Proceedings of 34th Annual Conference of International Association for Aquatic Animal Medicine* (ed. N Young), pp. 158–160. Hawaii.

Kennedy S, Lindstedt IJ, McAliskey MM, McConnell SA & McCullough SJ (1992). Herpesviral encephalitis in a harbour porpoise (*Phocoena phocoena*). *Journal of Zoo and Wildlife Medicine* **23**, 374–379.

Kennedy S (1998). Morbillivirus infections in aquatic mammals. *Journal of Comparative Pathology* **119**, 201–225.

Kinoshita RE, Mauroo N & Spielman D (2003). Management of a recent outbreak of melioidosis in marine mammals. In *Proceedings of 34th Annual Conference of International Association for Aquatic Animal Medicine* (ed. N Young), p. 31. Hawaii.

Lacave G, Eggermont M, Verslycke T, Brook F, Salbany A, Roque L & Kinoshita R (2004). Prediction from ultrasonographic measurements of the expected delivery date in two species of bottlenose dolphin (*Tursiops truncatus* and *Tursiops aduncus*). *Veterinary Record* **154**, 228–233.

McBain JF (1984). Sulfamethoxale toxicity in three killer whales. In *Proceedings of 16th Annual Conference of International Association for Aquatic Animal Medicine* (ed. N Young), p. 38. Hawaii.

McKinnie CJ & Dover SR (2003). Diagnosis and treatment of lingual carcinoma in an Atlantic bottlenose dolphin (*Tursiops truncatus*). In *Proceedings of 16th Annual Conference of International Association for Aquatic Animal Medicine* (ed. N Young), pp. 164–165. Hawaii.

Medway W & Geraci JR (1986). Clinical pathology of marine mammals. In *Zoo and Wild Animal Medicine* (ed. ME Fowler), pp. 791–797. WB Saunders: Philadelphia.

Meegan J, Romano T, Mazzaro L, Sidor I & Dunn JL (2006). Serosurveillance of wild bottlenose dolphins (*Tursiops truncatus*) and evaluation of both newly developed marine-specific and commercially available *Brucella abortus* serologic tests for the detection of antibodies to marine-origin *Brucella*. In *Proceedings of 37th Annual Conference of International Association for Aquatic Animal Medicine* (ed. W Van Bonn), pp. 65–67. Bahamas.

Menkhorst PW (2001). *A Field Guide to the Mammals of Australia*. OUP: Melbourne.

Miller WG, Adams LG, Ficht TA, Cheville NF, Payeur JP, Harley DR, House C & Ridgway SH (1999). Brucella-induced abortions and infection in bottlenose dolphins *(Tursiops truncatus)*. *Journal of Zoo and Wildlife Medicine* **30**, 100–110.

O'Brien JK & Robeck TR (2006). Development of sperm sexing and associated reproductive technology for sex preselection of captive bottlenose dolphins (*Tursiops truncatus*). *Reproduction, Fertility and Development* **18**, 319–329.

Perrin WF & Myrick AC (1980). Age determination of toothed whales and sirenians. In *Report of International Whaling Commission, Special Issue 3*, pp. 1–50. International Whaling Commission: Cambridge.

Queensland Wildlife Parks Association (unpub). Recommended minimum standards for exhibiting fauna in Queensland: marine mammals (eds D Spitall & P O'Callaghan).

Rector A, Lacave G, Mostmans S, Van Doorslaer K, Rehtanz M, Salbany A, Roque L, Ghim S, Jenson AB & Van Ranst M (2005). The genetic characterization of a novel close-to-root papillomavirus in a bottlenose dolphin: *Tursiops truncatus* papillomavirus type 1 (TTPV-1). In *Proceedings of Dolphin Neonatal and Reproduction Symposium* (ed. D Olson), p. 32. Indianapolis Zoo: Indianapolis, Indiana.

Reidarson TH (2003). Cetacea (whales, dolphins, porpoises). In *Zoo and Wild Animal Medicine* (eds ME Fowler & RE Miller), pp. 442–459. Elsevier: St Louis.

Ridgway SH, Green RF & Sweeney JC (1975). Mandibular anesthesia and tooth extraction in the bottlenosed dolphin. *Journal of Wildlife Diseases* **11**, 415–418.

Robeck TR, Steinman KJ, Yoshioka M, Jensen E, O'Brien JK, Katsumata E, Gili C, McBain JF, Sweeney J & Monfort SL (2005). Estrous cycle characterisation and artificial insemination using frozen-thawed spermatozoa in the bottlenose dolphin (*Tursiops truncatus*). *Reproduction* **129**, 659–674

Rogers T (2000). Cetaceans. In *Marine Wildlife, Proceedings 335*, pp. 277–289. Post Graduate Foundation in Veterinary Science: Sydney.

Shirai K & Sakai T (1997). Haematological findings in captive dolphins and whales. *Australian Veterinary Journal* **75**, 512–514.

Sweeney JC (1978). Infectious diseases. In *Zoo and Wild Animal Medicine* (ed. ME Fowler), p. 777. WB Saunders: Philadelphia.

Sweeney JC & Ridgway SH (1975). Common diseases of small cetaceans. *Journal of American Veterinary Medical Association* **167**, 533.

Sweeney JC, Reddy ML, Lipscomb TP, Bjorneby JM & Ridgway SH (2003). Respiratory tract. In *Handbook of Cetacean Cytology* (JC Sweeney, ML Reddy, TP Lipscomb, JM Bjorneby & SH Ridgway), pp. 11–16. Dolphin Quest: San Diego.

Terasawa F, Kitamura M, Fujimoto A & Hayama S (2002). Seasonal changes of blood composition in captive bottlenose dolphins. *Journal of Veterinary Medical Science* **64**, 1075–1078.

Townsend F (1999). Hand-rearing techniques for neonate cetaceans. In *Zoo and Wild Animal Medicine: Current Therapy 4* (eds ME Fowler & RE Miller), pp. 493–497. WB Saunders: Philadelphia.

Tsang KW, Kinoshita R, Rourke N, Yuen Q, Hu W & Lam WK (2002). Bronchoscopy of cetaceans. *Journal of Wildlife Diseases* **38**, 224–227.

Varela RA, Schwacke L, Fair PA & Bossart GD (2006). Effects of duration of capture and sample handling on critical care blood analytes in free-ranging bottlenose dolphins. *Journal of American Veterinary Medical Association* **229**, 1955–1961.

Wright KC, Griffin M, Lanter K & Salmon D (2005). The development and research of a commercially prepared fish substitute for captive marine birds and mammals. In *Proceedings of Dolphin Neonatal and Reproduction Symposium* (ed. D Olson), p. 46. Indianapolis Zoo: Indianapolis, Indiana.

19. Dugongs

Rupert Woods, Philip Ladds and David Blyde

1 TAXONOMY, DISTRIBUTION AND BIOLOGY

The Order Sirenia (manatees and dugongs) consists of two Families—the Trichechidae (manatees), which comprise four species, and the Dugongidae, the sole extant member of which is the dugong (*Dugong dugon*). The dugong is the most abundant and widespread of the Sirenia. They are classified as vulnerable to extinction in the World Conservation Union (IUCN) Red Data Book of Threatened Species (Bryden et al. 1998). The closest non-Sirenian relative of the dugong is the elephant (Marsh et al. 2000).

Dugongs are large herbivorous marine mammals. They inhabit tropical and subtropical coastal and inland waters in the Indo-West Pacific, between about 26–27° north and south of the equator. With relatively thin blubber layers compared to cetaceans and pinnipeds, they are intolerant of cold and are generally found in waters at and above 18°C. At the southern end of their distribution in Australia, dugongs may undertake seasonal movements that assist with thermoregulation (Bryden et al. 1998; Packard et al. 1984).

The most important dugong habitat in Australia is Torres Strait (Blanshard 2001), but dugongs are found from Moreton Bay in the east to Shark Bay in the west (Marsh et al. 2000). Aerial surveys have indicated that dugongs are the most numerous marine mammal in the in-shore waters of northern Australia. However, evidence suggests that dugong numbers have declined along most of the urbanised coast of eastern Queensland since the mid 1980s (Bryden et al. 1998). Habitat loss, drowning in gill or shark nets and traditional hunting have been implicated in the decline (Marsh 1997).

Dugongs are long-lived. The age-determination method developed for pinnipeds and toothed cetaceans has been adapted for dugongs, enabling their age to be determined from the number of growth layer groups in the tusks (Marsh 1980). The oldest dugong whose tusks have been examined for age determination was estimated as 73 yr old when she died (Marsh et al. 2000).

2 ANATOMY AND PHYSIOLOGY

Adult dugongs typically range from 2.4–3.0 m in length and 250–420 kg in weight. Females are slightly larger than males. There are reports of animals up to 3.31 m in length and weighing up to 600 kg (Blanshard 2001; Bryden et al. 1998). The relationship between body weight and length has been described by the equation:

$$y = (-34.251) - 14.976x + 55.218x^2$$

where y is the body weight (kg) and x is the body length from the tip of the snout to the notch in the tail fluke (in metres) (Spain & Heinsohn 1975). A selection of predicted body weights, together with an indication of the likely stage of maturity at certain body lengths, is presented in Table 19.1. Body length is an unsatisfactory indication of age except in young animals (Marsh et al. 1984a).

Table 19.1 Estimated body weight and likely stage of maturity of dugongs (*Dugong dugon*) based on body length

Length (m)	Estimated body weight (kg)	Likely stage of maturity
1.10	16	
1.20	27	1.2 m—newborn
1.25	33	
1.30	40	
1.40	53	
1.50	68	<1.5 m—suckling
1.60	83	
1.70	100	
1.75	109	
1.80	118	<1.8 m—probably suckling
1.90	137	
2.00	157	
2.10	178	
2.20	200	<2.2 m—immature
2.25	212	
2.30	223	
2.40	248	<2.4 m—probably immature
2.50	273	>2.5 m—mature
2.60	300	
2.70	328	
2.75	342	
2.80	357	
2.90	387	
3.00	418	
3.10	450	
3.20	483	
3.25	500	
3.30	518	

Estimated body weight predicted using the Spain and Heinsohn (1975) equation. Stage of maturity from Marsh (1997).
Source: Modified from Blanshard (2001).

Anatomy of the dugong has been reviewed by Marsh (1989) and a detailed guide to the necropsy of dugongs is available through the Australian Government Great Barrier Reef Marine Park Authority (Eros et al. 2007). The necropsy guide presents good diagrams and images of dugong anatomy and is available online at www.gbrmpa.gov.au.

Dugongs show a number of anatomical and physiological adaptations associated with life in a marine environment. They have fusiform bodies, fin-like fore limbs and horizontally expanded tail flukes, and they lack hind limbs. The flippers are short, the testes abdominal. The skull has a remarkably enlarged and sharply downturned premaxilla and a correspondingly elongated symphysial region of the mandible. The skeleton is extremely dense bone which presumably helps the animal overcome buoyancy problems when bottom-feeding in salt water. The feeding apparatus of the dugong reflects its role as a herbivore specialising on benthic seagrasses. The upper lip takes the form of a horseshoe-shaped disc or muzzle, which is a very complex and mobile structure (Marsh 1989).

Two pairs of upper incisors are present in juveniles. The deciduous incisors are small, do not erupt and are resorbed. Posterior growth of the permanent incisors (tusks) is similar in both sexes until about puberty, when the tusks are about 100 mm long. Female tusks continue to grow throughout life and may reach

Table 19.2 Physiological parameters for the Florida manatee (*Trichechus manatus latirostris*)

Parameter	Range
Respiratory rate	Variable, average 2–4 breaths/5 min in water, 3–15 breaths/5 min out of water
Heart rate	40–60 bpm (can be as low as 30 bpm with diving)
Body temperature	35.5–36°C

Source: Chittick & Walsh (2007).

180 mm. The tusks of males erupt after puberty and may reach about 150 mm, but never protrude more than a couple of centimetres. The erupted end is worn into a chisel shape. Apart from the tusks, the only functional teeth are the cheek teeth. Dugongs have a total of six cheek teeth in each jaw quadrant during their lifespan, but they are never all erupted and in occlusion at once. The dugong has a simple stomach and is considered to be a non-ruminant herbivore (Marsh 1989). It is a hindgut fermenter (Marsh et al. 2000).

There is a double ventricular apex to the heart and the lungs are long and extend posteriorly almost as far as the kidneys. The diaphragm is large and obliquely sloped. The main bronchus runs almost the entire length of the lung with only a few side branches. Even the most peripheral bronchioli possess cartilage within their walls (Marsh 1989).

Unrinsed seagrasses consumed by dugongs contain about 15 times more chloride and 30 times more sodium than most terrestrial pasture plants. Dugongs do not drink fresh water and presumably acquire their entire water intake from seagrass. Kidneys are elongated. The brain is relatively small (Marsh 1989).

Physiological parameters are not available for dugongs. Physiological parameters for Florida manatees (*Trichechus manatus latirostris*) are presented in Table 19.2 (Chittick & Walsh 2007).

3 REPRODUCTION

Although they can be long-lived (>70 yr), dugongs do not reach sexual maturity until 10–17 yr (females). Females are polyovular and polyoestrus and may undergo a number of sterile cycles before becoming pregnant (Marsh 1989). Gestation lasts approximately 13 mo. After the birth of a single calf there is an extended period of maternal investment and a prolonged intercalving interval (3–7 yr). Size at birth is variable with neonates ranging from about 1.10–1.25 m long and about 27–35 kg in weight. Calves begin to consume seagrass soon after birth but continue to suckle from their mothers for at least 18 mo (Bryden et al. 1998).

The dugong breeding season around Townsville or Cairns extends from August/September to December (Marsh et al. 1984a, b). Southern populations may lag a little behind their northern counterparts, with both mating activity and births concentrated in a few months at the end of each calendar year (Bryden et al. 1998).

The anal and genital apertures are contiguous in female dugongs. In males the genital aperture is distinctly separated from and anterior to the anus (Marsh et al. 2000).

4 HUSBANDRY

Some information is available on husbandry of manatees (Walsh & Bossart 1999) but very little information is available on dugongs.

4.1 Housing

Dugongs should be kept in natural sea water enclosures. Artificially produced salt water can be used to house dugongs but it is expensive and there may be problems with the availability of the constituents. Enclosures can be natural or artificial. If artificial, the enclosure walls and floor should be durable, watertight, non-porous, non-abrasive, non-toxic and easily cleaned and disinfected. Sharp projections and loose fittings should be avoided. Artificial pools have the advantage of being able to maintain water parameters better than natural pools. Natural pools, however, can be constructed much larger for less cost than artificial pools. The pools should be large enough to accommodate the natural behaviours of the animals and deep enough for the animals to dive and avoid the effects of ultraviolet radiation. Shade should be supplied and the temperature of the water should be maintained between 23°C and 30°C. If chemical treatment is used to maintain water quality and cleanliness, it should not cause harm to the animals (see Chapters 17 and 18 for further information on water quality). Minimum depth for long-term housing of one adult animal should be 4 m. Minimum length should be 11 m and minimum width should be 9 m. If more than one animal is to be housed, the length and the width should be increased by 50%.

4.2 Hospitalisation

Dugongs can be housed for short periods in small pools. The size of the animal determines the size of pool that it will require. Infant dugongs can be housed in aboveground pools as long as the water is turned over quickly enough to avoid build-up of bacteria, phosphates and ammonia. It is advantageous if the pool can be drained completely as this will allow capture and restraint of the animal far more easily than if it has to be caught in deep water.

4.3 Transport

Little published information is available on transport of dugongs. Manatees are generally secured in a stretcher and transported by truck on a 15 cm thick foam pad (Geraci & Lounsbury 1993). Dugongs should be kept wet and shaded from the sun. If possible they should be kept at 23–28°C during the transport. Temperatures outside this range must be avoided.

4.4 Individual marking and identification

Generally, individual dugongs are identified by their markings or other obvious physical differences. These differences may include the shape of the tail fin or any visible scars or notches. Microchips or passive integrated transponder (PIT) tags can be inserted IM into the epaxial muscles.

5 NUTRITION

5.1 Wild diets

Dugongs feed almost exclusively on several species of tropical and temperate seagrasses. Marine algae may be eaten if seagrass is scarce. Of the 10 genera of seagrasses occurring within the dugong's Australasian range, nine are known to be eaten (Lanyon 1991).

5.2 Captive diets

There is a reasonable amount of information on captive diets of manatees (Walsh & Bossart 1999), but very little information available on dugongs. A typical manatee diet may incorporate 75–85% leafy green vegetables such as romaine and leaf lettuce, cabbage, kale, hydroponic wheat, oats and sprouts; 15–20% dried forage such as timothy/alfalfa hay; and a small percentage of vegetables (e.g. carrots) and concentrates such as those designed for elephants. Lettuce by itself (romaine and green leaf) has been used for extended periods but has numerous drawbacks, including poor nutrient content and expense. Issues associated with captivity and nutrition include blockage of pool filtration associated with large amounts of particulate matter and fungal contamination of grains (Walsh & Bossart 1999).

Few dugongs have been maintained successfully in captivity. Ingredients of captive dugong diets include seagrass, cos lettuce, snowpea sprouts, cabbage, endive, wheatgrass and spinach. Dugongs feed from the ocean bed in the wild; efforts should be made to feed captive animals from the bottom of their enclosures. Food items can be anchored into a weighted frame that sinks to the bottom of the pool. Some captive dugongs may adapt to feeding from the surface. Advice should be sought from institutions that have held these animals.

5.3 Feeding sick animals

Dugongs have been reported to be notoriously difficult to maintain in captivity. If natural diets are used, animals will need to be hand-fed 40+ kg of seagrass a day. Long-term rehabilitation has been described as an expensive, labour-intensive and probably impractical option with little chance of success (Smith 1997).

It has been stated that, for manatees, diets should be designed with future release in mind and that food items vastly different from wild varieties (e.g. some fruits and vegetable) should not be used (Walsh & Bossart 1999). Nutritional supplementation has been administered by gastric tube to manatees (Walsh & Bossart 1999). About 2 L of gruel composed of monkey chow, lettuce and spinach was administered twice a day. A thin mixture is administered initially. Over time it is thickened, and may be delivered via stomach pump. Read Walsh and Bossart (1999) if supplementary feeding of dugongs is being considered.

6 RESTRAINT

Little information is available on restraint of dugongs.

6.1 Physical restraint

Geraci and Lounsbury (1993) give some good advice and useful diagrams regarding physical restraint in manatees. Techniques include manual restraint, nets and slings.

One method of capture involves surrounding the manatee with a net (30–35 m long and 4 m deep, of 10 cm stretch mesh) that is gradually drawn towards the shore until the animal is in water shallow enough for handlers to physically restrain it. A healthy animal may require 10–15 people (Geraci & Lounsbury 1993). Handlers must use extreme caution to avoid becoming entangled in the net and pulled into the water if the animal attempts to escape. Once in shallow water, the animal is manoeuvred onto a padded platform and secured with straps. A 3.5 m stretcher is usually appropriate. Once out of the water, manatees generally become calm and may need little restraint. Covering the eyes and wrapping a small piece of netting over the snout (do not obstruct breathing) helps to calm the animal. Attempts to assist unrestrained manatees are ill-advised and dangerous. One person cannot safely lift a manatee calf. Calves should be placed securely in a stretcher before removal from the water. Handlers must be prepared for sudden thrashing at all stages of handling (Geraci & Lounsbury 1993).

6.2 Chemical restraint

The unique anatomy, physiology and aquatic environment of sirenians make sedation and anaesthesia challenging (Chittick & Walsh 2007). Little information is available on chemical restraint of dugongs. However, a wide variety of anaesthetic agents and techniques have been used to restrain and anaesthetise Florida manatees (Chittick & Walsh 2007) and readers are encouraged to review that source prior to attempting chemical restraint of dugongs. There are anatomical and physiological differences between dugongs and manatees and it is unknown how applicable the techniques used in manatees may be to dugongs. An expert veterinary anaesthetist should be involved in any chemical restraint procedures.

Manatees are closely monitored during sedation and anaesthesia. Anaesthetic depth is determined by assessing blink response, jaw tone and resistance to limb movement (Chittick & Walsh 2007). Heart rates are calculated by listening to the heart with a stethoscope or by observing the pulsation of skin overlying the heart when animals are placed in dorsoventral recumbency. Electrocardiography is used for assessment of cardiac rate and rhythm. If oesophageal probes (e.g. an oesophageal stethoscope) are used, care should be taken to prevent insertion into the trachea. Use of a bronchoscope may assist with correct placement. Gingival capillary refill is generally less than or equal to 2 sec. Pulse oximetry readings are occasionally collected from the nasal septum, buccal mucosa or rostral gingival ridge of the lower jaw. Rectal reflectance pulse oximetry probes are placed in the nostril to obtain readings. Flexible rectal thermometers are difficult to place owing to high sphincter tone and fibrous faeces, and rectal temperatures read lower than core temperatures. Oral placement of thermometers alongside the caudal buccal mucosa give fairly accurate results, but thermometers can be damaged by the molars (Chittick & Walsh 2007).

6.2.1 Injectable agents

Little information is available on chemical restraint using injectable agents in dugongs. Reported drugs and dosages used in manatees have been reviewed by Chittick and Walsh (2007). The relatively low metabolic rate of sirenians may contribute to their apparent sensitivity to sedative and opioid drugs compared with terrestrial mammals. Diazepam (0.02–0.035 mg/kg IM) and midazolam (0.02–0.05 mg/kg IM) have been used extensively for both mild and pre-operative sedation. A dose of 0.045 mg/kg of midazolam IM will calm an excited manatee for approximately 60–90 min (Walsh & Bossart 1999). Pethidine (up to 1 mg/kg IM) has been used in conjunction with benzodiazepines for more fractious individuals or for procedures that may result in surgical discomfort. In captivity, after administration of these sedatives animals are often placed into a shallow pool to allow the agent to take effect and minimise the effects of restraint overriding the sedative (Chittick & Walsh 2007).

Local infusion of 2% lidocaine at the surgical site is effective, in conjunction with sedation, for analgesia in minor procedures.

Sedative or analgesic drugs are administered IM. Onset of effects after IM injection is approximately 15–20 min, although maximum effect may not occur until 25 min in some individuals. Drugs are administered into the gluteal muscles and long needles are required—18 G × 90 mm spinal needles for adults and 21 G × 3.75–5 cm needles for calves. Hand injection or a remote injection technique (Ryding 1982) could be used. If additional drugs or fluids need to be administered IV (see 7.1.1), this is most easily achieved after sedation (e.g. administration of analgesic such as butorphanol at 0.01–0.025 mg/kg IV).

6.2.2 Inhalation anaesthesia

Little information is available on inhalation anaesthesia in dugongs. Manatees are fasted for 24 hr prior to anaesthesia. Respirations are counted over 5 min to determine a normal rate because they are usually less than or equal to 1 breath/min. If an animal in the water shows signs of depression or inco-ordination after sedatives are administered, it is removed. In animals with pulmonary compromise, oxygen supplementation is often provided prior to administration of general anaesthesia (Chittick & Walsh 2007).

Occasionally manatees can be intubated under midazolam sedation alone, but others require induction with an inhalant administered by mask prior to intubation (Chittick & Walsh 2007). Mask administration is usually done with a modified 20 L clear plastic jug with the bottom removed. The edges are padded to allow a tighter fit over the animal's head. Breath-holding can occur and the patient may need to be stimulated by rocking side to side to facilitate breathing. Nasal intubation using a rhinoscope is preferred in manatees. Nasopharyngeal nematodes may complicate intubation and periodic suctioning of the endotracheal tube is required to remove mucus. A cuffed elongated foal tube of 8–14 mm internal diameter is recommended for Florida manatees, as well as a stylet to assist passage into the laryngeal opening. Because of the short length of the manatee trachea, care must be taken to avoid bronchial intubation. Once intubated, manatees are maintained with isoflurane levels similar to those used in domestic animals.

Mechanical ventilation is required to maintain adequate gas exchange and oxygenation. A detailed description of modifications to the standard mammalian approach to positive-pressure ventilation associated with anatomical and physiological traits of manatees is presented by Chittick and Walsh (2007). Considerations are similar to those required for other marine mammals. Mechanical ventilation does not mimic the sirenian respiratory cycle of inspiration closely followed by expiration. Ventilations are initially set at 6 breaths/min and in some cases peak inspiratory pressure must be set at about 20 mmHg to achieve lung expansion, oxygenation and carbon dioxide exchange. A slow inspiratory flow rate improves oxygenation while decreasing carbon dioxide accumulation.

Following recovery from anaesthesia, manatees are not placed in the water until they are appropriately responding to stimuli and breathing voluntarily. If possible, animals are placed in shallow water (<1 m deep) so they can surface to breathe by pushing off the bottom with their pectoral flippers (Chittick & Walsh 2007). Dopram has been used to stimulate breathing in manatees (Walsh & Bossart 1999).

7 CLINICAL PATHOLOGY AND NECROPSY

7.1 Haematology and biochemistry

7.1.1 Sample collection

Blood samples have been collected from the tail fluke of neonates, and from the pectoral flipper of immature animals (Blanshard 2001). Accessible veins in the fluke run roughly parallel to the leading edge but are located centrally within the fibrous connective tissue and their position is not apparent from the surface. The approach is vertical, using a needle marginally longer than half the maximum thickness of the fluke. A 25 G × 19 mm winged infusion set (butterfly needle) is useful. Insert the needle slowly to its full length, watching for any flash of blood up the tube. Withdraw the needle even more slowly if no blood is encountered, as sometimes blood will flow from a vessel that the needle presumably passed right through on insertion. If still unsuccessful in drawing blood from the initial attempt, withdraw the needle completely and reinsert a new needle at a slightly different location, continuing to search systematically along an imaginary line that transects the fluke.

Blanshard (2001) stated that increased thickness of the fluke and the potential danger of handlers being struck by a flailing tail probably make the pectoral flipper an easier and safer venipuncture site in bigger animals. A 21–23 G × 25 mm needle has been used for small calves and a 21 G × 38 mm for larger animals. Approaching from the palmar aspect of the flipper, the needle is inserted obliquely into the interosseous space between the radius and ulna and negative pressure is applied with a syringe as the needle is advanced. The anatomy of blood vessels in this area is not known.

Dugong blood has been reported as clotting very rapidly (Blanshard 2001). Blanshard (2001) recommended that the sample destined for haematological testing be transferred promptly into the anticoagulant tube and mixed gently for an extended period. A direct smear from the syringe hub or one made almost immediately after mixing the blood in the anticoagulant tube was recommended, to ensure that a differential white cell count could be obtained if the sample did ultimately clot.

7.1.2 Reference ranges and interpretation

Reference ranges are not available for dugong haematology and biochemistry. It has been suggested that comparison with haematology and biochemistry values from manatees may be helpful in interpreting any dugong results (Blanshard 2001). Haematology and biochemistry values from 12 healthy Florida manatees are presented in Table 19.3.

Blanshard (2001) stated that dugong neutrophils contain acidophilic specific granules termed 'pseudoeosinophils'. It was suggested that on histological sections a polymorphonuclear infiltrate could be misinterpreted as an eosinophilic response.

Table 19.3 Haematology and biochemistry values of 12 healthy Florida manatees (*Trichechus manatus latirostris*)

Parameters	Range
Haematology	
Hb (g/L)	103–120
RBC (10^{12}/L)	2.41–3.06
PCV (L/L)	0.32–0.40
MCV (fl)	121–135
MCH (pg)	37–43
MCHC (g/dL)	30–33
Platelets (10^9/L)	261–634
nRBC (100/WBC)	1–2
WBC (10^9/L)	4.0–11.7
Bands (%)	0
Neutrophils (%)	25–64
Lymphocytes (%)	21–77
Monocytes (%)	0–18
Eosinophils (%)	0
Basophils (%)	0–1
Biochemistry	
Glucose (mmol/L)	3.1–6.5
BUN (mmol/L)	4.6–11.4
Creatinine (mmol/L)	0.04–0.19
Total protein (g/L)	68–73
Albumin (g/L)	38–53
Globulin (g/L)	17–32
Total bilirubin (μmol/L)	0–1.7
ALP (U/L)	64–183
GGT (U/L)	39–64
ALT (U/L)	6–30
AST (U/L)	5–28
LDH (U/L)	94–372
CK (U/L)	79–302
Na (mmol/L)	142–157
K (mmol/L)	4.2–6.6
Cl (mmol/L)	90–103
CO_2 (mmol/L)	14–43
Ca (mmol/L)	2.52–3.04
P (mmol/L)	1–2.6
Fe (μmol/L)	10.6–35.6
Cholesterol (mmol/L)	2.8–8.5
Triglycerides (mmol/L)	0.4–1.6

Source: Walsh & Bossart (1999).

7.2 Necropsy

A detailed guide to the necropsy of dugongs is available through the Australian Government Great Barrier Reef Marine Park Authority (Eros et al. 2007). The guide discusses incident response procedures, necropsy techniques, specimen collection, preservation and transportation and determination of cause of death, and is well illustrated. An electronic version is available online at www.gbrmpa.gov.au and should be consulted prior to any necropsy being performed.

8 HAND-REARING

Dugongs <1.5 m in length will be suckling and those <1.8 m long are probably still suckling (Marsh 1997). Attempts have been made to hand-rear neonatal dugongs, but there is little published information on the process. Milk diets based on Di-Vetelact (Sharpe Laboratories) are used. Other ingredients such as coconut milk can be added to the Di-Vetelact to increase nutrient density. Hand-rearing is an extremely labour-intensive and prolonged exercise and should not be approached lightly. Sea World (Gold Coast, Qld, Australia) should be contacted for advice on dealing with a dependent dugong calf.

9 DISEASES

Though dugongs appear to be susceptible to a wide range of diseases (Campbell & Ladds 1981; Wilkie 2005) very little is known about the diseases. Marsh et al. (2000) reviewed the information then available and Blair (1981a, b) listed the parasites of Sirenia.

9.1 Dermatological diseases

Heavy barnacle infestation was observed as an incidental finding in a dugong that died of other causes. The barnacles were associated with 'impressive' indentations in the skin but the epidermis was intact and there was no inflammation (Reddacliff 1981; I Wilkie *pers. comm.*). Intra-epidermal cysts containing parasites morphologically resembling copepods have been reported in dugongs (Campbell & Ladds 1981). It is normal for dugongs to be extensively scarred, particularly on the dorsal surface. Parallel scars inflicted by the tusks of males are common on the skin of both sexes. Older 'scarback' individuals may also have a large area of unpigmented skin on the back above the pectoral fins (Marsh et al. 2000).

9.2 Gastrointestinal diseases

Constipation has been treated with mineral oil PO or enemas in manatees (Walsh & Bossart 1999). It has not been reported in dugongs. Salmonellosis has caused diarrhoea and death in at least one captive dugong (Elliott et al. 1981; see 9.6.2).

9.3 Lymphatic diseases

Other than some mild changes associated with parasitism (see 9.6.5) there are no reports of lymphatic disease in dugongs.

9.4 Respiratory diseases

Cochleotrema indicum (syn. *Pulmonicola pulmonalis*) (Blair 1981a; Blanshard 2001) was present in large numbers in the lung of a dugong with verminous bronchopneumonia, gastric nematodes and concurrent bacterial septicaemia. The animal was debilitated and was found swimming aimlessly in the shallows (Marsh et al. 2000).

9.5 Neurological and musculoskeletal diseases

Other than some mild changes associated with parasitism (see 9.6.5) there are no reports of neurological disease in dugongs. The majority of musculoskeletal disease is associated with vessel strikes (Wilkie 2005).

9.6 Infectious diseases

Other marine mammal groups are subject to dramatic population declines due to the impact of epizootic diseases. No epizootic outbreaks have been reported for dugongs (Marsh et al. 2000).

9.6.1 Viral diseases

There appear to be no published reports of viral infection in dugongs.

9.6.2 Bacterial diseases

Bacterial septicaemia was reported in wild dugongs examined in northern Queensland (Marsh et al. 2000) and Moreton Bay (I Wilkie *pers. comm.*), but the causal bacteria were not named. In the first study, rupture of intestinal abscesses and acute peritonitis (presumably due to bacterial infection) were observed in two other dugongs.

Salmonella lohbruegge was isolated from both kidney and liver of a dugong that died in captivity (Elliott et al. 1981). Prior to death, there was decreasing interest in surroundings, lethargy, weight loss, dehydration and severe diarrhoea.

9.6.3 Fungal diseases

There are no reports of fungal disease in dugongs.

9.6.4 Protozoal diseases

Infection of a dugong with *Toxoplasma* has been observed in Australia (I Wilkie *pers. comm.*). There was cachexia and multifocal hepatitis. The brain was not examined.

Cryptosporidium has been isolated from dugongs. DNA sequencing of one isolate revealed it to be a variant of *Cryptosporidium parvum*, which is a species that can infect humans (Blanshard 2001; Hill et al. 1997; Marsh et al. 2000; Morgan et al. 2000). It was not possible to ascertain whether *Cryptosporidium* contributed to illness in the animals from which it was isolated. It is also not known whether *Cryptosporidium* can complete its life-cycle in the dugong, whether the infection cycles within a herd and whether the animals can become infected from an external source (Marsh et al. 2000). There is, however, some evidence that the parasite may complete its life-cycle in this host and that infectious oocysts may be passed into the surrounding water (Blanshard 2001; Morgan et al. 2000).

9.6.5 Parasitic diseases

There are no reports of illness or lesions in dugongs due to cestode infestation. The nematodes *Heterocheilus tunicatus* and *Paradujardina halicoris* occur in the dugong (Arundel 1978) but neither appear to be the cause of overt disease. *P. halicoris* is found in the stomach, especially the gastric gland (Blair 1981a). Its presence in one study was associated with only a mild inflammatory response (Campbell & Ladds 1981). Large numbers of this nematode were present in the cardiac region of the stomach of a debilitated dugong that also had verminous pneumonia and bacterial septicaemia (Marsh et al. 2000).

At least 16 species of trematodes have been described in dugongs (Blair 1981a, b; Cribb 1998). In most cases these parasites, which occur in a variety of body locations, appear to be non-pathogenic. Illness or lesions have been observed in the following fluke infections of dugongs: *Lankatrema* spp. (Blair 1981a), *Fairdifex* (Blair 1981a, b), *Labicola elongata* (Blair 1979; Blanshard 2001), *Cochleotrema indicum* (Blair 1981a; Blanshard 2001), *Opisthotrema dujonis* and *Opisthotrema nasalis* (Budiarso et al. 1979). Undescribed trematodes have been associated with haemorrhages in the brain (B Hill *pers. comm.* to Blanshard 2001), lesions in the ovary (Marsh et al. 1984b), chronic pancreatitis with lymphoid hyperplasia and enteritis (Campbell & Ladds 1981).

9.7 Non-infectious diseases

9.7.1 Neoplasia

There are no reports of neoplasms in dugongs. Malignant lymphoma and papillomatosis have been described in manatees (Bossart et al. 2002; Hammer et al. 2005).

9.7.2 Toxins

Dugongs in northern Australia accumulate high levels of heavy metals with age (Denton et al. 1980; Gladstone 1996; Kemper et al. 1994). There is no evidence to suggest that this accumulation is particularly harmful to dugongs (Marsh et al. 2000). Lindane and dioxins have been identified in dugong livers and fat (Haynes et al. 1999; Marsh et al. 2000). The significance of these findings is unknown.

There are no records of dugongs being affected by oil spills in Australian waters (Marsh et al. 2000). Thirty-eight carcasses were recovered after the Nowruz oil spill in the Arabian Gulf in 1983 (Preen 1989).

There are no reports of marine algal biotoxins affecting dugongs.

9.7.3 Nutritional disorders

Other than inanition, there are no published reports of nutritional disorders in dugongs. Dugongs can synthesise ascorbic acid (Worthy 2001).

9.7.4 Strandings and entanglements

Little published information is available on the rescue, rehabilitation and release of dugongs. Their unique anatomy and physiology and their nutritional and husbandry requirements makes the captive care of dugongs difficult and very specialised. It requires considerable expertise and resources. If a dugong requires rescue and treatment, the authors recommend that the relevant state or territory wildlife agency and Sea World (Gold Coast, Qld, Australia) be contacted for advice.

Smith (1997) presented four management options for a live stranded or entangled dugong.

- Treatment and release—if entangled, the animal may simply require disentanglement from netting or other debris and can be released immediately. If stranded on a sandbar at low tide it may need to be moved into water or kept wet and shaded until high tide. The animal's condition should be assessed by a veterinarian before release. Advice should be sought from a dugong specialist on possible treatment. As with manatees (Geraci & Lounsbury 1993), careful and prolonged observation may be necessary to determine whether a dugong is actually in distress and requires help.
- Rehabilitation—the animal could be held for a period of treatment and rehabilitation, either at a temporary holding site or at a care facility such as a zoo or aquarium. However, dugongs have been reported as notoriously difficult to maintain in captivity. They need to be hand-fed >40 kg of seagrass a day. Long-term rehabilitation is an expensive, labour-intensive and probably impractical option with little chance of success.
- Relocation—in some states of Australia, e.g. New South Wales, water temperatures are generally

considered too cold for dugongs south of Sydney and, in winter, throughout the state. Animals that strand when water temperatures are too low may have a better chance of survival if relocated to the warmer waters of Queensland (if stranded on Australia's east coast), depending on their condition. Proposals for relocation should be assessed on a case-by-case basis. Approval for relocation must be obtained from the relevant wildlife authorities.

- Euthanasia—for dugongs that strand alive but are in very poor condition or severely injured the most humane option may be euthanasia. An animal should not be left to die. The preferred method of euthanasia is lethal injection, which must be administered by a veterinarian or under their direction. An acceptable alternative method is shooting, using a weapon of calibre such as .303.

Survival after stranding has not been determined for dugongs. Geraci and Lounsbury (1993) suggested that manatee calves captured as dependent young should not be released, since they have not learned important behaviours for avoiding boats and finding water refuges during cold weather.

9.7.5 Mortality during capture for research

There is a low risk associated with catching dugongs for research provided that the procedure is carried out by experienced operators (Marsh et al. 2000).

9.7.6 Acoustic pollution

Although to date there has been no documented evidence of marine seismic surveys being detrimental to populations of dugongs in Australia, there have been no detailed studies (Marsh et al. 2000). Vessel strikes are a major cause of mortality in Florida manatees, which may possess a limited low-frequency hearing sensitivity and therefore have difficulty detecting and locating boats at safe distances. The relevance of these results to dugongs is unknown because the anatomy of the dugong ear differs from that of the manatee.

9.7.7 Vessel strikes

There are relatively few records of dugong deaths due to vessel strikes in Australian waters (Marsh et al. 2000). Death due to collision is often sudden but may also be a result of a chronic long-term debilitation. Fresh open propeller wounds or skeg marks provide obvious clues. Propeller cuts made after death show no signs of haemorrhage or bruising to internal organs or musculature. Dead dugongs always float with the ventral aspect exposed and ante mortem wounds on this aspect are rare. Wounds responsible for death are usually located on the dorsal aspect. Chronic debilitation due to boat-induced injury usually involves infection. Even a minor externally healed propeller wound may be associated with significant internal pathology.

9.7.8 Incidental catch

Accidental entanglement in gill- and mesh-nets set by commercial fishers is considered a major but largely unquantified cause of dugong mortality in many countries. Shark nets set for human protection can be another source of dugong mortality. It is normal for dugongs to be extensively scarred, particularly on the dorsal surface. Parallel scars inflicted by the tusks of males are common on animals of both sexes. Because of this, net marks can be difficult to distinguish from other scars. Furthermore, if a dugong carcass is left to decay in the sun the skin will crack in a mosaic-type pattern that may resemble net entanglement marks (Marsh et al. 2000).

9.7.9 Hunting mortality

Dugong hunting is culturally significant to many Aboriginal and Torres Strait Islander communities in northern Australia (Marsh et al. 2000). Dugongs are caught for their meat and oil. Animals caught for meat are easy to identify. They usually have large pieces of meat missing from the carcass.

9.7.10 Other

In 2000, two dugongs from the Townsville region were necropsied. One dugong died from trauma—a 13 cm stingray barb had penetrated the abdominal wall and lodged within the small intestine. The other dugong appeared healthy and the cause of death was not determined. It was found washed up on a beach where, 2 d earlier, Cyclone Tessie crossed the Queensland coast (Marsh et al. 2000).

10 ZOONOSES

Cryptosporidiosis is a potential zoonotic disease when dealing with dugongs. Cryptosporidia are transmitted by the faecal/oral route, therefore handlers should pay attention to personal hygiene such as washing hands after contacting animals.

11 THERAPEUTICS

Little information is available on therapeutics in dugongs. Some information is available on therapeutics of manatees (Walsh & Bossart 1999).

11.1 Drug administration

Drugs are usually administered IM into the gluteal muscles, to avoid injury to organs under more cranial epaxial musculature. Long needles are required—18 G × 90 mm spinal needles for adults and 21 G × 3.75–5 cm needles for calves. Hand injection or a remote injection technique (Ryding 1982) could be used. In manatees the injection site is always surgically prepared so as to avoid contamination (Walsh & Bossart 1999). Should drugs or fluids need to be administered IV, this is most easily achieved after sedation (Chittick & Walsh 2007). There is no subcutaneous space, so administration of drugs via this route is not practical.

Antibiotics have been administered by stomach tube in manatees, or in food. However, Walsh and Bossart (1999) do not recommend this route for delivery of antibiotics because of associated loss of normal intestinal flora, diarrhoea and hypermotility that have been associated with their use. Oral medication, e.g. if included in monkey chow, may also be lost to the surrounding water (Walsh & Bossart 1999).

11.2 Drug use

In manatees, commonly used antibiotics include amikacin (7 mg/kg bid IM), penicillins (penicillin G and benzathine penicillin at 22 000 U/kg), cephalosporins such as ceftrioxone (22 mg/kg sid IM) and oxytetracycline at cattle doses (Walsh & Bossart 1999). Combination antibiotics such as amikacin with penicillins or amikacin with ceftrioxone have also been administered. Doses are based on those used in horses or cattle. Dexamethasone has been used at 0.25–0.5 mg/kg sid IM to stimulate the appetite and decrease inflammation.

11.3 Fluid therapy

Correcting or maintaining hydration status in neonatal dugongs has proved a challenge (Blanshard 2001). Although stomach tubing with a cut-down equine foal nasogastric tube was performed many times in two neonates, the ease with which the tube inadvertently entered the trachea (without any gag reflex from the animal) was considered very unnerving. Extraordinary care was required if the procedure was attempted.

Rectal administration of water to a dugong neonate using a round-ended feeding tube fitted on a 100 mL catheter-tipped syringe, and intraperitoneal administration of large volumes of balanced electrolyte solution in an emaciated 2 m animal have been used as alternative routes for rehydration. In the latter case, 19 G × 7 cm needles inserted to the hub against the ventral abdominal wall were barely long enough to reach the peritoneal cavity, and a small proportion of the 6 L of fluid given was deposited within the body wall (Blanshard 2001).

In manatees, the brachial vascular bundle is difficult to catheterise due to needle placement perpendicular to the arteriovenous plexus (Chittick & Walsh 2007); 18 G × 5 cm (adults) and 20 G × 3.75 cm (juveniles) needles have been used (Murphy 2003). This procedure is facilitated in manatees by prior sedation (Chittick & Walsh 2007). Water, administered at a rate of 2 L bid by stomach tube has been given to manatees to improve hydration (Walsh & Bossart 1999). Fluid was fed in by gravity, and about 75% of total volume given initially for the first two treatments to try to avoid reflux. Severely dehydrated manatees may require treatment tid. Electrolytes and PCV should be closely monitored.

12 SURGERY

Little information is available on surgical management in dugongs. Factors affecting surgery of other species of marine mammals are likely to apply (see Chapter 17). Minor procedures in manatees can be performed using a combination of sedation and infiltration of the site with local anaesthetic (Chittick & Walsh 2007).

ACKNOWLEDGMENTS

Very little information is available on dugong medicine and husbandry. Most of the unpublished information has been collated by Wendy Blanshard (Sea World, Qld, Australia). The authors thank Bruce Hill for reviewing the manuscript.

REFERENCES

Arundel JH (1978). Parasites and parasitic disease of Australian marine mammals. In *Fauna, Proceedings 36,* pp. 323–333. Post Graduate Committee in Veterinary Science: Sydney.

Blair D (1979). A new family of monosome flukes (Platyhelminths, Digenea) from the dugong, *Dugong dugon* (Muller). *Annales de Parasitologie (Paris)* **54**, 519–526.

Blair D (1981a). Helminth parasites of the dugong. In *The Dugong. Proceedings of a Symposium/workshop held at James Cook University*, 8–13 May 1979 (ed. H Marsh), pp. 275–285. James Cook University: Townsville.

Blair D (1981b). The monostome flukes (Digenea: Families Opisthotrematidae Poche and Rhabdiopoeidae Poche) parasitic in sirenians (Mammalia: Sirenia). *Australian Journal of Zoology Supplementary Series* **81**, 1–54.

Blanshard W (2001). Dugong strandings. In *Veterinary Conservation Biology: Wildlife Health and Management in Australasia—Marine Mammal Strandings. Proceedings of a Seminar/workshop held at Taronga Zoo*, 1 July 2001. Taronga Zoo: Sydney, Australia.

Bryden M, Marsh H & P Shaughnessy (1998). *Dugongs, Whales, Dolphins and Seals. A Guide to the Sea Mammals of Australasia*. Allen & Unwin: Sydney.

Bossart GD, Ewing RY, Lowe M, Sweat M, Decker SJ, Walsh CJ, Ghim SJ & Jenson AB (2002). Viral papillomatosis in Florida manatees (*Trichechus manatus latirostris*). *Experimental and Molecular Pathology* **72**, 37–48.

Budiarso IT, Palmieri JR, Imes GDJ, Allen JF & Lepes MM (1979). Two species of trematodes causing nasal lesions in dugongs. *Journal of Parasitology* **65**, 568.

Campbell RSF & Ladds PW (1981). Diseases of the dugong in northeastern Australia: a preliminary report. In *The Dugong: Proceedings of a Seminar/workshop held at James Cook University*, 8–13 May 1979 (ed. H Marsh). James Cook University: Australia.

Chittick EJ & Walsh MT (2007). Sirenians (manatees and dugongs). In *Zoo Animal and Wildlife Immobilization and Anesthesia* (eds G West, D Herd & N Caulkett), pp. 497–505. Blackwell Publishing: Iowa.

Cribb TH (1998). The diversity of the Digenea of Australian animals. *International Journal for Parasitology* **28**, 899–911.

Denton GRW, Marsh H, Heinsohn GE & Burdon-Jones C (1980). The unusual metal status of the dugong *Dugong dugon*. *Marine Biology* **57**, 201–219.

Elliot H, Thomas A, Ladds PW & Heinsohn GE (1981). A fatal case of salmonellosis in a dugong. *Journal of Wildlife Diseases* **17**, 203–208.

Eros C, Marsh H, Bonde R, O'Shea T, Beck C, Recchia C, Dobbs K, Turner M, Lemm S, Pears R & Bowater R (2007). *Procedures for the Salvage and Necropsy of the Dugong (Dugong dugon)* (2nd edn). Great Barrier Reef Marine Park Authority: Townsville.

Geraci JR & Lounsbury VJ (1993). *Marine Mammals Ashore: A Field Guide for Strandings*. Texas A&M Sea Grant Publications: Galveston.

Gladstone W (1996). *Trace Metals in Sediments, Indicator Organisms and Traditional Seafoods of the Torres Strait.* Great Barrier Reef Marine Park Authority: Townsville.

Hammer AS, Klausen B, Knold S, Dietz HH & Dutoit SJ (2005). Malignant lymphoma in a West Indian manatee (*Trichechus manatus*). *Journal of Wildlife Diseases* **41**, 834–838.

Haynes D, Muller J & McLauchlan MS (1999). Polychlorinated dibenzo-P-dioxins and dibenzofurans in Great Barrier Reef (Australia) dugongs (*Dugong dugon*). *Chemosphere* **38**, 255–262.

Hill BD, Fraser IR & Prior HC (1997). Cryptosporidium infection in a dugong (*Dugong dugon*). *Australian Veterinary Journal* **75**, 670–671.

Kemper C, Gibbs P, Obendorf D, Marvanek S & Lenghaus C (1994). A review of heavy metal and organochlorine DDT and PCBs levels in marine mammals in Australia. *Science of the Total Environment* **154**, 129–139.

Lanyon J (1991). The nutritional ecology of the dugong (*Dugong dugon*) in tropical north Queensland. PhD thesis. Monash University, Australia.

Marsh H (1980). Age determination of the dugong (*Dugong dugon* (Muller)) in northern Australia and its biological implications. *Report of the International Whaling Commission (Special Issue)* **3**, 181–201.

Marsh H (1989). Dugongidae. In *Fauna of Australia*. Mammalia, vol 1B. (eds DW Walton & BJ Richardson), pp. 1030–1038. AGPS: Canberra.

Marsh H (1997). *Dugongs*. Presented at the Stranded Marine Wildlife Workshop hosted by Queensland Department of Environment and Sea World Enterprises, 19–21 February 1997. Sea World Nara Resort: Main Beach, Queensland.

Marsh H, Heinsohn GE & Marsh LM (1984a). Breeding cycle, life history and population dynamics of the dugong, *Dugong dugon* (Sirenia: Dugongidae). *Australian Journal of Zoology* **32**, 767–788.

Marsh H, Heinsohn GE & Channells PW (1984b). Changes in the ovaries and uterus of the dugong, *Dugong dugon* (Sirenia: Dugongidae), with age and reproductive activity. *Australian Journal of Zoology* **32**, 743–788.

Marsh H, Eros C & Webb R (2000). Dugongs in health and disease. In *Marine Wildlife, Proceedings 335,* pp. 301–317. Post Graduate Foundation in Veterinary Science: Sydney.

Morgan U, Xiao L, Hill BD, O'Donoghue P, Limor J, Lal A & Thompson RCA (2000). Detection of the *Cryptosporidium parvum* 'Human' genotype in a dugong (*Dugong dugon*). *Journal of Parasitology* **86**, 1352–1354.

Murphy D (2003). Sirenia. In *Zoo and Wild Animal Medicine* (5th edn) (eds ME Fowler & RE Miller), pp. 476–482. Elsevier Science: St Louis.

Pachard JM, Rathbun GB, Domning DP, Best R & Anderson PK (1984). Sea cows and manatees In *All the World's Animals: Sea Mammals* (ed. D Macdonald), pp. 140–151. Torstar Books: New York.

Preen AR (1989). *Dugongs. Vol. 1: The Status and Conservation of Dugongs in the Arabian Region*. Technical Report. MEPA Coastal and Marine Management Series: Saudi Arabia.

Reddacliff GL (1981). Dermatoses: zoo animals. In *Proceedings 57,* pp. 407–417. Post Graduate Committee in Veterinary Science: Sydney.

Ryding FN (1982). Ketamine immobilization of southern elephant seals by a remote method. *British Antarctic Survey Bulletin* **57**, 21–26.

Spain AV & Heinsohn GE (1975). Size and weight allometry in a north Queensland population of *Dugong dugon* (Müller)(Mammalia: Sirenia). *Australian Journal of Zoology* **23**, 159–168.

Walsh MT & Bossart GD (1999). Manatee medicine. In *Zoo and Wild Animal Medicine: Current Therapy 4* (eds ME Fowler & RE Miller), pp. 507–516. WB Saunders: Philadelphia.

Wilkie I (2005). Necropsy findings in dugongs from Moreton Bay. In *Proceedings of the Australian Society for Veterinary Pathology*, p. 22. RSL Club, Cairns, Queensland.

Worthy GAJ (2001). Nutrition and energetics. In *Marine Mammal Medicine* (eds LA Dierauf & FMD Gulland), pp. 791–827. CRC Press: Boca Raton.

Appendix 1 Checklist of the mammals of Australia and its territories

Paul Andrew

This list includes extinct mammals that survived to European arrival and the few mammals that occur on Australian territories, such as Christmas Island and Lord Howe Island. Pelagic species have been included if they are known or are likely to have occurred in Australian waters.

1 TAXONOMY

Beyond the traditional Linnaean ranks, Subclasses, Suborders and (for dasyurids and murids where Families are large) Subfamilies have been added. It should be noted that this reference list contains no taxonomic novelties, and that no attempt has been made to resolve the taxonomic turmoil in bats and rodents.

The sequence of higher taxa follows Wilson and Reeder (2005). Species taxa are in alphabetical order within Family (Subfamily in dasyurids and murids). When there is a conflict in citation date between Strahan (1995) and Wilson and Reeder (2005), resolution is through Walton (1988). Walton (1988) is also followed for Geoffroy (not É. Geoffroy), Blainville (not de Blainville) and Tarragon (not de Tarragon).

With regard to species limits, variations between this list and Wilson and Reeder (2005) are referenced as follows.

- For split of *blythi* from *cristicauda* and use of *blythi* over *hillieri* see Woolley (2005).
- For split of *occidentalis* from *Pseudocheirus peregrinus* see Maxwell et al. (1996).
- For identity of *Pteropus banakrisi* from Moa Island see Helgen (2004).
- For split of *murrayi* from *Pipistrellus tenuis* see Churchill (1998: 67).
- For inclusion of Australian *adversus, australis* and *moluccarum* in *Myotis macropus* see Cooper et al. (2001).
- For *Balaenoptera omurai* see Wada et al. (2003).
- For *Orcaella heinsohni* see Beasley et al. (2005).
- For spelling of the species-group name in *Zyzomys palatalis* see Kitchener (1989).
- For the prehensile-tailed rat, which is technically without a name because the holotype has been re-identified (see Wilson & Reeder 2005: 1440), for the purposes of this list *Pogonomys mollipilosus* is retained.

2 COMMON NAMES

There are no novel common names in this list. The primary sources are Strahan (1995) and Wilson and Reeder (2005), with occasional names from Menkhorst and Knight (2004).

There is general favour in using names that are unique, descriptive, short and pronounceable. Whenever possible names that are parochial and convey little or no information, e.g. 'Monjon' and 'Rory's Pseudantechinus', are avoided. The use of scientific names as common names is also obviously problematic when species change genus. More importantly, common names and scientific names have different purposes; common names should reflect the phenotype, whereas scientific names reflect the genotype. Following this principle marsupial mole is used as a group—though not in the same Order as the Eurasian moles they are 'moles' by virtue of being subterranean.

There is a second issue with common names that has yet to attract much attention among Australian mammal names, that of common name construction. Duckworth and Pine (2003) discuss mammal common names at length, including construction etiquette, using Indochinese mammals as examples. Their guidelines are drawn from ornithology, for example, see Christidis and Boles (1994).

Perhaps the most trenchant issue here is how to deal with two-part group names. For example, within the recent Australian mammal literature we find freetail-bat (Strahan 1995), freetail bat (Menkhorst & Knight 2004) and free-tailed bat (Wilson & Reeder 2005).

The protocol of compounding group names is well-established in some Australian taxa, e.g. rock-wallaby, tree-kangaroo. This works because the two groups are well-known. Familiarity also makes acceptable the groups ringtail and brushtail, as used by Groves in Wilson and Reeder (2005). Elsewhere, retaining the adjectival form, where possible, keeps groups large for unfamiliar taxa, e.g. bent-winged bat (rather than bentwing-bat, but horseshoe-bat because the adjectival form is too clumsy as in horse-shoe-marked bat). The general rule adopted here is to compound group names comprising two nouns, but to retain two words when the first part of a group name is adjectival. This has the added advantage of being grammatical.

The application of a set of rules always highlights inconsistencies. For example, the use of cave-bat and forest-bat would likely leave the Species hard to find in an index (as would the groups elephant-seal, pilot-whale and sperm-whale), and pebble-mound and stick-nest are well-established names and only clumsily turned into adjectives. Any committee established to review Australasian mammal names should include consideration of how the name is to be written as well as spoken.

3 CHECKLIST OF THE MAMMALS OF AUSTRALIA AND ITS TERRITORIES

In this list the usual protocols have been followed, wherein the scientific name at each rank is followed by the name's author and the date of publication. The author and date are in parentheses if the taxon was first placed in a different Genus.

Class Mammalia Linnaeus, 1758

Subclass Prototheria Gill, 1872

ORDER MONOTREMATA Bonaparte, 1837

Family Tachyglossidae Gill, 1872 [echidnas]

Tachyglossus aculeatus (Shaw, 1792) Short-beaked echidna

Family Ornithorhynchidae Gray, 1825 [Platypus]

Ornithorhynchus anatinus (Shaw, 1799) Platypus

Subclass Marsupialia Iliger, 1811

ORDER NOTORYCTEMORPHIA Kirsch, 1977

Family Notoryctidae Ogilby, 1892 [marsupial moles]

Notoryctes caurinus Thomas, 1920 Northern marsupial mole

Notoryctes typhlops (Stirling, 1889) Southern marsupial mole

ORDER DASYUROMORPHIA Gill, 1872

Family Thylacinidae Bonaparte, 1838 [Thylacine]

Thylacinus cynocephalus (Harris, 1808) Thylacine

Family Myrmecobiidae Waterhouse, 1841 [Numbat]

Myrmecobius fasciatus Waterhouse, 1836 Numbat

Family Dasyuridae Goldfuss, 1820 [dasyurids]

Subfamily Dasyurinae Goldfuss, 1820

Antechinus adustus (Thomas, 1923) Rusty antechinus [tropical antechinus]

Antechinus agilis Dickman, Parnaby, Crowther & King, 1998 Agile antechinus

Antechinus bellus (Thomas, 1904) Fawn antechinus

Antechinus flavipes (Waterhouse, 1838) Yellow-footed antechinus

Antechinus godmani (Thomas, 1923) Atherton antechinus

Antechinus leo Van Dyck, 1980 Cinnamon antechinus

Antechinus minimus (Geoffroy, 1803) Swamp antechinus

Antechinus stuartii Macleay, 1841 Brown antechinus
Antechinus subtropicus Van Dyck & Crowther, 2000 Subtropical antechinus
Antechinus swainsonii (Waterhouse, 1840) Dusky antechinus
Dasycercus blythi (Waite, 1904) Brush-tailed mulgara [Ampurta]
Dasycercus cristicauda (Krefft, 1867) Crest-tailed mulgara
Dasykaluta rosamondae (Ride, 1964) Kaluta
Dasyuroides byrnei Spencer, 1896 Kowari
Dasyurus geoffroii Gould, 1841 Western quoll
Dasyurus hallucatus Gould, 1842 Northern quoll
Dasyurus maculatus (Kerr, 1792) Spot-tailed quoll [tiger quoll]
Dasyurus viverrinus (Shaw, 1800) Eastern quoll
Parantechinus apicalis (Gray, 1842) Dibbler
Phascogale calura Gould, 1844 Red-tailed phascogale
Phascogale tapoatafa (Meyer, 1793) Brush-tailed phascogale
Pseudantechinus bilarni (Johnson, 1954) Sandstone false antechinus [sandstone dibbler]
Pseudantechinus macdonnellensis (Spencer, 1895) Fat-tailed false antechinus
Pseudantechinus mimulus (Thomas, 1906) Alexandria false antechinus
Pseudantechinus ningbing Kitchener, 1988 Ningbing false antechinus
Pseudantechinus roryi Cooper, Aplin & Adams, 2000 Tan false antechinus
Pseudantechinus woolleyae Kitchener & Caputi, 1988 Woolley's false antechinus
Sarcophilus harrisii (Boitard, 1841) Tasmanian devil

Subfamily Sminthopsinae Archer, 1982

Antechinomys laniger (Gould, 1856) Kultarr
Ningaui ridei Archer, 1975 Wongai ningaui
Ningaui timealeyi Archer, 1975 Pilbara ningaui
Ningaui yvonnae Kitchener, Stoddart, & Henry, 1983 Southern ningaui
Planigale gilesi Aitken, 1972 Paucident planigale
Planigale ingrami (Thomas, 1906) Long-tailed planigale
Planigale maculata (Gould, 1851) Pygmy planigale [common planigale]
Planigale tenuirostris Troughton, 1928 Narrow-nosed planigale
Sminthopsis aitkeni Kitchener, Stoddart & Henry, 1984 Kangaroo Island dunnart
Sminthopsis archeri Van Dyck, 1986 Chestnut dunnart
Sminthopsis bindi Van Dyck, Woinarski & Press, 1994 Kakadu dunnart
Sminthopsis butleri Archer, 1979 Carpentarian dunnart [Butler's dunnart]
Sminthopsis crassicaudata (Gould, 1844) Fat-tailed dunnart
Sminthopsis dolichura Kitchener, Stoddart & Henry, 1984 Little long-tailed dunnart
Sminthopsis douglasi Archer, 1979 Julia Creek dunnart
Sminthopsis fuliginosus (Gould, 1852) Sooty dunnart
Sminthopsis gilberti Kitchener, Stoddart, & Henry, 1984 Gilbert's dunnart
Sminthopsis granulipes Troughton, 1932 White-tailed dunnart
Sminthopsis griseoventer Kitchener, Stoddart, & Henry, 1984 Grey-bellied dunnart
Sminthopsis hirtipes Thomas, 1898 Greater hairy-footed dunnart
Sminthopsis leucopus (Gray, 1842) White-footed dunnart
Sminthopsis longicaudata Spencer, 1909 Large long-tailed dunnart
Sminthopsis macroura (Gould, 1845) Stripe-faced dunnart
Sminthopsis murina (Waterhouse, 1838) Slender-tailed dunnart [common dunnart]
Sminthopsis ooldea Troughton, 1965 Ooldea dunnart
Sminthopsis psammophila Spencer, 1895 Sandhill dunnart
Sminthopsis virginiae (Tarragon, 1847) Red-cheeked dunnart
Sminthopsis youngsoni McKenzie & Archer, 1982 Lesser hairy-footed dunnart

ORDER PERAMELEMORPHIA Ameghino, 1889

Family Thylacomyidae Bensley, 1903 [bilbies]

Macrotis lagotis (Reid, 1837) Greater bilby
Macrotis leucura (Thomas, 1887) Lesser bilby

Family Chaeropodidae Gill, 1872 [Pig-footed bandicoot]

Chaeropus ecaudatus (Ogilby, 1838) Pig-footed bandicoot

Family Peramelidae Gray, 1825 [bandicoots]

Echymipera rufescens (Peters & Doria, 1875) Rufous spiny bandicoot

Isoodon auratus (Ramsay, 1887) Golden bandicoot
Isoodon macrourus (Gould, 1842) Northern brown bandicoot
Isoodon obesulus (Shaw, 1797) Southern brown bandicoot
Perameles bougainville Quoy & Gaimard, 1824 Western barred bandicoot
Perameles eremiana Spencer, 1897 Desert bandicoot
Perameles gunnii Gray, 1838 Eastern barred bandicoot
Perameles nasuta Geoffroy, 1804 Long-nosed bandicoot

ORDER DIPROTODONTIA Owen, 1866

SUBORDER VOMBATIFORMES Burnett, 1830

Family Phascolarctidae Owen, 1839 [Koala]
Phascolarctos cinereus (Goldfuss, 1817) Koala

Family Vombatidae Burnett, 1830 [wombats]
Lasiorhinus krefftii (Owen, 1872) Northern hairy-nosed wombat
Lasiorhinus latifrons (Owen, 1845) Southern hairy-nosed wombat
Vombatus ursinus (Shaw, 1800) Common wombat

SUBORDER PHALANGERIFORMES Szalay, 1982

Family Burramyidae Broom, 1898 [pygmy possums]
Burramys parvus Broom, 1896 Mountain pygmy possum
Cercartetus caudatus (Milne-Edwards, 1877) Long-tailed pygmy possum
Cercartetus concinnus (Gould, 1845) Western pygmy possum
Cercartetus lepidus (Thomas, 1888) Little pygmy possum [Tasmanian pygmy possum]
Cercartetus nanus (Desmarest, 1818) Eastern pygmy possum

Family Phalangeridae Thomas, 1888 [cuscuses and brushtails]
Phalanger mimicus Thomas, 1922 Southern common cuscus
Spilocuscus maculatus (Desmarest, 1818) Common spotted cuscus
Trichosurus arnhemensis Collett, 1897 Northern brushtail
Trichosurus caninus (Ogilby, 1836) Short-eared brushtail
Trichosurus cunninghami Lindenmayer, Dubach & Viggers, 2002 Mountain brushtail
Trichosurus johnstonii (Ramsay, 1888) Coppery brushtail
Trichosurus vulpecula (Kerr, 1792) Common brushtail
Wyulda squamicaudata Alexander, 1919 Scaly-tailed possum

Family Pseudocheiridae Winge, 1893 [ringtails and greater glider]
Hemibelideus lemuroides (Collett, 1884) Lemuroid ringtail
Petauroides volans (Kerr, 1792) Greater glider
Petropseudes dahli (Collett, 1895) Rock ringtail
Pseudocheirus occidentalis (Thomas, 1888) Western ringtail
Pseudocheirus peregrinus (Boddaert, 1785) Common ringtail
Pseudochirops archeri (Collett, 1884) Green ringtail
Pseudochirulus cinereus (Tate, 1945) Cinereous ringtail [Daintree River ringtail]
Pseudochirulus herbertensis (Collett, 1884) Herbert River ringtail

Family Petauridae Bonaparte, 1838 [possums and gliders]
Dactylopsila trivirgata Gray, 1858 Striped possum
Gymnobelideus leadbeateri McCoy, 1867 Leadbeater's possum
Petaurus australis Shaw, 1791 Yellow-bellied glider
Petaurus breviceps Waterhouse, 1839 Sugar glider
Petaurus gracilis (De Vis, 1883) Mahogany glider [ebony glider]
Petaurus norfolcensis (Kerr, 1792) Squirrel glider

Family Tarsipedidae Gervais & Verreaux, 1842 [Honey possum]
Tarsipes rostratus Gervais & Verreaux, 1842 Honey possum

Family Acrobatidae Aplin, 1987 [Feather-tailed glider]
Acrobates pygmaeus (Shaw, 1794) Feather-tailed glider

SUBORDER MACROPODIFORMES Ameghino, 1889

Family Hypsiprymnodontidae Collett, 1877 [Musky rat-kangaroo]
Hypsiprymnodon moschatus Ramsay, 1876 Musky rat-kangaroo

Family Potoroidae Gray, 1821 [bettongs and potoroos]
Aepyprymnus rufescens (Gray, 1837) Rufous bettong
Bettongia gaimardi (Desmarest, 1822) Eastern bettong

Bettongia lesueur (Quoy & Gaimard, 1824) Burrowing bettong
Bettongia penicillata Gray, 1837 Brush-tailed bettong
Bettongia tropica Wakefield, 1967 Northern bettong
Caloprymnus campestris (Gould, 1843) Desert rat-kangaroo
Potorous gilbertii (Gould, 1841) Gilbert's potoroo
Potorous longipes Seebeck & Johnston, 1980 Long-footed potoroo
Potorous platyops (Gould, 1844) Broad-faced potoroo
Potorous tridactylus (Kerr, 1792) Long-nosed potoroo

Family Macropodidae Gray, 1821 [kangaroos and wallabies]

Dendrolagus bennettianus De Vis, 1887 Bennett's tree-kangaroo
Dendrolagus lumholtzi Collett, 1884 Lumholtz's tree-kangaroo
Lagorchestes asomatus Finlayson, 1943 Central hare-wallaby [Lake Mackay hare-wallaby]
Lagorchestes conspicillatus Gould, 1842 Spectacled hare-wallaby
Lagorchestes hirsutus Gould, 1844 Rufous hare-wallaby
Lagorchestes leporides (Gould, 1841) Eastern hare-wallaby
Lagostrophus fasciatus (Péron & Lesueur, 1807) Banded hare-wallaby
Macropus agilis (Gould, 1842) Agile wallaby
Macropus antilopinus (Gould, 1842) Antilopine kangaroo [antilopine wallaroo]
Macropus bernardus Rothschild, 1904 Black wallaroo [Woodward's wallaroo]
Macropus dorsalis (Gray, 1837) Black-striped wallaby
Macropus eugenii (Desmarest, 1817) Tammar wallaby
Macropus fuliginosus (Desmarest, 1817) Western grey kangaroo
Macropus giganteus Shaw, 1790 Eastern grey kangaroo
Macropus greyi Waterhouse, 1845 Toolache wallaby
Macropus irma (Jourdan, 1837) Western brush wallaby
Macropus parma Waterhouse, 1846 Parma wallaby
Macropus parryi Bennett, 1835 Pretty-faced wallaby [whiptail wallaby]
Macropus robustus Gould, 1841 Common wallaroo
Macropus rufogriseus (Desmarest, 1817) Red-necked wallaby
Macropus rufus (Desmarest, 1822) Red kangaroo
Onychogalea fraenata (Gould, 1841) Bridled nail-tailed wallaby
Onychogalea lunata (Gould, 1841) Crescent nail-tailed wallaby
Onychogalea unguifera (Gould, 1841) Northern nail-tailed wallaby
Petrogale assimilis Ramsay, 1877 Allied rock-wallaby
Petrogale brachyotis (Gould, 1841) Short-eared rock-wallaby
Petrogale burbidgei Kitchener & Sanson, 1978 Monjon
Petrogale coenensis Eldridge & Close, 1992 Cape York rock-wallaby
Petrogale concinna Gould, 1842 Nabarlek
Petrogale godmani Thomas, 1923 Godman's rock-wallaby
Petrogale herberti Thomas, 1926 Herbert's rock-wallaby
Petrogale inornata Gould, 1842 Unadorned rock-wallaby
Petrogale lateralis Gould, 1842 Black-footed rock-wallaby [black-flanked rock-wallaby]
Petrogale mareeba Eldridge & Close, 1992 Mareeba rock-wallaby
Petrogale penicillata (Gray, 1825) Brush-tailed rock-wallaby
Petrogale persephone Maynes, 1982 Proserpine rock-wallaby
Petrogale purpureicollis Le Souef, 1924 Purple-necked rock-wallaby
Petrogale rothschildi Thomas, 1904 Rothschild's rock-wallaby
Petrogale sharmani Eldridge & Close, 1992 Mt Claro rock-wallaby [Sharman's rock-wallaby]
Petrogale xanthopus Gray, 1855 Yellow-footed rock-wallaby
Setonix brachyurus (Quoy & Gaimard, 1830) Quokka
Thylogale billardierii (Desmarest, 1822) Tasmanian pademelon
Thylogale stigmatica (Gould, 1860) Red-legged pademelon

Thylogale thetis (Lesson, 1827) Red-necked pademelon
Wallabia bicolor (Desmarest, 1804) Swamp wallaby

Subclass Eutheria Gill, 1872

ORDER SIRENIA Iliger, 1811

Family Dugongidae Gray, 1821
Dugong dugon (Müller, 1776) Dugong

ORDER SORICOMORPHIA Gregory, 1910

Family Soricidae Fischer von Waldheim, 1817 [shrews]
Crocidura trichura Dobson, 1889 Christmas Island shrew

ORDER CHIROPTERA Blumenbach, 1779

Family Pteropodidae Gray, 1821 [fruit-bats]
Dobsonia magna Thomas, 1905 Bare-backed fruit-bat
Macroglossus minimus (Geoffroy, 1810) Northern blossom-bat
Nyctimene cephalotes (Pallas, 1767) Pallas's tube-nosed bat
Nyctimene robinsoni Thomas, 1904 Eastern tube-nosed bat
Pteropus alecto Temminck, 1837 Black flying-fox
Pteropus brunneus Dobson, 1878 Percy Island flying-fox
Pteropus conspicillatus Gould, 1850 Spectacled flying-fox
Pteropus macrotis Peters, 1867 Big-eared flying-fox
Petropus melanotus Blyth, 1863 Black-eared flying-fox
Pteropus poliocephalus Temminck, 1825 Grey-headed flying-fox
Pteropus scapulatus Peters, 1862 Little red flying-fox
Syconycteris australis (Peters, 1867) Eastern blossom-bat [common blossom-bat]

Family Rhinolophidae Gray, 1825 [horseshoe-bats]
Rhinolophus megaphyllus Gray 1834 Eastern horseshoe-bat
Rhinolophus philippinensis Waterhouse, 1843 Large-eared horseshoe-bat

Family Hipposideridae Lydekker, 1891 [leaf-nosed bats]
Hipposideros ater Templeton, 1848 Dusky leaf-nosed bat
Hipposideros cervinus (Gould, 1854) Fawn leaf-nosed bat
Hipposideros diadema (Geoffroy, 1813) Diadem leaf-nosed bat
Hipposideros semoni Matschie, 1903 Semon's leaf-nosed bat
Hipposideros stenotis Thomas 1913 Northern leaf-nosed bat
Rhinonicteris aurantia (Gray, 1845) Orange leaf-nosed bat

Family Megadermatidae Allen, 1864 [Ghost bat]
Macroderma gigas (Dobson, 1880) Ghost bat

Family Emballonuridae Gervais, 1856 [Sheath-tailed bats]
Saccolaimus flaviventris (Peters, 1867) Yellow-bellied sheath-tailed bat
Saccolaimus mixtus Troughton, 1925 Papuan sheath-tailed bat [Troughton's pouched bat]
Saccolaimus saccolaimus (Temminck, 1838) Bare-rumped sheath-tailed bat
Taphozous australis Gould, 1854 Coastal sheath-tailed Bat [coastal tomb bat]
Taphozous georgianus Thomas, 1915 Common sheath-tailed bat [sharp-nosed tomb bat]
Taphozous hilli Kitchener, 1980 Hill's sheath-tailed bat
Taphozous kapalgensis McKean & Friend, 1979 Arnhem sheath-tailed bat
Taphozous troughtoni Tate, 1952 Troughton's sheath-tailed bat

Family Molossidae Gervais, 1856 [free-tailed bats]
Chaerephon jobensis (Miller, 1902) Northern free-tailed bat
Mormopterus beccarii Peters, 1881 Beccari's free-tailed bat
Mormopterus loriae (Thomas, 1897) Little free-tailed bat [Loria's mastiff bat]
Mormopterus norfolkensis (Gray, 1839) Eastern free-tailed bat
Mormopterus planiceps (Peters, 1866) Southern free-tailed bat
Tadarida australis (Gray, 1838) White-rumped free-tailed bat

Family Vespertilionidae Gray, 1821 [vespertilionid bats]

Subfamily Vespertilioninae Gray, 1821
Chalinolobus dwyeri Ryan, 1966 Large-eared pied bat
Chalinolobus gouldii (Gray, 1841) Gould's wattled bat
Chalinolobus morio (Gray, 1841) Chocolate wattled bat

Chalinolobus nigrogriseus (Gould, 1856) Hoary wattled bat
Chalinolobus picatus (Gould, 1852) Little pied bat
Falsistrellus mackenziei Kitchener, Caputi & Jones, 1986 Western false pipistrelle
Falsistrellus tasmaniensis (Gould, 1858) Eastern false pipistrelle
Kerivoula papuensis Dobson, 1878 Golden-tipped bat
Miniopterus australis (Tomes, 1858) Little bent-winged bat
Miniopterus schreibersii (Kuhl, 1817) Large bent-winged bat [Schreibers's long-fingered bat]
Murina florium Thomas, 1908 Flute-nosed bat
Myotis macropus Gould, 1855 Southern myotis
Nyctophilus arnhemensis Johnson, 1959 Arnhem long-eared bat
Nyctophilus bifax Thomas, 1915 Eastern long-eared bat
Nyctophilus geoffroyi Leach, 1821 Lesser long-eared bat
Nyctophilus gouldi Tomes, 1858 Gould's long-eared bat
Nyctophilus howensis McKean, 1975 Lord Howe Island long-eared bat
Nyctophilus timoriensis (Geoffroy, 1806) Greater long-eared bat
Nyctophilus walkeri Thomas, 1892 Pygmy long-eared bat
Pipistrellus adamsi Kitchener, Caputi & Jones, 1986 Cape York pipistrelle
Pipistrellus murrayi Andrews, 1900 Christmas Island pipistrelle
Pipistrellus westralis Koopman, 1984 Northern pipistrelle
Scoteanax rueppellii (Peters 1866) Greater broad-nosed bat
Scotorepens balstoni (Thomas, 1906) Inland broad-nosed bat
Scotorepens greyii (Gray, 1843) Little broad-nosed bat
Scotorepens orion (Troughton, 1937) Eastern broad-nosed bat
Scotorepens sanborni (Troughton, 1937) Northern broad-nosed bat
Vespadelus baverstocki (Kitchener, Jones & Caputi, 1987) Inland forest bat
Vespadelus caurinus (Thomas, 1914) Northern cave bat
Vespadelus darlingtoni (Allen, 1933) Large forest bat
Vespadelus douglasorum Kitchener, 1976 Yellow-lipped bat
Vespadelus finalysoni Kitchener, Jones & Caputi, 1987 Finlayson's bat
Vespadelus pumilus (Gray, 1841) Eastern forest bat
Vespadelus regulus (Thomas, 1906) Southern forest bat
Vespadelus troughtoni (Kitchener, Jones & Caputi, 1987) Eastern cave bat
Vespadelus vulturnus (Thomas, 1914) Little forest bat

ORDER CARNIVORA Bowdich, 1821

SUBORDER CANIFORMIA Kretzol, 1938

Family Canidae

Canis lupus
Canis lupus dingo Meyer, 1793, Dingo

Family Otariidae Gray, 1825 [eared seals]

Arctocephalus forsteri (Lesson, 1828) New Zealand fur-seal
Arctocephalus gazella (Peters, 1875) Antarctic fur-seal
Arctocephalus pusillus (Schreber, 1775) Australian fur-seal
Arctocephalus tropicalis (Gray, 1872) Subantarctic fur-seal
Neophoca cinerea (Péron, 1816) Australian sea-lion

Family Phocidae Gray, 1821 ['true' seals]

Hydrurga leptonyx (Blainville, 1820) Leopard seal
Leptonychotes weddellii (Lesson, 1826) Weddell seal
Lobodon carcinophaga (Hombron & Jacquinot, 1842) Crab-eater seal
Mirounga leonina (Linnaeus, 1758) Southern elephant seal
Ommatophoca rossii Gray, 1844 Ross seal

ORDER CETACEA Brisson, 1762

SUBORDER MYSTICETI Flower, 1864

Family Balaenidae Gray, 1821

Eubalaena australis (Desmoulins, 1822) Southern right whale

Family Balaenopteridae Gray, 1864

Balaenoptera acutorostrata Lacépède, 1804 Dwarf minke whale
Balaenoptera bonaerensis Burmeister, 1867 Antarctic minke whale
Balaenoptera borealis Lesson, 1828 Sei whale
Balaenoptera edeni Anderson, 1878 Bryde's whale
Balaenoptera musculus (Linnaeus, 1758) Blue whale
Balaenoptera omurai Wada, Olshi & Yamada, 2003 Omura's whale

Balaenoptera physalus (Linnaeus, 1758) Fin whale
Megaptera novaeangliae (Borowski, 1781) Humpbacked whale

Family Neobalaenidae Gray, 1873 [Pygmy right whale]

Caperea marginata (Gray, 1846) Pygmy right whale

SUBORDER ODONTOCETI Flower, 1869

Family Delphinidae Gray, 1821

Delphinus delphis Linnaeus, 1758 Short-beaked common dolphin
Feresa attenuata Gray, 1874 Pygmy killer whale
Globicephala macrorhynchus Gray, 1846 Short-finned pilot whale
Globicephala melas (Traill, 1809) Long-finned pilot whale
Grampus griseus (G. Cuvier, 1812) Risso's dolphin
Lagenodelphis hosei Fraser, 1956 Fraser's dolphin
Lagenorhynchus cruciger (Quoy & Gaimard, 1824) Hourglass dolphin
Lagenorhynchus obscurus (Gray, 1828) Dusky dolphin
Lissodelphis peronii (Lacépède, 1804) Southern right whale dolphin
Orcaella heinsohni Beasley, Robertson & Arnold, 2005 Australian snub-finned dolphin
Orcinus orca (Linnaeus, 1758) Common killer whale
Peponocephala electra (Gray, 1846) Melon-headed whale
Pseudorca crassidens (Owen, 1846) False killer whale
Sousa chinensis (Osbeck, 1765) Indo-pacific humpbacked dolphin
Stenella attenuata (Gray, 1846) Pantropical spotted dolphin
Stenella coeruleoalba (Meyen, 1833) Striped dolphin
Stenella longirostris (Gray, 1828) Spinner dolphin
Steno bredanensis (Lesson, 1828) Rough-toothed dolphin
Tursiops aduncus (Ehrenberg, 1833) Indo-pacific bottle-nosed dolphin
Tursiops truncatus (Montagu, 1821) Common bottle-nosed dolphin

Family Phocoenidae Gray, 1825 [porpoises]

Phocoena dioptrica Lahille, 1912 Spectacled porpoise

Family Physeteridae Gray, 1821 [sperm whales]

Kogia breviceps (Blainville, 1838) Pygmy sperm whale
Kogia sima (Owen, 1866) Dwarf sperm whale
Physeter catodon Linnaeus, 1758 Giant sperm whale

Family Ziphiidae Gray, 1865 [beaked whales]

Berardius arnuxii Duvernoy, 1851 Arnoux's beaked whale
Hyperoodon planifrons Flower, 1882 Southern bottle-nosed whale
Indopacetus pacificus (Longman, 1926) Longman's bottle-nosed whale
Mesoplodon bowdoini Andrews, 1908 Andrews's beaked whale
Mesoplodon densirostris (Blainville, 1817) Blainville's beaked whale
Mesoplodon ginkgodens Nishiwaki & Kamiya, 1958 Ginkgo-toothed beaked whale
Mesoplodon grayi Von Haast, 1876 Gray's beaked whale
Mesoplodon hectori (Gray, 1871) Hector's beaked whale
Mesoplodon layardii (Gray, 1865) Strap-toothed beaked whale
Mesoplodon mirus True, 1913 True's beaked whale
Tasmacetus shepherdi Oliver, 1937 Shepard's beaked whale
Ziphius cavirostris G. Cuvier, 1823 Cuvier's beaked whale

ORDER RODENTIA Bowdich, 1823

Family Muridae Gray, 1821 [rats and mice]

Subfamily Hydromyinae Alston, 1876

Conilurus albipes (Lichtenstein, 1829) White-footed rabbit-rat
Conilurus penicillatus (Gould, 1842) Brush-tailed rabbit-rat
Hydromys chrysogaster Geoffroy, 1804 Water rat
Leggadina forresti (Thomas, 1906) Forrest's mouse
Leggadina lakedownensis Watts, 1976 Lakeland Downs mouse
Leporillus apicalis (Gould, 1853) Lesser stick-nest rat
Leporillus conditor (Sturt, 1848) Greater stick-nest rat
Mastacomys fuscus Thomas, 1882 Broad-toothed rat
Melomys burtoni (Ramsay, 1887) Grassland melomys
Melomys capensis Tate, 1951 Cape York melomys
Melomys cervinipes (Gould, 1852) Fawn-footed melomys
Melomys rubicola Thomas, 1924 Bramble Cay melomys

Mesembriomys gouldii (Gray, 1843) Black-footed tree-rat
Mesembriomys macrurus (Peters, 1876) Golden-backed tree-rat
Notomys alexis Thomas, 1922 Spinifex hopping mouse
Notomys amplus Brazenor, 1936 Short-tailed hopping mouse
Notomys aquilo Thomas, 1921 Northern hopping mouse
Notomys cervinus (Gould, 1853) Fawn hopping mouse
Notomys fuscus (Jones, 1925) Dusky hopping mouse
Notomys longicaudatus (Gould, 1844) Long-tailed hopping mouse
Notomys macrotis Thomas, 1921 Big-eared hopping mouse
Notomys mitchellii (Ogilby, 1838) Mitchell's hopping mouse
Notomys mordax Thomas, 1922 Darling Downs hopping mouse
Pogonomys mollipilosus (Peters & Doria, 1881) Prehensile-tailed rat
Pseudomys albocinereus (Gould, 1845) Ash-grey mouse
Pseudomys apodemoides Finlayson, 1932 Silky mouse
Pseudomys australis Gray, 1832 Plains rat
Pseudomys bolami Troughton, 1932 Bolam's mouse
Pseudomys calabyi Kitchener & Humphreys, 1987 Kakadu pebble-mound mouse
Pseudomys chapmani Kitchener, 1980 Pilbara pebble-mound mouse
Pseudomys delicatulus (Gould, 1842) Delicate mouse
Pseudomys desertor Troughton, 1932 Desert mouse
Pseudomys fieldi (Waite, 1896) Shark Bay mouse
Pseudomys fumeus Brazenor, 1934 Smoky mouse
Pseudomys glaucus Thomas, 1910 Blue-grey mouse
Pseudomys gouldii (Waterhouse, 1839) Gould's mouse
Pseudomys gracilicaudatus (Gould, 1845) Eastern chestnut mouse
Pseudomys hermannsburgensis (Waite, 1896) Sandy inland mouse
Pseudomys higginsi (Trouessart, 1897) Long-tailed mouse
Pseudomys johnsoni Kitchener, 1985 Central pebble-mound mouse
Pseudomys laborifex Kitchener & Humphreys, 1986 Kimberley mouse
Pseudomys nanus (Gould, 1858) Western chestnut mouse
Pseudomys novaehollandiae (Waterhouse, 1843) New Holland mouse
Pseudomys occidentalis Tate, 1951 Western mouse
Pseudomys oralis Thomas, 1921 Hastings River mouse
Pseudomys patrius (Thomas & Dollman, 1909) Eastern pebble-mound mouse
Pseudomys pilligaensis Fox & Briscoe, 1980 Pilliga mouse
Pseudomys shortridgei (Thomas, 1907) Heath mouse
Uromys caudimaculatus (Krefft, 1867) Giant white-tailed rat
Uromys hadrourus (Winter, 1984) Masked white-tailed rat
Xeromys myoides Thomas, 1889 Water mouse [false water-rat]
Zyzomys argurus (Thomas, 1889) Common rock-rat
Zyzomys maini Kitchener, 1989 Arnhem Land rock-rat
Zyzomys palatalis Kitchener, 1989 Carpentarian rock-rat
Zyzomys pedunculatus (Waite, 1896) Central rock-rat
Zyzomys woodwardi (Thomas, 1909) Large rock-rat

Subfamily Murinae Murray, 1866

Rattus colletti (Thomas, 1904) Dusky rat
Rattus exulans (Peale, 1848) Pacific rat
Rattus fuscipes (Waterhouse, 1839) Bush rat
Rattus leucopus (Gray, 1867) Cape York rat
Rattus lutreolus (Gray, 1841) Swamp rat
Rattus macleari (Thomas, 1887) Maclear's rat
Rattus nativitatis (Thomas, 1889) Christmas Island rat
Rattus sordidus (Gould, 1858) Canefield rat
Rattus tunneyi (Thomas, 1904) Tunney's rat
Rattus villosissimus (Waite, 1898) Long-haired rat

ACKNOWLEDGEMENTS

I would like to thank Stephen Jackson, Stephen Van Dyck, Andrew Breed and Wendy Blanshard for their assistance.

REFERENCES

Beasley I, Robertson KM & Arnold P (2003). Description of a new dolphin, the Australian snubfin dolphin *Orcaella heinsohni* sp. nov. (Cetacea, Delphinidae). *Marine Mammal Science* **21**, 365–400.

Christidis L & Boles WE (1994). *The Taxonomy and Species of Birds of Australia and its Territories*. Royal Australasian Ornithologists Union Monograph 2. RAOU: Melbourne.

Churchill S (1998). *Australian Bats*. Reed New Holland: Sydney.

Cooper SJB, Day PR, Reardon TB & Schultz M (2001). Assessment of species boundaries in Australian Myotis (Chiroptera, Vespertilionidae) using mitochondrial DNA. *Journal of Mammalogy* **82**, 328–338.

Duckworth JW & Pine RH (2003). English names for a world list of mammals, exemplified by species of Indochina. *Mammal Review* **33**, 151–173.

Helgen KM (2004). On the identity of flying-foxes, genus Pteropus (Mammalia, Chiroptera), from islands in the Torres Strait, Australia. *Zootaxa* **780**, 1–14.

Kitchener DJ (1989). Taxonomic appraisal of Zyzomys (Rodentia, Muridae) with descriptions of two new species from the Northern Territory, Australia. *Records of Western Australian Museum* **14**, 331–373.

Maxwell S, Burbidge AA & Morris K (eds) (1996). *Action Plan for Australian Marsupials and Monotremes*. IUCN/SSC Australasian Marsupial and Monotreme Specialist Group: Canberra.

Menkhorst P & Knight F (2004). *Field Guide to the Mammals of Australia* (2nd edn). OUP: Melbourne.

Walton DW (1988). *Zoological Catalogue of Australia*, Vol. 5 Mammalia. Australian Biological Resources Study: Canberra.

Wilson DE & Reeder DM (eds) (2005). *Mammal Species of the World: A Taxonomic and Geographic Reference*. Johns Hopkins University Press: Baltimore.

Woolley PA (2005). The species of Dasycercus Peters, 1875 (Marsupialia, Dasyuridae). *Memoirs of Museum Victoria* **62**, 213–221.

Appendix 2 Guide to the identification of common parasites of Australian mammals

Dave Spratt, Ian Beveridge, Lee Skerratt and Richard Speare

Much of the original literature on helminth parasites of monotremes and marsupials may be found in Spratt et al. (1991).

1 COLLECTION PROCEDURES

1.1 Helminths (trematodes, nematodes, cestodes, acanthocephala)

Two types of collection procedures are used to collect parasites from carcasses: long and short. The division is based on the time taken to collect and preserve the parasites. In general, 'long' procedures refer to the optimal fixation procedures that will provide the best specimens for taxonomic studies. These procedures generally require more time, effort and resources than 'short' procedures, but maximise the chances of both collection and identification of any parasites present. 'Short' procedures are generally simpler, more applicable for use for specific organs or whole carcasses of small mammals and may be more suited for field situations.

All parasites collected for identification should be accompanied by a comprehensive label.

1.1.1 Long procedure

Examine body cavities, organs and subcutaneous tissues for parasites. Collect any parasites into a sterile container and wash briefly in 0.9% saline to remove mucus. Water can be used if saline is not available, however, if so then very brief washing is recommended. Helminths should be fixed with either 70% ethanol or 10% formalin. Ethanol is preferred as it enables genetic studies to be carried out. The ratio of volume of fixative to parasite should be at least 10:1. Hot fixation is preferred for trematodes (flukes) and nematodes (roundworms) to prevent contraction of specimens. For hot fixation, bring a small amount of solution to boil on a hotplate in a fume cupboard (good ventilation is essential), remove from heat and pour onto specimens in minimal amount of saline.

Alternatively, hot saline or hot water can be poured onto specimens and cold formalin or alcohol added immediately afterwards. As an alternative to hot fixation, large nematodes and trematodes can be fixed in a mixture of 95 parts of glacial acetic acid and 5 parts of concentrated formalin. This procedure is unsuitable for hair-like nematodes such as *Capillaria*. Care is needed when handling these chemicals as all are potentially hazardous (read their MSDS if you are unaware of their effects).

Specific procedures are as follows.

- Trematodes: fix in hot 70% ethanol or hot 10% formalin as outlined above.
- Nematodes: fix in hot 70% ethanol or hot 10% formalin as outlined above. In some mammals (particularly small ones), nematodes and trematodes may be tiny (0.3–1.0 mm) and tissues need to be examined under a stereo-microscope. In the field, opened intestines may be scraped with a microscope slide, the slide stirred in a bottle/beaker containing saline, the scraping allowed to sediment in the bottle/beaker, the sediment fixed in hot 70% ethanol or hot 10% formalin and sorted under a microscope on return to the laboratory.
- Cestodes and acanthocephala should be relaxed for 10–60 min (depending on size) in water (not saline) initially to prevent contraction, then fix in cold 70% ethanol or 10% formalin.

1.1.2 Short procedure

Short procedures are more suited to nematodes than other helminths. Make a large ventral incision in the carcass to ensure fixative enters thoracic and abdominal cavities and immerse carcass in cold 70% ethanol or 10% formalin. If fixing entire intestinal tract, make several small incisions in each of stomach, intestine, caecum and colon to allow penetration of fixative. Carcasses or organs may be frozen for indefinite periods but this procedure is useful only for nematodes and acanthocephala, not for trematodes and small cestodes as they are often damaged, preventing identification. Macropodoid stomachs can be dealt with in the field by slitting open the entire stomach, pouring boiling water onto contents then immediately adding cold 40% formaldehyde.

1.2 Arthropods

Arthropods should be fixed in cold 70% ethanol or submitted live (see below).

1.2.1 Ticks

If time and locality permit, collection of live ticks allows rearing of the less well-known immature stages (Fig. A2.1). Live engorged ticks should be wrapped in moist (dampened but not dripping) cottonwool and placed in a robust container with a few small ventilation holes in the lid. Unengorged specimens should be fixed in 70% ethanol.

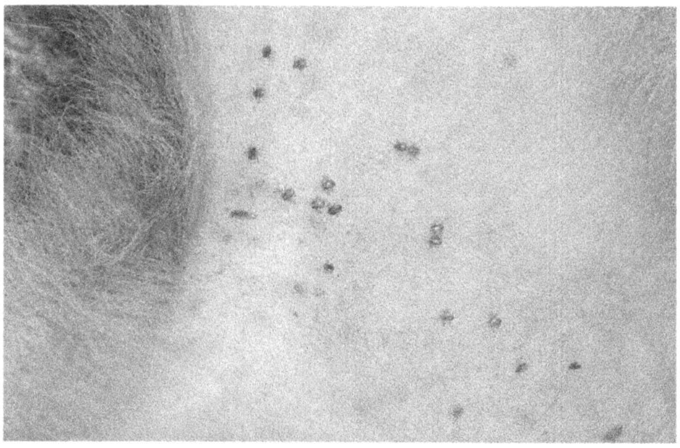

Figure A2.1 Collection of live ticks allows rearing of less well-known immature stages.

1.2.2 Fleas, hippoboscid flies, mites and lice

Fleas and hippoboscid flies are relatively large and extremely active, and require rapid collection into a small vial. To collect lice and mites, brush, comb or rub fur briskly over a sheet of white paper with a wad of cottonwool soaked in chloroform (Fig. A2.2a). The avenues for collecting mites are limited only by the ingenuity of the collector.

A low-power dissecting microscope or a good hand-lens are useful for finding smaller mites such as freshly attached chiggers (Fig. A2.2b), some of which live in minute pits in the skin. Mites attached to soft skin singly or in clusters may be poked off with a needle or sharpened matchstick, or a small piece of skin with mites attached may be snipped off. Hair follicles may be squeezed to evacuate mites that live within the skin but these are best obtained by skin scrapings with a scalpel blade (Fig. A2.3) or skin biopsy. The exoskeleton of mites contains chitin, which is impervious to potassium hydroxide (KOH) digestion. Skin scrapings and skin biopsies may be digested immediately or after formalin fixation in 10% KOH if mites are difficult to find and the specimen dissected under a stereo-microscope. Mites may also be found in the gastric mucosa of bats, the internal respiratory organs and particularly the nasal passages of mammals. Spinturnicid mites are large and slow-moving, common on the wings of bats and sometimes attached like ticks. Mites in particular should

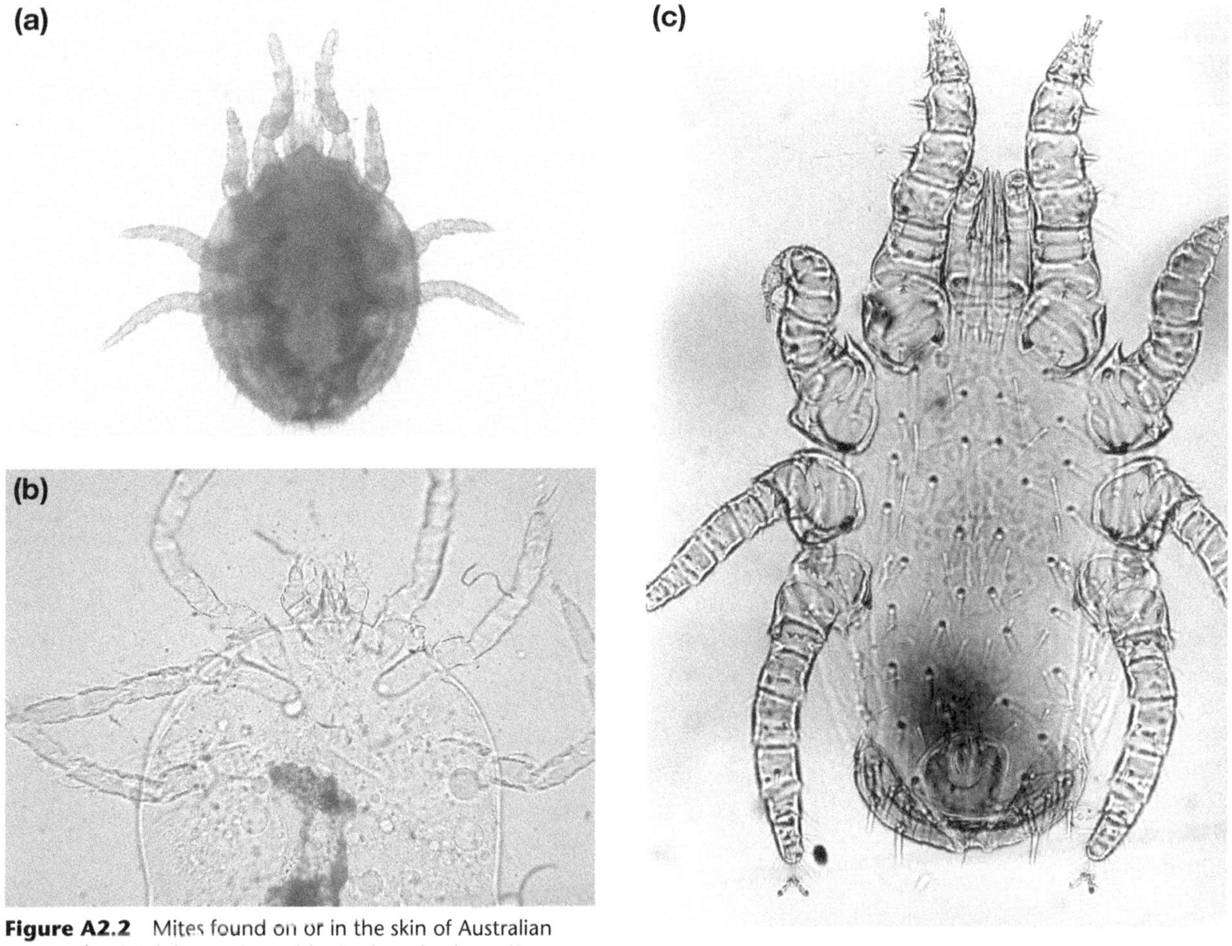

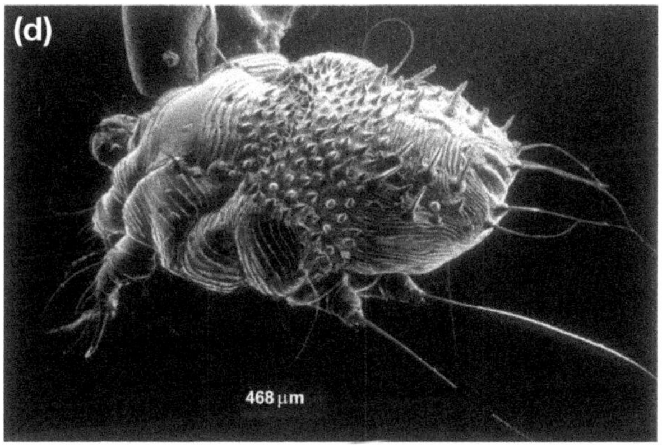

Figure A2.2 Mites found on or in the skin of Australian mammals. a) Adult prostigmatid mite from bush rat (*Rattus fuscipes*) (note 4 pairs of legs of adult mite in contrast to 3 pairs in larva). b) Larval trombiculid mite (chigger) from Proserpine rock-wallaby (*Petrogale persephone*) (note 3 pairs of legs in larva in contrast to 4 pairs in adult). c) *Thadeua serrata*, a mesostigmatid mite about 400 μm long found in discrete crateriform lesions in the inguinal region of several species of wallaby, may be associated with moderate to severe dermatitis. d) The most significant parasitic disease of common wombats (*Vombatus ursinus*) is sarcoptic mange caused by the introduced astigmatid mite, *Sarcoptes scabiei*.

be fixed in 70% ethanol and placed in small tubes (30–50 mm × 8–10 mm) because they are hard to find in large tubes and difficult to extract from minute tubes.

2 PARASITES IN FAECES

Parasites (particularly their eggs) are likely to be encountered in faeces. Techniques used for collection and examination of parasites include faecal mounts, faecal flotations, faecal sedimentation and the Baermann technique, all of which are described in detail in most standard parasitology texts. Due to a lack of knowledge of helminths and coccidia in native mammals, identification of their eggs and oocysts in faeces can be problematic.

2.1 Wet mounts

This technique relies on parasites being observed in the sample under a coverslip using a compound microscope.

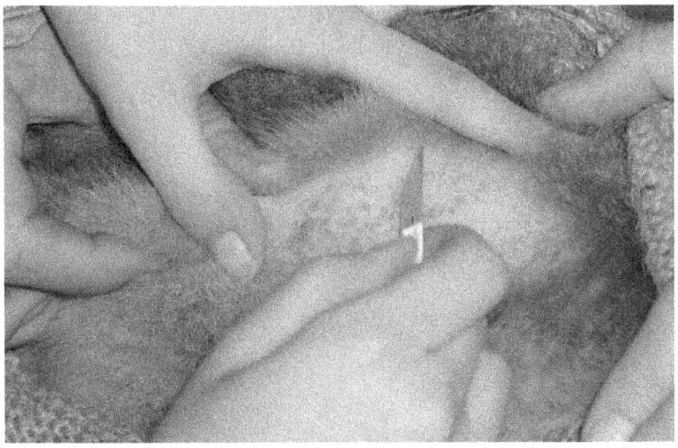

Figure A2.3 Mites living within the skin deep in hair follicles are best collected by scraping skin with a scalpel blade.

This technique is most useful for identification of mobile parasites such as some species of protozoa and nematode larvae.

2.2 Flotations

Many helminth eggs can be concentrated by flotation. This technique relies on use of a supersaturated solution, usually of magnesium sulphate or zinc sulphate, that lifts eggs to the surface of the column of liquid due to their lower specific gravity. Eggs are collected at the surface and are less obscured by faecal material. Nematode eggs are commonly encountered, but can frequently be identified only as 'strongylid' type eggs (Fig. A2.4b) or 'trichuroid' type eggs. Kangaroos in particular have numerous species of strongylid nematodes in their gastrointestinal tracts, but virtually none can be identified to genus by their eggs. The sole exception is the egg of *Globocephaloides* which is two-celled when passed in the faeces.

The use of larval cultures has not been investigated extensively. All cloacinid nematodes (a subfamily of strongylids) from which the third-stage larva has been described have three hook-like structures in the buccal capsule. They are particularly well-developed in species of *Labiostrongylus* (another strongylid). The larvae of species of *Strongyloides* are relatively small and present in freshly voided faeces. These may develop into infective larvae or free-living adults in larval culture. However, there are currently no known methods of distinguishing other genera.

Coccidian oocysts can be placed in potassium dichromate to allow sporulation. Sporulated oocysts of macropodoids (Fig. A2.4a) can be identified from diagrams and photographs published by Barker et al.

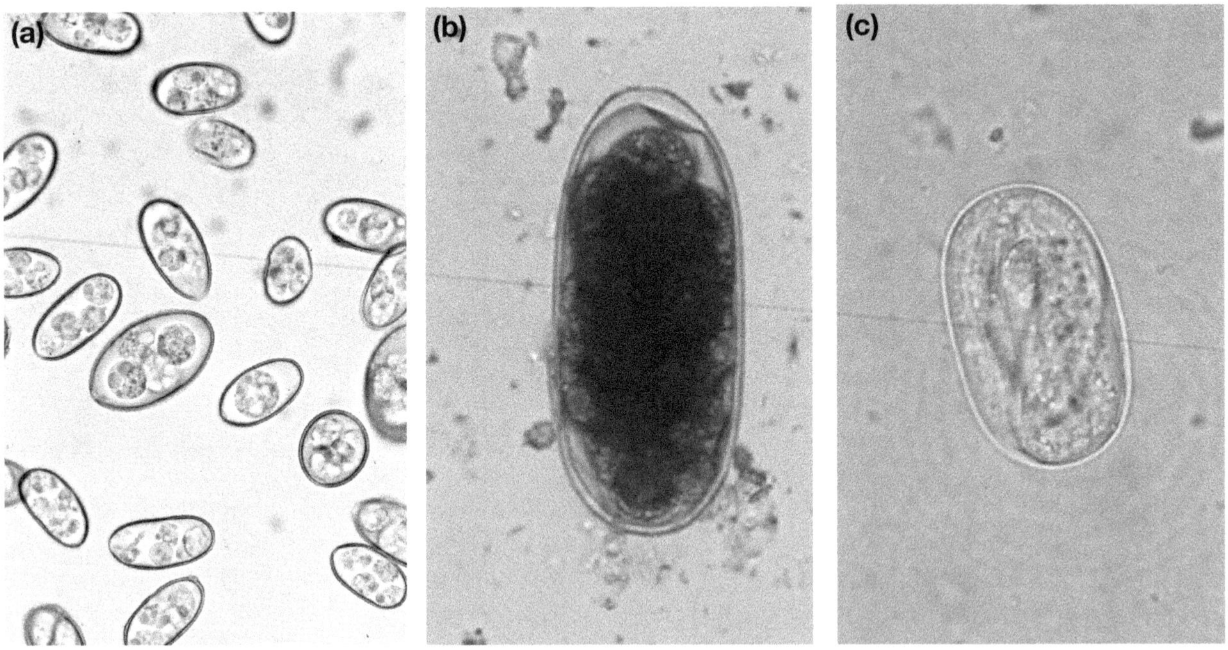

Figure A2.4 Oocysts and nematode eggs from the faeces of kangaroos. a) Oocysts of *Eimeria* spp. from the faeces of the eastern grey kangaroo (*Macropus giganteus*). b) Typical strongylid egg produced by stomach nematodes of the Subfamily Cloacininae, 120 μm long. c) Larvated egg of an unnamed species of *Strongyloides* found in the stomachs of kangaroos, 50 μm long.

(1972, 1988a, b, c, 1989), for wombats by Barker et al. (1979), for the echidna (*Tachyglossus aculeatus*) by Barker et al. (1985) and for the common brushtail (*Trichosurus vulpecula*) by O'Callaghan and O'Donoghue (2001). There are no identification guides to the coccidia of other native mammals.

2.3 Sedimentation

This technique relies on the sedimentation of trematode eggs in water due to their higher specific gravity. Due to low diversity in some hosts (mainly the diprotodonts), the species of trematode may be readily identified.

2.4 Baermann examination

The Baermann technique is used to concentrate live larvae. It relies on a faecal specimen containing larvae which are living and capable of movement. Faeces are placed on cheesecloth resting on a wire shelf in a funnel. A tube on the outlet is clamped and the funnel filled with warm water (30°C) to cover the faeces. After 6 hr water is collected from the tube into a centrifuge tube, spun and examined under a compound microscope for larvae. Examination of fresh faeces using this method is likely to yield larvae of *Strongyloides* (Fig. A2.4c) or of lungworms. Larvae of *Strongyloides* are unsheathed and have a straight tail that appears truncated at the tip, however, under oil immersion the tip has a V-shaped notch. Lungworm larvae are generally recognisable by their coiled or C-shape and kinked, notched or spiked tail. However, these larvae can be difficult to distinguish from the larvae of free-living nematodes if the sample has been collected from the ground.

3 LABELS

A label with host name (Latin binomial wherever possible), sex, approximate age (e.g. juvenile, adult), geographic locality, date, collector and the organ from which the parasite was collected should be enclosed in each container. Parasites without this information are of little value and are usually discarded when submitted to parasitologists and museums. Labels should be written firmly on good-quality paper in pencil or India ink, never in ballpoint pen.

4 PARASITES AND LESIONS ENCOUNTERED ON CLINICAL EXAMINATION OR AT NECROPSY

4.1 Platypuses (*Ornithorhynchus anatinus*)

4.1.1 Intestine

Three described trematodes and an undescribed liolopid trematode occur in the intestine of the platypus. One of these, *Mehlisia ornithorhynchi*, was reported associated with mild catarrhal enteritis (Whittington & Spratt 1989). The only nematode reported from the gastrointestinal tract, the trichostrongyloid *Tasmanema mundayi*, which normally occurs in the echidna, is considered a dubious host record (Spratt & Whittington 1989).

4.1.2 Subcutis

The catholic filarioid nematode *Cercopithifilaria johnstoni* occurs in the platypus as well as dasyurid, peramelid, petaurid and phalangerid marsupials and murid rodents, and possibly the echidna. Ixodid ticks are known intermediate hosts of this parasite (Spratt & Haycock 1988), thus *Ixodes ornithorhynchi*, the common tick of the platypus, is the likely vector in this host-parasite association.

4.1.3 Blood

The piroplasm *Theileria ornithorhynchi* occurs in erythrocytes and may also be found in leukocytes; it is considered non-pathogenic. The kinetoplastid flagellate protozoan *Trypanosoma binneyi* has been reported in the platypus but is not known to induce pathologic effects (Clark 2004).

4.1.4 Skin

Three unnamed nematodes have been described from the skin of the platypus. The pathological findings associated with them and arthropod parasites have been described (Whittington & Spratt 1989).

4.2 Echidnas (*Tachyglossus aculeatus*)

4.2.1 Intestine

The small intestine of echidnas in Australia is inhabited by a series of trichostrongyloid nematode species belonging to the genera *Nicollina* and *Tasmanema*. In Papua New Guinea, the long-beaked echidna (*Zaglossus bruijni*) is parasitised by a related genus, *Zaglonema*. These tiny red nematodes may occur in large numbers and have been associated with enteritis (McOrist & Smales 1986).

Large numbers of anoplocephalid cestodes may also be present. In most of eastern Australia the species present is *Linstowia echidnae*, but in north Queensland this species is replaced by *Echidnotaenia tachyglossi*. Large burdens may be present (>500) but appear not to be pathogenic.

Coccidiosis caused by a variety of species of *Eimeria* occurs in echidnas and the lesions have been described by Barker et al. (1985).

4.2.2 Body cavities and subcutis

Sparganosis caused by the plerocercoid of *Spirometra erinacei* is relatively common in the SC tissues of echidnas. The histopathological reaction to spargana is more marked than that found in eutherian mammals (Whittington et al. 1992). At least one of the cases reported by these authors was in fact a case of proliferative sparganosis, with massive numbers of small abnormal plerocercoids proliferating in body cavities (Beveridge et al. 1998).

4.2.3 Blood

The piroplasms *Babesia tachyglossi* and *Theileria tachyglossi* occur in erythrocytes, typically 2–8 organisms per cell in the former and 1 (occasionally 2–4) organism per cell in the latter (Clark 2004). Concurrent infection may occur but pathologic effects on the host are unknown.

4.2.4 Skin

Two species of ticks, *Aponomma concolor* and *A. tachyglossi*, are found on echidnas. While usually present in small numbers, heavy infestations of *A. concolor* may cause severe dermatitis (Beveridge 1993). Echidnas are commonly parasitised by a particularly large species of flea, *Bradiopsylla echidnae*, as well as several species of stickfast fleas of the genus *Echidnophaga* (Dunnet & Mardon 1974).

4.3 Macropods

4.3.1 Oesophagus

Species of the cloacinid nematode genera *Cyclostrongylus* and *Spirostrongylus* may be found in large numbers, coiled around the papillae in the oesophagus of wallabies of the Subgenus *Notamacropus* and the swamp wallaby (*Wallabia bicolor*). As far as known they have no pathogenic effects.

4.3.2 Stomach

The sacculated forestomach of most macropods contains large numbers of cloacinid nematodes. The commonest and most abundant genera are *Cloacina, Rugopharynx, Pharyngostrongylus, Parazoniolaimus* and *Labiostrongylus* (now divided into three genera, *Labiostrongylus, Labiosimplex* and *Labiomultiplex*) (Fig. A2.5a–f). Most are not known to be pathogenic. However, the third larval stages of *Parazoniolaimus collaris, Labiomultiplex eugenii* and *L. billardierii* penetrate lymphoid tissue at the junction of the squamous and glandular epithelia of the stomach and produce granulomata (Munday 1971; Smales 1977; Beveridge et al. 1985). In the wallaroo (*Macropus robustus*) large numbers of third-stage larvae of *Labiosimplex robustus* accumulate in nodules under the epithelial lining of the stomach. The third-stage larvae of *Rugopharynx rosemariae* penetrate the glandular mucosal surface of the stomachs of eastern and western grey kangaroos (*Macropus giganteus* and *M. fuliginosus*) to form numerous small crateriform lesions, often covering the surface of the stomach of juvenile kangaroos (Beveridge & Presidente 1978; Arundel et al. 1990). The tails of 1–3 white robust larval nematodes protrude from each lesion.

An undescribed species of *Strongyloides* occurs in the gastric mucosa of macropods, but only in regions with cardiac glandular mucosa. The nematodes are tiny and are not readily seen at necropsy, but in heavily infected kangaroos the gastric mucosa may appear erythematous. In the agile wallaby (*M. agilis*) a hyperplastic response occurs with the formation of isolated nodules of hyperplastic cardiac glands. Severe infections may be fatal (Winter 1954; Arundel et al. 1979; Speare et al. 1989).

The small pyloric area of the stomach of kangaroos is inhabited by the trichostrongyloid nematode genus *Filarinema*. These fine red nematodes are easily overlooked at necropsy, but in the red kangaroo (*M. rufus*) they may occur in large numbers. Encapsulated nematodes can be found in small elevated granulomata on the surface of the glandular epithelial lining of the forestomach (Arundel et al. 1979).

4.3.3 Small intestine

The duodenum of most wallabies is parasitised by species of the trichostrongyloid nematode genus *Austrostrongylus*. The nematodes are small, red and coiled spirally around villi. They are not known to be

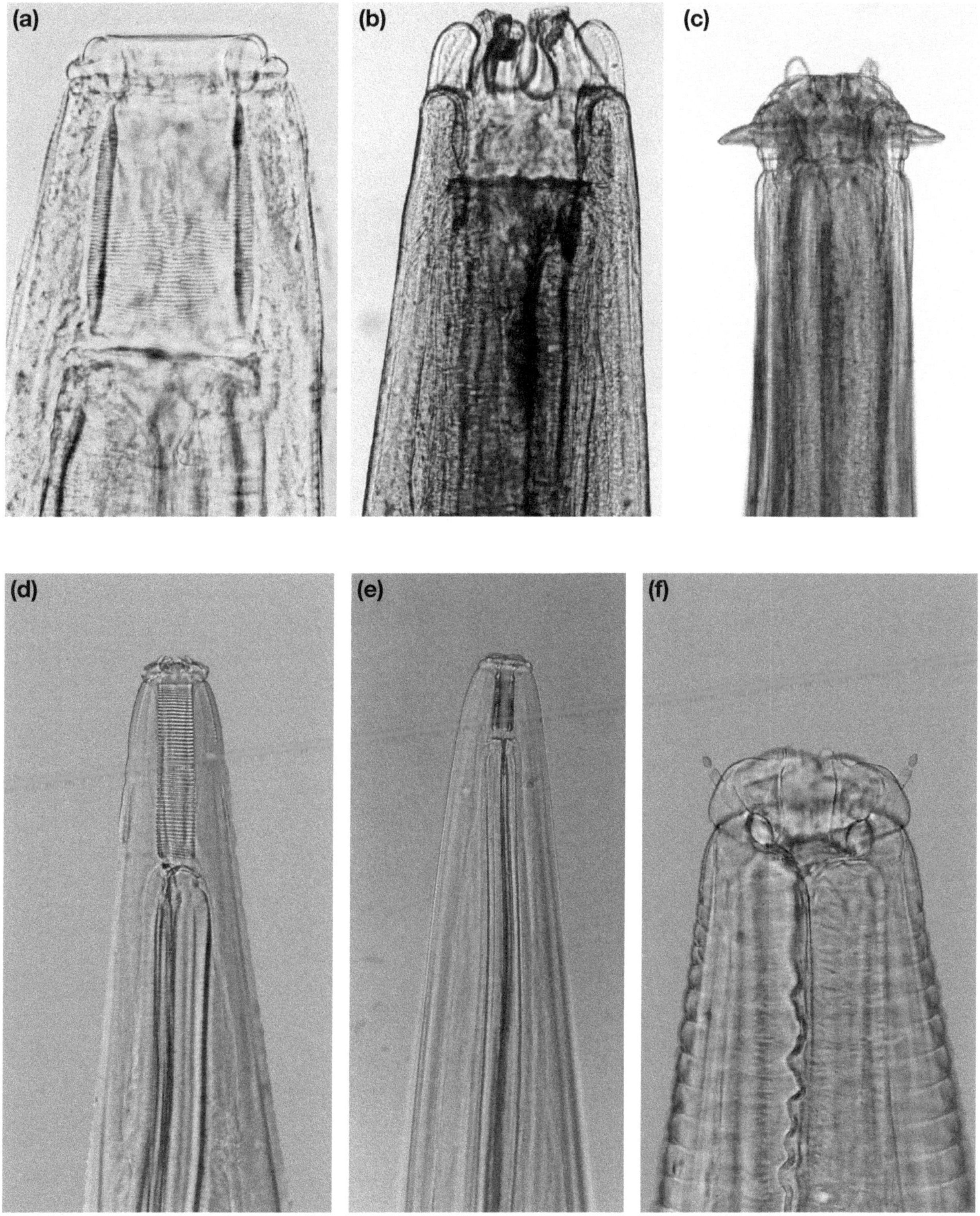

Figure A2.5 Anterior extremities of some common Genera and species of nematodes found in the stomachs of kangaroos and wallabies. All are strongylid nematodes and belong to the endemic Subfamily Cloacininae. a) *Rugopharynx rosemariae* from the eastern grey kangaroo, *Macropus giganteus*. b) *Labiostrongylus kungi* from the grey kangaroo. c) *Parazoniolaimus collaris* from the swamp wallaby, *Wallabia bicolor*. d) *Pharyngostrongylus kappa*. e) *Rugopharynx macropodis*. f) *Cloacina obtusa*, all are from the eastern grey kangaroo.

pathogenic. In contrast, a related trichostrongyloid genus, *Globocephaloides*, which is larger and red in colour and which also occurs in the duodenum, feeds on the host's blood and large numbers may kill juvenile kangaroos. *G. trifidospicularis* has been identified as a cause of anaemia and mortality in eastern grey kangaroos (Arundel et al. 1977) in south-eastern Australia and its counterpart in north-eastern Australia, *G. macropodis*, has been reported to kill juvenile eastern grey kangaroos maintained in a confined area with adult agile wallabies and pretty-faced wallabies (*M. parryi*) (Beveridge et al. 1984). The genus is readily identified by its large buccal capsule (Fig. A2.6a). A very similar genus, *Amphicephaloides* (Fig. A2.6b), occurs in pademelons of the genus *Thylogale* but has not been associated with anaemia or mortality.

Several species of anoplocephalid cestodes, all belonging to the genus *Progamotaenia*, occur in the intestines of macropods. In spite of their impressive size, they are not known to be pathogenic.

Lesions due to coccidia frequently occur in the small intestine of young macropods or in adults when animals are in stressful situations. Numerous white focal lesions may be evident on the external surface of the intestine and blood is frequently found in the lumen as a result of schizont rupture. The lesions have been described in detail by Barker et al. (1972).

4.3.4 Bile ducts and liver

Anoplocephalid cestodes of the genus *Progamotaenia* are common inhabitants of the bile ducts of kangaroos and wallabies. The most common species is *P. festiva*. Lesions are usually mild and consist of thickening of the bile duct wall, glandular hyperplasia and fibrosis (Presidente & Beveridge 1978). However, the most severe inflammatory response in macropods is seen in the northern nail-tail wallaby (*Onychogalea unguifera*) (Speare et al. 1989).

In south-eastern Australia, where macropod habitat overlaps with that of ruminants, macropods may be infected with *Fasciola hepatica*. Infections may be severe in small macropodids such as the Tasmanian pademelon (*Thylogale billardierii*) but flukes are reasonably well-tolerated by larger kangaroos. Substantial fibroplasia around bile ducts is seen (Spratt & Presidente 1981).

An eosinophilic vasculitis of portal venules caused by the muspiceoid nematode *Durikainema macropi* in terminal branches of the portal veins appears as pale yellow lines perpendicular to the distal edges of liver lobes in several species of macropods (Spratt & Speare 1982; Speare & Spratt 1982). Gross pathology of the liver in the advanced stage and histopathology of the liver in all stages is pathognomic. *D. macropi* causes an eosinophilic phlebitis of the terminal hepatic venules, often with obliteration, a feature which does not occur in any other known condition in macropods.

In the tammar wallaby (*M. eugenii*) hepatic lesions can be caused by a species of *Eimeria* (Munday 1988).

4.3.5 Caecum and colon

Hypodontus macropi (Fig. A2.6c) is a hookworm-like nematode which occurs in the caecum of wallabies, but in kangaroos its distribution extends into the terminal ileum. It feeds on the intestinal mucosa and can cause bleeding at sites of attachment. Its pathogenicity has not been determined, but heavy burdens have the potential to cause anaemia and hypoproteinaemia (Beveridge 1979). In eastern grey kangaroos a large nematode species, *Paramacropostrongylus toraliformis* (Fig. A2.6d), is found in the caecum. Larvae develop in nodular lesions found at the tip of the caecum. Species of *Macropostrongyloides* (Fig. A2.6e) also occur in the caecum or colon of several species of large kangaroos and appear to be blood feeders (Beveridge & Mawson 1978). Heavy burdens may therefore be pathogenic. In many species of kangaroos large numbers of pin-worms belonging to the genus *Macropoxyuris* (Fig. A2.6f) may be found in the colon. Even when present in very large numbers, these nematodes appear to have little effect. The trematode *Macropotrema pertinax* occurs in the caecum and colon of the agile wallaby (Blair et al. 1978) and causes a mild inflammatory response, and mucosal atrophy at attachment sites.

4.3.6 Lungs

Several species of lungworm belonging to the genus *Marsupostrongylus* occur in macropods. Most cause an interstitial pneumonia but infections are generally subclinical (McColl & Spratt 1982; Beveridge et al. 1985). A species of *Capillaria* caused verminous bronchitis, bronchiolitis and death in two colonies of captive rufous rat-kangaroos (*Aepyprymnus rufescens*), eastern bettongs (*Bettongia gaimardi*) and long-nosed potoroos (*Potorous tridactylus*) but has

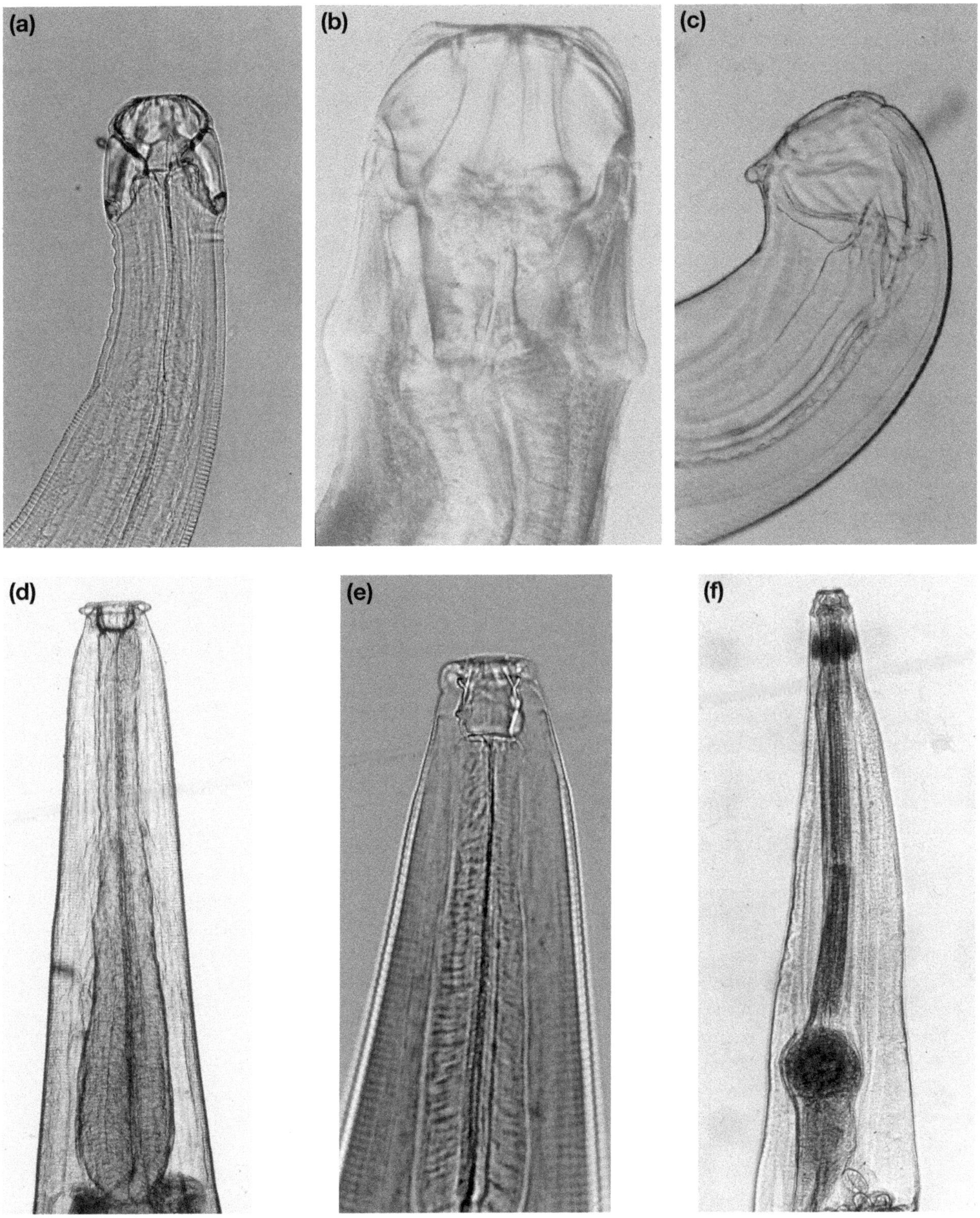

Figure A2.6 Anterior extremities of some of the common Genera of nematodes found in the small and large intestines of kangaroos and wallabies. a) *Globocephaloides trifidospicularis* from the duodenum of the grey kangaroo. b) *Amphicephaloides thylogale* from the duodenum of pademelons, *Thylogale* spp. c) *Hypodontus macropi* found in the ileum, caecum and colon of a wide range of macropodid species. d) *Paramacropostrongylus toraliformis* from the caecum of the grey kangaroo. e) *Macropostrongyloides baylisi* from the colon of the grey kangaroo. f) *Macropoxyuris* spp., a pin-worm from the colon of the grey kangaroo.

not been seen in wild animals (Love & Reddacliff 1992). The inflammatory cell exudate was dominated by lymphocytes, macrophages and plasma cells, but not eosinophils.

Cysts of *Echinococcus granulosus* are also common in the lungs of a variety of macropods, with the black-striped wallaby (*Macropus dorsalis*) and the swamp wallaby being major intermediate hosts in the sylvatic life-cycle of this parasite (Kumaratilake & Thompson 1982; Jenkins & Morris 2003; Jenkins 2006). Infections may be fatal in rock-wallabies (*Petrogale* spp.) and nail-tailed wallabies (*Onychogalea* spp.) (Johnson et al. 1998).

The kangaroo bot *Tracheomyia macropi* occurs attached firmly to the tracheal mucosa of red kangaroos and causes erosions, ulcerations and an inflammatory response (Mykytowycz 1963; Speare et al. 1989).

4.3.7 Heart

A large heartworm, *Breinlia ventricola*, has been found in red kangaroos in Western Australia (Spratt & Hobbs 2004).

Large numbers of this nematode are likely to be pathogenic in a manner similar to heartworm in dogs. Adults and immatures of *Durikainema macropi* have been found in the intracardiac venules and epicardial lymphatics of macropods (Speare & Spratt 1982; Spratt & Speare 1982) causing an epicardial lymphangitis. The associated pathological lesions are not known from any other disease in macropodids.

4.3.8 Body cavities and subcutaneous tissues

Several species of filarioid nematodes belonging to the genera *Breinlia* and *Johnstonema* occur in the pleural and peritoneal cavities and in SC and intermuscular connective tissues of kangaroos and wallabies (Spratt & Varughese 1975). The most significant of these, in terms of its impact on the game meat trade, is *Pelecitus* (formerly *Dirofilaria*) *roemeri* occurring in the SC and intermuscular connective tissues of the hind limbs, especially around the stifle joint, and occasionally in the SC fascia overlying shoulder, hips and abdomen (Spratt 1972). The large worms are trimmed out in meat destined for human consumption. Numerous species of tabanid flies or 'march flies' (Diptera: Tabanidae) act as intermediate host and vector of the nematode (Spratt 1974). *Pelecitus roemeri* is the only filarioid known from kangaroos and wallabies with sheathed microfilariae. These occur in the blood of some but not all species of kangaroo and wallaby hosts infected with this nematode.

4.3.9 Blood vessels and blood

Adults of *D. macropi* also occur in portal veins and venules of macropods and immatures were seen in the intradermal lymphatics of skin in the agile wallaby (Spratt & Speare 1982; Speare & Spratt 1982).

Theileria peramelis and *Babesia* spp. have been reported from the long-nosed potoroo and the Proserpine rock-wallaby respectively but piroplasms are not known from other potoroid and macropodid species (Clark et al. 2004). *Trypanosoma* spp. has been detected by PCR methods in the blood of an eastern grey kangaroo (Noyes et al. 1999).

4.3.10 Skin

Thadeua serrata is a minute mesostigmatid mite found in discrete crateriform lesions in the inguinal region of the red-necked wallaby (*M. rufogriseus*), Proserpine rock-wallaby, brush-tailed rock-wallaby (*P. penicillata*) and swamp wallaby and may be associated with moderate to severe dermatitis (Fig. A2.2c). Two additional species occur on wallabies, one of which occurs on the red-legged pademelon (*Thylogale stigmatica*), and two others on potoroos but associated pathology has not been reported.

Macropods are frequently infected with biting lice (Mallophaga). The principal genus encountered, *Heterodoxus*, containing numerous species, is found on large kangaroos and wallabies. The genus *Latumcephalum*, containing two species, is restricted to swamp wallabies while the genus *Paraboopia*, with a single species (*P. flava*) occurs on wallaroos. *Macrophilia* (four species) is restricted to pademelons (*Thylogale* spp.) while *Paraheterodoxus* (three species) occurs on bettongs and potoroos (von Kéler 1971). All belong to a single Family, the Boopidae. The complex array of species of *Heterodoxus* found on rock-wallabies has been included in attempts to elucidate the phylogeny of the various species of their rock-wallaby hosts (Barker & Close 1990).

Although originally termed wallaby flies, hippoboscid flies or 'keds' of the genera *Austrolfersia* and *Ortholfersia* occur commonly in the fur of kangaroos, wallabies, pademelons and rat-kangaroos. They are not pathogenic (Maa 1962). Unlike the sheep ked they are not wingless and frequently alight on humans handling animals in the wild or walking in the habitat of macropodoids.

4.4 Koalas (*Phascolarctos cinereus*)

4.4.1 Intestine

The small intestine of koalas is parasitised by a large anoplocephalid cestode, *Bertiella obesa*. While infections with anoplocephalids are usually not pathogenic, heavy infections with *B. obesa* have been associated with enteritis in koalas (Dickens 1976).

4.4.2 Lungs

Severe interstitial pneumonia has been associated with the lungworm *Marsupostrongylus longilarvatus*, a species which occurs more commonly in kangaroo and wallaby hosts (McColl & Spratt 1982).

4.4.3 Blood vessels

A muspiceoid nematode, *Durikainema phascolarcti*, occurs in large numbers in the pulmonary arteries of koalas, with nematodes occupying >95% of the diameter of pulmonary vessels in some transverse histological sections (Spratt & Gill 1998). First-stage larvae are seen in the blood and in histological sections of many organs.

4.4.4 Skin

Anaemia resulting from heavy tick infestation has been reported in the koala (Dickens 1976; Stone & Carrick 1991; Spencer & Canfield 1994). A heavy infestation of 25–50 ticks on a 5 kg koala may remove up to 125–250 mL of blood or 25–50% of static blood volume (Stone & Carrick 1991). The feeding process involves at least a five-fold concentration of the blood meal (Kemp et al. 1983).

4.5 Wombats

4.5.1 Small intestine

Strongyloides spearei is a tiny white nematode found in the intestinal epithelium of the common wombat (*Vombatus ursinus*) (Skerratt 1995). Although disease has not been described in wombats in the wild, the proclivity of species of the genus to cause disease in domestic animals and humans suggests that the species occurring in common wombats may be pathogenic. This was confirmed by a case of wasting and hypoproteinaemia in an immature common wombat in captivity in north Queensland, which presented with a heavy infection, cured by anthelmintic treatment (R Speare unpub).

The large anoplocephalid cestode *Paramoniezia johnstoni* occurs in both common and southern hairy-nosed wombats (*Lasiorhinus latifrons*) while a smaller species, *Phascolotaenia comani*, occurs only in the common wombat. In spite of their size, neither species is thought to be pathogenic.

Eimeria arundeli in common wombats has been associated with disease (Hum et al. 1991). Other species of *Eimeria* have been described from both common and hairy-nosed wombats (Barker et al. 1979) but have not been associated with disease.

4.5.2 Colon

The colon of common and southern hairy-nosed wombats is inhabited by large numbers (up to 6000) of strongyloid nematodes of the genera *Oesophagostomoides* and *Macropostrongyloides* (Fig. A2.7a, c). The genus *Phascolostrongylus* is restricted to common wombats (Fig. A2.7b). *Oesophagostomoides* and *Phascolostrongylus* both feed on gut contents and are probably not pathogenic. *Macropostrongyloides* appears to feed on blood and may therefore be a potential pathogen (Beveridge & Mawson 1978).

4.5.3 Lungs

A relatively robust lungworm, *Marsupostrongylus coulstoni*, occurs in the airways of the common wombat (Spratt 1979) but there is as yet no association with significant pulmonary disease (Skerratt 1998).

Hydatid cysts have been reported only in common wombats from Victoria (Jenkins 2006). The low prevalence or absence of hydatid infection in wombats from many areas implies that, unlike kangaroos and wallabies, they are an intermediate host of only local importance.

4.5.4 Liver

Common wombats infected with *Fasciola hepatica* tend to exhibit extreme fibrosis of the bile ducts and liver parenchyma through which flukes have migrated (Spratt & Presidente 1981). Although common wombats are

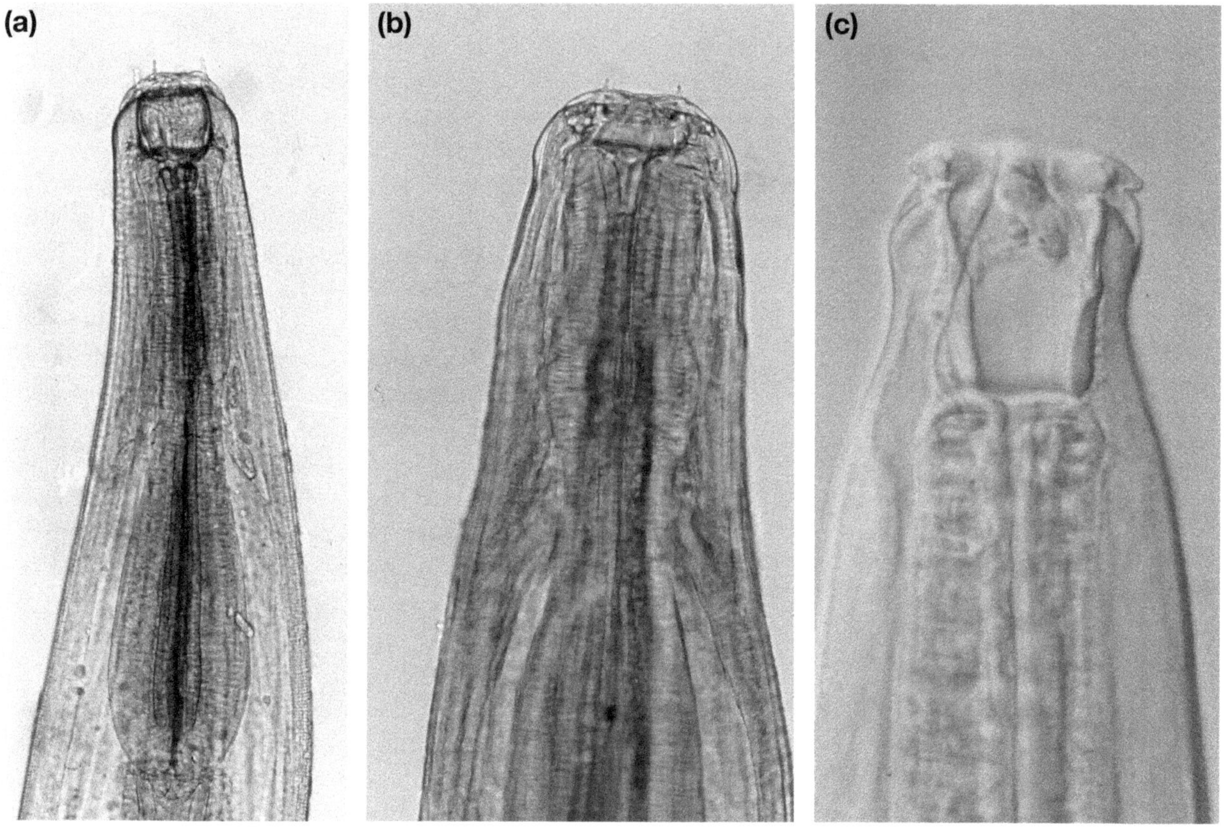

Figure A2.7 Anterior extremities of the three species of strongylid nematodes commonly found in the colons of wombats. a) *Oesophagostomoides* spp., found in both common wombat (*Vombatus ursinus*) and southern and northern hairy-nosed wombats (*Lasiorhinus latifrons* and *L. krefftii*). b) *Phascolostrongylus turleyi*, found only in the common wombat. c) *Macropostrongyloides lasiorhini*, found in both common and southern hairy-nosed wombats.

not likely to be a major reservoir host, the pathological reaction appears to be more extreme than in most other marsupials (PJ Presidente *pers. comm.*).

4.5.5 Blood

Trypanosoma spp. has been detected by PCR methods in the blood of a common wombat (Noyes et al. 1999) but the effect on the host is unknown.

4.5.6 Skin

The most significant parasitic disease of common wombats is sarcoptic mange due to the introduced parasite *Sarcoptes scabiei* (Skerratt et al. 1998, 2002) (Fig. A2.2d). It also occurs in southern hairy-nosed wombats but has not been reported from northern hairy-nosed wombats (*Lasiorhinus krefftii*) (Martin et al. 1998). Principal lesions observed are severe parakeratosis, alopecia, excoriation and fissuring of the parakeratotic crust and skin (Skerratt 2003a). In skin scrapings, mites are generally abundant. Eosinophils may be present in the dermis, suggesting that a Type I (immediate) hypersensitivity response may develop after a Type IV (delayed) hypersensitivity response and some immune tolerance may develop with severe infections (Skerratt 2003b).

The tick *Aponomma auruginans* is found frequently on common wombats. Tick burdens may be quite heavy and a significant reaction to tick bites may occur (Skerratt 1998). Other species of tick such as *Ixodes tasmani*, *I. cornuatus* and *I. victoriensis* are also found on common wombats.

Two species of *Lycopsylla* fleas occur only on wombats. Stick-fast fleas (*Echidnophaga* spp.) are common on wombats, particularly the southern hairy-nosed wombat, on which they occur in clusters along the vental abdomen. Five species occur on this host, three of them restricted to that host species. Heavy infections may cause mild dermatitis (Dunnet & Mardon 1974).

4.6 Dasyurids

4.6.1 Stomach

The stomachs of quolls (*Dasyurus* spp.) are parasitised by large spirurid nematodes of the genera *Cyathospirura* and *Cyclicospirura*. The nematodes often occur in nodules in the gastric epithelium. In antechinuses (*Antechinus* spp.) the spirurid nematode *Synhimantus* occurs in the lumen of the stomach, while *Antechiniella* has its anterior extremity embedded in elongate sleeve-like lesions which extend into the peritoneal cavity. Species of *Capillaria*, multi-intracellular inhabitants of the gastric mucosa, have been suggested as potential precipitating causes of the gastric ulceration commonly seen in males during the breeding season (Barker et al. 1978). Prevalence and intensity of helminth infection in the agile antechinus (*Antechinus agilis*) increased markedly in June–August leading up to the breeding season and die-off of males, and males were more heavily parasitised than females during this period (Beveridge & Barker 1976). However, it is unlikely that helminths were directly involved in the seasonal mortality of males. Deaths were related primarily to gastrointestinal haemorrhage, listeriosis and possibly babesiosis associated with an adrenocortical response to stress and reduced resistance to infection or latent disease (Barker et al. 1978).

4.6.2 Intestine

The small intestine of dasyurids is inhabited by a number of species of trichostrongyle nematodes of the genera *Dessetostrongylus, Patricialina* and *Woolleya*. None is known to be pathogenic. The Tasmanian devil (*Sarcophilus harrisii*) is frequently parasitised by large numbers of the cestode *Anoplotaenia dasyuri* (Gregory et al. 1975). The cestode is not known to be pathogenic and the larval stage occurs in the musculature of wallabies and potoroos (*Potorous* spp.). Tasmanian devils and quolls are also parasitised by species of the cestode genus *Dasyurotaenia* which embed their scoleces deeply into the mucosa of the small intestine, forming conspicuous elevations on the serosal surface. A sylvatic or wildlife cycle of *Trichinella pseudospiralis* occurs in Tasmanian devils, spot-tailed quolls (*Dasyurus maculatus*) and eastern quolls (*D. viverrinus*) in Tasmania, independent of domestic animals or humans (Obendorf et al. 1990). Minute (1–1.5 mm) adult males and females occur in the intestine and larvae in the skeletal muscles of these hosts. Transmission is dependent upon natural predator-prey associations with carnivores consuming infected animals as live prey or carrion. Carnivorous and carrion-feeding birds are also infected with *T. pseudospiralis* (Obendorf & Clarke 1992).

4.6.3 Body cavities and vessels

Larval stages of an ascarid nematode of pythons, *Ophidascaris robertsi*, occur commonly in the tissues and organs of small dasyurids. Their large size (7–8 cm) and activity may be highly debilitating (they sometimes protrude out through the skin), rendering the host susceptible to predation.

The filarioid nematode *Sprattia capilliformis* occurs below the liver in the superior mesenteric and portal veins of the northern quoll (*Dasyurus hallucatus*) and has sheathed microfilariae occurring in the blood.

4.6.4 Nasal cavity

The trichostrongyloid nematode *Nasistrongylus antechini* occurs in the nasal sinuses of the agile antechinus, brown antechinus (*A. stuartii*) and the dusky antechinus (*A. swainsonii*) causing erosion of the epithelium and necrosis of the turbinate bone (Beveridge & Barker 1975).

4.6.5 Blood

The haemogregarine apicomplexans *Hepatozoon dasyuroides* and *H. dasyuri* occur in erythrocytes of the kowari (*Dasycercus byrnei*) and the eastern quoll respectively (Clark et al. 2004). An unidentified species of *Babesia* occurs in the agile antechinus and its recrudescence in post-mating males may contribute to anaemia (Barker et al. 1978).

4.6.6 Skin

The flea *Uropsylla tasmanica*, occurring on eastern and spot-tailed quolls and the Tasmanian devil in Victoria and Tasmania, is unusual in that the larvae are parasitic on the host (Dunnet 1970). Eggs are cemented to the fur and when hatched the first-instar larvae use their large mandibles to burrow into the skin. The body of subsequent larval instars is modified with a narrow head and expanded trunk segments, permitting them to live in burrows in the skin with only the posterior segments and spiracles protruding. On maturation they drop to the ground and spin a cocoon.

Various dasyurid species are parasitised by biting lice belonging to the Family Boopidae. The genus *Boopia* occurs on species of *Dasyurus*, while the related genus *Phacogalia* is found on species of *Antechinus* (von Kéler 1971).

4.7 Possums and gliders

4.7.1 Stomach

Large spirurid nematodes belonging to the genera *Gongylonema, Protospirura, Physaloptera* and *Pseudophysaloptera* are occasionally encountered in the stomachs of possums and gliders but are not known to be pathogenic.

4.7.2 Small intestine

The duodenum of some possums is parasitised by small red trichostrongyloid nematodes of the genus *Paraustrostrongylus*. They are not known to be pathogenic. Species of *Parastrongyloides* also occur in the small intestine, but are tiny and not usually visible to the naked eye. Larvae occur in faeces and may be encountered if Baermann examinations are undertaken.

Cestodes of the anoplocephalid genus *Bertiella* are common in possums and gliders but are not known to be pathogenic.

4.7.3 Large intestine

In many species of possums and gliders there are few or no parasites in the large intestine. However, in the greater glider (*Petauroides volans*) oxyurid nematodes of the genera *Austroxyuris* and *Paraustroxyuris* may occur in millions, forming a thick layer over the mucosa. In spite of the large numbers of parasites, they do not appear to be pathogenic and may play a role in detoxification of the eucalypt leaf diet.

4.7.4 Lungs

Lungworms belonging to the genera *Marsupostrongylus* and *Filostrongylus* occur in the lungs of brushtails (*Trichosurus* spp.), the common ringtail (*Pseudocheirus peregrinus*) and the greater glider, associated with a variable degree of interstitial pneumonia.

4.7.5 Subcutis

The filarioid nematode *Cercopithifilaria johnstoni* has skin-inhabiting microfilariae and is transmitted by ticks of the genus *Ixodes*. It occurs in the greater glider and the common brushtail (*T. vulpecula*) (Spratt & Haycock 1988).

4.7.6 Blood vessels and blood

In contrast to the situation in the northern quoll, the filarioid nematode *Sprattia venacavincola* occurs above the liver in the vena cava and hepatic veins of brushtail possums (Spratt & Varughese 1975). It also has sheathed microfilariae in the blood.

A muspiceoid nematode, *Durikainema* sp., occurs in the pulmonary arteries of common brushtails in Tasmania and the northern brushtail (*T. arnhemensis*) in Darwin and is associated with focal pneumonitis (Obendorf et al. 1997).

Species of *Hepatozoon* occur in erythrocytes of the sugar glider (*Petaurus breviceps*), squirrel glider (*P. norfolcensis*), Herbert River ringtail (*Pseudochirulus herbertensis*) and common ringtail.

4.7.7 Skin

Ticks of the genera *Ixodes* and *Haemaphysalis* and fleas of the genera *Echidnophaga, Choristopsylla, Pygiopsylla* and *Acanthopsylla* all occur, but no lice are known from possums and gliders. Adult mites and larval trombiculid mites (chiggers) represent the most diverse ectoparasites of these hosts and the laelapid mite *Trichosurolaelaps crassipes*, when present in large numbers, may cause substantial irritation known as 'rump wear' in brushtails (Presidente 1984).

4.8 Bandicoots and bilbies

The parasites of bilbies (*Macrotis* spp.) are relatively unknown. The following information relates primarily to bandicoots (*Isoodon* spp. and *Perameles* spp.).

4.8.1 Stomach

Large spirurid nematodes of the genus *Physaloptera* occur commonly in the stomachs of bandicoots. Heavy burdens have been considered significant in the case of the eastern barred bandicoot (*Perameles gunnii*) (Lenghaus et al. 1990). The trichostrongyloid nematode *Peramelistrongylus* is also found commonly in the stomach but is not known to be pathogenic.

4 8.2 Small intestine

Trichostrongyloid nematodes of the genera *Mackerrastrongylus, Asymmetracantha* and *Beveridgiella* are common parasites of the duodenum of bandicoots. None is known to be pathogenic. Seuratoid nematodes of the genus *Linstowinema* occur in the intestines of bandicoots. The anterior ends of these nematodes are covered with large spines but in spite of their appearance they are not thought to be pathogenic. Numerous species of capillariid nematodes belonging to the genus *Eucoleus* occur in practically all epithelial tissues of bandicoots including the lips, scrotum, tongue and mucosal surfaces of the entire gastrointestinal tract.

Cestodes commonly encountered in bandicoots are anoplocephalids of the genus *Paralinstowia* and hymenolepidids of the genus *Potorolepis*. Again, none is considered pathogenic.

Acanthocephalans of the genus *Australiformis* are also found in the small intestine. They are large and there is a propensity for damage to the intestinal wall from the large hook-covered proboscis, but there are currently no reports of pathogenic effects.

4.8.3 Lungs

Lungworms of the genera *Filostrongylus* and *Marsupostrongylus* occur in the lungs of bandicoots and are associated with a variable degree of interstitial pneumonia (McColl & Spratt 1982). A highly characteristic pneumospirurid nematode, *Metathelazia naghiensis*, with a large buccal capsule with six prominent teeth, three pairs of large lips and irregular longitudinal cuticular ridges and tubercles over the body surface occurs in the southern brown bandicoot (*Isoodon obesulus*), eastern barred bandicoot and long-nosed bandicoot (*Perameles nasuta*) (Spratt 1980) but associated pathology has not been reported.

4.8.4 Blood

Trypanosoma thylacis occurs in the blood of the northern brown bandicoot (*I. macrourus*) and two undescribed species of *Trypanosoma* may occur in the blood of the southern brown and eastern barred bandicoots (Mackerras 1959; Bettiol et al. 1998). The haemogregarine *Hepatozoon peramelis* occurs in erythrocytes of the southern brown and long-nosed bandicoots and an undescribed species occurs in erythrocytes of the eastern barred bandicoot (Clark et al. 2004). The piroplasm *Babesia thylacis* occurs in erythrocytes of the southern brown bandicoot. *Theileria peramelis* occurs in erythrocytes and lymphocytes of the southern brown and long-nosed bandicoots and an undescribed species occurs in the northern brown bandicoot.

4.8.5 Skin

Ticks of the genera *Ixodes* and *Haemaphysalis* occur on bandicoots, species of the former genus acting as intermediate host and vector of the filarioid nematode *Cercopithifilaria johnstoni* which occurs SC in the two species of brown, the long-nosed and the eastern barred bandicoots of eastern Australia. These four species of bandicoot are considered the natural host of the paralysis tick *Ixodes holocyclus*. All three stages occur on these hosts although one or two stages may be found on a wide variety of monotreme, marsupial, rodent and avian hosts (Roberts 1970).

4.9 Bats

4.9.1 Intestine

The helminth parasites of Microchiropterans in Australia have been very poorly studied, with few records of parasites apart from trichostrongyles, dendrolecithid trematodes and one species of cestode. Megachiropterans appear to have few parasites, the most striking being the ascaridoid nematode *Toxocara pteropi*. Transmission appears to occur primarily by the transmammary route, such that infections are usually encountered in juvenile bats (Prociv 1985).

4.9.2 Blood

Piroplasms of the Genera *Hepatocystis* and *Polychromophilus* occur in erythrocytes of pteropid, hipposiderid and vespertilionid bats (Clark et al. 2004). *Trypanosoma pteropi* and *T. hipposideri* occur in pteropid and hipposiderid bats.

4.9.3 Skin

The paralysis tick *Ixodes holocyclus* is a significant pathogen in flying-fox (*Pteropus* spp.) colonies on the Atherton Tablelands, causing an ascending flaccid paralysis (Campbell et al. 2003). Bats fall from their roosts and die from progressive respiratory muscle paralysis, hypothermia (to which bats have a physiological propensity) or predation.

Spider-like bat flies (Nycteribiids) are commonly encountered crawling through the fur of bats and may transmit the piroplasms noted above but the effects of both the protozoans and the flies on bat hosts have not been studied.

4.10 Dingoes (*Canis lupus dingo*)

The parasites of dingoes are essentially similar to those of domestic dogs, with some notable exceptions.

4.10.1 Stomach

Species of the spirurid nematodes *Cyathospirura* and *Cylicospirura* are commonly found in nodules in dingo stomachs.

4.10.2 Intestine

Of greatest importance, because of its effect on intermediate hosts and risk to domestic animal and public health, is the cestode *Echinococcus granulosus*, which is hyperendemic in dingoes/wild dogs in south-eastern Australia (Jenkins & Morris 2003). Caution should be exercised in performing necropsies on dingoes because mature eggs of *E. granulosus* are infective directly from the tapeworm and ingestion of eggs may lead to development of hydatid cysts. Control methods for this parasite in domestic dogs have proved effective and it rarely occurs in them.

4.10.3 Trachea

The metastrongyloid nematode *Oslerus osleri* occurs commonly in nodules at the bifurcation of the trachea in dingoes, especially in south-eastern Australia, presumably because the suspected mode of direct transmission is regurgitative feeding of pups during weaning (Dunsmore & Spratt 1979). It has been reported in domestic dogs on only three occasions. These lesions may be severe and are likely to affect respiratory capacity.

4.11 Pinnipeds

4.11.1 Stomach

Large numbers of the ascaridoid nematode *Contracaecum ogmorhini* (also reported as *C. osculatum*) occur commonly in the stomachs of fur-seals (*Arctocephalus* spp.) and the Australian sea-lion (*Neophoca cinerea*). Heavy infections may be associated with gastric ulceration (Arundel 1978).

4.11.2 Small intestine

The small intestines of Australian pinnipeds often contain large numbers of the pseudophyllidean cestode *Diphyllobothrium arctocephalinum*. Although the cestodes may be more than 1 m long, they are not thought to be pathogenic.

A wide variety of species of *Diphyllobothrium* occur in Antarctic fur-seals (*Arctocephalus gazella*).

The hookworm parasite *Uncinaria hamiltoni* is found in the small intestine of the pups of the Australian sea-lion (*Neophoca cinerea*) and fur-seals (*Arctocephalus* spp.). It may cause anaemia and death in pups. It is probably transmitted trans-colostrally, based on experimental studies with related nematode species in seals in the northern hemisphere (Olsen & Lyons 1965).

Large numbers of acanthocephalans of the genus *Corynosoma* also occur in the intestines of pinnipeds. Since they attach themselves to the intestinal wall with an armed proboscis, there is some possibility that heavy infections may be pathogenic.

4.11.3 Lungs

The lungworm genus, *Parafilaroides*, occurs in Australian sea-lions and causes a significant interstitial pneumonia (Nicholson & Fanning 1982). *Parafilaroides hydrurgae* occurs in the leopard seal (*Hydrurga leptonyx*) although the pathology associated with this species is unknown.

4.11.4 Nasal cavities

The nasal cavities, trachea and bronchi of fur-seals and the Australian sea-lion are parasitised by mites of the genus *Orthohalarachne*. The mites are very large and are clearly visible to the naked eye. Nicholson and Fanning (1982) described lesions in the nasopharynx and turbinates of the Australian sea-lion associated with the mites. Seawright (1964) reported extensive alveolar collapse and excessive accumulation of mucocatarrhal exudate causing occlusion of small bronchi and bronchioles in the same host species infected with *O. attenuata*, resulting in death by asphyxiation.

4.11.5 Blubber and peritoneal cavity

Large cysts containing larval tapeworms are frequently found in the blubber and peritoneal cavity of pinnipeds. Currently referred to as *Monorygma grimaldii* and *Phyllobothrium delphini*, they are probably the larvae of unidentified tetraphyllidean cestodes occurring in the intestines of large sharks (Norman 1997).

4.12 Dugongs (*Dugong dugon*)

4.12.1 Stomach

The 12 cm ascarid nematode *Paradujardinia halicoris* occurs in large numbers in the stomach of the dugong, especially in the cardiac gland, but is not associated with a tissue response (Blair 1981a; Campbell & Ladds 1981). This is the only nematode parasite known from the dugong. Species of the opsthotrematid trematode

Lankatrema occur encapsulated in the wall of the stomach, cardiac gland and duodenum and may be responsible for enteritis (Blair 1981a; Campbell & Ladds 1981).

4.12.2 Caecum and large intestine

Paramphistome trematodes belonging to the genus *Indosolenorchis* occur in the caecum and first part of the large intestine of the dugong (Blair 1980) but their pathogenicity is unknown.

4.12.3 Lungs and associated airways

Four opisthotrematid trematodes representing the genera *Cochleotrema, Opisthotrema* and *Pulmonicola* occur in the lungs, eustachian tubes, middle ears and nasal passages of the dugong but their pathogenicity is unknown (Blair 1981b).

4.12.4 Skin

The monostome trematode *Labicola elongata* occurs commonly in abscesses in the upper lip of the dugong (Blair 1979).

REFERENCES

Arundel JH (1978). Parasites and parasitic diseases of Australian marine mammals. In *Fauna: Part B, Proceedings 36*, pp. 323–333. Post Graduate Committee in Veterinary Science: Sydney.

Arundel JH, Barker IK & Beveridge I (1977). Diseases of marsupials. In *The Biology of Marsupials* (eds B Stonehouse & D Gilmore), pp. 141–154. Macmillan: London.

Arundel JH, Beveridge I & Presidente PJA (1979). Parasites and pathological findings in enclosed and free-ranging populations of *Macropus rufus* (Desmarest) (Marsupialia) at Menindee, New South Wales. *Australian Wildlife Research* **6**, 361–379.

Arundel JH, Dempster KJ, Harrigan KE & Black R (1990). Epidemiological observations on the helminth parasites of *Macropus giganteus* Shaw in Victoria. *Australian Wildlife Research* **17**, 39–51.

Barker SC & Close RL (1990). Zoogeography and host associations of the *Heterodoxus octoseriatus* group and *H. ampullatus* (Phthiraptera: Boopidae) from rock-wallabies (Marsupialia: Petrogale). *International Journal for Parasitology* **20**, 1081–1087.

Barker IK, Harrigan KE & Dempster JK (1972). Coccidiosis in wild grey kangaroos. *International Journal for Parasitology* **2**, 187–192.

Barker IK, Beveridge I, Bradley AJ & Lee AK (1978). Observations on spontaneous stress-related mortality among males of the dasyurid marsupial *Antechinus stuartii* Macleay. *Australian Journal of Zoology* **26**, 435–447.

Barker IK, Munday BL & Presidente PJA (1979). Coccidia of wombats: correction of host–parasite relationships. *Eimeria wombati* (Gilruth and Bull, 1912) comb. nov. and *Eimeria ursini* Supperer, 1957 from the hairy-nosed wombat and *Eimeria arundeli* sp. n. from the common wombat. *Journal of Parasitology* **65**, 451–456.

Barker IK, Beveridge I & Munday BL (1985). Coccidia (*Eimeria tachyglossi* n. sp., *E. echidnae* n. sp. and *Octosporella hystrix* n. sp.) in the echidna, *Tachyglossus aculeatus* (Monotremata: Tachyglossidae). *Journal of Protozoology* **32**, 523–525.

Barker IK, O'Callaghan MG, Beveridge I & Close RL (1988a). Host parasite associations of *Eimeria* spp. (Apicomplexa: Eimeriidae) in rock wallabies, *Petrogale* spp. (Marsupialia: Macropodidae). *International Journal for Parasitology* **18**, 353–363.

Barker IK, O'Callaghan MG & Beveridge I (1988b). *Eimeria* spp. (Apicomplexa: Eimeriidae) parasitic in the rat kangaroos *Hypsiprymnodon moschatus, Potorous tridactylus, Aepyprymnus rufescens* and *Bettongia gaimardi* (Marsupialia: Potoroidae). *International Journal for Parasitology* **18**, 947–953.

Barker IK, O'Callaghan MG & Beveridge I (1988c). *Eimeria* spp. (Apicomplexa: Eimeriidae) parasitic in the genera *Setonix, Thylogale, Wallabia, Lagorchestes* and *Dendrolagus* (Marsupialia: Macropodidae). *International Journal for Parasitology* **18**, 955–962.

Barker IK, O'Callaghan MG & Beveridge I (1989). Host–parasite associations of *Eimeria* spp. (Apicomplexa: Eimeriidae) in kangaroos and wallabies of the genus *Macropus* (Marsupialia: Macropodidae). *International Journal for Parasitology* **19**, 241–263.

Bettiol SS, Jakes K, Le DD, Goldsmid JM & Hocking G (1998). First record of trypanosomes in Tasmanian bandicoots. *Journal of Parasitology* **84**, 538–541.

Beveridge I (1979). *Hypodontus macropi* Moennig, 1929, a hookworm-like parasite of macropodid marsupials. *Journal of Helminthology* **53**, 229–244.

Beveridge I (1993). Marsupial parasitic diseases. In *Zoo and Wild Animal Medicine: Current Therapy* (ed. ME Fowler), pp. 288–293. WB Saunders: Colorado.

Beveridge I & Barker IK (1975). *Nicollina antechini* sp. n. (Nematoda: Amidostomatidae) from the nasal cavity of the dasyurid marsupial, *Antechinus stuartii* Macleay, 1841, and associated pathology. *Journal of Parasitology* **61**, 489–493.

Beveridge I & Barker IK (1976). The parasites of *Antechinus stuartii* Macleay from Powelltown, Victoria, with observations on seasonal and sex-related variations in numbers of helminths. *Australian Journal of Zoology* **24**, 265–272.

Beveridge I & Mawson PM (1978). A taxonomic revision of the genera *Macropostrongyloides* Yamaguti and *Paramacropostrongylus* Johnston and Mawson (Nematoda: Trichonematidae) from Australian marsupials. *Australian Journal of Zoology* **26**, 763–787.

Beveridge I & Presidente PJA (1978). *Rugopharynx rosemariae* new species (Nematoda: Pharyngostrongylidae) from grey kangaroos (*Macropus giganteus* and *M. fuliginosus*) with life cycle stages and associated pathology. *International Journal for Parasitology* **8**, 379–387.

Beveridge I, Presidente PJA & Speare R (1985). Parasites and associated pathology of the swamp wallaby, *Wallabia bicolor* (Marsupialia). *Journal of Wildlife Diseases* **21**, 377–385.

Beveridge I, Speare R & Johnson PM (1984). New records of Globocephaloidinae (Nematoda) from Macropodidae in north Queensland. *Transactions of Royal Society of South Australia* **108**, 197–201.

Beveridge I, Friend SCE, Jeganathan N & Charles J (1998). Proliferative sparganosis in Australian dogs. *Australian Veterinary Journal* **76**, 757–759.

Blair D (1979). A new family of monostome flukes (Platyhelminths: Digenea) from the dugong (*Dugong dugon*) (Müller). *Annales de Parasitologie Humaine et Comparée* **54**, 519–526.

Blair D (1980). *Indosolenorchis hirudinaceus* Crusz, 1951 (Platyhelminthes: Digenea) from the dugong, *Dugong dugon* (Müller) (Mammalia: Sirenia). *Annales de Parasitologie humaine et comparée* **55**, 511–525.

Blair D (1981a). Helminth parasites of the dugong, their collection and preservation. In *The Dugong. Proceedings of Seminar/workshop at James Cook University*, 8–13 May 1979 (ed. H Marsh), pp. 275–285. James Cook University of North Queensland: Townsville.

Blair D (1981b). The monostome flukes (Digenea: Families Opisthotrematidae Poche and Rhabdiopoedae Poche) parasitic in sirenians (Mammalia: Sirenia). *Australian Journal of Zoology Supplementary Series* **81**, 1–54.

Blair D, Beveridge I & Speare R (1979). *Macropotrema pertinax* gen. et sp. nov. (Digenea: Paramphistomidae) from a wallaby, *Macropus agilis*, in northern Australia and associated pathology. *Annales de Parasitologie Humaine et Comparée* **6**, 585–592.

Campbell FE, Atwell RB & Smart I (2003). Effects of the paralysis tick, *Ixodes holocyclus*, on the electrocardiogram of the spectacled flying fox, *Pteropus conspicillatus*. *Australian Veterinary Journal* **81**, 328–331.

Campbell RSF & Ladds PW (1981). Diseases of the dugong in north-eastern Australia. In *The Dugong. Proceedings of Seminar/workshop at James Cook University*, 8–13 May 1979 (ed. H Marsh), pp. 176–181. James Cook University of North Queensland: Townsville.

Clark P, Adlard RD & Spratt DM (2004). Haemoparasites of Australian Mammals. In *Haematology of Australian Mammals* (ed. P Clark), pp. 147–162. CSIRO Publishing: Collingwood, Melbourne.

Dickens RK (1976). Koala in health and disease. In *Australian Fauna Management and Disease: Part A, Proceedings 29*, pp. 105–117. Post Graduate Committee in Veterinary Science: Sydney.

Dunnet GM (1970). Siphonaptera. In *The Insects of Australia*, pp. 647–655. MUP: Melbourne.

Dunnet GM & Mardon DK (1974). A monograph of Australian fleas (Siphonaptera). *Australian Journal of Zoology Supplementary Series* **30**, 1–273.

Dunsmore JD & Spratt DM (1979). The life history of *Filaroides osleri* in wild and domestic canids in Australia. *Veterinary Parasitology* **5**, 275–286.

Gregory GG, Munday BL, Beveridge I & Rickard MD (1975). Studies on *Anoplotaenia dasyuri* Beddard, 1911 (Cestoda: Taeniidae), a parasite of the Tasmanian devil: life cycle and epidemiology. *International Journal for Parasitology* **5**, 187–191.

Hum S, Barton NJ, Obendorf D & Barker IK (1991). Coccidiosis in common wombats (*Vombatus ursinus*). *Journal of Wildlife Diseases* **27**, 697–700.

Jenkins DJ (2006). *Echinococcus granulosus* in Australia, widespread and doing well! *Parasitology International* **55**, S203–S206.

Jenkins DJ & Morris B (2003). *Echinococcus granulosus* in wildlife around Kosciuszko National Park, south-eastern Australia. *Australian Veterinary Journal* **81**, 81–85.

Johnson PM, Speare R & Beveridge I (1998). Mortality in wild and captive rock wallabies and nailtail wallabies due to hydatid disease caused by *Echinococcus granulosus*. *Australian Mammalogy* **20**, 419–423.

Kemp DH, Stone BF & Binnington KC (1983). Tick attachment and feeding: the role of the mouthparts, feeding apparatus, salivary gland secretions and the host response. In *Physiology of Ticks* (eds FD Obenchain & R Galun), pp. 119–168. Pergamon: Oxford.

Kumaratilake LM & Thompson RCA (1982). Hydatidosis/ Echinococcosis in Australia. *Helminthological Abstracts* **51**, 233–252.

Lenghaus C, Obendorf DL & Wright FH (1990). Veterinary aspects of *Perameles gunnii* biology with special reference to species conservation. In *Management and Conservation of Small Populations. Proceedings of Conference in Melbourne, Australia* (edsTW Clark & JH Seebeck), pp. 89–108. Chicago Zoological Society: Brookfield, Illinois.

Love SCJ & Reddacliff GL (1992). Verminous bronchitis and bronchiolitis in potoroid marsupials associated with a new *Capillaria* sp. *Journal of Wildlife Diseases* **28**, 643–645.

Maa TC (1962). Notes on the Hippoboscidae (Diptera). *Pacific Insects* **4**, 583–614.

Mackerras MJ (1958). Catalogue of Australian mammals and their recorded internal parasites. Part I. Monotremes and marsupials. *Proceedings of Linnean Society of New South Wales* **53**, 101–125.

Martin RW, Handasyde K & Skerratt LF (1998). Current distribution of sarcoptic mange in wombats. *Australian Veterinary Journal* **76**, 4–7.

McColl KA & Spratt DM (1982). Parasitic pneumonia in a koala (*Phascolarctos cinereus*) from Victoria, Australia. *Journal of Wildlife Diseases* **18**, 511–512.

McOrist S & Smales L (1986). Morbidity and mortality of free-living and captive echidnas, *Tachyglossus aculeatus* (Shaw), in Australia. *Journal of Wildlife Diseases* **22**, 375–380.

Munday BL (1971). Gastric squamous tumour in the stomach of a pademelon (*Thylogale billardierii*) associated with an infestation with *Labiostrongylus* sp. larvae. *Journal of Wildlife Diseases* **7**, 125.

Munday BL (1988). Marsupial diseases. In *Australian Wildlife Proceedings 104*, pp. 299–357. Post Graduate Committee in Veterinary Science: Sydney.

Mykytowycz R (1963). Occurrence of bot-fly larvae *Tracheomyia macropi* Froggatt (Diptera: Oestridae) in wild red kangaroos, *Megaleia rufa* (Desmarest). *Proceedings of Linnean Society of New South Wales* **88**, 307–312.

Nicholson A & Fanning JC (1982). Parasites and associated pathology of the respiratory tract of the Australian sea lion, *Neophoca cinerea*. In *Wildlife Diseases of the Pacific Basin and Other Countries* (ed. ME Fowler), pp. 178–181. Fruitridge Printing: Sacramento.

Norman RJdeB (1997). Tetraphyllidean cysticerci in the peritoneal cavity of the common dolphin. *Journal of Wildlife Diseases* **33**, 891–895.

Noyes HA, Stevens JR, Teixeira M, Phelan J & Holz P. (1999). A nested PCR for the ssrRNA gene detects *Trypanosoma binneyi* in the platypus and *Trypanosoma* sp. in wombats and kangaroos in Australia. *International Journal for Parasitology* **29**, 331–339.

Obendorf DL & Clarke KP (1992). *Trichinella pseudospiralis* infections in free-living Tasmanian birds. *Journal of Helmintholological Society of Washington* **59**, 144–147.

Obendorf DL, Handlinger JH, Mason RW, Clarke KP, Forman AJ, Hooper PT, Smith SJ & Holdsworth M (1990). *Trichinella pseudospiralis* infection in Tasmanian wildlife. *Australian Veterinary Journal* **67**, 108–110.

Obendorf DL, Spratt D, Beveridge I, Presidente P & Coman B (1997). Parasites and diseases in Australian brushtail possums, *Trichosurus* spp. In *Biological Control of Possums* (ed. R Lynch), pp. 6–9. Royal Society of New Zealand Miscellaneous Series 45: Wellington.

O'Callaghan MG & O'Donoghue P (2001). A new species of *Eimeria* (Apicomplexa: Eimeriidae) from the brushtail possum, *Trichosurus vulpecula* (Diprotodontia: Phalangeridae). *Transactions of Royal Society of South Australia* **125**, 129–132.

Olsen OW & Lyons ET (1965). Life cycle of *Uncinaria lucasi* Stiles, 1901 (Nematoda: Ancylostomatidae) of fur seals, *Callorhinus ursinus* Linn., on the Pribilof Islands, Alaska. *Journal of Parasitology* **51**, 689–700.

Presidente PJA (1984). Parasites and diseases of brushtail possums (*Trichosurus* spp.). In *Possums and Gliders* (eds AP Smith & ID Hume), pp. 171–190. Australian Mammal Society/Surrey Beatty & Sons: Sydney.

Presidente PJA & Beveridge I (1978). Cholangitis associated with species of *Progamotaenia* (Cestoda: Anoplocephalidae) in the bile ducts of marsupials. *Journal of Wildlife Diseases* **14**, 371–377.

Prociv P (1985). Observations on the prevalence and possible infection source of *Toxocara pteropus* (Nematoda: Ascaridoidea) in Queensland flying-foxes. *Australian Mammalogy* **8**, 319–323.

Roberts FHS (1970). *Australian Ticks*. CSIRO: Melbourne.

Seawright AA (1964). Pulmonary acariasis in a Tasmanian fur seal. *Journal of Comparative Pathology* **74**, 97–100

Skerratt LF (1995). *Strongyloides spearei* n. sp. (Nematoda: Strongyloididae) from the common wombat *Vombatus ursinus* (Marsupialia: Vombatidae). *Systematic Parasitology* **32**, 81–89.

Skerratt LF (1998). Diseases and parasites of the common wombat *Vombatus ursinus* in the Healesville area of Victoria. In *Wombats* (eds RT Wells & PA Pridmore), pp. 317–328. Surrey Beatty & Sons: Sydney.

Skerratt LF (2003a). Clinical response of captive common wombats (*Vombatus ursinus*) infected with *Sarcoptes scabiei* var. *wombati*. *Journal of Wildlife Diseases* **39**, 179–192.

Skerratt LF (2003b). Cellular response in the dermis of common wombats (*Vombatus ursinus*) infected with *Sarcoptes scabiei* var. *wombati*. *Journal of Wildlife Diseases* **39**, 193–202.

Skerratt LF, Martin R & Handasyde K (1998). Sarcoptic mange in wombats. *Australian Veterinary Journal* **76**, 408–410.

Skerratt LF, Campbell NJH, Walton S, Kemp D, Murrell A & Barker SC (2002). The mitochondrial 12S gene is a suitable marker of populations of *Sarcoptes scabiei* from wombats, dogs and humans in Australia. *Parasitology Research* **88**, 376–379.

Smales LR (1977). The life history of *Labiostrongylus eugenii*, a nematode parasite of the Kangaroo Island wallaby (*Macropus eugenii*): the parasitic stages. *International Journal for Parasitology* **7**, 457–461.

Speare R & Spratt DM (1982). Pathology of a remarkable nematode parasitic in the portal veins of Macropodidae. In *Wildlife Diseases of the Pacific Basin and Other Countries* (ed. ME Fowler), pp. 113–117. Fruitridge Printing: Sacramento.

Speare R, Donovan JA, Thomas AD & Speare PJ (1989). Diseases of free-ranging Macropodidae. In *Kangaroos, Wallabies and Rat Kangaroos* (eds G Grigg, P Jarman & I Hume), pp. 705–734. Surrey Beatty & Sons: Sydney.

Spencer AJ & Canfield PJ (1994). Haematological characterization of heavy tick infestation in koalas, *Phascolarctos cinereus*. *Comparative Haematology International* **3**, 225–229.

Spratt DM (1972). Aspects of the life-history of *Dirofilaria roemeri* in naturally and experimentally infected kangaroos, wallaroos and wallabies. *International Journal for Parasitology* **2**, 139–156.

Spratt DM (1974). Comparative epidemiology of *Dirofilaria roemeri* infection in two regions of Queensland. *International Journal for Parasitology* **4**, 481–488.

Spratt DM (1979). A taxonomic revision of the lungworms (Nematoda: Metastrongyloidea) from Australian marsupials. *Australian Journal of Zoology Supplementary Series* **67**, 1–45.

Spratt DM (1980). *Metathelazia naghienesis* sp. n. (Nematoda: Pneumospiruridae) from the long-nosed bandicoot, *Perameles nasuta* (Marsupialia). *Journal of Parasitology* **66**, 1032–1035.

Spratt DM & Gill PA (1998). *Durikainema phascolarcti* n. sp. (Nematoda : Muspiceoidea: Robertdollfusidae) from the pulmonary arteries of the koala *Phascolarctos cinereus* with associated pathological changes. *Systematic Parasitology* **39**, 101–106.

Spratt DM & Haycock P (1988). Aspects of the life history of *Cercopithifilaria johnstoni* (Nematoda: Filarioidea). *International Journal for Parasitology* **18**, 1087–1092.

Spratt DM & Hobbs RP (2004). *Breinlia (Breinlia) ventricola* sp. nov. (Nematoda: Filarioidea) from the heart of the red kangaroo, *Macropus rufus*, in Western Australia. *Transactions of Royal Society of South Australia* **128**, 67–72.

Spratt DM & Presidente PJA (1981). Prevalence of *Fasciola hepatica* infection in native mammals in southeastern Australia. *Australian Journal of Experimental Biology and Medical Science* **59**, 713–721.

Spratt DM & Speare R (1982). *Durikainema macropi* gen. et sp. nov. (Muspiceoidea: Robertdollfusidae), a remarkable nematode from Macropodidae (Marsupialia). *Annals de Parasitologie Humaine et Comparée* **57**, 53–62.

Spratt DM & Varughese G (1975). A taxonomic revision of filarioid nematodes from Australasian marsupials. *Australian Journal of Zoology Supplementary Series* **35**, 1–99.

Spratt DM & Whittington RJ (1989). Nematode parasites from the skin of the platypus, *Ornithorhynchus anatinus* (Monotremata). *Systematic Parasitology* **13**, 223–229.

Spratt DM, Beveridge I & Walter EL (1991). A catalogue of Australasian monotremes and marsupials and their recorded helminth parasites. *Records of South Australian Museum Monologue Series* **1**, 1–105.

Stone BF & Carrick FN (1991). Some effects of ticks on koalas, especially tick paralysis. In *Koalas: Research for Management* (ed. G Gordon), pp. 65–69. World Koala Research: Brisbane.

von Kéler S (1971). A revision of the Australasian Boopidae (Insecta: Phthiraptera) with notes on the Trimenoponidae. *Australian Journal of Zoology Supplementary Series* **6**, 1–126.

Whittington RJ & Spratt DM (1989). Lesions associated with metazoan parasites of wild platypuses (*Ornithorhynchus anatinus*). *Journal of Wildlife Diseases* **25**, 521–526.

Whittington R, Middleton D, Spratt DM, Muntz F, Carmel B, Mc Cracken HE, Strakosch MR, Stephanson-Shaw J, Harper PAW & Hartley WJ (1992). Sparganosis in the monotremes *Tachyglossus aculeatus* and *Ornithorhynchus anatinus* in Australia. *Journal of Wildlife Diseases* **28**, 636–640.

Winter H (1958). Gastric strongyloidosis in kangaroos. *Australian Veterinary Journal* **34**, 118.

Appendix 3 Product sources

3M Pharmaceuticals Pty Ltd
9–15 Childers Rd, Thornleigh NSW 2120, Australia
Phone: +61 2 9875 6480
Fax: +61 2 9875 6415

Ambion, The RNA Company
21–30 Woodward St, Austin TX 78744-1832, USA
Phone: +1 512 651-0200
Fax: +1 512 651-0201
Website: <www.ambion.com/>

Abbott Australasia Pty Ltd
32–34 Lord St, Botany NSW 2019, Australia
Phone: +61 2 9384 9700
Fax: +61 2 9384 9800

Alcon Laboratories Australia Pty Ltd
Allambie Grove Park
25 Frenchs Forest Rd East, Frenchs Forest NSW 2086, Australia
Phone: +61 2 9452 9200
Fax: +61 2 9452 1568
Website: <http://www.alconlabs.com.au/>

Allergan Australia Pty Ltd
77 Ridge St, Gordon NSW 2072, Australia
Phone: +61 2 9498 0100
Fax: +61 2 9498 0290
Website: <http://www.allergan.com/>

Apex Laboratories Pty Ltd
61 Chivers Rd, Somersby NSW 2250, Australia
Phone: +61 2 4372 1661
Fax: +61 2 4372 1668
Website: <http://www.apexlabs.com.au/>

Animal Control Technologies
46–50 Freight Dve, Somerton Vic. 3062, Australia
Phone: +61 3 9308 9688
Fax: +61 3 9308 9622
Website: <http://www.animalcontrol.com.au/>

Aspen Pharmacare Australia Pty Ltd
34–36 Chandos St, St Leonards NSW 2065, Australia
Phone: +61 2 8436 8300
Fax: +61 2 9901 3540
Website: <http://www.aspenpharma.com.au/>

Ausrichter
2/21 Chester St, Camperdown NSW 2050, Australia
Phone: +61 2 9517 1166
Fax: +61 2 9516 5810
Website: <http://www.ausrichter.com>

Bacto Laboratories Pty Ltd
310–312 Elizabeth Dve, Liverpool NSW 2170, Australia
Phone: +61 2 9602 5499
Fax: +61 2 9823 9000
Website: <http://www.bacto.com.au/>

Bayer Diagnostics Business Group
875 Pacific Hwy, Pymble NSW 2073, Australia
Phone: +61 2 9391 6000
Fax: +61 2 9988 3311
Website: <http://www.bayer.com.au/>

Biolab Direct Australia
2 Clayton Rd, Clayton Vic. 3168, Australia
Phone: +61 3 9263 4300
Fax: +61 3 9562 9840
Website: <http://www.biolabgroup.com/>

Biolac
15 O'Shannassy St, Mount Pritchard NSW 2170, Australia
Phone: +61 2 9823 9874
Fax: +61 2 9823 9874
Website: <http://www.biolac.com.au/>

Boehringer Ingelheim Pty Ltd
85 Waterloo Rd, North Ryde NSW 2113, Australia
Phone: +61 2 8875 8800
Fax: +61 2 8875 8715
Website: <http://www.boehringer-ingelheim.com/>

Callington Haven
30 South St, Rydalmere NSW 2116, Australia
Phone: +61 2 9898 2788
Fax: +61 2 9684 4215
Website: <http://www.callingtonhaven.com/>

Catac Products UK Ltd
3–5 Chiltern Trading Estate
Earl Howe Rd, Holmer Green High Wycombe Buckinghamshire HP15 6QT, UK
Phone: +44 8453 707 040
Fax: +44 8706 207 041
Website: <http://www.catac.co.uk/>

Complan Foods
Trafalgar House
11 Waterloo Place, London SW1Y 4AU, UK
Phone: +44 8456 003 170
Website: <www.complanfoods.com/>

Copan Italia SPA
Via F. Perotti 10, 25125 Brescia, Italy
Phone: +39 030 268 7211
Fax: +39 030 268 7250
Website: <http://www.copaninnovation.com/>

Delvet Pty Ltd/Vetgroup Australia
B4, 26 Powers Rd, Seven Hills NSW 2147, Australia
Phone: +61 2 9624 8844
Fax: +61 2 9624 1544
Website: <http://www.delvetgroup.com.au/>

Fort Dodge Australia
Parkview Business Centre
Level 3, 1 Maitland Place, Baulkham Hills NSW 2153, Australia
Phone: +61 1800 257 178
Fax: +61 1800 252 976
Website: <http://www.wyeth.com/divisions/fort_dodge.asp>

Gelda Pharmaceutical
6320 Northwest Dve, Mississauga Ontario L4V 1J7, Canada
Phone: +1 905 673 9320
Fax: +1 905 673 8114
Website: <http://www.gelda.com>

GlaxoSmithKline Australia Pty Ltd
1061 Mountain Hwy, Boronia Vic. 3155, Australia
Phone: +61 3 9721 6000
Fax: +61 3 9729 5319
Website: <http://www.gsk.com.au>

Lacteeze (imported in Australia by Allergy Free Pty Ltd)
Shop 1008, Westfield Shoppingtown, Bondi Junction NSW 2022, Australia
Phone: +61 2 9387 2008
Tollfree phone: +61 1300 662 250
Fax: +61 2 9387 5790
Website: <http://www.allergyfree.com.au/>

Glen Forrest Stockfeeders (now trading as Specialty Feeds Pty Ltd)
3150 Great Eastern Hwy, Glen Forrest WA 6017, Australia
Phone: +61 8 9298 8111
Fax: +61 8 9298 8700
Website: <http://www.specialtyfeeds.com.au/>

Heinz Watties Australasia (HJ Heinz Co. Australia Lte)
Locked Bag 19057, Southbank Vic. 3006, Australia
Phone: +61 1800 037 058
Fax: +61 3 9861 5650
Website: <http://www.heinz.com.au/>

Hills Pet Nutrition Inc., Australia
PO Box 6433, Baulkham Hills BC NSW 2154, Australia
Phone: +61 1800 679 932
Website: http://www.hillspet.com.au

International Animal Health Products Pty Ltd
18 Healey Circuit, Huntingwood NSW 2148, Australia
Phone: +61 2 9672 7944
Fax: +61 2 9672 7988
Website: <http://www.iahp.com.au/>

Intervet Australia Pty Ltd
91–105 Harpin St, Bendigo East Vic. 3550, Australia
Phone: (free call) +61 1800 033 461
Fax: (free fax) +61 1800 817 414
Website: http://www.intervet.com.au/

Janssen-Cilag Pty Ltd
1–5 Khartoum Rd, North Ryde NSW 2113, Australia
Phone: +61 2 8875 3333
Website: <http://www.janssen-cilag.com.au/>

Jurox Pty Ltd
85 Gardiners Rd, Rutherford NSW 2320, Australia
Phone: +61 2 4931 8200
Fax: +61 2 4931 8222
Website: <http://www.jurox.com.au/>

Leiras Pharmaceuticals (Leiras Oy)
PO Box 415, Turku 20101, Finland
Phone: +358 2 333 2100
Fax: +358 2 333 2800
Website: <http://www.leiras.fi/>

Masterfoods Australia New Zealand Inc.
4 Corella Close, Wyong NSW 2259, Australia
Phone: +61 1800 816 016
Website: <http://www.masterfoods.com.au/>

Mavlab Pty Ltd
33 Rowland St, Slacks Creek Qld 4127, Australia
Phone: +61 7 3808 1399
Fax: +61 7 3808 4328
Website: <http://www.mavlab.com.au/>

Mazuri
PO Box 66812, St Louis Missouri 63166, USA
Phone: +1 1800 227 8941
Fax: +1 314 317 5276
Website: <www.mazuri.com>

Mead Johnson Nutritionals (Mead Johnson & Co.)
2400 West Lloyd Expressway, Evansville Indiana 47721-0001, USA
Phone: +1 812 429 5748
Fax: +1 812 429 5418
Website: <http://www.meadjohnson.com>

Mead Johnson (a division of Bristol-Myers Squibb Australia Pty Ltd)
556 Princes Hwy, Noble Park Vic. 3174, Australia
Tel: +61 3 9213 4000
Fax: +61 3 9701 1518
Website: <http://www.bmsa.com.au/>

Merial Australia Pty Ltd
Level 6, 79 George St, Parramatta NSW 2150, Australia
Phone: +61 2 9893 0000
Website: <http://au.merial.com/>

N. Stenning & Co. Pty Ltd
174 Parramatta Rd, Camperdown NSW 2050, Australia
Phone: +61 2 8594 9100
Fax: +61 2 8594 9199
Website: <http://www.nstenning.com.au/>

Novartis Pharmaceuticals Australia Pty Ltd
54 Waterloo Rd, North Ryde NSW 2113, Australia
Phone: +61 2 9805 3555
Fax: +61 2 9887 4551
Website: <http://www.novartis.com.au>

Nutricia Australia Pty Ltd
Talavera Corporate Centre
Level 4, Building D, 12–18 Talavera Rd, Macquarie Park NSW 2113, Australia
Phone: +61 1800 060 051
Website: <http://www.nutricia.com.au/>

Organon (NV Organon)
Molenstraat 110, 5342 CC Oss, Netherlands
Phone: +31 412 66 1222
Fax: +31 412 66 2617
Website: <http://www.organon.com/>

Paedpharm Pty Ltd
Suite 503, 35 Grafton St, Bondi Junction NSW 2022, Australia
Phone: +61 2 9300 1900
Fax: +61 2 9387 6654
Website: <http://www.paedpharm.com.au>

Parke Davis Pty Ltd
32 Cawarra Rd, North Caringbah NSW 2229, Australia
Phone: +61 2 9526 9500
Fax: +61 2 9525 6237

Parnell Laboratories (Aust.) Pty Ltd
6 Century Estate
476 Gardeners Rd, Alexandria NSW 2015, Australia
Phone: +61 2 9667 4411
Fax: +61 2 9667 4139
Website: <http://www.parnell.biz/>

Peptech Animal Health Pty Ltd
19–25 Khartoum Rd, Macquarie Park NSW 2113, Australia
Phone: +61 2 9870 8788
Fax: +61 2 9870 8787
Website: <http://www.peptech.com/>

Pet-Ag Inc.
255 Keyes Ave, Hampshire Illinois 60140, USA
Phone: +1 847 683 2288
Fax: +1 847 683 2343
Website: <http://www.petag.com/>

Pfizer Australia Pty Ltd
38–42 Wharf Rd, West Ryde NSW 2114, Australia
Phone: +61 2 9850 3333
Fax: +61 9850 3399
Website: <http://www.pfizer.com.au/>

Pfizer New Zealand Ltd
Level 3, Pfizer House
14 Normandy Rd, Mt Eden Auckland, New Zealand
Phone: +62 9 638 0000
Fax: +62 9 638 0021
Website: <http://www.pfizer.co.nz>

Pharmaderm Animal Health
60 Baylis Rd, Melville New York 11747, USA
Phone: +1 800 432 6673
Fax: +1 631 420 1572
Website: <http://www.pharmadermah.com/>

Roche Diagnostics Australia Pty Ltd
31 Victoria Ave, Castle Hill NSW 2154, Australia
Phone: +61 2 9860 2222
Fax: +61 2 9860 2111
Website: <http://www.rochediagnostics.com.au/>

Roche Products Pty Ltd
PO Box 255, Dee Why NSW 2099, Australia
Phone: +61 2 9454 9000
Fax: +61 2 9454 9152
Website: <http://www.roche-australia.com/>

Schering-Plough Animal Health
11 Gibbon Rd, Baulkham Hills NSW 2153, Australia
Phone: +61 2 9852 7200
Fax: +61 2 9852 7285
Website: <http://www.sp-animalhealth.com/>

Sharpe Laboratories Pty Ltd (Animal Health Division)
12 Hope St, Ermington NSW 2115, Australia
Phone: +61 2 9858 5622
Fax: +61 2 9858 5957

Sigma Pharmaceuticals Pty Ltd
96 Merrindale Dve, South Croydon Vic. 3136, Australia
Phone: +61 3 9839 2800
Fax: +61 3 9839 2801
Website: <http://www.sigmaco.com.au/>

Smith & Nephew Pty Ltd
315 Ferntree Gully Rd, Mt Waverley Vic. 3149, Australia
Phone: +61 3 8540 6506
Fax: +61 3 9544 5086
Website: <http://www.smith-nephew.com/>

Troy Laboratories Pty Ltd
98 Long St, Smithfield NSW 2164, Australia
Phone: +61 2 9604 6266
Fax: +61 2 9725 1772
Website: <http://www.troylab.com.au/>

Vetafarm
3 Bye St, Wagga Wagga NSW 2652, Australia
Tel: +61 2 6925 6222
Fax: +61 2 6925 6333
Website: <http://www.vetafarm.com.au/>

Vet-a-mix (Lloyd Laboratories)
PO Box 130, W Thomas Ave, Shenandoah Iowa 51601, USA
Phone: +1 712 246 4000
Fax: +1 712 246 5245
Website: <http://www.lloydinc.com/>

Veterinary Companies of Australia Pty Ltd
26 Binney Rd, Kings Park NSW 2148, Australia
Phone: +61 2 9679 5750
Fax: +61 2 9679 5766
Website: <http://www.vetcompanies.com.au/>

Virbac (Australia) Pty Ltd
15 Pritchard Place, Peakhurst NSW 2210, Australia
Phone: +61 2 9717 2000
Fax: +61 2 9533 1522
Website: <http://www.virbac.com.au/>

Wombaroo Food Products
PO Box 151, Glen Osmond SA 5064, Australia
Phone: +61 8 8391 1713
Fax: +61 8 8379 1339
Website: <http://www.wombaroo.com.au/>

Wyeth Nutritionals (Division of Wyeth Australia Pty Ltd)
Norwest Business Park
17–19 Solent Circuit, Baulkham Hills NSW 2153, Australia
Phone: +61 2 9761 8200
Fax: +61 2 9023 0022
Website: <http://www.wyeth.com.au/>

Appendix 4 Abbreviations

d	day
hr	hour
min	minute
mo	month
sec	second
wk	week
yr	year
IA	intra-articular
IM	intramuscular
IO	intraosseous
IP	intraperitoneally
IT	intratracheally
IV	intravenously
PO	orally
SC	subcutaneously
mL	millilitre
dL	decilitre
L	litre
kg	kilogram
g	gram
mg	milligram
μg	microgram
G	gauge
U	unit
IU	international units
sid	once a day
bid	twice a day
tid	three times a day
qid	four times a day
prn	as needed
q	every
ALT	alanine aminotransferase
ALP	alkaline phosphatase
AST	aspartate aminotransferase
BUN	blood urea nitrogen
Br	bilirubin

CBC	complete blood count
Chol	cholesterol
CK	creatine phosphokinase (creatine kinase)
GGT	gamma-glutamyl transferase
Hb	haemoglobin
HCT	haematocrit
ID	L-Iditol dehydrogenase (sorbitol dehydrogenase)
LDH	lactate dehydrogenase
MCH	mean corpuscular haemoglobin
MCHC	mean corpuscular haemoglobin concentration
MCV	mean corpuscular volume
PCV	packed cell volume
RBC	red blood cells
Trig	triglycerides
TSP	total serum protein
TPP	total plasma protein
WBC	white blood cells
GnRH	gonadotrophin-releasing hormone
FSH	follicle-stimulating hormone
HCG	human chorionic gonadotrophin
LH	luteinising hormone
CL	corpus luteum
BW	body weight
BY	back young
MAC	mean alveolar concentration
PY	pouch young
PEY	pouch emergent young
bpm	beats or breaths per minute
Ab	antibody
Ag	antigen
n	sample size
SD	standard deviation
CI	confidence interval

ACT	Australian Capital Territory	SA	South Australia
NSW	New South Wales	Tas.	Tasmania
NT	Northern Territory	Vic.	Victoria
Qld	Queensland	WA	Western Australia

Index

www.ingramcontent.com/pod-product-compliance
Ingram Content Group UK Ltd.
Pitfield, Milton Keynes, MK11 3LW, UK
UKHW060221061026
15625UKWH00007B/434

9 780643 091504